吴国雄先生简介

吴国雄 男，1943年3月生，籍贯广东，中国科学院大气物理研究所研究员，中国科学院院士。

主要著作有《副热带高压形成和变异的动力学问题》（国家自然科学基金研究专著/地球科学系列）（2002年）、《青藏高原与西北干旱区对气候灾害的影响》（2003年），以及《夏季副热带高压变化研究的新进展》（2006年）；在国内外期刊上发表论文400余篇。

教育背景、研究工作经历

1961—1966年	南京气象学院
1968—1978年	甘肃省张掖气象台、兰州西北中心气象台天气预报工程师
1974—1975年	北京大学地球物理系数值天气预报进修班
1978—1979年	中国科学技术大学研究生院
1980—1983年	英国伦敦大学帝国理工学院，获博士学位
1983—1984年	英国欧洲中期天气预报中心（ECMWF）访问科学家
1985年至今	中国科学院大气物理研究所研究员
1989—1991年	美国普林斯顿大学地球流体动力学实验室（GFDL）高级访问研究教授

学术团体任职情况

2022 年至今	国际科学理事会（ISC）中国国家委员会主席顾问
2018—2021 年	国际科学理事会（ISC）中国国家委员会主席
2012—2016 年	国际大地测量学与地球物理学联合会（IUGG）中国国家委员会主席
2011—2017 年	国际科学联盟（ICSU）中国国家委员会主席
2011—2014 年	国际科学联盟（ICSU）执行委员会（EB）委员
2007—2011 年	国际气象学和大气科学协会（IAMAS）主席
2005—2010 年	世界气候研究计划（WCRP）联合科学委员会（JSC）委员
2005—2008 年	国际"观测系统研究与可预报性试验"（THORPEX）计划科学顾问
2003—2007 年	国际气象学和大气科学协会（IAMAS）副主席
2003—2011 年	国际大地测量学与地球物理学联合会（IUGG）中国国家委员会副主席
2000—2005 年	全球能量水分交换试验（GEWEX）科学指导小组（SSG）成员
1998—2002 年	国际气候变化与可预报性研究（CLIVAR）科学指导小组成员
1998—2007 年	国际气象学和大气科学协会（IAMAS）中国国家委员会主席
1997—2003 年	国际气象学和大气科学协会（IAMAS）执行局成员
1993—2000 年	大气科学和地球流体力学数值模拟国家重点实验室（LASG）主任
1991—2000 年	国际气象和大气科学协会（IAMAS）动力气象委员会（ICDM）委员

获省级以上科技奖励和学术荣誉称号情况

2022 年	国际科学理事会（ISC）会士
2015 年	国际大地测量学与地球物理学联合会（IUGG）会士
2015 年	美国地球物理联合会（AGU）会士
2014 年	美国美华海洋大气学会（COAA）荣誉会士
2014 年	陈嘉庚地球科学奖
2012 年	英国皇家气象学会荣誉会员（Honorary Fellow of Royal Meteorological Society）
2008 年	何梁何利科学与技术进步奖——气象学奖
2007 年	国家自然科学奖二等奖（第 1 完成人）：海陆气相互作用及其对副热带高压和我国气候的影响
2004 年	科技部"国家重点实验室计划先进个人"奖
2001 年	全国优秀博士学位论文指导教师
1997 年	中国科学院科技进步奖二等奖（第 5 完成人）：灾害性气候预测方法及其应用研究
1992 年	中国科学院科技进步奖一等奖（第 8 完成人）：中期数值天气预报研究
1988 年	人事部"中青年有突出贡献专家"

少年成长　苦寒磨砺

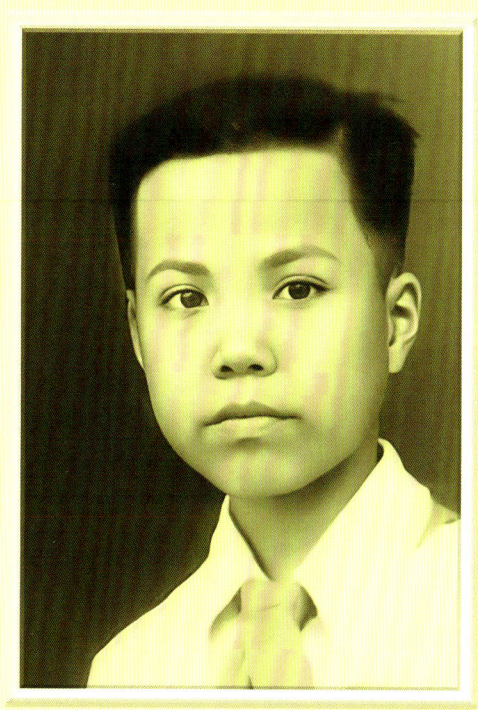

△ 1949—1955 年，在广东潮阳县棉城第二小学

△ 1955—1958 年，在广东潮阳县棉城中学

△ 1958—1961 年，在广东潮阳县第一高中

△ 1961 年，与同学在家乡广东潮阳县文光塔合影，右一为吴国雄

▲ 1961年，与同学在家乡文光公园合影，右一为吴国雄

青年求学　踔厉奋发

🔸 1965 年，南京气象学院气象系天气动力气象专业气 612 班在扬州瘦西湖五亭桥合影，后排左五为吴国雄，前排右三为其夫人刘还珠

🔸 1966 年，在南京气象学院

🔸 1968 年，离开南京气象学院赴张掖气象台前，吴国雄（右三）和夫人刘还珠（右一）与同学在天安门广场合影

🔺 1979年9月12日,赴英留学与同机部分留英研究生在中国驻英国大使馆教育处

🔺 1980—1983年期间,在英国伦敦泰晤士河南岸的雕塑前留影

🔺 1980年,在英国伦敦马克思墓前留影

1981年，与导师 J. Green（左）参观肯辛顿花园阿尔伯特纪念碑时合影

1981年，与英国伦敦帝国理工学院物理系大气物理专业研究生合影，后排左一为吴国雄

1982年，参观法国巴黎凡尔赛宫

1982年，参观英国伦敦白金汉宫

1983年，于英国伦敦帝国理工学院获得博士学位

1983年，在中国驻英国使馆，与章基嘉老师（右四）等合影，右一为吴国雄

矢志不渝，奉献身心

1970年，在甘肃张掖气象台，与台长陈万福（左）合影

1973年，离开张掖气象台赴兰州西北中心气象台工作前与同事合影，前排中为吴国雄，右二为其夫人刘还珠

1975年，北京大学数值预报进修班毕业留念，后排左二为吴国雄

🔺 1977年，在甘肃省气象局与薛纪善（左）合影

🔺 1984年，在欧洲中期天气预报中心（ECMWF）工作期间，和夫人刘还珠一起在巨型计算机CRAY_1A前与同事留影

🔺 1986年，在北京欢迎ECMWF主任Bengtsen博士访华留影

🔺 1986年夏，中国科学院大气物理研究所大气科学和地球流体力学数值模拟国家重点实验室（LASG）第一次国际研讨会部分学者合影。前排右三为所长、实验室主任曾庆存，右四为中科院院长卢嘉锡，右六为美国科学院院士Lindzen，右七为冯康，左二为杨大升，左一为李崇银；中排左一为吴国雄；后排左四为丑纪范，左五为麦文建，右六为陶诗言，右九为谢义炳

◀ 1990年，访问美国国家大气研究中心（NCAR）期间，吴国雄夫妇与Kasahara博士（中）合影

● 1994年，接待S. Manabe（右）和 J. Green（中）访问 LASG

● 1994年，接待 J. Green（左二）访问 LASG

● 1998年，参加在美国举行的中美大气科学合作工作会议，前排左一为吴国雄，右二为代表团团长、中国气象局局长邹竞蒙

● 1999年，国庆五十周年时，在天安门观礼台，与黄荣辉（左一）等合影

● 2001年，在北京主持第一届气候系统模式发展计划研讨会，会后形成决议上报国务院

● 2001年11月，主持的中国科学院知识创新工程重要方向项目"亚洲季风区海-陆-气相互作用对我国气候变化的影响"项目启动会

◐ 2002年，在北京的办公室

◐ 2005年，参加 LASG 成立 20周年国际学术研讨会

◐ 2005年8月11日，在北京国际气象学和大气科学协会（IAMAS）科学大会期间，与IAMAS秘书长 Roland List 等一起与学生志愿者合影

🔥🔥 2005年，在北京主持叶笃正先生（左图前左五，右图中）九十寿辰庆典

🔥 2005年，参加LASG成立20周年国际会议

🔥 2006年，在北京与同事合影，前排左一为吴国雄，前排中为在美国工作的学生刘辉

▲ 2007，在北京主持亚洲季风年（AMY2008）国际研讨会

▲▶ 2007年，在北京获国家自然科学奖时留影

● 2008 年，在郑州参加 973 项目会议

● 2008 年，在香港天文台，与台长林超英（右）合影

● 2009 年 11 月，接待 S. Manabe（前排右三）和 A. Oort（前排右二）

◐ 2009 年，主持 Manabe 访问 LASG 的学术报告

◐ 2010 年，在土耳其开会期间留影

◐ 2011 年，在北京主持热带对流年和亚洲季风年（YOTC&AMY）国际会议

● 2011年7月，在西藏林芝参观中国科学院青藏高原研究所观测站

● 2012年7月，LASG夏季国际研讨会休息期间的学术讨论

● 2011年，在杭州主持国家自然科学基金委员会第61期双清论坛

● 2012年，参加全球变化研究国家重大科学研究计划项目年会

● 2013年7月，IAMAS的五任主席和两位秘书长在达沃斯（Davos）合影

● 2013年，与周光召院长（右）在北京973项目专家组会议上合影

⬥ 2014年,课题组学术讨论

⬥ 2014年,在日本参加亚洲大洋洲地球科学学会(AOGS)年会

🔺 2016年，访问瑞典斯德哥尔摩大学

🔺 2016年，B. Hoskins（左四）、Bin Wang（王斌，右二）和 G. McBean（右三）访问中国科学院大气物理研究所

● 2018年，教师节留影

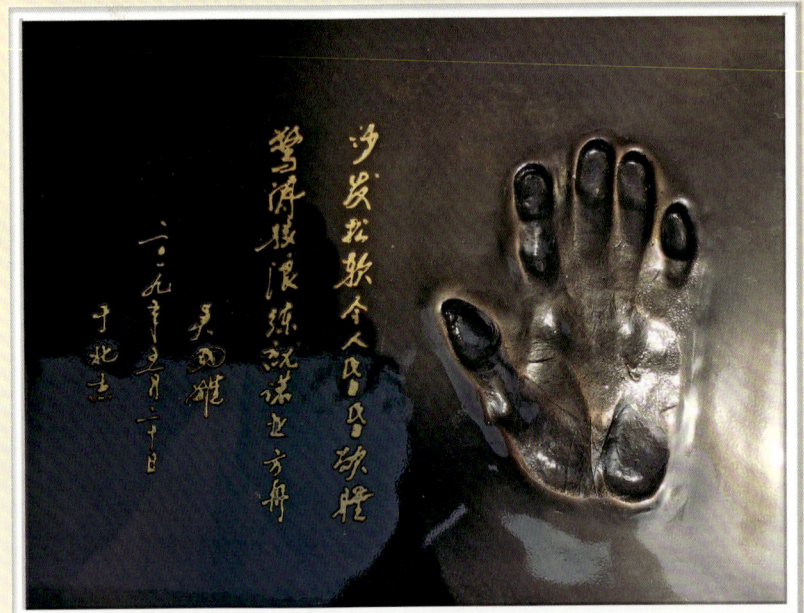

● 2019年，中关村院士墙留下手印

● 2020年，在北京录制西藏电视台《珠峰讲堂》，讲述青藏高原的天气气候影响

部分学生答辩留影

家庭美满　师友爱敬

○ 1970年，与夫人刘还珠在南京长江大桥

1972年，在张掖气象台工作（1968—1973年）期间与同事合影

○ 1984年，与ECMWF同事在英国雷丁（Reading）合影

○ 1986年，在北京拍摄的全家照

🔸 1986年，与叶笃正先生（中）和邹晓蕾（左）在鼓浪屿

🔸 1990年，与I. Held夫妇和G. Philander夫妇在普林斯顿（Princeton）家中

🔸 1986年，与夫人刘还珠在厦门

🔸 1990年，与夫人刘还珠在美国普林斯顿家中

● 1990年,与夫人刘还珠在美国普林斯顿

● 1991年,与 S. Manabe(后排右一)、J. Malmen 夫妇(前排左一、左二)在美国普林斯顿家中

● 1994年,J. Green(右一)来北京家中

△ 1994年，与陶诗言先生（中）和张智北（左）在台北

△ 1994年，与S. Manabe（右）访悬空寺

△ 1995年，和叶笃正先生（左）在香山

△ 1996年，与B. Hoskins（中）和刘雅章（左）参观圆明园

▶ 1996年，在北京郊游

◀▶ 2000年12月，课题组和家人在北京聚会

● 2004年，漫步泰山

● 2004年，庆贺叶笃正先生米寿（八十八岁），左起吴国雄、巢纪平、叶笃正、叶笃正夫人冯慧、丑纪范、符淙斌

● 2004年，与夫人刘还珠陪同叶笃正先生夫妇游北京植物园

● 2005年，吴国雄夫妇与叶笃正先生（右二）、陶诗言先生（左二）合影

● 2006年，吴国雄和夫人刘还珠在印度

● 2006年10月2日，在北京家中

● 2007年9月21日，金继明（前排左一）回国聚餐

2007年，与叶笃正夫妇（前排中）、巢纪平夫妇（前排右一和后排左一）、丑纪范夫妇（后排左二、三）和符淙斌夫妇（后排右一、二）聚会

2007年，于北京拍摄的全家福

⏶ 2008年，与儿子陪四哥（中）游览河南云台山

⏶ 2009年，参观尼亚加拉大瀑布

🔺 2010年，与家人一起郊游

🔺 2012年，与丑纪范先生（右）"比赛"

● 2011年,叶笃正先生九十五寿辰。后排右一、二为丑纪范夫妇

● 2012年,七十岁生日聚会

⬥ 2013年，在呼伦贝尔草原

⬥ 2016年，与夫人刘还珠在内蒙古赤峰

🔺 2016年12月，在美国地球物理联合会（AGU）年会期间师门聚会

🔺 2018年，AGU年会期间，张永生（前排左二）在马里兰家中款待老师和师门兄弟姐妹

副热带环流、亚洲季风以及青藏高原影响的若干动力学问题

——庆贺吴国雄先生八十寿辰

任荣彩 刘屹岷 毛江玉 段安民 等 编著

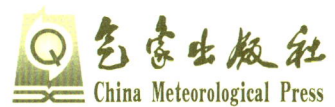

内 容 简 介

本书全面总结了吴国雄先生从事大气科学研究近60年以来与合作者一起，在大气环流动力学、副热带高压动力学、青藏高原动力学、亚洲季风动力学与位涡动力学等方面所作出的重要科学贡献。全书共分为6部分，其中，彩色图照部分通过吴国雄先生早年以来的学习、工作、科研交流以及生活图片，分门别类展示吴先生的日常工作缩影；正文第1~5部分每部分包括关于该部分主题的综述文章2~3篇，以及吴国雄先生相关的代表性论文原文4~6篇。本书适合从事大气科学工作的科研和业务工作者阅读，也可作为高等院校研究生的辅助教材。

图书在版编目（CIP）数据

副热带环流、亚洲季风以及青藏高原影响的若干动力学问题：庆贺吴国雄先生八十寿辰 / 任荣彩等编著. -- 北京：气象出版社，2023.1
ISBN 978-7-5029-7852-5

Ⅰ. ①副… Ⅱ. ①任… Ⅲ. ①大气科学－文集 Ⅳ. ①P4-53

中国版本图书馆CIP数据核字(2022)第220069号

副热带环流、亚洲季风以及青藏高原影响的若干动力学问题——庆贺吴国雄先生八十寿辰
Furedai Huanliu、Yazhou Jifeng yiji Qingzang Gaoyuan Yingxiang de Ruogan Donglixue Wenti
——Qinghe Wuguoxiong Xiansheng Bashi Shouchen

出版发行：气象出版社			
地　　址：北京市海淀区中关村南大街46号		邮政编码：100081	
电　　话：010-68407112（总编室）　010-68408042（发行部）			
网　　址：http://www.qxcbs.com		E-mail：qxcbs@cma.gov.cn	
责任编辑：黄红丽		终　审：吴晓鹏	
责任校对：张硕杰		责任技编：赵相宁	
封面设计：博雅锦			
印　　刷：北京地大彩印有限公司			
开　　本：889 mm×1194 mm　1/16		印　张：48	
字　　数：1550 千字		彩　插：18	
版　　次：2023年1月第1版		印　次：2023年1月第1次印刷	
定　　价：500.00 元			

本书如存在文字不清、漏印以及缺页、倒页、脱页等，请与本社发行部联系调换

丑纪范先生序言

欣逢吴国雄院士八十华诞，可喜可贺。他的学生们编著《副热带环流、亚洲季风以及青藏高原影响的若干动力学问题》一书，综述了吴院士在季风动力学、副热带动力学和青藏高原气候动力学等方面取得的学术成就和突出贡献，是一件很有意义的事情。

我和吴国雄院士都曾经在甘肃工作，相识多年，合作愉快。我们相继调入北京后，也一直保持着密切联系，特别是在副热带高压气候动力学的研究方面。众所周知，西太平洋副热带高压是与我国夏季雨带密切相关的重要环流系统。为探究副热带高压形成和变异的动力学问题，1995年起我们共同主持了两届国家自然科学基金委员会的重点项目"副热带高压带变异机理"和"夏季副热带高压变化及其影响天气气候异常的机理"。通过8年的系统研究，取得了很多重大理论性突破，纠正了如西太平洋副热带高压决定东亚降水异常等对副热带高压成因和影响的若干传统认识。作为这项研究的重要成果，还合作完成了国内第一部副热带高压动力学专著《副热带高压形成和变异的动力学问题》。需要说明的是，这项工作主要的贡献者，是吴国雄院士和他的学生们。

吴国雄院士建立了包括副热带非绝热加热和摩擦耗散在内的全型涡度方程，发展了热力适应理论，对副热带高压的形成机理提出了新的解释。他提出的"副热带四叶型加热"这一原创概念，阐明了副热带地区季风和沙漠共存的重要物理机制。他揭示了青藏高原引起的上升与大陆尺度"副热带四叶型加热"所产生的上升运动叠加，更好地解释了亚洲夏季风成为全球最强的季风系统、中亚干旱区成为全球最强的副热带内陆沙漠的机理。吴国雄院士还曾担任国际气象学和大气科学协会（IAMAS）主席，是连接大气科学领域中国学者和西方学者之间学术交流的国际著名气象学家，为推动中国的大气科学走向世界做出了重要贡献。

《副热带环流、亚洲季风以及青藏高原影响的若干动力学问题》即将出版，我很高兴为之作序。我相信，这本著作能够帮助广大读者特别是气象工作者更深入地了解吴国雄院士在天气和气候动力学领域的杰出成就，将会激励更多的年轻人勇攀大气科学新高峰。

丑纪范

2022年7月于北京

周秀骥先生序言

即将迎来我国著名气象学家、中国科学院院士吴国雄先生80华诞和从事大气科学研究60周年。吴国雄院士在季风动力学、副热带动力学和青藏高原动力学上取得了杰出的成就,对大气科学和气候学领域从广度和深度上都做出了重大贡献。

吴国雄院士于1966年毕业于南京气象学院。他深深酷爱大气科学专业,坚决要求从事气象预报工作,选择了环境艰苦的甘肃省张掖气象台。在干旱侵袭、风沙肆虐的张掖,他在满负荷的艰苦工作中还挤出时间复习大学课程和自学英语,把国际学术权威汤卜逊(Thompson)的著作《数值天气预报》翻译成了中文。1978年,吴国雄以第一名的优异成绩被叶笃正院士收为改革开放后第一届研究生,并顺利通过了出国研究生考试。但当他得知叶笃正先生也特别需要人帮忙恢复各项研究工作时,就向叶先生提出了拟放弃出国的想法。叶先生非常认真地说:"我们闭关自守了这么多年,国外的科学进展我们都不知道,你学成后再回国帮助我开展研究不迟。"这样他走出了国门,在求学和国外工作期间,西北农民的艰苦生存状况时常浮现在他的脑海中。很多年后,有人问当初怎么会舍弃一切回国,吴国雄不假思索地说:"走的时候就没想过有不回来的可能。"

吴国雄院士回国后潜心研究,取得了一系列开创性的成果,特别是在青藏高原天气和气候动力学上做出了重大贡献,受到国际和国内的广泛关注。他构建了在耗散系统中开展青藏高原气候动力学研究的理论框架,创立了青藏高原"热力适应理论""青藏高原感热气泵理论"。吴国雄院士提出的青藏高原感热气泵理论及加热激发垂直环流的模型,明确了高原斜坡感热热力效应在驱动亚洲季风和调节亚洲气候中的重要作用。他在中国科学院大气物理研究所大气科学和地球流体力学数值模拟国家重点实验室工作期间,曾担任两届实验室主任主持工作,为重点实验室持续取得优异成绩和优秀评估结果奉献精力,成绩显著。从2013年开始,我们合作组织为期10年的国家自然科学基金委员会青藏高原重大研究计划,他肩负专家组组长的重任,组织国内外优秀科研骨干,进一步推动了国内在青藏高原地-气耦合系统变化及其全球气候效应方面的研究。

在恭贺吴国雄院士80华诞之际,由他的学生们编著出版《副热带环流、亚洲季风以及青藏高原影响的若干动力学问题》一书,有非常重要的意义。该书的出版将使广大读者,特别是大气科学领域的专家学者们更多地了解吴国雄院士在大气科学研究中的学术成就,系统性地认识青藏高原气候动力学研究的历史进程和最新发展。更重要的是,该书的出版对年轻的大气科学工作者具有重要的激励作用,鼓励他们不畏困难、勇于探索、追求理想、奉献社会。

2022年6月于北京

吴国雄先生序言

皓月高悬碧空,清风频送花香;远听海啸起伏,近闻百虫奏鸣。在广东省潮阳县文光塔旁家里,院子中摆着一张洗得发白的小木桌,一盏煤油灯置于小圆桌中央。一家人如常围着小木桌:母亲和嫂嫂穿梭织网,我和四哥、姐姐则做作业、读课文——20世纪50年代童年的情景犹如发生在昨天一般。最愉快的时刻常常是做完作业后,哥哥给母亲奉上一盅工夫茶说:"妈妈,继续昨天的故事吧!"日复一日,妈妈给我们断断续续讲了三国、水浒、封神、说岳……年复一年,仁、义、礼、智、信,精忠报国,扶弱济贫等在我的脑海中潜移默化;呼风唤雨,腾云驾雾,宇宙深邃的奥妙在我幼小心灵里埋下了向往和求知的种子。冬去春来,雨雪雷电,旱涝灾害……千变万化的大自然愈来愈吸引着我,使我从小就迷上了变化莫测的气象科学。家里兄弟姐妹7人我排行最末,因家境贫苦,我们放学回家后常要帮家里织渔网贴补家用。虽然哥哥姐姐们聪明勤奋、品学兼优,但都为了减轻家庭负担而提前离家参加工作。一个难得的机会使我得以升读高中。记得那是初二的寒假,我一口气自学完初二和初三几何,完成了所有的初中几何作业。老师对我母亲说,这小孩要让他升学,如果有困难老师愿意提供帮助。全家人对我殷切期望,勒紧裤腰带送我上了高中。

我就读的广东省潮阳县第一高中是地区先进的高级中学,之前多年向许多院校输送了多种人才,唯独还没有人报考气象专业,认为气象是可望不可即的学科。1961年我高中毕业前全县举行统考,由于我的数、理、化及外语分数均获第一,学校鼓励我报考气象。于是我一口气报了三个气象专业的志愿。那一年正是国家困难时期,国民经济实行"调整、巩固、充实、提高"政策,中国科学技术大学取消气象专业,我被录取到南京气象学院,开始了五年的大学生涯。

人们常说上大学就学两样东西:一是打基础;二是学方法。对我来说,大学时期还是磨炼意志、树立艰苦奋斗人生观的五年。1959年初开始的"自然灾害困难时期"促使国家决心建立一所专门的气象学院。1960年招收的第一届学生是在南京大学气象系寄读的。1961年秋,当我从广东来到南京浦口,挑着行李,按着录取通知书的指引去寻找那让人憧憬的"依山傍水、风景秀丽"的"圣地"时,见到的却是满地泥泞、杂乱无章的建设工地。仅有的一幢小楼既是师生员工宿舍,又是全院唯一的办公场所,没有自来水,没有厕所。洗脸刷牙要到附近的池塘去。当时天旱,池塘半干,舀起一杯水尽是小虫,许多人都病了。上课是在没有完工的"服务楼",门、窗都还没有安装,雨水能淋湿半个教室。遍野的泥土地,旱天是"洋(扬)灰路",雨天变"水泥道",南方来的人数九寒天还只穿着单裤……大自然给我们这些"欲与天公试比高"的气象学者来了个下马威! 在那物质匮缺的年代,求知的愿望和为人民服务的决心都转化成无穷的精神力量。女同学把节余的粮票送给男同学;男子汉冬天到几里路外的盘城集湖面砸冰块,拉回来解决吃饭用水困难。大家利用课余时间拉煤渣铺路,平荒地植树、种蔬菜自给。周末学英雄、谈理想、树风格、讲情操……高昂的英雄气概使我们克服了一个个困难,温暖的兄弟情谊战胜了天寒地冻。建院的年代整个气象学院成了一座大熔炉! 直到两年后,教学楼盖起来了,最后一口井也终于打出水来了,大家才有了初步的生活和教学环境。记得自来水就要通进宿舍楼那一天,同学们集聚在洗脸间,打开所有的水龙头,等啊等啊,突然间,黄黄的泥水夹着铁锈哗啦哗啦地从一个个水龙头喷射出来,大家高呼着"水来了!""水来了!"欢呼雀跃。只有经历过寒冬的人才真正懂得太阳的温暖! 建院初期那艰苦的环境锤炼了"敢与天公试比高"的新气象学者,造就了他们克服困难的毅力,提高了他们的情商,成长为后来大气科学界的教授或高层次管理人才。"宝剑锋从磨砺出,梅花香自苦寒来!"五年的大学生活不仅让我们学到了基础知识和学习方法,更重要的是锤炼了我们战天斗地、克服困难的意志和毅力。

毕业那一年,我报考了中国科学院大气物理研究所的研究生。突然"文化大革命"爆发,接下来是近两年的漫长的毕业分配。恰逢各地在"精简机构",同学们都面临改行的威胁。我和未婚妻、同班同学刘还珠商量,向领导递上了决心书,要求到环境艰苦的甘肃省,因为那里是直接到气象台报到,可免遭改行的劫难。领导担心我的身体,怕吃不消西北的严寒。我没有动摇自己的决心,毅然来到位于河西走廊的张掖气象台,开始了十年漫长的西北经历。

当时整个张掖地区只有我们两名大学毕业的气象员。领导说,大学生应当搞最基础的工作,分配我们当观测员,我们愉快地接受,从最基本的观云测天开始。后来由于工作需要,才改当填图员、预报员。那里自学条件很差。为了执行我们事先商定好的五年自学计划,还珠为我订了三份英文刊物——《北京周报》、*Meteorological Magazine*(《气象杂志》)和 *Quarterly Journal of the Royal Meteorological Society*(《英国皇家气象学会季刊》)。就这样,把英语学习坚持下来了。领导号召大家向"九大""献礼"。我提出献两份礼,一份是设计能自行发电的自动风向风速观测仪,一份是把一本国际先进的、汤卜逊著的《数值天气预报》从英文翻为中文。领导很支持,特许免去我晚上的政治学习去搞"献礼"。我如期地交出了礼品,风向风速仪的图纸寄到了甘肃省气象局。书稿被南京气象学院院长章基嘉先生拿去编写教材。更为重要的是我们赢得了时间,学到了新知识。在张掖的五年,不仅完成了数学、物理、外语的自学计划,还涉足到数值预报的新领域,为日后的学习和科研打下了基础。

1969年中苏边界冲突后,气象部门受军队"双重领导"。1970年年初数九寒冬尚未结束,我和气象台另外七位同事被派送到祁连山下去开垦六十多亩荒地。在凛冽北风下拉着铁犁,抡着钉镐,硬是翻遍了半冻的荒地,种上了春小麦。后来购了黄牛、毛驴,白天干庄稼活,晚上八个人挤在小土炕上。人家听广播、打扑克,我又偎着马灯,除了读专业书外,还潜心研读南京中医学院编的《药理学》《病理学》。我看到农民看不起病,就利用自学的粗浅中医知识,给病人义务扎针、号脉、开处方,颇受欢迎。后来离开农村回气象台后,还有人到气象台找我免费看病。西北农村的艰苦生活使我永生难忘。

在张掖的五年也是我探索人生旅途、聚集精力于气象事业的重要起点。在那动荡的岁月和无人指引的环境中,我的爱好很多,一直苦苦探索。气象台新调来的台长丁兆林成了我的棋友。有一天,丁台长以前辈的身份认真地对我说:你这个人非常聪明,但又学医,又学无线电,又搞气象,长久下去会糟蹋才智,变成"万金油"。他劝我必须有所取舍。这次谈话使我醒悟,终止学医和无线电,从此一心一意钻研气象专业至今。一个人的成功往往有许多转折和机遇,往往需要许多人扶持和指点。正是一位基层领导作为过来人的指点使我在茫然探索中更坚定了前进的方向。

正当我们在张掖气象台的业务不断进步的时候,兰州的西北中心气象台却因为缺乏天气预报人员而苦恼。1973年夏天我和刘还珠被调往兰州,到西北中心气象台从事短期天气预报工作。在那里我们接触了经验丰富的老预报员和研究人员;使用的天气图也从东亚地区扩展到了全球,从地面和500 hPa增加到各个层次。我们把从前积累的知识应用到三维空间,视野顿然开阔。此后不久,北京大学与中央气象局联合举办数值预报进修班,幸运之神再次降临到我。由于我翻译过数值预报的书,被大家一致推选到北京大学学习。从1966年南京气象学院毕业到1974年再进北京大学校门整整八年,既具备了实践经验,又深知学习机会之难得。我们这批"老大学生"如饥似渴地学习,认真聆听谢义炳先生、杨大升先生、陈受钧先生等的授课。图书馆的教师说,你们一来图书馆又恢复了生气。我更是贪婪地吮吸着书本里的营养。第二年我带领高原组在中央气象局杜行远先生指导下进行五个月毕业实习。我们利用多时刻时间积分的思路和求高阶时间导数的方法,一口气设计了正压和三层斜压等三个短期预报模式,并实现了预报试验。杜行远先生对我们的工作深为赞誉。也正是这一阶段的经历,杜行远先生后来把我推荐给叶笃正先生,成为叶先生的研究生。从北京大学毕业回到兰州不久,我被任命为西北中心气象台天气预报工程师,组建了中期天气预报组和数值天气预报组。

1966年大学毕业时继续读研究生的理想虽然破灭了,但我坚信学习的机会一定会再来。1977年秋,《人民日报》发出中国科学院恢复研究生招生的消息,第二天我就报了名。1978年,经过初试、复试、面试,我成为中国科学院大气物理研究所叶笃正先生的研究生,我12年前未遂的理想终于实现了。接着又被选拔参加并通过出国留学研究生考试。同年秋天,我离开奋斗十年的大西北,踏上了新的征程。"十年磨一剑",机会总是属于有准备的人。

1979年秋,我到达英国,经过了短期英语培训,从1980年开始,我进入伦敦大学帝国理工学院物理系,师从格林(Green)博士进行气候动力学的研究,主攻大尺度地形和热源对大气环流的影响,证明了角动量、位涡及能量守恒在这种影响过程中的重要性。20世纪70年代初,德国的爱德曼(Edelmann)教授曾用正压模式进行数值试验,发现初始时刻的西风气流遇到山脉阻挡后会逐渐转变为东风气流。这与实际情况相去甚远,就像流水流经石头时怎么样也不会倒流回去一样。但是,不管爱德曼如何改变山脉形态始终无法改变这一不合理的结果,其中原因又不得而知,从而成了一个"爱德曼谜":谓之"正压地形波不稳定"现象。我的研究证明,这其实是角动量平衡的约束在正压模型中被歪曲所致,解开了"爱德曼谜"。在此基础上,进一步证明存在一个理论临界地形高度。当地形低于此临界高度时,其对气流的影响以爬坡效应为主;当地形高于此高度时,其作用以绕流效应为主。我的论文还阐述了地形和热源外强迫作用及涡度输送的内强迫作用,以及它们之间的非线性效应在大气环流平均态形成中的重要作用,该成果在美国《大气科学杂志》发表时,被美国科学院狄金森(Dickinson)院士评价为杰出的、具有指导性的工作。

1983年7月我获得了伦敦大学物理学博士学位。那时正值欧洲中期天气预报中心(ECMWF)致力于研究地形效应,我被邀请到该中心进行数值预报试验研究。ECMWF是在1976年开始筹建的,1979年投入业务运行,1980年其中期预报准确率就超过各国,跃居世界第一位至今。其成功的经验在于按照"气象之父"——Rossby的教导,把科学家高度组织起来,从而集中了大家的智慧,取得了高速发展。在这个世界顶尖的研究中心工作一年多,我接触了许多世界一流的科学家,学到了一流的管理经验,为我后来从事国家重点实验室管理工作提供了借鉴。

20世纪80年代,美国就在全世界广招人才,英国的博士毕业生可以轻易进入美国职场,部分中国留英学生也已经渡洋西去。但仍有相当一部分还是主动回国为科学的春天建功立业。记得在我即将取得博士学位的那一个圣诞节,十多名留英的学生团聚在伦敦我的住所,使馆的领导也参加了我们的联欢。我们憧憬着祖国科学的春天,尽管深知回国后待遇很低,工作条件很差,我们还是相互鼓励,下定决心学成后回国为祖国的科学发展贡献力量。1984年末我告别了生活五年多的英国,与还珠一起回到祖国,全家定居北京。还珠在国家气象中心工作,我则回到中国科学院大气物理研究所。1985年起我被聘为研究员,从事天气气候动力学研究,从此有了相对稳定的研究环境。生活和工作条件的确比在英国下降了许多。我和还珠每月都是86元工资,五人挤在集体宿舍中,在集体食堂用餐。白天协助所长进行开放实验室(1985年建立)工作,晚上才是自己的研究时间。1987年因为劳累我得了心脏病。但是每当想起在西北严寒的冬天,衣不遮体的孩子们蜷缩在土墙下晒太阳的情景;想起在天府之国遇到满头白发的老太太携着小孙女,拿着公社介绍信要饭而又羞于张口的凄凉景象,我自不敢苟生,希望用自己所学为实现国强民富做出点滴贡献,我和同事们默默无闻地奋斗着。

1989年,我应聘到美国普林斯顿大学任高级访问研究教授(Senior Visiting Research Professor)。我工作的地球流体力学实验室(GFDL)有许多国际大气科学界的权威,如诺贝尔奖获得者、"温室气体效应之父"Manabe、"中期数值天气预报之父"Miyakoda、台风专家Kurihara、动力气象专家Held、资料专家Oort、ENSO专家Philander、海洋模拟专家Bryan等。有机会与他们一起切磋和生活,受益匪浅。GFDL是一个研究中心,其组织形式与业务中心ECMWF又不相同。开放、流动、联合和竞争是其发展的根本保障,也正是我国国家重点实验室建设的形式。我和还珠在这样的环境中工作,不仅在业务和科研成果上收益很大,而且还学到了现代科研管理的先进方法。

1990年,美国总统发布命令,为所有在美国的中国人签署绿卡。鉴于GFDL老一代科学家已近退休年龄,中年科学家很少,其主任Malhman博士找我谈话,挽留我们在GFDL工作,并主动要为我们办绿卡。我感谢了他的好意,也对他说,他的好朋友叶笃正先生是我的老师,他之所以得到中、美和世界大气科学界的敬仰(叶笃正先生是年被选为美国气象学会名誉会员),不仅因为他是一位伟大的科学家,更重要的是因为他是一位爱国者(叶先生1951年从美国回国定居),为发展中国和世界的大气科学做出了重要贡献。我告诉他,中国还很穷,中国的大气科学还落后,我吮吸着穷苦大众的乳汁长大,必须将所学首先奉献给他们。Malhman 充分理解我们的选择，在 1991 年夏天我们结束工作即将起程回国之际，为我们举行了送别午餐，预祝我们在发展中国大气科学上做出新贡献。

1991 年夏天，我和还珠又回到祖国，回到各自的单位。从 1993 年起，我开始接任大气科学和地球流体力学数值模拟国家重点实验室（LASG）主任。我除了把自己所学奉献给酷爱的事业，又尝试着把西方先进的管理方法引进实验室的运行机制中。酸甜苦辣皆有，挑战与机遇并存。其实，不管是科研和管理，我们都是在创新的路上与同事们不断地探索，庆幸者的我们毕竟迈出了第一步。

1984 年底从 ECMWF 回国时恰逢国内"科学之春"时期，1985 年成立"国家自然科学基金委员会"，1986 年我得到第一届自然科学基金大气科学学科最大的一项资助（5 万元），开始与年轻学者一起涉足前沿科学研究，包括波-流相互作用、E-P 通量和余差环流、平均经圈环流维持机制以及大气运动的多平衡态等，为年轻学者如饥似渴探索自然奥秘的精神深深感动。20 世纪 90 年代初，我们继续探索印度洋海-气相互作用与 ENSO 的联系及其对气候的影响。20 世纪 90 年代中期，根据副热带高压系统在天气、气候及其变化中的重要性和对其认知的不足，我们的研究团队和中国气象局丑纪范先生的团队紧密合作，共同主持了两期有关副热带高压系统形成与变异的重点基金项目，揭示了外部热力强迫和内部动力强迫在西太平洋副热带高压和南亚高压的形成与变异中的不同作用；证明副热带高压和降水之间存在着相互作用等，把基础研究和业务应用联系起来。丑纪范先生深邃的思维和"不记人小过，不念人旧恶，不发人隐私"的宽阔胸怀令我终生难忘。21 世纪初，为了推动我国大气科学和海洋科学的交叉联合，受中国科学院地学部和国家自然科学基金委员会地学部的委托，我和国家海洋局的苏纪兰先生相继组织了两期"双清论坛"，并组织了中国科学院知识创新工程重要方向项目"亚洲季风区海-陆-气相互作用对我国气候变化的影响"和科学技术部 973 项目"亚印太交汇区海-陆-气相互作用对亚洲季风的影响"。通过大家的共同努力，推动了海洋科学和大气科学的交融发展。苏纪兰先生严谨的科学态度和乐观开朗的性格给我们留下了深刻的印象。为了进一步认识青藏高原大地形对我国和全球的天气气候的影响，2012 年中国气象局制定了加强青藏高原观测体系的宏伟计划。2013 年由富有战略思想的周秀骥先生发起、我和徐祥德等共同组织的国家自然科学基金委员会为期十年的重大研究计划"青藏高原地-气耦合系统变化及其全球气候效应"，总共资助了 9 个集成项目、33 个重点项目和 45 个培育项目。通过大家的共同努力，我们对青藏高原地-气系统的耦合过程的认识、对相关的资料处理和数值模拟水平都有了显著提高；发现了许多青藏高原影响区域和全球的大气和海洋环流以及气候的新事实，并揭示了相关的影响机制。取得的成果充分显示了重大科学研究计划通过把科学家组织起来开展综合性和交叉性的重大科学研究的重要性。在组织这个重大科学计划的同时，在中国科学院和国家自然科学基金委员会的联合支持下，我和同事以及学生们一起又挑战另一个前沿课题"地表位涡制造及其天气气候效应"，发现青藏高原地表位涡制造影响局地和下游环流和气候的机制，得到青藏高原地表位涡通量和位涡制造影响天气气候变化的新指标，享受到获得新成果的乐趣。

回味自英国归来后的生活和工作，深深体会到，当一个人的志愿和爱好融合到人民的事业之中，犹如滴水回归大海，虽历经惊涛骇浪，但是其乐无穷！

感谢我的同事对我的研究工作进行归纳和剖析。其实科学研究也是后浪推前浪，永无止境。期望本书的出版能对广大年轻学者有所裨益。

吴国雄

2022 年 6 月于北京

本文修改自 1999 年本人为北京出版社出版的《中国中青年院士文集》的《自序》

前言——编者的话

半个多世纪以来,吴国雄先生始终矢志不渝,全身心地投入到所热爱的大气科学事业,取得了世人瞩目的科研成就。先生的研究成果涵盖了大气动力学、气象学、气候学以及全球变化等诸多领域,包括大气动力学的理论和气候系统模式发展、天气气候预测应用研究以及现代气候动力学的热点科学问题等。时值先生80寿诞,学生们萌发了将先生的科研成果进行系统性整理、概括撰文成册的想法,倡议得到了先生的允准和支持。随即,由工作在国内外多个科研院所及业务部门的学生们组成了撰稿团队,分成13个撰写小组分别负责5个研究方向的成果概括和相关科研进展综述,并定书名为《副热带环流、亚洲季风以及青藏高原影响的若干动力学问题》,现已经完稿,即将交付出版社。希望通过本书的出版,使更多的大气科学科研和业务工作者全面系统地了解吴国雄先生的科研成就,铭记先生对科学事业的卓越贡献,同时,本书也可作为大气科学相关专业高等院校研究生的辅助教材。

根据吴国雄先生400多篇科研论文以及数部科学专著所涉及的科学问题,本书分别对5个科学主题进行概括、总结和展示。每个部分首先通过2~3篇综述性文章系统地概括该主题的研究进展以及吴国雄先生的主要科学贡献,然后通过4~7篇代表性论文的全文,进一步展示先生的科学贡献。本书还展示了先生早年以来的一些学习、工作、科研交流和生活图片,作为先生日常工作和生活的缩影,供读者鉴赏。

其中,第1部分"大气环流与数值模拟的若干基本问题"概括了先生在大气环流动力学、海-气相互作用以及数值模式系统发展方面的核心成果,3篇综述文章分别由虞越越等、任荣彩等和包庆等执笔撰写。第2部分"副热带高压的若干基本问题"概括了先生在副热带环流动力学方面的核心贡献,2篇综述文章分别由刘屹岷等和张琼等执笔撰写。第3部分"青藏高原气候效应的若干基本问题"概括了先生在青藏高原动力学方面的核心成就,2篇综述文章由段安民等、何编等执笔撰写。第4部分"亚洲季风的若干基本问题"概述了先生在季风动力学方面的核心科学贡献,3篇综述文章分别由毛江玉等、刘伯奇等和万日金等执笔撰写。第5部分"位涡及位涡源的若干基本问题"概括总结了先生在位涡动力学理论和应用方面的科学贡献,3篇综述文章分别由崔晓鹏等、何编等和马婷婷等执笔撰写。所展示的大部分图片由吴国雄先生提供,由刘屹岷、任荣彩和王军等筛选整理。

本书得以顺利出版,首先感谢吴国雄先生的允准、大力帮助和指导!其次感谢各位执笔人以及所有参与写作或提供图表材料的同志们!感谢刘博文等13位研究生整理录入30篇代表性论文!并感谢中国科学院大气物理研究所大气科学和地球流体力学数值模拟国家重点实验室(LASG)的支持!最后,感谢气象出版社黄红丽等各位编辑的大力帮助和支持。本书出版还得到了广东省基础与应用基础研究重大项目(2020B0301030004)和国家自然科学基金项目(91937302、42288101、42075052)的联合资助。

任荣彩　刘屹岷　毛江玉　段安民　等
2022年6月于北京

目 录

吴国雄先生简历

吴国雄先生工作生活图片集锦

丑纪范先生序言

周秀骥先生序言

吴国雄先生序言

前言——编者的话

第 1 部分 大气环流与数值模拟的若干基本问题

综述 1 各种大气经向环流理论模型及其相互关系 ………………………………………………
…………………… 虞越越，任荣彩，叶万恒，林琳，陈怡然，张伶俐，罗彦恒，杨漉茹，柴佳颖(3)

综述 2 ENSO 与印度洋相互作用的事实和过程 …………………… 任荣彩，宋歌，孙舒悦，李倩，孟文(17)

综述 3 气候系统模式 FGOALS 发展历程 …… 包庆，何编，王晓聪，李剑东，周林炯，吴小飞，李矜霄(25)

代表性论文 1 A GCM Simulation of the Relationship between Tropical-Storm Formation and ENSO
………………………………………………………… WU Guoxiong, LAU Ngar-Cheung(44)

代表性论文 2 赤道印度洋-太平洋地区海气系统的齿轮式耦合和 ENSO 事件 I. 资料分析 …………
………………………………………………………………………………… 吴国雄，孟文(69)

代表性论文 3 大气热力强迫和动力强迫的调配及平均经圈环流的仿真模拟 ……… 吴国雄，蔡雅萍(78)

代表性论文 4 A Further Study of the Surface Zonal Flow Predicted by an Eddy Flux
Parametrization Scheme ……………………………… WU Guoxiong, WHITE Andrew Arthur(87)

第 2 部分 副热带高压的若干基本问题

综述 4 副热带高压的形成和年际变异 ………………………………………… 刘屹岷，刘平，任荣彩(105)

综述 5 南亚高压的双模态和年际及年代际变化 ………………………………… 张琼，刘屹岷，任荣彩(122)

代表性论文 5 影响大气涡度发展的若干热力过程——I. 热力适应过程 ……… 吴国雄，刘屹岷(131)

代表性论文 6 Summertime Quadruplet Heating Pattern in the Subtropics and the Associated
Atmospheric Circulation ……………………………………… WU Guoxiong, LIU Yimin(146)

代表性论文 7 Relation between the Subtropical Anticyclone and Diabatic Heating …………
………………………………………………………… LIU Yimin, WU Guoxiong, REN Rongcai(154)

代表性论文 8 Multi-Scale Forcing and the Formation of Subtropical Desert and Monsoon ……………
………………… WU Guoxiong, LIU Yimin, ZHU Xiaying, LI Weiping, REN Rongcai,
DUAN Anmin, LIANG Xiaoyun(176)

代表性论文 9 The Bimodality of the 100 hPa South Asia High and Its Relationship to the Climate
Anomaly over East Asia in Summer ………… ZHANG Qiong, WU Guoxiong, QIAN Yongfu(196)

代表性论文 10　Genesis of the South Asian High and Its Impact on the Asian Summer Monsoon Onset
　………………………………………… LIU Boqi, WU Guoxiong, MAO Jiangyu, HE Jinhai(209)

代表性论文 11　Location and Variation of the Summertime Upper-Troposphere Temperature Maximum over South Asia ………… WU Guoxiong, HE Bian, LIU Yimin, BAO Qing, REN Rongcai(228)

第 3 部分　青藏高原气候效应的若干基本问题

综述 6　青藏高原对亚洲气候格局的影响 …… 段安民,王在志,张萍,王子谦,卓海峰,刘新,李伟平(255)

综述 7　青藏高原对亚洲夏季风的关键调制作用 ………………… 何编,刘屹岷,包庆,梁潇云,毛江玉(278)

代表性论文 12　青藏高原感热气泵和亚洲夏季风 …… 吴国雄,李伟平,郭华,刘辉,薛纪善,王在志(297)

代表性论文 13　Tibetan Plateau Forcing and the Timing of the Monsoon Onset over South Asia and the South China Sea ………………………………………… WU Guoxiong, ZHANG Yongsheng(307)

代表性论文 14　The Influence of the Mechanical and Thermal Forcing of the Tibetan Plateau on the Asian Climate ………… WU Guoxiong, LIU Yimin, WANG Tongmei, WAN Rijin, LIU Xin, LI Weiping, WANG Zaizhi, ZHANG Qiong, DUAN Anmin, LIANG Xiaoyun(327)

代表性论文 15　Astronomical and Hydrological Perspective of Mountain Impacts on the Asian Summer Monsoon ………………………… HE Bian, WU Guoxiong, LIU Yimin, BAO Qing(350)

代表性论文 16　Weakening Trend in the Atmospheric Heat Source over the Tibetan Plateau during Recent Decades. Part I: Observations ……………… DUAN Anmin, WU Guoxiong(365)

代表性论文 17　The Role of Air-Sea Interactions in Regulating the Thermal Effect of the Tibetan-Iranian Plateau on the Asian Summer Monsoon ……………………………………………
　………………………………………… HE Bian, LIU Yimin, WANG Ziqian, BAO Qing(385)

第 4 部分　亚洲季风的若干基本问题

综述 8　亚洲季风的气候特征和本质 ……………………………………………… 刘伯奇,毛江玉(413)

综述 9　亚洲夏季风爆发的阶段性及其物理机制 ……………………… 毛江玉,刘伯奇,颜京辉(423)

综述 10　华南春雨的气候成因和季节内振荡 ……………………… 万日金,潘蔚娟,毛江玉(435)

代表性论文 18　Air-Sea Interaction and Formation of the Asian Summer Monsoon Onset Vortex over the Bay of Bengal ………………………………………………
　………… WU Guoxiong, GUAN Yue, LIU Yimin, YAN Jinghui, MAO Jiangyu(446)

代表性论文 19　Interannual Variability in the Onset of the Summer Monsoon over the Eastern Bay of Bengal ……………………………………… MAO Jiangyu, WU Guoxiong(471)

代表性论文 20　Thermal Controls on the Asian Summer Monsoon ………………………………
　………… WU Guoxiong, LIU Yimin, HE Bian, BAO Qing, DUAN Anmin, JIN Feifei(489)

代表性论文 21　Roles of Forced and Inertially Unstable Convection Development in the Onset Process of Indian Summer Monsoon ……………………… WU Guoxiong, LIU Boqi(502)

代表性论文 22　Asian Summer Monsoon Onset Barrier and Its Formation Mechanism ……………
　………… LIU Boqi, LIU Yimin, WU Guoxiong, YAN Jinghui, HE Jinhai, REN Suling(521)

代表性论文 23　The Nature of the Thermal Forcing of the Asian Summer Monsoon ……………
　………………………………………… WU Guoxiong, LIU Yimin, HE Bian(543)

第5部分 位涡及位涡源的若干基本问题

综述 11　倾斜涡度发展理论和全型涡度方程及应用 ················· 崔晓鹏,姚秀萍,郑永骏(555)

综述 12　地表位涡制造、位涡环流特征及在气候动力学中的应用 ············· 何编,生宸,谢永坤(565)

综述 13　垂直运动方程以及青藏高原位涡强迫的天气效应 ···
················· 马婷婷,刘屹岷,毛江玉,张冠舜,汤艺琼(579)

代表性论文 24　湿位涡和倾斜涡度发展 ················· 吴国雄,蔡雅萍,唐晓菁(590)

代表性论文 25　影响大气涡度发展的若干热力过程——Ⅱ. 全型涡度方程 ·······················
················· 吴国雄,刘还珠,刘屹岷,刘平(605)

代表性论文 26　Eurasian Cooling Linked with Arctic Warming: Insights from PV Dynamics ···········
················· XIE Yongkun, WU Guoxiong, LIU Yimin, HUANG Jianping(620)

代表性论文 27　PV-Q Perspective of Cyclogenesis and Vertical Velocity Development Downstream of
the Tibetan Plateau ················· WU Guoxiong, MA Tingting, LIU Yimin, JIANG Zhihong(642)

代表性论文 28　Characteristics of the Potential Vorticity and Its Budget in the Surface Layer over
the Tibetan Plateau ················· SHENG Chen, WU Guoxiong, TANG Yiqiong, HE Bian,
XIE Yongkun, MATingting, MA Ting, LI Jinxiao, BAO Qing, LIU Yimin(671)

代表性论文 29　Quantification of Seasonal and Interannual Variations of the Tibetan Plateau
Surface Thermodynamic Forcing Based on the Potential Vorticity ·······················
················· HE Bian, SHENG Chen, WU Guoxiong, LIU Yimin, TANG Yiqiong(691)

代表性论文 30　Linkage between Cross-Equatorial Potential Vorticity Flux and Surface Air
Temperature over the Mid-High Latitudes of Eurasia during Boreal Spring ···············
················· SHENG Chen, WU Guoxiong, HE Bian, LIU Yimin, MA Tingting(703)

附录　吴国雄先生论文、著作列表 ················· (725)

第1部分

大气环流与数值模拟的若干基本问题

综述 1

各种大气经向环流理论模型及其相互关系

虞越越[1]，任荣彩[2]，叶万恒[1]，林琳[1]，陈怡然[1]，张伶俐[1]，罗彦恒[1]，杨滟茹[1]，柴佳颖[1]

(1 南京信息工程大学大气科学学院/气象灾害教育部重点实验室(KLME)/气象与环境联合研究中心(ILCEC)/气象灾害预报预警与评估省部共建协同创新中心(CIC-FEMD)，南京 210044；2 中国科学院大气物理研究所，北京 100029)

摘要：大气经向环流是完成地球-大气系统角动量、热量和水分的输送与收支平衡以及各种能量间转换的重要机制，与天气和气候的形成密切关联。文中介绍了大气经向环流的观测事实，回顾了各种大气经向环流理论模型及其相互关系。以气压作为垂直坐标的大气环流模型包括欧拉平均环流、变形欧拉平均环流等模型。欧拉平均环流呈三圈结构，低纬度和高纬度地区的直接环流圈（哈得来环流和极地环流）由非绝热加热的非均匀性驱动，中纬度间接环流（费雷尔环流）则由涡动热量和动量输送驱动。从欧拉平均环流分离出的变形欧拉平均环流可近似认为是拉格朗日观点下的平均环流，其呈单圈结构，由非绝热加热和波动通量散度共同驱动，可表征等熵空气块的运动轨迹。此外，以位温或相当位温作为垂直坐标的大气环流模型——大气经向质量环流和大气经向角动量环流模型，同样是呈半球尺度的单圈环流型，可直观表示冷/暖、干/湿空气的经向传输，自然区分绝热动力过程与非绝热加热过程对大气质量、能量和角动量传输的贡献，已成为理解大气环流和天气气候演变的重要工具。

1 引言

大气经向环流是完成地球-大气系统角动量、热量和水分的输送与收支平衡以及各种能量间转换的重要机制，同时又是这些物理量输送、收支和转换的重要结果。环流往往引导不同性质的气团活动，并影响锋、气旋和反气旋等天气系统的产生和移动，与天气和气候的形成紧密相关。因此，研究大气经向环流的特征及其形成、维持、变化和作用，对提高天气预报和气候预测水平具有重要意义。本文将从大气经向环流的观测事实出发，回顾各种大气经向环流的理论模型及其相互之间的关系，尝试阐明其对实际大气经向环流的再现能力和适用情境，包括以气压作为垂直坐标的大气经向环流模型，如欧拉平均环流和变形欧拉平均环流，以及以位温或相当位温作为垂直坐标的大气经向质量环流和大气经向角动量环流。

2 大气经向环流的观测事实

南、北半球近地面往往呈现"三风四带"环流特征，即三个纬向风带——极地东风带、中纬盛行西风带、低纬东风带以及四个纬向气压带——极地高压带、副极地低压带、副热带高压带、赤道低压带，且高低气压带呈交叉排列（图1）。具体地，在低纬度地区，近地面为稳定的信风带，北半球为东北信风，南半球为东南

通讯作者：虞越越，yuyy@nuist.edu.cn
资助项目：国家自然科学基金项目 42075052

信风。南、北半球两支信风交汇的地带称为赤道辐合带或热带辐合带、赤道低压区或赤道槽或赤道无风带。中纬度地区盛行西风,其在北半球很不稳定,常伴有波动、气旋、锋面等活动,夏季近地面西风很弱;而南半球西风带风速较北半球更强,风向更稳定。南北纬30°附近为副热带高压带。高纬度地区在冬、夏均为较浅薄的弱东风,称为极地东风带,其在北半球主要分布在北大西洋低压和北太平洋低压向极一侧,多变化,而在南半球主要分布在南极洲沿海的印度洋地区,较稳定。副极地低气压带位于南北纬60°附近,极地高压带位于南北两极地区。近地面风带与气压带均具有显著的季节变化:从冬到夏,南/北半球风带与气压带南/北移,其中信风带范围扩展,强度增强,盛行西风带和极地东风带强度减弱;从夏到冬,南/北半球风带与气压带北/南移,其中信风带范围缩小,强度减弱,盛行西风带和极地东风带强度增强。

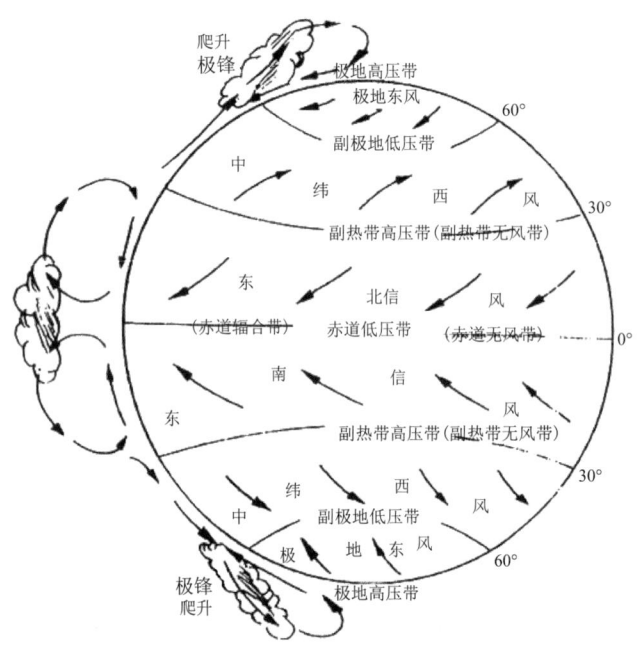

图1 全球大气环流示意图(摘自卜永芳和韩明娟(1987))

对流层中高层以大范围西风为主,仅低纬度地区为东风带。在中、高纬度地区出现行星锋区及与之相应的急流中心。具体地,在低纬度地区,高层为东风,其在对流层内所跨纬距随高度增加而变窄。从冬到夏,东风带北移,范围扩大,强度增大,与近地面信风带的季节变化特征类似。中纬度地区,自地面到对流层顶均盛行西风,其所跨纬距随高度增加而变宽,强度也随高度增加而增强;在副热带高压北缘对流层高层(~200 hPa)存在副热带西风急流,往往与其下的副热带锋区紧密相连。中高纬度地区,对流层高层(~300 hPa)存在极锋急流,位于极锋上空,急流中心附近风速一般为45~55 m·s^{-1},甚至可达105 m·s^{-1}。副热带西风急流和极锋急流的中心强度和位置均存在显著的季节变化:副热带西风急流轴在南/北半球冬季位于南/北纬20°~30°,夏季位置南/北移,高度稍有降低;极锋急流在南/北半球冬季的平均位置是南/北纬40°~60°,夏季南/北移至60°~70°N;冬季强于夏季。

1929年德国科学家Brewer和Dobson通过观测发现,平流层存在一个半球尺度的经向环流,后被称为BD环流(Brewer-Dobson circulation),又称热带外气泵(Dobson et al., 1929;Brewer, 1949)。BD环流主要包括从热带对流层进入平流层的上升运动、平流层中的向极运动以及中高纬度地区的下沉运动。BD环流的发现成功解释了平流层水汽含量较低以及高纬度臭氧浓度比热带地区更高的原因,随后的观测数据进一步证实了这个结果(Appenzeller et al., 1996;Bonazzola and Haynes, 2004;Rind et al., 2001;Cordero et al., 2002)。关于BD环流的形成机制,普遍认可的一个驱动因子是大气行星波的发展和耗散(Scott and Polvani, 2004)。由于对流层行星波强迫在南、北半球和不同季节的差异,BD环流在北半球强于南半球,冬季强于夏季(田文寿 等, 2020)。BD环流对臭氧及其他痕量气体的分布特征、混合作用及其气候效应具有重要影响(Stohl et al., 2003;Sigmond et al., 2004;Mclandress et al., 2010)。

3 气压坐标下的大气经向环流理论模型

3.1 欧拉平均环流(Eulerian-mean circulation)

欧拉平均环流是以气压为垂直坐标轴的一种理想大气环流模型。欧拉平均环流的结构特征如图2所示,它主要由三个经向子环流圈组成:哈得来环流、费雷尔环流以及极地环流。其中位于低纬度的哈得来环流与高纬度的极地环流是直接热力环流圈,即气流在较暖处上升,在对流层上部向较冷处流动,在较冷处下沉,在低层由冷处回流向暖处,从而构成闭合环流圈;位于中纬度地区的费雷尔环流则是一个间接环流圈,与直接环流圈方向相反。欧拉平均环流取决于四个驱动因子:非绝热加热、纬向外强迫、涡动热通量散度的经向差异以及涡动动量通量散度的垂直差异。以下将围绕这四个因子,以哈得来环流和费雷尔环流为例介绍直接和间接环流圈的可能形成机制。

3.1.1 哈得来环流

哈得来环流最早被认为是一个半球尺度的单圈环流,与科里奥利力的偏转效应共同作用形成了低纬度的信风带(Hadley,1735)。普遍认为哈得来环流的形成主要与太阳辐射的纬度差异有关。太阳辐射引起的非均匀非绝热加热可以驱动一个纬向对称经圈环流,在热源(热带)处为上升支、热汇(中高纬度)处为下沉支。然而,观测表明这种热力直接环流主要存在于热带地区,无法横跨半球。这是因为哈得来环流具有斜压不稳定性,即在斜压波动发展旺盛的热带外地区,由于斜压波伴随的热量和动量输送对环流具有显著的调控作用,哈得来环流会迅速消失。

吴国雄(1988a)通过对热带地区的热量和角动量的定量收支分析,说明了哈得来环流的形成和维持机制。具体地,他指出热带地区的非绝热加热至关重要,可与哈得来环流垂直上升支的绝热冷却作用相平衡。Wu和Cubalsch(1985)利用欧洲中期天气预报中心高分辨业务谱模式通过数值模拟研究也证明,厄尔尼诺(El Nino)持续的赤道东太平洋海温距平所导致的非绝热加热异常,与哈得来环流之间存在正反馈,可显著加强哈得来环流。角动量收支分析(White,1949;Priestley,1951;叶笃正 等,1955;Kung,1968;Newton,1971a;吴国雄,1988a)表明,热带近地层为角动量源区,近地表角动量的制造主要由向赤道的水平支惯性矩来平衡,其上自由大气中向极地的水平支惯性矩则由自由大气中涡动角动量通量的辐散来平衡。图2是根据上述成果归纳的三圈环流中热量和角动量收支的主要贡献示意。

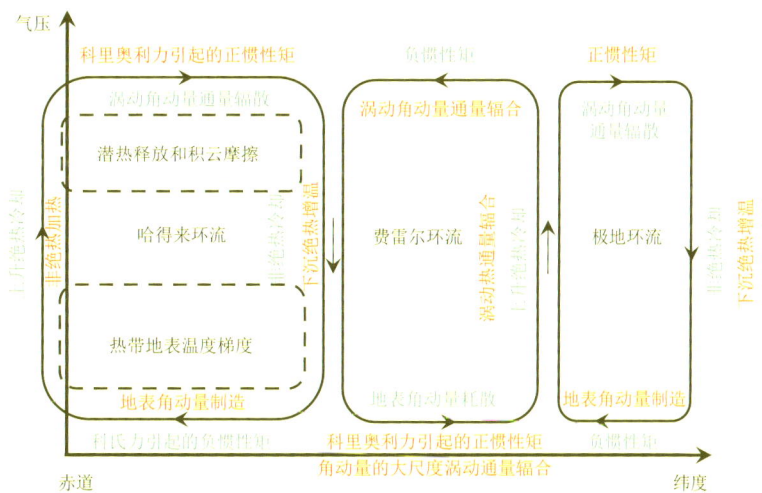

图2 基于吴国雄(1988a)全球角动量收支及热量收支分析结果归纳的三圈环流主要贡献示意
(红色字样表示正收支,蓝色表示负收支)

吴国雄和刘还珠(1987)和吴国雄(1988b)还发现,低纬度的哈得来环流在垂直方向上存在双圈结构,这也随后被 Hoskins(1989)的计算所证实。该双圈结构只出现在冬半球,并从冬到夏呈现规律性的变化(图略)。Schineider 和 Linzen(1977)、吴国雄和蔡雅萍(1994)认为,哈得来环流的双圈结构与直接环流的多种驱动机制有关:近地层直接环流由热带地表温度梯度所激发,而高层的直接环流则主要由热带深对流导致的潜热释放和积云摩擦所驱动。

3.1.2 费雷尔环流

与哈得来环流不同,位于中纬度地区的费雷尔环流是斜压波驱动的结果。即非绝热加热作用较小,涡动热通量和涡动动量通量散度起主导作用。从热力角度来看,热通量基本是向极地的,在温差最显著的对流层低层南、北纬50°附近,热通量达到极大值,其靠极地一侧有涡动热通量辐合,其靠赤道一侧有涡动热通量辐散。为平衡涡动热通量引起的热量差异,在其靠极地一侧需要冷却,于是形成绝热上升运动,而在其靠赤道一侧需要加热,于是形成绝热下沉运动,形成一个间接环流圈。这一机制被吴国雄(1988a)的热量收支分析所证明(图2)。热带外地区的涡动热通量辐合(散)与费雷尔环流的绝热冷却(加热)大小相当,控制着中高纬度地区的热量收支平衡。北半球,行星波和天气波对感热的涡动输送同等重要,而在南半球,天气波支配着感热的涡动输送。后来,张韬等(2002)利用 GOALS-5 全球海-陆-气耦合模式,说明了 ENSO 对大气经向环流及其涡动热通量的影响,发现 ENSO 暖事件发生时定常波动的经向涡动热通量变化,是北半球对流层热带外地区温度异常的主要原因,而瞬变波的影响起抵消作用,证实了北半球行星波的重要作用。

从动量角度来看,涡动动量通量方向基本也是向极地的,其在副热带急流所在的对流层顶南、北纬30°附近达到极大值,在其靠极地一侧均有涡动动量通量辐合。高层远比低层显著。为平衡涡动动量通量引起的纬向动量和角动量差异,北半球/南半球30°以北/以南需要有向东的纬向平均科里奥利力,即在高层有纬向平均的向赤道水平支。而在低层,有用于平衡地表角动量耗散的向极地回流支。吴国雄(1988a)的收支分析结果的确表明,在中纬度自由大气中,涡动角动量通量的辐合确实是由向赤道的水平支惯性矩来平衡(图2)。有研究(White,1949;Priestley,1951;叶笃正 等,1955;Kung, 1968;Newton,1971a)发现,北半球行星波(波数1~3)对西风角动量的涡动输送与天气波(波数4~9)输送相当,而在南半球是天气波更为主导;行星尺度输送的最大值在上对流层和平流层,天气尺度输送的最大值中心局限于对流层顶附近。在中纬度近地层,角动量的耗散则主要由极地的水平支惯性矩来平衡,但纬向角动量的大尺度涡动通量辐合对地表纬向流抵抗地表摩擦得以维持也十分重要(Lorenz,1967;Kidson et al.,1969;Oort et al.,1971,1974;Newton,1971b,1972),因此地表纬向风的模拟水平可用于评估统计气候模型中大尺度涡动通量参数化方案的优劣(Schneider et al.,1974;Saltzman,1978;Wu and White,1986)。

总之,气压坐标下大气经圈环流的形成不仅与二维空间中的纬向对称环流有关,还与三维空间中的辐射与动力过程相互作用紧密相关。低纬度和高纬度的直接环流圈主要由辐射引起的热带-极地温度差驱动;而这一温度差同时又使得中纬度地区的纬向平均有效位能达到最高,有利于斜压波的发展,斜压波伴随向极的涡动热量输送可平衡辐射带来的热力差异,同时斜压波伴随的涡动动量输送可将有效位能转化为动能,进而平衡近地摩擦对大气动量的耗散。

欧拉平均环流理论框架的弱点在于,其基于准地转近似,因此涡动和纬向平均流对热量和动量收支的作用总是接近相互抵消,很难定量表征涡动强迫的净作用。连续方程中的涡动和纬向平均流相关项也几乎抵消,示踪物传输计算也较难在欧拉平均环流框架下进行。此外,因为气压坐标下的地转分量纬向平均后为0,欧拉平均环流反映的仅是非地转分量;也就是说,欧拉平均环流的特征较难直观地表达包含地转分量的实际大气经向环流。实际上,中纬度地区费雷尔环流结构与实际可以观测到的环流特征并不相符。

3.2 变形欧拉平均环流(transformed Eulerian-mean circulation,TEM)

由于涡动热通量辐合与绝热冷却效应总是接近相互抵消,非绝热加热是很小的剩余项,而对于大气环流中的一个特定气块而言，只有非绝热加热才能改变其热量收支，从而达到新的热平衡。于是，Andrews 和 McIntyre(1976)将气压坐标下的纬向平均经向运动速度和垂直运动速度分别减去涡动热通量相关项和涡动动量通量散度对应的绝热运动部分，得到了与非绝热加热过程相关的运动部分。这一剩余运动部分组成的经向环流叫作变形欧拉平均环流，也称为剩余环流（$\tilde{v} = [v] - (\Theta_p^{-1}[v^* \theta^*])_p$，$\tilde{\omega} = [\omega] + (a\cos\varphi)^{-1}(\Theta_p^{-1}[v^* \theta^*]\cos\varphi)_\varphi$）。变形欧拉平均环流可近似认为是拉格朗日观点下的平均环流，可表征等熵空气块的运动轨迹。变形欧拉平均环流在平流层的部分通常被用于表征平流层 BD 环流，研究平流层的物质输送问题。

变形欧拉平均环流的冬季气候平均分布特征如图 3 所示。与欧拉平均环流的三圈特征不同，变形欧拉平均环流是一个半球尺度的单圈环流，即在热带地区有非绝热加热上升运动，对流层中高层向极运动，在高纬度地区非绝热冷却下沉，在对流层低层向赤道回流。变形欧拉平均环流的强度和范围具有显著的季节变化：从冬季到夏季，北半球单圈环流的南边界北移至赤道北侧，范围缩小，强度减小；从夏季到冬季，北半球单圈环流的南边界南移至赤道南侧，范围扩展，强度增强。

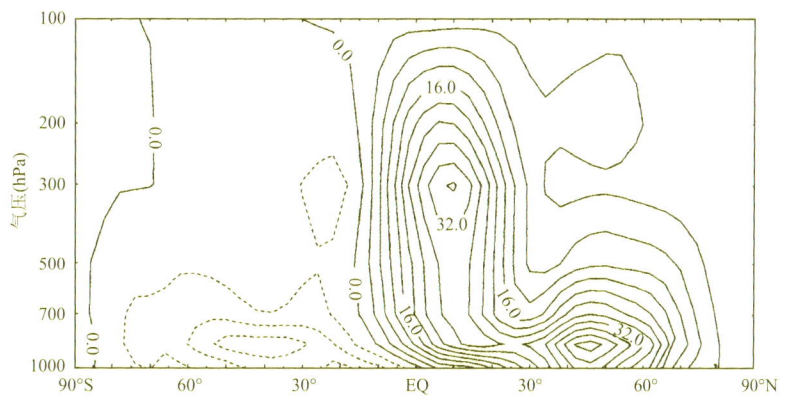

图 3　以冬季平均的质量流函数表征的冬季变形欧拉平均环流（或剩余环流）

（等值线，单位：10^2 kg·s^{-1}）(摘自 Holton(2004))

变形欧拉平均环流的结构和强度只取决于两个因子——非绝热加热和准地转近似下的 Eliassen-Palm(E-P)波通量散度项。E-P 通量（$E = (E(\varphi), E(p)) = a\cos\varphi(-[u^* v^*], f_0 \Theta_p^{-1}[v^* \theta^*])$）散度项可表征涡动通量散度与热通量散度以特定的组合形式影响大气经向环流。E-P 通量辐散可以看作是一个波动驱动的向东纬向力，E-P 通量辐合则可看作是一个波动驱动的向西纬向力（$[u]_t - f_0 \tilde{v} = (a\cos\varphi)^{-1} \nabla \cdot E + [\zeta]$）。E-P 通量辐合可以解释高层西风急流减速（吴国雄 等，1990b）。Charney 和 Drazin(1961)、Dickinson(1969)和 Holton(1975)、Edmon 等(1980)、吴国雄和陈彪(1988,1990a)等一系列工作进一步证明了 E-P 通量散度项对纬向流的决定作用，提出了无加速理论。无论是二维场合(Charney and Drazin, 1961)或是三维场合（吴国雄 等，1988，1990a），无论是干模型或是湿模型(Stone and Salustri, 1984；吴国雄 等，1988)，无加速理论均被证明是成立的。此外，E-P 通量辐合辐散还可以通过"向下控制原则"在中纬度驱动经向变形欧拉平均环流(Haynes et al., 1991；Holton et al., 1995；Haynes, 2005)。变形欧拉平均环流水平支的分布与 E-P 通量及其散度一一对应。北半球的 E-P 通量及其散度均比南半球强得多。其在近地面 42°N 有辐散中心，主要由天气波贡献；在中纬度 500 hPa 附近有辐合中心，主要由行星波贡献。行星尺度的 E-P 通量及其辐合冬季强于夏季，北半球强于南半球。也就是说北半球行星波的波流相互作用要比南半球强得多。天气尺度波的 E-P 通量散度则在南半球比行星波强，在北半球与行星波相当。E-P 剖面和变形欧拉平均环流是诊断波流相互作用、波动传播和地转位涡输送的有效工具。

吴国雄等(1988)进一步给出了干、湿模型下的 E-P 剖面以及变形欧拉平均环流(图略)，结果表明湿模型中的变形欧拉平均环流与干模型中类似，都是一个单圈环流。两者在高层高度相似，但在低层湿环流更强，中纬度地区的水平支也更加连续。这是因为当把潜热由外强迫源转为与涡动水汽输送相联系的内强迫源后,水汽凝结效应使 E-P 通量及其散度在对流层下部增强了,尤以天气尺度为甚。且无论是行星尺度波段或是天气尺度波段,水汽作用均使低层的辐散中心向极地漂移,使得副热带地区出现明显的辐合。因而水汽涡动的输送过程是热带、副热带低层西风减速的重要因素。需要注意的是,与湿环流的垂直支相对应的加热场不包括非绝热凝结潜热释放,因此不宜用湿模型下的剩余环流去研究热带及近地面与相变加热相联系的一些热力特征。基于上述结果,在表 1 中列出了干、湿模型中变形欧拉平均环流的异同点。

表 1 干模型和湿模型下变形欧拉平均环流的异同

	干模型	湿模型
基本假设	准地转、球面 p 坐标中的动量方程、热成风关系、连续方程和热力方程	除干模型所有方程外,有水汽守恒方程(纬向平均)
方程组表达式	$\begin{cases} [u]_t - f_0 \tilde{v} = (a\cos\varphi)^{-1} \nabla \cdot \boldsymbol{E} + [\zeta] \\ f_0 [u]_p - a^{-1} R^* [\theta]_\varphi = 0 \\ (a\cos\varphi)^{-1} (\tilde{v}\cos\varphi)_\varphi + (\tilde{\omega})_p = 0 \\ [\theta]_t + \Theta_p \tilde{\omega} = [Q_m] \end{cases}$	$\begin{cases} [u]_t - f_0 \tilde{v}_m = (a\cos\varphi)^{-1} \nabla \cdot \boldsymbol{E}_m + [\zeta] \\ f_0 [u]_p - a^{-1} R^* [\theta]_\varphi = 0 \\ (a\cos\varphi)^{-1} (\tilde{v}_m\cos\varphi)_\varphi + (\tilde{\omega})_{mp} = 0 \\ [\theta_e]_t + \Theta_{ep} \tilde{\omega}_m + \sigma'[\omega] = [Q_d] \\ [q]_t + q_{spm}\tilde{\omega} = [S-C] + M \\ \theta_e = \theta + A(p)q \end{cases}$
	湿模型中考虑了水汽的作用,即将凝结热从干模型的外部强迫源 $[Q_m]$ 中转化为湿系统中的内部强迫源。	
E-P 通量	$\boldsymbol{E} = (E(\varphi), E(p))$ $= a\cos\varphi(-[u^*v^*], f_0 \Theta_p^{-1}[v^*\theta^*])$	$\boldsymbol{E}_m = (E_m(\varphi), E_m(p))$ $= a\cos\varphi(-[v^*u^*], f_0 \Theta_{ep}^{-1}[v^*\theta_e^*])$
E-P 通量散度	准地转位涡的涡动输送	除涡动输送外,还包括了大尺度涡动的水汽输送对基本态的内强迫效应
P 坐标系下剩余速度	$\tilde{v} = [v] - (\Theta_p^{-1}[v^*\theta^*])_p$ $\tilde{\omega} = [\omega] + (a\cos\varphi)^{-1}(\Theta_p^{-1}[v^*\theta^*]\cos\varphi)_\varphi$	$\tilde{v}_m = [v] - ([v^*\theta_e^*]\Theta_{ep}^{-1})_p$ $\tilde{\omega}_m = [\omega] + (a\cos\varphi)^{-1}([v^*\theta_e^*]\cos\varphi\Theta_{ep}^{-1})_\varphi$
流函数	$\begin{cases} 2\pi a g^{-1}\cos\tilde{\omega} = -a^{-1}\tilde{\chi}_\varphi \\ 2\pi a g^{-1}\cos\varphi \tilde{v} = \tilde{\chi}_p \end{cases}$	$\begin{cases} 2\pi a g^{-1}\cos\tilde{\omega}_m = -a^{-2}\tilde{\chi}_{m\varphi} \\ 2\pi a g^{-1}\cos\varphi \tilde{v}_m = \tilde{\chi}_{mp} \end{cases}$
无加速理论	成立,即当 $[\zeta] = [Q_m] = \nabla \cdot \boldsymbol{E} \equiv 0$ 时,$[u]_t = [\theta]_t = 0$	成立,即当 $[\zeta] = [Q_d] = \nabla \cdot \boldsymbol{E}_m \equiv 0$ 时,且 $M = [C-S]$ 时,存在解 $\tilde{v}_m = \tilde{\omega}_m = [u]_t = [\theta_e]_t = [q]_t = 0$

需要指出的是,虽然变形欧拉平均环流可以直接表征气块运动,但其本质上是从气压坐标下的三圈环流中人为分离出来的数学环流。它的计算取决于纬向平均速度流和纬向偏差场,它们与气候平均和气候异常一样对基准场有依赖性。另外,变形欧拉平均环流的定义决定了气柱内空气的无辐散性(垂直积分为零),无法反映某区域内大气质量的变化。

4 位温(相当位温)坐标下的大气环流理论模型

4.1 等熵大气经向质量环流

除了大气低层静力不稳定区域以外,大气的位温随高度单调升高,因此位温(或熵)可以作为垂直坐标用于研究大气环流的分布。基于绝热条件下空气位温的守恒性质,穿越等熵面的垂直运动必须由非绝热加热或冷却引起,沿等熵面的运动仅与绝热过程有关,这与变形欧拉平均环流有异曲同工之处(Pauluis et al.,2011)。基于位温坐标,Townsend 和 Johnson(1985)、Johnson(1989)、Held 和 Schneider(1999)等提出了大气经向质量环流的理论框架,其可近似为拉格朗日环流,等熵质量环流相对于气压坐标下的经向环流具有独特性和优越性。利用等熵/位温坐标定义的经向质量环流可以直接、客观地量化各纬度带内的冷、暖空气的经向输送,避免了变形欧拉平均环流在这方面的局限。在做沿等熵面的纬向平均或纬向积分时,质量环流考虑了沿纬圈各经度的空气块质量的差异,包含地转和非地转分量的共同作用,这完全不同于纯粹的欧拉平均经圈速度环流。亦即,对于处于地转平衡的大气,欧拉平均环流的经向速度为零,而对于经向质量环流,由于各纬度带经向速度所携带的空气质量不同,在纬向平均速度为零的情形下依然可以存在穿越纬圈的质量的南北净输送。总之,全球大气经向质量环流对大气经向物质输送的描述比欧拉平均经向环流更客观、更直接,也弥补了变形欧拉平均环流理论的一些局限,因此,已逐渐成为研究大气质量传输、理解大气环流运作机理的重要理论工具。表 2 列出了等熵大气质量环流与气压坐标下经向环流的优劣对比。

表 2 气压坐标下的经向环流和位温坐标下的大气经向环流对比

	气压坐标下的经向环流	位温坐标下的经向质量环流
环流性质	等压层内质量不变(定义),纬向平均经向环流仅为速度环流。$\bar{v} = \frac{1}{2\pi \bar{m}} \int_0^{2\pi} vm \, d\lambda = \frac{1}{2\pi} \int_0^{2\pi} v \, d\lambda = \bar{v}$	等熵层内质量随时空变化,纬向平均经向质量环流不等同速度环流。$\hat{v} = \frac{1}{2\pi \bar{m}_\theta} \int_0^{2\pi} v m_\theta \, d\lambda \neq \frac{1}{2\pi} \int_0^{2\pi} v \, d\lambda = \bar{v}$
体现模态	只包含非地转流,中纬度很小。地转波动引起的经向环流部分须由波动输送项分离出的剩余环流才能表示。	包含非地转流和地转流。波动部分(主要在中纬度)对纬向平均经向环流的贡献可直接算出。
高、低纬度联系	三圈,热带和极区不能由经向环流相连。	单圈,将热带和极区连接起来。
时空可加性	速度环流在时、空上可平均,但不可累加(除非密度不随时空变化)。	质量环流在时、空上既可平均又可加(物理意义等同)。

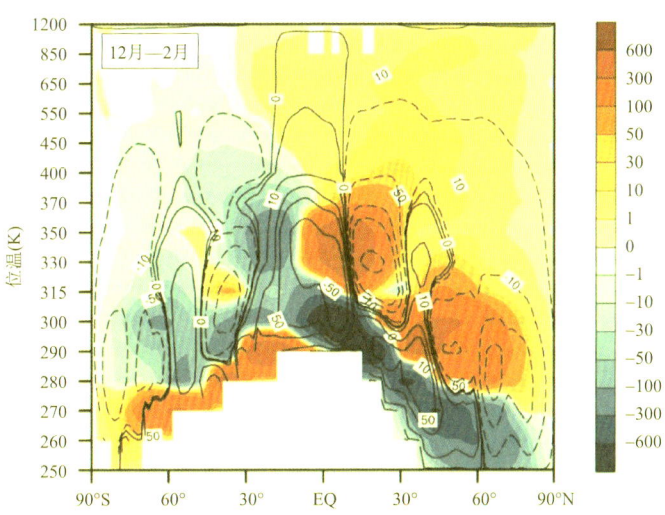

图 4 1979—2011 年冬季平均的气候平均纬圈积分的经向质量通量(填色,单位:$10^8 \text{ kg} \cdot \text{s}^{-1}$)、垂直质量通量(等值线,单位:$10^8 \text{ kg} \cdot \text{s}^{-1}$)的垂直(位温)-经向分布(摘自虞越越(2015))

等熵大气经向质量环流的结构与变形欧拉平均环流类似,是一个半球尺度的单圈环流(图4),由热带地区的上升支、高层由赤道向极地的暖支、中高纬度地区的下沉支以及低层由极区向赤道的冷支组成。Cai和Shin(2014)进一步强调,在半球尺度单圈环流内部还存在三个子环流圈:平流层质量环流、对流层热带外质量环流以及对流层热带质量环流,三者有机结合在一起。其中,平流层质量环流支与平流层的变形欧拉平均环流特征相似,可以近似表征BD环流。等熵大气经向质量环流的季节变化与变形欧拉平均环流类似。

等熵大气经向质量环流的形成和维持与非绝热加热及斜压不稳定波动紧密相关(Johnson,1989)。中纬度地区斜压不稳定的罗斯贝波活动决定了热带外等熵大气经向质量环流形成和维持的机制,如图5所示。斜压不稳定波动往往伴随温度场落后于高度场的温压配置(图5a)。在高度槽的后部,温度槽(冷)使等熵面上凸,更多/少空气质量位于低/高层;而在高度槽的前部,温度脊(暖)使得等熵面下凹,使更多/少的空气质量位于高/低层。高层,槽前南风引起的向极质量输送大于槽后北风引起的向赤道质量输送,从而形成净的向极地质量输送(向极地暖支);而在低层,槽前南风向极地的质量输送则小于槽后北风引起的向赤道质量输送,从而形成净的向赤道质量输送(向赤道冷支)。因此,某等熵面上的波动振幅越大、西倾越明显,则该等熵面以上的层次将有更多的向极地输送的大气质量,同时该等熵面以下将有更多向赤道输送的大气质量。

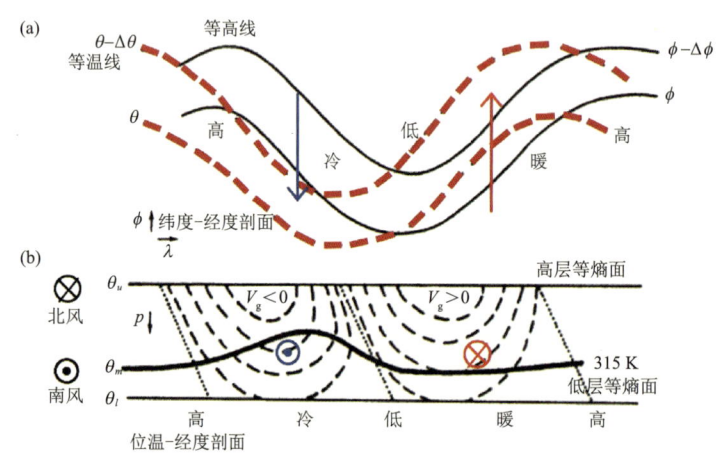

图5　斜压不稳定波动对应的位势高度、位势温度以及经向风场及其引起大气经向质量输送机制示意。此处为由低层等熵面 θ_l、中层等熵面 θ_m 以及高层等熵面 θ_u 组成的两层模型。其中(a)为经度-纬度剖面;(b)为与其对应的经度-垂直剖面,经向准地转风以 V_g 符号在图中标出,斜直虚线分隔槽与脊(摘自Johnson(1989))

等熵大气经向质量环流的高层暖支表征低纬度暖空气向极输送的强度,主导着平流层北半球环状模(NAM)和平流层极涡的强度变化(Yu et al.,2018a),而低层冷支则表征了极区冷空气向南输送或冷空气南侵的强度及其路径,其与中纬度的寒潮爆发和极端低温的发生紧密相关(Iwasaki and Mochizuki,2012;Iwasaki et al.,2014;Yu et al.,2015a,2015b,2015c;Liu and Chen,2021a;Liu et al.,2021b)。等熵大气经向质量环流向极暖支和向赤道冷支的耦合变化,反映了平流层-对流层动力耦合的过程,关注大气质量环流的异常变化,是理解平流层向下影响对流层寒潮低温的重要途径。伴随平流层极涡振荡或NAM事件的向下影响以及异常信号的系统性经向传播,可由等熵大气经向质量环流向极暖支和向赤道冷支的先后同位相变化来解释。例如,平流层爆发性增温(SSW)发生前的大气行星波活动及经向交换加强,首先表现为经向质量环流暖支自上而下逐层异常加强,反映在环流异常场上即为动力和热力异常信号的同时向下和系统性向极传播(Cai and Ren,2006,2007;Ren and Cai,2006,2007,2008)。换言之,大多数情况下平流层向下传播的暖异常信号实际并不能下传到对流层。Yu和Ren(2019)的研究进一步指出,大气经向质量环流高层向极地暖支和低层向赤道冷支不同的耦合变化,决定了平流层-对流层环流动力耦合的特征,进一步决定了平流层环流异常的天气气候效应(Yu et al.,2018b,2018c)。此外,平流层北半球环状模(NAM)与地面北极涛动(AO)的同位相关系也可用等熵大气质量环流理论解释。对质量与温度异常的定量分析表明,在 AO 负位相的冬季,等熵大气经向质量环流往往异常偏强,早冬高层暖支将更多的暖空气向极区输送,导致平流层出现异常暖高压(NAM 负位相),平流层暖支的加强略早于低层向赤道冷支,使得整层气柱质量盈余,低层极区也出现高压(AO 负位相),形成极区看似"正压"的结构(Cai and Ren,2007;Yu et al.,2014)。但该环流模型也存在一些不足:在计算过程中,较难对等熵质量进行空间尺度分解,只能对经向风进行空间尺度分解,因此较难在严格意义上分离出不同尺度波动的质量输送;基于该环流框架下的质量收支平衡受资料时、空分辨率的影响显著。

4.2　湿等熵大气经向质量环流

在等熵大气经向质量环流的基础上,Paulius 等(2008,2010)进一步提出以相当位温或湿熵为垂直坐标定义纬向平均质量环流,称为"湿等熵经向质量环流"。该环流框架对研究降水较为丰富的中纬度地区及夏半球的经向环流特征更具优越性。湿质量环流与质量环流的基本特征类似,空气在近赤道雨带区域上升至对流层高层;之后在高层一部分空气在副热带下沉,另一部分空气在天气波动的作用下穿越风暴轴继续向极地输送,在极地下沉至近地层并向赤道方向回流(图 6a)。在此基础上 Liang 等(2021)进一步改进了湿等熵大气质量环流计算方案,分离了干、湿环流分量。对比干支(图 6b)、湿支(图 6c)与总的质量环流(图 6a)可见,热带外的质量环流高层暖支与湿环流支的质量输送量级相当,使得中高纬度地区湿质量环流的强度是等熵质量环流的 2 倍。这也与湿模型中的变形欧拉平均环流特征类似,但可应用于所有纬度和所有情形下。

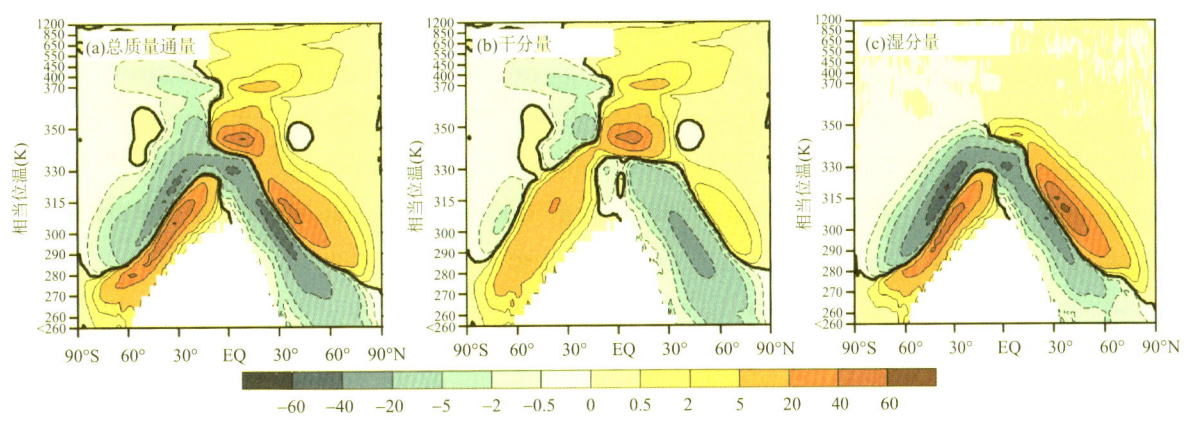

图 6　1979—2019 冬季平均(a)等熵大气经向质量通量及其(b)干、(c)湿分量(10^9 kg·s^{-1})(摘自 Liang 等(2021))

由于与潜热输送紧密相关,湿质量环流除在湿空气的质量输送描述方面具有优越性外,还有助于更全面地描述湿大气中的能量输送情况。湿等熵大气经向质量环流框架常被用于对流、降水系统的机理解释和定量研究。Laliberté 等(2012,2013)的工作表明,湿环流支中暖湿空气的上升主要由深对流和斜升对流系统引起,并进一步加强系统性降水。Paulius 和 Mrowiec(2013)将对流过程引起的垂直质量、能量和熵的输送归因于高能高熵的空气块上升以及低能低熵空气块下沉的共同作用。此外,还有研究着眼于湿质量环流年代际以及更长时间尺度的变化对水汽分布及其次生现象的影响。例如:Laliberté 和 Kushner(2014)考查了中纬度沿湿等熵面的水汽输送对近年来夏季北极对流层热容量的年代际和长期变率的贡献;Laliberté 和 Paulius(2010)通过燃料使用较为均衡(A1B)情景下 20 世纪和 21 世纪的数值模拟试验指出,经向热量输送的减少以及大气层结的加强使得质量环流强度随着全球变暖有减弱趋势,但湿质量环流和质量环流之差(即湿环流支)在冬季有显著的加强趋势。Liang 等(2021)指出,湿等熵大气质量环流的干湿分支在指示极地异常高温事件和中纬度异常低温事件方面存在显著差异。

4.3 等熵大气经向角动量环流

Johnson(1989)、Shin(2012)以及Cai和Shin(2014)通过考查质量输送所携带的角动量输送(大气角动量质量环流,图7),同时研究了全球角动量输送特征以及高层副热带西风带和低层赤道东风带的形成和维持机制。他们指出,伴随大气高层质量的向极地输送(等熵大气质量环流暖支)有角动量向极地输送,而伴随低层质量的向赤道输送(等熵大气质量环流冷支)有角动量向赤道输送。就长期平均而言,质量环流无辐散,而角动量却由赤道向极地单调减少,因而在高层有角动量辐合、低层有角动量辐散。但这会导致西风急流无限加强,而地面中纬度西风带减弱、热带东风带无限加强,与实际不符。进一步研究发现,大气长波的西倾结构有利于高层进入极区的高角动量气块将角动量向下传播,表现为在热带外地区往往存在气压矩负值大值区。伴随斜压波的角动量向下传播,一方面保证了高层西风急流不至于过强,另一方面,使得低层空气得到角动量。对于热带外地区,由高层下传的角动量强于经向角动量辐散,使得地面西风带得以维持;而热带地区,由于大尺度波动较弱,表现为由经向角动量辐散所主导的热带东风带,这就为地球主要风带的形成和维持提供了合理的机制解释。但该模型也存在一定的不足,如:环流强度不与纬向平均纬向风完全对应,还受到空气质量变化的影响;且与等熵大气经向角动量环流一样,基于该环流框架下的角动量收支受时、空分辨率的影响显著。

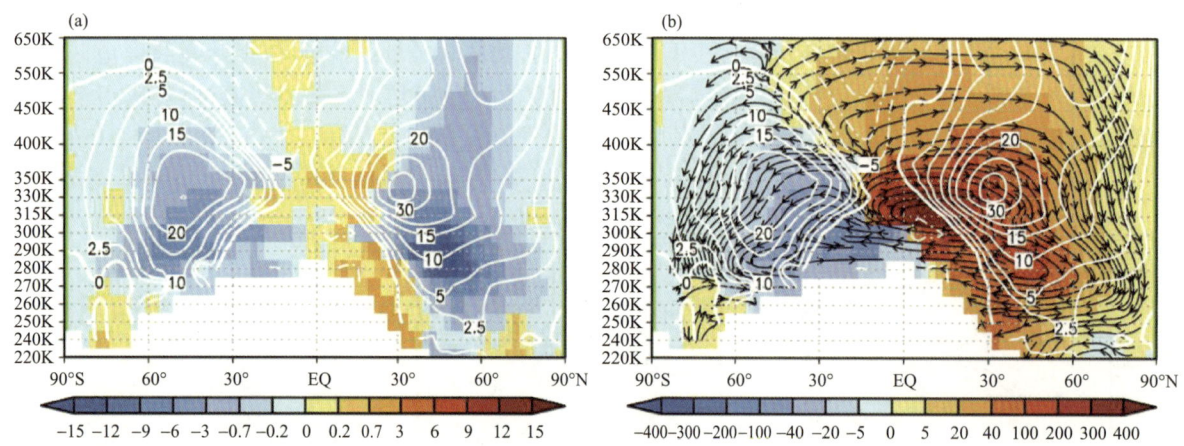

图7 1979—2010冬季平均(a)纬向平均气压矩(填色,单位:Hadley),负/正值表示向下/上的西风角动量输送;(b)等熵大气经向角动量环流(质量流函数:流线;垂直积分的角动量质量通量:填色;单位:Hadley)以及纬向平均纬向风(等值线,单位:m·s^{-1})(摘自Cai和Shin(2014))

5 总结与讨论

介绍了大气经向环流的观测事实,回顾了各种大气经向环流理论模型及其相互关系,包括欧拉平均环流、变形欧拉平均环流、等(湿)熵大气经向质量环流以及等熵大气经向角动量环流。以气压作为垂直坐标的大气环流模型包括欧拉平均环流、变形欧拉平均环流。欧拉平均环流呈现三圈结构,低纬度和高纬度的直接环流圈(哈得来环流和极地环流)由非绝热加热的非均匀性驱动,中纬间接环流(费雷尔环流)则由涡动热量和动量输送驱动。从欧拉平均环流分离出的变形欧拉平均环流可近似认为是拉格朗日观点下的平均环流,可表征等熵空气块的运动轨迹,呈单圈结构,由非绝热加热和波动通量散度驱动。变形欧拉平均环流又被进一步分为了干、湿模型。近年来,以位温或相当位温作为垂直坐标的大气环流模型(等(湿)熵大气经向质量环流以及等熵大气经向角动量环流)陆续建立。在这一新框架下,冷/暖、干/湿空气自然分离,其单圈传输特征得以直观描述;绝热动力过程与非绝热加热过程自然分离,体现其分别在空气质量、能量、角动量传输中的作用,为直观定量地描述大气经向环流观测事实、解释其中的关键物理过程提供了新视角和新工具。表3概括了文中所回顾的几种大气经向环流的特征、对比以及相互关系。

表3 各种大气经向环流理论模型的总结与对比

理论模型名称		垂直坐标	环流性质特性	环流结构	优越性	不足
欧拉平均环流			速度环流。仅包含非地转流的作用	三圈,热带和极区不能由经向环流相连	能清楚再现热力驱动的直接环流,可较好地解释地面气压和风带	1)在准地转的中纬度地区,费雷尔环流结构在实际观测中并不明显存在; 2)较难定量表征涡动强迫的净作用; 3)较难计算示踪物传输
变形欧拉平均环流	干模型	气压	速度环流。分离出与非绝热加热过程相关的垂直运动部分以及涡动强迫的水平运动	半球尺度单圈,热带和极区直接相连,但在中纬度环流较弱	1)能再现从热源到热汇的单圈环流特征,与能量传输方向一致; 2)涡动强迫作用可直接算出; 3)可表征等(湿)熵空气块的运动轨迹;	1)本质上是从欧拉平均环流中人为分离出来的数学环流。取决于纬向平均速度流和纬向偏差场,对基准场有依赖性; 2)气柱内空气具有无辐散性,较难反映某区域内的质量收支; 3)湿模型垂直支对应的加热场不包括凝结潜热,不宜用于热带及近地面与相变加热相关的热力特征研究
	湿模型		速度环流。考虑水汽守恒,将凝结潜热加入到内部非绝热强迫得到相关的水平运动	半球尺度单圈,中纬度经向环流更加连续	4)湿模型能更好地描述降水丰富的中纬度地区或夏半球的经向环流特征	
等熵大气经向质量环流		位温	质量通量环流。包含非地转流和地转流	半球尺度单圈,但在中纬度环流较弱	1)质量环流在时空上既可平均又可加(物理意义等同); 2)更加全面地表征等熵空气的经向运动,不仅包含涡动强迫部分,也包含非地转流的作用; 3)以及变形欧拉环流的大部分优点	1)较难对等熵质量进行空间尺度分解,因此较难严格分离出不同尺度波动的作用; 2)基于该环流框架下的质量收支受时空分辨率的显著影响
等湿熵大气经向质量环流		相当位温		半球尺度单圈,中纬度经向环流更加连续		
等熵大气经向角动量环流		位温	角动量通量环流	半球尺度单圈	1)可表征伴随等熵空气块运动的角动量输送; 2)对高低层主要风带的形成和维持提供了合理定量的机制解释	1)环流强度不与纬向平均纬向风完全对应,受到空气质量变化的影响; 2)基于该环流框架下的角动量收支受时空分辨率的显著影响

除文中介绍的大气经向环流理论模型之外,还有以能量等作为垂直坐标的模型,Paulius 等(2011)和 Kjellsson 等(2014)对以多种热力学变量作为垂直坐标的经向质量环流特征做了系统性的考查和比较,加深了对大气经向环流理论的理解。

参考文献

卜永芳,韩明娟,1987.气象学与气候学基础[M].北京:高等教育出版社:93.
田文寿,黄金龙,邹镕,等,2020.平流层大气环流的典型系统及变化特征综述[J].气象科学,40(5):629-631.
吴国雄,1988a.平均经圈环流在大气角动量和感热收支中的作用[J].大气科学,12(1):6-17.
吴国雄,1988b.关于大气平均经圈环流的一种计算方案[J].中国科学B,4:442-450.
吴国雄,刘还珠,1987.时间平均全球大气环流统计图集[M].北京:气象出版社:212.
吴国雄,陈彪,1988.不同波数域中的干湿E-P剖面和余差环流[J].大气科学特刊,12(s1):94-106.
吴国雄,陈彪,1990a.原始方程系统中的无加速定理.Ⅱ.纬向平均温度的变化[J].大气科学,14(2):143-154.

吴国雄,陈彪,1990b.E-P 剖面的年变化和对流层西风加速[J].中国科学 B,7:775-784.
吴国雄,蔡雅萍,1994.大气热力强迫和动力强迫的调配及平均经圈环流的仿真模拟[J].气象学报,52(2):138-148.
叶笃正,杨大升,1955.北半球大气中角动量的年变化和它的输送机构[J].气象学报,26(4):281-294.
虞越越,2015.大气经向质量环流与北半球冬季寒潮爆发[D].北京:中国科学院大气物理研究所.
张韬,吴国雄,郭裕福,2002.一个具有高分辨率海洋分量的海气耦合模式[J].应用气象学报,13(6):688-695.
ANDREWS D G, MCINTYRE M E, 1976. Planetary waves in horizontal and vertical shear: The generalized Eliassen-Palm relation and the mean zonal acceleration[J]. J Atmos Sci, 33(11): 2031-2048.
APPENZELLER C, HOLTON J R, ROSENLOF K H, 1996. Seasonal variation of mass transport across the tropopause [J]. Geophys Res,101(D10): 15071-15078.
BONAZZOLA M, HAYNES P H, 2004. A trajectory-based study of the tropical tropopause region[J]. Geophys Res, 109: D20112.
BREWER A W, 1949. Evidence for a world circulation provided by the measurements of helium and water vapour distribution in the stratosphere[J]. Quart J Roy Meteor Soc, 75: 351-363.
CAI M, REN R C, 2006. 40−70 day meridional propagation of global circulation anomalies[J]. Geophys Res Lett, 33: L06818.
CAI M, REN R C, 2007. Meridional and downward propagation of atmospheric circulation anomalies. Part I: Northern Hemisphere cold season variability[J]. J Atmos Sci, 64: 1880-1901.
CAI M, SHIN C S, 2014. A total flow perspective of atmospheric mass and angular momentum circulations: Boreal winter mean state[J]. J Atmos Sci, 71: 2244-2263.
CHARNEY J G, DRAZIN P G, 1961. Propagation of the planetary-scale disturbance from the lower into the upper atmosphere[J]. J Geophys Res, 66: 83-109.
CORDERO E, NEWMAN P A, WEAVER C, et al, 2002. Stratospheric Dynamics and Transport of Ozone and Other Tracer Gases. Chapter 6: Stratospheric Ozone An Electronic Text [M]. Greenbelt, Maryland, USA: NASA, GSFC.
DICKINSON R E, 1969. Theory of planetary wave-zonal flow interaction[J]. J Atmos Sci, 26(1): 73-81.
DOBSON G M B, HARRISON D N, LAWRENCE J, 1929. Measurements of the amount of ozone in the Earth's atmosphere and its relation to other geophysical conditions[J]. Proc R Soc, 122: 456-486.
EDMON H J, HOSKINS B J, MCINTYRE M E, 1980. Eliassen-Palm cross sections for the troposphere[J]. J Atmos Sci, 37(12): 2600-2616.
HADLEY G R, 1735. Ⅵ. Concerning the cause of the general trade-winds[J]. Philosophical Transactions of the Royal Society of London, 39.
HAYNES P H, 2005. Stratospheric dynamics[J]. Annu Rev Fluid Mech, 37: 263-293.
HAYNES P H, MARKS C J, MCINTYRE M E, et al, 1991. On the "downward control" of extratropical diabatic circulations by eddy-induced mean zonal forces[J]. Atmos Sci, 48: 651-678.
HELD I M, SCHNEIDER T, 1999. The surface branch of the zonally averaged mass transport circulation in the troposphere[J]. J Atmos Sci, 56: 1688-1697.
HOLTON J R, 1975. The dynamic meteorology of the stratosphere and mesosphere[J]. Meteorological Monographs, 15(37): 218.
HOLTON J R, 2004. An Introduction to Dynamic Meteorology[M]. Seattle, WA, USA: Academic Press: 327.
HOLTON J R, HAYNES P H, MCINTYRE M E, et al, 1995. Stratosphere-troposphere exchange[J]. Rev Geophys, 33: 403-439.
HOSKINS B J, 1989. Diagnostic of the global atmospheric circulation, based on ECMWF analyses 1979−1989[M]. Seattle, WA, USA: Reading University: 217.
IWASAKI T, MOCHIZUKI Y, 2012. Mass-weighted isentropic zonal mean equatorward flow in the northern hemispheric winter[J]. SOLA, 8: 115-118.
IWASAKI T, SHOJI T, KANNO Y, et al, 2014. Isentropic analysis of polar cold air mass streams in the northern hemispheric winter[J]. J Atmos Sci, 71: 2230-2243.
JOHNSON D R, 1989. The forcing and maintenance of global monsoonal circulations: An isentropic analysis[J]. Adv Geo-

phys, 31: 43-316.

KIDSON J W, VINCENT D G, NEWWELL R E, 1969. Observational studies of the general circulation of the tropics: Long term mean values[J]. Quart J Roy Meteor Soc, 95: 258-287.

KJELLSSON J, DOOS K, LALIBERTÉ F B, et al, 2014. The atmospheric general circulation in thermodynamical coordinates[J]. J Atmos Sci, 71: 916-928.

KUNG E C, 1968. On the momentum exchange between the atmosphere and earth over the Northern Hemisphere[J]. Mon Wea Rev, 96: 337-341.

LALIBERTÉ F, PAULUIS O, 2010. Winter intensification of the moist branch of the circulation in simulations of 21st century climate[J]. Geophyis Res Lett, 37: L20707.

LALIBERTÉ F, SHAW T, PAULUIS O, 2012. Moist recirculation and water vapor transport on dry isentropes[J]. J Atmos Sci, 69: 875-890.

LALIBERTÉ F, SHAW T, PAULUIS O, 2013. A Theory for the lower-tropospheric structure of the moist isentropic circulation[J]. J Atmos Sci, 70: 843-854.

LALIBERTÉ F, KUSHNER P J, 2014. Midlatitude moisture contribution to recent Arctic tropospheric summertime variability[J]. J Climate, 27: 5693-5707.

LIANG R X, YU Y Y, SHI C H, et al, 2021. Roles of wet and dry components of moist isentropic mass circulation in changing the extratropical surface temperature in the northern hemispheric winter [J]. Geophys Res Lett, 48: e2020GL091587.

LIU Q, CHEN G X, 2021a. Zonal shift in the cold airmass stream of the east Asian winter monsoon[J]. Environ Res Lett, 16: 124028.

LIU Q, CHEN G X, WANG L, et al, 2021b. Southward cold airmass flux associated with the east Asian winter monsoon: Diversity and impacts[J]. J Climate, 34(8): 3239-3254.

LORENZ E, 1967. The nature and theory of the general circulation of the atmosphere[M]. Geneva: World Meteorological Organization monograph, Rep Prog Phys: 161.

MCLANDRESS C, JONSSON A I, PLUMME D A, et al, 2010. Separating the dynamical effects of climate change and ozone depletion. Part I: Southern Hemisphere stratosphere[J]. Climate, 23: 5002-5020.

NEWTON C W, 1971a. Mountain torques in the global angular momentum balance[J]. J Atmos Sci, 28: 623-628.

NEWTON C W, 1971b. Global angular momentum balance: Earth torques and atmospheric fluxes[J]. J Atmos Sci, 28: 1329-1341.

NEWTON C W, 1972. Southern Hemisphere general circulation in relation to global energy and momentum balance requirements[J]. Meteor Monog, 13: 215-246.

OORT A H, RASMUSSON E M, 1971. Atmospheric Circulation Statistics[M]. Washington, DC: US Government Printing office.

OORT A H, BOWMAN H D, 1974. A study of the mountain torque and its interannual variations in the Northern Hemisphere[J]. J Atmos Sci, 31: 1974-1982.

PRIESTLEY C H B, 1951. A survey of the stress between the ocean and atmosphere[J]. Australian J Sci Res, 4(3): 315-328.

PAULUIS O, CZAJA A, KORTY R, 2008. The global atmospheric circulation on moist isentropes[J]. Science, 321: 1075-1078.

PAULUIS O, CZAJA A, KORTY R, 2010. The global atmospheric circulation in moist isentropic coordinates[J]. J Climate, 23: 3077-3093.

PAULUIS O, SHAW T, LALIBERTÉÉ F, 2011. A statistical generalization of the transformed Eulerian-mean circulation for an arbitrary vertical coordinate system[J]. J Atmos Sci, 68: 1766-1783.

PAULUIS O, MROWIEC A A, 2013. Isentropic analysis of convective motions[J]. J Atmos Sci, 70: 3673-3688.

REN R C, CAI M, 2006. Polar vortex oscillation viewed in an isentropic potential vorticity coordinate[J]. Adv Atmos Sci, 23: 884-890.

REN R C, CAI M, 2007. Meridional and vertical out-of-phase relationships of temperature anomalies associated with the northern annular mode variability[J]. Geophys Res Lett, 34: L07704.

REN R C, CAI M, 2008. Meridional and downward propagation of atmospheric circulation anomalies. Part II: Southern Hemisphere cold season variability[J]. J Atmos Sci, 65: 2343-2359.

RIND D, CHANDLER M, LONERGAN P, et al, 2001. Climate change and the middle atmosphere: 5. Paleostratosphere in cold and warm climates[J]. Geophys Res, 106: 20195-20212.

SALTZMAN B, 1978. A survey of statistical-dynamical models of the terrestrial climate[J]. Adv Geophys, 20: 183-304.

SCHINEIDER E L, LINZEN R S, 1977. Axially symmetric steady-state models of the basic state for instability and climate studies[J]. J Atmos Sci, 34: 263-279.

SCHNEIDER S H, DICKINSON R E, 1974. Climate Modeling[J]. Rev Geophys, 12(3): 447-493.

SCOTT R K, POLVANI L M, 2004. Stratospheric control of upward wave flux near the tropopause[J]. Geophys Res Lett, 31: L02115.

SIGMOND M, SIEGMUND P C, MANZINI E, et al, 2004: A simulation of the separate climate effects of middle-atmospheric and tropospheric CO_2 doubling[J]. Climate, 17: 2352-2367.

SHIN C S, 2012. A hybrid Lagrangian/Eulerian View of the Global Atmospheric Mass Circulation: Seasonal Cycle [D]. Tallahassee, FL, USA: Florida State University.

STOHL A, COAUTHORS, 2003. A backward modeling study of intercontinental pollution transport using aircraft measurements[J]. Geophys Res,108: 4370.

STONE P H, SALUSTRI G, 1984. Generalization of the quasi-geostrophic Eliassen-Palm flux to include eddy forcing of condensation heating[J]. J Atmos Sci, 41: 3527-3536.

TOWNSEND R D, JOHNSON D R, 1985. A diagnostic study of the isentropic zonally averaged mass circulation during the first GARP global experiment[J]. J Atmos Sci, 42: 1565-1579.

WHITE R M, 1949. The role of mountains in the angular momentum balance of the atmosphere[J]. J Meteor, 6: 353-355.

WU G X, CUBALSH U, 1985. The impact of the El Niño anomaly on the mean meridional circulation as simulated by a high resolution model[J]. ECMWF Tech Memo, 105: 1-25.

WU G X, WHITE A A, 1986. A further study of the surface zonal flow predicted by an eddy flux parametrization scheme [J]. Quart J Roy Meteor Soc, 112(474): 1041-1056.

YU Y Y, REN R C, HU J G, et al, 2014. A mass budget analysis on the interannual variability of the polar surface pressure in the winter season[J]. J Atmos Sci, 71: 3539-3553.

YU Y Y, CAI M, REN R C, et al, 2015a. Relationship between warm air mass transport into upper polar atmosphere and cold air outbreaks in winter[J]. J Atmos Sci, 72: 349-368.

YU Y Y, REN R C, CAI M, 2015b. Dynamical linkage between cold air outbreaks and intensity variations of the meridional mass circulation[J]. J Atmos Sci, 72: 3214-3232.

YU Y Y, REN R C, CAI M, 2015c. Comparison of the mass circulation and AO indices as indicators of cold air outbreaks in northern winter[J]. Geophys Res Lett, 42: 2442-2448.

YU Y Y, CAI M, REN R C, 2018a. A stochastic model with a low-frequency amplification feedback for the stratospheric Northern Annular Mode[J]. Climate Dynam, 50: 3757-3773.

YU Y Y, CAI M, REN R C, et al, 2018b. A closer look at the relationships between meridional mass circulation pulses in the stratosphere and cold air outbreak patterns in northern hemispheric winter[J]. Climate Dynam, 51: 3125-3143.

YU Y Y, CAI M, SHI C H, REN R C, 2018c. On the linkage among strong stratospheric mass circulation, stratospheric sudden warming, and cold weather events[J]. Mon Wea Rev, 146: 2717-2739.

YU Y, REN R, 2019. Understanding the variation of stratosphere-troposphere coupling during stratospheric Northern Annular Mode events from a mass circulation perspective[J]. Clim Dyn, 53: 5141-5164.

综述 2

ENSO 与印度洋相互作用的事实和过程

任荣彩[1]，宋歌[1,2]，孙舒悦[3]，李倩[4]，孟文[5]

(1 中国科学院大气物理研究所，北京 100029；2 中国科学院大学，北京 100049；3 国家气象中心，北京 100081；
4 美国麻省理工学院，MA 02139 美国；5 美国国家海洋大气局环境预报中心(NCAR)，MD 20740 美国)

摘要：文中回顾了关于 ENSO 与印度洋相互作用事实和过程的研究结果，包括 ENSO 与印度洋海温(SST)变化第一主导模态——海盆一致(Indian Ocean basin，IOB)模态以及与第二主导模态——印度洋偶极子(Indian Ocean dipole，IOD)的相互作用。一方面 ENSO 在其发展期和衰减期，可通过激发"大气桥"过程，分别影响当年秋—冬季的 IOD 事件和次年春—夏季的 IOB 事件异常，另一方面，印度洋 IOB SST 异常，可通过激发印度洋上空纬向季风环流，并与太平洋上空的沃克(Walker)环流形成"齿轮式"耦合，进而影响太平洋表层的纬向风、对流活动以及暖水堆积，从而促使东太平洋 ENSO 事件的发生；秋—冬季 IOD 事件的发生，同样可以反过来通过触发太平洋纬向风异常而影响暖水堆积，促使次年相反位相的 ENSO 事件发生。最新研究还指出，印度洋次表层 IOD(Sub-IOD)事件可独立于表层 IOD 事件在冬—春季发生，并主要受到冬—春成熟的中太平洋 ENSO 事件的影响。

1 引言

发生在赤道太平洋的 ENSO(El Nino-Southern Oscillation)代表着热带太平洋海-气相互作用的主导变率，是全球气候年际尺度变率的主要强迫源，对亚洲乃至全球各个季风系统都有重要影响。印度洋是亚洲季风环流系统的一个交汇地，海温异常与亚洲区域气候异常的联系更为紧密也更直接，被誉为传递 ENSO 影响的"电容器"(Xie et al.，2009)。印度洋海温的变化在多时间尺度上受到 ENSO 的调制和影响。如印度洋海温的主导模态——海盆一致(IOB)模，受 ENSO 所激发的"大气桥"过程影响(Lau et al.，1996；Klein et al.，1999；Wang，2002)，总是跟随 ENSO 峰值滞后几个月达到峰值(Chiu et al.，1983；Du et al.，2009；Li et al.，2012)。印度洋海温变化的第二个主导模态——东西偶极子模态多在秋季发生，与 ENSO 发展阶段异常沃克环流所引起的低层赤道东风的触发作用有关(Saji et al.，1999；Webster et al.，1999)。另外，也有诸多的证据表明，尽管热带印度洋的海温变率要小于太平洋的海-气耦合模态——ENSO，但其亦可反过来影响 ENSO 事件的发生、发展和维持(Wu et al.，2004；Annamalai et al.，2005；Kug et al.，2006；Xie et al.，2009；Wang，2019)。全面深入理解印度洋与 ENSO 在多种时间尺度上的相互作用关系和相互作用过程，不仅是认识热带海洋间相互影响的重要方面，也是理解全球气候系统年际尺度以上变率，特别是把握亚洲季风异常机理的关键途径之一。

本文将针对 ENSO 与印度洋的相互影响，回顾近 20 多年以来国内外的主要研究进展。将主要关注

通讯作者：任荣彩，rrc@lasg.iap.an.cn
资助项目：国家重点研发专项 2019YFA0606700 第一课题

两者之间通过大气过程(不包括海洋交换过程)所发生的相互作用,特别是印度洋两个主要模态(IOB 和 IOD)与 ENSO 的相互联系及相互作用。

2 ENSO 与印度洋海盆(IOB)SST 异常

2.1 ENSO 影响 IOB SST

2.1.1 年际尺度

ENSO 事件发生时的海温异常,可在热带印度洋-太平洋上空激发异常的 Walker 环流,通过一系列的"大气桥"过程,影响印度洋海盆(IOB)的海温异常(Lau et al.,1996;Klein et al.,1999;Alexander et al.,2002)。所谓"大气桥",是指赤道太平洋的 SST 异常与印度洋的 SST 异常通过大气的桥梁作用联系起来。具体地,异常 Walker 环流的下沉支可在印度尼西亚和西太平洋地区引起异常的海平面气压场,进而通过云辐射强迫、感热、风-蒸发过程以及海洋动力等过程,导致 El Niño/La Niña 年的冬—春季 IOB SST 的一致上升/下降(Chiu et al.,1983;Cadet,1985;Klein et al.,1999;Alexander et al.,2002;Annamalai et al.,2005)(见图1),月平均的 IOB 海温异常总是在 El Niño/La Niña 成熟位相后的次年春季达到极值,其与 ENSO 的最大正相关出现在 ENSO 超前 4~5 个月时(Li et al.,2012;孙舒悦 等,2015)。更进一步地,ENSO 引发的"大气桥"所导致的海表面热通量异常主要贡献北印度洋的 SST 变化(Klein et al.,1999;Murtugudde et al.,1999),而西南印度洋海温的异常主要受"大气桥"所激发的自东向西传播的下沉 Rossby 波动力过程的影响(Huang et al.,2002;Xie et al.,2002)。

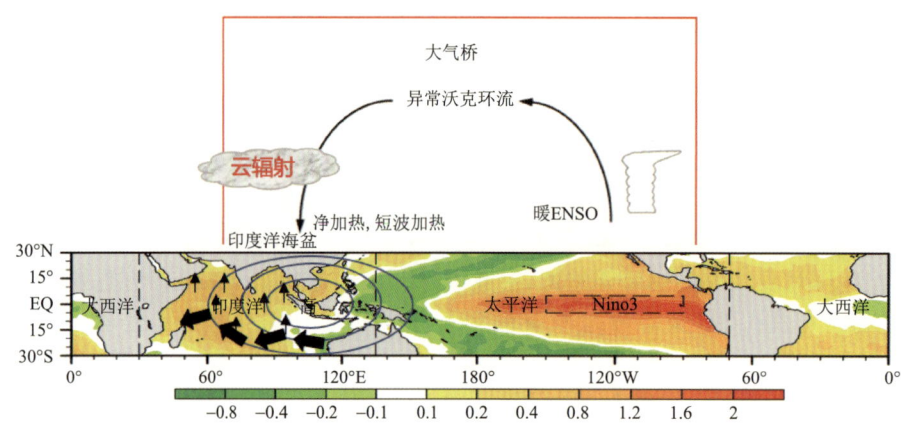

图1 ENSO 影响印度洋 IOB SST 的"大气桥"过程示意

(下图中的细箭头代表风-蒸发反馈过程,粗箭头代表西传海洋 Rossby 波过程,阴影为热带海温异常,蓝色线条为地面高压系统示意)

印度洋 IOB 海温对 ENSO 强迫的这种滞后响应,还受到 ENSO 衰减快慢的调制。近年来的研究证明,当 ENSO 事件衰减异常偏慢,即 4 月 Niño3 指数的强度大于所有正或负事件的 Niño3 指数的平均值时,由于 ENSO"大气桥"过程的持续时间更长,IOB SST 的响应会持续到夏季,且响应强度会在夏季更强(Ren et al.,2016)。由于夏季印度洋的海温异常与南亚季风的密切关系,后来的研究的确发现,ENSO 与孟加拉湾季风爆发早晚密切关系,正是由这些衰减偏慢的 ENSO 事件所决定的。亦即,当去除所有衰减偏慢的 ENSO 事件后,ENSO 与季风爆发的关系就不再显著了(Sun et al.,2017)。这说明 ENSO 衰减的快慢,不仅仅影响 IOB SST 的异常,还关系到 ENSO 是否能有效地影响南亚夏季风的异常。

2.1.2 年代际尺度

IOB SST 除了在年际时间尺度上与 ENSO 存在密切关联外,在年代际时间尺度上与 ENSO 及其影响的年代际变化也存在关联。首先,尽管在年际尺度上 IOB SST 与 ENSO 密切关联,但 IOB SST 最显著的变化是其显著的长期上升趋势。根据 1950—2009 年英国的 HadISST1 数据估计,IOB SST 上升趋势达到 0.0419 ℃/10 a;而 ENSO 并不存在显著的长期上升趋势,IOB SST 的长期上升被认为是全球变暖的区域响应(Li et al.,2012)。然而,去除长期趋势后,南印度洋(Allan et al.,1995)和北印度洋(Chowdary et al.,2012)分别被发现存在显著的年代际变化特征,并且是与 ENSO-IOB 关系的年代际变化有关(Cole et al.,2000)。特别是发生在 20 世纪 70 年代前后的 IOB SST 的变化,是过去 140 年以来最为显著的年代际变化(Li et al.,2012)。Xie 等(2010)指出,20 世纪 70 年代前后 ENSO 对 IOB 的效应有显著的增强;Li 等(2012)的研究指出,20 世纪 70 年代前后 IOB SST 最强的年代际变化出现在夏季,并不是出现在 ENSO 效应最强的春季,表现为 SST 的显著上升。那么,这种夏季的变化是否还是与 ENSO 有关呢?

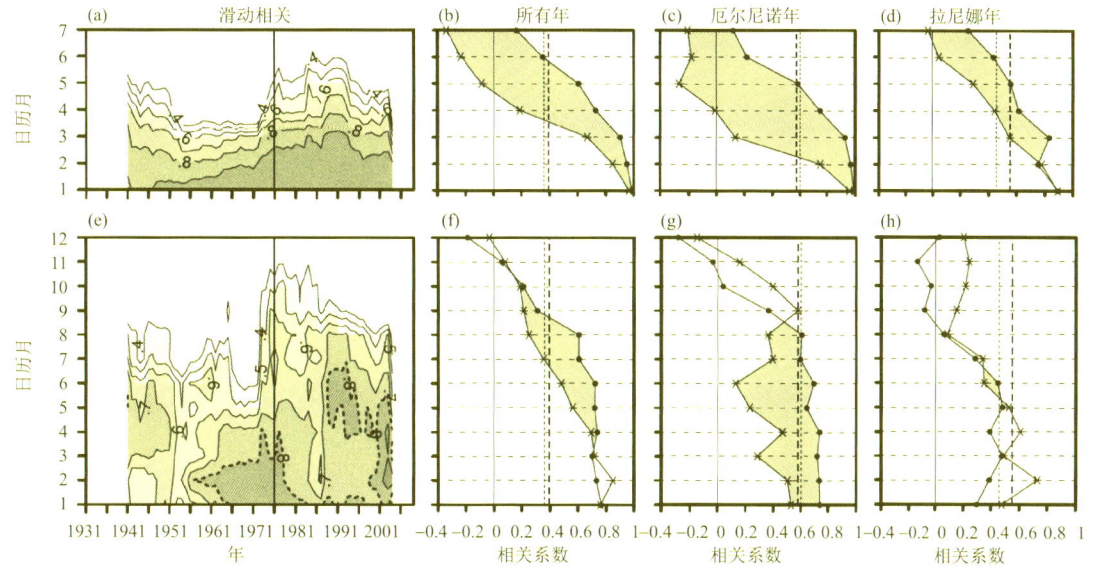

图 2 (a)每 21 年滑动的冬季(NDJ)Nino3 指数的滞后自相关;以及 1946—1976 年阶段(*)和 1977—2002 年阶段(●)的(b)所有 ENSO 年,(c)仅 El Nino 年,和(d)仅 La Nina 年的 Nino3 指数的滞后自相关;(e)—(h)与(a)—(b)类似,但为 Nino3 指数与各月 IOB SST 指数的滞后交叉相关。所有分图的纵坐标代表 ENSO 衰减年的月份,(a,e)中的横坐标为实际年,其他图横坐标为相关系数(摘自 Li 等(2012))

Li 等(2012)的结果表明,上述变化与 ENSO 事件衰减快慢的年代际变化有关。平均而言,由于 ENSO 事件在 20 世纪 70 年代之后的衰减显著变慢(图 2a,2b),导致其在印度洋的最强效应由 20 世纪 70 年代前的春季延迟到了 20 世纪 70 年代后的夏季(图 2e,2f),因此印度洋 SST 最显著的年代际变化发生在夏季。进一步地,上述 ENSO 延迟衰减的情形,主要以 El Nino 的延迟衰减所主导,20 世纪 70 年代前后 La Nina 事件的衰减并无显著变化(图 2c,2d),因此 ENSO 影响夏季印度洋 SST 的年代际变化表现的是年代际的变暖。实际上,后来的研究也证实,不管是 El Nino(引起 IOB 变暖)或是 La Nina(引起 IOB 变冷)事件,由于伴随其延迟衰减,其所激发的一系列影响 IOB 的"大气桥"过程都会相应延迟,它们的衰减快慢的确关系到当年夏季印度洋 IOB 是否会出现显著的 SST 异常;只是发生延迟衰减的 La Nina 事件的个数在 20 世纪 70 年代前后是相当的,而所有发生延迟衰减的 El Nino 事件都出现在了 20 世纪 70 年代之后,因此引起 IOB SST 的年代际显著上升(Ren et al.,2016)。这一结果表明,ENSO 的衰减快慢,可在夏季印度洋产生显著的效应,20 世纪 70 年代前后 El Nino 和 La Nina 延迟衰减事件个数的不对称,是导致夏季 IOB SST 年代际上升的主要原因。

2.2 印-太"齿轮式"耦合(GIP)过程指示 ENSO 事件发生

在国际上关于 ENSO 影响印度洋 IOB 的"大气桥"机制的成果发表前夕,国内已经出现了类似的研究成果。吴国雄等(1998)的研究指出,ENSO 期间赤道中东太平洋 SST 与印度洋 IOB SST 的显著正相关,与太平洋上空异常的纬向-垂直 Walker 环流圈与赤道印度洋上空反方向的异常纬向-垂直季风环流圈的齿轮式(印-太齿轮,gearing between the Indian and Pacific Ocean,GIP)耦合密切相关。具体地,暖 ENSO 时对应顺时针的季风环流圈与逆时针的 Walker 环流圈的啮合(反向啮合),冷 ENSO 时相反(正向啮合)。重要的是,季风环流圈一般早于 Walker 环流圈形成,两者的 GIP 啮合点也首先在印度洋上空形成;然后在海-气相互作用正反馈机制(暖 SST 上空大气低层风辐合,西侧为西风异常,东侧为东风异常,西风异常又会进一步加强 SST 暖异常的发展;对于冷 SST 异常的发展则相反)的作用下,季风环流圈逐步扩大,Walker 环流圈逐步缩小,GIP 啮合点逐步向东传播进入太平洋,低空的异常西(东)风进而引起东部地区 SST 上升(下降)以及对流发展(抑制),SST 异常进一步向东发展至中东太平洋,最终可导致暖(冷)ENSO 事件的爆发(图 3)。由此说明,ENSO 事件的起始扰动有可能可追溯到印度洋上空的异常纬向-垂直季风环流圈以及 GIP 啮合点的出现,这强调了印度洋环流异常对 ENSO 的可能触发作用。

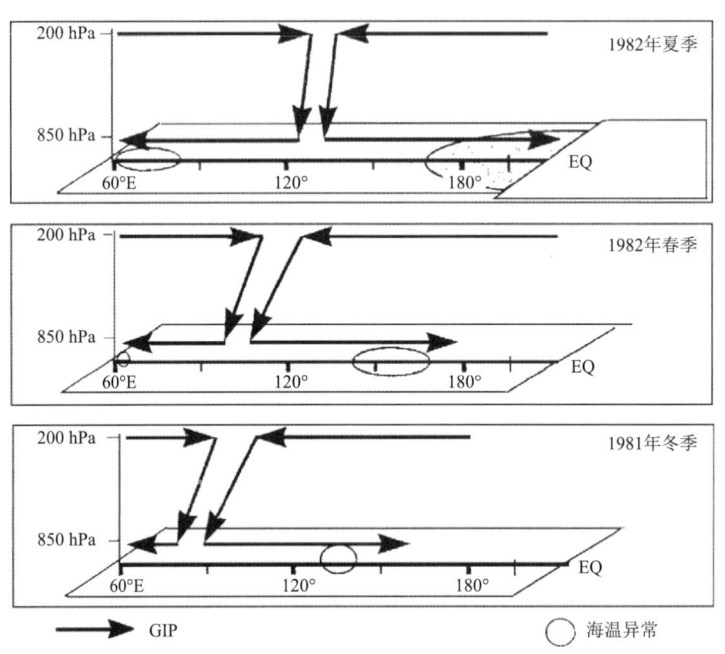

图 3　1982—1983 年 El Niño 事件及其前期,赤道印度洋季风垂直环流圈与太平洋 Walker 垂直环流圈反向啮合的"印-太齿轮"以及相关的海温异常从印度洋向太平洋传播的示意图(摘自吴国雄等(1998))

后续孟文等(2000)在气候模式的模拟结果中又进一步证实了 ENSO 事件中印-太 GIP 耦合以及相关的海气相互作用正反馈过程的存在,而且发现,只要在赤道太平洋或赤道印度洋两者中的任何一区域的海表,加入异常的纬向风应力异常信号,就可以触发 GIP 耦合过程,并以此作为桥梁,影响到另一区域的海-气相互作用和 SST 异常。这说明 GIP 耦合过程不仅代表着印度洋影响 ENSO 发展的过程,也反映着 ENSO 发展过程中通过异常 Walker 环流而影响印度洋 SST 的"大气桥"过程。尽管印度洋 SST 变率相对较小,但 ENSO 与印度洋之间的确是存在着相互影响和相互作用。关于这一点,目前仍然是国内外相关研究的热点之一。

与"印-太齿轮"影响过程类似,近年来的研究结果的确发现,印度洋 IOB 增暖可激发大气 Kelvin 波向东传播,进而可对赤道西太平洋地区的东风异常有加强作用,当印度洋 IOB 海温异常超前 2~3 个月时,其与赤道西太平洋纬向风异常有最大的负相关(Wang,2019)。然而,同时也有证据说明,引起 ENSO 的赤道纬向风异常主要是局地海洋过程作用的结果。印度洋 IOB SST 异常主要在 ENSO 的衰减期，有助于西太平洋的纬向风异常以及西太平洋副热带高压的维持，称为印度洋的"电容器"效应(Wang，2019；Xie et al.，2009)。

最新的研究将印-太齿轮耦合机制的触发，与夏季青藏高原上空的由感热负异常所激发的"负感热斜压模态"联系起来，指出该斜压模态的出现可在印度洋上空强迫出类似的反向季风环流圈，以及与 Walker 环流的反向啮合，从而可以通过海-气耦合过程以及啮合点的东传，在随后的冬季触发 El Nino 事件的发生(Yu et al.，2022)。这再次证实了印-太齿轮耦合机制是印度洋异常信号指示 ENSO 事件发生的重要机制。

3 ENSO 与印度洋偶极子(Indian Ocean dipole，IOD)相互作用

3.1 ENSO 影响 IOD 过程的复杂性

ENSO 对印度洋的另一个重要影响是在 ENSO 事件的发展位相显著调制印度洋海温的第二主导变化模态——东、西反位相的海温异常模态(IOD)。正是由于 ENSO 的主导调制作用，IOD 具有很强的季节锁相特性，通常在秋季成熟，冬季衰亡(Behera et al.，1999；Saji et al.，1999；Webster et al.，1999)。关于 IOD 事件期间的海-气耦合过程，大多数研究认为是正在发展的 ENSO 事件，可通过"大气桥"过程触发赤道印度洋上的纬向风异常，进而触发海洋波动传播并通过 Bjerknes 反馈过程引起 IOD 事件的发生(Klein et al.，1999；Shinoda et al.，2004；Huang et al.，2007；Wang，2019)。也有研究指出，IOD 是赤道印度洋中固有的海-气耦合过程的产物，其对印度洋海温的季节—年际尺度变率有其独立的作用(Webster et al.，1999；Saji et al.，2003)。而 ENSO 仅在与 IOD 同时发生时可影响 IOD 事件的周期、强度和形成过程(Behera et al.，2006)。除 ENSO 之外，南亚季风异常、南印度洋偶极子以及海洋中的印尼贯穿流等都可以触发 IOD 事件的发生(Ashok et al.，2003；Annamalai et al.，2004；Song et al.，2004；Behera et al.，2008；Zhang et al.，2020)。对过去 60 年 IOD 事件的统计发现，的确存在 IOD 事件独立于 ENSO 发生的情形(Song et al.，2022)。

最新的研究指出，次表层也存在海温偶极子(subsurface Indian Ocean dipole，Sub-IOD)，而且是赤道印度洋次表层海温变化的 EOF 第一主导模态，具有 15～20 个月的振荡周期，被认为是赤道印度洋局地海-气相互作用形成的固有的东—西振荡变化，可能为表层 IOD 的准两年振荡提供了基础(Sayantani et al.，2015；Song et al.，2021)。大多数 Sub-IOD 伴随 IOD 在夏—秋季发生，但相当一部分 Sub-IOD 可在冬—春季独立于 IOD 发生，此时发生的 Sub-IOD 是受到冬—春季成熟的且海温中心明显偏西的中太平洋 ENSO 的远程影响(Song et al.，2022)。这表明中太平洋 ENSO 对印度洋 Sub-IOD 存在影响。最新的研究证据也指出，前一年发生的中部型 El Nino，也更有利于当年强的正 IOD 事件发生(Yang et al.，2021)。不同类型 ENSO 对印度洋 IOD 振荡的不同影响，说明了 ENSO 影响的多样性，也使得表层 IOD 与 Sub-IOD 的关系更为复杂，这些都有待进一步研究。

3.2 IOD 对 ENSO 事件的可能影响

基于 IOD 指数超前 ENSO 指数 1 年的最大负相关，Izumo 等(2014)指出负(正)的 IOD 事件有助于 El Nino(La Nina)事件的发生。例如，负 IOD 事件中印度洋东南部的正 SSTA 在秋天达到峰值，可在西太平洋产生东风异常，从而有助于西太平洋暖水的累积；在随后的冬季，当 IOD 异常迅速消失时，会导致太平洋上东风异常的突然崩溃，进而可导致 El Nino 事件的发展。巢清尘等(2001)指出，在热带西太平洋西风异常爆发前，赤道东印度洋的西风异常已经出现并维持，然后快速地向东扩展到热带西太平洋，有利于西太平洋西风爆发，是 El Nino 发生不可缺少的条件之一。此外，Wang 等(2021)发现，当 El Nino 事件开始发展的时间相对较早时，印度洋正 IOD 事件以及大西洋 Nina 的共同作用，可成为极端 El Nino 发生的

必要条件之一。

总之，ENSO对IOD有显著的调制作用，但仍有独立于ENSO的IOD事件，不同类型ENSO影响的差异更增加了ENSO与IOD关系的复杂性。IOD可能在ENSO的发展位相对太平洋的纬向风产生影响，还可能对极端El Niño事件的发生有贡献，但仍需要更多的证据。此外，在更长的年代际时间尺度上，目前的证据初步指出了太平洋的年代际振荡（PDO）和多年代际振荡（IPO）可能对IOD的东极有影响（Dong et al.，2016；Krishnamurthy et al.，2016），但也有证据说明印度洋次表层海洋波动过程的变化起了主要作用（Ashok et al.，2004），目前还缺少定论。

4　总结与讨论

文中回顾了太平洋ENSO与印度洋两个主要模态之间，通过大气过程相互作用的主要研究成果，在说明了ENSO通过"大气桥"过程影响印度洋IOB的重要事实后，重点阐述了吴国雄等（1998）提出的印度洋海表温度异常通过印度洋季风环流圈与太平洋上空的Walker环流圈的"齿轮式啮合"影响ENSO爆发的成果。在回顾了ENSO对IOD的显著调制作用的同时，指出了ENSO对IOD的多重影响及其复杂性，也回顾了近年来有关IOD影响ENSO爆发的事实。本文也指出了一些尚不清楚的相关科学问题。总之，本文的回顾说明，太平洋ENSO与印度洋的主导模态之间是存在双向的相互作用，期望本文的回顾将有助于行业界较全面地了解太平洋与印度洋的相互影响过程，期望对认识热带海洋间相互影响、理解全球气候系统变化以及把握亚洲季风异常机理有所裨益。

参考文献

巢清尘，巢纪平，2001. 热带西太平洋和东印度洋对ENSO发展的影响[J]. 自然科学进展，12：63-70.

孟文，吴国雄，2000. 赤道印度洋-太平洋地区海气系统的齿轮式耦合和ENSO事件Ⅱ. 数值模拟[J]. 大气科学（1）：15-25.

孙舒悦，任荣彩，2015. ENSO与印度洋海盆海温多尺度相互作用及其对气候影响的研究进展[J]. 气象科技，43(5)：8.

吴国雄，孟文，1998. 赤道印度洋-太平洋地区海气系统的齿轮式耦合和ENSO事件：Ⅰ. 资料分析[J]. 大气科学，24（4）：15-25.

ALEXANDER M, BLADÉ I, NEWMAN M, et al, 2002. The atmospheric bridge: The influence of ENSO teleconnections on air-sea interaction over the global oceans[J]. J Climate, 15：2205-2231.

ALLAN R J, LINDESAY J A, REASON C J C, 1995. Multidecadal variability in the climate system over the Indian Ocean region during the austral summer[J]. J Climate, 8：1853-1873.

ANNAMALAI H, MURTUGUDDE R, 2004. Role of the Indian Ocean in regional climate variability[J]. Washington DC American Geophysical Union Geophysical Monograph Series，147：213.

ANNAMALAI H, XIE S P, MCCREARY J P, et al, 2005. Impact of Indian Ocean sea surface temperature on developing El Niño[J]. J Climate, 18：302-319.

ASHOK K, GUAN Z, YAMAGATA T, 2003. A look at the relationship between the ENSO and the Indian Ocean dipole[J]. J Meteor Soc Japan, 81：41-56.

ASHOK K, CHAN W L, MOTOI T, et al, 2004. Decadal variability of the Indian Ocean dipole[J]. Geophys Res Lett, 31.

BEHERA S K, KRISHNAN R, YAMAGATA T, 1999. Unusual ocean-atmosphere conditions in the tropical Indian Ocean during 1994[J]. Geophys Res Lett, 26：3001-3004.

BEHERA S K, LUO J J, MASSON S, et al, 2006. A CGCM study on the interaction between IOD and ENSO[J]. J Climate, 19：1688-1705.

BEHERA S K, LUO J J, YAMAGATA T, 2008. Unusual IOD event of 2007[J]. Geophys Res Lett, 35.

CADET D L, 1985. The Southern Oscillation over the Indian Ocean[J]. Journal of Climatology, 5：189-212.

CHIU L S, NEWELL R E, 1983. Variations of zonal mean sea surface temperature and large-scale air-sea interaction[J]. Quart J Roy Meteor Soc, 109：153-168.

CHOWDARY J S, XIE S P, TOKINAGA H, et al, 2012. Inter-decadal variations in ENSO teleconnection to the Indo-Western Pacific for 1870—2007[J]. J Climate, 25: 1722-1744.

COLE J E, DUNBAR R B, MCCLANAHAN T R, et al, 2000. Tropical Pacific forcing of decadal SST variability in the western Indian Ocean over the past two centuries[J]. Science, 287: 617-619.

DONG L, ZHOU T, DAI A, et al, 2016. The footprint of the inter-decadal Pacific oscillation in Indian Ocean sea surface temperatures[J]. Scientific Reports, 6: 21251.

DU Y, XIE S P, HUANG G, et al, 2009. Role of air-sea interaction in the long persistence of El Nino-induced north Indian Ocean warming[J]. J Climate, 22: 2023-2038.

HUANG B, KINTER J L, 2002. Interannual variability in the tropical Indian Ocean[J]. J Geophys Res: Oceans, 107 (C11):3199, doi: 10.1029/2001JC0012789.

HUANG B, SHUKLA J, 2007. Mechanisms for the interannual variability in the tropical Indian Ocean. Part II: Regional processes[J]. J Climate, 20: 2937-2960.

IZUMO T, LENGAIGNE M, VIALARD J, et al, 2014. Influence of Indian Ocean dipole and Pacific recharge on following year's El Niño: Inter-decadal robustness[J]. Climate Dyn, 42: 291-310.

KLEIN S A, SODEN B J, LAU N C, 1999. Remote sea surface temperature variations during ENSO: Evidence for a tropical atmospheric bridge[J]. J Climate, 12: 917-932.

KRISHNAMURTHY L, KRISHNAMURTHY V, 2016. Decadal and interannual variability of the Indian Ocean SST[J]. Climate Dyn, 46: 57-70.

KUG J S, KANG I S, 2006. Interactive feedback between ENSO and the Indian Ocean[J]. J Climate, 19: 1784-1801.

LAU N C, NATH M J, 1996. The role of the "atmospheric bridge" in linking tropical Pacific ENSO events to extratropical SST anomalies[J]. J Climate, 9: 2036-2057.

LI Q, REN R C, CAI M, et al, 2012. Attribution of the summer warming since 1970s in Indian Ocean basin to the inter-decadal change in the seasonal timing of El Nino decay phase[J]. Geophys Res Lett, 39.

MURTUGUDDE R, BUSALACCHI A J, 1999. Interannual variability of the dynamics and thermodynamics of the tropical Indian Ocean[J]. J Climate, 12: 2300-2326.

REN R, SUN S, YANG Y, et al, 2016. Summer SST anomalies in the Indian Ocean and the seasonal timing of ENSO decay phase[J]. Climate Dyn, 47: 1827-1844.

SAJI N H, GOSWAMI B N, VINAYACHANDRAN P, et al, 1999. A dipole mode in the tropical Indian Ocean[J]. Nature, 401: 360-363.

SAJI N H, YAMAGATA T, 2003. Structure of SST and surface wind variability during Indian Ocean dipole mode events: Coads observations[J]. J Climate, 16: 2735-2751.

SAYANTANI O, GNANASEELAN C, 2015. Tropical Indian Ocean subsurface temperature variability and the forcing mechanisms[J]. Climate Dyn, 44: 2447-2462.

SHINODA T, HENDON H H, ALEXANDER M A, 2004. Surface and subsurface dipole variability in the Indian Ocean and its relation with ENSO[J]. Deep Sea Research Part I: Oceanographic Research Papers, 51: 619-635.

SONG G, HUANG B, REN R, et al, 2021. Basin-wide connections of upper-ocean temperature variability in the equatorial Indian Ocean[J]. J Climate: 1-50.

SONG G, REN R C, 2022. Linking the subsurface Indian Ocean dipole to central Pacific ENSO[J]. Geophys Res Lett, 49: e2021GL096263.

SONG Q, GORDON A L, 2004. Significance of the vertical profile of the indonesian through flow transport to the Indian Ocean[J]. Geophys Res Lett, 31.

SUN S, REN R, WU G, 2017. Onset of the Bay of Bengal summer monsoon and the seasonal timing of ENSO's decay phase[J]. International Journal of Climatology, 37: 4938-4948.

WANG C, 2019. Three-ocean interactions and climate variability: A review and perspective[J]. Climate Dyn, 53: 5119-5136.

WANG C, 2002. Atmospheric circulation cells associated with the El Nino-Southern Oscillation[J]. J Climate, 15: 399-419.

WANG J Z, WANG C, 2021. Joint boost to super El Niño from the indian and Atlantic Oceans[J]. J Climate, 34: 4937-4954.

WEBSTER P, MOORE A, LOSCHNIGG J, et al, 1999. Coupled ocean-atmosphere dynamics in the Indian Ocean during 1997—1998[J]. Nature, 401: 356-360.

WU R, KIRTMAN B P, 2004. Understanding the impacts of the Indian Ocean on ENSO variability in a coupled GCM[J]. J Climate, 17: 4019-4031.

XIE S P, ANNAMALAI H, SCHOTT F A, et al, 2002. Structure and mechanisms of south Indian Ocean climate variability[J]. J Climate, 15: 864-878.

XIE S P, HU K, HAFNER J, et al, 2009. Indian Ocean capacitor effect on Indo-Western Pacific climate during the summer following El Niño[J]. J Climate, 22: 730-747.

XIE S P, DU Y, HUANG G, et al, 2010. Decadal shift in El Niño influences on Indo-Western Pacific and east Asian climate in the 1970s[J]. J Climate, 23: 3352-3368.

YANG K, CAI W, HUANG G, et al, 2021. Is preconditioning effect on strong positive Indian Ocean dipole by a preceding central Pacific El Niño deterministic? [J]. Geophys Res Lett, 48: e2020GL092223.

YU W, LIU Y, XU L, et al, 2022. Potential impact of spring thermal forcing over the Tibetan Plateau on the following winter El Niño-Southern Oscillation[J]. Geophys Res Lett, 49: e2021GL097234.

ZHANG L Y, DU Y, CAI W, et al, 2020. Triggering the Indian Ocean dipole from the Southern Hemisphere[J]. Geophys Res Lett, 47: e2020GL088648.

综述 3

气候系统模式 FGOALS 发展历程

包庆[1]，何编[1]，王晓聪[1]，李剑东[1]，周林炯[2]，吴小飞[3]，李矜霄[1]

(1 中国科学院大气物理研究所，北京 100029；2 美国普林斯顿大学地球流体力学实验室，新泽西州 美国；
3 成都信息工程大学，成都 610225)

摘要：文中回顾了中国科学院大气物理研究所大气科学和地球流体力学数值模拟国家重点实验室(LASG)大气环流模式 F/SAMIL 及其耦合版本 GOALS 和 FGOALS 模式的发展历程。20 世纪 90 年代初，吴国雄课题研究组将谱模式引进到 LASG，在动力框架和物理过程方面做了系统的改进和发展。LASG 数值模拟课题组建立了中国第一个全球海洋-大气-陆地气候系统模式——GOALS 系统，提供给科学家开展青藏高原和亚洲季风等方面的气候模拟研究。2002 年，通过国家自然科学基金委员会，吴国雄等科学家递交了"中国应加快发展自己的气候系统模式"的信息专报，受到国务院有关领导的高度重视并亲笔批示，为中国自主研发气候系统模式起到了重要推动作用。从此，中国的科研院所和高校迎来了气候系统模式发展的黄金时段。随着中国计算条件的逐步改善，大气模式和耦合版本不断发展、完善，LASG 大气谱模式水平分辨率从最初 R15 提升到 R42，垂直分辨率由 9 层提高到 26 层，物理参数化方案在辐射过程、积云对流、边界层和陆面过程等进一步发展与完善，耦合模式 FGOALS 也实现了可插拔模块间的直接耦合。近年来，LASG 新一代大气模式采用了有限体积球立方网格，发展了显式对流降水方案，实现了分辨率提升到 25 km 的目标，耦合模式 FGOALS-f3 作为分辨率最高的三个气候系统模式之一参与了 CMIP6 国际高分辨率模式比较计划、全球季风模拟比较计划、北极放大效应模拟比较计划和青藏高原相关次季节预测计划。同时，FGOALS 模式参与了世界气象组织次季节-季节预报的业务化计划(S2S)。展望未来，下一代 LASG 模式将提升至千米级分辨率，在物理过程和动力框架两方面共同挑战"灰区分辨率"，实现千米级尺度自适应，不仅将在国际耦合模式比较计划中继续崭露头角，还将在青藏高原数值模拟、天气-气候一体化的无缝隙预测等领域中发挥巨大作用，在探索科学问题的同时积极服务国家的重大需求和国防安全。

1 引言

气候系统模式适用于理解过去、现在和未来气候变化情景下多圈层相互作用的物理规律，包括全球能量和水循环规律、重大天气和气候灾害的发生机制。气候系统模式可广泛用于极端天气和气候事件的多尺度预测，不仅服务于气象相关的防灾、减灾，而且为水资源调配、外交谈判、粮食安全和国防安全等领域提供科学依据。

中国科学家认识到建立可持续发展的具有自主知识产权的全球气候系统模式的必要性和紧迫性。中

通讯作者：包庆，baoqing@mail.iap.ac.cn
资助项目：国家自然科学基金项目 4217050404

国科学院大气物理研究所大气科学和地球流体力学数值模拟国家重点实验室(LASG)是中国最早开展气候系统模式研发的机构,率先在20世纪90年代发展了中国第一个包括全球大气、海洋、陆地和海冰的物理气候系统模式GOALS(吴国雄 等,1997;Zhang et al.,2000)(图1),参与了国际模式比较和合作,开展了青藏高原和亚洲季风等方面的气候模拟研究(Wu et al.,1998),受到了国内外同行的重视和好评。LASG在气候系统模式发展和应用领域取得了丰硕的成果,引起了普林斯顿大学地球流体力学实验室(GFDL)世界知名模式专家Syukuro Manabe(2021年诺贝尔物理学奖获得者)的关注,他分别于1993、2005、2009年三次访问中国科学院大气物理研究所(简称大气所),与LASG科学家进行了广泛而又深入的交流。

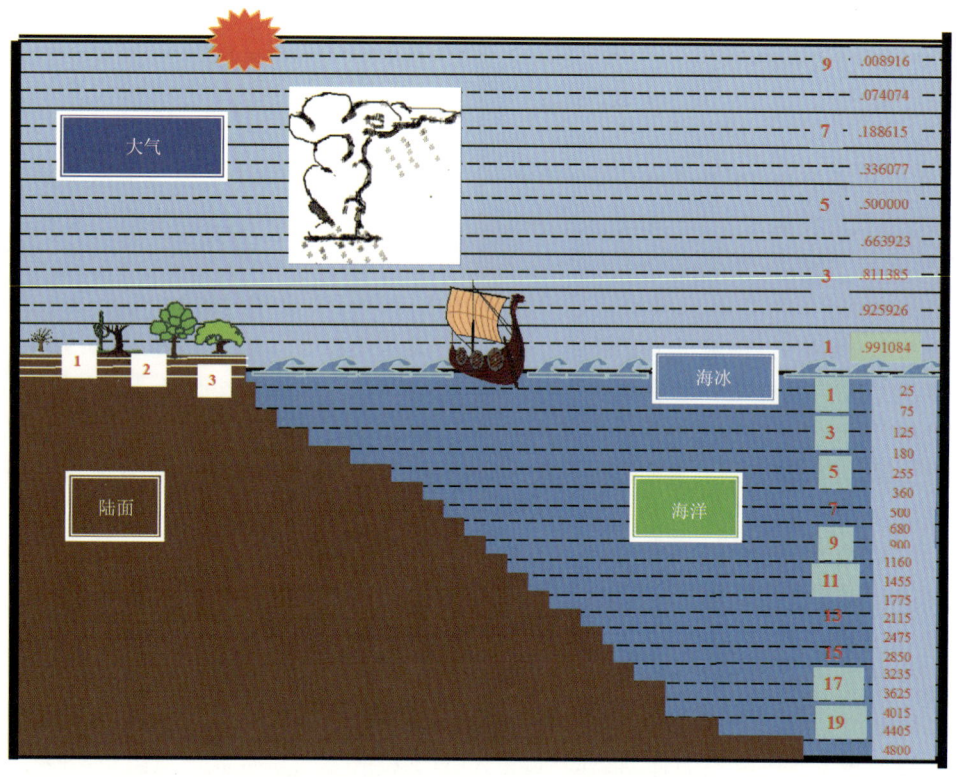

图1 中国第一个大气-海洋-陆面耦合的气候系统模式GOALS示意(吴国雄 等,1997,图1)

气候系统模式的发展是一项巨大的系统工程,仅LASG的科学家难以长期保持中国气候系统模式的先进性和可持续发展。在广泛听取国内科学家建议的基础上,2001年7月起,围绕中国应发展自己的气候系统模式的重要性和政策建议,由国家自然科学基金委员会地球科学部组织大气所吴国雄、王斌、宇如聪、俞永强、张学洪和郭裕福等专家学者和基金管理人员共同编写建议书。经过近一年时间的准备,先后五次修改、凝练和完善,最终完成"中国应加快发展自己的气候系统模式"建议书,于2002年7月上报中共中央和国务院。建议书受到国务院相关领导同志的高度重视并亲笔批示,有关部委共同协商落实。"中国应加快发展自己的气候系统模式"建议书为中国研发气候系统模式起到了重要的推动作用。从此,中国的科研院所和高校迎来了气候系统模式发展的黄金时段。

2 大气环流模式F/SAMIL发展历程

2.1 大气模式动力框架发展历程

LASG大气环流模式F/SAMIL的动力框架发展经历了两个阶段。20世纪90年代初,吴国雄将大气环流谱模式R15L9引进到LASG(球谐函数采用菱形15波截断,水平分辨率相当于7.5°经度×4.5°纬度,垂直方向共分9层),该大气模式是在澳大利亚谱模式的基础上发展起来的(吴国雄 等,1997)。在动力框架方面,吴国雄、刘辉和李伟平等(Liu et al.,1994;Wu et al.,1996)为减少截断误差和负地形影响,采用了曾庆存(1963)提出的参考大气扣除方法(SAS),取重力内波速度为常数,参考温度廓线可写成气压的解析表示。对流层,参考大气温度廓线与观测接近;平流层,则不能反映温度的升高。因此采用欧洲中心5年平均温度廓线作为模式的参考大气,同时对反照率方案做了改进。Wu等(1996)通过对R15L9模式10年的AMIP积分结果的分析指出:SAS方案有助于减少模式在格陵兰高地、落基山脉、青藏高原、安第斯山脉和南极高地等高大山脉周围的扰动误差,模拟结果更近于观测(图2)。吴统文、刘平和王在志等(Wu et al.,2003;王在志 等,2005a)将水平分辨率由R15提升到R42(水平分辨率相当于2.81°经度×1.66°纬度),并开展了耦合气候态的模拟评估(吴统文 等,2004);王在志(2005b)将垂直方向从9层的σ坐标系提高到26层σp混合坐标系,采用半隐式时间积分方案,按照模块化和标准化要求对R42L26模式系统的结构进行调整,便于进一步的应用开发。王鹏飞等(2006)对辐射模块进行OMP并行运算处理,应用消息传递技术(MPI)完成了R42L26分辨率模式的并行化。R42L26模式系统被正式命名为Spectral Atmosphere Model of IAP LASG,简称SAMIL(周天军 等,2005;包庆 等,2006)。至此,大气模式SAMIL的分辨率、计算效率和模块化水平达到了国际主流气候模式的水平。

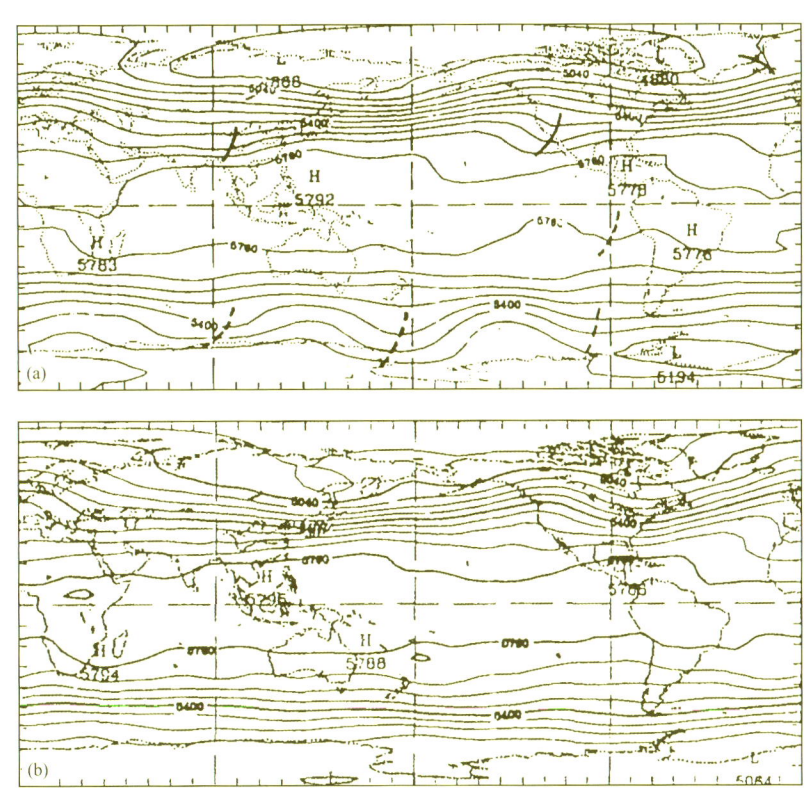

图2 Wu等(AAS,1996 图2)通过对R15L9模式采用SAS方案前、后的国际大气环流模式比较计划(AMIP)积分结果

高分辨率大气环流模式的发展过程是大气环流模式F/SAMIL发展经历的第二个阶段。2010年,LASG数值模拟课题组与GFDL的Shian-Jiann Lin研究团队合作,逐步将球立方有限体积动力框架FV3引入大气所气候系统模式中,采用有限体积球立方网格的新一代大气环流模式被命名为大气所LASG有限体积大气环流模式(finite volume atmosphere model of IAP LASG,简称为FAMIL)(周林炯,2015)。FV3动力框架为基于立方球剖分的全球准均匀网格,在计算效率、计算精度、局地加密和可扩展性等方面性能突出,被美国国家环境预报中心(NCEP)遴选为新一代天气预报系统的动力内核。FV3支持全球聚焦变网格,变网格的FV3在聚焦区可平滑过渡到更高分辨率,实现特定区域的加密功能。为了在LASG模式中引入FV3动力框架,首先在大气环流模式SAMIL的R42L26分辨率中接入FV3动力框架的半拉

格朗日通量形式平流方案(flux-form semi-Lagrangian transport scheme，缩写FFSL)。通过理想试验和干框架试验表明FFSL平流方案能够实现模式本身的保形和正定特性。与传统谱分量合成相比，FFSL方案克服了"负水汽问题"，降水频次得到改善(Wang X C et al.，2013)。Zhou等(2012)在国家超级计算天津中心的"天河一号"超级计算机上，成功进行了全球12.5 km和6.25 km超高分辨率的大气环流模式FAMIL万核计算规模的性能测试。结果表明，单独运行的大气模式FAMIL具有优越的并行计算性能和并行I/O性能。Li等(2017)在国家超级计算广州中心的"天河二号"超级计算机上，基于耦合器开展第二代大气环流模式FAMIL2测试，指出耦合系统模式在CPU使用率、CPU节点间信息传输等待时间、代码向量化、Gflops平均值、Gflops峰值五个方面表现优异，耦合系统具有良好的可扩展性。

2.2 大气模式物理过程参数方案发展历程

2.2.1 对流参数化发展

由于大气环流模式网格尺度大约在100 km，不能直接解析湍流、积云等小尺度天气过程，气候模式一般通过参数化间接表示次网格过程对平均网格的贡献。物理过程参数化是模式模拟及预测的最大不确定性来源，其中尤以对流-云-辐射的参数化过程最为复杂，大气环流模式的整体性能很大程度上受制于上述湿物理过程的参数化水平。LASG大气模式发展至今，先后引进替换了Manabe等(1965)、Tiedtke(1989)、Zhang和McFarlane(1995)等对流参数化方案，并开展了对流方案在LASG大气模式中的模拟性能评估(宋晓良，2005；刘琨 等，2010；Wang et al.，2011)。尽管已有的研究工作丰富了人们对积云对流问题的理解，积云参数化的研究也取得了一定进展，但由于积云对流本身所涉及的多尺度相互作用的复杂性，积云参数化的研究进展依然缓慢，其中闭合假设和积云模型的不确定性最大(Arakawa，2004)。夹卷和积云上升速度是积云模型中的重要组成部分，诸如气溶胶活化、衡量物输送等过程都与之有关。此外，积云垂直运动还便于准确地定出云顶高度。近年来，得益于大涡模拟和场地试验数据，夹卷率和积云垂直动量在理论和参数化领域都取得明显进展(Siebesma and Cuijpers，1995；de Roode et al.，2012；Wang and Zhang，2014)。

尺度自适应参数化方案是数值模式研发领域中的难点，也是热点。LASG大气模式组自主研发显式对流降水方案(resolved convective precipitation，简写RCP，软著号2017R11S036935)，实现积云对流降水显式化，将传统的积云对流降水用单参数云微物理方程改写，然后分别计算它们的云微物理属性，减少传统积云对流参数化方案中由于对流效果的平均化和强烈依靠对流参数准确性带来的误差。RCP方案在25~100 km的水平分辨率下基本实现了尺度自适应，无需调参。针对第六次国际耦合模式比较计划(CMIP6)模式的评估结果中，采用RCP方案的FAMIL2模式和其耦合版本FGOALS-f3显著增强了热带大气季节内振荡(MJO)的模拟能力(图3)，减少了双赤道辐合带(Double_ITCZ)误差，提高了厄尔尼诺-南方涛动(ENSO)的模拟能力。RCP方案准确再现极端降水强度和日变化特征，并且提高了对热带气旋(台风)模拟能力(He et al.，2019；Li et al.，2019；Bao et al.，2020a，2020b；Ahn et al.，2020；Li et al.，2021b)。

2.2.2 云参数化发展

云作为气候系统中水循环的重要成分，覆盖了地球约2/3的面积，在地气系统的辐射收支中起着重要作用。一方面，云通过反射太阳短波辐射，对地-气系统起到降温作用；另一方面，云通过吸收地表和云下大气放射的红外长波辐射，并以云顶较低的温度向外发射长波辐射，对地-气系统起到保温作用。由于云的水平尺度通常在100 m~10 km，远小于模式的水平网格距(~100 km)，因此云的参数化必须考虑湿度次网格分布，也就是所谓的统计云参数化方案。采用的PDF分布模型大致有均一分布、正态分布、贝塔分布等(Bony and Emanuel，2001；Tompkins，2002)。利用云分辨模式数据，Wang等(2015)分析了不同PDF分布对云模拟结果的影响。对云辐射传输过程来说，仅仅有云量的垂直分布是不够的，还需要知道

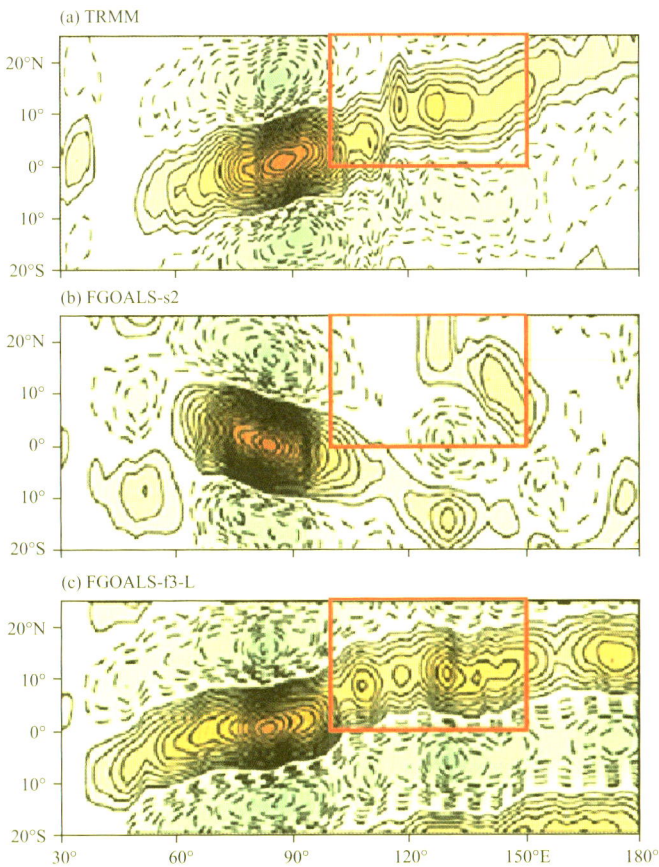

图 3　气候系统模式 FGOALS-s2 和 FGOALS-f3-L 对 MJO 的模拟性能。(a)TRMM 卫星观测,(b)参与 CMIP5 试验的 FGOALS-s2 模式结果,(c)参与 CMIP6 试验的 FGOALS f3-L 模式结果(Ahn et al.,2020,GRL,图 1)

云在垂直方向上的几何结构,以及云内水凝物的水平非均匀性。忽略云的次网格结构会对辐射和云微物理过程带来显著偏差(Fowler et al.,1996;Cairns et al.,2000;Gu and Liou,2001)。近年来,LASG 大气模式团队发展了一个次网格云生成器,可以产生满足多种约束条件的云样本。采用次网格云生成器并结合蒙特卡洛抽样,可有效提高模式对云辐射过程的模拟能力(Wang et al.,2021)。

2.2.3　辐射参数化发展

LASG 大气模式陆续采用了多个大气辐射参数化方案。GOALS 模式早期低分辨率版本 R15L9 采用的是 k 分布辐射方案,中等分辨率 R42L26 版本先后采用了英国气象局的 Edwards 辐射方案的改进版本 SES1 和 SES2,其中 SES2 版本先后参加了 CMIP3/CMIP5 气候模式比较计划;FAMIL 模式采用 RRT-MG 辐射方案。下面主要介绍 SES2 和 RRTMG 辐射方案。

SAMIL R42L26 采用的大气辐射方案最初是由 Edwards 等(1996)基于二流近似辐射传输方程提出的。澳大利亚气象局 Sun 等(1999a,1999b)对该辐射方案在气体吸收、云辐射参数化方面做了很多改进,模拟效果进一步得到改进,命名为 SES1 和 SES2。SES1 是 SAMIL R42L26 分辨率版本最初采用的大气辐射方案。但是,SES1 仅考虑了水汽、臭氧和 CO_2 的气体吸收作用,没有考虑 O_2、CH_4、N_2O 和 CFC 等大气中均匀混合气体的影响;也没有考虑对辐射过程和气候变化有重要影响的气溶胶辐射效应,同时 SES1 辐射方案的计算效率也较低。针对这些问题,Sun(2005)对 SES1 做了进一步改进,改进后的 SES2 辐射方案称为 SES2。SES2 增加了辐射计算的谱分辨率,其中短波波段由 4 个增为 9 个,长波波段由 7 个增为 8 个,进而引入 O_2、CH_4、N_2O 和 CFC 等 7 种均匀混合吸收性气体。和 SES1 一样,对气体吸收仍采用相关 k 分布方法处理,但 SES1 中相关 k 分布系数是根据 19 个参考气压和 3 个参考温度计算,而 SES2

中的参考气压和温度分别增加到59个和5个,从而得到了更为准确的气体吸收系数。谱分辨率增加后,SES2还对光谱积分过程进行了合理优化,提高了计算效率。此外,SES2方案还考虑了气溶胶辐射参数化,从而能够研究气溶胶的直接辐射效应。SES2是FGOALS-s版本所用的辐射方案。SES2采用了较新的气体吸收系数,从而明显增强了对气体吸收的模拟能力,在对流活跃区域和赤道辐合带增强了对水汽吸收的模拟,使得上述区域晴空OLR和入射地表短波辐射与观测的偏差有所减小。在SES2中引入CH_4和N_2O等均匀混合温室气体后,不仅在对流活跃区域,在全球范围内模拟的晴空和云天OLR正偏差明显减小,地表入射短波的正偏差也较原SES1辐射方案结果有所减小。此外,大气顶出射的短波、大气向下的长波和大气吸收短波辐射与观测的偏差均有不同程度的减小,优于SES1的结果。

环流模式快速辐射传输模式(the rapid radiative transfer model for GCMs,简写为RRTMG)是全球多个大气研究和气候模式研究团队采用的通用大气辐射方案(Clough et al.,2005),该方案基于相关k分布方案,其气候模式版本有14个短波和16个长波波段,可以有效处理计算温室气体、云和气溶胶的大气辐射效应。2014年起,大气模式FAMIL采用了RRTMG大气辐射方案,并为该方案重新设置了温室气体和气溶胶输入场。

2.3 大气模式评估

SAMIL和FAMIL系列模式研发以来,先后进行了一系列模式评估与诊断应用研究,涉及动力框架以及辐射、云降水对流和陆面等诸多物理过程,内容众多。限于篇幅,这里侧重总结天气与气候研究中最为基本而重要的能量平衡与水循环方面的内容,同时介绍新一代FAMIL模式对MJO、极端降水、降水日变化、热带气旋等极端天气气候事件的模拟技巧。

2.3.1 能量收支的误差

经过大气辐射过程、积云对流和诊断云等物理参数化过程调整,FGOALS-s版本(SAMIL)模拟的全球辐射通量的年平均结果与观测的偏差大幅度减小(图4)。其中大气顶能量收支的年变化及其平均值与观测更为接近。在积云对流方案调整基础上,通过对诊断云物理方案的进一步调整,SAMIL对云物理量模拟更为合理,赤道辐合带等区域在很大程度上克服了单一积云对流物理过程调整引起的云宏观和微观属性不匹配问题,能模拟出夏季气候平均辐射通量的全球分布特征,尤其在东亚区域有较好的模拟能力。研究还表明,在热带和副热带对流活跃区域SAMIL中积云对流过程偏差对辐射通量的模拟偏差有很大影响,而模式中较为简单的诊断云方案也会将云宏观物理量模拟偏差带入云微观量模拟中,也是主要偏差源之一。因此,急需进一步改进FGOALS大气模式的云物理和对流参数化过程。

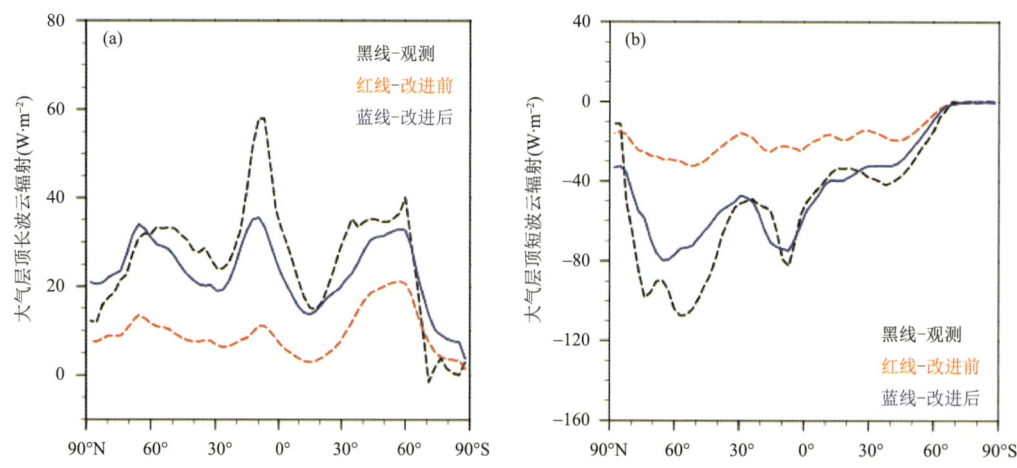

图4 FGOALS-s版本大气模式SAMIL模拟的全球多年平均的大气顶(a)长波和(b)短波云辐射效应。红线和蓝线分别为物理方案调整前、后模式结果(图来自李剑东 等,2010)

FAMIL 的第一个版本 FAMIL1 具有灵活的水平分辨率伸缩性。FAMIL1 最高的实用分辨率可到全球 25 km。FAMIL1 可以在当时世界上最快的超级计算机天河 1A 上进行大规模并行运算，是 IAP/LASG 第三代全球海洋-大气-陆地耦合系统模式 FGOALS 的大气分量，参与了第 6 次国际耦合模式评估计划 CMIP6。FAMIL1 能够准确再现能量和水分平衡特征，通过与观测和再分析资料进行比较，FAMIL1 很好地模拟了大气顶部和地表的辐射通量以及地表的潜热、感热通量。同时，FAMIL1 还很好地模拟了能量平衡的季节变化和地理分布。多年平均降水蒸发误差是 10^{-5} 量级。比较能量平衡和水分平衡的关系时发现，FAMIL1 对长波辐射通量、降水蒸发和大气垂直运动之间的联系与再分析资料非常一致（图 5）。

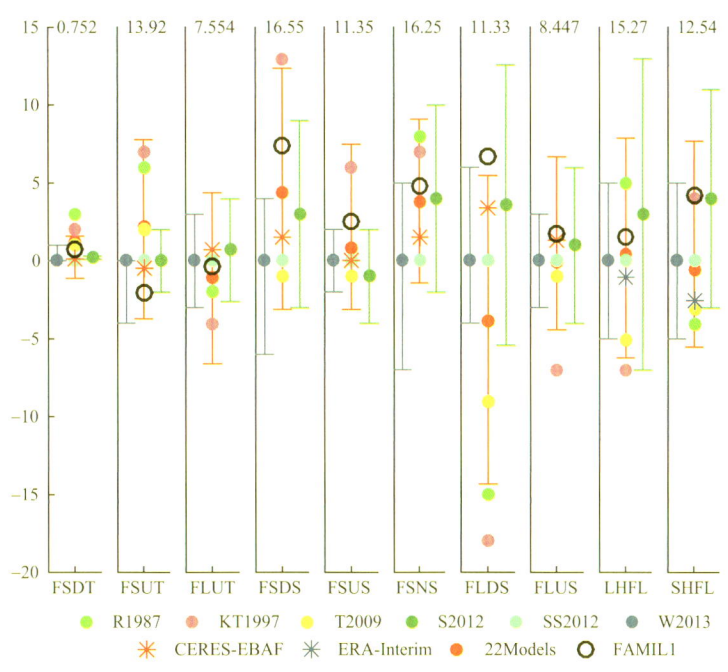

图 5　FAMIL1 全球年平均大气顶部和地表的能量平衡与不同观测、再分析资料、不同时期的经典文献和 22 个 CMIP6 模式的比较（图引自 Zhou et al.，2015）

气溶胶引起的辐射强迫是当前气候变化和归因研究的热点科学问题，基于全球模式可研究其大时、空尺度的变化特征。为此，LASG 大气模式研发团队选用有代表性的气溶胶折射指数资料和米散射程序，综合考虑了辐射波长、气溶胶粒径和相对湿度等影响因子，获得黑炭、有机碳、硫酸盐、硝酸盐、沙尘和海盐气溶胶光学属性的离线计算结果，进而结合到大气模式中；同时为模式引入了多种云反照率效应计算方案。基于大气模式 SAMIL 计算的 21 世纪最初 10 年全球平均的人为气溶胶直接和间接辐射强迫量值与 IPCC AR4 及 AR5 多模式结果相当（图 6）；由此为大气模式建立了一套完整的气溶胶辐射气候效应参数

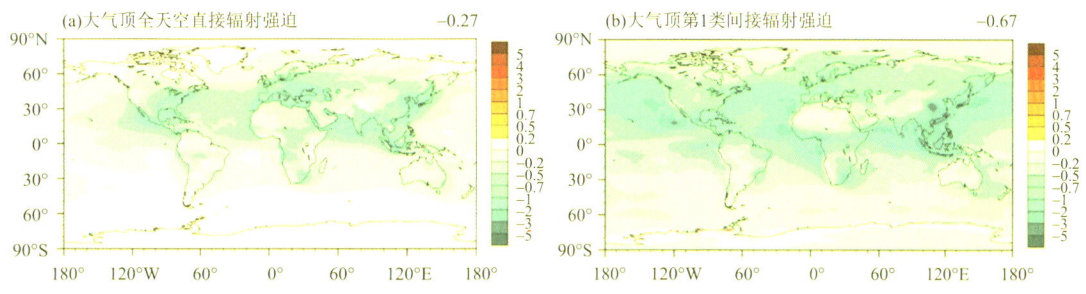

图 6　FGOALS-s 大气模式计算的全球年平均大气顶部人为气溶胶(a)直接和(b)第 1 类间接辐射强迫
（时段为 21 世纪最初 10 年，相对于工业革命前的 19 世纪 50 年代；单位：W·m^{-2}）

化模块,为辐射强迫、人为气候归因等研究打下坚实的模式基础。LASG 大气模式团队进一步利用 CMIP5 试验资料,计算了全球人为气溶胶辐射强迫的历史变化和不同 RCP 排放情景下的未来变化。结果表明:东亚已超过欧美成为北半球人为气溶胶辐射强迫最强的区域,东亚气溶胶辐射强迫的增强使得北半球气溶胶辐射强迫高值区从 20 世纪 80 年代起自欧美所在高纬度地区开始向低纬度南移,影响全球的能量收支和温度变化。与欧美区域相比,除高气溶胶含量的影响而外,东亚区域较强的夏季低层水汽会增加亲水性气溶胶的吸湿增长,强化其光学厚度及直接辐射强迫。

2.3.2 极端天气气候事件模拟技巧

由极端降水引发的灾害占中国自然灾害的 70%,准确的模拟和预估极端降水事件对于防灾减灾至关重要,目前的气候模式对于极端降水的模拟仍是充满挑战性的问题。FAMIL2 模式缓解了青藏高原南坡虚假降水问题,模拟青藏高原地区的日降水概率密度分布与高分辨率卫星资料一致(Bao et al.,2020a);Bao 等(2020b)表明,新一代 FAMIL2 大气模式能够准确再现亚洲夏季风的降水日变化峰值时间,特别是再现了中国陆地夜间降水峰值(又称"巴山夜雨")的特点。从日降水强度和降水频次的概率密度函数及极端降水天数的空间分布方面分析发现:四种再分析资料和大多数 CMIP 模式均不能抓住 150 mm·d^{-1} 以上的极端降水,并且低估了极端降水的频次。FAMIL2 模式由于对水汽和垂直加热率的模拟更接近真实,提高了对中国东部极端降水的模拟能力(He S et al.,2019)。

热带大气季节内振荡(MJO)是大气环流的重要系统,MJO 的活动对天气和气候有明显影响。偏弱的 MJO 是气候模拟中国际公认的难题,大气模式 FAMIL2 和全耦合版本 FGOALS-f3,准确再现了 MJO 东传的强度和速度特征。Jiang 等(2015)评估了 CMIP5 所有模式模拟 MJO 的性能,分析指出大气所前一代模式(FGOALS-s2)模拟 MJO 东传明显偏弱,Ahn 等(2020)评估结果表明,相较于 CMIP5 模式版本,采用 RCP 方案的 CMIP6 大气所 FGOALS-f3 模式模拟的 MJO 东传强度增强、速度合理,因此 MJO 的模拟性能获得了显著提升。

在 100 km 标准分辨率下,国际主流气候模式模拟台风个数偏少、模拟能力不足。以美国 NCAR CAM 最新谱元模式版本为例,对照试验中西北太平洋地区仅能模拟出约每年 9 个台风(Zarzycki et al.,2015),相较于 IBtrack 观测数据中西北太平洋平均每年 77 个台风,误差达 88%。FAMIL 研发团队的评估指出,通过采用显式对流降水方案 RCP 改进了 FAMIL2 模式中对流耦合赤道波群和 MJO 的模拟能力,使用 100 km 分辨率能够再现全球热带气旋的生成位置、移动路径、季节循环、年际变率等特征。分辨率 100 km 的 FAMIL2 模拟出西北太平洋热带气旋爆发个数为 76 个/年,与观测结果一致,优于百千米国际主流模式(以 CAM 最新的谱元版本为例)9 个/年的结果。此外,FAMIL2 有能力模拟出热带气旋爆发个数的年际变率,在西北太平洋、北大西洋的热带气旋爆发个数时间相关系数分别是 0.51 和 0.49,超过了显著性 95% 的检验标准(Li et al.,2019;He B et al.,2019;Bao et al.,2020b)。

3 陆面过程模式发展历程

3.1 SSIB 陆面方案的改进及其在 LASG 模式中的应用

SSIB 陆面方案是 SIB(simple biosphere model)的简化版本,主要用于陆-气耦合模式中的陆面过程模拟(Sellers et al.,1986;Xue et al.,1991)。Liu 和 Wu(1997)改进了 SSIB 陆面方案并将其与 LASG 大气环流谱模式 R15L9 进行双向耦合。因此,SSIB 也成为了 LASG 第一代海-陆-气耦合模式 GOALS 的陆面分量(吴国雄 等,1997)。SSIB 包含 3 层土壤和 1 层植被,考虑了 11 种不同的植被类型;土壤按深度可划分为表层(2 cm)、根层(10 cm)和深层土壤(1 m);预报量包括植被对降水的截留量、地表雪盖量、植被冠层、土壤表层和土壤深层的温度以及 3 层土壤含水量等 8 个预报量。Liu 和 Wu(1997)基于耦合 SSIB 的 R15L9 模式进行了季风爆发和夏季环流模拟,发现 SSIB-R15L9 耦合模式可以有效改进 7 月气候态和季风季节推进特征的模拟。孙岚等(2000)进一步评估表明,耦合 SSIB 后的 R15L9 模式能够显著改善全球陆地水平衡状况,其模拟的降水分布更接近实况。

3.2 CLM 陆面方案的改进及其在 LASG 模式中的应用

随着模式发展,GOALS 模式的大气分辨率由 R15L9 逐步提升至 R42L26,并正式命名为 SAMIL 模式(王在志 等,2005a,2005b)。为了适应大气模式分辨率提高和物理过程完善的需求,包庆等(2006)首次将美国国家大气研究中心(NCAR)通用陆面模式 CLM 2.0 与 SAMIL-R42L26 进行耦合。相对于 SSIB 耦合模式版本,CLM 耦合模式版本对地表感热、温度、降水率、潜热通量和海平面气压场的模拟能力大幅度提高,尤其是夏季表面感热通量场方面,使亚洲北部和东南部、格陵兰岛以及北美洲大部分地区的数值从 100 $W \cdot m^{-2}$ 降低至约 60 $W \cdot m^{-2}$,与 NCEP 再分析资料基本一致。耦合 CLM2.0 的 SAMIL-R42L26 不但能确保陆地表面能量收支平衡,还能较好地模拟亚洲季风区地表感热和潜热的演变趋势。此后,SAMIL 模式及其海-气耦合版本 FGOALS-s 的陆面分量始终以 CLM 为主,但模式版本不断提升,如 FGOAL-s2 中陆面模式版本由 CLM 2.0 更新至 CLM 3.0 (Bao et al.,2013)。近年来,LASG 采用有限体积球立方网格发展了新一代大气模式 FAMIL 1.0 及其耦合版本 FGOALS-f 1.0(Zhou et al.,2015),其陆面模式版本为 CLM3.5(Subin et al.,2011),而其更新版本 FAMIL 2.0/FGOALS-f 2.0 的陆面分量模式则进一步升级至 CLM 4(Oleson et al.,2010;包庆 等,2019)。

此外,为了改进 FGOALS-f 模式对青藏高原及周边地区陆-气相互作用的模拟性能,FAMIL 模式研发团队进一步将陆面分量模式升级至 CLM 4.5,并进行了一系列改进和优化。首先,为了准确模拟青藏高原湖泊的物理过程,研发团队一方面通过将海洋模型中的垂直混合 K 廓线(KPP)方案耦合到 CLM 湖泊模型中,以更加真实地模拟湖泊水体垂直混合过程,提高对湖表温度及湖温廓线的模拟结果(Zhang et al.,2019);同时又通过改进湖冰内部的热导率、表面反照率、粗糙度和空气密度的计算方案,并加入湖水盐分对结冰点的影响和表面升华过程对冰厚的影响以提高 CLM 对湖冰的模拟精度,从而更加充分地描述湖冰的冻结与融化过程(Zhu et al.,2020)。然后,针对青藏高原地区高寒草甸下垫面,建立了高寒草地根系生物量数据集和改进了 CLM 4.5 中高原含根土壤水热属性算法,有效提高了土壤含水量和土壤温度的模拟性能(苏有琦 等,2020)。最后,为了准确模拟青藏高原复杂地形对地表入射短波的吸收、遮挡和吸收作用,模式研发团队在 CLM 4.5 增加了三维地形辐射参数化方案。该三维地形参数化方案将入射短波辐射分解为五个分量:直接辐射通量、散射辐射通量、直接-反射辐射通量、散射-反射辐射通量以及多次散射-反射辐射通量,进而实现对复杂地形区地表短波辐射过程的准确计算(Liou et al.,2007;Lee et al.,2013)。耦合三维参数化方案后,FGOALS-f2 对青藏高原地表短波辐射、气温和降水的模拟性能具有明显提升(念国魁,2020)。

3.3 雪盖模型发展及其在 LASG 模式中的应用

积雪下垫面是一个非常重要的陆面类型,积雪表面反照率和内部热量、能量传输对天气系统有很大的影响。考虑到当时 LASG 大气环流谱模式的陆面过程 SSIB 中的积雪参数化方案过于简单,无法刻画积雪的重要特征。孙菽芬等(1999)参考多种积雪模型的优点,并进行补充和发展,形成了一个物理过程更加真实的一维积雪模型。该积雪模型改用内能替代温度作为预报量,克服了相变过程中直接预报温度所造成的计算误差,有效改进了相变过程的预报;通过积雪压实方程(Anderson,1976)考虑了雪花形态破坏过程和压实过程而引起的密度变化;优化设计了雪盖的光学特性、热力学特性及水流流动特性等参数化方案。经对比,该一维简化模型的模拟结果不但与 Jordan 复杂积雪模型(Jordan,1991)几乎有同等精度,而且与俄罗斯 Yershov 站实测数据吻合十分理想。

为了进一步理解不同复杂程度积雪模型对气候模拟的影响,Jin 等(1999a)对一维积雪模型 SNTHERM (U. S. Army Cold Regions Research Engineering Laboratory model)、三层简化积雪模型 SAST(snow-atmosphere-soil transfer)和 BATS(biosphere-atmosphere transfer scheme)的积雪分量模型

等三个方案进行了对比研究。结果表明,物理过程最复杂的 SNTHERM 模型的模拟结果和观测结果吻合度最高,尤其是在日变化模拟方面优势明显。物理过程最简单的 BATS 模型的模拟精确度足够满足季节尺度的模拟需求,但由于其物理过程简单,无法刻画积雪的垂直变化和日变化物理属性,因此可能对天气和水文模拟产生影响。而在合理的垂直分层设计下,中等复杂程度的 SAST 模型的模拟结果与复杂模型 SNTHERM 精确度相当。基于评估结果,Jin 等(1999b)进一步基于 SNTHERM 研发了可用大气环流的三层积雪模型。同时雪模型还参考了 BATS 的物理参数化方案,将其可适用范围扩展至植被覆盖区域。Sun 等(2001)将三层积雪模型 SAST 与陆面过程 SSIB 耦合,利用俄罗斯及法国的观测数据对改进后的 SSIB 模型进行了检验,改进后 SSIB 能较好再现观测情况。

3.4 动态植被模型在 LASG 模式中的应用

植被-大气相互作用模式 AVIM(atmosphere vegetation interaction model)是 Ji(1995)在一维地表过程模式 LPM (Ji and Hu,1989)的基础上加入植被生理过程发展而来,包括了完整的植被物理过程和生物过程。为了实现 GOALS 模式植被分布的动态模拟,Zeng 等(2008a,2008b)将 SAMIL-R42L9 模式与 AVIM 进行了双向耦合,首次实现了 SAMIL 模式对叶面积指数(LAI)和净初级生产力(NPP)的动态模拟。然后,曾红玲等(2010)基于耦合 AVIM 后的 SAMIL-R42L9 模式,探讨了全球植被分布对气候和大气环流的潜在影响,指出植被通过改变地表特征参数进而改变地表通量和能量收支,且这种影响可以从地表向上延伸至对流层的中高层,并可进一步增强三圈环流,从而使现有的气候和植被分布更加稳定。

为了进一步实现模式对碳循环过程的模拟,Wang J 等(2013)将动态植被模型 VEGAS 2.0 引入到 FGOALS-s2。VEGAS 2.0 能够对陆地植被的时、空演变特征进行动态模拟(Zeng et al.,2004,2005)。耦合 VEGAS 2.0 后,FGOALS-s2 所模拟的全球陆地初级生产总值(GPP)和 NPP 均接近观测值,并在热带和季风区体现出良好的季节变化特征。

4 耦合模式发展历程

4.1 气候系统模式 GOALS

LASG 国家重点实验室耦合气候系统模式经历了三个主要的阶段,第一阶段包括两个主要的耦合模式版本的研发。LASG 数值模拟团队率先在 20 世纪 90 年代发展了中国第一个包括全球大气、海洋、陆地和海冰的物理气候系统模式 GOALS,GOALS 耦合模式是数值模拟团队集体的研究成果,吴国雄、张学洪和石广玉课题研究小组对模式的发展起核心作用(吴国雄 等,1997;Zhang et al.,2000)(图 2),气候系统模式 GOALS 的大气分量采用 15 波菱形截断,垂直方向分 9 层(R15L9),水平分辨率相当于 7.5°经度×4.5°纬度。耦合模式 GOALS 的陆面分量模型为 SSiB,SSiB 模型包括 3 层土壤和 1 层植被以及 11 种植被类型。耦合模式 GOALS 的海洋分量模式是一个基于原始方程的 20 层海洋环流模式,其水平分辨率为 5°经度×5°纬度,垂直方向分 20 层,其中 1000 m 以上有 9 层。GOALS 模式的北边界取在 72°N,除了北极被处理为一个岛外,模式系统包括了整个北冰洋。GOALS 模式中建立了一个简单的热力学海冰模式,人为地增加了南极附近海表盐强迫。GOALS 模式采用通量订正的方法控制耦合模式常见的气候漂移问题,能够合理再现大气中水循环的收支情况。在年际变化尺度上,GOALS 海-气耦合模式能够较为准确模拟出类似 ENSO 的海温异常事件。GOALS 模式参与了国际模式比较计划(Kang et al.,2002),被用于青藏高原和亚洲季风领域的数值模拟研究(Wu et al.,1998;李伟平 等,2001;刘屹岷 等,1999;Zhang et al.,2000;刘平 等,2000;Wu et al.,2003;Liang et al.,2005),受到了国内外同行的重视和好评。21 世纪初,LASG 数值模拟团队更新了气候模式系统 GOALS 的版本,实现了全球大气环流谱模式(R42L9)与海洋环流模式(T63L30)在 40°S~40°N 的开洋面上海-气通量交换的完全耦合,大气和海洋分量模式分辨率得到提升。新一代 GOALS 模式系统稳定积分了 40 年,基本上不存在明显的气候漂移问题。通过对比新一代 GOALS 模式对热带和副热带地区海温、海表风应力、洋面净通量和降水等的气候平均态的模拟结果与实测资料，表明模式再现了当今气候的主要特征（吴统文 等，2004）。

4.2 气候系统模式 FGOALS-s

气候系统模式 FGOALS-s 是 LASG 耦合模式发展的第二阶段的成果，包含三个耦合模式版本：FGOALS-s1.0、FGOALS-s1.1 和 FGOALS-s2。2004 年，气候系统模式 FGOALS 1.0 由 LASG 发布。FGOALS 1.0 的组成包括两个大气分量模块：谱模式版本大气环流谱模式 SAMIL 和其耦合版本 FGOALS-s1.0；另一个是格点版本 GAMIL（Wang et al.，2004），其相关耦合版本称为 FGOALS-g1.0。海洋分量模块是 LASG IAP common ocean model（缩写为 LICOM）（刘海龙 等，2004）。陆面、海冰和耦合器来自于 NCAR CCSM2。与 GOALS 相比，FGOALS-s1.0 各分量模式的分辨率有显著提高（王在志 等，2005a），主要的物理过程也有大量改进（周天军 等，2005）。FGOALS-s1.0 采用了先进的计算技术，包括气候模式耦合器（CPL5）、应用消息传递接口（MPI）和标准化的 NETCDF 数据格式的输入/输出（I/O）。然而，FGOALS-s1.0 在模拟过程中也发现存在一些偏差，如在大气层顶能量平衡、地表及海表的温度、云量以及亚洲季风区的降水分布。与 1.0 版本相比，FGOALS-s1.1 大气分量模式物理过程的改进主要包括：Edwards-Slingo 辐射方案（Edwards and Slingo，1996）、质量通量的积云对流参数化方案 Tiedtke（Tiedtke，1989）和诊断云参数化方案。在 Edwards-Slingo 辐射方案中，短波带的分辨率从 4 波段扩展到 9 波段，长波带从 7 波段扩展到 8 波段，吸收气体的种类增加：引入了 O_2、CH_4、N_2O 和 CFC 等温室气体。另外，辐射模型可以考虑 5 类气溶胶的直接效应，这 5 类气溶胶包括黑炭、灰炭、硫酸盐、海盐和沙尘（Sun，2005；Li et al.，2009）。质量通量的积云对流参数化方案选择 Tiedkte 的初始版本，调整了浅对流的横向对流卷入/卷出率，为激发深对流设定了相对湿度临界值，并考虑云水转换成雨水的效率（刘屹岷 等，2007）。诊断云方案中，低云临界值改为由不同的陆面、海洋和雪/冰盖区来控制。此外，FGOALS-s1.1 大气分量 SAMIL 模式的源代码已经升级为 Fortran90 标准，应用了先进的模块化技术（包括动态内存、module 和 interface 的使用等），节省了计算所需的内存。FGOALS-s1.1 模式与 1.0 版本相比，不仅实现了模式层顶 TOA 能量平衡，而且改进了 1.0 版本中冷舌过度西伸的问题，同时在印度洋和西太平洋上 FGOALS-s1.1 模式模拟的热带 SST 更加接近观测。此外，FGOALS-s1.1 能够模拟出东亚夏季风年际变率第一主要模态的特征和其与 El Nino 衰减位相的关系，为以后开展东亚季风预测打下坚实基础（Bao et al.，2010）。

为了参与 IPCC 第 5 次评估报告有关数值试验（第五次国际耦合模式比较计划，CMIP5），LASG 数值模拟团队进一步完善耦合模式各个模块，不仅对大气、海洋和耦合器进一步发展，而且增加了动态植被过程和海洋生物化学过程，形成了 FGOALS-s2 地球系统模式（Wang et al.，2014）。FGOALS-s2 的大气分量版本是 SAMIL2.4.7（Bao et al.，2013），垂直方向仍然采用混合坐标，分为 26 层，从地面到 2.19 hPa；水平分辨率为 42 波菱形截断，物理过程方面通过改进积云参数化中闭合假设和夹卷率、完善非局地边界层方案和层云方案等工作提高模式性能。FGOALS-s2 的海洋分量是 LICOM2（Liu et al.，2012），垂直方向分为 30 层，水平分辨率赤道地区 0.5°经度×0.5°纬度，非赤道地区 1°经度×1°纬度。耦合系统的陆面分量模式为 CLM3（包庆 等，2006）。FGOALS-s2 海冰模式的热力学分量采用的是 CCSM 海冰模式版本 5（CSIM5）（Briegleb et al.，2004）。耦合器升级为无通量订正的 NCAR 第六代耦合器。FGOALS-s2 对大气中重要变率事件如 PNA 和 NAO 遥相关型、ENSO 的非周期特征和振幅的模拟性能达到国际主流模式水平。采用改进后的耦合模式 FGOALS-s2，完成了 IPCC 第 5 次评估报告要求的所有核心数值试验、部分优先做和选做试验，截至 2022 年 3 月，大气所发布的数据集下载总量达 424 TB。IPCC 第 5 次评估报告第 9 章图 7.36 对参加 CMIP5 所有模式 ENSO 评估表明，FGOALS-s2 模拟 ENSO 的技巧名列前茅。

4.3 气候系统模式 FGOALS-f

气候系统模式 FGOALS-f 是 LASG 耦合模式发展的第三阶段的成果，耦合模式包含了三个主要的版本：FGOALS-f1，FGOALS-f2 和 FGOALS-f3。大气环流模式 FAMIL1 接入 NCAR 第七代耦合器 CPL7，形成了气候系统模式 FGOALS-f1 版本，FGOALS-f1 可长期稳定运行的分辨率为 C48(200 km)(Zhou et al.，2015)。FGOALS-f2 和 FGOALS-f3(简写为 FGOALS-f2/3)为 FGOALS-f1 的升级版本，是 LASG 研发的新一代海-陆-气-冰耦合的气候系统模式，其大气分量是有限体积大气环流模式 FAMIL2。FAMIL2 中采用了自主研发的显式对流降水方案 RCP，对台风、小时级极端降水、降水日循环模拟性能突出。FGOALS-f2/3 成功解决了高分辨率下耦合模式不稳定的问题，耦合系统模式以 C96(100 km)可以稳定运行 500 模式年以上(He et al.，2019；Bao et al.，2020b；Li et al.，2021a，2021b)；FGOALS-f2/3 的海洋分量为 LICOM3(Lin et al.，2020)，海冰分量为海冰模式 CICE4.0(Hunke et al.，2010)，陆面分量为通用陆面模式 CLM4.0(Oleson et al.，2010)，分量模型经由耦合器 CPL7 实现耦合。在耦合模式比较计划的第六阶段(CMIP6)，FGOALS-f3 大气分量 FAMIL2 的标准分辨率为 100 km，模式版本为 FGOALS-f3-L；高分辨率为 25 km，模式版本为 FGOALS-f3-H。与 FGOALS-f3 相比，FGOALS-f2 海洋分量为 POP2 海洋模式(Smith et al.，2002)，其余分量模型与 FGOALS-f3 模式一致。气候系统模式 FGOALS-f2 版本主要应用于次季节、季节、年际和年代际等多时间尺度的无缝隙天气气候预测，以及青藏高原相关次季节预测计划 LS4P。FGOALS-f3 为参与 IPCC 第 6 次评估报告有关数值试验的模式版本，其中 FGOALS-f3-H 实现了分辨率提升到 25 km 的目标，FGOALS-f3 也作为分辨率最高的三个气候系统模式之一参与国际高分辨率模式比较计划 CMIP6 HighResMIP，进入了全球高分辨率气候模式发展的前列，另外两家耦合模式分别为英国 HadGCM3-GC31-HM、意大利 CMCC-CM2-VHR4。美国海洋大气局太平洋实验室 Yann Planton 博士发表在 BAMS 的评估论文指出(Planton et al.，2021)，FGOALS-f3-L 模式对 Double_ITCZ 和 ENSO 的模拟性能优越，热带太平洋核心区的海温误差(RMSE)仅为 0.04 ℃·℃$^{-1}$，优于 CMIP6 模式的平均误差 0.16 ℃·℃$^{-1}$。印度热带气象研究所 Halder 等(2021)评估指出，在 27 个 CMIP6 模式中，FGOALS-f3-L 模式对印度洋年代际变率模拟的评分技巧最高，对年际变率的模拟评分技巧排名第二。截至 2022 年 3 月，FGOALS-f3-L 模式参加 AMIP、CMIP 和 GMMIP 等子计划的数据集被下载量达到 216 TB，成功下载 NetCDF 文件个数达 88 万个，高分辨率数据集被下载总量达 270 TB，成功下载 NetCDF 文件个数达 29 万个。Bao 等(2020b)指出，水平分辨率的提高能够显著改进模式在日变化和极端降水方面的模拟性能。对比全球 100 km 中等分辨率 FGOALS-f2/3 模式，全球 25 km 的 FGOALS-f2/3 能够更好地再现全球降水的日变化主要特征，特别是能够准确模拟亚洲夏季风的降水日变化峰值时间。在台风模拟方面，FGOALS-f2/3 在高分辨率和中等分辨率下均能够解析台风活动；在全球台风个数、生成位置、移动路径和生命周期等台风关键指标方面性能突出；随着模式水平分辨率的提升，台风模拟的负偏差显著减少，最大风速、螺旋雨带和台风眼等台风的关键精细化结构特征较中等分辨率版本显著提高(Li et al.，2021b)。在 GMMIP 试验中，利用 CMIP6 版本开展了青藏高原地形动力和热力敏感性试验，结果表明，高原热力强迫作用对亚洲夏季风在亚洲大陆上的形成具有决定性作用。与 CMIP5 模式版本的结果相比，高原对季风降水的影响在西太平洋上有一定的不确定性(图7)，这需要进一步改进和发展模式来减少这方面的模拟误差(He et al.，2020)。此外，FGOALS-f3-L 还参加了北极海冰比较计划，开展了大样本集合试验来模拟北极海冰减少对全球气候变化的影响(He et al.，2021)。多模式集合的结果表明，北极海冰减少是导致冬季欧亚大陆东部变冷趋势出现的重要原因(Smith et al.，2022)。

4.4 耦合模式评估

FGOALS 模式研发以来，对全球降水、温度和环流等主要气候特征进行了系统性的评估。随着模式分辨率和物理参数化过程的不断改进，模式对气候平均态的模拟性能也随之不断提升。这里以时间顺序，简述 FGOALS 模式对气候平均态的模拟状况。

2003 年，FGOALS 模式实现了全球大气环流谱模式(R42L9)与海洋环流模式(T63L30)在 40°S~40°N 洋面上海-气通量交换的完全耦合。该模式系统积分 40 年后基本上不存在明显的气候漂移。对模式稳定 30 年平均气候态分析结果表明，相较于观测，FGOALS 模式基本上再现了热带、副热带地区海温、海表风

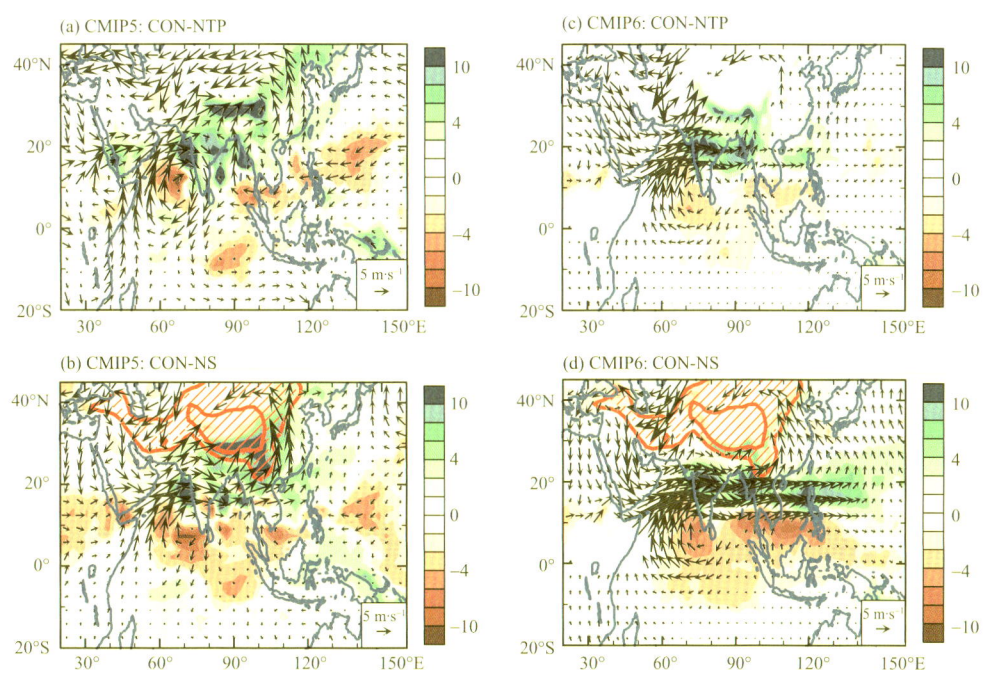

图 7 气候态夏季降水(阴影,mm·d^{-1})和 850 hPa 风场(矢量,m·s^{-1})异常。(a)CMIP5 版本的参考试验减去无青藏伊朗高原地形试验;(b)CMIP5 版本的参考试验减去无青藏伊朗高原感热加热试验;(c)、(d)分别与(a)、(b)同,但为 CMIP6 版本的结果

应力、洋面净通量和降水等的气候平均态主要特征,对热带气候平均态已具有一定的模拟能力。不过与观测相比,区域性差异明显存在,比如沿赤道西太平洋/暖池区和靠近南美沿岸的东太平洋海域以及印度洋海表温度明显偏高约 2 K,所模拟的赤道东太平洋海温冷舌西伸明显,造成赤道中太平洋海温明显偏低等偏差。这些模拟误差,与模式中海表风应力和洋面所得到或释放的净热通量有密切的关系。海表温度的模拟误差反过来也影响到对降水的模拟。2004 年,利用改进后的 FGOALS 模式 R42L9 分辨率版本,采用观测的海温和海冰进行了 20 年(1979—1998 年)稳定积分的 AMIP 试验,重点分析了模式对亚洲季风区气候的模拟性能。模式能够抓住印度和东亚季风降水的主要区域和环流分布型。与观测相比,模拟的高原夏季加热偏强,由此导致偏强的东亚夏季风和偏弱的印度季风,同时西太平洋副高过于西伸。2013 年,基于改进后的 FGOALS-f 谱模式版本(FGOALS-s2),参加了第 5 次全球耦合模式比较计划(CMIP5)。新版本 FGOALS-s2 成功克服了气候漂移问题,可基本再现全球的海表温度、降水和大气环流特征。该模式可以准确地再现赤道太平洋海表温度的年循环、亚洲夏季风环流和降水中心。此外,模式可以真实地再现 ENSO 的准规则周期、PDO 空间分布和 20 世纪季风降水的历史变化。模式模拟的全球平均温度略高于观测,可能与没有考虑气溶胶的间接效应有关。2019 年,新版本的 FGOALS 有限体积球立方网格版本 FGOALS f3 参加了第 6 次全球耦合模式比较计划的标准大气模式试验(1979—2014 年),共提交了 3 组试验的逐日和逐月模式数据。FGOALS-f3 可以抓住全球大气环流和降水的主要空间特征,对 MJO 传播、热带气旋和极端降水有较高的模拟能力。

4.5 耦合模式应用

耦合模式应用方面,LASG 基于 FGOALS-f2 模式建立天气-气候一体化的集合无缝隙预测系统,简称为 FGOALS-f2 无缝隙预测系统。FGOALS-f2 预测系统对 MJO、热带气旋(台风)、厄尔尼诺、北极海冰等的预测技巧达到国际同类模式预测系统先进水平。FGOALS-f2 无缝隙预测系统的预测结果多次用于国家级会商,参与单位组织的咨询报告撰写,在中国气象局国家气候中心、自然资源部国家海洋环境预报中心、水利部信息中心等多家国家级和省级业务平台实现应用,服务于国家需求和国防安全。FGOALS-f2

无缝隙预测系统加入世界气象组织(WMO)次季节-季节(S2S)预测计划,向全球用户提供实时预测。在国际 S2S 参与模式系统中,FGOALS-f2 系统的实时预报时效性最高,重要指标的预测水平达到国际同类模式水平:MJO 预报技巧可达 30 天,接近欧洲中心 33 天的 MJO 预报技巧(Lim et al.,2018;Jiang et al.,2020)。FGOALS-f2 系统加入国际 S2S 计划,以高时效性和高预报技巧多次助力一带一路地区防灾减灾,提升了中国在 S2S 领域的国际影响力。

5 气候系统模式 FGOALS 未来展望

5.1 动力框架和分辨率未来展望

高分辨率大气环流模式是大气模式发展的重要趋势之一,高分辨率的模式在云、降水、极端天气事件等方面的模拟能力较低分辨率模式有明显优势。截至 2022 年,欧洲中期数值预报中心(ECMWF)的第 5 代季节预测模式分辨率已经提升到全球 36 km,天气预报模式分辨率为 9 km;2019 年 6 月,美国海洋大气局(NOAA)通过美国下一代全球预报系统研制计划(NGGPS)遴选了有限体积球立方动力框架 FV3 为新一代全球预报系统的核心,并将 GFS 天气预报模式正式更新为 GFSv15 系统,目前 GFS 预报系统已经升级到 GFSv16,分辨率为 12.5 km。FGOALS 参与 CMIP6 的最高水平分辨率为 C384(~25 km),垂直层数为 32 层。在未来的模式研发中,FGOALS 模式将进入天气-气候一体化模式的研发周期,计划将模式水平分辨率提升到 C768(12.5 km)。在 CMIP7 周期,进一步升级到非静力的动力框架,优化模式动力网格的计算性能及与国产芯片深度耦合,发展新一代大气环流模式 FAMIL3。在完成水平 12.5 km 垂直 64 层版本的 CMIP7 气候系统模式 FGOALS-f4 研发后,开展全球千米级模拟的尝试,提升模式动力框架在超高分辨率条件下的准确性和稳定性,对模式动力框架性能进行全面评估检验和改进。

采用全球聚焦拉伸网格方法,是提高中国模式水平分辨率的重要方法。FGOALS-f4 模式仍将采用球立方网格有限体积动力内核 FV3,通过动力框架施密特转换,实现变网格功能。在 C768 分辨率下,均匀网格分辨率为全球 12.5 km,2.5 倍聚焦情况下的聚焦区域分辨率能达到 5 km,并囊括青藏高原及其周边地区(图8)。8 倍聚焦情况下(C768_r8),聚焦区的分辨率为 12.5 km/8≈1.56 km。未来将基于该变网格动力框架展开计算性能和稳定性等方面的进一步研究。

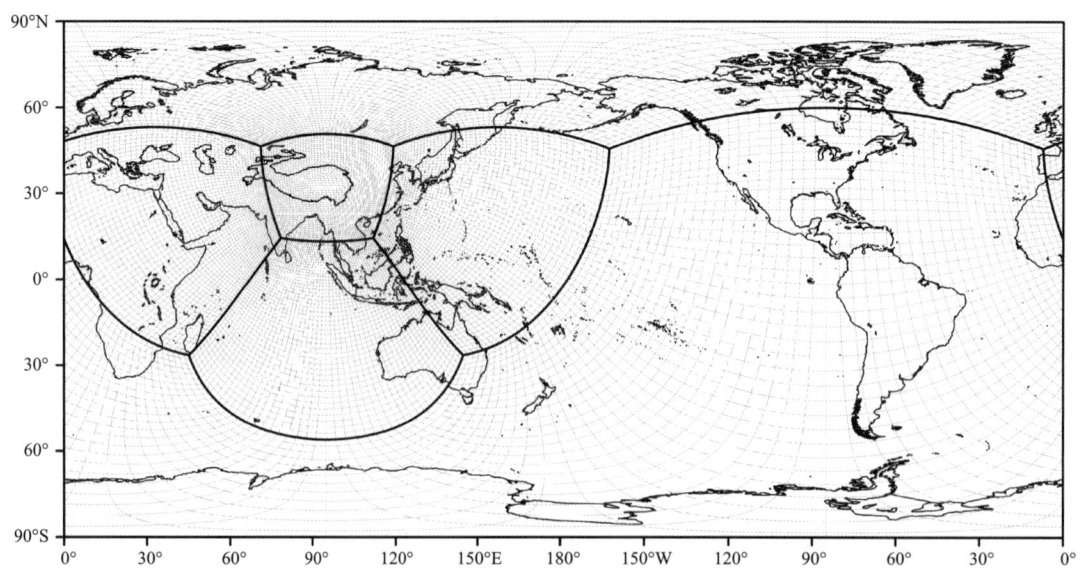

图 8 聚焦青藏高原地区的球立方有限体积动力框架的网格单元示意(刘安岭绘制)。示意图的网格分辨率选取 C768,拉伸率设为 2.5,聚焦区分辨率 5 km,聚焦对面区分辨率约 30 km,拉伸区的分辨率从 5 km 逐渐过渡到 30 km,聚焦中心为(95°E,32°N),示意图采用 16 格距间隔绘制,避免原始网格线过度密集的问题

5.2 物理过程发展未来展望

大气模式分辨率提高后,关键物理过程参数化方案的改进和完善成为国外各个模式研发中心提高模式模拟性能的重要途径。与高分辨率大气模式匹配的关键物理过程的改进研究主要体现在积云对流物理过程参数化、三维云辐射传输过程、边界层物理过程参数化、显式云微物理过程参数化(包含云-气溶胶相互作用过程)等方面。FGOALS 模式在原有模式研发基础上,将发展多尺度自适应参数化方案。完善显式对流降水方案,研发尺度自适应边界层过程,挑战"灰区"分辨率,建立全球聚焦千米级的气候系统模式 FGOALS-f4。在千米级变分辨率下,分析云微物理和边界层内各物理量在对流系统和边界层内中分布情况,结合云模拟器开展诊断分析,将模拟结果与卫星资料开展对比工作。根据比较结果,使用变网格框架测试不同分辨率的模拟性能,重点关注模式在亚洲季风区的技巧,通过极端降水的日变化特征、复杂地形区极端降水的再现能力等方面,全面评估模式对极端天气气候事件和亚洲季风系统的模拟技巧。

5.3 初始化模块未来展望

模式初始化过程将影响预测系统天气-气候多时间尺度的预测准确性。在近 10 年的模式研发周期中,LASG 数值模拟组发展了一套基于牛顿松弛迭代和时间滞后扰动的模式初始化方案能够较为高效地同化多种全球再分析资料,并产生多个有效预测样本。然而,现有初始化方法不能同化多源观测数据,如站点 GPS 探空、卫星观测数据和雷达观测数据等。在今后的研究中,LASG 数值模拟组将完善初始化方案,基于国产芯片和超算运行系统,在球立方有限体积动力内核上研发国产卫星和雷达观测资料同化系统,建立再分析数据系统,并最终实现提升天气-气候多尺度预报技巧的目标。

5.4 模式应用未来展望

LASG 数值模拟课题组将开发新一代高分辨率大气模式及其耦合系统模式,参加第七次国际耦合模式比较计划(CMIP7)。新一代高分辨率大气模式的发展目标是实现全球 12.5 km 的分辨率,改进与分辨率相适应的物理过程参数化方案,保证高分辨率模式长期积分稳定,提高模式对极端天气气候事件(例如:台风、MJO、青藏高原和东亚季风区的极端强降水)的模拟技巧。

青藏高原地形复杂,目前国内模式分辨率不足以解析高原复杂地形,会导致青藏高原地区大气环流、下垫面物质和能量交换过程模拟得不准确。实现青藏高原地区精细化的模拟和预测,提高气候预估能力和次季节-季节预测技巧。LASG 数值模拟课题组将进一步研发全球变网格聚焦的大气环流模式及其耦合版本,实现青藏高原地区千米级精细化的模拟,为科学家进一步理解青藏高原天气和气候效应提供更加准确的科学工具和数据。

LASG 数值模拟课题组还将依托新版本高分辨率模式及变网格版本建立千米级热带气旋多尺度模拟/预测系统。发展适合台风天气预报/短期气候预测的初始化方案,结合尺度自适应的参数化方案,开展西北太平洋热带气旋的精细化模拟和预测。进一步理解西北太平洋热带气旋的产生和发展机制,提升热带气旋的预测技巧,促进热带气旋预测业务与下游应用产业之间的联动。

在不久的将来,我们将利用新一代高分辨率耦合模式系统,开展天气-气候一体化的集合无缝隙预测,以更高质量和精细化的无缝隙预测产品助力中国及至全球防灾减灾和可持续发展,进一步提高 LASG 模式的国内外影响力。

参考文献

包庆,刘屹岷,周天军,等.2006.LASG/IAP 大气环流谱模式对陆面过程的敏感性试验[J].大气科学,30(6):1077-1090.
包庆,吴小飞,李矜霄,等.2019.2018–2019 年秋冬季厄尔尼诺和印度洋偶极子的预测[J].科学通报,64:73-78.
李剑东,刘屹岷,吴国雄,2010.积云对流和云物理过程调整对气候模拟的影响[J].大气科学,34(5):891-904.
李伟平,吴国雄,刘屹岷,等.2001.青藏高原表面过程对夏季青藏高压的影响——数值实验[J].大气科学,25:809-816.

刘海龙,俞永强,李薇,等,2004. LASG/IAP 气候系统海洋模式(LICOM 1.0)参考手册[M]. 北京:科学出版社:1-128.
刘琨,刘屹岷,吴国雄,2010. SAMIL 模式中 Tiedtke 积云对流方案对热带降水模拟的影响[J]. 大气科学,34(1):163-174.
刘平,吴国雄,2000. 1998 年夏季长江流域降水异常研究——热带环流Ⅱ:数值试验[C]//孙安健,吴国雄,李永康. 严重旱涝与低温的诊断分析和预测方法研究. 北京:气象出版社:72-77.
刘屹岷,吴国雄,刘辉,1999. 谱模式中负地形的处理与东亚副热带气候的模拟[J]. 大气科学,23(6):652-662.
刘屹岷,刘琨,吴国雄,2007. 积云对流参数化方案对大气含水量及降水的影响[J]. 大气科学,32(6):1201-1211.
念国魁,2020. 地形辐射效应对青藏高原及亚洲季风影响的数值模拟研究[D]. 北京:中国科学院大气物理研究所.
宋晓良,2005. 两种质量通量型积云参数化方案在气候模拟中的评估分析研究[D]. 北京:中国科学院大气物理研究所.
苏有琦,张宇,宋敏红,等,2020. 基于实测土壤属性 CLM4.5 对青藏高原高寒草甸模拟性能的评估[J]. 高原气象,39(6):1295-1308.
孙岚,吴国雄,孙菽芬,2000. 陆面过程对气候影响的数值模拟:SSIB 与 IAP/LASG L9R15 AGCM 耦合机器模式性能[J]. 气象学报,58(2):179-193.
孙菽芬,金继明,吴国雄,1999. 用于 GCM 耦合的积雪模型的设计[J]. 气象学报,57(3):293-300.
王鹏飞,黄刚,2006. 数值模式预报时效对计算精度和时间步长的依赖关系[J]. 气候与环境研究,11:395-403.
王在志,吴国雄,刘平,等,2005a. 全球海-陆-气耦合模式大气模式分量的发展及其气候模拟性能Ⅰ——水平分辨率的影响[J]. 热带气象学报,21(3):225-237.
王在志,宇如聪,王鹏飞,等,2005b. 全球海-陆-气耦合模式大气模式分量的发展及其气候模拟性能Ⅱ——垂直分辨率的提高及其影响[J]. 热带气象学报,21(3):238-247.
吴国雄,张学洪,刘辉,等,1997. LASG 全球海洋-大气-陆面系统模式(GOALS/LASG)及其模拟研究[J]. 应用气象学报,8(增刊):15-28.
吴统文,吴国雄,王在志,等,2004. GOALS/LASG 模式对气候平均态的模拟[J]. 气象学报,62(1):20-30.
曾红玲,季劲钧,吴国雄,2010. 全球植被分布对气候影响的数值模拟[J]. 大气科学,34(1):1-11.
曾庆存,1963. 大气运动的特征参数和动力方程[J]. 气象学报,33:472-483.
周林炯,2005. 大气环流模式的研发及对能量平衡和水分收支的模拟[D]. 北京:中国科学院大气物理研究所.
周天军,宇如聪,王在志,等,2005. 亚洲季风区海陆气相互作用对我国气候变化的影响(第四卷):大气环流模式 SAMIL 及其耦合模式 FGOALS-s[M]. 北京:气象出版社.
AHN M S, KIM D, KANG D, et al, 2020. MJO propagation across the Maritime Continent: Are CMIP6 models better than CMIP5 models? [J]. Geophys Res Lett, 47: e2020GL087250.
ANDERSON E A, 1976. A Point Energy and Mass Balance Model of a Snow Cover. Office of Hydrology[M]. National Weather Service, Silver Spring, MD, NOAA Technical Report NWS, 19: 1-150.
ARAKAWA A, 2004. The cumulus parameterization problem: Past, present, and future[J]. J Climate, 17(13): 2493-2525.
BAO Q, WU G X, LIU Y M, et al, 2010. An introduction to the coupled model FGOALS1.1-s and its performance in East Asia[J]. Adv Atmos Sci, 27(5): 1131-1142.
BAO Q, LIN P F, ZHOU T J, et al, 2013. The flexible global ocean-atmosphere-land system model, spectral version 2: FGOALS-s2[J]. Adv Atmos Sci, 30(3): 561-576.
BAO Q, LI J, 2020a. Progress in climate modeling of precipitation over the Tibetan Plateau[J]. Natl Sci Rev, 7(3): 486-487.
BAO Q, LIU Y M, WU G X, et al, 2020b. CAS FGOALS-f3-H and CAS FGOALS-f3-L outputs for the high-resolution model intercomparison project simulation of CMIP6[J]. Atmos Oceanic Sci Lett, 13(6): 576-581.
BONY S, EMANUEL K A, 2001. A parameterization of the cloudiness associated with cumulus convection: Evaluation using TOGA COARE data[J]. J Atmos Sci, 58(21): 3158-3183.
BRIEGLEB B P, BITZ C M, HUNKE E C, et al, 2004. Scientific Description of the Sea Ice Component in the Community Climate System Model, Version Three[R]. NCAR Tech. Note NCAR/TN-463+STR, 70.
CAIRNS B, LACIS A, CARLSON B, 2000. Absorption within inhomogeneous clouds and its parameterization in general circulation models[J]. J Atmos Sci, 57: 700-714.
CLOUGH S A, SHEPHARD M W, MLAWER E J, et al, 2005. Atmospheric radiative transfer modeling: A summary of

the AER codes[J]. J Quant Spectrosc Ra, 91: 233-244.

DE ROODE S R, SIEBESMA A P, JONKER H J J, et al, 2012. Parameterization of the vertical velocity equation for shallow cumulus clouds[J]. Mon Weather Rev, 140(8): 2424-2436.

EDWARDS J M, SLINGO A, 1996. A studies with a flexible new radiation code. I: Choosing a configuration for a large-scale model[J]. Q J R Meteor Soc, 122: 689-720.

FOWLER L, RANDALL D, RUTLEDGE S, 1996. Liquid and ice cloud microphysics in the CSU general circulation model. Part I: Model description and simulated microphysical processes[J]. J Climate, 9: 489-529.

GU Y, LIOU K, 2001. Radiation parameterization for three-dimensional inhomogeneous cirrus clouds: Application to climate models[J]. J Climate, 14: 2443-2457.

HALDER S, PAREKH A, CHOWDARY J S, et al, 2021. Assessment of CMIP6 models' skill for tropical Indian Ocean sea surface temperature variability[J]. Int J Climatol, 41: 2568-2588.

HE B, BAO Q, WANG X C, et al, 2019. CAS FGOALS-f3-L model datasets for CMIP6 historical atmospheric model intercomparison project simulation[J]. Adv Atmos Sci, 36(8): 771-778.

HE B, LIU Y M, WU G X, et al, 2020. CAS FGOALS f3-L model datasets for CMIP6 GMMIP Tier-1 and Tier-3 experiments[J]. Adv Atmos Sci, 37(1): 18-28.

HE B, ZHANG X Q, DUAN A M, et al, 2021. CAS FGOALS-f3-L large-ensemble simulations for the CMIP6 polar amplification model intercomparison project[J]. Adv Atmos Sci, 38(6): 1028-1049.

HE S, YANG J, BAO Q, et al, 2019. Fidelity of the observational/reanalysis datasets and global climate models in representation of extreme precipitation in East China[J]. J Climate, 32: 195-212.

HUNKE E C, LIPSCOMB W H, 2010. CICE: The Los Alamos Sea Ice Model Documentation and Software User's Manual Version 4.1[M]. Tech Rep LA-CC-06-012, 675.

JI J, 1995. A climate-vegetation interaction model: Simulating physical and biological processes at the surface[J]. J Biogeogr, 22: 445-451.

JI J, HU Y, 1989. A simple land surface process model for use in climate studies[J]. Acta Meteorol Sin, 3: 342-351.

JIANG X, WALISER D E, XAVIER P K, et al, 2015. Vertical structure and physical processes of the Madden-Julian oscillation: Exploring key model physics in climate simulations[J]. J Geophys Res Atmos, 120: 4718-4748.

JIANG X, ÁNGEL F ADAMES, KIM D, et al, 2020. Fifty years of research on the Madden-Julian oscillation: Recent progress, challenges, and perspectives[J]. J Geophys Res Atmos, 125: e2019JD030911.

JIN J M, GAO X G, WU G X, 1999a. Comparative analysis of physically based snowmelt models for climate simulations[J]. J Climate, 12: 2643-2657.

JIN J M, GAO X G, WU G X, 1999b. One-dimentional snow water and energy balance model for vegetated surfaces[J]. Hydrol Process, 13: 2467-2482.

JORDAN R A, 1991. One dimensional temperature model for a snow cover[J]. CRREL Special Report, 91(1b): 1-48.

KANG I S, JIN K, WANG B, et al, 2002. Intercomparison of the climatological variations of Asian summer monsoon precipitation simulated by 10 GCMs[J]. Clim Dynam, 19(5-6): 383-395.

LEE W L, LIOU K N, WANG C C, 2013. Impact of 3-D topography on surface radiation budget over the Tibetan Plateau[J]. Theor Appl Climatol, 113: 95-103.

LI J, LIU Y, WU G, 2009. Cloud radiative forcing in Asian monsoon region simulated by IPCC AR4 AMIP models[J]. Adv Atmos Sci, 26(5): 923-939.

LI J, BAO Q, LIU Y, et al, 2017. Evaluation of the computational performance of the finite-volume atmospheric model of the IAP/LASG (FAMIL) on a high-performance computer[J]. Atmos Oceanic Sci Lett, 10(4): 329-336.

LI J, BAO Q, LIU Y, et al, 2019. Evaluation of FAMIL2 in simulating the climatology and seasonal-to-interannual variability of tropical cyclone characteristics[J]. J Adv Model Earth Syst, 11(4): 1117-1136.

LI J, BAO Q, LIU Y, et al, 2021a. Dynamical seasonal prediction of tropical cyclone activity using the FGOALS f2 ensemble prediction system[J]. Weather and Forecasting, 36(5): 1759-1778.

LI J, BAO Q, LIU Y, et al, 2021b. Effect of horizontal resolution on the simulation of tropical cyclones in the Chinese Academy of Sciences FGOALS-f3 climate system model[J]. Geosci Model Dev, 14(10): 6113-6133.

LIANG X, LIU Y, WU G, 2005. The role of land-sea distribution in the formation of the Asian summer monsoon[J]. Geophys Res Lett, 32(3): L03708.

LIM Y, SON S, KIM D, 2018. MJO prediction skill of the subseasonal-to-seasonal prediction models[J]. J Climate, 31(10): 4075-4094.

LIN P F, YU Z P, LIU H L, et al, 2020. LICOM model datasets for CMIP6 ocean model intercomparison project (OMIP)[J]. Adv Atmos Sci, 37(3): 239-249.

LIOU K N, LEE W L, HALL A, 2007. Radiative transfer in mountains: Application to the Tibetan Plateau[J]. Geophys Res Lett, 34: L23809.

LIU H, ZHAO Y C, WU G X, 1994. Characteristics of a spectral climate model (R15L9) with reference atmospheric reduction[C]. Beijing: Abstract of PRC/USA 5th Workshop on Climate studies: 31.

LIU H, WU G, 1997. Impacts of land surface on climate of July and onset of summer monsoon: A study with an AGCM plus SSiB[J]. Adv Atmos Sci, 14(3): 289-308.

LIU H, LIN P, YU Y, et al, 2012. The baseline evaluation of LASG/IAP climate system ocean model (LICOM) version 2[J]. J Meteor Res, 26(3): 318-329.

MANABE S, SMAGORINSKY J, STRICKLER R F, 1965. Simulated climatology of a general circulation model with a hydrologic cycle[J]. Mon Weather Rev, 93(12): 769-798.

OLESON K W, LAWRENCE D M, BONAN G B, et al, 2010. Technical Description of Version 4.0 of the Community Land Model (CLM)[R]. NCAR Technical Note NCAR/TN-478+STR, doi: 10.5065/D6FB50WZ.

PLANTON Y Y, GUILYARDI E, WITTENBERG A T, 2021. Evaluating climate models with the CLIVAR 2020 ENSO metrics package[J]. Bull Amer Meteor Soc, 102: E193-E217.

SELLERS P J, MINTZ Y, SUD Y C, et al, 1986. A simple biosphere model (SiB) for use within general circulation models[J]. J Atmos Sci, 43: 505-531.

SIEBESMA A, CUIJPERS J, 1995. Evaluation of parametric assumptions for shallow cumulus convection[J]. J Atmos Sci, 52(6): 650-666.

SMITH D M, EADE R, ANDREWS M B, et al, 2022. Robust but weak winter atmospheric circulation response to future Arctic sea ice loss[J]. Nature Communications, 13(1): 1-15.

SMITH R, GENT P, 2002. Reference Manual for the Parallel Ocean Program (POP)[R]. Los Alamos unclassified report LA-UR-02-2484.

SUBIN Z M, RILEY W J, JIN J, et al, 2011. Ecosystem feedbacks to climate change in California: Development, testing, and analysis using a coupled regional atmosphere and land surface model (WRF3-CLM3.5)[J]. Earth Interactions, 15: 1-38.

SUN S, XUE Y, 2001. Implementing a new snow scheme in simplified simple biosphere model[J]. Adv Atmos Sci, 18(3): 335-354.

SUN Z, 2005. Parameterizations of radiation and cloud optical properties[R]. BMRC Research Report: 1-6.

SUN Z, RIKUS L, 1999a. Improved application of ESFT to inhomogeneous atmosphere[J]. J Geophys Res Atmos, 104(D6): 6291-6304.

SUN Z, RIKUS L, 1999b. Parameterization of effective radius of cirrus clouds and its verification against observations[J]. Q J R Meteor Soc, 125: 3037-3056.

TIEDTKE M, 1989. A comprehensive mass flux scheme for cumulus parameterization in large-scale models[J]. Mon Weath Rev, 117(8): 1779-1800.

TOMPKINS A M, 2002. A prognostic parameterization for the subgrid-scale variability of water vapor and clouds in large-scale models and its use to diagnose cloud cover[J]. J Atmos Sci, 59(12): 1917-1942.

WANG J, BAO Q, ZENG N, et al, 2013. Earth system model FGOALS-s2: Coupling a dynamic global vegetation and terrestrial carbon model with the physical climate[J]. Adv Atmos Sci, 30(6): 1549-1559.

WANG X C, BAO Q, LIU K, et al, 2011. Features of rainfall and latent heating structure simulated by two convective parameterization schemes[J]. Sci China: Earth Sci, 54: 1779-1788.

WANG X C, LIU Y M, WU G X, et al, 2013. The application of flux-form semi-Lagrangian transport scheme in a spectral atmosphere model[J]. Adv Atmos Sci, 30: 89-100.

WANG X C, ZHANG M H, 2014. Vertical velocity in shallow convection for different plume types[J]. J Adv Model Earth Syst, 6(2):478-489.

WANG X C, LIU Y M, BAO Q, et al, 2015. Comparisons of GCM cloud cover parameterizations with cloud-resolving model explicit simulations[J]. Sci China: Earth Sci, 58: 604-614.

WANG X C, MIAO H, LIU Y M, et al, 2021. Dependence of cloud radiation on cloud overlap, horizontal inhomogeneity, and vertical alignment in stratiform and convective regions[J]. Atmospheric Research, 249:105358.

WANG Z Z, WU G X, WU T W, et al, 2004. Simulation of Asian monsoon seasonal variations with climate model R42L9/LASG[J]. Adv Atmos Sci, 21: 879-889.

WU G, LIU H, ZHAO Y, et al, 1996. A nine-layer atmospheric general circulation model and its performance[J]. Adv Atmos Sci, 13: 1-8.

WU G, ZHANG Y, 1998. Tibetan Plateau forcing and timing of the monsoon onset in South Asia and Southern China Sea [J]. Mon Weath Rev, 126(4): 913-927.

WU T W, LIU P, WANG Z Z, et al, 2003. The performance of atmospheric component model R42L9 of GOALS/LASG [J]. Adv Atmos Sci, 20: 726-742.

XUE Y K, SELLERS P J, KINTER II J L, et al, 1991. A simplified biosphere model for global climate studies[J]. J Climate, 4: 345-364.

ZARZYCKI C M, JABLONOWSKI C, 2015. Experimental tropical cyclone forecasts using a variable-resolution global model[J]. Mon Weath Rev, 143: 4012-4037.

ZENG H, JI J, WU G, 2008a. An updated coupled model for land-atmosphere interaction. Part II: Simulations of biological processes[J]. Adv Atmos Sci, 25(4): 632-640.

ZENG H, WANG Z, JI J, et al, 2008b. An updated coupled model for land-atmosphere interaction. Part I: Simulations of physical processes[J]. Adv Atmos Sci,25(4): 619-630.

ZENG N, QIAN H, MUNOZ E, et al, 2004. How strong is carbon cycle-climate feedback under global warming? [J]. Geophys Res Lett, 31: L20203.

ZENG N, MARIOTTI A, WETZEL P, 2005. Terrestrial mechanisms of interannual CO_2 variability[J]. Global Biogeochemi Cy, 19: GB1016.

ZHANG G, MCFARLANE N, 1995. Sensitivity of climate simulations to the parameterization of cumulus convection in the Canadian climate center general-circulation model[J]. Atmosphere Ocean, 33(3): 407-446.

ZHANG Q, JIN J, WANG X, et al, 2019. Improving lake mixing process simulations in the community land model by using k profile parameterization[J]. Hydrol Earth Syst Sci, 23: 4969-4982.

ZHANG X H, SHI G, LIU H, et al, 2000. IAP Global Ocean-Atmosphere-Land System Model[M]. Beijing: Science Press: 252.

ZHOU L, LIU Y, BAO Q, et al, 2012. Computational performance of the high-resolution atmospheric model FAMIL[J]. Atmos Oceanic Sci Lett, 5: 355-359.

ZHOU L, BAO Q, LIU Y, et al, 2015. Global energy and water balance: Characteristics from finite-volume atmospheric model of the IAP/LASG (FAMIL 1)[J]. J Adv Model Earth Syst, 7(1): 1-20.

ZHU L, JIN J, LIU Y, 2020. Modeling the effects of lakes in the Tibetan Plateau on diurnal variations of regional climate and their seasonality[J]. J Hydrometeor, 21: 2523-2536.

A GCM Simulation of the Relationship between Tropical-Storm Formation and ENSO

WU Guoxiong[1], LAU Ngar-Cheung[2]

(1 Program in Atmospheric and Oceanic Sciences, Princeton University, Princeton, New Jersey;
2 Geophysical Fluid Dynamics Laboratory/NOAA, Princeton University, Princeton, New Jersey)

Abstract: A low-resolution Geophysical Fluid Dynamics Laboratory (GFDL) general circulation model has been integrated for 15 years. In the course of this experiment, the observed month-to-month sea surface temperature (SST) variations in the tropical Pacific Ocean were incorporated in the lower boundary condition. The output from this model run was used to investigate the influence of El Niño Southern Oscillation (ENSO) events on the variability of tropical-storm formation.

Criteria for detecting tropical cyclogenesis and tropical-storm formation were developed for the model. Tropical storms appearing in the model atmosphere exhibit many typhoonlike characteristics: Strong cyclonic vorticity and convergence in the lower troposphere, strong anticyclonic vorticity and divergence near the tropopause, and intense precipitation. It is demonstrated that, despite its coarse resolution, the model is capable of reproducing the observed geographical distribution and seasonal variation of tropical-storm formation.

The relationship between simulated tropical-storm formation and ENSO was explored using correlation statistics, composite fields for the warm and cold phases of ENSO, and individual case studies. Significant correlations were found between eastern equatorial Pacific SST anomalies and tropical-storm formation over the Western North Pacific, Western South Pacific, and Western North Atlantic. In these areas, below-normal frequency of tropical-storm formation was simulated in warm El Niño years, whereas more tropical storms occurred in La Niña years. The correlation between tropical-storm formation and equatorial SST changes is particularly high for fluctuations on time scales of less than 3—4 years. During the boreal summer months (June—October), there exists a seesaw in the frequency of tropical-storm formation between Western and central North Pacific: while more tropical storms were generated over Western North Pacific during La Niña years, less tropical storms were detected over central North Pacific. The reverse situation prevails in El Niño years. Over the Indian Ocean, the relationship between storm formation and ENSO exhibits a seasonal dependence.

本文原载于 *Monthly Weather Review*, 1992, 120: 958-983

1 Introduction

During the last few decades, our knowledge of the structure, formation, movement, and prediction of tropical storms (abbreviated as TS hereafter) has expanded considerably. The progress in the research related to TS has been facilitated by the availability of satellite observations, as well as the dedication of numerous diagnostic and modeling efforts to this problem (for example, see Chen et al., 1979; Anthes, 1982; Wang et al., 1987). It is now known that the formation of TS requires not only warm sea surface temperature (SST) and strong conditional instability in the lower troposphere but also favorable conditions in the environmental flow. During El Nino years, the difference in sea level pressure between the southeastern tropical Pacific and western Pacific reaches a minimum, and the tropical circulation undergoes dramatic changes (Rasmusson et al., 1982; Rasmusson et al., 1983b; Philander, 1989). The dynamic and thermodynamic variations associated with ENSO exert a notable influence on TS activity. Reductions in TS formation during El Nino years over Western North Pacific, Western South Pacific, and Western North Atlantic have been reported in a number of observational studies (Pan, 1982; Nicholls, 1984; Gray, 1984a; Chan, 1985, 1990; Dong, 1988; Li, 1988). However, in view of the multiple factors contributing to the formation of TS, it is difficult to isolate the impact of ENSO-related SST anomalies on TS activity by using observational data alone. Alternate approaches to this problem appear to be necessary.

The first attempt to simulate typhoon-type vortices by large-scale general circulation models (GCMs) was made by Manabe et al. (1970) at the Geophysical Fluid Dynamics Laboratory (GFDL). By using a global moist model with a horizontal grid spacing of 417 km, Manabe and his collaborators found good agreement between the observed and simulated sites of TS formation (see their Fig. 5.12). The capability of higher-resolution models to predict TS formation as well as to simulate the geographical and seasonal distributions of storm activity has also been demonstrated by Bengtsson et al. (1982). The ability of GFDL models to simulate the formation of TS was further assessed by the comprehensive study of Broccoli and Manabe (1990). Using multiyear integrations of GCMs with different resolutions, they found that the model-simulated TS resemble the corresponding observed systems in their thermodynamic structure. Moreover, the geographical distribution of the annual frequency of TS in their experiments is similar to the observed distribution.

Concurrently, the advances in our understanding of ENSO (for example, see Bjerknes, 1966, 1969; Rown-tree, 1972; Philander, 1989) have motivated many numerical experiments aimed at simulating this family of phenomena (e.g., Blackmon et al., 1983; Shukla et al., 1983b; Palmer et al., 1984; Wu et al., 1987). The time-varying nature of the tropical Pacific SST forcing as well as the seasonality of the model response were incorporated explicitly in a series of GCM experiments reported by Lau (1985). The model analyzed in the latter study has nine levels in the vertical and uses a rhomboidal truncation at 15 wavenumbers. Variations in the tropical Pacific SST between $30°N$ and $30°S$ were updated every day using temporal interpolations from monthly observations for the 1962—1976 period. In other parts of the World Ocean, climatological monthly mean SST data were used. By using this experimental design, Lau was able to reproduce the temporal evolution of individual ENSO episodes occurring in the 15-yr period.

Since the GFDL model is capable of producing TS-like features, and since the same model can simulate the essential characteristics of ENSO events with the incorporation of time-varying SST forcing in the lower boundary condition, it would be feasible to investigate the relationship between TS formation and ENSO by analyzing the appropriate model output from the experiments described by Lau (1985). Considering that the tropical Pacific SST changes provide the only external forcing on interannual time scales in that integration, the model data should reveal any correlation between TS formation and ENSO more clearly than observational data. The objective of the present study is to investigate the nature of such relationships in the model atmosphere. Suitable criteria for detecting TS formation are developed for the model in section 2. In section 3, the temporal and spatial characteristics of the simulated TS are examined. Composites of TS over the Western North Pacific are computed, and the typical structure of the simulated storms is described. The seasonal variation and geographical distribution of the frequency of TS formation in the model atmosphere are also documented. The relationships between ENSO and the genesis of TS over different oceans are then studied in section 4. Section 5 is devoted to the understanding of these relationships through the use of seasonal or monthly composites of selected fields for La Niña and El Niño events. Conclusions from this study and discussions on further work relevant to this subject are given in section 6.

2 Criteria for detecting tropical-storm formation

In the present model study, the daily output from the experiment described by Lau (1985) was analyzed. The meteorological fields examined include geopotential height at 200, 500, and 1000 hPa, temperature and horizontal wind at 200 and 950 hPa, water vapor mixing ratio at 950 hPa, vertical velocity at 500 hPa, and precipitation. The horizontal grid spacings of these model data are 7.5° of longitude and 4.5° of latitude. The following simple criteria have been used to identify cyclone systems at grid points L in the latitude zone between 40.5°S and 40.5°N:

• At 1000 hPa, the geopotential height at L must be at a minimum relative to the eight surrounding points; the 950 hPa circulation at L must be both cyclonic and convergent; and the 500 hPa vertical motion at L must be directed upward.

• The thickness between 200 and 1000 hPa over L must be at a maximum relative to the four neighboring points and must exceed (by 60 m) the corresponding mean value of the grid points lying within 1500 km to the west and east.

• The wind speed at 950 hPa must exceed gale-force strength (i.e., 17.2 m s^{-1}) at one or more of the nine grid points in the proximity of L.

The systems satisfying these criteria will henceforth be referred to as cyclones. Since emphasis is placed on storm genesis, any cyclone that is located within a distance of 7.5° of longitude or 9° of latitude from a cyclone identified one day earlier will be regarded as the same cyclone having migrated from its previous location and will be excluded from this census.

The criteria listed above were applied on a day-to-day basis to the model output. During the 15-yr integration, there are altogether 10011 surface cyclones detected in the domain under investigation. Figure 1a shows the geographical distribution of total number of cyclone formation at individual grid points. The latitudinal distribution of the frequency of cyclone formation is depicted in Fig. 1b. These results indicate that a substantial portion of the detected cyclones are located in extratropics, mainly in the winter season. Local maxima are also found near the equator and in subtropical regions. In the northern subtropics between 13.5° and 31.5°N, the longitudinal distribution of the frequency of cyclogenesis (Fig. 1c) is characterized by local maxima over western Africa, Saudi Arabia, Bay of Bengal, Western North

Pacific (WNP), and Western North Atlantic (WNA) to the east coast of the United States. The frequencies in these regions are particularly high in summer.

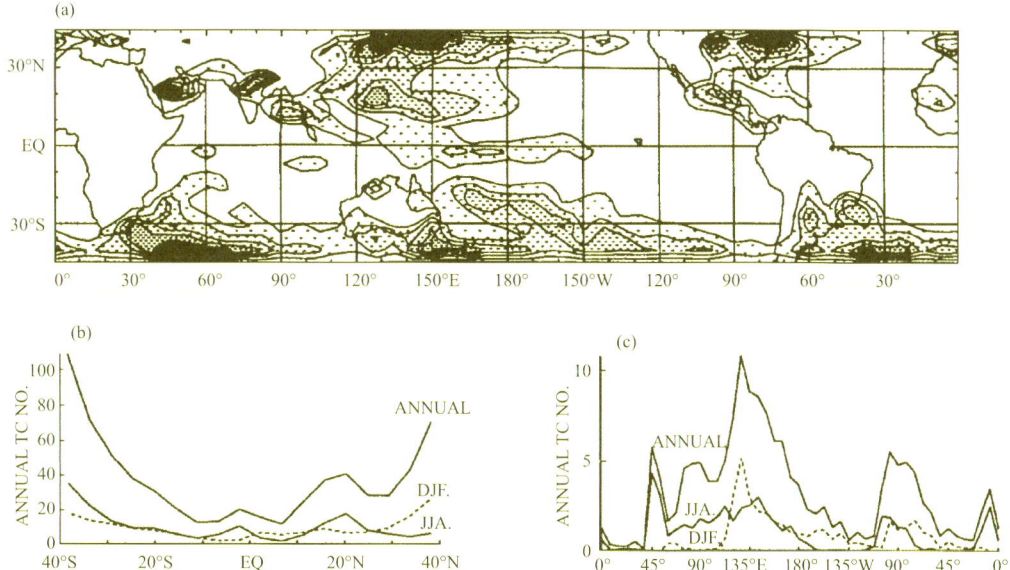

Fig. 1 (a) Distribution of the frequency of model-simulated cyclone formation at each 4.5° latitude × 7.5° longitude grid box. The values shown here correspond to the total number of occurrences of cyclone formation during the entire 15-yr integration. Annual frequency may hence be obtained by dividing these values by 15. Contour interval in 10. The contours plotted correspond to values of 10, 20, 30, and 40. Stippling with progressively higher densities is used to depict regions of frequent cyclone formation. Regions with more than 50 occurrences are indicated by solid black. (b) Latitudinal distribution of annual and seasonal cyclone frequencies summed over all longitudes. (c) Longitudinal distribution of annual and seasonal cyclone frequencies between 13.5° and 31.5°N

Analysis of the seasonal variation of the simulated disturbances shows that the cyclones over western Africa, Saudi Arabia, and the Bay of Bengal mostly develop in summer months, whereas the systems over WNP and WNA occur every season. In order to differentiate TS from other types of disturbances, composite charts of the detected cyclone systems have been constructed for different areas and different seasons by the following procedure: the center of each of these cyclones was assigned as the origin of a common coordinate system, with the abscissa and ordinate corresponding to longitudinal and latitudinal displacement from the cyclone center, respectively. The model data within 30° of longitude (i.e., four grid points in the zonal direction) and 31.5° of latitude (i.e., seven grid points in the meridional direction) from each cyclone center were translated to this common frame of reference. Composite plots were then obtained by averaging the data over all cases in the common coordinate system. Some of the composite results are presented in Fig. 2.

At 38.25°N in January, the surface cyclone is typically associated with a cyclonic wave that is characterized by temperature perturbations lagging behind geopotential height perturbations (figures not shown) and by strong westerlies at 200 hPa (Fig. 2a). The composite pattern of 200 hPa wind for the WNP region in January (Fig. 2b) resembles that pertaining to 38.25°N. These systems should therefore be classified as baroclinic disturbances. On the other hand, the composite for WNP in July is very different from its wintertime counterpart and bears many distinct TS characteristics, with a geopotential low at 500 hPa, an anticyclone at 200 hPa, and very weak wind speed directly above the surface low (refer to

Fig. 3).

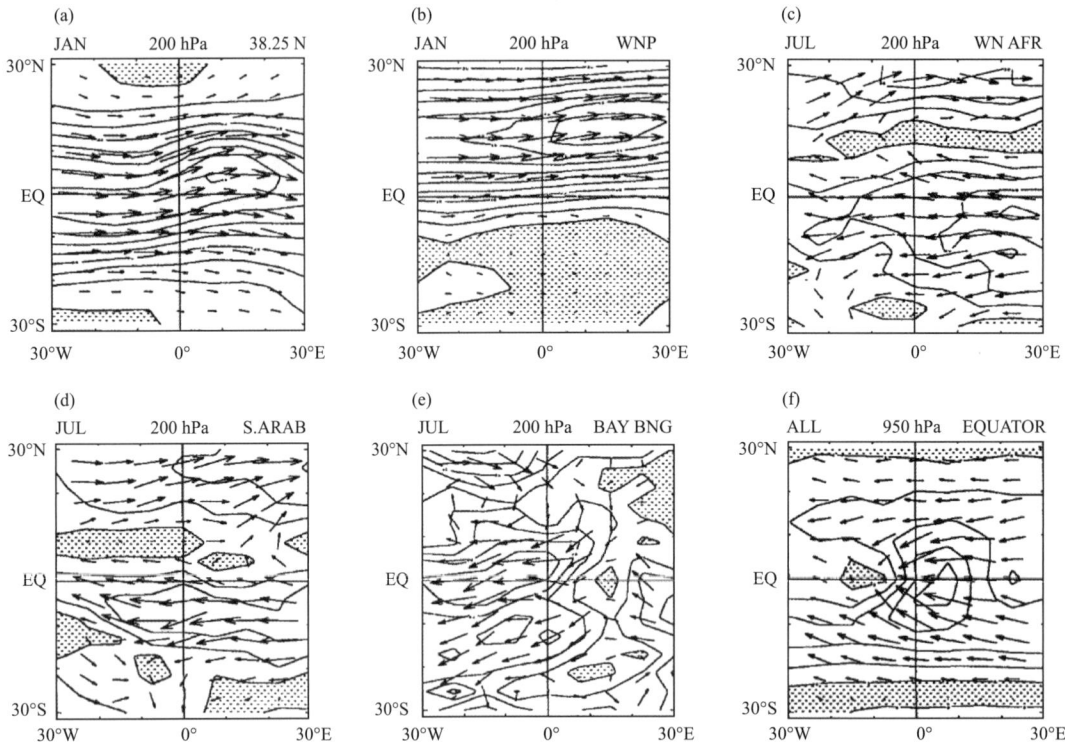

Fig. 2 Composites of velocity vectors at 200 hPa [(a)—(e)] and at 950 hPa [(f)] for cyclones identified in different seasons and different areas (see text for details of composite procedure). Contours indicate wind speed at intervals of 5 m s^{-1} at 200 hPa and 2 m s^{-1} at 950 hPa. Stippling indicates wind speeds less than 5 m s^{-1} at 200 hPa and 2 m s^{-1} at 950 hPa. Panel (a) is for 38.25°N in January; Panel (b) is for WNP in January; Panel (c) is for Western North Africa in July; Panel (d) is for southern Saudi Arabia in July; Panel (e) is for the Bay of Bengal in July; and Panel (f) is for the equatorial belt between 4.5°S and 4.5°N and for all seasons

The July composites for western Africa, Saudi Arabia, and the Bay of Bengal (Figs. 2c—2e) are all characterized by strong 200 hPa easterlies directly above the surface-low center. Composites of water vapor mixing ratio at 950 hPa for western Africa and Saudi Arabia (figures not shown) indicate very dry conditions. These results suggest that the perturbations over western Africa are cyclonic systems developing along the North African dry baroclinic zone, as noted by Carlson (1969) and Burpee (1974).

In the equatorial zone, the disturbances mainly occur to the east of the "maritime continent" in the western Pacific (Fig. 1a). The composite pattern of the 950 hPa wind field (Fig. 2f) shows that most of the equatorial perturbations do not resemble TS, since they are characterized by surface easterlies on both the northern and southern flanks of the disturbance center. This rather unique flow pattern allows the equatorial systems to be distinguished from other disturbances.

It is evident from the composite patterns shown in Fig. 2 that the cyclonic systems generated in the model include not only TS, but also baroclinic cyclones, continental dry cyclones, and equatorial easterly perturbations. On the basis of this evidence, the following additional constraints were applied to each of the cyclones detected earlier so as to retain only those systems with TS characteristics:

1) The 200 hPa zonal wind above the cyclone center L must not exceed 5 m s^{-1} (westerly).

2) The 950 hPa relative humidity at L must exceed 70%.

3) The 950 hPa zonal wind component must not be easterly at both the grid point located 4.5° to the

north and the point located 4.5° to the south of L.

Experimentation with different thresholds indicates that the results are not sensitive to the particular cutoff values chosen above. The impact of the additional constraints on the distribution of the frequency of cyclogenesis has been evaluated by comparing the pattern in Fig. 1 with the corresponding patterns obtained by incorporating the new constraints one at a time (not shown). As expected, the first constraint serves to exclude baroclinic perturbations. The second constraint effectively removes tropical continental dry disturbances, particularly those located over western Africa and Saudi Arabia. The third constraint removes those equatorial easterly disturbances possessing the structure noted in Fig. 2f. As will be demonstrated in the next section, the application of these additional constraints makes retention of only those typhoonlike systems possible, resulting in a realistic seasonal and spatial distribution of the frequency of TS formation.

The selection criteria developed in the present study may be compared with those implemented in Broccoli and Manabe (1990), who have analyzed various versions of essentially the same GCM. Both studies require the surface pressure at the storm center to be at a minimum and the wind speed in the vicinity of the storm to exceed gale-force strength. Broccoli and Manabe (1990) excluded most of the baroclinic cyclones by considering only grid points equatorward of ~30° latitude, and by conducting the search only during a 6-month hurricane season, which corresponds to the warm half of the annual cycle. They have also eliminated the continental dry cyclones from their census by surveying only the oceanic grid points. The differences between the two detection procedures notwith-standing, the gross spatial characteristics of the frequency of TS occurrence as reported in both studies are in fair agreement with the observations (see Broccoli et al., 1990; Table 1; and section 3 of this paper).

3 Model climatology of TS formation

In order to ascertain that the criteria and constraints developed in the preceding section are appropriate for identifying circulation systems with TS characteristics, the composite TS structure in the WNP area will be examined in this section. The geographical and seasonal variations of simulated TS formation will also be documented. The model results presented here will be compared with observational data.

While assessing the realism of the model results presented here, the fact that the cyclones are generated by a GCM with very low spatial resolution (i.e., rhomboidal truncation at only 15 wavenumbers) should be kept in mind. It is anticipated that this model is not capable of reproducing the fine details of the mesoscale cyclone structure. Therefore, this study will concentrate on those aspects of the tropical storms with spatial scales longer than that resolvable by the model grid. The prediction experiments performed by Krishnamurti et al. (1989) offer ample evidence linking increased model resolution with improved forecast skill of TS formation, movement, and structure. Hence, the findings reported here need to be further validated by analyzing model runs with higher resolution. It is, however, noteworthy that the regional climatology of TS occurrence in the 15-wavenumber model is not notably different from that in a 30-wavenumber model (see Broccoli et al., 1990).

a. Composite TS structure over WNP

Altogether, 128 storms were identified over WNP (0°—45°N, 101.25°—168.75°E) in the summer months of July, August, and September in the 15-yr model integration. The composites of the horizontal wind vector, relative vorticity, and divergence fields at 200 and 950 hPa, and the geopotential height fields at 200 and 1000 hPa are shown in Fig. 3. The same composite procedure described in the previous section has been used to construct these patterns. At the lower level there is a distinct region of convergence surrounding the storm center, with a radial extent of about 1000 km (Fig. 3f), as well as a core region of cyclonic relative vorticity surrounded by anticyclonic vorticity (Fig. 3d). The horizontal circulation exhibits a notable degree of asymmetry about the storm center, with southerlies in the eastern sector being much stronger than the northerlies to the west (Fig. 3b). The polarities of the composite vorticity and divergence fields at the upper level are opposite to those at 950 hPa. The 200 hPa circulation over the storm is characterized by anticyclonic vorticity (Fig. 3c), divergence (Fig. 3e), and wind vectors spiraling out of the vortex center (Fig. 3a). A low geopotential center appears at 1000 hPa (Fig. 3h), with gradients to the west being noticeably weaker than those to the east and north. The asymmetric structure of the low-level depression is similar to that observed in the Northern Hemisphere (Shea et al., 1973) and is related to the ambient circulation in the Western North Pacific. In that region there exists a semipermanent subtropical high pressure center to the northeast of the site of storm formation, whereas a quasi-stationary monsoon low system prevails to the west. This spatial configuration of the mean-flow environment agrees well with observations (Ding et al., 1981). The average depth of the simulated TS in terms of D value for the geopotential height field at 1000 hPa (which is defined as the difference between the height or pressure value at the low center and the mean value at the two grid points located at 1500 km to the east and west of the center, see Frank, 1977) is -70 m (or -9 hPa), about half of the corresponding observed amplitude as reported by Frank. The low center becomes weaker with increasing height: the D value at 500 hPa is only -24 m, or about -2 hPa. At 200 hPa (Fig. 3g), a high geopotential center with a D value of $+26$ m appears. In reality, the sign reversal in the geopotential height field for a TS over WNP usually occurs above 200 hPa (Frank, 1977). The present results suggest that the vertical extent of the simulated storm is shallower than that observed. The low-

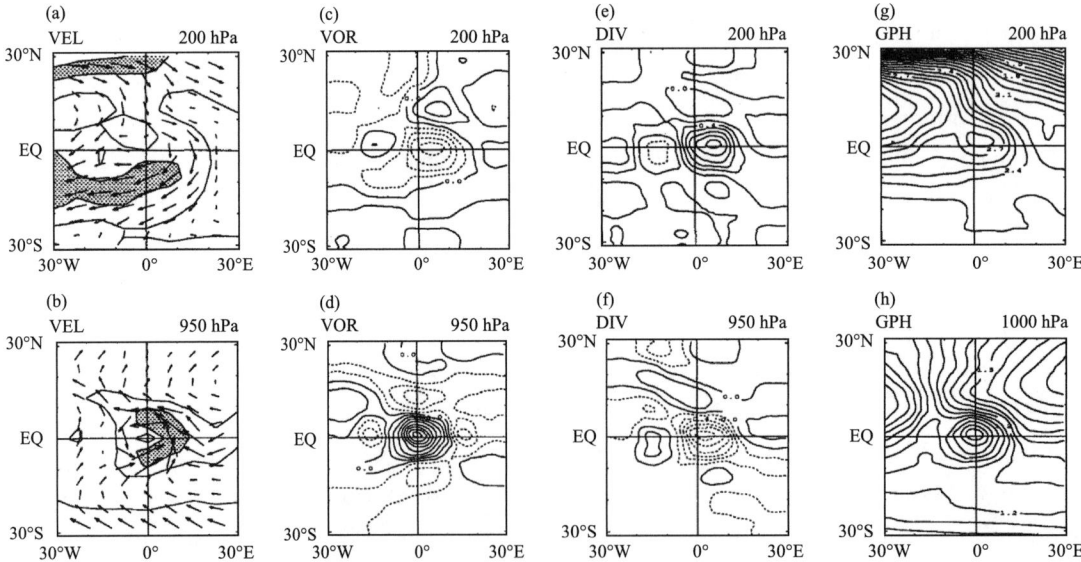

Fig. 3 Composite patterns of the model data at 200 hPa (upper panels), 950 hPa [(b), (d), and (f)], and 1000 hPa (h) for (a) and (b) horizontal wind, contour interval in 5 m s^{-1}, stippling indicates wind speed larger than 10 m s^{-1}; (c) and (d) relative vorticity, contour interval in 5×10^{-6} s^{-1}, dashed contours indicate negative vorticity; (e) and (f) divergence, contour interval in 2×10^{-6} s^{-1}, dashed contours indicate convergence; and (g) and (h) geopotential height, the contour values at 200 hPa have been subtracted by a constant value of 12200 m, contour interval in 10 m

er level of the TS is also characterized by high moisture content, strong upward motion, and heavy precipitation with rates reaching 34 mm day^{-1} (figures not shown). Numerical experiments on typhoon formation with a fine-mesh model by Kurihara and Tuleya (1981) show that maximum rainfall rates of 48—72 mm day^{-1} can be produced on the first day of storm formation.

The amplitudes of the model patterns displayed in Fig. 3 are generally lower than those documented by Frank (1977) for observed cyclones, thus implying that the present low-resolution model tends to underestimate the storm intensity. However, it is also evident from the composites that the TS generated in the present model exhibit considerable qualitative similarities to their observed counterparts.

b. Regional distribution of TS

Figure 4a shows the geographical distribution of the frequency of TS formation at individual grid points during the 15-yr model integration. The pattern in Fig. 4a bears some resemblance to the corresponding distribution in the real atmosphere (Fig. 4b), as presented by Gray (1979). For both the observed and simulated atmospheres, TS mainly develops over the oceans. There are more TS generated in the Northern Hemisphere than in the Southern Hemisphere. The Bay of Bengal and the region east of the Philippines over the western Pacific Ocean are the most favorable regions for TS formation. The occurrence of TS formation near the Brazilian coast in the model atmosphere does not appear to be supported by the available observations. Other sites of storm formation include WNA, Western South Pacific, and the southern Indian Ocean. No storm occurs over the Eastern South Pacific and Eastern Atlantic, where the circulation is dominated by subtropical high pressure centers, with prevalent large-scale subsidence.

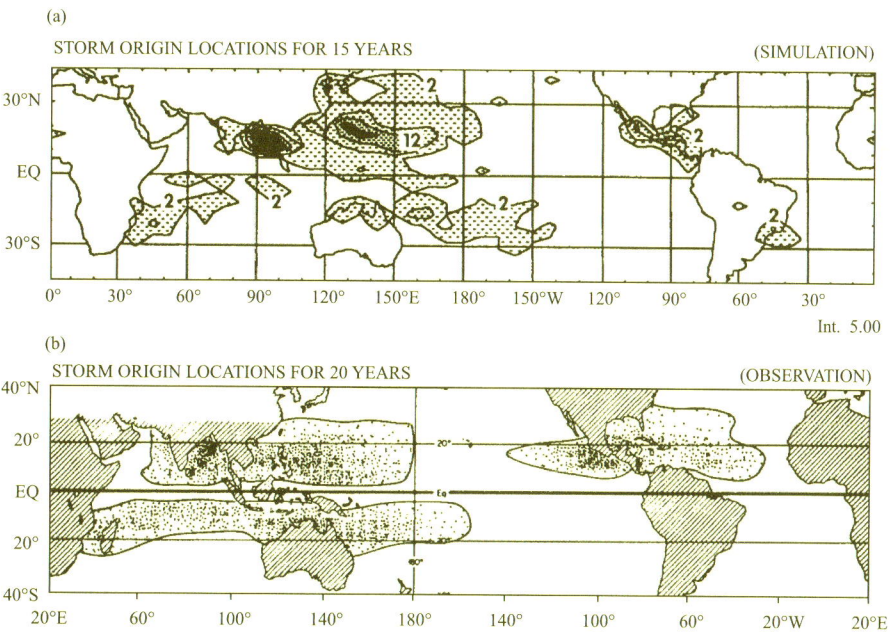

Fig. 4 Geographical distribution of the formation of tropical storms in the (a) simulated and (b) observed atmospheres. The model pattern depicts the total number of occurrence of storm formation during the entire 15-yr integration. Values with the ranges of 2—7, 7—12, and 12—17 are depicted by light, medium, and dense stippling, respectively. Regions with more than 17 occurrences are indicated by solid black. The observed pattern is reproduced from Gray (1979) and depicts the locations of individual storm cases during a 20-yr period

Over the southern central Pacific, the area of enhanced TS formation in the model extends farther to the east. This eastward shift is consistent with the corresponding displacement of the South Pacific convergence zone (SPCZ) in the model from the observed position (see Lau, 1985, Fig. 3). The TS formation over WNA in the model is less evident than that in the observations. This discrepancy may be related to the fact that the simulated 200 hPa trough in the Atlantic sector is displaced too far west of its observed position (see Lau, 1985, Fig. 4b). Accordingly, the climatological North Atlantic tropical anticyclone at 200 hPa in the model also shifts westward, thus inhibiting the formation of storms in the central Atlantic.

The overall agreement in the geographical distribution of TS between model and observation is encouraging. To further evaluate the model capability in simulating the spatial dependence of storm formation, the global ocean was divided into ten regions, that is, northern and southern Indian Ocean (NIO, 33.75°—101.25°E; SIO, 18.75°—123.75°E), WNP (101.25°—168.75°E), Western South Pacific (WSP, 123.75°E—131.25°W), central North Pacific (CNP, 168.75°E—138.75°W), Eastern North and South Pacific (ENP, 138.75°W to the west coast of the United States; ESP, 131.25°—63.75°W), WNA (east coast of the United States to 26.25°W), eastern North Atlantic (ENA, 26.25°—3.75°W), and South Atlantic (SAT, 63.75°W—18.75°E). The latitudinal extent of these regions is from the equator to 45°N and 45°S for the northern and southern regions, respectively. The mean annual frequency of TS formation in each ocean region was evatluated, and the results are compared to the corresponding observations (Gray, 1968) in Table 1. In view of the systematically weaker TS intensities simulated in this low-resolution model (see section 3a), it may not be too meaningful to compare the actual number of observed and modeled storms satisfying a similar set of selection criteria. Instead, the frequency of TS formation in a given region has been expressed as a fraction (in percent) of the total number of TS occurring throughout the globe. These relative frequencies have been computed separately for the model and observed atmospheres and are also shown in Table 1. The high percentages of storm events over WNP and WSP and the near absence of storms over SAT are reproduced in the model statistics. The percentages of simulated TS formation over WNP and WSP are somewhat too high. The percentages over other oceans, especially over ENP and WNA, are lower than the observations. It should be pointed out that, due to inaccuracies in the earlier part of the observational record, the number of observed hurricane occurrences presented in Table 1 may be lower than reality (see Gray, 1977; Anthes, 1982). The inclusion of satellite observations indicates that the annual number of hurricane occurrences is approximately 80 (Frank, 1987).

In Table 2, the percentage of the total TS occurring in different latitude belts in the model is compared to that in the real atmosphere (Gray, 1968). In both the model and observed atmospheres, most TS are generated in the subtropical latitude belt, with only about one-eighth of all TS forming to the north of the Tropic of Cancer.

Table 1 Simulated and observed mean annual frequency of tropical-storm formation in different regions. The observations are from Gray (1968)

Area		Modeled		Observed	
Ocean	As defined by Gray	Number	Percentage	Number	Percentage
NIO	Ⅲ + Ⅳ	8	11	8	13
WNP + CNP	Ⅱ	31	44	22	36
ENP	Ⅰ	5	7	10	16
WNA	Ⅷ	3	4	7	11
ENA		1	1		
SIO	Ⅴ	8	11	6	10
WSP	Ⅵ + Ⅶ	13	18	9	14
SAT		2	3	0	0
Total		71	99	62	100

Table 2 Fraction of tropical storms occurring in different latitude belts. Storm formation in all months of the year has been considered. The observations are from Gray (1968). The model results are based on the ratios of the number of storms detected in various zones to the total number (see data given in parentheses)

Modeled	Observed	
$9°-18°$N and $9°-18°$S	$9°-22.5°$N and $9°-22.5°$S	$10°-20°$N and $1°-20°$S
48% (511/1063)	64% (683/1063)	65%
>22.5°N 14% (154/1063)		>22°N 13%

c. Seasonal variation of TS

In Fig. 5, the seasonal variations of the maximum and minimum frequency of TS formation over WNP and CNP within the 15-yr integration period are shown together with the corresponding observations (Chen et al., 1979). In the observed and model atmospheres, both maximum and minimum frequencies attain peak values in the summer months, whereas no TS is detectable from November to May in some years.

The hemispheric averages of the simulated frequency of TS formation for individual calendar months are presented in Fig. 6a. Note that the time axis for the Northern Hemisphere values is shifted with respect to the Southern Hemisphere data by 6 months, so as to align the two sets of statistics to the same phase of the seasonal cycle. In both hemispheres, maximum TS formation occurs in summer, and minimum TS formation is simulated in late winter or early spring. This gross feature is similar to the observations shown in Fig. 6b. In the winter half year, the model produces more TS than the real atmosphere. As a result, the amplitude of seasonal variation in TS frequency in the model is weaker than that inferred from the observations.

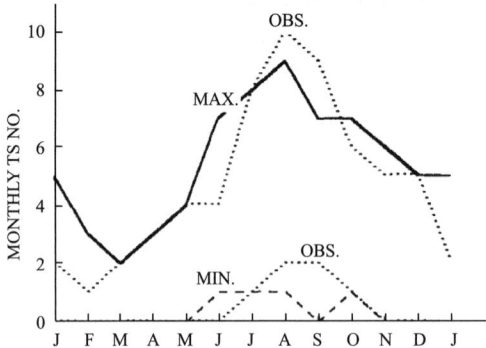

Fig. 5 Seasonal variation of the highest (solid curve) and lowest (dashed curve) frequency of storm formation over the western and central North Pacific for individual calendar months in the 15-yr simulation. The corresponding observed values are indicated by dotted curves

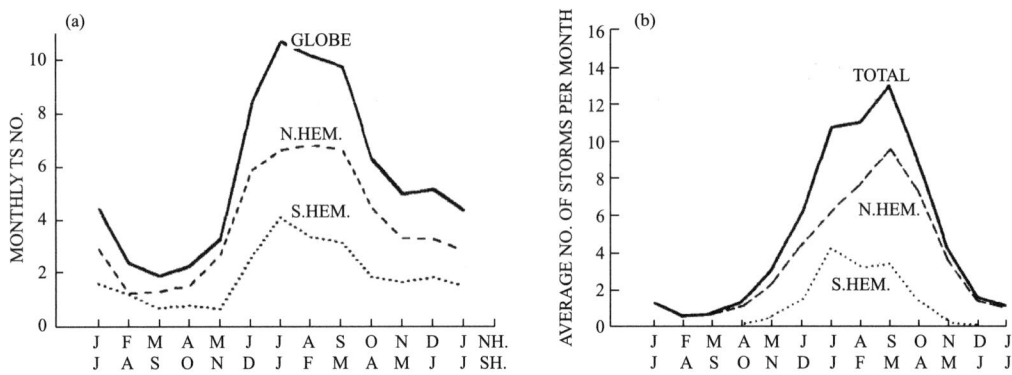

Fig. 6 Seasonal variation of the mean frequency of tropical-storm formation in individual calendar months, for the entire globe (solid curve), Northern Hemisphere (dashed curve), and Southern Hemisphere (dotted curve). The results are presented separately for the (a) model and (b) observed atmospheres. Note the 6-month shift in the time axes used for presenting the data in the two hemispheres

4 Modulation of frequency of TS formation by ENSO events

a. Definition of El Niño and La Niña

In order to delineate the impact of ENSO-related variability on TS formation, the SST average over an appropriate portion of the equatorial Pacific will be used as an ENSO index. In Fig. 7, two time series of monthly and regional mean SST anomaly are presented. The anomaly for each month is defined as the deviation of the monthly mean from the 15-yr average of the corresponding calendar month. The dashed curve is based on area averages for the central equatorial Pacific (4.5°S—4.5°N, 176.25°E—138.75°W). The solid curve denotes the SST anomaly in the eastern equatorial Pacific (9°S—0°, 138.75°—86.25°W). The gross features of these two curves are rather similar, with the exception of 1976, during which the wanning is confined to the vicinity of the Peru-Ecuador coast. In order to conform with other observational ENSO studies (e.g., Angell, 1981; Rassmuson et al., 1982), the eastern equatorial Pacific SST anomaly is used here to define the warm events. Altogether five warm ENSO events, that is, 1963, 1965, 1968/1969, 1972, and 1976 (shown by stippling in Fig. 7) can be identified. For the cold La Niña events, the SST anomalies in both the eastern and central equatorial Pacific are considered.

Four cold episodes, that is, 1964, 1967, 1970/1971, and 1973—1975, can be identified during the 15-yr period. In the following discussion, the SST anomaly over the eastern equatorial Pacific will be used as an index of ENSO.

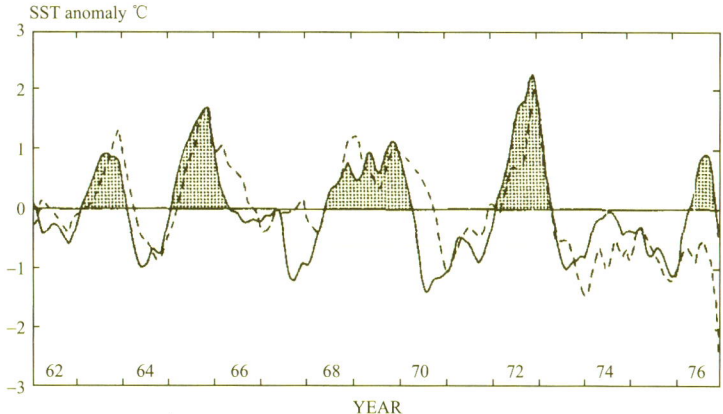

Fig. 7　Variation of observed SST anomalies in the central equatorial Pacific (dashed curve) and in the eastern equatorial Pacific (solid curve). Stippling denotes El Niño episodes determined using the SST data for the eastern equatorial Pacific

b. Relationship between frequency of TS formation and ENSO

The relationship between global frequency of TS formation and ENSO will be assessed first. The model frequency of TS formation in individual years is obtained by the census described in section 2 and correlated with the August values of the SST anomaly in the eastern equatorial Pacific. The August SST anomaly is chosen because this month corresponds to the developing phase of most El Niño events in the 1962—1976 period (Rasmusson et al., 1982; Lau, 1985). Assuming that the data values for individual years are mutually independent, the 95% and 99% significance limits for the correlation coefficients are estimated to be 0.51 and 0.64, respectively. Preliminary analyses show that both high- and low-frequency variations exist in TS frequency and in SST anomaly. Since this study's main emphasis is on the events with ENSO time scale, the fluctuations with that time scale have been retained by a high-pass filter. Following Holloway (1958), a five-point Gaussian filter has been designed to differentiate ENSO events occurring on time scales less than approximately 5 years from the much more slowly varying variations. The frequency response of this filter is presented in Fig. 8. This filter has been applied to the time series of both TS frequency and SST anomaly.

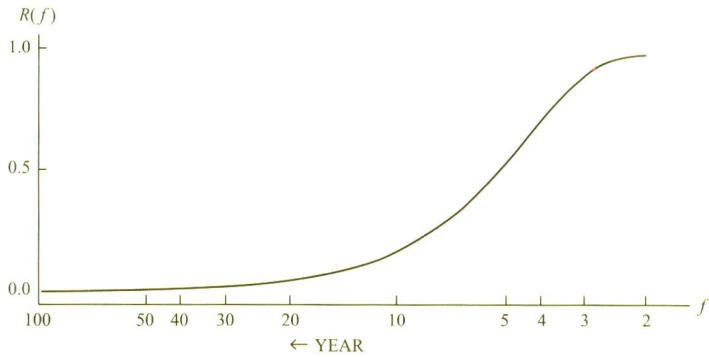

Fig. 8　Frequency response of the high-pass filter used in this study

Figure 9 shows the time series of SST anomaly in the eastern equatorial Pacific and the annual mean frequency of TS formation for all oceans. The correlation coefficients for the unfiltered (Fig. 9a) and filtered (Fig. 9b) cases are −0.65 and −0.67, respectively, and are both above the 95% significance limit. The frequency of TS formation is below normal in most warm events and above normal in most cold events. This relationship is more evident in the high-pass-filtered data shown in Fig. 9b. Correlation analyses have also been performed for individual oceans and different seasons of the calendar year. Some of the results obtained from the filtered data are shown in Table 3. With the exception of the values for ENP, ENA, SIO and NIO, and the summertime value for WSP, the correlation coefficients are mostly negative, that is, less (more) TS are generated in El Niño (La Niña) years. Within the period of May-November, the negative correlations between ENSO and TS formation over WNP, WSP, and WNA are all above the 95% confidence level. Figures 10 and 11 present the variations of frequency of TS formation in individual oceans in relation to the eastern equatorial Pacific SST. The model results summarized here may be compared with observational analyses for the following individual regions.

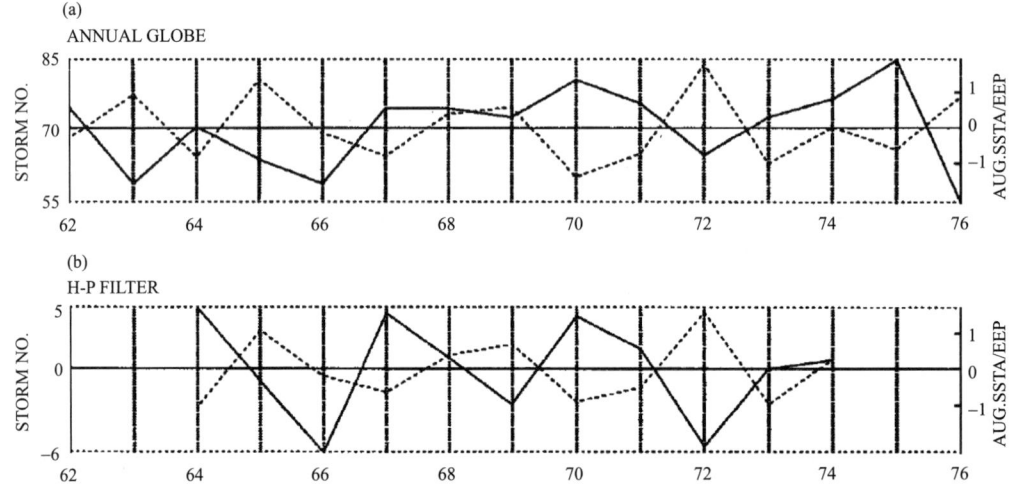

Fig. 9 (a) Variation in the annual frequency of simulated tropical-storm formation over the globe (solid curve) and observed August SST anomaly in the eastern equatorial Pacific (dashed curve), (b) As in (a), but for high-pass-filtered storm frequency and SST indices

Table 3 Correlation coefficients between August SST anomaly in the eastern equatorial Pacific and frequency of tropical-storm formation in different seasons and different oceans. The time series have been processed through a high-pass filter. Italics denotes values exceeding the 95% significance level

	Globe	WNP	WSP	ENP	WNA	ENA	SAT	NIO	SIO
December-April	−0.04	−0.88	0.21	0.27	0.23	0.00	−0.17	−0.37	0.38
May-November	−0.61	−0.70	−0.62	0.21	−0.69	0.19	−0.15	0.57	0.04
Annual	−0.67	−0.78	−0.11	0.30	−0.55	0.19	−0.20	0.54	0.23

1) WNP (Fig. 10a) and CNP (Fig. 10b)

The strongest negative correlation between TS formation in the model and ENSO is found in WNP. Except for the period between January and March, the negative correlation coefficients for other seasons are all above the 95% confidence level. Correlation analyses have been performed for the summer period between June and October in different longitudinal domains in the region (Table 4). Negative correlations are obtained in every longitudinal domain west of 168.75°E, with the strongest negative correlation coefficient reaching −0.89 for the region between 146.25° and 168.75°E. On the contrary, significant positive correlation is found over the central North Pacific (168.75°E−138.75°W). Therefore, a seesaw in TS formation between WNP and CNP occurs in the model in response to ENSO: during El Niño years, while WNP experiences less TS formation, more TS are generated over CNP. The reverse situation holds for La Niña years.

Using observational data of the 1948−1982 period, Chan (1985) found that the spectra of both TS frequency in the WNP and the Southern Oscillation index peak in the 3−3.5-yr frequency band, and that the cross spectrum of these two indices has significant coherence on this time scale. Chan (1990) also found that the number of TS is below normal in the western part of WNP during El Niño years. At the same time, more TS tend to develop in the eastern part of that region. Based on data for the 1900−1979 period, Li (1988) reported a significant correlation between ENSO and typhoon occurrence in WNP. His results indicate that the mean annual number of typhoon occurrence in WNP (excluding South China Sea) increases from 21.4 for El Niño years to 26.2 for La Niña years. Dong (1988) also found negative and positive correlations between TS frequency and eastern Pacific SST anomaly for the regions located

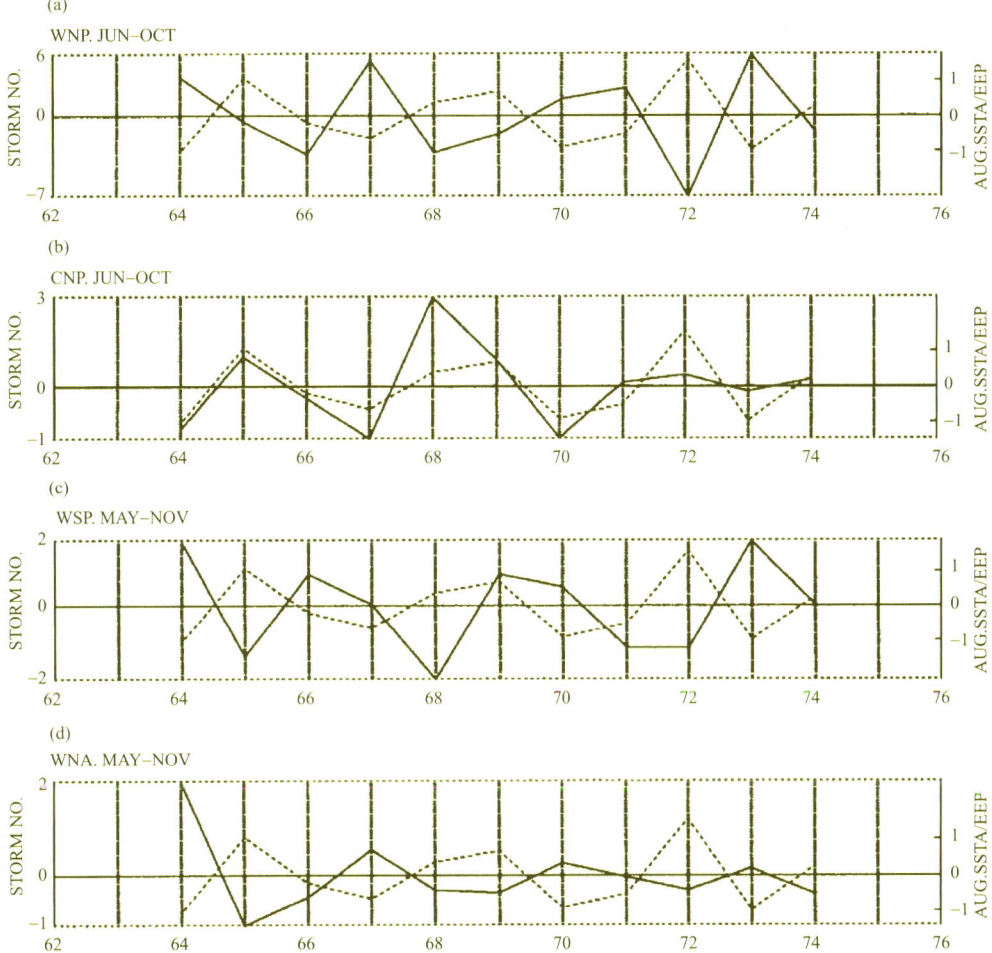

Fig. 10 Variation in the frequency of simulated tropical-storm formation in summer months over different oceans (solid curves) and observed August SST anomaly in the eastern equatorial Pacific (dashed curves). The storm frequency and SST indices have been processed through a high-pass filter. Panel (a) is for WNP between June and October; Panel (b) is for CNP between June and October; Panel (c) is for WSP between May and November; and Panel (d) is for WNA between May and November

west and east of 160°E, respectively. These results are well represented in the model. On the other hand, by only considering strong El Niño events in the 1955—1979 period, Ramage and Hori (1981) did not find any significant correlation between storm frequency and ENSO. However, it is worth noting that not all El Niño and La Niña events in this period were considered in Ramage and Hori's study. Hence, it is difficult to reconcile their findings with the present results.

2) WSP(Fig. 10c)

A significant negative correlation coefficient (-0.62) between the model TS formation and ENSO in this region is found only for the period between May and November. Based on observational data for the 1950—1974 period, Nicholls (1979) reported a negative correlation between the Darwin June—August sea level pressure and the TS frequency over the "cyclone area" (5°—35°S, 105°—165°E). On analyzing the data for 1913—1937 and 1964—1982, Nicholls (1984) found that the number of TS over this particular area exhibits negative correlations with the eastern Pacific SST and Darwin pressure from well before the start of the cyclone season to well into the season (October to April). This relationship is particularly strong in the early part of the cyclone season (October—December). The correlations are much weaker during the middle and late cyclone season (December—April). These observational results are well represented by the model as well. It is evident from Fig. 10c and Table 3 that the correlation coefficient in the southern winter and spring months is significantly negative, whereas that in the southern summer months becomes weakly positive.

3) WNA (Fig. 10d)

The negative correlation between the variations of the model TS frequency in the Western North Atlantic in summer months and the eastern equatorial Pacific SST anomaly is significant and comparable to that in WNP. In the real atmosphere, the relation between hurricane occurrence over the Atlantic and ENSO events seems to be more complicated (see Gray 1984a, 1984b; and Gray et al. 1987). However, the latter studies do show a negative correlation between TS frequency and ENSO.

4) NIO and SIO(Fig. 11)

From the unfiltered model results, no significant correlation between TS formation and ENSO can be identified in the Indian Ocean in any period of the calendar year. This model finding is in agreement with observational results (see Mandal,1989). However, the filtered data show that the TS variations in the NIO region may be correlated with the SST index. The correlations exhibit a strong seasonal dependence (Table 3). Weakly negative correlation (-0.44) is found in the winter months of January to March (Fig. 11a), whereas significant positive correlation ($+0.65$) is found in the summer months of June—October (Fig. 11b), when the monsoon trough migrates over the Bay of Bengal.

Table 4 Correlation coefficients between August SST anomaly in the eastern equatorial Pacific and frequency of tropical-storm formation during the summer months (June—October) in different longitudinal domains of the western and central North Pacific. The time series have been processed through a high-pass filter. Italic and bold numbers denote values exceeding the 95% and 99% significance levels, respectively

Longitude	101.25°—123.75°E	123.75°—146.25°E	123.75°—168.75°E	146.25°—168.75°E	168.75°E—138.75°W
Correlation coefficient	-0.40	-0.61	-0.78	-0.89	$+0.63$

In the last decade, many efforts have contributed to expanding our observational knowledge of the relationship between Indian monsoon rainfall and ENSO events (e.g. Angell,1981). Strong negative correlations between Indian monsoon rainfall and warm ENSO events have been reported, for example,

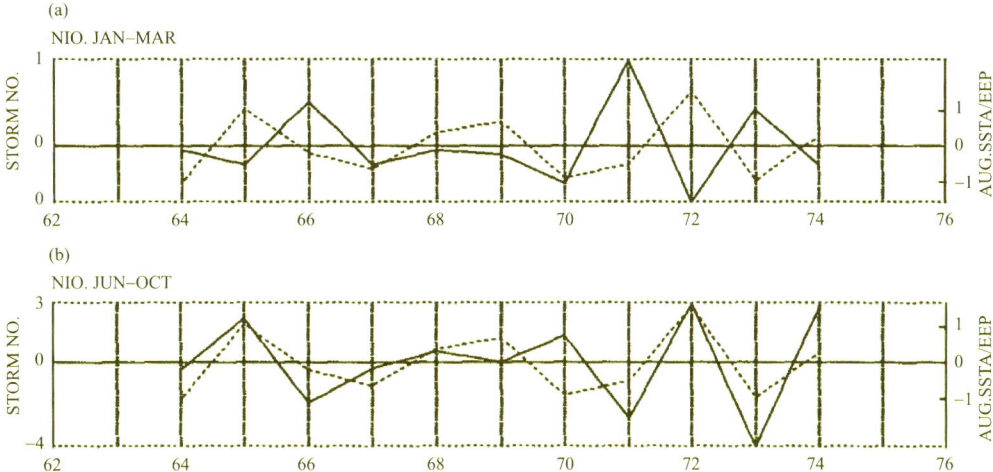

Fig. 11 Variation in the frequency of simulated tropical-storm formation over north Indian Ocean (solid curves) and observed August SST anomaly in the eastern equatorial Pacific (dashed curves). The storm frequency and SST indices have been processed through a high pass filter. Panel (a) shows January—March, and Panel (b) shows June—October

by Shukla and Paolino (1983a), and Rasmusson and Carpenter (1983a). These authors also noted that the precipitation anomaly over Sri Lanka and extreme southern India is different from the anomaly in the interior of the Indian subcontinent. For instance, above-normal precipitation in the former regions were observed during the autumn of the warm episode. As most of the TS in NIO occur in the Bay of Bengal (see Fig. 4), the monsoon rainfall over much of the Indian subcontinent might not be the direct result of such disturbances. In fact, Ding and Reiter (1983) found that when the WNP region experiences fewer typhoons, the rainfall over continental India is well below normal, whereas the monsoon rainfall over the Bay of Bengal itself is above normal. In other words, there seems to exist a positive relationship between monsoon rainfall over the Bay of Bengal and ENSO. Assuming that a large fraction of monsoon rainfall there is due to the development of TS systems, the positive correlation between ENSO and summertime TS formation over NIO, as found in the model, may not be unreasonable. However, further investigations of this relationship are evidently required.

5) Other oceans

The correlations between ENSO and TS formation over the South Atlantic and eastern North Atlantic are not significant. Over ESP, there is no TS formation both in the model and in observations. Over ENP, fewer TS are found in La Niña years (figure not shown). There seems to be a weak positive correlation between TS formation over ENP and the SST index. However, the correlation coefficients for different months and for the whole year are all below the 95% significance level.

To summarize the geographical dependence of the relationship between TS formation and ENSO, the data for TS formation in seven La Niña and six El Niño years have been extracted separately from the model record. The distributions of the mean frequency of TS formation for the cold and warm years are shown in Figs. 12a and 12b, respectively. For comparison with Fig. 4, the values shown here are the corresponding annual mean values at each grid point multiplied by 15. The difference between these two patterns is shown in Fig. 12c. To suppress small-scale features, a nine-gridpoint space smoother (Holloway, 1958) has been applied to the difference field in Fig. 12c. The results indicate that, although the spatial patterns of TS formation for the warm and cold years are similar to each other, the frequency of TS formation over different oceans does change noticeably. For the whole globe, the annual frequency of

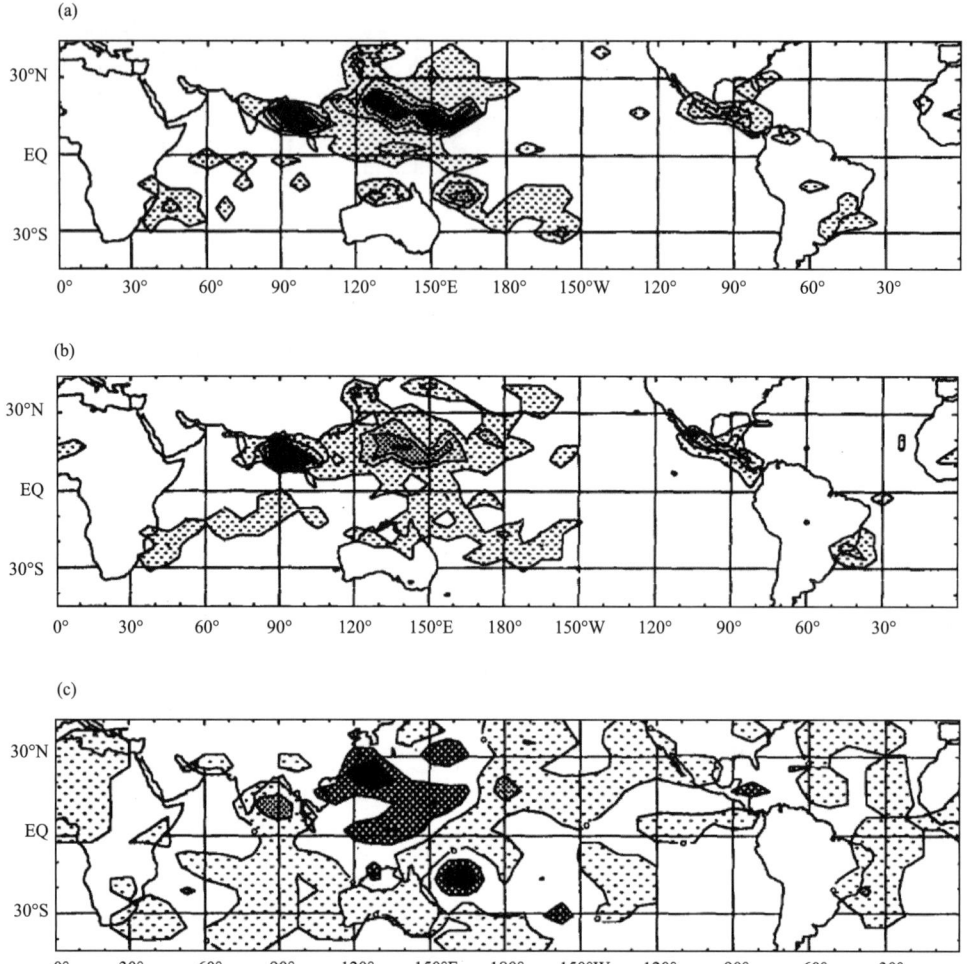

Fig. 12 Geographical distribution of the frequency of tropical-storm formation in the composite La Niña years (a) and El Niño years (b), and the difference between the two composites (c). For ease of comparison with Fig. 4, the values shown here are expressed in numbers of occurrences per 15 year. In panels (a) and (b), values within the ranges of 2−7, 7−12, and 12−17 are depicted by light, medium, and dense stippling, respectively. Regions with more than 17 occurrences are indicated by solid black. A nine-point space smoother has been used to construct panel (c). For the latter panel, light and medium stippling denote values of 0 to −2.5 and values less than −2.5, respectively; whereas values of 2.5−5 and values larger than 5 are indicated by crosshatching and solid black, respectively. Positive values in (c) indicate more TS formation during La Niña years

TS formation is reduced from 77 in La Niña years to 65 in El Niño years. The decrease in frequency of TS formation in La Niña years occurs mainly over the Bay of Bengal and CNP and also over SIO and ENP. The increase in TS formation during La Niña years is discernible over WNP, WSP, and WNA. The variation of TS formation over WNP between the warm and cold episodes of ENSO is the strongest both in magnitude and areal extent. The polarity of the change in frequency of TS formation over WNP is opposite to that over CNP, thus confirming the existence of the seesaw in TS formation between WNP and CNP.

To further analyze the impact of ENSO on TS formation in individual months, the seasonal variations of the composite frequency of storm formation for warm and cold events are presented in Fig. 13. Results are shown only for those maritime areas where the frequency of TS formation in the period of

May—November is significantly correlated with ENSO (Table 3). Over WNP (Fig. 13a), more TS are generated in La Niña years than in El Niño years in the periods of April to September and December to February. Observational analyses (Li,1988) show that the strongest negative correlations between the frequency of TS formation and Pacific SST occur from July to November. Over WSP (Fig. 13b), the largest differences between the El Niño and La Niña years are found from September to December and also from April to June. The former period is in the early part of the Australian cyclone season (October—April). This result is in agreement with that inferred from the observations (Nicholls, 1979).

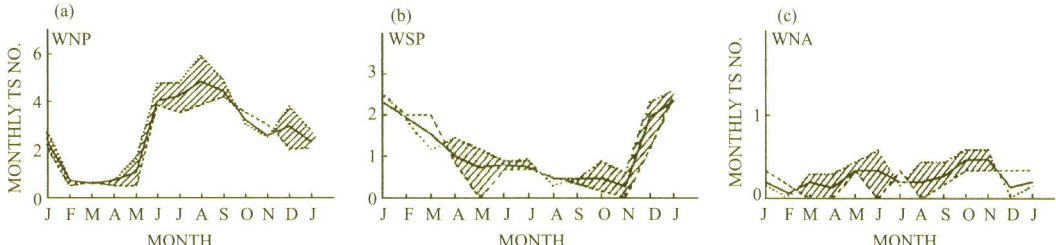

Fig. 13 Seasonal variation of the composite frequency of storm formation during individual calendar months in (a) WNP, (b) WSP, and (c) WNA. The composite values for warm El Niño and cold La Niña years are depicted by the dashed and dotted curves, respectively. The 15-yr mean frequency is depicted by the solid curves. Stippling indicates those months in which the number of tropical storms in La Niña years is more than that in El Niño years

5 ENSO composites of the large-scale flow

As discussed in section 1, TS are generated under certain favorable conditions in the local SST field as well as the ambient circulation pattern. In the present paper, only the influence of the large-scale flow field on TS formation will be investigated. The direct impact of different types of SST distribution on the atmospheric circulation and TS statistics will be considered in future experiments (see discussion in section 6). This section is devoted to a comparison of composite circulation charts for warm and cold phases of ENSO. The wind field has been decomposed into rotational and divergent components by computing the stream function and velocity potential, respectively. The results presented here are based on the boreal summer months (July—September). Results for the annual mean are very similar to the summer composites.

Figure 14 shows the difference between the La Niña and El Niño composites of simulated streamfunction and rotational wind at 950 and 200 hPa. At 200 hPa (Fig. 14a), two anomalous large-scale cyclones are found to straddle the equator over the central Pacific. Anticyclonic vorticity prevails over the remaining areas. Strong anomalous westerlies along the equator are simulated between 150°E and 120°W, while the rest of the equatorial zone is dominated by anomalous easterlies. The 950 hPa difference field (Fig. 14b) is characterized by a spatial pattern similar to that for the 200 hPa level, except for a sign reversal. The patterns shown here are reminiscent of the analytic solution of Gill (1980) for the atmospheric response to tropical heating symmetric about the equator. The anomalous anticyclonic vorticity at 950 hPa over the central Pacific as well as the Bay of Bengal would tend to inhibit storm development during La Niña years. On the other hand, the anomalous cyclonic vorticity at 950 hPa and anticyclonic vorticity at 200 hPa over much of the remaining maritime area should favor storm development during the cold events.

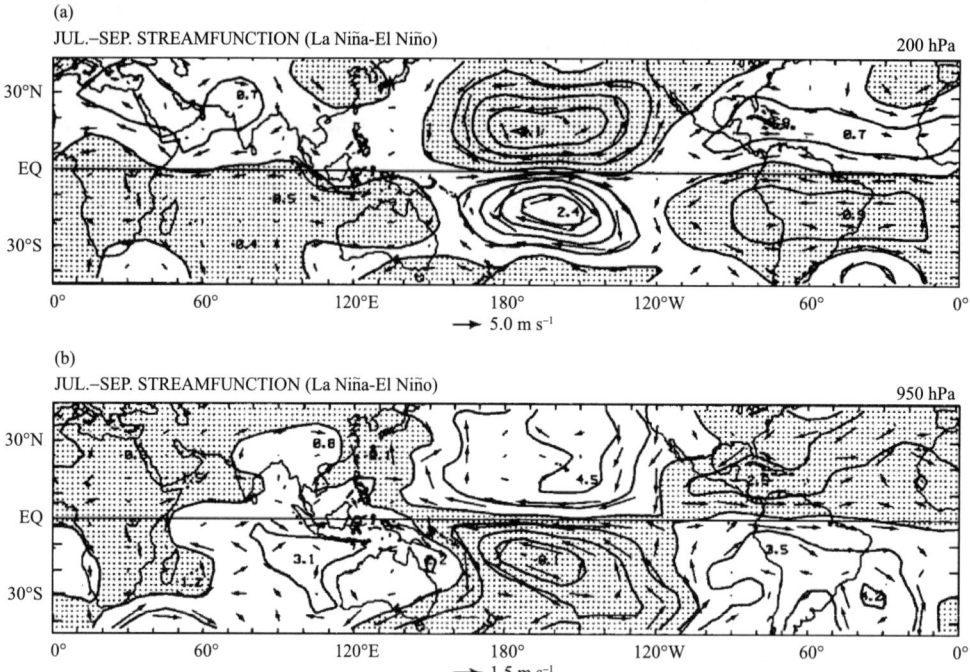

Fig. 14 Distribution of the difference between the La Niña and El Niño composites of simulated streamfunction and rotational wind vector at (a) 200 hPa, contour interval for stream function: $5 \times 10^6 \mathrm{m}^2 \mathrm{s}^{-1}$ and (b) 950 hPa, contour interval: $2 \times 10^6 \mathrm{m}^2 \mathrm{s}^{-1}$, for the July—September period. The scale for the wind vectors is given at the bottom of each panel. Stippling indicates negative streamfunction

The difference charts of velocity potential and divergence wind, as shown in Fig. 15, are also characterized by a sign reversal between the lower and upper levels. The eastern Pacific comes under the influence of anomalous low-level divergence and upper-level convergence during La Niña. The anomalous low-level convergence over the Caribbean Sea, WNP, WSP, and the Indian subcontinent, together with anomalous divergence at 200 hPa over the same sites, are in favor of local TS development during the cold events.

Considering the streamfunction and velocity potential composites together, it is seen that the WNP, WSP, and Caribbean regions are associated with anomalous convergence and cyclonic vorticity at the low level and with divergence and anticyclonic vorticity at the upper level during La Niña. These regions coincide with sites of above-normal frequency of TS formation during the cold events. On the other hand, the Bay of Bengal is associated with anomalous low-level anticyclonic vorticity during the same period. Below-normal storm frequency might then be expected during the cold events.

The seesaw of TS formation between WNP and CNP found in the last section can also be interpreted in light of the composite streamfunction and velocity potential patterns. In La Niña years at 950 hPa, while anomalous cyclonic vorticity and convergence prevail in the WNP region, CNP is dominated by anomalous anticyclone vorticity and divergence. These circulation anomalies tend to enhance TS formation in WNP and suppress it in CNP.

Over the WNP region, an above-normal number of simulated storms (28 per year) is generated in the La Niña year of 1974, whereas the least frequent simulated storm development (18 per year) occurs in the El Niño year of 1972. In order to understand the difference in large-scale flow between the summers of these two years, the west-east cross section of the August mean wind field along 11.25°N is shown in Fig. 16. This latitude corresponds to the southern edge of the region of maximum storm fre-

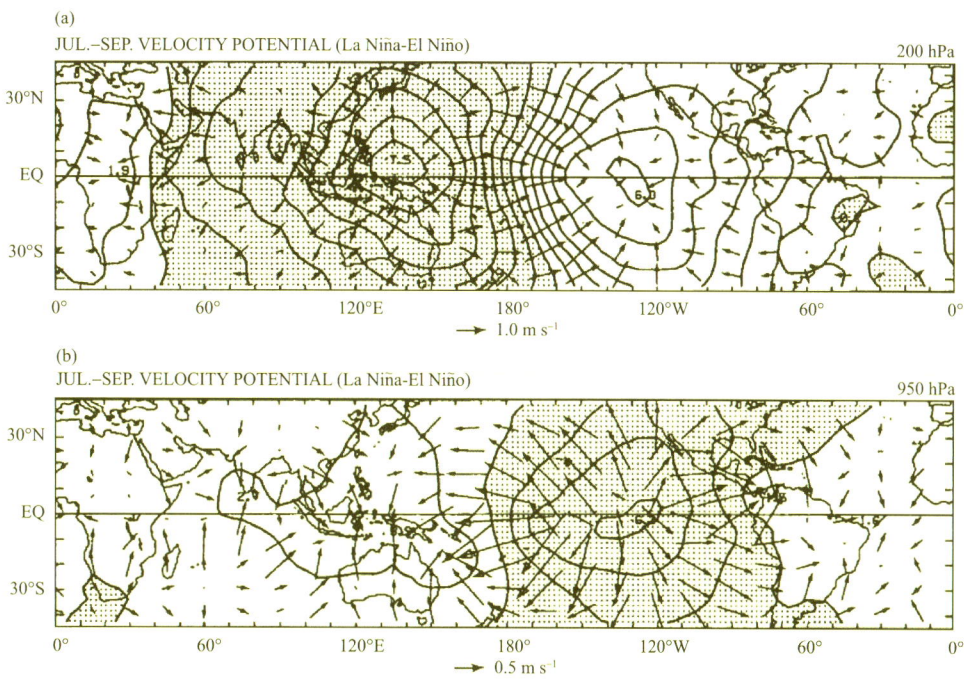

Fig. 15 As in Fig. 14, but for velocity potential and divergent wind vector. Contour intervals for the velocity potential are $1 \times 10^6 \, m^2 \, s^{-1}$ and $2 \times 10^6 \, m^2 \, s^{-1}$ for the 200 and 950 hPa levels, respectively

quency over WNP (see Fig. 4a). In 1974 (Fig. 16b), strong monsoon westerlies extend eastward from the Indian Ocean to WNP. These westerlies meet with the strong easterly trades at 150°E, and strong low-level convergence and rising motion are simulated at this longitude. The enhanced cyclonic vorticity and convergence associated with the deepening of the east Asia monsoon trough in this La Niña year are therefore conducive to storm formation. In the El Niño year of 1972 (Fig. 16a), as the warm SST anomaly migrates eastward, the strong rising center over the western Pacific is also shifted toward the dateline. At the same time, the monsoon westerlies over South Asia are weakened and terminated at the Gulf of Siam (105°E), thus resulting in a second rising center there. In the WNP region west of 160°E, no apparent low-level convergence or large-scale ascent is discernible. This environment is not favorable for storm development in the WNP region (Frank, 1987).

Figure 17 shows the north-south cross sections of the wind field for 1972 and 1974 along 142.5°E, which corresponds to the longitude of maximum frequency of storm formation. In the La Niña year of 1974 (Fig. 17b), cross-equatorial southerlies prevail over the low latitudes, resulting in rising motion over the entire 10°—30°N zone. On the other hand, in the El Niño year of 1972 (Fig. 17a), the cross-equatorial southerlies meet with the northerlies at about 12°N in the lower troposphere. However, the ascent associated with this convergence is weak, and upper-layer sinking is simulated over the same region, so that the meridional flow turns toward the equator near 700 hPa, and the strongest ascent in the upper troposphere occurs at 0°—10°S.

The features revealed by the above case study are also discernible from the composites shown in Figs. 14 and 15. During La Niña years, intensified near-equatorial easterlies at 950 hPa prevail over much of the Pacific. At the same time, enhanced surface westerlies occur in the equatorial zone extending from the western Indian Ocean to the Indonesian archipelago. These model results agree with those from observational analyses performed by Angell (1981) and Rasmusson and Carpenter (1982).

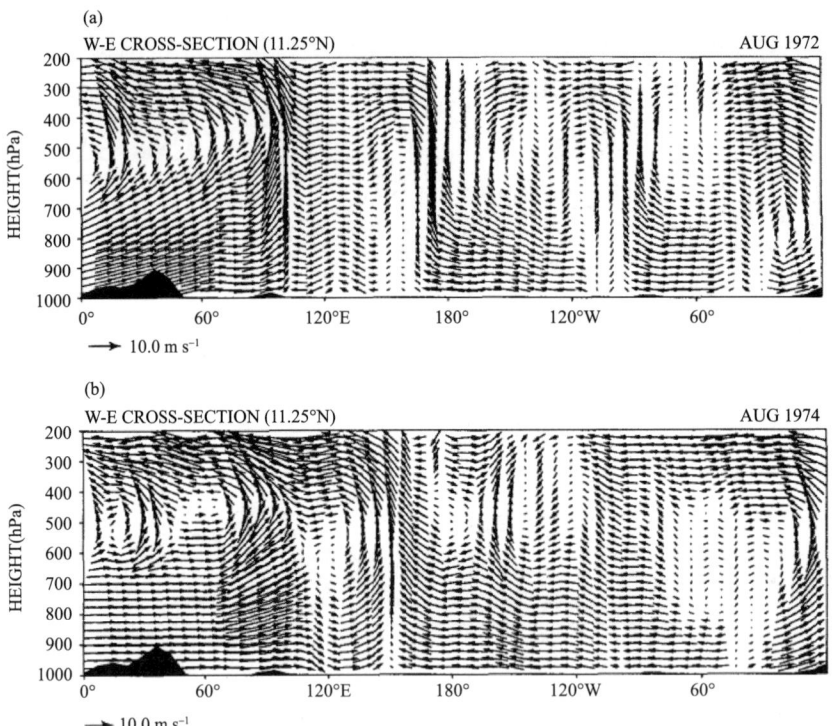

Fig. 16 Longitude-pressure distribution of the circulation at 11.25°N in (a) August 1972 and (b) August 1974. For better visualization of the circulation pattern, the vertical component of the wind vector has been multiplied by a factor of 2000 before plotting. The scale for the resultant wind vector thus obtained is given at the lower left-hand corner of each panel

It is known from various observational studies that, during El Niño years, the warm waters of the western tropical Pacific migrate eastward, leading to the shifting of the heating center toward the date line (Lau et al., 1985, 1986; Sardeshmukh et al., 1985). The forced circulation in the equatorial zonal plane associated with this heating center will also be displaced eastward, resulting in the intensification of surface easterlies over the Americas and of surface anticyclonic vorticity over WNA. The upper-level outflow from the heating area also enhanced the equatorial upper westerlies in the central American region (Arkin, 1982), so that the 200 hPa anticyclonic vorticity over WNA is weakened. These developments in the observed ambient flow structure in the North Atlantic sector are accompanied by a reduction in TS activity in this region during El Niño and are reproduced well in the model (see Figs. 14—17). It seems that the interannual variation of TS activity in the WNA is mostly related to the longitudinal displacement of the Walker circulation during El Niño and La Niña events.

6 Conclusions and discussions

In spite of the coarse resolution of the GCM used for this study, it is demonstrated that typhoonlike events do occur in the course of the experiment. The large-scale structure of the TS appearing in the model atmosphere bears considerable resemblance to the observed characteristics. The model disturbances examined here have a smaller vertical extent and generally have lower intensities than the corresponding features in the observed atmosphere or in simulations using models with much finer meshes.

The geographical distribution and seasonal variation of TS formation simulated by the model are

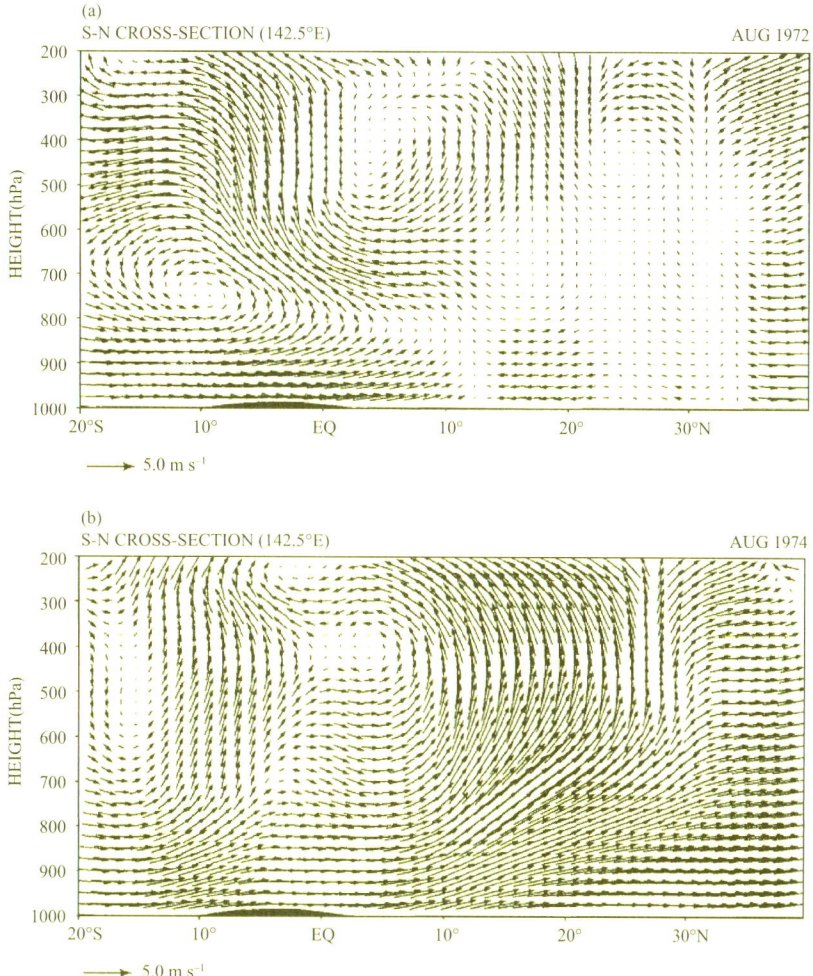

Fig. 17 Latitude-pressure distribution of the circulation at 142.5°E, in (a) August 1972 and (b) August 1974. For better visualization of the circulation pattern, the vertical component of the wind vector has been multiplied by a factor of 750 before plotting. The scale for the resultant wind vector thus obtained is given at the lower left-hand corner of each panel

similar to the corresponding observational results. The most favorable region for TS formation is the WNP region east of the Philippines. Other sites of enhanced storm activity are WSP, the eastern part of the Bay of Bengal, WNA, and SIO. Most of the TS occur in the warm half of the seasonal cycle.

Over the WNA, WNP, and WSP regions, more TS are simulated in La Niña years than in El Nino years. This relationship is particularly evident in the summer months in WNP and WNA, and in spring and early summer in WSP. These model results agree with observations. In the Indian Ocean, the correlation between TS activity and ENSO seems to be seasonally dependent. In El Niño years, above-normal storm activity is simulated in summer, and below-normal activity occurs in winter. The reverse situation applies to La Niña years. The latter results remain to be substantiated by more detailed investigations of observational data.

In La Niña years, the Walker circulation shifts westward, resulting in stronger rising motion over the Indonesian archipelago. At the same time, the monsoon trough and the associated westerlies extending from the Indian Ocean to the western Pacific are intensified. The strengthened convergence and cyclonic vorticity along the monsoon trough are accompanied by an enhancement of TS activity over WNP and WSP. In El Niño years, as the convective heating over the equatorial Pacific shifts eastward, above-normal low-level convergence and upper-level divergence occur over the eastern Pacific. In the South American sector, low-level equatorial easterlies and high-level westerlies are simulated, resulting in the weakening of low-level cyclonic vorticity and convergence, and upper-level anticyclonic vorticity and divergence over the tropical WNA. The frequency of TS formation in WNA during warm events is accordingly lowered.

In connection with ENSO, there exists a seesaw in TS formation frequency between WNP and CNP. This can also be explained by the anomalous tropical circulation. In La Niña years, low-level equatorial easterlies over the central Pacific and equatorial westerlies over the eastern Indian Ocean and the western Pacific Ocean are intensified. Enhanced low-level convergence and cyclonic vorticity occur over WNP, while a strong anticyclone associated with divergence dominates CNP. The polarity of the anomalous circulation at the upper level is opposite to that at the lower level. Therefore, during La Niña years, more TS tend to develop over WNP, whereas their appearance over CNP is suppressed.

The present study reveals the existence of significant correlations between TS formation over several maritime areas and ENSO. These relationships are then interpreted in terms of the changes in the large-scale tropical circulation during warm and cold events. However, the mechanisms contributing to such correlations require further study. Numerical experiments by Keshavamurty (1982) have shown that the tropical atmospheric response to equatorial SST anomalies is sensitive to the location of the anomaly relative to the ascending and descending branches of the Walker circulation. The results from the latter experiments and the evidence provided in the present study suggest that the geographical location and spatial pattern of the imposed SST anomaly may have a significant impact on the frequency of TS formation. To better understand the mechanisms linking ENSO events to TS formation, as well as the sensitivity of TS formation to various spatial configurations of the SST anomalies, several additional experiments incorporating different types of SST anomaly patterns have recently been launched at GFDL. Analysis of these new integrations will hopefully shed further insights on the role of perturbations at various ocean sites in altering the variability of TS occurrence.

References

ANGELL J K, 1981. Comparison of variations in atmospheric quantities with sea surface temperature variations in the equatorial eastern Pacific[J]. Mon Wea Rev, 109: 230-243.

ANTHES R A, 1982. Tropical Cyclones, Their Evolution, Structure, and Effects[M]. Boston: American Meteorological Society: 208.

ARKIN P A, 1982. The relationship between interannual variability in the 200 hPa tropical wind field and the Southern Oscillation[J]. Mon Wea Rev, 110: 1393-1404.

BENGTSSON L, BOTTGER H, KANAMITSU M, 1982. Simulation of hurricane-type vortices in a general circulation model[J]. Tellus, 34: 440-457.

BJERKNES J, 1966. A possible response of the atmospheric Hadley circulation to equatorial anomalies of ocean temperature [J]. Tellus, 18: 820-829.

BJERKNES J A B, 1969. Atmospheric teleconnections from the equatorial Pacific[J]. Mon Wea Rev, 97: 163-172.

BLACKMON M L, GEISLER J E, PITCHER E J, 1983. A general circulation model study of January climate anomaly patterns associated with interannual variation of equatorial Pacific sea surface temperatures[J]. J Atmos Sci, 40: 1410-1425.

BROCCOLI A J, MANABE S, 1990. Can existing climate models be used to study anthropogenic changes in tropical cyclone climate? [J]. Geophys Res Lett, 17: 1917-1920.

BURPEE R W, 1974. Characteristics of North African easterly waves during the summer of 1968 and 1969[J]. J Atmos

Sci, 31:1556-1570.

CARLSON T N, 1969. Synoptic histories of three African disturbances that developed into Atlantic hurricanes[J]. Mon Wea Rev, 97: 256-276.

CHAN J C L, 1985. Tropical cyclone activity in the Northwest Pacific in relation to the El Nino/Southern Oscillation phenomenon[J]. Mon Wea Rev, 113: 599-606.

CHAN J C L, 1990. The influence of sea-surface temperatures on tropical cyclone activity in the Western North Pacific[R]. Abstracts of the Int. TOGA Scientific Conf. Honolulu, HI.

CHEN L, DING Y H, 1979. On West Pacific Typhoon[M]. Beijing: Science Press:491.

DING Y H, REITER E R, 1981. Some conditions influencing the variability in typhoon formation over the west Pacific Ocean[J]. Arch Meteor Geophys Bioklim, A30: 327-342.

DING Y H, REITER E R, 1983. Large-scale hemispheric teleconnections with the frequency of tropical cyclone formation over the Northwest Pacific and North Atlantic Ocean[J]. Arch Meteor Geophys Bioklim, A32: 311-337.

DONG K, 1988. El Nino and tropical cyclone frequency in the Australian region and the Northwest Pacific[J]. Aust Meteor Mag, 36:219-255.

FRANK W M, 1987. The structure and energetic of the tropical cyclone. Paper Ⅰ: Storm structure[J]. Mon Wea Rev, 105: 1119-1135.

GILL A E, 1980. Some simple solutions for heat-induced tropical circulation[J]. Quart J Roy Meteor Soc, 106: 447-462.

GRAY W M, 1968. Global view of the origin of tropical disturbances and storms[J]. Mon Wea Rev, 96: 669-700.

GRAY W M, 1977. Tropical cyclone genesis in the Western North Pacific[J]. J Meteor Soc Japan, 55: 465-481.

GRAY W M, 1979. Hurricanes: Their formation, structure and likely role in the tropical circulation: Meteorology over the Tropical Oceans[J]. Roy Meteor Soc: 155-218.

GRAY W M, 1984a. Atlantic seasonal hurricane frequency: Part Ⅰ: El Nino and 30 hPa quasi-biennial oscillation influences [J]. Mon Wea Rev, 112: 1649-1668.

GRAY W M, 1984b. Atlantic seasonal hurricane frequency. Part Ⅱ: Forecasting its variability[J]. Mon Wea Rev, 112: 1669-1683.

GRAY W M, MIELKE P, BERRY K, 1987. Statistical analysis of Gray's Atlantic seasonal hurricane forecastscheme[R]. 17th Conf on Hurricanes and Tropical Meteorology. Miami, Amer Meteor Soc: 227-230.

HOLLOWAY J L, 1958. Smoothing and Filtering of Time Series and Space Fields[J]. Advances in Geophysics, 4: 351-389.

KESHAVAMURTY R N, 1982. Response of the atmosphere to sea surface temperature anomalies over the equatorial Pacific and the teleconnections of the Southern Oscillation[J]. J Atmos Sci,39:1241-1259.

KRISHNAMURTI T N, OOSTERHOF D, DIGNON N, 1989. Hurricane prediction with a high resolution global model[J]. Mon Wea Rev, 117: 631-669.

KURIHARA Y, TULEYA R E, 1981. A numerical simulation study on the genesis of a tropical storm[J]. Mon Wea Rev, 109: 1629-1653.

LAU K M, CHAN P H, 1985. Aspects of the 40—50 day oscillation during the northern winter as inferred from outgoing longwave radiation[J]. Mon Wea Rev, 113: 1889-1909.

LAU K M, CHAN P H, 1986. The 40—50 day oscillation and the El Nino/ Southern Oscillation: A new perspective[J]. Bull Amer Meteor Soc, 67: 533-534.

LAU N C, 1985. Modeling the seasonal dependence of the atmospheric response to observed El Nino in 1962—1976[J]. Mon Wea Rev, 113: 1970-1996.

LI C Y, 1988. Actions of typhoon over the western Pacific (including the South China Sea) and El Nino[J]. Adv Atmos Sci, 5: 107-115.

MANABE S, HOLLOWAY J L JR, STONE H M, 1970. Tropical cyclone in a time-integration of a global model of the atmosphere[J]. J Atmos Sci, 27: 580-613.

MANDAL G S, 1989. Low frequency oscillations and seasonal variability of tropical cyclones in north Indian Ocean[R]. Second Int. Workshop on Tropical Cyclones (IWTC-II), Manilla, WMO/ TP No. 319, WMO/CAS/OFDA: 341-354.

NICHOLLS N, 1979. A possible method for predicting seasonal tropical cyclone activity in the Australian region[J]. Mon Wea Rev, 107: 1221-1224.

NICHOLLS N, 1984. The Southern Oscillation, sea-surface temperature, and interannual fluctuations in Australian tropical cyclone activity[J]. J Climatol, 4: 661-670.

PALMER T N, MANSFIELD D A, 1984. Response of two atmospheric general circulation models to sea-surface temperature anomalies in the tropical East and West Pacific[J]. Nature, 310: 483-485.

PAN Y H, 1982. The effect of the thermal state of equatorial eastern Pacific on the frequency of typhoon over western Pacific[J]. Acta Meteorol Sin, 40: 24-34.

PHILANDER S G H, 1989. El Niño, La Niña, and the Southern Oscillation[M]. New York: Academic Press: 293.

RAMAGE C S, HORI A M, 1981. Meteorological aspects of El Niño[J]. Mon Wea Rev, 109: 1827-1835.

RASMUSSON E M, CARPENTER T H, 1982. Variations in tropical sea surface temperature and surface wind fields associated with the Southern Oscillation/El Niño[J]. Mon Wea Rev, 110: 354-384.

RASMUSSON E M, CARPENTER T H, 1983a. The relation between eastern equatorial Pacific sea surface temperatures and rainfall over India and Sri Lanka[J]. Mon Wea Rev, 111(3): 517-528.

RASMUSSON E M, WALLACE J M, 1983b. Meteorological aspects of the El Niño/ South Oscillation[J]. Science, 222: 1195-1202.

ROWNTREE P R, 1972. The influence of tropical East Pacific Ocean temperature on the atmosphere[J]. Quart J Roy Meteor Soc, 98: 290-321.

SARDESHMUKH P D, HOSKINS B J, 1985. Vorticity balances in the tropics during the 1982—1983 El Niño-Southern Oscillation event[J]. Quart J Roy Meteor Soc, 111: 261-278.

SHEA D J, GRAY W M, 1973. The hurricane's inner core region: Ⅰ. Symmetric and asymmetric structure[J]. J Atmos Sci, 30: 1544-1564.

SHUKLA J, PAOLINO D A, 1983a. The Southern Oscillation and long-range forecasting of the summer monsoon rainfall over India[J]. Mon Wea Rev, 111: 1830-1837.

SHUKLA J, WALLACE J M, 1983b. Numerical simulation of the atmospheric response to equatorial Pacific sea surface temperature anomalies[J]. J Atmos Sci, 40: 1363-1392.

WANG Z L, FEI L, 1987. Handbook of Typhoon Forecasts[M]. Beijing: China Meteorology Press: 360.

WU G X, CUBASCH U, 1987. The impact of El Niño anomaly on mean meridional circulation and transfer properties of the atmosphere[J]. Sci Sin B, 30: 533-545.

代表性论文 2

赤道印度洋-太平洋地区海气系统的齿轮式耦合和 ENSO 事件
Ⅰ. 资料分析

吴国雄,孟文

(中国科学院大气物理研究所大气科学和地球流体力学数值模拟国家重点实验室,北京 100029)

摘要:利用历史观测数据,研究了印度洋海表温度(SST)的季节变化特征,证实赤道印度洋和东太平洋 SST 年际变化有显著的正相关,指出这种正相关是由于沿赤道印度洋上空纬向季风环流和太平洋上空 Walker 环流之间显著的耦合造成的。这两个异常的纬向环流圈之间的耦合形式看起来很像是存在于赤道印度洋和太平洋上空的一对齿轮(简写为 GIP),当一个作顺时向变化时,另一个则作反时向变化。文中还证明 ENSO 事件与 GIP 的年际异常存在很好的对应关系,暖事件时 GIP 为反向运转;冷事件时 GIP 为正向运转;异常的 GIP 的啮合点位于印尼群岛附近。对 20 世纪 80 年代以来的 ENSO 事件的分析表明,每次事件前期异常的 GIP 的啮合点首先出现在印度洋上空,然后逐渐传入太平洋,引起 GIP 东侧的大气纬向风 u 和 SST 同时发生异常变化。当这种风场和 SST 的异常变化发展东传到达赤道中东太平洋时,导致 ENSO 事件最终出现。本文由此指出印度洋上空纬向环流的异常可以通过印度洋和太平洋上空大气系统的齿轮式耦合去影响赤道中东太平洋的海-气相互作用并触发 ENSO 事件发生。

关键词:异常 Walker 环流;异常纬向季风环流;齿轮式耦合

1 引言

自 Bjerkness(1969)提出 ENSO 的概念以来,人们对发生在热带太平洋的海气系统中的这一强讯号进行了广泛研究。Rasmusson 等(1982)对 20 世纪 80 年代以前 ENSO 的总体统计行为进行分析,概括了其发生、发展、维持和消弱等生命史各阶段的特征。Philander(1990)对太平洋范围大气和海洋系统的气候特点和海气相互联系作了全面的总结。巢纪平(1993)则从大气和海洋的热力和动力过程,以及海洋和大气的耦合波等角度去研究与 ENSO 相关的动力学。McCreary 等(1991)、Neelin 等(1992)等还用简单概念模式、中等简化模式和耦合 GCM 去研究 ENSO。这些模式的模拟结果都表明,海气耦合会激发出不稳定的低频耦合模态,并用这种不稳定慢波去解释热带海气系统中的年际变化。然而模式中的年际变化并不等于 ENSO 循环。ENSO 事件的发生存在显著的时空非均一性(2~9 年周期)。西传的 El Nino 与赤道中东太平洋海温的季节变化有锁相联系,而东传的 El Nino 夏季首先在西太平洋出现正的 SST 异常,然后向东传播。迄今的耦合模式都未能成功地模拟这些个例特征。如果把全球海气系统当成一个整体，那么 El Niño 的发生除了热带太平洋的海气相互作用外，看来还应该有来自热带太平洋以外的激发源。

与 ENSO 事件紧密联系的一个系统是亚洲季风系统。其实早在 20 世纪 30 年代 Walker 等(1932)就发现南方涛动与印度的季风降水关系密切。这一结论也为文献 Rasmusson 等(1982)的资料分析所证实。许多作者(Barnett,1983;Meehl,1987)建立了印度季风和 ENSO 之间风场和海平面气压场相关联的模型。Webster 等(1992)提出了亚洲季风和 ENSO 之间存在"选择性相互作用"的概念，Yasunari(1990)以及 Ju 等(1995)则发现两者之间能相互调节。Villwock 等(1994)以及 Tourre 等(1995,1997)分别分析了印度洋和西太平洋地区的 SST、海洋表层 400 m 中的热贮量以及纬向和经向表面风应力的变化，证明它们主要表现为 ENSO 的特征，其变化的主要形式与 ENSO 同相。吴国雄等(1995)在分析赤道中部印度洋和赤道东部太平洋在 1979—1988 年期间的月平均 SST 变化时发现，两者之间的正相关竟高达+0.76，远高于 99% 的信度水准(0.24)。如此显著的同时性遥相关是不可能用海洋内部的过程去解释的，而必须借助于大气环流的特征及海气相互作用予以解释。本文的目的之一就是揭示上述遥相关的机制，通过分析亚洲季风纬向环流和 Walker 环流的耦合效应，揭示印度洋地区季风环流的异常导致 ENSO 事件发生的事实。第 2 节分析赤道印度洋地区 SST 变化的基本特征；第 3 节分析印度洋和太平洋沿赤道地带纬向变化的相关特征，证明两者之间以"齿轮组合方式"紧密地耦合在一起；第 4 节分析 ENSO 事件与上述"印太齿轮式耦合(GIP)"的联系；第 5 节证明 20 世纪 80 年代以来的 ENSO 事件均可追溯到印度洋上空的激发作用。若干结论和讨论在第 6 节给出。

本文所用资料为"国际大气模式比较计划(AMIP)"所提供的 1979 年 1 月—1992 年 12 月的 SST 资料，分辨率为 2°×2°，欧洲中期天气预报中心(ECMWF)所提供的同期 850 hPa 和 200 hPa 的风场格点资料，以及 NCAR 提供的同期向外长波辐射(OLR)格点资料。

2 印度洋-太平洋赤道地区 SST 的变化特征

图 1 给出印度洋沿赤道地区(6°S~6°N)多年平均的 SST 的季节变化。3 条曲线分别代表西部(45°~70°E)，中部(70°~85°E)，和东部(85°~100°E)区域的演变状况。最高 SST 在各区域均出现在 4 月份；西部区域在 11 月份出现次高值。最低 SST 在中、西部出现在 8 月份；在东部出现在 11 月份。SST 的年较差在西部最强，达 2.5 ℃；在中部次之，为 1.3 ℃；在东部最弱，只有 1.0 ℃。一般东部 SST 比西部暖和。最大的东西海温差出现在 8 月和 1 月，各为 1.5 ℃ 和 0.9 ℃。另一个重要特征是西部存在显著的半年波，其讯号在中部仍可观测到，在东部已消失。看来，太阳高度角在赤道上空的半年周期至少不应是 SST 半年波的主要原因。注意到 SST 半年波的最高(4 月)和最低(8 月)分别对应于与索马里急流相联系的印度洋赤道强西风的建立前期和最盛时期。从 5 月份开始，70°E 以西的赤道印度洋地区表面向上的潜热通量和感热通量均可达 150 W·m^{-2} 甚至 200 W·m^{-2} 以上(Wu et al., 1998)。由此推测，沿赤道西风急流上巨大的感热通量和潜热通量损失是印度洋西部 SST 在 4—8 月急速下降的主要原因。

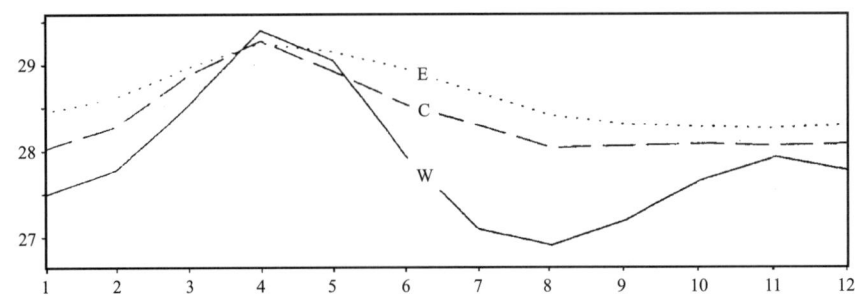

图 1 沿赤道(6°S~6°N)印度洋不同区域上，1980—1992 年平均的 SST 的季节变化

单位：℃；W：西部区域(45°~70°E)，C：中部区域(70°~85°E)，E：东部区域(85°~100°E)

注意到在图 1 中，印度洋东西海区中心相距仅有 35°，而 1 月和 8 月的 SST 差别竟达 1～1.5 ℃。这一东西向海温梯度虽略低于赤道东太平洋的，但却比赤道西太平洋的高约 1 倍。这表明至少在北半球的隆冬和盛夏，赤道印度洋的纬向平流过程对 SST 的局地变化也是非常重要的。图中各海区从 1—4 月的升温对应着东风应力；而从 4—8 月的降温对应着西风应力，正说明了纬向平流过程在印度洋 SST 年际变化中的重要作用。

把每一格点上 12 个月的月均 SST 与其年均值比较可计算该点对该年 SST 的方差；由多年该方差的平均可计算气候平均年变率 F_A。由每一格点上的月均 SST 与该月对多年平均 SST 比较可计算该点在该月多年平均的方差；对 12 个月的方差求平均可计算气候平均年际变率 F_I。通过分析 $F=F_I/F_A$ 的空间分布，可以了解 SST 的年际变化和季节变化的相对强弱。其结果由图 2 表示。由图可见，全球 F 值大于 0.4 的海区基本位于 20°S～20°N 之间的近赤道带。也就是说，年际汛号较强的区域位于近赤道海区。在厄瓜多尔和哥伦比亚沿岸以及赤道中太平洋 160°E～130°W 区域，$F>1$，年际变化讯号超过季节变化。在印度洋季风区，季节变化占优，尤其在索马里沿岸和阿拉伯海，季节变化讯号远超过年际变化。因此，为集中研究年际变化规律，在下面的分析中，季节变化的讯号均予扣除。

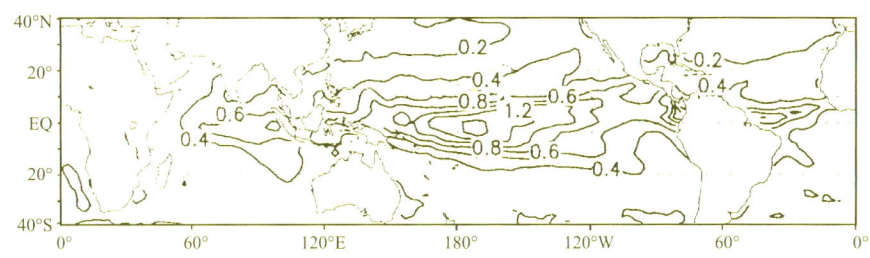

图 2　1980—1992 年观测到的平均的 SST 的年际变率与季节变率之比

3　赤道印度洋-太平洋的海气系统中的齿轮式组合

图 3 为 1979 年 1 月—1992 年 12 月共 168 个月中赤道东太平洋(150°～90°W，10°S～10°N)和赤道中印度洋(60°～90°E，10°S～10°N)SST 的年际变化。两海区的 SST 变化出现异常高的正相关，其相关系数竟高达+0.52，远远高于 99% 的信度水准(+0.20)。由于该两海区之间受马来半岛和印度尼西亚等岛屿所隔，且海洋中的过程为慢过程，它们间 SST 异常如此高的同时正相关是不能用海洋内部过程去解释的。一种可能的假设是两海区海温之间的变化通过大气风应力场的密切相关被联接起来。本文的工作就是利用观测资料去检验这一假设。

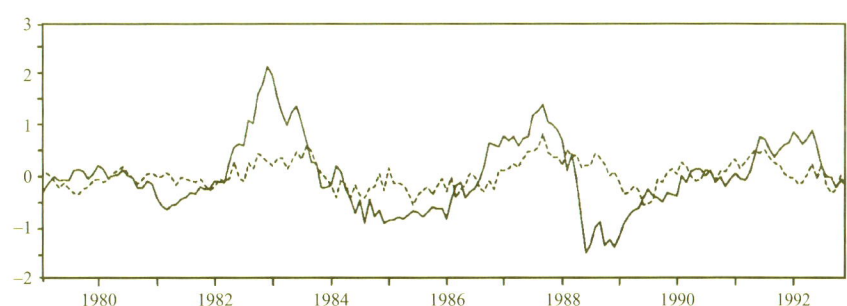

图 3　1979 年 1 月—1992 年 12 月共 168 个月中赤道东太平洋海区(实线：150°～90°W，10°S～10°N)和赤道中部印度洋洋区(虚线：60°～90°E，10°S～10°N)SST 对气候月平均离差(SSTA)的时间演变。单位为 ℃

沿赤道的对称环流在太平洋地区以 Walker 环流为主要特征，在印度洋地区以季风纬向环流为主要特征。Webster 等(1992)的研究表明，200 hPa 和 850 hPa 高度上的纬向风分量 u 的相对强弱是该两环流的很好的表征。图 4 示出了印度洋地区及中东太平洋地区上空纬向风分量的年际变化。由图可见,无论是在 200 hPa 还是在 850 hPa,两地区 u 的年际异常基本都是反号,负相关系数在上层为 -0.24,在低层为 -0.44,均超过了 99% 的置信水准(0.20)。这一结果表明,当印度洋上季风环流偏弱时,太平洋上的 Walker 环流偏弱,反之亦然。据此,我们可以把 Walker 环流和季风纬向环流想象为啮合在一起的一对"齿轮":当一个齿轮作顺时针向转动时,另一个齿轮则作反时针向转动,将其简称为"印-太齿轮(gearing between the Indian and Pacific Ocean,GIP)"。也就是说,季风纬向环流的异常可导致 Walker 环流的异常,而 Walker 环流的异常也可导致季风纬向环流的异常。还由于 Walker 环流的变化与 ENSO 现象紧密相连,那么,可以推测,季风纬向环流的异常与 ENSO 应当是互为因果的。在下面章节中将对此作进一步分析。

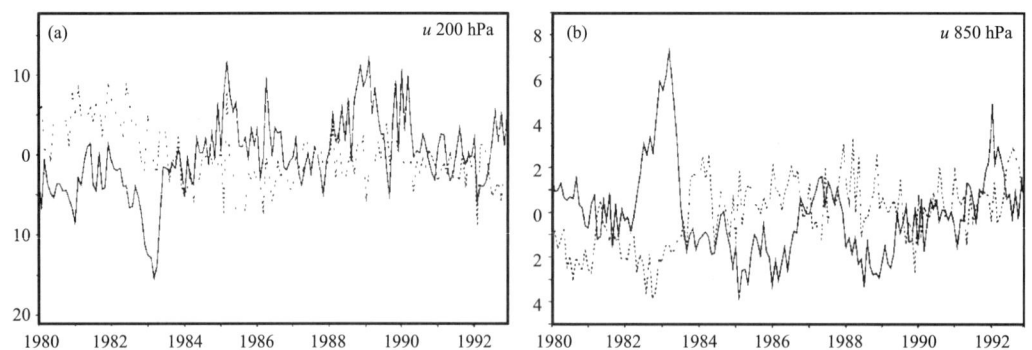

图 4　1980—1992 年沿赤道带(2.5°S~2.5°N)、印度洋(40°~100°E,虚线)及中东太平洋(180°~90°W,实线)上空 200 hPa(a)和 850 hPa(b)高度处,逐月平均纬向风对该时段相应月份平均纬向风的偏差的时间演变。单位为 m·s^{-1}

4　GIP 与 ENSO 型 SSTA 的联系

对 1979 年 1 月—1992 年 12 月的月均 SST 资料滤去季节变化,并进行标准化,然后进行 EOF 分析,其第一分量的方差贡献达 28.1%,其空间型和时间特征向量如图 5 所示。它是一显著的 ENSO 型:1982—1983 年、1986—1987 年,以及 1991—1992 年 3 次 El Niño 事件及 1984—1985 年和 1988—1989 年的 La Niña 事件都表现得很清楚。

把图 5b 所示的 ENSO 特征向量与 200 hPa 和 850 hPa 上同期的流函数场 ψ、速度位势 χ,以及纬向风 u 求线性回归分析,可得到与 ENSO 型相关的各物理量场,其结果由图 6 表示。χ 和 u 场具有以赤道为准对称的特征,而 ψ 场则有以赤道为反对称的特征。在赤道中东太平洋的暖 SST 上空,850 hPa 的一对气旋式环流和 200 hPa 的反气旋式环流横跨赤道两侧(图 6a),在西太平洋及印度洋(850 hPa 上为印度洋东部)环流性质正相反。在速度势 χ 场上(图 6b),850 hPa 的低位势中心和 200 hPa 上的高位势中心均位于印尼群岛附近。因此,在高低空沿赤道的旋转风分量和无旋风分量具有相似的分布。在其共同作用下,赤道中东太平洋低空为西风异常,高空为东风异常;而印度洋和西太平洋低空为东风异常,高空为西风异常(图 6c)。换言之,印-太齿轮组转动方向与 ENSO 密切相关。如果我们定义 Walker 环流和季风纬向环流加强的方向为齿轮组的正方向,则上述的分析表明:El Niño 期间,齿轮组反向转动;La Niña 期间,齿轮组正向转动。

图 7 为向外长波辐射的相应的回归场分布。它表明,在 El Niño 期间,印尼群岛附近的对流显著减弱(OLR 增加 9 W·m^{-2} 以上);赤道中东太平洋对流加强,尤其在 170°W 附件对流增加明显(OLR 降低达 15 W·m^{-2} 以上);印度洋地区的对流也有所加强。因此可确定,齿轮组的啮合点位于印尼群岛附近。

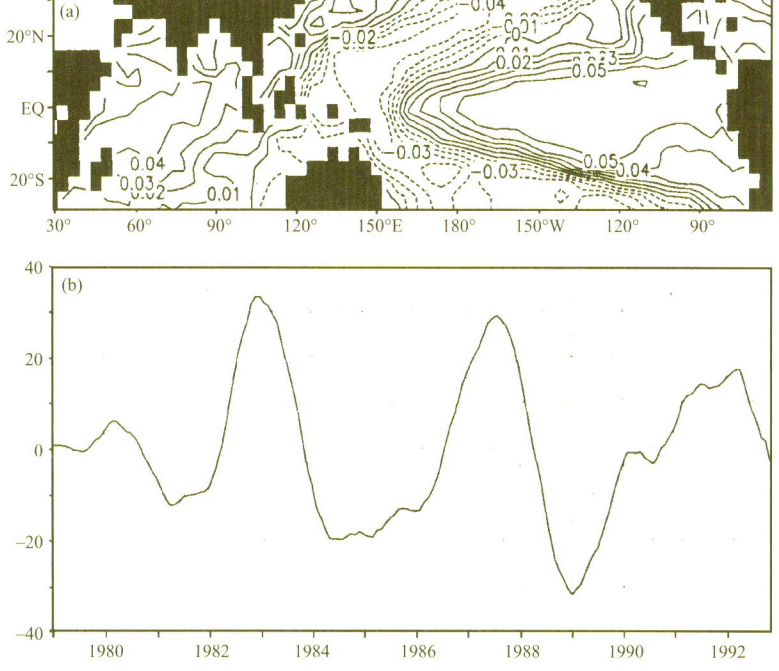

图 5 利用 1979 年 1 月—1992 年 12 月的月均标准化 SST 资料进行 EOF 分析得到的第一特征向量
(a)空间型;(b)时间特征向量

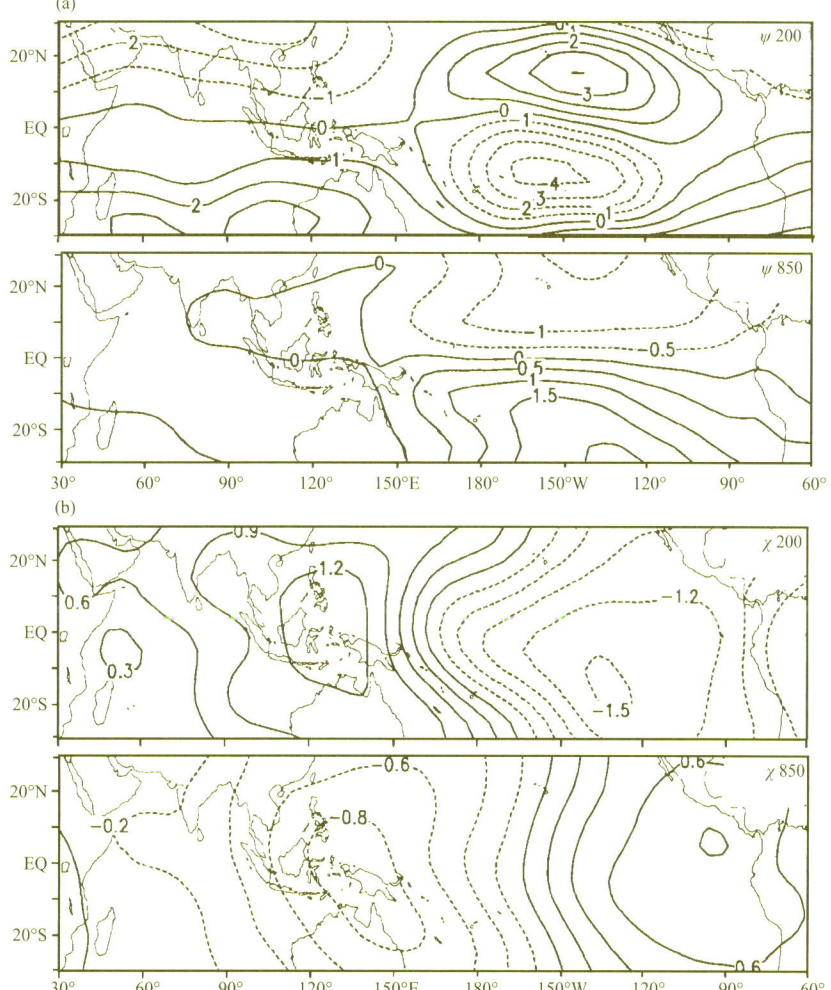

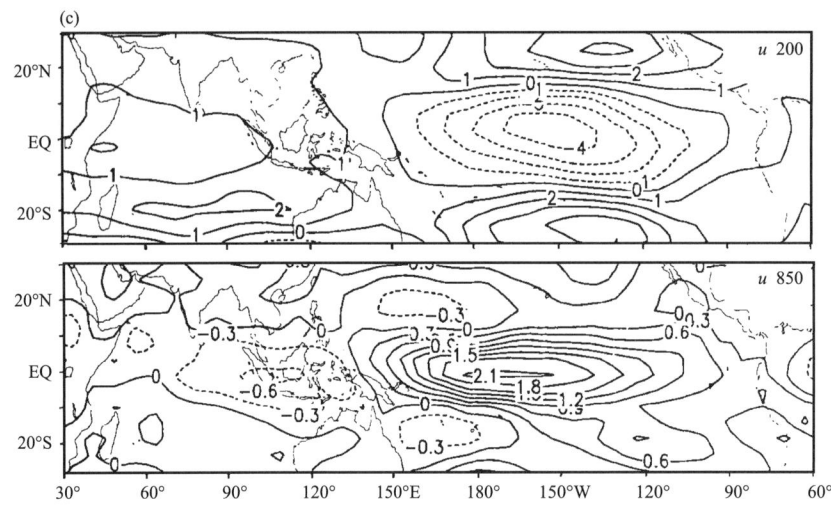

图 6 图 5b 中表征 ENSO 事件的时间特征向量与 200 hPa（上图）和 850 hPa（下图）气象要素的回归场分布。
(a)流函数场 ψ，单位为 $10^6 \ m^2 \cdot s^{-1}$；(b)速度位势场 χ，单位为 $10^6 \ m^2 \cdot s^{-1}$；(c)纬向风速 u，单位为 $m \cdot s^{-1}$

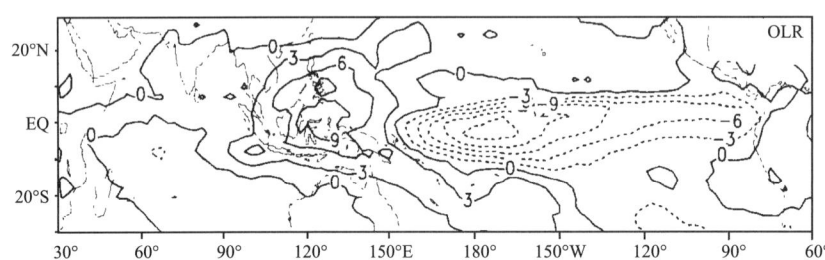

图 7 图 5b 中表征 ENSO 事件的时间特征向量与向外长波辐射 OLR 的回归场的分布，单位为 $W \cdot m^{-2}$

5 来自印度洋的 ENSO 触发机制

图 8 中给出了沿赤道带(2.2°S~2.2°N)SST、850 hPa 和 200 hPa 纬向风，以及向外长波辐射等异常量在 1980—1992 年期间的演变。从 SST 的演变（图 8a）可以清楚地看到 1982—1983 年、1986—1987 年，以及 1991—1992 年的 3 次 El Niño 事件和 1984—1985 年，以及 1988—1989 年的两次 La Niña 事件。它们与图 5 的结果完全吻合。从图 8a 也可以看出赤道中东太平洋的 SST 异常与印度洋的基本同号。此外，尽管早期征兆可追溯到西太平洋，但 ENSO 事件基本上是中东太平洋中的现象。由于中印半岛及印尼群岛的阻隔，没有发现 SST 异常从印度洋向太平洋传播而激发 ENSO 的迹象。

然而，沿赤道的高低空异常纬向风的情况则与 SST 的状况不同。从图 8b 和图 8c 可以看到图 6、7 所揭示的在 ENSO 事件期间的 GIP 现象，即 Walker 环流和季风环流同时增强或减弱。除此以外，还可清楚地看出：

(1)在 1980—1992 年，与每次 ENSO 事件相伴随的中东太平洋低空纬向风的异常均可追溯到印度洋的源地。即高低空反号的纬向风异常首先出现在印度洋，然后向东传播进入太平洋，强度增大，并伴有 SST 异常发生。

(2)由于 0 风速线表征着 GIP 的啮合点，图 8b 和图 8c 表明，每次 ENSO 事件中 GIP 的啮合首先出现在印度洋，然后向东传播。在此期间，季风环流圈扩大，Walker 环流圈缩小。

(3)与此期间 3 次 El Niño 事件相伴随，OLR 的负异常也首先出现在印度洋上空，然后向东传播（图 8d）。其中心一般位于 GIP 的啮合和 SSTA 之间：其西部为反向 GIP 的啮合点，东部为正的 SSTA。换言

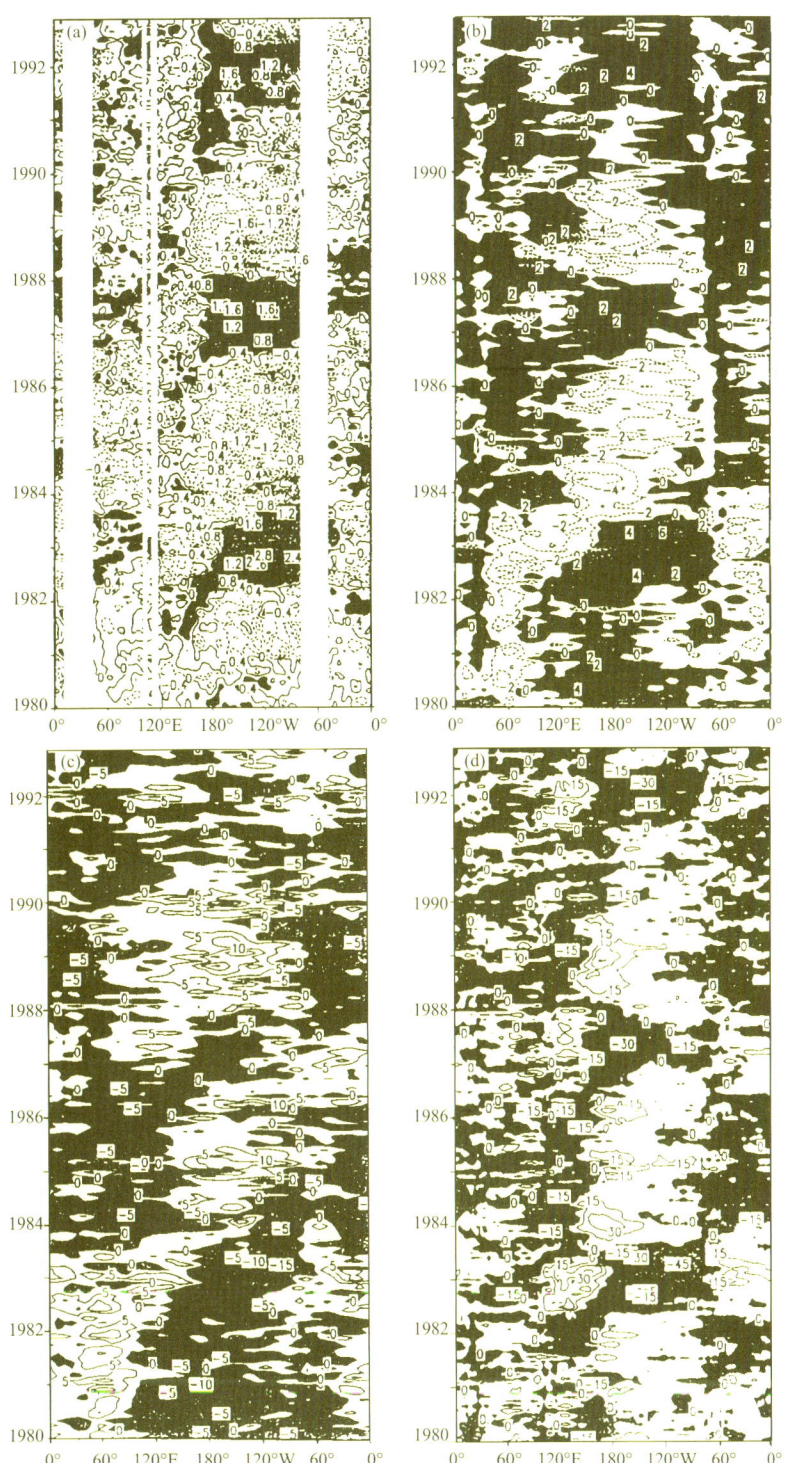

图 8 1980—1992 年沿赤道(2.2°S～2.2°N)要素场年际异常随时间的演变
(a)SST(℃);(b)850 hPa 纬向风(m·s^{-1});(c)200 hPa 纬向风(m·s^{-1});(d)向外长波辐射 OLR(W·m^{-2})

之,对流的加强出现在 850 hPa 异常西风区中。

以 1982—1983 年的 El Nino 事件为例(图 9)。1981 年冬天,在印度上空出现弱小的反向齿轮组(图 9a),其东部低空的西风和 OLR 异常已进入西部太平洋,在那里出现小范围的弱的正 SST 异常。到 1982 年初(图 9b),异常季风环流圈已扩展到整个印度洋,啮合点位于 120°E 附近;正 SST 异常也东扩到日期变更线附近。到 1982 年夏天(图 9c),异常季风环流圈继续东扩,GIP 的啮合点已位于西太平洋;整个赤道东太平洋已出现很强的正 SST,El Niño 已进入强盛阶段。由此看来,ENSO 事件的起始扰动可以是源于印度洋上空的纬向风异常。

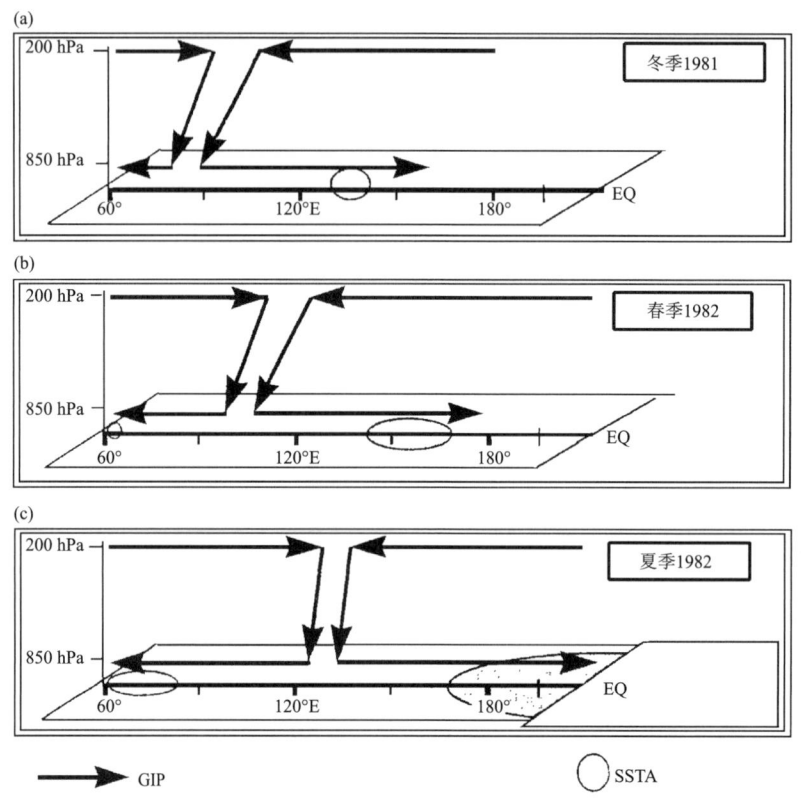

图 9　1982—1983 年 El Niño 期间及其前期异常风"齿轮啮合现象"从印度洋向太平洋传播及与 SST 异常配置的示意图

6　讨论和结论

通过对观测资料的分析发现,赤道印度洋 SST 在西南季风强盛期(4—8 月)的下降,在东北季风强盛期(12—4 月)的上升,以及中西部 SST 半年波的存在与沿赤道东西向 SST 差异较大、风应力作用较强有关。证明赤道中东太平洋和赤道中部印度洋的 SST 之间显著的正相关是由于在太平洋上空的 Walker 环流和印度洋上空的季风环流存在齿轮式的强耦合造成的。当这种太平洋-印度洋齿轮耦合为正向时,Walker 环流和季风纬向环流场均加强,东太平洋和印度洋 SST 为负异常。反之,当 GIP 为反向时,Walker 环流和季风环流均减弱,东太平洋和印度洋 SST 为正异常。

通过对 ENSO 型 SST 的时间特征向量与高低空流场、速度势场、纬向风场以及 OLR 同时相关特征进行分析,我们发现 ENSO 与 GIP 有很紧密的联系:El Niño 对应着反向的 GIP,而 La Niña 对应着正向的 GIP,GIP 的啮合点位于印尼群岛附近。

通过对 1980—1992 年沿赤道 SST 及 850 hPa 和 200 hPa 纬向风时间演变的分析表明,尽管与 ENSO 相联系的 SST 异常主要是太平洋现象,但与之相联系的沿赤道异常大气环流却显著的源于印度洋地区。GIP 的啮合点首先出现在印度洋,然后逐渐向东传入太平洋。与之对应这时赤道太平洋 SST 出现异常,并随着 GIP 的东传,SST 异常也向东扩展并发展成 ENSO 事件。换言之,从大气环流的角度来说,印度洋地区季风环流的异常可能是 ENSO 事件形成的一种触发机制。在本研究的第 II 部分,我们将通过使用耦合气候模式对这一结论进行检验。

参考文献

巢纪平,1993. 厄尔尼诺和南方涛动动力学[M]. 北京:气象出版社:309.

吴国雄,孙凤英,王晓春,1995. 降水对热带海表温度异常的邻域响应,Ⅱ. 资料分析[J]. 大气科学,19(6):663-676.

BARNETT T P, 1983. Interaction of the monsoon and Pacific trade wind system at interannual time scales-Part Ⅰ: The equatorial zone[J]. Mon Wea Rev, 111:756-773.

BJERKNESS J, 1969. Atmospheric teleconnections from the equatorial Pacific[J]. Mon Wea Rev, 97:163-172.

JU J, SLINGO J M, 1995. The Asian summer monsoon and ENSO[J]. Quart J Roy Meteor Soc, 121:1133.

MCCREARY J P, ANDERSON D L T, 1991. An overview of coupled ocean-atmosphere models of El Nino and the Southern Oscillation[J]. J Geophy Res, 96(supplemeant): 3125-3150.

MEEHL G A, 1987. The annual cycle and interannual variability in the tropical Indian and Pacific Ocean regions[J]. Mon Wea Rev, 115:27-50.

NEELIN J D, LATIF M, ALLAART M A F, et al, 1992. Tropical air-sea interaction in general circulation models[J]. Climate Dynamics, 7:73-104.

PHILANDER S G H, 1990. El Nino, La Nina and the Southern Oscillation[M]. San Diego: Academic Press: 287.

RASMUSSON E M, CARPENTER T H, 1982. Variations in tropical sea surface temperature and surface wind fields associated with the Southern Oscillation/El Nino[J]. Mon Wea Rev, 110:354-384.

TOURRE Y M, WHITE W B, 1995. ENSO signals in global upper-ocean temperature[J]. J Phys Oceanogr, 25: 1317-1332.

TOURRE Y M, WHITE W B, 1997. Evolution of the ENSO signal over the Indo-Pacific domain[J]. J Phys Oceanogr, 27: 683-696.

VILLWOCK A, LATIF M, 1994. Indian Ocean response to ENSO[R]. Internationl conf. on monsoon variability and prediction. Vol. Ⅱ, WCRP-84. Geneva, Switzerland. 530-537.

WALKER G T, BLISS E W, 1932. World Weather V[J]. Mem R Meteor Soc, 4:53-84.

WEBSTER P T, YANG S, 1992. Monsoon and ENSO: Selectively interactive systems[J]. Quart J Roy Meteor Soc, 118: 877-926.

WU G X, ZHANG Y S, 1998. Tibetan Plateau forcing and the Asian monsoon onset over South Asia and Southern China Sea[J]. Mon Wea Rev, 126 (4): 913-927.

YASUNARI T, 1990. Impact of Indian monsoon on the coupled atmosphere/ocean system in the Tropical Pacific[J]. Meteor Atmos Phys, 44:29-41.

> 代表性论文 3

大气热力强迫和动力强迫的调配及平均经圈环流的仿真模拟

吴国雄，蔡雅萍

(大气科学和地球流体力学国家重点实验室(LASG),中国科学院大气物理研究所,北京 100029)

摘要：通过研究平均经圈环流(MMC)及其所受的内外强迫作用的相互配置,指出对 MMC 的热力和动力强迫满足确定的调配率。这一调配率受大气内在的斜压性、静力稳定度及绝对涡度制约。

利用辐射加热和凝结加热参数化方案,结合欧洲中期天气预报中心(ECMWF)的分析资料,对 1 月份平均经圈环流进行数值仿真模拟。结果表明,热带对流加热可以形成双层 Hadley 环流结构；涡动动量输送对双 Hadley 环流的形成也有一定影响。中高纬度的 MMC 则主要由外动量强迫及大气的动量和热量输送特征决定。

关键词：平均经圈环流；热力强迫；动力强迫

1 引言

平均经圈环流(MMC)作为大气环流成员之一,与平均纬圈环流、水平环流、西风带及温度分布等有不可分割的联系。MMC 作为一种次级环流,只有当大气的静力平衡和地转平衡状态被破坏后才被激发出来。它引起大气产生相应调整,重建新的静力平衡和地转平衡状态(Eady,1950)。叶笃正和朱抱真(1958)指出,大气平均状态中的三个经圈环流是地球自转、非均匀加热、涡动输送过程和摩擦的共同作用所形成和维持的。对全球角动量收支和感热收支的分析(吴国雄 等,1988)则表明,MMC 的水平支的惯性矩平衡着角动量在近地表的制造和在自由大气中的输送；而其垂直支的绝热变温效应则平衡着大气中的非绝热加热和涡动热输送。

在对 ECMWF 1979—1984 年平均资料的分析中,吴国雄和刘还珠(1987)发现,低纬度的 Hadley 环流在垂直方向上存在双圈结构,这也为后来 Hoskins 等(1989)的计算所证实。由于热带平均经向风的分析对初值形成方案十分敏感(Hollingsworth and Cats,1981),Trenberth 和 Olsen(1988)曾认为,Hadley 环流的这种双层结构可能与分析误差有关。然而,双 Hadley 环流结构只出现在冬半球,而且从冬到夏呈现规律性的变化(参见文献吴国雄等(1987)20～21 页图),却不能用分析误差去解释。事实上,Schneider 和 Linzen(1977)的数值试验表明,热带地表的温度梯度可激发近地层的直接环流,而高层的潜热释放和积云摩擦可造成高层的直接环流。因此,Hadley 环流的双层结构可能是存在的。但是他们的理想化的模拟结果与资料分析的差异较大,地表温度梯度所激发出的环流只出现在 800 hPa 以下。本文试图用数值分析和数值仿真模拟相结合的方法去研究 MMC 的形成和维持,由此揭示出双 Hadley 环流生成的可能机

本文原载于《气象学报》,1994,52(2)：138-148

制。在第 2 节中,我们首先讨论 MMC 及其源汇项的配置。第 3 节着重讨论潜热释放和辐射冷却在 Hadley 环流形成中的作用。第 4 节比较各种内外强迫作用在 MMC 维持中的贡献。若干结论在第 5 节给出。

2 平均经圈环流的源汇分布

纬向平均的动量方程和热力方程可分别表示为:

$$[u]_t = \xi[v] - [u]_p[\omega] - (a\cos^2\varphi)^{-1}[u^*v^*\cos^2\varphi]_\varphi - [u^*\omega^*]_p + S \tag{1}$$

$$[\theta]_t = -a^{-1}[\theta]_\varphi[v] - [\theta]_p[\omega] - (a\cos\varphi)^{-1}[v^*\theta^*\cos\varphi]_\varphi - [\theta^*\omega^*]_p + Q \tag{2}$$

式中,$\xi = f - (a\cos\varphi)^{-1}[u\cos\varphi]_\varphi$ 为纬向平均绝对涡度,"[ ]"和"*"分别表示纬向平均及对纬向平均的偏差。等号右端第 1、2 项代表纬向平均经圈环流对纬向平均量的贡献,而第 3 至第 5 项分别代表涡动的水平和垂直输送等内强迫及外强迫源对纬向平均量的贡献。把这些强迫项用下述符号表示:

$$\begin{cases} F_1 = -(a\cos^2\varphi)^{-1}f[u^*v^*\cos^2\varphi]_\varphi = fx_1 \\ F_2 = -f[u^*\omega^*]_p = fx_2 \\ F_3 = fS \\ F = \sum_{i=1}^{3}F_i = fx = f(x_1 + x_2 + S) \end{cases} \tag{3}$$

$$\begin{cases} H_1 = -\alpha(a\cos\varphi)^{-1}[v^*\theta^*\cos\varphi]_\varphi \\ H_2 = -\alpha[\omega^*\theta^*]_p \\ H_3 = \alpha Q \\ H = \sum_{i=1}^{3}H_i \end{cases} \tag{4}$$

式中,$\alpha = (p/p_0)^\kappa R/(ap)$。定义静力稳定度参数 A、斜压性参数 B、和惯性稳定度参数 C 如下:

$$\begin{cases} A = -(a^2\rho\cos\varphi)^{-1}[\ln\theta]_p \\ B = (a^2\rho\cos\varphi)^{-1}[\ln\theta]_\varphi \\ C = (\cos\varphi)^{-1}f\xi \end{cases} \tag{5}$$

再根据纬向平均连续方程:

$$(a\cos\varphi)^{-1}[v\cos\varphi]_\varphi + [\omega]_p = 0 \tag{6}$$

定义流函数 ψ:

$$\begin{cases} [v] = (\cos\varphi)^{-1}\psi_p \\ [\omega] = -(a\cos\varphi)^{-1}\psi_\varphi \end{cases} \tag{7}$$

最后利用纬向平均地转关系:

$$[f + u\tan\varphi/a][u] = -a^{-1}[\Phi]_\varphi \tag{8}$$

静力关系:

$$[\Phi]_p = -[RT/p] \tag{9}$$

及位温定义:

$$[\theta] = [T][p_0/p]^\kappa \tag{10}$$

可得热成风关系:

$$[f + 2u\tan\varphi/a][u]_p = \alpha[\theta]_\varphi \tag{11}$$

根据关系式(3)—(11),可把动量方程(1)和热量方程(2)分别写成:

$$f[u]_t = C\psi_p + \delta B\psi_\varphi + F \tag{12}$$

$$\alpha[\theta]_t = -B\psi_p - A\psi_\varphi + H \tag{13}$$

式中,$\delta = f/\mathscr{F} = f/[f + 2a^{-1}u\tan\varphi] \approx 1$。在定常状态,式(12)、(13)构成联立方程组:

$$\begin{cases} B\psi_\varphi + C\psi_p = -F \\ A\psi_\varphi + B\psi_p = H \end{cases} \tag{14}$$

其系数行列式为：

$$\Delta = B^2 - AC = [\rho g^2]^{-1} ([u]_z \mathscr{F})^2 (1 - Ri f\xi\mathscr{F}^2) \tag{15}$$

或者

$$\Delta = B^2 - AC = -[\rho g a^2 \cos^2\varphi]^{-1} f P_E$$

式中，P_E 为 p 坐标下的 Ertel 位涡：

$$P_E \equiv -g(f\boldsymbol{k} + \nabla\times\boldsymbol{v}) \cdot \nabla\theta$$

Ri 为 Richardson 数：

$$Ri \equiv N^2 ([u]_z)^{-2}$$

在地球大气中，$Ri \gg 1$，且 $f\xi\mathscr{F}^2 \approx 1$，因此有：

$$\Delta = B^2 - AC < 0 \tag{16}$$

或等价的

$$fP_E > 0 \tag{16}'$$

式(16)′意味着大气是对称稳定的。于是从式(14)可得解：

$$\begin{cases} \psi_\varphi = -a\cos\varphi[\omega] = -\dfrac{1}{\Delta}(CH + BF) \\ \psi_p = a\cos\varphi[v] = \dfrac{1}{\Delta}(AF + BH) \end{cases} \tag{17}$$

当 $F = H \equiv 0$，即式(14)为齐次时，由于 $\Delta \neq 0$，根据达朗贝尔法则，MMC 只有零解，即：

$$[v] = [\omega] \equiv 0, (当 F = H = 0 时) \tag{18}$$

这意味着：

(1)平均经圈环流是由大气的动量源或热量源的外强迫作用及内部涡动动量输送和热量输送的内强迫作用造成的。它的强度和分布受到三个参数的影响，即大气的静力稳定度、惯性稳定度和斜压性。当没有强迫作用或内外源处于平衡状态时，MMC 不可能存在。

(2)在经圈环流中心，$\psi_\varphi = \psi_p = 0$。这时由式(17)有：

$$\begin{cases} BF + CH = 0 \\ AF + BH = 0 \end{cases} \quad (在 MMC 中心处) \tag{19}$$

既然 $\Delta = B^2 - AC \neq 0$，该线性齐次方程组只有零解：

$$F = H = 0$$

也就是说，在 MMC 中心处，作用于大气的外部动量源和热量源分别为涡动动量输送和热量输送所平衡。

(3)对于 MMC 上、下水平支上 $[\omega] = 0$ 的点，由式(14)有：

$$[v] = \frac{H}{\cos\varphi B} = -\frac{x}{\xi} \quad (当[\omega] = 0 时) \tag{20}$$

在北半球，$\xi > 0$，$B < 0$，因此有：

$$[v] < 0 \quad (当 x > 0，或 H > 0 时)$$
$$[v] > 0 \quad (当 x < 0，或 H < 0 时)$$

在南半球，$\xi < 0$，$B > 0$，因此有：

$$[v] > 0 \quad (当 x > 0，或 H > 0 时)$$
$$[v] < 0 \quad (当 x < 0，或 H < 0 时)$$

也就是说，动量源(汇)和(或)热量源(汇)区对应着向赤道(向极地)的水平非地转运动。

(4)对于 MMC 纯上升支及下沉支上 $[v] = 0$ 的点，由式(14)有

$$[\omega] = -\frac{1}{a\cos\varphi}\psi_\varphi = \frac{1}{a\cos\varphi}\frac{fx}{B} = -\frac{1}{a\cos\varphi}\frac{H}{A} \quad (当[v] = 0 时) \tag{21}$$

由于在两半球均有 $f/B<0$,因此:

$$-[\omega]>0 \quad (当 x>0 \text{ 或 } H>0)$$
$$-[\omega]<0 \quad (当 x<0 \text{ 或 } H<0)$$

也就是说,动量和(或)热量的源(汇)区对应着上升(下沉)运动。

综上所述,纬向平均动量和(或)热量的源汇分布与 MMC 分布之对应关系可用图 1a 表述如下:在一个直接环流的上升支和近地层对应着动量和(或)热量的源;在其下沉支和对流层顶附近对应着汇。在一个间接环流的上升支和对流层顶附近对应着动量和(或)热量的源;在其下沉支和近地层对应着汇。当热带存在双 Hadley 环流时,则源汇分布应如图 1b 所示。这时,热带必存在双层的动量和热量的源和汇。

值得注意的是,在满足式(20)、(21)的那些点上,对 MMC 的热力强迫和动力强迫之间存在着确定的比例关系,它们受大气内在的斜压性,静力稳定度及惯性稳定度所调制。换言之,对 MMC 的动力强迫和热力强迫存在下述调配关系:在 $[\omega]=0$ 的点上,动量源与热量源之比正比于绝对涡度与斜压性之比;在 $[v]=0$ 的点上,动量源和 f 的积与热量源之比正比于斜压性与静力稳定度之比,或者说正比于位温面相对于等压面水平倾角的正切。

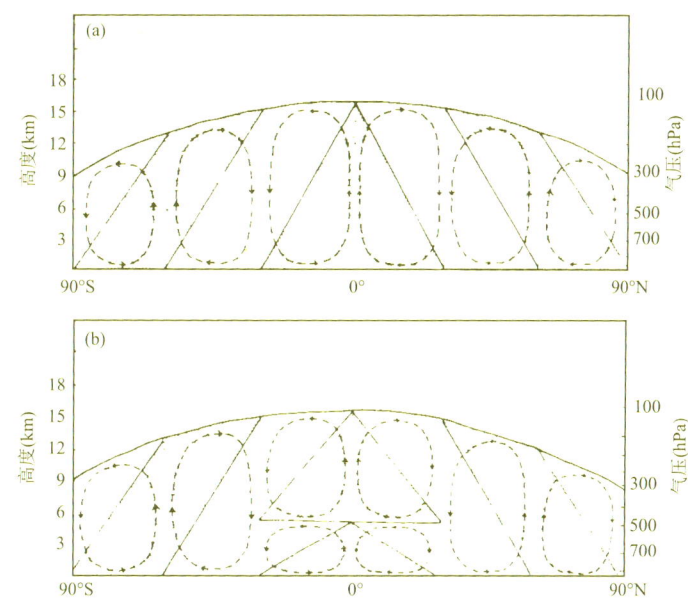

图 1 平均经圈环流和动量、热量强迫源(阴影区)和汇(无阴影区)的相对配置
((a)典型的三圈环流情况;(b)热带 Hadley 环流存在双层结构时)

3 热源强迫与 Hadley 环流

应用热成风关系(11),从动量方程(12)和热量方程(13)可导出经圈环流满足的方程为:

$$(A\psi_\varphi)_\varphi + 2B\psi_{\varphi p} + (C\psi_p)_p + B_p\psi_\varphi + B_\varphi\psi_p$$
$$= (F_1 + F_2 + F_3)_p + (H_1 + H_2 + H_3)_\varphi = F_p + H_\varphi \quad (22)$$

在导出上式时利用了近似式 $|[u]|/(a\Omega\cos\varphi)\ll 1$。该偏微分方程的性质由 $\Delta=B^2-AC$ 的符号(或 Richardson 数的大小)决定。由于大气一般具有对称稳定性($fP_E>0$),从式(16)或式(16)′知,式(22)是一个椭圆型方程。由于齐次椭圆型方程不能在内点取得极大值,又由于对全球积分而言 ψ 的边值近似为零,于是当式(22)中 $F=H\equiv 0$ 时,有解 $\psi\equiv 0$。这同样表示,当没有强迫作用或当内外源在各处均处于平衡状态时,大气中不可能存在平均经圈环流。此外,由于式(22)是一个线性方程,因此 MMC 可以看成是各种强迫作用产生的环流之和。

为求式(22)的数值解,在垂直方向把大气分为40层,间隔为25 hPa,在水平方向从南极到北极取网格点间隔为3°,共得61×41个网格点。在两极及 $p=0$ 处,ψ 边值取为零。在 $p=1000$ hPa处,由式(17)求得 ψ_φ 后,再由中央差分法解得 ψ 的下边界值。我们这里要研究1月份的环流,网格点上所需要的 $[u]$、$[T]$、$[u^*v^*]$、$[u^*\omega^*]$、$[v^*T^*]$、$[\omega^*T^*]$ 和外动量源 F_3 及热源 H_3 均利用吴国雄等(1987)所提供的相应资料内插得到。它们是从ECMWF 13层的五年(1979年9月—1984年8月)平均1月份资料计得的。把这些格点值代入式(22),便可由超张驰法求解该椭圆型方程,得到MMC的分布。

不考虑其他因子作用时,把上述非绝热加热 H_3 的分布代入式(22),求解得到的MMC如图2所示。非绝热加热在中高纬度的作用较小,在低纬度所激发的强大的直接环流中心分别位于250 hPa和800 hPa,强度分别为 14×10^3 kg·s^{-3} 和 12×10^3 kg·s^{-3}。由于感热加热所激发的环流很弱,且位于900 hPa以下(见Schineider等(1977)),可见这种热带环流主要应该由潜热加热和辐射加热造成。

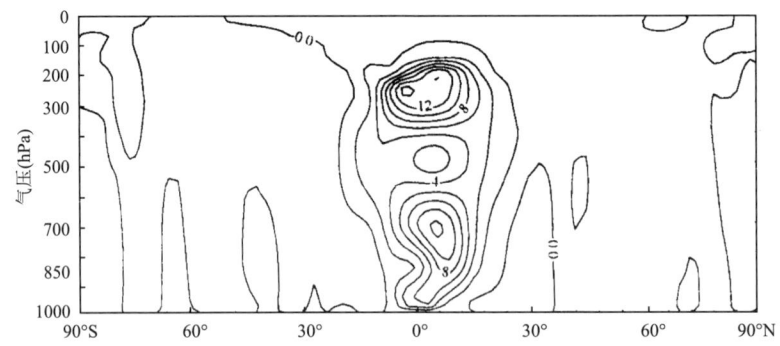

图2 1月份大气的外部热源 H_3 所激发的平均经圈环流

(等值线间隔为 2×10^3 kg·s^{-3}。外部热源 H_3 取自吴国雄等(1987),由ECMWF五年资料从热力学方程求得)

为研究辐射加热和热带对流活动对平均经圈环流的影响,我们引入如下的辐射加热参数化方案:

$$Q_{\text{rad}} = \left(\frac{p_0}{p}\right)^\kappa \frac{T_e(\varphi,p)-[T(\varphi,p)]}{\tau(p)} \tag{23}$$

式中,$[T(\varphi,p)]$ 为1月份纬向平均温度分布,$\tau(p)$ 为 p 坐标中的辐射加热时间尺度,

$$\tau(p) = \begin{cases} 30 \text{ d} & (p>300 \text{ hPa}) \\ (0.01p-1)\times15 \text{ d} & (200 \text{ hPa} \leqslant p \leqslant 300 \text{ hPa}) \\ 15 \text{ d} & (p<200 \text{ hPa}) \end{cases}$$

$T_e(\varphi,p)$ 为辐射平衡温度:

$$T_e(\varphi,p) = 0.5\left[T(\varphi)+\Gamma H_0\ln\frac{p}{p_0}+\left|T(\varphi)+\Gamma H_0\ln\frac{p}{p_0}\right|\right]+213$$

式中,Γ 为干绝热递减率,假定为常数,H_0 为标准大气的高度,$[T(\varphi)+213]$ 为理想化的1000 hPa辐射平衡温度。假定1月份 $T(\varphi)$ 在15°S取极大值,则可令:

$$T(\varphi) = 25\{1-2\sin^2[0.86(\varphi+15)]\}+60$$

上述辐射平衡温度 T_e 的取值给出均一的大气上界温度分布(constant skin temperature)以保证大气上界纬向风有界。式(23)表明,辐射冷却正比于温度对辐射平衡状态的偏差值。Rodgers和Walshaw(1966)证明,由式(23)所表达的辐射方案对地球大气而言是相当精确的。

本研究中对凝结加热取如下参数化形式:

$$Q_{\text{con}} = \frac{1}{c_p}\left(\frac{p_0}{p}\right)^\kappa \tilde{Q}(p)\exp\left[-\left(\frac{\varphi-\varphi_0}{\text{d}\varphi}\right)^2\right] \tag{24}$$

式中,φ_0 为凝结加热中心纬度,$\text{d}\varphi$ 模拟热带降水带的高斯半宽。$\tilde{Q}(p)$ 模拟凝结加热的垂直分布:

$$\widetilde{Q}(p) = \begin{cases} C_Q(p-p_t)/(p_1-p_t) & p_t \leqslant p \leqslant p_1 \\ C_Q & p_1 \leqslant p \leqslant p_2 \\ C_Q(p_b-p)/(p_b-p_2) & p_2 \leqslant p \leqslant p_b \end{cases} \quad (25)$$

它满足：

$$\int_{p_t}^{p_b} \widetilde{Q}(p) \frac{\mathrm{d}p}{g} = L\rho_{\text{water}} P_r \quad (26)$$

式中，p_t 和 p_b 分别为凝结层顶和底的高度，而 p_1 和 p_2 各为均匀凝结层的顶和底的高度；L 为凝结潜热，P_r 为降水率，ρ_{water} 为水的密度。

积云摩擦参数化取简单的通量形式：

$$F_{\text{con}} = -g \frac{\partial}{\partial p}[M_c(u-u_c)] \quad (27)$$

式中，M_c 为积云质量通量，它通过下述关系：

$$Q_{\text{con}} = \omega_c \frac{\partial \theta}{\partial p} = -gM_c \frac{\partial \theta}{\partial p} \quad (28)$$

由 Q_{con} 求得。

由观测的年降水量从式(25)、(26)可算出 C_Q，通过调节 $\mathrm{d}\varphi$ 可模拟热带降水的分布。假定潜热的释放通过浅薄对流和深厚对流完成，两者之比为8:7，则可取如下表参数。

	p_t (hPa)	p_1 (hPa)	p_2 (hPa)	p_b (hPa)	P_r (m·a^{-1})
深对流层	100	200	300	500	0.8
浅对流层	500	800	850	900	0.7

为模拟冬季状况，在式(24)中，取 $\varphi_0=6°\text{S}$，$\mathrm{d}\varphi=3°$。把上述参数代入式(23)、(24)便可得非绝热加热的分布。由此模拟得到的平均经圈环流如图3所示。积云对流在加热区南北侧均强迫出直接经圈环流(图3a)。然而南侧环流较弱，中心在 250 hPa 附近。北侧环流较强，在近赤道 250 hPa 和 850 hPa 附近各激发出强度为 $13×10^3$ kg·s^{-3} 和 $11×10^3$ kg·s^{-3} 的环流中心。高低纬度的辐射加热差异在低纬激发出的直接环流中心在 500 hPa 附近，但强度仅为凝结加热所激发的 1/3（图略）。因此积云加热和辐射加热共同激发的环流（图3b）基本上与图3a相似。比较图3和图2可见，由式(23)和(24)的参数化方案所描述的加热率能较好地模拟大气中的纬向平均加热，它们所强迫的 MMC 十分相似。

4 纬向平均经圈环流的模拟

由辐射加热和凝结加热所激发的 MMC 不能产生足够的热输送以平衡高低纬的加热差异，由此造成的行星大气之南北温差和西风急流会变得十分巨大，导致正压和斜压不稳定扰动的发展。此外，地形和海陆热力差异也能产生大气扰动。扰动的发展有效地减少南北温差。扰动的动量和热量输送则破坏大气的地转平衡和静力平衡，从而激发出非地转次级环流，以使大气恢复新的平衡关系（Eady, 1950; Green, 1970）。为分析扰动输送对 MMC 的激发作用，我们把从吴国雄等(1987)得到的 (F_1+F_2) 及 (H_1+H_2) 分别代入模型(22)中，算得的 MMC 如图4所示。这种涡动量强迫和热量强迫各在南北半球激发出三圈环流。由涡动动量强迫产生的环流中心一般位于 500 hPa 左右；且中纬度的间接环流比热带直接环流要强 1～3 倍。值得注意的是，在夏半球 Hadley 环流呈现为完整的单一环流圈，而在冬半球在 250 hPa 和 650 hPa 附近却各有一个强度约为 $2×10^3$ kg·s^{-3} 的中心。由于南北半球1月份涡动输送的最大差异是行星尺度涡动的输送，由此推知冬半球上层 Hadley 环流的存在与冬半年行星尺度涡动发展及其角动量输送特征有一定关系。涡动热量输送所激发出的环流的中心的纬度位置在南半球与动量输送所激发出的环流的位置大致吻合（25°S 和 45°S）；在北半球前者比后者北移了 10° 左右。

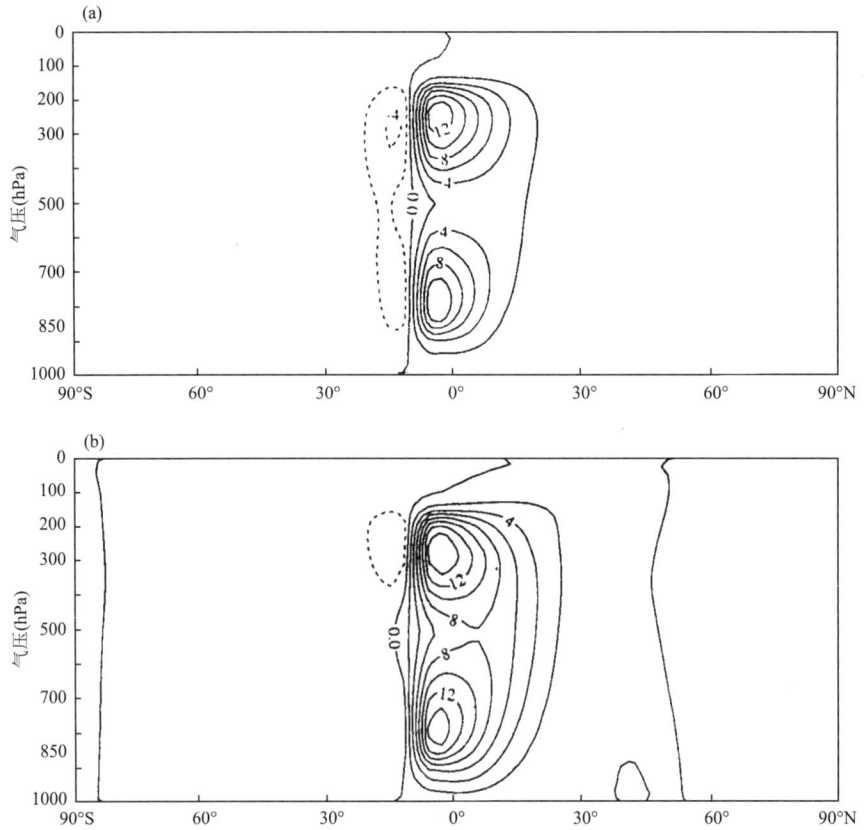

图 3 用凝结加热参数化方案(24)和辐射加热参数化方案(23)模拟得到的 1 月平均经圈环流(等值线间隔为 2×10^3 kg·s^{-3}。(a)凝结加热所激发的平均经圈环流;(b)凝结加热和辐射加热共同激发的平均经圈环流)

分析表明,积云摩擦的动量强迫效应诱发的 MMC 较弱(图略)。这里直接用吴国雄等(1987)所提供的外动量源 F_3 来研究其对 MMC 的贡献。图 5 所展示的即为 F_3 所激发的 MMC。其中心多集中在对流层下层。除南极环流和北半球 Ferrel 环流各出现在 750 hPa 和 800 hPa 外,其余各中心均出现在 900 hPa。其中北半球低层的 Hadley 环流强度达 15×10^3 kg·s^{-3}。根据第 2 节讨论,激发图 5 所示的 MMC 的外动量源汇主要集中在近地层,且与地面风带分布有很好的对应关系。换言之,摩擦力矩应是激发这类近地层环流的主要因子。值得注意的是北半球热带地区在 300 hPa 附近还存在着一个弱小的直接环流,强度为 3×10^3 kg·s^{-3} 左右,比动量内强迫产生的略强。在此两直接环流间为一间接环流。看来,热带上层 Hadley 环流的形成也与对流层上层存在动量源有关。

综上所述,250 hPa 附近的上层 Hadley 环流中心主要由非绝热加热造成,但也受外动量强迫和涡动动量输送的影响。此三者贡献的比例约为 15∶3∶2。

如式(23)和(24)所表示的非绝热加热方案与本节由资料计算得的所有强迫作用(H_1、H_2 及 F)加在一起代入式(22),则由其激发出的 MMC(图 6a)在两半球均呈明显三圈结构,与用 1 月份实测 $[v]$ 和 $[w]$ 所计算得到的 MMC(图 6b,取自吴国雄等(1987))十分一致。在模拟和分析中,北半球的 Hadley 环流都具有明显的双层结构。模拟结果还表明,在下层 Hadley 环流的形成中,动力强迫和热力强迫的贡献相当。

5 结论

MMC 是由大气的外动量和(或)热量强迫以及内输送过程所激发出来的。纬向平均的动力强迫(F)

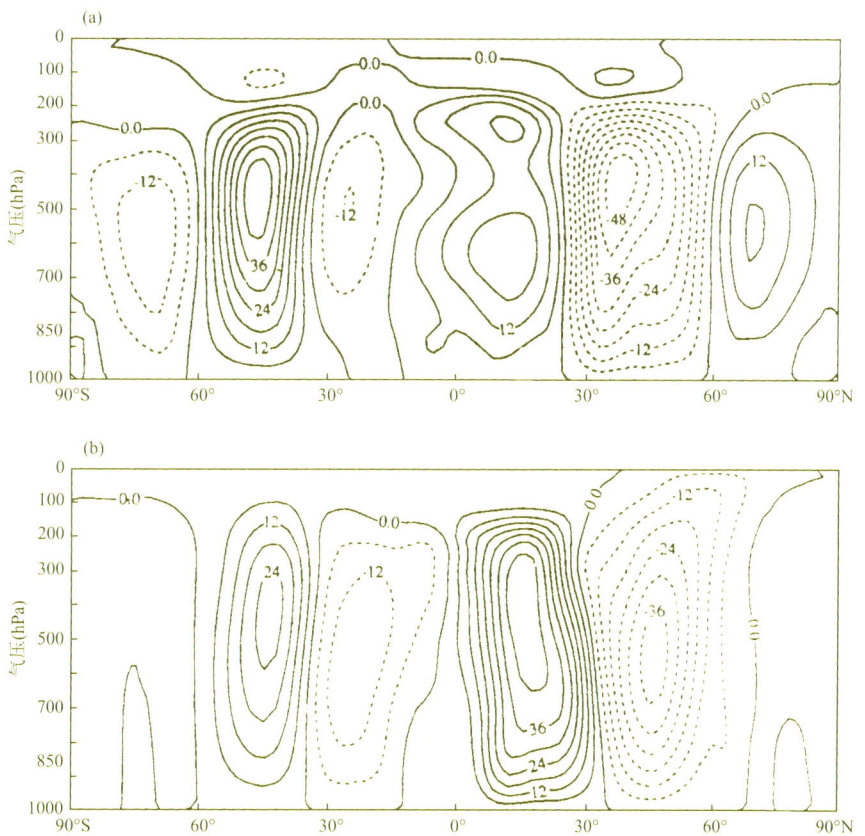

图 4　1 月份大气波动输送过程所强迫的平均经圈环流(等值线间隔为 6×10^2 kg·s^{-3}。(a)水平和垂直涡动动量输送所激发的平均经圈环流；(b)水平和垂直涡动感热输送所激发的平均经圈环流)

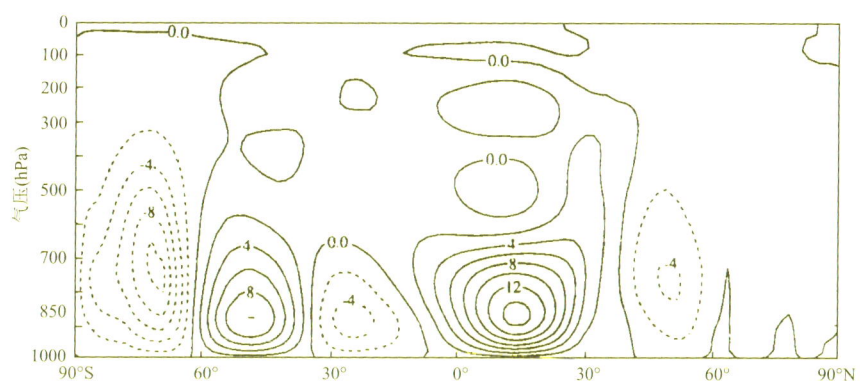

图 5　1 月份大气的外动量源(F_3)所激发的平均经圈环流(等值线间隔为 2×10^2 kg·s^{-3}。外部动量源 F_3 由 ECMWF 资料从 u 动量方程求得(吴国雄 等,1987))

和热力强迫(H)之间存在确定的比例关系。在垂直运动$[\omega]$为零的点上,动力强迫与热力强迫之比正比于绝对涡度与斜压性之比。在水平运动$[v]$为零的点上,动力强迫与科氏参数之积和热力强迫之比正比于斜压性和静力稳定度之比。

热带深对流导致的潜热释放可在对流层上部形成热源并导致上层 Hadley 环流形成。外动量源及涡动输送过程对上层 Hadley 环流的形成也有贡献,但作用比前者弱。在下层 Hadley 环流的形成中,热力强迫和动力强迫的贡献相当。在中纬度间接环流的形成中,非绝热加热作用较小。外动量强迫和涡动的动量和热量输送过程均起着重要的作用。

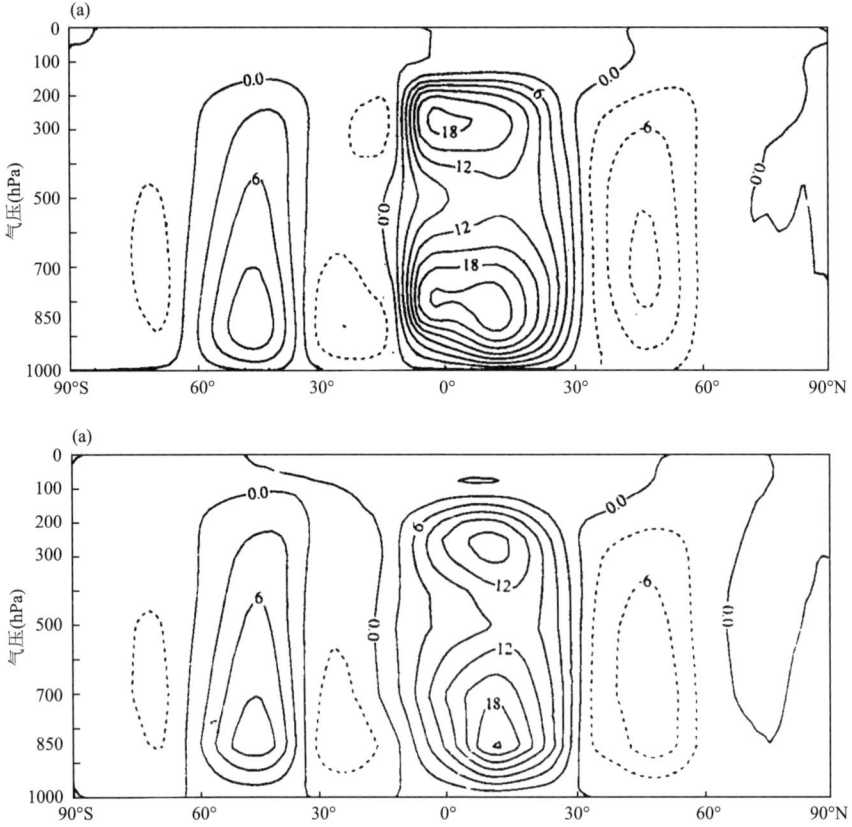

图6 1月份平均经圈环流的分布(等值线间隔为 3×10^3 kg·s^{-3}。(a)用辐射、凝结加热参数化方案及其他内外强迫模拟得到的平均经圈环流(详见正文);(b)由纬向平均风 $[v]$ 和 $[\omega]$ 计得的平均经圈环流(吴国雄 等,1987))

参考文献

叶笃正,朱抱真,1958. 大气环流的若干基本问题[M]. 北京:科学出版社:156.

吴国雄,TIBALDI S,1988. 平均经圈环流在大气角动量和感热收支中的作用[J]. 大气科学,12(1):8-17.

吴国雄,刘还珠,1987. 全球大气环流时间平均统计图集[M]. 北京:气象出版社:212.

EADY E T,1950. The cause of the general circulation of atmosphere[J]. Roy Meteor Soc Cent Proc:156-172.

GREEN J S A,1970. Transfer properties of the large-scale eddies and the general circulation of the atmosphere[J]. Quart J Roy Meteor Soc,96:157-185.

HOSKINS B J, HSU H H, JAMES I N, et al,1989. Diagnostics of the Global Atmospheric Circulation Based on ECMWF Analyses 1979—1989[R]. WMO/TD-326.

HOLLINGSWORTH A, CATS G,1981. Initialization in the tropics[C]. ECMWF workshop on "Tropical meteorology and its effects on medium range weather prediction at middle latitudes". ECMWF, England:Reading,105-142.

RODGERS C D, WALSHAW C D,1966. The computation of infrared cooling in planetary atmospheres[J]. Quart J Roy Meteor Soc,92:67-92.

SCHINEIDER E K, LINZEN R S,1977. Axially symmetric steady-state models of the basic state for instability and climate studies. Part Ⅰ:Linearized calculations[J]. J Atmos Sci,34:263-279.

TRENBERTH K E, OLSON J G,1988. ECMWF global analyses 1979—1986:Circulation statistics and data evaluation [R]. NCAR Technical Note, NCAR/TN-300+STR,94.

A Further Study of the Surface Zonal Flow Predicted by an Eddy Flux Parametrization Scheme

WU Guoxiong[1], WHITE Andrew Arthur[2]

(1 Institute of Atmospheric Physics, Chinese Academy of Sciences, Peking, People's Republic of China; 2 Meteorological Office, Bracknell)

Abstract: The problem of calculating the surface zonal flow using Green's large-scale eddy flux parametrizations in spherical geometry is re-examined. In a 2-level, quasi-geostrophic (QG1) model, low-latitude easterlies and mid-latitude westerlies of reasonable intensity can be obtained by assuming a more realistic baroclinicity than that applied in a study reported by White in 1977. This result also depends on the use of a more realistic value for the latitude average of a certain transfer coefficient; but no detailed treatment of its spatial variation is found to be necessary. Some further solutions are obtained using a strictly inconsistent formulation in which the Coriolis parameter is allowed its true latitude variation in all terms (in contrast to the QG1 model). In line with the conclusions of White's earlier study it seems in this case that detailed specification of the relevant transfer coefficient's spatial variation would be needed in order to produce a realistic surface zonal flow. Regarding the surface flow problem as a test of Green's parametrization scheme, we conclude that the latter performs quite well if QG1 approximations are applied to the Coriolis parameter. However, further refinement of the scheme is evidently required for good performance when these approximations are not applied.

1 Introduction

One of the most important aspects of the tropospheric general circulation is the zonal average surface zonal flow U_s. It is well established that large-scale eddy fluxes of zonal angular momentum are mainly responsible for the maintenance of U_s against the effects of frictional interaction with the earth's surface—see Lorenz (1967), Kidson et al. (1969) and Oort and Rasmusson (1971). Figure 1 shows a typical latitude profile of U_s and the angular momentum and vorticity transfers required to support it. Clearly, the eddy motion does not act simply to transfer angular momentum down the angular velocity gradient or vorticity down the absolute vorticity gradient (Lorenz, 1967; Starr, 1968; Read, 1986). Thus the surface zonal flow affords a stringent test of large-scale eddy parametrization schemes of the type used in statistical climate models (see Schneider and Dickinson 1974; Saltzman 1978).

Green (1970)—herein referred to as G70—proposed a mixing-type parametrization scheme which attempted to overcome the difficulties mentioned above. He suggested that the large-scale transient eddy motion acts to mix, in a readily idealizable fashion, only those quantities which are closely conserved in the quasi-geostrophic eddy evolution. Such quantities include potential temperature and potential vorticity, but not angular momentum or vorticity (neither of which obeys a Lagrangian conservation law in three-dimensional baroclinic motion). Green proposed simple expressions for the eddy fluxes of conserved quantities in terms of the relevant mean gradients and eddy diffusivities—called transfer coefficients.

By assuming a reasonable form for the zonal mean temperature field, Green used his potential vorticity mixing model to calculate a theoretical surface zonal flow variation in a β-plane formulation. His results were realistic: westerlies in middle latitudes and easterlies in extreme latitudes, with reasonable maximum values. In an extension of Green's study, White (1977) (herein denoted as W77) showed that the various assumptions of Green's analysis were uncritical so long as a realistic β effect was present. Examination of the problem in a quasi-geostrophic spherical polar formulation, however, led to less encouraging results. It was found that geometrical effects (including the variation of absolute vorticity gradients on the sphere) conspired to make the calculated surface flow much more sensitive to the assumed behaviour of a certain transfer coefficient than was the case in the β-plane formulation. In particular, W77 concluded that fairly detailed treatment of the spatial variation of the transfer coefficient was necessary in order to obtain a surface zonal flow of realistic strength and having polar easterlies. The required treatment might indeed go beyond what is readily justifiable within the framework of Green's scheme.

The present paper re-examines some of the conclusions of W77, clarifying certain aspects of the spherical polar problem and drawing attention to some overlooked features. Our study is partly prompted by results obtained by Vallis (1982). He found realistic surface zonal flows in a 3-level, zonal average climate model framed in spherical geometry. Eddy fluxes were represented by Green's parametrizations but without the detailed treatment of spatial variations that W77's results suggested would be necessary.

Following Marshall (1981), a 2-level model is used here instead of the continuous functional representation adopted in W77. This considerably simplifies the analysis, especially as regards the application of a crucial integral constraint on the solution.

In section 2 the derivation of the differential equation for the surface zonal flow U_s (as a function of latitude) is outlined. Some useful approximate analytical solutions are considered in section 3, and numerical solutions of the full equation are presented in section 4. The Phillips (1963) type 1 replacements, whereby the Coriolis parameter f is represented by a mid-latitude value, are applied for the most part, but the effects of allowing full latitude variation of f are also investigated. A summary and conclusions are in given in section 5.

Zonal mean models based on Green's parametrizations have been used by Sela and Win-Nielsen (1971), Win-Nielsen and Fuenzalida (1975), Ohring and Adler (1978) and White and Green (1984) as well as Vallis (1982). Marshall (1981) applied the technique in an ocean circulation model. Extensions to three dimensions have been made by White and Green (1982) and Shutts (1983). The relation of mixing theories such as Green's to the various functional extremization approaches of Bretherton and Haidvogel (1976) and Paltridge (see Paltridge, 1981) has been elucidated by Shutts (1981).

The applicability of diffusive parametrizations has of course been widely questioned. Lorenz (1979) found that a diffusion law related observed eddy heat fluxes to zonal mean temperature gradients realisti-

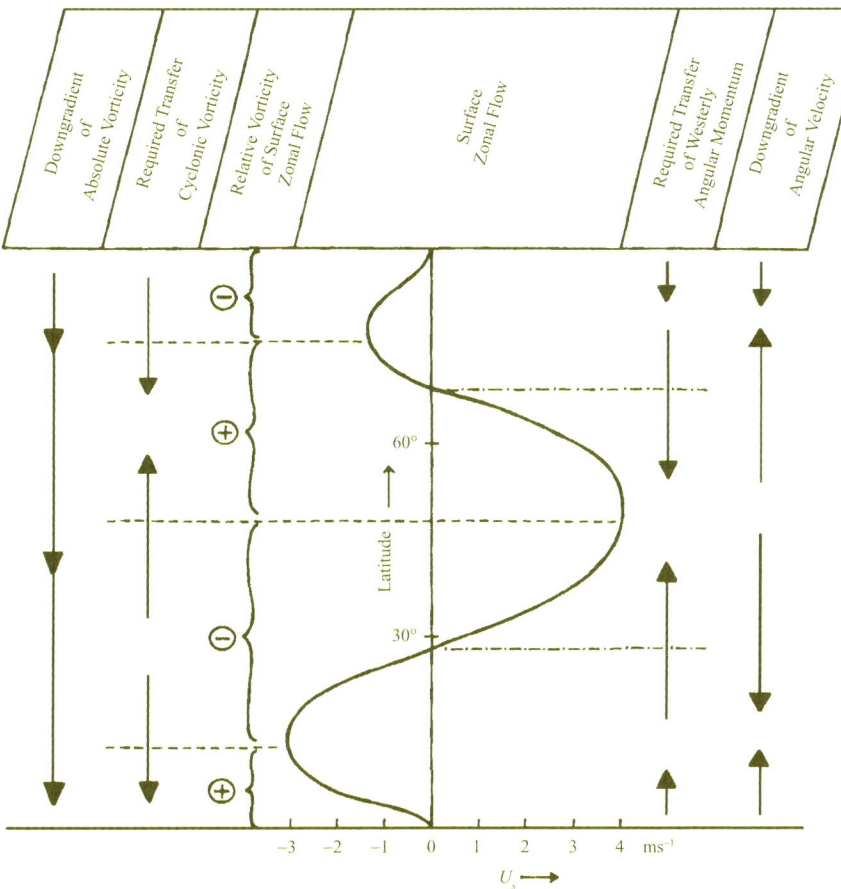

Figure 1 An idealized latitude variation of surface zonal flow U_s. Maximum values are typical of observed tropospheric behaviour. Also shown are the dynamical transfers required to support U_s, against the effects of surface friction, and some relevant zonal mean quantities. (The gradients of absolute vorticity and angular velocity indicated are typical of the height-averaged troposphere.) Based on a diagram given in unpublished lecture notes by J. S. A. Green

cally only at the largest space-scales and on long time-scales. However, in his data analysis Lorenz examined only the simplest law, in which the single diffusion coefficient is assumed constant. Studies of the eddy potential vorticity flux in relation to zonal mean fields (Wiin-Nielsen and Sela, 1971; Edmon et al., 1980; Pfeffer, 1981; Karoly, 1982a; Schmitz and Dethloff, 1984) reveal predominantly down-gradient transfers. Small regions of up-gradient flux do occur, especially near the tropopause, but typical tropospheric behaviour on seasonal and longer time-scales appears qualitatively consistent with the notion of potential vorticity diffusion.

We consider that dynamical parametrizations of the type proposed by Green offer a sufficiently reasonable idealization of the behaviour of real eddy motion to justify careful study of their implications—the more so because no superior scheme has, to our knowledge, been put forward.

2 Equation for the surface flow in a 2-level model

The dynamical basis of W77 was a type 1 quasi-geostrophic (QG1) model having continuous vertical structure and a standard representation of static compressibility. Here we use a 2-level QG1 model of incompressible flow (although a crude adjustment for the effects of static compressibility will be introduced in section 4).

The vorticity equation is applied at levels 1 and 2 (height $z = H/4$, $3H/4$) and the thermodynamic equation at level 2 ($z = H/2$). A smooth, rigid boundary is located at level 4 ($z = H$) and a rough, rigid boundary at level 0 ($z = 0$). The potential vorticity equations are

$$(\partial/\partial t + \mathbf{v}_i \cdot \nabla)q_i = Q_i \qquad i = 1, 3 \tag{1}$$

in which $\mathbf{v}$ is the non-divergent geostrophic flow and Q is a (small) source term whose form need not be quoted. The potential vorticities $q_i (i=1, 3)$ are given by

$$q_i = \zeta_i + f + \mu_i^2 (\psi_3 - \psi_1) \tag{2}$$

where ψ is the streamfunction corresponding to $\mathbf{v}$; $\zeta = \nabla^2 \psi$ is the relative vorticity; $\mu_1^2 = -\mu_3^2 = 4f_0^2/gBH^2$, $(gB)^{1/2}$ being the buoyancy frequency at level 2 and f_0 a typical mid-latitude value of $f = 2\omega\sin\theta$ (ω = earth's rotation rate, θ = latitude).

Upon assuming that the zonal average zonal stress vanishes at level 4, it follows from the QG1 momentum equation (see W77, Eq. (43)) applied at levels 1 and 3 that

$$\tau_s = \frac{1}{2}\rho_0 H(\overline{v_1'\zeta_1'} + \overline{v_3'\zeta_3'}) = \frac{1}{2}\rho_0 H(\overline{v_1'q_1'} + \overline{v_3'q_3'}) \tag{3}$$

in the steady state. Here τ_s, $(=\overline{\tau}(z=0))$ is the zonal average zonal stress, v is the poleward component of $\mathbf{v}$ and ρ_0 is the mean fluid density. The overbar and prime signify zonal average and deviation (eddy) components respectively. According to Eq. (3), τ_s is simply related to the poleward eddy fluxes of relative vorticity ζ and potential vorticity q.

Following G70 and W77, the eddy fluxes of q are parametrized in terms of the zonal mean gradients of potential vorticity as

$$\overline{v_i'q_i'} = -(K_i/R)\partial \overline{q_i}/\partial \theta \qquad i = 1, 3 \tag{4}$$

in which K_1, K_3 are transfer coefficients and R is the mean radius of the earth. From Eqs. (2), (3) and (4) may be derived a differential equation relating the surface zonal flow U_s to τ_s, and the mid-level baroclinicity b ($=-(1/R)\partial \overline{\phi}_2/\partial\theta, \phi = \log$ (potential temperature)):

$$\frac{d^2}{d\sigma^2}[U_s (1-\sigma^2)^{1/2}] - \frac{R^2 \tau_s}{\rho_0 HK (1-\sigma^2)^{1/2}} = 2\omega R - gaH \frac{d^2}{d\sigma^2}\left[\frac{b}{f_0}(1-\sigma^2)^{1/2}\right] - \frac{\gamma f_0 R^2 b}{BH (1-\sigma^2)^{1/2}} \tag{5}$$

Here $\sigma = \sin\theta$ and the quantities K, a and γ are given by

$$K = \frac{1}{2}(1+p)K_1, a = (1+3p)/4(1+p), \gamma = 2(1-p)/(1+p) \tag{6}$$

with

$$p = K_3/K_1 \tag{7}$$

Equation (5) is formally similar to Eq. (45) of W77. Our assumptions in solving it are the following.

(a) The baroclinicity function

Various analytical forms are chosen for the baroclinicity b (which would of course be determined internally in a full climate model). Each is expressible as

$$b = (\varepsilon\Delta\phi/R)F(\theta) \tag{8}$$

where $F(\theta)$ is the un-normalized functional form, and the constant ε is chosen so that

$$R\int_0^{\pi/2} b \, d\theta = \Delta\phi \tag{9}$$

Thus $\Delta\phi$ is the difference in log (potential temperature) between pole and equator at level 2 ($z=H/2$). $\Delta\phi$ is related to the corresponding pole to equator potential temperature difference, ΔT_p, by $\Delta\phi \approx \Delta T_p/\hat{T}_p$ (where $\hat{T}_p$ is a constant reference value). In deriving the numerical solutions presented in section

4 we take $\Delta\phi = 0.154$ in all cases. This corresponds to $\Delta T_p \simeq 40$ °C — a typical Northern Hemisphere (NH) winter value.

(b) Latitude variation of K_1, K_3 and K

We assume that K_1 and K_3 have the same latitude variation. This is equivalent to assuming separability in latitude and height of the relevant transfer coefficient, and has the convenient consequence that the ratio p (see Eq. (7) and section 2(d)) is independent of latitude. Separability of the transfer coefficient is in broad accord with the results of Wiin-Nielsen and Sela's (1971) diagnostic study (see their Figs. 10 and 12).

The actual latitude variation of K_1, K_3 is a matter of some uncertainty. From a consideration of baroclinic wave dynamics, G70 argued that K (viewed as a correlation between meridional velocity components and meridional particle displacements) should attain a maximum value in middle latitudes, where the baroclinicity is greatest, and should decrease to very small values in extreme latitudes, where the baroclinicity is small. For a β plane extending between latitudes $y=0$ and $y=L$, variation as $\sin^2(\pi y/L)$ appears appropriate. Indeed, G70 formalized this expected behaviour in terms of an integrative hypothesis'. W77 found that this formalization could not reasonably be extended to the spherical polar problem (see W77, sections 2 and 4). However, variations of the form $\cos^r\theta \sin^s\theta$ (where r and s are small integers) seem to us intuitively reasonable, and the consequences of assuming such forms will be re-examined in section 4(b). We take

$$K = K_m G(\sigma) \tag{10}$$

Where

$$K_m = 2.5 \times 10^7 \Delta\phi \text{ m}^2\text{s}^{-1} \tag{11}$$

is the mid-latitude maximum value of K. Thus the structure function $G(\sigma)$ attains its maximum value of unity where $K = K_m$. Equation (11) is similar to the form adopted by G70, p. 175, but the constant factor is smaller. This accounts partly for the fact that G70 considered potential temperature differences between 20° and 70°N; but the choice (11) has been made primarily to ensure that, when $\Delta\phi \approx 0.15$, K_m takes a value similar to those observed in the NH winter troposphere ($\approx 4 \times 10^6$ m^{2}s^{-1}; see Wiin-Nielsen and Sela (1971) and Sela and Wiin-Nielsen (1971)). In W77, this value was adopted for the latitude average of K, thus giving an overestimate by a factor of about two.

(c) Surface stress law

The surface stress τ_s is assumed expressible in terms of the surface zonal flow U_s. As in G70 and W77 we take

$$\tau_s = \rho_0 k U_s \tag{12}$$

where k is a friction-layer parameter. Equation (5) can then be re-written in terms of the dependent variable $U = U_s(1-\sigma^2)^{1/2}$ as

$$\frac{d^2 U}{d\sigma^2} - \frac{\nu^2 U}{(1-\sigma^2)} = 2\omega R - gaH \frac{d^2}{d\sigma^2}\left[\frac{b}{f_0}(1-\sigma^2)^{1/2}\right] - \frac{\gamma f_0 R^2 b}{BH(1-\sigma^2)^{1/2}} \tag{13}$$

in which

$$\nu^2 = kR^2/HK \tag{14}$$

For the friction-layer parameter k we adopt the value 2×10^{-2} ms^{-1}, independent of latitude. This is somewhat larger than the value assumed in W77, but is consistent with the surface zonal stress and zonal flow determinations reviewed by White (1974).

With the above values of k, K_m and $\Delta\phi$, and $R = 6.4 \times 10^6$ m, $H = 10^4$ m, we obtain

$$\nu^2 = 22/G(\sigma) \tag{15}$$

where $G(\sigma)$ is the structure function introduced in Eq. (10).

(d) Boundary conditions and surface torque constraint

As in W77, boundary conditions on U_s are applied as

$$U_s = 0 \text{ at } \sigma = 0, 1 \tag{16}$$

and the constraint

$$\int_0^1 U_s (1-\sigma^2)^{1/2} d\sigma = \int_0^1 U d\sigma = 0 \tag{17}$$

is also imposed. Constraint (17) represents the vanishing of the total surface torque between solid earth and atmosphere in the steady state, and it is obeyed by requiring the constant p (see Eqs. (6) and (7)) to take the appropriate value. Thus p is determined by the model, and does not have to be specified externally.

Our 2-level model is more consistent than the continuous vertical structure formulation used in W77. The quantities a and γ are here both determined via Eqs. (6) once p has been found by applying constraint (17); whereas in W77 constraint (17) was obeyed by varying only the equivalent of γ, and the equivalent of a was specified externally. In fact, the term in a is of minor importance in tropospheric parameter ranges, and W77's procedure amounts to only a small quantitative inconsistency. The most important new aspects of this study are investigations of the effects of different choices for the baroclinicity b and the latitude structure of the transfer coefficient K (as it appears in the quantity ν^2—see Eqs. (14) and (15)). We also examine how results are changed when f_0 in Eq. (13) is replaced by $f = 2\omega \sin\theta = 2\omega\sigma$.

Equation (13) is solved numerically by dividing the interval $x = [0, 1]$ into N_I equal intervals and using the Gaussian algorithm, or a recurrence method, to obtain the discrete solution which obeys conditions (16) and (17). For further details see Wu (1983). We take $N_I = 18$; higher resolutions lead to only very small changes. In section 3, however, we consider some useful analytical solutions of the problem.

3 Analytical solutions

If the term in τ_s is neglected, Eq. (5) can be written as

$$\frac{d^2}{d\sigma^2}\left[\left(U_s + \frac{gaH}{f_0}b\right)(1-\sigma^2)^{1/2}\right] = 2\omega R - \frac{\gamma f_0 R^2 b}{BH(1-\sigma^2)^{1/2}} \tag{18}$$

Analytical solution of Eq. (18) using the procedure outlined in section 2 is straightforward. With $f_0 = \sqrt{2} \cdot \omega$, and a particular choice of baroclinicity, $(U_s/\omega R)$ depends only on the non-dimensional parameters

$$\delta_1 = \Delta\phi/(BH) \tag{19}$$

$$\delta_2 = gH\Delta\phi/(\omega^2 R^2) \tag{20}$$

The parameter δ_1 is the ratio of the potential temperature differences between pole and equator, and between bottom ($z = 0$) and top ($z = H$) of the model troposphere. δ_2 is a Rossby number based on the thermal wind shear and the absolute velocity of rotation ωR of a point fixed on the earth at the equator. Typical tropospheric values of δ_1 and δ_2 are 1 and 0.05.

Solutions in this zero-stress approximation should be regarded as representing the limit of very weak surface stress, for the constraint (17) cannot be justified if τ_s vanishes completely. As was discussed in W77, such solutions are useful guides to the forms of solutions of the complete equation (5) although they invariably overestimate extremal values. Here, some solutions of Eq. (18) will be presented to indicate the consequences of making various changes in the conditions applied in solving Eq. (5). No algebraic results will be quoted, and all physical interpretation will be deferred until section 4. (Note, however, that our choices for the baroclinicity in this section are dictated largely by a desire to avoid algebraically complicated results. Our objective is to demonstrate qualitative effects only.)

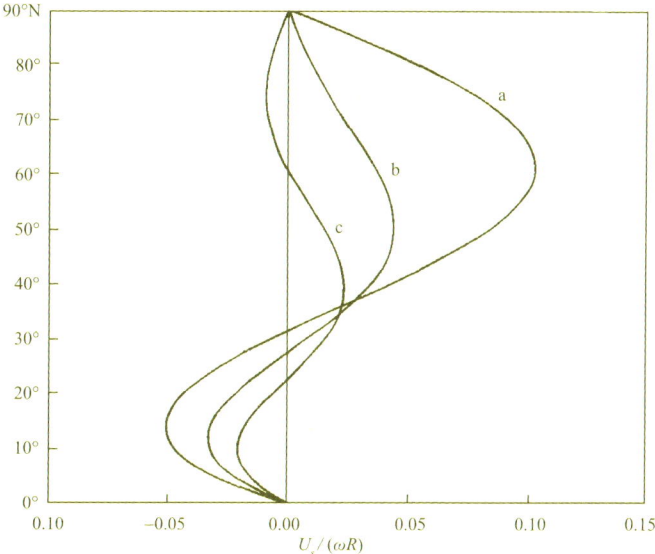

Figure 2 Latitude variations of surface zonal flow (expressed as $U_s/\omega R$) obtained by solving Eq. (18) analytically, assuming various forms for the baroclinicity. Curves (a), (b), (c) correspond respectively to the baroclinicities given by Eqs. (21), (22), (23)—whose latitudes of maximum horizontal temperature gradient are in decreasing order

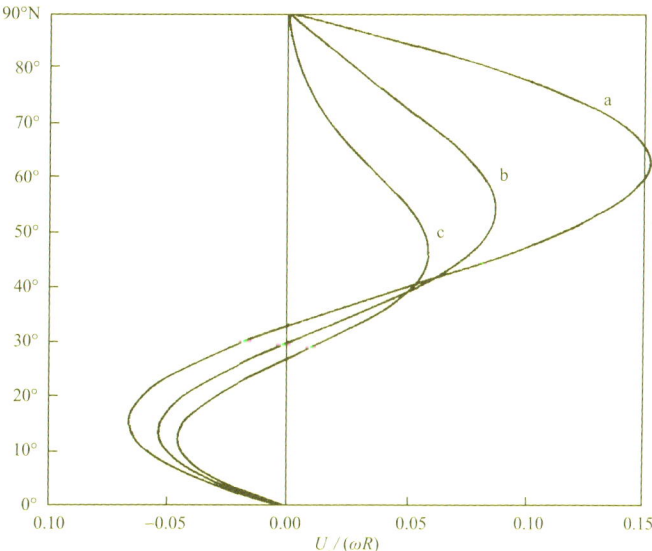

Figure 3 As Figure 2, but showing the results obtained when $f_0 = \sqrt{2} \cdot \omega$ (in Eq. (18)) is assigned the true latitude variation of the Coriolis parameter

An important feature of any assumed baroclinicity is the latitude θ_{max} of its maximum. Figure 2 shows the variations of $U_s/\omega R$ obtained by solving Eq. (18) when $\delta_1 = 1, \delta_2 = 0.05$ and

$$b = (3\Delta\phi/R) \sin^2\theta\cos\theta \qquad \theta_{max} = 54°44' \qquad (21)$$

$$b = (15\Delta\phi/2R) \sin^2\theta \cos^3\theta \qquad \theta_{max} = 39°14' \qquad (22)$$

$$b = (105\Delta\phi/8R) \sin^2\theta \cos^5\theta \qquad \theta_{max} = 32°18' \qquad (23)$$

Moving the latitude of maximum baroclinicity equatorwards clearly has two effects: (i) to reduce extremal values; and (ii) ultimately to produce polar easterlies. Qualitatively similar conclusions may be reached by representing the baroclinicity by a δ function centred at $\theta = \theta_{max}$ (White 1974).

Figure 3 shows the variations of $U_s/\omega R$ obtained by solving Eq. (18) with the same choices of baroclinicity, Eqs. (21)—(23), and the same values of δ_1 and δ_2, but with $f_0 = \sqrt{2} \cdot \omega$ replaced by $f = 2\omega\sigma$ (the true latitude variation of the Coriolis parameter). It is evident that equatorward movement of θ_{max} leads to marked reductions of extremal values—as in the previous case—but polar easterlies do not appear even when $\theta_{max} = 32°18'$. Comparison of corresponding curves in Fig. 2 and 3 suggests another important result: putting $f = 2\omega\sigma$ instead of $f = f_0 = \sqrt{2} \cdot \omega$ in Eq. (18) leads to marked increases of extremal values.

4 Numerical solutions

Interpretation of results is facilitated by writing Eq. (5) in terms of τ_s

$$\tau_s = -\frac{\rho_0 HK (1-\sigma^2)^{1/2}}{R^2} \left\{ 2\omega R - gaH \frac{d^2}{d\sigma^2}\left[\frac{b}{f_0}(1-\sigma^2)^{1/2}\right] - \frac{d^2}{d\sigma^2}[U_s(1-\sigma^2)^{1/2}] - \frac{\gamma f_0 R^2 b}{BH(1-\sigma^2)^{1/2}} \right\} \qquad (24)$$

The terms on the right-hand side represent various components of the height-integrated potential vorticity flux, as expressed by Green's transfer equations (see Eqs. (3) and (4)). The first three terms are essentially barotropic in character: they represent components of the height-integrated absolute vorticity gradient. The planetary vorticity gradient term is usually the largest of these barotropic terms (and the first of the second derivative terms reinforces it over wide ranges of latitude given a realistic baroclinicity, b). Thus the net effect of the barotropic terms is to give negative values of τ_s, and hence easterly surface flow. The remaining r.h.s. term in Eq. (24) represents baroclinic effects. It tends to give positive values of τ_s, and hence westerly surface flow, if $\gamma > 0$ and $b > 0$. Since the total stress torque over the Northern Hemisphere must vanish in the present model in the steady state (see Eq. (17)), the quantity γ—here assumed independent of latitude—must be positive when the temperature on level surfaces increases monotonically from pole to equator ($b \geqslant 0$). Thus surface westerlies are most likely to occur in middle latitudes, where b attains a maximum; see G70. In low latitudes surface easterlies are expected. In high latitudes the dominance of barotropic and baroclinic effects is unclear: all the barotropic terms are subject to the factor $(1-\sigma^2)^{1/2}$, but this factor cancels from the baroclinic term. According to Eqs. (6) and (7), positive γ corresponds to a bias of the transfer coefficients K_i towards the lower level of the 2-level atmosphere, and thus to the steering level of baroclinic waves being below $z = H/2$.

(a) Dependence of solutions on the form of the baroclinicity

Let us neglect the latitude variation of the transfer coefficients K_i. Taking $G(\sigma) = 1$ in Eq. (15) we then obtain $\nu^2 = 22$. Curve (a) in Fig. 4 shows the solution $U_s(\theta)$ which results in this case when

$$b = (2\Delta\phi/R)\sin\theta\cos\theta \tag{25}$$

There are no polar easterlies, and easterly and westerly maxima are 8.7 and 12.5 ms^{-1} respectively. As noted in W77 for a similar solution (see Fig. 10 of that paper), these maxima are much larger than observed tropospheric values (see, for example, Fig. 1, above). However, the baroclinicity given by Eq. (25) attains a maximum at $\theta = 45°$ and is symmetric about that latitude. In the Northern Hemisphere winter, the baroclinic zone is located between 20 and 50°N; see Fig. 5(a). The form

$$b = (3\Delta\phi/R)\sin\theta\cos^2\theta \tag{26}$$

has a maximum close to 35°N, and better represents the observed NH winter variation (see Fig. 5(b)). The analytical results of section 3 suggest that the form Eq. (26) will give reduced surface flow intensities, as compared with the form Eq. (25), because it has maximum baroclinicity nearer the equator. Figure 4, curve (b), shows the solution $U_s(\theta)$ obtained by solving Eq. (13) with $\nu^2 = 22$, as before, but with b represented by Eq. (26). Marked reductions of intensity are indeed seen: Maximum easterlies and westerlies are 6-9 and 7.3 ms^{-1} respectively. In physical terms, the equatorward movement of the baroclinic zone serves to cancel out a considerable part of the barotropic vorticity transfer in low/middle latitudes, and requires a weaker baroclinic vorticity transfer in high latitudes than before. Thus both easterlies and westerlies are weakened.

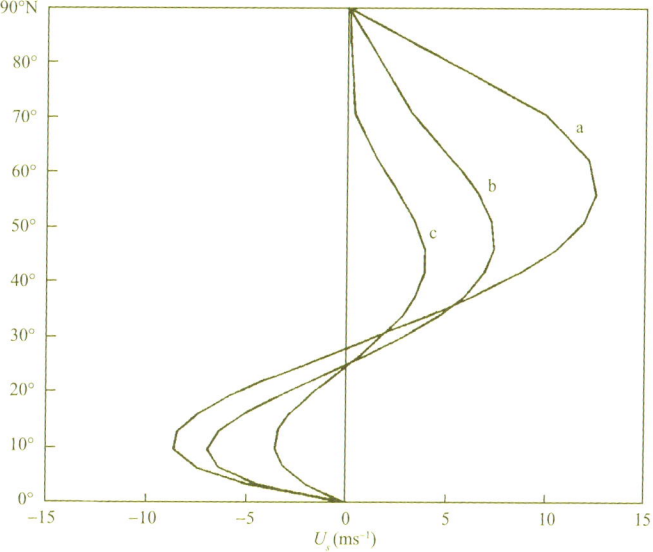

Figure 4 Latitude variations of surface zonal flow U_s obtained by solving Eq. (13) numerically under various assumptions. Curve(a): $\nu^2 = 22$ (independent of latitude), baroclinicity given by Eq. (25); curve (b): $\nu^2 = 22$, baroclinicity given by Eq. (26); curve (c): as for curve (b), but with latitude variation of ν^2 permitted according to Eq. (31) (which includes an adjustment to allow for the effect of static compressibility). $\Delta\phi = 0.154$ in each case

The effect of moving the baroclinic zone equatorward was not noted in W77. In fact, it was incorrectly stated (p. 111) that "solutions ⋯ are not sensitive to the choice of b". Here we have found significant reductions in extremal values when the form Eq. (26), rather than Eq. (25), is used to represent b. A further example—involving a greater change in b—will be considered in section 4(c).

(b) Dependence of solutions on the latitude variation of the transfer coefficient

In W77 it was found that considerable reductions in extremal values of U_s could be obtained by allowing for the latitude variation of the transfer coefficient K. The assumed variation was of the form $\sigma^2(1$

$-\sigma^2$), and the latitudinal mean value of K was held constant in the comparison. This is a reasonable adjustment (see G70), but, as noted in section 2(b), the mean value adopted was equal to observed tropospheric mid-latitude maxima ($\approx 4 \times 10^6$ m^2 s^{-1}) and was thus too high by a factor of two.

Here we investigate the effect of applying the structure function

$$G(\sigma) = [(3\sqrt{3})/2]\sigma(1-\sigma^2) \tag{27}$$

to the K_i (see Eq. (10)). Thus the K_i retain their previous values at the maximum latitude ($35°16'$) only, and are decreased at all other latitudes. Figure 6, curve (b), shows $U_s(\theta)$ obtained by solving Eq. (13) with b given by Eq. (25) and $\nu^2 = 22/G(\sigma)$; curve (a) shows the corresponding solution when $\nu^2 = 22$ (given above as curve (a) in Fig. 4). Easterly and westerly maxima are reduced to 6.1 and 8.8 ms^{-1}. These values represent considerable reductions (and are comparable to those found in section 4(a)). Even greater reductions are obtained if K is assumed to vary as $\sigma(1-\sigma^2)$, as in W77. If the grounds for comparison are those adopted here, curve A of W77's Fig. 10 and curve B of W77's Fig. 11 correspond to one another; a reduction of easterly/westerly maxima from 9/13 ms^{-1} to 4/7 ms^{-1} is seen.

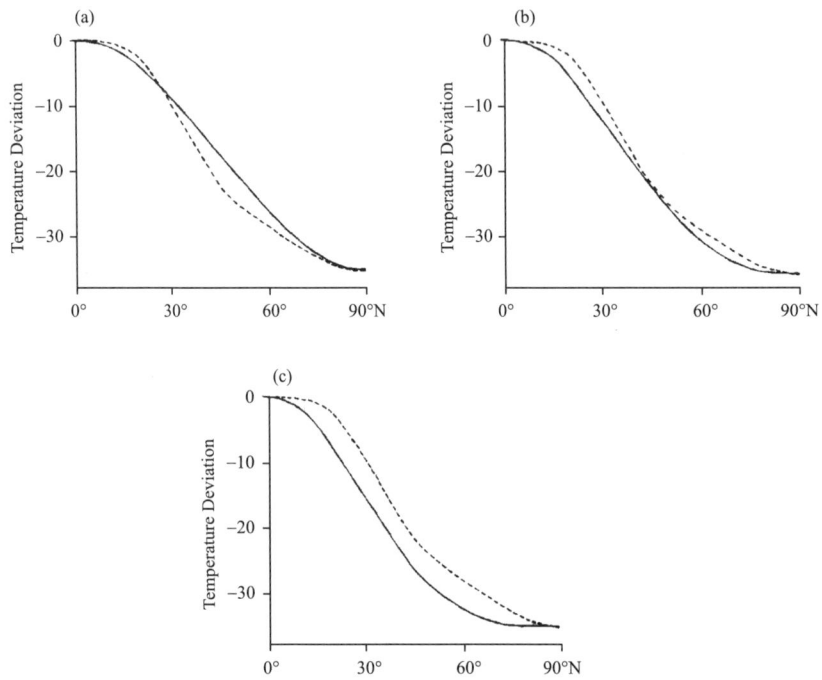

Figure 5 (a)—(c). Comparison of observed latitude variation of zonal mean temperature at 500 hPa in January with three analytic forms. Dotted lines: observed variation (for $\theta < 75°$N, from Oort and Rasmusson (1971); for $75° \leqslant \theta \leqslant 85°$N, calculated from data of Lejenas and Madden (1982) (P. Nalpanis, private communication)). Full lines: variations implied by (a) Eq. (25), (b) Eq. (26), (c) Eq. (32); the amplitude factor has been chosen in each case to give the same extreme temperature difference as the observed value. All temperatures are defined relative to an arbitrary zero at the equator

(c) Effects of accounting for static compressibility

Static compressibility has been neglected so far in this study. It was represented in W77, however, and may easily be accounted for in our 2-level formulation. With $\rho = \rho_i$, at level i, the counterpart of Eq. (13) turns out to be

$$\frac{d^2 U}{d\sigma^2} - \frac{\mu^2 U}{(1-\sigma^2)} = 2\omega R - g\hat{a}H \frac{d^2}{d\sigma^2}\left[\frac{b}{f_0}(1-\sigma^2)^{1/2}\right] - \frac{\hat{\gamma}f_0 R^2 b}{BH(1-\sigma^2)^{1/2}} \tag{28}$$

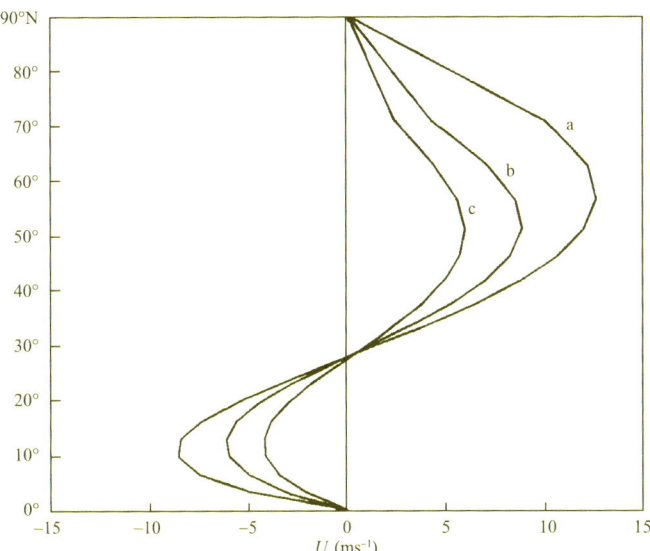

Figure 6 Latitude variations of surface zonal flow U_s obtained by solving Eq. (13) numerically under various assumptions. Curve (a): $\nu^2 = 22$ (independent of latitude), baroclinicity given by Eq. (25); curve (b): $\nu^2 = 22/G(\sigma)$, with $G(\sigma)$ given by Eq. (27), baroclinicity given by Eq. (25); curve (c): as for curve (b), but with ν^2 scaled so as to allow crudely for the effects of static compressibility (see Eq. (31)). $\Delta\phi = 0.154$ in each case

Here

with

$$\left.\begin{aligned} \hat{a} &= (1 + 3\hat{p})/4(1 + \hat{p}) \\ \hat{\gamma} &= 2(\rho_2 - \rho_3 \hat{p})/\rho_1(1 + \hat{p}) \\ \mu^2 &= \rho_s k R^2 / \bar{\rho}^z H \tilde{K} \\ \tilde{K} &= (\rho_1 K_1 + \rho_3 K_3)/2\bar{\rho}^z \\ \rho_s &= \rho(z = 0) \\ \bar{\rho}^z &= (\rho_1 + \rho_3)/2 \\ \hat{p} &= \rho_3 K_3 / \rho_1 K_1 \end{aligned}\right\} \quad (29)$$

and

Equation (28) can be solved numerically by the same methods as was Eq. (13). (An isothermal reference atmosphere, with a reasonable scale height, may be assumed in order to determine the densities ρ_i.) It is found that the solutions differ from those obtained by solving Eq. (13), in corresponding cases, almost entirely through the differences between the quantities μ^2 and ν^2. Since $\tilde{K}$ (see Eq. (29)) is a mass-weighted average of K_1, and K_3, it is equivalent to K, as given by Eq. (6), in the incompressible case. Thus

$$\mu^2 = \rho_s \nu^2 / \bar{\rho}^z \quad (30)$$

This difference can be allowed for by solving Eq. (13) with the term in ν^2 suitably scaled up. A convenient way of doing this is to determine K_1, rather than K, from Eq. (10); the effect is to increase ν^2 by a factor of $2/(1 + \hat{p})$, which (somewhat fortuitously) is almost equal to typical values of $\rho_s/\bar{\rho}^z$. To take account of static compressibility we shall in the remainder of this paper follow the procedure of replacing the term ν^2 in Eq. (13) by $2\nu^2/(1 + \hat{p})$. Figure 6, curve (c), shows the solution obtained when the case previously examined in section 4(b), with baroclinicity given by Eq. (25), is treated in this way: the term in ν^2 in Eq. (13) now takes the form

$$-44U/[(1+p)G(\sigma)(1-\sigma^2)] \tag{31}$$

(with $G(\sigma)$ given by Eq. (27), as before). Westerly and easterly maxima are now reduced to 5.9 and 4.2 ms^{-1} respectively. Figure 4, curve (c), shows the solution obtained using the adjustment Eq. (31) when b is represented by Eq. (26) (the more realistic form). Extremal values are now 4.0 and -3.5 ms^{-1}. These are similar to typical observed values (see Fig. 1), though weak westerlies (<0.4 ms^{-1}), rather than easterlies, here occur poleward of 70°. The extremal values are similar also to those obtained by Vallis (1982) in a 3-level, pressure coordinate, zonal average climate model which used Green's (1970) parametrization scheme—see his Fig. 5. In Vallis's model, the baroclinicity was generated internally, a fairly detailed treatment of diabatic effects being incorporated. Full latitude variation of the Coriolis parameter was allowed in the explicit formulation, but QG1 approximations were applied in the eddy flux parametrizations (and thus largely determined the surface zonal flow).

Since Vallis obtained weak surface easterlies (<1 ms^{-1}) poleward of 80°, we have derived some further solutions of Eq. (7) to investigate the circumstances in which polar easterlies may occur in our model. The arguments given at the beginning of this section suggest that polar easterlies will appear when the assumed baroclinicity is sufficiently small in high latitudes that the barotropic terms in Eq. (23) become dominant there. The form

$$b = (4\Delta\phi/R)\sin\theta\cos^3\theta \tag{32}$$

has a maximum at 30°N (see Fig. 5(c)). Figure 7 shows the solution obtained using Eq. (32) when static compressibility is allowed for as in Eq. (31). A broad band of weak polar easterlies—extending as far south as 60°N and with a maximum of about 0.8 ms^{-1}—is produced, and easterly/westerly maxima are both about 3.0 ms^{-1}. The polar easterlies are thus much more extensive than in Vallis's result. Although the assumed form of b is somewhat unrealistic (see Fig. 5(c)), we feel that the solution shown in Fig. 7, as compared with curve (c) of Fig. 4, is sufficient to demonstrate the sensitivity of the high latitude surface zonal flow to changes in the form of the baroclinicity (and hence to other modelling assumptions). This sensitivity is suggested also by the analytical solutions presented in section 3. Thus we consider that the differences between our results and Vallis's are of little significance, given the differences in formulation between the two models.

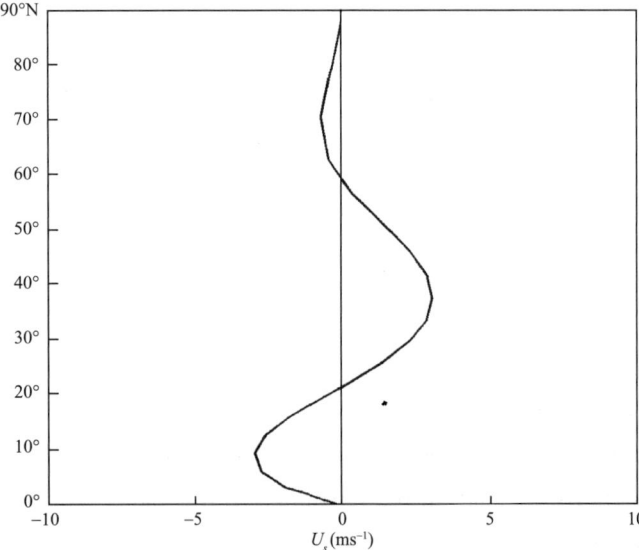

Figure 7 Latitude variation of surface zonal wind U_s btained by solving Eq. (13) numerically when the baroclinicity is given by Eq. (32). Latitude variation of v^2 as in Eq. (31) (static compressibility allowed for). $\Delta\phi = 0.154$

(d) Effects of allowing latitudinal variation of the Coriolis parameter

In the QG1 model the Coriolis parameter f appears as the mid-latitude value f_0 (here taken as $\sqrt{2} \cdot \omega$) in the vortex stretching term of the vorticity equation and also in the thermal wind equation. Thus f_0 appears in the second and third terms on the r. h. s. of Eq. (13). Although allowing $f = 2\omega\sigma$ instead of $f = f_0$ in these terms is strictly inconsistent because of the consequent loss of conservation properties in the basic formulation (see W77, p. 115), it seems reasonable to expect that such variations should give a more accurate representation of behaviour on the sphere. (Similar variable f replacements have been made by Edmon et al. (1980), Karoly (1982b) and Andrews (1983), amongst others, in various studies of large-scale atmospheric motion.) To investigate this possibility we have solved Eq. (13) with and without the indicated replacements. With $f = 2\omega\sigma$, the r. h. s. of Eq. (13) becomes

$$2\omega R - \frac{gaH}{2\omega}\frac{d^2}{d\sigma^2}\left[\frac{b(1-\sigma^2)^{1/2}}{\sigma}\right] - \frac{2\gamma\omega R^2 b\sigma}{BH(1-\sigma^2)^{1/2}} \tag{33}$$

We consider the case in which

$$b = (15\Delta\phi/2R)\sin^2\theta\cos^3\theta \tag{34}$$

and

$$\nu^2 = 44/[(1+p)G(\sigma)] \tag{35}$$

(with $G(\sigma)$ as given by Eq. (27)). In Fig. 8, curve (a) is the result obtained when $f = f_0$; curve (b) is the result when $f = 2\omega\sigma$ only in the γ term in Eq. (33); curve (c) is that when $f = 2\omega\sigma$ in both the γ and a terms. Clearly, allowing the full latitude variation of the Coriolis parameter considerably increases extremal values, and polar easterlies do not appear; Easterly and westerly maxima in the $f = f_0$ case are 4.9 and 6.0 ms^{-1}, increasing to 7.8 and 11.6 ms^{-1} in the $f = 2\omega\sigma$ case. These results are in accord with the indications of the analytical solutions presented in section 3. To obtain reasonable surface zonal flows it is evidently necessary to assume variations of the K_i which are far more peaked in middle latitudes (as first suggested by White (1974)). Physical interpretation is straightforward. The effect of allowing $f =$

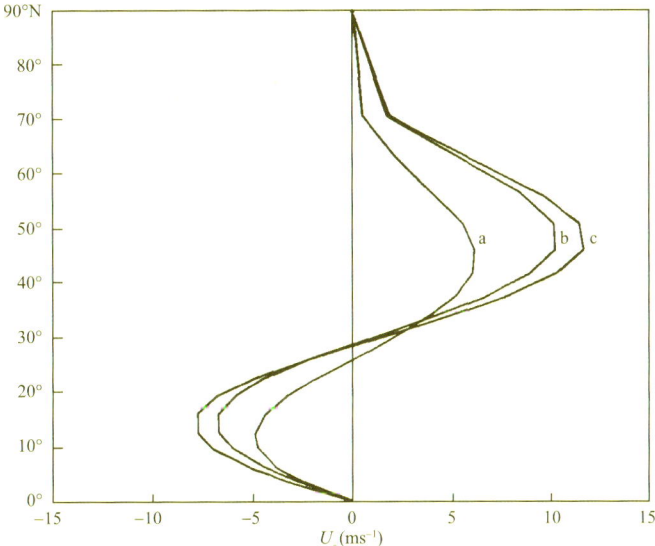

Figure 8 Effect of latitude variation of the Coriolis parameter f on the calculated surface zonal flow $U_s(\theta)$. Solutions obtained numerically, baroclinicity given by Eq. (34), ν^2 given by Eq. (35) (static compressibility allowed for). Curve (a) shows the solution obtained from Eq. (13), in which the usual $f = f_0$ replacements of the QG1 formulation are made; curve (c) shows the solution obtained when $f = 2\omega\sigma$ is retained throughout; curve (b) shows an intermediate case. See text for further details. $\Delta\phi = 0.154$ in each case

$2\omega\sigma$ is to decrease baroclinic vorticity transfers equatorward of 45° and to increase them poleward of 45°. Hence both easterly and westerly surface wind maxima are markedly increased, and the tendency for polar easterlies to appear is less than before.

5 Concluding remarks

In the atmosphere, mean meridional circulations and stationary eddies, in addition to transient eddies, transfer angular momentum meridionally. The stationary eddies transfer only one quarter to one third of the total angular momentum flux (see Oort and Rasmusson, 1971), and some of this stationary eddy transfer must be contributed by essentially transient eddies which are localized in longitude by the stationary long-wave pattern. The theory developed in G70 should therefore account for nearly all the eddy flux of angular momentum. In middle and high latitudes the flux by meridional circulations is negligible in the height average (which is the proper consideration in the surface flow problem). Meridional circulations are quite important in the height-integrated angular momentum balance in very low latitudes (<10°—see Kidson el al. (1969)); but White (1974) found that the calculated surface zonal flow was only slightly changed when the contribution of the meridional circulation was allowed for in solving the relevant equation (equivalent to our Eq. (5)).

The present study has confirmed W77's finding that representing the expected smallness of the transfer coefficient K in extreme latitudes reduces the intensity of the calculated surface zonal flow. However, because we have specified a more realistic latitude average of K, predicted numerical values of $U_s(\theta)$ are here somewhat less than those found in W77.

Use of a more realistic representation of the zonal average baroclinicity than in W77 also contributes a noticeable reduction of calculated surface zonal flow intensities. Indeed, the surface zonal flow has a realistic intensity when both of the above effects are suitably represented and the latitude variation of the Coriolis parameter is approximated in the prescribed manner of the QG1 formulation. Because the zone of subtropical easterlies is somewhat reduced, the implied latitudinal extent of the Hadley cell and equatorward penetration of the Ferrel cell are also more realistic. These results are consistent with those obtained by Vallis (1982) using similar QG1 parametrizations in a zonal average climate model which differs from ours in several respects. However, Vallis obtained weak easterlies in high latitudes where we obtain weak westerlies. It appears that the balance which determines the high-latitude surface flow in Green's scheme is very delicate; polar easterlies or polar westerlies may be produced depending on details of model formulation and behaviour.

Contrary to the implications of W77's results, we conclude therefore that Green's parametrization scheme can produce realistic surface flows on the sphere so long as QG1 dynamics are assumed. Nevertheless, and as stated in W77, allowing the full latitude variation of the Coriolis parameter leads to a considerable increase in the intensity of the calculated surface zonal flow, and the tendency for polar easterlies to appear is much reduced. The formulation used is then not strictly consistent dynamically (because it does not imply the good integral conservation properties of the QG1 model) but it may be considered to be more accurate than QG1. In this case, Green's parametrization scheme, as at present formulated, does not pass the surface flow test. The production of a realistic surface zonal flow would evidently require a more detailed specification of the latitude and height variation of the relevant transfer coefficient—to an extent which the theory is not yet equipped to justify. It is a challenge for future work to establish good theoretical grounds for the required spatial variations.

References

ANDREWS D G, 1983. A conservation law for small-amplitude quasi-geostrophic disturbances on a zonally asymmetric basic flow[J]. J Atmos Sci, 40: 85-90.

BRETHERTON F P, HAIDVOGEL D B, 1976. Two-dimensional turbulence above topography[J]. J Fluid Mech, 78: 129-154.

EDMON H J, HOSKINS B J, MCINTYRE M E, 1980. Eliassen-Palm cross-sections for the troposphere[J]. J Atmos Sci, 37: 2600-2616.

GREEN J S A, 1970. Transfer properties of the large-scale eddies and the general circulation of the atmosphere[J]. Quart J R Met Soc, 96: 157-185.

KAROLY D J, 1982a. Eliassen-Palm cross-sections for the Northern and Southern Hemispheres[J]. J Atmos Sci, 39: 178-182.

KAROLY D J, 1982b. Atmospheric vacillations in a general circulation model. III: Analysis using transformed Eulerian-mean diagnostics[J]. IBID, 39: 2916-2922.

KIDSON J W, VINCENT D G, NEWELL R E, 1969. Observational studies of the general circulation of the tropics: Long term mean values[J]. Quart J R Met Soc, 95: 258-287.

LEJENAS H, MADDEN R A, 1982. The annual variation of the large-scale 500 hPa and sea level pressure fields[R]. University of Stockholm.

LORENZ E N, 1967. The Nature and Theory of the General Circulation of the Atmosphere[M]. Geneva: WMO.

LORENZ E N, 1979. Forced and free variations of weather and climate[J]. J Atmos Sci, 36: 1367-1376.

MARSHALL J C, 1981. On the parametrization of geostrophic eddies in the ocean[J]. J Phys Oceanog, 11: 257-271.

OHRING G, ADLER S, 1978. Some experiments with a zonally averaged climate model[J]. J Atmos Sci, 35: 186-205.

OORT A H, RASMUSSON E M, 1971. Atmospheric Circulation Statistics[M]. NOAA Professional Paper: 5.

PALTRIDGE G W, 1981. Thermodynamic dissipation and the global climate system[J]. Quart J R Met Soc, 107: 531-547.

PFEFFER R L, 1981. Wave-mean flow interactions in the atmosphere[J]. J Atmos Sci, 38: 1340-1359.

PHILLIPS N A, 1963. Geostrophic motion[J]. Reviews of Geophysics, 1: 123-176.

READ P L, 1986. Super-rotation and diffusion of axial angular momentum. II: A review of quasi-axisymmetric models of planetary atmospheres[J]. Quart J R Met Soc, 112: 253-272.

SALTZMAN B, 1978. A survey of statistical-dynamical models of the terrestrial climate[J]. Advances in Geophysics, 20: 183-304.

SCHMITZ G, DETHLOFF K, 1984. Interpretation of quasi-geostrophic potential vorticity fluxes on the basis of climatological data[J]. Z Meteor, 34: 159-165.

SCHNEIDER S H, DICKINSON R E, 1974. Climate modeling[J]. Rev Geophys Space Phys, 12: 447-493.

SELA J, WIIN NIELSEN A, 1971. Simulation of the atmospheric energy cycle[J]. Mon Wea Rev, 99: 460-468.

SHUTTS G J, 1981. Maximum entropy production states in quasi-geostrophic dynamical models[J]. Quart J R Met Soc, 107: 503-520.

SHUTTS G J, 1983. Parameterization of travelling weather systems in a simple model of large-scale atmospheric flow[J]. Advances in Geophysics, 25: 117-172.

STARR V P, 1968. The Physics of Negative Viscosity Phenomena[M]. New York: McGraw-Hill.

VALLIS G K, 1982. A statistical-dynamical climate model with a simple hydrological cycle[J]. Tellus, 34: 211-227.

WHITE A A, 1974. Large scale momentum transfer in a general circulation model of the atmosphere[J]. University of London.

WHITE A A, 1977. The surface flow in a statistical climate model: A test of a parameterization of large-scale momentum fluxes[J]. Quart J R Met Soc, 103: 93-119.

WHITE A A, GREEN J S A, 1982. A non-linear atmospheric long wave model incorporating parametrizations of transient baroclinic eddies[J]. IBID, 108: 55-85.

WHITE A A, GREEN J S A, 1984. Transfer coefficient eddy flux parametrizations in a simple model of the zonal average atmospheric circulation[J]. IBID, 110: 1035-1052.

WIIN-NIELSEN A, SELA J, 1971. On the transport of quasi-geostrophic potential vorticity[J]. Mon Wea Rev, 99: 447-459.

WIIN-NIELSEN A, FUENZALIDA H, 1975. On the simulation of the axisymmetric circulation of the atmosphere[J]. Tellus, 27:199-214.

WU G X, 1983. The influence of large-scale orography upon the general circulation of the atmosphere[D]. London: University of London.

第2部分

副热带高压的若干基本问题

综述 4

副热带高压的形成和年际变异

刘屹岷[1], 刘平[2], 任荣彩[1]

(1 中国科学院大气物理研究所,北京 100029;2 School of Marine and Atmospheric Sciences, Stony Brook University, Stony Brook, USA, NY 11794-5000)

摘要:在南、北半球的副热带地区,存在气压场或位势高度场的相对高值带,即所谓的"副热带高压带"。太阳辐射、地球自转和海-陆-气相互作用等外部强迫过程使得副热带高压带断裂为若干个区域高压中心,与副热带天气系统的运动和水汽传输及其邻近地区的天气和气候紧密联系。文中重点回顾了 20 世纪 90 年代以来吴国雄研究团队开创的副热带高压(副高)形成和变异动力学研究,包括副高的特征、副高与垂直运动的关系、热力适应、LOSECO 四叶型加热与副高的形成。副热带沙漠与季风孪生现象和印度洋海温异常对西太副高的影响,阐述了降水与西太平洋副高之间是一种相互影响的关系,必须通过大气外强迫的异常去理解夏季的降水异常。最后讨论了需要进一步深入研究的问题。

关键词:副热带高压;Hadley 环流下沉支;热力适应;海-陆-气相互作用

1 引言

副热带是一条环绕全球的干旱带,集中了世界上绝大部分的沙漠区域。在南、北半球副热带地区,存在着气压场或位势高度场的相对高值带,即为所谓的"副热带高压带",简称"副高"。沿着副热带干旱带,又存在全球雨量最为丰沛的季风区,如东亚季风区等。副高变异是影响东亚持续天气气候异常的主要环流因子。

早在 20 世纪 60 年代,中国气象学家陶诗言、黄士松就西太副高和南亚高压的变化对中国天气的影响进行了一系列研究(如黄士松和余志豪,1962;陶诗言,1963),研究成果已成为中国天气和短期气候预测的指南。在 20 世纪 90 年代中期之前,大气科学家特别是中国学者,对副高的研究总体上可分为两大类:一类是研究副高自身结构、活动规律及其对天气气候的作用;另一类是研究影响副高活动的因子。虽然在副高天气影响研究上成果丰硕,但经典的 Hadley 环流下沉等解释副高形成的理论仍存在明显的局限,它不能解释副高的时空变化特征(Hoskins,1996)。同时,由于副热带能量关系比其他纬度带复杂,能量方程不能简化,与发展比较成熟的"中高纬大气环流动力学理论"和"热带海-气相互作用及 ENSO 动力学"相比而言,副热带动力学研究显得较为薄弱。

在这种背景下,从国家第九个"五年计划"期间开始,吴国雄院士和丑纪范院士连续主持了国家自然科学基金委员会"副热带高压带变异机理"和"夏季副热带高压变化及其影响天气气候异常的机理"重点项目;随后吴国雄院士又接连主持了 3 届基金委创新群体以及中国科学院重要方向项目、科技部 973 等重大

通讯作者:刘屹岷,lym@lasg.iap.ac.cn

计划和项目,推动了副高变异机理及其影响的一系列研究,取得了一系列重大理论突破,刷新了对副高成因的若干传统认识,并完成了国内第一部副热带高压动力学专著《副热带高压形成和变异的动力学问题》(吴国雄 等,2002)。文中回顾了近二十年来吴国雄研究团队在副高气候动力学方面的若干研究进展,主要内容涉及副高的形成以及年际变化等,最后给出总结与对未来的展望。

2 副高的三维结构特征

本节利用再分析资料,阐述副热带高压的气候学特征。通过引入"经向偏差",揭示了纬向平均副热带高压带的垂直结构及其季节和年际变化特征(刘平 等,2000)。

2.1 位势高度的经向偏差

早期对副高带的研究基本上是定性的,而对于其强度、范围以及在各种时间意义尺度上的变率研究很少,其中原因主要在于没能找到使得各个层次上副高带之间可以相互比较的基准。对于对流层的任意等压面上,纬向平均的位势高度和风场之间的地转平衡关系几乎存在于除赤道外的所有纬度上,如图1所示。

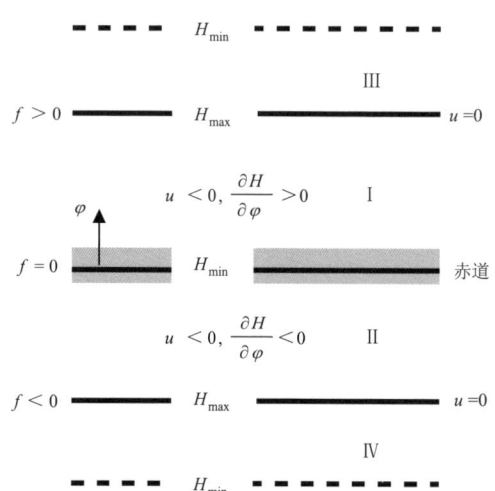

图1 位势高度经向偏差示意图,其中 f、u、φ、H 分别为科里奥利参数、纬向风、纬度和位势高度,箭头指向北(引自刘平 等,2000)

由于纬向风在Ⅰ、Ⅱ区域为东风带($u<0$),而Ⅲ、Ⅳ区域为西风带($u>0$),在Ⅰ、Ⅲ和Ⅱ、Ⅳ之间有纬向风0线存在。根据地转风关系,位势高度在Ⅰ、Ⅱ区域随纬度增大,而在Ⅲ、Ⅳ区域随纬度减小。因此,在Ⅰ、Ⅲ,Ⅱ、Ⅳ交界面上有位势高度极大值存在。由此可见,位势高度极大值与 $u=0$ 线重合。同时可以得到,靠近赤道的小范围内有位势高度极小值存在(图1中灰色阴影)。

刘平等(2000)定义了位势高度经向偏差,即在某一等压面上,以赤道上位势高度值为基准,求取各个纬度与之的偏差。这种"经向偏差"把副热带高压带范围确定了下来,"经向偏差"的两条极大值线与纬向风0线重合,同为副热带高压带的脊线。对流层中各个等压面上均存在图1所示的关系,因此这种方法可以将各个层次上位势高度的经向偏差进行比较,即副热带高压带在各个层次上的脊线位置、强度乃至范围有了可比性。

2.2 位势高度垂直结构的季节和年际变化

图2为赤道上气候平均的位势高度经向偏差在1、7月的分布。1月(图2a),两半球该偏差均随高度

增高,强度减小,范围缩小,且向赤道靠拢。在 850 hPa 以上的层次,南半球的强度略大于北半球;南、北半球副高带的强度有明显的季节变化,尤其是北半球。南半球(图 2 的左半部分),在低层(850 hPa 以下),7 月(图 2b)略强于 1 月;而在 850 hPa 以上的层次,1 月明显较 7 月强,即南半球的夏季副热带高压带存在于较其冬季更高的层次。而北半球,即图 2 的右半部分。副热带高压带的强度在 1 月随高度是减弱的,但 7 月却有明显的不同。除了在 800 hPa 以下的层次上有随高度减弱的特征外,在 800 hPa 以上的层次随高度增大。在 800~300 hPa 的层次上增幅较弱,而 300 hPa 以上增幅很明显。另外,850 hPa 以下的层次,1 月的经向偏差明显大于 7 月;但在 700 hPa 以上的层次,7 月明显强于 1 月。总体来看,南、北半球的副高带在对流层中、高层总是夏半年强于冬半年;而在低层(850 hPa 以下)则相反,冬半年略强于夏半年。副热带高压如此复杂的季节变化是以往的分析所未预料到的,其原因与夏季存在更强的纬向非对称非绝热加热有关(Wu and Liu,2003;Liu et al.,2001,2004a,2004b)。

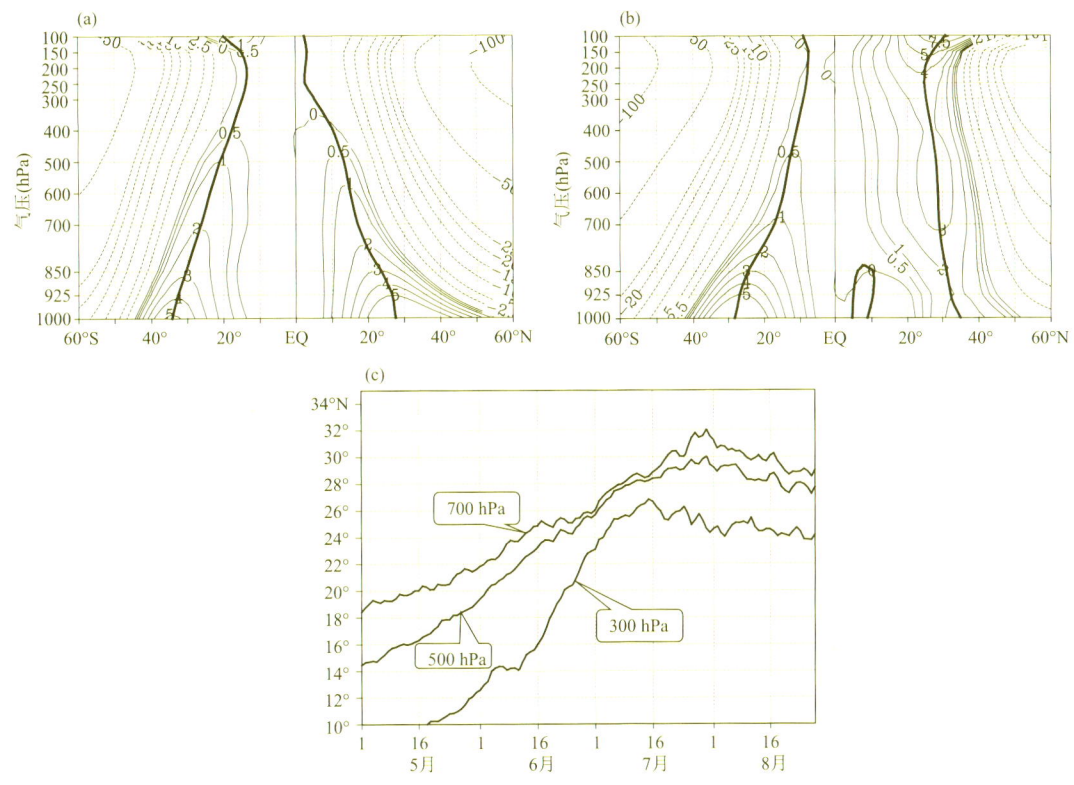

图 2 1980—1997 气候态 1 月(a)、7 月(b)平均经向偏差的纬度-高度剖面,粗实线为纬向平均纬向风 0 线。
(c)北半球平均副高带脊线([u]=0)在三个等压面上的逐日演变(修改自刘平 等,2000)

多年平均月平均纬向风 0 线(图 2c),即副高带的脊线位置在南、北半球均表现出明显的季节变化,从 3 月到 8 月,位置从南向北;而从 9 月到 2 月,位置从北向南。在整个过程中脊线位置基本平行,说明副热带高压带的季节移动是整层次同步的(刘平 等,2000)。

副热带高压带在北半球夏季无论脊线位置或是强度均有独特的特征。气候平均的 5 月 1 日—8 月 31 日 700、500、300 hPa 上副高带脊线的演变曲线表明(图 2c),在 700 和 500 hPa 上,6 月下旬副高带脊线位置变化不大,而在前、后的变化都比较显著,直到 8 月初达到最北。而 300 hPa 上,6 月上旬副高带脊线几乎稳定在 14°N 附近,前、后的变化较明显,尤其是 6 月中旬以后,比其下面两个层次的变化显著更快。虽然这 3 条曲线的变化并非完全一致,但均表现出两次明显的跳跃式演变。这种整体性和同步性在副高带的年际变化中也有所体现,但与季节变化不同的是低层的变化较高层大,表明副高带位置的季节变化和年际变化不同。因此,对于副高带的研究,从整体上入手是很必要的。

3 副高与垂直运动的关系

3.1 纬向平均副高

副高带是沿副热带地区气压最高的环状带。利用地转关系可以定义副高位置

$$\begin{cases} u = 0 \\ \dfrac{\partial u}{\partial y} \begin{cases} > 0 \text{ 北半球} \\ < 0 \text{ 南半球} \end{cases} \end{cases} \tag{1}$$

副高的强度可用经向质量通量的堆积或辐合来估计,即

$$\text{Int.} = \partial(f\rho u)/\partial y = f\rho \partial u/\partial y + \rho u \beta \tag{2}$$

在等压坐标下,该强度指数正比于位势高度本身,因此与上述位势高度的经向偏差的定义一致(吴国雄等,2004)。

吴国雄等(2004)利用再分析资料研究了定常条件下二维和三维空间上副热带高压与垂直运动的关系,比较了 Hadley 环流和副高动力学的差异。自由大气中,当纬向气流(u)被南北向的加热差异激发出来以后,地转的作用使其偏转而出现经向质量通量。在地转关系的约束下,沿着副高脊线的这种经向质量通量的辐合,必然被气压梯度力所产生的质量通量的辐散所平衡,因此副高出现在东西风的交界面上(图 3a)。

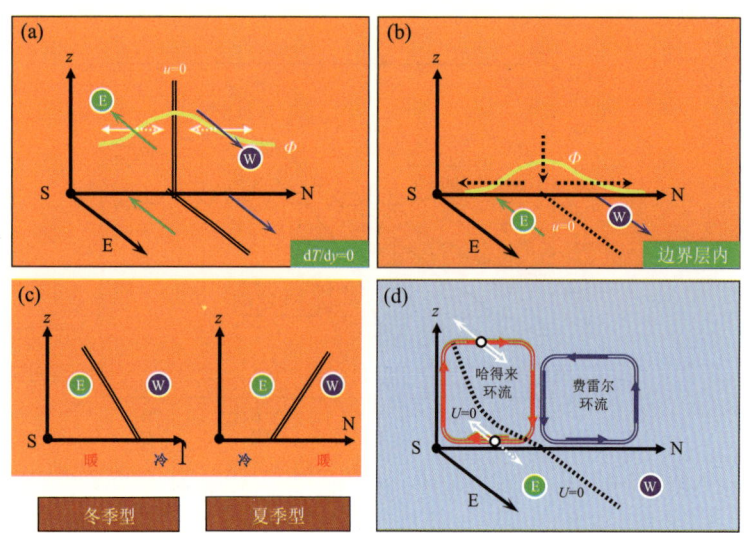

图 3 纬向平均副热带高压和经向 Hadley 环流圈维持机制的动力学示意图。(a) 沿南北方向副高脊线上,地转偏向力($-f\rho u$,虚线箭头)引起的经向质量辐合与由气压梯度力($-\nabla \phi$,实线箭头)引起的辐散相平衡;(b) 行星边界层内,穿越等压线由副高向外的气流辐散与边界层顶的进入边界层的下沉运动相平衡;(c) 按照热成风关系,副高脊线随高度上升向暖区倾斜,冬季形成西风切变,夏季在亚洲季风区形成东风切变;(d) 与 Hadley 环流水平支相关联的惯性力矩(空心箭头)与地面摩擦引起的角动量的产生(虚线箭头)以及高空从热带向中纬度的角动量的辐散(实线箭头)相平衡(引自吴国雄 等,2004)

3.2 纬向平均副高与 Hadley 环流

传统观点将副高的形成归于辐射冷却和 Hadley 环流下沉支的作用。然而,在北半球冬季 Hadley 环流最强时,副热带高压反而最弱;夏季 Hadley 环流最弱时,副高反而最强。因此,用传统论点解释副高的存在似乎不恰当。吴国雄等(2004)的结果表明,由于地球自转,在副热带地区出现了最大的经向质量通量

的辐合,导致纬向平均副热带高压的形成(式(2))。在热成风关系的制约下,除北半球夏季外,副高脊线随高度上升向赤道倾斜(图 2、图 3d、图 4)。而 Hadley 环流的下沉支从对流层顶垂直地延伸到行星边界层(图 3d、图 4),因此副热带高压脊线与该下沉支在自由大气中位置分离,年际变化反相;在行星边界层中摩擦耗散作用使两者重合,年际变化同相,但垂直运动对副高的形成不起作用。

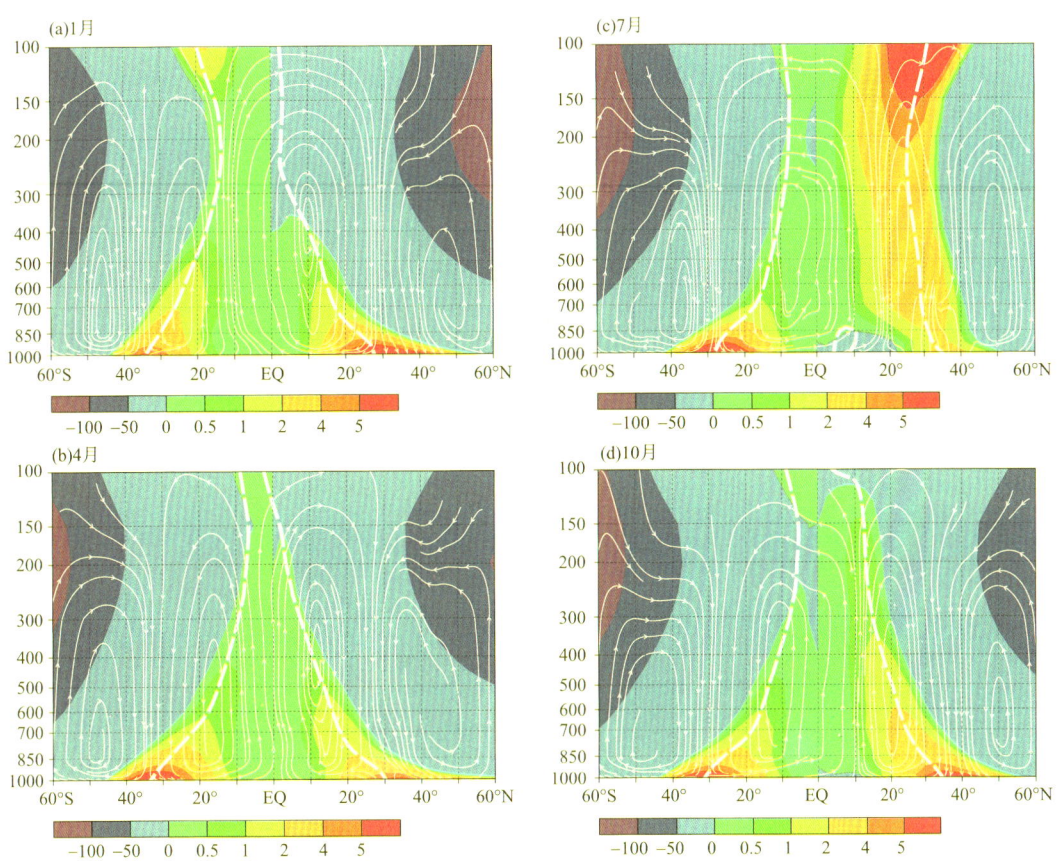

图 4　1980—1997 年月平均纬向平均位势高度对赤道的经向偏差(阴影)、纬向平均 $u=0$ 线(粗虚线)以及纬向平均经向环流(带箭头线)的纬度-高度剖面((a)1 月;(b)4 月;(c)7 月;(d)10 月,单位:dagpm) (引自吴国雄 等,2004)

3.3　三维空间

三维空间上,沿行星边界层的副高脊线处为下沉运动,副热带大洋东部强烈的下沉运动对应着强烈的向赤道气流。自由大气中,副高与垂直下沉运动的分布位置存在差异。500 hPa 上,大洋上副高东部大气下沉,西部大气上升(图 5)(刘屹岷和吴国雄,2000)。一方面,这与等熵面的北高南低的倾斜分布在动力学上是一致的;另一方面还表明,副高的形成也与非绝热加热密切相关。以上分析说明,不论是在自由大气还是在边界层中,不能简单地把下沉运动看成是副高形成的原因。结合涡度方程和连续方程可得

$$\frac{d\zeta}{dt} \propto (1-\kappa)(f+\zeta)\frac{\omega}{p} - (f+\zeta)\frac{1}{\theta}\frac{d\theta}{dt} \tag{3}$$

因此,下沉本身只引起正涡度发展,只有位温增加才能导致负涡度的发展。由此看来,简单地用下沉运动本身去解释涡度的发展和副高的形成存在片面性。西太平洋副高(西太副高)下面为弱的辐散和下沉运动,有利于局地湿度降低,这是西太副高区为晴天的原因(刘屹岷和吴国雄,2000;任荣彩 等,2007)。

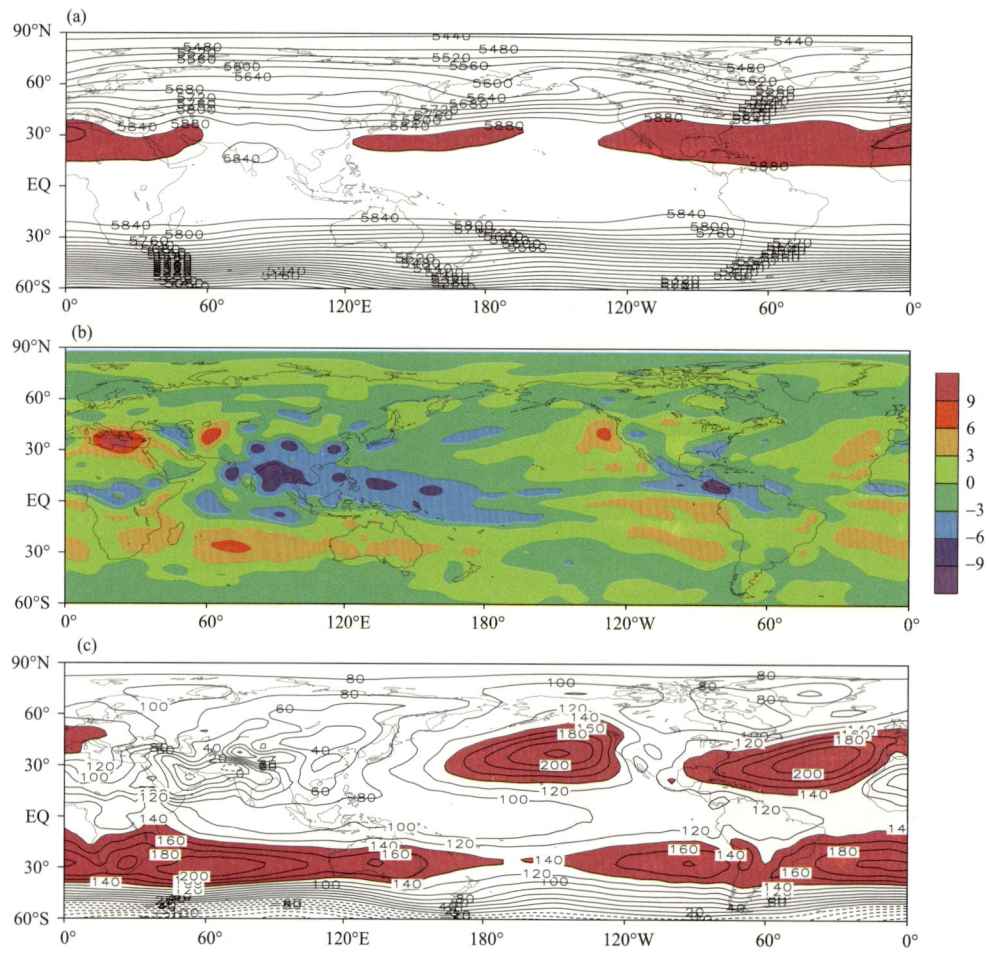

图 5 由 NCEP/NCAR 1980—1995 年资料计算得到的 7 月月平均 (a) 500 hPa 位势高度场；(b) 500 hPa 垂直速度 $\omega \times 100$；(c) 1000 hPa 位势高度场。单位：(a)和(c)是 gpm，(b)中的 ω 是 $Pa \cdot s^{-1}$（引自刘屹岷和吴国雄，2000）

4 热力适应

正如 Eady(1950)所阐明的，降水不仅是大气环流的结果，也是大气环流形成的原因之一。"季风降水是形成西太副高和南亚高压的原因"这一观点主要是受到 Webster(1972)和 Egger(1978)的两组数值试验的启发。他们分别用两层大气环流模式(750 hPa/250 hPa 和 800 hPa/400 hPa)，以加热驱动模式，得到了与观测相似的西太副高和南亚高压。由于引进的加热主要是季风降水的凝结潜热加热，因此他们得出潜热加热是形成夏季环流的主要驱动因子。Hoskins(1996)把季风加热驱动沙漠的思想予以发展，认为大陆的季风降水通过 Rossby 波强迫的下沉运动造成其西部洋面高压。Chen 等(2001)在一个线性位涡模型中引入正弦型的垂直加热廓线，也模拟出了海面副高和对流层顶附近的高压，从而推断深对流加热是夏季副高形成的原因。然而这些结果排除了夏季陆面上强大表面感热的影响，仅是环流对对流加热的斯维尔德鲁普(Sverdrup)响应(吴国雄 等，2008)。

4.1 热力适应和过流

吴国雄和刘屹岷(2000a)利用位涡理论，讨论了非绝热加热导致的大气动力特征的变化，阐明了大气动力过程向外加热适应的原理。指出在加热随高度上升增加的低层，正的位涡制造被地面负的摩擦涡源、侧边界的负位涡输入及上边界正位涡的输出所平衡；在加热随高度上升减少的上层，负的位涡制造被下边界正位涡的输入及侧边界负位涡的输出所平衡。在气柱垂直积分模型中,摩擦施加于气柱的负位涡除了被气柱中正位涡的制造所抵消外,所余的负位涡在高层沿侧边界向外输送。

当大气柱受热时,低层将产生正涡度和气旋性环流,高层则产生负涡度和反气旋性环流(图6a)。由经典涡度方程可导出(吴国雄 等,2008;Wu et al.,2009)

$$\frac{\partial^2 w}{\partial z^2} = f^{-1}\frac{\partial}{\partial z}(\zeta'_t + \mathbf{V}\cdot\nabla\zeta + \beta v) \tag{4}$$

在定常运动中采用正则模的解,则沿气旋/反气旋系统中心的东西方向上,有

$$w \propto -\beta f^{-1}\frac{\partial v}{\partial z} \tag{5}$$

于是在加热中心的东侧有上升运动发展,在其西侧则盛行下沉运动(图6a)。物理上,上述垂直运动的分布是大气维持涡度平衡所必须的。这是因为在这种定常地转框架下有

$$\beta v + f \nabla\cdot\mathbf{V} = 0 \tag{6}$$

偏南风对应的负地转涡度平流必须由局地辐合的正涡度制造所平衡;而偏北风对应的正地转涡度平流必须由局地辐散的负涡度制造所平衡。由大气质量守恒性可知,加热区东部低空辐合/高空辐散产生上升运动,而其西部低空辐散/高空辐合则产生下沉运动。

当大气柱被冷却时,低空出现反气旋,高空出现气旋(图6b)。同样的道理,这时下沉运动在东部出现,上升运动在西部出现。

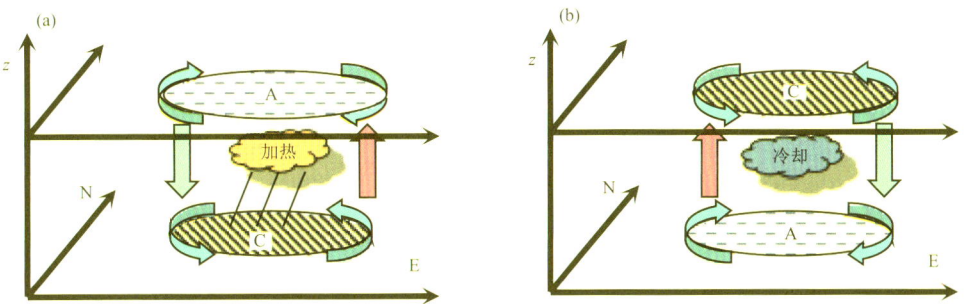

图6 示意图显示了在夏季副热带地区,旋转地球上的大气对外部加热的适应。(a)非绝热加热在地表附近产生气旋式环流C,在对流层上部产生反气旋式环流A,空气在环流的东部上升,在西部下沉;(b)非绝热冷却在地表附近产生反气旋环流,在对流层上部产生气旋式环流,空气下沉于环流的东部,上升在西部(引自Wu et al.,2009)

吴国雄和刘屹岷(2000a)还讨论了过流问题:在非绝热加热区域之上,加热为0。根据热力学方程,垂直运动速度应当为0。然而,在实际大气中,加热区之上往往存在着垂直运动,这种现象被称为"过流"(吴国雄和刘屹岷,2000)。这是因为在加热区上方均存在的深厚层次,导致加热区大气在下层辐合上升,在上层辐散,同时在周边地区引起上层辐合下沉和下层的辐散。大气加热通过这一"补偿效应"影响着邻域的大气运动。

4.2 垂直非均匀加热与能量频散

在不考虑大气内部热力结构的作用和地面摩擦作用时,位涡方程可写为

$$\frac{\partial \zeta_{\theta_z}}{\partial t} + \mathbf{V}\cdot\nabla\zeta_{\theta_z} + \beta v\theta_z = (1-\kappa)(f+\zeta)\theta_z\frac{\omega}{p} - (f+\zeta)\theta_z\frac{Q}{\theta} + \\ (f+\zeta)\frac{\partial Q}{\partial z} - \frac{\partial v}{\partial z}\frac{\partial Q}{\partial x} + \frac{\partial u}{\partial z}\frac{\partial Q}{\partial y} \tag{7}$$

对于长时间尺度的演变,涡度的局地变化可略,方程(7)左端为位涡方程中的平流项和β项。右端分别是垂直运动项、热源项、加热垂直变化项以及加热的纬向和经向变化项。根据尺度分析(参见吴国雄 等,1999,2002),在热源区加热的垂直变化项比其他项大一个量级,而资料分析及数值试验的结果(刘屹岷

等,1999a)表明强烈的季风潜热加热区中平流项很弱。因此最大加热的下方有

$$v \propto \frac{f+\zeta}{\beta\theta_z}\frac{\partial Q}{\partial z} > 0 \qquad \theta_z \neq 0 \tag{8}$$

这即表明,热源中心下方能够出现偏南风,使气旋式环流出现在热源西侧,反气旋式环流出现在热源东侧。在最大潜热加热中心上方,则有

$$v \propto \frac{f+\zeta}{\beta\theta_z}\frac{\partial Q}{\partial z} < 0 \qquad \theta_z \neq 0 \tag{9}$$

因此,热源区上方出现偏北风,反气旋式环流出现在热源西侧,气旋式环流出现在热源东侧。上述过程即是环流对深对流加热的基本热力适应(图7a)。数值试验的结果与上述理论分析一致(Liu et al., 2001;刘屹岷 等,2001)(图7、图8、图9)。

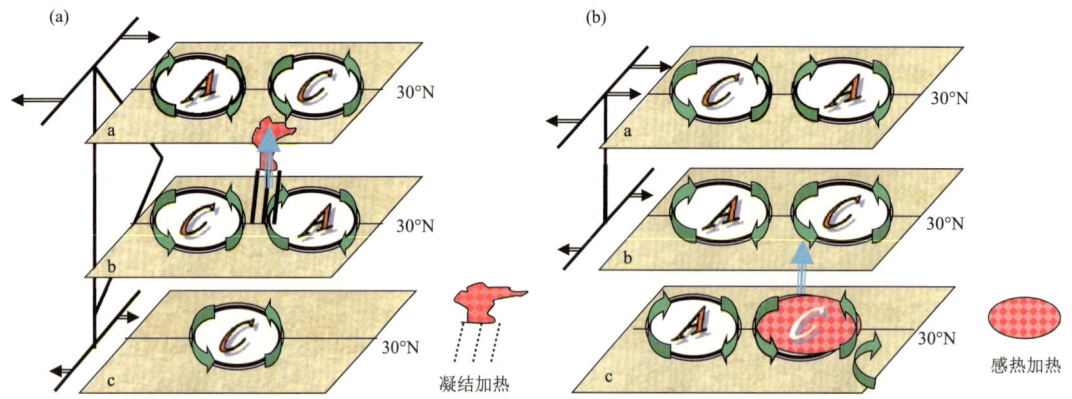

图7 深对流加热(a)和感热加热(b)对副热带环流影响示意(引自吴国雄 等,2002)

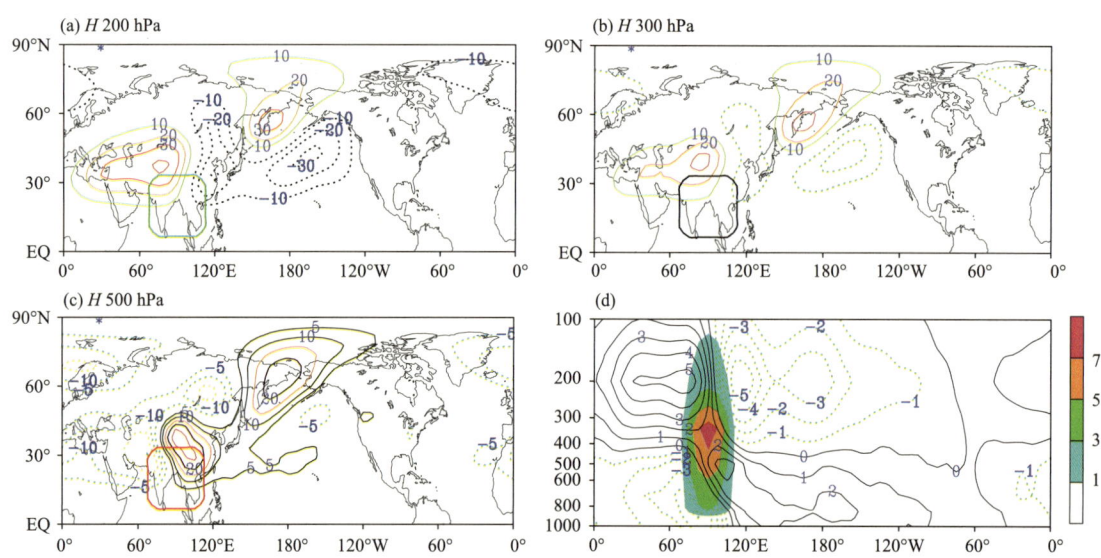

图8 理想热源数值试验积分第13~24个月平均的位势高度的纬向偏差的空间分布。(a)200 hPa;(b)300 hPa;(c)500 hPa;(d)沿30°N的垂直剖面。单位:dagpm。(a)—(c)粗线界定最大加热层(=0.336)上加热为0的区域(引自 Liu et al., 2001)

表面感热加热(SE)以近地层加热为主。这时,除了近地面层外,根据式(7),有

$$\frac{\partial \zeta_z}{\partial t} \propto \frac{(f+\zeta_z)}{\theta_z}\frac{\partial Q^{SE}}{\partial z} < 0 \tag{10}$$

这就是说,SE加热产生了反气旋涡源。在时间尺度很短时,SE加热区上空将出现深厚的反气旋。对准定常态有

$$\mathbf{V}\cdot\nabla\zeta_z+\beta v\propto\frac{(f+\zeta_z)}{\theta_z}\frac{\partial Q^{SE}}{\partial z}<0 \tag{11}$$

由于副热带地区低空为东、西风交界处，$u\approx 0$，水平平流作用很小。根据式(11)，有

$$\beta v<0 \tag{12}$$

可以看到，在 SE 的作用下，纬向均匀的副热带高压带便出现断裂：在 β 效应作用下，SE 区上空出现强的北风($v<0$)，其西侧为反气旋，东侧为气旋(图 7b)。又由于在南暖北冷的背景温度场中，西风随高度上升增强，使副热带高空受西风控制，$u>0$。根据式(11)

$$u\frac{\partial\zeta}{\partial x}<0 \tag{13}$$

那里的平流作用使均匀风带出现波动，在 SE 区的下游出现反气旋式环流。根据热力适应原理，在近地层则有气旋式涡度生成(图 7b 下部)。

盛夏副热带大陆西岸有很强的感热加热(感热通量 >100 W·m^{-2})。副热带低层出现北太平洋和北大西洋副高，陆面上为低气压；高层的反气旋则出现在陆地上，大洋上空为低槽；隆冬强烈的感热加热出现在大陆东岸沿海地区。这使低空大洋上出现气旋，大陆上出现反气旋；对流层高层大洋上为脊，陆地上为槽，与图 7b 的情况吻合(刘屹岷 等，1999b；Wu et al.，2009)。由此可推论，表面感热加热对冬、夏气候基本态的形成十分重要。由于表面感热通量 100 W·m^{-2} 所产生的涡源的量级为 $(10^{-10}$ s$^{-2})$，它在 24 小时内即可强迫出强大的副热带反气旋系统。足见感热加热对大气环流的形成非常重要。

4.3 水平非均匀加热与能量频散

值得注意的是，在图 8 中热源区以北，虽然加热为 0，但高、低层均出现反气旋性环流异常，在中、高纬度地区也呈现出相当正压的 Rossby 波列特征。这显然已经不能用垂直非均匀加热予以解释。事实上，造成这种现象的原因是水平非均匀加热效应(图 9)。

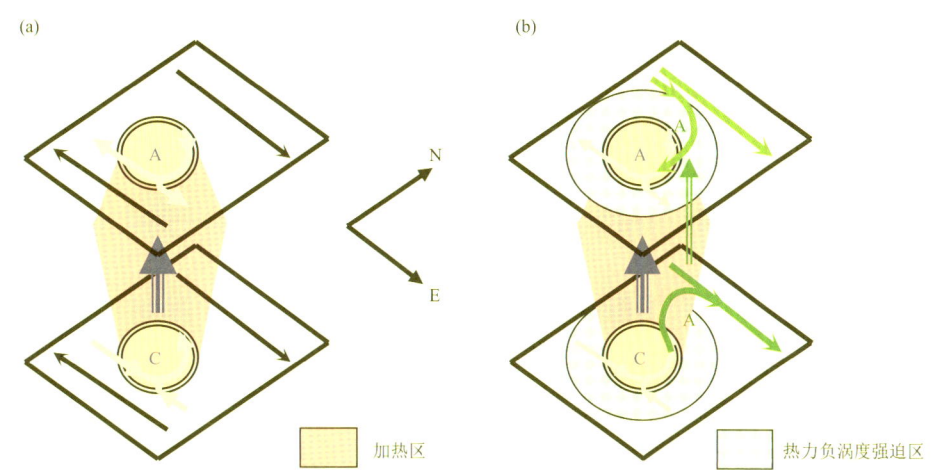

图 9　副热带环流对空间非均匀加热的热力适应示意。(a)垂直非均匀加热在加热区内产生低层的辐合和气旋式环流及高层的辐散和反气旋环流；(b)水平非均匀加热在加热区内外产生负涡度强迫，使加热区北面低层辐合、高层辐散，出现上升运动及深厚的反气旋中心(引自刘屹岷 等，2001)

对方程(7)中各项的计算表明，方程右端中，在加热区内，加热的垂直变化项比其他项大一个量级，而在加热区以外为 0；加热区外其余项可略，但其中加热的经向变化项在加热区北部及其以北地区的高、低层均产生负的涡度强迫，比加热的垂直变化项小 1~2 个量级。图 8 的数值试验还指出，在加热区内出现上层辐散、下层辐合，是上述环流的次级响应；另一个辐散(高层)及辐合(低层)中心出现在加热区以北，由此产生的垂直上升运动及其造成的负涡度强迫在加热区内最大，在加热区以北达到第二个极值。加热的经向变化项和 ω 项的共同作用使加热区北部边界附近及以北地区涡度为负、位势高度为正(图 9)(刘屹岷

等,2001)。

5 LOSECO 四叶型加热与副高形成

5.1 LOSECO 四叶型加热形成

在两半球的副热带地区,冬季海洋为热源,陆地为冷源;夏季则与之相反。根据第 4 节中热力适应原理可以推知,夏季副热带地区的上升运动应当出现在大陆东部和海洋西部,而下沉运动应当发生在大陆西部和大洋东部。在沿岸地区还存在次级环流。沿着副热带的海岸线特别是大陆西部的海岸线,散度风从海面吹向大陆。与其相联系的局地垂直纬圈环流在大陆的东、西岸为上升运动,在海洋的东、西部为下沉运动。这种局地次级环流叠加在如图 6 所示的大尺度上升运动上,使大陆/海洋东部观测到的总的上升/下沉运动强于海洋/大陆西部的上升/下沉运动(图 10)。

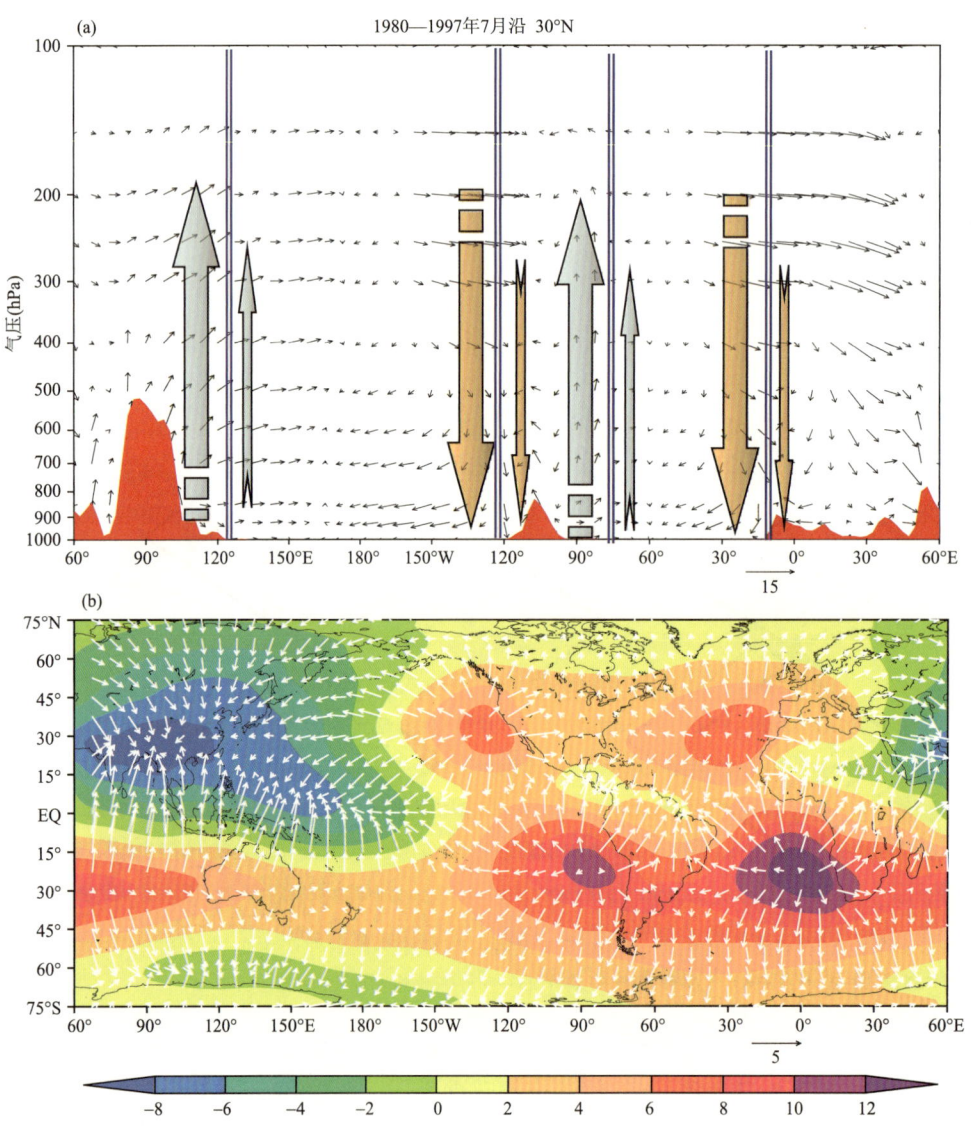

图 10 (a)1980—1997 年 7 月平均沿 30°N 纬圈环流和(b)1000 hPa 速度势(阴影,单位:10^6 m²·s⁻¹)及散度风(箭矢,单位:m·s⁻¹)的分布。(a)中的双垂线为海陆分界线,垂直箭头示意垂直运动的分布(引自吴国雄 等,2008)

图 11 给出了 1980—1997 年平均的北半球副热带 7 月平均各要素场的分布。海陆分布不仅决定着副热带垂直运动的分布，还能调节陆面和洋面上的物理过程，使副热带近海海面在冬季对大气的感热加热增强，在夏季成为大气感热(SE)的汇，从而加剧了冬、夏季沿大陆东部海岸的海陆热力差异。夏季大陆东岸的陆面加热利于对流发展，增强了大尺度的上升运动；相反，其邻近洋面上的感热冷却利于近洋面稳定度的增加，减弱了大尺度的上升运动。于是夏季副热带大陆的东部对流强烈发展，加热以对流凝结加热(CO)为主。位于邻近西太平洋地区和西大西洋地区的 D 点则不同：尽管由于位于洋面上，辐射冷却(LO)占主导地位，但大尺度海洋冷却区的西部有上升运动发展，那里还存在另一种主加热(CO)。另外由于存在辐散风驱动的近海次级环流(图 9b)，以及海陆分布导致近海海表对大气的冷却作用，这里的 CO 加热强度远不及陆面上 C 点的 CO 加热。因此，该处的加热呈现双主加热(D)的分布：第一主加热为整层长波辐射冷却(LO)，第二主加热为凝结加热(CO)。图 11b 给出各加热中心 LO、SE、CO(简称为 L、S、C) 和 D 的位置。

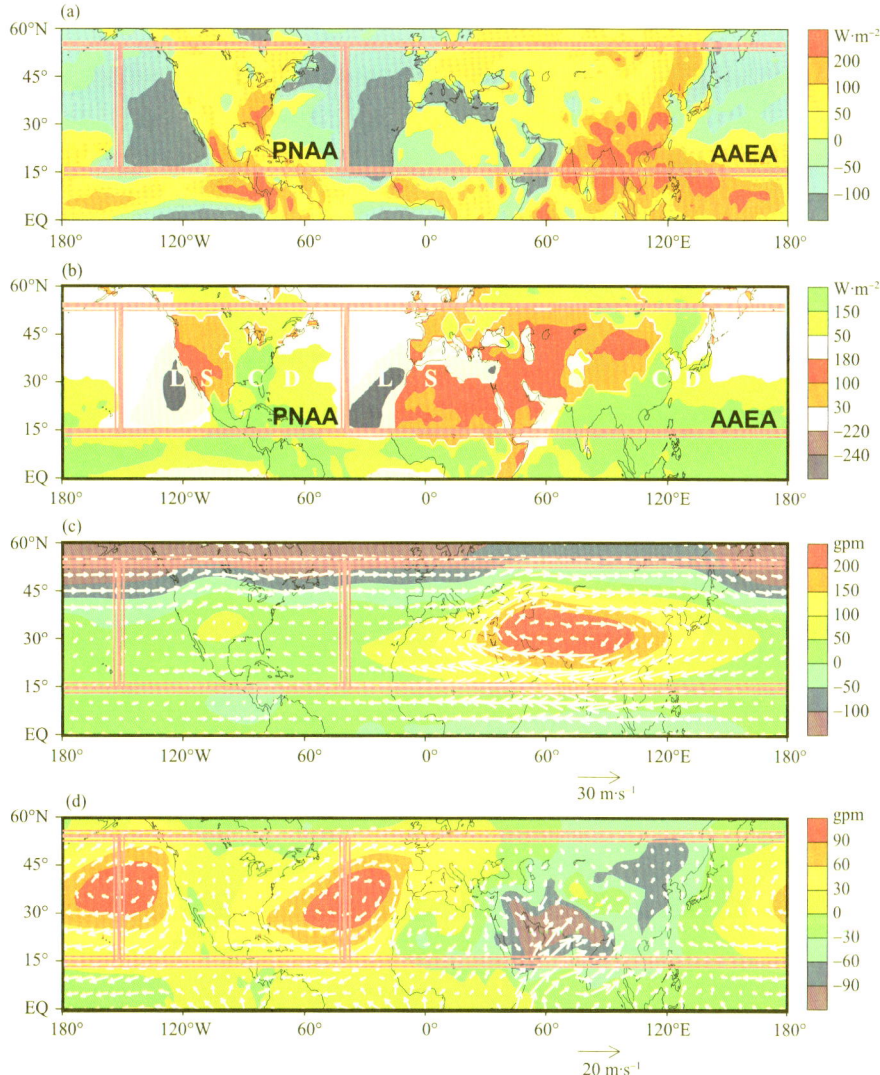

图 11 1980—1997 年平均北半球副热带 7 月平均各要素场的分布。(a) 气柱总加热率(W·m^{-2})；(b) <−220 W·m^{-2} 的长波辐射冷却(LO，蓝色)，>30 W·m^{-2} 的感热加热(SE，橙色和红色)及 >50 W·m^{-2} 的凝结加热(CO，绿色)；(c) 100 hPa 和 (d) 1000 hPa 的对赤道平均的纬向偏差风场(矢量)和高度场(阴影，单位：gpm) (引自 Wu and Liu, 2003；Liu et al., 2004b)

5.2 沿大陆西岸加热的自组织特征

夏季大洋上空大气大尺度的冷却(图 11a)及沿大陆西岸局地的次级环流(图 10b)造成大洋东部有强烈的下沉运动发展(图 10a),进而引发强烈的下沉升温。沿岸海域强大的北风应力产生显著的离岸海流,又导致冷海水上翻。于是在行星边界层出现下冷上暖的稳定层结,成为副热带低层云最多的区域,云顶出现强烈的长波辐射冷却(LO)。在边界层内加热随高度上升减小,稳定度增大,成为负位涡的源地。在定常态,这种加热所致的负涡度制造必须由大的正地转涡度平流所平衡,从而加剧了该处大尺度环流的北风,副高环流得以加强。该区域大气的总加热廓线以 LO 加热廓线为主要特征。

在大陆西部沿海,大尺度的下沉运动(图 10a)虽被沿大陆西岸的次级环流(图 10b)所减弱,但盛行的下沉运动使空气湿度降低,天空干燥少云,地面温度高。边界层中来自北方的大尺度环流中的冷空气加大了陆-气温差,因此,这里有很强的地表感热加热(SE)在边界层内加热随高度上升减弱,稳定度也降低。感热加热引起的负涡度制造被偏北气流的正地转涡度输送平衡,使大气呈现定常态。该区域的总加热廓线接近 SE 的加热廓线。

综上所述,由于海陆分布及其对多圈层相互作用的调控,使得夏季副热带大陆及其毗邻的海域的主要加热分布被有规则地组织起来,自西向东呈现出西岸近海的长波辐射冷却(LO)-大陆西部的表面感热加热(SE)-大陆东部的对流凝结加热(CO)-东岸近海的双主加热(D)分布,称为 LOSECOD 四叶型加热。

5.3 "四叶型加热"拼图和夏季副热带高压的形成

Liu 等(2004b)根据副热带的海陆分布,把全球划分为五大区域,大陆位于每一区域的中心。每一区域的加热分布都呈现 LOSECOD 四叶型加热型,而夏季副热带的加热分布呈现为有趣的马赛克(mosaic)形的 LOSECOD 拼图(图 11)。

四叶型中每一叶加热的垂直廓线均不相同,根据式(8)—(9)和式(12)—(13),其在高、低空所激发的环流也不一样。在副热带地区冷却覆盖着大洋东部全部以及西部的偏中纬地区。每一洋盆东部的冷却率都超过 $-100~\text{W}\cdot\text{m}^{-2}$。与此冷却相对应的区域在 5 个海面中都无一例外地呈现出海面的反气旋环流和高层的气旋式环流。另一个更为明显的特征是,在每一洋盆东部长波辐射冷却(LO)最强的地方,都存在表面强的向赤道气流和高层的向极气流。这是因为那里的主加热是 LO,且最大的冷却出现在边界层顶,在近地层 $\partial Q/\partial z \ll 0$,而在自由大气中 $\partial Q/\partial z > 0$。

图 11c、d 给出夏季对流层高、低层位势高度的纬向偏差。1000 hPa 上的低压出现在陆面上,其西侧存在强大的向赤道气流;高压则出现在海洋上,其中心偏向大洋东部。在 200 hPa 上,高压出现在大陆上空,大槽位于海洋上空,南半球高压强度均超过 30 gpm,北美上空的高压超过 60 gpm,而横跨非洲和欧亚大陆的南亚高压强度则超过 120 gpm。由于表面感热加热主要出现在 800 hPa 以下,且在大陆西部加热强烈,那里 $\partial Q/\partial z \ll 0$,因而对应着强大的向赤道表面气流。垂直积分深对流凝结加热(CO)的主要加热区位于两大陆的东岸和两大洋的西岸。在所有这些加热中心处,都可以观测到低空的向极气流和高空的较强的向赤道气流。于是在低空气旋出现在 CO 中心的西部,反气旋出现在 CO 中心的东部。在高空气旋环流在 CO 中心东部,而反气旋环流出现在其西部。其中 LO 冷却强化了 SE 引起的副高分布,并使近海面反气旋中心偏向大洋东部(图 11d)。在四叶型加热的共同作用下,整个副热带高低空系统于是呈现出纬向非轴对称分布(Wu and Liu, 2003; Liu et al., 2004b; Wu et al., 2009)。

5.4 副热带沙漠与季风的形成

Wu 等(2009)研究了夏季副热带地区存在的三种类型的大气强迫,提出了通过行星涡度和温度的经向平流在非绝热加热和涡度制造之间形成正反馈的新机制,以解释副热带地区每一个大陆西部沙漠和东部季风共同存在的现象(图 12)。首先,陆地上的大陆尺度加热和海洋上的冷却导致大陆东部和西部的空气上升,海洋东部和西部的空气下沉。其次,沿海地区局部尺度的海风强迫增强了海洋东部上空的空气下沉和大陆东部上空的上升。这导致在每个大陆和相邻海洋形成了夏季副热带 LOSECOD 四叶型加热。这种四叶型对应于大陆西部的干燥气候和东部的湿润气候。第三，区域尺度的地形抬升加热在地形以东产生向极上升气流，在地形以西产生向赤道的下沉气流。其中青藏高原位于欧亚大陆东部。高原强迫的环流与大陆尺度强迫产生的环流在东亚地区一致，在非洲-欧亚大陆上形成了最强的季风和内陆及西部大陆最大的沙漠。

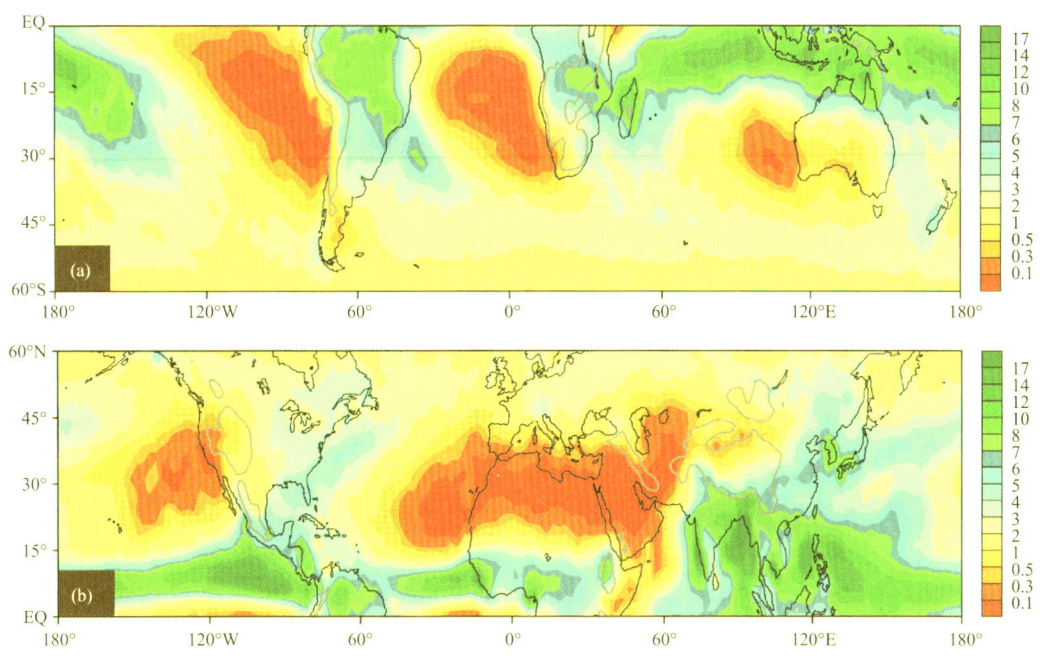

图 12　观测的多年平均 1 月和 7 月降水分布。单位 mm·d^{-1}（引自 Wu et al.，2009）

Wu 等（2009）提出了一种新的非绝热加热与涡度产生之间的正反馈机制，该正反馈通过热量和行星涡度的经向传输而产生。该机制被用来解释副热带沙漠和季风的形成。海洋东部的强低层长波辐射冷却和大陆西部的强地表感热产生负涡度，该负涡度被来自高纬度的正行星涡度平流平衡。在海洋东部的向赤道气流产生了低的海面温度和稳定的层结，进而导致低层云的形成和强大的局地长波辐射冷却的维持。大陆西部的向赤道气流携带着寒冷、干燥的空气，从而增强了当地的感热以及下层土壤的水分释放。这些因素导致沙漠区气候干燥。而在大陆东部，凝结加热在对流层下部产生正涡度，这与来自低纬度经向气流的负行星涡度平流相平衡。该气流带来温暖潮湿的空气，从而增强了与降雨相关的局部对流不稳定和凝结加热。这些因素产生了潮湿的季风气候。

6　印度洋海温异常对西太副高的影响

20 世纪 90 年代及之前，很多学者就 ENSO 与中国降水的联系开展了大量的研究，发现赤道中东太平洋的海表温度异常（SSTA）与中国东部的降水存在一定的相关（叶笃正 等，1996）。但是相关并不等于响应，逻辑学上相关的两种现象常常是通过第三媒介产生因果联系的。吴国雄和刘还珠（1995）的数值试验结果表明，赤道中东部正 SSTA 在其西侧引发一对跨赤道的气旋对和沿着赤道出现西风异常，但降水异常只出现在正 SSTA 区的西侧，呈现出邻域响应的特征，对中国的天气没有直接的影响。吴国雄等（2000b）发现与 ENSO 相关联的印度洋海温异常通过二级热力适应机制影响中国的天气和西太副高的异常。

6.1 印度洋海温影响的二级热力适应

吴国雄等(2000b)的研究证明,由于大气环流对热带、副热带正 SSTA 表层异常感热加热的热力适应,在其东部出现南风及对流降水;进而由于大气对该对流凝结加热的适应,在其东部出现副高异常,从而证明热带和副热带的 SSTA 可以通过"两级热力适应"去影响副高的异常。他们利用 IAP/LASG GOALS 全球气候系统模式,在北印度洋强加 1.5 K 的 SST 异常,进行一组往复 7 月的干、湿大气的对比试验。模拟的降水对大气的潜热加热只出现在湿大气试验中,但不容许出现在干大气中。结果表明,干大气的响应从一开始就局限在 SST 异常区附近(图 13a),在定常态为显著的非对称加热的 Gill 型环流,湿大气的响应则从一开始就呈现感热加热和潜热加热两种不同的分布。对感热加热的响应一直停留在 SST 异常区附近,与干大气情况相似。对潜热加热的响应从第一天开始就随着降水区的东移而向东扩展,从第三天开始异常东南风已到达中国东部海岸,其定常态的响应如图 13b 所示。降水异常区从中国东南沿海向西南延伸到赤道印度洋(吴国雄 等,2000b)。这些结果与图 7a 所示的潜热加热和高、低空副高的分布一致。

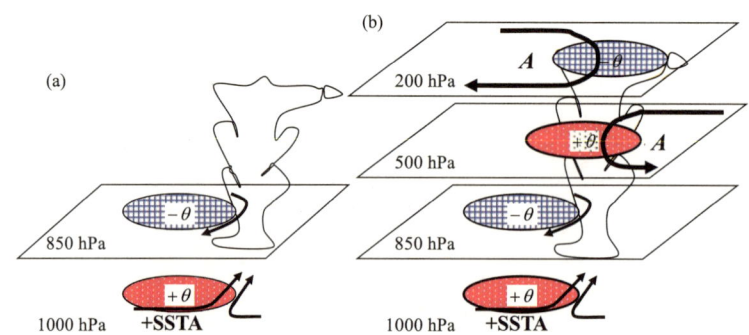

图 13 印度洋 SSTA 激发西太副高异常的"两级热力适应"的概念模型(引自吴国雄 等,2000b)

6.2 印度洋海温年际变化异常及其对西太副高的影响

刘平(1999)把 Lindzen-Nigam 关于热带平均海表温度影响平均低空环流的理论拓展到 SSTA 的影响中,证明是由于 SSTA 的"空间分布型"影响着低纬度地区对流降水及副高的异常。Liu X L 等(2020)将 2~9 年滤波后的冬季高原西部降水分为两部分:与印度洋一致变暖(或 ENSO)有关部分(图 14)和与之无关部分。其中,与印度洋一致变暖有关部分是将印度洋 SSTA 指数回归到高原西部降水上,该部分降水与高原西部降水序列的相关系数为 0.62,方差贡献为 38.1%。其影响的物理过程主要通过两个途径:其一是印度洋地区低层反气旋的水汽输送,其二是引起高层副热带西风急流的增强和南移,从而形成有利于垂直运动的散度场配置和天气尺度扰动的增强、中纬度槽脊活动的南移。

7 总结和讨论

传统上认为西太副高支配着中国夏季的降水,两者似乎是一种主从的关系,因此常常通过预报副高的异常去进行降水异常的预测。本研究指出,西太副高在 500 hPa 上对应着大范围的上升运动,而非下沉运动,因此在三维空间上,用下沉运动去解释副高形成的传统理论存在片面性。相关副热带高压动力学的主要成果进展在吴国雄等(2002,2003,2008)、Wu 等(2004)、Liu 和 Wu(2004a)、Liu Y M 等(2020)和刘屹岷等(2020)等的回顾文章中有系统总结。本文概述了如下几个方面的成果:

(1)揭示了副热带高压脊线与 Hadley 环流下沉支在自由大气中位置分离,年际变化反相的特征;在行星边界层中摩擦耗散作用使两者重合,年际变化同相的特征;也证明了副高的形成源于空气质量通量的堆积或辐合,垂直运动对副热带高压的形成不起作用。

(2)推导出了包括了大气热力结构的内强迫项以及加热和摩擦的外强迫项的全型垂直涡度倾向方程,

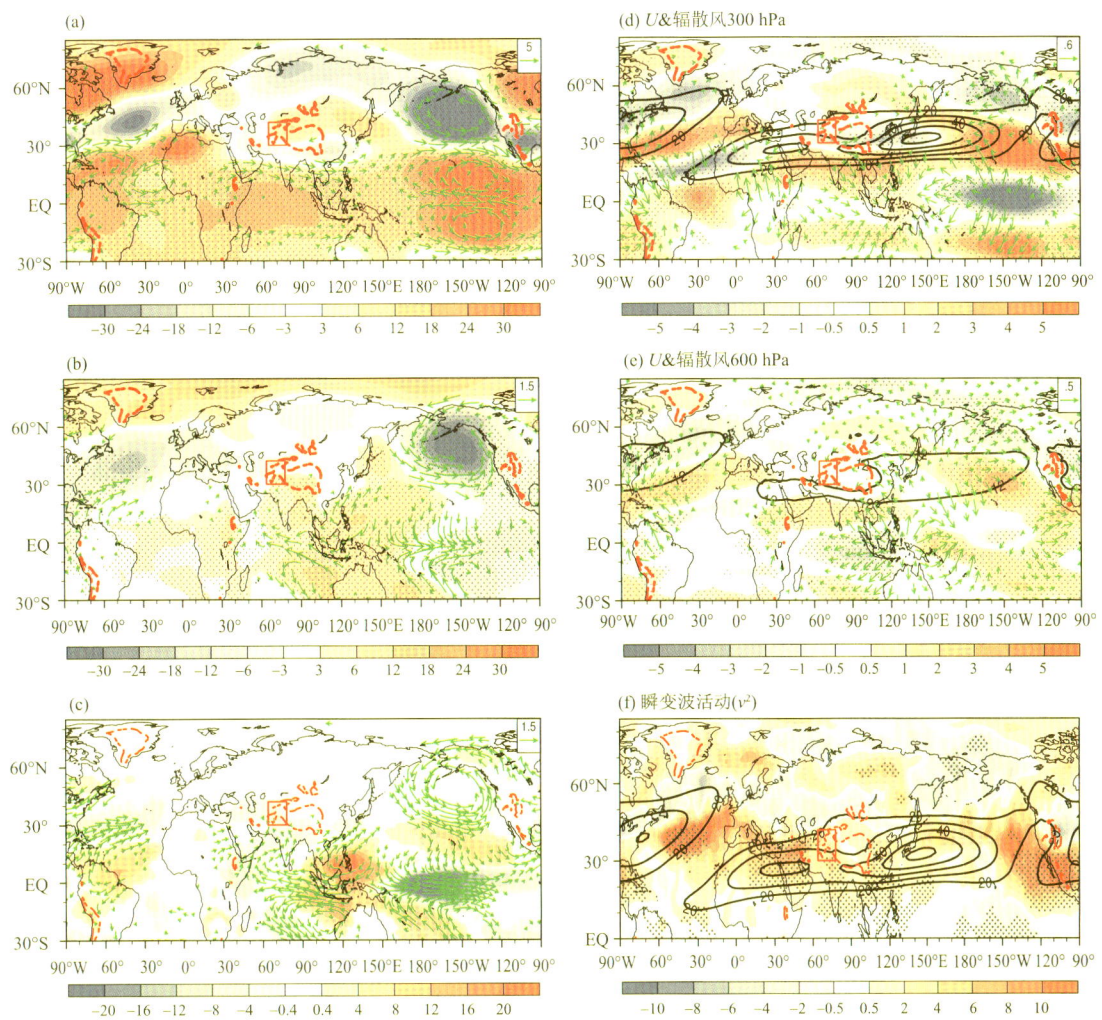

图 14　与印度洋变暖(ENSO)有关的降水对应的环流。回归的(a)300 hPa 位势高度场(阴影;单位:gpm)和风场(矢量;单位:m·s^{-1});(b)同(a),但为 1000 hPa;(c)向外长波辐射 OLR(阴影,单位:W·m^{-2})和 850 hPa 风场(矢量;单位:m·s^{-1});(d)300 hPa 纬向风(阴影;单位:m·s^{-1})和辐散风(矢量;单位:m·s^{-1});(e)同(d),但为 600 hPa;(f)300 hPa 瞬变波活动(阴影;单位:m^2·s^{-2})。打点和矢量表示均通过 95% 的显著性检验。(d)—(f)中黑色等值线表示副热带西风急流,其中(d)和(f)中表示风速大于 20 m·s^{-1},(e)中表示风速大于 12 m·s^{-1}(引自 Liu X L et al.,2020)

用于研究副高的形成机制。

(3)提出了"热力适应"理论,揭示了外热源激发大气环流异常的机理,提出了"过流"的概念。

(4)建立副热带热力涡度制造和行星涡度平流间的 Sverdrup 平衡关系,揭示副热带加热对大气经向风场的影响,建立副热带"四叶型加热"拼图及其影响副高形成和形态变异的模型,揭示不同形式外热源激发环流异常及副高单体形成的机理,证明地球自转效应及外部加热的垂直分布形态是大气中副高基本形态的决定性因子。

(5)明确夏季副热带地区存在的三种类型的大气强迫,提出了通过行星涡度和温度的经向平流在非绝热加热和涡度产生之间产生正反馈的新机制,证明季风和沙漠是多尺度强迫的孪生现象。

(6)提出了"两级热力适应"理论及拓展的"Lindzen-Nigam"理论,揭示了热带太平洋、印度洋海表面温度异常的"空间分布型"影响着低纬度地区对流降水及副高的异常。

(7)指出西太平洋副高存在十分清楚的年际变化,与印度洋海温异常密切相关,影响了亚洲气候异常;并阐述了影响的物理过程和机理。

上述研究指出了降水与西太副高之间是一种相互作用的关系,必须通过大气外热源的异常去理解夏季的降水异常。尽管关于副高的动力研究已取得一定的进展,但由于副高不同时间尺度的变异不仅涉及圈层相互作用和云-辐射反馈过程,还与大气内部的非线性过程密切相关。因此,尚需持续不断的探索,进一步完善模式,提高对副高变化的模拟能力,以揭示副高变异与中国天气和气候变化相互作用的机制,为提高天气气候预报准确率寻求新的途径。

参考文献

黄士松,余志豪,1962. 副热带高压结构及其同大气环流有关若干问题的研究[J]. 气象学报,31(4):339-359.

刘平,1999. 副高带、西太副高年际变化特征及其与海表温度异常的联系[D]. 北京:中国科学院大气物理研究所.

刘平,吴国雄,李伟平,等,2000. 副热带高压带的三维结构特征[J]. 大气科学,24(5):577-584.

刘屹岷,吴国雄,刘辉,等,1999a. 空间非均匀加热对副热带高压带形成和变异的影响. Ⅲ:凝结潜热加热与南亚高压及西太平洋副高[J]. 气象学报,57(5):525-538.

刘屹岷,刘辉,刘平,等,1999b. 空间非均匀加热对副热带高压带形成和变异的影响. Ⅱ:陆面感热加热与东太平洋副高[J]. 气象学报,57(4):385-396.

刘屹岷,吴国雄,2000. 副热带高压研究回顾及对几个基本问题的再认识[J]. 气象学报,58(4):500-512.

刘屹岷,吴国雄,宇如聪,等,2001. 热力适应、过流、频散和副高Ⅱ. 水平非均匀加热与能量频散[J]. 大气科学,24(4):317-328.

刘屹岷,姜继兰,何编,2020. 副热带高压气候动力学研究回顾[J]. 气象科学,40(5):585-595.

任荣彩,刘屹岷,吴国雄,2007. 1998年7月南亚高压影响西太平洋副热带高压短期变异的过程和机制[J]. 气象学报,65(2):183-197.

陶诗言,1963. 中国夏季副热带天气系统若干问题的研究[M]. 北京:科学出版社.

吴国雄,刘还珠,1995. 降水对热带海表温度异常的邻域响应Ⅰ:数值模拟[J]. 大气科学,19(4):421-434.

吴国雄,刘屹岷,刘平,1999. 空间非均匀加热对副热带高压形成和变异的影响——Ⅰ:尺度分析[J]. 气象学报,57(3):257-263.

吴国雄,刘屹岷,2000a. 热力适应、过流、频散和副高Ⅰ. 热力适应和过流[J]. 大气科学,24(4):433-446.

吴国雄,刘平,刘屹岷,等,2000b. 印度洋海温异常对西太副高的影响——大气中的两级热力适应[J]. 气象学报,58(5):513-522.

吴国雄,丑纪范,刘屹岷,等,2002. 副热带高压形成和变异的动力学问题[M]. 北京:科学出版社:314.

吴国雄,丑纪范,刘屹岷,等,2003. 副热带高压研究进展及展望[J]. 大气科学,27(4),503-517.

吴国雄,刘屹岷,任荣彩,等,2004. 定常态副热带高压与垂直运动的关系[J]. 气象学报,62(5):587-597.

吴国雄,刘屹岷,宇婧婧,等,2008. 海陆分布对海气相互作用的调控和副热带高压的形成[J]. 大气科学,32(4):720-740.

叶笃正,黄荣辉,1996. 长江黄河流域旱涝规律和成因研究[M]. 济南:山东科学技术出版社:384.

CHEN P, HOERLING M P, DOLE R M, 2001. The origin of the subtropical anticyclones[J]. Journal of the Atmospheric sciences, 58:1827-1835.

EADY E T, 1950. The cause of general circulation of atmosphere[J]. Centen Proc Roy Meteor Soc:156-172.

EGGER J, 1978. On the theory of planetary standing waves:July[J]. Beitr Phys Atmos, 51:1-14.

HOSKINS B J, 1996. On the existence and strength of the summer subtropical anticyclones:Bernhard Haurwitz memorial lecture [J]. Bulletin of the American Meteorological Society, 77(6):1287-1292.

LIU X L, LIU Y M, WANG X C, et al, 2020. Large-scale dynamics and moisture sources of the precipitation over the western Tibetan Plateau in boreal winter[J]. Journal of Geophysical Research Atmospheres, 125:e2019JD032133.

LIU Y M, WU G X, LIU H, et al, 2001. Condensation heating of the Asian summer monsoon and the subtropical anticyclone in the eastern hemisphere[J]. Climate Dynamics, 17(4):327-338.

LIU Y M, WU G X, 2004a. Progress in the study on the formation of the summertime subtropical anticyclone[J]. Advances in Atmospheric Sciences, 21(3):322-342.

LIU Y M, WU G X, REN R C, 2004b. Relationship between the subtropical anticyclone and diabatic heating[J]. Journal of Climate, 17:682-698.

LIU Y M, LU M M, YANG H J, et al, 2020. Land-atmosphere-ocean coupling associated with the Tibetan Plateau and its climate impacts[J]. National Science Review, 7: 534-552.

WEBSTER P J, 1972. Response of the tropical atmosphere to local, steady forcing[J]. Monthly Weather Review, 100: 518-541.

WU G X, LIU Y M, 2003. Summertime quadruplet heating pattern in the subtropics and the associated atmospheric circulation[J]. Geophysical Research Letters, 30: 1201-1204.

WU G X, LIU Y M, LIU P, 2004. Formation of the Summertime Subtropical Anticyclone. East Asian Monsoon[M]. Singapore: World Scientific: 499-544.

WU G X, LIU Y, ZHU X, et al, 2009. Multi-scale forcing and the formation of subtropical desert and monsoon[J]. Annales Geophysicae, 27: 3631-3644.

综述 5

南亚高压的双模态和年际及年代际变化

张琼[1]，刘屹岷[2]，任荣彩[2]

（1 瑞典斯德哥尔摩大学，斯德哥尔摩 10691；2 中国科学院大气物理研究所，北京 100029）

摘要：本文总结回顾了关于南亚高压在气候统计意义上的双模态分布型，探讨了其对应的动力热力结构以及南亚高压的年际和年代际变化。资料诊断和数值模拟试验均证明南亚高压具有"趋热性"，其双模态的维持主要取决于青藏高原以及周围地区上空大气的热力状况。数值试验结果表明青藏高原夏季的强加热所激发的纬向非对称斜压不稳定的发展，会导致南亚高压的双模态转换的准双周振荡。与高压中心位置东—西振荡相联系的南亚高压双模态，是南亚高压与青藏高原热源异常变化在准双周时间尺度上密切耦合的结果。青藏高原热源异常具体表现在伊朗高原和青藏高原的感热加热的相互作用，高原感热加热强弱是导致南亚高压双模态维持和转换，调节亚洲副热带季风区的水汽辐合以及东亚气候异常的主导因子。南亚高压的年际变化与东亚季风凝结潜热释放密切相关，而青藏高原感热的年代际减弱引发南亚高压和西太平洋副高减弱，贡献于中国东部"南涝北旱"降水异常分布。

1 引言

夏季，在对流层上层和平流层下层，青藏高原及其周边地区存在一个大尺度高压系统——南亚高压。南亚高压是平流层低层除极涡外最强、最稳定的环流系统，盛夏时节南亚高压的水平尺度可延伸至半球范围。Flohn(1957)提出，这个巨大反气旋的形成是青藏高原大气热源抬升的结果。南亚高压的东西振荡是盛夏高层副热带环流的一个明显特征。陶诗言和朱福康(1964)最早指出夏季南亚高压有围绕青藏高原作往返振荡的趋势。罗四维等(1982)利用多年历史天气图资料以 100°E 为界划分了东部型、西部型和带状型南亚高压，该分型被广泛地应用于天气学分析中，并明确地将东、西部型高压的相互转换称为南亚高压的东西振荡，这种振荡基本上是一种中期天气过程，具有明显的天气学意义。后来朱抱真和宋正山(1984)通过资料分析依据南亚高压在不同的经度位置对高原地区的降水和积云对流以及印度季风的影响，划分了伊朗型高压以及以 90°E 为界的东型青藏高压和西型青藏高压。张琼等（张琼和吴国雄，2001；Zhang et al.，2002）通过对 NCEP/NCAR 的月平均和候平均资料的统计分析发现，夏季南亚高压的一个主要气候特征是南亚高压中心在经度位置上的双模态分布：青藏模态和伊朗模态。他们指出，气候意义上的南亚高压双模态与以往天气学意义上的南亚高压的东西振荡无论从时空尺度还是维持的物理机制均有显著的不同，青藏模态以上升运动和非绝热加热为特征，而伊朗模态则以下沉运动和绝热加热为特征。南亚高压双模态的重要性还在于它与东亚地区大面积洪涝灾害的发生密切相关（张琼和吴国雄，2001）。青藏模态对应着孟加拉湾地区和青藏高原南部、中国南海，以及长江流域至日本南部地区降水偏多，印度和朝鲜半岛地区的降水偏少；而伊朗模态对应着相反的降水异常分布。

通讯作者：张琼，qiong.zhang@natgeo.su.se

因此,理解对流层高层南亚高压双模态的形成和转换机制,以及其和对流层中低层副热带高压之间关系,有助于理解东亚区域气候异常的机制。

2 南亚高压双模态及其热力动力结构

图 1a 显示,从多年气候统计来看,夏季南亚高压中心主要集中在青藏高原和伊朗高原上空,而很少出现在 70°~80°E, Zhang 等(2002)依此将南亚高压划分为青藏高原模态和伊朗高压模态。图 1b 和 1c 分别是合成的青藏高压和伊朗高压的环流场,其直观且明显的特征就是南亚高压中心位于青藏高原和伊朗高原这两个大地形上空。

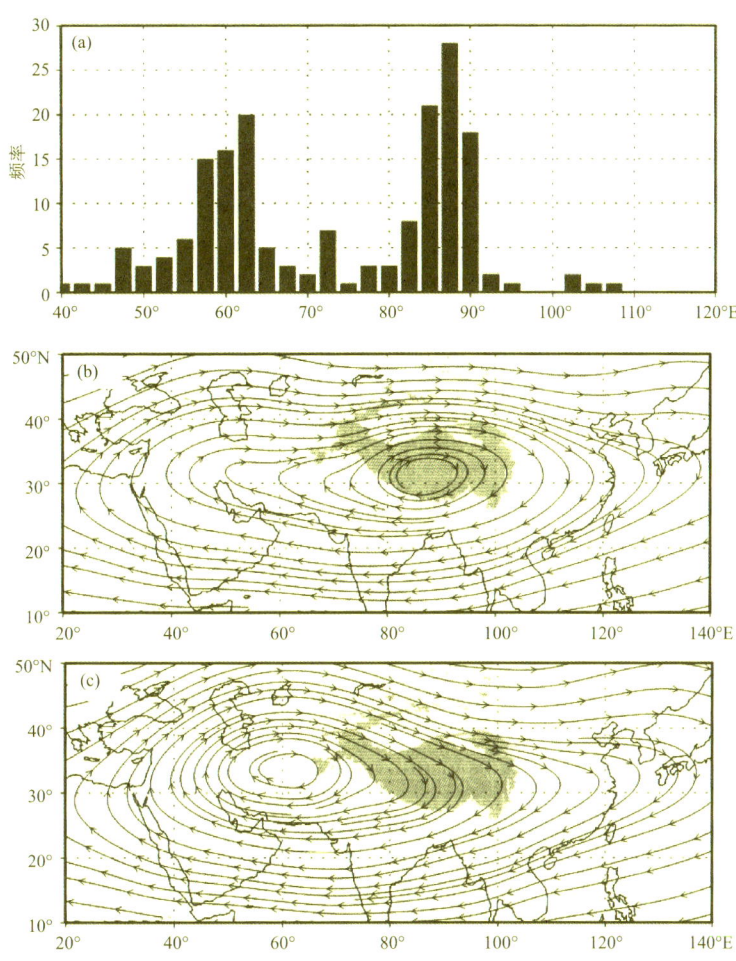

图 1　1980—1994 年盛夏 7,8 月共 180 个候的 100 hPa 南亚高压中心频数分布(a),以及合成的 100 hPa 青藏高压(b)和伊朗高压(c)环流图。180 个候中共有 77 个青藏高压型,62 个伊朗高压型(引自 Zhang et al.,2002)

从图 2 可以看到青藏高压和伊朗高压在垂直结构上的异同点。其共同的特征体现在位温的垂直异常结构(图 2b,d),在 100 hPa 处高压中心是冷中心,而 200 hPa 以下是暖气柱,即高层的高压中心总是处于暖气柱上空,体现了其"趋热性"。不同的是,青藏高原上空的暖气柱对应强的上升气流(图 2a),而伊朗高原上空的暖气柱对应下沉气流(图 2c)。对热力学方程各项的诊断发现表明,青藏高压上空气柱的加热主要来自于强的非绝热加热,尤其是近地层加热,上升冷却对此有补偿作用。在伊朗高原上空对流层低层的加热来自于地表非绝热加热,但对流层中高层气柱的增暖主要由绝热下沉增温造成。因此,南亚高压双模态的维持主要取决于青藏高原以及周围地区上空大气的热力状况。

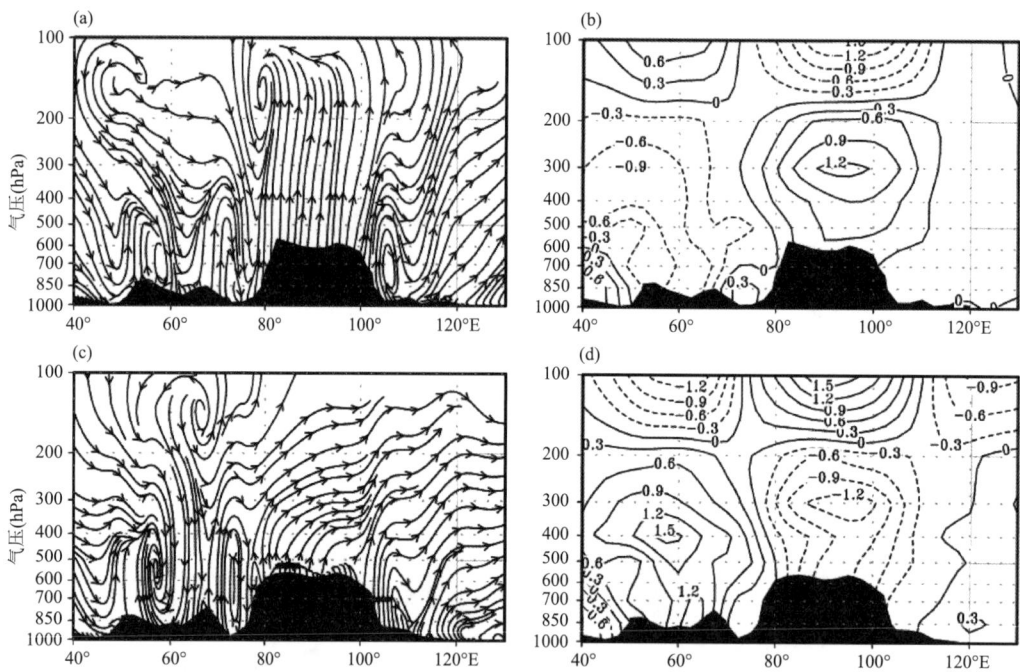

图 2 对应于 77 个青藏高压型(a),(b)和 62 个伊朗高压型(c),(d)的垂直环流和位温异常的沿 30°N 的气压-经度剖面合成图,(a)和(c)为垂直纬圈环流,(b)和(d)为位温异常(引自 Zhang et al.,2002)

3 夏季高原加热和南亚高压准双周振荡

Liu 等(2007)证明青藏高原夏季的强加热能激发纬向非对称斜压不稳定发展,产生南亚高压的东/西部型双模态及准双周振荡。他们利用原始方程模式 IGM1 进行数值试验,图 3 给出在初始场取纬向对称环流,通过在真实青藏高原地形和高原加热强迫下的 300 hPa 等压面上的纬向偏差流函数分布。可见高原上超过临界强度的加热能引发非对称斜压不稳定,该加热导致高原上空出现一个位涡最小值。当加热足够强时,能导致位涡纬向梯度激增,平流作用使反气旋以东的位涡增加,以西的位涡减小;反气旋向西部上空移动,直至被非定常的位涡平流所阻尼。这一过程大约为两周,在流场上表现为高原上空的反气旋中心处于东西交替转换的准双周振荡。用理想的高原地形和加热强迫及纬向对称初始环流得到了类似的结果(Liu et al.,2007;吴国雄 等,2008)。

以上数值试验结果与观测统计结果一致(Liu et al., 2004,2013;Zhang et al., 2002;Ren et al., 2019)。Ren 等(2019)对南亚高压准双周振荡过程的合成分析表明,在青藏高原热源异常偏强时,南亚高压中心多离开青藏高原而在伊朗高原上空,即呈南亚高压的伊朗高原模态;反之,在青藏高原热源异常偏弱时,南亚高压中心则反而多位于青藏高原上空,即呈南亚高压的青藏高原模态。以青藏高原热源异常强峰值位相为例,南亚高压的伊朗高原模态,对应着异常偏强的青藏高原地面气压负异常、低空辐合、上升运动和高空辐散,与此相联系的是伊朗高原上空异常偏强的高空辐合和补偿性下沉运动。这种动力下沉造成的异常偏强的绝热加热效应,可加强伊朗高原上空的南亚高压中心。同时,由于在准双周振荡过程中青藏高原上的地面低压的异常强度更强、对流层上升运动更加异常深厚,高空的大气非绝热加热异常中心发生在更高的层次(约 150 hPa),这使得在南亚高压所在高度(约 200 hPa)仍为低压异常,而在更高的平流层为高压异常,从而南亚高压多呈伊朗高原模态(图 4,P11)。在青藏高原热源异常偏弱位相时情形相反,即伴随弱的热源、青藏高原上空的非绝热加热中心高度相对偏低,在南亚高压所在高度为高压异常(而伊朗高原上的绝热加热异常偏弱),南亚高压多呈青藏高原模态(图 4,P4)。由此表明,与高压中心位置东—西振荡相联系的南亚高压双模态,是南亚高压与青藏高原热源异常变化在准双周时间尺度上密切耦合的结果。

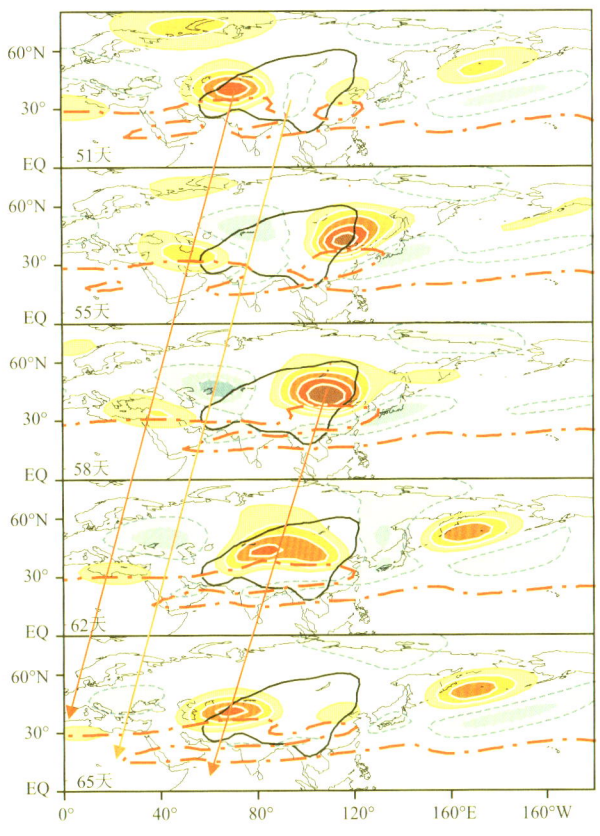

图3　IGM1试验中在青藏高原地形和高原加热强迫的300 hPa等压面上的纬向偏差流函数分布(单位：$10^7\ m^2\cdot s^{-1}$)，红色点划线为$u=0$线(引自吴国雄 等，2008)

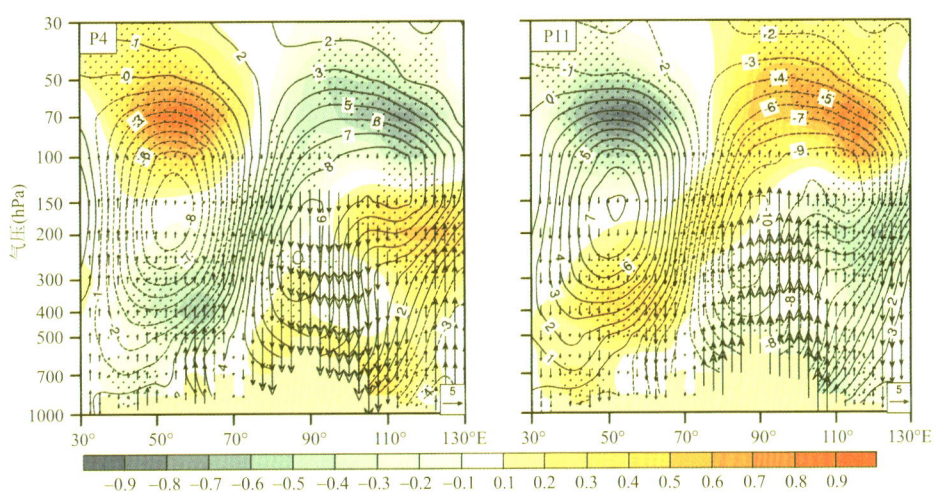

图4　基于青藏高原热源异常偏弱和偏强峰值位相合成的南亚高压所在纬度(27.5°~37.5°N)位势高度(gpm，等值线)、位温(K，阴影)以及垂直运动($10^{-3}\ Pa\cdot s^{-1}$，箭头)的经向-高度分布(引自Ren et al.，2019)

为了深入理解青藏高原热源的主要控制因子以及伊朗高原热力结构与青藏高原热源之间的相互作用，Wu 等(2016)和Liu等(2017)基于区域气候系统模式WRF开展了青藏高原感热和潜热的敏感性试验，结果表明夏季青藏高原和伊朗感热加热存在相互影响和反馈，形成了观测到的伊朗高原感热加热→青藏高原感热加热和凝结潜热释放→大气垂直环流之间的准平衡耦合系统(图5，Liu et al.，2017)，从而影响大气环流。青藏高原上的感热-潜热相互反馈在这个耦合系统中起主要作用,正是由于夏季青藏-伊朗高原感热气泵的影响,改变了其上空的温度和环流结构,有利于副热带季风型经圈环流的发展,从而给亚洲副热带季风提供了大范围上升运动的背景条件(Liu et al.,2020)。两大高原的感热加热对其他地区的影响有相互加强也有相互抵消,伊朗高原和青藏高原的感热加热的共同作用对亚洲副热带季风区的水汽辐合作出最主要的贡献。

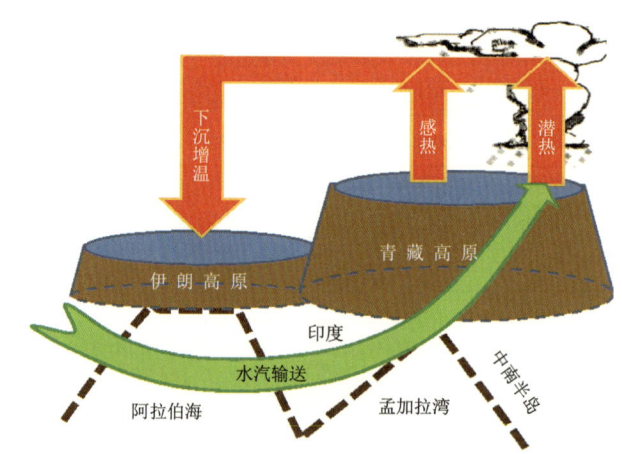

图 5 青藏高原、伊朗高原的热力强迫以及南亚的水汽输送所构成的一个相互反馈的耦合系统示意图(引自 Liu et al.,2017)

4 南亚高压东西振荡及其对西太副高及东亚降水的影响

陶诗言和朱福康(1964)指出,夏季南亚高压与 500 hPa 西太平洋副热带高压的进退有紧密联系,二者有相向和相背而行的趋势。相向而行是指南亚高压东部型在建立的过程中,西太副高也有一次西伸北上过程。反之,相背而行是指南亚高压西部型在建立的过程中,我国大陆东部的西太副高常向东南撤。张琼和吴国雄(2001)通过再分析资料进一步拓宽了这一对应关系,指出对应于南亚高压的双模态的分布,500 hPa 环流图上不仅西太平洋副热带高压的东西分布存在差异,伊朗副高的东西分布也存在差异,导致整个亚澳季风区出现大范围的气候异常。

概言之,盛夏季节副热带大陆的加热使大气低层出现低压,高层出现高压,于是大陆东部受上升运动控制,西部受下沉运动控制。高原强烈的地表加热也使高原上空出现浅薄的表层低压和深厚的中上层高压,因此高原及其东侧为上升运动,西侧为下沉运动。副热带的环流因而表现为在洲际尺度的热力环流上叠加同位相的高原热力环流,从而加剧东亚的夏季风气候及中亚的干热气候。

如上所述,南亚高压中心的青藏高原和伊朗高原双模态特征,是与青藏高原上空大气热源强度的准双周振荡紧密耦合导致的高压中心位置的东—西振荡。除了与双模态相联系的东—西振荡,南亚高压的东缘还存在规律性的向东亚地区东伸或西退至青藏高原,表现为另一种形式的东西振荡。任荣彩等(2007)对 1998 年长江流域"二度梅"过程的分析和模拟发现,南亚高压向东亚上空的异常东伸,可通过高空负涡度平流的动力强迫,诱导低层西太副高的北上发展,对应 1998 年梅雨过程的异常中断,而后南亚高压西退回到高原上空时西太副高减弱南退,造成了"二度梅"的开始。Ren 等(2015)的研究指出,南亚高压东缘的东伸、西退过程具有显著的 10~30 天周期,南亚高压的东伸(西退)过程对应着长江—黄河中下游地区降水偏多(偏少),中国南部地区降水偏少(偏多)。

祝传栋等(2022)研究了南亚高压这两类东—西振荡的联系以及它们对亚洲地区环流和天气影响的差异。结果表明,南亚高压中心的双模态东—西振荡位相可显著影响其东缘东伸/西退的发生及其幅度。尽管在南亚高压中心呈青藏高原和伊朗高原模态时,均可以出现南亚高压东缘的向东亚东伸,但青藏高原模态下发生东伸的频率明显高于伊朗高原模态；在伊朗高原模态时则更容易出现南亚高压东缘的西退。而且，在青藏高原模态下发生的南亚高压东缘东伸的幅度也比伊朗高原模态时更大。南亚高压中心的双模态东—西振荡主要与印度北部及整个青藏高原地区的降水异常型密切联系，并与异常降水有关的热力和动力作用变化相耦合。而南亚高压东缘的东伸/西退则通过引起西太副高的西进/东退，与东亚地区偶极子型的降水异常（青藏高原中东部、长江与黄河之间的中下游地区的降水异常与长江以南地区的相反）相联系。此外，当南亚高压中心为青藏高原模态且东缘发生东伸时，与南亚高压中心为伊朗高原模态且东缘发生西退时，青藏高原西部与中东部的降水异常总是呈显著反位相变化。

5　南亚高压年际-年代际变化及其对洪涝的影响

年际尺度上，南亚高压强度异常表现为其西、东部的位势高度（温度）异常在欧亚大陆上空呈现符号基本一致的变化模态（图6），与高原东部到长江中下游一带的季风潜热释放的异常关系密切（Zhang et al., 2016）。诊断及数值试验揭示该潜热加热影响南亚高压的机理为，潜热的垂直分布在对流层上层是随高度递减的，使得东亚副热带地区上层出现反气旋性环流异常，南亚高压东部加强、东伸；还会激发一西传Rossby波，在中亚地区形成反气旋性异常，造成地中海东岸高层大气的南风暖平流和伊朗高原上空的下沉增温等绝热加热，进而出现局地降水负异常和近地面感热作用的非绝热加热增强，它们共同使得南亚高压西部的位势高度升高（图7）。在20世纪后半期，中国南部地区降水出现增加趋势，伴随着高空北风切变的增强和其西侧南亚高压和对流层上层暖中心的增强（Wu et al., 2015）。结果表明，大气环流对降水异常的反馈是区域气候空间分布型变化的一个重要贡献因素。

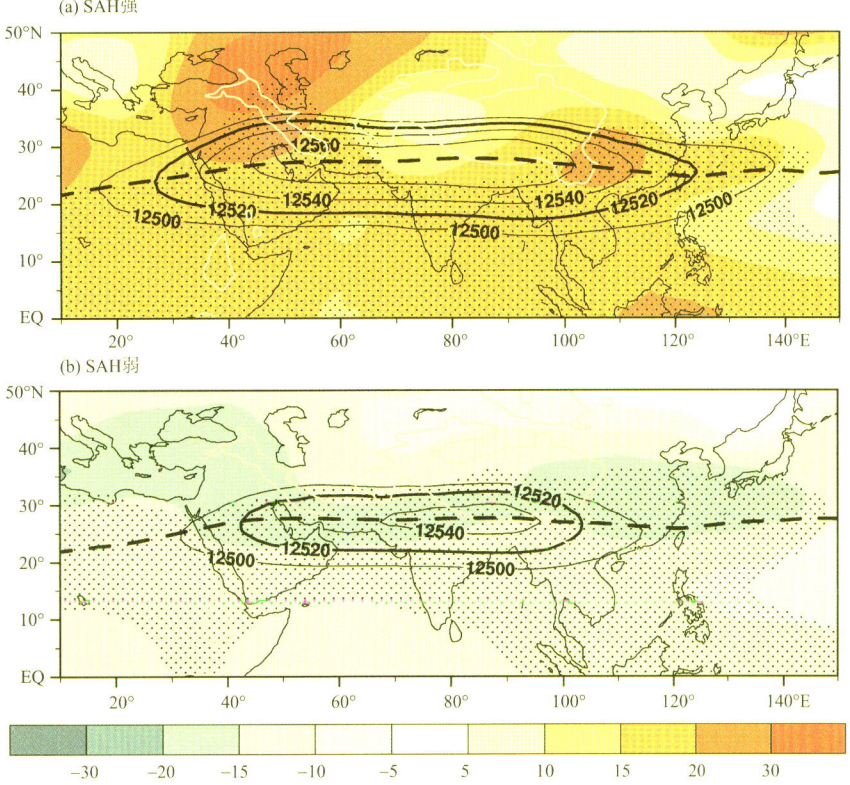

图6　根据南亚高压指数（SAHI）合成的200 hPa位势高度场（等值线）和对气候平均的偏差（填色）：(a)SAHI正异常年；(b)SAHI负异常年，单位为gpm。打点区域表示位势高度异常达到95%信度检验的区域，加粗的黑色等值线为12520 gpm，黑色粗断线表示南亚高压强、弱年的脊线，灰色等值线是地形1500 m等高线（引自Zhang et al., 2016）

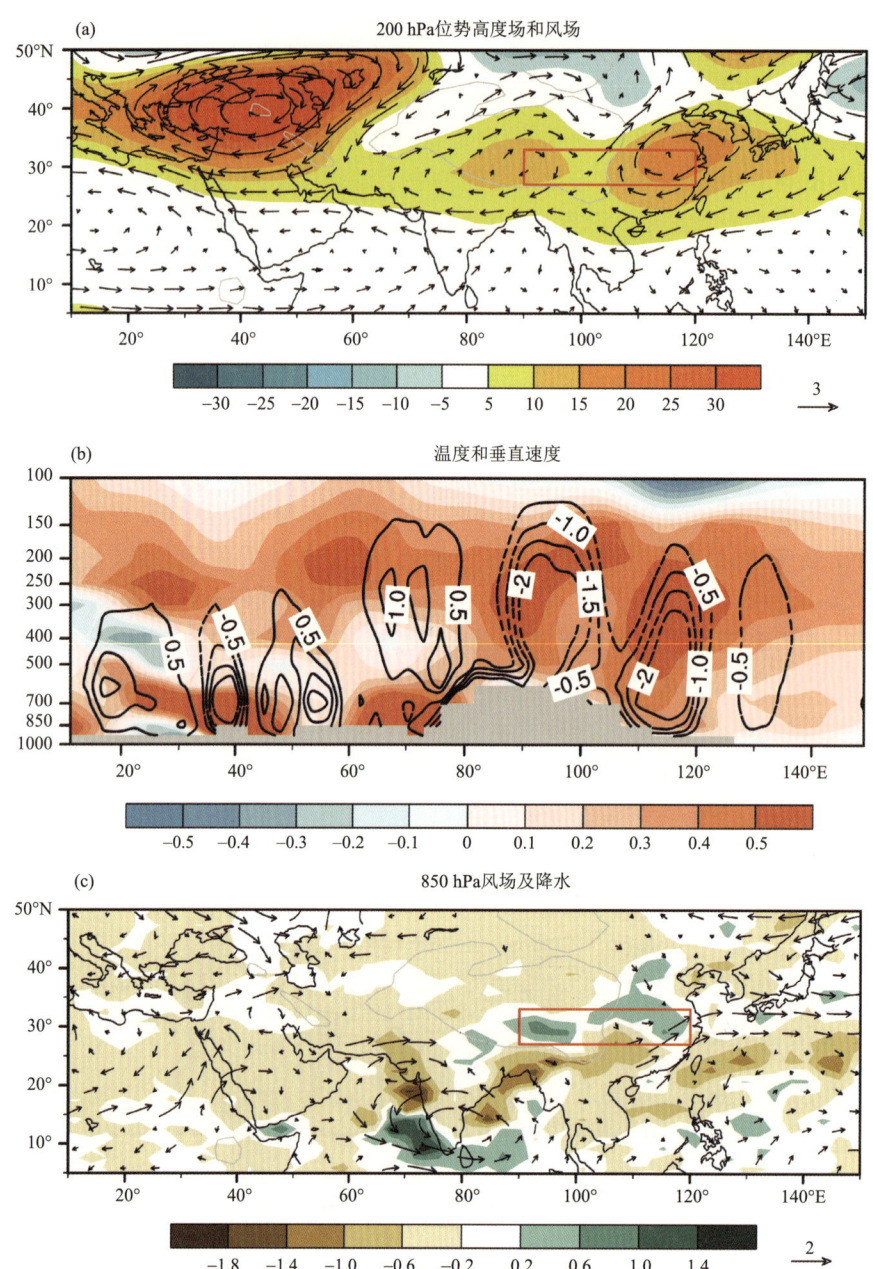

图 7 大气环流模式中环流对高原东部到长江中下游地区异常非绝热加热的响应（敏感性试验与对照试验之差）。填色分别为位势高度场(a,单位：gpm),温度(b,单位：K)和降水(c,单位：mm·d^{-1})；矢量为风场(单位：m·s^{-1})，垂直速度单位是 0.01 Pa·s^{-1}（引自 Zhang et al., 2016）

青藏高原上的夏季感热在 20 世纪末之前呈现出减弱的年代际变化。Duan 等(2011,2013)的诊断指出青藏高原增暖是造成中国南涝北旱(长江涝、黄河旱)年代际气候异常的重要影响因子之一。Liu 等(2012)的诊断和数值试验进一步从位涡理论证明，青藏高原地表加热是激发低空气旋式环流和高空反气旋式环流的重要原因。自 20 世纪 70 年代中到 21 世纪初，青藏高原地表风速减弱导致感热加热减弱，"感热气泵效应"受到抑制。环绕青藏高原的低空气旋式环流因此减弱，导致东亚地区偏南风减弱，雨带因此滞留在江南造成了南涝北旱(华南涝、华北旱)的空间分布特征(图8)。

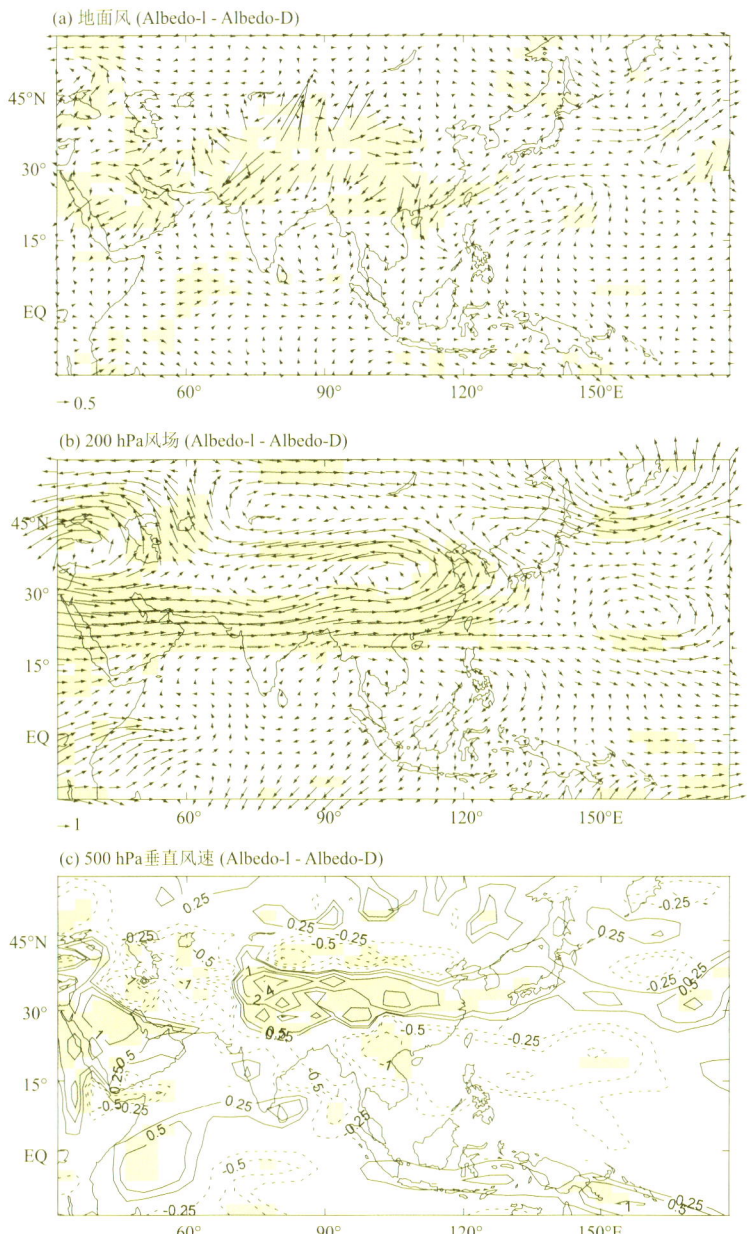

图 8 模拟的夏季平均风场(m·s^{-1})和垂直速度(Pa·s^{-1})对减弱的高原感热的响应。
(a)表面风；(b)200 hPa 风场和 (c)500 hPa 垂直风速(引自 Liu et al.，2012)

6 总结

  文中回顾了夏季南亚高压的双模态特征的气候统计诊断分析以及数值试验的主要研究成果。气候统计诊断分析表明南亚高压的形成和维持是由于其"趋热性"，数值试验结果证明了南亚高压双模态的维持机制是由于青藏高原和伊朗高原表面的感热加热作用。青藏高原热源异常变化的准双周时间尺度导致了高压中心双模态之间的转换呈现准双周振荡。南亚高压双模态的东西振荡带动 500 hPa 西太平洋副热带高压的东西进退，影响东亚地区降水分布异常。南亚高压存在十分清楚的年际和年代际变化，东亚季风区的凝结潜热加热的年际变化和青藏高原表面感热加热的年代际变化与此相对应，影响了亚洲气候异常。

参考文献

罗四维,钱正安,王谦谦,1982. 夏季 100 hPa 南亚高压与我国东部旱涝关系的天气气候研究[J]. 高原气象,1(2):1-10.

任荣彩,刘屹岷,吴国雄,2007. 1998 年 7 月南亚高压影响西太平洋副热带高压短期变异的过程和机制[J]. 气象学报,65(2):183-197.

陶诗言,朱福康,1964. 夏季亚洲南部 100 hPa 流型的变化及其与太平洋副热带高压进退的关系[J]. 气象学报,34(4):385-395.

吴国雄,刘屹岷,宇婧婧,等,2008. 海陆分布对海气相互作用的调控和副热带高压的形成[J]. 大气科学,32(4):720-740.

张琼,吴国雄,2001. 长江流域大范围旱涝与南亚高压的关系[J]. 气象学报,59(5):569-577.

朱抱真,宋正山,1984. 青藏高压的形成过程和准周期振荡——观测事实的分析[C]//青藏高原科学试验文集(一). 北京:科学出版社:303-313.

祝传栋,任荣彩,2022. 夏季南亚高压两类东—西振荡过程的联系及其天气效应对比[J]. 大气科学. doi:10.3878/j.issn.1006-9895.2106.21075.

DUAN A M, LI F, WANG M R, et al, 2011. Persistent weakening trend in the spring sensible heat source over the Tibetan Plateau and its impact on the Asian summer monsoon[J]. Journal of Climate, 24(21):5671-5682.

DUAN A M, WANG M R, LEI Y H, et al, 2013. Trends in summer rainfall over China associated with the Tibetan Plateau sensible heat source during 1980—2008[J]. Journal of Climate, 26(1):261-275.

FLOHN H, 1957. Large-scale aspects of the "summer monsoon" in South and East Asia [J]. Journal of the Meteorological Society of Japan, 35A:180-186.

LIU B, WU G, MAO J, et al, 2013. Genesis of the South Asian high and its impact on the Asian summer monsoon onset [J]. J Climate, 26:2976-2991.

LIU Y, WU G, REN R, 2004. Relation between the subtropical anticyclone and diabatic heating[J]. J Climate, 17:682-698.

LIU Y, HOSKINS B J, BLACKBURN M, 2007. Impact of Tibetan orography and heating on the summer flow over Asia [J]. J Met Soc Japan, 85B:1-19.

LIU Y M, WU G X, HONG J L, et al, 2012. Revisiting Asian monsoon formation and change associated with Tibetan Plateau forcing:Ⅱ. Change[J]. Climate Dynamics, 39 (5):1183-1195.

LIU Y, WANG Z, ZHUO H, et al, 2017. Two types of summertime heating over the Asian large-scale orography and excitation of potential-vorticity forcing. II. Sensible heating over Tibetan-Iranian Plateau[J]. Sci China:Earth Sci, 60(4):733-744.

LIU Y M, LU M M, YANG H J, et al, 2020. Land-atmosphere-ocean coupling associated with the Tibetan Plateau and its climate impacts[J]. National Science Review, 7:534-552.

REN R, ZHU C, CAI M, 2019. Linking quasi-biweekly variability of the South Asian high to atmospheric heating over Tibetan Plateau in summer[J]. Climate Dyn,53(5-6):3419-3429.

REN X, YANG D, YANG X, 2015. Characteristics and mechanism of the subseasonal eastward extension of the South Asian high[J]. J Climate, 28 (17):6799-6822.

WU G X, HE B, LIU Y M, et al, 2015. Location and variation of the summertime upper troposphere temperature maximum over South Asia[J]. Climate Dynamics, 45:1-18.

WU G, ZHUO H, WANG Z, et al, 2016. Two types of summertime heating over the Asian large-scale orography and excitation of potential-vorticity forcing. Ⅰ. Over Tibetan Plateau[J]. Sci China:Earth Sci, 59(10):1996-2008.

ZHANG Q, WU G, QIAN Y, 2002. The bimodality of the 100 hPa South Asia high and its relationship to the climate anomaly over East Asia in summer[J]. J Meteorol Soc Japan, 80:733-744.

ZHANG P F, LIU Y M, HE B, 2016. Impact of East Asian summer monsoon heating on the interannual variation of the South Asian high[J]. Journal of Climate, 29 (1):159-173.

代表性论文 5

影响大气涡度发展的若干热力过程
—— I. 热力适应过程

吴国雄,刘屹岷

(中国科学院大气物理研究所,北京 100029)

  副热带高压系统是沿着两半球副热带分布的高值气压系统或高值位势高度系统。在流场分布上表现为反气旋。因此,副热带高压有较大的负值相对垂直涡度。其变化规律受涡度方程所约束。因此通过垂直涡度倾向方程(以下简称涡度方程或 ζ-方程)以研究副热带高压的活动规律及变异成因是行之有效的办法。由于涡度的变异受到大气中内、外热力过程的显著影响,我们除了介绍传统的涡度方程外,还将讨论能描述热力过程和动力过程共同作用的新的涡度方程。

  为了研究内、外热力强迫影响副高变异的机制,我们首先引进一个热力学参数 Burger 数,把副热带大气运动特性与低纬度和高纬度的大气运动特性区分开来。然后利用位势涡度(简称位涡)综合描述气块动力和热力结构的特点,以揭示大气中热力适应的原理。这一热力适应原理说明了为什么在地球表面的热力强迫能影响上部对流层的环流。

1 副热带大气运动的特征

展开大气运动的热力学方程:

$$\begin{cases} \dfrac{D\theta}{Dt} = Q \\ \dfrac{D}{Dt} = \dfrac{\partial}{\partial t} + \mathbf{V} \cdot \nabla \end{cases} \tag{1}$$

得到:

$$\dfrac{\partial \theta}{\partial t} + u\dfrac{\partial \theta}{\partial x} + v\dfrac{\partial \theta}{\partial y} + \dfrac{\theta}{g}N^2 w = Q \tag{2}$$

式中,Q 为加热率,$\mathbf{V}$ 和 ∇ 为三维风矢量和梯度算子,$N^2 = \dfrac{g}{\theta}\dfrac{\partial \theta}{\partial z}$,$N$ 为布伦特-维赛拉(Brunt-Vaisala)振荡频率,其他符号为气象常用。对于准定常问题可略去局地变化项,再利用热成风关系:

$$\dfrac{\partial u}{\partial z} = -\dfrac{g}{f\theta}\dfrac{\partial \theta}{\partial y}, \dfrac{\partial v}{\partial z} = \dfrac{g}{f\theta}\dfrac{\partial \theta}{\partial x} \tag{3}$$

式(2)可写为:

$$u\dfrac{\partial v}{\partial z} - v\dfrac{\partial u}{\partial z} + \dfrac{N^2}{f}w = \dfrac{g}{f\theta}Q = Q^* \tag{4}$$
$\quad\quad\quad\quad$ (a) $\quad$ (b) $\quad\quad$ (c)

本文原载于《副热带高压形成和变异的动力学问题》第四章.北京:科学出版社,2002

对于一给定的纬向分布流场 $u=u(z)$,外界的加热 Q^* 将激发出纬向非对称风场 v,其量值大小 v^* 取决于(4)式左端各项的相对贡献。这时有下述三种可能:

$$\begin{cases} (a) \text{ 当纬向平流为主时}, v^* \sim \dfrac{H_Q}{u}Q^*; \\ (b) \text{ 当经向平流为主时}, u^* \sim \dfrac{H_u}{u}Q^*; \\ (c) \text{ 当垂直对流为主时}, v^* \sim \dfrac{f^2}{N^2 \beta H_Q}Q^* \end{cases}$$

式中,H_Q 为加热的特征垂直尺度,H_u 为风场变动特征垂直尺度,在导得结论(c)时,利用了地转涡度关系 $\beta v^* \sim f\dfrac{\partial w^*}{\partial z}$。把上述关系式代回(4)式,可以估计各项的相对重要性:

$$v^* - \frac{H_Q}{H_u}v^* + \frac{N^2 H_Q^2}{f^2}\left(\frac{\beta}{u}\right)v^* \sim \frac{H_Q}{u}Q^* = \Re$$

由于 $L=(u/\beta)^{1/2}$ 为定常 Rossby 波波长,如引进 Burger 数

$$B = N^2 H_Q^2/(f^2 L^2) \tag{5}$$

则上式成为:

$$v^* - \frac{H_Q}{H_u}v^* + Bv^* \sim \Re \tag{6}$$
$$\quad\text{(a)} \quad\quad \text{(b)} \quad\quad \text{(c)}$$

(6)式即为与(4)式相对应的相对尺度分析方程。式中的 Burger 数 B 就是 Held(1978)在讨论大气斜压不稳定时用到的重要参数 $\gamma = f^2 u/(\beta N^2 H_Q H)$ 的倒数(Hoskins and Karoly,1981)。参照 Hoskins(1987)的讨论,取下述参数值可计得各纬带 B 参数的值(表1)。

表1 各纬带 B 参数的值

φ	$N(\text{s}^{-1})$	$f(\text{s}^{-1})$	$H(\text{km})$	$L(\text{km})$	B
10°N	1.1×10^{-2}	$f_{10°}$	15	2000	11
30°N	1.1×10^{-2}	$f_{30°}$	15	2500	0.91
50°N	1.1×10^{-2}	$f_{50°}$	12	3000	0.16

根据上表给出的 Burger 数在不同纬带的值,可以得出如下结论:由于 Burger 数在低纬很大,在高纬很小,在副热带接近 1,因此为平衡外界热源,在热带主要依靠绝热上升冷却;在高纬度主要依靠水平冷平流;在副热带,水平平流和绝热冷却的作用相当。由此看来,副热带动力和热力学要比低纬和中高纬的更为复杂。下面就热带和中高纬两种极端情况讨论大气在准定常热源强迫下的运动特征。

1.1 热带地区的斜压结构

热带地区 $B \approx 11$。于是热力方程(2)可简写为:

$$N^2 w = Q^* f \tag{7}$$

即非绝热加热必须由上升运动的绝热冷却来平衡。把上式代入简化的地转涡度方程:

$$\frac{\mathrm{D}_h \zeta}{\mathrm{D}t} = f\frac{\partial w}{\partial z} + \text{其他项} \tag{8}$$

得到加热随高度增加有气旋式涡度增长;反之,有反气旋式涡度增长。于是热带地区大气热源(例如深对流系统)的上下方出现性质相反的流场,显现出"斜压结构"。这一定性结论由于使用了地转假设,不是十分严格。本章将在第3节用位涡理论从热力适应的角度对此进行严格讨论。

1.2 中高纬度的相当正压结构

在中高纬度 $B \approx 0.16$。则热力方程(2)可简写为:

$$u\frac{\partial \theta}{\partial x} + v\frac{\partial \theta}{\partial y} = Q \tag{9}$$

利用热成风关系式(3)可得:

$$\frac{\partial}{\partial z}\left(\frac{v}{u}\right) = \frac{g}{f\theta_0 u^2}Q \tag{10}$$

上式表明,仅当存在局地加热的情形,风向才随高度改变。在没有局地热源的情形,风向随高度不变,显示出"相当正压"的结构。这就是 Smagorinsky(1953)指出的"平流极限"。

1.3 副热带运动的特征

在副热带地区, $B \approx 1$。因此,热力方程中的平流项和绝热上升项均重要,应针对不同问题进行具体分析。

本节的讨论表明,我们可以用 Burger 数的大小来区分大气中的运动特征。在低纬热带地区, $B \gg 1$,加热主要为垂直运动的绝热冷却所平衡;大气环流具有斜压的结构;高层流场和低层流场位相相反。在中高纬度地区, $B \ll 1$,加热主要为水平冷平流所平衡;大气环流具有相当正压结构;在远离热源的区域,高低层流场位相相同。在副热带地区, $B \approx 1$,外部加热为水平冷平流和垂直上升冷却所共同平衡。因此副热带大气的运动学和动力学要比纯热带的或纯中高纬的更为复杂壮观。

2 位涡方程

大气的热力学方程可以写成拉格朗日形式,位温是一个守恒量,便于分析和研究。但大气的动力学方程及由之导得的涡度方程不具备这种性质,动量和涡度都是不守恒的量,给分析和研究带来诸多不便。本节将仿照 Ertel(1942)的研究,将支配大气运动的动量方程、热力方程、连续方程、状态方程和水汽方程结合为一体,导出具有综合特征的位涡方程,并获取具有守恒的物理量:湿位涡和干位涡(吴国雄 等,1995)。

对欧拉形式的动量方程:

$$\frac{\partial \boldsymbol{V}}{\partial t} + \nabla\left(\frac{V^2}{2} + \varphi\right) - \boldsymbol{V} \times \boldsymbol{\zeta}_a = -\alpha \nabla p + \boldsymbol{F}_v \tag{11}$$

求矢量积($\nabla \times$),可得涡度方程:

$$\frac{\partial}{\partial t}\boldsymbol{\zeta}_a - \nabla \times (\boldsymbol{V} \times \boldsymbol{\zeta}_a) = \nabla p \times \nabla \alpha + \boldsymbol{F}_\zeta \tag{12}$$

式中, $\boldsymbol{V}$ 为三维风矢, φ 为外作用力的势函数, $\boldsymbol{F}_v$ 为表面摩擦力, $\boldsymbol{F}_\zeta = \nabla \times \boldsymbol{F}_v$ 为涡度摩擦耗散, $\boldsymbol{\zeta}_a$ 为绝对涡度。设 $\boldsymbol{\Omega}$ 为地球自转角速度,则有:

$$\boldsymbol{\zeta}_a = \nabla \times \boldsymbol{V} + 2\boldsymbol{\Omega} \tag{13}$$

对三维涡度方程(12)求标量积($\boldsymbol{k} \cdot$),便得到经典的垂直涡度倾向方程:

$$\frac{\mathrm{D}}{\mathrm{D}t}(f + \zeta_z) + (f + \zeta_z)\nabla_h \cdot \boldsymbol{V} = N_z + \left(\frac{\partial u}{\partial z}\frac{\partial w}{\partial y} - \frac{\partial v}{\partial z}\frac{\partial w}{\partial x}\right) + \boldsymbol{k} \cdot \boldsymbol{F}_\zeta \tag{14}$$

式中, $N_z = \boldsymbol{k} \cdot \nabla p \times \nabla \alpha$ 为水平力管数,右端第二项为倾斜项。上述各式表明,单位质量的动量 $\boldsymbol{V}$、三维绝对涡度 $\boldsymbol{\zeta}_a$ 和垂直涡度 ζ_z 均不是守恒量。

考虑了潜热释放及其他形式非绝热加热 Q_d 的热力方程可表示为:

$$c_p \frac{T}{\theta} \frac{\mathrm{D}\theta}{\mathrm{D}t} = -L\frac{\mathrm{D}q}{\mathrm{D}t} + Q_d \tag{15}$$

式中,符号均为气象常用。由相当位温定义:

$$\theta_e = \theta \exp\left(\frac{Lq}{c_p T}\right) \tag{16}$$

代入式(15),并略去高阶小量,得:

$$\frac{\mathrm{D}\theta_e}{\mathrm{D}t} = \left(\frac{\partial}{\partial t} + \mathbf{V}\cdot\nabla\right)\theta_e = \frac{\theta_e}{c_p T}Q_d \equiv Q \tag{17}$$

用 $\nabla\theta_e$ 点乘式(12),并利用矢量关系:

$$\nabla\theta_e \cdot \nabla\times(\mathbf{V}\times\boldsymbol{\zeta}_a) = -\nabla\cdot[\nabla\theta_e\times(\mathbf{V}\times\boldsymbol{\zeta}_a)]$$

$$\nabla\theta_e \times(\mathbf{V}\times\boldsymbol{\zeta}_a) = \mathbf{V}(\boldsymbol{\zeta}_a\cdot\nabla\theta_e) + \boldsymbol{\zeta}_a\left(\frac{\partial\theta_e}{\partial t} - Q\right)$$

得:

$$\left(\frac{\partial}{\partial t}+\mathbf{V}\cdot\nabla\right)(\boldsymbol{\zeta}_a\cdot\nabla\theta_e) + (\boldsymbol{\zeta}_a\cdot\nabla\theta_e)\nabla\cdot\mathbf{V} = (\nabla p\times\nabla\alpha)\cdot\nabla\theta_e + \nabla\theta_e\cdot\mathbf{F}_\zeta + \boldsymbol{\zeta}_a\cdot\nabla Q \tag{18}$$

其次,用比容 α 乘上式两边并利用连续方程:

$$\frac{\mathrm{D}\alpha}{\mathrm{D}t} - \alpha\nabla\cdot\mathbf{V} = 0 \tag{19}$$

便可得方程:

$$\frac{\mathrm{D}P_m}{\mathrm{D}t} = \alpha(\nabla p\times\nabla\alpha)\cdot\nabla\theta_e + \alpha\nabla\theta_e\cdot\mathbf{F}_\zeta + \alpha\boldsymbol{\zeta}_a\cdot\nabla Q \tag{20}$$

式中,

$$P_m = \alpha\boldsymbol{\zeta}_a\cdot\nabla\theta_e \tag{21}$$

为湿空气位势涡度,或简称湿位涡(MPV)。它等于单位质量气块的绝对涡度在 $\nabla\theta_e$ 方向的投影与 $|\nabla\theta_e|$ 的积。式(20)则为精确形式的湿位涡方程。对干空气而言,$q=0$。这时式(20)蜕化为干空气的位涡方程,而湿位涡式(21)则蜕化为 Ertel 位涡:

$$\frac{\mathrm{D}P}{\mathrm{D}t} = \alpha\nabla\theta\cdot\mathbf{F}_\zeta + \alpha\boldsymbol{\zeta}_a\cdot\nabla Q \tag{22}$$

和

$$P = \alpha\boldsymbol{\zeta}_a\cdot\nabla\theta \tag{23}$$

这是因为位温是多元大气的变量,$\theta = \theta(\alpha,\beta)$,因此

$$(\nabla p\times\nabla\alpha)\cdot\nabla\theta \equiv 0$$

由式(22)知,在绝热无摩擦大气中 Ertel 位涡 p 是一个守恒量。

下面研究湿位涡的守恒特性。无摩擦、湿绝热大气满足:

$$\mathbf{F}_\zeta = 0, Q = 0$$

这时湿位涡方程(20)成为:

$$\frac{\mathrm{d}P_m}{\mathrm{d}t} = \alpha(\nabla p\times\nabla\alpha)\cdot\nabla\theta_e \qquad (\mathbf{F}_\zeta = 0, Q = 0) \tag{24}$$

由于压容力管项($\nabla p\times\nabla\alpha$)对单位横截面的积分等于气压梯度力沿其周界所做的功,根据式(13),该做功量引起横截面上绝对涡度的变化。因此,式(24)的物理意义就易于理解:湿位涡的个别变化等于单位质量气块因气压梯度力做功所引起的绝对涡度个别变化在 $\nabla\theta_e$ 方向上的投影与 $|\nabla\theta_e|$ 的积。尽管式(24)右端在形式上与 Bennetts 和 Hoskins(1979)在 Bousinesq 近似下所导得的形式略有差别,但易于证明,它们在本义上是一致的。

设状态为 (p_0, T_0, q) 的未饱和空气受扰动沿干绝热上升至抬升凝结高度(LCL)达到饱和态 (p, T, q)。此后($p \leqslant p_c$)上升气块便处于饱和状态,因此有:

$$\theta_e(p_0, T_0, q) = \theta_e[p, T, q(T)] = \theta_e(p, \alpha) \qquad \text{当 } p \leqslant p_c \tag{25}$$

这就是说,当 $p \leqslant p_c$ 时,等 θ_e 面与等 p 面和等 α 面的相交轴平行,于是 $\alpha(\nabla p\times\nabla\alpha)\cdot\nabla\theta_e \equiv 0$。因此,对饱和空气(这时 $p = p_c, T = T_d$)或未饱和空气受抬升至凝结高度以上,有:

$$\frac{\mathrm{d}P_m}{\mathrm{d}t} \equiv 0 \qquad (p \leqslant p_c) \tag{26}$$

或

$$P_m = \alpha \boldsymbol{\zeta}_a \cdot \nabla \theta_e \equiv 常数 \quad (p \leqslant p_c) \tag{27}$$

换言之，在绝热无摩擦的饱和空气中，湿位涡是守恒的。这时

$$\theta_e = \theta(p_c, T_c) \exp\left(\frac{Lq}{c_p T_c}\right)$$

另由 θ 的守恒性知 $\theta(p_c, T_c) = \theta(p_0, T_0)$，所以有

$$\theta_e = \theta(p_0, T_0) \exp\left(\frac{Lq}{c_p T_c}\right) \tag{28}$$

式中，T_c 的计算较为繁琐。由于在温度一对数压力图上等 q 线的斜率很大，T_c 与露点温度 T_d 接近。因此在分析中可近似地用 T_d 代替 T_c，也就是说可以由马格拉斯公式从 q 和 p 直接计算 T_c。

简析位涡方程(22)、(20)或(26)发现，位涡方程具有如下的显著优点：

a. 它是由支配大气运动的 7 个基本方程综合得到的，具有动力和热力的内部一致性。质块的位涡 P 或 P_m，以及导致位涡变化的外摩擦和加热效应，均与大气的动力和热力结构有关。因此，位涡的变化是气块所受的动力和热力作用的综合结果。

b. 位涡 P 是一个守恒量。对于饱和湿空气，P_m 也是一个守恒量。其变化方程具有拉格朗日的特征，能简化所讨论的物理过程。

因此，在下面关于副高动力学的研究中，我们将采用(22)式及其导得的方程为基本出发方程。

3 热力适应和过流

为了认识位涡及位涡方程在研究副高中的可能作用，这里首先分析位涡的特征，及其与大气运动的涡度、环流的关系。然后利用位涡方程分析大气运动如何向外界强迫适应，即热力适应问题。在此基础上认识热带、副热带加热在副高形成中的可能作用。

3.1 位涡、涡度和环流

根据位涡的定义(23)，为简化讨论，仿 Hoskins(1991)采用 θ 坐标，由于 $\nabla_\theta \theta \equiv 0$，有：

$$P = \frac{1}{\rho} \boldsymbol{\zeta}_a \cdot \boldsymbol{n} \frac{\partial \theta}{\partial z} \tag{29}$$

式中，$\boldsymbol{n}$ 为沿 θ 递增方向的单位矢量，Δz 为两 θ 面之间的厚度。在 θ 面上的单位面积元为 $ds = dxdy$，对 (29)式实施面积分可得：

$$\frac{1}{S} \int P dm = \int_{\theta_1}^{\theta_2} \zeta_{an} d\theta \tag{30}$$

式中，S 为 θ 面上的积分面积，ζ_{an} 为 S 面上沿 $\boldsymbol{n}$ 方向的平均绝对涡度。式(30)表明单位面积上以质量为权重的位涡积分等于以 θ 为权重的绝对涡度的积分。由于式(30)是从一般定义式(23)出发，与是否存在加热或/和摩擦等外过程无关。因此具有普适性，是质块运动过程必须遵循的基本关系。由于质块运动中质量 $dm = \rho dxdydz$ 守恒。因此式(30)的物理含义可解释如下：

a. 在绝热无摩擦场合，θ 和 P 守恒。因此绝对涡度的改变必须伴有面积 S 的改变。S 缩小，绝对涡度增加；S 增加，绝对涡度减小。

b. 对截面积 S 不变的流柱而言，以质量为权重的位涡的增加(减少)必伴有以 θ 为权重的绝对涡度的增加(减少)。

根据斯托克斯(Stokes)定理，面积 S 法线方向上涡度的面积分等于沿其边界 Γ 上环流的积分，即：

$$\int_S \boldsymbol{\zeta}_a \cdot \boldsymbol{n} ds = \oint_\Gamma \boldsymbol{V}_r \cdot d\boldsymbol{l} = C_a \tag{31}$$

式中，$\boldsymbol{V}_r$ 为沿边界 Γ 的速度，C_a 为沿 Γ 的绝对环流。则(29)式在 S 上的积分还可写为：

$$\int P dm = \int_{\theta_1}^{\theta_2} C_a d\theta \tag{32}$$

于是得到与 Hoskins(1991)一致的绝对环流表达式。式(32)与(30)是等价的。它指出,以质量为权重的位涡之和等于以 θ 为权重的绝对环流。

考虑静力稳定的静止大气中有三个水平分布的等熵面,在区域 S 中,非绝热加热 Q 在下层($\theta_1-\theta_2$)中由零单调增加到 Q_0,在上层($\theta_2-\theta_3$)中由 Q_0 递减至零(见图1a)。设 θ_1 面即为地面。这时在气柱中将发生如下两个过程:

a. 热力过程:θ_2 面上的加热使该处的 θ 面由 A 点降至 A_1 点(图1b)。于是由定义(29),静力稳定度和位涡 P 在下层增加,上层减少。同时低层变薄,质量减少;高层增厚,质量增加。在流场不变的情形,质量的变化抵偿了 P 的变化,使式(30)成立。

b. 动力过程:由于 Burger 数很大,由 $N^2W \approx Q$(见第1节中讨论),质点 A 受热上升。于是底层辐合,上层辐散(见图1c)。在纯动力过程中,根据 θ 坐标下的连续方程 $\frac{\partial}{\partial t}\left(\frac{\partial p}{\partial \theta}\right) = -\nabla_\theta \cdot \left[\mathbf{V}\left(\frac{\partial p}{\partial \theta}\right)\right]$,$p$ 是气压,底层辐合流使质量增加,高空辐散流使质量减少。这时(30)式左端 P 和 $\mathrm{d}m$ 在低层增大,高层减少。于是由式(30)右端知,低层激发出气旋式涡度,高层出现反气旋涡度增长,流场因而向加热场适应。这里的讨论和结论与 Hoskins(1991)从环流关系进行的讨论和结论相一致,只是在讨论动力过程向热力过程适应时用了更严格的证明。

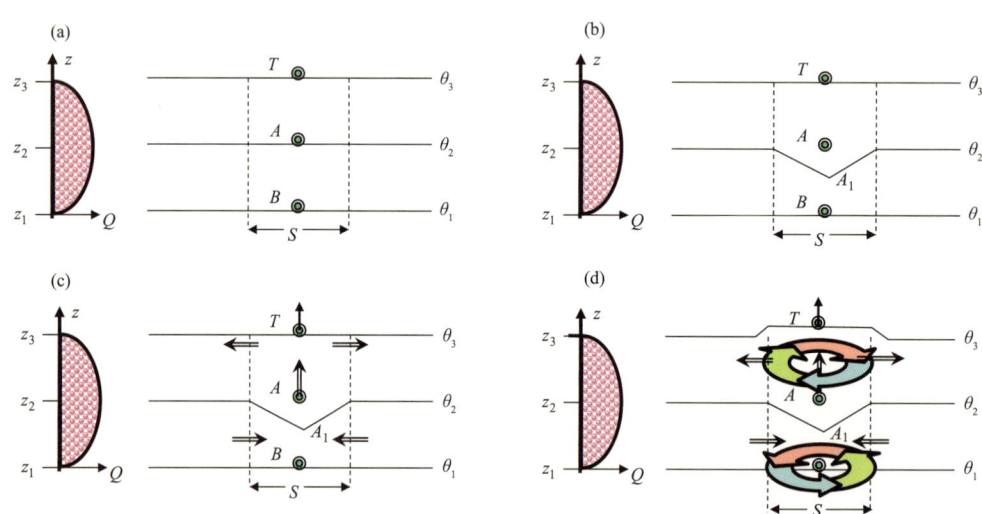

图1 大气环流(箭矢)对外部加热 Q 的热力适应发展过程示意图。图中左侧表示区域 S 中 Q 随高度(等熵面)的分布;(a)起始态,T、A 和 B 表示起始位于 θ_3、θ_2 和 θ_1 上的质点;(b)加热的热力效率使气柱 θ 位温升高,θ_2 面下凹;位涡 P 在下层增加,上层减少;(c)加热使质点 A 上升,其动力效应使下层空气辐合,上层辐散,导致质量 m 在下层增加,上层减小;(d)在(b)和(c)的共同作用下,根据关系式(30),下层出现气旋环流,高层出现反气旋环流

3.2 大气中的热力适应

大气运动在时间和空间上都具有多尺度特征。对于绝热无摩擦大气,不同时空尺度运动学研究中的一个共同课题是流场和气压场之间的互相调整。自从 Rossby(1937a,1937b)和 Oboukhov(1949)相继提出这一"地转适应"问题以来。我国学者对此进行了系统的研究。叶笃正(Yeh T C,1957)首先证明对于大尺度大气运动,流场向气压场适应;对于小尺度运动则是气压场向流场适应。曾庆存(1963a)证明,判定上述不同尺度适应特征的临界尺度即为 Rossby 变形半径。曾庆存(1963b,1963c)还进一步证明大气运动在时间尺度上也可分为快的适应过程和慢的演变过程。叶笃正和李麦村(1965,1980)指出,各种尺度大气运动的发展都可分为三个阶段:首先是非常迅速的适应阶段,它使不平衡的运动迅速恢复平衡;第二阶段为发展阶段,大气运动中观测到的大的变化都发生在这一阶段;第三阶段为准平衡阶段,一个慢变的过程。巢纪平(1980)对于上述的多时间尺度特征给出了简单的物理解释。叶笃正和巢纪平(1998)定义了

下述四个时间尺度：
$$T_1 = L/C, T_2 = f_0^{-1}, T_3 = (\beta L)^{-1}, T_4 = LV^{-1}$$
式中，L 为准地转平衡附近的特征空间尺度，V 为背景流场速度，$C = (gH)^{1/2}$ 为重力波波速。他们证明当运动的水平特征尺度接近 Rossby 变形半径时，有：
$$T_1 \approx T_2 < T_3 < T_4$$
即大尺度动力过程至少存在三个特征时间尺度。

在具有外热源和耗散的情形，则大气的适应问题不再仅是一个运动学问题，而且是一个动力学问题。由于问题的斜压特征，必须引进热力学方程予以研究。巢纪平和许有丰(1961)应用一个两层线性模式以研究热源作用下高度场的中长期变化，计算表明，高度场对加热场的适应极其迅速，至第 10 天左右，高度场和加热场之间的相对形势已趋于准定常。张可苏(1980)利用一个不包含平流过程的高度线性化模型，以研究有热源和耗散情况下的大气适应过程，结果表明，大气运动同时具有向"外部热源"和向"地转运动"适应的性质。在定常"外源"作用下系统的垂直结构完全由热源分布决定；热源强度向上减弱有利于生成热高压，向上增强有利于生成热低压。

严格地说，在外热源强迫下的大气运动中，热量及涡度输送在大气准平衡状态的重建中是非常重要的，而且是高度非线性的。因此，必须利用非线性模型对适应过程中大气的动力、热量状况进行研究。丑纪范(1995a,1995b)以及李建平和丑纪范(1998)利用非线性理论研究了大气系统的全局行为，指出，当外源固定不变或变化非常缓慢时，系统存在两种特征时间尺度：向吸引子适应的快过程和在吸引子上演变的慢过程。当外源变化时则存在第三种时间尺度，即宏观状态随外参数变化而演变的更为缓慢的过程。如第 2 节中所讨论的，Ertel 位涡 $P(=\alpha\boldsymbol{\zeta}_a \cdot \nabla\theta)$ 综合了气块的动力和热力特征，且 P 的变化方程具有拉格朗日方程特征，用于进行适应过程研究利于揭示内在的物理机制。因此，可利用位涡方程以分析大气中的热力适应过程，分析适应过程中的适应阶段和发展阶段的动力学问题（吴国雄和刘屹岷，2000）。在后面的章节中，我们还将以热力适应理论分析准平衡阶段的频散过程和副热带对流凝结强迫在副高形成中的作用。从而利用热力适应理论把热带、副热带大气运动的发展、频散及副高的形成作为一个整体联系起来。

下面我们首先导出总位涡方程，并用其分别讨论加热中心下层和上层的热力适应问题，证明斜压位涡及对称不稳定在维持位涡准平衡中的重要性。接着提出"过流"(overshooting)的概念，证明其在热力适应中的必然性和重要作用。然后利用数值试验分析赤道深对流凝结潜热加热和近地面感热加热中的热力适应特征。

利用连续方程：
$$\frac{\partial \rho}{\partial t} + \nabla \cdot \boldsymbol{V}\rho = 0$$

可以把平流形式的全微分算子：
$$\frac{DA}{Dt} = \left(\frac{\partial}{\partial t} + \boldsymbol{V} \cdot \nabla\right)A$$

改写为通量形式的全微分算子：
$$\frac{d}{dt}(\rho A) = \left(\frac{\partial}{\partial t} + \nabla \cdot \boldsymbol{V}\right)\rho A = \rho \frac{DA}{Dt} \tag{33}$$

则位涡方程(22)可改写成：
$$\frac{dW}{dt} = \boldsymbol{F}_\zeta \cdot \nabla\theta + \boldsymbol{\zeta}_a \cdot \nabla Q \tag{34}$$

式中，
$$W = \alpha P = \boldsymbol{\zeta}_a \cdot \nabla\theta \tag{35}$$

为气块的总位涡。把 W 分解为垂直(W_v)和水平(W_h)两部分：
$$\begin{cases} W = W_v + W_h \\ W_v = (f + \zeta_z)\theta_z \\ W_h = \nabla \times \boldsymbol{V} \cdot \nabla_h \theta \end{cases} \tag{36}$$

可见 W_v 为天气分析中常用的位涡,它是惯性稳定度参数 $(f+\zeta_z)$ 和静力稳定度参数的乘积,可定义为 W 的正压分量。而 W_h 是与风的垂直切变及水平位温梯度相联系的,在地转关系成立时,W_h 正比于 $(-|\nabla_h \theta|^2 < 0)$(Wu and Liu, 1998),可定义为 W 的斜压分量。

取大气大尺度运动的特征值为:

$$\begin{cases} \Delta z \sim 10^3 \text{ m}, \Delta x, \Delta y \sim 10^6 \text{ m}, \Delta \theta \sim 10 \text{ K}, \Delta V \sim 10 \text{ m} \cdot \text{s}^{-1}, \\ w \sim 10^{-3} \text{ m} \cdot \text{s}^{-1}, \Delta F \sim 10^2 \text{ W} \cdot \text{m}^{-2} \end{cases} \quad (37)$$

则由热通量 F 可估得加热率:

$$Q = -\frac{\theta}{c_p T \rho} \frac{\partial F}{\partial z} \sim 10^{-4} \text{ K} \cdot \text{s}^{-1}$$

由此知:

$$W = W_v + W_h = (f+\zeta)\frac{\partial \theta}{\partial z} + \nabla \times \boldsymbol{V} \cdot \nabla_h \theta \sim 10^{-6} \text{ K} \cdot \text{s}^{-1} \cdot \text{m}^{-1}(1+10^{-1}) \quad (38)$$

如定义 TU 为一总位涡单位:

$$1 \text{ TU} = 10^{-6} \text{ K} \cdot \text{s}^{-1} \cdot \text{m}^{-1}$$

则

$$W = W_v + W_h \sim \text{TU} + 10^{-1} \text{ TU}$$

也就是说在"一般场合",当 θ 面的水平倾斜很小(10^{-3})时,W 的"正压"分量 W_v 要比"斜压"分量 W_h 大一个量级。在"特殊场合",当 θ 面的水平倾角很大($10^{-2} \sim 10^{-1}$)时,W 的斜压分量将与正压分量相当甚或更大。

(1)下层大气中的热力适应

把式(34)左端展开,可估得各项尺度如下:

$$\underset{(a)}{\frac{\partial W}{\partial t}} + \underset{(b)}{\nabla_h \cdot \boldsymbol{V} W} + \underset{(c)}{\frac{\partial}{\partial z} w W} = \underset{(d)}{\boldsymbol{F}_\zeta \cdot \nabla \theta} + \underset{(e)}{\boldsymbol{\zeta}_a \cdot \nabla Q} \quad (39)$$

$$\qquad 10^{-11} \qquad 10^{-12} \qquad 10^{-11} \qquad 10^{-11}$$

也就是说,在一般场合总位涡的水平通量辐散要比其垂直通量辐散大约一个量级,或说总位涡的垂直通量辐散不足以平衡非绝热的位涡制造。在图1的 $(z_1 \sim z_2)$ 层中 $\frac{\partial Q}{\partial z} > 0, \boldsymbol{\zeta}_a \cdot \nabla Q > 0$,根据式(39),在时间尺度小于1天时,局地位涡将增长 $\left(\frac{\partial W}{\partial t} > 0\right)$,出现低层气旋环流(图1a)。当时间尺度很大,即准定常场合,$\frac{\partial W}{\partial t} \to 0$,这时在区域 S 的内点上,式(39)中(b)、(c)两项分别为(−)和(+),均不能平衡加热的位涡制造。因此,在边界层上,必须有:

$$\boldsymbol{F}_\zeta \cdot \nabla \theta \cong -\boldsymbol{\zeta}_a \cdot \nabla Q$$

即:

$$F_z \cong -(f+\zeta_a)\frac{\partial Q}{\partial z}\bigg/\frac{\partial \theta}{\partial z} < 0 \quad (40)$$

换言之,$\frac{\partial Q}{\partial z} > 0$ 所制造的位涡必为摩擦项所平衡。因之加热随高度的增加导致表面产生气旋环流和负的摩擦力矩。

由式(40)估得的摩擦耗散率高达 10^{-9} s^{-1}。换言之,边界层内 $100 \text{ W} \cdot \text{m}^{-2}$ 的加热在2~3小时内即可产生强达 10^{-5} s^{-1} 的表面气旋环流。然而这一表面气旋不能无限增长,这是因为一方面在区域 S 中的加热使 θ 面在 S 中下凹,因而沿边界 θ 面变得很倾斜,$\nabla_h \theta$ 指向 S 区域内部;另一方面,表面气旋环流的出现形成风的垂直切变,由此产生的水平涡度沿边界指向 S 区域的外部,位涡平衡的性质于是将改变。此时沿边界:

$$W_h = \nabla \times \boldsymbol{V} \cdot \nabla_h \theta < 0 \tag{41}$$

如取式(37)中的$(\Delta x, \Delta y)$为$10^4 \sim 10^5$ m,则有：

$$W_v \sim 10^0 \text{ TU}, W_h \sim (10^0 - 10^1) \text{ TU}$$

因之有：

$$|W_h| > |W_v| \tag{42}$$

及：

$$W = W_v + W_h < 0 \tag{43}$$

即沿S的边界Γ出现了对称不稳定结构。由式(41)—(43)知,式(39)左端的大项为正值,位涡的平衡性质成为：

$$\nabla_h \cdot \boldsymbol{V}W \approx \boldsymbol{F}_\xi \cdot \nabla\theta + \boldsymbol{\zeta}_a \cdot \nabla Q$$
$$\text{(b)} > 0 \quad \text{(d)} < 0 \quad \text{(e)} > 0 \tag{44}$$

在区域S上从z_1到z_2对式(39)积分,并利用高斯定理：

$$\int_S \nabla_h \cdot (\boldsymbol{V}A) \mathrm{d}s = \oint_\Gamma (A\boldsymbol{V}) \cdot \boldsymbol{n} \mathrm{d}l \tag{45}$$

可得：

$$\int_{z_1}^{z_2} \mathrm{d}z \oint_\Gamma W\boldsymbol{V} \cdot \boldsymbol{n} \mathrm{d}l + \int_S W_2 w_2 \mathrm{d}s = \int_{z_1}^{z_2} \int_S \left[F_\xi \theta_z + (f + \zeta_z)\frac{\partial Q}{\partial z}\right] \mathrm{d}s \mathrm{d}z$$
$$\text{(b)} > 0 \quad\quad \text{(c)} > 0 \quad\quad \text{(d)} < 0 \quad\quad \text{(e)} > 0 \tag{46}$$

式中,Γ为区域S的边界线,反时向为正,$\boldsymbol{n}$为Γ外法向上的单位矢量。由于低层辐合把周围W小的流体源源不断地输入区域S中($\boldsymbol{V}_\Gamma \cdot \boldsymbol{n} < 0$);而$z_2$界面上的上升运动又把下层制造的正涡度向高层输送,再根据式(43),可知式(39)中(b)—(e)各项的符号为(+)、(+)、(−)、(+)。这就是说,低层加热所制造的正涡度除了被摩擦耗散外,还通过上界涡度的"逸散"及侧边界交换的"稀释"作用而保持平衡。通过式(41)—(46)的讨论表明,正是由于加热使沿边界θ面的倾斜及斜压位涡W_h的加强,导致位涡的稀释,才使得加热区域中位涡的平衡得以维持。

(2) 自由大气中的热力适应

在$z_2 \sim z_3$层,$\frac{\partial Q}{\partial z} < 0$(图1),有负涡度制造及反气旋式环流形成。在该层中摩擦项可略。在定常状态根据式(39)的尺度分析,负涡度的制造必须被水平总位涡W的辐散项所平衡。由于该层中$\nabla_h \theta$仍指向S区域内部,而上层反气旋环流所产生的垂直风切变——即水平涡度指向S区域外部,因此式(41)—(43)中关于正压和斜压总位涡沿边界分布的分析仍然成立。换言之,沿上层的边界Γ也出现对称不稳定结构,其位涡平衡的性质成为：

$$\nabla_h \cdot \boldsymbol{V}W \approx \boldsymbol{\zeta}_a \cdot \nabla Q$$
$$\text{(b)} < 0 \quad \text{(e)} < 0 \tag{47}$$

在区域S上从$z_2 \sim z_3$对(39)式积分可得：

$$\int_{z_2}^{z_3} \mathrm{d}z \oint_\Gamma W\boldsymbol{V} \cdot \boldsymbol{n} \mathrm{d}l - \int_S W_2 w_2 \mathrm{d}s = \int_{z_2}^{z_3} \int_S (f + \zeta_z)\frac{\partial Q}{\partial z} \mathrm{d}s \mathrm{d}z$$
$$\text{(b)} < 0 \quad\quad \text{(c)} < 0 \quad\quad \text{(e)} < 0 \tag{48}$$

因此,上层加热所制造的负位涡度一方面被底层上传的正位涡所"稀释",但更主要的,必为沿边界Γ的负位涡的"散逸"所平衡。因此负的斜压位涡沿边界向外"散逸"是保持位涡平衡的主渠道。

(3) 整层大气中的热力适应

如果在S上从$z_1 \sim z_3$对式(39)积分,则定常的位涡平衡关系可以由式(46)和(48)之和予以分析。这时通过界面θ_2的位涡输送项(c)消失;在求垂直积分时加热项中的大项$\left(f\frac{\partial Q}{\partial z}\right)$为零,唯余小项$\zeta_z \frac{\partial Q}{\partial z}$的积分。换言之,在对式(39)从$z_1 \sim z_3$的积分中,(b)、(d)为大项。因此,表面摩擦制造的负位涡(式(46)中的

d 项<0)必须靠上层大气中负位涡的"逸散"(式(48)中 b 项<0)才能建立平衡。这时,沿着气柱上层的侧边界负位涡的逸散(式(48)b 项)必须大于下层侧边界正涡度的"稀释"。

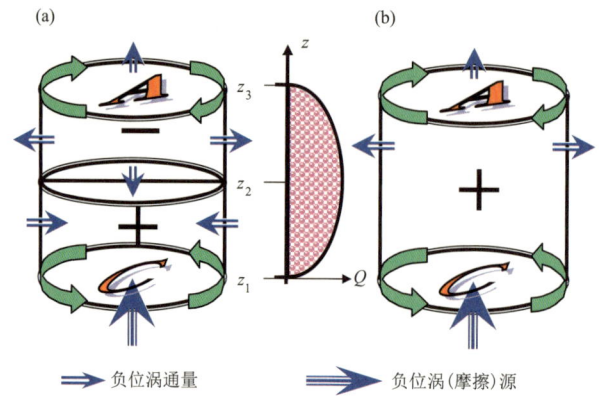

图 2　在大气热力适应过程中总位涡 W 收支示意图。(+)、(−)表示总位涡的源和汇。箭矢表示总位涡负通量的方向。(a)在分层模式中,下层加热所致的总位涡源被摩擦汇及边界的负通量平衡;上层加热所致的总位涡汇被侧边界的向外逸散、通过过流向上层以上自由大气的输出,及通过下边界向上的正位涡通量所平衡;(b)在整层气柱模式中,底边界摩擦所产生的负 W 源除了抵消加热产生的正位涡 $(\zeta_z \partial Q/\partial z)$ 外,主要通过侧边界的外逸及上边界过流的上传而建立平衡

综上所述,我们可以把热力适应过程中位涡的平衡机制用图 2 示意如下:在分层模型中(图 2a),在加热随高度增加的低层,正的位涡制造被地面负的摩擦涡源、侧边界的负位涡输入及上边界正位涡的输出所平衡;在加热随高度减少的上层,负的位涡制造被下边界正位涡的输入及侧边界负位涡的输出所平衡。在气柱垂直积分模型中(图 2b),摩擦施加于气柱的负位涡除了被气柱中正位涡的制造 $\left(\int_{z_1}^{z_3} \zeta_z \frac{\partial Q}{\partial z} \mathrm{d}z\right)$ 所抵消外,所余的负位涡在高层沿侧边界向外输送。

3.3　热力适应中的"过流"问题

在 θ_3 层上非绝热加热为零。在没有水平平流作用的情形,由热力方程推知,垂直运动应为零。我们这里定义的"过流"或说"上冲",就是指 θ_3 层及以上层次仍然存在上升运动及辐散(见图 1、2)。这里我们首先证明"过流"是非绝热加热的必然伴随现象。

在图 1 中,取 $\theta > \theta_3$ 的任意一层 θ_4,这时由于在大气中存在扩散过程,在 $\theta_2 \sim \theta_4$ 层中,$\frac{\partial Q}{\partial z} < 0$ 有负涡度制造。因此,图 1 以及上一小节关于 $z_2 \sim z_3$ 层的自由大气中热力适应的讨论也适于 $\theta_2 \sim \theta_4$ 层中。于是在 $\theta_3 \sim \theta_4$ 层也存在反气旋环流以及上升和水平辐散。由此可知,θ_3 层上的过流是必然现象。还由于 θ_4 是任取的,因此过流现象可以出现在较厚的层次中。在真实大气和模式大气中,由于存在扩散,在加热区上方也存在 $\frac{\partial Q}{\partial z} < 0$ 的深厚层次。

过流首先增加了 z_2 层以上反气旋涡度向外散逸($\int \mathrm{d}z \oint W \mathbf{V} \cdot \mathbf{n} \mathrm{d}l < 0$)的垂直厚度,更多地补偿了摩擦制造的负涡度。

其次,过流的绝热冷却作用使 θ_3 面上凸(图 1d),于是使 $z_1 \sim z_3$ 层中 θ_z 减少。从位涡的观点看,整层气柱中静力稳定度的减少使气柱中正压位涡 W_v 减少,沿边界的对称不稳定于是更迅速出现,从而加速了高低层位涡平衡的建立(见(46)、(47)两式)。

除此之外,加热区大气在下层的辐合、上升及在上层的辐散还在周边地区引起上层辐合、下沉、及下层的辐散。大气加热通过这一"补偿效应"影响着邻域的大气运动。

过流是大气中的重要过程。一个明显的例子是赤道和热带的大气结构,那里的对流云一般可伸展到

12 km 左右。而对流层顶却高达 16 km(约为 390 K)。从 200 hPa 向上，θ 面向上凸起，在对流层上部及平流层下部形成比中纬度还要冷的低温中心。而在 200 hPa 以下，θ 面下凹显著。热带的这一热力适应及过流特征对全球气候的形成有十分重要的作用。在下一章中我们将看到，在过流的作用下副热带地区地表面的感热加热能够导致对流层上层副热带高压的生成。

4 热力适应的数值模拟

为了考察大气流场对外热源的适应过程，我们利用气候系统模式 LASG/IAP GOALS(吴国雄 等，1997)进行数值模拟。该模式的大气部分为 9 层 15 波菱形截断的全球大气环流模式(Wu et al.，1996)，垂直向采用 σ-坐标。海洋部分为 20 层全球海洋环流模式(Zhang et al.，1996)。陆地模式为 SSiB 模式(Liu and Wu，1997;Xue et al.，1991)。辐射过程采用 Shi(1981)的 K-分布方案。在试验中，设定全球为海洋；海洋模式关闭，以给定的 7 月纬向平均海表温度分布作为大气的下边界。太阳高度角固定为 7 月 15 日的值。其他所有外强迫(包括 CO_2、气溶胶、云量等)及大气状态均取 7 月纬向平均值。起始风场和水平温度梯度 $\nabla_h T$ 设定为零。为了简化讨论，以及便于与已有研究成果比较，外热源的中心在试验中被置于赤道上；加热范围 S 界定为 11°S～11°N，165°E～157°W；加热场在水平面上呈轴对称分布，从中心向边界递减。

4.1 大气流场对深对流凝结加热的适应

在 S 区域上给出恒定的深对流凝结加热，中心强度为 5 ℃·d^{-1}，1 ℃·d^{-1} 的热源区域如图 3 粗实线所示。加热的垂直廓线见图 4 中的阴影部分。除此之外，模式中其他的非绝热加热，包括感热加热及其他潜热加热等，均设为零。图 3 给出积分第 1 天和第 3 天低层 $\sigma=0.991$ 和高层 $\sigma=0.189$ 层上风场的分布，显示出典型的 Gill 型响应(Gill，1980)。沿赤道向东的 Kelvin 波以大约 30°·d^{-1} 的相速传播，关于赤道对称。向西的 Rossby 波相速为 12°·d^{-1}。这也与 Matsuno(1966)的结果类似。这种环流的响应解释了赤

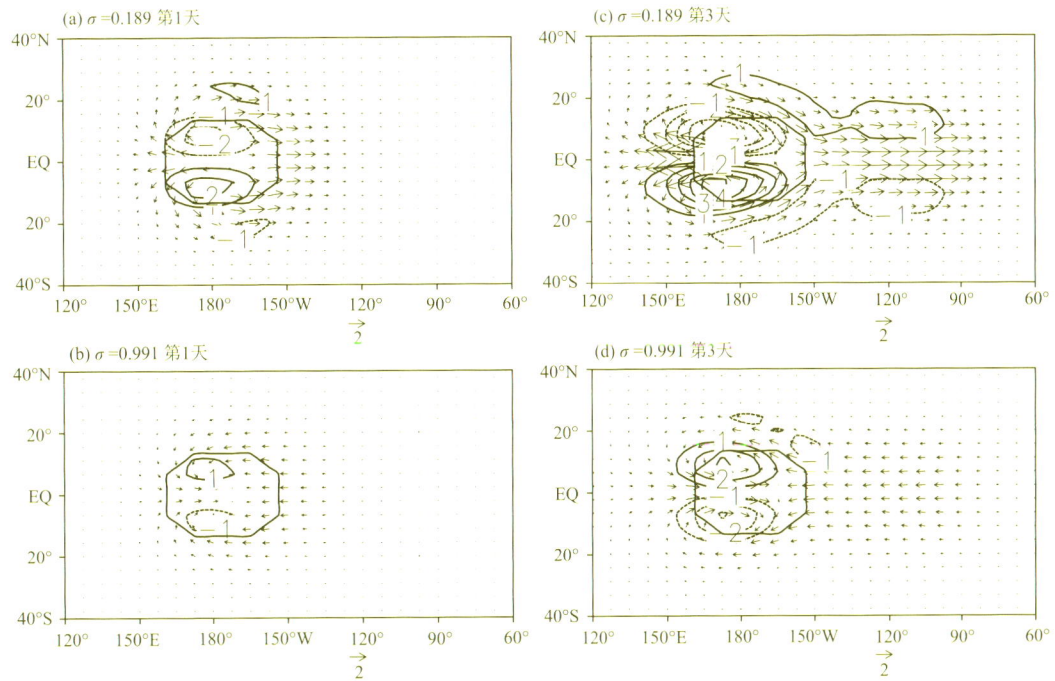

图 3 在深对流凝结加热的热力适应数值试验中第 1 天((a)、(b))和第 3 天((c)、(d))在上层 $\sigma=0.189$((a)、(c))和下层 $\sigma=0.991$((b)、(d))的流场和涡度场(单位 10^{-6} s^{-1}，实线为正，虚线为负)的分布，粗线界定最大加热层($\sigma=0.336$)上加热大于 1 K·d^{-1} 的区域

道地区对流层高、低层纬向风和经向风的相对大小,以及对称波型以行星尺度存在的趋势。图3的结果表明,该模式能合理地模拟出大气流场对外热源强迫的响应。它还表明加热在低层造成正的涡度和水平辐合,高层与之相反。其邻域出现较弱的补偿式流场。

图4是本试验中沿赤道位温和风矢量(纬向风 u 和垂直速度 ω)的垂直剖面的演变。凝结潜热加热使等θ面(如284 K等值线)下凹,气块穿过等θ面辐合上升。在最大加热层(336 hPa)以上向外辐散。近地层$Q=0$,等θ面基本保持水平。200 hPa上,出现了明显的"过流":那里有明显的上升运动、辐散及冷中心。

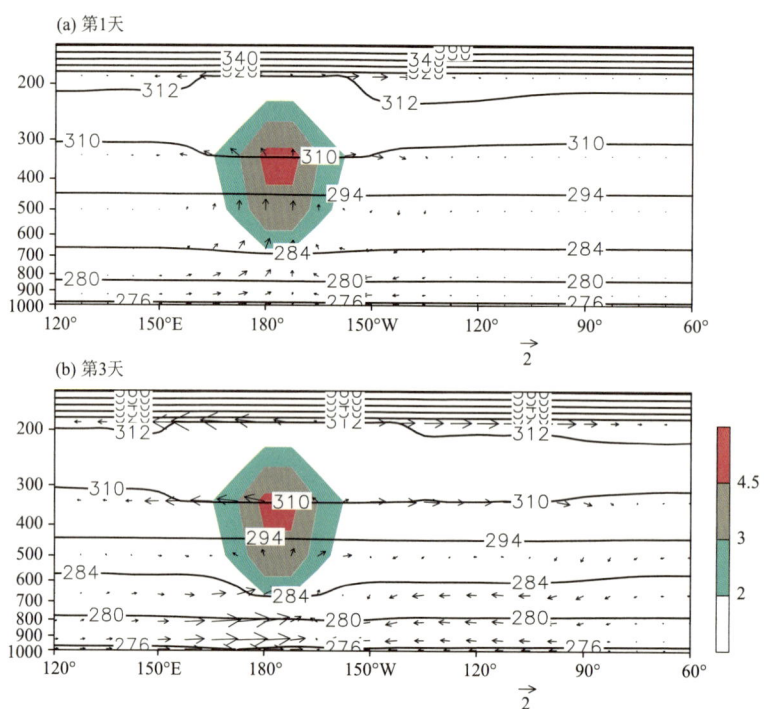

图4 在深对流凝结加热的热力适应数值试验中沿赤道的垂直剖面上等位温线(K)和流场(单位:u,m·s^{-1};ω,-40 Pa·s^{-1})的分布。阴影区表示加热强度。(a)第1天;(b)第3天

4.2 大气流场对表面感热加热的适应

将理想化的表面感热置于图5所示的赤道地区,中线位于176°W,强度是100 W·m^{-2}。1 W·m^{-2}的热源区域如图5中粗实线所示。为突出环流对感热源的响应并避免较小尺度的扰动,除设定的热源外,将模式中的凝结加热和其他表面感热去掉。

图5a和5d给出积分第1天和第3天热源中心上空温度垂直扩散项的曲线,表现了感热加热的垂直变化。最大的加热位于近地层的920 hPa附近,向上迅速减小,在800 hPa附近已接近0。这是因为感热加热是由边界层中的小尺度湍流运动所造成,因此其扩散被限制在对流层低层。

图5c和5f给出积分第1和第3天$\sigma=0.991$层上风场的响应。显示加热在低层造成正的涡度和水平辐合,热源区中西部赤道南北两侧各有一对称于赤道的气旋式环流。图5b和5e给出$\sigma=0.664$层上的响应流场。该层已是感热影响范围的上层。与低层形势相反,在热源区上方为水平辐散和对称于赤道的反气旋式环流。

图6是本试验中沿赤道位温和风矢量(纬向风 u 和垂直速度 ω)的垂直剖面的演变。加热后的第一天,热源区的等θ面即开始下凹,气块穿过等θ面辐合上升。因感热加热的垂直层次较低,在$\sigma=0.664$层上已是辐散。辐合辐散及垂直上升运动随着时间增强。

上述的低层辐合和气旋式环流,高层辐散和反气旋式环流结构与热力适应原理是完全一致的。感热加热引起θ面下降空气上升。因表面加热的垂直梯度很大,造成低层θ面与地面相切。由于赤道地区不

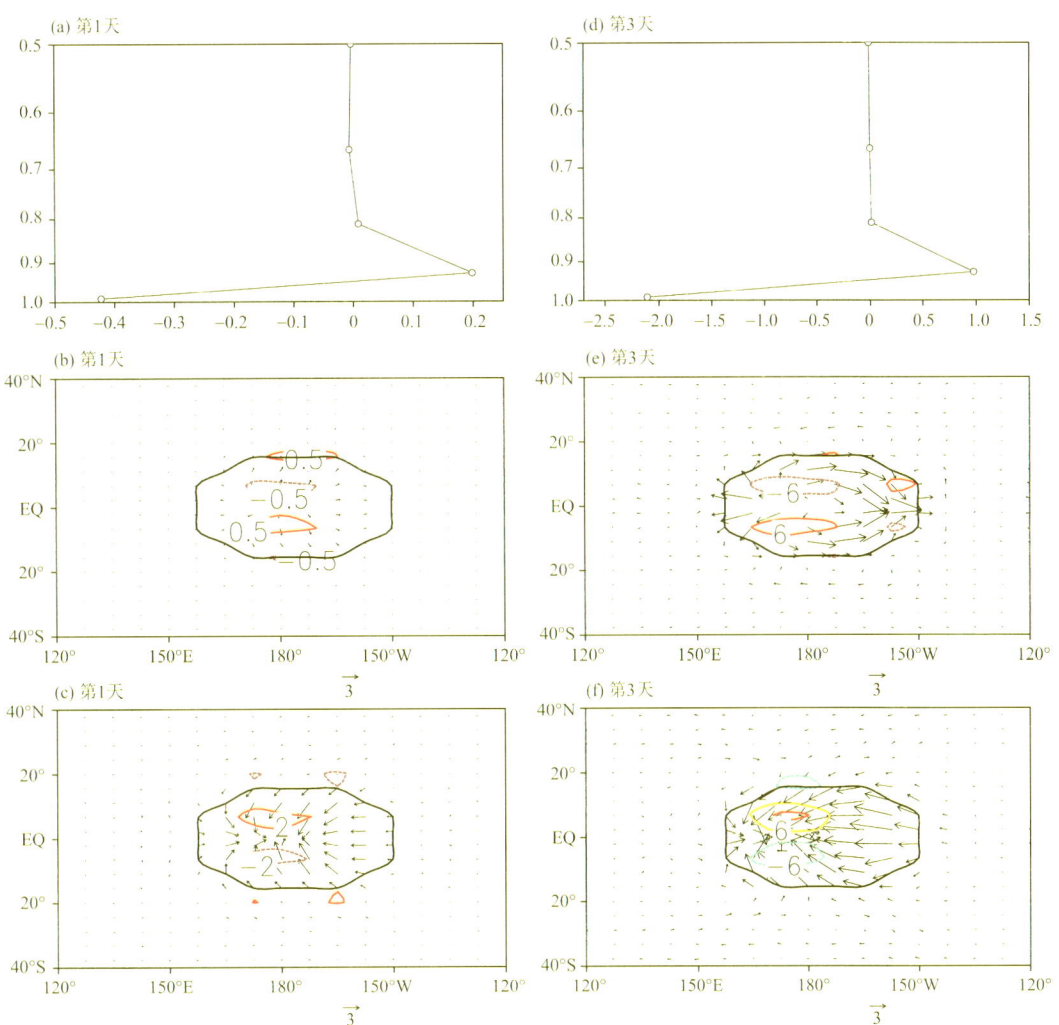

图5 在表面感热加热的热力适应数值试验中第1天((a)、(b)、(c))和第3天((d)、(e)、(f))沿赤道的加热率垂直廓线(a)和(d),单位:1 K·d^{-1},以及在上层 $\sigma=0.664$((b)、(e))和下层 $\sigma=0.991$((c)、(f))的流场和涡度场(单位:10^{-6} s^{-1},实线为正,虚线为负)的分布。粗线界定加热大于 1 W·m^{-2} 的区域

存在辐散流场,因此加热只影响周围地区。尤其值得注意的是,尽管表面感热加热源基本只出现在800 hPa 以下的边界层中,$\sigma=0.664$ 以上层次已没有感热加热,但过流作用使 500 hPa 出现冷中心及水平辐散。加热的西边界(165°E)和东边界(157°W)285 K 的 θ 面已明显倾斜。由此可见,过流是大气热力适应过程中使近地层加热能影响其上层次环流的重要机制。

在图5中,除赤道外,涡度的量级为 10^{-6} s^{-1}。根据图5和图6,可以估算总位涡 W 各分量沿加热区边界的量级如下:

$$(f+\zeta) \sim (10^{-6} \sim 10^{-5}) \text{ s}^{-1}, \quad \frac{\Delta\theta}{\Delta z} \sim 10^{-3} \sim 10^{-4} \text{ K·m}^{-1},$$

$$\left(\frac{\Delta u}{\Delta z}, \frac{\Delta v}{\Delta z}\right) \sim 10^{-2} \text{ s}^{-1}, \quad \left(\frac{\Delta\theta}{\Delta x}, \frac{\Delta\theta}{\Delta y}\right) \sim 5 \times 10^{-6} \text{ K·m}^{-1}$$

可见即使对水平尺度达 4×10^6 m、平均加热为 64 W·m^{-2}(200/π)的赤道表层加热,沿着加热区的边界其所激发的水平涡度(风的垂直切变)也比垂直涡度大3个量级以上。根据(38)式,可估算沿边界总位涡的正压分量 W_v 和斜压分量 W_h 的量级为:

$$W_v = (f+\zeta)\theta_z \sim (10^{-2} \sim 10^{-4}) \text{ TU}$$
$$W_h = \nabla \times \mathbf{V} \cdot \nabla_h \theta \sim (10^{-1} \sim 10^{-2}) \text{ TU}$$

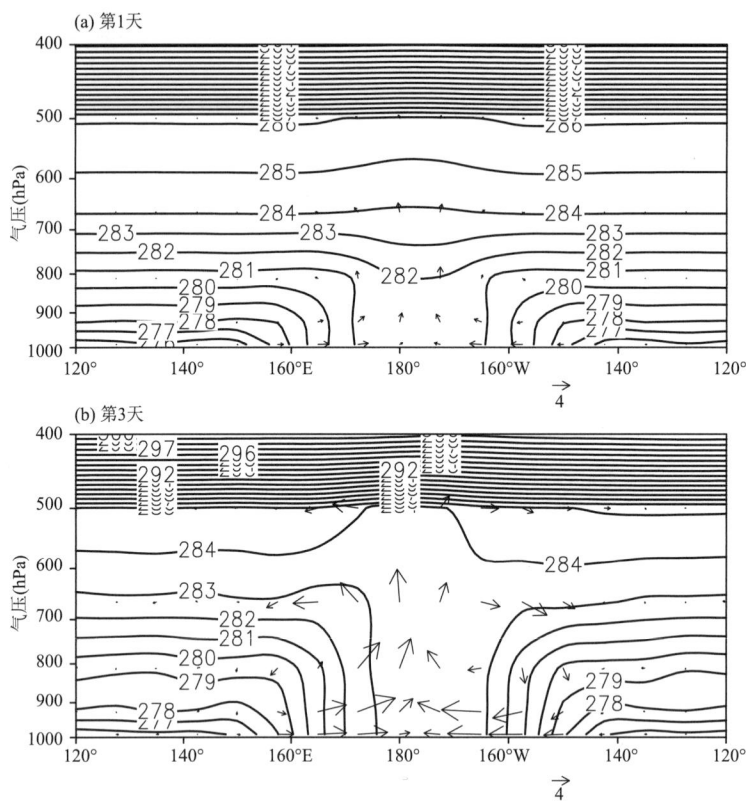

图 6 在表面感热加热的热力适应数值试验中沿赤道的垂直剖面上等位温线(K)和
流场(单位:u,m·s^{-1};ω,-40 Pa·s^{-1})的分布;(a)第 1 天;(b)第 3 天

还由于 $W_h<0$,因此式(42)和(43)成立,沿边界 Γ 出现了对称不稳定($W<0$)。因此也证明在热力适应过程中,由式(46)给出的下层大气总位涡平衡,由式(48)表示的上层大气的总位涡平衡,以及由图 2 描述的气柱中总位涡平衡的分析是正确的。此外,图 5 和图 6 还表明,向热源区低空的辐合和高空的辐散还在加热区周边引起低空辐散和高空辐合。因此,周边地区在低空有反气旋涡度形成;在高空有气旋涡度形成。另外,上升区的东西两侧有下沉运动,形成类似 Walker 环流的纬向直接热力环流圈,最强上升运动位于热源中心 176°W。这些结果表明赤道感热加热不仅影响加热区上空的环流,还在邻域强迫出补偿性环流。

在上一节及本节中,我们利用位涡定义和位涡方程讨论了大气流场向非绝热加热适应的热力适应原理。证明热力适应过程中近地面出现气旋式环流,加热层顶出现过流现象。还证明在此过程中地表摩擦施加给气柱的负涡度源必须在上层大气柱向外逸散,成为自由大气中的负涡源。当这一负涡源出现在西风带时,还将通过频散作用影响中高纬大气的环流。在后面的章节中,我们将看到,在热带及副热带的外强迫加热影响副热带高压的变异过程中,热力适应起着很大的作用。

参考文献

巢纪平,1980.关于"大气各类运动的多时间尺度特征"的讨论[C]//中央气象局气象科学研究所.第二次全国数值天气预报会议论文集.北京:科学出版社:193-195.

巢纪平,许有丰,1961.二层线性模式里长期天气过程的一些计算[C]//顾震潮,等.动力气象论文集.北京:科学出版社:90-95.

李建平,丑纪范,1998.大气动力学方程组的定性理论及其应用[J].大气科学,22(4):443-453.

丑纪范,1995a.大气动力学方程组的全局分析[J].北京气象学院学报,1:1-12.

丑纪范,1995b.大气动力学的若干进展和趋势[C]//现代大气科学的前沿与展望.北京:气象出版社:71-75.

吴国雄,蔡雅萍,唐晓菁,1995.湿位涡和倾斜涡度发展[J].气象学报,53:387-405.

吴国雄,张学洪,刘辉,等,1997. LASG 全球海洋-大气-陆面模式(GOALS/LASG)及其模拟研究[J]. 应用气象学报,8(增刊):15-28.

吴国雄,刘屹岷,2000. 热力适应、过流、频散和副高 I. 热力适应和过流[J]. 大气科学,24:433-446.

叶笃正,李麦村,1965. 大气运动中的适应问题[M]. 北京:科学出版社.

叶笃正,李麦村,1980. 大气各类运动的多时间尺度特征[C]//中央气象局气象科学研究所. 第二次全国数值天气预报会议论文集. 北京:科学出版社:181-192.

叶笃正,巢纪平,1998. 论大气运动的多时态特征-适应发展和准定常演变[J]. 大气科学,22(4):385-398.

张可苏,1980. 在有热源和耗散情况下的大气适应过程[J]. 大气科学,4(3):199-211.

曾庆存,1963a. 初始扰动结构对适应过程的影响及观测风场的应用[J]. 气象学报,33(1):37-50.

曾庆存,1963b. 大气中的适应过程和发展过程,(一)[J]. 气象学报,33(2):163-174.

曾庆存,1963c. 大气中的适应过程和发展过程,(二)[J]. 气象学报,33(3):281-289.

BENNETTS D A, HOSKINS B J, 1979. Conditional symmetric instability-a possible explanation for frontal rainbands[J]. Quarterly Journal of the Royal Meteorological Society, 105: 945-962.

ERTEL H, 1942. Ein neuer hydrodynamische wirbdsatz[J]. Meteorology Z Braunschweig, 59: 277-281.

GILL A E, 1980. Some simple solutions for heat-induced tropical circulation[J]. Quarterly Journal of the Royal Meteorological Society, 106:447-662.

HELD I, 1978. The vertical scale of an unstable baroclinic wave and its importance for eddy heat flux parameterizations[J]. Journal of the Atmospheric Sciences, 35: 572-576.

HOSKINS B J, 1987. Diagnosis of forced and free variability in the atmosphere[C]//CATTLE H. Atmospheric and Oceanic Variability. Bracknell, UK: James Glaisher House. 57-73.

HOSKINS B J, 1991. Towards a PV- view of the general circulation[J]. Tellus, 43AB: 27-35.

HOSKINS B J, KAROLY D J, 1981. The steady linear response of a spherical atmosphere to thermal and orographic forcing[J]. Journal of the Atmospheric Sciences, 38: 1179-1196.

LIU H, WU G X, 1997. Impacts of land surface on climate of July and onset of summer monsoon: A study with an AGCM plus SSiB[J]. Advances in Atmospheric Sciences, 14(3): 289-308.

MATSUNO T, 1966. Quasi-geostrophic motions in the equatorial area[J]. Journal of the Meteorological Society of Japan, 44: 25-42.

OBOUKHOV A M, 1949. The Problem of Geostrophic Adaptation[M]. Izuestioa of Academy of Science USSR, Ser. Geography and Geophysics, 13: 281-289.

ROSSBY C G, 1937a. On the mutual adjustment of pressure and velocity distribution in certain simple current system, I [J]. Journal Marine Research, 1: 15-28.

ROSSBY C G, 1937b. On the mutual adjustment of pressure and velocity distribution in certain simple current system, II [J]. Journal Marine Research, 2: 239-263.

SHI G Y, 1981. An accurate calculation and the infrared transmission function of the atmospheric constituents[D]. Tohoku: Tohoku University of Japan.

SMAGORINSKY J, 1953. The dynamical influence of large scale heat sources and sinks on the quasi-stationary mean motions of the atmosphere[J]. Quarterly Journal of the Royal Meteorological Society, 79: 342-366.

WU G X, LIU H, ZHAO Y C, et al, 1996. A nine-layer atmospheric general circulation model and its performance[J]. Advances in Atmospheric Sciences, 13:1-18.

WU G X, LIU H, 1998. Vertical vorticity development owing to down-sliding at slantwise isentropic surface[J]. Dynamics of Atmospheres and Oceans, 27(1-4):715-743.

XUE Y K, SELLERS P J, KINTER J L, et al, 1991. A simplified biosphere model for global climate studies[J]. Journal of Climate, 4: 345-364.

YEH T C, 1957. On the formation of quasi-geostrophic motion in the atmosphere[J]. Journal of the Meteorological Society of Japan, 75: 130-137.

ZHANG X H, CHEN K M, JIN X Z, et al, 1996. Simulation of thermohaline circulation with a twenty layer oceanic general circulation model[J]. Theoretical and Applied Climatology, 55: 65-88.

代表性论文 6

Summertime Quadruplet Heating Pattern in the Subtropics and the Associated Atmospheric Circulation

WU Guoxiong, LIU Yimin

(State Key Laboratory of Atmospheric Sciences and Geophysical Fluid Dynamics (LASG),
Institute of Atmospheric Physics (IAP), Chinese Academy of Sciences, Beijing 100029, China)

Abstract: A quadruple heating pattern is found over each subtropical continent and its adjacent oceans in summer based on data diagnosis. The ocean region to the west is characterized by strong longwave radiative cooling (LO); the western and eastern portions of the continent are dominated by sensible heating (SE) and condensation heating (CO), respectively; and the ocean region to the east is characterized by double dominant heating (D), with LO prevailing CO. These compose a LOSECOD heating quadruplet. Its general feature is heating over the continent and cooling over the oceans. A distinct circulation pattern accompanies this heating pattern: in the upper troposphere, anticyclonic circulation over the continent is accompanied by cyclonic circulations over the oceans on its western and eastern sides; near the surface, cyclonic circulation over the continent is accompanied by anticyclonic circulations over the oceans on both sides. This circulation feature is interpreted as the atmospheric thermal adaptation to the quadruplet heating. It is further demonstrated that the global summer subtropical heating and circulation may be viewed as "mosaics" of such a quadruplet heating and circulation patterns, respectively.

1 Introduction

Many studies (Queney, 1948; Charney and Eliassen, 1949; Bolin, 1950; Yeh, 1950; Rodwell and Hoskins, 2001) have shown the importance of large-scale mountains in the formation of atmospheric circulation in winter. However, the circulations in summer subtropics seem to be more related with thermal forcing, and the formation mechanism is more complicated compared with other latitudes (Hoskins, 1987). In recent years, the impacts of monsoon condensation heating on the formation of subtropical anticyclones have been reported (Hoskins, 1996; Wu, 1999a, 1999b; Liu et al., 1999, 2001; Rodwell and Hoskins, 2001; Chen et al., 2001). All of these studies show that the surface anticyclones forced by monsoon heating alone are too weak compared to observations. Liu et al. (2001, Figure 8) further showed in a numerical experiment with monsoon condensation heating as the sole external forcing, that

本文原载于 *Geophysical Research Letters*, 2003, 30(5): 1-5

the induced South Asian high in the upper troposphere is also too weak, whereas the middle tropospheric subtropical anticyclones over western oceans are too strong. They then demonstrated that the introduction of land surface sensible heating to the simulation brought them close to the observed strength. On the other hand, as Rodwell and Hoskins (2001, Figure 8) showed, the local cooling over eastern oceans does significantly enhance the in situ subsidence and subtropical anticyclones over the oceans. All of these results indicate that different diabatic heatings play different roles in the formation of subtropical anticyclones (Liu and Wu, 2000), and should be considered in synthesis. In this regard, the present study employs the reanalysis data of the National Centers for Environmental Prediction/ National Center for Atmospheric Research (NCEP/NCAR) (Kalney et al., 1996) from 1980 to 1997 to demonstrate the distributions of individual as well as total diabatic heating against circulations in the summer subtropics.

2 A Quadruple Heating Pattern and the Associated Circulation Pattern

The July-mean column-integrated heating over the eastern North Pacific, North American and the western North Atlantic (PNAA, 150°—40°W) is presented in Figure 1a. Its prominent feature is cooling over ocean and heating over continent in the subtropics. Figure 1b demonstrates the distributions of the intense longwave radiative cooling (LO) and dominant local heating in the area. Because over the oceans, LO usually overwhelms other heatings, only those LO stronger than -220 Wm^{-2} are shown in the figure so that the secondary heating can be observed. To confirm the dominance of different heatings in different locations, the vertical profiles of LO, diffusive sensible heating (SE), deep convective heating (CO), and total heating (TH) at four typical sites L (30°N, 122°W), S (30°N, 108°W), C (30°N, 80°W) and D (30°N, 60°W) within different heating lobes as shown in Figure 1b are presented in Figure 1c, respectively.

By comparing Figure 1a with Figure 1b, the PNAA area can be divided into four sub regions according to their heating features. Over the eastern Pacific, the strong negative TH (a) is due to LO with an intensity exceeding -220 Wm^{-2} (b). Figure 1c (left) demonstrates that over this LO lobe SE is weak, no obvious CO is observed, and strong LO of -6.5 Kd^{-1} appears in the low layer between $\sigma=0.85$ and 0.95. In Figure 1a, the two heating maximums over western and eastern North America coincide with the SE and CO centers shown in Figure 1b, and possess maximum heating of more than 150 Wm^{-2} and 200 Wm^{-2}, respectively. Over the SE lobe, CO is secondary (Figure 1c, middle left) and SE dominates in the lower troposphere with a maximum of about 6 Kd^{-1} near the surface. Over the CO lobe (middle right), the huge CO of more than 3 Kd^{-1} in the upper troposphere is the main feature. Over the Atlantic to the east, a condensation heating of 50 to 150 Wm^{-2} exists to the west of 45°W (Figure 1b). Figure 1c shows that at site D, CO together with shortwave radiative heating (figure not shown) in the layer between $\sigma=0.2$ and 0.6 exceed LO, whereas above and below this layer they are weaker than LO. The column-integrated heating in this lobe is then negative (Figure 1a). This feature distinguishes lobe D from lobe CO, although significant latent heating and wet climate exist in both lobes. As a whole, the shapes of TH profile over the LO, SE, CO, and D lobes (Figure 1c) are determined respectively by LO, SE, CO, and double dominant heating (D). Therefore, the atmospheric diabatic heating over PNAA demonstrates a quadruplet that is organized as LO over the ocean to the west, SE over the western continent, CO over the eastern continent and D over the ocean to the east, forming as a "LOSECOD" quadruplet, and with heating over land and cooling over ocean in general.

The July-mean wind field and geopotential height deviation (colored) from its equatorial zonal mean

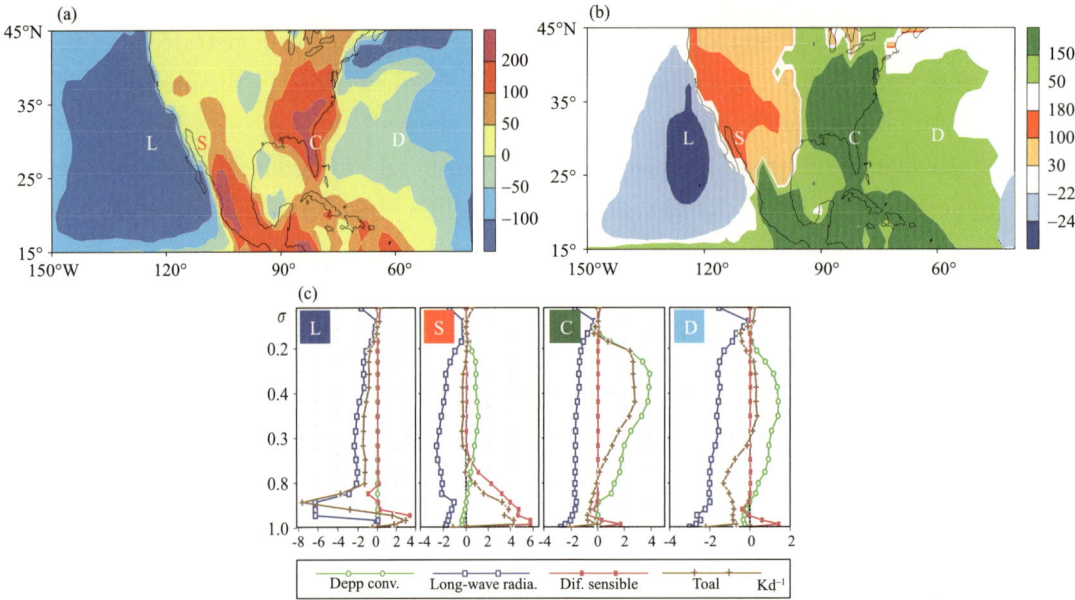

Figure 1 July-mean distribution of column-integrated diabatic heating over the PNAA region. (a) Total heating. (b) Main local heating, blue color denotes longwave radiative cooling (LO); yellow and red, sensible heating (SE); and green, condensation heating (CO). (c) Vertical profiles of different heatings at the marked locations L, S, C, and D representing, respectively, those over the LO, SE, CO, and D lobes. Units: Wm^{-2} in (a) and (b), but Kd^{-1} in (c)

at 100 hPa and 1000 hPa over the PNAA region are shown in Figure 2. The continental area where the SE and CO heating prevail is characterized by the existence of surface cyclonic and upper tropospheric anticyclonic circulations. In contrast, the oceanic area where LO prevails is characterized by the existence of surface anticyclonic and upper tropospheric cyclonic circulations. This can be understood through the PV-θ view, according to which a heating (cooling) can produce lower layer cyclonic (anticyclonic) vorticity and upper layer anticyclonic (cyclonic) vorticity (Hoskins, 1991; Wu and Liu, 2000). However, such anticyclonic circulations near the surface (Figures 2b and 3d) are strongly asymmetric about their central meridional axis. That is, over the LO lobe, the equatorward flow is strongly developed, whereas over the D lobe, a band of southwesterly wind extends northeastward from the Florida Peninsula, just in coordination with the band of deep condensation heating over western Atlantic (Figure 1b). Furthermore, the meridional winds over the LO and SE lobes to the west are equatorward near the surface but poleward in the upper troposphere, whereas they are poleward near the surface but equatorward in the upper troposphere over the CO and D lobes to the east. The correspondence between the profile of heating Q and meridional wind v can be interpreted by employing the following Sverdrup balance (Wu et al., 1999b; Liu et al., 2001):

$$\beta v \approx (f+\zeta)\theta_z^{-1}\partial Q/\partial z$$

The decrease (increase) with height of a heating Q produces negative (positive) vorticity. In a steady state and in the absence of zonal advection, this must be balanced by positive (negative) planetary-vorticity advection that is brought in by meridional winds from high (low) latitudes. Since the TH in the LO and SE lobes decreases with height rapidly in the lower layer, but increases with height in the deep upper layer (Figure 1c), the in situ strong near-surface equatorward flow and weaker upper-layer poleward flow should be generated, as demonstrated in Figure 2. A similar argument applies in the CO and D lobes. In summary, the circulation pattern shown in Figure 2 is primarily forced by the LOSEC-OD quadruple heating pattern.

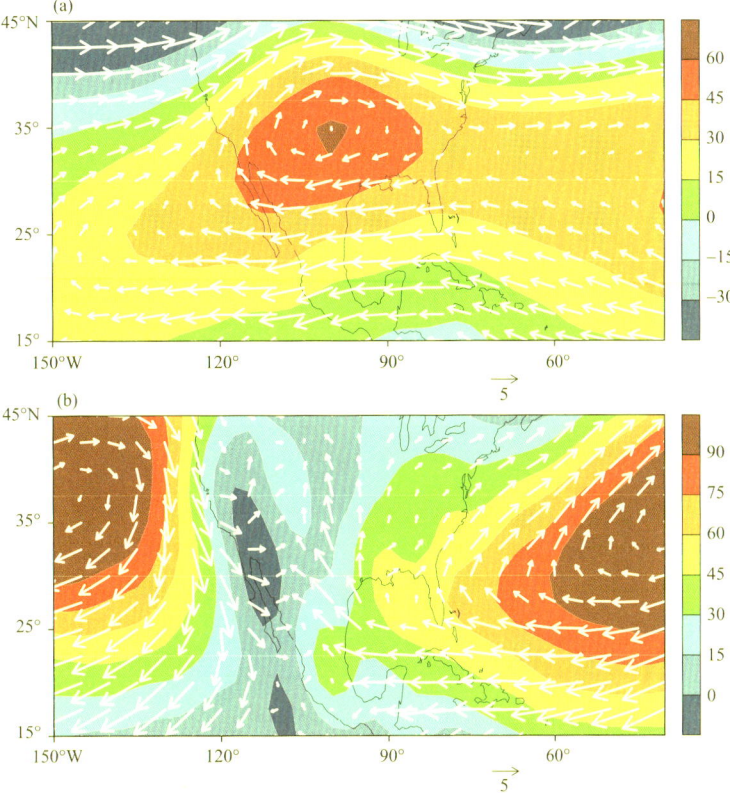

Figure 2 July-mean circulations at (a) 100 hPa and (b) 1000 hPa over the PNAA region. Coloured shading indicates the deviation of geopotential height from the equatorial zonal-mean in units of geopotential meters (gpm). Arrows represent horizontal winds. In correspondence with the LOSECOD quadruplet heating, the circulations in the region also possess a unique pattern: the lower tropospheric cyclone is over the continent and accompanied by anticyclonic circulations over the oceans on its western and eastern sides, whereas in the upper troposphere, the anticyclone is over the continent and accompanied by cyclonic circulations over the oceans on its two sides

3 Mosaics of the Heating Quadruplet and Circulation Pattern

In the northern subtropics there are two big continents. Besides the PNAA region, the rest is defined as the Atlantic-Africa-Eurasia-Pacific region, or AAEP. It will be shown elsewhere that the heating quadruplet is mainly forced by the land distribution. Therefore the lateral boundaries are chosen at the longitudes over the oceans where SH is negligible and the surface meridional wind components vanish. The distribution of the July-mean column-integrated heating and main local heating over the entire northern subtropics are respectively presented in Figures 3a and 3b. They agree with the distributions of the apparent heat source Q_1, apparent moisture sink Q_2 (Yanai et al., 1973), and outgoing longwave radiation calculated from the same data source from 1980 to 1994 (Yanai and Tomita, 1998). The vertical profiles of the heating in each lobe are similar to those presented in Figure 1c and shown by Yanai et al. (1992) and Yanai and Tomita (1998), and are thus not shown here. Despite the huge area occupied by AAEP, a heating quadruplet similar to the one over PNAA can be identified. A strong LO lobe is located over Eastern North Atlantic. A SE lobe occupies the vast western and central AAEP area. A very intense CO lobe is found over the East Asian monsoon region. Although a deep condensation heating of more than 50 Wm^{-2} extends from the East China Sea to Japan (Figure 3b), radiative cooling overwhelms the heating over western Pacific (Figure 3a).

The July-mean circulations of the Northern Hemisphere at 100 and 1000 hPa are shown in Figures 3c and 3d, respectively. Along the subtropics over AAEP, the circulation pattern as demonstrated in Figure 2 can also be detected, and is well coordinated with the LOSECOD quadruplet. At 1000 hPa, a cyclonic low prevails in the SE and CO lobes. At 100 hPa, in correlation with the vast longitude-span of the SE and CO lobes, the anticyclone covers the whole AAEP domain with a deviation height about three times as strong as its counterpart over PNAA. This is because in the absence of advection, the intensity of the geopotential height of a forced cyclone/anticyclone is proportional to the strength and the squared zonal half-length of the forcing. Furthermore, when the circulation patterns over AAEP and PNAA are

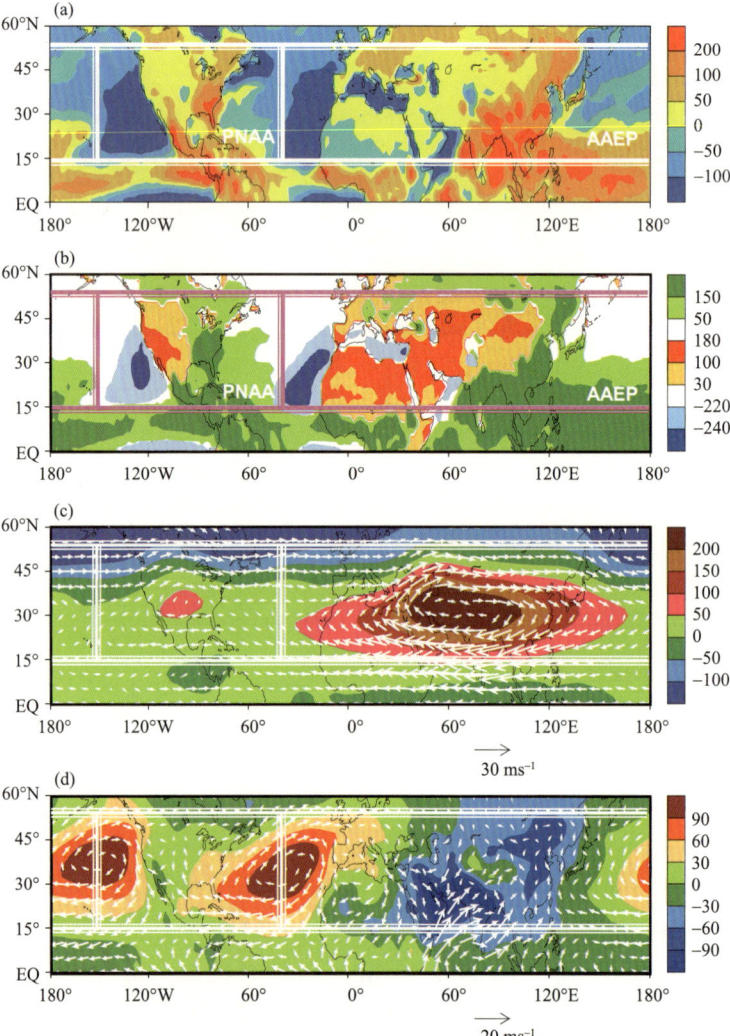

Figure 3 July-mean distributions in the Northern Hemisphere of the column-integrated (a) total heating and (b) main local heating in units of Wm^{-2}, and the wind vector and deviation of geopotential height from the equatorial zonal-mean at (c) 100 hPa and (d) 1000 hPa with units of gpm. The heating distributions (a and b) demonstrate a mosaic of the quadruplet LOSECOD heating patterns, and the circulations (c and d) also demonstrate a mosaic of the circulation patterns as shown in Figure 2

placed side by side, the two troughs at 100 hPa and the two strong subtropical anticyclones at 1000 hPa appear just at the joined edges. It becomes apparent that for each strong oceanic surface subtropical anticyclone, its eastern part is substantially affected by radiative cooling and continental sensible heating, whereas its western part is to a great extent affected by radiative cooling as well as condensation heating associated with the summer monsoon.

In the southern subtropics there are three continents. By selecting the three longitudes of 110°W, 20°W, and 90°E as boundaries, the southern subtropics can be divided into three regions: the Pacific-South America-Atlantic (PSAA), Atlantic-South Africa-Indian Ocean (ASAI), and Indian Ocean-Australia-Pacific (IAUP) regions, as shown in Figure 4. The distribution of the January-mean TH is presented in Figure 4a, and the main local heating, in Figure 4b. The LOSECOD quadruplet can also be identified between 15° and 35°S in each region, with the central heating lobes SE and CO located over the continent, and radiative cooling over the oceans.

The January-mean circulations of the Southern Hemisphere at 100 and 1000 hPa are shown in Figures 4c and 4d, respectively. Again, the circulation pattern as seen in Figure 2 is prominent over each continent. The longitude spans of the individual heating patterns in the southern subtropics differ very little from each other. Therefore, the intensities of the three upper tropospheric anticyclones (greater than 30 gpm) are similar (Figure 4c). When the three circulation patterns are tiled side by side, troughs

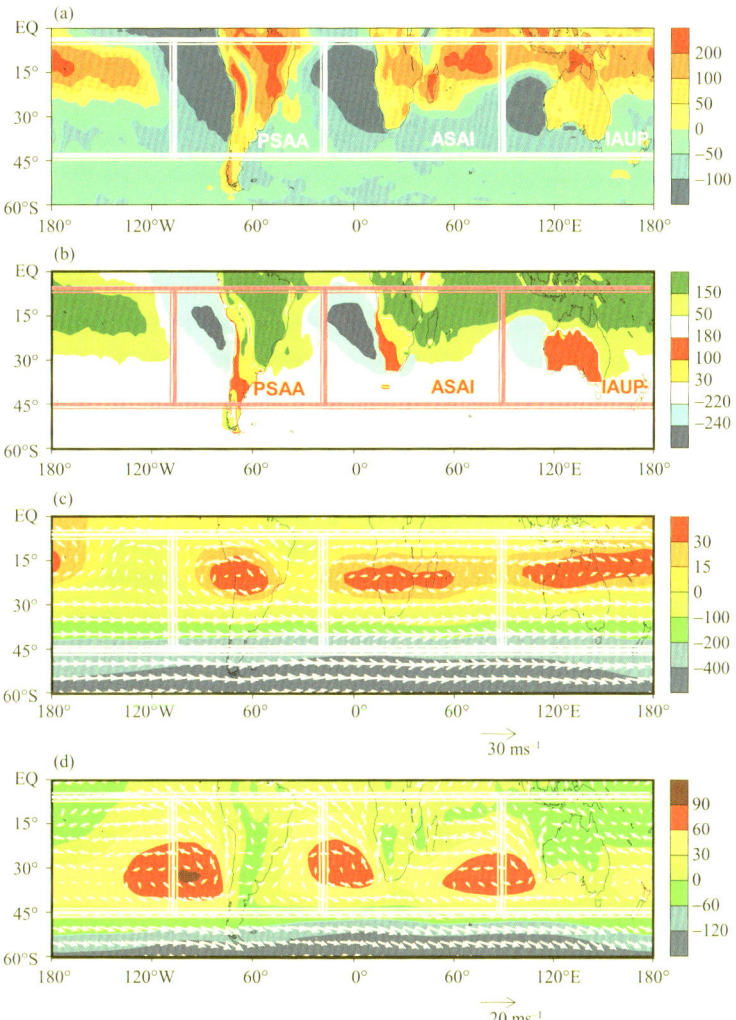

Figure 4 Same as in Figure 3 for the January-mean distributions in the Southern Hemisphere

in the upper troposphere and anticyclones near the surface are found at the three joined edges, as is the case in the Northern Hemisphere.

4 Conclusions and Discussions

We have shown that over each continent and its adjacent oceans in summer subtropics, there exist a heating quadruplet LOSECOD and an associated circulation pattern with surface cyclonic and upper-layer anticyclonic circulations over the continent but surface anticyclonic and upper-layer cyclonic circulations over the oceans, and the summer subtropical circulations can be considered as a mosaic of such circulation patterns. In this mosaic, the subtropical anticyclones in the upper troposphere appear over the centers of the circulation patterns, whereas near the surface they appear at the edges of the two adjacent patterns. Because the zonal sizes within the quadruplet are comparable in all regions except AAEP, the mosaic intensities are similar over the other regions but very strong over AAEP. Consequently, over the AAEP region are observed the most pronounced upper tropospheric anticyclone (the South Asian high), the most vast and severe deserts (the Sahara, Takla Makan and Gobi deserts etc.), and the most intense summer monsoon (Asian summer monsoon).

The observed climate distribution over the whole globe is very complicated. It depends not only on the land-sea distribution but also on the earth's orography, the interactions between different climate subsystems, and different circulation patterns in different latitudes, and so forth. Although as a first approximation, the circulation pattern shown in Figure 2 can be considered as the thermal adaptation of the atmosphere to the LOSECOD quadruplet heating through the PV-θ view and the Sverdrup balance, many dynamical aspects of the formation of the subtropical circulation remain unclear. For instance, how does the local meridional circulation, including the Hadley cell, interact with the subtropical circulation? To what extent do the heating patterns themselves depend on the atmospheric circulation pattern? How do zonal advection and mountain forcing affect their configurations? A more important issue concerns why the heating is organized in such a quadruplet pattern. In any case, an understanding of the peculiar LOSECOD quadruplet heating and the associated circulations in the summer subtropics will help us in the study of climate variability and predictability.

References

BOLIN B, 1950. On the influence of the earth's orography on the westerlies[J]. Tellus, 2: 184-195.

CHARNEY J G, ELIASSEN A, 1949. A numerical method for predicting the perturbations of the middle latitude westerlies [J]. Tellus, 1: 38-54.

CHEN P, HOERLING M P, DOLE R M, 2001. The origin of the subtropical anticyclones[J]. Journal of the Atmospheric Sciences, 58: 1827-1835.

HOSKINS B J, 1987. Diagnosis of forced and free variability in the atmosphere[M]//CATTLE H. Atmospheric and Oceanic Variability. Bracknell, UK: James Glaisher House: 57-73.

HOSKINS B J, 1991. Towards a PV-θ view of the general circulation[J]. Tellus, 43AB: 27-35.

HOSKINS B J, 1996. On the existence and strength of the summer subtropical anticyclones[J]. Bulletin of the American Meteorological Society. 77: 1287-1292.

KALNAY E, KANAMITSU M, KISTLER R, et al, 1996. The NCEP/NCAR 40-year reanalysis project[J]. Bulletin of the American Meteorological Society, 77: 437-471.

LIU Y M, WU G X, LIU H, et al, 1999. Impacts of spatially inhomogeneous heating on the formation and variation of subtropical anticyclones, Ⅲ, Condensation heating and the subtropical anticyclones over South Asia and Western North Pacific

[J]. Acta Meteorological Sinica, 57: 525-538.

LIU Y M, WU G X, 2000. On the studies of subtropical anticyclones: A review[J]. Acta Meteorologica Sinica, 58: 500-512.

LIU Y M, WU G X, LIU H, et al, 2001. Condensation heating of the Asian summer monsoon and the subtropical anticyclones in the eastern hemisphere[J]. Climate Dynamics, 17: 327-338.

QUENEY P, 1948. The problem of air flow over mountains: A summary of theoretical studies[J]. Bulletin of the American Meteorological Society, 29: 16-26.

RODWELL M R, HOSKINS B J, 2001. Subtropical anticyclones and summer monsoon[J]. Journal of Climate, 14: 3192-3211.

WU G X, LIU Y M, 1999a. Basic dynamics and experiments of subtropical anticyclone[R]. paper presented at XXII General Assembly, International Union of Geodesy and Geophysics, Birmingham, UK.

WU G X, LIU Y M, LIU P, 1999b. Impacts of spatially inhomogeneous heating on the formation and variation of subtropical anticyclones, I. Scale analysis[J]. Acta Meteorologica Sinica, 57: 257-263.

WU G X, LIU Y M, 2000. Thermal adaptation, overshooting, dispersion and subtropical anticyclone, I. Thermal adaptation and overshooting[J]. Chinese Journal of Atmospheric Sciences, 24: 433-446.

YANAI M, ESBENSEN S, CHU J H, 1973. Determination of bulk properties of tropical cloud clusters from large-scale heat and moisture budgets[J]. Journal of the Atmospheric Sciences, 30: 611-627.

YANAI M, LI C, SONG Z, 1992. Seasonal heating of the Tibetan Plateau and its effects on the evolution of the Asian summer monsoon[J]. Journal of the Meteorological Society of Japan, 70: 319-351.

YANAI M, TOMITA T, 1998. Seasonal and interannual variability of atmospheric heat sources and moisture sinks as determined from NCEP-NCAR reanalysis[J]. Journal of Climate, 11: 463-482.

YEH T C, 1950. The circulation of the high troposphere over China in the winter of 1945—1946[J]. Tellus, 2: 173-183.

Relation between the Subtropical Anticyclone and Diabatic Heating

LIU Yimin, WU Guoxiong, REN Rongcai

(State Key Laboratory of Numerical Modeling for Atmospheric Sciences and Geophysical Fluid Dynamics, Institute of Atmospheric Physics, Chinese Academy of Sciences, Beijing 100029, China)

Abstract: Monthly mean reanalysis data and numerical experiments based on a climate model are employed to investigate the relative impacts of different types of diabatic heating and their synthetic effects on the formation of the summertime subtropical anticyclones. Results show that the strong land surface sensible heating (SE) on the west and condensation heating (CO) on the east over each continent generate cyclones in the lower layers and anticyclones in the upper layers, whereas radiative cooling over oceans generates the lower-layer anticyclone and upper-layer cyclone circulations. Such circulation patterns are interpreted in terms of the atmospheric adaptation to diabatic heating through a potential vorticity-potential temperature view. A Sverdrup balance is used to explain the zonally asymmetric configuration of the surface subtropical anticyclones. The strong deep CO that is maximized in the upper troposphere over the eastern continent and the adjacent ocean is accompanied by upper-tropospheric equatorward flow and weaker lower-tropospheric poleward flow, whereas the very strong longwave radiative cooling (LO) that is maximized near the top of the planetary boundary layer over the eastern ocean is accompanied by strong surface equatorward flow and weaker upper-layer poleward flow. The center of the surface subtropical anticyclone is then shifted toward the eastern ocean, and its zonal asymmetry is induced. This study concludes that in the summer subtropics over each continent and its adjacent oceans LO, SE, CO, and a double-dominant heating (D) from west to east compose a LOSECOD heating quadruplet. A specific zonal asymmetric circulation pattern is then formed in response to the LOSECOD quadruplet heating. The global summer subtropical heating and circulation can then be viewed as "mosaics" of such quadruplet heating and circulation patterns, respectively.

1 Introduction

In winter, westerlies dominate in the midlatitudes and subtropics in the free troposphere, and mountain forcing plays an important role in the formation of the circulation patterns in these areas

(Charney and Elliason, 1949; Bolin, 1950; Yeh, 1950; Rodwell and Hoskins, 2001). In the summer subtropics on the other hand, the westerlies are weak, and thermal forcing becomes more important in influencing the circulation pattern (Hoskins, 1987). Webster (1972) used a two-layer linear model to show that latent heating is an important local source in maintaining the stationary circulations in low latitudes. Based on another two-layer linear model, Egger (1978) attempted to separate the contributions of sensible heating and latent heating from the maintenance of the summertime subtropical anticyclones and reached a similar conclusion. However due to the coarse vertical resolution, their models greatly underestimated the impacts of the sensible heating in the planetary boundary layer, and this conclusion needs to be reexamined.

During the past several years, the impacts of monsoon condensation heating on the formation of the subtropical anticyclone have been reported on by different studies (Hoskins, 1996; Liu et al., 1999b, 2001; Rodwell and Hoskins, 2001). All of these results demonstrate that the surface anticyclones forced by monsoon heating alone are too weak compared to observations. In contrast, Chen et al. (2001) reported that latent heating can produce a strong surface subtropical anticyclone in a linear quasigeostrophic model. Rodwell and Hoskins (2001) used numerical experiments to show that the longwave radiative cooling (LO) over the eastern oceans significantly enhances the local descending motion and the oceanic subtropical anticyclones. Wu et al. (1999) demonstrated that the zonal advection of vorticity near the ridgeline of the subtropical anticyclone is weak and the vorticity equation can be simplified to a simple Sverdrup balance. Using this balance, Wu and Liu (2000) and Liu et al. (1999a, 2001) found that both the land surface sensible heating and deep condensation heating are important in breaking the symmetric subtropical anticyclone into isolated systems. These studies used either an individual heating or the column-integrated heating. While significant results have been obtained, many details remain unknown. For instance, the surface subtropical anticyclones over the North Pacific and North Atlantic appear as "triangular shaped", and their centers are biased to the east. Chen et al. (2001) interpreted such a "tilting" as a result of the meridional shear of the zonal wind. However the simulated anticyclone following this line was presented as a parallelogram with its center biased westward, quite unlike the observations. All of these imply that the synthetic diabatic effects on the maintenance of the summertime subtropical anticyclone are still unclear.

In this study, the reanalysis of the National Centers for Environment Prediction-National Center for Atmospheric Research (NCEP-NCAR; Kalnay et al., 1996) from 1980 to 1997 and a climate model are employed to investigate the separated and synthetic impacts of different thermal forcings on the formation of the summertime subtropical anticyclones. The diabatic heating in the reanalysis is not an observation, but the product of a GCM that depends on physical parameterization schemes (Newman et al., 2000). Comparisons between the NCEP heating and Q_1 from European Centre for Medium-Range Weather Forecasts (ECMWF) data (Nigam et al., 2000; Rodwell and Hoskins, 2001) and Tropical Ocean Global Atmosphere (TOGA) data (Lin and Johnson, 1996) show that both the horizontal distribution of the column-integrated heating and the vertical heating profile are similar. Duan (2003, personal communication) also made comparisons between the NCEP-NCAR reanalysis and two other datasets. One is for observations from July 1993 to March 1999 from six Automatic Weather Stations (AWSs) over the Tibetan Plateau (Li et al., 2001). The other is the Global Energy and Water Cycle Experiment (GEWEX) Asian Monsoon Experiment (GAME) Intesive Observing Period (IOP) reanalysis from 1 April to 31 October 1998 completed by the Japan Meteorological Research Institute and the Japan Meteorological Agency (unpublished). Results show that the surface sensible heat flux and latent heat flux provided by NCEP-NCAR agree with the AWS data, and there is no significant difference in the variation and magnitude between the NCEP-NCAR and GAME-IOP daily datasets. This then validates the usage of the NCEP-NCAR reanalysis for the present study, although caution is required since both the AWS and GAME-IOP data have limited spatial and temporal coverage. Section 2 shows the observed distributions of the subtropical circulation against diabatic heating in July for the Northern Hemisphere and in January for the Southern Hemisphere. In section 3, the relevant dynamics concerning the atmospheric responses to different kinds of heating are reviewed briefly, and the factors in generating the zonally asymmetric subtropical anticyclones are explored. Numerical experiments are then designed in section 4 to compare with observations and to verify the hypothesis presented in section 3. In section 5, all of these different kinds of forcings are put together to get their synthetic impacts on the formation of the summertime subtropical anticyclone. Conclusions and discussion are presented in section 6.

2 Diabatic heating and circulation in summer subtropics

a. Column-integrated total heating and surface sensible heating

The distributions of the column-integrated total diabatic heating (TH) retrieved from the NCEP-NCAR reanalysis for January and July are presented in Fig. 1. Along the summer subtropics, except in lower latitudes, heating is usually over land and cooling is over ocean. The distributions of the zonal deviation of geopotential height at different levels are shown in Fig. 2. In both the Southern and Northern Hemispheres, the surface lows are located over the subtropical continents, with strong equatorward flow existing along, and to the west of, the western coastal regions (Figs. 2b and 2d), whereas the surface subtropical anticyclone centers are located in the eastern parts of the oceans, with magnitudes of more than 30 gpm in the Southern Hemisphere and 90 gpm in the Northern Hemisphere. At 200 hPa (Figs. 2a and 2c), the subtropical anticyclones are observed over continents, whereas troughs are located over oceans. The magnitude of the anticyclones in the Southern Hemisphere is more than 30 gpm. The North American high is more than 60 gpm, whereas the South Asian high is stronger than 120 gpm and spans a vast longitudinal domain in the subtropics ranging from North Africa to the western Pacific. It is more than twice as strong as its counterpart over North America. Comparing Fig. 2 to Fig. 1, a prominent feature is then obtained: the positive TH over each subtropical continent in the summer hemisphere is accompanied by the surface cyclone and the upper-layer anticyclones, whereas the negative TH over each subtropical ocean sector is accompanied by the surface anticyclone and the upper-layer cyclonic circulation. Such a coordination between TH and the circulation pattern can be well understood by using

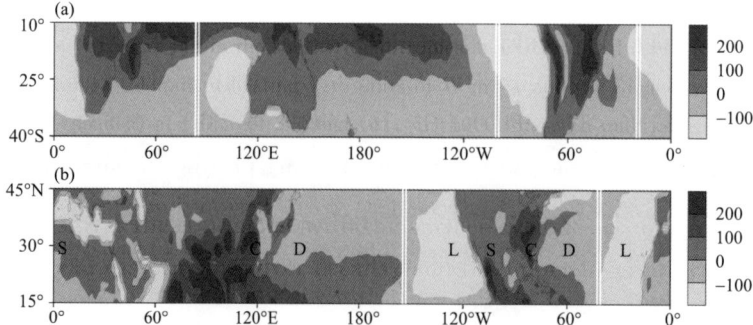

Fig. 1 Monthly mean column-integrated total heating in (a) Jan in the southern subtropics and (b) Jul in the northern subtropics. Units are W m^{-2}

the potential vorticity-potential temperature (PV-θ) view proposed by Hoskins (1991); namely, heating (cooling) generates lower-layer cyclonic (anticyclonic) circulation and upper-layer anticyclonic (cyclonic) circulation. Therefore, as a first approximation, the summer subtropical circulation as presented in Fig. 2 can be interpreted as the adaptation of atmospheric circulation to the diabatic heating (Wu and Liu, 2000) along the subtropics.

Fig. 2 Monthly mean distributions of the zonal deviation of the geopotential height (units are gpm) and of sensible heat flux at the surface (shading, units are W m^{-2}) at (a), (c) 200 and (b), (d) 1000 hPa, for (a), (b) Jan and (c), (d) Jul

The distributions along summer subtropics of the surface sensible heat flux are also shown in Fig. 2. In January in the Southern Hemisphere (Figs. 2a and 2b), it is more than 100 W m^{-2} over Australia, and over the western coasts of South Africa and South America. In July in the northern subtropics (Figs. 2c and 2d), the area of more than 100 W m^{-2} covers western North America, North Africa, and western and middle Asia. A prominent feature is then obtained: strong surface sensible heating (SE) exists over each subtropical continent in the summer hemisphere, particularly over the western coasts. Comparing Fig. 2 with Fig. 1, we see that except for the eastern part, the positive TH over the western and central parts of each continent in the summer subtropics mainly result from the in situ SE. In addition, the centers of either the surface cyclones or the upper-layer anticyclones over continents are located over the SE areas. All of these imply the significance of the continental SE in the maintenance of the summertime subtropical circulations.

b. Deep condensation heating (CO)

The distributions in summer months of deep condensation heating (CO) and the associated zonal deviation winds in the upper and lower troposphere are presented in Fig. 3. In January (Figs. 3a and 3b), the three subtropical heating regions are located, respectively, from eastern Africa to about 80°E, from eastern Australia to 120°W, and from the eastern coast of Brazil to 25°W. Strong CO along the western coast of South America is also observed due to the Andes. At 850 hPa (Fig. 3b), poleward flow dominates the region of convective heating except along the eastern coast of Australia where weak equatorward flows exist due to the local orographic forcing. The strong CO centers of more than 250 W m^{-2} over Madagascar and along the eastern coast of Brazil in the subtropics and over the eastern side of the date line in the Tropics are all accompanied by poleward flows of more than 4 m s^{-1}. Apparent subtropical cyclone and anticyclone circulations are therefore observed, respectively, to the west and east of these strong CO centers. In the upper troposphere (Fig. 3a), equatorward flow prevails over these regions of deep condensation heating in the subtropics, and anticyclonic and cyclonic circulations are observed, respectively, to the west and east of these heating centers.

In July (Figs. 3c and 3d), CO of more than 50 W m^{-2} also exists over areas from the eastern continents to the western oceans along the subtropics. While the area of more than 100 W m^{-2} over the western Atlantic is more confined to the coastal region, it appears over the whole western Pacific and extends beyond the date line. Strong CO of more than 250 W m^{-2} is observed over the southeastern continental United States and eastern China. As in the Southern Hemisphere, equatorward flow in the upper troposphere (Fig. 3c) and poleward flow in the lower troposphere (Fig. 3d) prevail over these two strong heating areas. The flows are more pronounced and organized particularly over the Asian monsoon area. As a result, prominent anticyclone circulations are found to the west of the strong CO regions in the upper troposphere, and to their east in the lower troposphere.

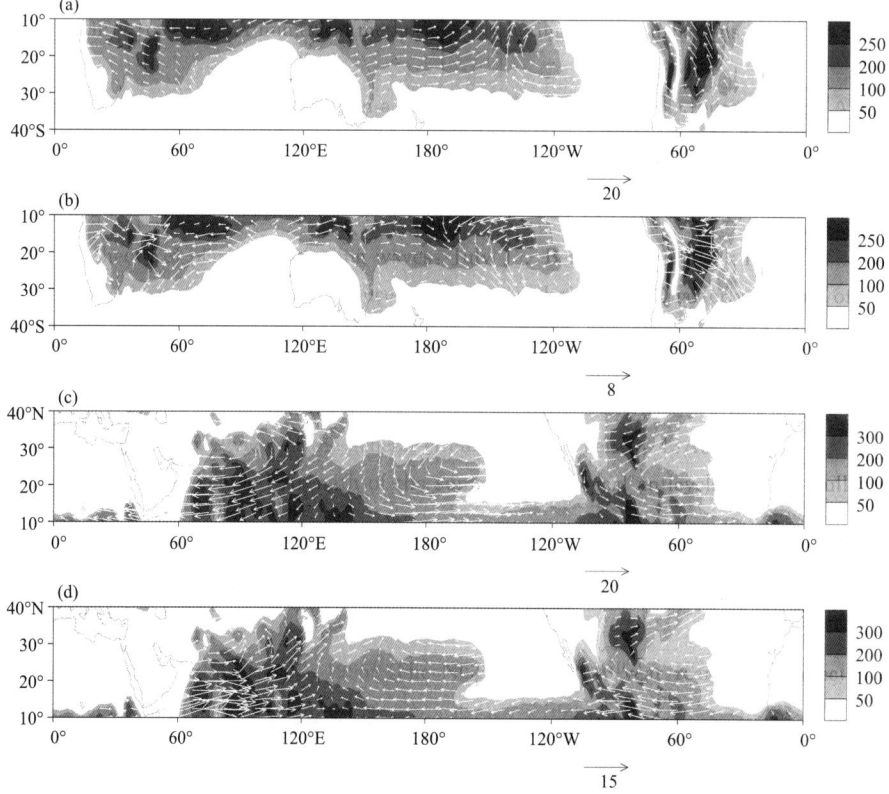

Fig. 3 Monthly mean distributions of the zonal deviation wind (vector; units are m s^{-1}) and of convective condensation heating (shading; units are W m^{-2}) at (a), (c) 200 and (b), (d) 850 hPa, for (a), (b) Jan and (c), (d) Jul

c. Radiative cooling

The distributions in summer months of the columnintegrated cooling (copied from Fig. 1) and the associated zonal deviation winds in the upper and lower troposphere are presented in Fig. 4. Cooling covers the eastern oceans and the poleward side of the western oceans in the subtropics. In particular, it is stronger than −100 W m⁻² over the eastern coastal region of each ocean basin. The zonal deviation circulations, in association with such cooling, bear common features over each of the five subtropical ocean basins; namely, the surface anticyclonic and upper-layer cyclonic circulations appear in the cooling region over oceans, as discussed in section 2a and emphasized by Rodwell and Hoskins (2001). Another remarkable feature revealed in Fig. 4 is that strong surface equatorward flow and upper-layer poleward flow appear over the eastern offshore ocean region where radiative cooling becomes the main feature of the atmospheric heating. Such a relationship between the meridional flow and radiative cooling can also be detected over the Mediterranean Sea. The only exception is observed along the western coastal region of the Arabian Sea, where southerlies develop in the lower troposphere due to the strong suction of the elevated heating over the Tibetan Plateau in summer (Wu et al., 1997a; Ye and Wu, 1998).

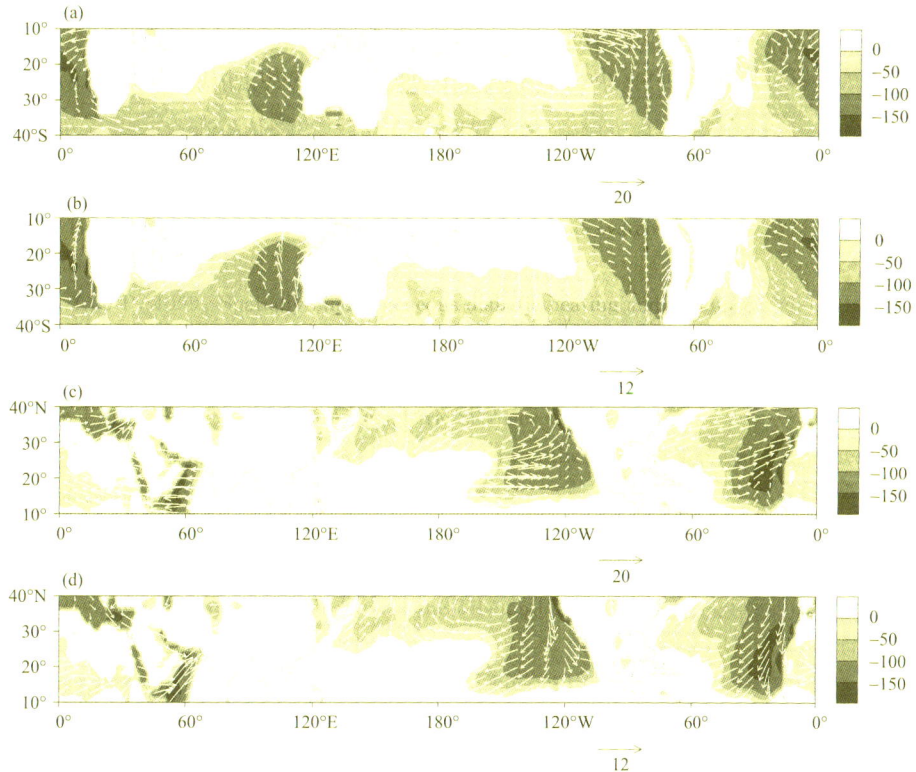

Fig. 4　Monthly mean distributions of the zonal deviation wind (vector; units are m s⁻¹), and of column-integrated cooling (shading; units are W m⁻²) at (a), (c) 200 and (b), (d) 1000 hPa, for (a), (b) Jan and (c), (d) Jul

3　Dynamics relevant to the maintenance of the subtropical anticyclone

Although the theory of the thermal adaptation of atmospheric circulation to diabatic heating presented above can be used to explain the general distributions of the summertime subtropical anticyclone, it cannot explain the asymmetric configurations of the subtropical anticyclone in several detailed aspects. For instance, strong meridional flows either in the upper or in the lower troposphere along the subtropics are usually associated with strong condensation heating (Fig. 3) or radiative cooling (Fig. 4); the equatorward flow in the region along, and to the west of, the western coast of each continent near the earth's surface is much developed; and the center of the surface anticyclone over ocean is biased eastward (Fig. 2), etc. All these cannot be simply explained by the thermal adaptation theory. A theoretical study by Gill (1980) shows that the atmospheric response to a heating source or sink located in low latitudes exhibits an asymmetric Rossby wave pattern. In the Northern Hemisphere, southwesterlies (northeasterlies) are generated to the east of the heating source (sink). Although this theory can be used to interpret the tilting of the subtropical anticyclones, it cannot explain the eastward bias of the surface subtropical anticyclone centers.

To help understand these asymmetries, based on the NCEP-NCAR reanalysis, the July mean heating profiles at different locations along the ridgeline of the subtropical anticyclone in the Northern Hemisphere are present in Fig. 5. Those in the southern subtropics are similar and are not shown here. To confirm the dominance of different types of heating in different locations, the vertical profiles of CO, SE, LO, and TH (including other types of heating) at typical sites are presented. Figure 5a is for the area based in the Western Hemisphere, whereas Fig. 5b is for the area based in the Eastern Hemisphere. The four sites L (30°N, 122°W), S (30°N, 108°W), C (30°N, 80°W), and D (30°N, 60°W) selected for plotting in Fig. 5a and the four sites L (30°N, 24°W), S (30°N, 0°), C (30°N, 120°E), and D (30°N, 145°E) selected for plotting in Fig. 5b are located, respectively, within different heating lobes as marked in Fig. 1b (and also in Fig. 13).

At the L sites (left panels) over the eastern Pacific and Atlantic, weak SE is near the surface, and no apparent CO is observed. Strong LO values of 26.5 K day^{-1} over the eastern Pacific and Atlantic appear in the lower layer between 0.5 and 1.5 km ($\sigma \sim 0.85$ and 0.95) above sea level. The profile of TH then follows the LO profile. These regions are therefore defined as the LO heating lobes. At the S sites (middle-left panels) over western North America (Fig. 5a) and North Africa (Fig. 5b), CO is weak, and SE dominates the lower troposphere with a maximum of about 6 K day^{-1} near the surface. The TH profile in these areas then follows the SE profile. These regions are therefore defined as the SE heating lobes. At the C sites (middle-right panels) along the eastern coasts of North America (Fig. 5a) and China (Fig. 5b), the thick CO with a maximum of about 4 K day^{-1} in the upper troposphere is the main feature, and the TH profile follows the CO profile. These regions are therefore defined as the CO heating lobes. At the D sites (right panels) over the western Atlantic (Fig. 5a) and western Pacific (Fig. 5b), CO is less than 2 K day^{-1}. The CO together with other types of heating (figure not shown) exceed the longwave radiative cooling in the layer between $\sigma = 0.2$ and 0.6 over the western Atlantic and are near $\sigma = 0.2$ over the eastern Pacific, but are weaker than the cooling above and below these layers. Thus the in situ column-integrated TH is negative as shown in Fig. 1, and the TH profile in these regions is determined mainly by the double-dominant heating (D), that is, LO and CO. These regions are therefore defined as the D heating lobes. The negative TH in lobe D contributes to the occurrence of the lower- (upper-) layer anticyclonic (cyclonic) circulation and distinguishes it from lobe CO, whereas the existence of the secondary dominant heating CO in this lobe results in the development of the lower-(upper-) layer poleward (equatorward) flow wet climate and distinguishes the lobe from the LO lobe in which the climate is rather dry.

Results from Fig. 5 show that the TH profiles vary from one location to the other, and follow the profiles of LO at L, SE at S, CO at C, and the double-dominant heating at D, respectively. They compose a LOSECOD heating quadruplet over each continent and its adjacent oceans (Wu and Liu, 2003).

Let us now consider how the vertical differential heating can influence the atmospheric circulations. In the subtropics, particularly along the ridgeline of the subtropical anticyclone, both the vorticity advection and transient processes are weak (Wu et al., 1999; Rodwell and Hoskins, 2001; Liu et al., 2001), and vorticity equation can be simplified to the Sverdrup balance:

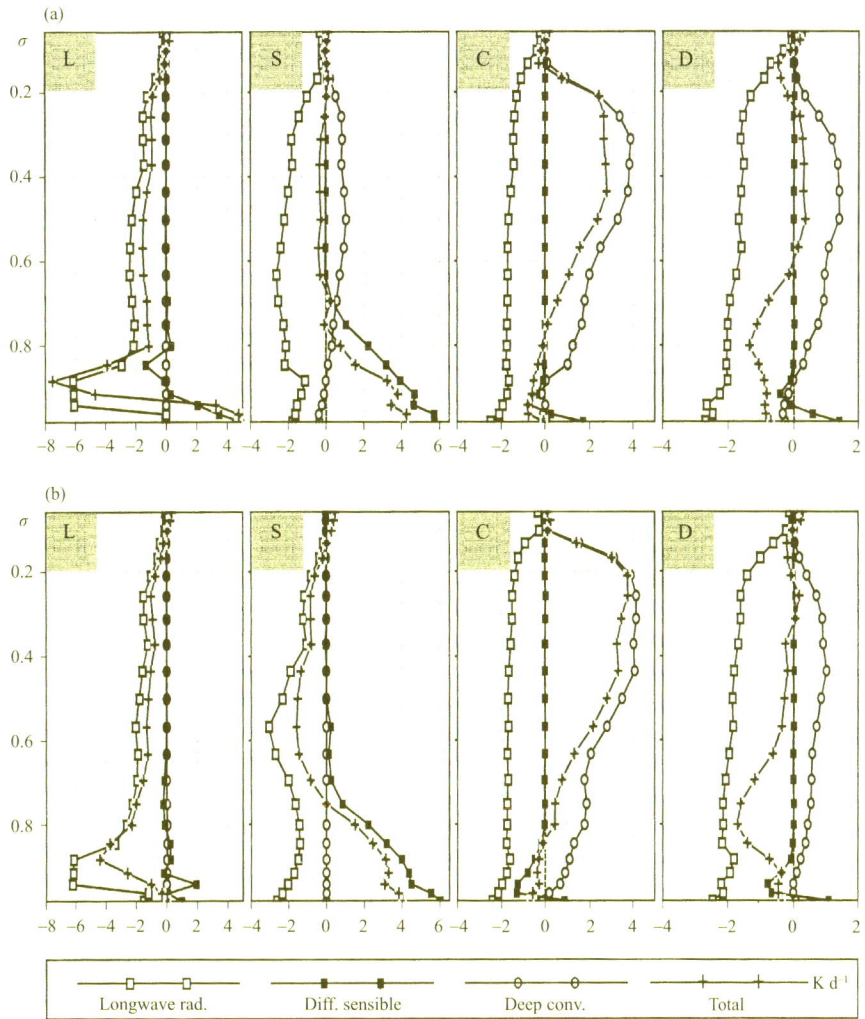

Fig. 5 Vertical profiles of various Jul mean heating at sites L, S, C, and D as indicated in Figs. 1b and 13 in the (a) western and (b) eastern Northern Hemisphere. Shown are longwave radiative cooling (LO, open square), diffusive sensible heating (SE, filed square), deep condensation heating (CO, open circle), and total heating (TH, cross, including all other heating). Units are K day^{-1}

$$\beta v \approx \theta_z^{-1}(f+\zeta)Q_z \quad (\theta_z \neq 0) \tag{1}$$

This implies that in the absence of horizontal advection and at a steady state, the increase (decrease) of the relative vorticity due to the vertical differential heating should be compensated for by the planetary vorticity (Q_z) advection brought in through the meridional winds from low (high) latitudes. Therefore the thermally forced circulation along the subtropics depends strongly on the vertical profile of the heating. Since f is positive in the Northern Hemisphere but negative in the Southern Hemisphere, in a statically stable atmosphere ($\theta_z > 0$) a heating that increases with altitude will generate poleward flow, whereas a heating that decreases with altitude will produce equatorward flow. This then provides another base for understanding the formation of the asymmetric configuration of the summertime subtropical anticyclone, and can be summarized below by the schematic diagrams presented in Fig. 6.

a. Surface sensible heating

During summer along the subtropics, the land surface sensible heat flux over the western continents usually exceeds 100 W m^{-2}, which amounts to a heating rate (Q) of 10^{-5} K s^{-1}. For a large-scale at-

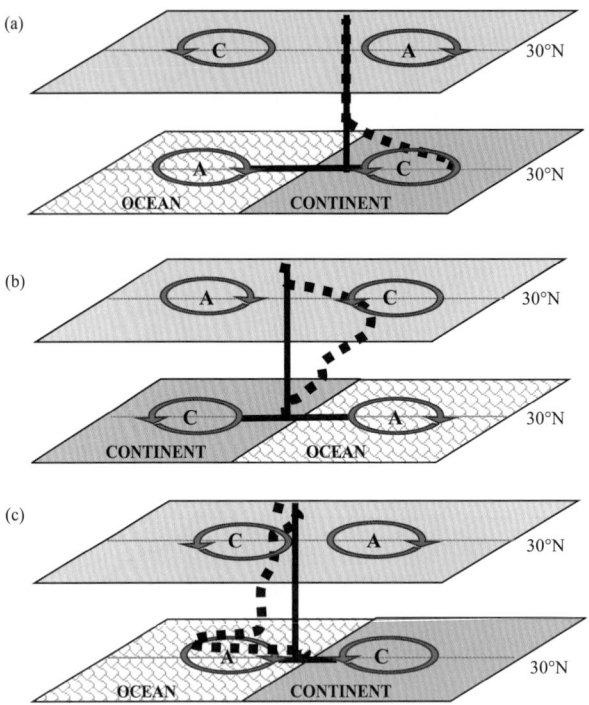

Fig. 6 Schematic diagram indicating the summertime subtropical atmospheric response to the vertically differential diabatic forcing of (a) sensible heating, (b) deep condensation heating, and (c) longwave radiative cooling. Here, A denotes an anticyclone, C denotes a cyclone, and the profiles at the centers of each panel indicate the dominating heating

mospheric system such as the subtropical anticyclone, the order of magnitude for θ_z is estimated as $O(\theta_z)$ $\sim 10^{-2}$ K m^{-1}, and that for the forcing term on the right-hand side of (1) is estimated as 10^{-10} s^{-2}. Then the forced equatorward flow in the lower layers is estimated as $1-10$ m s^{-1}. This means that, in response to a surface sensible heat flux of 100 W m^{-2}, the equatorward winds of several meters per second will be forced over the heating region in the lower layers with a thickness of about 1 km (Fig. 6a). Therefore equatorward flow on the western side of the surface cyclone should be stronger than the poleward flow on its eastern side, resulting in the westward bias of the cyclone center over the continent (Fig. 2). The intensified equatorward flow along the western coast of the continent brings colder air and favors the enhancement of sensible heat release from the warm land surface. This then explains why the strongest sensible heating in the summer subtropics appears along the western coast of each continent (Fig. 2).

b. Deep convective condensation heating

Along the subtropics, maximum CO usually occurs at a height ($z = Z_M$) between 300 and 400 hPa where the heating rate can be several degrees per day. This, following the thermal wind balance, results in a strong westerly to its north and an easterly to its south above the heating region. Thus, even in the upper troposphere in the subtropics, vorticity advection above a deep convection area is usually small. By using (1), the diabatic term $\theta_z^{-1}(f+\zeta)Q_z$ is estimated to be $+10^{-10}$ s^{-2} below Z_M, but -10^{-9} s^{-2} above this level. Therefore a poleward flow of several meters per second is forced below Z_M, while a slightly stronger equatorward flow is forced in the upper troposphere. Such a mechanism then explains why in Fig. 3 the equatorward flow in the upper troposphere and the poleward flow in the lower troposphere are in accordance with the strong deep condensation heating in the subtropics. These forced strong meridional winds then contribute toward strengthening the formation of the subtropical anticyclone to the west of the deep convection region in the upper troposphere, and to its east in the lower troposphere, as schematically shown in Fig. 6b and observed in Fig. 3.

c. Longwave radiative cooling

The maximum LO is usually below 850 hPa with an intensity of about -6 K day^{-1}. Following a similar argument, such radiative cooling can produce a vorticity forcing of $+10^{-10}$ s^{-2} in the upper troposphere and -10^{-9} s^{-2} in the lower troposphere. Poleward flow of several meters per second in the upper layer and stronger equatorward flow in the lower troposphere are then forced. Such radiation-induced meridional flows reinforce those circulation patterns generated by sensible heating (Fig. 6a) and favor the formation of the subtropical anticyclone to the west of the radiative cooling in the lower troposphere, but to its east in the upper troposphere, as shown in Fig. 6c. We may therefore propose that the strong negative vorticity forcing near the surface due to LO over the eastern oceans contributes to the eastward shift of the surface oceanic subtropical anticyclone, resulting in the asymmetric configuration of the surface subtropical anticyclones over oceans.

4 Numerical experiments

To verify further the atmospheric response to external thermal forcing, numerical experiments are designed by employing the Global Ocean-Atmosphere-Land System (GOALS) model developed at the Institute of Atmospheric Physics/State Key Laboratory of Numerical Modeling for Atmospheric Sciences and Geophysical Fluid Dynamics (IAP/LASG; Wu et al., 1997b; Zhang et al., 2000; Table 1). Its atmospheric component is a spectral general circulation model that possesses nine vertical levels in s coordinates and is rhomboidally truncated at wavenumber 15 in the horizontal. The oceanic component is a gridpoint model with horizontal resolution of 4° latitude ×5° longitude and 20 vertical layers (Zhang et al., 1996). The land surface processes are represented by the Simplified Simple Biosphere (SSiB) model (Xue et al., 1991), which has been implemented in the atmospheric component (Liu and Wu, 1997). The GOALS climate model can simulate the mean climate reasonably well and has been used in climate studies for different purposes (Houghton et al., 2001; Kang et al., 2002).

In this study the ocean component of the model is switched off, and the required sea surface temperature (SST) and sea ice are prescribed by using the climate mean observation data of 1979—1988 developed for the Atmospheric Model Intercomparison Project (AMIP). Because the adjustment period measured by the global surface energy balance is about half a year for each experiment, all the experiments in this section are integrated for 12 model years, and the July means calculated from the last 10 years are taken for comparisons. The normal integration is defined as a control run (CON) as shown in Table 1. A pair of GCM experiments (SH-1 and SH-2) is designed to study the contributions of sensible heating and radiative cooling to the formation of the summertime subtropical anticyclones. Another group of sensitivity runs is also designed for the boreal summer as perpetual July experiments. In these runs, the solar zenith angle is fixed at the value corresponding to 15 July, the SST assumes its zonal means, and the initial fields are taken from the July mean zonal states of the multiyear integration of the GOALS model. All the experiments are integrated for 24 months, and the results from the last 12 months are extracted for analysis. To show the responses of the atmospheric circulation to a prescribed heating, only those zonal deviation fields are plotted. All the perpetual July experiments designed for the present study are also presented in Table 1 and are described in detail in the following corresponding sections.

The distributions in CON of the July means are demonstrated in Fig. 7. The simulated distribution of the surface sensible heat flux (shading) agrees in general with that of the NCEP-NCAR reanalysis (Figs. 2c and 2d), although the magnitude over eastern North America is too high. Surface cyclones and upper-layer anticyclones are produced over continents, whereas surface anticyclones and upper-layer cyclones are located over oceans. Despite the coarse resolution of the model, it is able to produce the large-scale features of the subtropical circulations.

Table 1 Experiments of different external forcings (see text for detailed description)

GCM experiments		Perpetual July experiments	
SH-1	SE alone	SH-0P	idealized SH in an aqua planet
SH-2	SE+LO	LH-0P	idealized CO in an aqua planet
CON	SE+LO+CO +orography	LH-1P	CO+LO in an aqua planet
		LH-2P	CO+LO+orography
		SH-2P	SE+LO+orography
		CON-P	SE+LO+CO+orography

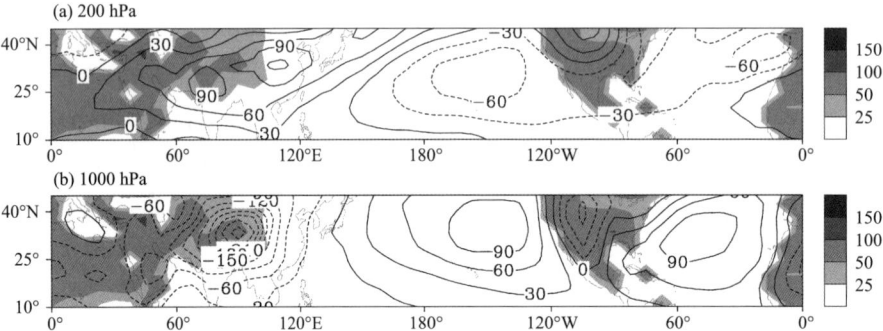

Fig. 7 Jul mean distributions in CON of the zonal deviation of geopotential height (units are gpm) and of the surface sensible heat flux (shading; units are W m^{-2}) at (a) 200 and (b) 1000 hPa

a. Idealized thermal forcing

1) SURFACE SENSIBLE HEATING

Perpetual July and an aquaplanet are assumed for the experiment. To mimic the SE distributions over North America and over North Africa and the subtropical Asia, surface sensible heating is imposed in the two regions from 0° to 105°E and from 90° to 120°W, and between 24.4° and 46.6°N, with a heating maximum of 150 W m^{-2} located along the western boundaries and decreasing in sinusoidal form to zero along the eastern boundaries, as depicted in Fig. 8c. The cloud distribution is prescribed as zonal symmetric and is obtained from CON. The model-induced condensation heating and sensible heating are eliminated from the thermodynamic equation and are not allowed to warm the atmosphere. This experiment is labeled SH-0P. Figure 8 shows that such SE produces anticyclone highs of 50 gpm at 500 hPa (Fig. 8a) and strong surface lows of 80—100 gpm over the two heating regions (Fig. 8b). The two lows at 500 hPa and the two surface highs of more than 40 gpm are generated over the two "oceans" between the two heating areas. The vertical cross section along 30°N (Fig. 8c) shows that the main atmospheric response over the heating region is the generation of cyclone circulation below 700 hPa and anticyclone circulation above it, as is anticipated from the thermal adaptation theory. In lower layers, strong northeries are generated over the western heating region (Fig. 8b). It is important to note that with the imposed surface sensible heating alone, the generated pattern of geopotential height deviation at 1000 hPa already captures the main features in the observation (Fig. 2d). This implies the importance of land surface sensible heating in the formation of the subtropical anticyclone in summer at least in the lower layers.

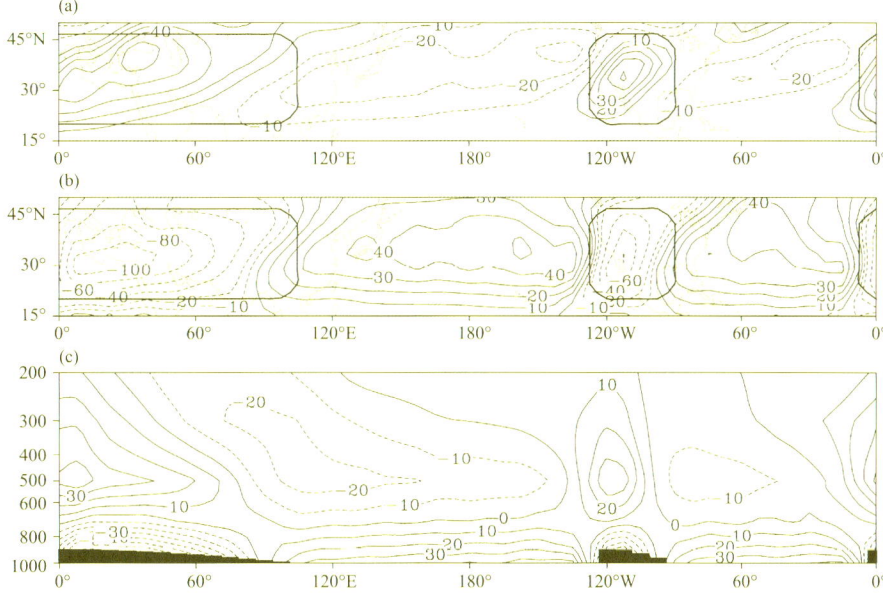

Fig. 8 Zonal deviation of geopotential height (units are gpm) at (a) 500 and (b) 1000 hPa, and (c) its vertical cross section at 30°N in the idealized perpetual Jul experiment SH-0P. The heavy curves in (a) and (b) bound the region where the heating is more than 1 W m^{-2}. Shading in (c) indicates the surface sensible heating region

2) DEEP CONVECTIVE CONDENSATION HEATING

The strongest convective heating in July is observed over the northern Bay of Bengal. Following the NCEP-NCAR reanalysis, a localized deep convective condensation heating is prescribed over a region from 6.7° to 33.3°N and from 67.5° to 112.5°E in an aquaplanet. A maximum heating center of 8 K day^{-1} is located at 20°N, 90°E where the basic zonal flow U vanishes at the height $\sigma=0.336$ (Fig. 9c). The heating then decreases from the center outward in sinusoidal form to zero at the boundaries in both the longitudinal and meridional directions. This perpetual July experiment is labeled LH-0P. The results shown in Fig. 9 demonstrate that over the heating region a southerly is forced at 850 hPa below the maximum heating (Fig. 9b), whereas a northerly is forced at 200 hPa above the maximum heating (Fig. 9a). Figure 9c shows the vertical cross section of the zonal deviation of geopotential height along 30°N where the initial U is about 4 m s^{-1}. Above the level of maximum heating, positive and negative geopotential heights are generated, respectively, to the west and east of the heating; whereas below this level, the pattern is out of phase with that in the upper layer. These are the results anticipated from the Sverdrup balance (1), which are in good agreement with the observations presented in Fig. 3. However, the anticyclone centers in both the upper and lower layers are too close to the heating (Fig. 9c) compared to observations. Comparison of Fig. 9c with Fig. 8c shows that over the heating region, the surface sensible heating tends to generate an out-of-phase vertical pattern of the geopotential height, whereas the deep convective heating tends to produce meridional winds in the free atmosphere subject to the Sverdrup relation (1) that was defined as a "local mode" by Chen (2001) and a surface cyclone in response to the heating. The main features presented in Fig. 9c are also captured in the analytical solutions of Chen based on a linear quasigeostrophic model (QG) and under a 5 m s^{-1} basic flow (his Fig. 7e). However, two significant differences exist between the GCM and the linear QG outputs. First, the linear advection in the

QG moves the local mode eastward, resulting in a striking out-of-phase vertical structure in streamfunction above the heating region. On the contrary, in the GCM as shown in Fig. 9c, despite the existence in the initial fields of the basic westerly zonal flow (4 m s^{-1}) along 30°N, such a local mode is over the heating region, just like the resting basic flow solutions in Chen (his Fig. 7a). This is because in a GCM, the strong subtropical CO heating along the weak zonal westerly basic flow can result in a strong westerly to the north and an easterly to the south of the heating region, and can produce a resting zonal flow across the heating region. Such a mechanism of the reaction of the basic flow to the heating is absent in the linear model, and the model solutions become very sensitive to the prescribed basic flow as was reported. Second, the maximum amplitudes of the streamfunction in the QG solutions appear near the top (14 km) and at the lower boundary of the heating (Figs. 5 and 7 in Chen). This is because the heating profile given as a function of $\sin(\alpha z)$ with a $\alpha = \pi/(14 \text{ km})$ in his study produces a vorticity forcing proportional to $\cos(\alpha z)$ that is maximized at the upper and lower boundaries. As a result a strong surface anticyclone is forced in the local mode. However, in the GCM, a strong subtropical anticyclone does not appear at the surface. Instead, the maximum amplitudes appear just to the west and east of the heating center at, respectively, $\sigma = 0.189$ and 0.500, one level above and below the maximum heating level ($\sigma = 0.336$). Due to the strong CO heating, the static stability close to the heating center becomes very small, and the vorticity forcing in the Sverdrup balance (1) is maximized just below and above the heating center. Such a mechanism of the heating-adjusted static stability and vorticity forcing is also absent in the prescribed forcing in the QG model, and the local mode in a resting basic flow exhibits an antisymmetric pattern about the heating center (Fig. 5a in Chen). We will show that the GCM results presented in LH-0P (Fig. 9) are more robust not only in the following experiments, but also in other studies based on different models.

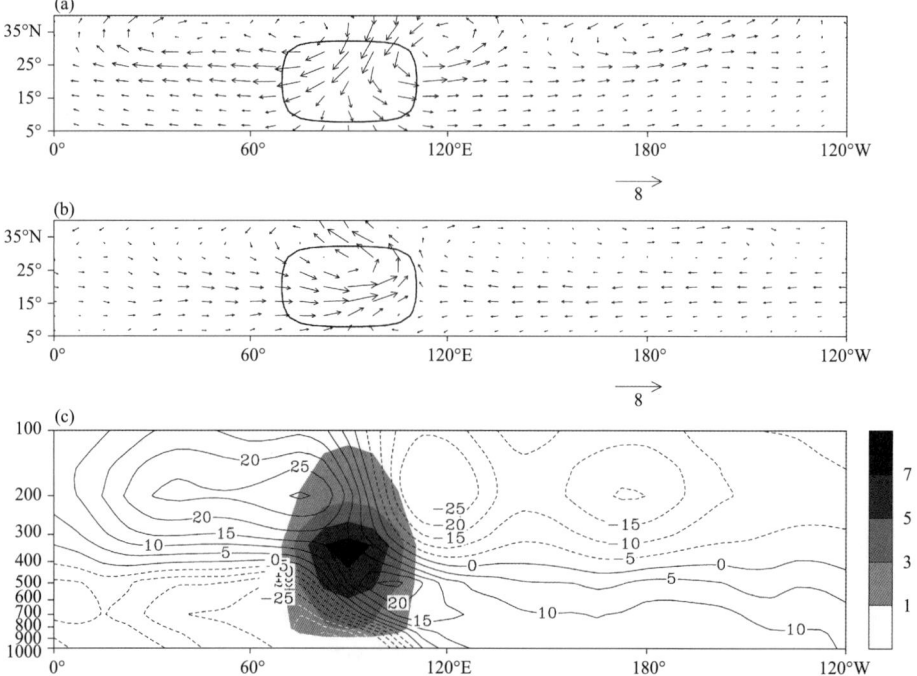

Fig. 9 Zonal deviation of horizontal wind (units are m s^{-1}) at (a) 200 and (b) 850 hPa, and (c) the vertical cross section of the geopotential height at 30°N (units are gpm) in the idealized perpetual Jul experiment LH-0P. The heavy curves bound the region where the heating is more than 1 K day^{-1} at $\sigma = 0.336$. Shading indicates the vertical profile of the heating (units are K day^{-1})

b. Sensitivity experiments

1) SENSIBLE HEATING AND RADIATIVE COOLING

Two sensitivity experiments are designed. The first experiment is similar to CON but with the removal of surface sensible heating in the thermodynamic equation. The difference between CON and this first run then approximately represents the atmospheric response to surface sensible heating alone, and is labeled SH-1. The second experiment is also similar to CON but with the removal of latent heating in the thermodynamic equation. The results are labeled SH-2 and can be considered as the atmospheric response to surface sensible heating and radiative cooling (Table 1). The results from these two experiment sets are presented in Fig. 10. The distributions of the zonal deviation of geopotential height in SH-1 (Figs. 10a and 10b) agree in general with those in CON (Fig. 7), and the intensities of the upper-and lower-layer anticyclones are close to their counterparts in CON. These imply the significance of the surface sensible heating over land in the formation of the summertime subtropical circulations. However, the centers of the surface subtropical anticyclones and the upperlayer cyclones over the North Pacific and North Atlantic shift too far to the west compared to those in CON. In particular, the surface subtropical anticyclone over the North Pacific looks rather symmetric about the date line. These may be attributed to the lack of radiative cooling as was proposed in the preceding section. In SH-2 in which both the effects of the surface sensible heating and the radiative cooling are presented, the zonal asymmetry of these oceanic systems appears, and the simulated subtropical circulations (Figs. 10c and 10d) are closer to those in CON. The anticyclone center over the North Pacific is located at 145°W both in SH-2 and in CON, in agreement with the observation (Fig. 2d). Results from the above experiments then prove that it is mainly the radiative cooling over the eastern oceans that changes the symmetric surface anticyclones over the oceans to asymmetric, and shifts their centers eastward toward the western coasts of the continents. The main discrepancy between SH-2 and CON exists over the western oceans. This may be attributed to the exclusion of condensation impacts in SH-2.

2) CONDENSATION HEATING

The July mean rainfall distribution in CON is presented in Fig. 11a. It is close to the distribution of the condensation heating in the NCEP-NCAR reanalysis (Figs. 3c and 3d). To investigate the impacts of such a condensation heating on the circulation, a pair of the aforementioned perpetual July experiments is assigned. The first experiment, called LH-1P (Table 1), assumes an aquaplanet and uses the three-dimensional distribution of the condensation heating that exists in CON and is consistent with the rainfall distribution presented in Fig. 11a. The second experiment is the same as LH-1P but with the inclusion of orography and the replacement of the zonal mean SST by the observed SST, and defined as LH-2P (Table 1). Results from LH-1P are shown in Figs. 11b and 11c. In response to the latent heating from eastern China to the date line along the subtropics, the in situ surface southwesterlies and strong upper-layer northeasterlies appear. A strong upper-layer anticyclone is formed to the west of the heating, while the surface anticyclone is formed to its east, as expected from the Sverdrup balance (1). The strong rainfall over Latin America is accompanied by a surface low and an upperlayer high, respectively. It is worthwhile to compare these with the corresponding results of other studies. Using a linear two-layer model, Webster (1972) introduced a latent heating distribution for June—July—August (JJA) as a prescribed diabatic heating to the integration, which was based on the estimates of cloudiness and also possesses a maximum over China and the western Pacific along the boreal subtropics (his Fig. 2c). Strong northeasterlies at 250 hPa and southwesterlies at 750 hPa were generated over this heating area, and a strong upper-layer subtropical anticyclone from Japan to eastern North Africa and a weaker low-layer subtropical

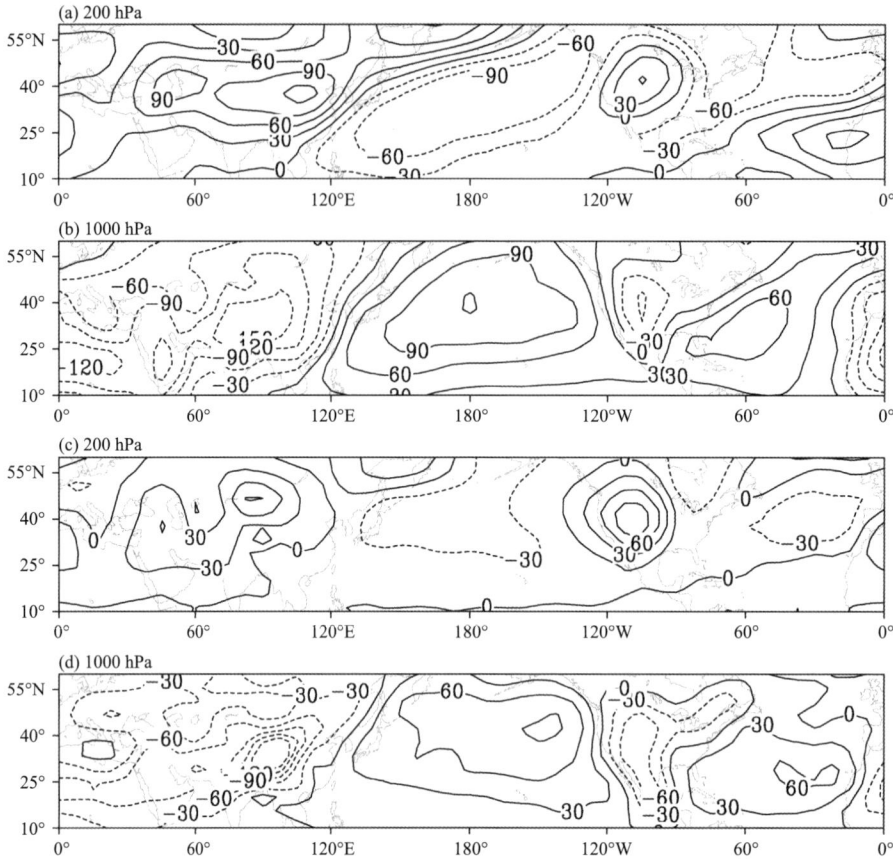

Fig. 10 Jul mean zonal deviations of geopotential height (units are gpm) at (a), (c) 200 and (b), (d) 1000 hPa calculated from the GCM experiments (a), (b) SH-1 and (c), (d) SH-2

anticyclone over the western Pacific were produced (his Fig. 15). Similar features are also presented in the linear model outputs of Lin (1983, his Fig. 9). They agree well with those shown in Fig. 9 and Figs. 11b and 11c.

In all these experiments, the center of the upper-layer anticyclone is located over eastern China, too far to the east compared to the observation (Fig. 2c). This is in agreement with the results of LH-0P (Fig. 9), in which the center of the upper-layer anticyclone is just to the west of the heating center. In LH-2P, in which the orography distribution is included, the locations of both the surface cyclone and upper-layer anticyclone are shifted westward.

3) COMPARISON BETWEEN SENSIBLE HEATING AND CONDENSATION HEATING

For this purpose, another pair of perpetual July experiments is designed. The first is similar to LH-2P but with the inclusion of sensible heating, and labeled CONP. The second is the same as CON-P but with the removal of the latent heating in the thermodynamic equation and labeled SH-2P (Table 1). The CON-P and SH-2P runs correspond to, respectively, the full GCM experiments CON and SH-2 described before except for the usage of the perpetual July setting. The distributions of the July mean geopotential height field in the perpetual runs are similar to those in the corresponding GCM experiments (figures not shown), and the usage of the perpetual July experiment for the present purpose is validated. The cross sections along 30°N of the zonal deviation of the geopotential height for these experiments are shown in Fig. 12. The vertical out-of-phase feature in SH-2P (Fig. 12a) is similar to that in the idealized experiment SH-0P (Fig. 8c), and enhanced due to the inclusion of orography. In LH-2P (Fig. 12b),

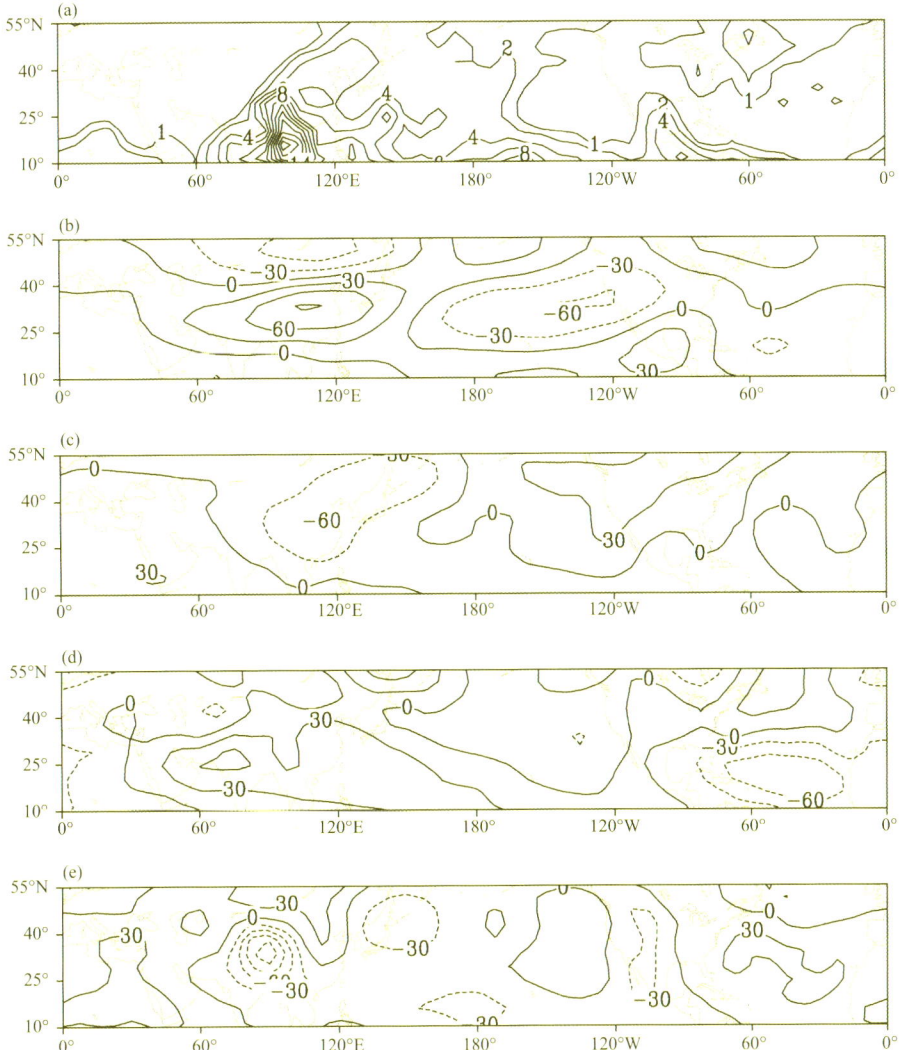

Fig. 11 (a) Jul mean precipitation in CON (units are mm day^{-1}), and the zonal deviations of geopotential height (units are gpm) at (b), (d) 200 and (c), (e) 1000 hPa calculated from the perpetual Jul experiments (b), (c) LH-1P and (d), (e) LH-2P

the Sverdrup balance pattern that was obtained from the idealized run LH-0P (Fig. 9c) is prominent over either the Asian monsoon region or the North American monsoon region: The local maximum positive geopotential height is observed to the west of the maximum heating in the upper layer and to its east in the lower layer, not at the surface. In addition, a surface cyclone exists just below the heating.

It is interesting to compare the results of SH-2P and LH-2P with those of CON-P. The vertical out-of-phase feature in CON-P is well simulated in SH-2P particularly over the oceans. Surface continental cyclones and oceanic anticyclones in CON-P are also presented well in SH-2P. The magnitude of the surface anticyclone over the Pacific in SH-2P is over 80 gpm and accounts for about 80% of its counterpart in CON-P (>100 gpm), and that over the Atlantic is more than 60 gpm and accounts for more than 70% of its counterpart in CON-P (>80 gpm). The contributions to these surface anticyclones from condensation heating become secondary. In these aspects, sensible heating is more important than latent heating in maintaining the circulations. Over the two monsoon regions, however, the geopotential height fields in CON-P bear the strong impacts of the Sverdrup balance, which is driven by condensation heating. As

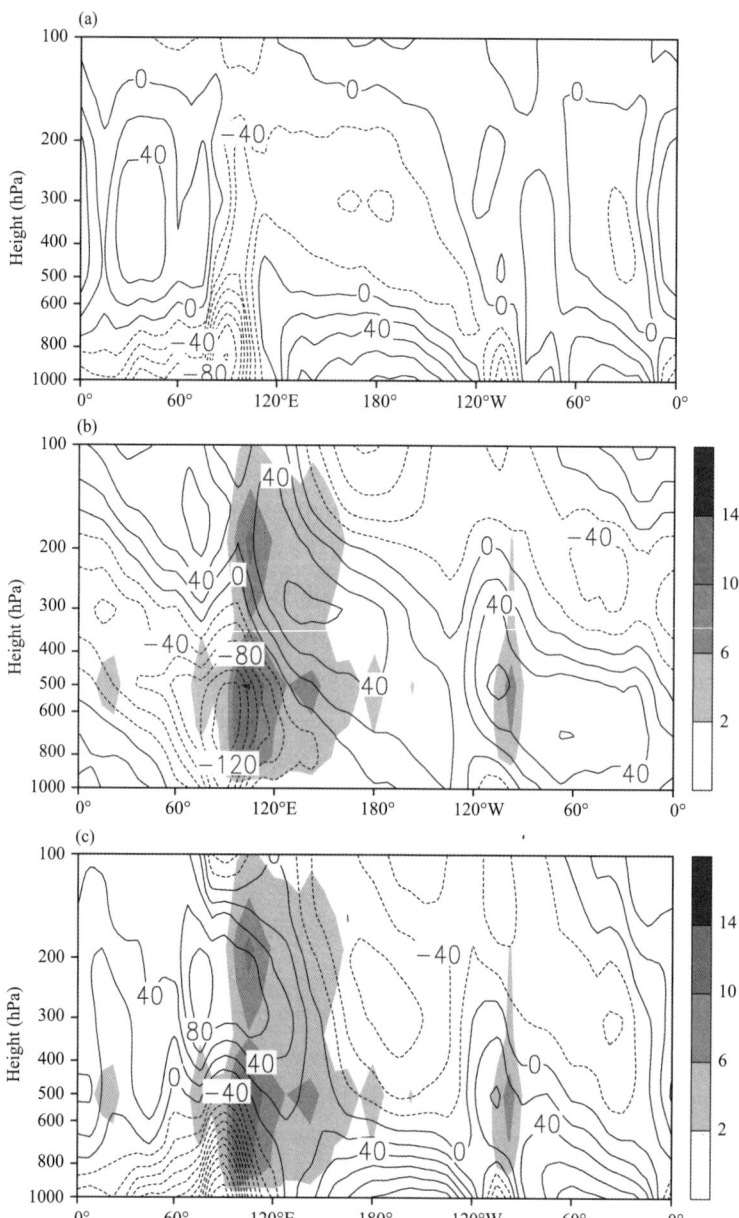

Fig. 12 Cross sections of the zonal deviation of geopotential height (units are gpm) along 30°N in perpetual Jul experiments (a) SH-2P, (b) LH-2P, and (c) CON-P. Shading indicates the Jul distribution of precipitation in CON, which is taken as a latent heating source in the experiment

a result, a geopotential height center of 60 gpm at 200 hPa over Asia and another center of similar strength at 500 hPa over North America produced by condensation heating in LH-2P are rather close to their counterparts in CON-P, and the subtropical anticyclone at 500 hPa over western Pacific in CON-P owes its existence mainly to the condensation heating. It is important to note that the sensible heating and condensation heating act together to strengthen the anticyclone over Asia and the troughs over oceans in the upper troposphere, as well as the oceanic subtropical anticyclones and the huge Asian cyclone near the surface.

It becomes clear that the sensible heating and condensation heating in summer play different roles in maintaining the subtropical circulations. Together with radiative cooling, they are all important in forming the summertime subtropical anticyclones. Although the magnitude of the latent heating associated with the summer monsoon is stronger than the sensible heating, the sensible heating over the land surface is more fundamental in maintaining the summertime subtropical anticyclones near the surface. It also contributes significantly to the vertical out-of-phase features of the subtropical circulation.

The above conclusion contradicts those drawn by Webster (1972) and Egger (1978), who attempted to separate the contributions of sensible and latent heating to the maintenance of the summertime subtropical anticyclones and stressed the dominant importance of latent heating. However, as shown in Fig. 5, a substantial part of sensible heating is concentrated below 800 hPa. Such very low-layer heating exerts profound impacts on the circulations not only in the lower layers but also in the upper layers as demonstrated in Figs. 8, 10, and 12. However, the sensible heating above 800 hPa is negligible compared with other heating over all heating lobes (Fig. 5). Therefore it may not be appropriate to evaluate the impacts of sensible heating on the atmospheric circulation by only using the kinds of two-layer circulation models in which the lower layer is at either 800 or 750 hPa. Chen et al. (2001) introduced two latent heating sources, respectively, over Asia and North America to the aforementioned linear QG model (Chen, 2001) to study the atmospheric responses to the summertime diabatic heating, and interpreted the surface oceanic subtropical anticyclones in the Northern Hemisphere as a remote response of Rossby waves forced by the large-scale heating source over Asia, and the circulations over the two continents as a local response to monsoon latent heat release in the midtroposphere. As discussed before, the lack of the mechanism in the QG model of the heating-adjusted static stability and vorticity forcing can lead to an exaggerated surface response. In addition, the observed monsoon rainfall is located from eastern China to the western Pacific and from eastern North America to the western Atlantic (Fig. 3), whereas the imposed latent heating sources for their experiment were located between 50° and 130°E and between 116° and 94°W, too far to the west of the observations. These differences make it difficult to compare the current study with their results.

5 Synthetic thermal forcing and distribution of the subtropical anticyclone

The above analysis implies that the land-sea distribution in the summer subtropics is crucial in forming the distributions of the subtropical anticyclone, and the roles of the different kinds of diabatic heating across each continent and its adjacent oceans should be considered synthetically. To verify this further, the distributions of the circulation pattern against the local dominant thermal forcing along the subtropics are presented in Fig. 13 based, again, on the NCEP-NCAR reanalysis. The subtropical area in the two hemispheres is divided into five subareas, each having one continent located at its center. The distributions of the local dominant heating/cooling are presented in Figs. 13a and 13d. Those SEs stronger than 30 W m^{-2} are shaded in orange and red, and COs stronger than 50 W m^{-2} are in green. Because the radiative cooling over oceans is very strong, only those LO stronger than -220 W m^{-2} are shown in blue so that the secondary dominant heating (CO) over the western oceans can be presented. For the purposes of comparison, the labels L, S, C, and D marked in Fig. 1b are also marked in Fig. 13. Following Wu and Liu (2003), the labels L, S, C, and D are located, respectively, in the lobes over which the LO, SE, CO, and D heatings prevail (Fig. 5). The LOSECOD heating quadruplet is observed over each subarea, although its zonal scale varies from one subarea to another. In coordination with each heating quadruplet, the circulations in the upper troposphere (Figs. 13b and 13e) and in the lower troposphere (Figs. 13c and 13f) exhibit a specific pattern: the surface cyclone and upper-layer anticyclone are located

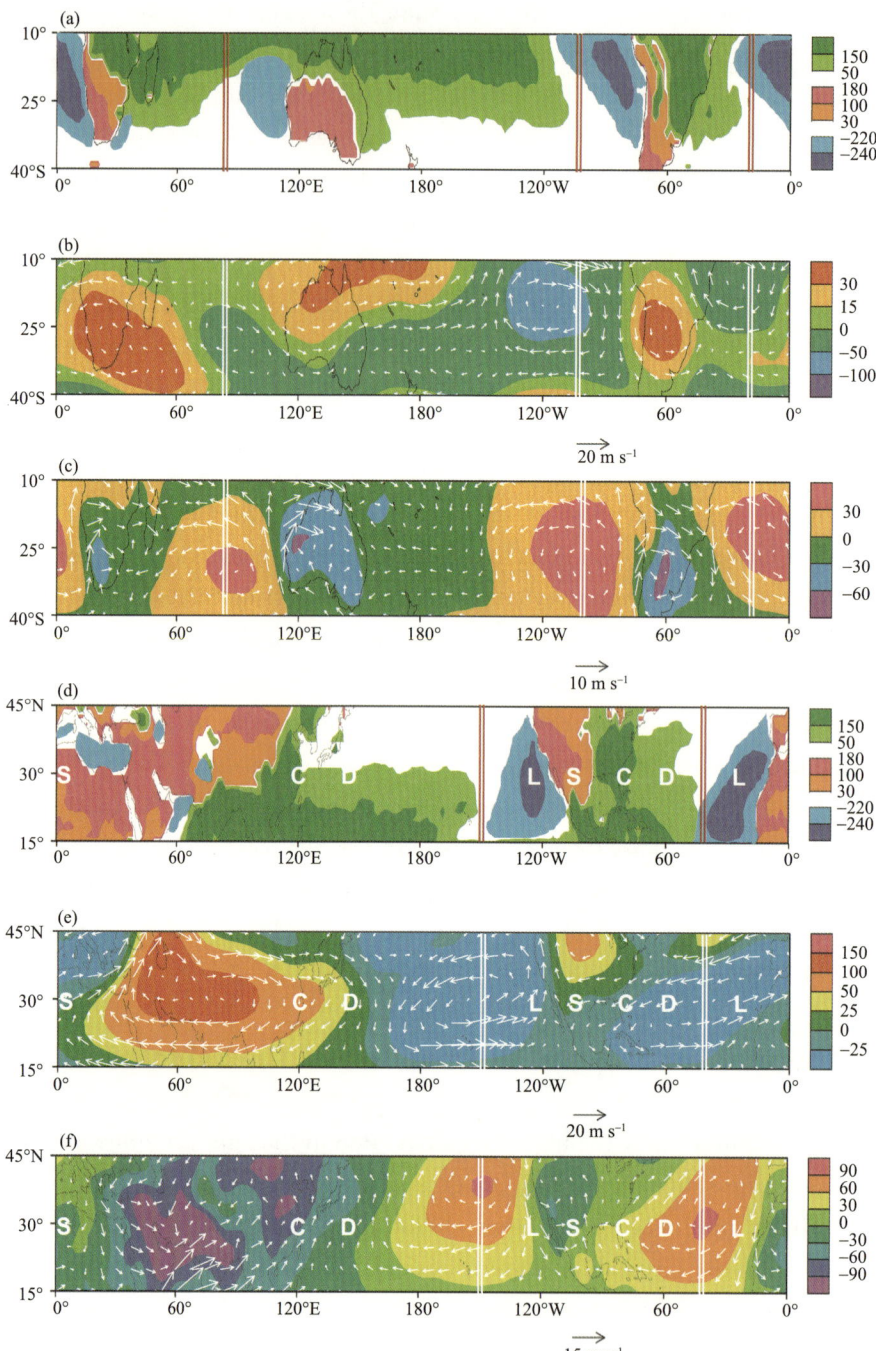

Fig. 13 Distributions in (a)—(c) Jan in the southern subtropics and (d)—(f) Jul in the northern subtropics of the (a), (d) different kinds of dominant heating with units of W m^{-2}, and the zonal deviations of wind (arrow) and geopotential height (shaded) at (b), (e) 200 and (c), (f) 1000 hPa with units of gpm. The sites L, S, C, and D as shown in Fig. 1b indicate, respectively, the locations in the LO, SE, CO, and D heating lobes

over the SE and CO lobes over the continent, with the surface cyclone accompanied with anticyclonic circulations on its western and eastern sides over oceans, and the upper-layer anticyclone accompanied with cyclonic circulations on the both sides over oceans. Because the amplitude of the forced circulation depends on the horizontal scale and intensity of the forcing, in response to the larger-scale SE and CO lobes over Asia the upper-layer Asian anticyclone is much stronger than its counterparts over other continents.

In addition, in response to the decrease/increase with height of the dominant heating in the lower/upper layers, the strong surface equatorward flow and upper-layer poleward flow are found over the LO and SE lobes in the west. Whereas in response to the increase/decrease with height of the dominant heating in the lower/upper layers, the surface poleward flow and stronger upper-layer equatorward flow are observed over the CO and D lobes in the east. The circulation pattern, therefore, presents a zonally asymmetric configuration. Furthermore, when the circulation patterns over the five subareas are tiled side-by-side, the lower-layer oceanic subtropical anticyclone and the upper-layer trough appear at the edges of the two adjacent subareas, whereas the upper-layer continental subtropical anticyclone and lower-layer cyclone appear at the center of each subarea. Therefore the global general circulation in the summer subtropics can be viewed as a mosaic of the unique circulation pattern over each subarea.

6 Conclusions and discussions

It has been shown that the atmospheric thermal adaptation to the sensible and condensation heating over land and radiative cooling over ocean are essential in producing subtropical anticyclones over continents in the upper troposphere and over oceans in the lower troposphere. The surface sensible heating over the land surface is fundamental in maintaining the summertime subtropical anticyclones near the surface. It also contributes significantly to the vertical out-of-phase features of the subtropical circulation. Because the isentropic surfaces in the subtropics in the lower-troposphere slope upward toward higher latitudes, adiabatic flows slide downward on the eastern side of the lower-tropospheric anticyclone, but upward on its western side. Because the horizontal advection of vorticity is neglectable along the anticyclonic ridgeline, the atmospheric vertical motion there is mainly determined by the vertical shear of the meridional winds at a steady state. Therefore the thermally adapted meridional flows cause descending motion over the eastern ocean and western continent, and ascending motion over the western ocean and eastern continent, just in phase with the adiabatic ascent/descent.

As pointed out by Rodwell and Hoskins (2001), the descending motion over the eastern ocean increases the air temperature and reduces its relative humidity above the cold ocean surface, stabilizing the air and forming low stratus clouds in the planetary boundary layer, in favor of the in situ radiative cooling near the top of the planetary boundary layer and the development of diabatic descending motion. On the other hand, the ascent over the western ocean decreases the air temperature and increases its relative humidity, in favor of the development of convection and diabatic ascent. The maximum radiative cooling over the eastern ocean occurs below the level $\sigma = 0.8$, and the maximum condensation heating over the western ocean occurs near $\sigma = 0.3$ (Fig. 5). Following the Sverdrup balance (1), stronger equatorward flow and anticyclone vorticity are produced on the eastern sides of the surface anticyclone over oceans and the upper-layer anticyclone over continents, whereas weaker poleward flow and cyclone vorticity are produced on the eastern sides of the upper-layer troughs over oceans and the surface cyclones over continents. Such asymmetric meridional winds induced by condensation heating and radiative cooling then strengthen the circulation pattern along the subtropics, which was produced by the sensible heating over land, and deform its configuration both in the upper and lower troposphere. Surface anticyclone centers are shifted eastward, whereas the upper-tropospheric troughs tilt from the east in high latitudes to the west in low latitudes. The pattern of asymmetric subtropical circulation is then formed. Therefore, we can reach the conclusion that different types of heating play different roles and all are important in the formation of the summertime subtropical anticyclones. Their maintenance is closely associated with the land-sea distribution and can be interpreted in terms of the atmospheric adaptation to diabatic heating by using the PV-q view and the Sverdrup balance (1). The global summer subtropical heating and circulation may be viewed as "mosaics" of the LOSECOD heating quadruplet and circulation patterns, respectively.

Although the primary relation between diabatic heating and the formation of the subtropical anticyclone has been investigated in this study, many questions still remain unclear or have been left untouched. For instance, how does a unique heating quadruplet exist over each continent area in summer? Why are the radiative cooling and vertical descending motion so strong and so well coupled over the eastern oceans? What is the role of air-sea interaction in this coupling? Besides these, some more practical problems concern the variability of the subtropical anticyclone at synoptic scales. In such circumstances, not only external forcing, but also internal forcing, including the interaction of circulations in different latitudes and between different weather systems, are both important. All these problems are significant for further understanding the nature and variations of the subtropical anticyclones and need to be investigated. We look forward to seeing more new insights on these points in the near future.

References

BOLIN B, 1950. On the influence of the earth's orography on the westerlies[J]. Tellus, 2: 184-195.

CHARNEY J G, ELIASSEN A, 1949. A numerical method for predicting the perturbations of the middle latitude westerlies [J]. Tellus, 1: 38-54.

CHEN P, 2001. Thermally forced stationary waves in a quasigeostrophic system[J]. Journal of the Atmospheric Sciences, 58: 1585-1594.

CHEN P, HOERLING M P, DOLE R M, 2001. The origin of the subtropical anticyclones[J]. Journal of the Atmospheric Sciences, 58: 1827-1835.

EGGER J, 1978. On the theory of planetary standing waves: July[J]. Beiträge zur Physik Der Atmosphäre, 51: 1-14.

GILL A E, 1980. Some simple solutions for heat-induced tropical circulation[J]. Quarterly Journal of the Royal Meteorological Society, 106: 447-662.

HOSKINS B J, 1987. Diagnosis of forced and free variability in the atmosphere[M]//CATTLE H. Atmospheric and Oceanic Variability. James Glaisher House: 57-73.

HOSKINS B J, 1991. Towards a PV-θ view of the general circulation[J]. Tellus, 43AB: 27-35.

HOSKINS B J, 1996. On the existence and strength of the summer subtropical anticyclones[J]. Bulletin of the American Meteorological Society, 77: 1287-1292.

HOUGHTON J T, DING Y, GRIGGS D J, et al, 2001. The Scientific Basis. Contribution of Working Group I to the Third Assessment Report of the Intergovernmental Panel on Climate Change[M]. Cambridge: Cambridge University Press: 471-523.

KALNAY E, KANAMITSU M, KISTLER R, 1996. The NCEP/NCAR 40-year reanalysis project[J]. Bulletin of the American Meteorological Society, 77: 437-471.

KANG I S, JIN K, LAU K M, et al, 2002. Intercomparison of atmospheric GCM simulated anomalies associated with the 1997/98 El Niño[J]. Journal of Climate, 15: 2791-2805.

LI G P, DUAN T Y, HAGINOYA S, et al, 2001. Estimates of the bulk transfer coefficients and surface fluxes over the Tibetan Plateau using AWS data[J]. Journal of the Meteorological Society of Japan, 79(2): 625-635.

LIN B D, 1983. The behavior of stationary waves and the summer monsoon[J]. Journal of the Atmospheric Sciences, 40: 1163-1177.

LIN X, JOHNSON R H, 1996. Heating, moistening, and rainfall over the western Pacific warm pool during TOGA COARE[J]. Journal of the Atmospheric Sciences, 53: 3367-3383.

LIU H, WU G X, 1997. Impacts of land surface on climate of July and onset of summer monsoon: A study with an AGCM plus SSiB[J]. Advances in Atmospheric Sciences, 14(3): 289-308.

LIU Y M, LIU H, LIU P, et al, 1999a. Spatially inhomogeneous diabatic heating and its impacts on the formation and variation of subtropical anticyclone. Ⅱ. Land-surface sensible heating and the subtropical anticyclone over the northeast Pacific and North America[J]. Acta Meteorologica Sinica, 57(4): 385-396.

LIU Y M, WU G X, LIU H, et al, 1999b. Spatially inhomogeneous diabatic heating and its impacts on the formation and variation of subtropical anticyclone. Ⅲ. Convective condensation heating and the subtropical anticyclone over South Asia and Northwest Pacific[J]. Acta Meteorologica Sinica, 57(5): 525-538.

LIU Y M, WU G X, LIU H, et al, 2001. Dynamical effects of condensation heating on the subtropical anticyclones in the eastern hemisphere[J]. Climate Dynamics, 17: 327-338.

NEWMAN M, SARDESHMUKH P D, BERGMAN J W, 2000. An assessment of the NCEP, NASA, and ECMWF reanalyses over the tropical west Pacific Warm Pool[J]. Bulletin of the American Meteorological Society, 81: 41-48.

NIGAM S, CHUNG C, DEWEAVER E, 2000. ENSO Diabatic Heating in ECMWF and NCEP-NCAR Reanalyses, and NCAR CCM3 Simulation[J]. Journal of Climate, 13: 3152-3171.

RODWELL M R, HOSKINS B J, 2001. Subtropical anticyclones and monsoons[J]. Journal of Climate, 14: 3192-3211.

WEBSTER P J, 1972. Response of the tropical atmosphere to local, steady forcing[J]. Monthly Weather Review, 100: 518-541.

WU G X, LI W P, GUO H, et al, 1997a. The sensible heat driven air-pump over the Tibetan Plateau and the Asian summer monsoon[C]// YE D Z. Collection in the Memory of Dr. Zhao Jiuzhang. Beijing: Science Press: 116-126.

WU G X, ZHANG X H, 1997b. The LASG global ocean-atmosphere-land system model GOALS/LASG and its simulation study[J]. App Meteor, 8(spec):15-28.

WU G X, LIU Y M, LIU P, 1999. Spatially inhomogeneous diabatic heating and its impacts on the formation and variation of subtropical anticyclone. Ⅰ. Scale analysis[J]. Acta Meteorologica Sinica, 57: 257-263.

WU G X, LIU Y M, 2000. Thermal adaptation, overshooting, dispersion and subtropical anticyclone. Ⅰ. Thermal adaptation and overshooting[J]. Chinese Journal of Atmospheric Sciences, 24: 433-446.

WU G X, LIU Y M, 2003. Summertime quadruplet heating pattern in the subtropics and the associated atmospheric circulation[J]. Geophysical Research Letters, 30(5): 1201-1204.

XUE Y K, SELLERS P J, KINTER J L, et al, 1991. A simplified biosphere model for global climate studies[J]. Journal of Climate, 4: 345-364.

YE D Z, WU G X, 1998. The role of the heat source of the Tibetan Plateau in the general circulation[J]. Meteorology and Atmospheric Physics, 67: 181-198.

YEH T C, 1950. The circulation of the high troposphere over China in the winter of 1945—1946[J]. Tellus, 2: 173-183.

ZHANG X H, CHEN K M, JIN X Z, et al, 1996. Simulation of thermohaline circulation with a twenty-layer oceanic general circulation model[J]. Theoretical and Applied Climatology, 55: 65-88.

ZHANG X H, SHI G Y, LIU H, et al, 2000. IAP Global Ocean-Atmosphere-Land System Model[M]. Beijing: Science Press:252.

Multi-Scale Forcing and the Formation of Subtropical Desert and Monsoon

WU Guoxiong[1], LIU Yimin[1], ZHU Xiaying[1], LI Weiping[2], REN Rongcai[1], DUAN Anmin[1], LIANG Xiaoyun[2]

(1 State Key Laboratory of Numerical Modeling for Atmospheric Sciences and Geophysical Fluid Dynamics (LASG), Institute of Atmospheric Physics (IAP), Chinese Academy of Sciences, Beijing 100029, China; 2 National Climate Center, China Meteorological Administration, Beijing 100081 China)

Abstract: This study investigates three types of atmospheric forcing across the summertime subtropics that are shown to contribute in various ways to the occurrence of dry and wet climates in the subtropics. To explain the formation of desert over the western parts of continents and monsoon over the eastern parts, we propose a new mechanism of positive feedback between diabatic heating and vorticity generation that occurs via meridional advection of planetary vorticity and temperature. Monsoon and desert are demonstrated to coexist as twin features of multi-scale forcing, as follows.

First, continent-scale heating over land and cooling over ocean induce the ascent of air over the eastern parts of continents and western parts of oceans, and descent over eastern parts of oceans and western parts of continents. Second, local-scale sea-breeze forcing along coastal regions enhances air descent over eastern parts of oceans and ascent over eastern parts of continents. This leads to the formation of the well-defined summertime subtropical LOSECOD quadruplet-heating pattern across each continent and adjacent oceans, with long-wave radiative cooling (LO) over eastern parts of oceans, sensible heating (SE) over western parts of continents, condensation heating (CO) over eastern parts of continents, and double dominant heating (D: LO+CO) over western parts of oceans. Such a quadruplet heating pattern corresponds to a dry climate over the western parts of continents and a wet climate over eastern parts. Third, regional-scale orographic-uplift-heating generates poleward ascending flow to the east of orography and equatorward descending flow to the west.

The Tibetan Plateau (TP) is located over the eastern Eurasian continent. The TP-forced circulation pattern is in phase with that produced by continental-scale forcing, and the strongest monsoon and largest deserts are formed over the Afro-Eurasian Continent. In contrast, the Rockies and the Andes are located over the western parts of their respective continents, and orography-induced ascent is separated from ascent due to continental-scale forcing. Accordingly, the deserts and monsoon climate over these continents are not as strongly developed as those over the Eurasian Continent.

A new mechanism of positive feedback between diabatic heating and vorticity generation, which occurs via meridional transfer of heat and planetary vorticity, is proposed as a means of explaining the formation of subtropical desert and monsoon. Strong low-level longwave radiative cooling over eastern parts of oceans and strong surface sensible heating on western parts of continents generate negative vorticity that is balanced by positive planetary vorticity advection from high latitudes. The equatorward flow generated over eastern parts of oceans produces cold sea-surface temperature and stable stratification, leading in turn to the formation of low stratus clouds and the maintenance of strong in situ longwave radiative cooling. The equatorward flow over western parts of continents carries cold, dry air, thereby enhancing local sensible heating as well as moisture release from the underlying soil. These factors result in a dry desert climate. Over the eastern parts of continents, condensation heating generates positive vorticity in the lower troposphere, which is balanced by negative planetary vorticity advection of the meridional flow from low latitudes. The flow brings warm and moist air, thereby enhancing local convective instability and condensation heating associated with rainfall. These factors produce a wet monsoonal climate. Overall, our results demonstrate that subtropical desert and monsoon coexist as a consequence of multi-scale forcing along the subtropics.

Keywords: Meteorology and atmospheric dynamics (Climatology; General circulation; Precipitation)

1 Introduction

Compared with the large number of studies on the formation of monsoons (e.g., Webster et al., 1998; Wang, 2006), few have investigated the dynamics of desert formation. Prior to the 1970s, the formation of desert had long been attributed to the descending arm of the Hadley Cell along the subtropics: its adiabatic heating was thought to reduce the atmospheric relative humidity, while the associated horizontal divergence near the surface suppressed the formation of cloud, resulting in subtropical aridity. However, severe subtropical arid climates occur during summer when the Hadley Cell is weakest, whereas such aridity is weak during winter when the Hadley Cell is strongest. In addition, many moist, monsoonal areas also exist within the subtropics. Therefore, the climate in such areas cannot be readily explained in terms of the traditional view (i.e., the Hadley Cell).

Charney (1975) proposed a local biosphere-albedo-atmosphere feedback mechanism to explain the enhancement of the Sahel Desert, arguing that over-grazing in arid areas increases surface albedo, and that radiative heat loss over desert areas is compensated by adiabatic descent, which in return results in a decrease in relative humidity and leads to desertification. Such a localized mechanism ignores the influence of advection, and is applicable only to tropical areas (e.g., Bounoua and Krishnamurti, 1991), where the Burger Number is large (Hoskins, 1986).

Most deserts are found in the subtropics, where the Burger Number is close to one. In these areas, the advection of heat is comparable with adiabatic heating in balancing external heating, and cannot be ignored when considering the formation of desert. Yang et al. (1992) proposed a linkage between aridity and convection via a closed overturning "Walker Cell". Using trajectory analysis, Rodwell and Hoskins (1996) found that descent over the eastern Sahara, Mediterranean, and Kyzylkum Desert are mainly of mid-latitude origin (their Fig. 9) rather than resulting from Walker-type circulation. The authors proposed a monsoon-desert mechanism for desertification whereby remote diabatic heating in the Asian monsoon region induces a Rossby-wave pattern to the west, resulting in clear sky and creating a local diabatic enhancement "which effectively doubles the strength of descent".

Across the subtropics, zonal advection of vorticity or temperature is weak. In the present study, after investigating the characteristics of atmospheric forcing across the summer subtropics at different spatial scales, a new mechanism is proposed for the enhancement of dry and moist climate, involving a positive feedback between diabatic heating and vorticity generation via meridional advection of planetary vorticity and temperature. With reference to this mechanism, desert over the western parts of continents and monsoon over the eastern parts of continents are demonstrated to represent coexisting twin features of a single system.

The remainder of the manuscript is organized as follows. In Section 2, we briefly introduce the model and data used in the present analysis. Large-scale continental forcing and local-scale coastal sea-breeze forcing in summer are analyzed in Sections 3 and 4, respectively, and are used as a basis upon which to explain the formation of summertime quadruplet heating across each subtropical continent and its adjoining oceans. New, local positive feedback mechanisms for the formation of monsoon and desert are presented in Section 5, and the impacts of regional-scale mountain forcing on subtropical arid and wet climates are considered in Section 6. Section 7 provides a summary of the main findings.

2 Data and Model

We employed the reanalysis of the National Centers for Environment Prediction/National Center for Atmospheric Research (NCEP/NCAR; Kalnay et al., 1996) and Xie-Arkin data (Xie and Arkin, 1996) from 1980 to 1997 to investigate the separate and combined impacts of different thermal forcings on the formation of monsoon and desert. Diabatic heating in the NECP/NCAR reanalysis is not observation data, but the product of a general circulation model (GCM) that depends on physical parameterization schemes (Newman et al., 2000). Comparisons among NCEP heating, Q_1 from ECMWF data (Nigam et al., 2000; Rodwell and Hoskins, 2001), and TOGA data (Lin and Johnson, 1996) reveal similar horizontal distributions of column-integrated heating and the vertical heating profile.

Duan and Wu (2005) compared NCEP reanalysis data and two other data sets: observations from six Automatic Weather Stations upon the Tibetan Plateau (abbreviated as TP hereafter) from July 1993 to March 1999 (Li et al., 2001), and GAME-IOP reanalysis data from 1 April to 31 October 1998, as compiled by the Japan Meteorological Research Institute and the Japanese Meteorological Agency (Yamazaki et al., 2000a, 2000b; see also http://game.suiri.tsukuba.ac.jp/cdroms/CD-ROM.html). The comparison revealed that the surface sensible heat flux and latent heat flux provided by NCEP/NCAR agree with the AWS data, and there is no significant difference in spatial/temporal variations and magnitude between the NCEP/NCAR and GAME-IOP daily data sets. This finding validates the use of the NCEP/NCAR reanalysis in the present study, although a degree of caution is required because AWS and GAME-IOP data have limited spatial/temporal coverage.

Numerical experiments were performed using the Global Ocean-Atmosphere-Land System model developed at LASG (IAP/LASG GOALS; Wu et al., 1997a; Wu T W et al., 2003). The atmospheric component is a spectral GCM with nine vertical levels in σ-coordinates and that is rhomboidally truncated at wave number 42 in the horizontal. Land-surface processes are represented by the SSiB model (Xue et al., 1991), which has previously been implemented in the atmospheric component (Liu and Wu, 1997). The GOALS climate model can simulate the mean climate reasonably well and has been used in climate studies for a variety of purposes (IPCC, 2001; Kang et al., 2002).

In this study, the required sea surface temperature (SST) and sea ice are prescribed based on climate mean observation data for the period 1979—1988, as compiled for the Atmospheric Model Intercomparison Program (AMIP). Because the adjustment period measured by the global mean surface temperature is about 6 months for each experiment (data not shown), all the experiments in this study are integrated for 12 model years, and the monthly means for the last 10 years are calculated for comparison.

3 Land-Sea Distribution and Continental-Scale LOSECOD Heating

The global distributions of column-integrated total diabatic heating in January and July are shown in Figs. 1a and 1b, respectively. The total diabatic heating includes diffusive sensible heating (SE), condensation latent heating (CO), long-wave radiative cooling (LO), and short-wave radiative heating (SH). The atmosphere along the equator and in near-equator tropical areas is a strong heating source except over the eastern Pacific and Atlantic, where it becomes a permanent heat sink in association with the existence of in situ cold tongues of SST. There is generally no apparent seasonal change in these tropical areas. In contrast to the tropics, a marked seasonal change in total diabatic heating occurs in the extra-tropics. In January (Fig. 1a), during the Northern Hemisphere winter, atmospheric heating is positive over oceans, but negative over continents. The Southern Hemisphere is in summer during January, and the atmosphere is characterized by heating over continents and cooling over oceans. In the northern extra-tropics during July (Fig. 1b), atmospheric heating occurs over continents, whereas cooling occurs over oceans. The Southern Hemisphere is in winter at this time, and atmospheric heating over continents is negative, while weak atmospheric heating is observed over mid-latitude oceans. In summary, the total heating in the summer extra-tropics is positive over continents, but negative over oceans:

$$\int_0^{P_0} Q \frac{\mathrm{d}p}{g} > 0, \quad \text{over continents} \tag{1}$$

$$\int_0^{P_0} Q \frac{\mathrm{d}p}{g} < 0, \quad \text{over oceans} \tag{2}$$

where Q represents total heating ($=$SE+CO+LO+SH), P_0 is surface pressure, and g is acceleration due to gravity.

According to the PV-θ view (Hoskins, 1991) and the theory of thermal adaptation (Wu and Liu, 2000; Liu et al., 2001), the atmosphere is expected to respond to summertime continental heating by the generation of cyclonic circulation in the lower troposphere and anticyclonic circulation in the upper troposphere (Fig. 2a), whereas it responds to cooling over oceans by the generation of anticyclonic circulation near the surface and cyclonic circulation in the upper troposphere (Fig. 2b). Because in summertime the zonal flow and vorticity advection across the subtropics are weak, the quasi-steady state geostrophic vorticity equation,

$$u \frac{\partial \zeta}{\partial x} + v \frac{\partial \zeta}{\partial y} = -\beta v - f \nabla \cdot \mathbf{V}$$

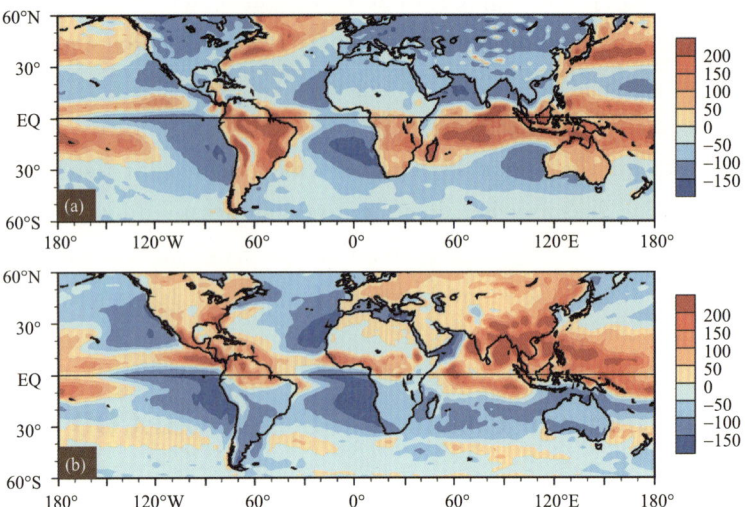

Fig. 1 Distributions of column-integrated total heating averaged over the period 1980—1997 for January (a) and July (b). Unit is Wm^{-2}

can be simplified to

$$0 \approx -\beta v - f \nabla \cdot \boldsymbol{V} \tag{3}$$

Under the first-order approximation $0 \approx \nabla \cdot \boldsymbol{V} + \frac{\partial w}{\partial z}$, Eq. (3) can be rewritten as

$$\frac{\partial w}{\partial z} \approx \frac{\beta}{f} v$$

By assuming a characteristic height H and a normal mode distribution of w with height, we have

$$w \propto -H^2 \frac{\partial^2 w}{\partial z^2} \propto -\frac{\beta H^2}{f} \cdot \frac{\partial v}{\partial z} \tag{4}$$

According to Eq. (3), on the eastern side of the heating region and the western side of the cooling region, the negative advection of planetary vorticity brought about by poleward flow in the lower troposphere should be balanced by in situ positive vorticity generated in response to horizontal convergence, while the positive advection of planetary vorticity brought about by equatorward flow in the upper troposphere should be balanced by in situ negative vorticity generated in response to horizontal divergence. Under the constraint of fluid continuity, ascending motion should develop on the eastern side of the heating region over continent and the western side of the cooling region over ocean. In other words, the ascending motion over the eastern continent and the western ocean is forced by the negative vertical shear of the in situ meridional wind, as described in Eq. (4). In contrast, the vertical shear of the in situ meridional wind is positive on the western side of the heating region over continent and the eastern side of the cooling region over ocean. According to Eq. (4), descending motion should develop in these regions, as shown schematically in Fig. 2.

Figure 3 shows the July mean vertical cross-section of streamline (u, w) along 30°N, in which the vertical velocity has been amplified by a factor of 100. The general features in the cross-section show a good fit with the thermal adaptation described above: ascending motions develop over the eastern parts of the North American and Eurasian continents and over the western parts of the North Pacific and North Atlantic, whereas descending motions develop over the western parts of the North American and Eurasian continents and over the eastern parts of North Pacific and North Atlantic.

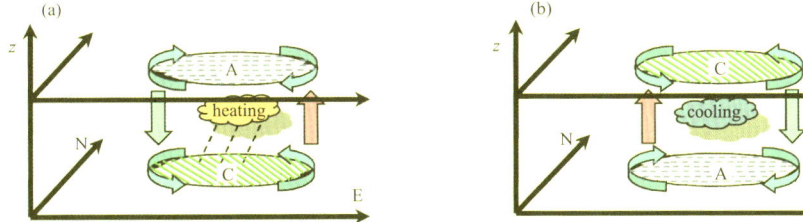

Fig. 2 Schematic diagram showing atmospheric adaptation on the rotating earth to external heating in the summertime subtropics: (a) Diabatic heating generates cyclonic circulation (C) near the surface and anticyclone circulation (A) in the upper troposphere, with air ascent developing over the eastern part of the circulation and descent over the western part. (b) Diabatic cooling generates anticyclonic circulation near the surface and cyclone circulation in the upper troposphere, with air descent developing over the eastern part of the circulation and ascent over the western part

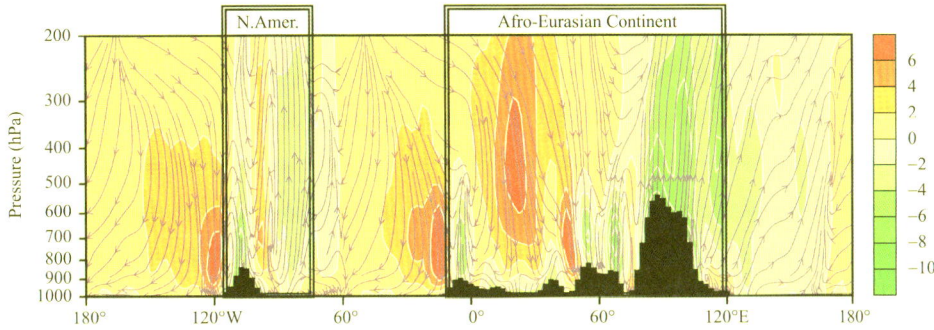

Fig. 3 Vertical cross-section of subtropical circulation along 30°N (vertical velocity has been amplified by a factor of 100). The boxes indicate the locations of the North American and Afro-Eurasian continents. Blue shading indicates orography, and color shading indicates the intensity of the p-vertical velocity ω (interval, 2 Pa s^{-1}), as shown in the color scale to the right

4 Local-Scale Sea Breeze and Heating Pattern across the Summertime Subtropics

The color shading in Fig. 3 shows the magnitude of the vertical p-velocity ω along 30°N. Although ascending motions develop over both eastern parts of continents and western parts of oceans, the rising over land is apparently stronger than over the sea to the east. Similarly, the descending motion over the eastern ocean is much stronger than that over the neighboring land to the east. This feature can also be detected from the distributions of heating/cooling in the summertime subtropics (see Fig. 1): atmospheric heating over the eastern subtropical continents is always stronger than that over the neighboring ocean to the east, and atmospheric cooling over the eastern subtropical oceans is always the strongest along the subtropical zones.

To examine the origin of the above pattern, Fig. 4 shows the velocity potential χ and the corresponding divergent wind component at 1000 hPa. In the northern extra-tropics in January (Fig. 4a), the velocity potential is generally low over land and high over ocean. Air flows from land to ocean, occurring as a land-breeze except over the eastern subtropical North Atlantic. In July (Fig. 4b) in the Northern Hemisphere, the velocity potential is low over oceans and high over continents. During this month, a strong sea breeze dominates coastal regions in the summertime extra-tropics. This seasonal contrast in the occurrences of land and sea breezes is also evident in the Southern Hemisphere, although not as prominent as that in the Northern Hemisphere.

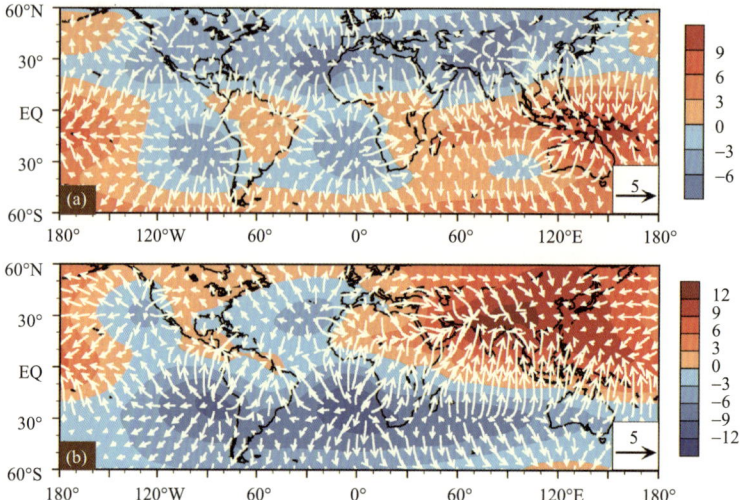

Fig. 4 Distributions of the monthly mean velocity potential (shading, unit is 10^{-6} m^{2}s^{-1}) and divergent wind (arrows, ms^{-1}) at 1000 hPa in January (a) and July (b)

In summary, summertime sea breezes along subtropical coastal regions form a localized circulation, with ascending motion over coastal land and descending motion over the adjacent coastal sea. When such a localized sea-breeze circulation is superimposed on the continental-scale circulation induced by the land-sea distribution (see Fig. 2), the ascending motion over the eastern continent is enhanced, while the ascending motion over the neighboring western ocean region, to the east of the continent, is suppressed to some extent. Similarly, the descending motion over the eastern ocean region is intensified, while the descending motion over the neighboring western continent is weakened. These findings explain the asymmetry in the observed distribution of vertical velocity (Fig. 3).

The distribution of vertical motion associated with continental-scale forcing and local sea breeze is closely related to in situ heating/cooling. Figure 5 shows various vertical heating profiles at different sites across the North American and Eurasian continents along 30°N in July. The locations of the selected sites are as follows: LO (122°W, 30°N), SE (108°W, 30°N), CO (80°W, 30°N), and D (60°W, 30°N) for North America (Fig. 5a), and LO (24°W, 30°N), SE (0, 30°N), CO (120°E, 30°N), and D (145°E, 30°N) for Afro-Eurasia (Fig. 5b). These sites are located within different heating lobes (see Fig. 6b), and are representative of the heating characteristics of the surrounding areas. Site LO is located over the eastern ocean, where the column-integrated total heating is negative. The strong atmospheric sinking in the region (Fig. 3) indicates strong atmospheric cooling. The characteristics of heating in this region can be described as follows:

$$\text{At Site LO,} \begin{cases} \int_0^{p_0} Q \frac{\mathrm{d}p}{g} < 0 \\ \mathbf{V} \cdot \nabla \theta < 0, \quad \text{in the free atmosphere} \end{cases} \quad (5)$$

In these LO regions, condensation heating is very weak, and surface sensible heating is negligible (Fig. 5, left-hand panels). These regions are located to the east of surface oceanic anticyclones that are produced by summertime atmospheric cooling over oceans, as discussed in Section 3, and are marked by strong equatorward surface wind-drag. Because the SST in these regions is cold due to the upwelling of cold sea water in association with surface equatorward flow, and because the atmosphere is subject to adiabatic warming due to strong descent, atmospheric stratification in the boundary layer is stable, and low-level stratus cloud is formed persistently (data not shown). The total heating profile is therefore dominated by long-wave radiative cooling, with a maximum cooling of about 6 ℃ per day at about 900 hPa, close to the top of the low-level stratus cloud.

Site SE is located over the western continent, where column-integrated total heating is positive (Fig. 1b). Weak atmospheric sinking in this region (Fig. 3) indicates cooling in the free atmosphere. The characteristics of heating in this region can be described as follows:

$$\text{At Site SE}, \begin{cases} \int_0^{p_0} Q \frac{\mathrm{d}p}{g} > 0 \\ \mathbf{V} \cdot \nabla \theta < 0, \quad \text{in the free atmosphere} \end{cases} \quad (6)$$

Therefore, atmospheric heating should be confined to the boundary layer, and the total heating profile in this region follows the diffusive sensible heating profile (Fig. 5, panels located second from left).

Site CO is located over the eastern continent, where column-integrated total heating is strongly positive. The atmosphere ascends intensely in this region (Fig. 3), indicating strong heating in the free atmosphere. The characteristics of heating in this region can be described as follows:

$$\text{At Site CO}, \begin{cases} \int_0^{p_0} Q \frac{\mathrm{d}p}{g} > 0 \\ \mathbf{V} \cdot \nabla \theta > 0, \quad \text{in the free atmosphere} \end{cases} \quad (7)$$

Therefore, the total heating profile in this region is strongly consistent with the profile of deep condensation heating (Fig. 5, panels located third from left).

Site D is located over the western ocean, where the column-integrated total heating is negative (Fig. 1b). The atmosphere also ascends in this region, albeit much more weakly than at Site CO (Fig. 3). This finding indicates the existence of condensation heating in the free atmosphere. The characteristics of heating in this region can be described as follows:

$$\text{At Site D}, \begin{cases} \int_0^{p_0} Q \frac{\mathrm{d}p}{g} < 0 \\ \mathbf{V} \cdot \nabla \theta > 0, \quad \text{in the free atmosphere} \end{cases} \quad (8)$$

Therefore, the total heating profile in this region is determined by two dominant heating profiles: long-wave radiative cooling (LO) and condensation heating (CO) (Fig. 5, right-hand panels).

When these dominant heating features are linked side by side, it is then possible to explain the formation of the specific summertime heating pattern identified by Wu and Liu (2003, Figs. 3b and 4b) and Liu et al. (2004). As demonstrated in Figs. 6b and 7b, in both hemispheres the dominant summertime subtropical heating, from west to east across each continent and adjacent oceans, is well organized into a specific pattern: long-wave radiative cooling (LO) to the west of the continent, surface sensible heating (SE) in the western part of the continent, deep condensation heating (CO) in the eastern part, and double dominant heating (D=LO+CO) to the east. Consequently, the summertime subtropical diabatic heating shows a mosaic of the LOSECOD quadruplet heating pattern.

The summertime zonal flow along the subtropics is weak, and the atmospheric adaptation to external heating is manifest as the following Sverdrup vorticity balance (Wu and Liu, 2000; Liu et al., 2001)

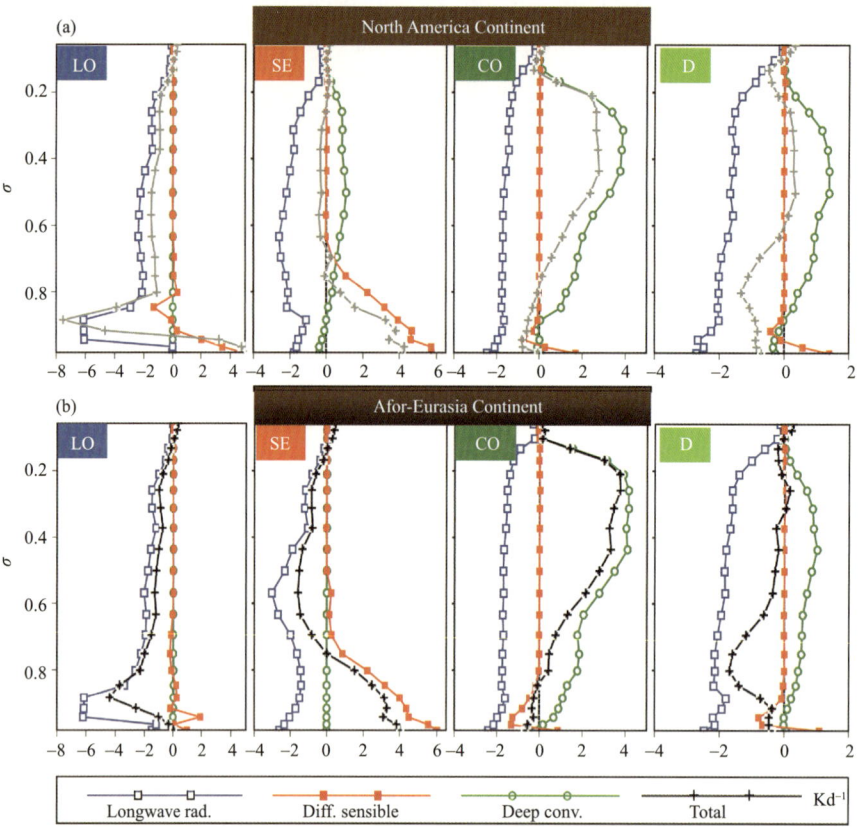

Fig. 5 Vertical heating profiles in sigma (σ) coordinates for locations LO, SE, CO, and D (see Fig. 6b for locations) across (a) North America and adjacent oceans, and (b) the Afro-Eurasian Continent and adjacent oceans (after Wu and Liu, 2003; Liu et al., 2004). The lines with open squares, filled squares, open circles, and "plus" signs represent long-wave radiative cooling (LO), diffusive sensible heating (SE), deep convective condensation heating (CO), and total heating, respectively. Unit is Kd^{-1}

or, to distinguish this equation from the oceanic Sverdrup balance, thermal vorticity balance (TVB):

$$\beta v \approx \theta_z^{-1}(f+\zeta)\frac{\partial Q}{\partial z} \tag{9}$$

This equation states that the generation of vorticity due to summertime diabatic heating in the subtropics is balanced mainly by the meridional transfer of planetary vorticity. Thus, increased heating with height produces poleward flow, whereas decreased heating with height generates equatorward flow. According to the vertical distributions of heating profiles shown in Fig. 5, the total heating profiles over the eastern ocean and western continent decrease with height in the lower troposphere, and increase with height in the upper troposphere. Therefore, in boreal summer, strong near-surface northerlies and upper-layer southerlies are expected to be generated over the eastern ocean and western continent. Over the eastern continent and western ocean areas, the total heating increases with increasing height in the lower troposphere, and shows a rapid decrease with height in the upper troposphere. Therefore, near-surface southerlies and strong upper-layer northerlies are expected to be generated over these areas. The same patterns apply to austral summer; consequently, anticyclonic circulation appears in the upper troposphere over continents, with tilting troughs over oceans (Figs. 6a and 7a). Near the surface, cyclone circulation is intensified over land, while anticyclone circulation is enhanced over ocean, especially over the eastern ocean (Figs. 6c and 7c).

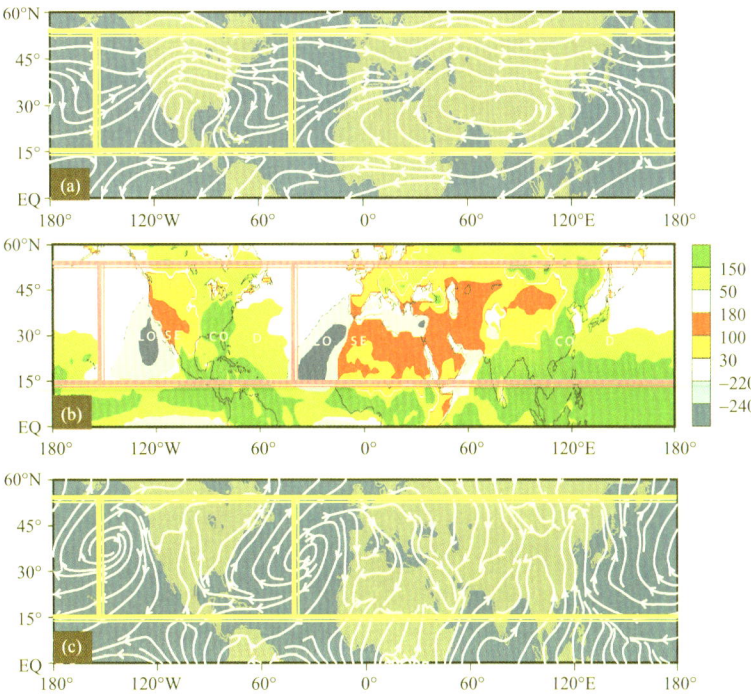

Fig. 6 Distributions of July mean quadruplet LOSECOD dominant heating (b, unit is Wm^{-2}; adopted from Fig. 3b of Wu and Liu, 2003), and circulation at 200 hPa (a) and 1000 hPa (c). Each grid along the subtropics has one continent in the center and is bounded by the longitudes at which the centers of sea-surface anticyclones are located. The abbreviations LO, SE, CO, and D in Panel (b) indicate the geographic locations at which the vertical profiles of heating are plotted in Fig. 5

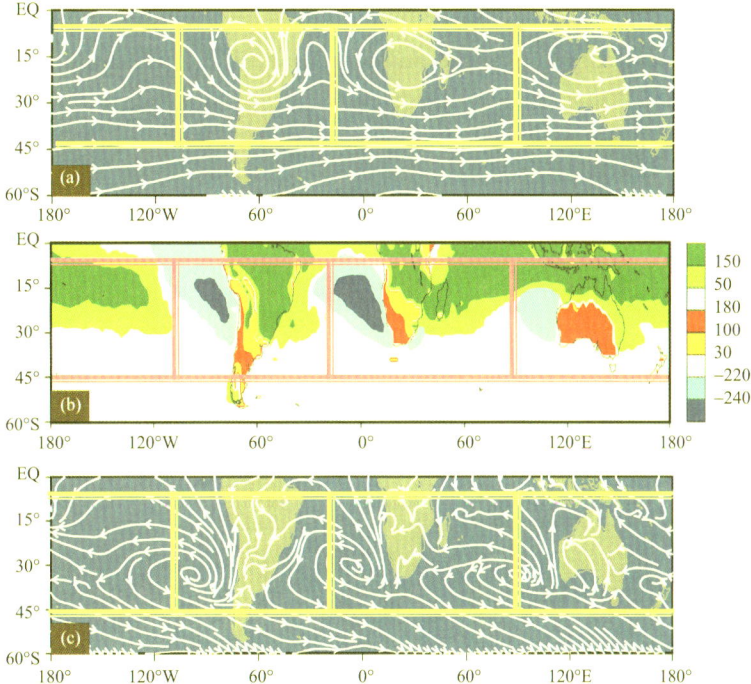

Fig. 7 Distributions of January mean quadruplet LOSECOD dominant heating (b, unit is Wm^{-2}; adopted from Fig. 4b of Wu and Liu, 2003), and circulation at 200 hPa (a) and 1000 hPa (c). Each grid along the subtropics has one continent in the center and is bounded by the longitudes at which the centers of sea-surface anticyclones are located

5 Local Positive Feedback Mechanism and Development of Dry and Wet Climate

In nature, the atmospheric heating source is not just the cause of atmospheric motion; it is also a consequence of the general circulation (Eady, 1950). Atmospheric ascent over the eastern continent and descent over its western part, as induced by continental-scale and sea-breeze coastal-scale forcing, favor the occurrence of a wet climate event in the east and dry climate event in the west. However, the formation of desert and monsoon requires the persistence of such events, requiring in turn a positive feedback mechanism.

Over the eastern ocean and western continent, the increase in radiative cooling with height in the LO region and the decrease in sensible heating with height in the SE region (Fig. 5) generate negative vorticity in the lower troposphere. In the absence of zonal advection and at a steady state, this negative vorticity generation is balanced mainly by positive planetary vorticity advection from higher latitudes, resulting in strong meridional wind from higher latitudes (Eq. 9). As discussed above, over the eastern parts of oceans, this strong equatorward meridional wind causes strong ocean upwelling and cold SST. The wind also enhances in situ stable stratification and sustains low-level stratus cloud and long-wave radiative cooling at the cloud top, thereby leading to the further generation of negative vorticity. Accordingly, a positive feedback is established between longwave radiative cooling and negative vorticity forcing.

Over the western continent, this meridional flow brings cold and dry air from high latitudes into the area, resulting in an increased difference in temperature and moisture between the land surface and overlying air, and enhancing the surface sensible heat flux and evaporation. Strong negative vorticity is continuously generated and the soil becomes drier. Thus, the meridional transfer of heat and vorticity results in a positive feedback between negative vorticity generation and surface sensible heat transfer, and the soil becomes progressively drier. This positive feedback process continues until it is balanced by surface friction (which is ignored in Eq. (9)). Dry, arid, and even desert-like climates are formed over the western continent, as shown in Figs. 6 and 7 (and also in Fig. 12, which is discussed later in the text).

Over the eastern continent and adjacent western ocean areas, the increase in condensation heating with height (Fig. 5) generates positive vorticity. In the absence of zonal advection and at a steady state, this positive vorticity generation is balanced mainly by negative planetary vorticity advection from lower latitudes, resulting in strong meridional wind from lower latitudes Eq. (9). This poleward meridional flow brings warm and moist air into the region, and creates convective instability, enabling convection to develop. Condensation heating is then sustained, and positive vorticity continues to be generated. Thus, the meridional transfer of heat and vorticity generates a positive feedback between positive vorticity generation and condensation heating, until equilibrated by frictional dissipation. Wet and even monsoon-like climates are formed over the eastern continent, as shown in Figs. 6, 7, and 12.

6 Regional-Scale Orographic Forcing and Formation of Monsoon and Desert

Figures 1 and 3 reveal significant differences between East Asia and North and South America in the intensity of vertical motion and the horizontal extent of convective condensation heating. The ascending air over East Asia covers the area east of 75°E, and becomes intense east of 90°E (Fig. 3). In contrast, the ascent of air over North and South America is relatively weak, and the regions of ascending air produced by mountain forcing and continental forcing are separated by descending air.

Such a phenomenon can be seen more clearly from the horizontal distribution of vertical velocity at 500 hPa, as shown in Fig. 8. Within the subtropics in Eurasia during July (Fig. 8a), atmospheric descent dominates the western part of the continent except over the Iran High, where orography-forced rising occurs. The centers of strong descent near 20°E and 60°E correspond to a Mediterranean-type persistent dry summertime climate and the Kyzylkum Desert, respectively; the strong ascent dominates the vast eastern continent from 75°E eastwards. Intense rising centers of greater than -8×10^{-2} Pas^{-1} over the TP and East China are linked together. In contrast, the two rising centers over North America (Fig. 8b) are separated by a region of descent: one ascending center of -4×10^{-2} Pas^{-1} is located over North Florida, and another center with similar intensity is located over the Rocky Mountains.

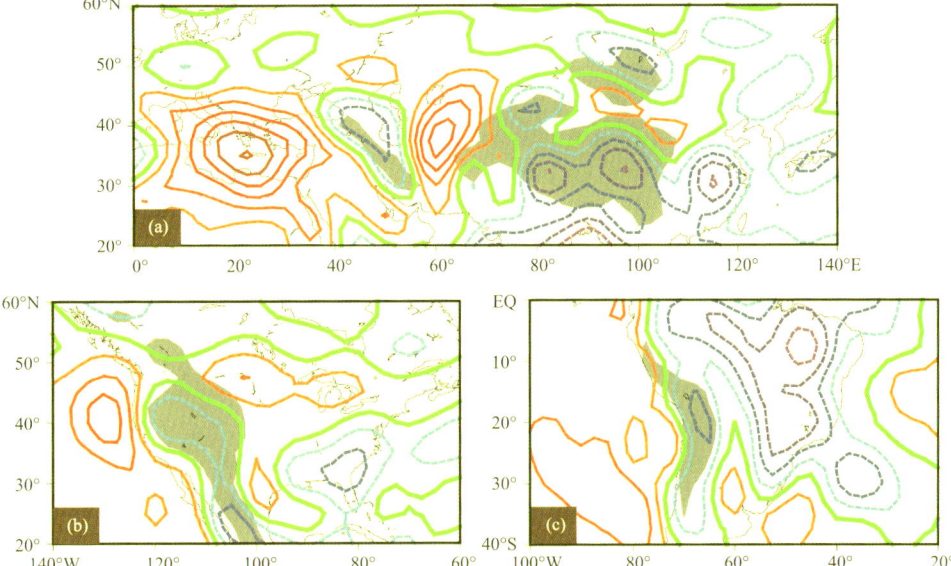

Fig. 8 Distributions of "vertical velocity" w at 500 hPa over (a) the Afro-Eurasia Continent in July, (b) North America in July, and (c) South America in January. Contour interval is 2×10^{-2} Pas^{-1}. Solid and dashed lines indicate air descent and ascent, respectively; the green line denotes the zero isoline, and gray shading indicates topography above 1500 m

Similar features are observed in South America. Within the subtropics in South America during January (Fig. 8c), one ascending center is located over east Brazil, and another over the Andes, separated by descent over central South America. Furthermore, because the dominant heating over the eastern continent is convective condensation heating (Fig. 5), the horizontal extent of diabatic heating in these areas is consistent with convective condensation heating induced by monsoon or monsoon like rainfall. As shown in Fig. 1, the heating area of >100 Wm^{-2} over East Asia extends northeastward to Northeast China. In contrast, comparable areas in other continents are confined to subtropical regions.

The above findings indicate that the East Asian monsoon is the strongest subtropical monsoon in the world, as also reported in many previous studies (e.g., Tao and Chen, 1987; Yanai and Li, 1994; Yanai and Wu, 2006). Other studies (Yeh et al., 1957; Wu et al., 1997b) have suggested that the strength of the East Asian monsoon is due to the existence of the TP. To further investigate the formation of the strong East Asian monsoon, we perform idealized numerical experiments to demonstrate how the location of the plateau affects the intensity and extent of desert and the East Asian monsoon.

Figure 9 shows the July mean vertical heating profiles averaged over the western part of the TP (78°—90°E, 29°—38°N) and the eastern part (90°—105°E, 27°—40°N). The magnitude of large-scale near-surface condensation heating over the eastern part of the plateau (4.5 Kd^{-1}) is slightly higher than that over the western part (3.5 Kd^{-1}), and the magnitude of longwave radiative cooling over the western part is slightly higher than that over the eastern part; however, all other types of heating are similar over the two parts. The profile of total heating is positive throughout the entire troposphere, with a striking maximum of about 10 Kd^{-1} near the surface. The atmospheric adaptation to such uplift heating is expected to generate strong surface cyclonic circulation and upper-layer anticyclonic circulation. The superimposition of these orography-induced local circulations on continental-scale circulations is likely to affect the climate of the surrounding area.

To verify the above hypothesis, we performed idealized aqua-planet numerical experiments based on the GOALS SAMIL GCM, which is described in Section 2. An idealized Afro-Eurasian continent, as outlined by the thick dashed lines in Figs. 10 and 11, is embedded into the aqua planet. The use of these types of aqua-planet sensitivity experiments can simplify the complicated dynamical diagnosis and accentuate the physics under consideration. Furthermore, because the outcomes of such experiments can be linked to the corresponding theoretical understanding, ensemble evaluations are generally not required. The SST and sea ice are the zonal averaged AMIP products. The seasonal variable land-surface albedo and solar radiation are prescribed using the zonal averaged products of the long-term integration of the GOALS SAMIL GCM. The surface temperature over land is forecasted in the model. The model is integrated for 12 years, and the outcomes from the last 10 years are extracted for the current analysis.

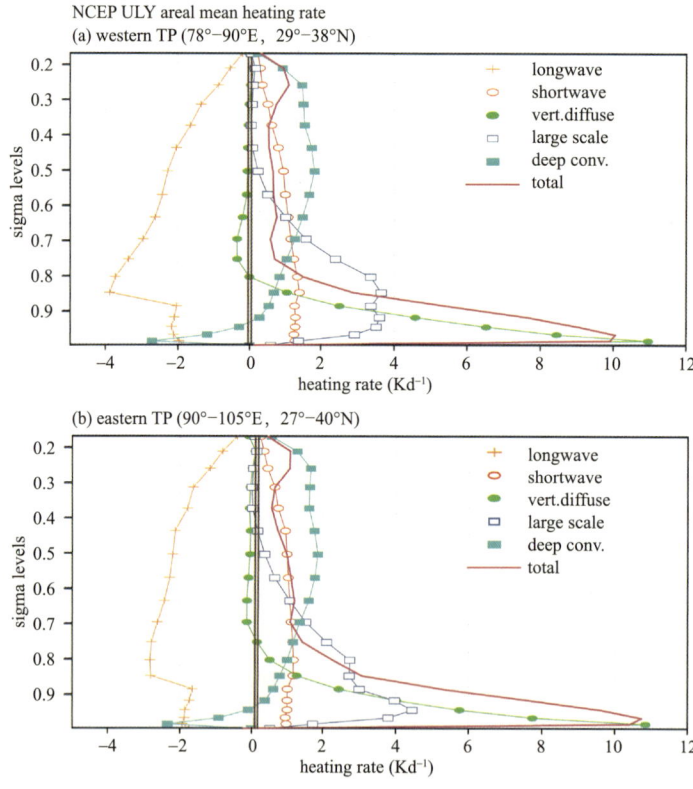

Fig. 9 July mean vertical profiles in sigma (s-) coordinates of various heatings over (a) the western Tibetan Plateau (78°—90°E, 29°—38°N) and (b) the eastern Tibetan Plateau (90°—105°E, 27°—40°N). Lines with "plus" signs, open circles, filled circles, open squares, and filled squares represent long-wave radiative cooling, short-wave radiative heating, vertical diffusive heating, large-scale condensation heating, and convective condensation heating, respectively. Red solid lines indicate total heating. Unit is Kd^{-1}

We performed three experiments, as briefly described in Table 1. The first experiment (NOMT) possesses only the land-sea distribution, mimicking the existence of the Afro-Eurasian Continent. In the second experiment (MT90), an idealized ellipsoidal "TP" is placed on the continent. Its maximum height is 5000 m, and its center is located at 87.5°E, 32.5°N, mimicking the actual plateau. The third experiment (MT60) is the same as MT90, except that the center of the "TP" is moved to 60.0°E, 32.5°N, about 30° to the west of its actual location.

Table 1 Design of the Aqua-Planet Sensitivity Experiments Using an Idealized Afro-Eurasian Continent

	Center of Mountain	Mountain Shape	Mountain Height	Purpose of the Experiment
NOMT	/	/	/	Assess how circulation and rainfall respond to the land-sea distribution
MT90	(87.5°E, 32.5°N)	Ellipsoid	5 km	Assess how circulation and rainfall respond to a mountain located on the east of the continent
MT60	(60.0°E, 32.5°N)			Assess how circulation and rainfall respond to a mountain located on the west of the continent

Figure 10 shows the simulated July mean precipitation and circulation at 925 hPa. The area with a July−January difference in wind direction of >120° is marked by the thick red lines in Figs. 10a−c. This area, together with the area with a July mean rainfall of more than 6 mmd^{-1}, indicates the occurrence of monsoon. In the no-mountain experiment (NOMT; Fig. 10a) along the subtropical area, the summertime heating over land and cooling over ocean (figures not shown, refer to Fig. 1) generate near-surface cyclonic circulation over continent and anticyclonic circulation over ocean. An intense inter-tropical convergence zone (ITCZ) appears along the equator in the Pacific, and remarkable tropical monsoon rainfall occurs over South Asia and the Indochina Peninsula; however, the model fails to represent the West Africa monsoon or the westerly along the western coast, as most of the features that contribute to these systems (e.g., the shape of coastline, topography, and the cold tongue in the eastern equatorial Atlantic) are not included in the idealized model. Nevertheless, weak East Asian monsoon rainfall of 6−8 mmd^{-1} is observed over the southeastern corner of the Afro-Eurasian continent.

In the MT60 experiment, in which the TP is superimposed over the western central continent (Fig. 10b), all the above features that appear in the NOMT run (e.g., continental cyclonic circulation, ITCZ, tropical monsoons, and the weak East Asian monsoon) are almost unchanged. However, another cyclonic circulation is produced surrounding the plateau, and significant rainfall (>8 mmd^{-1}) appears over the southeastern part of the plateau. This plateau-forced rainfall is geographically separated from the weak East Asian monsoon associated with continental-scale forcing.

In the MT90 experiment, in which the TP is located at its present-day position, with its center near 90°E (Fig. 10c), we still observe most of the features produced by continental forcing (see Fig. 10a). However, because the plateau has been moved eastward in this experiment, the cyclonic circulation surrounding the plateau, as shown in Fig. 10b, is also shifted eastward. Consequently, the rainfall forced by the plateau is merged into the rainfall band produced by continental-scale forcing, and they enhance each other. The East Asian monsoon now extends northward toward North China.

Figure 10d shows the differences between MT90 and NOMT in terms of daily rainfall and circulation at 925 hPa. The introduction of the "TP" intensifies not only the ITCZ, but also the East Asian monsoon rainfall, from the eastern part of the plateau to East China and Japan and Korea (shading), but results in reduced rainfall over South Asia (dotted contours). This occurs because the plateau generates a strong northerly to its west and southwest, but induces a strong southerly to its east and northeast, as shown by the difference flow (Fig. 10d).

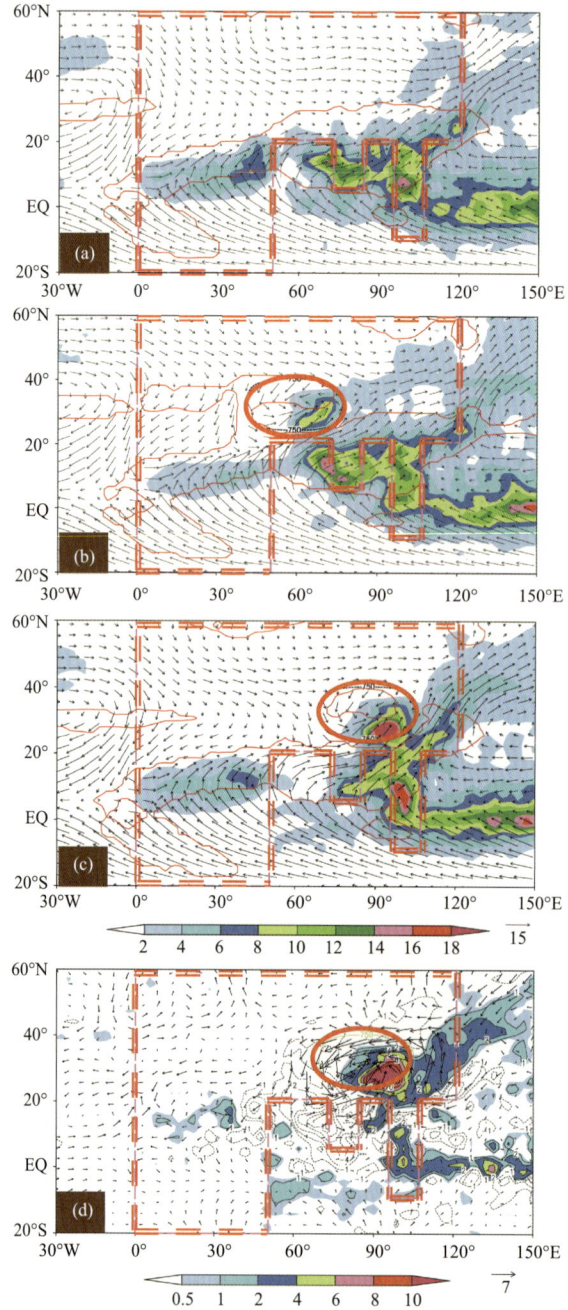

Fig. 10 Distributions of the 10-year July mean rainfall (color shading, unit is mmd^{-1}) and wind at 925 hPa (arrows, ms^{-1}) in the experiments NOMT (a), MT60 (b), and MT90 (c). The differences between MT90 and NOMT are shown in (d). The red lines in (a)—(c) mark the area where the change in wind direction between July and January exceeds 120°. The thick dashed line denotes the idealized coastline, and the thick ellipse in (b)—(d) indicates land above 750 m

To further illustrate the influence of the location of the TP on the formation of monsoon and desert, Fig. 11 shows the distributions of rainfall, meridional winds in the lower and upper troposphere, and a zonal vertical cross-section of subtropical circulation along 32.5°N, as generated from experiments MT60 and MT90. In MT60 (Figs. 11a—c), there exist two anticyclonic circulations at 200 hPa (Fig. 11a) and two related cyclonic circulations at 850 hPa (Fig. 11b), corresponding to continental-scale forcing and plateau forcing, respectively. Subtropical rainfall possesses two centers: one over the eastern plateau and the other over East Asia. The vertical cross-section of circulation (Fig. 11c) shows two ascending branches over the Afro-Eurasian continent: one over the eastern plateau and another over East Asia, in good correspondence with the rainfall regions. Of note, a moderate air-descending region occurs below 300 hPa in the troposphere, between the two ascending branches. This feature is similar to that observed along the subtropics over North America and South America (see Figs. 3, 8b, and 8c).

In contrast, experiment MT90 (Figs. 11d—f) produces only one anticyclonic circulation at 200 hPa (Fig. 11d) and one cyclonic circulation at 850 hPa (Fig. 11e), with relatively intense meridional winds and rainfall over the eastern part of the continent. In the vertical cross-section of circulation (Fig. 11f), there exists only one ascending region, spanning from the eastern plateau to Western Pacific, in good correspondence with the rainfall distribution over East Asia (see Figs. 11d and 11e) and in good agreement with the observed distribution of vertical motion in the area (Figs. 3 and 8a).

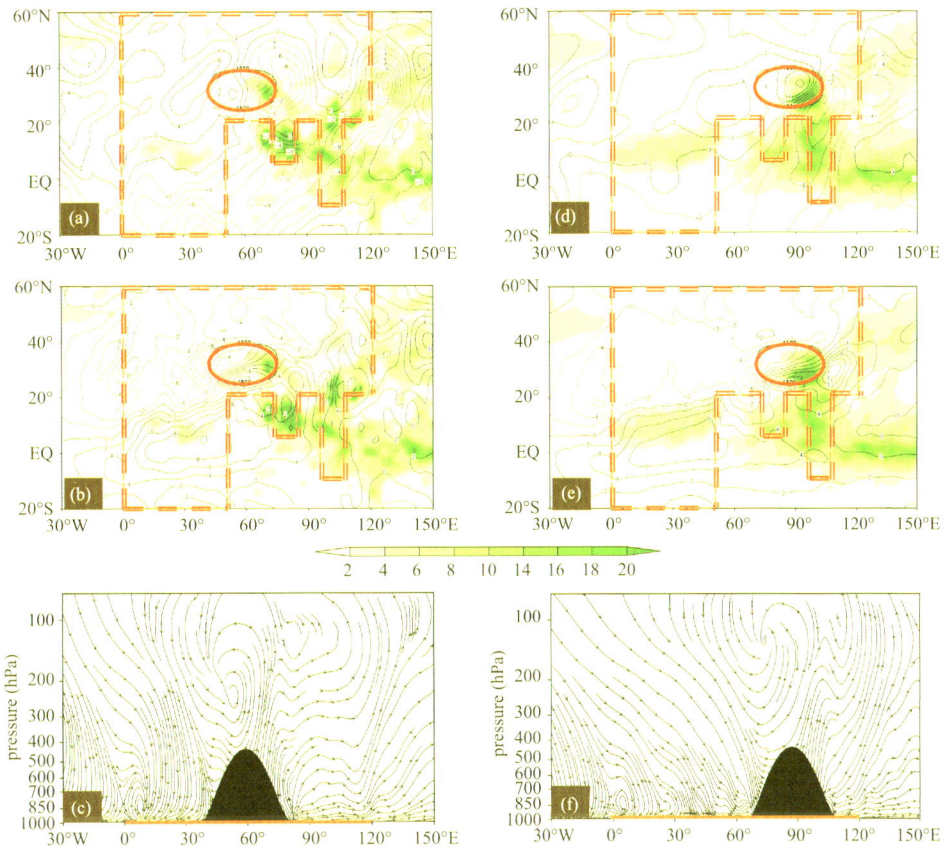

Fig. 11 Distributions of the 10-year July mean rainfall (a, b, d, and e; shading, unit is mm d^{-1}), meridional wind (solid contours (positive values) denote southerly wind, and dotted contours (negative values) denote northerly wind; interval is 2 ms^{-1}) at 200 hPa (a and d) and 850 hPa (b and e), and corresponding vertical cross-sections of circulation along 32.5°N (c and f; black area represents the plateau) in the experiments MT60 (a—c) and MT90 (d—e). Thick dashed lines represent the idealized coastline, and ellipses indicate land with an elevation of 1500 m

The results of the sensitivity experiments (Figs. 10 and 11) confirm our hypothesis that the location of the TP in the eastern part of the Afro-Eurasian continent results in an intensification of the monsoon climate to its east and the desert climate to its west. As shown in Fig. 12b, the area with more than 5 mmd^{-1} of July mean rainfall over East Asia has reached Northeast China, and the rainfall in South China exceeds 10 mmd^{-1}. The East Asian monsoon is therefore the strongest subtropical monsoon in the world, in terms of both spatial coverage and intensity. To the west of the TP, the arid and desert areas with rainfall less than 2 mmd^{-1} extend from about 70°E westward to North Africa, forming the largest and driest desert area in the world. In contrast, the Rockies and Andes are located on the western continent. As shown in Figs. 12a and 12b, the monsoon-like rainfall over the eastern continent and the arid climate over the western continent in North and South America are not as pronounced as those over subtropical Asia. Because the mountain ranges in Australia and Southern Africa are not as high as those in other continents, the monsoon and desert climates in these areas are somewhat atypical, although greater summer rainfall does occur over the eastern continent and a drier area occurs over the western continent.

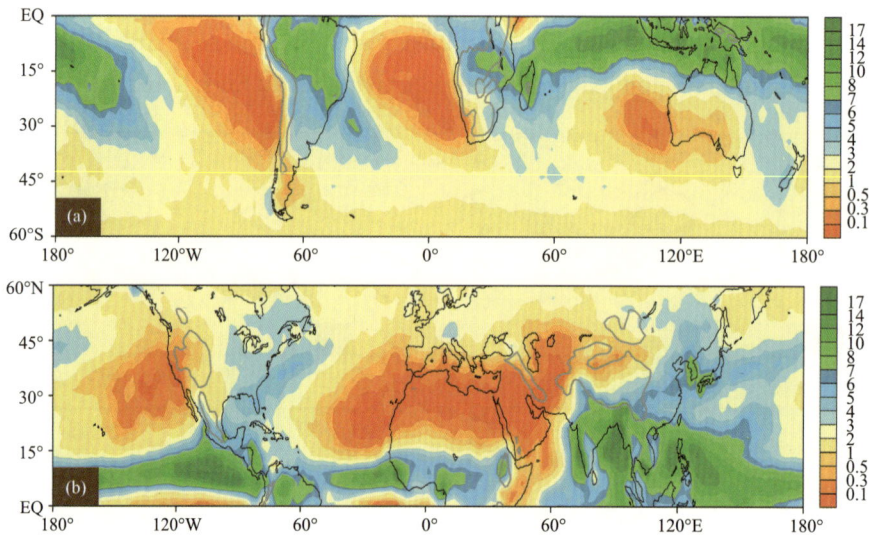

Fig. 12　Distributions of monthly mean rainfall (mmd^{-1}) averaged over the period 1980—1997 for (a) January and (b) July, and grey lines indicate topography with an elevation of 1000 m in (a) and 1500 m in (b)

7　Summary and Discussion

The climate system is an open dissipative system that is subjected to various kinds of forcing. It behaves differently in response to different types of external forcing, and the response is complicated by feedback between external forcing and climate. In this study, we investigated three kinds of forcing (i.e., large-scale continental forcing, sea-breeze local forcing along coastal regions, and regional-scale plateau forcing) and demonstrated their contributions to the formation of monsoon and desert in the summertime subtropics from various perspectives. First, continental-scale heating over land and cooling over ocean produce near-surface cyclonic circulation over continent and anticyclonic circulation over ocean, and reversed circulations in the upper troposphere. Subject to the vorticity balance and continuity constrains, air ascends over the eastern continent and western ocean, and descends over the eastern ocean and western continent. Second, the sea breeze along coastal regions enhances the rising of air over the eastern continent to the west of the coast, and suppresses the rising over the western ocean to its east. The sea breeze also enhances the descent of air over the eastern ocean to the west of the western coast, and weakens the descent over the western continent to its east. Thus, as a consequence of the land-sea distribution, the dominant diabatic heating becomes well organized across each subtropical continent and its adjacent oceans in summer, and appears as a LOSECOD quadruplet-heating pattern.

Meridional transfer of heat and planetary vorticity give rise to positive feedback between vorticity generation and diabatic heating (longwave radiative cooling on the eastern parts of oceans, surface sensible heating on the western parts of continents, and condensation heating on the eastern parts of continents and western parts of oceans); the feedback continues until it is balanced by frictional dissipation. A warm, moist poleward flow develops in the lower troposphere over the eastern continent, resulting in abundant rainfall and a monsoon-like climate. On the western side of the continent, a cold, dry equatorward flow develops near the surface, soil moisture decreases, and an arid or even desert climate is developed. It should be noted that the divergent wind along the coastal line associated with sea-breeze forcing (as shown in Fig. 4b) is not independent of the rotational wind produced by continental forcing (Figs 6 and 7); instead, they are related by formula (3), and both are formed in response to the contrasting heat capacities of land and ocean.

It is also worthwhile to compare the present results with those of previous studies. Chen et al. (2001) employed a linear quasi-geostrophic vorticity equation model to study the atmospheric response to a prescribed convective latent heating with a sinusoidal vertical profile. For a given constant basic flow of $U=0$ or $U<0$, the atmospheric response is simply a THB type, as presented in Eq. (9) and as obtained by Liu et al. (2001). When the basic flow becomes positive (i.e., $U=5$ or 10 ms^{-1}), the response to the latent heating is wave-like, and a weak surface anticyclone appears downstream. These findings were used to explain the formation of the summertime subtropical anticyclone over the ocean surface. However, the intensity of the downstream anticyclone is too weak to explain the summertime surface anticyclone. More importantly, the near-surface zonal flow is weak in the summertime subtropics, and the dynamics developed by Chen et al. (2001) show a better fit to middle and high latitudes, where westerlies are dominant, rather than the subtropics.

Rodwell and Hoskins (2001) recognized the " 'duality' between the monsoon condensation heating and the low-level subtropical circulation in the sense that either one would be very different without the other." The authors speculated that desert is formed as a remote Rossby-wave response to the west of subtropical summertime monsoon heating, with interaction with mid-latitude westerlies, resulting in a region of adiabatic descent. However, the strongest descent in response to forcing is close to the western part of the forcing (Fig. 8; see also Gill, 1980); in contrast, the observed deserts on the western parts of continents are located far from the monsoon region on the eastern continent. In fact, based on their standard model integration and trajectory analysis (Fig. 9 in Rodwell and Hoskins, 1996), all the trajectories along the subtropical westerly descend strongly over the Mediterranean Sea, and most of the trajectories that originated from the Asian monsoon region and along the tropical easterlies begin to ascend before reaching the Southern Hemisphere. There occurs limited interaction between the westward and eastward trajectories because the former travel above 300 hPa, whereas the latter travel below 400 hPa, despite their similar original heights. In fact, Figs. 6a and 8a show that the strong descent centers located over the Mediterranean and mid-Asia occur below the upper troposphere westerlies, in agreement with the trajectory analysis of Rodwell and Hoskins mentioned above. Figures 3, 8a, and 12 show that the strong descent and deserts over North Africa-Mediterranean and mid-Asia are closely related to the continental-scale forcing and the forcing of the Iran Plateau and the TP.

In terms of the third type of forcing, the uplifted heating of regional-scale plateau forcing in summer generates a pronounced Rossby-wave-type circulation with poleward flow and atmospheric ascent to the east, and equatorward flow and descent to the west of the plateau. The TP is located within the eastern Eurasian continent. Atmospheric circulation produced by the TP is in phase with that of continental-scale and sea-breeze forcing. The strongest monsoon and deserts are therefore formed over the African and Eurasian continents. In contrast, the Rockies and Andes are located over the western parts of continents; although the descending motion over the oceanic regions to the west of North and South America are strongly intensified, the ascending regions over the continents due to orographic forcing are separated from those due to continental forcing. The desert and monsoon climates over these continents are not as strong as those over the Eurasian Continent.

In summary, the observed summertime subtropical climate pattern, with abundant rainfall or monsoon on the east of the continent and aridity or desert on the west, is formed in response to the combined effects of continental-scale forcing, local-scale coastal sea-breeze forcing, and regional-scale mountain forcing, as well as the positive feedback between diabatic heating and heating-induced vorticity generation in the lower troposphere, which occurs via the meridional transfer of heat and planetary vorticity. Therefore, monsoons and deserts in the summertime subtropics coexist as twin features of the climate system produced by multi-scale circulations in association with various types of external forcing.

References

BOUNOUA L, KRISHNAMURTI T N, 1991. Thermal dynamic budget of the 5 day wave over the Sahara Desert during summer[J]. Meteorology and Atmospheric Physics, 47(1):1-25.

CHARNEY J G, 1975. Dynamics of deserts and drought in Sahel[J]. Quarterly Journal of the Royal Meteorological Society, 101(428):193-202.

CHEN P, HOERLING M P, DOLE R M, 2001. The origin of the subtropical anticyclones[J]. Journal of the Atmospheric Sciences, 58(13):1827-1835.

DUAN A M, WU G X, 2005. Role of the Tibetan Plateau thermal forcing in the summer climate patterns over subtropical Asia[J]. Climate Dynamics, 24(7-8):793-807.

EADY E T, 1950. The cause of general circulation of atmosphere[J]. Roy Meteor Soc Cent Proc, 76:156-172.

GILL A E, 1980. Some simple solutions for heat-induced tropical circulation[J]. Quarterly Journal of the Royal Meteorological Society, 106(449):447-462.

HOSKINS B J, 1986. Diagnosis of forced and free variability in the atmosphere[C]//CATTLE H. Atmospheric and Oceanic Variability. Bracknell : Royal Meteorological Society:57-63.

HOSKINS B J, 1991. Towards a PV-θ view of the general-circulation[J]. Tellus Series a-Dynamic Meteorology and Oceanography, 43AB:27-35.

IPCC, 2001. Chap. 8, Climate Change: The Scientific Basis[C]//HOUGHTON J T, DING Y H, GRIGGS D G, et al. Contribution of Working Group I to the Third Assessment Report of the Intergovernmental Panel on Climate Change. Cambridge and New York: Cambridge University Press: 881.

KALNAY E, KANAMITSU M, KISTLER R, et al, 1996. The NCEP/NCAR 40-year reanalysis project[J]. Bulletin of the American Meteorological Society, 77(3): 437-471.

KANG I S, JINA K, LAU K M, et al, 2002. Intercomparison of atmospheric GCM simulated anomalies associated with the 1997/98 El Nino[J]. Journal of Climate, 15(19):2791-2805.

LI G P, DUAN T Y, HAGINDYA S, 2001. Estimates of the bulk transfer coefficients and surface fluxes over the Tibetan Plateau using AWS data[J]. Journal of the Meteorological Society of Japan, 79(2):625-635.

LIN X, JOHNSON R H, 1996. Heating, moistening, and rainfall over the western Pacific warm pool during TOGA COARE[J]. Journal of the Atmospheric Sciences, 53(22):3367-3383.

LIU H, WU G X, 1997. Impacts of land surface on climate of July and onset of summer monsoon: A study with and AGCM plus SSiB[J]. Advances in Atmospheric Sciences, 14(3):289-308.

LIU Y M, WU X G, LIU H, et al, 2001. Condensation heating of the Asian summer monsoon and the subtropical anticyclone in the eastern hemisphere[J]. Climate Dynamics, 17(4):327-308.

LIU Y M, WU G X, REN R C, 2004. Relationship between the subtropical anticyclone and diabatic heating[J]. Journal of Climate, 17(4):682-698.

NEWMAN M, SARDESHMUKH P D, BERGMAN J W, 2000. An assessment of the NCEP, NASA, and ECMWF reanalyses over the tropical west Pacific warm pool[J]. Bulletin of the American Meteorological Society, 81(1):41-48.

NIGAM S, CHUNG C, DEWEAVER E, 2000. ENSO diabatic heating in ECMWF and NCEP-NCAR reanalyses, and NCAR CCM3 simulation[J]. Journal of Climate, 13(17):3152-3171.

RODWELL M J, HOSKINS B J, 1996. Monsoons and the dynamics of deserts[J]. Quarterly Journal of the Royal Meteorological Society, 122(534):1385-1404.

RODWELL M J, HOSKINS B J, 2001. Subtropical anticyclones and summer monsoons[J]. Journal of Climate, 14(15):3192-3211.

TAO S Y, CHEN L, 1987. A review of recent research on the East Asian summer monsoon in China[C]//CHANG C P, KRISHNAMURTI T N. Monsoon Meteorology. Oxford: Oxford University Press: 60-92.

WANG B, 2006. The Asian Monsoon[M]. Chichester: Springer: 787.

WEBSTER P J, MAGANA V O, PALMER T N, 1998. Monsoons: Processes, predictability, and the prospects for prediction[J]. Journal of Geophysical Research, 103(C7):14451-14510.

WU G X, ZHANG X H, LIU H, et al, 1997a. The LASG global ocean-atmosphere-land system model GOALS/LASG and its simulation study[J]. Journal of Applied Meteorological Science, 8(spec.):15-28.

WU G X, LIU Y M, HE B, et al, 1997b. Sensible heat driven air-pump over the Tibetan Plateau and its impacts on the Asian summer monsoon[C]//YE D Z. Collections on the Memory of Zhao Jiuzhang. Beijing: Science Press: 116-126.

WU G X, LIU Y M, 2000. Thermal adaptation, overshooting, dispersion and subtropical anticyclone. I. Thermal adaptation and overshooting[J]. Chinese Journal of Atmospheric Sciences, 24:433-446.

WU G X, LIU Y M, 2003. Summertime quadruplet heating pattern in the subtropics and the associated atmospheric circulation[J]. Geophysical Research Letters, 30(5):1201. doi:10.1029/2002GL016209.

WU T W, LIU P, WANG Z Z, et al, 2003. The performance of atmospheric component model R42L9 of GOALS/LASG[J]. Advances in Atmospheric Sciences, 20(5):726-742.

XIE P P, ARKIN P A, 1996. Analyses of global monthly precipitation using gauge observations, satellite estimates, and numerical model predictions[J]. Journal of Climate, 9(4):840-858.

XUE Y, SELLERS P J, KINTER J L, et al, 1991. A simplified biosphere model for global climate studies[J]. Journal of Climate, 4(3):345-364.

YAMAZAKI N, 2000a. Current status of GAME reanalysis project and some preliminary results-heavy precipitation on July 22, 1998 and diurnal variations of precipitable water[C]. Proc. Int. GAME/HUBEX Workshop. Sapporo, Japan: GAME project: 24-29.

YAMAZAKI N, 2000b. Release of GAME reanalysis data[J]. Tenki, 47: 659-664.

YANAI M, LI C F, 1994. Mechanism of heating and the boundary-layer over the Tibetan Plateau[J]. Monthly Weather Review, 122(2):305-323.

YANAI M, WU G X, 2006. Effects of the Tibetan Plateau[C]//WANG B. The Asian Monsoon. Berlin: Springer: 513-549.

YANG S, WEBSTER P J, DONG M, 1992. Longitudinal heating gradient: Another possible factor influencing the intensity of the Asian summer monsoon circulation[J]. Advances in Atmospheric Sciences, 9(4):397-410.

YEH T C, LO S W, CHU P C, 1957. The wind structure and heat balance in the lower troposphere over Tibetan Plateau and its surroundings [J]. Acta Meteorologica Sinica, 28: 108-121.

The Bimodality of the 100 hPa South Asia High and Its Relationship to the Climate Anomaly over East Asia in Summer

ZHANG Qiong[1], WU Guoxiong[1], QIAN Yongfu[2]

(1 State Key Laboratory of Atmospheric Sciences and Geophysical Fluid Dynamics, Institute of Atmospheric Physics, Chinese Academy of Sciences, Beijing 100029, China; 2 Department of Atmospheric Sciences, Nanjing University, Nanjing 210093, China)

Abstract: The NCEP/NCAR pentad mean reanalysis data from 1980 to 1994 are employed to examine the activities of the 100 hPa South Asia high (SAH) during the boreal summer. The results show that there exists bimodality in the longitude location of the SAH. According to the two preferred regions for the SAH, the SAH is classified into the Tibetan Mode (TM) and the Iranian Mode (IM), respectively. The studies on the maintenance mechanism, both from circulation structure and thermal structure, manifest that the SAH has the feature of warm preference. The diagnosis based on the thermodynamic equation further reveals that the TM is closely related to the diabatic heating of the Tibetan Plateau, whereas the IM is more associated with the adiabatic heating in the free atmosphere, as well as the diabatic heating near the surface.

The statistical composites, and case studies, corresponding to the two modes show that the SAH bimodality is strongly related to the climate anomalies over East Asia. In the case of TM, the southerly airflow over the subtropics and the northerly airflow over the middle-high latitude at 850 hPa are enhanced, forming a convergence zone along 30°N and resulting in increased rainfall extending from the south Japan, Korea Peninsula, and Yangtze-Yellow river valley of China to the Tibetan Plateau. In the case of IM, at 850 hPa the mid-latitude East Asia is dominated by a huge anomalous anticyclone, thus results in the decreased rainfall over the area.

1 Introduction

In summer in the upper troposphere, and lower stratosphere, there exists a large scale high pressure system (Tibetan high) over the Tibetan Plateau and its surrounding area, which is referred to the

South Asia high (SAH) in this paper hereafter. Flohn (1960) suggested that the formation of the huge anticyclone is a result of the elevated heating of the Tibetan Plateau to the atmosphere. Based on the analyses of the International Geophysical Year (IGY) data, it has been recognized that the SAH is the strongest and the steadiest circulation system at the 100 hPa level besides the polar vortex (Mason and Anderson, 1963). Tao and Zhu (1964) reported that the longitudinal shifting of the SAH leads that of the 500 hPa Subtropical Anticyclone over Western Pacific (SAWP) for a few days, so that the activity of the SAH could be used as an index for the short-medium range weather forecasts in Asia. The longitudinal shifting of the SAH towards or apart from the Tibetan Plateau is then suggested as an east—west oscillation during the midsummer (commonly from July to August in the Northern Hemisphere) (Tao and Zhu, 1964). Based on the daily 100 hPa synoptic charts published by the China Meteorological Observatory, Luo et al. (1982) further classified the SAH into three circulation patterns: the eastern pattern; western pattern; and belt-type pattern. The maintaining period for one pattern is about 10—13 days, which is close to the 13—15 days oscillation period of the stream function over the Tibetan Plateau revealed by Krishnamurti et al. (1973).

It is well known that the SAWP at 500 hPa level is a substantially important system influencing the summer rainfall variability over East Asia. Since there exist difficulties in forecasting the activity of SAWP (Liu and Wu, 2000), the relationship between the east—west oscillation of the 100 hPa SAH, and the activities of SAWP, provides some new clues on the short-term prediction of the flood and drought occurring in this region in summer. Thus, the studies on the effects of the east—west oscillation of the SAH on summer precipitation have drawn intensive attentions. For example, during the Qinghai-Xizang Plateau Meteorology Experiment (QXPMEX) in 1979, many synoptic analyses were conducted based on the observation data, and most results have been applied to routine forecasts (Zhao et al., 1978; Zhu et al., 1980; Luo et al., 1982; Chen et al., 1983; Xu, 1983; Yang et al., 1984).

Besides the synoptic analyses mentioned above, the mechanism responsible for the east—west oscillation of the SAH was also studied. The arguments fall into two basic categories concerning whether thermal effect, or dynamical forcing of the Tibetan Plateau is more important for the oscillation. The thermal forcing proponents emphasize the huge sensible heating of the Tibetan Plateau, which causes the adjustment of the subtropical circulation, together with the effects of the latent heating over the eastern China plain, resulting in the east—east oscillation (Qian, 1978; Liu et al., 1987). The dynamical forcing proponents think much of the interaction between the circulation systems (Institute of the Central Meteorological Bureau of China, 1978), they suggested that the adjustments of the nearby circulation induced by the Plateau cause a kind of forcing oscillation that is different from the self-oscillation due to its thermal forcing. The case study shows that such a forcing oscillation may be related to the eccentric effect of the Northern Hemisphere polar vortex. Some of the possible mechanisms have also been suggested based on the dishpan experiments (Zhang et al., 1977), which show that an obstacle can independently produce a kind of oscillation without thermal effect. However, such an oscillation is confined to a small range, and the center of the SAH can hardly move out of the Tibetan Plateau. In contrast, a distinct large range of oscillation occurs, when the external circulation effect is included into the dishpan experiment.

Recently, we examined the climatic characteristics of the SAH by employing the 40-year NCEP/NCAR reanalysis. It is found that the SAH exhibits as bimodality in longitude location during the northern summer, one location is over the Tibetan Plateau, and another over the Iranian Plateau. Such bimodality is different from the east—west synoptic oscillation as suggested by Tao and Zhu (1964). The importance of the SAH bimodality also lies on its close relation with the large area occurrence of flood and drought (Zhang and Wu, 2001). Therefore, it is necessary to study further the features of the bimodality, as well as the mechanism of the alternation from one mode to another.

In this work, we present the SAH bimodality by using NCEP/NCAR pentad mean reanalysis data. Some diagnostic methods described in section two are applied to study its thermal, and dynamical structure. Section three presents the description of the bimodality. The thermal and dynamical features of the SAH are addressed in section four. Section five presents the association of the bimodality with the climate anomaly in Asia. The possible connection between the SAH bimodality, and the climate anomaly over Asia, is described in section six via case study. Section seven lists the main summary of this study.

2 Data and methods

2.1 Data

The pentad mean data for this study comes from the NCEP/NCAR daily reanalysis set (Kalnay et al., 1996), covering the 15-yr period from 1980 to 1994. The reanalyses were obtained by assimilating past data into a frozen state-of-the-art analysis/forecast model system. The database was enhanced with many sources of observations that were not available in real time operations, and the products can be regarded as one of the most complete, physically consistent meteorological datasets. To establish the pentad mean reanalysis, the daily mean data are recomputed by arithmetical mean. For the purpose of convenience, six pentads are uniformly distributed in each month, a pentad value is usually obtained from the corresponding 5-day mean, for those 31-day months, the sixth pentad values are obtained from the corresponding 6-day mean, and for a 28-day or 29-day February, the sixth pentad value is obtained from 3- or 4-day mean.

Considering the planetary scale of the SAH during summer, the domain of 40°—140°E, 20°S—60°N is chosen to capture the main activities of the SAH. The geopotential height, wind and air temperature data at the standard pressure levels (1000, 850, 700, 500, 400, 300, 200 and 100 hPa) are analyzed objectively on a 2.5°×2.5° lat-lon resolution. Besides, the NCEP surface temperature are analyzed on a 192×94 Gaussian grid.

The pentad mean rainfall data on a 2.5°×2.5° lat-lon grid are obtained from the Climate Prediction Center Merged Analysis of Precipitation (CMAP) dataset (Xie and Arkin, 1997). We used the version derived by merging rain gauge observation with rainfall estimates inferred from various satellite observations, but without numerical model outputs. Over land the data is mainly based on information from rain gauge observations, while over the ocean they primarily use satellite estimates made with several different algorithms based on outgoing longwave radiation, and scattering and emission of microwave radiation.

The anomalies throughout this study are referred to the departures from their corresponding 15-yr means.

2.2 Diagnostic methods

The thermal structure of the bimodality is examined based on the thermodynamic equation

$$\frac{\partial T}{\partial t} = -\boldsymbol{V} \cdot \nabla T - \left(\frac{P}{P_0}\right)^{\kappa} \omega \frac{\partial \theta}{\partial p} + \frac{Q_1}{C_p} \tag{1}$$

where T is temperature, θ potential temperature, $\mathbf{V}$ horizontal wind, ω vertical p-velocity, $\kappa = R/C_p$, Q_1 diabatic heating, C_p the specific heat at constant pressure of dry air, $p_0 = 1000$ hPa.

The velocity ω in (1) has been obtained kinematically by integrating the mass continuity equation (He et al., 1987)

$$D + \frac{\partial \omega}{\partial p} = \frac{1}{a\cos\phi}\left[\frac{\partial u}{\partial \lambda} + \frac{\partial}{\partial \phi}(v\cos\phi)\right] + \frac{\partial \omega}{\partial p} = 0 \tag{2}$$

with the surface boundary condition

$$\omega = \omega_S = -g\rho_S\left(\frac{u_S}{a\cos\phi}\frac{\partial h}{\partial \lambda} + \frac{v_S}{a}\frac{\partial h}{\partial \phi}\right)$$
$$\text{at } p = p_S \tag{3}$$

In (2) and (3) u and v are the zonal and meridional components of horizontal wind $\mathbf{V}$, respectively, D horizontal divergence, a mean earth radius, ϕ latitude, λ longitude, p pressure, g acceleration of gravity, ρ density, and h terrain height. The suffix S denotes the surface value.

Assuming that the motion is approximately adiabatic in the layer between 100 and 200 hPa, we impose the additional condition near the tropopause

$$\omega = \omega_T = -\left(\frac{\partial \theta}{\partial t} + \mathbf{v}\cdot\nabla\theta\right)\bigg/\left(\frac{\partial \theta}{\partial p}\right)$$
$$\text{at } p = p_T = 150 \text{ hPa} \tag{4}$$

The original estimates of the horizontal divergence D_0 are corrected by adding

$$D' = \frac{\omega_T - \omega_S - \int_{p_T}^{p_S} D_0 \, dp}{p_S - p_T} \tag{5}$$

Then $D = D_0 + D'$ is used to obtain ω from (2).

3 Description of the bimodality

As shown in Fig. 1, in the multi-year mean 100 hPa geopotential height field, the midsummer mean SAH which is centered over the Tibetan Plateau, and its neighborhood exhibits as a huge system covering most part of the subtropics in the Northern Hemisphere. Another high system lies over the Rocky Mountains with a much smaller scale. Such a distribution may imply that in summer the high system in the upper level is related to the large scale topography, and their thermal features may be deduced from such a morphological character.

Figure 1b and Fig. 1c respectively, show the zonal section of the mean vertical circulation along 30°N and meridional section along 90°E. The ascending motions related to the thermal effect of the highlands are observed in the area of the SAH. The strong ascending motion over the central Tibetan Plateau is up to 100 hPa (Fig. 1b). Due to the appearance of the topographic trapped wave, the ascending motions on the two sides of the Tibetan Plateau, are weakened to some extent. It is found that the ascending motion over the Iranian Plateau is limited only in the lower troposphere, and the descending motions are in dominant in the upper troposphere. In the meridional section (Fig. 1c), it is observed that a strong summer monsoon circulation expands over a large area extending from the Tibetan Plateau to the equator, again illustrating the considerable thermal effects of the Tibetan Plateau in summer. The above mean vertical circulations suggest that the maintenance of the SAH over the Tibetan Plateau, and its neighborhood, is mainly related to the thermal effects of the huge highland during summer.

Generally, the location of the SAH is described by its ridge line, and the center. At 100 hPa easter-

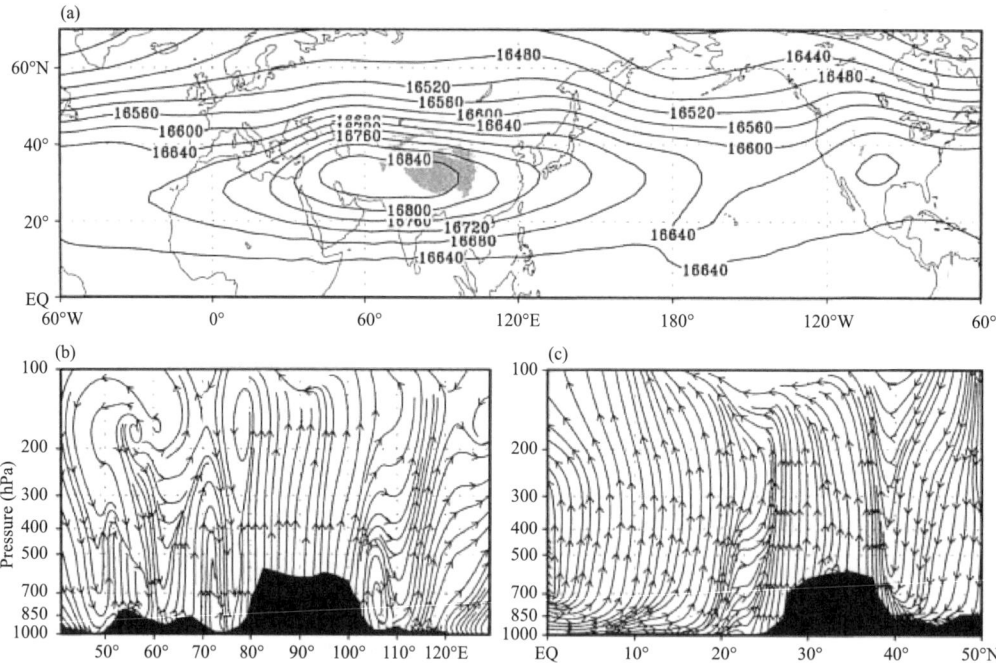

Fig. 1 (a) July—August mean 100 hPa geopotential height from the NCEP/NCAR pentad mean reanalysis for 1980—1994, unit: gpm. The topography greater than 3000 m are shaded. (b) The pressure-longitude section for the mean vertical circulation along 30°N. (c) The pressure-latitude section along 90°E. ω is amplified 150 times when plotting

lies prevail in low latitudes, and westerlies prevail in mid-latitudes. The ridge line of the SAH is then defined as the interface between the easterlies and westerlies, and its center is found at a grid point where the geopotential height along the ridge line is the greatest. Thereby the location of the SAH center can be described by the longitude and latitude coordinates of the point.

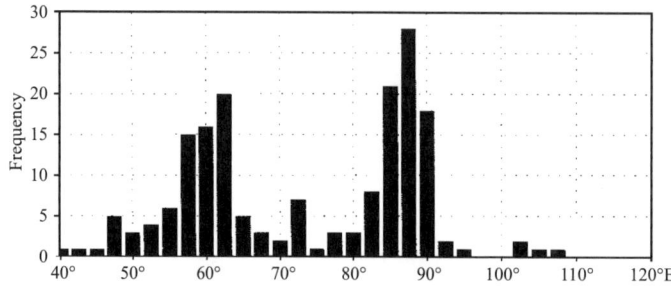

Fig. 2 The longitude-frequency distribution of the SAH major center during midsummer from the pentad mean data. For 15 summers there are totally 180 pentads involved in the statistics

When the activities of the SAH in summer are examined in detail by the individual pentad data, it is found that its longitude location exhibits as bimodality. Figure 2 shows the statistical frequency of the longitude location of the SAH in summer. The period considered for the statistics is July and August, which means there are 12 pentads in each midsummer, totally 180 pentads during 1980—1994. Quite similar to that of the monthly mean data (not shown), the SAH centers possess two preferable locations corresponding to the location of the Tibetan Plateau to the east and Iranian Plateau to the west, but scarcely appear near 70°—80°E, where the centered region of the climate mean SAH is located (Fig. 1a). According to its preferable location, the SAH can be classified into the Tibetan Mode (TM), and the

Iranian Mode (IM), respectively. Those center longitudes lie between 82.5°—92.5°E, are defined as the TM, and the center longitudes that lie between 55°—65°E, are defined as the IM. Statistically, among the total 180 pentads there are 77 pentads for TM (42.8%), and 62 pentads for IM (34.4%).

The 100 hPa streamline composites for the two modes are shown in Fig. 3. Besides the remarkable planetary feature of the SAH, its location reflects the preference to the highlands. Moreover, it is observed that the IM lies more northward than the TM.

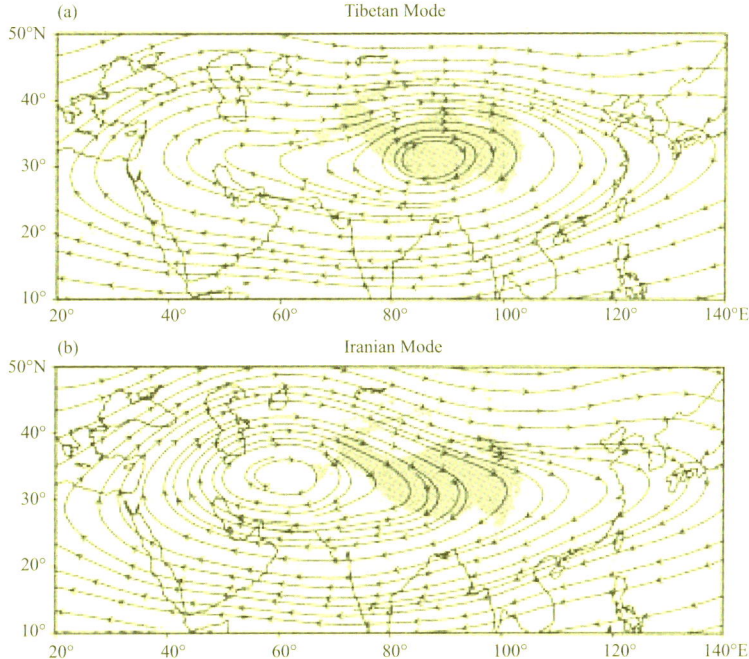

Fig. 3 The 100 hPa streamline composite corresponding to the TM (a) and the IM (b). 77 TM cases, and 62 IM cases are included in 1980—1994 July—August pentads

4 Thermal and dynamical structure of the bimodality

4.1 Vertical structure

To understand the thermal and dynamical features of the SAH bimodality, more composites for the two modes are presented in the following.

Figure 4 shows the pressure-longitude sections composed for the two modes along the ridge line latitude. Here we present the anomalies of the vertical circulation and the potential temperature to address the dynamical, and thermal structure of the TM and IM. It can be found that the TM corresponds to the anomalous strong ascending motion over the central Tibetan Plateau (Fig. 4a), which enhancing the original ascending motion over the region in Fig. 1b, and accompanied by the anomalous warm column (Fig. 4c). In the case of the IM, the anomalous ascending motions are observed over the Iranian Plateau (Fig. 4b), which weakens the original descending motion over the region in Fig. 1c, and also accompanied by the anomalous warm column. It is also found that contrasting to the tropospheric warm column, the SAH center at 100 hPa exhibits as a cold one (Figs. 4c, d). The above vertical structures show that in both the TM and the IM case, the cold SAH center is corresponding to the anomalous ascending motions (Figs. 4a, b) and the anomalous warm column (Figs. 4c, d). In the other words, it indicates that the SAH has the feature of warm preference. It may suggest that the anomalous warm column is related to the SAH center via the anomalous ascending motion in the column. The anomalous ascending motions enhance the divergence at upper level, thus increase the negative vorticity at the region, and result in the maintenance of the anticyclone center over the region. Such a warm preference feature of the SAH is also demonstrated by the meridional section composites (not shown).

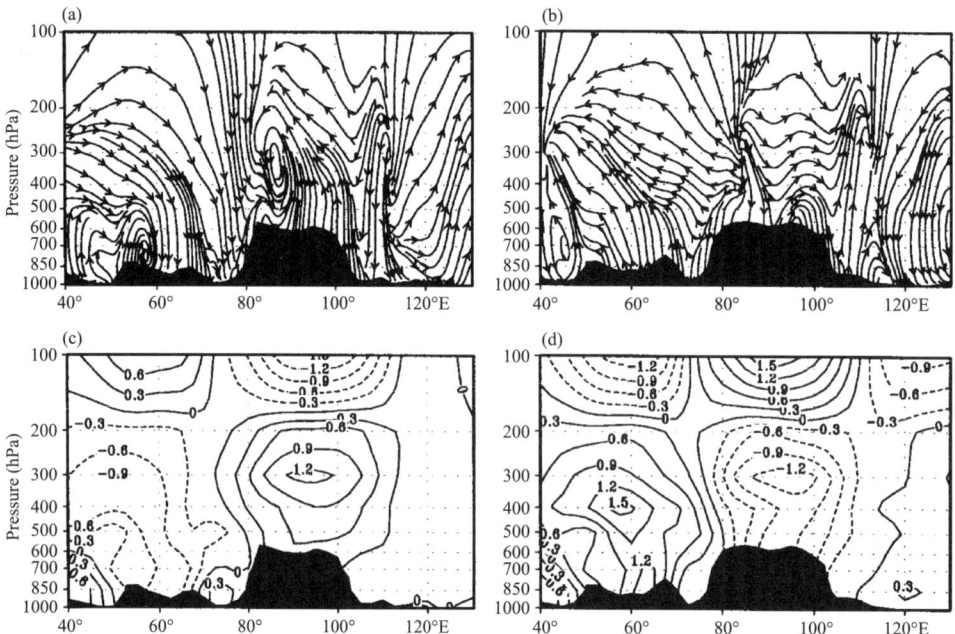

Fig. 4 The pressure-longitude cross sections composed for the TM ((a) and (c), along 30°N) and the IM ((b) and (d), along 32.5°N). (a) and (b) are vertical circulation anomalies, ω is amplified 150 times when plotting; (c) and (d) are potential temperature anomalies, unit is K. 77 TM cases, and 62 IM cases are included in 1980—1994 July—August pentads

4.2 The thermodynamic equation diagnosis

To further examine the thermal characteristics of the SAH bimodality and the possible reason of the SAH's warm preference, the terms of the thermodynamic equation (1) are calculated. Figure 5 shows the pressure-longitude cross sections of the terms of Eq. (1) along the SAH ridge line latitude for the TM. It is found that the obvious warming occurs over the Tibetan Plateau up to 200 hPa. Figure 5d indicates that such a warming results from the diabatic heating over the Tibetan Plateau, while the strong cooling due to ascent over the Plateau shown in Fig. 5c compensates the diabatic heating to a large extent. The horizontal advection shown in Fig. 5b contributes only a small part to the warming. Therefore, it indicates that the contribution of the diabatic heating of the Tibetan Plateau, is much important to the maintenance of the TM.

Figure 6 shows the similar pressure-longitude cross sections for the IM. The warming center over the Iranian Plateau is located in the lower and middle troposphere (Fig. 6a), acting as a warm background for the IM, and also illustrating the warm preference of the SAH. Similar to the case of the TM, such a warming over the Iranian Plateau in the lower troposphere is mainly due to the in situ surface heating (Fig. 6d). However, unlike the case shown in Fig. 5d, the warming in the middle troposphere centered at 400 hPa over the Iranian Plateau (Fig. 6a) is mainly due to the in situ adiabatic heating (Fig. 6c), compensated by the horizontal advection (Fig. 6b) and diabatic cooling (Fig. 6d).

In summary, it is indicated that the center of the SAH tends to stay over the warm air column. For the TM, such a warming is mainly due to the diabatic heating over the Tibetan Plateau; whereas for the IM, besides the diabatic heating in the lower troposphere, the adiabatic heating associated with descent in the middle troposphere over the Iranian Plateau is more important. Therefore, the maintenance of the SAH over the certain region mostly depends on the thermal effect of the atmosphere over the region.

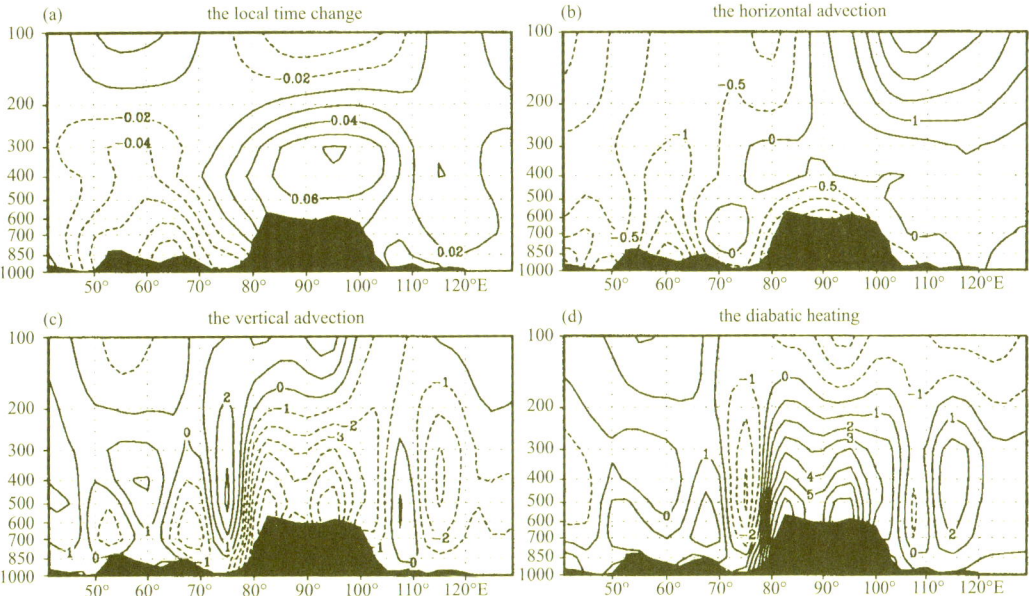

Fig. 5 The pressure-longitude cross sections composed for the terms in Eq. (1) along 30°N for the TM, unit is Kday^{-1}. (a) Local time change, (b) horizontal advection, (c) vertical advection, and (d) diabatic heating. 77 TM cases, and 62 IM cases are included in 1980—1994 July—August pentads

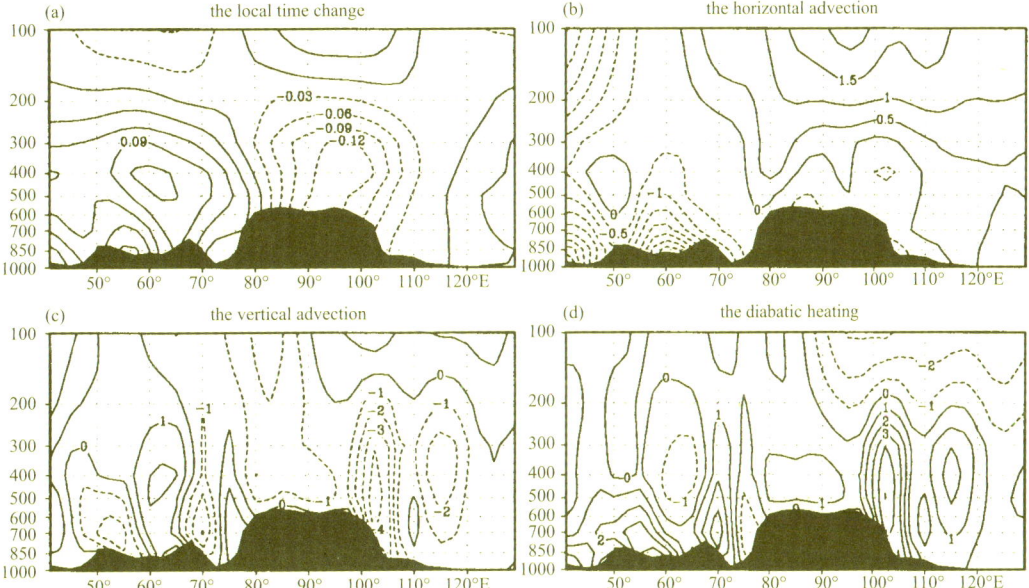

Fig. 6 Pressure-longitude cross sections composed for the terms in Eq. (1) along 32.5°N for the IM, unit is Kday^{-1}. (a) Local time change, (b) horizontal advection, (c) vertical advection, and (d) diabatic heating. 77 TM cases, and 62 IM cases are included in 1980—1994 July—August pentads

5 SAH bimodality and climate anomaly over Asia

To contrast the climate anomaly over Asia between the two SAH modes, we present composite maps of surface temperature, 850 hPa wind, and rainfall for 77 TM cases minus 62 IM cases. Because of the quite close case numbers of the two modes, in total 180 pentads, the anomalies associated with the TM, and the IM, tend to have opposite polarities. Thus, the composite TM-minus-IM will be simply referred to as dominant TM anomalies.

The differences of the surface temperature for the two modes show a distinct east—west contrast corresponding to the SAH center (Fig. 7a). It further confirms the warm preference of the SAH center as shown in Figs. 4c—d. The positive surface temperature anomalies are mostly associated with the enhanced near surface heat flux. Therefore, supporting the results shown in Fig. 5 and Fig. 6, it suggests that the diabatic heating over the Tibetan Plateau and the Iranian Plateau play an important role for the maintenance of the SAH over the two highlands.

As shown in Fig. 7b, a prominent cyclonic anomaly is found over middle latitude East Asia in 850 hPa wind difference between the TM and IM cases. The southerly airflows to the southeast of the cyclonic anomaly and the northerly airflows to the northwest of the cyclonic anomaly are significantly intensified. Such a wind anomaly brings about a convergence zone near 30°N along the lower reaches of the Yangtze River Valley, to south Japan. As a result, the large area differences of the rainfall for the two modes are found at the north and south side of 30°N (Fig. 7c). In the case of the TM, an enhanced rainfall belt extends from the northeast China, Korea Peninsula, and south Japan to the Yangtze-Yellow river valley, and the Tibetan Plateau. The composite rainfall difference attains a 95% significance level in most parts of these regions. To the south of 30°N, the rainfall increases in the case of the IM, however, only a few regions show the significant difference between the two modes. It suggests that the SAH bimodality is more related to rainfall anomalies over East Asia. It also noticed that significant differences are found near the Iranian Plateau at 850 hPa wind, whereas the rainfall over the area between the two modes has no much differences due to the rainless climate over this desert area.

6 Discussion

The statistical results show that the climate anomalies over East Asia are significantly related to the SAH bimodality. To illustrate how the 100 hPa SAH bimodality is connected to the rainfall anomaly over East Asia, two cases, July 1982 and July 1994 are chosen to have a comparison. Figure 8a shows that, in July 1982, the increased rainfall is distributed along a belt from south Japan, Korea Peninsula and Yangtze-Yellow river valley of China, to the Tibetan Plateau and India. Such a rainfall anomaly pattern corresponds with TM at 100 hPa, westward-extended SAWP at 500 hPa (Fig. 8b), and an anomalous anticyclone over subtropics at 850 hPa (Fig. 8c). Thus, the intensified southerly airflow in the subtropics, and northerly airflow in mid-high latitude, bring about the convergence zone near 30°N, as a result, the rainfall over this region is enhanced.

In contrast, in the case of July 1994, the rainfall over East Asia at mid-latitude such as Japan, Korea and Yangtze-Yellow river valley of China, is decreased (Fig. 8d). As shown in Fig. 8e, in this case, 100 hPa SAH presents as IM and 500 hPa SAWP weakens and withdraws eastward. At 850 hPa, a dominant anomalous anticyclone locates over the mid-latitude East Asia, resulting in decreased rainfall over

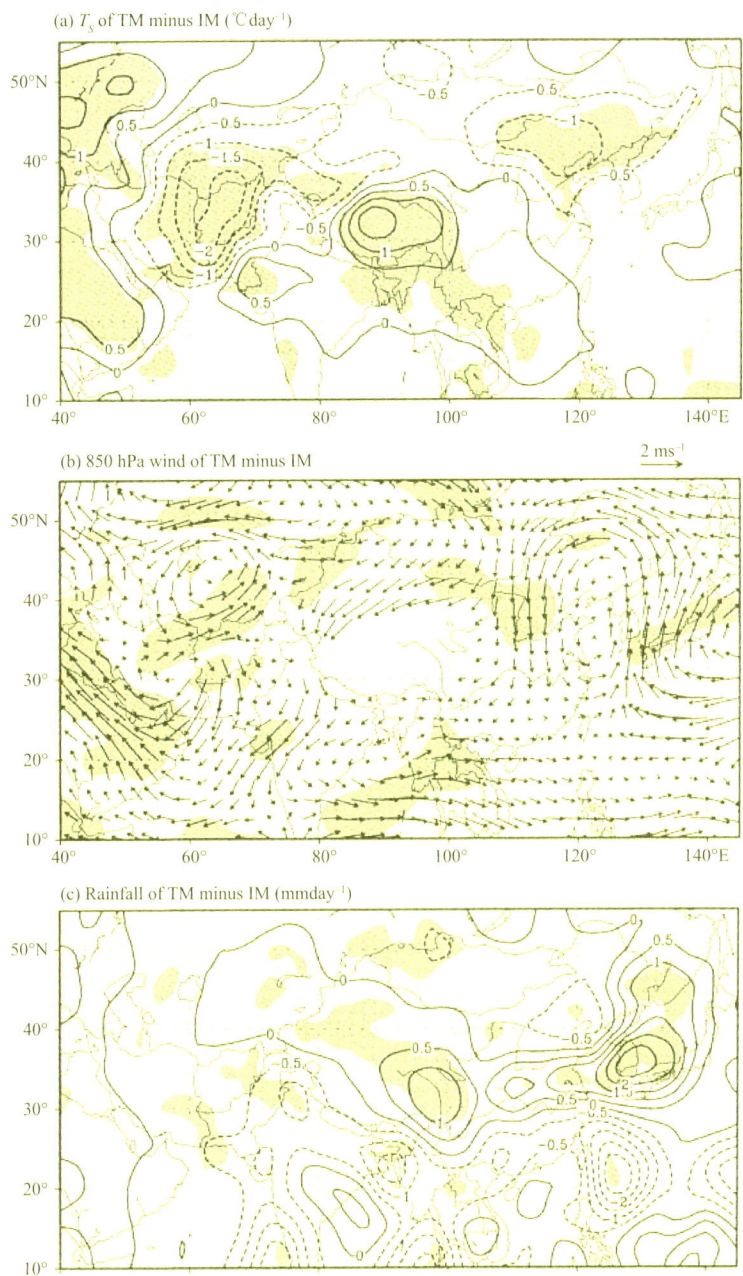

Fig. 7 Composite difference of (a) surface temperature, (b) 850 hPa wind and (c) rainfall between the TM and IM. Shading denote region of difference at 95% significance level. The wind vectors where the topography greater than 3000 m are set in default in (b). 77 TM cases, and 62 IM cases are included in 1980—1994 July—August pentads

the area.

The results from the above case study are consistent with the statistical results in Fig. 7 over middle-high latitudes, especially over East Asia. However, some contradictions between the case study, and composite analysis are found over the tropics, both in rainfall and wind anomalies. It may suggest that the climate anomalies over East Asia are strongly related to the SAH bimodality, whereas the climate anomalies over the tropics may be more related to the tropical atmosphere anomaly.

In the above two case studies, the relationship between the 100 hPa SAH and the 500 hPa SAWP is similar to the viewpoint of Tao and Zhu (1964). That is, following the eastward (westward) shifting of the 100 hPa SAH, the 500 hPa SAWP behaves as westward extending (eastward withdrawing). Furthermore, it is noticed that the 500 hPa Iranian subtropical high also has the extending, and withdrawing behavior, following the shifting of the 100 hPa SAH (Figs. 8b and e). Such a relationship between the SAH bimodality and 500 hPa subtropical system is much important and helpful to short-term weather forecasting. However, the interaction process between the upper and middle-low level subtropical system remains a key issue and need to be further studied.

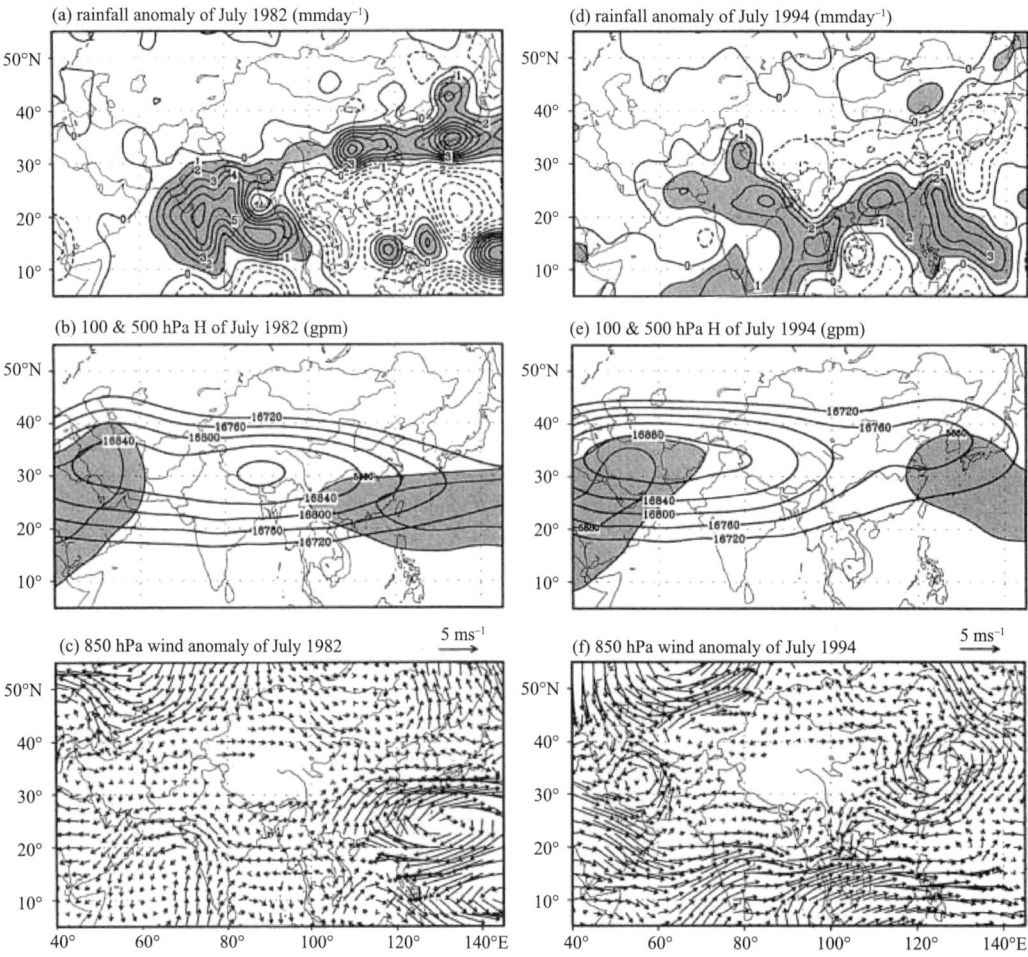

Fig. 8 Rainfall anomaly and atmospheric circulation structure for July 1982 ((a), (b) and (c)) and for July 1994 ((d), (e) and (f)). (a) and (d), rainfall anomaly, those anomalies greater than 1 mmday^{-1} are shaded. (b) and (e), 100 and 500 hPa geopotential height distribution, those greater than 5860 gpm at 500 hPa are in shading, the contour interval for 100 hPa height is 40 gpm and only those greater than 16720 gpm are plotted. (c) and (f), 850 hPa wind anomaly, the vectors where the topography greater than 3000 m are set in default

7 Summary

The pentad mean reanalysis data from NCEP/NCAR reanalysis are employed to examine the activities of the SAH during summer. The diagnoses show that there exists bimodality in the longitude location of the SAH. According to the two regions where the SAH preferring to stay, the SAH is classified into the Tibetan Mode, and the Iranian Mode, respectively. The studies on the maintenance mechanism both from circulation structure and thermal structure manifest that the SAH has the feature of warm preference. The diagnosis on the thermodynamic equation further reveals that the TM is closely related to the diabatic heating of the Tibetan Plateau, whereas the IM is more associated with the adiabatic heating in the free atmosphere as well as the diabatic heating near the in situ surface. The composite of the surface temperature for the two SAH modes suggests that the sensible heat over the Tibetan Plateau and Iranian Plateau may play an important role in the maintenance of the SAH over the two highlands.

The statistical analysis and the case study show that SAH bimodality is strongly related to the climate anomaly over East Asia. The TM corresponds with an increased rainfall belt extending from the south Japan, Korea and Yangtze-Yellow river valley to the Tibetan Plateau. In the case of IM, a decreased rainfall pattern is found over the above area.

The SAH bimodality presented in this study is substantially different to the previous east—west oscillation of the SAH on some aspects. First, the time scale is different: the bimodality distribution is examined based on climatological data, whereas the east—west oscillation is often referred as a synoptic process; second, the spatial scale is also different, the shifting scope of the bimodality extends much larger than that of the east—west oscillation, as shown in Fig. 2 and Fig. 3, the TM is located over the Tibetan Plateau centered at about 90°E, and the IM is near the Iranian Plateau centered at about 60°E. However, when the concept of east—west oscillation is first presented by Tao and Zhu (1964), it is suggested as the longitudinal shifting of the SAH towards and apart from the Tibetan Plateau. Their study emphasized the influence of the SAH on summer rainfall in China, and did not put much attention to far westward extension of the SAH to the Iranian Plateau. Third, the maintenance and variation mechanism of the bimodality and east—west oscillation differs considerably. Although the structure, and feature of the TM and IM are different, their preference to a certain region presents as a relatively steady status. The mechanism of the east—west oscillation as discussed in the introduction remains uncertain. In a parallel research, we further analyze the activity and feature of the Tibetan Mode in more details, and have found that all the TM cases can also be divided into two types: Eastern TM and western TM types. The shifting between these two types and the spatial span of such shifting are more similar to the aforementioned west-east oscillation of the SAH. The conjecture about the east—west oscillation of the SAH on a climate time scale will be published in a separate paper.

References

CHEN G Y, LIAO Q S, 1983. The relationship of 100 hPa South Asian high and summer precipitation in China. Meteorological Science Technique Symposium, NO. 6[C]. Beijing: China Meteorological Press: 123.

FLOHN H, 1960. Recent investigation on the mechanism of the "summer monsoon" of southern and eastern Asia[C]. Proc Symp Monsoon of the World. New Delhi: Hind Union Press: 75-88.

HE H, MCGINNIS J W, SONG Z, et al, 1987. Onset of the Asian summer monsoon in 1979 and the effect of the Tibetan Plateau[J]. Mon Wea Rev, 115: 1966-1995.

Institute of the Central Meteorological Bureau, 1978. The preliminary discussion on the east—west oscillation of the South Asian high[C]//Symposium paper collections on Tibetan Plateau Meteorological (1975—1976). Beijing: Science Press: 148-157.

KALNAY E, KANAMITSU M, KISTLER R, et al, 1996. The NCEP/NCAR 40-year reanalysis project[J]. Bull Amer Meteor Soc, 77: 437-471.

KRISHNAMURTI T N, DAGGUPATY S M, FEIN J, et al, 1973. Tibetan high and upper tropospheric tropical circulation during Northern summer[J]. Bul Amer Meteor Soc, 54: 1234-1249.

LIU F M, WEI S H, 1987. The East—West Oscillation of the 100 hPa South Asia High and Its Forecasting. The Effects of the Tibetan Plateau on Weather of China during Summer[M]. Beijing:Science Press:111-117.

LIU Y M,WU G X, 2000. Reviews on the study of the subtropical anticyclone and new insights on some fundamental problems[J]. Acta Meteorological Sinica, 58:500-512.

LUO S W,QIAN Z A,WANG Q Q, 1982. The study for 100 hPa South Asian high and its association with climate of East China[J]. Plateau Meteorology, 1: 1-10.

MASON R B,ANDERSON C E, 1963. The development and decay of the 100 hPa summertime anticyclone over Southern Asia[J]. Mon Wea Rev, 91:3-12.

QIAN Y F, 1978. The calculation method of the heating field and the effects of the heating during a Tibetan High process [C]// Symposium paper collections on Tibetan Plateau Meteorological (1975—1976). Beijing:Science Press:199-212.

TAO S Y,ZHU F K, 1964. The variation of 100 hPa circulation over South Asia in summer and its association with march and withdraw of West Pacific Subtropical high[J]. Acta Meteorological Sinica, 34:385-395.

XIE P,ARKIN P A, 1997. Global precipitation: A 17-year monthly analysis based on gauge observations, satellite estimates, and numerical model outputs[J]. Bull Amer Meteor Soc, 78: 2539-2558.

XU B N, 1983. Some statistical characters of the South Asia high during early summer and its relation to weather of northeast Guizhou Province. Meteorological Science Technique Symposium, No. 6[M]. Beijing: China Meteorological Press: 108-115.

YANG Z,YANG H, 1984. The relation between the South Asia high and precipitation of Qinghai Province during flood period. Qinghai-Xizang Plateau Meteorology Experiment Symposium[M]. Beijing: Science Press:159-171.

ZHANG K S,YE D Z, 1977. Simulation of the movements of the Tibetan high and the application on summer prediction[J]. Scientia Sinica, 4: 360-368.

ZHANG Q,WU G X, 2001. The large area flood and drought over Yangtze River Valley and its relation to the South Asia high[J]. Acta Meteorological Sinica, 59(5): 569-577.

ZHAO W,ZHU F K, 1978. Midsummer precipitation over south Xinjiang Province and the South Asia high[J]. Meteorology, No. 9.

ZHU F K,LU L Y,CHEN X J,et al, 1980. The South Asia High[M]. Beijing: Science Press: 95.

Genesis of the South Asian High and Its Impact on the Asian Summer Monsoon Onset

LIU Boqi[1,2], WU Guoxiong[2], MAO Jiangyu[2], HE Jinhai[1]

(1 Key Laboratory of Meteorological Disaster of Ministry of Education, Nanjing University of Information Science and Technology, Nanjing 210044, China; 2 State Key Laboratory of Numerical Modeling for Atmospheric Sciences and Geophysical Fluid Dynamics (LASG), Institute of Atmospheric Physics, Chinese Academy of Sciences, Beijing 100029, China)

Abstract: The formation of the South Asian high (SAH) in spring and its impacts on the Asian summer monsoon onset are studied using daily 40-yr ECMWF Re-Analysis (ERA-40) data together with a climate-mean composite technique and potential vorticity-diabatic heating (PV-Q) analysis. Results demonstrate that, about two weeks before the Asian summer monsoon onset, a burst of convection over the southern Philippines produces a negative vorticity source to its north. The SAH in the upper troposphere over the South China Sea is then generated as an atmospheric response to this negative vorticity forcing with the streamline field manifesting a Gill-type pattern. Afterward the persistent rainfall over the northern Indochinese peninsula causes the SAH move westward toward the peninsula. Consequently, a trumpet-shaped flow field is formed to its southwest, resulting in divergence-pumping and atmospheric ascent just over the southeastern Bay of Bengal (BOB).

Near the surface, as a surface anticyclone is formed over the northern BOB, an SST warm pool is generated in the central-eastern BOB. This, together with SAH pumping, triggers the formation of a monsoon onset vortex (MOV) with strong surface southwesterly developed over the BOB. Enhanced air-sea interaction promotes the further development and northward migration of the MOV. Consequently, the wintertime zonal-orientated subtropical anticyclone belt in the lower troposphere splits, abundant water vapor is transported directly from the BOB to the subtropical continent, and heavy rainfall ensues; the atmospheric circulation changes from winter to summer conditions over the BOB, and Asian summer monsoon onset occurs.

1 Introduction

The Asian summer monsoon (ASM) onset is an important part of the seasonal transition from winter to summer. Many studies (Wu and Zhang, 1998; Mao et al., 2002a; Wang and LinHo, 2002; Mao and Wu, 2007) have pointed out that the ASM onset starts over the eastern Bay of Bengal (BOB) area, followed by onset over the South China Sea (SCS) and finally by onset over India. Krishnamurti et al. (1981) identified a type of low-level vortex known as the onset vortex, which forms during the Indian summer monsoon onset at about 10°N in the BOB and eastern Arabian Sea (Ananthakrishnan et al., 1968). As the vortex moves northward and develops, the Indian summer monsoon is established (Vinayachandran et al., 2007). Features of the vortex, such as its structure and rainfall distribution, have been studied previously (Pisharoty and Asnani, 1957; Rao and Rajamani, 1970; Sharma and Srinivasan, 1971). Atmospheric barotropic instability associated with horizontal wind shear, together with diabatic heating, is thought to be responsible for the vortex formation (Krishnamurti et al., 1981; Krishnamurti, 1981, 1985; Krishnamurti and Ramanathan, 1982; Mao and Wu, 2011). Mak and Kao (1982) showed that vertical wind shear and baroclinic instability are also important for the vortex development. In addition to these internal dynamics, Joseph (1990) proposed that the vortex is also influenced by the local sea surface temperature (SST) because, if the SST is above 28 ℃, and if other factors are favorable, organized deep convection will develop in the tropical atmosphere (Gray, 1968, 1975, 1998; Harr and Chan, 2005). Usually, a warm pool (SST>30.5 ℃) appears over the Arabian Sea about 1 week before the Indian summer monsoon onset. The warm SST leads to the formation of a vortex that affects the Indian summer monsoon onset over Kerala in early June (Joseph, 1990; Shenoi et al., 1999, 2005).

The monsoon onset vortex (MOV) also exists over the BOB in early May during the BOB summer monsoon onset (Lau et al., 1998; He and Shi, 2002; Liu et al., 2002; Yan, 2005). Case studies show that while the onset vortex is moving northward over the BOB region, the wintertime zonally oriented subtropical anticyclone belt splits over the BOB with the onset of the Asian summer monsoon (Mao et al., 2002b). Vinaychandran et al. (2007) studied the structure of the onset vortex and showed that the warmest SST is in the eastern BOB before the summer monsoon onset. Wu et al. (2011, 2012) emphasized the key role of air-sea interaction in the formation of the BOB MOV. They found that the strong surface sensible heating resulted from the BOB warm pool in spring leads to vortex formation. Previous research on the MOV has been confined to case studies. The current study will focus on climate-mean characteristics of the MOV formation and development in order to improve our understanding of the BOB summer monsoon onset process.

In addition to the lower tropospheric circulation associated with the vortex, the upper tropospheric circulation over the BOB also undergoes noticeable changes in the seasonal transition from winter to summer. Mao et al. (2004) pointed out that this seasonal transition starts with the occurrence of northward tilting of the subtropical ridge surface and a positive meridional temperature gradient (MTG) in the upper troposphere over the BOB. The ridgelines always tilt toward the warmer region with increasing height, and the moment when this tilt becomes perpendicular (i.e., $\partial T/\partial y = 0$) is defined as the monsoon onset time in the region. Based on such a concept, a series of the onset dates of the BOB summer monsoon has been produced (Mao and Wu, 2007).

The SAH is another characteristic synoptic system over the Asian continent in boreal summer (Mason and Anderson, 1963), which forms near the tropopause over the Indochinese peninsula before the BOB monsoon onset (He et al., 2006; Liu et al., 2009, 2012). The SAH is a thermal high pressure system, whose intensity depends on the distribution of heterogeneous diabatic heating over the Tibetan Plateau and the Asian summer monsoon area (Wu and Liu, 2003; Wu et al., 2007). Its seasonal evolution and variation in boreal summer also depend strongly on changes in atmospheric heating, especially the thermal effect of the Tibetan Plateau (Flohn, 1960; Krishnamurti and Rodgers, 1970; Krishnamurti, 1971a, 1971b; Liu et al., 2001; Zhang et al., 2002; Wu et al., 2002; Qian et al., 2002; Duan and

Wu, 2005; Duan et al., 2008). Data diagnoses (Reiter and Gao, 1982; Zhu et al., 1986; Chen et al., 1991) demonstrate that the SAH first appears in spring over the SCS just prior to the ASM onset. During the BOB monsoon onset, its center moves northwestward and intensifies over the Indochinese peninsula. Finally, after the ASM onset, it settles over the southern flank of the Tibetan Plateau.

It is still not clear how the SAH is generated in spring, and how its development in the upper troposphere can help the MOV development, which has previously been thought to arise solely through air-sea interaction and energy conversions over the BOB, and contribute to the ASM onset. We address these questions in the current study. The remainder of this paper is organized as follows. Section 2 describes the data and method applied in this study. Section 3 investigates the variation in the SAH during the BOB summer monsoon onset, emphasizing the mechanism for the SAH formation and evolution. Section 4 describes the climate characteristics of the BOB MOV, focusing on the coupling between the upper and lower synoptic systems, and summarizing the three-dimensional (3D) features of the BOB summer monsoon onset process. Finally, a discussion and summary are provided in section 5.

2 Data and methodology

The datasets used in this study are as follows: (1) The 40-yr European Centre for Medium-Range Weather Forecast (ECMWF) Re-Analysis data (ERA-40) (Uppala et al., 2005) data are for the period 1979—2001, including 3D wind field, geopotential height, air temperature, relative humidity, precipitation and surface heating flux, with a horizontal resolution of $2.5° \times 2.5°$. The variables are arranged on standard isobaric surfaces from 1000 hPa to 1 hPa, excluding the surface values. For consistency, the rainfall records within the ERA-40 dataset are directly applied in the present study. Except for excessive precipitation in the tropics, the ERA-40 rainfall data are shown to be reliable in comparison with other precipitation data derived from satellites, such as the 3B42 products of the Tropical Rainfall Measuring Mission (TRMM) and daily precipitation records of the Global Precipitation Climatology Project (GPCP) (Voisin et al., 2008; Chan and Nigam, 2009). (2) Daily global sea-surface winds with a high horizontal resolution of $0.25° \times 0.25°$ supplied by the National Oceanic and Atmospheric Administration (NOAA) (Zhang et al., 2006) and National Climatic Data Center (NCDC) (Smith and Reynolds, 2004). In this dataset the wind speeds are generated by blending observations from six satellites, and the wind directions from the National Center for Environment Prediction/Department of Energy Global-Reanalysis 2 (NCEP-2) dataset (Kanamitsu et al., 2002) are interpolated into the blended speed grids. (3) Daily optimum interpolation SST (OISST) analysis data (AVHRR products) are from the NOAA Satellite and Information Service with the same resolution as the sea surface winds. (4) Daily sea surface sensible heating flux with a horizontal resolution of $1° \times 1°$ obtained from the Objectively Analyzed Air-Sea Fluxes (OAFlux) for the global oceans, sponsored by the Woods Hole Oceanographic Institution (WHOI) Cooperative Institute for Climate and Ocean Research (CICOR).

In our composite study of the monsoon onset processes, it is important to have a robust definition of the onset date. As mentioned above, the earliest ASM onset is the BOB monsoon onset. In this study, we will focus on the BOB monsoon onset. Following Mao and Wu (2007), the BOB onset date is defined as the day when the following criteria are satisfied: (1) the area-averaged upper-tropospheric (200—500 hPa) MTG over the eastern BOB (5°—15°N, 90°—100°E) changes from negative to positive and (2) the MTG remains positive for more than 10 days. Table 1 lists the resultant BOB summer monsoon onset dates from 1979 to 2001. The mean onset date of the BOB summer monsoon is 2 May. The WCMWF Interim Re-Analysis (ERA-Interim) dataset (Dee et al., 2011) for the period 1979—2010 has been employed to verify what is shown in Table 1, and the results are quite similar. Other criteria for the monsoon onset are also used for comparison purpose, including the area-averaged 850 hPa zonal wind (U850) and outgoing longwave radiation (OLR) reaching threshold values of 0 m s^{-1} and 230 W m^{-2}, respectively. The correlation coefficients of the MTG with U850 and with OLR are 0.94 and 0.67, respectively, with both above the 99% statistical significance level (0.481), indicating that the onset date defined by the sign-change of MTG is reliable. The marked interannual variability of the BOB summer monsoon onset date requires us to use a composite analysis technique described below to obtain climate-means of monsoon onset processes. For each individual year the onset date is defined as the zero day (D0), with the dates before (after) the onset date being labeled as negative (positive) days. All the variables are chosen from -30 days (D-30) to $+30$ days (D$+30$) relative to the BOB summer monsoon onset date each year. The climate states of a particular variable for each of the 61 days (D-30 to D$+30$) during the summer monsoon onset are thus calculated by arithmetically averaging the data over each corresponding day for the period 1979—2001.

Table 1 The BOB summer monsoon onset date for years from 1979 to 2001. "E" and "L" mark, respectively, the monsoon onset occurring after an El Nino and La Nina event

Year	Date	Year	Date
1979	09 May	1991	22 Apr
1980	11 May	1992(E)	06 May
1981	07 May	1993	19 May
1982	22 Apr	1994	01 May
1983(E)	13 May	1995(E)	07 May
1984	11 Apr	1996(L)	27 Apr
1985(L)	16 Apr	1997	16 May
1986	09 May	1998(E)	13 May
1987	03 May	1999(L)	09 Apr
1988	02 May	2000(L)	25 Apr
1989	30 Apr	2001(L)	29 Apr
1990	14 May		

3 Evolution of the SAH before the BOB summer monsoon onset

Figure 1 shows the evolution of the atmospheric circulation at 150 hPa during the BOB summer monsoon onset. On D-15 and earlier (Fig. 1a), South Asia is dominated by a large anticyclone with its ridge axis extending from the equatorial central Pacific to the north Indian Ocean. There is no closed SAH at this time. By D-12 (Fig. 1b), as northerly develops over the eastern SCS between 10° and 20°N, a closed anticyclone (viz., the SAH) appears in the upper troposphere over the western SCS and the southeastern Indochinese peninsula. The SAH strengthens and moves westward from the SCS to the Indochinese peninsula from D-9 to D-6 (Figs. 1c and 1d); meanwhile the ridgeline remains around 10°N with little northward migration. The SAH center starts to shift to the north of 10°N from D-6 to D0, with the streamline field to its south exhibiting a "trumpet" shape and strong divergence over the southeastern BOB (Fig. 1e). When the BOB summer monsoon onset occurs on D0 the SAH shifts northward with its ridgeline situated near 15°N. The strong upper-level divergence persists over the southeastern BOB (Fig. 1f). After the summer monsoon builds up over the BOB, the SAH continues to develop and move farther northwestward (Fig. 1g). The center of the SAH is located near Yangon in Burma on D+6 (Fig. 1h), and the strong trumpet-shaped upper-layer divergence on its southwest persists and moves slowly eastward. The above results imply that the SAH evolution may be closely associated with the onset process of the BOB summer monsoon. We therefore consider below the SAH formation process prior to the BOB summer monsoon onset.

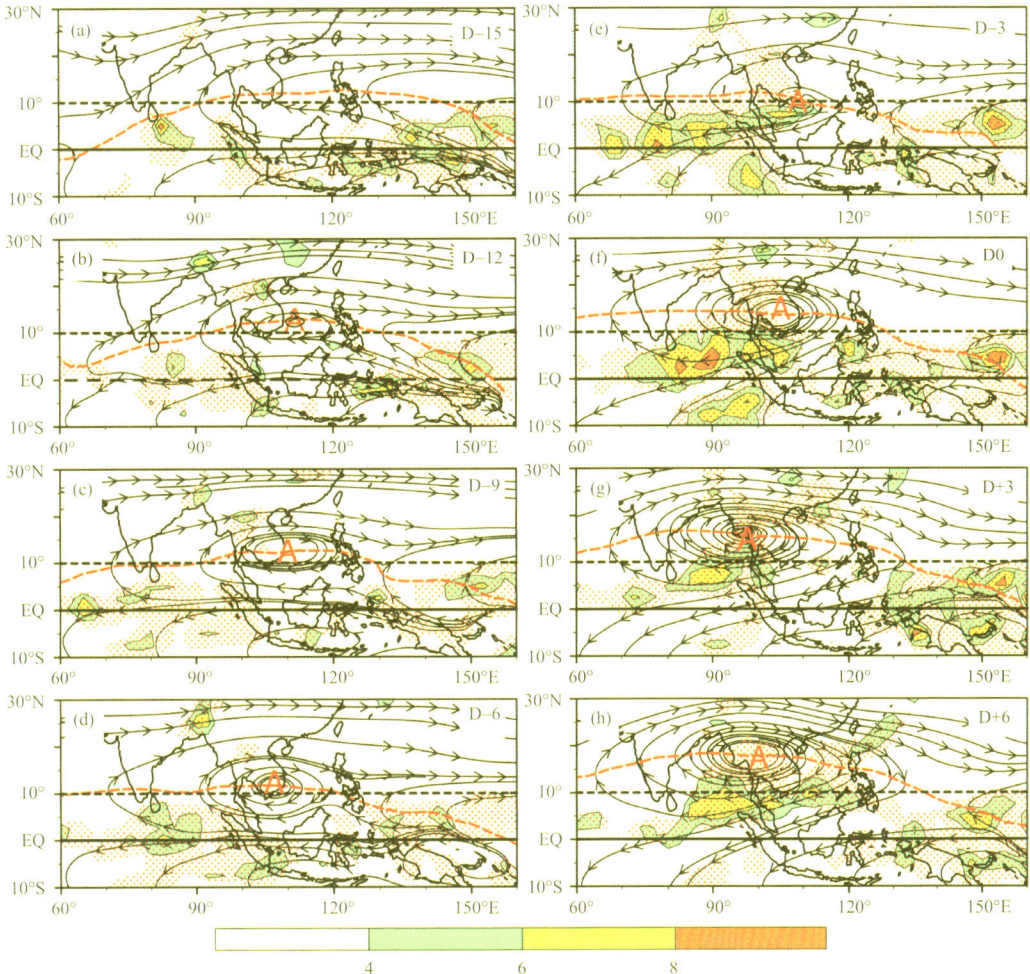

Fig. 1 Daily evolution of the 150 hPa streamline and divergent field (shading, units: 10^{-6} s^{-1}), and the diabatic heating averaged from 500 hPa to 200 hPa (red stipple denotes >1.5 K day^{-1}) during the BOB summer monsoon onset period: (a) D−15, (b) D−12, (c) D−9, (d) D−6, (e) D−3, (f) D0, (g) D+3, and (h) D+6. The letter "A" denotes the anticyclone center, and the ridgeline is plotted as a red dashed line

a. Formation of the SAH

Figure 2 shows the geopotential height deviation from the zonal mean across South Asia in the pre-monsoon phase, averaged over 5°−15°N. On D−19 and D−16 in the upper troposphere, there is only one center above 150 hPa but two weak centers at 200 hPa, one over the SCS and the other over the ocean to the east of the Philippines (Figs. 2a and 2b). On D−13, the western high-pressure center (i.e., the SAH) intensifies (Figs. 1b and 2c), while the eastern one weakens and disappears. Thus the SAH formation presents as a substitution of the original anticyclone above the equatorial western Pacific by the western anticyclone over the western SCS and the southeastern Indochinese eninsula.

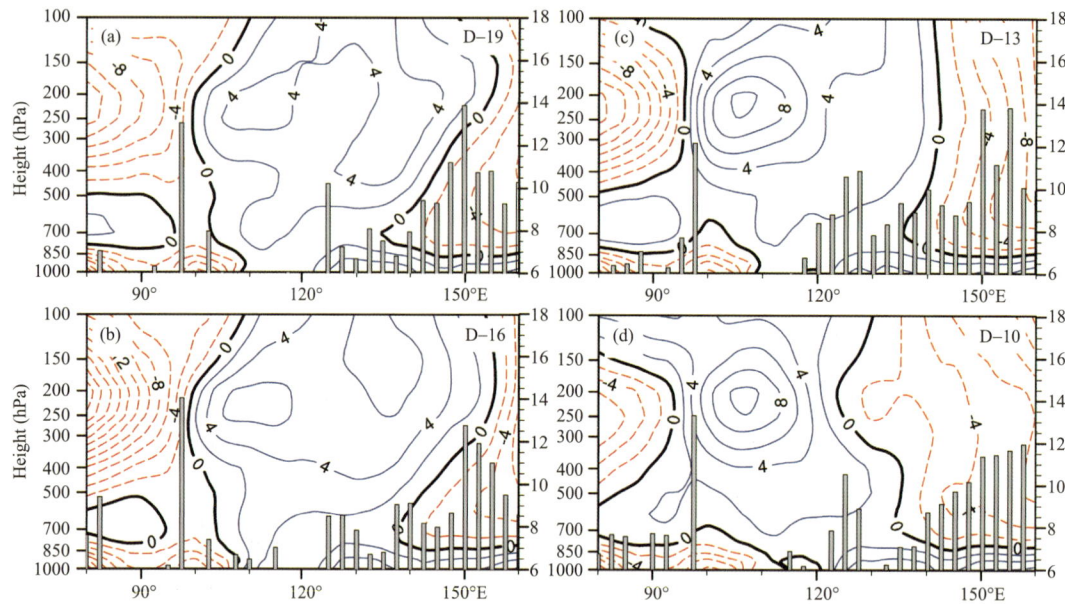

Fig. 2　Pressure-longitude cross-section (averaged over 5°—15°N) of the deviation of geopotential height (contours, units: gpm) from the zonal mean: (a) D—19, (b) D—16, (c) D—13, (d) D—10. The bar represents the longitudinal profile of ERA-40 precipitation averaged over 5°—10°N (units: mm d^{-1}), with its magnitude marked on the left vertical axis

Note that before the SAH formation phase from D—19 to D—16 (Figs. 2a and 2b), the main convection is over the western Pacific, with little rainfall over the SCS and the southeastern Indochinese peninsula (100°—120°E). On D—13 and D—10 (Figs. 2c and 2d), significant rainfall of more than 9 mm d^{-1} occurs over the southern Philippines, but also little precipitation occurs over SCS. During this period, a northerly wind also develops in the upper troposphere over the eastern SCS (Figs. 1b and 1c), implying a potential association between the convection over the southern Philippines and the development of northerlies over the eastern SCS, which contributes to the generation of the SAH.

This relationship is investigated using the pressure-latitude cross section of local meridional circulation and diabatic heating averaged over the Philippines (120°—130°E) on D—19 and D—13 (Figs. 3a and 3b, respectively). Prior to SAH formation on D—19 (Fig. 3a), downward motion controls the southern Philippines region (around 10°N), with upper-layer southerlies above 500 hPa and lower-layer northerlies below. There is no apparent diabatic heating in the region. On D—13 (Fig. 3b), a strong deep convective condensation heating of more than 2.5 K d^{-1} develops over the southern Philippines, and prevailing northerlies in the upper layers are established to the north of the heating region.

Following the Ertel potential vorticity equation (Ertel, 1942; Hoskins et al., 1985), the vertical vorticity equation can be expressed as (Wu and Liu, 1998; Wu et al., 1999; Liu et al., 1999, 2001, 2004):

$$\frac{\partial \zeta}{\partial t} + \mathbf{V} \cdot \nabla \zeta + \beta v = -(f + \zeta) \nabla \cdot \mathbf{V} + \frac{f + \zeta}{\theta_z} \frac{\partial Q}{\partial z} + S \qquad (1)$$

in which the internal forcing term associated with atmospheric baroclinicity has been ignored because the current study focuses on the SAH that is in the upper troposphere in tropics, where the isentropic surfaces are horizontally located. On the right hand side of Eq. (1), the three terms represent the effects of three dimensional divergence, and the vertical and horizontal gradients of diabatic heating, respectively.

Scale analysis indicates that in the tropics the vorticity generation due to diabatic heating usually prevails over the atmospheric compressibility, and the first term on the right hand side of Eq. (1) can be ignored as well. Within the heating region the vertical heating gradient ($\partial Q/\partial z$) dominates the forcing, whereas at the border of the diabatic heating, the horizontal heating gradient (i. e. S) plays a more important role (Wu and Liu, 1998; Liu et al., 2001). The vorticity source

$$S = -\frac{1}{\theta_z}\frac{\partial v}{\partial z}\frac{\partial Q}{\partial x} + \frac{1}{\theta_z}\frac{\partial u}{\partial z}\frac{\partial Q}{\partial y} = -\frac{g}{f\theta\theta_z}(\nabla_h \theta \cdot \nabla_h Q) \tag{2}$$

represents an external forcing caused by the correlation between the horizontal gradients of atmospheric temperature and diabatic heating. Here, θ_z and Q are the static stability and diabatic heating source, respectively; ∇_h denotes horizontal gradient; and the other terms follow convention in meteorology. Of note is that Q is calculated with the scheme deduced by Yanai et al. (1973), which takes the form

$$Q = Q_1 = C_p\left[\frac{\partial T}{\partial t} + \mathbf{V} \cdot \nabla_h T + \left(\frac{p}{p_0}\right)^{R/C_p}\omega\frac{\partial \theta}{\partial p}\right] \tag{3}$$

Here T, C_p and R are the air temperature, the specific heat of dry air at constant pressure and the gas constant for dry air, respectively. Atmospheric warming by latent heat release within tropical convection usually produces a warm core and an inward horizontal temperature gradient around the heating border. To the north of the heating, this gradient and the environmental MTG are positively correlated with the horizontal heating gradient. A negative vorticity source is then produced on the northern rim of the heating (Liu et al., 2001).

When convection develops over the southern Philippines on D−13, the strong local diabatic heating produces a considerable negative vorticity source ($<-2\times10^{-11}$ s^{-2}) in the upper troposphere to its north (Fig. 3c). The S center over the Philippines is at 400−300 hPa. Accompanied by the occurrence of the negative vorticity source is a remarkable reversal of the upper tropospheric circulation to its north: as shown in Figs. 3a and 3b in the latitude domain to the north of 10°N, the dominant southerly on D−19 is replaced by prevailing northerly on D−13. Because the northerly is produced in the upper troposphere along the longitude span between 120° and 130°E which is located over the east of the SCS, the development of the northerly can help the isolation of an upper-layer anticyclone circulation over the SCS from the main anticyclone system over the western Pacific, in favor of the genesis of the SAH.

It should be noted that in Table 1, the average onset date is 2 May, while the earliest onset date was 9 April in 1999 and the latest was 19 May in 1993; both are nearly 20 days away from the mean. The atmospheric state during those extreme years can be closer to mid-late winter or midsummer, which may cause the composite results different substantially. To clarify the question, a standard deviation σ of the monsoon onset day is calculated from the data listed in Table 1, and the thresholds of $\pm 0.7\ \sigma$ are chosen to select the early and late monsoon onset years. Five Early years (1982, 1984, 1985, 1991, and 1999) and six Late years (1980, 1983, 1990, 1993, 1997, and 1998) were then selected from the total 23 years, and three groups of onsets (All, Early and Late) were composed. The evolutions for the Early and Late onset groups of diabatic heating, circulation and relevant negative vorticity source S near the Philippines are displayed in Figure 3 in comparison with the composite (All case). As in the composite case, the diabatic heating centers in the Early case (Figs. 3d and 3e) and Late case (Figs. 3g and 3h) both move northward to the southern Philippines from D−19 to D−13. In association with the northward displacement of the diabatic heating, a negative vorticity source ($S<0$) appears in the mid-upper troposphere near 10°N in either the Early case (Fig. 3f) or the Late case (Fig. 3i). In each case in response to this negative vorticity forcing, the prevailing southerly in the upper troposphere over the latitude band 10°—20°N on D—19 (Figs. 3d and 3g) has been reversed to dominant northerly on D—13 (Figs. 3e and 3h) as in the All composite, which is in favor of the anticyclone vorticity formation over the SCS to its west. In view of the similar features of the different monsoon onset groups in the evolutions of diabatic heating, the occurrence of negative vorticity source near the southern Philippines, and the wind reversal over the eastern SCS, it is reasonably justified that the composite results presented in Figures 3a—c are of representative, even for both early and late monsoon onset groups.

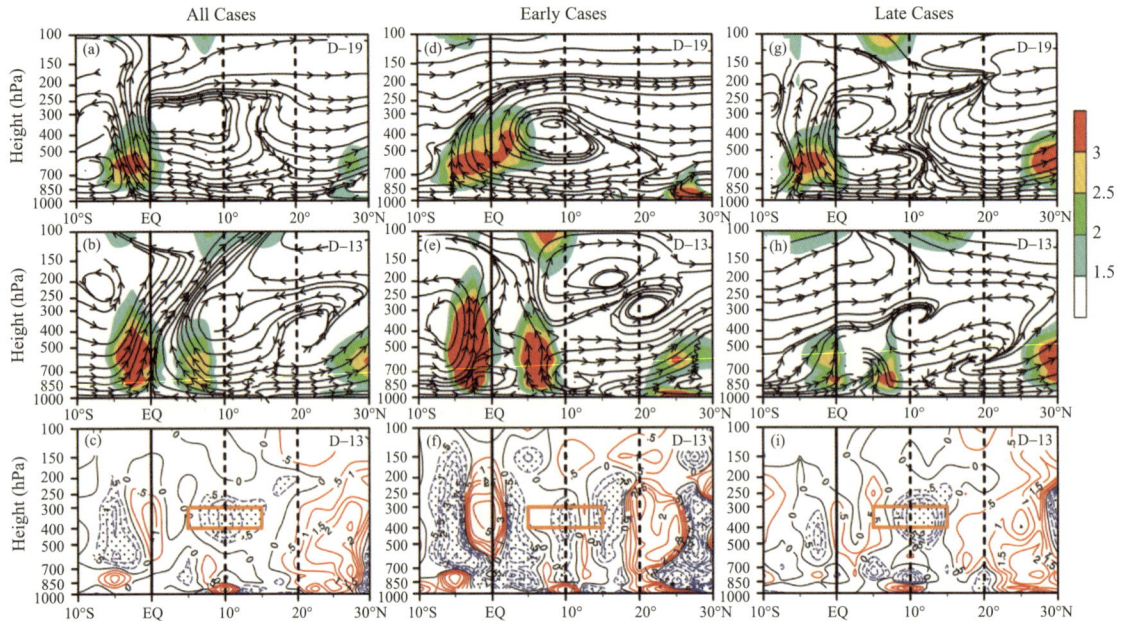

Fig. 3 120°—130° averaged Pressure-latitude cross section of Q_1 (shaded, units: K day^{-1}) and meridional circulation on D—19 (a, d, g) and D—13 (b, e, h); and of vorticity source S (c, f, i) on D—13 (units: 10^{-11} s^{-2}, values stronger than -0.5 are stippled). The left column (a, b, c) is for the all-case composing, while the middle column (d, e, f) for the early group, and the late group is in the right column (g, h, i). The orange boxes in (c), (f) and (i) denote the area with local maximum S

Figure 4 presents the 400—300 hPa averaged horizontal distributions of S, Q_1 and streamline field on D—13 (Fig. 4a), also shown are the time series from D—19 to D+5 of S, $\mathbf{V} \cdot \nabla \zeta, \beta v$, and ($\mathbf{V} \cdot \nabla \zeta + \beta v$) (Fig. 4b) averaged over the region (7.5°—12.5°N, 120°—125°E), an area of the south Philippines. It is evident that part of the negative vorticity source over the south Philippines is balanced by horizontal advection of relative vorticity due to the prevailing easterly across the source and by the horizontal advection of planetary vorticity due to the local northerly (Fig. 4a). Consequently, the streamline field exhibits a Gill-type response to the negative vorticity forcing source over the Philippines (Gill, 1980), with the SAH establishing over the western SCS and the southeastern Indochinese peninsula. Afterward, although the stream-field and the associated vorticity advection in the upper troposphere vary, the negative vorticity source over the Philippines remains stable at about -1.0×10^{-11} s^{-2} until monsoon onset (Fig. 4b). The mean value of S is one order of magnitude larger than the one of ($\mathbf{V} \cdot \nabla \zeta + \beta v$) from D—13 to D0. Furthermore, the other forcing terms in (1) are small enough to be ignored. The stability and persistence of this near equatorial negative vorticity forcing source support the Gill-type atmospheric response to the negative vorticity forcing in the establishment of the SAH.

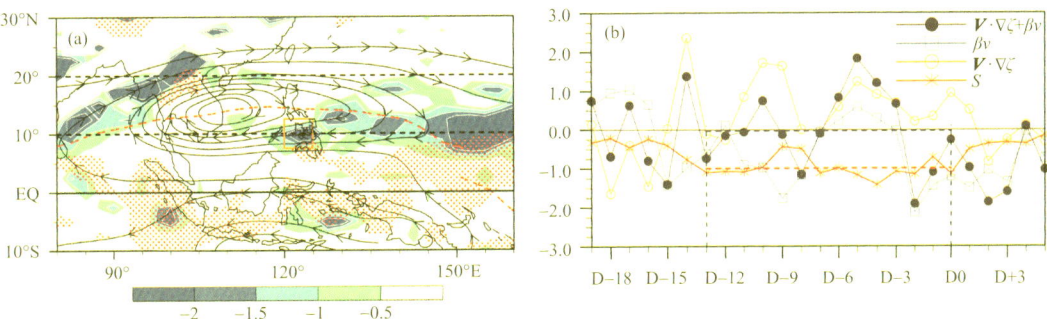

Fig. 4 (a) The 400-300 hPa averaged horizontal distributions of S (shading, interval is $0.5 \times 10^{-11}\ s^{-2}$), Q_1 (red stipple indicates >1.5 K day^{-1}) and streamline field on D-13, and (b) Time series of S, $\mathbf{V} \cdot \nabla \zeta$, βv, $(\mathbf{V} \cdot \nabla \zeta + \beta v)$ and over the region (7.5°-12.5°N, 120°-125°E) shown by the orange box in (a). The red and purple bold dash lines in (b) represent the mean value of S and $(\mathbf{V} \cdot \nabla \zeta + \beta v)$ from D-13 to D0, respectively

b. Strengthening of the SAH in the premonsoon period

After D-13, the center of the SAH migrates gradually westward toward the Indochinese peninsula as its intensity increases (Fig. 1). Figure 5 shows the D-5 to D-1 averaged precipitation and circulation in the monsoon area. Over the northern Indochinese peninsula around 20°N, a rainfall center is isolated from the convection near the equator before the onset date (Fig. 5a). The condensation heating released by the rainfall center can produce negative relative vorticity and contribute to SAH strengthening in the upper troposphere.

During this period, there is strong surface sensible heating over the Indian peninsula, which contributes substantially to the generation of a surface cyclone over land and anticyclones over the Arabian Sea and the northern BOB (Fig. 5b), accompanied by significant local descent (Fig. 5c). As the descending air reaches the sea surface, a divergent westerly develops with strong divergence over the ocean but convergence over the land (Fig. 5d). There are two water vapor paths surrounding the Indochinese peninsula: the southwesterly path over the northwestern BOB brings moisture to the Indochinese peninsula, while the southeasterly carries water vapor to the land from the SCS and west Pacific. The moisture moving from the ocean to the Indochinese peninsula is then transported northward by the strong lower-layer southerly (Fig. 5e). Moisture then converges near the northern Indochinese peninsula, contributing to the occurrence of the rainfall maximum there (Fig. 5a). Condensation heating due to the rainfall thus contributes to the development of the southerly in the lower layers and the strengthening of the SAH in the upper troposphere during the premonsoon period.

4 BOB summer monsoon onset process

As mentioned above, the SAH forms and develops before the BOB summer monsoon onset. Before we investigate the SAH impacts on the monsoon onset, we look at the onset processes. Based on case studies of the ASM onset in 1998 and 2003, Wu et al. (2011, 2012) have reported that air-sea interaction plays an important role in the MOV formation during the BOB summer monsoon onset. In this section the composite technique is employed again to study the climate-mean characteristics of the MOV formation and its connection with the ASM onset.

a. Generation of the MOV

Both the atmospheric circulation and the thermal conditions on the sea surface change distinctly dur-

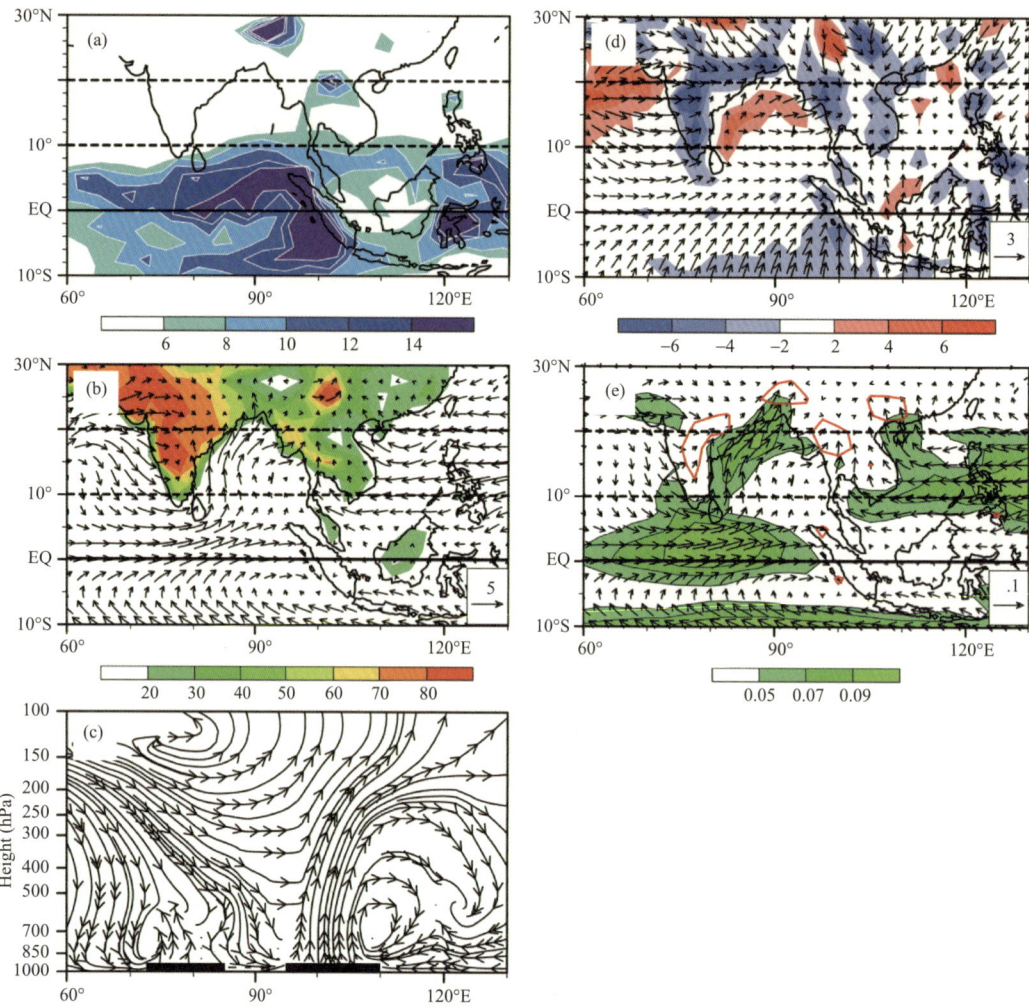

Fig. 5 The D−5 to D−1 averaged distributions of (a) precipitation (mm d^{-1}), (b) surface sensible heating (shading, W m^{-2}) and 10 m wind (vectors, m s^{-1}), (c) local zonal circulation (averaged over 10°−20°N), (d) 10 m divergent wind (vectors, m s^{-1}) and its divergence (shading, 10^{-6} s^{-1}), and (e) 925 hPa moisture transport (vectors, with its maximum shaded, g m kg^{-1} s^{-1}) and its maximum convergence (red contour is −0.1 g kg^{-1} s^{-1})

ing the BOB summer monsoon onset. Before the onset, strong sensible heating (>12 W m^{-2}) is observed beneath the strong equatorial surface westerly over the southern BOB (Figs. 6a−c). At 700 hPa a distinct local cyclonic circulation pattern over the tropical BOB develops in response to this sensible heating (Figs. 7a−c). The SST in the southeastern BOB is relatively colder than the one in the central BOB (Fig. 9). One possible reason is that, over the southern BOB, the stronger surface westerly prevents the local SST from rapidly rising up through the wind-evaporation-SST feedback (Xie and Philander, 1994). Furthermore, the near-equatorial cloud related to local rainfall could also inhibit the solar radiation from warming the SST directly. Over the northern BOB, a distinct anticyclone is situated near the surface (Figs. 6a−c) and a continuous ridgeline of the subtropical anticyclone is located at 700 hPa with a weak Indo-Burma trough over the eastern BOB (Figs. 7a−c). These features are associated with downward motion (Fig. 5c) with diabatic cooling (Figs. 8a−c) over the northern BOB. The resultant clear sky and weak surface wind conditions favor increasing solar radiation and decreasing energy release from the ocean, as reported by Wu et al. (2011, 2012), and a warm pool (SST>30 ℃) is formed between 10°N and 15°N over the central BOB before the summer monsoon onset (Figs. 9a—c). This short-lived warm pool in spring only occurs over the central eastern BOB, because before the BOB summer monsoon onset the prevailing southwesterly winds along the northwestern BOB (Figs. 6a—c) cause upwelling in the region due to the offshore currents, and a buildup of cold sea surface water (Figs. 9a—c). A similar situation exists over the northeastern BOB because of the prevailing surface northerly winds.

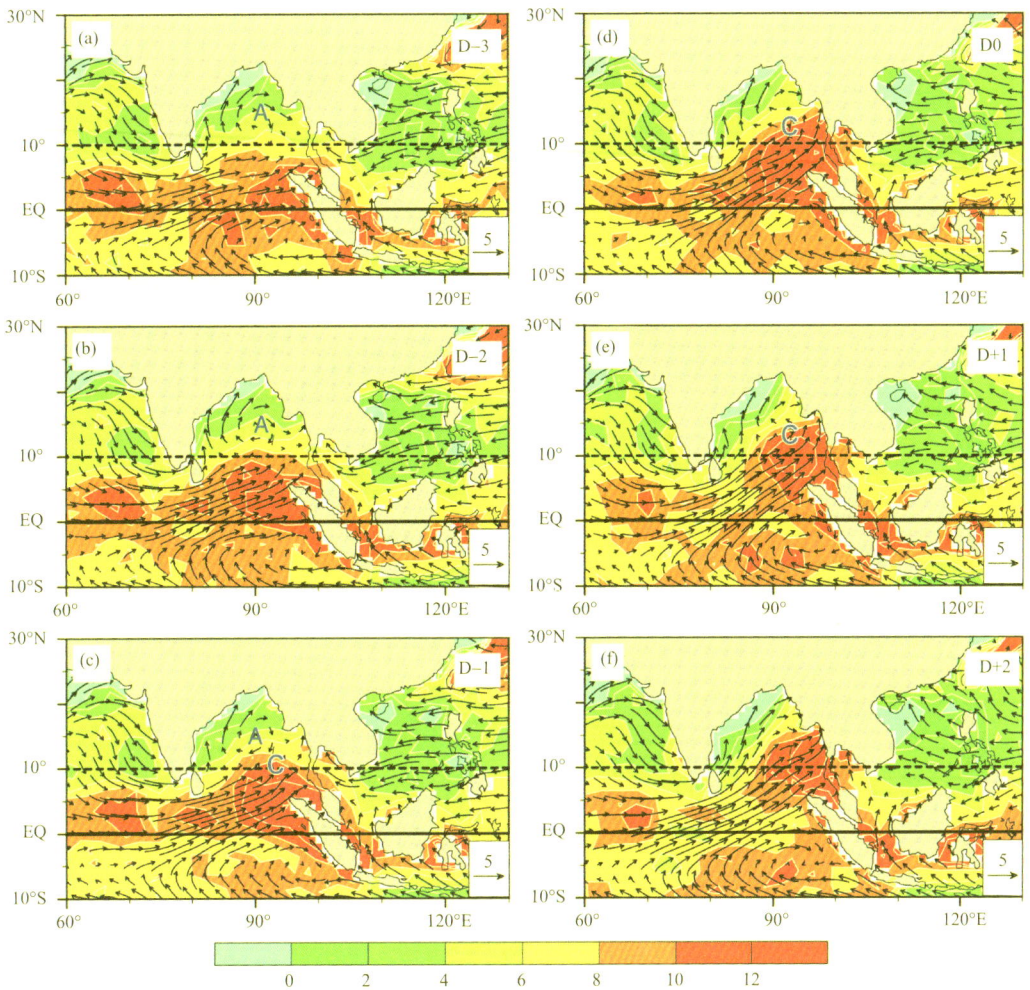

Fig. 6　Daily evolution of sea surface sensible heating (shading, unit is W m^{-2}) and sea surface wind (vectors, m s^{-1}) from D−3 to D+2 for (a) to (f), respectively. "A" and "C" denote the anticyclone and cyclone centers, respectively

After the summer monsoon onset, the southwesterly sweeps over most of the central and eastern BOB, and induces onshore currents along the eastern coast (Figs. 6d—f), with local downwelling maintaining warm SST there. This, together with the increased local surface wind, enhances the local surface sensible heating. As the atmosphere is heated in an already warm area over the ocean warm pool, atmospheric available potential energy is generated; cyclonic circulation or even a MOV develops in the region. The positive feedback due to local air-sea interaction supports further development and northward displacement of the MOV even after the onset date (Fig. 6e).

At 700 hPa, the cyclone near Sri Lanka also moves northward (Fig. 7d). As the MOV develops, the cyclone merges with the Indo-Burma trough and the subtropical high ridgeline splits. Consequently the equatorial westerly over the southern BOB is linked directly with the subtropical westerly over the central Indochinese peninsula and the southern coast of China, and abundant water vapor is transported to the Indochinese peninsula, accompanied by heavy rain over the northeastern BOB and the western Indochinese peninsula (Figs. 7e, 7f, and 8d-f). The ASM onset commences.

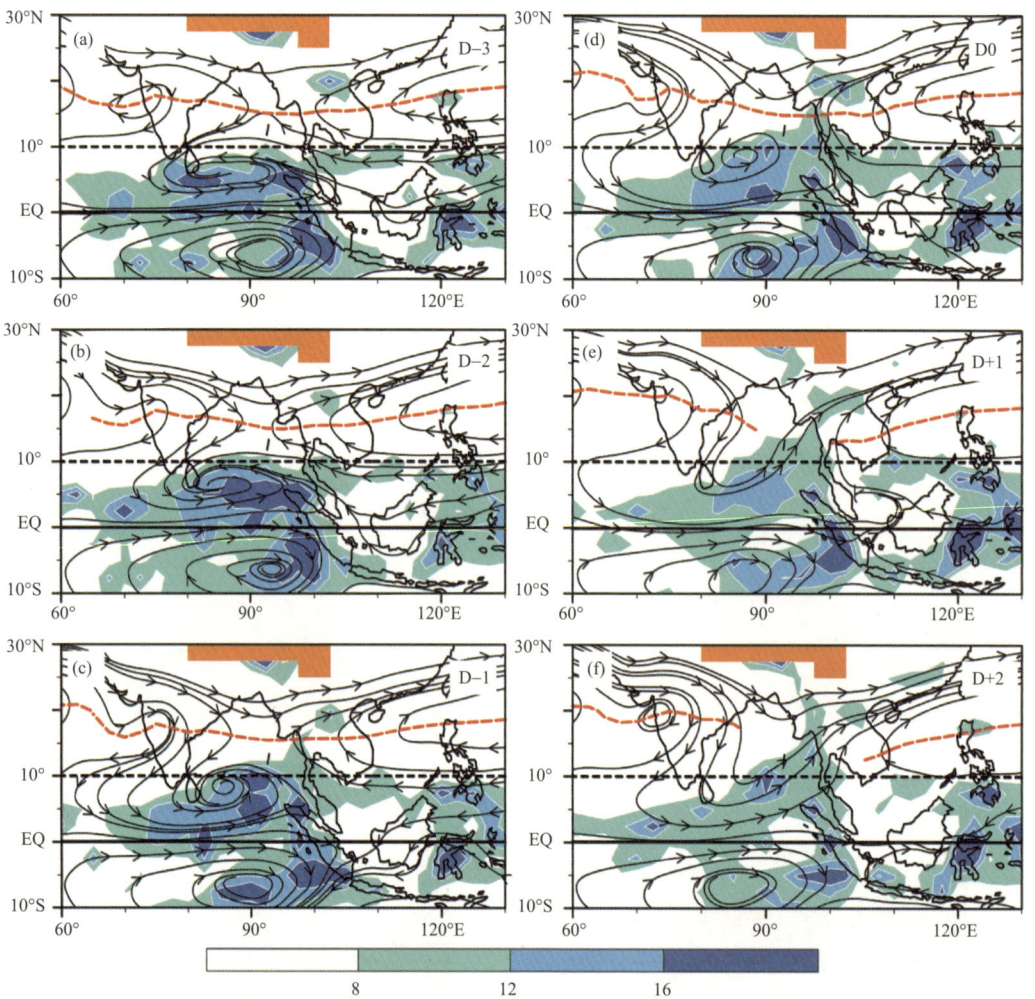

Fig. 7 Daily evolution of 700 hPa streamline and ERA-40 rainfall (shading, mm d^{-1}) from D−3 to D+2 for (a) to (f), respectively. Topography and ridgeline are marked by orange area and red dashed line, respectively

b. Evolution of the SAH

During the BOB summer monsoon onset the SAH also develops significantly, as shown in Figure 1 (right column). On D−3 the anticyclone center is still over the southeast of the Indochinese peninsula with its ridgeline at 10°N over the BOB (Fig. 1e), and the divergence on its southwest over the southern BOB is weak. From D−3 to D0 (Fig. 1f), the ridgeline shifts rapidly from 10°N to 15°N, and the upper divergence to the southwest of the SAH center strengthens significantly, reaching a maximum value of more than 8×10^{-6} s^{-1} over the southern BOB. After the ASM onset (Figs. 1g and 1h), the SAH strengthens and its center moves farther northward, and the trumpet-shaped strong upper-layer divergence on its southwest persists and moves slowly eastward.

This evolution of the SAH in the upper troposphere strongly influences the MOV generation near the surface and the ASM onset. Figure 10 shows the 3-D structure of the atmospheric circulation during the MOV formation process. On D−3 (Fig. 10a), a belt of weak positive surface relative vorticity lies over the southern BOB, with a band of surface wind cyclonic shear just to the south of the surface anticyclone over the northern BOB. Meanwhile, the SAH ridgeline is located near 10°N, with persistent up-

Fig. 8 Daily evolution of the 10°—20°N averaged pressure-longitude cross-section of apparent heating source Q_1 (K d^{-1}, values of >4 K d^{-1} are shaded) from D−3 to D+2 for (a) to (f), respectively

per tropospheric divergence of more than 4×10^{-6} s^{-1} to its south. While the SAH is moving northward from D−3 to D0 (Figs. 1e and 1f), the upper-layer divergence becomes more organized and stronger (> 8×10^{-6} s^{-1}) over the southern BOB (Fig. 10b). The strong ascending motion at 600 hPa is better correlated with the maximum of upper layer divergence than with the large value of the lower cyclonic vorticity over the southern BOB. This implies a close coupling between the upper-layer divergence and mid-layer air ascent; that is, a pronounced pumping effect above the southern part of the BOB during the gestation of the MOV. On D−1 (Fig. 10c), the very strong upper-layer divergence, the mid-troposphere ascent and the strong surface vorticity are all well coupled; they become perpendicular near 10°N where the MOV appears (Fig. 6c). As mentioned above, the SST in the southeastern BOB is relatively colder compared to the central BOB (Fig. 9); therefore, this strong pumping effect can be considered as a trigger for the formation of the MOV. After the BOB summer monsoon onset, the SAH ridgeline remains near 15°N over the northern BOB (Figs. 10e and 10f). The MOV continues to move northeastward into the warm pool area over the northeastern BOB where surface winds intensify and sea surface sensible heating strengthens rapidly (Figs. 6d—f). The positive feedback due to air-sea interaction further promotes development of the MOV after the BOB summer monsoon onset. Moreover, the positive meridional gradient of SST over the BOB could facilitate northward advance of the MOV. Deep convection then commences over the northeastern BOB and Indochinese peninsula as the MOV merges into the Indo-Burma

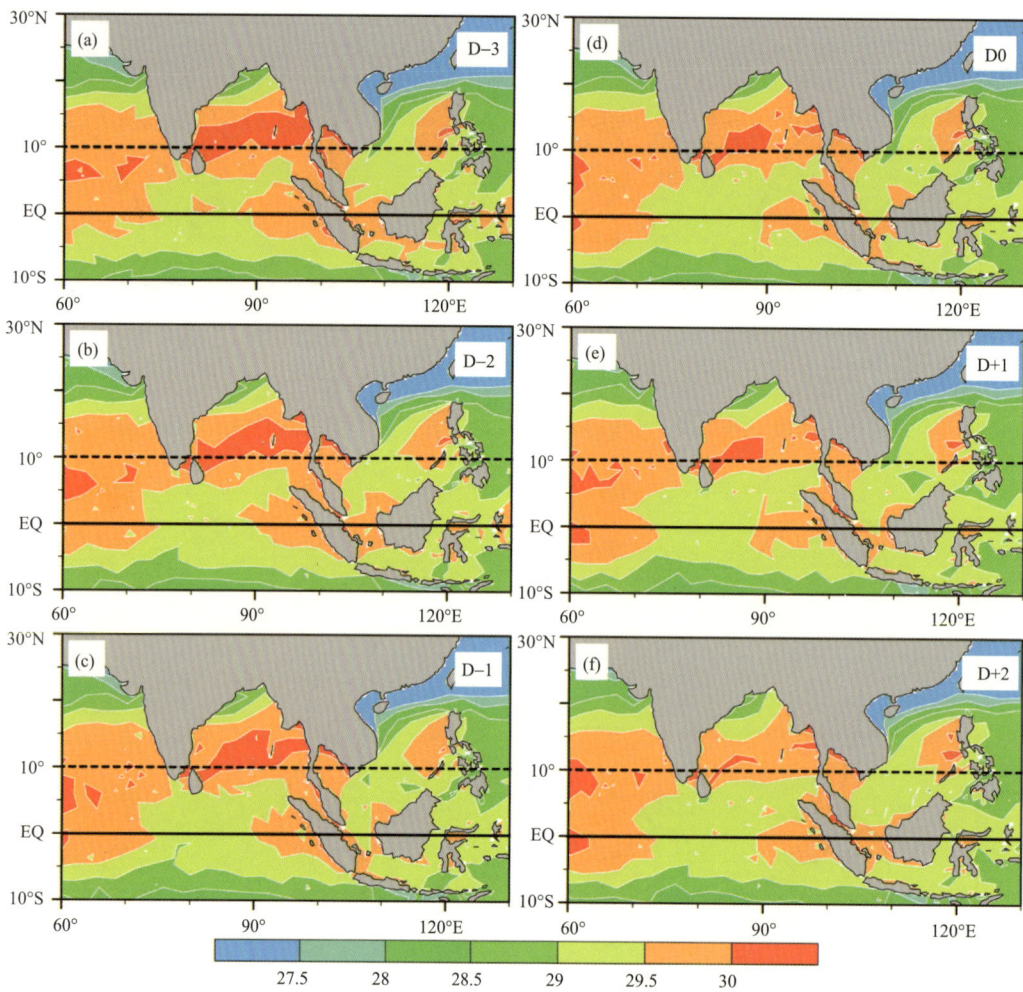

Fig. 9 Daily evolution of sea surface temperature (unit: °C) from D−3 to D+2 for (a) to (f), respectively

trough, accompanied by the splitting of the subtropical anticyclone ridgeline (Figs. 7e and 7f). The strong condensation heating released by the deep monsoon convection sustains the SAH over the peninsula (Figs. 1g and 1h). However, although the upper-layer divergence is still maintained over the southern BOB, it is decoupled from the atmospheric ascent in the middle layer and the vortex near the surface (Figs. 10e and 10f). Consequently, the surface vortex ceases to develop and finally disappears on D+2 (Fig. 6f).

5 Summary and discussion

Composite analysis based on the ERA-40 dataset has been used to study the climate-mean characteristics of the generation of the SAH, the Asian summer monsoon onset process, and the impacts of the SAH on the BOB MOV formation and monsoon onset. In the East Asian and western Pacific areas as season evolves from winter to spring, the center of the lower tropospheric anticyclone shifts gradually from eastern China to the western Pacific. By April and early May, the moist tropical southeasterly that originates from the western Pacific "warm pool" region merges into the subtropical easterly on the south of the anticyclone, then hits the warm land of the south Philippines and triggers local convective instability. Convection and the related diabatic heating strengthen noticeably over the southern Philippines

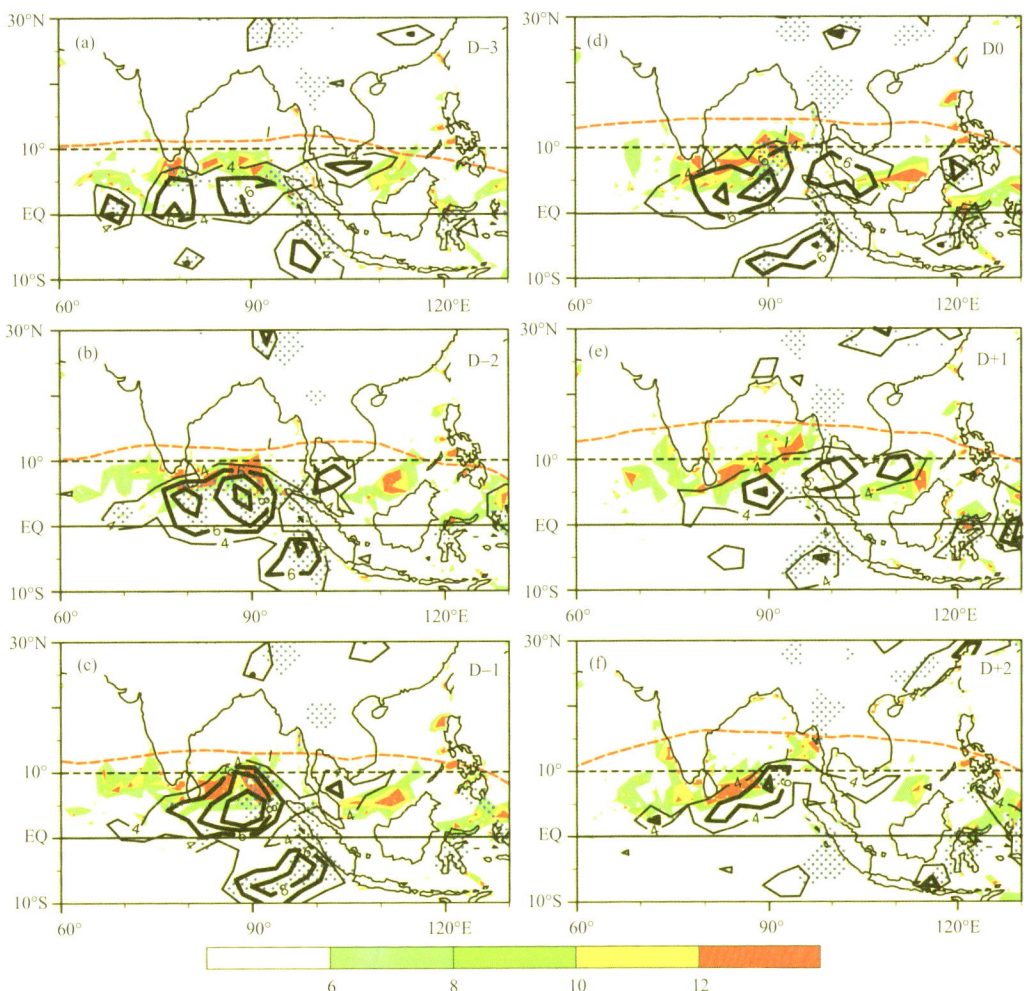

Fig. 10 Daily evolution of 150 hPa divergence (black solid contours, interval is 2×10^{-6} s^{-1}, thicker line indicates values of >6) and ridgeline of the SAH (red dashed line), 600 hPa ascending motion (blue stippled area, only values smaller than -0.08 Pa s^{-1} are plotted), and sea surface positive relative vorticity (shading, 10^{-6} s^{-1}) from D$-$3 to D$+$2 for (a) to (f), respectively

about a fortnight before the ASM onset and persist in the region until the ASM onset. A persistent negative vorticity source (S<0) in the mid-upper troposphere is induced poleward of the convection center as illustrated schematically in Figure 11. The SAH is thus generated to the northwest of the persistent negative vorticity source S and the resultant streamline field manifests itself as a Gill-type response (Gill, 1980). Meanwhile, an abnormal cyclone develops in the upper troposphere over the equatorial western Pacific, causing the original anticyclone to weaken and eventually disappear. Thus, both the development of an anticyclone over the western SCS and southeastern Indochinese peninsula and the dampening of the original anticyclone over the equatorial west Pacific contribute to the SAH formation.

In the lower troposphere, strong sensible heating over the Indian peninsula produces a surface anticyclone and downward flow over the northern BOB prior to the BOB monsoon onset. Strong divergent wind from the surface anticyclone brings abundant moisture into the Indochinese peninsula. The surface southerlies then transport the vapor to the north where heavy rainfall develops. The latent heat release associated with precipitation then contributes to the westward movement and intensification of the SAH after it is generated.

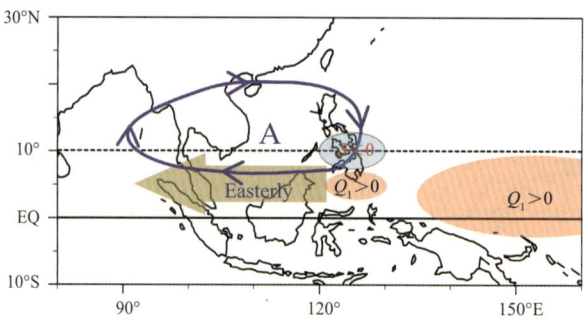

Fig. 11 Schematic diagram presenting the mechanism responsible for the SAH (marked by "A") formation: persistent convective diabatic heating in spring over the south Philippines produces a sustained negative vorticity source over the philippines, and the South Asian high is generated as a Gill-type atmospheric response to the negative forcing source. The marked area over the equatorial west Pacific denotes the diabatic heating which is isolated from the one over the south Philippines

The development of the surface anticyclone over the northern BOB in spring provides favorable conditions for the formation of a short-lived warm pool over the central BOB, and contributes to the genesis of the MOV over the BOB just before the monsoon onset. In the upper troposphere, owing to the development of the SAH, the upper-layer divergence on the southwestern side of the SAH increases over the southern BOB and is then coupled with stronger ascending motion in the middle troposphere. Such a pumping effect triggers the genesis of the MOV near the surface and intensifies the surface southwesterly over the northeastern BOB. Subsequently, the development of onshore ocean currents and downwelling near the coast maintains the local warm SST and the surface sensible heating is enhanced over the eastern BOB. As a consequence, the MOV develops further during and even after the monsoon onset. At 700 hPa, in tandem with the northeastward development of the surface MOV, the tropical cyclone moves northward and merges into the Indo-Burma trough and the original zonally oriented subtropical high over the northern BOB splits. Thus, the southwesterly in front of the trough brings abundant moisture to the Indochinese peninsula where deep convection builds up. Torrential rainfall develops over the eastern BOB and western Indochinese peninsula, and the ASM onset thus commences over the eastern BOB. In short, the ASM onset occurs firstly over the eastern BOB because of a close coupling between the upper-tropospheric pumping because of the development of the SAH and the near surface air-sea interaction on the BOB. The pumping effect due to the strengthening of the SAH triggers the MOV genesis, and the local air-sea interaction plays a key role in the MOV development, which contributes to the BOB summer monsoon onset.

Our results, which are based on a climate-mean analysis, demonstrate the impact of coupling between air-sea interaction and SAH evolution on the BOB MOV formation and development. However, the MOV formation process can be very different from year to year because of the pronounced interannual variability of the BOB summer monsoon onset. In the ocean, the warm pool formation is affected by both the solar radiation and the mixed-layer depth; the earlier (later) the summer monsoon onset, the shorter (longer) the warming time of SST by solar radiation before the monsoon onset and so the colder (warmer) the SST. It is also true that the shallower the mixed-layer depth, the easier it is for a warm pool to form. Thus, the influence of air-sea interaction is different from year to year. In the atmosphere, the major factors that affect the SAH generation and evolution, including the condensation heating over the southern Philippines and the moisture transport and the sensible heating over the Indochinese peninsula, also possess distinct individual variability. In addition, as shown in Table 1 there is close relationship between the onset time of BOB summer monsoon and ENSO events; that is, early onset after the La Nina but late onset after the El Nino events. The work of Wu and Zhang (1998) has also demonstrated that the ASM onset is strongly modulated by the intraseasonal oscillations from the south, east and northwest directions. How the ENSO events and ISOs affect the interannual variation of the Asian monsoon onset is still unclear. Therefore, further studies on the interannual variability of the coupling between the SAH evolution and air-sea interaction are required for a comprehensive understanding of the dynamical processes of the Asian summer monsoon onset.

References

ANANTHAKRISHNAN R, SRINIVASANV, RAMAKRISHNAN A R, 1968. Synoptic features associated with onset of southwest monsoon over Kerala, Forcasting Manual[R]. FMU Technical Rep. IV-18.2. Pune, India: India Meteorological Department.

CHAN S C, NIGAM S, 2009. Residual Diagnosis of Diabatic Heating from ERA-40 and NCEP Reanalyses: Intercomparisons with TRMM[J]. Journal of Climate, 22(2):414-428.

CHEN L X, ZHU Q G, LUO H B, et al, 1991. East Monsoon[M]. Beijing: China Meteorological Press.

DEE D P, UPPALA S M, SIMMONS A J, et al, 2011. The ERA-Interim reanalysis: Configuration and performance of the data assimilation system[J]. Quarterly Journal of the Royal Meteorological Society, 137(656):553-597.

DUAN A M, WU G X, 2005. Role of the Tibetan Plateau thermal forcing in the summer climate patterns over subtropical Asia[J]. Climate Dynamics, 24(7-8):793-807.

DUAN A, WU G X, LIANG X Y, 2008. Influence of the Tibetan Plateau on the summer climate patterns over Asia in the IAP/LASG SAMIL model[J]. Advances in Atmospheric Sciences, 25(4):518-528.

ERTEL H, 1942. Ein neuer hydrodynamische wirbdsatz[J]. Meteorologische Zeitschrift Braunschweig, 59: 33-49.

FLOHN H, 1960. Recent investigation on the mechanism of the "summer monsoon" of Southern and Eastern Asia[C]. Monsoon of the World. New Delhi: Hind Union Press: 75-88.

GILL A E, 1980. Some simple solutions for heat-induced tropical circulation[J]. Quarterly Journal of the Royal Meteorological Society, 106(449):447-462.

GRAY W M, 1968. Global view of origin of tropical disturbances and storms[J]. Monthly Weather Review, 96(10): 669-700.

GRAY W M, 1975. Tropical Cyclone Genesis[R]. Atmospheric Science Paper NO. 234. Fort Collins, Colorado: Department of Atmospheric Science, Colorado State University.

GRAY W M, 1998. The formation of tropical cyclones[J]. Meteorology and Atmospheric Physics, 67(1-4):37-69.

HARR P A, CHAN J C L, 2005. Monsoon Impacts on Tropical Cyclone Variability[C]//CHANG C P, WANG B, LAN N C G. The Global Monsoon System: Research and Forecast. Hangzhou: Secretariat of the World Meteorological Organization: 512-542.

HE J H, MIN W, SHI X, et al, 2002. Splitting and eastward withdrawal of the subtropical high belt during the onset of the South China Sea summer monsoon and their possible mechanism[J]. J Nanjing Univ (Nat Sci), 38(3):318-330.

HE J H, WEN M, WANG L J, et al, 2006. Characteristics of the onset of the Asian summer monsoon and the importance of Asian-Australian "land bridge"[J]. Advances in Atmospheric Sciences, 23(6):951-963.

HOSKINS B J, MCINTYRE M E, ROBERTSON A W, 1985. On the use and significance of isentropic potential vorticity maps[J]. Quarterly Journal of the Royal Meteorological Society, 111(470):877-946.

JOSEPH P V, 1990. Warm pool over the Indian Ocean and monsoon onset[J]. Tropical Ocean-Atmosphere Newsletter, 53: 1-5.

KANAMITSU M, EBISUZAKI W, WOOLLEN J, et al, 2002. NCEP-DOE AMIP-II reanalysis (R-2)[J]. Bulletin of the American Meteorological Society, 83(11):1631-1643.

KRISHNAMURTI T N, 1971a. Observational study of the tropical upper tropospheric motion field during the Northern

Hemisphere summer[J]. Journal of Applied Meteorology and Climatology, 10(6):1066-1096.

KRISHNAMURTI T N, 1971b. Tropical east-west circulations during northern summer[J]. Journal of the Atmospheric Sciences, 28(8):1342-1347.

KRISHNAMURTI T N, 1981. Cooling of the Arabian Sea and the onset-vortex during 1979[C]. Recent progress in the equatorial oceanography: A report of the final meeting of SCOR WORKING GROUP 47. Venice, Italy:1-12. Available from Nova University, Ocean Science Center, Dania, FL 33004.

KRISHNAMURTI T N, 1985. Summer monsoon experiment—a review[J]. Monthly Weather Review, 113(9):1590-1626.

KRISHNAMURTI T N, RODGERS E, 1970. 200 hPa wind field June, July, August 1967[C]. Rept No 70-2. Dept of Meteor, Florida State University, Tallahassee: 115.

KRISHNAMURTI T N, ARDANUY P, RAMANATHAN Y, et al, 1981. On the onset vortex of the summer monsoon [J]. Monthly Weather Review, 109(2):344-363.

KRISHNAMURTI T N, RAMANATHAN Y, 1982. Sensitivity of the monsoon onset to differential heating[J]. Journal of the Atmospheric Sciences, 39(6):1290-1306.

LAU K M, WU H T, YANG S, 1998. Hydrologic processes associated with the first transition of the Asian summer monsoon: A pilot satellite study[J]. Bulletin of the American Meteorological Society, 79(9):1871-1882.

LIU B Q, HE J H, WANG L J, 2009. Characteristics of the South Asia high establishment processes above the Indo-China Peninsula from April to May and their possible mechanism[J]. Chinese Journal of Atmospheric Sciences, 33(6):1319-1332.

LIU B Q, HE J H, WANG L J, 2012. On a possible mechanism for Southern Asian convection influencing the South Asian high establishment during winter to summer transition[J]. Journal of Tropical Meteorology, 18(4):473-484.

LIU Y M, WU G X, LIU H, et al, 1999. The effect of spatially nonuniform heating on the formation and variation of subtropical high part III: Condensation heating and South Asian high and western Pacific subtropical high[J]. Acta Meteorologica Sinica, 57(5):525-538.

LIU Y M, WU G X, LIU H, et al, 2001. Condensation heating of the Asian summer monsoon and the subtropical anticyclone in the eastern hemisphere[J]. Climate Dynamics, 17(4):327-338.

LIU Y M, CHAN J C L, MAO J Y, et al, 2002. The role of Bay of Bengal convection in the onset of the 1998 South China Sea summer monsoon[J]. Monthly Weather Review, 130(11):2731-2744.

LIU Y M, WU G X, REN R C, 2004. Relationship between the subtropical anticyclone and diabatic heating[J]. Journal of Climate, 17(4):682-698.

MAK M, KAO C Y J, 1982. An instability study of the onset-vortex of the southwest monsoon, 1979[J]. Tellus, 34(4):358-368.

MAO J Y, WU G X, LIU Y M, 2002a. Study on modal variation of subtropical high and its mechanism during seasonal transition part I: Climatological features of subtropical high structure[J]. Acta Meteorologica Sinica, 60(4):400-408.

MAO J Y, WU G X, LIU Y M, 2002b. Study on modal variation of subtropical high and its mechanism during seasonal transition part II: Seasonal transition index over Asian monsoon region[J]. Acta Meteorologica Sinica, 60(4):409-420.

MAO J Y, CHAN J C L, 2004. Relationship between the onset of the South China Sea summer monsoon and the structure of the Asian subtropical anticyclone[J]. Journal of the Meteorological Society of Japan, 82(3):845-859.

MAO J Y, WU G X, 2007. Interannual variability in the onset of the summer monsoon over the eastern Bay of Bengal[J]. Theoretical and Applied Climatology, 89(3-4):155-170.

MAO J Y, WU G X, 2011. Barotropic process contributing to the formation and growth of tropical cyclone Nargis[J]. Advances in Atmospheric Sciences, 28(3):483-491.

MASON R B, ANDERSON C E, 1963. The development and decay of the 100 hPa. summertime anticyclone over Southern Asia[J]. Monthly Weather Review, 91(1): 3-12.

PISHAROTY P R, ASNANI G A, 1957. Rainfall around monsoon depressions over India[J]. Mausam, 8(1): 15-20.

QIAN Y F, ZHAO Q, YAO Y H, et al, 2002. Seasonal variation and heat preference of the South Asia high[J]. Advances in Atmospheric Sciences, 19(5):821-836.

RAO K V, RAJAMANI S, 1970. Diagnostic study of a monsoon depression by geostrophic baroclinic model[J]. Indian

Journal of Meteorology and Geophysics, 21: 187-194.

REITER E R, GAO D Y. 1982. Heating of the Tibet Plateau and movements of the South Asian high during Spring[J]. Monthly Weather Review, 110(11):1694-1711.

SHARMA M C, SRINIVASAN V. 1971. Centres of monsoon depressions as seen in satellite pictures[J]. Indian Journal of Meteorology and Geophysics, 22(3): 357-360.

SHENOI S S C, SHANKAR D, SHETYE S R. 1999. On the sea surface temperature high in the Lakshadweep Sea before the onset of the southwest monsoon[J]. Journal of Geophysical Research-Oceans, 104(C7):15703-15712.

SHENOI S C, SHANKAR D, GOPALKRISHNA V V, et al. 2005. Role of ocean in the genesis and annihilation of the core of the warm pool in the southeastern Arabian Sea[J]. Mausam, 56(1): 147-160.

SMITH T M, REYNOLDS R W. 2004. Improved extended reconstruction of SST (1854−1997)[J]. Journal of Climate, 17(12):2466-2477.

UPPALA S M, KALLBERG P, SIMMONS A J, et al. 2005. The ERA-40 re-analysis[J]. Quarterly Journal of the Royal Meteorological Society, 131(612):2961-3012.

VINAYACHANDRAN P N, SHANKAR D, KURIAN J, et al. 2007. Arabian Sea mini warm pool and the monsoon onset vortex[J]. Current Science, 93(2):203-214.

VOISIN N, WOOD A W, LETTENMAIER D. 2008. Evaluation of precipitation products for global hydrological prediction [J]. Journal of Hydrometeorology, 9(3):388-407.

WANG B, LINHO. 2002. Rainy season of the Asian-Pacific summer monsoon[J]. Journal of Climate, 15(4):386-398.

WU G X, LIU H Z. 1998. Vertical vorticity development owing to down sliding at slantwise isentropic surface[J]. Dynamics of Atmospheres and Oceans, 27(1-4):715-743.

WU G X, ZHANG Y S. 1998. Tibetan Plateau forcing and the timing of the monsoon onset over South Asia and the South China Sea[J]. Monthly Weather Review, 126(4):913-927.

WU G X, LIU Y M, LIU P. 1999. The effect of spatially nonuniform heating on the formation and variation of subtropical high Part Ⅰ: Scale analysis[J]. Acta Meteorologica Sinica, 57(3): 257-263.

WU G X, CHOU J F, LIU Y M. 2002. Dynamics of the formation and variation in Subtropical Anticyclones[M]. Beijing: Science Press: 314.

WU G X, LIU Y M. 2003. Summertime quadruplet heating pattern in the subtropics and the associated atmospheric circulation[J]. Geophysical Research Letters, 30(5).

WU G X, LIU Y M, ZHANG Q, et al. 2007. The influence of mechanical and thermal forcing by the Tibetan Plateau on Asian climate[J]. Journal of Hydrometeorology, 8(4):770-789.

WU G X, GUAN Y, WANG T M, et al. 2011. Vortex genesis over the Bay of Bengal in spring and its role in the onset of the Asian summer monsoon[J]. Science China-Earth Sciences, 54(1):1-9.

WU G X, GUAN Y, LIU Y M, et al. 2012. Air-sea interaction and formation of the Asian summer monsoon onset vortex over the Bay of Bengal[J]. Climate Dynamics, 38(1-2):261-279.

XIE S P, PHILANDER S G H. 1994. A coupled ocean-atomsphere model of relevance to the ITCZ in the eastern Pacific [J]. Tellus Series A-Dynamic Meteorology and Oceanography, 46(4):340-350.

YAN J H. 2005. Asian summer monsoon onset and advancing process and the variation of the subtropical high[D]. Beijing: Graduate University of Chinese Academy of Sciences.

YANAI M, ESBENSEN S, CHU J H. 1973. Determination of bulk properties of tropical cloud clusters from large-scale heat and moisture budgets[J]. Journal of the Atmospheric Sciences, 30(4):611-627.

ZHANG H M, BATES J J, REYNOLDS R W, et al. 2006. Assessment of composite global sampling: Sea surface wind speed[J]. Geophysical Research Letters, 33(17).

ZHANG Q, WU G X, QIAN Y F. 2002. The bimodality of the 100 hPa South Asia high and its relationship to the climate anomaly over East Asia in summer[J]. Journal of the Meteorological Society of Japan, 80(4):733-744.

ZHU Q G, HE J H, WANG P X. 1986. A study of circulation differences between East-Asian and Indian summer monsoons with their interaction[J]. Advances in Atmospheric Sciences, 3(4): 466-477.

Location and Variation of the Summertime Upper-Troposphere Temperature Maximum over South Asia

WU Guoxiong[1], HE Bian[1,2], LIU Yimin[1], BAO Qing[1], REN Rongcai[1]

(1 State Key Lab of Atmospheric Sciences and Geophysical Fluid Dynamics (LASG), Institute of Atmospheric Physics, Chinese Academy of Sciences, Beijing 100029, China; 2 State Key Laboratory of Loess and Quaternary Geology, Institute of Earth Environment, Chinese Academy of Sciences, Xian 710061, China)

Abstract: The upper-troposphere-temperature-maximum (UTTM) over South Asia is a pronounced feature in the Northern Hemisphere summer. Its formation mechanism is still unclear. This study shows that the latitude location of the upper-tropospheric warm-center (T) coincides with the subtropical anticyclone, and its longitude location is determined by the zonal distribution of vertical gradient of heating/cooling ($Q_z = \partial Q / \partial z$), which is different from the Gill's model. Since both convective heating and radiation cooling decrease with height in the upper troposphere, the heating/cooling generates vertical northerly/southerly shear, leading to a warm/cold center being developed between heating in the east/west and cooling in the west/east. The location of the UTTM coincides with the South Asian high (SAH) and is between a radiation cooling in the west and the Asian-monsoon convection heating source in the east. The UTTM is sensitive to this convective heating: increased heating in the source region in a general circulation model causes intensification of both the SAH and UTTM, and imposing periodic convective heating there results in oscillations in the SAH, UTTM, and vertical motion to the west with the same period. Diagnoses of reanalysis indicate that such an inherent subtropical T-Q_z relation is significant at interannual timescale. During the end of the twentieth century, rainfall increase over South China is accompanied by an increasing northerly flow aloft and intensification in the SAH and UTTM to the west. Results demonstrate that the feedback of atmospheric circulation to rainfall anomalies is an important contributor to the regional climate anomaly pattern.

Keywords: upper-troposphere temperature maximum (UTTM); feedback between rainfall and circulation; temperature-vertical heating gradient (T-Q_z) relation

1 Introduction

In the boreal summer half of the year there exists in the upper troposphere of the Eurasian subtropics a huge high pressure system centered over South Asia southwest of the Tibetan Plateau (TP), which was formerly known as the Tibetan High (Ye and Gao, 1979) but is now referred to more frequently as the South Asian high (SAH). The center of the climate mean SAH first appears in April over the South China Sea (SCS) before the onset of the Asian summer monsoon (ASM) (Liu B Q et al., 2012, 2013); it then gets strengthened over the Indochina Peninsula during the AMS onset, which continues for more than one month from the end of April/early May over the Bay of Bengal (BOB) region until the end of May/early June over North India (Wu and Zhang, 1998; Wang and LinHo, 2002); and finally it settles over the climate mean region until autumn. The SAH is in fact the strongest and steadiest circulation system besides the polar vortex in the upper troposphere of the Northern Hemisphere (Mason and Anderson, 1963). The geographic location of the SAH's center varies at different timescales ranging from a few days (Tao and Zhu, 1964) to quasi-biweekly (Krishnamurti et al., 1973; Luo and Yanai, 1983, 1984; Liu et al., 2007; Peng et al., 2014) and even longer (Zhang et al., 2002). The activities of the SAH are closely linked to local weather variation and have been used as an index for short- to medium-range weather forecasts in Asia (Ye and Gao, 1979).

How the gigantic SAH is formed and maintained has long been a subject of interest to meteorologists. Strong summer heating over the TP was once considered a major contributor to its formation (Flohn, 1957; Yeh et al., 1957; Yanai et al., 1992), and the strong latent heat release associated with the ASM was later shown to have important impacts on its maintenance (Chen et al., 1985; Liu Y M et al., 2001, 2013). Wu and Liu (2003) and Wu et al. (2009) found that the summertime circulation configuration in the subtropics is a consequence of the atmospheric response to an organized continental-scale quadruple diabatic heating (LOngwave radiation cooling, surface SEnsible heating, COndensation heating, and Double dominant heating; called LOSECOD), together with local-scale sea breeze forcing and regional-scale mountain forcing. In this regard, the huge and strong SAH is due mainly to the large longitudinal span of the Eurasian continent and the large size and high elevation of the Iranian Plateau and the TP along the subtropics.

Another spectacular phenomenon in the upper troposphere of the subtropics in summer is a remarkable warm temperature maximum. Using a 14-year dataset, Li and Yanai (1996) analyzed the horizontal distribution of the mean temperature in the 200—500 hPa layer in summer and found that "a huge warm air mass is centered on South Asia with the maximum temperature ($\geqslant -22$ ℃) over the southern TP". Though there are numerous studies on the formation and variation of the SAH, there are a lot fewer studies on the formation of this upper-troposphere-temperature-maximum (UTTM). Tamura et al. (2010) found an abrupt temperature increase in the upper troposphere southwest of the TP during the early ASM monsoon onset from late April to mid-June, and suggested that this was due to local adiabatic warming that was closely associated with the anomalous Hadley circulation. Several proposals have attributed the formation of the UTTM to various external thermal forcing. Early investigations emphasized the importance of the thermal forcing of the TP (e.g., Yeh et al., 1957; Flohn, 1957; Hahn and Manabe, 1975; Ye and Gao, 1979). Based on ECMWF TOGA analyses on a 2.5°×2.5°grid from 1985 to 1992, Yanai and Li (1996) demonstrated that in the boreal summer months the maximum vertically integrated heat source $<Q_1>$, moisture sink $<Q_2>$, and minimum OLR flux are all located in a region stretching from the southern TP to the northern BOB, indicating the importance of monsoon condensation heating in the development of the warm center.

Another proposal concerning the location of the UTTM is based on local vertical coupling: the precipitating convection over North India thermodynamically couples the upper troposphere to the air of the highest entropy subcloud layer so that the "thermal forcing of continental India is important in setting the location and amplitude of the UTTM" (Boos and Kuang, 2010, 2013). However, the statistical-equilibrium theory (Emanuel, 1991; Emanuel et al., 1994) links the change in the subcloud-layer entropy θ_e to the change in the thickness (virtual temperature) of the convective layer bounded between the free and lifted convection levels, rather than the entropy and temperature themselves. The thickness was defined from the convective cloud base to the cloud top extending through most of the troposphere, which is between 150 and 800 hPa over North India in the observation (Fig. 2 in Boos and Kuang, 2010), and is different from the UTTM layer, which is in the upper troposphere between 400 and 200 hPa. Furthermore, convective diabatic heating in the tropics is substantially balanced by adiabatic cooling as well as longwave radiation cooling (Sardeshmukh and Hoskins, 1985; Rodwell and Hoskins, 2001). Therefore, can the convective heating over North India locally couple the UTTM to the high subcloud entropy? If not, what is the mechanism responsible for its formation and variation? These are challenging issues in monsoon dynamics. In this study, we will investigate the general configuration of the SAH and UTTM and analyze their relation with external diabatic heating and the associated mechanism based on data diagnosis, general circulation numerical experiments and dynamic analysis. The rest of the text is organized as follows: Section 2 presents the latitude locations of large-scale tropical monsoon systems, including the SAH, UTTM, surface entropy, and monsoon convection. The intrinsic coherent structure among these systems is analyzed based on the geostrophic and thermal wind balances and axisymmetric monsoon dynamics. The longitude location of the UTTM in relation with the vertical gradient of diabatic heating, the so-called T-Q_Z ($Q_z = \partial Q/\partial z$) relation, and the associated mechanism are discussed in Section 3. Such a newly established T-Q_Z relation is compared to the Gill's (1980) model in Section 4. In Section 5, time dependent numerical sensitivity experiments are used to verify the rendered mechanism. The implication of the results for the observed regional climate change is explored in Section 6. Discussion and conclusion are presented in Section 7.

2 Latitude location of the UTTM

The observational monthly datasets used in this study are derived from ERA40 datasets (Uppala et al., 2005). The summer climate equilibrium state is indicated by climate variables averaged for the June—July—August (JJA) period from 1979 to 2002. For precipitation, the PREC/L data (Chen et al., 2002) are applied.

In summer in the upper troposphere over the Asian monsoon area, a zonal belt of high geopotential is located between 15° and 35°N, and its magnitude increases with height (Fig. 1a). At 400 hPa, regions higher than 760 dagpm (green) appear over the western Pacific and above Saudi Arabia, while a weak subtropical ridge just emerges in the region to the south of the TP and north of the BOB. At 200 hPa, the unique SAH becomes prominent, with the 1249 dagpm contour covering the entire Eurasian subtropics. The area encircled by 1254 dagpm is seated from the Arabian Peninsula to the southeastern TP, with its ridgeline located along 26°—27°N and its center shifted westward compared to the weak ridge at 400 hPa. A warm region of the 200—400 hPa mass-weighted mean temperature is located along the latitudes between 20° and 35°N, forming the UTTM, with its elliptical center warmer than 246 K, ranging from 60° to 100°E above Afghanistan, Pakistan, North India, and the southwestern TP. Here we take the average temperature from 200 to 400 hPa, not 500 hPa, to obtain the upper troposphere mean temperature, because 500 hPa is close to the surface of the TP and the temperature distribution at this level is strongly affected by the ground surface heating of the TP (Fig. 5 of Yanai et al., 1992). While the ridgeline of the SAH tilts northward slightly with increasing height, the elliptic long axis of the UTTM coincides well with the upper-layer SAH axis, in agreement with the fact that the axis of subtropical anticyclones always tilts toward warmer regions with increasing height (Wu et al., 2002).

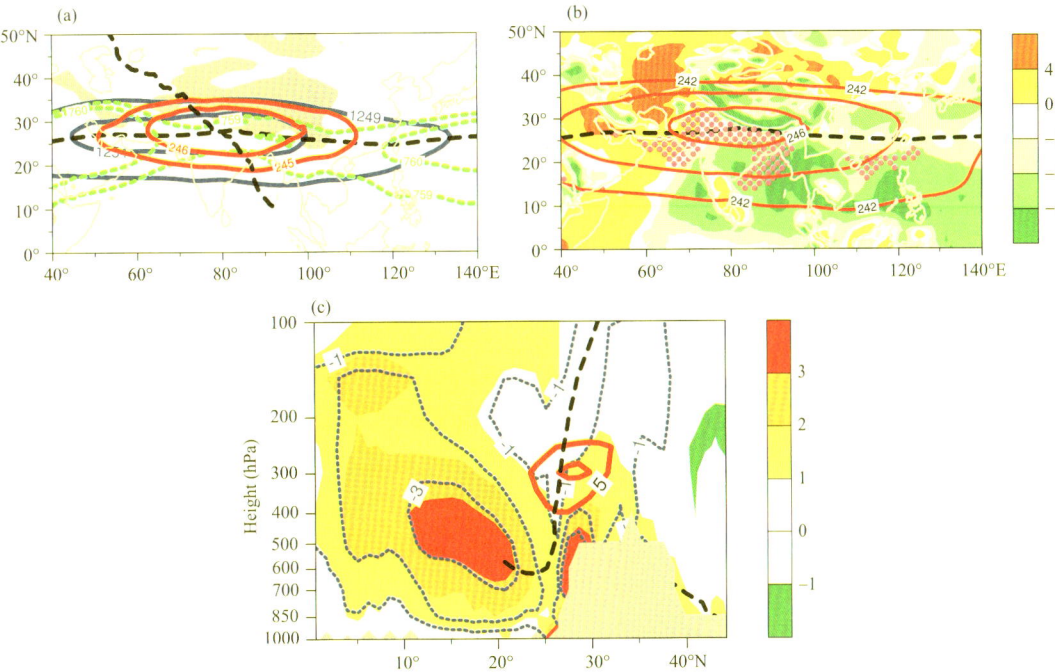

Fig. 1　The JJA mean distributions of a geopotential height (unit: dagpm) at 200 hPa (blue solid) and 400 hPa (green dashed), 200—400 hPa mass-weighted mean temperature (red solid, unit: K), and zero zonal and meridional wind contours at 200 hPa (black dashed); (b) 500 hPa vertical velocity (shading, unit: hPa s^{-1}), 200—400 hPa mass-weighted mean temperature (red contour, unit: K), surface entropy≥356 K (purple stippled; unit: K), and contour of $u=0$ at 300 hPa (black dashed); and (c) 60°—100°E mean diabatic heating Q_1/c_p (shading, unit: K d^{-1}) and adiabatic heating (blue dotted contour, unit: K d^{-1}), ridgeline (black dashed line), and temperature deviation from the (40°—160°E, 0°—50°N) area mean (red contour, interval: 5 ℃)

Can the formation of this UTTM be attributed directly to the vertical coupling of the local convection with high surface entropy over North India? For this, the spatial pattern of UTTM should be in good coordination with high surface entropy and strong vertical motion. However, this is not the case in the subtropics, as demonstrated in Fig. 1b: the climate-mean maximum air ascent (green shading) is observed over the southern slope of the TP and South China in the subtropics, and over the eastern Arabian Sea (AS), the and eastern BOB and western Indochina Peninsula region, and SCS in the tropics; while the high surface entropy (purple stippling) appears mainly over the northern AS, northwestern BOB, and northern SCS in the tropical oceans, and over northern and northeastern India. It is interesting to see that over the ocean region of the AS, BOB, and SCS, high surface entropy is located just to the north of strong ascent (rainfall), in good agreement with the axisymmetric theory (Privé and Plumb, 2007a). However, over the northern Indian subcontinent, high surface entropy is accompanied by descent (orange shading) over the west, but ascent over the east, with the high surface entropy appearing between the two strong ascents located respectively to its north and south, presenting a surface circulation-driven east-west asymmetry of the monsoon by suppressing moist convection to the west, while encouraging rainfall in the east (Xie and Saiki, 1999; Wu et al., 2009). Within the 246 K UTTM contour, while the northeastern part is characterized by strong ascent and lower entropy, its southwestern portion features descent and high entropy. In the existence of such a highly asymmetric geometry over North India and the southwestern Tibetan Plateau, the limitation of the axisymmetric theory is evident, as was discussed by Prive and Plumb (2007b).

To illustrate the general features more clearly, a latitude-pressure cross section of the diabatic heating and adiabatic cooling, averaged over the longitudinal span of the UTTM (60°—100°E), is calculated and presented in Fig. 1c. The main diabatic heating Q (shading, calculated from (10)) is over the tropics around 10°—20°N and over the southern slope of the TP, corresponding respectively to the southern and northern branches of the South Asian monsoon (SAM; Wu et al., 2012a). These two heating maxima are basically compensated for by local adiabatic cooling (dotted curve, the ω term in (10)) associated with local air ascent, as was demonstrated by Rodwell and Hoskins, 2001). On the other hand, the UTTM is seated in a region just between the two maximum heating centers, and the vertical motion is weak in this region.

Shown in Fig. 2 is the allocation of the 200—400 hPa mass weighted air temperature (red), 200 hPa zonal velocity (black) and surface θ_{se} (blue) in different sectors in South Asia. It reveals the following common features in all sectors: the maximum surface entropy is associated with easterly shear in the upper troposphere, the maximum surface entropy is located to the south of the UTTM, and the UTTM is seated by the ridge line of subtropical anticyclone where u vanishes. The overlapping of the UTTM and the ridgeline is subject to the following geostrophic balance, hydrostatic balance, and thermal wind relation:

$$fu = -\partial\phi/\partial y \tag{1}$$

$$\partial\phi/\partial\ln p = -RT \tag{2}$$

$$\frac{\partial u}{\partial \ln p} = \frac{R}{f}\left(\frac{\partial T}{\partial y}\right) \tag{3}$$

where f is the Coriolis parameter, u is zonal wind, ϕ is geopotential height, y is the meridional coordinate, and p, T, and R denote pressure, temperature, and the gas constant for dry air, respectively. Then higher geopotential along the SAH in the subtropics implies the occurrence of a westerly to its north and an easterly to its south, with vanishing zonal wind along its axis (1). The higher geopotential along the ridgeline also implies a thicker layer beneath corresponding to a warmer layer below (2). On the northern side of the SAH the increasing westerly flow with height corresponds to a warmer region situated to its south, whereas on the southern side of the SAH, the increasing easterly flow with height corresponds to a warmer region situated to its north (3). In other words, the UTTM must be located along with the SAH in the upper troposphere. As demonstrated in Fig. 1a, the area of the 200—400 hPa mass-weighted temperature warmer than 245 K is elliptical and spans from 50°E to 110°E, with its long axis coincident with the ridgeline of the subtropical anticyclone at 200 hPa.

Why should the SAH and the UTTM be located in subtropics? How the high surface entropy, the SAH and the UTTM are allocated in such a manner as shown in Fig. 2? A series of study (Schneider and Lindzen, 1977; Schneider, 1977, 1987; Held and Hou, 1980) on the dynamics of the Hadley circulation demonstrated that in response to an axisymmetric diabatic heating, the atmospheric circulation adopts two distinct regimes: the thermal equilibrium (TE) regime in extra-tropics and the angular momentum conservation (AMC) regime in the tropics. In the TE regime the planetary vorticity is large and the Rossby radius of deformation is small, a weak forcing cannot generate meridional circulation. Whereas in the AMC regime the planetary vorticity is small and the Rossby radius of deformation is large, a rela-

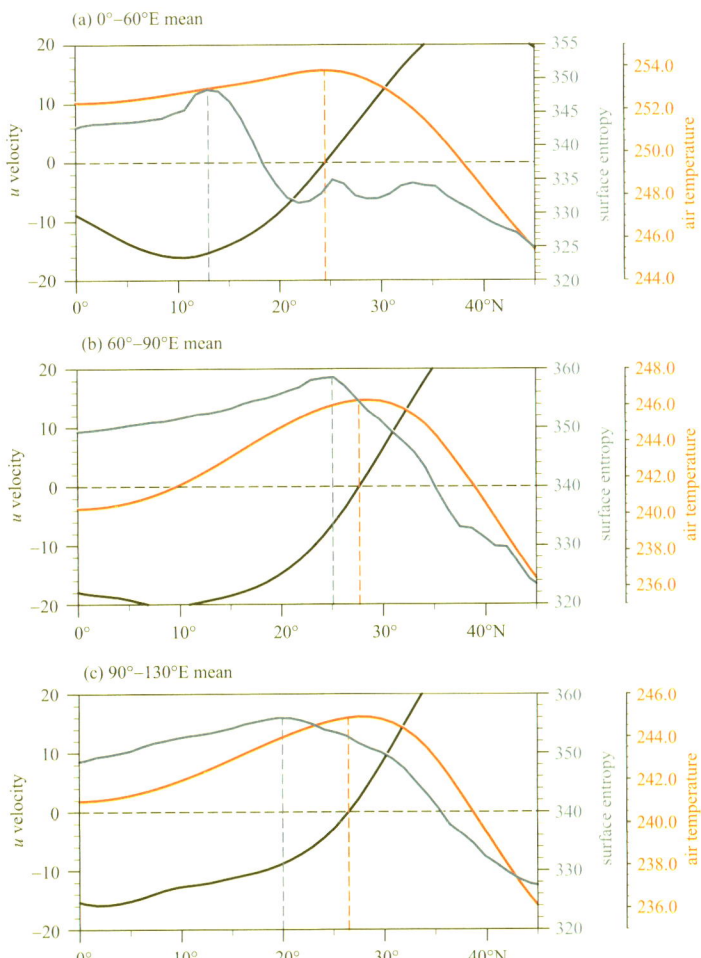

Fig. 2　ERA40 July-mean profiles of surface θ_{SE} (K, blue line), 200—400 hPa mass weighted mean air temperature (K, red line), and 200 hPa zonal velocity u (m s^{-1}, black line) for (a) 0°—60°E mean, (b) 60°—90°E mean, and (c) 90°—130°E mean. The black dashed line indicates $u=0$

tively weak forcing may overcome the in situ planetary vorticity and produce a meridional circulation. Plumb and Hou (1992) further studied the axisymmetric atmospheric response to an off-equator external forcing centered at 25°N and identified the regime transition from the TE type to the AMC type. They found that for a stronger forcing the relative vorticity becomes significant and the absolute vorticity may vanish, then the atmospheric response takes the AMC form, forming a thermally forced tropical monsoonal flow. Corresponding to this monsoonal meridional circulation with a cross-equatorial southerly in the lower troposphere and a reversed northerly in the upper troposphere, a vertical easterly shear presents in tropics. Such a background is in favor of the development of large scale ascent since absolute vorticity advection increases with height ($w \propto \partial[-\mathbf{V} \cdot \nabla(f+\zeta)]/\partial z$). Consequently convection can develop where the surface entropy is high. The cumulus friction of the tropical convection in return can further strengthen the easterly shear with height in the upper troposphere (Schneider and Lindzen, 1976, 1977). Furthermore, the easterly in the upper troposphere implies a temperature/geopotential height increase with increasing latitude. On the contrary in extra-tropics, the temperature is in local TE and decreases polewards together with decreasing geopotential height. Consequently, a horizontal temperature/geopotential-height gradient reversal appears in the subtropics where the UTTM and SAH are located, just to the north of the surface high entropy and tropical convection.

3 Longitude location of the UTTM and vertical gradient of diabatic heating

In extra-tropics the zonal distribution of temperature is connected with the vertical shear of meridional wind v, subject to the following thermal wind relation

$$\frac{\partial v}{\partial \ln p} = \frac{R}{f}\left(\frac{\partial T}{\partial x}\right) \tag{4}$$

Since the vertical shear of meridional wind from 400 to 200 hPa is negative to the east of the SAH center and positive to its west, according to (4) the maximum temperature in the 400—200 hPa layer should be located near the SAH center, as shown in Fig. 1a. The above analyses thus show that the center of the UTTM is coincident with the center of the SAH aloft. In other words, the SAH is a warm entity in nature. Now the problem becomes where in the subtropics the SAH and UTTM should be located.

Since the latitude location of the UTTM coincides with the SAH, and since in the subtropics $u \approx 0$ and $\partial \zeta / \partial y \approx 0$, the relative vorticity advection is negligible there. At a steady state and in a frictionless atmosphere the vorticity generation in the subtropics due to diabatic heating should be balanced by planetary vorticity advection (Liu et al., 2001):

$$\begin{aligned}\beta v &\approx (f+\zeta)\theta_z^{-1}(\partial Q/\partial z) - g(fT\theta_z)^{-1}(\partial T/\partial x)(\partial Q/\partial x) \\ &- g(fT\theta_z)^{-1}(\partial T/\partial y)(\partial Q/\partial y), \quad \theta_z \neq 0, \mathbf{V}\cdot\nabla\zeta \to 0\end{aligned} \tag{5}$$

where f is the Coriolis parameter, $\beta = df/dy$, $\theta_z = \partial\theta/\partial z$, ζ is relative vorticity, θ is potential temperature, g is acceleration of gravity, and Q is the diabatic heating. Assuming the following variable scales in the upper troposphere (Wu et al., 1999):

$$\theta_z \sim 10^{-2}\ \text{K m}^{-1},\ \Delta T \sim 10\ \text{K},\ (\Delta x, \Delta y) \sim 10^6\ \text{m},\ \Delta z \sim 10^3\ \text{m},\ Q \sim 10^{-5}\ \text{K s}^{-1}$$

Then the orders of magnitude for the three terms on the right-hand side of (5) are, respectively, $10^{-10}\ \text{s}^{-2}$, $10^{-11}\ \text{s}^{-2}$, and $10^{-11}\ \text{s}^{-2}$. This implies that the vorticity generation induced by horizontally inhomogeneous heating is at least one order of magnitude smaller than that due to the vertical heating gradient. Thus, the relation between meridional wind and the vertical gradient of diabatic heating Q can be well expressed as the following Sverdrup balance (Liu et al., 2001):

$$\beta v \approx (f+\zeta)\theta_z^{-1}(\partial Q/\partial z),\ \theta_z^{-1} \neq 0, \mathbf{V}\cdot\nabla\zeta \to 0 \tag{6}$$

where

$$z = -H \cdot \ln p$$

in which H is the scale height of the atmosphere. Taking the derivative of (4) with respect to x and substituting (6) into the resultant equation leads to:

$$\begin{cases}\dfrac{\partial^2 T}{\partial x^2} \approx \gamma \dfrac{\partial}{\partial x}\left(\dfrac{\partial^2 Q}{\partial z^2}\right) & \mathbf{V}\cdot\nabla\zeta \to 0 \\ \gamma = f(f+\zeta)H/(R\beta\theta_z) & \theta_z \neq 0\end{cases} \tag{7}$$

Since the domain under consideration is in the upper troposphere in the subtropics, in deriving (7) the local vertical variation of $[(f+\zeta)\theta_z^{-1}] \approx (f\theta_z^{-1})$ is neglected. Thus the parameter γ can be considered a constant along the x axis. Assume normal mode distributions for both diabatic heating Q and deviation (from zonal mean) temperature T:

$$\begin{cases} Q = Q(x)\cos\left(\dfrac{\pi z}{H_Q}\right) \\ T = T(x)\cos\left(\dfrac{\pi z}{H_Q}\right) \\ T(x) = T_0\cos\left(\dfrac{\pi x}{L}\right) \end{cases} \tag{8}$$

Here, H_Q is the characteristic height of diabatic heating Q, T_0 is the amplitude of temperature, and L is the characteristic horizontal scale of the temperature center. Substituting (8) into (7) leads to the following T-Q_Z relation between the temperature and the vertical gradient of diabatic heating Q

$$\begin{cases} T(x) \approx \gamma L^2 H_Q^{-2} \partial Q/\partial x = \lambda \partial Q/\partial x \qquad \mathbf{V} \cdot \nabla \zeta \to 0 \\ \lambda = \gamma L^2 H_Q^2 \end{cases} \quad (9)$$

Solution (9) indicates that in the subtropics, the maximum temperature is on the west/east of diabatic heating/cooling Q for a quarter of a wavelength, and the minimum temperature is on the west/east of diabatic cooling/heating Q for a quarter of a wavelength. The stronger the heating and the larger of the longitudinal scale of the temperature, the stronger the deviated temperature center. Figure 3 shows the climate mean July-distributions of the 200—400 hPa averaged air temperature T and diabatic heating Q in the Northern Hemisphere. The calculation of Q in this study is from the expression of apparent heat source (Yanai et al., 1973) as follows:

$$Q = \frac{Q_1}{c_p} = \left(\frac{p}{p_0}\right)^\kappa \left(\frac{\partial \theta}{\partial t} + \mathbf{V} \cdot \nabla \theta + \omega \frac{\partial \theta}{\partial p}\right) \quad (10)$$

where Q_1 is the apparent heat source, c_p is the specific heat of air at constant pressure, p_0 is 1000 hPa, $\kappa = R/c_p$, and ω is vertical velocity in the p-coordinate. Figure 3 shows that two warm centers are located over the Eurasian (Fig. 3a) and North American (Fig. 3b) continents, with their long axes being coincident with the local ridgelines of the subtropical anticyclone. Diabatic heating and cooling are respectively located to their east and west, and the much stronger warm center over Eurasia than over North America is accompanied with much stronger heating and larger longitude span, in agreement with the T-Q_Z relation (9) just developed. Two cold centers are located over Northern Pacific and Northern Atlantic (Fig. 3a). Due to the existence of the upper tropospheric trough over ocean in summer, the ridgeline is broken over the eastern ocean. Even so, the cold trough over ocean is accompanied with heating in its west and long wave radiation cooling in its east, closely subject to the T-Q_Z relation.

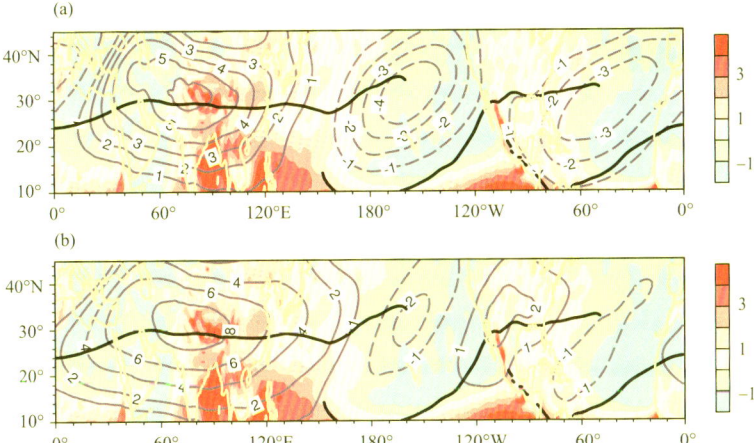

Fig. 3 1979—1998 July-mean 200—400 hPa averaged air temperature deviation (K, blue contour) from the zonal mean (a) and from the 180°—360°E mean (b), and diabatic heating (K d^{-1}, shading). Heavy black curve indicates where $u=0$, and $\partial u/\partial y > 0$

Let us focus on the UTTM. Fig. 4a shows the distribution of diabatic heating Q for the upper troposphere averaged between 400 and 200 hPa. Along the latitude belt of 26°—28°N where the ridgeline of SAH is located, the positive/negative heating is located to the east/west of 70°E, with the heating maximum appearing near 90°E. The heating in the east is due to monsoon condensation, and the cooling in the west is longwave radiation cooling, in good agreement with the summertime LOSECOD heating pattern in the subtropics (Wu et al., 2009). The corresponding temperature distribution along 26°—28°N calculated based on (9) demonstrates a positive phase to the east of the cooling and west of the heating maximum, spanning between 60° and 90°E (Fig. 4b), in good correspondence with the observed 246 K contour of the UTTM as shown in Fig. 1a. The vertical-longitude cross section of the temperature distribution averaged along the latitude belt of 26°—28°N as presented in Fig. 4c demonstrates that despite the unevenness of the heating near 90°E, the calculated temperature (contour) and the temperature retrieved from the ERA40 reanalysis (shading) are generally in phase and both exhibit westward tilting with increasing height. In the upper layer between 300 and 120 hPa, the UTTM denoted by the positive temperature deviation is located between 50° and 110°E in both the calculation and reanalysis. The calculated temperature anomalous center is 1.5—2.0 °C, which is also close to the reanalysis (>1.5 °C).

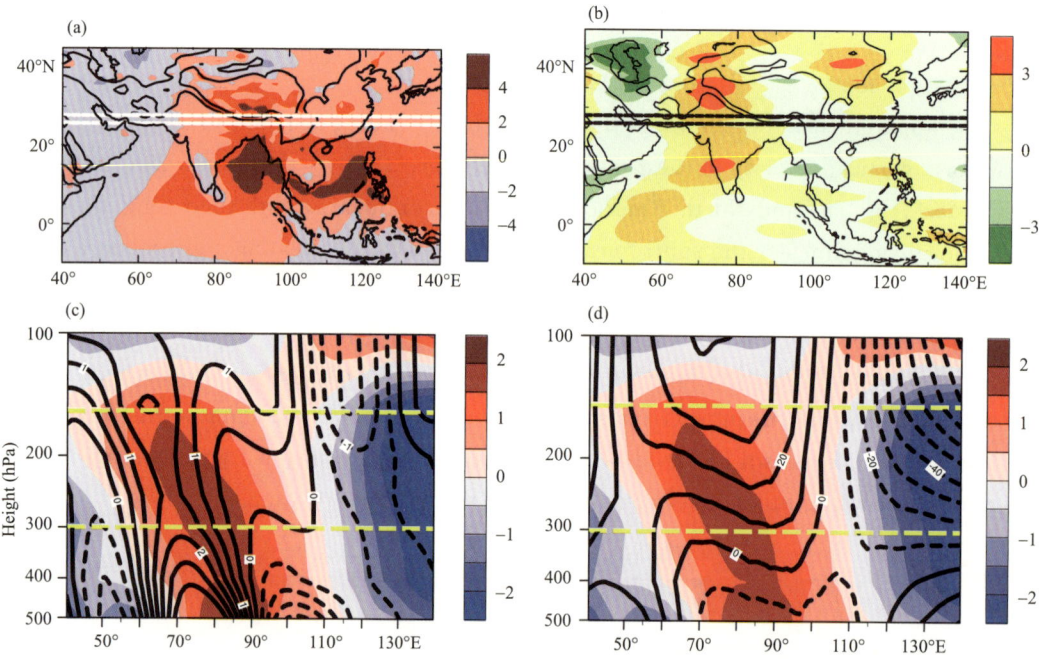

Fig. 4 The JJA mean distributions of the 200—400 hPa mass-weighted mean (a) diabatic heating rate and (b) the calculated temperature $\lambda \partial Q(x)/\partial x$; and the 26°—28°N mean cross sections of the deviations from their 40°—140°E means of (c) calculated temperature $\lambda \partial Q(x)/\partial x$ (contour) and temperature in reanalysis (shading) and (d) temperature (shading) and geopotential height (contour) in reanalysis. Units are K d^{-1} in (a), K for temperature in (b) to (d), and dagpm for geopotential height in (d)

There is a phase difference between the calculated temperature and the temperature obtained from reanalysis in the layer below 300 hPa. Particularly at 500 hPa, the "observed" warm maximum is over the TP centered at 90°E, whereas the calculated warm maximum is located to its west, centered near 75°E. This may be due to the strong impact of surface sensible heating, since diffusive surface heating can directly warm up the in situ atmosphere (Luo and Yanai, 1984) so that Q and T are in phase there, indicating the limitation of the T-Q_z relation in the lower troposphere. Such phase shifting decreases with increasing height. At 300 hPa the calculated UTTM center is located at 70°E, which is shifted westward by about 10° of longitude compared to the reanalysis; at levels above 200 hPa it agrees well with the reanalysis. The distributions in the reanalysis of geopotential height and temperature deviations from their means at 40°—140°E (Fig. 4d) demonstrate that in coordination with the warm center, the deviation of geopotential height is positive in the upper layers and negative in the lower layers, indicating increased air column thickness. While the warm center presents westward tilting, the SAH axis also tilts westward with height. The SAH becomes well defined above 350 hPa, just above the warm center.

The primary cause of cooling being located to the west of heating over Eurasia is due to the continental-scale thermal forcing along the subtropics (Wu, et al., 2009): in summer the atmospheric heating source is over continent while heating sink is over ocean. Along the subtropics, a heating generates air ascent (decent) in the east (west); whereas a cooling generates air descent (ascent) in the east (west). Consequently, in the upper troposphere the thermal forcing in summer generates air ascent to the east (west) and descent to the west (east) over continent (ocean), corresponding to the configuration with condensation heating being located to the east of long wave cooling over continent but to its west over ocean as demonstrated in Fig. 3. It is worthwhile to mention that, while the convective heating on the east of the UTTM is strong and uneven, the cooling on its west is weak and relatively uniform (Fig. 4a). This may due to the fact that, part of the longwave radiation comes from the response to the monsoonal convective heating to its east (Rodwell and Hoskins, 1996, 2001) and imply that the monsoon latent heating in the east is more active in forcing the UTTM.

The underpinning mechanism can be understood by the schematic diagram shown in Fig. 5. Along the subtropical latitude belt where the SAH is seated, huge condensation heating occurs mainly over and to the east of the southern TP (blue upward arrow), with its maximum located in the middle troposphere close to 400 hPa. Positive/negative vorticity is thus generated in the lower/upper troposphere below/above the heating maximum (right-hand side of (6)). Because the relative vorticity advection in subtropics is weak, negative/positive planetary vorticity advection in the lower/upper troposphere is required to maintain the atmosphere at a steady state (left-hand side of (6)), resulting in the local development of a southerly/northerly (black arrow) in the lower/upper troposphere (Wu et al., 1999; Liu et al., 1999, 2001). Under the thermal wind constrain (4), such a vertical northerly shear over the monsoon convection region should be accompanied with a warm center on its west and a cold center on its east. Therefore the dynamics of the T-Q_z relation can be interpreted as follows: In summer along the subtropics where the relative vorticity advection is weak, the UTTM should be generated to the west of the strong Asian monsoon convection so that the vertical northerly shear of the meridional wind induced by the convective heating is balanced by the zonal temperature gradient, and the SAH is produced aloft. Consequently, the Coriolis force ($fv<0$, orange arrow at 200 hPa) associated with the northerly flow in the upper troposphere above the convective heating is balanced by the eastward pressure gradient force ($-\partial\phi/\partial x>0$), and a new geostrophic and hydrostatic equilibrium state in the subtropics can be achieved.

A similar argument can be applied to the contributions of cooling (purple downward arrow) in the west to the UTTM. Over the subtropical western continent in summer, atmospheric heating is characterized as strong surface sensible heating in the lower troposphere but longwave radiation cooling in the upper troposphere (Wu and Liu, 2003; Wu et al., 2009). Because surface sensible heating and longwave radiation cooling both decrease with height, according to (6) vertical southerly shear is generated in the west of the UTTM. Following the thermal wind constraint (4), the vertically averaged air temperature in the east of the southerly shear region should be warmer than that in the west, and the UTTM should develop in the eastern end of the cooling, as depicted in Fig. 5.

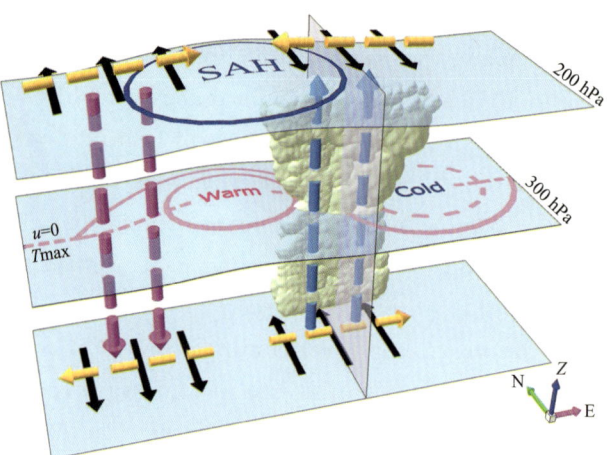

Fig. 5 Schematic diagram of the T-Q_z mechanism contributing to the longitudinal location of the upper-troposphere temperature maximum (UTTM): Strong monsoon convective latent heating along the subtropics (blue arrow) results in the local development of a vertical northerly shear (black arrow), and induces an eastward decreasing temperature gradient over the heated layer in the upper troposphere, forming the UTTM and the aloft SAH to the west of the heating. The vertical southerly shear over the western Eurasian subtropics, which is due to strong surface sensible heating and longwave radiation cooling in the upper troposphere, contributes to the occurrence of the UTTM and SAH on the eastern end of the cooling. The induced Coriolis force (fv, orange arrow) is in geostrophic balance with the pressure gradient force. Refer to text for details

4 Comparison of the T-Q_z model with the Gill's Model

It should be pointed out that either the Gill's model or the T-Q_z model does not apply in middle and high latitudes. This is because in higher latitudes the Burger number ($B = N^2 H_Q^2 / f^2 L^2$, in which $N = \left(\frac{g}{\theta} \frac{\partial \theta}{\partial z}\right)^{1/2}$ is the Brunt-Vaisala frequency, H_Q is a characteristic height of heating, and $L = (u/\beta)^{1/2}$ is the wave length of stationary Rossby wave) is small ($B \approx 10^{-1}$) (Hoskins, 1987), and the thermodynamic equation takes an advection-limit approximation (Smagorinsky, 1953). Warm temperature is thus formed downstream of a heating center. Whereas in subtropics and tropics the Burger number is much larger ($B \approx 10^0 - 10^1$), where temperature advection becomes secondary and the relation between heating and temperature distribution is more complicated. In the following we will make comparisons between the Gill's model and the T-Q_z model just developed.

4.1 Consistency of the Gill's model with the potential vorticity theory

The Gill's model (Gill, 1980) is a shallow-water equation model in the tropics and at the surface, which uses the equatorial beta-plane approximation and takes the following non-dimensional form:

$$\varepsilon u - \frac{y}{2} v = -\frac{\partial p}{\partial x} \tag{11.1}$$

$$\varepsilon v + \frac{y}{2} u = -\frac{\partial p}{\partial y} \tag{11.2}$$

$$\varepsilon p + \frac{\partial u}{\partial x} + \frac{\partial v}{\partial y} = -Q \tag{11.3}$$

In this model, (x, y) are non-dimensional distance with x eastwards and y measured northwards from the equator. The parameter ε represents Rayleigh friction as well as Newtonian cooling, and p is surface pressure. Q is defined as a "heating rate". The Gill's model has been successfully used in interpreting the tropical dynamics particularly in the lower troposphere. Can this model be directly used to study the subtropical dynamics in the upper troposphere? From (11) the following steady-state vorticity equation can be reached:

$$\frac{1}{2}v = \frac{y}{2}Q + \frac{y}{2}\varepsilon p - \varepsilon \zeta \tag{12}$$

Comparing (12) with (6), one can find that Q in (11.3) is actually presenting a vorticity forcing source rather than a heating itself. Defining Q in (11.3) as a "heating rate" will lead to inconsistency with the potential vorticity (PV) dynamics if the vertical gradient of diabatic heating is of opposite sign with the heating itself. This is because, following the PV theorem (Ertel, 1942; Hoskins et al., 1985; Wu and Liu, 2000), a vertical vorticity source is proportional to the vertical heating gradient $\partial Q/\partial z$. For a heating that is increasing with height ($\partial Q/\partial z > 0$), the static stability of the heated layer is increased and positive vorticity is generated. Thus the vorticity forcing possesses the same sign as the heating, and southerly is induced in the heating region with low pressure located in the west of the heating following either (12) or (6). However, if the heating is decreasing with height ($\partial Q/\partial z < 0$), a negative vorticity is produced in the heating region, because the static stability of the heated layer is decreased, and a northerly is required to transport positive planetary vorticity for maintaining a steady state according to the PV relation (6). High pressure (warm temperature) thus develops in the west of the heating. Whereas in the Gill's model (11), the prescribed heating still produces southerly (12) with low pressure (cold temperature) developing in its west, which is opposite to the observation as shown in Fig. 3 and results in inconsistency with the PV theory. Similar argument can be applied to the case when a cooling source ($Q<0$) is decreasing with height ($\partial Q/\partial z > 0$). Since both the vertically decreasing convective heating and vertically decreasing longwave radiation cooling occur in the upper troposphere, it means that defining Q as a heating rate may hamper the application of the Gill's model to the upper troposphere.

4.2 Applicability of the Gill's model in the upper troposphere

The above inconsistency can be overcome if the original "heating rate Q" in the Gill's model is replaced by a vorticity forcing source that is proportional to the vertical heating gradient $\partial Q/\partial z$, and the modified Gill's model becomes equivalent to the following form:

$$\begin{cases} \varepsilon u - \dfrac{y}{2}v = -\dfrac{\partial p}{\partial x} & (13.1) \\ \varepsilon v + \dfrac{y}{2}u = -\dfrac{\partial p}{\partial y} & (13.2) \\ \dfrac{1}{2}v = \dfrac{y}{2}\dfrac{\partial Q}{\partial z} + \dfrac{y}{2}\varepsilon p - \varepsilon \zeta & (13.3) \end{cases}$$

In which (13.3) is the corresponding modified vorticity equation, which then becomes consistent with the PV dynamics (6). For a prescribed forcing $\partial Q/\partial z$, there are three unknowns, i.e., u, v, and p. In such case, prescribing any inadequate solution for any unknown may lead to an incompatible set of equation. For instance, one cannot set $u=0$ then employ this system to study the subtropical anticyclone system along the ridgeline because by doing so the equation set becomes a set of three equations with two unknowns. In other words, caution is needed when employing the original Gill's model to study the subtropical dynamics.

Another limitation in applying Gill's model in the upper troposphere is the existence of dissipation. In the frictionless T-Q_z model only meridional wind is generated over the heating (cooling) region in the subtropics, and circulation and temperature centers are produced right along the same latitude of the heating region. Due to the existence of friction in the Gill's model, southwesterly (northeasterly) is generated over the heating (cooling) region, and cyclone (anticyclone) circulation is produced to the northwest of heating (cooling) source region. The strong friction and the extra Newtonian cooling impact ($y\varepsilon p/2$) may generate too much positive vorticity (13.3) and too strong wind (13.1 and 13.3) in higher latitudes, and causing bias of the circulation response in higher latitude in the upper troposphere.

It is interesting to demonstrate that if the Rayleigh friction and Newtonian cooling are further removed from the modified Gill's model (13), the system becomes:

$$\begin{cases} -\dfrac{y}{2}v = -\dfrac{\partial p}{\partial x} & (14.1) \\ \dfrac{y}{2}u = -\dfrac{\partial p}{\partial y} & (14.2) \\ \dfrac{1}{2}v = \dfrac{y}{2}\dfrac{\partial Q}{\partial z} & (14.3) \end{cases}$$

Such a non-dissipative system is thus equivalent to our hydrostatic and Sverdrup vorticity balanced system (formula (4) and (6)). It can now be employed to investigate the subtropical dynamics even for a prescribed solution of $u=0$. In such circumstances, the solution becomes

$$\begin{cases} u = -\dfrac{2}{y}\dfrac{\partial p}{\partial y} & (15.1) \\ v = y\dfrac{\partial Q}{\partial z} & (15.2) \\ p(x) = p(x_0) + \int \dfrac{y^2}{2}\dfrac{\partial Q}{\partial z}dx & (15.3) \end{cases}$$

In the upper troposphere, the deep convective heating and longwave radiation cooling both decrease with increasing height. Solution (15) then indicates the development of northerly over a heating region and southerly over a cooling region, with high pressure (also warm temperature) being located between heating in the east and cooling in the west and low pressure (also cold temperature) being located between heating in the west and cooling in the east. These conclusions are in agreement with the solution (9) we just obtained based on the PV framework.

The above discussions demonstrate that for the original Gill's surface model to be consistent with the potential vorticity dynamics and applicable in the upper troposphere, two modifications are needed. One is to replace the "heating rate" (Q) in the original model version of the continuity equation by a vorticity forcing that is proportional to the vertical heating gradient ($\partial Q/\partial z$); and another is to remove or substantially reduce the momentum friction and Newtonian cooling.

5 Numerical experiments

To verify the proposed T-Q_z mechanism which is developed for a steady state atmosphere and see whether the change in the UTTM over North India can be attributed to the change in the monsoon latent heating to its east, a set of numerical experiments is designed using the time-dependent Spectral Atmospheric Model developed at IAP/LASG (SAMIL; Wu et al., 2003; Bao et al., 2010), which can reasonably simulate the East Asian monsoon (Wu et al., 2012a; Bao et al., 2013). A 10-year integration of the SAMIL is defined as the control run CON. Because the ridgeline of the SAH in CON is located around 30°N, which deviates from its counterpart in reanalysis (Fig. 1a) by about 3—4 degrees northward, a box over the Asian monsoon region bounded by 90°—120°E and 28°—32°N along the modeled SAH axis is selected as the forcing source region S for designing a sensitivity experiment SEN, as indicated in Fig. 6a. In this SEN experiment every setting is the same as in the CON run except that in the thermodynamic equation, an extra convective heating, which is characterized by a cosine vertical heating profile with a maximum of 3 ℃ per day located at 500 hPa, is added in the box region during JJA. Both the CON and SEN experiments are integrated for 10 years, and the results from the last 5 years are extracted for the comparison studies.

This imposed convective heating results in increased rainfall of more than 1 mm per day in the source region and decreased rainfall of more than 1 mm per day to the west of the forcing (Fig. 6a). Consequently, the local northerly at 200 hPa is enhanced, contributing to the intensification of the SAH to the west. This is similar to the result obtained by Schneider (1987), who used a steady, nonlinear, inviscid, shallow-water equation model to study the low-latitude upper tropospheric circulation response to zonal asymmetric forcing and found that the upper tropospheric divergent northerly is generated over the forcing region (his Fig. 1b). The time means of UTTM in July in the two experiments are presented in Fig. 6b. It shows that due to the enhanced heating in the box region, the areas encircled by the 244, 245, and 246 K isotherms in the SEN experiment (dashed contours) are all enlarged compared to their counterparts in the CON run (solid contours). The 28°—32°N averaged longitude-height cross-section (Fig. 6c) demonstrates the existence of an anticlockwise secondary circulation with strong air ascent over the source region, where intensified heating is imposed and air descends to the west of the heating region, which presents a Rossby wave response to the forcing as revealed in Rodwell and Hoskins (2001). Below 300 hPa, the air gets warmer by more than 0.6 ℃ between 500 and 300 hPa to the east of 90°E due to direct convective heating. In the upper troposphere air warming of more than 0.4 ℃ appears to the west of 90°E. All these results are in good agreement with the analytical solution (9) and the data diagnosis presented in Fig. 4, and justify the proposed T-Q_Z mechanism as presented in Fig. 5.

The response of surface entropy to the imposed extra monsoon latent heating is demonstrated in Fig. 6d. Despite the surface cooling in the east and warming in the west (Fig. 6c), the surface entropy is increased to the east of 90°E but decreased to its west, with the maximum located over the northern tip of India. This suggests a moistening of surface air in the east and drying in the west, in good coordination with the increased rainfall and air ascent in the east and decreased rainfall and air descent in the west.

Another sensitivity experiment PER is a perpetual July experiment that is based on the same SAMIL AGCM as used for the CON and SEN runs. However, the solar azimuth is fixed at 15 July throughout the integration, and a periodic thermal forcing term

$$\left(\frac{\partial T}{\partial t}\right)_{osc} = Q\cos(\omega t) \tag{16}$$

is added to the thermodynamic equation in the same box region with the same vertical heating profile as described for SEN. An amplitude of the maximum heating rate $Q=5$ K day^{-1} at 500 hPa and a frequency $w = (60 \text{ day})^{-1}$ are assigned. Then the PER experiment is integrated for four years, and the results from the last three years (36 months) are retrieved for the following diagnosis. A wavelet power analysis and the corresponding significance diagnosis are applied to the 200 hPa geopotential height in the two experiments CON and PER in the forcing source region S (90°—120°E, 28°—32°N) and the response region R (70°—90°E, 28°—32°N). In the CON run, neither the forcing region S nor the response region R

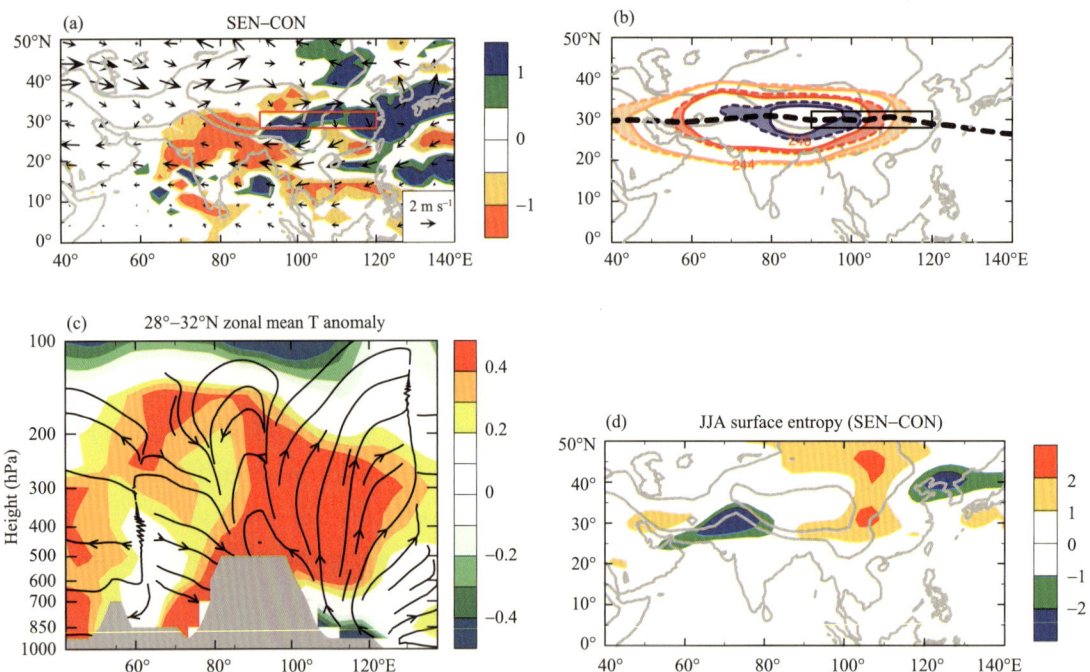

Fig. 6 Distributions of the JJA mean of (a) differences between the forcing experiment SEN and control experiment CON of rainfall (shading, unit: mm day^{-1}) and circulation at 200 hPa (arrow); (b) 200–400 hPa mass-weighted temperature in the SEN run (dashed curve) and CON run (solid curve); and the differences between SEN and CON of (c) 28°–32°N zonal mean circulation (streamlines, vertical motion has been amplified by a factor of 500) and temperature (shading, unit: K); and (d) surface entropy (unit: K). The heavy dashed black curve in (b) denotes the ridgeline of the SAH at 200 hPa. The box in (a) and (b) indicates where the extra convective heating with a maximum of 3 K day^{-1} at 500 hPa is imposed in the SEN experiment

shows any apparent periodic oscillation (figures not show). However, in the PER run corresponding to the imposed periodic forcing, the wavelet power analysis at the 200 hPa geopotential height demonstrates that a significant 60 day oscillation signal exists not only in the forcing region S (Fig. 7a and b), but also in the response region R (Figs. 7c and d). Figure 8a presents the evolution of the normalized heating at 200 hPa in the forcing region S. The normalized geopotential height at 200 hPa, the temperature averaged between 400 and 200 hPa, and the vertical velocity w at 300 hPa in the response region R are also shown. In the S region the heating evolution demonstrates a restricted 60 day oscillation. In the response region R, the atmosphere demonstrates a well-defined and inherent 60 day oscillation too: in response to the positive/negative forcing in convective heating in S, in the R region there appear positive/negative geopotential height at 200 hPa, a warm/cold air center in the upper troposphere between 400 and 200 hPa, and air descent/ascent at 300 hPa. The correlation coefficients between heating in the S region in the east and the geopotential height, temperature, and vertical motion in the R region in the west reach 0.80, 0.67, and 0.57, respectively, exceeding the 99% significance level (0.542). While the vertical air motion in the west responds almost concurrently to the forcing in the east, there is a delay in the response of temperature and geopotential height in the R region compared to the forcing in the S region (Fig. 8b). The correlation coefficients between the forcing and the responding temperatures with a delay of 2, 4, 6, 8, and 10 days are, respectively, 0.63, 0.67, 0.68, 0.66, and 0.61. These results highlight the significance of the circulation response to the monsoon rainfall: an oscillating rainfall anomaly of the ASM in return can force a corresponding oscillation in the atmospheric circulation.

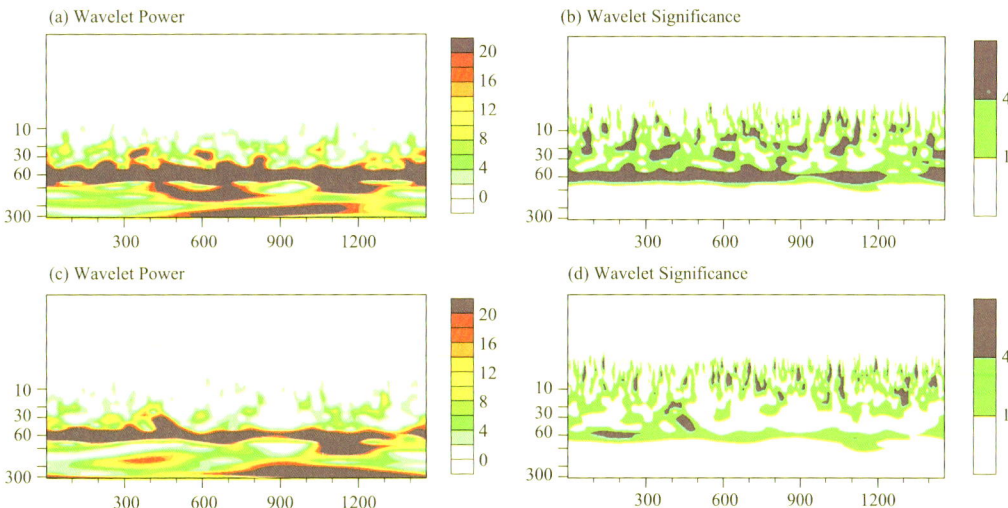

Fig. 7 Distributions in the PER experiment of wavelet power (a, c) and significance (b, d) for the 200 hPa geopotential height in the forcing source region S (90°—120°E, 28°—32°N) (a, b) and in the response region R (70°—90°E, 28°—32°N) (c, d); abscissa is for integration day and coordinate is for period (unit: day)

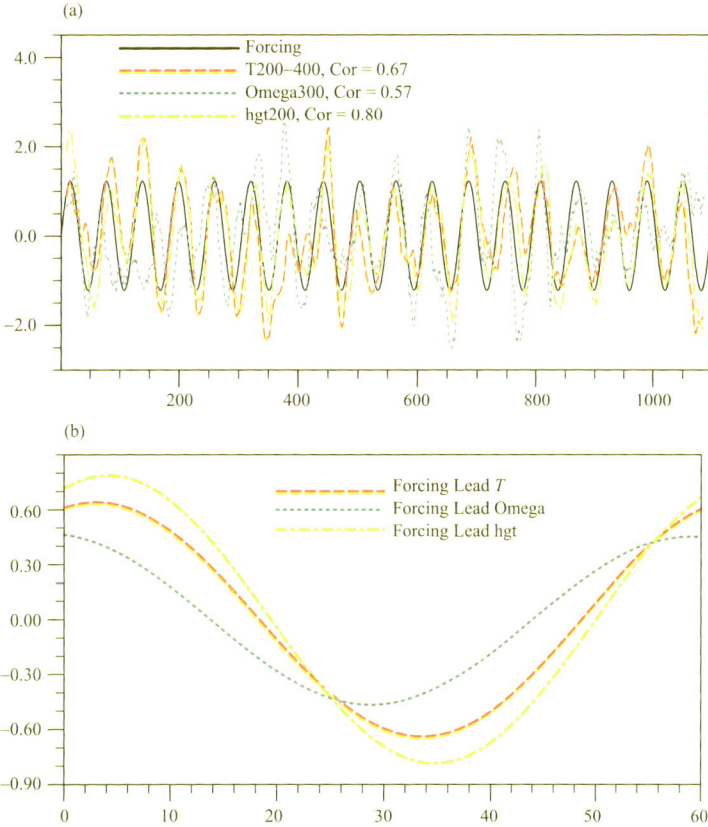

Fig. 8 Evolutions in the PER experiment of the normalized 200—500 hPa mass-weighted mean heating in the forcing source region S (black solid curve), and the geopotential height at 200 hPa (green dashed-dotted curve), the 200—400 hPa mass-weighted temperature (red dashed curve), and pressure vertical motion at 300 hPa (blue dotted curve) in the response region R (a); and the corresponding time-lag correlations between the forcing in region S and those in the response region R (b). A 20-day running mean has been used on the original data to filter out high-frequency noise

To see the characteristics of the atmospheric circulation response to the exerted convective heating in the source region, we produce the composite crosssections averaged between 28° and 32°N of the zonal circulation and temperature deviation from their (40°—140°E) means for the warm phase, mean state, cold phase, and the difference between the warm and cold phases. The response structures for the delays of 2, 4, 6, 8, and 10 days with respect to the forcing are similar to those of the simultaneous response. The composite results of the simultaneous response are presented in Fig. 9. The T-Q_Z mechanism and the associated secondary circulation are prominent in all the zonal cross sections along the ridgeline of the SAH. In all phases of the oscillation, the area to the east of 80°E is characterized by air ascent, warm in the lower troposphere and cold in the upper troposphere. These are typical features of monsoon convection. The warmth in the lower layer is caused by the convective latent heating, and the cold in the upper layer is a consequence of the overshooting of deep convection (Wu and Liu, 2000). The upper boundary of the 2 ℃ warm region in the source region is located at 200 hPa in the mean case (Fig. 9b) but is higher up in the warm phase (Fig. 9a) and lower down in the cold phase (Fig. 9c). In contrast, the area to the west of 80°E is characterized by air descent, warm in the upper troposphere and cold in the mid troposphere. The cold in the lower layer must be due mainly to radiation cooling, and the warmth in the upper layer corresponds to the UTTM as discussed above. The intensity of the UTTM is strongest in the warm phase (Fig. 9a) and weakest in the cold phase (Fig. 9c). This implies that the circulation variation in the western region is tightly linked to the convective forcing in the source region to the east. As presented in Fig. 9d, the configuration of the difference is pretty similar to the schematic diagram (Fig. 5) inferred from theoretical analysis. Because the forced diabatic heating was imposed to the east of 90°E, the separation between air ascent and descent is located by 90°E as well, again in good coordination with the results obtained by Rodwell and Hoskins (2001). Results from these experiments depict a well-developed T-Q_Z diagram and show that enhanced convective heating in the forcing region not only causes local tropospheric warming, but also forces a UTTM to its west.

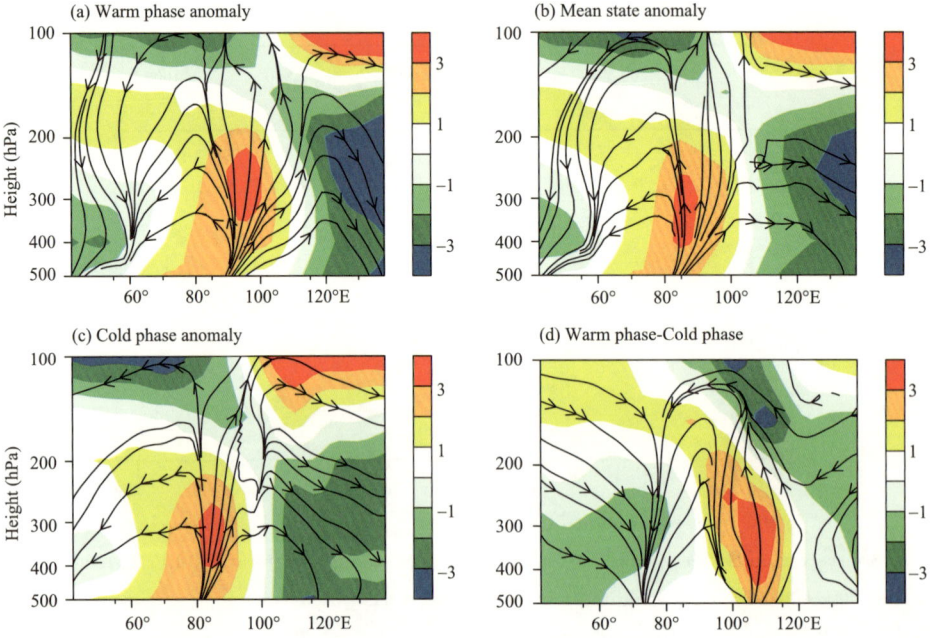

Fig. 9 28°—32°N mean cross sections of zonal circulation (streamline) and temperature deviations (shading, unit: ℃) from the corresponding zonal means in 40°—140°E produced from the PER experiment in (a) warm phase, (b) mean state, (c) cold phase, and (d) difference between (a) and (c)

6 Implication for regional climate change

During the last two decades of the twentieth century, summertime rainfall over South China has increased, due mainly to the weakening of TP thermal forcing (Duan and Wu, 2008; Duan et al., 2011, 2013; Liu Y M et al., 2012). According to the above T-Q_z mechanism, the intensification of the monsoon rainfall in South China should exert significant influence on regional climate change, at least in the Asian monsoon area. Figure 10 shows the differences between the periods (1991—2000) and (1981—1990) of the JJA mean precipitation (a), geopotential height at 200 hPa (b), and the UTTM (c). Along the latitude band of 24°—30°N where the SAH is seated (b), there is increased precipitation to the east of 90°E over the southeastern TP and southern China, but decreased precipitation to the west of 90°E over northeastern India in the last decade of the twentieth century compared to the previous decade (Fig. 10a). During the same period, the SAH is enhanced: the areas encircled by the 12540, 12520, and 12500 dagpm contours at 200 hPa are all increased (Fig. 10b). At the same time, the intensity of the UTTM increases, and its coverage expands remarkably (Fig. 10c). The area encircled by 246 K is about double, and the areas encircled by 245 K and 244 K are also enlarged, in good correspondence with the increased precipitation over the area to the east of 90°E.

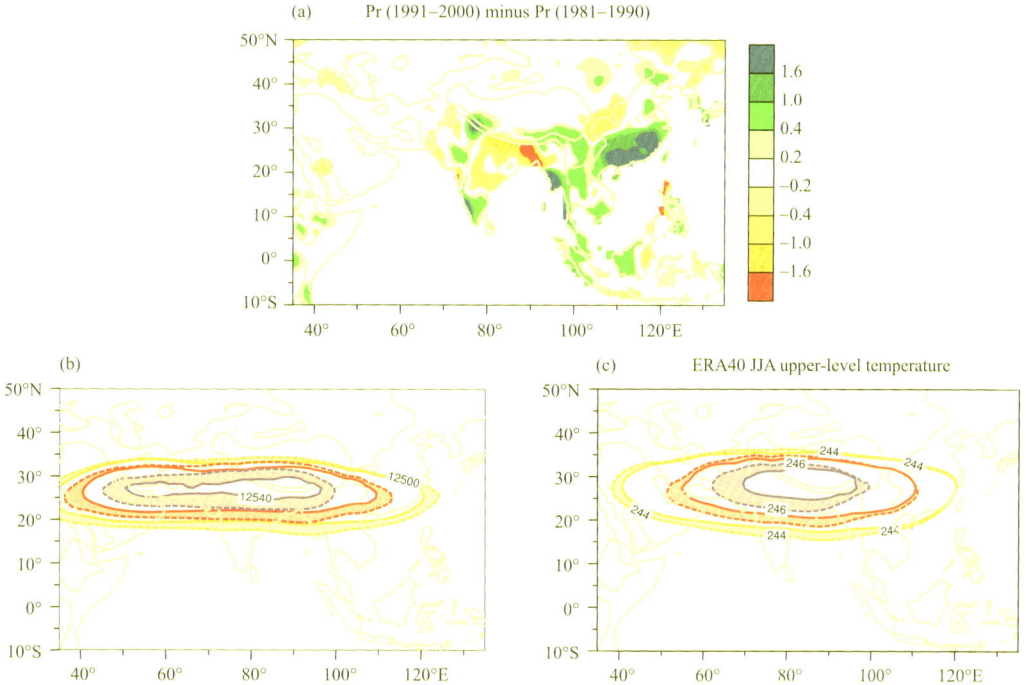

Fig. 10 Decadal changes in JJA mean climate between the periods (1991—2000) and (1981—1990) of (a) precipitation based on the PREC/L dataset (unit: mm d^{-1}), (b) 200 hPa geopotential height (unit: dagpm) and (c) the 200—400 hPa mass-weighted temperature (unit: K) based on ERA40 reanalysis. The solid and dashed curves in (b) and (c) denote, respectively, the 1981—1990 and 1991—2000 means

Figure 11 shows the climate changes in different elements in the ASM regions. For this purpose, the evolutions from 1981 to 2002 of JJA mean climate elements are produced in different monsoon regions, and a 5-year running mean smoother is applied to the original dataset to weaken the interannual signals. A careful scrutiny of these time series reveals that their changes are very well organized indeed

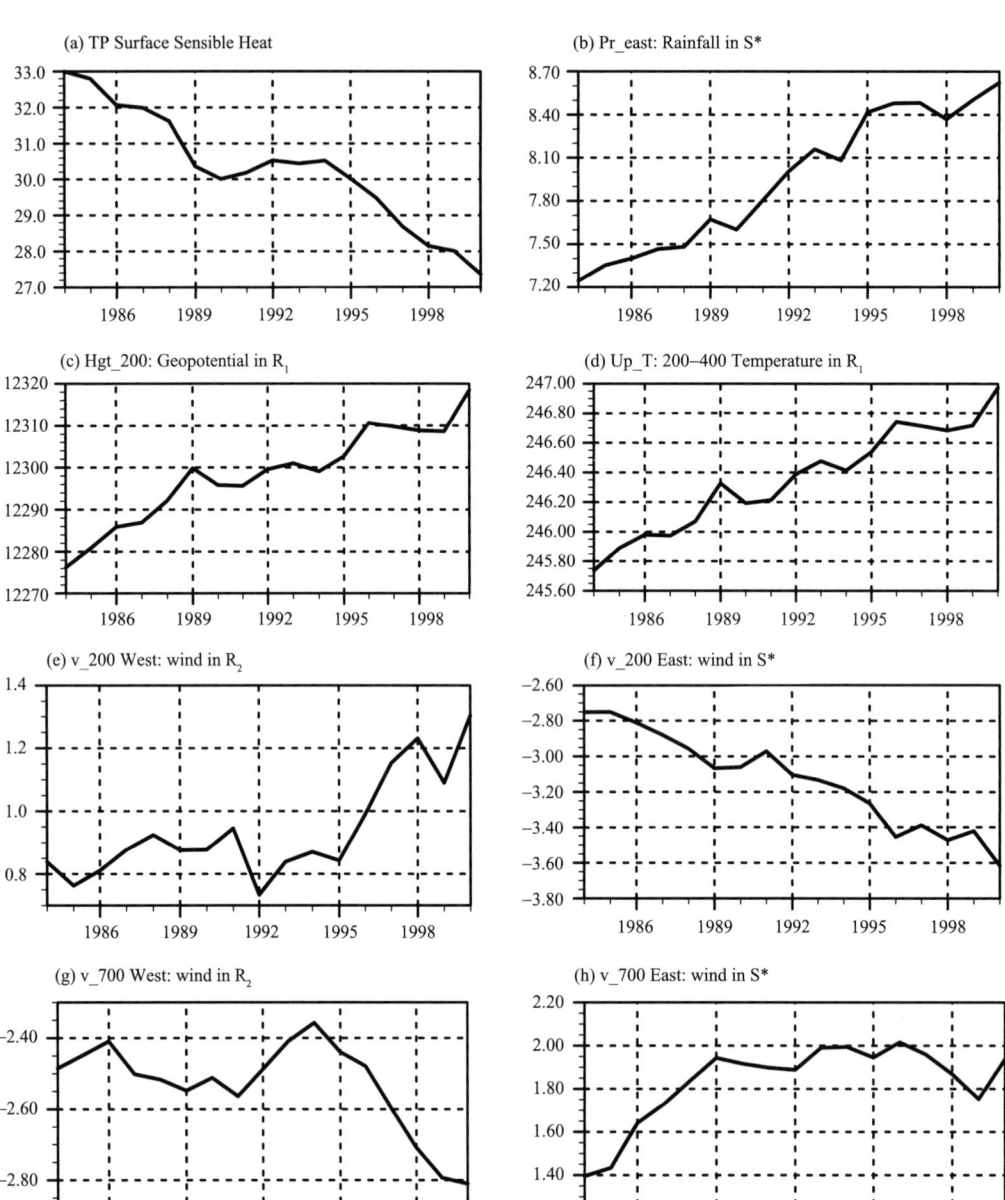

Fig. 11 Evolution from 1984 to 2000 of the JJA mean based on the ERA40 reanalysis of (a) surface sensible heat flux on the Tibetan Plateau region (80°—100°E, 26°—36°N unit: W m^{-2}), (b) precipitation in the forcing source region S* (Pr_east: 90°—120°E, 24°—28°N, unit: mm d^{-1}), (c) 200 hPa geopotential height in the response region R$_1$ (hgt_200: 70°—90°E, 24°—28°N unit: gpm), (d) 200—400 hPa mass-weighted temperature in R$_1$ (Up_T, unit: K), 200 hPa meridional wind (e) in the response region R$_2$ (V_200 West: 50°—80°E, 24°—28°N) and (f) in S* (V_200 East, unit: m s^{-1}), and 700 hPa meridional wind (g) in R$_2$ (V_700 West, Unit: m s^{-1}) and (h) in S* (V_700 East, unit: m s^{-1})

and can be interpreted using the T-Q_Z mechanism just developed. During the period 1984 to 2000, the sensible heating averaged over the TP (80°—100°E, 26°—36°N) decreases from 33 to 27 W m^{-2} (Fig. 11a). Because the TP surface sensible heating is an important driving force of the East ASM (Liu Y M et al., 2012; Wu and Zhang, 1998; Wu et al., 2007, 2012a, 2012b), the weakening of the TP forcing should result in a weakened low-layer southerly over East Asia, with intensified precipitation being confined to South China. Consequently, the monsoon rainfall in the source region S* (Pr east: 90°—120°E, 24°—28°N) increases during the same period (Fig. 11b). The source region S* defined here for reanalysis is shifted four degrees southward relative to the aforementioned source region S for modeling diagnoses because the ridgeline of the SAH is located by 26°—27°N in reanalysis, but by 30°N in the model, so there is a southward shifting of three to four degrees. As discussed above, the increasing latent heat release in the source region S* causes the intensifying SAH (hgt_200, Fig. 11c) and UTTM averaged between 200 and 400 hPa (Up_T, Fig. 11d) to its west in the response region R_1 (70°—90°E, 24°—28°N). According to (6), the intensification of the monsoon in the source region should also cause a local increase in the northerly at 200 hPa (V_200 East, Fig. 11f) and the southerly at 700 hPa (V_700 East, Fig. 11h). Further westward in the response region R_2 (50°—80°E, 24°—28°N), the strengthening SAH and the UTTM should result in an increasing southerly at 200 hPa (V_200 West, Fig. 11e) and a northerly at 700 hPa (V_700 West, Fig. 11g) there, and the reduction of local rainfall as shown in Fig. 10a.

The applicability of the mechanism sketched in Fig. 5 to local interannual climate variability is also verified by calculating the corresponding correlation coefficients. For this purpose, the original datasets from ERA40 JJA mean climate elements are used, and the results are presented in Table 1. The rainfall in the source region S* (Pr_east) is highly and significantly correlated with the UTTM (0.97), the SAH (0.94) in Region R_1, and the meridional wind at 200 hPa in the source region S* (−0.95) and in the response region R_2 (0.67). These indicate that at an interannual timescale, anomalous monsoon rainfall can affect regional atmospheric circulation as well.

Table 1 Correlation coefficients at the interannual timescale among those variables presented in Fig. 11: 200—400 hPa mass-weighted temperature (Up_T) in the response region R_1 (70°—90°E, 24°—28°N), 200 hPa geopotential height in R_1 (Hgt_200), precipitation in the forcing source region S* (Pr_east: 90°—120°E, 24°—28°N), and the 200 hPa meridional wind in S* (V_200_East) and in the response region R_2 (V_200_West: 50°—80°E, 24°—28°N)

Correlation coefficient	UP_T	Hgt_200	Pr_east	V_200_east	V_200_west
Up_T	1.0				
Hgt_200	0.99	1.0			
Pr_east	0.97	0.94	1.0		
V_200_east	−0.98	−0.97	−0.95	1.0	
V_200_west	0.72	0.73	0.67	−0.79	1.0

The 95% and 99% significance levels are, respectively, 0.51 and 0.64.

These results based on reanalysis data thus demonstrate that the different regional climate change or variation signals detected in different regions in Asia are dynamically well organized into a $T\text{-}Q_Z$ pattern as presented by the schematic diagram shown in Fig. 5.

7 Discussion and conclusion

The classical dynamics has demonstrated that in response to an axisymmetric diabatic heating, the atmospheric circulation adopts two distinct regimes: The TE regime in extra-tropics and the AMC regime in the tropics, and in the upper troposphere the maximum temperature and geopotential height should be located in subtropics. It is demonstrated in this study that the longitude location of the summertime upper-tropospheric maximum temperature is a consequence of the atmospheric circulation response to the atmospheric diabatic heating along the subtropics, and presents a $T\text{-}Q_Z$ relation: a warm temperature center is located between a monsoon heating in its east and a longwave radiation cooling in its west; whereas a cold temperature center is located between a monsoon heating in its west and a longwave radiation cooling in its east. A T-Q_Z mechanism is thus delineated for the longitude location of the UTTM over North India: In summertime in the subtropics where the relative vorticity advection is weak, strong latent heat released by the Asian monsoon can generate a strong vertical gradient of diabatic heating, resulting in a vertical northerly shear in situ. The UTTM and the associated upper tropospheric anticyclone SAH aloft thus develop to the west of the latent heating so that the eastward zonal-temperature decrease matches the vertical northerly shear in the monsoon region and the thermal wind balance is maintained. Over western Eurasia, strong surface sensible heating and longwave radiation cooling in the free atmosphere produce vertical southerly shear, and the UTTM develops on the eastern end of the radiation cooling region. Thus, the UTTM is generated to the west of monsoon convective heating and to the east of radiation cooling.

It is shown that the T-Q_Z model developed here is based on the potential vorticity (PV) theory and is different from the original Gill's model. In order to keep consistency of the Gill's model with the PV theory, it is demonstrated that the originally defined "heating rate" Q in the continuity equation of the Gill's model should be replaced by a vorticity forcing which is proportional to the vertical gradient of diabatic heating, i.e., $\partial Q/\partial z$. Furthermore, if the Rayleigh friction and Newtonian cooling in the Gill's model are removed, the system becomes identical to the PV system employed in this study and can be used to study the upper tropospheric dynamics.

Results from numerical sensitivity experiments show that enhancing the convective heating to the east of 90°E along the SAH ridgeline can intensify the SAH and UTTM to its west, and imposing a periodic convective latent heating in this source region can also induce periodic oscillations in the geopotential height, vertical air descent, and temperature to the west with the same frequency and a similar phase. Diagnoses based on the ERA40 reanalysis data also show that at either an interannual or decadal scale, the change and variation of the Asian summer rainfall in the forcing region are significantly correlated with the change and variation of the SAH, UTTM, and vertical motion in the response region to the west. All these results justify our conclusion that the appearance and variation of the summertime upper tropospheric warm center situated over North India can to a large extent be attributed to the strong convective latent heating associated with the Asian monsoon rainfall over the area southeast of the TP and over South China, and also to the longwave radiation cooling over the western Eurasia.

The T-Q_Z mechanism identified in this study highlights the inherent structure of the Asian monsoon system and the significant monsoon feedback on the circulation. However, the monsoon system is an open-dissipative system. When its variability is studied, the impacts of external forcing such as that of the ocean, the land-sea thermal contrast, and the thermal forcing of the Iranian Plateau and TP should be taken into account.

The current study focuses on the allocation of the upper tropospheric warm center and the monsoon latent heating along the subtropics without considering the whole structure and configuration of the UTTM. Because the summertime large-scale circulation in the subtropics is a consequence of multiscale forcing—including continental-scale LOSECOD thermal forcing, regional-scale mountain forcing, and local-scale sea-breeze forcing—and because the UTTM is closely tied to the SAH, which links the westerly in the mid-latitudes to the easterly in the tropics, a full understanding of the formation and variation of the UTTM will require great effort and deserves further study.

References

BAO Q, WU G X, LIU Y M, et al, 2010. An introduction to the coupled model FGOALS1.1-s and its performance in East

Asia[J]. Advances in Atmospheric Sciences, 27(5):1131-1142.

BAO Q, LIU P F, ZHOU T J, et al, 2013. The Flexible Global Ocean-Atmosphere-Land system model, Spectral Version 2: FGOALS-s2[J]. Advances in Atmospheric Sciences, 30(3):561-576.

BOOS W R, KUANG Z M, 2010. Dominant control of the South Asian monsoon by orographic insulation versus plateau heating[J]. Nature, 463(7278):218-233.

BOOS W R, KUANG Z M, 2013. Sensitivity of the South Asian monsoon to elevated and non-elevated heating[J]. Scientific Reports, 3(1):1-4.

CHEN L X, DUAN T, LI W, 1985. The variation of the atmospheric heat source and the budget of atmospheric energy on the Qinghai-Xizang Plateau during summer 1979[J]. Acta Meteorologica Sinica, 43: 1-12.

CHEN M Y, ARKIN P A, 2002. Global land precipitation: A 50-yr monthly analysis based on gauge observations[J]. Journal of Hydrometeorology, 3(3):249-266.

DUAN A M, WU G X, 2008. Weakening trend in the atmospheric heat source over the Tibetan Plateau during recent decades. Part Ⅰ: observations[J]. Journal of Climate, 21(13):3149-3164.

DUAN A M, LI F, WANG M R, et al, 2011. Persistent weakening trend in the spring sensible heat source over the Tibetan Plateau and its impact on the Asian summer monsoon[J]. Journal of Climate, 24(21):5671-5682.

DUAN A M, WANG M R, LEI Y H, et al, 2013. Trends in summer rainfall over China associated with the Tibetan Plateau sensible heat source during 1980—2008[J]. Journal of Climate, 26(1):261-275.

EMANUEL K A, 1991. A scheme for representing cumulus convection in large-scale models[J]. Journal of the Atmospheric Sciences, 48(21):2313-2335.

EMANUEL K A, NEELIN J D, BRETHERTON C S, 1994. On large-scale circulations in convecting atmospheres[J]. Quarterly Journal of the Royal Meteorological Society, 120(519):1111-1143.

ERTEL H, 1942. Ein neuer hydrodynamische wirbdsatz[J]. Meteorology Z Braunschweig, 59: 277-281.

FLOHN H, 1957. Large-scale aspects of the "summer monsoon" in South and East Asia[J]. Journal of the Meteorological Society of Japan, 35A:180-186.

GILL A E, 1980. Some simple solutions for heat-induced tropical circulation[J]. Quarterly Journal of the Royal Meteorological Society, 106(449):447-462.

HAHN D G, MANABE S, 1975. Role of mountains in South Asian monsoon circulation[J]. Journal of the Atmospheric Sciences, 32(8):1515-1541.

HELD I M, HOU A Y, 1980. Nonlinear axially-symmetric circulations in a nearly inviscid atmosphere[J]. Journal of the Atmospheric Sciences, 37(3):515-533.

HOSKINS B J, 1987. Diagnosis of forced and free variability in the atmosphere[C]//CATTLE H. Atmospheric and Oceanic Variability. Bracknell : James Glaisher House: 57-73.

HOSKINS B J, MCINTYRE M E, ROBERTSON A W, 1985. On the use and significance of isentropic potential vorticity maps[J]. Quarterly Journal of the Royal Meteorological Society, 111(470):877-946.

KRISHNAMURTI T N, DAGGUPATY S M, FEIN J, et al, 1973. Tibetan high and upper tropospheric tropical circulation during northern summer[J]. Bulletin of the American Meteorological Society, 54(12):1234-1249.

LI C F, YANAI M, 1996. The onset and interannual variability of the Asian summer monsoon in relation to land sea thermal contrast[J]. Journal of Climate, 9(2):358-375.

LIU B Q, HE J H, WANG L J, 2012. On a possible mechanism for southern Asian convection influencing the South Asian high establishment during winter to summer transition[J]. Journal of Tropical Meteorology, 18(4):473-484.

LIU B Q, WU G X, MAO J Y, et al, 2013. Genesis of the South Asian high and its impact on the Asian summer monsoon onset[J]. Journal of Climate, 26(9):2976-2991.

LIU Y M, LIU Y M, LIU P, 1999. The effect of spatially nonuniform heating on the formation and variation of subtropical high part Ⅲ: Condensation heating and South Asian high and western Pacific subtropical high[J]. Acta Meteorologica Sinica, 57(5):525-538.

LIU Y M, WU G X, LIU H, et al, 2001. Condensation heating of the Asian summer monsoon and the subtropical anticyclone in the eastern hemisphere[J]. Climate Dynamics, 17(4):327-338.

LIU Y M, BRIAN H, MICHAEL B, 2007. Impact of Tibetan orography and heating on the summer flow over Asia[J]. Journal of the Meteorological Society of Japan (Special 125th Anniversary Issue), 85B:1-19.

LIU Y M, HONG J, DONG B, et al, 2012. Revisiting Asian monsoon formation and change associated with Tibetan Plateau forcing: II. Change[J]. Climate Dynamics, 39(5):1183-1195.

LIU Y M, HU J, HE B, et al, 2013. Seasonal evolution of subtropical anticyclones in the climate system model FGOALS-s2[J]. Advances in Atmospheric Sciences, 30(3):593-606.

LUO H B, YANAI M, 1983. The large-scale circulation and heat sources over the Tibetan Plateau and surrounding areas during the early summer of 1979. Part I: Precipitation and kinematic analyses[J]. Monthly Weather Review, 111(5): 922-944.

LUO H B, YANAI M, 1984. The large-scale circulation and heat sources over the Tibetan Plateau and surrounding areas during the early summer of 1979. Part II: Heat and moisture budgets[J]. Monthly Weather Review, 112(5):966-989.

MASON R B, ANDERSON C E, 1963. The development and decay of the 100 hPa. summertime anticyclone over southern Asia[J]. Monthly Weather Review, 91(1): 3-12.

PENG Y P, HE J H, CHEN L X, et al, 2014. A study on the characteristics and effect of the low-frequency oscillation of the atmospheric heat source over the eastern Tibetan Plateau[J]. Journal of Tropical Meteorology, 20(1):17-25.

PLUMB R A, HOU A Y, 1992. The response of a zonally symmetric atmosphere to subtropical forcing: Threshold behavior[J]. Journal of the Atmospheric Sciences, 49(19):1790-1799.

PRIVE N C, PLUMB R A, 2007a. Monsoon dynamics with interactive forcing. Part I: Axisymmetric studies[J]. Journal of the Atmospheric Sciences, 64(5):1417-1430.

PRIVE N C, PLUMB R A, 2007b. Monsoon dynamics with interactive forcing. Part II: Impact of eddies and asymmetric geometries[J]. Journal of the Atmospheric Sciences, 64(5):1431-1442.

RODWELL M J, HOSKINS B J, 1996. Monsoons and the dynamics of deserts[J]. Quarterly Journal of the Royal Meteorological Society, 122(534):1385-1404.

RODWELL M J, HOSKINS B J, 2001. Subtropical anticyclones and summer monsoons[J]. Journal of Climate, 14(15): 3192-3211.

SARDESHMUKH P D, HOSKINS B J, 1985. Vorticity balances in the tropics during 1982—1983 El Nino-Southern Oscillation event [J]. Quarterly Journal of the Royal Meteorological Society, 111(468):261-278.

SCHNEIDER E K, 1977. Axially symmetric steady-state models of the basic state for instability and climate studies. Part II: Nonlinear calculations[J]. Journal of the Atmospheric Sciences, 34(2):280-296.

SCHNEIDER E K, 1987. A simplified model of the modified Hadley circulation[J]. Journal of the Atmospheric Sciences, 44(22):3311-3328.

SCHNEIDER E K, LINDZEN R S, 1976. A discussion of the parameterization of momentum exchange by cumulus convection[J]. Journal of Geophysical Research-Oceans and Atmospheres, 81(18):3158-3160.

SCHNEIDER E K, LINDZEN R S, 1977. Axially symmetric steady-state models of the basic state for instability and climate studies. Part I: Linearized calculations[J]. Journal of the Atmospheric Sciences, 34(2):263-279.

SMAGORINSKY J, 1953. The dynamical influence of large scale heat sources and sinks on the quasi-stationary mean motions of the atmosphere[J]. Quarterly Journal of the Royal Meteorological Society, 79(341):342-366.

TAMURA T, TANIGUCHI K, KOIKE T, 2010. Mechanism of upper tropospheric warming around the Tibetan Plateau at the onset phase of the Asian summer monsoon[J]. Journal of Geophysical Research-Atmospheres, 115: D02106. doi:10.1029/2008JD011678.

TAO S Y, ZHU F K, 1964. The variation of 100 hPa circulation over South Asia in summer and its association with arch and withdraw of west Pacific subtropical high[J]. Acta Meteorological Sinica, 34: 385-395.

UPPALA S M, KÅLLBERG P W, SIMMONS A J, et al, 2005. The ERA-40 re-analysis[J]. Quarterly Journal of the Royal Meteorological Society, 131(612):2961-3012.

WANG B, LINHO, 2002. Rainy season of the Asian-Pacific summer monsoon[J]. Journal of Climate, 15(4):386-398.

WU G X, ZHANG Y S, 1998. Tibetan Plateau forcing and the timing of the monsoon onset over South Asia and the South China Sea[J]. Monthly Weather Review, 126(4):913-927.

WU G X, LIU Y M, LIU P, 1999. Impacts of spatial differential heating on the formation and variation of the subtropical anticyclone. I. scale analysis[J]. Acta Meteorologica Sinica, 57(3): 257-263.

WU G X, LIU Y M, 2000. Thermal adaptation, overshooting, dispersion and subtropical anticyclone Part I: Thermal adaptation and overshooting[J]. Journal of the Atmospheric Sciences, 24: 433-446.

WU G X LIU Y M, LIU P, et al, 2002. Relation between the zonal mean subtropical anticyclone and the sinking arm of the Hadley cell[J]. Acta Meteorologica Sinica, 61: 635-636.

WU G X, LIU Y M, 2003. Summertime quadruplet heating pattern in the subtropics and the associated atmospheric circulation[J]. Geophysical Research Letters, 30(5).

WU G X, ZHANG Q, DUAN A M, et al, 2007. The influence of mechanical and thermal forcing by the Tibetan Plateau on Asian climate[J]. Journal of Hydrometeorology, 8(4):770-789.

WU G X, LIU Y, ZHU X, et al, 2009. Multi-scale forcing and the formation of subtropical desert and monsoon[J]. Annales Geophysicae, 27(9):3631-3644.

WU G X, LIU Y M, HE B, et al, 2012a. Thermal Controls on the Asian summer monsoon[J]. Scientific Reports, 2:404.

WU G X, DONG P, LIU Y, et al, 2012b. Revisiting Asian monsoon formation and change associated with Tibetan Plateau forcing: I. Formation[J]. Climate Dynamics, 39(5):1169-1181.

WU T W, LIU P, WANG Z Z, et al, 2003. The performance of atmospheric component model R42L9 of GOALS/LASG [J]. Advances in Atmospheric Sciences, 20(5):726-742.

XIE S P, SAIKI N, 1999. Abrupt onset and slow seasonal evolution of summer monsoon in an idealized GCM simulation [J]. Journal of the Meteorological Society of Japan, 77(4):949-968.

YANAI M, ESBENSEN S, CHU J H, 1973. Determination of bulk properties of tropical cloud clusters from large-scale heat and moisture budgets[J]. Journal of the Atmospheric Sciences, 30(4):611-627.

YANAI M H, LI C F, SONG Z S, 1992. Seasonal heating of the Tibetan Plateau and its effects on the evolution of the Asian summer monsoon[J]. Journal of the Meteorological Society of Japan, 70(1B):319-351.

YANAI M, LI C F, 1996. Seasonal and interannual variability of atmospheric heating[C]. Atlanta, Ga: 8th Conference on Air-Sea Interaction/Conference on the Global Ocean-Atmosphere-Land System (GOALS): 102-106.

YE D Z, GAO Y X, 1979. Meteorology of the Qinghai-Xizang Plateau[M]. Beijing: Chinese Science Press.

YEH T C, LO S W, CHU P C, 1957. The wind structure and heat balance in the lower troposphere over Tibetan Plateau and its surroundings [J]. Acta Meteorologica Sinica, 28: 108-121.

ZHANG Q, WU G X, QIAN Y F, 2002. The bimodality of the 100 hPa South Asia high and its relationship to the climate anomaly over East Asia in summer[J]. Journal of the Meteorological Society of Japan, 80(4):733-744.

第3部分

青藏高原气候效应的若干基本问题

综述 6

青藏高原对亚洲气候格局的影响

段安民[1,2,3]，王在志[4,5]，张萍[2,3]，王子谦[6]，卓海峰[2]，刘新[7]，李伟平[4,5]

(1 厦门大学海洋与地球学院近海海洋环境科学国家重点实验室，厦门 361102；2 中国科学院大气物理研究所，北京 100029；3 中国科学院大学地球与行星科学学院，北京 100049；4 中国气象局地球系统数值预报中心，北京 100081；5 灾害天气国家重点实验室，北京 100081；6 中山大学大气科学学院，广州 510275；7 中国科学院青藏高原研究所，北京 100101)

摘要：吴国雄院士团队自 20 世纪 90 年代以来一系列开创性的工作将青藏高原(以下简称高原)热力强迫的研究推向了一个新的高度。值此庆祝吴国雄院士诞辰 80 周年之际，文中回顾了青藏高原热力学和动力学的历史发展过程，并基于吴国雄院士团队近几十年的研究成果，系统性地回顾了高原的热力特征、高原加热的基本理论框架，以及高原热力强迫对亚洲气候格局的影响。

1 引言

青藏高原(以下简称高原)平均海拔在 4000 m 以上，横跨欧亚大陆，总面积超过 250 万平方千米，素有除南极和北极以外的"地球第三极"之称，是世界上海拔最高、面积最大、地形最复杂的高原，其热力和动力作用对亚洲乃至全球大气环流和气候系统都有着极其重要的影响。

高原大地形对大气的动力和热力强迫作用是同时存在的，不同背景环流下二者相对重要性不同。冬季，高原以机械强迫为主，进入高原的西风急流形成以高原为轴，南侧气旋、北侧反气旋的"偶极子"偏差环流，有助于华南早春连阴雨的形成。夏季，高原以热力强迫为主，使得周边大气从低空向高原辐合而上升，犹如一部巨型"气泵"屹立在欧亚大陆上空，且主要受到高原地表感热的驱动。因此，青藏高原这种热力效应被定义为"感热驱动气泵(TP-SHAP)"。TP-SHAP 是导致亚洲大气环流从冬到夏演变中出现季节突变的重要原因，也是亚洲夏季风爆发和维持的重要因素。

春、夏季(4 月—9 月)青藏高原为大气热源，其中春季热源以地表感热为主，而夏季以凝结潜热为主；10 月高原由热源转为热汇，冬季成为冷源。夏季副热带最强的大气热源出现在高原上空，其中感热加热是夏季高原上空近地层非绝热加热的基本形式，而潜热加热最大值出现在对流层低层并占主导地位。冬季高原为"弱冷源"，且与高原西侧至东南角凝结潜热释放有关。

高原热力强迫影响大气环流的基本原理为"热力适应"。大气环流对高原加热的响应呈现出低层气旋、高层反气旋的结构，并在加热区东侧上升、西侧下沉。高原引起的上升(下沉)与大陆尺度四叶型热力强迫(LOSECOD)(Wu and Liu，2003)所产生的上升(下沉)运动叠加，使得东亚夏季风(中亚干旱区)成为全球最强的季风系统(干旱区)之一。此外，青藏高原是大气运动重要的负涡度源，能够激发 Rossby 波列向下游传播，并影响亚洲乃至全球的大气环流异常。

高原热源异常能够影响亚洲夏季风的年际及年代际变率。在年际尺度上，春季高原感热正异常往往

通讯作者：段安民，amduan@xmu.edu.cn
资助项目：国家自然科学基金重大研究计划战略研究项目 92037000；国家自然科学基金重点项目 42030602

会导致随后夏季江淮流域降雨偏多；此外，高原热源与海表温度异常相互作用，并能够协同影响东亚夏季风雨带。在全球变暖的背景下，高原感热在20世纪80年代—20世纪末呈现年代际的减弱趋势，是东亚季风区出现南涝北旱的重要原因之一。

20世纪90年代，吴国雄院士提出了青藏高原感热气泵（TP-SHAP）的概念，这一开创性的研究工作为今后关于高原热力影响的研究奠定了基础，自此青藏高原热力强迫的研究也被推向了一个新的高度。值此庆祝吴国雄院士诞辰80周年之际，文中主要基于吴国雄院士团队近几十年的研究成果，回顾了青藏高原动力强迫和热力强迫的历史发展过程、青藏高原加热影响大气环流的基本理论框架，并重点回顾了青藏高原热力强迫对亚洲气候格局的影响。

2 青藏高原动力、热力作用的研究历史

2.1 高原动力和热力强迫作用的提出

青藏高原气象学的系统性研究始于20世纪40年代末。早期人们主要关注高原大地形对大气环流的机械强迫作用。Queney（1948）引入三个临界波数来区分不同类型的一维地形波，系统性地总结了地形爬坡作用。随后，Bolin（1950）和Yeh（1950）在二维平面上研究了青藏高原大地形对西风急流的南北两支绕流作用，并表明这种分支是形成东亚大槽以及印缅槽的重要原因。周晓平等（1958）进一步研究了地形坡度对Rossby波波速的影响，并指出冬季青藏高原南、北两支绕流在下游的汇合导致东亚急流的形成。Wu（1984）进一步研究表明，大气对地形动力作用的行星尺度响应存在一个临界高度（约1 km），当地形低于该临界高度时，则气流以爬坡为主，而当地形高于该临界高度时，则以绕流为主。实际大气中气流遇到青藏高原时，绕流和爬坡是可以同时发生的，这就表明气流不仅受到地形机械强迫作用，还与地形抬升加热有关。

早在20世纪50年代，中国学者叶笃正等（1957）和德国学者Flohn（1957）先、后分别利用当时仅有的地面观测和探空资料发现青藏高原在夏季是一个巨大的大气热源。与此同时，朱抱真（1957a，1957b）指出考虑了热源和地形的共同作用后所得的扰动流型才更接近实况，并由此提出热源和地形作用通过动力过程在大气环流中相互制约的统一性。20世纪80年代以来，随着观测资料、再分析资料、数值模拟以及动力理论的深入发展，关于青藏高原影响的研究逐渐从"是什么"进入到回答"为什么"的动力学研究阶段。Held（1983）通过理论研究指出，在基本气流很强时地形的机械作用比热力作用重要；而在基本气流很弱时则地形的热力作用更为重要。也就是说，高原大地形对大气的动力和热力作用是同时存在的，只是不同背景环流下二者相对重要性可能会不一样。自此以后，以青藏高原的机械和热力强迫作用为核心的青藏高原气象学成为天气气候研究领域的一个热点科学问题。

2.2 青藏高原动力强迫和热力强迫的相对重要性

Held等（1990，2002）利用线性模型研究了大气对地形与热力强迫的响应。研究表明，地形强迫和加热强迫对大气定常波的形成都非常重要，但其相对重要性受到西风基本气流的调控。当西风气流很强时地形对大气环流的机械强迫作用是主要的，而当西风气流很弱时地形对大气环流的热力强迫作用更重要。冬季沿青藏高原的西风很强，夏季沿青藏高原的西风很小。因此可以推断，冬季青藏高原对大气环流的影响以机械强迫为主，而夏季青藏高原对大气环流的影响则以热力强迫为主。

王同美等（2008）利用美国国家环境预报中心/美国国家大气研究中心（NCEP/NCAR）的再分析资料研究了不同季节高原的热、动力作用对周围环流的影响。图1展示了冬、夏季高原周围850 hPa纬向偏差流场和位温的分布。他们发现高原对西风带的动力作用在冬季最强，使得高原周围形成以高原为轴，南侧气旋、北侧反气旋的非对称的"偶极子"（TP-dipole mode）偏差环流（图1a），流线在中亚上空辐散，而在中国东部辐合。它比传统认识的爬坡、绕流的影响范围大得多，其范围遍及东亚的高、低纬度。高纬度地区的反气旋环流使其以西的暖空气向北平流，而以东的冷空气向南平流。因此，亚洲高纬度地区的等温线由

西北向东南倾斜,导致亚洲中纬度地区内陆气温比沿海气温暖,位温场的西侧(50°E)、东侧(130°E)在40°N有10 K的温差,在50°N位温差达14 K。而高原南侧的气旋性环流使得干空气向南输送至南亚次大陆,并将湿空气向北输送至中南半岛和华南地区,其有利于在亚洲夏季风爆发开始前形成印度干旱和中南半岛湿润的气候,这也使得印度半岛感热加强而中南半岛的潜热加强;此外,高原东侧南北气流的辐合则有利于华南早春连阴雨(persistent rains in early spring,简称 PRES)的形成(万日金 等,2006,2008a,2008b;Wu et al.,2007;王同美 等,2008;Wang and Wu,2007;Wu et al.,2015;吴国雄 等,2016a)。青藏高原对华南春雨的影响将在第四章进行系统详细的回顾。

随着西风带的季节性北移,夏季青藏高原成为一个巨大的热源,热力作用激发了亚洲副热带地区巨大的气旋性环流(图1b)。图1这种大尺度的偏差环流形势在早期的数值试验中就已经被发现了(吴国雄 等,2005),数值试验结果表明,相比于大尺度海陆热力差异对偏差环流的影响,青藏高原才是其形成的主要因素。进一步的分析发现,"青藏高原感热气泵(TP-SHAP)"是导致亚洲大气环流从冬到夏演变中出现季节突变的重要原因,也是亚洲夏季风爆发和维持的重要因素。关于TP-SHAP的提出、形成与维持的相关研究将在本文第4节进行详细的回顾和总结,此处不再赘述。

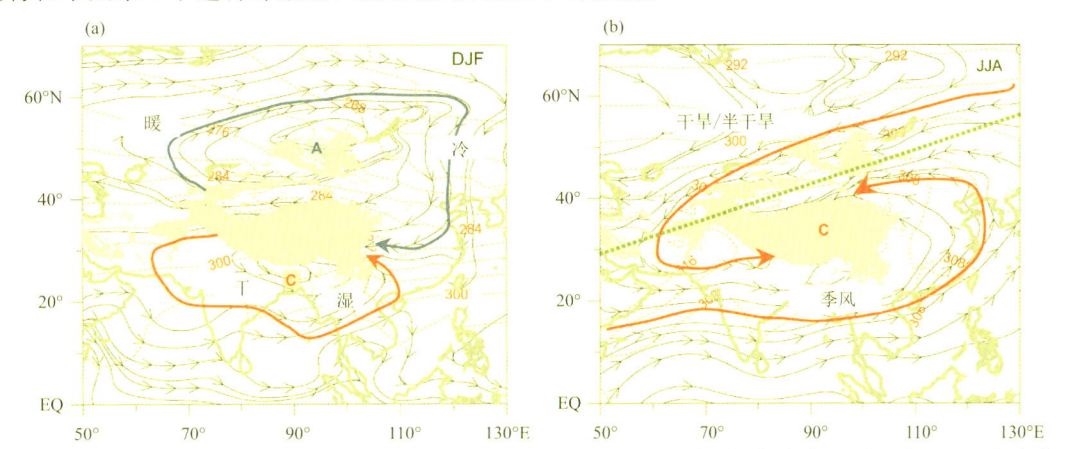

图1 (a)冬季(12月—2月)、(b)夏季(6月—8月)850 hPa纬向偏差流场(灰色实线)和位温场(K,红色虚线)。阴影表示海拔高度超过1800 m的地形(摘自 Wu et al.,2015a)

3 青藏高原大气热源的特征

3.1 高原大气热源(汇)的定义

大气热源(汇)是一个反映空气柱热量收支的物理量,当某地区大气柱内净能量收支为正时,则称此地区大气为热源;净能量收支为负,则称为冷源(叶笃正和高由禧,1979)。其表达式为(Duan et al.,2008,2018,2022):

$$AHS = SH + LH + NR$$

式中,AHS 表示大气总热源(atmosphere heat source),SH 表示地表感热(surface sensible heat),LH 表示大气中的凝结潜热释放(latent heat release),NR 表示大气柱的净辐射通量(net radiation flux)。接下来介绍各成分的具体算法。

青藏高原地表感热通量采用总体动力学公式进行计算:

$$SH = C_p \rho C_{DH} V_0 (T_s - T_a)$$

式中,$\rho = 0.8$ kg·m^{-3},为空气密度;$C_p = 1005$ J·kg^{-1}·K^{-1},为干空气定压比热;C_{DH} 为热量拖曳系数(无量纲量);V 为10 m风速(单位:m·s^{-1});T_s 为地表温度(单位:℃);T_a 为2 m空气温度(单位:℃)。$C_{DH} = 4 \times 10^{-3}$ (Duan et al.,2008)。

凝结潜热则根据降水数据进行估计：
$$LH = P_r L_w \rho$$
式中，P_r 为降水率，凝结潜热系数 $L_w = 2.5 \times 10^6$ J·kg^{-1}，ρ 为水的密度。

净辐射通量的计算如下：
$$NR = R_\infty - R_0 = (S_\infty^\uparrow - S_\infty^\downarrow) - F_\infty - (S_0^\downarrow - S_0^\uparrow) - (F_0^\downarrow - F_0^\uparrow)$$
式中，R_∞ 与 R_0 分别表示大气层顶（top of atmosphere，TOA）与地表的净辐射通量，∞与0分别表示大气层顶和地表，S 和 F 分别表示短波、长波辐射通量，↓和↑表示辐射通量的传输方向。

然而，高原热源研究面临的最大困难就是资料问题。目前高原观测站仍然较少，且可用的卫星数据时间序列较短，不同再分析资料中的高原热源误差较大，这些原因导致准确定量地估算高原热源仍旧困难（竺夏英 等，2012；Duan et al.，2014）。

3.2 高原大气热源的年循环特征

气候平均而言，高原上垂直积分的总加热每年有两次冷、热源的转换，冬季为冷源，4月转变为热源，4—9月均为加热大气，且7月达到最大；而在10月转为冷源，10月—次年3月均为冷却大气，且12月达到最小（赵平 等，2001；周秀骥 等，2009）。

王同美等（2008）利用 NCEP 加热率资料计算了高原热源的年循环。图2给出了1月、4月、7月和10月高原区域月平均加热率的垂直分布。图3为在1、4、7、10月高原区域平均加热率的垂直分布，冬季高原加热表现为冷源，3月由于地面感热增强低层开始出现加热，4月大气柱的总加热转为热源主要是因为低层感热的迅速增长，6、7、8月感热均维持较大值，季风爆发后潜热加热作用也显著增强，在7月出现峰值，这是感热和潜热释放共同作用的结果。夏季的潜热加热最大高度偏低，在500 hPa 左右。10月的状况与3月类似，低层因感热加热而成为弱的加热。

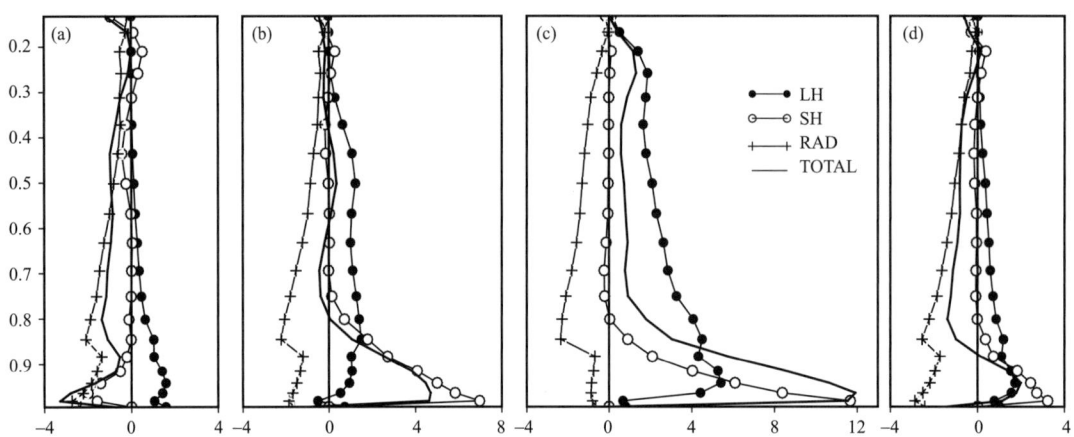

图2 高原区域（80°～100°E，27.5°～37.5°N）平均的非绝热加热率垂直廓线（单位：K·d^{-1}）。LH：潜热加热，SH：感热加热，RAD：辐射冷却，TOTAL：总的非绝热加热。(a)1月；(b)4月；(c)7月；(d)10月（摘自 Wu et al.，2007，JH）

随着气象观测站的增加和再分析资料的发展和更新，已有研究基于多种资料（站点观测资料，MERRA2、JRA55、ERA-interim、ERA5 等再分析资料）计算得到的青藏高原大气热源年循环特征与以上结果基本一致，但强度、峰值出现的时间等略有差别（Duan et al.，2008，2018，2022；段安民 等，2022）。Zhao 等（2018）使用最新的青藏高原观测资料和卫星数据研究了青藏高原上空大气总加热源 AHS 的各分量的年循环和月方差（图3）。结果表明，在春季特别是在3月和4月，感热 SH（50～70 W·m^{-2}）比潜热 LH（5～20 W·m^{-2}）大得多时，SH 在 AHS 中占主导地位。5月，SH 达到全年的峰值。然而，随着雨季的到来，LH 在5月和6月迅速增加，并在5月下旬和6月初超过 SH。LH 在7月与总热源一样达到顶峰，而且整个夏季都超过了 SH，因此成为 AHS 最重要的组成部分。净辐射通量全年表现为降温效应，但春季和夏季($-40\sim-70$ W·m^{-2})的冷却效果小于秋季和冬季($-80\sim-110$ W·m^{-2})。春季 SH 的方差大于 LH 的方差，5 月的数值最大。夏季 LH 的方差大于 SH，8 月达到 300 W^2·m^{-4} 的峰值。

综上所述，春、夏季青藏高原为大气热源，其中春季高原地表感热为大气热源的主要分量并在 5 月达到感热峰值，夏季凝结潜热占据高原热源主要成分并在 7 月出现潜热峰值，10 月高原由热源转为冷源，冬季成为冷源。

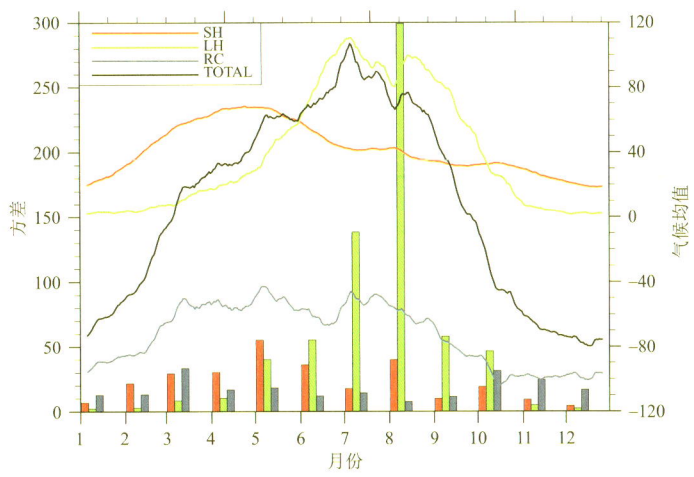

图 3　由 73 个台站的平均观测数据计得的青藏高原上空大气热源每个分量的气候平均值（实线；右纵坐标，单位：W·m^{-2}）和月方差（柱条，左纵坐标），红色表示 SH，绿色表示 LH，蓝色表示辐射加热 RC，黑色表示它们的总和（摘自 Zhao et al.，2018）

3.3　夏季高原热源特征

（1）夏季高原热源气候态特征

夏季青藏高原为一个强大的热源，其对大气的加热作用应该包括两部分，其一是高原自身的抬升加热，其二是其东侧被加强的季风降水所释放的潜热加热（吴国雄 等，2004，2005）。图 4 是利用 NCEP/NCAR 资料得到的 1980—1999 年 7 月沿 32.5°N 亚洲大陆平均的非绝热加热及其各分量的高度-经度剖面，垂直方向取 δ 坐标。感热加热（图 4a）集中分布在地表，在高原上强度最大，强度最大可达到 10 K·d^{-1}，并随高度增加而迅速减弱，到 500 hPa 附近（高原上 σ=0.8）消失。高原上空的潜热加热几乎存在于整个对流层，但与中国东部地区不同，其最大加热层次不是位于对流层中高层，而是出现在对流层低层，反映夏季高原上空旺盛的浅积云对流活动（图 4b）。同时，由于高原上空辐射冷却（图 4c）的强度不足以平衡感热或潜热加热的任何一项，导致高原上空的气柱成为强大的大气热源（图 4d）。高原西侧亚洲内陆地区降水稀少，潜热加热几乎为零，但近地层感热和中高层的辐射冷却造成这里近地层为较弱的大气热源，其上为大气冷源（热汇）。高原东侧东亚季风区上空的大气热源几乎完全取决于深对流加热，中低层的辐射冷却强度只有大陆中西部的一半左右，感热加热也远小于大陆中西部地区。因此，就自由大气而言，夏季副热带亚洲大陆西部为大气热汇，东部为大气热源，最强的大气热源出现在高原上空。感热加热是夏季高原上空近地层非绝热加热的基本形式，潜热加热在中高层则占主要地位。

（2）夏季高原热源的年际变化特征

本小节主要以 7 月高原热源的时空分布来介绍夏季高原热源的异常特征。段安民（2003a，2003b）使用旋转经验正交函数（REOF）展开分析了 7 月的高原区域大气热源强度主要空间分布型和相应的时间系数（旋转主成分 RPC）。图 5 给出了以高原为主体的一矩形区域（75°~100°E，23.809°~40.952°N）共 150 个格点 1958—1999 年 7 月大气热源强度的前 4 个 REOF。从中可见前 4 个型的累积方差贡献才达到总方差的三分之一，REOF1 方差贡献还不足 10%，这意味着 7 月高原加热的局地性非常显著。第一 REOF（图 5a）突出了高原东北部的大气加热异常，从中可看出以高原东北部为中心的大气加热异常的空间尺度最大。第二 REOF(图 5b)加热中心位于高原西南地区,第三 REOF(图 5c)和第四 REOF(图 5d)则分别表示以克什米尔和高原东南部为中心的大气热源空间型。4 个空间型共表示出总方差的 33%,每型表示的局地方差均达 80% 以上。

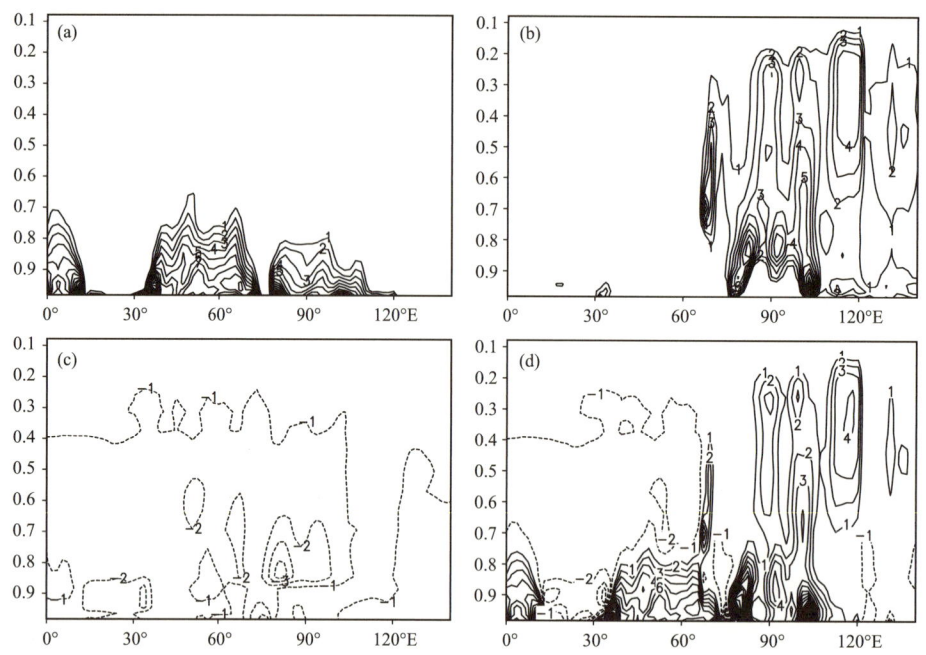

图 4　气候平均的 7 月沿 32.5°N 的非绝热加热垂直廓线。(a)感热加热;(b)潜热加热;(c)辐射冷却;(d)总的非绝热加热;等值线间隔为 1 K·d^{-1},垂直风向为 σ 坐标,σ=0.9 和 σ=0.1 在高原上空分别相当于 540 hPa 和 60 hPa (摘自吴国雄 等,2004)

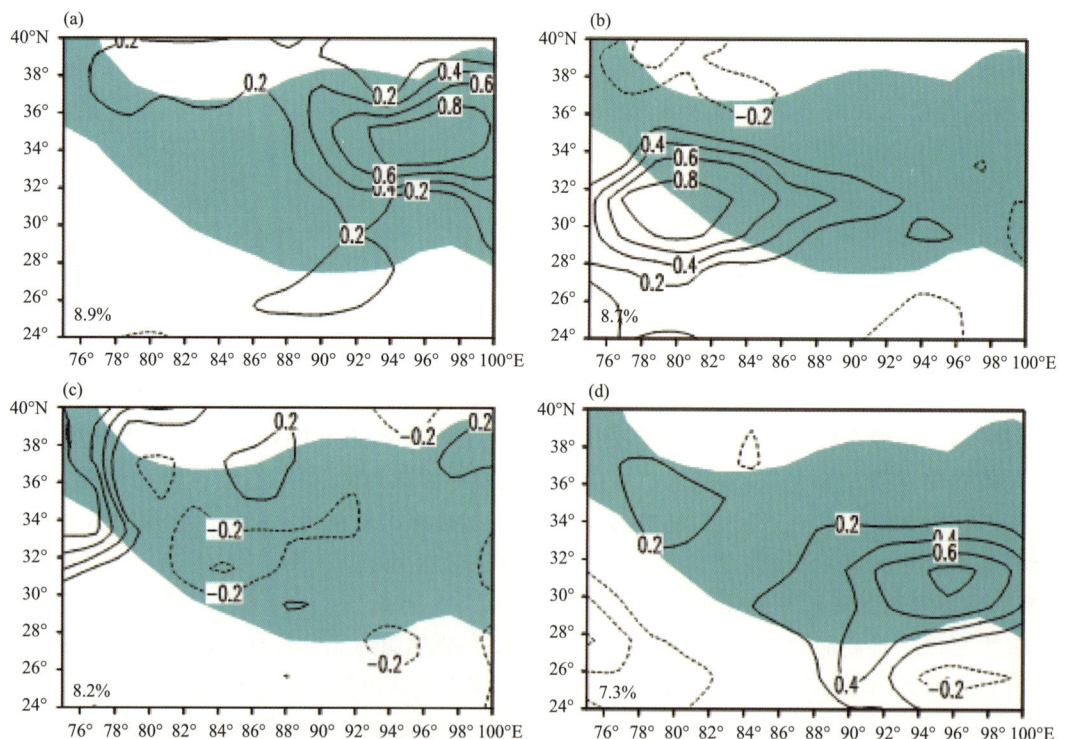

图 5　7 月青藏高原大气热源 H 的前 4 个 REOF((a)—(d)分别为 REOF1、REOF2、REOF3 和 REOF4,等值线是大气热源强度与该空间型的时间系数的相关系数,图中左下角的黑体字是各 REOF 的解释方差百分比,阴影部分为 3000 m 以上的地形)(摘自段安民,2003a)

为了认识 7 月份高原大气热源各主要空间型的年际和年代际的变化特征,他们进一步用小波分析方法对前 4 个 RPC 的时间演变特征进行分析。结果表明它们均存在 2~4 年的周期成分,并且这一周期成分主要出现在 20 世纪 70、80 年代。此外,高原上空的大气热源还有较显著的年代际变化特征,如 20 世纪 50—60 年代的 6~12 年周期。

各 RPC 与 7 月东亚中、低空水平风场、水汽通量场以及降水场的相关系数表明,高原大气加热不同的空间分布型所对应的东亚大气环流明显不同,因此相应的降水分布也不相同。其中,华北地区的降水与高原东部加热型有显著负相关,与克什米尔加热型有显著正相关;西南南部、华南地区和菲律宾等地的降水与克什米尔加热型有显著负相关,而日本的降水则与之有显著正相关。因此,这也说明,在研究 7 月高原热状况与同期东亚大气环流和降水的关系时,应注意高原加热的局地特征。仅用高原主体面积平均的加热指数来研究高原的热状况对气候的影响存在一定的局限性。关于夏季高原热源对东亚夏季风的影响将在本文第 5 小节进行详细的综述。

3.4 冬季高原热源特征

(1) 冬季高原热源气候态特征

与高原夏季这种强大的热源特征相比,高原在冬季并没有加强陆地上空的冷源,反而减弱了陆地上的冷却效应,相对于同纬度地区呈现出"弱冷源"特征(这与吴国雄老师早期的工作略有不同:那时认为高原冬季冷却大气、夏季加热大气)。宇婧婧等(2009,2011a)利用 ERA40、JRA25 及 NCEP2 三套再分析资料计算了冬季高原上空垂直积分的非绝热加热(图6)。注意到三套资料中,虽然高原主体大部分地区上空非绝热加热为负,但与周围其他大陆上空的热汇相比,负值偏小,即呈现出一个"弱冷源"特征,这一结果与通常人们的设想不同。进一步分析冬季高原"弱冷源"的原因发现,与周围大陆相比,高原上空长波辐射冷却是偏弱的,且潜热释放加热在高原西侧至东南角是相对较大的,从而造成冬季高原主体偏西为弱冷源,高原西侧和东南角有非绝热加热大值中心。这一结果也被 LASG 自主研发的全球大气环流谱模式 SAMIL 设计的理想高原隆起的水球试验进一步验证。

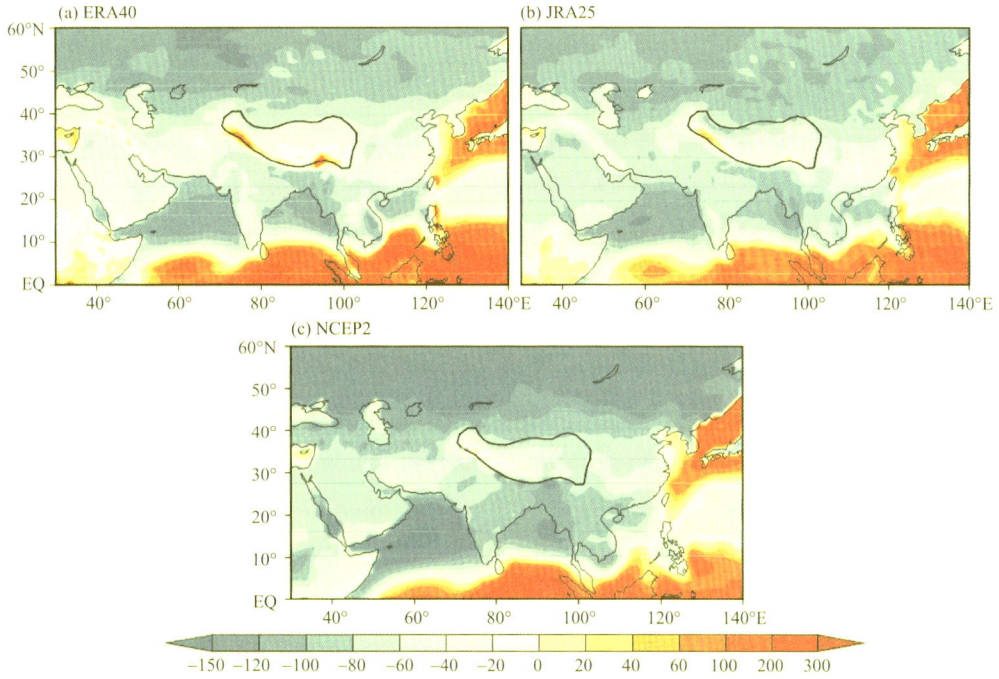

图 6 冬季(DJF)气候平均大气柱垂直积分的非绝热加热的分布(粗实线表示 3000 m 高度)。
(a) ERA40;(b) JRA25;(c) NCEP2,单位:W·m^{-2}(摘自宇婧婧,2009)

通过在数值试验中改变高原的位置,发现处于西风带中的高原以动力作用为主,使其周围形成以高原为轴,南侧气旋、北侧反气旋的"偶极子"型偏差环流。此时高原西侧因地形阻挡作用产生的上升运动造成凝结潜热释放,以及高原南侧气旋性环流为高原南侧带来充沛水汽,导致高原南侧潜热增加,总的非绝热加热增加。最终造成冬季高原上空"弱冷源"的热力特征。这体现了高原动力作用影响其热力特征的结果。

(2)冬季高原热源的年际变化特征

宇婧婧等(2011b)在研究冬季青藏高原热源特征的基础上进一步研究了冬季高原上空非绝热加热的年际变化特征及其与北半球大气环流场异常的联系。他们首先对冬季1月份高原上空非绝热加热的年际变率进行了EOF分析,发现,三套再分析加热资料的EOF第一模态均为最主要的模态(ERA40的EOF1方差贡献为48.9%,JRA25的EOF1方差贡献为39.6%,NCEP2的EOF1方差贡献为31.2%),显著超过第二模态的分布。且这三种资料EOF1模态的分布(图7)均显示,冬季高原非绝热加热的变化主要集中在高原西侧至东南侧地区。结合上述关于气候平均的非绝热加热分布可以得出,高原非绝热加热年际异常大值区与气候平均的非绝热加热大值区基本分布一致,均在高原西侧至东南侧地区。而此非绝热加热大值区正是高原西侧至东南侧的潜热大值区,表明高原1月非绝热加热异常主要表现为西侧至东南侧的潜热变化。冬季(12月、1月、2月)三个月的非绝热加热异常的主要特征基本一致。

进一步通过资料诊断及数值试验发现,冬季高原的这种非绝热加热异常与整个北半球的异常环流模态密切相关,北半球从西北到东南的相当正压遥相关波列可能是造成高原冬季非绝热加热异常的主要原因。

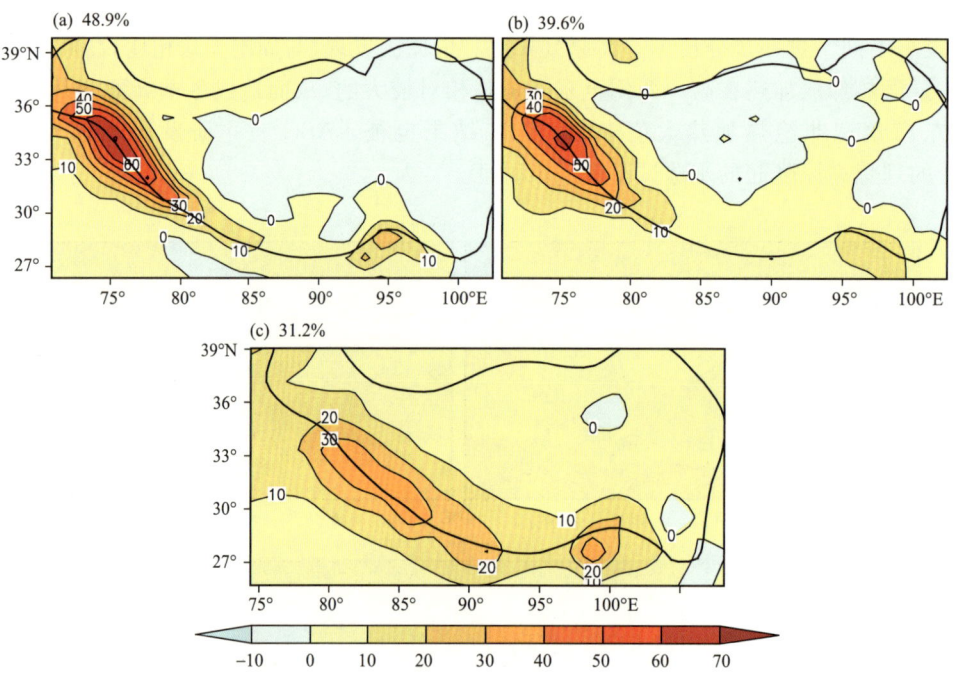

图7 1月青藏高原上空大气柱垂直积分的非绝热加热异常的主要模态(EOF1)(粗实线表示3000 m高度)。(a)ERA40;(b)JRA25;(c)NCEP2,单位:W·m^{-2}(摘自宇婧婧 等,2011b)

4 青藏高原感热气泵的形成与维持

利用多套再分析资料(NCEP1、NCEP2、MERRA2;吴国雄 等,1997a;Wu et al.,2007)以及GOALS(global ocean-atmosphere-land system)气候模式的大气环流分量模式SAMIL(spectral atmospheric model of the IAP/LASG)的数值模拟结果(Wu et al.,1996)发现,在夏季,青藏高原上空有明显的等熵面

下凹的特征,表明此时高原为一热源;高原低空四周的大气被高原"抽吸"上升,并在对流层上部向外"排放"(图8a、b)。而冬季,暖中心主要位于热带海洋上,高原主体及南坡上空为显著的冷却作用,这使得青藏高原上空的大气在冬季下沉并向高原低空四周"排放"(图8c、d)。由于巨大的青藏高原位于欧亚大陆中东部(70°~110°E)的副热带地区,这种周而复始的抽吸-排放作用和其所致的大范围的大气上升-下沉犹如一部巨型"气泵"屹立在欧亚大陆中、东部的副热带地区上空,有效地调控了亚洲大气环流和气候季节变化,显著地影响着亚洲冬季风和夏季风的形成与变化。

数值试验结果还表明,巨大的青藏高原气泵是由其地表感热加热所驱动的,即青藏高原"感热驱动气泵"(吴国雄 等,1997a;Wu et al.,2012a)。此外,TP-SHAP 在夏季使高原成为大气的重要负涡度源,它维持着高空的南亚高压,还通过激发 Rossby 波影响着北半球的大气环流和气候(吴国雄 等,1997a,2000a,2002,2004,2016a,2016b,2018;吴国雄,2004;刘新 等,2001;Duan et al.,2005;Wu et al,2006,2007,2016;王同美 等,2009)。进一步的研究表明,这种热抽吸现象也存在于伊朗高原,伊朗高原的感热气泵效应大大减少了北非的降水量,增加了阿拉伯海和印度次大陆北部的降水量,有效地促进了南亚夏季风的发展。青藏高原和伊朗高原(TIP)的这种抽吸效应构成了青藏-伊朗高原感热气泵(TIP-SHAP),不仅增强了东亚季风,还将南亚雨带向北移动,并有助于印度季风北支的形成(Wu et al.,2012)。因此,将青藏高原和伊朗高原视为一个统一体并研究它们对亚洲夏季风(Asian summer monsoon,简称 ASM)的联合影响更符合实况。

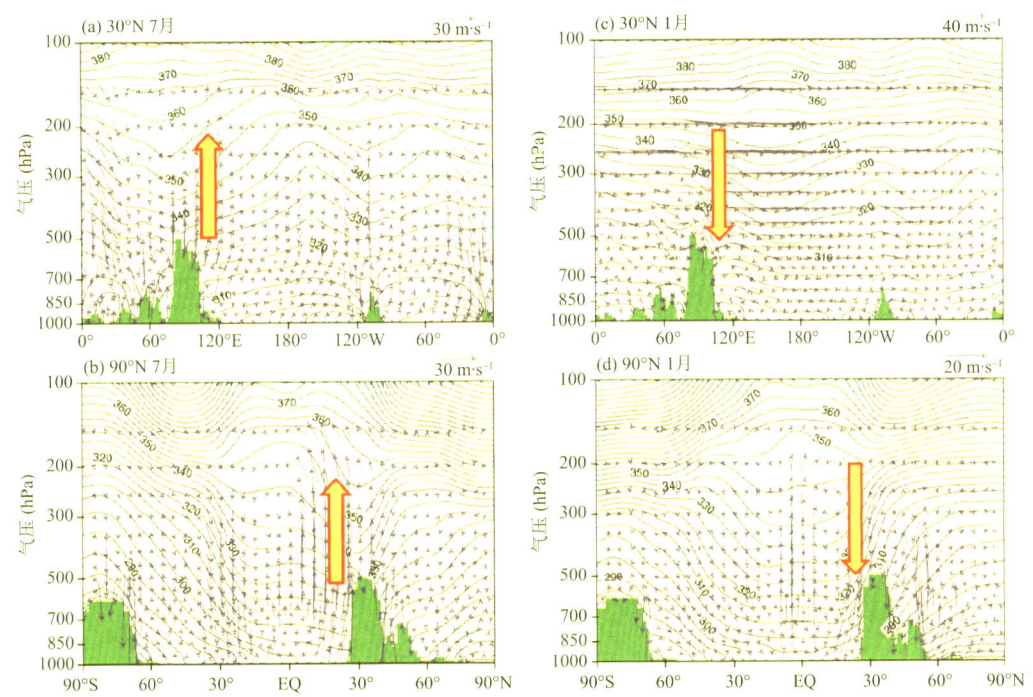

图8 1986—1995 年(a,b)7月和(c,d)1月平均的位温(等值线,单位:K)和垂直环流(箭头,单位:m·s⁻¹)
沿 30°N(a,c)、90°E(b,d)的剖面分布(摘自吴国雄 等,2018)

4.1 热力适应与感热气泵的形成

为讨论大气环流对非绝热加热的适应问题,吴国雄等(2000a)提出了热力适应理论,采用 Ertel 位涡(Ertel,1942;Hoskins,1991)及位涡方程来描述加热场和环流场涡度的变化关系。

Ertel 位涡的定义是:$P = \alpha \zeta_a \cdot \nabla \theta$

式中,$\alpha = 1/\rho$ 是比容,ρ 是流体质块的密度,ζ_a 是三维绝对涡度矢量,θ 是位温。位涡是以比容为权重的三维绝对涡度与位温梯度的标量积。由 Ertel 位涡的定义可知,用这样一个物理量可以综合描述流体质块的动力和热力两方面特征。令 $W=\rho P$，位涡方程可表示为：

$$\underbrace{\frac{\partial W}{\partial t}}_{(a)} + \underbrace{\nabla_h \cdot \mathbf{V}W}_{(b)} + \underbrace{\frac{\partial}{\partial z}wW}_{(c)} = \underbrace{\boldsymbol{\zeta}_a \cdot \nabla\dot{\theta}}_{(d)} + \underbrace{\mathbf{F}_\zeta \cdot \nabla\theta}_{(e)}$$

在大气下层（边界层）中，根据大气大尺度运动中各物理量的特征尺度进行尺度分析（吴国雄 等，1999）。可以得到上述方程中各项的特征尺度为：(b) 10^{-11}；(c) 10^{-12}；(d) 10^{-11}；(e) 10^{-11}。从尺度分析的结果可以看出，总位涡的水平通量散度(b)要比其垂直通量散度(c)约大一个量级，也就是说非绝热加热(d)和摩擦项制造的位涡(e)将主要通过水平位涡通量辐散(b)来平衡。

为了说明青藏高原加热引起的大气环流变化的适应过程机理。我们给出了以下示意图：

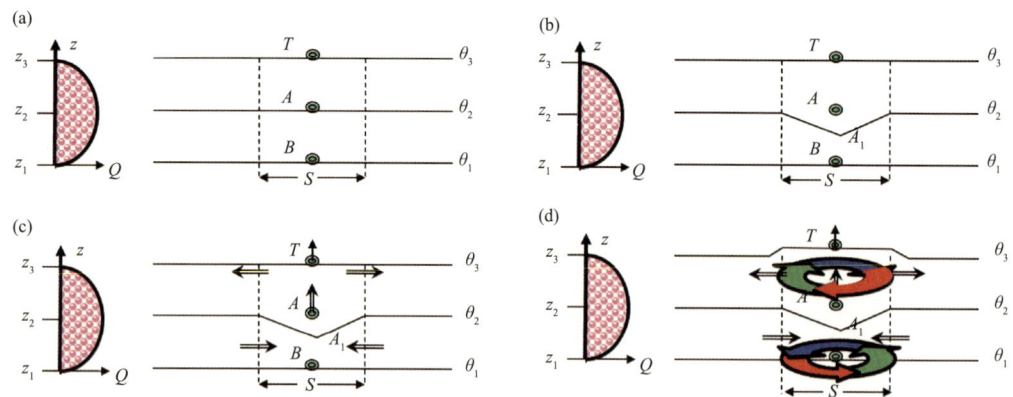

图 9 图中大气环流（箭矢）对外部加热 Q 的热力适应的发展过程示意图，图中左侧表示区域 S 中 Q 随高度（等熵面）的分布，T、A 和 B 表示起始于 θ_3、θ_2 和 θ_1 上的质点。(a)起始态；(b)热力效应使气柱位温升高；(c)动力效应使下层空气辐合，上层辐散；(d)热力适应使下层出现气旋环流，高层出现反气旋环流；顶部出现冷中心"过流"（改自吴国雄和刘屹岷，2000a，图 1）

由示意图可以看出大气边界层中（图 9a；$z_1 \sim z_2$ 层）有 $\frac{\partial \dot{\theta}}{\partial z}>0$，因而有 $\boldsymbol{\zeta}_a \cdot \nabla\dot{\theta} > 0$。当时间尺度较短时，有 $\left(\frac{\partial W}{\partial t}>0\right)$，也就是局地位涡将增加，使低层出现气旋式环流和上升运动。当时间尺度很大时，有 $\left(\frac{\partial W}{\partial t}\to 0\right)$，且下层辐合使(b)项小于 0，不能平衡加热的位涡制造。虽然(c)项为正，但量级太小，也不足以平衡加热制造的位涡。因此必然有(d)与(e)的平衡，也就是 $\mathbf{F}_\zeta \cdot \nabla\theta \approx -\boldsymbol{\zeta}_a \cdot \nabla\dot{\theta}$。由此可知边界层中加热制造的正位涡主要被摩擦耗散所平衡。

自由大气（图 9b；$z_2 \sim z_3$ 层）中有 $\frac{\partial \dot{\theta}}{\partial z}<0$，因而有 $\boldsymbol{\zeta}_a \cdot \nabla\dot{\theta}<0$。即有负涡度制造和形成反气旋式环流。而且在自由大气中摩擦力较小，(e)可以忽略。对较长时间尺度，并略去局地变化项，则根据位涡方程可以得到在自由大气中加热制造的负位涡(d)主要被水平位涡通量辐散项(b)项平衡。因此加热在大气上层制造的负位涡必然要通过向四周的散逸来维持平衡。

对于加热区上空整个气柱而言，摩擦施加于气柱的负位涡除了被气柱中正位涡的制造所抵消外，所余的负位涡在高层沿侧边界向外输送。

在气柱的上部 θ_3 层（图 9c）上非绝热加热为零，但由于热力适应的结果，在 θ_3 层及其以上层次中仍然存在上升运动及反气旋环流，也就是吴国雄等(2000a)所定义的"过流"（overshooting）。可以证明"过流"是非绝热加热的必然伴随现象。

过流是大气热力适应中的重要过程，它的重要性在于：首先，过流增加了 z_2 层以上反气旋涡度向外散逸($\iiint \nabla \cdot V W drdydz < 0$)的垂直厚度，更多地补偿了摩擦制造的负涡度。其次，过流的绝热冷却作用使 θ_3 面上凸(图 9d)，于是使 $z_1 \sim z_3$ 中 θ_e 减小。

夏季青藏高原地区上空由于高原的加热作用，在 150 hPa 以下 θ 面明显下凹，反映出高原上空大气的位温高于同纬度其他地区。相反，在 150 hPa 以上高空 θ 面却明显的向上凸起，表现出高原大气上层位温小于同纬度其他地区。并且可以看到青藏高原上空在 150 hPa 高度，尽管加热已经很小但仍然有上升运动存在。青藏高原上空大气的热力和动力特征，是热力适应和过流理论的典型例证。高原大气加热特征对邻近地区大气环流及气候具有十分重要的意义。

从位涡的观点来看，整层气柱中静力稳定度的降低使气柱中正压位涡减少，于是沿边界的对称不稳定迅速地出现，从而加速了高低层位涡的平衡。另外，加热区大气在下层辐合、上升，而在上层辐散并引起周边地区的大气上层辐合、下沉以及下层辐散。通过这种"补偿效应"，加热区的热力作用影响着其周边地区的大气运动(吴国雄 等，2018)。

基于热力适应理论，由副热带非绝热加热引起的垂直运动的特征环流形势如图 10。大气环流对热源强迫的响应在对流层低层产生气旋性环流，而在高层产生反气旋环流，与此同时，加热区的东侧伴随上升运动，而其西侧产生下沉运动(图 10a)。夏季青藏-伊朗高原大地形能够产生强烈的抬升加热，因此能够在其上空观察到这种垂直运动模态(图 10b)。

由于青藏高原位于欧亚大陆东部，高原引起的上升(下沉)与大陆尺度的热力强迫所产生的上升(下沉)运动同步，因此使得东亚夏季风成为全球最强的季风系统之一，而中亚干旱成为全球最显著的干旱区之一(Wu et al.，2003；Duan et al.，2005)。此外，由于青藏高原是大气运动重要的负涡度源，高原北侧处在西风带中，这一负涡度源能够激发出 Rossby 波列，通过能量频散影响着亚洲乃至北半球的大气环流异常(刘新 等，2002；吴国雄 等，2004，2005；吴国雄，2004；Duan et al.，2005)。关于高原热力强迫对亚洲气候格局的具体影响将在本文第 5 节进行详细的回顾。

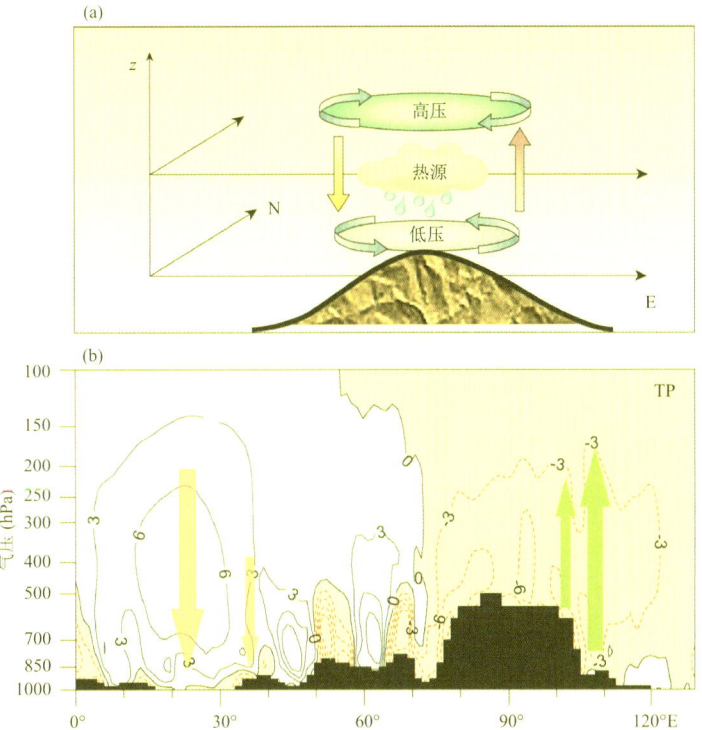

图 10 (a)夏季副热带地区的大气对外部加热响应的示意图。热源东(西)侧空气上升(下降)，近地表(对流层高层)产生异常气旋(反气旋)性环流；(b)基于 ERA-interim 再分析资料得到的 1979—2010 年 7 月平均沿 32.5°N 的垂直速度的经度-气压截面(单位：hPa·s^{-1})(摘自 Wu et al.，2015)

4.2 角动量守恒与大范围上升运动

青藏高原感热气泵作用的特殊性除了体现在高耸的青藏高原表面抬升加热影响外,还与其所处的地理位置和范围有关。根据平均经圈环流理论,当热源位于赤道附近时可激发出典型的 Hadley 经圈环流,即在热带上升、副热带下沉。而青藏高原感热加热位于远离赤道的副热带区,这样的热源地理位置对经圈环流和垂直运动会产生什么影响?根据 Plumb 和 Hou(1992)对热源离开赤道时的大气环流响应所做的理论分析,如果热源强度达到一定强度,会激发出角动量守恒型(AMC)环流。如果热源强度不够大,产生的负涡度不足以影响行星涡度时,就不会有 AMC 型经圈环流响应,而是相对强度较弱的热力平衡型环流响应。在热带和副热带区由于绝对涡度通常较小,相对容易激发出 AMC 型经圈环流。因此对青藏高原的感热加热,在青藏高原上空会使得绝对涡度减少,从而会激发出 AMC 型环流响应,即在高原上空形成大范围的上升运动。

吴国雄等(2016b)通过数值模拟研究了夏季青藏高原上高于 2 km 主体上的感热加热特征,并分析了高原感热对局地季风经圈环流和垂直运动的影响,结果表明高原主体感热加热在近对流层顶处(接近 100~150 hPa)出现了最小位涡强迫中心,从而有利于激发出 AMC 型环流,并在整个青藏高原上空及其南侧形成强烈上升运动。这与 Plumb 和 Hou(1992)的理论分析相吻合。

因此,青藏高原感热加热通过激发高原上空大范围的上升运动,对高原周边大气形成了显著的抽吸效应。另外,Plumb 和 Hou(1992)中 AMC 型经圈环流形成的判据还与热源的宽度或范围有关:在相同热源强度下,热源范围越小,越容易激发出 AMC 型经圈环流,也即与热源强度的经向梯度有关。青藏高原陡峭地形的特点使得感热气泵的侧边界效应更加显著(参见 Wu et al., 2007)。该效应为亚洲季风首先在孟加拉湾地区爆发提供了有利的环流条件。

对于同样位于副热带的伊朗高原感热影响,刘屹岷等(2017)基于 WRF 模式进行了有无感热的模拟试验,结果显示,伊朗高原的感热也能激发到达对流层顶的上升运动,也可形成类似 SHAP 的抽吸效应。但上升强度和范围都不如青藏高原感热的响应,因此这也表明青藏高原独特地理位置和分布特点,使得感热加热效应具有其特殊性。

青藏高原感热加热通过激发 AMC 型经圈环流,形成抽吸效应,加大了地面风速,这样进一步加大了地面感热强度,使得感热会出现大范围突然增长。对观测资料的分析表明(Wu et al., 1998),尽管从晚冬开始,在伊朗、阿富汗等区域出现大于 100 W·m^{-2} 的感热加热,但一直局限在局地范围内,直到 4 月青藏高原上出现大于 100 W·m^{-2} 的感热加热后,到 5 月初,强大的感热加热区快速扩大到了从地中海到高原的广大地区。这样的发展过程对感热的影响形成了正反馈作用,SHAP 也因而得以维持。

5 青藏高原热源对亚洲气候格局的影响

5.1 高原热源对大尺度、局地气候的影响

(1)高原热源对亚洲大尺度环流的影响

根据热力适应理论,青藏高原和伊朗高原上空的非绝热加热会激发出一个近地层的浅薄低压和中高空深厚的南亚高压,使得高原及其东部产生上升运动,西部下沉运动。图 10 表示 7 月平均沿 32.5°N 的亚洲大陆经向风分量和垂直速度的气压-经度剖面。从该图可看到无论是伊朗高原上空还是青藏高原上空都有明显的表层浅薄的气旋式环流及高空深厚的反气旋式环流。两个反气旋中心分别位于 60°E 和 90°E 处,对应着 2.2 节提到的南亚高压双模态。Wu 等(2003)指出,由于旋转地球上的海陆分布,夏季副热带大陆及其邻近海域上空大气的主要加热呈现为 LOSECOD 四叶型加热(图 11):大陆西部以西的洋面上空以长波辐射冷却(LO-)为主;大陆西部上空以表面感热加热(SE-)为主;大陆东部上空以深对流凝结潜热加热(CO-)为主;而大陆东部以东的洋面上空则存在双主加热(D-),即长波辐射冷却和深对流凝结潜热加

热。根据热力适应,该 LOSECOD 四叶型加热所激发出的大陆尺度环流在低层为气旋式,在高层为反气旋式。因此,大陆东岸有强烈上升的潮湿气流,而大陆西部洋面对着为强烈下沉的干冷气流。这导致沿副热带的高压呈非轴对称的形态,使得副热带大陆东部潮湿多雨,西部干旱少雨(Wu et al.,2003,2009;Liu et al.,2004;吴国雄 等,2004,2005)。

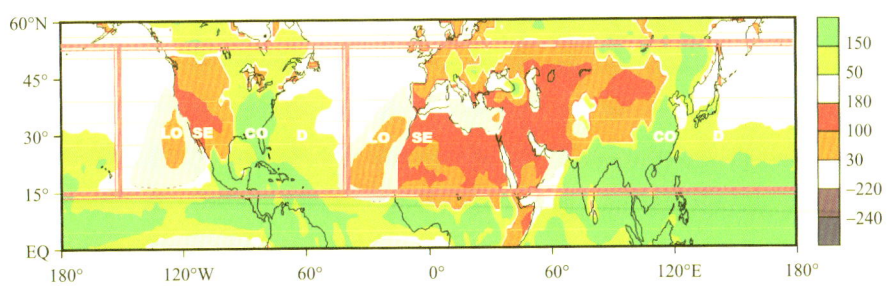

图 11　夏季沿着副热带的非绝热加热,即为 5 个四叶型加热型的拼图(改自吴国雄 等,2005)

在图 12a 中可明显看到,在 110°E 以东及 30°E 以西的大陆东、西两侧对流层中高层和对流层低层的风向相反:30°E 以西低空盛行北风,高空盛行南风,而 110°E 以东情况正好相反,低空南风,高空北风。也就是说夏季亚洲副热带地区的环流可以简单地看成是伊朗高原和青藏高原加热所激发的热力环流嵌套在亚洲大陆尺度的热力环流上而形成的。这两种尺度的热力环流都是以低空的气旋式环流和高空的反气旋式环流为主要特征。低空北风和高空南风最大风速中心出现在 20°E 附近亚洲大陆西边界,次级中心出现在 60°E 附近高原西侧。另一方面,低空偏南风和高空偏北风中心则出现在 115°E 附近的大陆东边界,次级中心出现在 100°E 附近高原东侧(图 12a)。由于夏季沿该副热带涡度平流项相对为小项可以略去,定常的涡度收支方程于是可简化为:

$$\beta v + (f + \zeta)\nabla \cdot \mathbf{V} \approx 0$$

因此热力环流的东侧低空 $\nabla \cdot \mathbf{V} < 0$,为气流辐合区;高空 $\nabla \cdot \mathbf{V} > 0$,为气流辐散区。低层抽吸和高层辐散效应导致这里上升运动强烈(图 12b),偏南暖湿气流在这里辐合上升,伴有较多的降水。同理,热力环流西侧低空为偏北干冷气流与辐散,高空为偏南风和辐合,为下沉运动区,造成这里气温高,降水少,气候干燥。因此图 12b 中的上升运动分布与图 12a 中的环流分布有很好的对应关系:对应着扎格罗斯山脉(50°E),苏莱曼山脉(70°E)及青藏高原各有三组下沉-上升运动,且最强的中心均在近地层;而 30°E 以西及 110°E 以东的广大地区则分别有大范围的下沉和上升运动,且中心均在自由大气中。这种大陆尺度的上升/下沉运动与地形强迫的上升/下沉运动因同相叠加而变得很强。因此使得北非和中亚成为全球的显

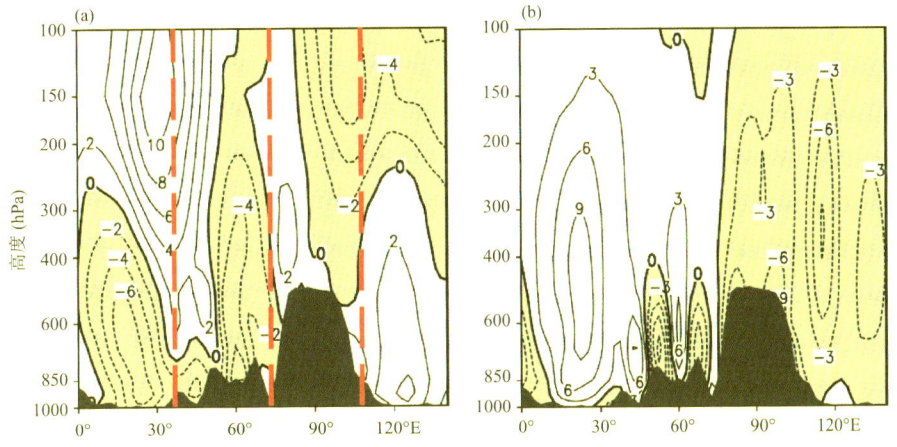

图 12　气候平均的 7 月沿 32.5°N 的经向风剖面(a)和垂直速度剖面(b)。等值线间隔分别为 2 m·s^{-1} 和 2×10^{-2} Pa·s^{-1},图中阴影部分代表地形(摘自 Duan et al.,2005)

著干旱区之一,而东亚夏季风成为全球最激烈的季风系统之一。

青藏高原对气候的影响与落基山脉和安第斯山的情况既有共同之处,但也存在差异。作为大气的抬升热源,每一座山脉都激发出浅薄的低层气旋式环流和深厚的中、高层反气旋式环流。因此,落基山脉和安第斯山也可以激发出其上空及其东部的上升运动和西部的下沉运动,但落基山脉和安第斯山均位于大陆的西部。其激发的上升运动与大陆东部大陆尺度的上升运动并不连成一片,中间为下沉运动所分隔。因此,北美和南美的季风气候并没有亚洲季风那么明显。相反,落基山和安第斯山所激发的下沉运动加强了其西部的大陆尺度的下沉运动,使副热带东太平洋地区成为全球显著的下沉运动区。冷洋面上强烈的下沉增温形成稳定的层结和层状低云,使副热带太平洋东部成为强大的长波辐射冷却中心(吴国雄 等,2005;Duan et al.,2005)。

(2)高原热源对局地气候的调节作用

青藏高原地形复杂,其激发的大气波动对局地气候型起着调节作用。图13给出了7月平均亚洲地区500 hPa的流场和垂直速度场、沿32.5°N的降水纬向分布、位温,以及由垂直速度表示的大气定常波纬向分布。500 hPa上亚洲大陆副热带地区自西向东共有4个上升运动中心(图13a),分别位于高原西南部85°E附近、高原东南部100°E附近、中国东部115°E附近以及日本南部130°E附近,每个上升中心相距约15°,其位置对应于降水峰值区域(图13b)和波动的波峰(图13c)。段安民(2003)证明地表加热激发出的波动在垂直方向是减幅的,而深对流激发的波动最大振幅在自由大气中。从图4和图13c可以看出,伊朗高原和青藏高原地表有很强的感热加热,其所激发的扰动的振幅随高度迅速减小。而在东亚地区,最大加热在对流层中上层,波动的最大振幅也在对流层上层。可见,夏季与青藏高原相联系的夏季加热分布对高原及其下游地区的局地气候分布型也有重要的调节作用。

5.2 高原热源影响亚洲夏季风的形成

亚洲夏季风是全球最强和最为复杂的季风,它不仅与亚洲和太平洋、印度洋构成的大尺度海陆分布格局有关,也与亚洲南部次大陆尺度的海陆交错分布以及青藏高原大地形的存在有密切联系。吴国雄等(2005)利用GOALS气候模式系统来模拟青藏高原对亚洲气候格局的影响,试验表明,亚洲夏季风是海陆分布及青藏高原热力强迫共同作用的结果,且在亚洲季风形成中海陆分布的作用是第一位的,高原的作用是第二位的,若没有青藏高原的存在,就不会形成"现代"的亚洲季风。夏季青藏高原及其抬升加热对大气环流的影响加强了欧亚大陆尺度的加热对大气环流的影响,在中亚干旱和东亚季风中起着放大器的作用。

Wu等(2007,2012a)进一步利用水球试验探究了青藏高原热力强迫影响亚洲夏季风的物理过程。本文仅总结了TIP-SHAP影响亚洲夏季风的结构(图14),关于青藏高原对亚洲夏季风的关键调制作用将在本章另文进行详细论述。首先,TIP-SHAP对周边大气具有局地抬升作用,使得周围大量空气向高原辐合。此外,TIP-SHAP引起的低层巨大的气旋式环流能够将大量水汽从印度洋输送到南亚和东亚地区。其中,水汽输送带可分为三大支路:热带第一个分支横越阿拉伯海、孟加拉湾及中国南海,引起强降水,形成南亚夏季风的南部分支,这主要是由于海陆热力对比所致;由于受到TIP-SHAP的作用,第二支路向北穿过印度北部和孟加拉湾向高原的南坡方向流动,形成南亚夏季风的北支路;第三支输送水汽维持东亚季风,其同时受到海陆热力对比和青藏高原的热力强迫。同时,TIP-SHAP通过改变高原上空温度和环流结构,有利于激发副热带季风型经圈环流,从而在副热带地区形成大范围上升运动。因此由海陆热力差异及TIP-SHAP所产生的上述三个作用促成了亚洲夏季风的最终形成。

5.3 高原热源年际变率对东亚夏季风的影响

(1)高原热源对东亚夏季风的直接影响

Luo等(1983,1984)指出1979年夏季风爆发期间,高原的加热作用对高原及其周边地区低空环流演变、大尺度降水、水汽和垂直运动场均有较大影响。此后,罗会邦和陈蓉(1995)研究表明当高原夏季热源

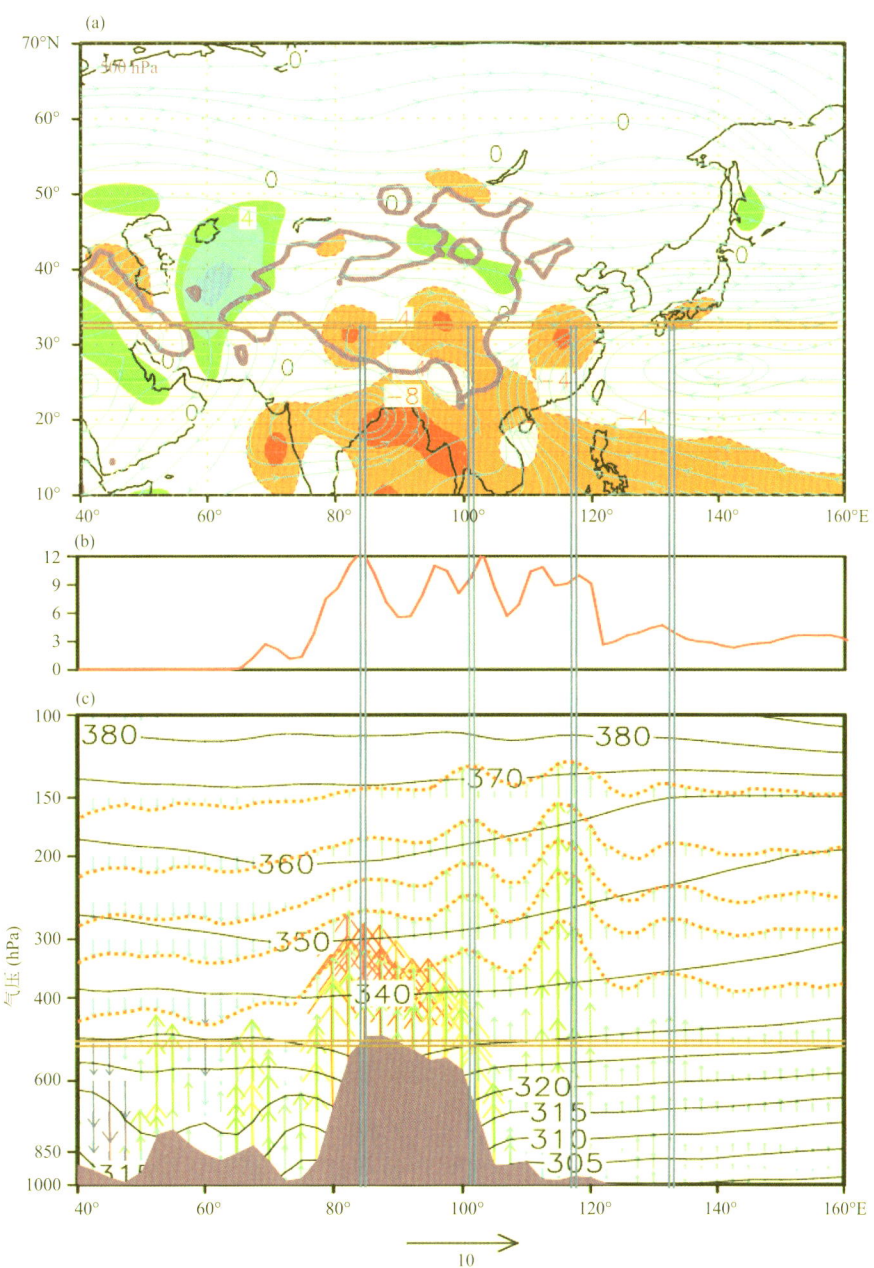

图 13 7 月气候平均的(a)500 hPa 流场和垂直运动场,(b) 沿 32.5°N 的降水(纵坐标,mm·d^{-1}),以及(c) 位温(K,实线)和由垂直速度(-100 Pa·s^{-1},箭矢)表示的大气定常波的纬向分布,阴影部分表示地形(摘自吴国雄等,2004)

增强时,长江上游和淮河流域降水增多,而华南地区降水减少。为了进一步研究高原热源对东亚夏季风(East Asian summer monsoon,简称 EASM)的预报意义,赵平等(2001)发现春季高原热源与夏季长江流域降水存在明显正相关,并认为高原春季 4 月的热源(感热为主)对于随后的夏季江淮地区、华南及华北地区的降水均有一定的指示意义。此后的大量研究也得到了类似的结论,春季高原热源(感热为主)对于随后的夏季江淮地区的降水均有一定的预报意义(段安民,2003;Duan et al.,2005,2020;Wang et al.,2014)。由于季风爆发前感热加热是高原大气热源的主要成分,并持续至夏季,进而造成夏季中国东部江淮流域低层辐合和高层辐散,强烈的水汽辐合和旺盛的对流活动造成局地降水增加。

Wang 等(2014)利用观测资料及 WRF 模式进行个例分析,进一步探讨了春季高原热源影响夏季江淮主雨带降水的物理过程。研究表明,春季高原非绝热加热及局地环流之间的正反馈使得春季高原感热异

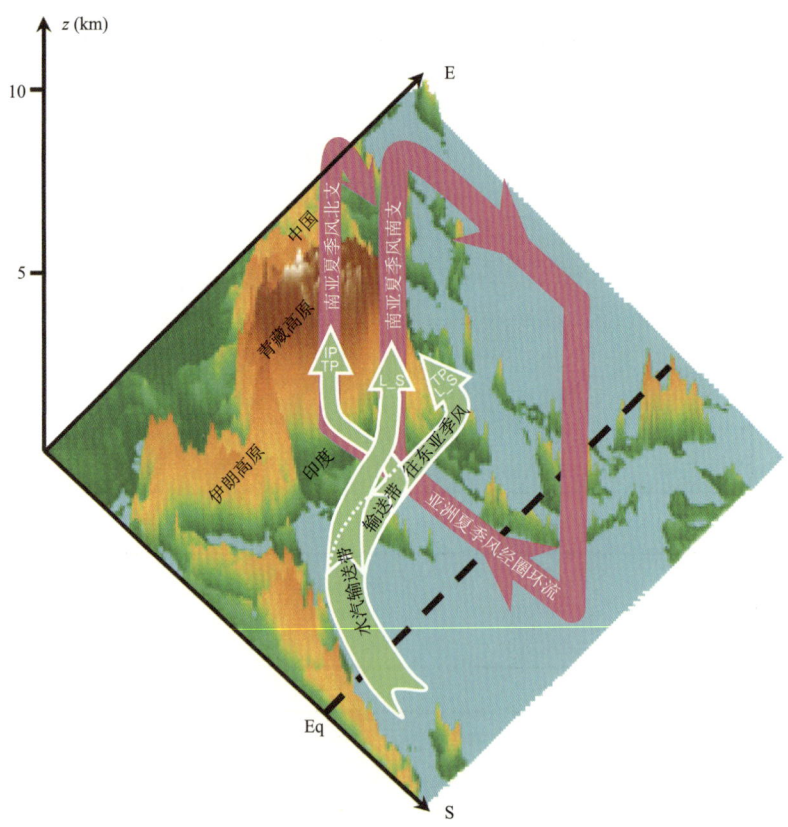

图 14 青藏高原加热影响亚洲夏季风形成的整体结构示意图。对于南支季风,受到热带地区海陆热力对比的影响,沿输送带的水汽主要被抬升;北支的水汽从传送带向山麓和坡地抬升并产生强降水,其主要受 TIP-SHAP 控制;其余水汽向东北方向输送以维持东亚夏季风的形成,并同时受到陆海热对比和高原热强迫的控制(摘自 Wu et al.,2015)

常持续至夏季,进而引发大尺度波通量的东传使得异常南北气流在江淮流域交汇并逐渐形成一条显著的水汽辐合带;春季高原感热偏强还会造成向东输送的暖平流增强,该异常的暖平流会促使大气的上升运动加强,从而造成淮河流域降水增多;此外,春季高原感热偏强造成夏季高原地区更频繁的低涡活动频发,使得下游淮河流域对流天气的发生发展,从而造成江淮流域的总降水量以及极端降水的频次均偏多。即春季高原感热增强持续至夏季,可通过定常波的传播、暖平流及高原低涡的向东输送等方式造成江淮流域降雨增多。

然而,此研究的结论是基于个例分析得到的,以上影响过程在某些年份可能是起作用的,但并不见得大多数年份都是如此。因此 Sun 等(2019)基于观测资料及 CESM1.2.0 的海气耦合模式对于春季高原热源如何影响夏季江淮降雨的这个问题重新进行了探讨。研究表明,春季高原热源正异常通过产生 Rossby 波列并向下游传播,在西北太平洋上空形成一个相当正压的异常反气旋,从而影响西北太平洋的海表及次表层海温异常,即春季高原热源的信号存储在了西北太平洋中(图15),进而影响东亚夏季风雨带位置和强度。

(2)高原热源与海洋对东亚夏季风的协同影响

虽然青藏高原及其热状况被认为是影响区域及全球大气和海洋环流异常的强迫源,但青藏高原强迫本身也应该被认为是大气和海洋环流的结果。这是因为青藏高原的热源和机械强迫是由风、温度、湿度和地面径流等各种大气变量决定的。揭示青藏高原强迫的表现形式有助于阐明高原强迫在海-陆-气相互作用框架中的作用。近几十年来,青藏高原与海洋的相互作用及其对亚洲季风的协同作用日益受到人们的关注。

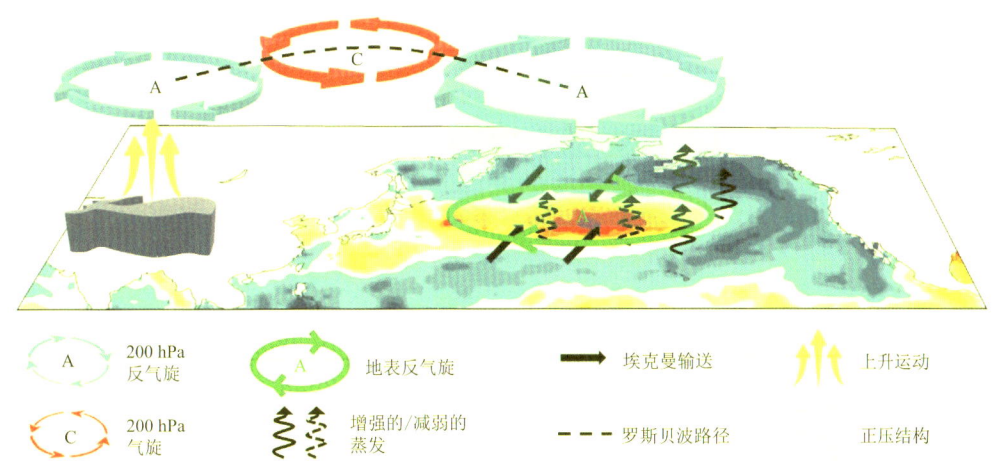

图 15　春季高原感热影响太平洋海表温度异常的机制图（摘自 Liu Y M et al.，2020）

　　Hu 等（2015）利用数据诊断及耦合的数值模式探讨了青藏高原夏季热源及印度洋海温一致模态（Indian Ocean basin mode，简称 IOBM）对东亚夏季风年际变化的相对重要性。结果表明，IOBM 主要影响东亚夏季低层环流。而对于东亚夏季降水而言，高原热力强迫比印度洋海温更重要。其原因是高原地处东亚季风区上游的副热带地区，夏季偏强的高原大气热源能在东亚北部激发出相当正压结构的气旋性异常环流，从而带来北风异常；另一方面，高原热源偏强还加强华南的低空偏南风急流，南、北气流在东亚季风区主雨带位置附近辐合，形成大范围的降水正异常。总之，IOBM 正位相和高原热源增强都有利于东亚夏季风降水增加。在此基础上，刘森峰等（2017）发展了同时考虑春季高原地面热源和印度洋海温预测 EASM 年际变率的方法，这一方法预测的准确性高于以往单因子的预报方法。

　　众所周知，在 ENSO 发展年与衰减年，EASM 会产生不同的响应。EASM 主雨带降水在衰减年夏季显著偏强，而在发展年夏季却相反（Liu and Duan，2017）。Liu S F 等（2020）认为，在夏季全球海表温度异常（sea surface temperature anomaly，简称 SSTA）准四年振荡的不同位相下，高原热力作用对 SSTA 产生的不同响应能够进一步对东亚夏季风的主雨带降水产生反馈，从而造成 EASM 主雨带的不对称响应（图 16）。发展年夏季高原偏弱的热源响应所产生的环流响应减弱了 EASM 气候态的季风环流系统，并造成主雨带降水减少；而在衰减年夏季，高原热源及 ENSO 对 EASM 主雨带的影响是一致的，同时加强了西北太平洋反气旋，从而造成主雨带降水增加。

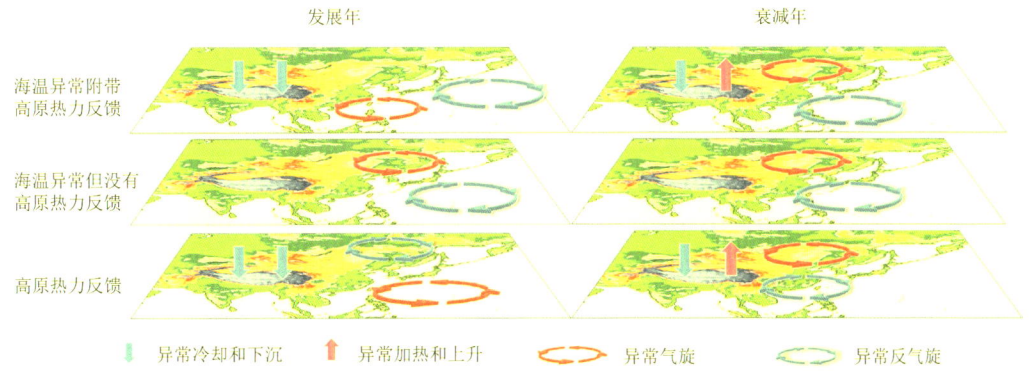

图 16　低层环流在有/无青藏高原热力反馈效应时对全球海表温度异常的响应示意图（摘自 Liu S F et al.，2020）

5.4 高原热源年代际变率对东亚夏季风的影响

很多研究表明,20世纪下半叶人为温室气体排放导致了高原上空的气候变暖,且其影响程度可能强于全球的其他区域(Duan et al.,2006)。与此相反,自20世纪80年代到20世纪末,高原感热及高原上空总热源却呈现出明显的减弱趋势,以春季最为明显。春季高原感热(Tibetan Plateau sensible heating,简称 TPSH)减弱主要是由于全球变暖背景下,不同纬度间增暖幅度不同所导致。欧亚大陆东部中高纬度增暖幅度远强于副热带增暖幅度,其导致整个副热带的经向温度和气压梯度减弱,从而减弱东亚西风急流及地表西风,造成20世纪80年代之后 TPSH 出现显著的减弱趋势(Duan et al.,2008,2009,2011,2013;Yang et al.,2009;Wang et al.,2012;Zhu et al.,2012)。

TPSH 的这种年代际变率是如何影响季风的呢?资料诊断及数值模拟都表明减弱的 TPSH 与亚洲季风之间存在正相关关系。TPSH 减弱导致亚洲季风环流减弱,其主要表现为对流层低层的异常反气旋、对流层高层的异常气旋以及减弱的西太平洋副热带高压。对流层低层异常的环流导致华南地区的偏南风减弱、华南地区异常水汽辐合而华北地区异常水汽辐散,从而造成东亚季风区呈现南涝北旱的现象(图17)(Duan et al.,2011,2013;Liu et al.,2012)。

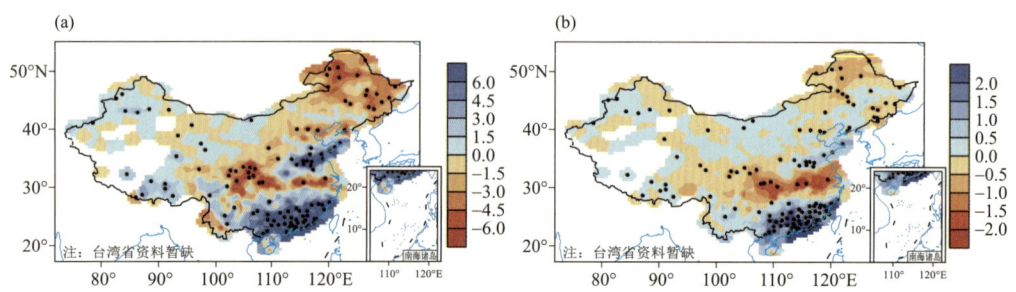

图17 1980—2008 年夏季(a)中国降水变化趋势(mm/a)及(b)中国降水对春季高原感热第一主成分的偏回归(移除高原积雪的影响)。打点表示超过90%置信水平(摘自 Duan et al.,2013)

6 总结和展望

6.1 总结

青藏高原(以下简称高原)是除了南极和北极以外的"世界第三极",它的机械强迫和独特的热力效应对区域甚至全球气候都具有重要的影响。吴国雄院士团队自20世纪90年代以来一系列开创性的工作将青藏高原热力强迫的研究推向了一个新的高度。

高原大地形对大气的动力和热力强迫作用是同时存在的,只是不同背景环流下二者相对重要性可能会不一样。冬季,青藏高原的西风很强,其对大气环流的影响以机械强迫为主。进入高原的西风急流形成以高原为轴,南侧气旋、北侧反气旋的非对称的"偶极子"(TP-dipole mode)偏差环流,其导致了华南早春连绵阴雨(PRES)的形成。夏季,青藏高原的西风很弱,其对大气环流的影响则以热力强迫为主。它激发的东亚地区巨大的气旋性环流,能够将高原周边的大气从低空向上进行周而复始的抽吸上升,犹如一部巨型"气泵"屹立在欧亚大陆中、东部的副热带地区上空,并主要受到高原地表感热的驱动。因此,青藏高原被定义为"感热驱动气泵(TP-SHAP)"。TP-SHAP 是导致亚洲大气环流从冬到夏演变中出现季节突变的重要原因,也是亚洲夏季风爆发和维持的重要因素。

春、夏季(4月—9月)青藏高原为大气热源,其中春季高原地表感热为大气热源的主要分量并在5月达到峰值,夏季凝结潜热占据高原热源的主要成分并在7月出现峰值,10月高原由热源转为冷源,冬季成为冷源。夏季副热带亚洲大陆西部为大气热汇,东部为大气热源,最强的大气热源出现在高原上空。感热加热是夏季高原上空近地层非绝热加热的基本形式,潜热加热最大值出现在对流层低层,占高原热源的主要地位。7月高原异常加热的局地性非常显著,其所对应的东亚大气环流及降水异常也明显不同。冬季青藏高原呈"弱冷源"特征,这与高原西侧至东南角是相对较大的潜热释放加热有关;此外,冬季高原非绝热加热异常的大值区也位于其西侧至东南侧,并与整个北半球的异常环流模态密切相关。

青藏高原热力强迫影响大气环流的主要途径为"热力适应"。大气环流对高原加热的响应呈现出低层气旋、高层反气旋的结构,并在加热区东侧伴随上升运动,西侧为下沉运动。高原引起的上升(下沉)与大陆尺度LOSECOD四叶型热力强迫所产生的上升(下沉)运动同步,因此使得东亚夏季风成为全球最强的季风系统之一,而中亚干旱成为全球最显著的干旱区之一。此外,由于青藏高原是大气运动重要的负涡度源,高原北侧处在西风带中,这一负涡度源能够激发出Rossby波列,通过能量频散影响着亚洲乃至北半球的大气环流异常。

青藏高原热源除了在调节亚洲夏季风的形成中扮演重要角色,其也能够在年际和年代际时间尺度上对亚洲夏季风产生影响。在年际尺度上,春季高原感热正异常往往会导致夏季江淮流域降雨增多;此外,高原热源及SSTA之间存在相互作用,并能够协同影响东亚夏季风雨带。在全球变暖的背景下,高原感热在20世纪80年代—20世纪末出现了年代际的减弱趋势,其使得东亚季风区呈现南涝北旱的现象。

6.2 展望

本文主要回顾了高原大气热源的气候效应,而目前我们面临的挑战是这种高原热源是如何形成的,其主要的驱动因子是什么? 高原热源与大气环流之间两者互为因果,这使得对高原热源变异机理的探究难度变得更大。关于高原热状况年代际(Zhou et al.,2019;Gao et al.,2020;Sun et al.,2020;Liu et al.,2021)和年际变率(Cui et al.,2014;Liu et al.,2015;Chen et al.,2017;Jiang et al.,2016;Wang et al.,2016,2018;Zhao et al.,2018;Liu S F et al.,2020;Hu et al.,2021;Zhang et al.,2021)的驱动因子,已有一些工作对其进行了探讨。段安民等(2022)近期对该科学问题进行了系统的综述回顾,并将不同月份的高原热源分成了四类,进一步探究了每一类高原热源异常的海气驱动因子。尽管高原热源变异机理已经取得一定的研究进展,然而目前影响高原热源的不同海气信号的定量贡献还值得进一步探讨。

基于以上大量研究成果,不可否认的是,青藏高原热力强迫在亚洲乃至全球气候的形成和变化都扮演着举足轻重的角色。然而青藏高原的动力和热力作用的相对重要性目前仍然存在很多分歧。Boos等(2010,2013)认为对南亚季风的形成而言,青藏高原的热力作用并不重要,重要的是高耸的喜马拉雅山对来自北方干冷空气的机械隔离作用,使得印度低空高能量地区得以与高空的南亚暖中心通过对流发展耦合起来,从而维持印度季风。事实上,Wu等(2012a)发现,夏季并不存在从青藏高原北部的副热带向南亚入侵的干冷空气;且他们的试验中喜马拉雅山脉(HIM)斜坡上的地表感热加热仍然存在,这并不能因此否认高原热力强迫的主导地位。如果在HIM试验的基础上把喜马拉雅山斜坡上的感热加热去除,尽管喜马拉雅山脉仍然存在,但此时南亚北部的季风降水消失,这与Wu等(2007)水球试验中斜坡没有加热导致局部的抽吸抬升作用消失的结果是一致的。以上分歧也因此引发了异常激烈的"季风之争"(Qiu,2013)。近期,Son等(2019)基于GFDL大气环流模式定量探讨了青藏高原动力及热力作用对东亚夏季风的贡献。研究表明山脉动力作用与东亚夏季风65%的降水相关,而抬升的热源和海陆热力差异却仅分别占了15%,山体拖曳也占据了少于5%的东亚季风降水。因此他们认为青藏高原的动力作用被确定为影响东亚夏季风的主要强迫源。他们认为高原上游西风激发的准静止正压Rossby波通过影响下游槽脊经向位置的变化,从而影响东亚夏季雨带的演变(Son et al.,2020)。值得注意的是,在他们的试验设计中,消除高原热源加热只是通过增加辐射通量,使其与感热、潜热抵消,从而使高原总热源为零。然而实际大气中,高原热源各分量之间始终存在复杂的相互反馈过程,这会导致高原热源的变化。因此,Son等人的这种对消除高原热力作用的方法是否合理还有待商榷。

最后,尽管已有不少学者分别从理论、观测和数值模拟等方面研究了高原机械和热力作用对大气环流和天气气候的影响,但不可否认的是,在自然界中,青藏-伊朗高原的地形及其地表感热的存在是不可分割的。高原大地形是一直存在并且短时间内并不发生变化,而大气环流却总是在变化中,因此,大地形的热力作用在调节气候系统的演变中扮演了至关重要的角色。今后需要从协同论的角度出发,在理论研究和资料分析中统一考虑高原机械和热力强迫的综合作用。

参考文献

段安民,2003.青藏高原热力和动力强迫对东亚气候格局的影响[D].北京:中国科学院大气物理研究所.
段安民,吴国雄,2003.7月青藏高原大气热源空间型及其与东亚大气环流和降水的相关研究[J].气象学报(4):447-456.
段安民,张萍,2022.青藏高原大气热源年际变率及其驱动因子[J].大气科学,46(2):455-472.
刘新,吴国雄,李伟平,2001.夏季青藏高原加热和大尺度流场的热力适应[J].自然科学进展,11(1):33-39.
刘新,吴国雄,刘屹岷,等,2002.青藏高原加热与亚洲环流季节变化和夏季风爆发[J].大气科学,6:781-793.
刘屹岷,王子谦,卓海峰,等,2017.夏季亚洲大地形双加热及近对流层顶位涡强迫的激发Ⅱ:伊朗高原-青藏高原感热加热[J].中国科学:地球科学,47(3):354-366.
罗会邦,陈蓉,1995.夏半年青藏高原东部大气热源异常对环流和降水的影响[J].气象科学,15(4):94-102.
刘森峰,段安民,2017.基于青藏高原春季感热异常信号的中国东部夏季降水统计预测模型[J].气象学报,75(6):903-916.
万日金,吴国雄,2006.江南春雨的气候成因机制研究[J].中国科学D辑,36(10):936-950.
万日金,吴国雄,2008a.江南春雨的时空分布[J].气象学报,66(3):310-319.
万日金,王同美,吴国雄,2008b.江南春雨和南海副热带高压的时间演变及其与东亚夏季风环流和降水的关系[J].气象学报,66(5):800-807.
王同美,吴国雄,万日金,2008.青藏高原的热力和动力作用对亚洲季风区环流的影响[J].高原气象,27(1):1-9.
王同美,吴国雄,宇婧婧,2009.春季青藏高原加热异常对亚洲热带环流和季风爆发的影响[J].热带气象学报,25(B12):92-102.
吴国雄,2004.我国青藏高原气候动力学研究的近期进展[J].第四纪研究,24(1):1-9.
吴国雄,李伟平,郭华,等,1997a.青藏高原感热气泵和亚洲夏季风[C]//叶笃正.赵九章纪念文集.北京:科学出版社:116-126.
吴国雄,张学洪,刘辉,等,1997b.LASG全球海洋-大气-陆面系统模式(GOALS/LASG)及其模拟研究[J].应用气象学报,S1:16-29.
吴国雄,刘屹岷,刘平,1999.空间非均匀加热对副热带高压形成和变异的影响——Ⅰ:尺度分析[J].气象学报,57(3):257-263.
吴国雄,刘屹岷,2000a.热力适应、过流、频散和副高 Ⅰ.热力适应和过流[J].大气科学,24(4):433-446.
吴国雄,刘平,刘屹岷,等,2000b.印度洋海温异常对西太副高的影响——大气中的两级热力适应[J].气象学报,58(5):513-522.
吴国雄,刘新,张琼,等,2002.青藏高原抬升加热气候效应研究的新进展[J].气候与环境研究,7(2):184-201.
吴国雄,毛江玉,段安民,等,2004.青藏高原影响亚洲夏季气候研究的最新进展[J].气象学报,62(5):528-540.
吴国雄,刘屹岷,刘新,等,2005.青藏高原加热如何影响亚洲夏季的气候格局[J].大气科学,24(1):1-10.
吴国雄,何编,刘屹岷,等,2016a.青藏高原和亚洲夏季风动力学研究的新进展[J].大气科学,40(1):22-32.
吴国雄,卓海峰,王子谦,等,2016b.夏季亚洲大地形双加热及近对流层顶位涡强迫的激发(Ⅰ):青藏高原主体加热[J].中国科学:地球科学,46(9):1209-1222.
吴国雄,刘屹岷,何编,等,2018.青藏高原感热气泵影响亚洲夏季风的机制[J].大气科学,42(3):488-504.
叶笃正,罗四维,朱抱真,1957.西藏高原及其附近的流场结构和对流层大气的热量平衡[J].气象学报,28(2):108-121.
叶笃正,高由禧,周明煜,等,1979.青藏高原气象学[M].北京:科学出版社.
宇婧婧,2009.冬季青藏高原热状况及其与北半球大气环流的联系[D].北京:中国科学院大气物理研究所.
宇婧婧,刘屹岷,吴国雄,2011a.冬季青藏高原上空热状况的分析——Ⅰ气候平均[J].气象学报,69(1):79-88.
宇婧婧,刘屹岷,吴国雄,2011b.冬季青藏高原上空热状况的分析——Ⅱ年际变化[J].气象学报,69(1):89-98.
赵平,陈隆勋,2001.35年来青藏高原大气热源气候特征及其与中国降水的关系[J].中国科学,31(4):327-332.
周晓平,顾震潮,1958.大地形对于高空行星波传播的影响[J].气象学报,29(2):99-103.
周秀骥,赵平,陈军明,等,2009.青藏高原热力作用对北半球气候影响的研究[J].中国科学:地球科学,39(11):1473-1486.

朱抱真,1957a. 大尺度热源-热汇和地形对西风带的常定扰动（一）[J]. 气象学报,28(2):122-140.

朱抱真,1957b. 大尺度热源-热汇和地形对西风带的常定扰动（二）[J]. 气象学报,28(3):198-224.

竺夏英,刘屹岷,吴国雄,2012. 夏季青藏高原多种地表感热通量资料的评估[J]. 中国科学:地球科学,42(7):1104-1112.

BOLIN B, 1950. On the influence of the earth's orography on the general character of the westerlies[J]. Tellus, 2(3): 184-195.

BOOS W R, KUANG Z M, 2010. Dominant control of the South Asian monsoon by orographic insulation versus plateau heating[J]. Nature, 463(7278): 218-222.

BOOS W K, KUANG Z M, 2013. Sensitivity of the South Asian monsoon to elevated and non-elevated heating[J]. Sci Rep, 3: 1192.

CHEN X Y, YOU Q L, 2017. Effect of Indian Ocean SST on Tibetan Plateau precipitation in the early rainy season[J]. J Climate, 30: 8973-8985.

CUI Y F, DUAN A M, LIU Y M, et al, 2014. Interannual variability of the spring atmospheric heat source over the Tibetan Plateau forced by the North Atlantic SSTA[J]. Climate Dynamics, 45(5-6):1617-1634.

DUAN A M, WU G X, 2005. Role of the Tibetan Plateau thermal forcing in the summer climate patterns over subtropical Asia[J]. Climate Dynamics, 24: 793-807.

DUAN A M, WU G X, ZHANG Q, et al, 2006. New proofs of the recent climate warming over the Tibetan Plateau as a result of the increasing greenhouse gases emissions[J]. China Sci Bull, 51: 1396-1400.

DUAN A M, WU G X, 2008. Weakening trend in the atmospheric heat source over the Tibetan Plateau during recent decades. Part Ⅰ: Observations[J]. J Climate, 21: 3149-3164.

DUAN A M, WU G X, 2009. Weakening trend in the atmospheric heat source over the Tibetan Plateau during recent decades. Part Ⅱ: Connection with climate warming[J]. J Climate, 22: 4197-4212.

DUAN A M, LI F, WANG M R, et al, 2011. Persistent weakening trend in the spring sensible heat source over the Tibetan Plateau and its impact on the Asian summer monsoon[J]. J Climate, 24: 5671-5682.

DUAN A M, WANG M R, LEI Y H, et al, 2013. Trends in summer rainfall over China associated with the Tibetan Plateau sensible heat source during 1980—2008[J]. J Climate, 26: 261-275.

DUAN A M, WANG M R, XIAO Z X, 2014. Uncertainties in quantitatively estimating the atmospheric heat source over the Tibetan Plateau[J]. Atmos Oceanic Sci Lett, 7(1): 28-33.

DUAN A M, LIU S F, ZHAO Y, et al, 2018. Atmospheric heat source/sink dataset over the Tibetan Plateau based on satellite and routine meteorological observations[J]. Big Earth Data, 2(2):179-189.

DUAN A M, HU D, HU W T, et al, 2020. Precursor effect of the Tibetan Plateau heating anomaly on the seasonal march of the East Asian summer monsoon precipitation[J]. Journal of Geophysical Research: Atmospheres, 125: e2020JD032948.

DUAN A M, LIU S F, HU W T, et al, 2022. Long-term daily dataset of surface sensible heat flux and latent heat release over the Tibetan Plateau based on routine meteorological observations[J]. Big Earth Data:1-12.

ERTEL H,1942. Ein neuer hydrodynamischer wirbelsatz[J]. Meteorologische Zeitschrift, 59: 271-281.

FLOHN H, 1957. Large-scale aspects of the "summer monsoon" in South and East Asia[J]. Journal of the Meteorological Society of Japan, 35A: 180-186.

GAO K L, DUAN A M, CHEN D L, 2020. Interdecadal summer warming of the Tibetan Plateau potentially regulated by a sea surface temperature anomaly in the Labrador Sea[J]. Int J Climatol, 41 (Suppl 1): E2633-E2643.

HELD I M, 1983. Stationary and Quasi-stationary Eddies in the Extratropical Troposphere: Theory [M]. London: Academic Press: 127-168.

HELD I M, TING M, 1990. Orographic versus thermal forcing of stationary waves: The importance of the mean low-level wind[J]. Journal of Atmospheric Sciences, 47(4): 495-500.

HELD I M, TING M, 2002. Northern winter stationary waves: Theory and modeling[J]. Journal of the Atmospheric Sciences, 47: 495-500.

HOSKINS B J, 1991. Towards a PV-θ view of the general circulation [J]. Tellus, 43 (4): 27-35.

HU J, DUAN A M, 2015. Relative contributions of the Tibetan Plateau thermal forcing and the Indian Ocean sea surface temperature basin mode to the interannual variability of the East Asian summer monsoon[J]. Climate Dynamics, 45: 2697-2711.

HU S, ZHOU T, WU B, 2021. Impact of developing ENSO on Tibetan Plateau summer rainfall[J]. J Climate, 34(9): 3385-3400.

JIANG X W, LI Y Q, YANG S, et al, 2016. Interannual variation of summer atmospheric heat source over the Tibetan Plateau and the role of convection around the western maritime continent[J]. J Climate, 29(1): 121-138.

LIU H C, DUAN K Q, LI M, et al, 2015. Impact of the North Atlantic oscillation on the dipole oscillation of summer precipitation over the central and eastern Tibetan Plateau[J]. Int J Climatol, 35: 4539-4546.

LIU S F, DUAN A M, 2017. Impacts of the global sea surface temperature anomaly on the evolution of circulation and precipitation in East Asia on a quasi-quadrennial cycle[J]. Climate Dynamics, 51: 4077-4094.

LIU S F, DUAN A M, WU G X, 2020. Asymmetrical response of the East Asian summer monsoon to the quadrennial oscillation of global sea surface temperature associated with the Tibetan Plateau thermal feedback[J]. Journal of Geophysical Research: Atmospheres, 125, e2019JD032129.

LIU Y M, WU G X, REN R C, 2004. Relationship between the subtropical anticyclone and diabatic heating[J]. J Climate, 17: 682-698.

LIU Y M, WU G X, HONG J L, et al, 2012. Revisiting Asian monsoon formation and change associated with Tibetan Plateau forcing: II. Change[J]. Climate Dynamics, 39(5): 1183-1195.

LIU Y M, LU M M, YANG H J, et al, 2020. Land-atmosphere-ocean coupling associated with the Tibetan Plateau and its climate impacts[J]. National Science Review, 7(3): 534-552.

LIU Y, CHEN H, LI H, et al, 2021. What induces the interdecadal shift of the dipole patterns of summer precipitation trends over the Tibetan Plateau? [J]. Int J Climatol, 41: 5159-5177.

LUO H B, YANAI B M, 1983. The lager-scale circulation and heat sources over Tibetan Plateau and surrounding areas during early summer of 1979, Part I: Precipitation and kinetic analysis[J]. Mon Wea Rev, 111: 922-944.

LUO H, YANAI B M, 1984. The large-scale circulation and heat sources over the Tibetan Plateau and surrounding areas during the early summer of 1979. Part II: Heat and moisture budgets[J]. Mon Wea Rev, 112: 966-989.

PLUMB R A, HOU A Y, 1992. The response of a zonally symmetric atmosphere to subtropical thermal forcing: Threshold behavior[J]. J Atmos Sci, 49(19): 1790-1799.

QIU J, 2013. Monsoon Melee[J]. Science, 340(6139): 1400-1401.

QUENEY P, 1948. The Problem of air flow over mountains a summary of theoretical studies [J]. Bull Amer Meteor Soc, 29(1): 16-26.

SON J H, SEO K H, WANG B, 2019. Dynamical control of the Tibetan Plateau on the East Asian summer monsoon[J]. Geophysical Research Letters, 46(5): 7672-7679.

SON J H, SEO K H, WANG B, 2020. How does the Tibetan Plateau dynamically affect downstream monsoon precipitation? [J]. Geophysical Research Letters, 47(23).

SUN J, YANG K, GUO W, et al, 2020. Why Has the inner Tibetan Plateau become wetter since the mid-1990s? [J]. J Climate, 33(19): 8507-8522.

WANG M R, ZHOU S W, DUAN A M, 2012. Trend in the atmospheric heat source over the central and eastern Tibetan Plateau during recent decades: Comparison of observations and reanalysis[J]. China Sci Bull, 57: 548-557.

WANG R, WU G, 2007. Mechanism of the spring persistent rains over southeastern China[J]. Sci China Ser D, 50: 130-144.

WANG Z Q, DUAN A M, WU G X, 2014. Time lagged impact of spring sensible heat over the Tibetan Plateau on the summer rainfall anomaly in East China: Case studies using the WRF model[J]. Climate Dynamics, 40: 1-14.

WANG Z Q, DUAN A M, YANG S, et al, 2016. Atmospheric moisture budget and its regulation on the variability of summer precipitation over the Tibetan Plateau[J]. J Geophys Res Atmos, 122(2): 614-630.

WANG Z Q, YANG S, LAU N, et al, 2018. Teleconnection between summer NAO and east China rainfall variations: A bridge effect of the Tibetan Plateau[J]. J Climate, 31(16): 6433-6444.

WU G X, 1984. The nonlinear response of the atmosphere to large-scale mechanical and thermal forcing [J]. J Atmos Sci,

41(16): 2456-2476.

WU G X, LIU H, ZHAO Y C, et al, 1996. A nine-layer atmospheric general circulation model and its performance[J]. Adv Atmos Sci, 13: 1-18.

WU G X, ZHANG Y S, 1998. Tibetan Plateau forcing and the timing of the monsoon onset over South Asia and the South China Sea[J]. Monthly Weather Review, 126: 913-927.

WU G X, LIU Y M, 2003. Summer quadruplet heating pattern in the subtropics and the associated atmospheric circulation [J]. Geophys Res Lett, 30: 1-4.

WU G X, MAO J Y, DUAN A M, et al, 2006. Current progresses in study of impacts of the Tibetan Plateau on Asian summer climate[J]. Acta Meteor Sinica, 20(2): 144-158.

WU G X, LIU Y M, ZHANG Q, et al, 2007. The influence of mechanical and thermal forcing by the Tibetan Plateau on Asian climate[J]. Journal of Hydrometeorology, 8(4): 770-789.

WU G X, LIU Y M, ZHU X, et al, 2009. Multi-scale forcing and the formation of subtropical desert and monsoon[J]. Ann Geophys, 27: 3631-3644.

WU G X, LIU Y M, HE B, et al, 2012a. Thermal controls on the Asian summer monsoon[J]. Nature Sci Rep, 2(404): 1-7.

WU G X, GUAN Y, LIU Y M, et al, 2012b. Air-sea interaction and formation of the Asian summer monsoon onset vortex over the Bay of Bengal[J]. Climate Dyn, 38: 261-279.

WU G X, DUAN A M, LIU Y M, et al, 2015. Tibetan Plateau climate dynamics: Recent research progress and outlook [J]. National Science Review, 2(1): 100-116.

WU G X, LIU Y, 2016. Impacts of the Tibetan Plateau on Asian climate[J]. Meteorological Monographs, 56: 7.1-7.29.

YANG K, QIN J, GUO X F, et al, 2009. Method development for estimating sensible heat flux over the Tibetan Plateau from CMA data[J]. J Appl Meteor, 48: 2474-2486.

YEH T C, 1950. The circulation of the high troposphere over China in the winter of 1945−1946[J]. Tellus, 2(3): 173-183.

ZHANG P, DUAN A M, 2021. Dipole mode of the precipitation anomaly over the Tibetan Plateau in mid-autumn associated with tropical Pacific-Indian Ocean sea surface temperature anomaly: Role of convection over the northern maritime continent[J]. Journal of Geophysical Research Atmospheres, 126: e2021JD034675.

ZHAO Y, DUAN A M, WU G X, 2018. Interannual variability of late-spring circulation and diabatic heating over the Tibetan Plateau associated with Indian Ocean forcing[J]. Advances in Atmospheric Sciences, 35: 927-941.

ZHOU C, ZHAO P, CHEN J, 2019. The interdecadal change of summer water vapor over the Tibetan Plateau and associated mechanisms[J]. J Climate, 32(13): 4103-4119.

ZHU X Y, LIU Y M, WU G X, 2012. An assessment of summer sensible heat flux on the Tibetan Plateau from eight datasets[J]. Sci China Earth Sci, 55: 779-786.

> 综述 7

青藏高原对亚洲夏季风的关键调制作用

何编[1,2]，刘屹岷[1,2]，包庆[1,2]，梁潇云[3]，毛江玉[1,2]

(1 中国科学院大气物理研究所，北京 100029；2 中国科学院大学，北京 100049；
3 中国气象局地球系统数值预报中心，北京 100081)

摘要：青藏高原气候动力学是国际前沿的热点科学问题。近几十年来，虽然青藏高原热力、动力强迫对季风的影响已有诸多研究，但由于亚洲夏季风系统的复杂性，高原对季风影响的相对重要性仍存争议。从全型涡度方程和位涡理论以及辐射-对流平衡理论出发，吴国雄院士带领其团队近十几年来在青藏高原影响亚洲夏季风机理这一科学问题上做出了诸多创新性成果。文中主要回顾了他们基于上述气候动力学理论，并结合自主研发的数值模式 SAMIL 及其耦合版本 FGOALS-s2 开展的一系列研究，包括亚洲大地形及海陆分布对季风影响的重要性、青藏高原热动力强迫对南亚夏季风水分循环以及海陆气相互作用的调制和青藏高原热力强迫对东亚夏季风影响的物理机制几个方面的进展，为进一步深入理解青藏高原气候动力学问题提供一些参考。

关键词：青藏高原；亚洲夏季风；地形强迫；数值模拟；海陆气相互作用

1 引言

值吴国雄院士 80 岁寿辰之际，文中回顾了吴院士带领他的研究团队，在青藏高原气候动力学方面近 10 多年的主要研究进展。从 20 世纪 50 年代以来，青藏高原作为全球最大的地形，其对天气和气候的影响被人们开始重视起来。早期的研究主要关注高原大地形对西风气流的阻挡作用(Queney，1948；Charney and Eliassen，1949；Bolin，1950；Yeh，1950；顾震潮，1951)。而从 1957 年开始，Flohn 和叶笃正先生分别在独立工作的情况下，发现青藏高原在夏季是一个巨大的热源，是它的热力效应对周围大气具有调控作用(Flohn，1957；叶笃正 等，1957)，青藏高原热力学的研究开始变得热起来。在国内，叶笃正先生带领他的团队逐步开展了青藏高原热源的诊断分析及影响的相关工作(叶笃正 等，1979)。吴国雄先生当时正是叶笃正先生团队成员，他在青藏高原影响亚洲季风上开展了诸多研究。吴国雄等(1997)提出了青藏高原感热气泵理论(TP-SHAP)，他们指出高原在春、夏季表面的感热输送造成低层气流向高原地区的辐合，形成夏季高原上空强烈的上升运动，犹如一个巨大的气泵调节着周围低层环流的季节性演变。夏季，由于高原表面感热加热的作用，低层气流和大量的水汽在高原附近辐合，随着强烈的上升运动带入到对流层高层，向周围地区辐散，调解着南亚、孟加拉湾、中国南海和西北太平洋季风的活动。如果没有这种感热加热，高原表面的空气将沿着等熵面做绝热运动，而不会在垂直方向上作穿越等熵面对大气进行加热(吴国雄 等，2002，2005)。

在 2010 年左右，国际上一些学者对青藏高原热力强迫影响亚洲夏季风形成提出了不同观点(Boos and Kuang，2010；Molnar et al.，2010)，他们认为高原表面的感热加热不重要，而是其南侧喜马拉雅山

通讯作者：何编，heb@lasg.iap.ac.cn

的隔断作用导致暖湿气流在印度北部堆积,触发局地湿对流,驱动季风环流。他们基于 Emanuel(1994)提出的辐射-对流平衡理论,并结合仅保留喜马拉雅山地形的敏感性试验试图证明上述观点。然而,辐射-对流平衡假设以局地的观点来理解湿对流的发展对大气的驱动作用在热带是非常合理的,但是在副热带地区,由于大尺度环流需要满足准地转平衡约束,局地对流的扰动受外界影响因素比较复杂,南亚季风区高原南侧的强大湿对流与青藏高原的热力强迫之间的关系是不清楚的。针对上述问题,吴国雄院士团队同时从位涡理论和辐射-对流平衡理论出发,并结合自主研发的大气模式 SAMIL2 及其耦合版本 FGOALS-s2 模拟的基础上,对亚洲大地形的动力、热力作用如何影响亚洲夏季风做了较为全面和深入的研究。本文的回顾主要分为以下 6 个小节,第 2 小节回顾了青藏-伊朗高原和海陆热力差异对亚洲夏季风影响的相对重要性,第 3 小节回顾了青藏-伊朗高原热力、动力强迫对南亚夏季风的影响,第 4 小节回顾了青藏-伊朗高原影响南亚夏季风的能量和水循环及其中的反馈过程,第 5 小节回顾了青藏高原热力强迫对东亚夏季风影响的研究,最后第 6 小节给出了全文总结和展望。

2 青藏-伊朗高原和海陆热力差异对亚洲夏季风影响的相对贡献

基于中国科学院大气物理研究所自主研发的大气环流模式 SAMIL-R42L9（Wu et al.，2003),我们开展了一系列数值试验,从单纯的水球试验开始,逐步加入海陆分布和青藏高原地形,用于理解从简单强迫到多尺度地形强迫对亚洲夏季风形成的影响,进一步明确海陆分布和青藏高原大地形对亚洲夏季风形成的相对贡献。在试验设计上,首先将模式中地球表面覆盖上水,也就是大气模式下垫面强迫仅考虑海洋,定义为水球试验(AQU)作为参考试验。AQU 试验使用的海表温度(SST)是来自于第二次大气模式比较计划(AMIP-II)提供的纬向平均气候态海表温度,具有季节变化。将来自 AQU 的风、温度、湿度和地表气压的纬向平均值作为其他敏感性试验的大气初始值。为了提取亚洲海陆分布和大尺度地形强迫对季风的影响的主要特征,这里简化了观测中的海陆分布和地形强迫,用理想地形替代,以突出他们的主要作用和尽可能地避免一些模式模拟误差带来的影响。海陆分布敏感性试验中,主要考虑了三种不同几何形状的陆地分布类型。第一个试验(MID),增加一个位于 $0°\sim120°E$ 和 $30°\sim90°N$ 矩形的大陆,用以研究中高纬度陆-海热力对比对环流的影响。第二个试验(SUB),将第一个试验中的中纬度大陆南部边界向南延伸 $10°$ 进入副热带,模拟欧亚大陆的主体部分。第三个试验(TRO)引入了($0°\sim50°E$，$35°S\sim20°N$)、($75°\sim85°E$，$5°\sim20°N$)和($95°\sim105°E$，$9°S\sim20°N$)三个矩形的热带陆地,用以代表非洲、印度和中南半岛的热带次大陆,并与欧亚大陆主体合并,形成"非洲-欧亚大陆"。所有试验积分 10 年,使用最后 8 年稳定态的平均进行分析。

2.1 副热带亚洲大陆的作用

当整个地球表面全是海洋,在太阳辐射和地球自转的作用下,南、北半球大气分别存在简单的经向三圈环流(图 1a),沿赤道为热带强降水区(ITCZ),南北半球 $30°\sim60°$ 之间各存在一个温带次强降水区(图 1b)。为了定量表征季风,这里采用冬、夏盛行风向偏转大于 $120°$,且伴有明显的降水变化来定义季风区(拉梅奇,1978)。在 AQU 试验中,三圈环流随季节有南北的移动,行星风带的季节移动,在南北半球对流层低层 $30°$ 和 $60°$ 位置分别造成两条狭窄的冬、夏盛行风向近乎相反的纬向区域带(图 1c)。$30°$ 的区域带对应着副热带高压带的中心区,那里风速很小且区域带内冬、夏降水的差异不明显(<2 mm)。位于 $60°$ 附近的那条冬、夏风向近乎相反的区域带内的冬夏降水差异更小,这说明 AQU 试验条件下不存在季风现象。MID 试验结果显示,当亚洲大陆南界仅达到 $30°N$ 时,虽然在大陆的南部边沿存在冬夏风向近乎相反的转向,但是由于此时大陆南部位于温带次强雨带区,风向的转换并没有带来冬夏降水明显的差异,也没有季风现象存在。当温带中纬度大陆南伸到 $20°N$ 时,在大陆东南边缘区域产生了冬夏风向的近乎反转(夏季为西南气流,冬季为东北气流),且该地区冬季为干冷气候,夏季为暖湿,冬夏存在因为风向的转换而带来的较明显干湿差异。虽然大陆东南边缘的副热带夏季风降水还比较弱,该雨带与沿赤道的热带

强降水区是分离共存(图 2a),但是 SUB 试验表明,深入到副热带地区的大尺度陆地的存在是亚洲季风形成的前提和根本。

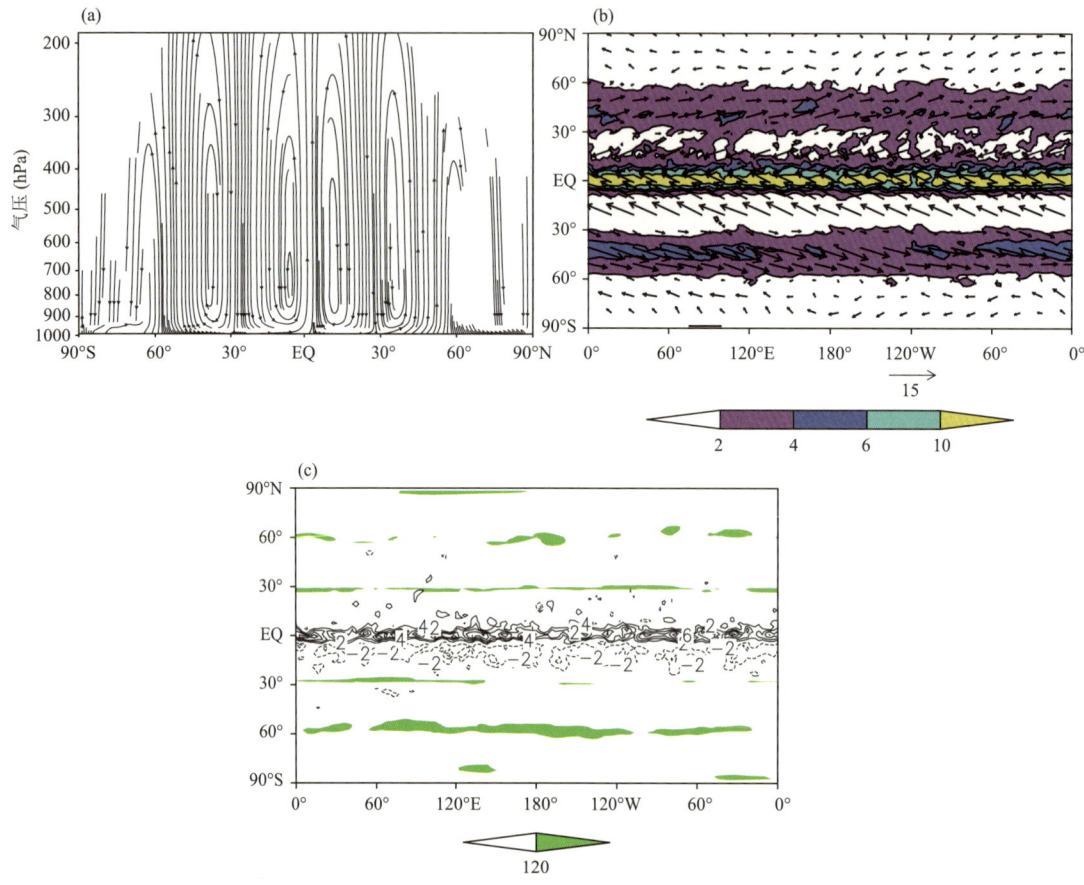

图 1　SAMIL-R42L9 水球试验模拟的(a)7 月纬向平均的经圈环流,(b)地面风场(矢量,m·s^{-1})和降水(填色,mm·d^{-1}),(c)7 月与 1 月的降水差异(等值线,间隔 2 mm·d^{-1})和地面风向的偏转(填色区为转向大于 120°的地区)(摘自梁潇云 等,2006)

2.2　热带亚洲次尺度陆地的作用

当热带亚洲次尺度陆地存在时,热带次尺度陆地通过"两级热力适应"机制,使得热带陆地上空对流层低层的偏南气流得到维持,引导过赤道气流出现,南北半球空气进行质量交换,南亚次大陆上空的热带夏季风形成。热带次尺度陆地同时存在且海陆交错分布时(TRO),热带陆地之间的热力强迫起到互相加强的作用,南亚次大陆和中南半岛上空的热带夏季风得到加强(图 2b)。从 TRO-SUB 的差异图上来看(图 2c),夏季热带和副热带陆地的热力强迫共同作用使副热带大陆低层气旋式环流加强,导致西南风将大量水汽从热带输送到 20°N 以北,给大陆东南部带来丰沛的夏季风降水。以上结果说明在现代亚洲夏季风形成中,海陆分布的作用是最基本的,也是第一位的。

2.3　青藏高原的作用

(1)青藏高原在亚洲季风形成中的作用

然而,在上述试验中,仅考虑海陆分布产生的大陆东南部的夏季降水还是较观测偏弱得多。我们进一步探讨了亚洲大陆上青藏高原大地形强迫对亚洲夏季风的影响。在 SUB 试验设计的基础上,我们添加了一个中心位于(87.5°E,32.5°N),中心高度为 5000 m 的理想的青藏高原(SUB-TP),试验中的其他条件同 SUB 试验。根据位涡理论,夏季青藏高原作为一个强大的抬升热源,其热力强迫作用可在对流层低层形

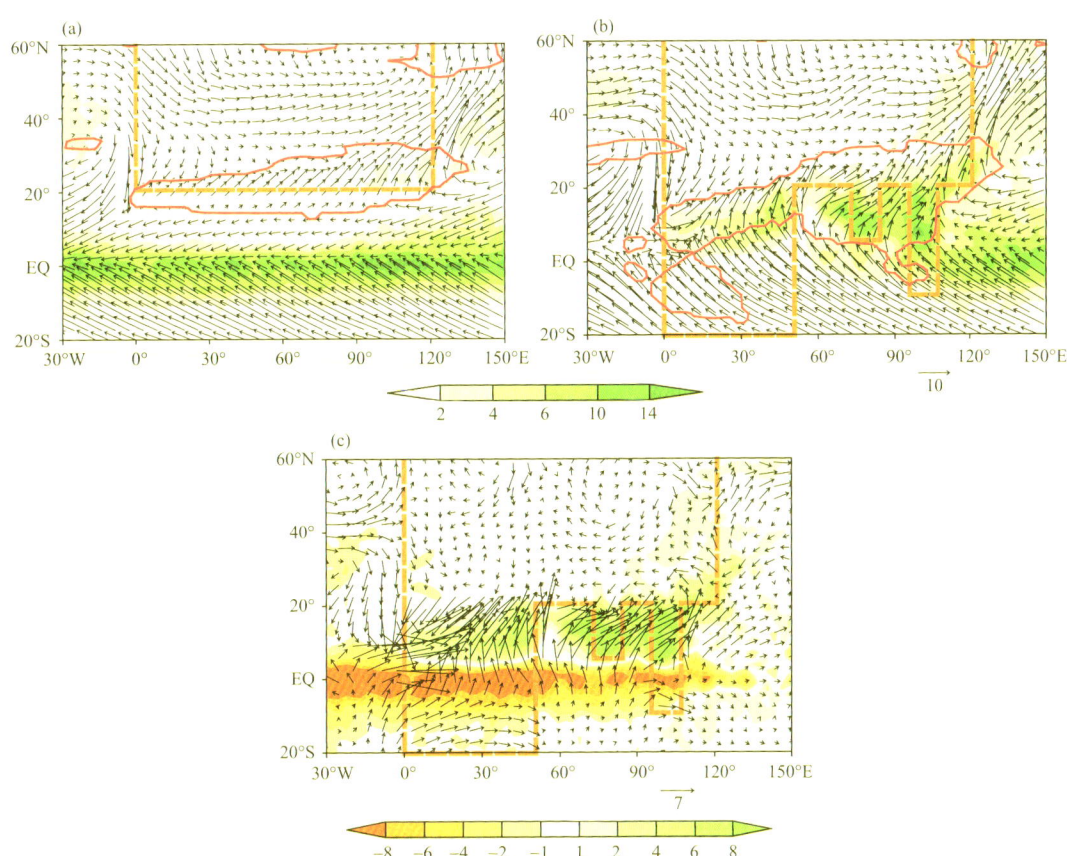

图 2 SAMIL-R42L9 理想试验模拟的 7 月地面风场(矢量,m·s^{-1})和降水(填色,mm·d^{-1})。(a)SUB 试验,(b)TRO 试验,(c)TRO-SUB。图中橘黄色粗虚线表示理想的陆地边界,红色曲线所包围的区域为 7 月和 1 月地面风转向大于 120°的地区(摘自梁潇云 等,2006)

成一个西偏北、东偏南的气旋性环流。同时青藏高原的存在还增加了海陆的热力差异,使陆地东面海洋上的副热带高压加强。从而在高原东南部形成强的西南气流,加强了东亚副热带陆地上空的西南气流,造成大陆夏季风降水增加。但是,副热带陆地以南的热带地区盛夏虽然部分地区有弱的西南气流出现,但无夏季降水伴随,故无热带季风的产生(图 3a)。夏季,青藏高原作为一个强热源,其热力作用可以影响到对流层上层,高原的强感热加热通过热力适应,在高原上空的对流层高层引起一个强的反气旋式异常环流。这个反气旋式异常环流大大加强了南亚陆地上空高层的反气旋式环流并使其北移到青藏高原的南部。由此可见,夏季对流层上层南亚高压的形成受到青藏高原的显著影响。虽然青藏高原的存在,大大加强了对流层低层的气旋式环流和对流层高层的反气旋环流,但是无论是低层还是高层,都没有越赤道气流的形成,南、北半球无质量的交换,而且大陆东南部的夏季风降水较弱。

我们又进一步在 TRO 试验设计的基础上,添加同样的一个理想的青藏高原地形(TRO-TP)。当热带次尺度陆地和青藏高原均存在时,亚洲大陆东南部夏季风偏南气流和降水大大加强,与观测比较接近(图 3b)。SUB-TP 和 TRO-TP 试验表明,在亚洲季风的形成中海陆分布的作用是第一的,没有热带次尺度陆地的存在,热带季风不能形成。青藏高原大地形在亚洲季风形成中的作用是第二位的,它对亚洲季风起着调整的作用,并使得亚洲季风向高层发展,季风系统更加深厚。由于大地形的机械阻挡和热力共同作用,青藏高原的西侧对流层低层无论冬夏均是偏北气流,从而高原西南侧的印度季风被减弱,季风区向南收缩。夏季青藏高原的热力作用引起的对流层低层的气旋式环流,加强了高原东南侧的孟加拉湾和中南半岛地区的热带夏季风和东侧的东亚副热带夏季风和降水,并使大陆副热带季风区向北和向东扩展。

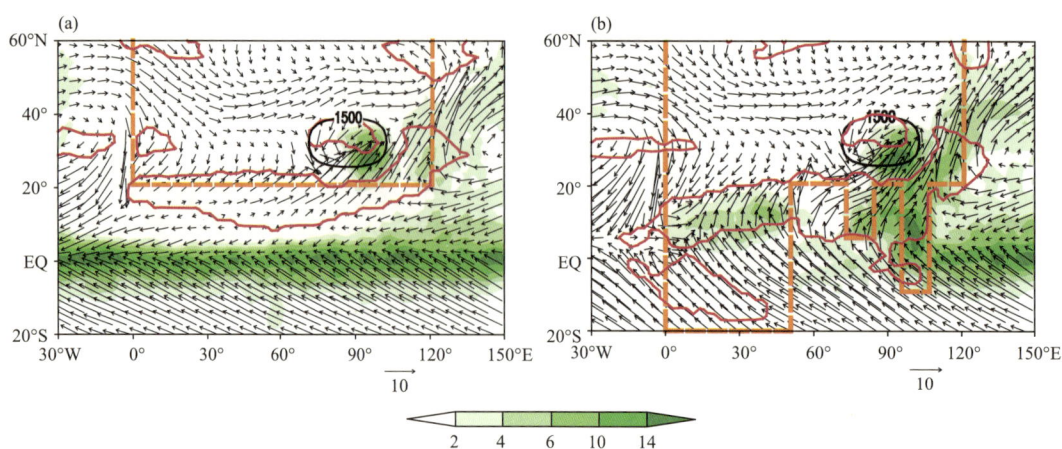

图 3　SAMIL-R42L9 理想试验模拟的 7 月地面风场（矢量，m·s^{-1}）和降水（填色，mm·d^{-1}）。(a)SUB-TP 试验，(b)TRO-TP 试验。图中橘黄色粗虚线表示理想的陆地边界，黑色椭圆曲线为 1500 m 的地形等高线，红色曲线所包围的区域为 7 月和 1 月地面风转向大于 120°的地区（摘自梁潇云 等，2006）

(2)青藏高原对于亚洲夏季风爆发地点的影响

吴国雄和张永生(1998,1999)的研究发现亚洲夏季风于 4 月底到 5 月初首先在孟加拉湾东-中南半岛西南地区爆发，然后于 5 月中旬和 6 月上旬分别出现在南海和印度半岛地区，呈三个分阶段爆发的特征。而且亚洲季风爆发的阶段性特征与青藏高原的感热加热变化一致。由此可见，青藏高原的热力和动力强迫对亚洲夏季风的爆发有着重要的作用。为了进一步研究青藏高原大地形在亚洲季风环流形成中的作用，以及它对亚洲夏季风爆发过程的影响，我们设计了一个青藏高原中心位于(60°E，32°N)，其他条件同 TRO-TP 的试验(TRO-TP60)。

图 4 是 TRO-TP 和 TRO-TP60 两个试验中亚洲夏季风的进程。虽然是理想试验，但是在 TRO-TP 试验中亚洲夏季风爆发的阶段性还是很明显的：首先是出现在孟加拉湾地区，其次是中南半岛东部和中国南海地区北部，最后是印度半岛以及阿拉伯海地区，这与实际气候平均的亚洲西南夏季风的爆发过程是比较一致的（毛江玉 等，2002）。在 TRO-TP60 试验中，亚洲夏季风的爆发进程则为：首先是印度西南夏季风爆发，紧接着孟加拉湾西南夏季风爆发，大约 20 天之后南海夏季风爆发。对比两个试验结果发现，青藏高原的西移引起印度夏季风的爆发提前两周左右，孟加拉湾地区夏季风的爆发推迟两周左右。这说明青藏高原的动力热力作用造成亚洲夏季风的爆发呈现出阶段性，青藏高原对亚洲夏季风爆发地点的"锚定"作用。无论是 TRO-TP 试验还是 TRO-TP60 试验中，南海夏季风均偏弱很多，也许在南海夏季风的形成中青藏高原和中南半岛的作用远远没有南海海温增暖或其他因子的作用重要。

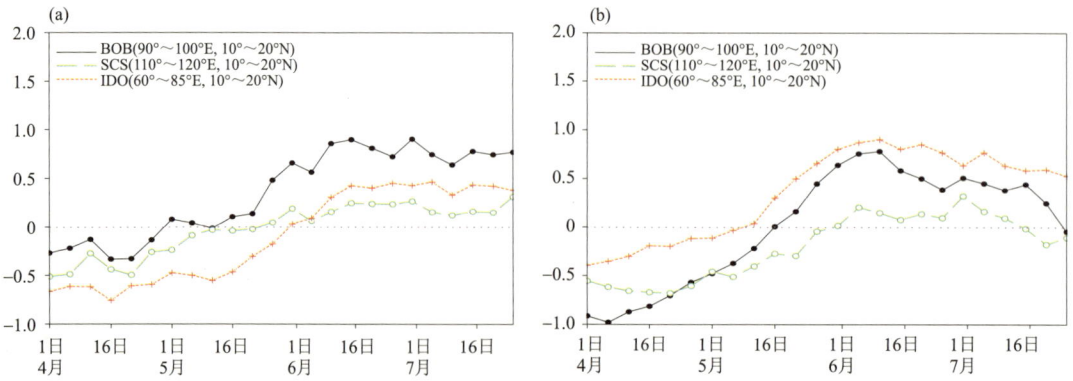

图 4　SAMIL-R42L9 理想试验中亚洲季风区对流层上层(500～200 hPa)平均温度经向梯度($\frac{\partial T}{\partial y}$)的演变(10^{-3} K·km^{-1})。(a) TRO-TP 试验，(b)TRO-TP60 试验（改自梁潇云 等，2006）

随着模式性能的提高,我们基于改进的 SAMIL2 版本(Bao et al.,2013)模式进一步探究了关于青藏-伊朗高原和大尺度海陆热力差异对南亚夏季风影响的相对贡献。与基于 SAMIL-R42L9(Wu et al.,2003)开展的水球试验的设计类似(梁潇云 等,2006;Wu et al.,2007),这里开展的参考试验(CON)和一系列敏感性试验的海温和海冰强迫同样是气候态仅包括季节变化的 AMIP II 数据资料,所有的外强迫场(温室气体、太阳常数、臭氧、气溶胶)都固定为气候态的值。所有试验积分 7 年,取后 5 年夏季平均作为分析对象。CON 试验中地形强迫为真实世界地形(图 5a);NMT 试验(图 5c)中将全球的地形全部设为 0,只包含海陆分布的差异;由于我们希望进一步考察青藏高原和伊朗高原对南亚季风的影响,因此,我们还设计了 L_S 试验,它与 NMT 的不同点是将青藏高原和伊朗高原的地形设为 0(图 5d),但是全球其他地区地形保持不变。

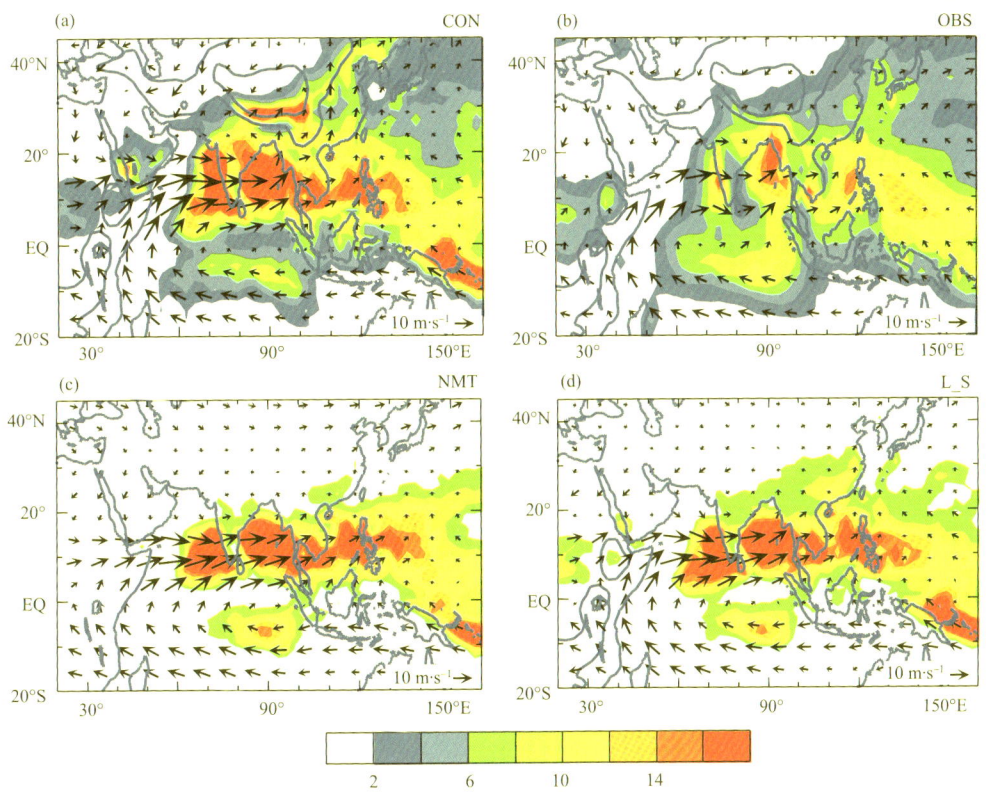

图 5 亚洲大地形和海陆分布对亚洲夏季风形成的影响。SAMIL2 模拟的气候态夏季(6—8 月)平均的降水(填色,mm·d^{-1})和 850hPa 风场(矢量,m·s^{-1})分布。(a)参考试验 CON,(b)观测降水(GPCP)和再分析资料的风场(NCEP2),(c)NMT 试验,全球地形高度设为 0,(d)L_S 试验,仅青藏-伊朗高原地形设为 0(引自 Wu et al.,2012)

我们利用控制实验 CON、无地形试验 NMT 以及没有青藏-伊朗高原试验 L_S 来讨论海陆热力差异对南亚季风形成和维持的影响,同时讨论高原的相对重要性。图 5a 给出的是 CON 试验模拟的南亚夏季风和降水的空间分布,和观测资料对比(图 5b),虽然模式在降水量级的模拟上偏强,但是把握住了印度洋、孟加拉湾、南海几个强降水中心,高原南侧以及印度北部地区的降水也跟观测比较接近,此外,模式能够模拟出东亚季风区的降水,强降水雨带略微偏北。这个分布相比旧版本 SAMIL-R42L9 基于理想化水球试验的结果更加接近观测。一方面是由于模式性能的提高,另一方面也是由于地形强迫更加接近真实情况。进一步从 NMT 的模拟结果上看(图 5c),模式模拟出的 20°N 以南的热带地区降水和 CON 试验的结果非常接近,仍然保持了几个极值降水中心;但是最明显的变化是副热带地区 20°N 以北的南亚季风以及东亚季风减弱很多。而在不包含青藏-伊朗高原地形的试验中(L_S 试验,图 5d),我们得到的结果与 NMT 试验比较接近,印度大陆以及东亚地区的降水相比 NMT 略微增强。同时我们注意到,由于 L_S 试验中包括了非洲东部的山脉,它们的动力以及热力强迫作用可以使得索马里急流存在并维持,进一步增强了南亚夏季风。因此,对比 Wu 等(2007)中水球试验的结果,可以看出在不同模式类似的试验设计中,海陆分布对南亚夏季风形成的影响是第一位的,而青藏-伊朗高原的存在是亚洲夏季风进一步向北推进并且在副热带大陆造成强降水的至关重要的因素。

3 青藏-伊朗高原不同区域热力、动力强迫对南亚夏季风分布型的影响

在模式中去除高原地形得到的季风环流和降水的响应是同时去除了青藏-伊朗高原动力作用和表面加热作用的共同结果。不能单纯地归因于青藏高原的纯动力作用。我们进一步开展了分离地形热力强迫和动力强迫的敏感性试验,用于比较青藏-伊朗高原动力作用和抬升加热作用对南亚夏季风形成的影响。为了避免在无高原表面感热加热试验中引入额外的冷源,我们采用在保持地表能量平衡不变的情况下,让相应地区地形高度大于 500 m 以上的感热加热无法加热大气,也就是在大气热力学方程中地表温度的垂直耗散项设为 0 这种方法来开展无地形的感热加热试验。该试验的结果是地形的纯动力强迫的影响,而参考试验与无感热试验的差异则为地形抬升加热对大气的影响。

3.1 青藏-伊朗高原动力作用的影响

首先我们在图 6a 给出了 CON 试验减去 L_S 试验的低层风场异常和降水场异常的空间分布;这个结果是排除了海陆热力差异的作用的影响外,受到青藏-伊朗高原热力、动力强迫作用共同影响的结果。从图中我们可以看出比较明显的特征是在整个亚洲大陆上空风场上出现一个气旋式异常并且伴有降水的增加,特别是在青藏高原南侧地区有比较强的降水异常;而在 20°N 以南热带海洋上,特别是在印度大陆西侧的阿拉伯海降水有所减少,出现了一个较强的反气旋式异常。我们进一步分析高原不同区域的动力和热力强迫作用对这个异常场的相对贡献。

我们在 L_S 试验的基础上,仅考虑青藏-伊朗高原的地形强迫,但不考虑其加热作用(IPTP_M 试验),如图 6b 所示,IPTP_M 试验模拟出的季风环流以及降水场的特征和 L_S 试验非常接近,说明青藏-伊朗高原的动力隔断作用对亚洲夏季风降水在 20°N 以北印度大陆上贡献并不明显;同样的,我们做了单独伊朗高原(图 6c)和青藏高原(图 6d)的动力隔断试验,它们的模拟结果也反映出了类似特征,亚洲季风降水主要位于 20°N 以南的大陆和海洋上,而无法到达印度北部以及高原南侧及东亚部分地区。另外,在环流场的模拟上没有一个试验可以模拟出图 6a 中高原附近的气旋式环流场以及印度大陆西侧的反气旋式环流。这说明,在不考虑高原抬升加热的情况下,仅仅由于大地形的动力作用是无法在亚洲大陆内陆产生大尺度范围内的辐合抬升运动的。

实际上,已有研究已经指出,大地形周围环流场的形成与维持跟本身的水平尺度、垂直高度以及加热状况有关。Bolin(1950)和 Yeh(1950)从理论上证明了青藏高原的大地形最显著的作用是在冬季将西风气流分成两支,并且可以影响到下游环流系统;Wu(1984)的工作进一步证明了大气运动在地形的热力和动力强迫下的响应是非线性的,从能量守恒以及角动量守恒的观点出发,得到了气流遇阻挡作用存在一个临界高度,当超过这个高度的时候气流以绕流为主,而低于这个高度的时候气流以爬流为主,这个高度通常小于 1000 m 左右,对于我们所研究的青藏、伊朗这两个高原来说,青藏高原的尺度远远超过了这个临界高度,也就意味着,气流不依赖于外力而通过自身动力爬上高原形成降水是不可能的。

3.2 青藏-伊朗高原热力作用的影响

为了进一步研究青藏-伊朗高原大地形的表面感热加热对南亚夏季风的影响,我们开展了三组地形的热力敏感性试验,在不改变地形高度的情况下分别将伊朗高原(IP_NS)、青藏高原(TP_NS)和这两者(IPTP_NS)的感热加热设为 0,并用 CON 试验的结果减去无感热试验的结果得到(IP_SH),(TP_SH)和(IPTP_SH)如图 7 所示。在 IP_SH(图 7a)中,伊朗高原的热力强迫作用导致了高原西部产生了一个气旋

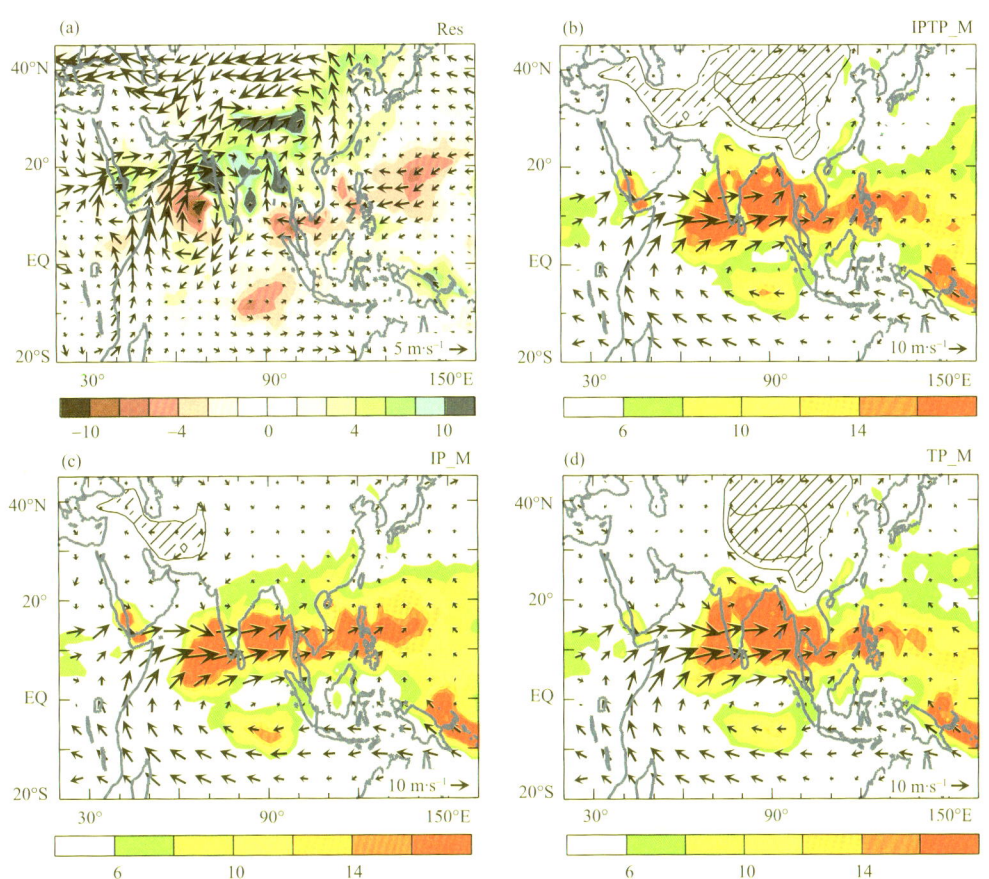

图 6 青藏-伊朗高原动力强迫对亚洲夏季风的影响。SAMIL2 模拟的气候态夏季(6—8 月)平均的降水(填色，mm·d^{-1})和 850 hPa 风场(矢量，m·s^{-1})分布。(a) CON 试验减去 L_S 试验，(b)IPTP_M 试验，青藏-伊朗高原有地形无感热加热，(c) IP_M 试验，伊朗高原有地形无感热加热，(d)TP_M，青藏高原有地形无感热加热 (引自 Wu et al. , 2012)

式环流异常，与图 6a 中伊朗高原和阿拉伯半岛上的环流型相接近。在降水的异常场上，热带印度洋和西北太平洋的降水有所减少，而亚洲大陆 100°E 的地区降水增加，主要位于印度大陆西侧，巴基斯坦以及高原南侧斜坡地区。在 IP_SH 试验中伊朗高原热力强迫导致的印度大陆降水增加和图 6a 中印度地区降水的变化较为一致，说明了伊朗高原的抬升加热作用是在造成以及维持南亚夏季风北部的降水上有很重要的作用。

仅考虑青藏高原表面感热加热的试验结果上(图 7b)，高原的热力强迫同样在高原附近造成了一个气旋式环流场的出现，但是水平尺度上比伊朗高原的影响大很多，占据主要作用。同时，降水在 80°E 以西有所减少，而在增加 80°E 以东，特别是孟加拉湾地区，高原南侧的斜坡以及东亚地区有所增加。这个降水场异常的空间分布和图 6a 中高原附近以及南亚、东亚地区的降水分布类似，表明了高原表面感热加热在影响亚洲夏季风在南亚大陆上的降水以及东亚地区降水上占据了主导地位。

最后，IPTP_SH 试验的结果表明(图 7c)，青藏-伊朗高原作为一个整体，它们的表面加热造成的降水和环流异常与图 6a 十分接近。这说明，青藏-伊朗高原的抬升加热作用是导致亚洲夏季风在南亚大陆，青藏高原南侧及东侧形成强大季风降水的关键因素。从位涡动力学的理论上去理解，青藏-伊朗高原的表面加热之所以对季风的调制作用这么重要是因为它可以激发高原周边低层的气旋式环流异常，这个环流异常可以看作是 PV 对地形加热的一种响应，由于地表的加热作用，高原周围的等熵面下凹并和高原地表相交，产生正的 PV 异常，在大尺度静力稳定的条件下必然出现正涡度异常，对应气旋式环流异常。

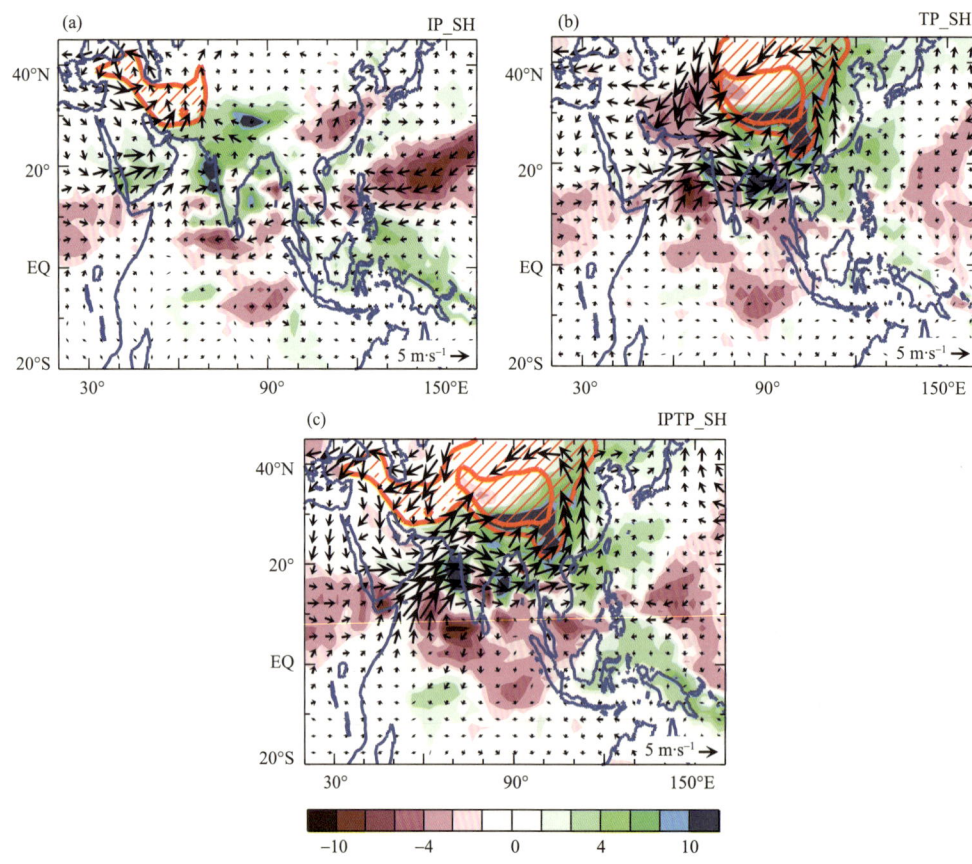

图 7 青藏-伊朗高原热力强迫对亚洲夏季风的影响。SAMIL2 模拟的气候态夏季(6—8 月)平均的降水(填色,mm·d^{-1})和 850 hPa 风场(矢量,m·s^{-1})分布。(a) IP_SH 试验(CON 试验减去 IP_M 试验),(b)TP_SH 试验(CON 试验减去 TP_M 试验),(c) IPTP_SH 试验(CON 试验减去 IPTP_M 试验)(引自 Wu et al.,2012)

3.3 喜马拉雅山斜坡抬升加热的影响

Wu 等(2007)基于水球试验发现了理想梯形高原表面加热在高原平台和侧面的斜坡对大气的影响具有不同作用,其斜坡的感热加热效应对局地垂直运动具有更重要的影响。然而由于模型和试验设计本身的问题,高原斜坡加热效应对季风环流和降水的影响并没有进一步开展研究。在 Boos 和 Kuang(2010)的工作中,他们开展了仅保留喜马拉雅山地形强迫的数值试验,发现南亚夏季风整体变化不大,因此归结于是喜马拉雅山的隔断作用而不是高原整体的加热作用对南亚夏季风的形成具有主导作用。实际上,他们的试验仅考虑了地形本身的变化,虽然去除了高原平台以及东侧的大部分地区,但是高原南侧的斜坡依然存在,南侧的感热加热依旧会对大气运动产生影响。为了说明这一影响,我们同样进行了地形以及无感热的敏感性试验。基于 Boos 和 Kuang(2010)的试验设计,第一个试验我们保留了喜马拉雅山的地形(HIM,图 8c),另一个试验中,将其斜坡上的感热加热去除(HIM_M,图 8d)。模拟的结果表明,HIM 试验模拟出了和 CON 试验(图 8a)类似的南亚季风空间分布特征,在印度北部以及喜马拉雅山的南侧斜坡上出现了大量的降水,表面流场有爬上高原南侧斜坡的运动;然而,在没有感热加热的试验 HIM_M(图 8d)中,从赤道地区输送过来的水汽无法爬上喜马拉雅山的斜坡,而是分成东西两支,沿着地形等高线绕过喜马拉雅山。西侧的分支在印度半岛西部形成了一个气旋,增加了南亚季风的降水,而使得印度北部降水减少。绕流的形成正是因为在没有斜坡地形的加热作用及外力强迫的情况下,气块只能沿等熵面做绝热运动,而无法穿越等熵线爬上高原。

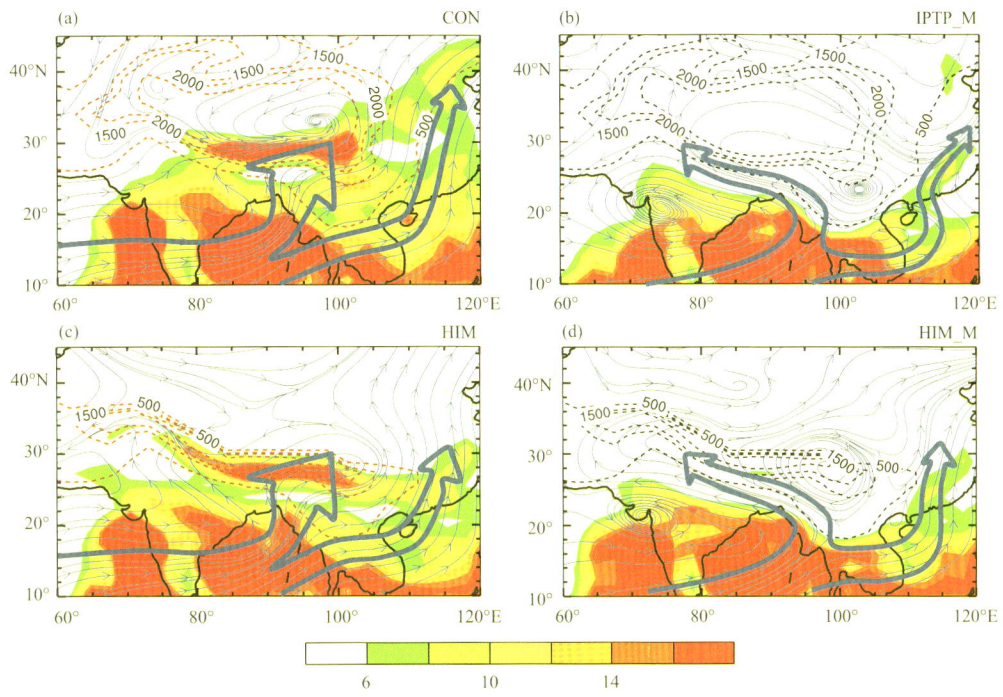

图8 山脉扰流和爬流对夏季降水的相对贡献。SAMIL2 模拟的气候态夏季(6—8 月)平均的降水(填色,mm·d^{-1})和 $\sigma=0.89$ 的流场分布。(a) CON 试验,(b) IPTP_M 试验,(c) HIM 试验,(d) HIM_M 试验。深蓝色箭头表示气流绕过或者爬上喜马拉雅山脉。黑虚线为等高线(引自 Wu et al.,2012)

4　青藏-伊朗高原影响南亚夏季风的能量和水分循环及其中的反馈过程

在第 3 节的研究中,我们主要通过数值模拟的方式,结合位涡理论研究了青藏-伊朗高原热力、动力强迫对亚洲夏季风系统的低层环流和降水异常的调制作用。对于季风系统而言,湿对流的激发和对环流的相应的反馈是季风系统影响周围大气乃至全球气候的关键过程。而以往的研究多侧重于高原地形强迫对大尺度环流的影响过程,对季风区湿对流的影响尚不清楚。例如,Boos 和 Kuang(2010)提出喜马拉雅山隔断了北方冷空气的入侵,使得印度北部暖湿气流堆积,形成近地面相当位温的极大值,是激发南亚地区湿对流的重要驱动力。但是这种隔断作用是否存在,暖湿气流的堆积是否受高原的热力调控是不清楚的。基于 Emanuel(1994)提出的辐射-对流准平衡理论,我们进一步研究了青藏-伊朗高原的表面热力强迫作用对南亚季风区湿对流触发的关键调制作用。在此基础之上,进一步分析了高原热力强迫对印度洋海气相互作用的影响及其中的反馈过程。初步探索了青藏-伊朗高原影响亚洲夏季风的能量和水分循环中所起的作用。

4.1　青藏-伊朗高原动力和热力强迫对南亚季风区湿过程影响的相对重要性

为了探讨青藏高原是否对北半球夏季高纬地区的冷空气南下具有阻挡作用,我们从天文辐射的角度分析了太阳辐射驱动的北半球夏季经地面经向风与高原地形之间的联系。图 9a 给出的是北半球夏至日,太阳辐射直射北回归线的日照时长示意图。从大气层顶来说,副热带地区 $A(30°N)$ 和热带地区 $B(17°N)$ 受到的太阳辐射强度相等,但是由于地轴的倾斜,北半球地区的日照时长随着纬度的增加是增加的。简单计算可知 A_1 点比 B_1 点时长多约 1 个小时。因此,A_1 点受到的总辐量量比 B_1 点受到的辐射要高约 6%(图 9b),这与观测到的近地表辐射收支在相同纬度带上的差异是一致的(图略)。因此,辐射收支的差异决定了在北半球夏季近地面很难有冷空气从高纬度向南入侵到低纬度热带地区,跟有无高原地形的阻挡没有必然联系。基于 CERES(clouds and the earth's radiant energy system)卫星观测资料,我们在图 9c 中给出了印度季风区 75°~100°E 平均的大气层顶太阳辐射年循环特征。可以看出,太阳辐射的最大值在 5—7 月的北半球高纬度 30°~50°N 达到最大值。对副热带 20°~30°N 而言,太阳辐射从 1 月开始逐渐增强,到夏季 7 月份达到极值后逐渐减弱。

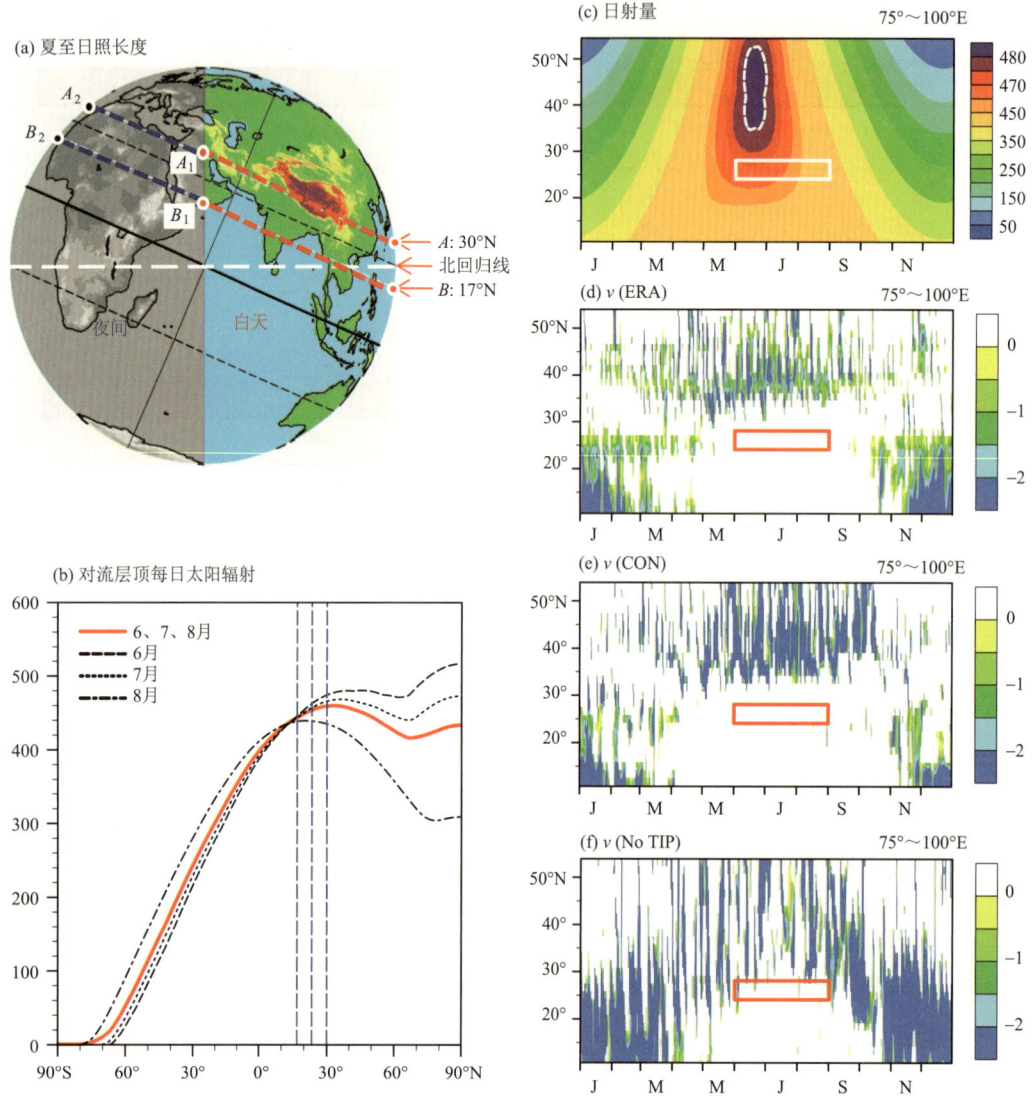

图 9 (a)北半球夏至日大气层顶太阳辐射和日照时长示意图,(b)夏季纬向平均的大气层顶太阳辐射(DSR,W·m^{-2}),(c) 75°~100°E 平均的 DSR(白虚线代表 480 W·m^{-2} 线)的年循环特征,以及 75°~100°E 平均的平均的近地面经向风,(d) ERA-interim 的结果,(e) CON 试验的结果,(f) NoTIP 试验的结果。(c)—(f)中的正方形表示 JJA 期间的 SASM 区域(24°~28°N)。(b)中的三条虚线分别表示 30°N、23.5°N 和 17°N 的纬度(引自 He et al.,2015)

这种年循环特征也反映在近地面冷空气的活动上。我们计算了 ERA-interim 再分析资料中近地表经向风的逐日年循环特征(图 9d),其中负值(填色)表示北风。可以看出,从北方南下的冷空气活动具有明显的季节变化,和太阳辐射的年循环特征类似,在冬季,即使有青藏高原大地形的存在,也还有副热带地区的北风向南入侵到热带地区。而在夏季,北方冷空气的活动局限于 30°N 以北的地区。我们进一步在数值模拟中开展了有无地形对冷空气活动影响的模拟。参考试验(CON)与第 3 节相同,是一个标准的 AMIP 试验。其模拟出的经向风季节循环(图 9d)与 ERA-interim 再分析资料非常接近。而在无青藏-伊朗高原地形的试验中(NoTIP,图 9f),经向风的模拟同样具有与参考试验中较为类似的季节变化,北半球高纬度冷空气的活动局限于30°N以北的地区。这说明在夏季,青藏高原的存在不是阻挡北方冷空气南下的关键因素,那么印度北部地区近地面激发湿对流的关键热力因子,相当位温θ_e极大值的形成也跟高原的阻挡作用没有必然联系。

我们进一步通过分析有无高原和有无高原表面加热的试验研究了青藏-伊朗高原热动力强迫对夏季印度北部相当位温极大值形成的可能影响。从相当位温θ_e的表达式(1)可知,近地面θ_e的变化主要取决于近地面位温θ_{sur}本身,以及水汽q_{sur}的影响。

$$\theta_e = \theta \exp\left(\frac{Lq}{C_p T}\right) \tag{1}$$

从再分析资料中可以看出,近地面位温的分布(图10a)大值区主要位于欧亚大陆,特别是青藏高原地区达到极大值。而近地面水汽q_{sur}的分布(图10b)大值区主要位于副热带和热带的海洋上,在青藏高原南侧斜坡也存在极大值,这和近地面相当位温的分布(图10c)有较高的相似度。这说明印度北部地区θ_e的变化很有可能主要受水汽的变化主导。从湿对流发展的过程上来说,我们在图10d中给出了青藏高原南侧斜坡附近四个点(图10c,蓝色点)平均的环境温度层结曲线及气块沿湿绝热线上升的曲线。在气块达到抬升凝结高度(LCL)后由于自身浮力上升,在气块温度比环境高的情况下具有对流有效位能(CAPE),容易产生对流降水。为了进一步明确高原动力、热力作用对这个过程的影响。我们分析了参考试验CON,无地形试验NoTIP,以及无感热试验TIP_NS对这一过程的模拟情况。参考试验CON基本模拟出了相当位温θ_e和降水(图10g)的大尺度分布特征,相对应的位温(图10e)和表面水汽和环流(图10f)也与再分析资料非常接近。

从湿对流发展的过程上看(图10h),参考试验的结果总体上与再分析资料相似,只是CAPE的值要高于再分析资料,因此相应模拟出来的高原地区斜坡降水要强于观测。当把高原地形去掉时(图10i),很明显地表位温有所减少,但是其分布仍然是位于欧亚大陆上。从水汽和环流的变化上来看(图10j),孟加拉湾地区向北的季风比参考试验明显减弱,因此输送的水汽有所减少,这导致近地面θ_e(图10k)相比参考试验在印度北部地区有明显降低,这必然会影响局地对流的触发。从层结曲线上看(图10l),LCL有明显的升高,并且湿绝热线位于环境温度曲线的右方,说明气块无法靠自身浮力产生对流,有效位能为0。从降水场上也明显可以看出,在印度北部地区,强降水无法发生。无高原表面感热加热的结果(图11a,b,c)与无地形试验的结果非常相似,说明青藏-伊朗高原的感热加热对近地面相当位温θ_e极大值在印度北部形成具有至关重要的作用。并且这种作用主要是通过影响周围的水汽输送作为最主要的途径。为了证明这一点,我们给出了CON试验减去TIP_NS试验的相关变量分布,可以看出表面位温的增加主要出现在青藏高原地区(图11d),而水汽的增加主要位于青藏高原南侧和东侧斜坡地区,伴随低层气旋式的环流异常(图11e)。这个分布与相当位温的分布高度相似(图11f)。我们还定量计算了高原热力强迫作用下近地面位温和水汽变化对相当位温影响的相对重要性(式2,3)。

$$\frac{\Delta \theta_e}{\theta_e} = \frac{L}{C_p} \frac{\Delta q_{sur}}{T_{sur}} + \frac{\Delta \theta_{sur}}{\theta_{sur}} \tag{2}$$

$$\left|\frac{\Delta \theta_e}{\theta_e}\right| = 2.5 \times 10^3 \left|\frac{\Delta q_{sur}}{T_{sur}}\right| + \left|\frac{\Delta \theta_{sur}}{\theta_{sur}}\right| \tag{3}$$
$$\qquad\qquad\quad \text{E} \qquad\qquad\quad \text{M} \qquad\qquad\quad \text{T}$$

其中"| |"表示无量纲变量。图11g—i分别给出了(3)中三个项(T、M、E)对于CON和TIP_NS的差值的分布。很明显,除了青藏高原地区外,在印度北部和东亚地区,相当位温的变化的主要贡献项是水汽项。

除此之外,局部地表蒸发是大气水汽的另一个来源。我们计算了印度北部地区的蒸发和降水的区域平均值。结果表明,无论是再分析资料,还是在参考试验中,局部蒸发都远小于降水。这意味着,即使所有从地面蒸发的地表水蒸气都在局地凝结并掉落,也只会占当地降水的三分之一左右(图略)。这说明,由高原热力强迫引起的印度洋水汽输送对印度北部和东亚地区的相当位温极大值、湿对流和大陆季风的形成有重要贡献。还要说明的是,我们还在"干大气"模式框架中开展了有无高原表面感热加热对环流影响的模拟(图略)。结果表明,即使没有湿对流过程对大气的反馈作用,高原表面的感热加热依然会在低层造成

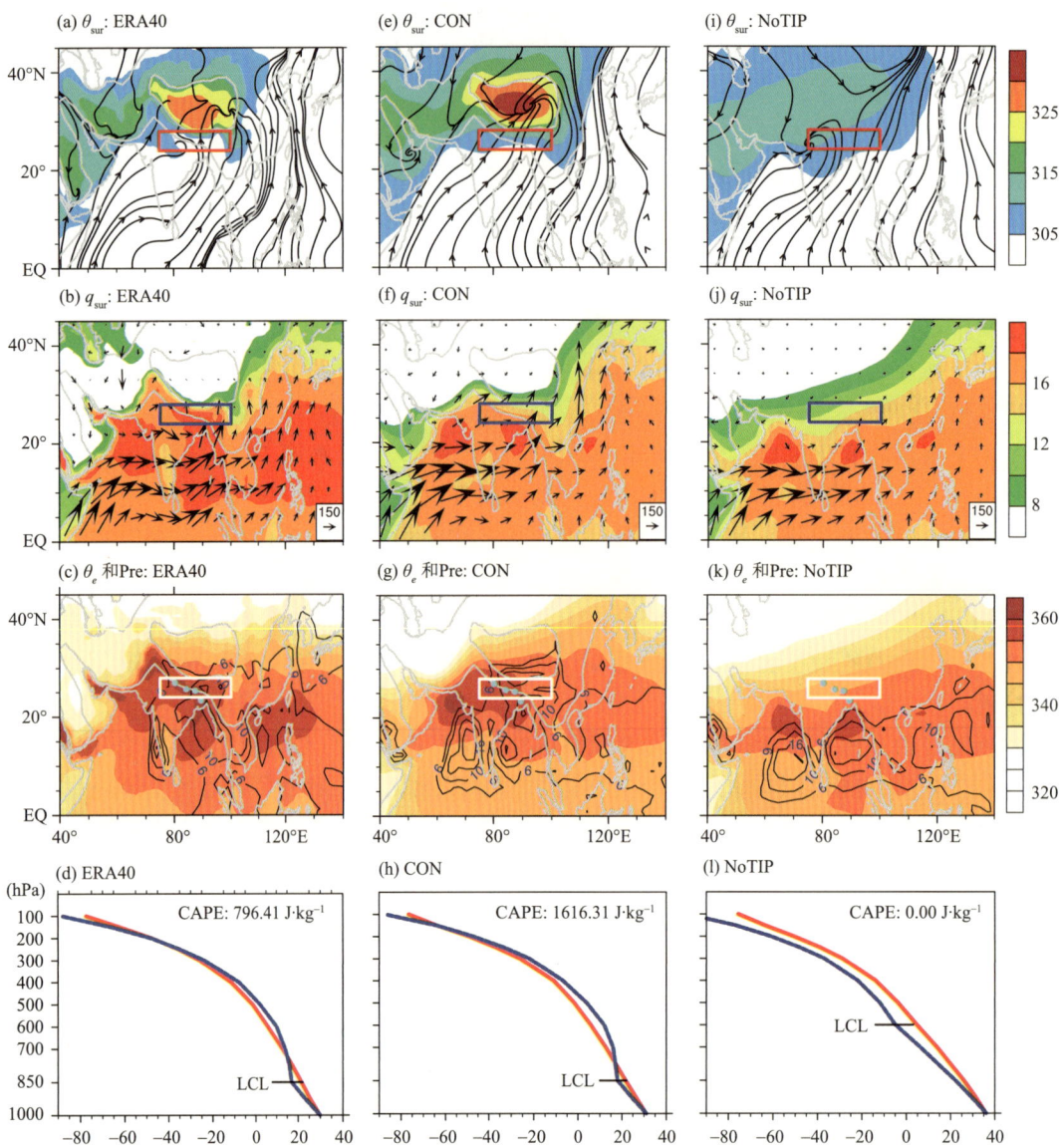

图 10 夏季平均的表面位温 θ_{sur}(K)和流场(第一行),q_{sur}(g·kg^{-1})和 850 hPa 水汽通量(矢量,kg·m^{-1}·s^{-1})(第二行),θ_e(K)和降水(等值线,mm·d^{-1})(第三行)。第四行给出了印度北部四个淡蓝色格点平均的环境温度层结曲线(K,红色)和气块沿湿绝热线上升的温度(K,蓝色),其中(d)由 ERA-40 资料计算,(h)由 CON 试验计算,(l)由 NoTIP 试验计算。长方形表示区域(24°~28°N, 75°~100°E)(引自 He et al., 2015)

气旋式环流异常,从印度洋向北吹向高原南侧斜坡地区。这说明高原感热加热对大陆地区季风环流和降水形成的影响是第一位的。

4.2 海气相互作用对青藏-伊朗高原热力强迫的调制

上面两节回顾的研究都是在观测的海温强迫下,利用大气模式模拟的结果。我们知道亚洲季风是一个海陆气耦合的复杂气候系统,青藏高原作为亚洲最大的地形强迫,对周围的海气相互作用过程必然会有影响,这种影响进一步会反馈给季风系统。因此,非常有必要研究青藏高原的热动力强迫作用如何影响周围海洋,并分析其中的反馈机制。我们利用气候系统模式 FGOALS-s2 开展了有无青藏-伊朗高原表面加热的两组试验,并同其大气模式分量模拟的有无青藏-伊朗高原表面加热的两组试验开展比较(图 12)。其中 CON 试验是标准的 AMIP 试验,TIPNS 试验是在 AMIP 试验中将青藏-伊朗高原表面加热去除。

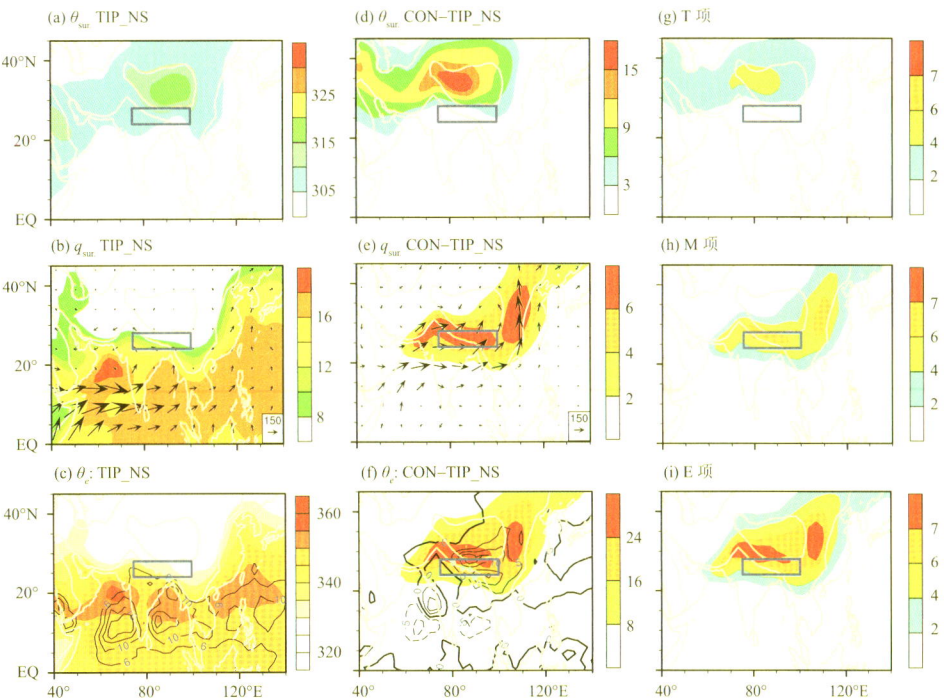

图 11 夏季平均的 θ_{sur} (K)（第一行），q_{sur} (g·kg^{-1}) 和 850 hPa 水汽通量（矢量，kg·m^{-1}·s^{-1})（第二行），θ_e (K) 和降水（等值线，mm·d^{-1})（第三行）。其中 (a)—(c) 是 TIP_NS 的结果，(d)—(f) 是 CON—TIP_NS 试验的结果，(g)—(i) 分别是 T，M，E 项（引自 He et al.，2015）

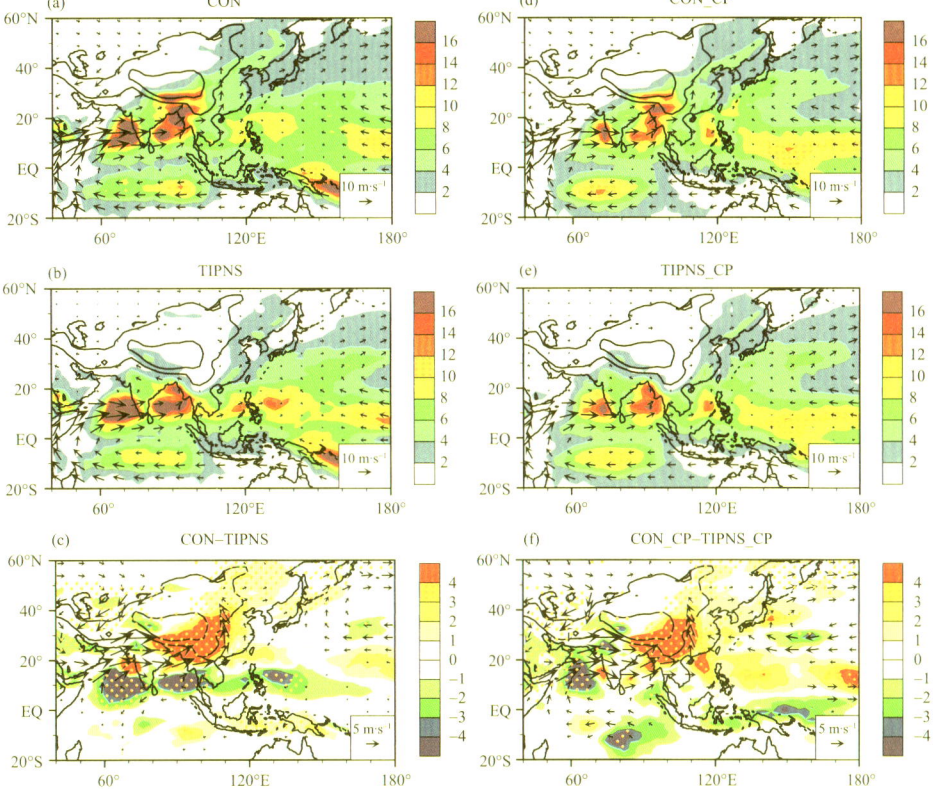

图 12 FGOALS-s2 模拟的夏季气候平均降水（填色，mm·d^{-1}）和 850 hPa 风场（矢量，m·s^{-1}）。(a)CON 试验，(b)TIPNS 试验，(c)CON－TIPNS 试验，(d) CON_CP 试验，(e)TIPNS_CP 试验，(f)CON_CP－TIPNS_CP 试验（引自 He et al.，2019）

CON_CP 是海气耦合试验，TIPNS_CP 是在耦合模式中将青藏-伊朗高原表面加热去除。从图 8a、b 中我们可以看出，两组试验都能较好的模拟出亚洲夏季风的基本特征。值得注意的是，这里 CON 试验模拟的季风降水要比在 Wu 等（2012）要更加接近观测，这也是模式对流方案优化后的结果，有利于增加敏感性试验结果的可靠性。当把高原表面感热去除后，两组试验（图 12b，e）在亚洲大陆地区模拟的降水都有所减少，特别是青藏高原及南侧斜坡地区的强降水消失，说明高原热力作用对大陆季风调控的重要性。另一方面也说明，无论考不考虑海气耦合过程，高原热力强迫对大陆地区季风降水和环流的影响变化不大。从参考试验和无感热试验的差异场（图 12c，f）比较可知。高原热力强迫对海洋地区降水的影响具有显著不同。在海温固定的试验中，高原热力强迫导致降水在热带印度洋和太平洋显著减少，而在海气耦合的试验中，降水主要在阿拉伯海、热带印度洋和太平洋地区有所减少，而在西太平洋地区反而有所增加。这跟海洋对高原热力强迫调控的响应密切相关。

为了理解海气相互作用对高原热力强迫的响应及反馈，我们比较了高原热力强迫在有无海气耦合试验中的差异，利用（CON_CP－TIPNS_CP）－（CON－TIPNS）的结果，我们分析了海气耦合对高原热力强迫作用的间接影响（图 13）。从降水和 850 hPa 环流的响应上来看，较为明显的是降水在青藏高原的北印度洋减少，在热带印度洋增加，而在南印度洋减少的经向三极空间型。这与 AMIP 试验中，青藏-伊朗高原直接热力强迫造成的降水（图 12c）是相反的。环流场上在欧亚大陆围绕高原出现一个强大的反气旋式环流异常，而在印度洋 10°N 附近存在一个较小的气旋式异常，在南印度洋则为反气旋，这种环流型和降水异常密切相关。总体上来说，海气耦合倾向于减弱高原的热力强迫作用，使得从印度洋向亚洲内陆输送的水汽减少，从而导致高原南侧和北印度降水减少，相应的热带印度洋降水增加。从地表温度的响应来说（图 13b），在陆地上，由于高原热力作用的减弱，高原南侧和印度大陆降水减少，云量减少，到达地表的太阳辐射增加，因此温度升高。而在印度洋上，海温的变化则是高原直接强迫的结果。它使得北印度洋海温减少，而热带和南印度洋海温升高。基于混合层热量收支方程我们发现，导致印度洋海温降低的重要原因是地表净短波的减少，以及海洋混合层冷海水的水平输送，和印尼西海岸附近的上翻流。而热带印度洋海温升高的主要原因是海表的潜热通量减少，南半球印度洋海温升高的原因则是地表净辐射的增强。我们进一步给出了大气垂直环流对海气相互作用的间接响应（图 13c）。结果表明，与降水和环流空间型对应的 10°N 存在一个上升运动，同时伴随向南北两侧的下沉运动，是造成高原南侧和北印度洋以及南印度洋降水减少的直接原因。我们最后在图 14 中给出了青藏-伊朗高原热力强迫调控印度洋海气相互作用及其反馈的示意图。高原强迫的直接影响导致一个气旋环流异常在高原附近对流层低层生成，而其间接效应在高原附近形成一个反气旋式环流异常，并且伴随高原南侧的"气旋-反气旋"偶极型环流异常（细红箭头），与垂直方向上的经向环流异常耦合（粗红箭头），与高原的直接强迫作用相抵抗。在印度洋海温和海流的响应上，高原的热力强迫会造成热带印度洋表层自东向西的海流，同时混合层出现自西向东的海流。这样在纬向形成一个次级环流，使得印度洋西部下沉增暖，东部上翻冷却。

5 青藏高原表面热力强迫对东亚夏季风的调制作用

在第 3、4 节主要介绍了吴国雄院士团队在青藏高原热动力强迫调制南亚夏季风活动上的研究进展，本节将简单介绍吴院士团队在高原热力强迫调节东亚夏季风上的数值模拟研究工作。在 2007 年的研究工作中（包庆，2007），他们利用在 AMIP 数值试验中将青藏高原上的地表反照率增加和减少 50% 的方法来研究青藏高原异常变暖和变冷对东亚大气环流的影响。地表反照率减少则有利于地表吸收更多的太阳辐射，高原变暖。相反地，地表反照率增加则青藏高原表面变冷。高原增暖试验（alb_w）减去高原变冷试验（alb_c）的结果作为环流对青藏高原热力强迫的响应来进行分析。结果表明，青藏高原地区地表面温度有很强的正异常，高原大部分地区增暖幅度均超过 2 K。这正是由于高原地表反照率改变，引起地表面热力状况发生变化的结果。另外，在长江中下游地区、孟加拉湾东北部和印度半岛孟买以东大部分地区地表面温度出现明显负异常（图 15a）。从高层的环流来看（图 15b），青藏高原上空存在强大的反气旋性环流异

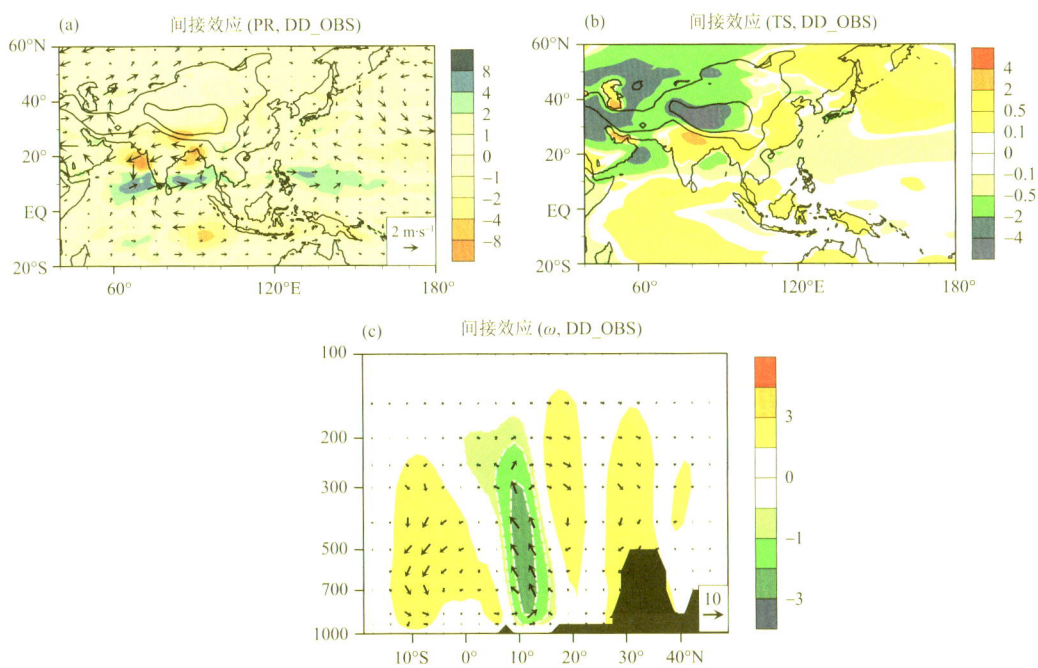

图13 海气相互作用对青藏伊朗高原热力强迫影响亚洲夏季风的间接调控作用(DD_OBS=(CON_CP－TIPNS_CP)－(CON－TIPNS))。(a)夏季平均的降水(填色，mm·d^{-1})和850 hPa风场(矢量，m·s^{-1})异常场。(b)表面温度(K)，(c) 75°～110°E平均的垂直速度(阴影，10^{-2} Pa·s^{-1})和经向环流(矢量，v,m·s^{-1}，-3ω，10^{-2} Pa·s^{-1})(引自 He et al.，2019)

常，这表明南亚高压(简称 SAH)强度增强。青藏高原北侧西风急流和南侧东风急流风速增大、位置北移。同时，东亚地区高层的东北风异常增大。另外，SAH 下游西北太平洋上空存在明显的气旋性环流异常。而低层的环流场来说(图 15d)，对流层低层东亚包括西北太平洋地区有很强的反气旋性环流异常，表明西太副高增强。在强大的西北太平洋副热带高压的西北侧，西南季风异常增强，携带大量暖湿气流由孟加拉湾和中国南海途经我国华南地区和朝鲜半岛一直到达日本北部地区。上升区以东出现补偿性弱下沉，对应于低层反气旋性环流中东部的北风气流和降水减弱。该下沉运动要比由于热力强迫引发的上升运动弱很多(图 7b)。

基于副热带环流对非绝热加热响应的 Sverdrup 平衡关系，他们对高原热力强迫影响下游环流的物理过程进行了解释(图 16)：青藏高原陆面加热增暖激发深对流加热引起对流层低层和高层环流变化的初始倾向；由于环流对高原上热源的适应过程，高原上空低层出现气旋式环流，高层为反气旋环流。伴随气旋环流东侧的南风偏差，高原以东地区降水增多，导致东亚大陆地区由于凝结潜热释放增加而出现温度的正差异。平衡态时，中国东部的潜热加热使高层北风增强，反气旋环流位于高层加热区西边；低层在加热区南风增强，加强其东部的反气旋环流，西太平洋副高表现为加强西伸。

6 总结和展望

文中回顾了吴国雄院士带领的研究团队近十几年来在青藏高原影响亚洲夏季风研究上的科研进展。他们通过运用位涡理论和辐射-对流准平衡理论开展分析，并结合自主研发的大气环流模式 SAMIL 及其海气耦合版本 FGOALS-s2 开展了一系列数值试验，揭示了青藏-伊朗高原作为亚洲最大的地形，其热力强迫对亚洲夏季风在亚洲大陆的形成与维持具有至关重要的作用。高原的表面加热是一种地形的抬升加热，因此其动力作用也是非常重要的。但是，在夏季，高原的动力作用主要体现在降低抬升凝结高度，使得局地的湿对流更加容易激发，而不是隔断南北冷暖空气交汇。印度洋向亚洲大陆低层的水汽输送仍然是

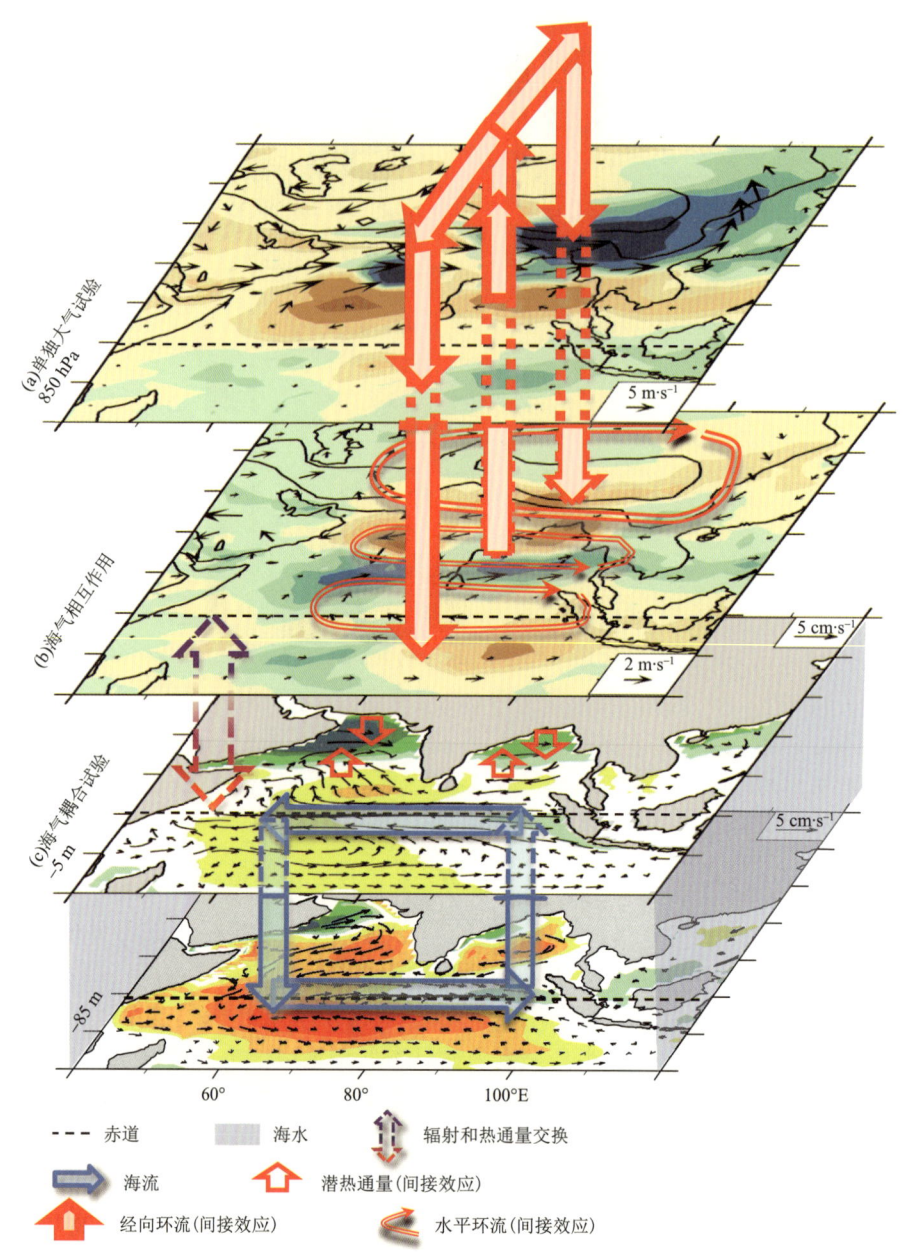

图 14 青藏伊朗高原热力强迫调节海气相互作用对亚洲夏季风影响的示意图。高原强迫的直接影响导致一个气旋环流异常在高原附近对流层低层生成,而其间接效应在高原附近形成一个反气旋式环流异常,并且伴随高原南侧的"气旋-反气旋"偶极型环流异常(细红箭头),与垂直方向上的经向环流异常耦合(粗红箭头),与高原的直接强迫作用相抵抗(引自 He et al., 2019)

高原的热力作用主导的。此外,由于亚洲季风系统的复杂性,和数值模式本身的局限,我们对青藏高原影响亚洲夏季风认识仍旧存在不足之处。比如,目前气候模式水平分辨率较粗,无法有效解析青藏高原陡峭地形,另一方面物理过程参数化有待进一步完善,这都将导致青藏高原地形的动力作用和热力作用模拟存在偏差。从最新的耦合模式比较计划 CMIP6 的结果上看,相比较于 CMIP5 的结果,大多数模式对亚洲夏季风的模拟仍旧存在系统性误差,这也必然导致我们对调控亚洲夏季风内部变率过程的认识存在局限性。因此,进一步发展气候系统模式,开展相关的模拟和理论研究,是深化青藏高原气候动力学认识的关键途径。

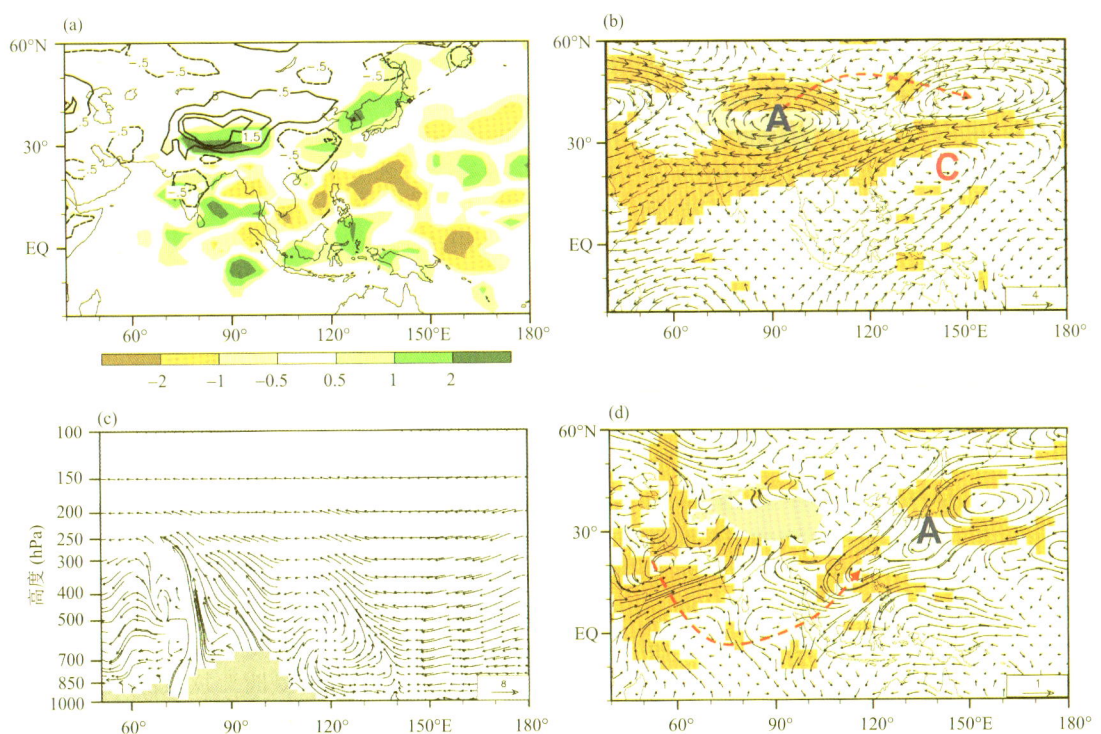

图 15 青藏高原增暖(高原反照率×0.5减去反照率×1.5)的敏感性试验结果。气候态夏季平均的(a)表面温度(等值线,℃)和降水(阴影,mm·d^{-1})异常,(b)200 hPa 风场异常(m·s^{-1}),(c)沿33°N纬圈剖面的环流场(纬向风速单位 m·s^{-1},垂直速度单位 hPa·d^{-1}),(d)850 hPa 风场异常(m·s^{-1})。其中(b,d)中黄色填色为通过95%显著性检验区域(引自包庆,2007)

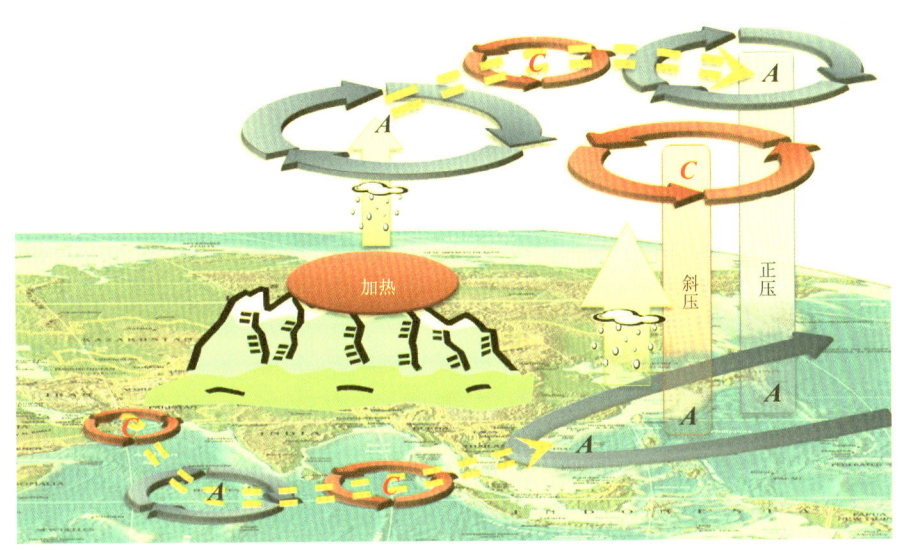

图 16 青藏高原热力强迫影响东亚夏季环流和降水示意图(引自包庆,2007)

参考文献

包庆,2007. 青藏高原气候动力学的数值模拟研究[D]. 北京:中国科学院研究生院(中国科学院大气物理研究所).
顾震潮,1951. 西藏高原对东亚大气环流的动力影响和它的重要性[J]. 中国科学,2:283-303.
拉梅奇 C S,1978. 季风气象学[M]. 北京:科学出版社:1-190.

梁潇云,刘屹岷,吴国雄,2006. 热带、副热带海陆分布与青藏高原在亚洲夏季风形成中的作用[J]. 地球物理学报(4): 983-992.

毛江玉,吴国雄,刘屹岷,2002. 季节转换期间副热带高压带形态变异及其机制的研究Ⅰ:副热带高压结构的气候学特征[J]. 气象学报,60(4):400-408.

吴国雄,李伟平,郭华,等,1997. 青藏高原感热气泵和亚洲夏季风[C]//叶笃正. 赵九章纪念文集. 北京:科学出版社: 116-126.

吴国雄,张永生,1998. 青藏高原的热力强迫和机械强迫作用以及亚洲季风的爆发Ⅱ.爆发时间[J]. 大气科学,1:51-61.

吴国雄,张永生,1999. 青藏高原的热力强迫和机械强迫作用以及亚洲季风的爆发Ⅰ.爆发地点[J]. 大气科学,6:825-838.

吴国雄,刘新,张琼,等,2002. 青藏高原抬升加热气候效应研究的新进展[J]. 气候与环境研究,7:184-201.

吴国雄,刘屹岷,刘新,等,2005. 青藏高原加热如何影响亚洲夏季的气候格局[J]. 大气科学,29(1):47-56.

叶笃正,罗四维,朱抱真,1957. 西藏高原及其附近的流场结构和对流层大气的热量平衡[J]. 气象学报,28(2):108-121.

叶笃正,高由禧,1979. 青藏高原气象学[M]. 北京:科学出版社:278.

BAO Q,LIN P,ZHOU T,et al,2013. The flexible global ocean-atmosphere-land system model version: FGOALS-s2[J]. Adv Atmos Sci,30(3):561-576. doi:10.1007/s00376-012-2113-9.

BOLIN B,1950. On the influence of the earth's orography on the general character of the westerlies[J]. Tellus,2: 184-195.

BOOS W R,KUANG Z,2010. Dominant control of the South Asian monsoon by orographic insulation versus plateau heating[J]. Nature,463:218-222.

CHARNEY J G, ELIASSEN A,1949. A numerical method for predicting the perturbation of the middle latitude westerlies[J]. Tellus,1:38-55.

EMANUEL K A,NEELIN J D,BRETHERTON C S,1994. On large-scale circulations in convecting atmosphere[J]. Quart J Roy Meteor Soc,120:1111-1143.

FLOHN H,1957. Large-scale aspects of the "summer monsoon" in South and East Asia[J]. J Meteor Soc Japan,35A:180-186.

HE B,GUO X W,LIU Y M,et al,2015. Astronomical and hydrological perspective of mountain impacts on the Asian summer monsoon[J]. Sci Rep,5:17586. doi:10.1038/srep17586 (2015).

HE B,LIU Y,WU G,et al,2019. The role of air-sea interactions in regulating the thermal effect of the Tibetan-Iranian Plateau on the Asian summer monsoon[J]. Climate Dynamics,52(7-8):4227-4245. doi:https://doi.org/10.1007/s00382-018-4377-y.

MOLNAR P,BOOS W R,BATTISTI D S,2010. Orographic controls on climate and paleoclimate of Asia:Thermal and mechanical roles for the Tibetan Plateau[J]. Annual Review of Earth and Planetary Sciences,38:77-102.

QUENEY P,1948. The problem of air flow over mountains: A summary of theoretical studies[J]. Bull Amer Meteorol Soc,29:16-29.

WU G X,1984. The nonlinear response of the atmosphere to large-scale mechanical and thermal forcing[J]. J Atmos Sci,41(16):2456-2476.

WU G X, LIU Y M, ZHANG Q, et al,2007. The influence of the mechanical and thermal forcing of the Tibetan Plateau on the Asian climate[J]. J Hydrometeorol,8:770-789.

WU G X,LIU Y M,HE B,et al,2012. Thermal controls on the Asian summer monsoon[J]. Sci Rep,2(404):1-7. doi:10.1038/srep00404.

WU T W, LIU P, WANG Z Z,et al,2003. The performance of atmospheric component model R42L9 of GOALS/LASG[J]. Adv Atmos Sci,20:726-742.

YEH T C,1950. The circulation of the high troposphere over China in the winter of 1945—1946[J]. Tellus,2(3):173-183.

代表性论文 12

青藏高原感热气泵和亚洲夏季风

吴国雄[1],李伟平[1],郭华[1],刘辉[1],薛纪善[2],王在志[2]

(1 大气科学和地球流体力学国家重点实验室·中国科学院大气物理研究所,北京 100029;
2 广东省气象局热带气象研究所,广州 510080)

摘要:通过对 GOALS-LASG 气候模式的数值试验,发现高原上空气柱在冬季的下沉运动和在夏季的上升运动起着一个"气泵"的作用。证明该"气泵"主要是由青藏高原与大气的感热交换所驱动的,并被定义为青藏高原"感热气泵"(SHAP)。SHAP 的有效工作导致高原地区由冬到夏大气环流的突变及南亚副热带高压的突然北跳,并维持着亚洲季风期。盛夏季节,SHAP 对低空暖湿大气的动力抽吸作用维持了高原地区的降水并调节着东亚和南亚的气候。它在高层的动力排放作用和表面感热源在高原上空形成的深厚负涡度源作用下形成盛夏高原上空强大的副热带高压;并通过能量频散,影响着全球的高空气候。

关键词:青藏高原感热气泵;能量频散;Rossby 波列;南亚副热带高压

1 引言

亚洲季风区大气环流由冬到夏的季节变化具有突变特征(Ye et al.,1959;Matsumoto,1992;Murakami and Matsumoto,1994)。它通常出现在 5—6 月,与亚洲季风的爆发紧密相关(Krishnamurti et al.,1985;Hirasawa et al.,1995)。Hahn 和 Manabe(1975)以及朱抱真(Wu et al.,1996b)的大气环流数值试验表明,亚洲季风的存在是以高原的存在为前提的。在不包含高原的试验中亚洲季风消失,印度洋上的雨带和辐合带接近赤道,与太平洋上空的赤道辐合带(ITCZ)相似。更多的研究表明季风是大气对海陆热力差异变化的响应(Murakami and Ding,1982;Johnson et al.,1987;Luo and Yanai,1983,1984;He et al.,1987;陈隆勋 等,1991)。尤其是高原对大气的抬升加热对亚洲季风的形成和维持起着重大的作用(Staff Members,1957;Flohn,1957,1969;叶笃正和高由禧,1979;Ye,1981,1982;Li and Yanai,1996)。由此可推测在上述所报道的大气环流数值模拟中,在无地形试验中亚洲季风的消失并不单纯是由于高原机械强迫效应的消失,更主要的可能是抬升的高原加热被排除的结果。本文的试验就是为了证实这一推论而设计的。

这里所用的模式是 LASG 的全球海-气-陆耦合气候模式 GOALS/LASG(吴国雄 等,1997),它包含有 9 层大气、20 层海洋、3 层土壤、1 层植被,以及 2 层海冰。在大气模式中包含有辐射、扩散及干-湿对流调整的参数化方案。下垫面过程包括动量、潜热和感热的相互交换。模式还设计有各种开关,以控制各圈层的相互作用。在本文的研究中,海洋和土壤两部分的开关被关闭;陆表植被分布由观测场提供;海温和海冰分布由国际"大气模式比较计划"(AMIP)所提供的 1979—1988 年的观测值内插到模式的高斯网格上得到。模式分辨率为菱形截断 15 波。模式垂直坐标为 σ-坐标;最低层 $\sigma=0.991$,离地面约 70 m。

本文原载于《赵九章纪念文集》,北京:科学出版社,1997:116-126

本研究设计了两个试验。第一个试验为控制试验(CON)，它就是用 AMIP 提供的海温和海冰资料，从 1979 年 1 月 1 日积分到 1988 年 12 月 31 日共 10 年得到的。其模拟的平均气候状况可见(Wu et al.,1996a)。第二个试验为高原无感热加热试验(NSH)，它与 CON 的唯一差异是在青藏高原高度高于 3 km 的格点上在积分热力学方程的每一时间步中令地表对大气的感热加热为零，其余过程均与 CON 一致。这两个试验的气候平均态可通过计算各自试验结果的 10 年平均得到，也就是说，每个试验的月均量包含有 10 个样本量。通过分析用 NSH 的结果减去 CON 的相应结果的差异分布，可认识高原感热加热对气候状态的影响。

2 青藏高原感热气泵及其对低空环流的影响

对 CON 中 $\sigma=0.991$ 面上 10 年平均的月均流场再进行 12 个月平均，得到年均流场。然后用 1 月和 7 月的月均流场减去该年均流场，可得到 1 月和 7 月的偏差流场，结果如图 1 所示。这种处理方法能滤除中、短期天气过程及年际变化的讯号，而保留了影响气候型的稳定因子。比较图 1a 和 1b 发现，1 月和 7 月的偏差流场在全球中低纬几乎是处处反向。中纬度的反向流场反映着 1 月份北半球阿留申低压，冰岛低压，及 7 月份南半球南美和澳大利亚西南海域低压等半永久性大气活动中心的作用。低纬度的反向流场反映着主要季风系统的作用。仔细对照发现，流场总是从冬半球的大陆辐散流向海洋，然后跨越赤道进入夏半球，最后向夏半球陆地辐合，反映着海陆热力差异支配着季风环流的特征。从图中还可看出沿副热带分布的在冬夏季反号的主要辐合-散中心在北半球位于北非-阿拉伯半岛，青藏高原，北太平洋中部，及拉丁美洲；在南半球位于南部非洲，澳大利亚，及南美玻利维亚高原。低纬度的季风系统就是往返于两半球这些辐合/散中心之间的主要风系。根据这些南北中心之间的组合，从图 1 中我们可以大致划分下述五大季风系统：南美-北非季风系，南美-拉丁美洲的美洲季风系；澳大利亚-北太平洋中部的澳大利亚-西太平洋季风系；澳大利亚-青藏高原的东亚季风系；以及南部非洲-青藏高原间的南亚-南非季风系。

图 1 中另一明显的特征是青藏高原上空大尺度垂直运动的变化。在 $\sigma=0.811$ 的高度上，高原南侧 1 月份出现全球唯一的强度大于 8×10^{-4} hPa·s^{-1} 的偏差下沉运动中心，7 月份出现全球唯一的强度大于 8×10^{-4} hPa·s^{-1} 的偏差上升运动中心。计算表明(Wu et al.,1996b)，模式中高原表面 1 月份对大气的冷却率为 50 W·m^{-2}；7 月份对大气的加热率为 85 W·m^{-2}。前者使高原上空气柱冷却达 1 ℃·d^{-1}；后者则以近 2 ℃·d^{-1} 的速率加热其上空大气。因此，高原地表与大气的感热交换使其上的大气在冬季冷却下沉，像一个排气泵一样向低层大气源源不断地输送质量；在夏季受热上升，像一个吸气泵一样把低层大气吸向高空。图 1 表明，青藏高原的这一"感热气泵"不仅调节着北太平洋上空半永久性活动中心与高原地区质量交换的方向，还控制着除了南非-北美和南美-拉丁美洲季风区以外全球另外三大季风区的质量交换方向。由于 1 月和 7 月是气压场相对稳定的月份，还由于求平均的时段足够长，局地变化项很小，则图 1 的平均偏差流场可以近似地看成平均偏差轨迹场。于是高原的"感热气泵"在 1 月份把大陆空气沿下列三条途径向南半球排放：第一条从高原腹地沿阿拉伯海—索马里沿岸—经南印度洋流到南部非洲辐合；第二条从华南经南海—孟加拉湾—经赤道中印度洋到达澳大利亚；第三条从蒙古—我国东北—经过日本到达中部北太平洋，然后穿越赤道进入南太平洋，与流向澳大利亚的气流汇合。7 月份这三股气流又沿着相反的方向向高原腹地辐合。在 1 月份它们以东北季风形式从北半球跨越赤道；在 7 月份它们又从南半球跨越赤道以西南季风的形式影响亚洲和西太平洋地区。从这一意义上，高原"感热气泵"和"南非泵""澳大利亚泵"形成低层大气运动的三个驱动源，主宰着从西太平洋，东亚和南亚到东南非洲，南印度洋和澳大利亚的大气环流季节变化。

在 NSH 试验中，我们把 3 km 高度以上的"感热气泵"(SHAP)关闭，则高原上空 7 月份的上升运动将大为削弱。从图 2a 可以看出，在(NSH−CON)的差异散度场中，高原上空的低层为辐散；在其周围则为异常的辐合，其中以华南、中南半岛及印度的差异辐合为甚。差异涡度场(图 2b)表现为高原地区的反气旋涡度和沿华南、中南半岛到印度的气旋性涡度。在斯堪特纳维亚半岛和加勒比海还有两个较弱的正差

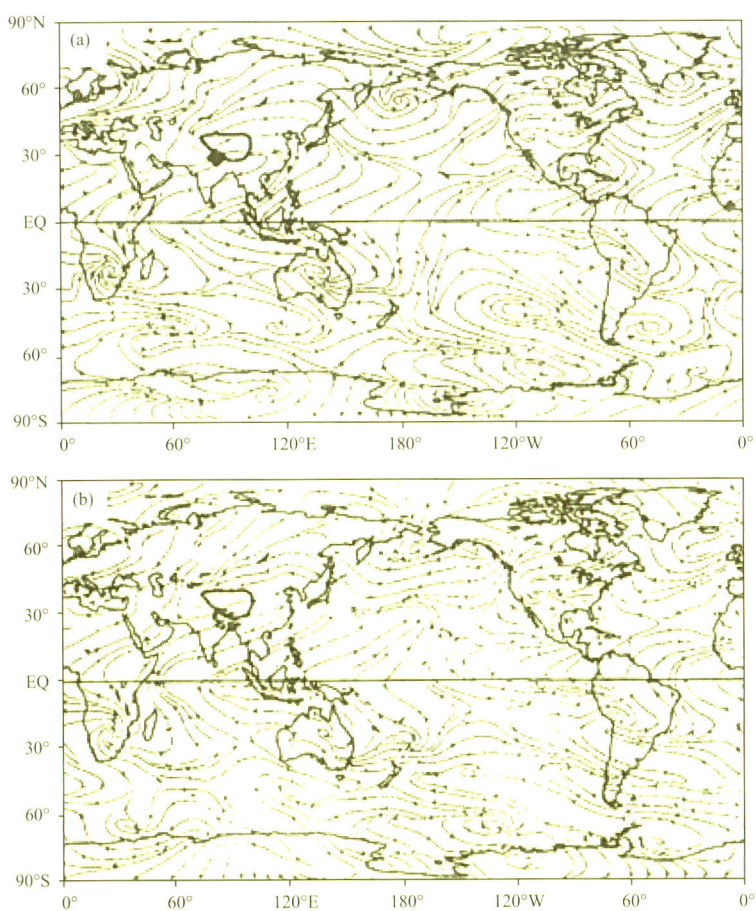

图 1 在 CON 试验中，$\sigma=0.991$ 面上月气候平均流场与年气候平均流场的偏差分布（细实线），以及 $\sigma=0.811$ 面上月平均垂直运动 ω 与其年平均的偏差（阴影区）等值线间隔为 2×10^{-4} hPa·s^{-1} 的分布。粗实线为青藏高原高于 3 km 的等高线 (a) 1月，ω 只给出大于 8×10^{-4} hPa·s^{-1} 的区域；(b) 7月，ω 只给出小于 -8×10^{-4} hPa·s^{-1} 的区域

异涡度区，其成因不详。于是降水在高原及华北地区减少；沿华南、中南半岛及印度一带增加（图2c）。与加勒比海的正差异涡度中心相对应，那里的降水也增加。总的来说，SHAP的关闭对低空环流的影响以高原及其南侧、东侧的地区最为显著，表现出"邻域响应"的主要特征。

3 感热气泵对高空环流的影响

图 3 示出两种试验中 7 月份的 200 hPa 流场及其差异分布。CON 中的 200 hPa 流场（图3a）在北半球中低纬地区具有显著的 2 波特征。两个反气旋中心分别位于青藏高原上空及墨西哥附近；两大槽则分别位于两大洋上空并自东北向西南方向倾斜。跨赤道进入南半球的东北气流位于西起东大西洋，经非洲，印度洋，直至西太平洋的广大区域。这些特征与观测到的北半球夏季 200 hPa 的流场分布（Peixoto and Oort, 1992）非常相似。在 NSH 试验中（图3b）乌拉尔地区有弱脊发展；原位于东北亚的槽消失；白令海峡及巴芬湾上空出现低槽。最显著的变化是原位于高原上空的南亚高压（青藏高压）消失，反气旋中心出现在华南至南海上空。原来位于西太平洋地区的跨赤道气流也消失。东亚大气环流的总体特征与观测到的早春的大气环流形势更为接近。由此看来，夏季 200 hPa 青藏高压的维持是与青藏高原对大气的感热加热分不开的。图 3c 为 (NSH－CON) 试验所得的差异场分布。该图的显著特征是高原上空有强烈的气旋式流场；以它为中心出现了一组波数为 2 的罗斯贝波波列。向下游的一支，其反气旋式辐散中心正好叠加

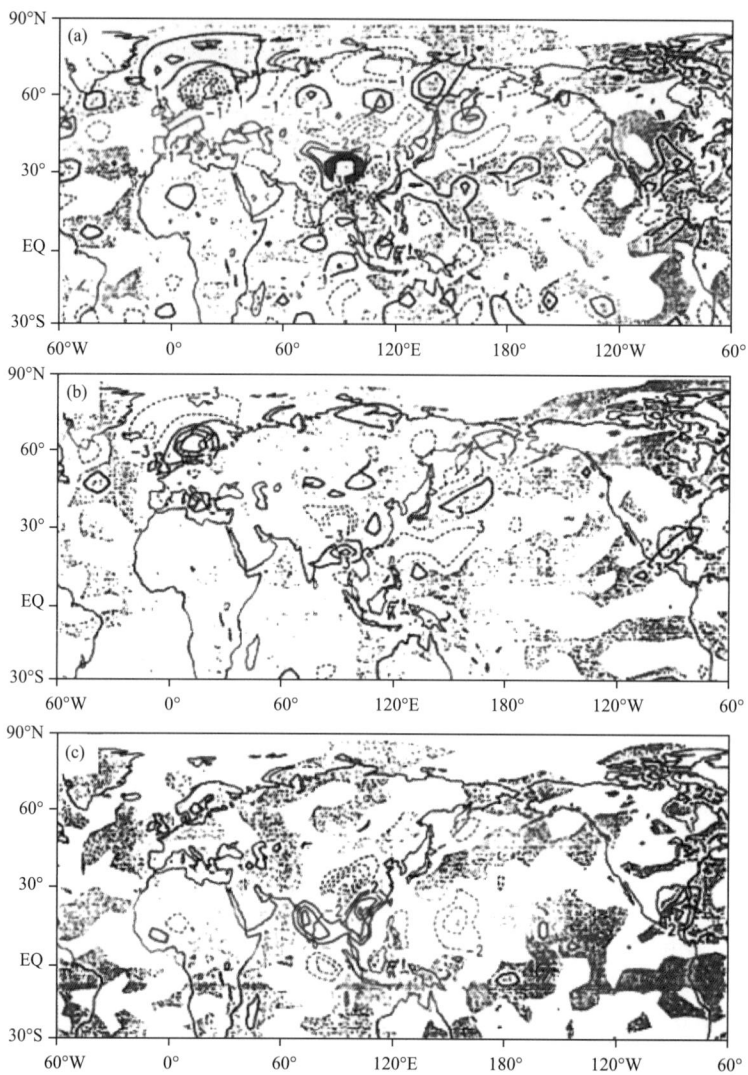

图 2 在青藏高原感热气泵关闭后，7月份 $\sigma=0.991$ 面上 NSH 试验与 CON 试验的差异场分布。(加阴影区表示正值区)(a)差异散度场(等值线间隔为 1×10^{-6} s^{-1})；(b)差异涡度场(等值线间隔为 3×10^{-6} s^{-1})；(c)差异降水场(等值线间隔为 2 mm)

在原东北亚槽区(图 3a)上并使该槽消失(图 3b)；此后该波列折向东南，并沿着太平洋洋中槽向低纬延伸的西风区向东南方向传入热带，跨越赤道，并在东太平洋进入南半球。上游的一支从高原向西北伸展，反气旋式辐散中心位于乌拉尔地区，使那里出现浅脊(图 3b)；位于英国北部北海海域的气旋式辐合中心叠加在原来的浅脊南部(图 3a)，并使之消失。由于定常波能量只能向下游频散，上述高原上游波列在平均气候态中出现，这说明由高原辐合涡源所激发的向上游传播的瞬变波列具有相似的波长和特征。除了上述的主波列外，由高原向下游传播的波列在太平洋北部出现分支。向北的一支经阿拉斯加后折向南，在巴芬湾上空出现强烈的气旋式差异流场，导致在 NSH 试验中巴芬湾上空出现低槽。此后，该分支波列向东南方向经中北大西洋传到非洲西岸佛得角北部，受副高轴($u=0$)及其南部的东风气流所阻而消失。

与低空环流的邻域性特征相比，高原感热气泵对高层大气环流的影响几乎是全球性的。它所激发的 Rossby 波列不仅影响了北半球的槽脊分布，还通过赤道东太平洋的"西风通道"把扰动传进了南半球。

为了更进一步了解影响大气环流的高原 SHAP 的特征，我们在图 4 中给出 7 月份 200 hPa 涡度、温度、西风及垂直环流的变化。200 hPa 最大的涡度差异出现在高原上空，强度大于 $+2\times10^{-5}$ s^{-1}。在其周围沿华南，中南半岛到南亚有一个反号的负涡度异常区。上述分布形式与图 2 所示的低空环流的"邻域响

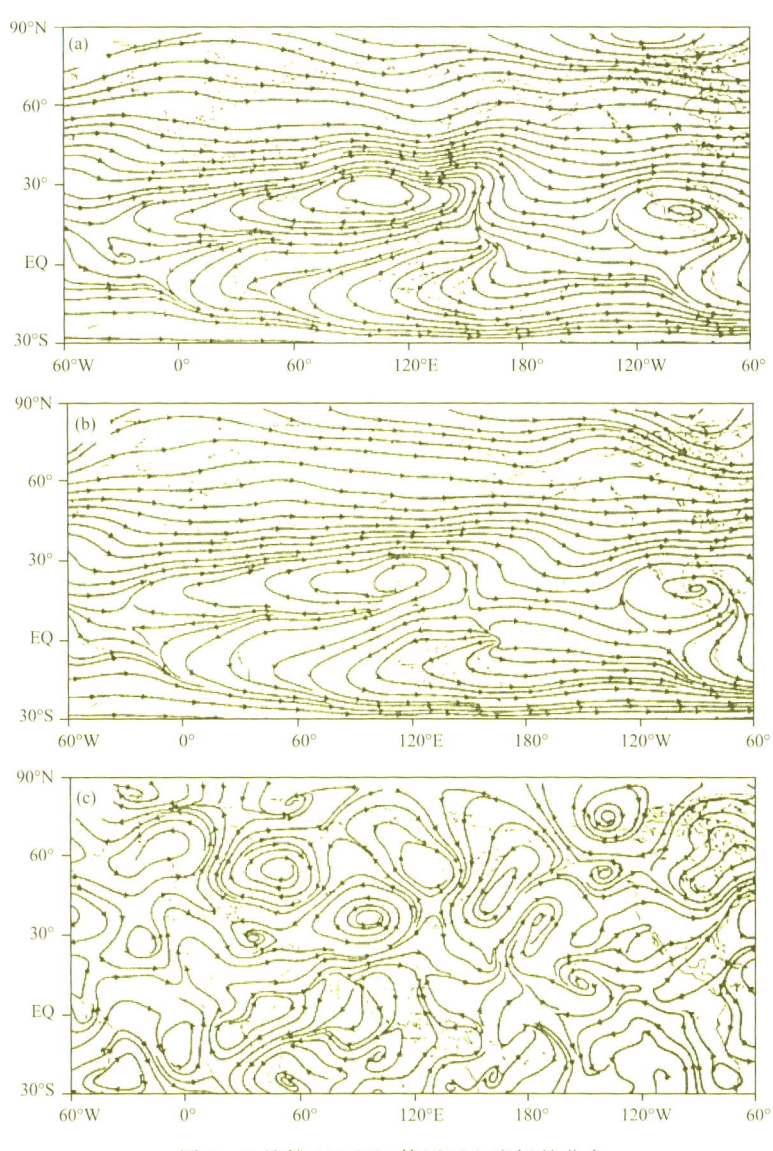

图 3 7 月份 200 hPa 等压面上流场的分布
(a)控制试验(CON);(b)无感热气泵试验(NSH);(c)NSH 与 CON 之差

应"特征相对应:高空的辐合和正涡度偏差区与低空的辐散和负涡度偏差区对应,降水减少;反之亦然。在中高纬地区,高原上空的正涡度差异则以 Rossby 波列的形式向全球传播,其路径与流场差异场的情况(图 3c)完全一致。在沿 33°N 的剖面上,温度差异场以邻域响应最为显著(图 4b)。尽管高原 SHAP 的关闭影响着全球的气温变化,除了大西洋上空增暖外,对流层其他地方的大气均变冷,然而,剧烈的温度变化发生在高原毗邻区域上空。从 60°~120°E 的范围内,对流层降温均在 1 ℃以上;最大降温在高原地面处达 6 ℃。相反,在平流层,高原上空气温升高最大值达 5 ℃以上。这意味着高原上空的对流层顶是下降的。差异纬圈环流在高原上空是下沉的。这与图 2 和图 3 中所示高原上空的差异辐合及低空的差异辐散的特征一致。图 4c 所示为沿 90°E 的西风差异分布。主要差别出现在 10°~55°N 的范围内。在高原所处范围内,500 hPa 以下南侧为东风差异,北侧为弱的西风差异;500 hPa 以上刚好相反。因此在高原上空大气中,低层有反气旋式涡度差异(图 2b),高层有气旋式涡度差异(图 3c)。图 4c 还表明高原上空有下沉差异环流,其中心在高原南侧,与图 1b 所示的偏差上升中心的位置吻合。与其对应有高层辐合异常和低层辐散异常,此外,孟加拉湾北部低空有气旋式涡度差异及辐合差异。这与图 2 所示的降水差异分布也对应得很好。

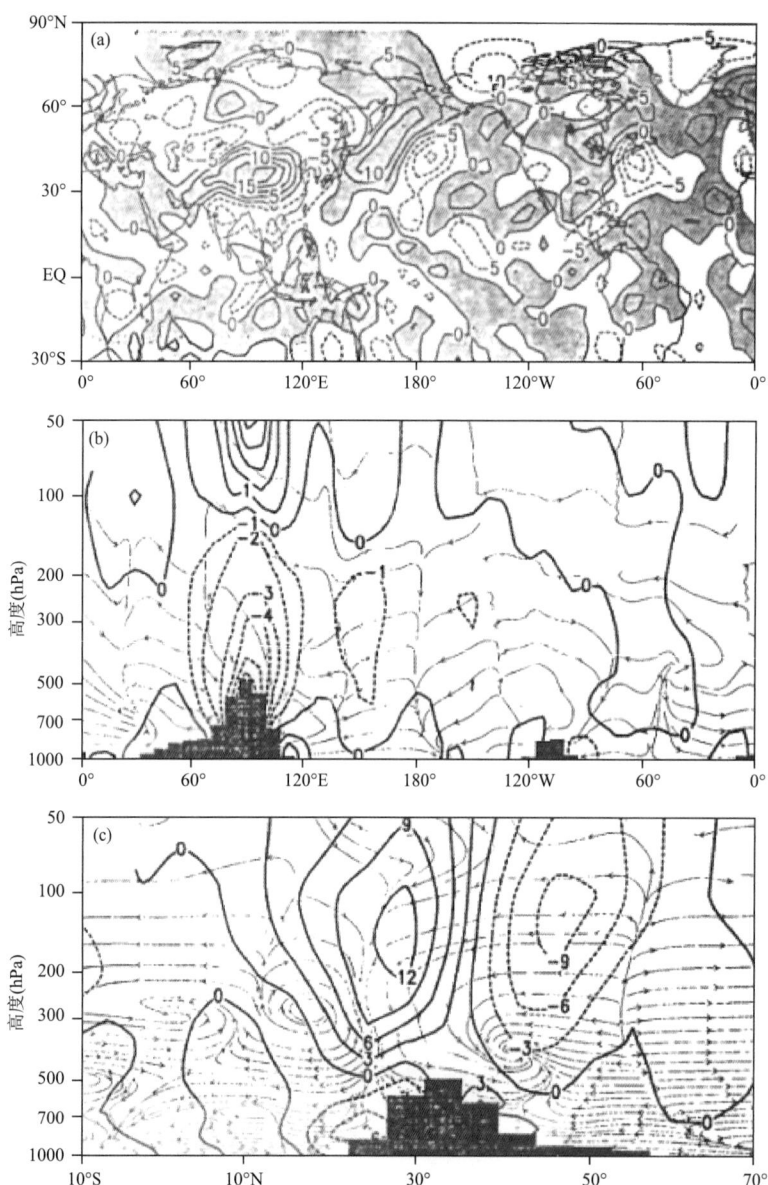

图 4 在青藏高原感热气泵关闭后，NSH 试验与 CON 试验的差异分布 (a) 200 hPa 涡度场（等值线间隔为 $5\times10^{-6}\ s^{-1}$），阴影区为正值区；(b) 沿 33.33°N 的温度剖面（等值线间隔为 1℃）和纬圈环流；(c) 沿 90°E 的西风剖面（等值线间隔为 $2\ m\cdot s^{-1}$）和经圈环流

从以上的分析可以得出如下结论：高原 SHAP 关闭对 7 月份大气环流的影响是通过下述两种机制起作用的。第一、它导致高层大气辐合，在高原上空下沉，然后在低层向外辐散，引起邻域的降水异常；第二、它导致低层大气失热强烈冷却。由于涡度的局地变化正比于加热率随高度的变化，这种低层失热将产生正的涡源，使除了近地层外，高原上空气旋式涡度显著增加。它以 Rossby 波的形式向外频散，影响着全球的气候。

4 高原感热气泵对季节变化和季风期的影响

叶笃正等（Ye et al., 1959）的研究指出，东亚大气环流的季节变化在 5—6 月具有突变的特征。图 5a 为 CON 试验中 300 hPa 纬向西风分量 u 沿 90°E 剖面的季节变化。1—3 月，东风主要在赤道附近维持；中纬度急流稳定在 35°N 附近。东西风分界线（即副高轴位置）在 5 月下旬以前一直维持在 10°N 附近。

在5月底6月初,东西风界线突然从10°N北跳到27°N附近。西风急流也北撤至45°N附近。模式成功地模拟了副高轴在季节转换期间的突发性北跳。在高原感热气泵关闭掉的试验中(图5b)副高轴从5月位于10°N附近到7月份达到最北的20°N左右,历时近两个月,变化也非常缓慢。由此看到,高原感热气泵是导致高原附近大气环流在季节转化中发生突变的重要原因。

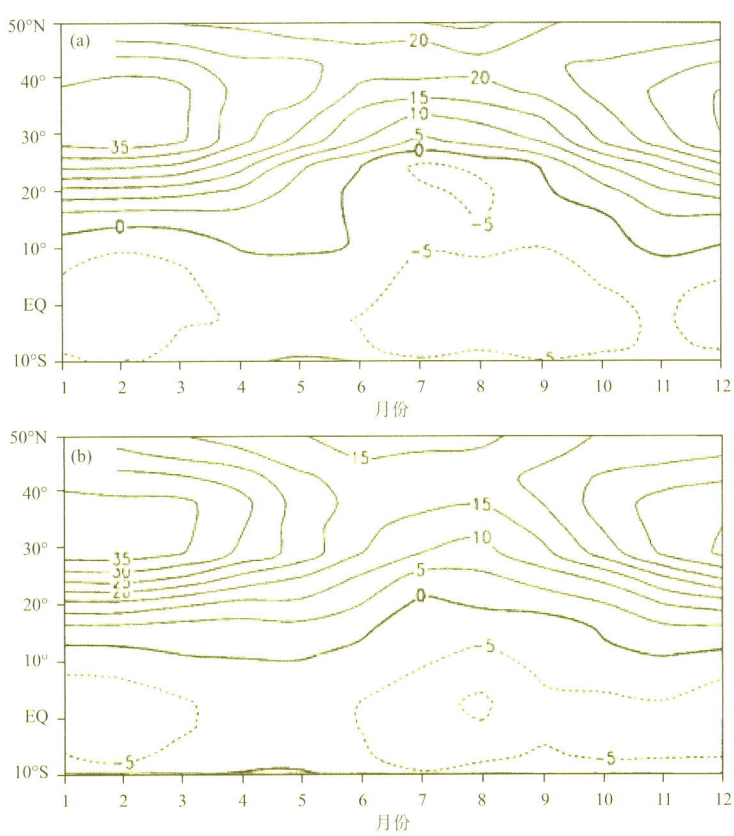

图5 沿9°E的剖面上,月均300 hPa纬向西风(m·s^{-1})及副高轴线(等值线 $u=0$)随季节的变化
(a)CON试验;(b)NSH试验

Li 和 Yanai(1996)曾用1979—1992年共14年平均的上部对流层(200~500 hPa)沿30°N和5°N的温差为因子去研究季节的变化。指出200~500 hPa高度上30°N和5°N的温差正值区可以作为亚洲夏季风的一个度量。我们使用NCAR-NCEP所提供的1982—1995年的月均再分析资料进行同样的计算,其结果如图6a所示。一般说来,热带的温度比副热带高。但在亚洲季风区夏季风期间,沿30°N的温度却高于5°N的。根据图6a,亚洲夏季风期从5月中旬开始至9月结束。我们对CON试验的10年平均的资料进行类似计算,结果如图6b所示。所模拟的夏季风期从5月中旬持续到9月中旬,与观测结果相当接近。在NSH试验中,季风期在6月上旬才开始,到8月中旬已结束,季风期比CON的缩短了一半。而且持续最强的经度也偏离高原,移至120°E海岸线附近,与200 hPa反气旋中心所在的经度(图3b)一致。从图6的分析可以看到,高原感热气泵对亚洲季风的维持是极为重要的。当这一气泵停止工作或工作低效时,亚洲季风期便会大大缩短,从而导致气候异常。

5 讨论和结论

在设计本文试验时,为了使敏感试验NSH的结果不致与实际大气环流相去太远,仍保留有亚洲季风区的基本环流特征,但同时又能揭示高原感热气泵的重要作用,我们只去除了青藏高原高于3 km处对大气的感热加热。高原低于3 km处对大气的感热加热仍然保留。严格地说,我们只是令高原的感热气泵

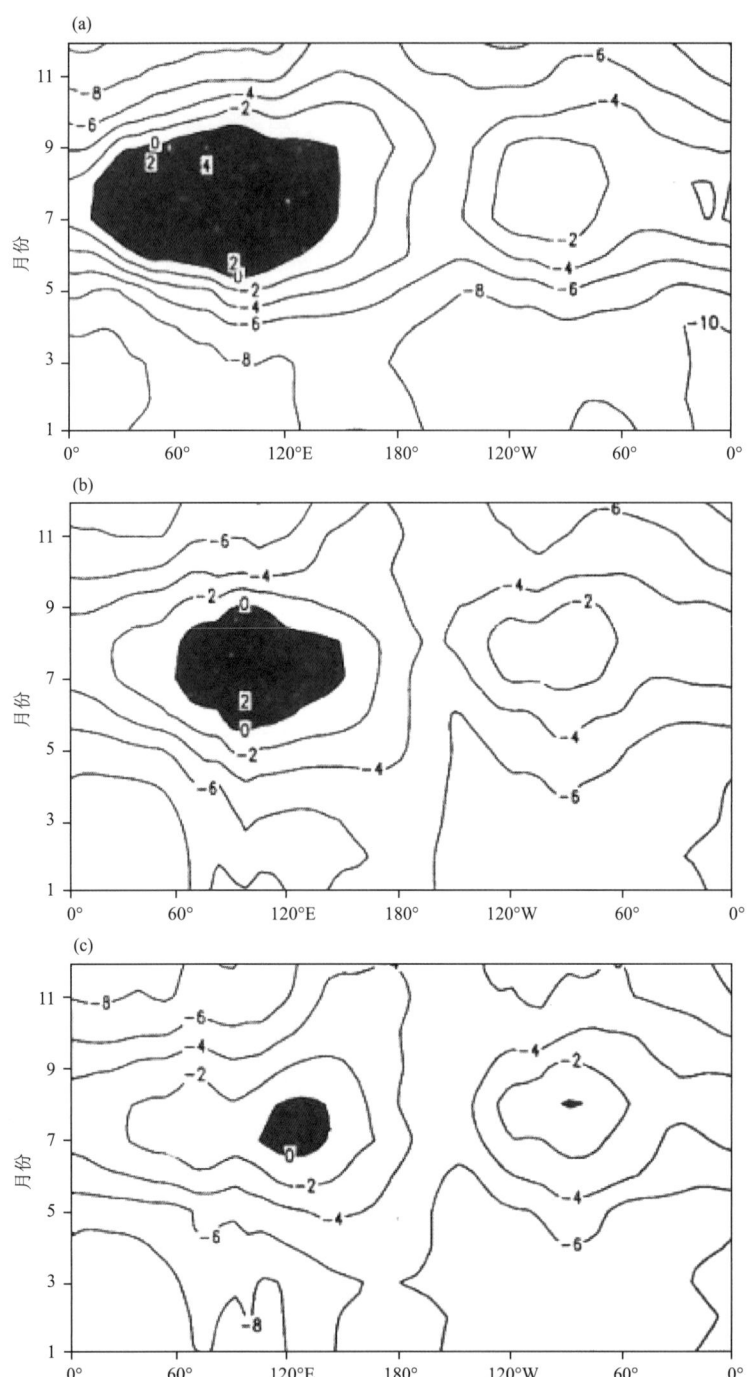

图 6 200 到 500 hPa 平均气温在 30°N 和 5°N 之间的温差沿经度分布的季节变化(单位为℃),阴影区为正值区,表征季风期;(a)观测到的(1982—1995 年);(b)CON 试验;(c)NSH 试验

处于低效的工作状态。即便如此,我们还是清楚地看到了高原感热气泵对东亚夏季环流及气候的重要控制作用。

高原感热气泵是高原地表与大气之间强烈的感热交换所驱动的气泵。由于这一热交换发生在数千米高的中低对流层中,因此该气泵能有效地工作并调节着大气有规律的季节变化。冬季高原地表的冷却作用把高层大气吸向高原地区,在那里下沉并向高原低空四周的区域排放。夏季高原地表的加热作用则把高原周围的低层大气吸向高原,在那里上升并在上部对流层向外辐散。高原感热气泵如此周而复始的抽吸和排放大气的作用有效地调控着亚洲的冬季风和夏季风。

通过设计 NSH 试验,让高原感热气泵处于低效的工作状态,并与气泵完全开放的 CON 试验相比较,可透视高原 SHAP 对全球气候的贡献。我们发现,在夏季高原 SHAP 对大气环流的影响是通过下述两种机制起作用的。动力上,它吸进低空的周边暖湿大气,在把其泵向上层的过程中释出降水后在对流层上层向外排放。热力上,出现强烈的负涡度源。它维持着高原上空的夏季南亚高压。高原上空对流层高层的辐散气流和负涡度源通过 Rossby 波向上下游频散,影响着北半球各地的气候。大西洋上空的 Rossby 波列在南传中在非洲西岸受副高轴($u=0$)所阻而消失。太平洋上空的 Rossby 波列则穿越赤道东太平洋上空槽底的西风气流进入南半球。因此,高原夏季的感热气泵能够影响世界范围的气候状况。

本研究还指出,高原的感热气泵是导致高原附近的环流在自冬到夏的季节演变中呈现突变的重要原因。也是维持亚洲季风的极为重要的因素。

参考文献

陈隆勋,朱乾根,罗会邦,等,1991. 东亚季风[M]. 北京:气象出版社.

吴国雄,张学洪,刘辉,等,1997. LASG 全球海洋-大气-陆面系统模式(GOALS/LASG)及其模拟研究[J]. 应用气象学报,8(suppl):15-28.

叶笃正,高由禧,1979. 青藏高原气象学[M]. 北京:科学出版社.

FLOHN H,1957. Large-scale aspects of the "summer monsoon" in South and East Asia[J]. J Meteor Soc Japan, 35A:180-186.

FLOHN H,1969. Contributions to a meteorology of the Tibetan highlands[J]. Atmos Sci, 130.

HAHN D G, MANABE S,1975. The role of mountain in the South Asia monsoon[L]. J Atmos Sci, 32:1515-1541.

HE H, MCGINNIS J W, SONG Z S, et al,1987. Onset of the Asian summer monsoon in 1979 and the effect of the Tibetan Plateau[J]. Mon Wea Rev, 115:1966-1994.

HIRASAWA N, KATO K, TAKEDA T,1995. Abrupt change in the characteristics of the cloud zone in subtropical East Asia around the middle of May[J]. J Meteor Soc Japan, 73 (2):221-239.

JOHNSON D R, YANAI M, SCHOAK T K,1987. Global and regional distributions of atmospheric heat sources and sinks during the GWE[M] // CHANG C P, KRISHNAMURTI T N. Monsoon Meteorology. Oxford: Oxford University Press:271-297.

KRISHNAMURTI T N, JAYAKAUMAR P K, SHENG J, et al,1985. Divergent circulation on the 30 to 50 day time scale[J]. J Atmos Sci, 42:364-375.

LI C F, YANAI M,1996. The onset and interannual variability of the Asian summer monsoon in relation to land-sea thermal contrast[J]. J Climate, 9:358-375.

LUO H, YANAI M,1983. The large-scale circulation and heat sources over the Tibetan Plateau and surrounding areas during the early summer of 1979. Part Ⅰ: Precipitation and kinematic analyses[J]. Mon Wea Rev, 111:922-944.

LUO H, YANAI M,1984. The large-scale circulation and heat sources over the Tibetan Plateau and surrounding areas during the early summer of 1979. Part Ⅱ: Heat and moisture budgets[J]. Mon Wea Rev, 112:966-989.

MATSUMOTO J,1992. The seasonal changes in Asian and Australian monsoon regions[J]. J Meteor Soc Japan, 70:257-273.

MURAKAMI T, DING Y H,1982. Wind and temperature changes over Eurasia during the early summer of 1979[J]. J Meteor Soc Japan, 60:183-196.

MURAKAMI T, MATSUMOTO J,1994. Summer monsoon over the Asian continent and Western North Pacific[J]. J Meteor Soc Japan, 72:719-745.

PEIXOTO J, OORT A,1992. Physics of Climate[M]. New York: Springer AIP Press.

Staff Members of the Section of Synoptic and Dynamic Meteorology, Institute of Geophysics and Meteorology, Academia Sinica,1957. On the general circulation over eastern Asia (Ⅰ)[J]. Tellus, 9:432-446.

WU G X, LIU H, ZHAO Y C, et al,1996a. A nine-layer atmospheric general circulation model and its performance[J]. Adv Atmos Sci, 13(1):1-18.

WU G X, ZHU B Z, GAO D Y,1996b. The impact of Tibetan Plateau on local and regional climate[M] // Institute of Atmospheric Physics, CAS. Atmospheric Circulation to Global Change. Beijing: China Meteorological Press:425-440.

YE T Z, 1981. Some characteristics of the summer circulation over the Qinghai-Xizang (Tibet) Plateau and its neighbourhood[J]. Bull Amer Meteor Soc, 62:14-19.

YE T Z, 1982. Some aspects of the thermal influences of the Qinghai-Tibetan Plateau on the atmospheric circulation[J]. Arch Meteor Geophys Bioklim, A31:205-220.

YE T Z, TAO S Y, LI M C, 1959. The abrupt change of circulation over the Northern Hemisphere during June and October[J]. The Atmosphere and the Sea in Motion: 249-267.

Tibetan Plateau Forcing and the Timing of the Monsoon Onset over South Asia and the South China Sea

WU Guoxiong, ZHANG Yongsheng

(State Key Laboratory of Atmospheric Sciences and Geophysical Fluid Dynamics, Institute of Atmospheric Physics, Chinese Academy of Sciences, Beijing 100029, China)

Abstract: Observations were employed to study the thermal characteristics of the Tibetan Plateau and its neighboring regions, and their impacts on the onset of the Asian monsoon in 1989. Special attention was paid to the diagnosis of the temporal and spatial distributions of surface sensible and latent heat fluxes. Results show that the whole procedure of the outbreak of the Asian monsoon onset is composed of three consequential stages. The first is the monsoon onset over the eastern coast of the Bay of Bengal (BOB) in early May. It is followed by the onset of the East Asian monsoon over the South China Sea (SCS) by 20 May, then the onset of the South Asian monsoon over India by 10 June. It was shown that the onset of the BOB monsoon is directly linked to the thermal as well as mechanical forcing of the Tibetan Plateau. It then generates a favorable environment for the SCS monsoon onset. Afterward, as the whole flow pattern in tropical Asia shifts westward, the onset of the South Asian monsoon occurs.

Finally, the timing of the onset of the Asian monsoon in 1989 was explored. It was shown that the onset of the Asian monsoon occurs when the warm or rising phase of different low-frequency oscillations reach the "East Asian monsoon area" (EAMA) concurrently. These include the warm phase of the eastward propagating two- to three-week oscillation (TTO) of the upper-layer temperature in middle latitudes, the rising phase of the northward propagating Madden-Julian oscillation of the southern tropical divergence, and the rising phase of the westward propagating TTO of the western Pacific divergence. It was concluded that the timing of the Asian monsoon onset is determined when the favorable phases of different low-frequency oscillations are locked over the EAMA.

1 Introduction

The seasonal transition from winter to summer in the Far East and South Asia is characterized by the abrupt changes in general circulation and weather patterns in the region (Ye et al., 1959; Matsumoto, 1992; Murakami and Matsumoto, 1994). This usually occurs in May and June in association with the onset of the Asian monsoon (Krishnamurti et al., 1985; Hirasawa et al., 1995). It has long been recognized that climate mean outbreak of the Asian summer monsoon starts first in the South China Sea (SCS) region in early and middle May, then propagates westward gradually. In early and middle June, it reaches the South Asian subcontinent region. The onset of the South Asian monsoon then occurs (Tao and Chen, 1987; Chang and Chen, 1995).

Both the onsets of the East Asian and South Asian monsoons are consequences of the atmospheric response to the changes in the contrast of thermal heating between land and ocean (Murakami and Ding, 1982; Johnson et al., 1987; Luo and Yanai, 1983, 1984; He et al., 1987; Chen et al., 1991). Particularly, the elevated heating of the Tibetan Plateau to the atmosphere plays a fundamental role in the formation and maintenance of the summer circulation at least over Asia (Staff Members of Academia Sinica, 1957; Flohn, 1957; Flohn and Reiter, 1969; Ye and Gao, 1979; Ye, 1981, 1982; Luo and Yanai, 1983, 1984; Chen et al., 1985; He et al., 1987; Li and Yanai, 1996). This has been further elucidated and confirmed by a series of numerical experiments (Hahn and Manabe, 1975). In the case in which orography is excluded, an Asian monsoon does not appear, and the convergence and rainy belt over the Indian Ocean is located near the equator just like the intertropical convergence zone (ITCZ) over the Pacific. Their experiments remind us that for the formation of the Asian monsoon, the differential heating between land and ocean is only a part of the story, and the important mechanism must lie in the influences of the Tibetan Plateau.

It has long been puzzled why the monsoon onset occurs earlier in the SCS region than over the Indian region, and how this East Asian monsoon onset is linked to the forcing of the Plateau. Chang and Chen (1995) briefly reviewed several hypotheses that had been proposed to explain the SCS monsoon onset considering the plateau forcing, and suggested that the SCS monsoon onset is triggered by the approach of a midlatitude trough-front system. Since in late spring and early summer, most of such trough systems, which intrude the SCS region, are associated with the India-Burma trough, then, how the Tibetan Plateau affects the formation of the India-Burma trough requires further investigation.

Another important subject in the literature of monsoon study that needs to be clarified further concerns the timing of monsoon onset. Chen and Chang (1980) and Krishnamurti et al. (1981) found that the onset of the East Asian monsoon is associated with the generation, development, and movement of the so-called onset vortex or "monsoon low" in the lower troposphere. The low-frequency oscillation (LFO) of either the Madden and Julian (1971, 1972) mode (MJO) or two- to three-week mode (TTO) has been considered as an important mechanism that modulates the monsoon activities (Tao et al., 1963; Murakami, 1976; Krishnamurti and Bhalme, 1976; Yasunari, 1979, 1980; Krishnamurti and Subrahmanyam, 1982; Krishnamurti et al., 1985; Lorenc, 1984; Murakami et al., 1986; Chen, 1987; Nakazawa, 1992; Chen and Chen, 1995). Their studies remind us that when the timing of monsoon onset is studied, in addition to the plateau impacts, attention should also be paid to the atmospheric motions with different frequencies.

In this paper we employ observational study for the year 1989, and try to get some new insights into the problems concerning how the plateau forcing is associated with the Asian monsoon onset and what the causes are of the location and timing of the early monsoon onset over Asia. The data used include the European Centre for Medium-Range Weather Forecasts (ECMWF) analyzed data, the ECMWF Tropical Ocean Global Atmosphere (TOGA) complementary data, the National Centers for Environmental Prediction (NCEP)-National Meteorological Center (NMC) outgoing longwave radiation (OLR) data, and the Chinese rainfall data collected from 336 observation stations and archived at the Data Center of the Institute of Atmospheric Physics. The OLR data is resolved at a network of 50 longitude by 5° latitude. The ECMWF data is a twice-daily (0000 and 1200 UTC) objective analyzed grid data with a resolution of 2.5° longitude by 2.5° latitude. The ECMWF TOGA complementary data are contended in the ECMWF extended FGGE level-Ⅲ dataset, which is also twice-daily grid data but with a finer resolution of 1.125° latitude by 1.125° longitude. It includes surface wind stress, latent heat flux, sensible heat flux, net radiation, etc. The data were used as initial values of the ECMWF assimilation system for routine medium-range weather forecasts. It compensates the sparse data coverage over the plateau area and over the oceans. From the performance of the ECMWF forecasts and from the comparison of its analyses with those of Ye and Gao (1979) by using the station-based observation, we found the dataset, in general, is good in fidelity and is appropriate for the present research.

In sections 2 and 3, the seasonal variations of differential heating between land and ocean in the Asian monsoon region are examined. The significance of the surface sensible heat flux of the Tibetan Plateau in the seasonal transition of the general circulation over the monsoon area is also considered. The focus in section 4 is on how the huge sensible heating of the Tibetan Plateau in late spring and early summer in 1989 leads to the onset of the Asian monsoon in the SCS region. The importance of both the thermal and mechanical forcing of the plateau is investigated. In section 5, efforts are made to understand the timing of the Asian monsoon onsets at various locations by considering the interactions of different low-frequency oscillations during the seasonal transition period. Some conclusions and discussions are given in section 6.

2 The thermal features in the boreal tropical and subtropical regions

During the seasonal transition from winter to summer in 1989, the evolutions of surface latent heat flux in different latitude zones are shown in Fig. 1. At the bottom of each panel is shown the distribution of the mean orography meridionally averaged over the latitude domain. In the Tropics (Fig. 1a), the latent heat flux from ocean surface is much larger than that from land surface. Over the ocean surface, the intensity of most of the maximum centers exceeds 200 W m^{-2}. On the contrary, over the African continent it is basically below 50 W m^{-2}. From late May to late June, there are three maximum perturbation centers of more than 200 W m^{-2} appearing over the western Pacific and propagating westward. The first and second centers correspond to typhoons 8903 and 8905, respectively, and the third one appeared already in late June.

The evolution of surface latent heat flux in the sub-tropical zone (27.5°—37.5°N), in which the Tibetan Plateau is located, is shown in Fig. 1b. In the area west of 80°E, the land surface latent heat flux is basically below 50 W m^{-2}, similar to what occurs in the Tropics. Whereas in the area east of 80°E, its seasonal transition is remarkable. Before early May, as the offshore cyclone disturbances develop near the east China coast and propagate eastward one after another, periodically intensified latent heat fluxes are observed over the western Pacific. At the same time the flux over land surface is weaker. After the middle of May, the aforementioned cyclogenesis disappears, and the surface latent heat flux over the western Pacific becomes very weak. The maximum surface latent heat flux is now observed over the plateau. The contrast in surface latent heat flux between land surface and sea surface is reversed. This of course will exert considerable impacts on the general circulation of the atmosphere. However, since the substantial expansion and increase of surface latent heat flux over the plateau and over east China occur in May onward, in accordance with the rainy season there, when we are studying those persistent processes that have occurred over the plateau for some period and contributed to the Asian monsoon onset, we may therefore disregard the influence of surface latent heat flux.

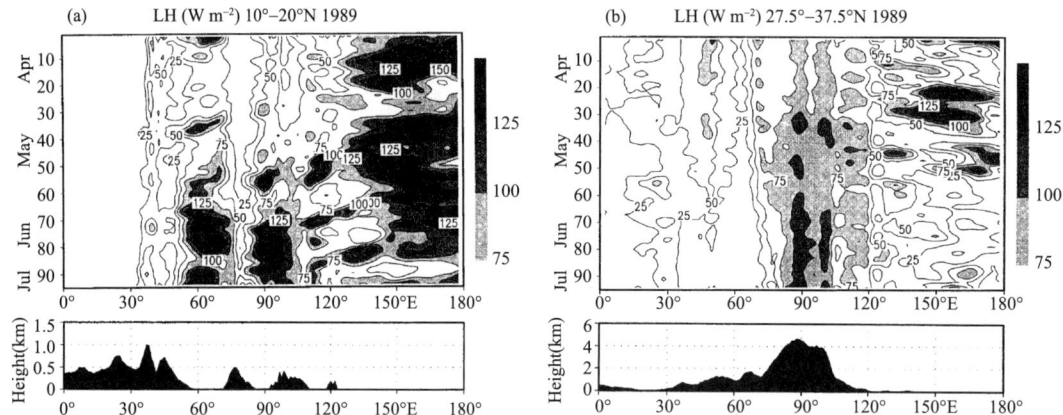

Fig. 1 The evolution of longitudinal distribution of surface latent heat flux during April to July 1989, averaged over $10°-20°N$ (a), and $27.5°-37.5°N$ (b). Interval is 25 W m^{-2}. Stippling indicates the area of more than 75 W m^{-2}. The numbers along the vertical coordinate denote the days counted from 1 April. At the bottom of each panel is the meridional mean orographic height (km) averaged over the corresponding latitude zone

The evolution of surface sensible heat flux from November 1988 to June 1989 in the same subtropical zone is shown in Fig. 2. Over the western Pacific and before the spring equinox, the offshore cyclone disturbances accompanied by large surface sensible heat flux develop one after another and propagate eastward. After the equinox, as the surface air temperature along the east China coastal region is getting warmer and closer to the nearby sea surface temperature, those cyclone disturbances observed in Fig. 1b do not generate large surface sensible heat flux when they move from continent area to the offshore ocean area after the equinox. The western Pacific then becomes tranquil. Some regions are even controlled by downward flux. Although the sensible heat flux to the west of 30°E remains small, prominent changes occur over the massive Asian continental area during the period. In the winter months, downward flux are observed over the western part of the Tibetan Plateau and along the coast area of eastern China. From late winter onward, upward flux of more than 100 W m^{-2} appears at first in Iran, Afghanistan, and Pamirs, and is intensified in April. At the beginning of May, this strong positive sensible heat flux has already covered the vast areas from the Mediterranean Sea to the Tibetan Plateau, with maximum intensity of more than 200 W m^{-2} located over the western part of the plateau. Along the eastern coast of the Asian continent, the reversal of the gradient of surface sensible heat flux occurs by the spring equinox. After the equinox, the thermal contrast between land and sea has changed completely. It is important that, unlike the seasonal change of latent heat flux, the strong surface sensible heating over the plateau and middle-east Asian region appears much earlier and persists afterward. Therefore, the surface sensible heating of the plateau must have played some roles in the seasonal transition at least over the surrounding regions and in the onset of the Asian monsoon. This has been discussed by many other authors and will be further investigated below.

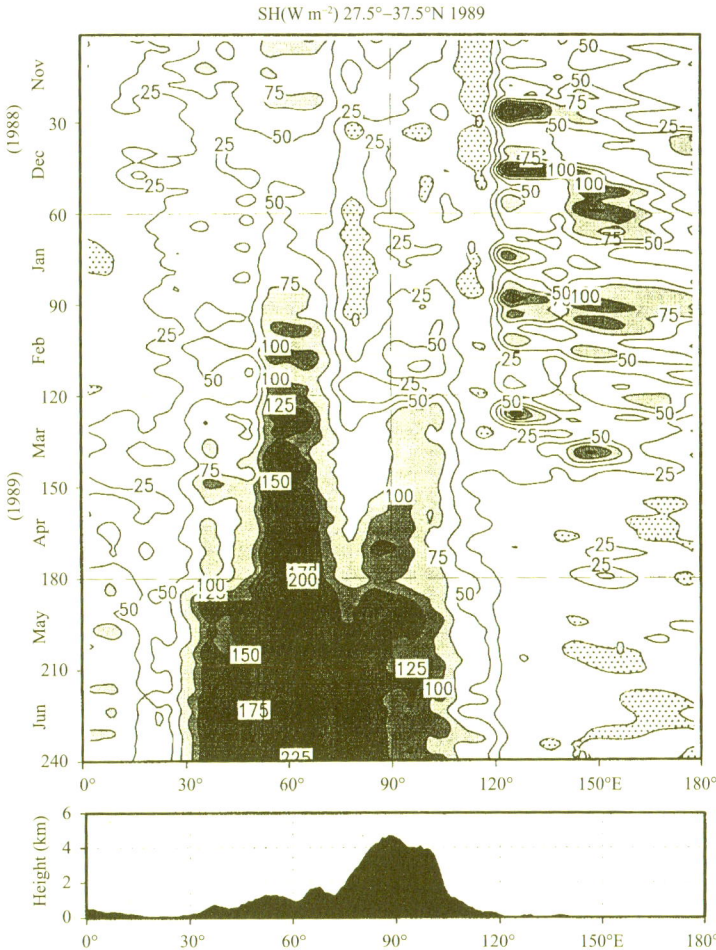

Fig. 2 The evolution of longitudinal distribution of surface sensible heat flux during November 1988 to June 1989, meridionally averaged between 27.5° and 37.5°N. Interval is 25 W m^{-2}. Light stippling indicates the area of more than 75 W m^{-2}, and dotted stippling, negative values. The numbers along the vertical coordinate denote the days counted from 1 November. The bottom panel shows the meridional mean orographic height (km) averaged between 27.5° and 37.5°N

3 Sensible heating of the Tibetan Plateau

During the onset of the East Asian monsoon, heavy rainfall occurs in the SCS and extends to the coast area of southern China. Figure 3 shows the latitude-time distribution of 10 day total rainfall in 1989, which is averaged over the 110°—125°E domain. It can be seen from the figure that after the middle of May, heavy rain of more than 75 mm occurs in the area south of 30°N. In addition, according to daily observations, low OLR of less than 220 W m^{-2} and strong southerlies at 850 hPa in the SCS region of (10°—20°N, 110°—120°E) appears by 20 May. Then 20 May can be chosen as the onset day of the SCS monsoon in 1989. This choice is close to that of 15—20 May by Chen et al. (1996) based upon the analysis of temperature of blackbody at cloud top (T_{BB}), and that of 10—15 May by Q. Ye (1995, personal communication) based upon the minimum OLR data analysis. On 10 June, the heavy rainband jumps northward to the latitude domain 27°—30°N, and dry period starts in southern China. In the following sections, we will show this is the time when the South Asian monsoon onset occurs.

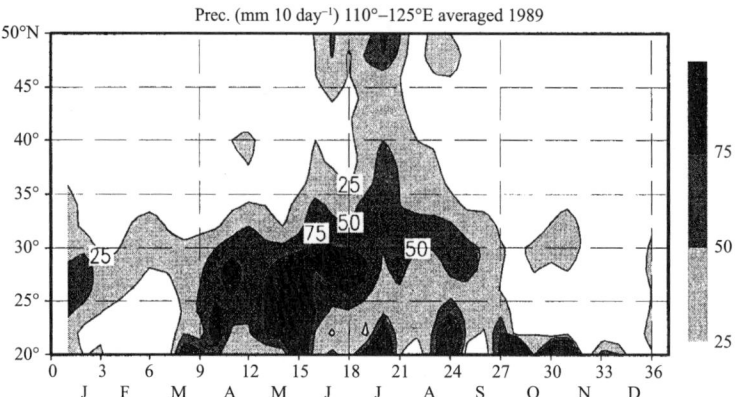

Fig. 3 Meridional movement of the 10 day total rainfall in 1989 averaged over the longitude domain 110°—125°E. Interval is 25 mm. Shading varies from very heavy, heavy, to light, indicating respectively, regions of more than 75 mm, between 75 and 50 mm, and between 50 and 25 mm

To understand the thermal aspects of the plateau in association with the Asian summer monsoon onset, the evolutions of the daily mean sensible heat flux at the surface (SH) and the temperature at 300 hPa (T) averaged over the Tibetan Plateau area (27.5°—37.5°N, 80°—100°E) in the transition period are shown in Fig. 4. The temperature at 300 hPa is representative of the average temperature of the tropospheric column over the plateau. Three spells of abrupt increase in T occur at the ends of late April, middle May, and early June, respectively. The amounts of temperature increase during the abrupt warming spells are 6°, 7°, and 4 ℃ over 5, 6, and 3 days, respectively. Enhanced sensible heating appears about 10 days before the first and second spells. The structure of the three-spell abrupt warming is an important characteristic phenomenon. We will discuss this further when we investigate the monsoon onset and low-frequency oscillations of the year in the following context.

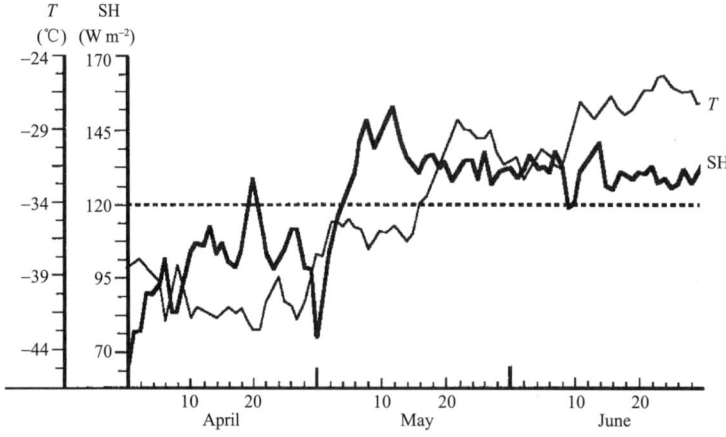

Fig. 4 The evolutions during April—June of surface sensible heat flux SH (W m^{-2}), and 300 hPa temperature T (℃), averaged over the Tibetan Plateau area (27.5°—37.5°N, 80°—100°E)

To further elucidate the importance of surface sensible heating in the increase of column temperature, let T_o be the local change of temperature of the unit column between 200 and 500 hPa; T_v, the heating rate due to temperature advection; and T_s, the heating rate due to surface sensible heat flux SH. Then we can employ observation data and the thermal dynamic equation to estimate the relative importance of surface sensible heating and horizontal advection. The results for the plateau area (27.5°—37.5°N, 80°—

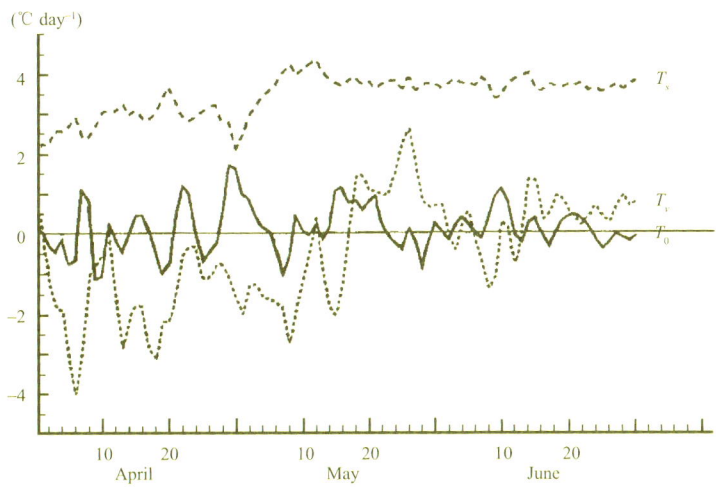

Fig. 5　The evolutions during April—June 1989 of the 200—500 hPa mean temperature advection (T_v) and local change (T_a), and the surface sensible heating averaged over the Tibetan Plateau (27.5°—37.5°N, 80°—100°E) and measured in degrees Celsius per day

100°E) are shown in Fig. 5. It becomes obvious that during the seasonal transition, the surface-sensible heating over the plateau prevails over advection in the local temperature change. From the beginning of April to the end of June and over the plateau, the increase of column temperature due to surface sensible heating gradually increases from 2 ℃ to 4 ℃ per day, while that due to advection is secondary and even negative. It is interesting to notice that before the onset of the SCS monsoon (20 May), advection over the plateau plays a role in cooling the atmosphere. This implies that the outflow over the eastern flank of the plateau is warmer than the inflow over its western flank, and the warm temperature ridge must be located to the east of the plateau. This is indeed the case of observed temperature distribution (as will be shown in Fig. 6b). Astonishingly, just 3 days before the onset, the advection T_v changes its sign. In conjunction with sensible heating, it contributes to the strong atmospheric abrupt warming over the plateau as identified in Fig. 4. This important feature reminds us that not only the surface sensible heating of the plateau, but also the atmospheric circulation pattern are important factors in contributing to the onset of the Asian monsoon.

The surface sensible heat flux over the plateau warms the atmosphere at an elevation of about 5 km, while the surrounding air is far above the boundary underneath. Such elevated heating of the plateau has profound impacts on the seasonal transition in Asian regions. As a matter of fact, from the beginning of May onward, the difference in the meridional mean 200 hPa temperature between the subtropics from 27.5° to 37.5°N and the Tropics from the equator to 10°N, becomes positive in the monsoon region while it remains negative in other longitude regions (figures not shown). The general feature is the same as what was reported by Li and Yanai (1996), and in favor of the development and maintenance of the South Asian high over the plateau and the Asian monsoon onset.

4　Thermal and mechanical forcing of the Tibetan Plateau and the Asia monsoon onset

One of the essential questions concerning the Asian monsoon onset is why the monsoon onset occurs earlier in the SCS region than in the Indian region? In this section we try to explore the possible mechanism linking the Asian monsoon onset to the plateau forcing.

As shown in Fig. 4, the second abrupt warming of the atmosphere over the plateau starts in the middle of May. To present the basic thermal and dynamical characteristics of the circulation before the SCS onset, we specified the period May 5—11 as the "pre-SCS onset" period since during this period the column temperature over the plateau and its local change are representative and remain relatively stable (Figs. 4 and 5). The distributions of surface sensible heat flux and 200 hPa temperature averaged over this period are shown in Fig. 6. The sensible heat flux is below 100 W m^{-2} only over the Pamirs and Kunlun Shan mountain ranges, and over the southeast flank of the plateau. From the remainder of the plateau, large SH flux of more than 100 W m^{-2} is pumped to the atmosphere. Particularly over the central and northeastern parts of the plateau the sensible heat flux exceeds 150 W m^{-2}. Compared with the western Pacific where the surface sensible heat flux is less than 25 W m^{-2} (refer to Fig. 2), the plateau does look like a huge stove. However, the warmest column is by no means over the plateau. Due to the strong advection, the warm temperature ridge in each layer of the upper troposphere is shifted to the downwind side of the sensible heat source over the plateau. As shown in Fig. 6b, the apparent east—west gradient of temperature at 200 hPa is just over the plateau. Along 30°N, a cold trough of −56 ℃ is at 60°E, while a warm center of −52 ℃ is near 115°E, just to the north of the SCS. Over the plateau region, the column over its eastern part is warmer than that over its western part. Since the warm column over the plateau in summer acts as a huge chimney that sucks the lower-tropospheric air from the surroundings (Wu et al., 1996), and since the air column over the eastern plateau is warmer than that over its western part, the lower-tropospheric inflow along the eastern boundary of the plateau should be stronger than the inflow along its western boundary, as will be shown in Fig. 8a. This means that convection should be easier to develop along the eastern coast of the Bay of Bengal (BOB), in favor of the earlier monsoon onset in the eastern region.

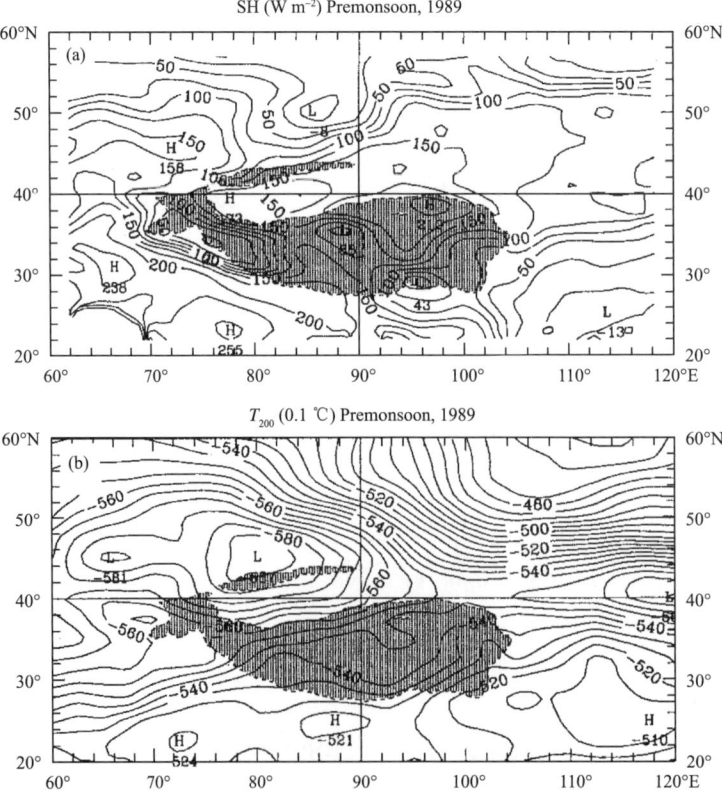

Fig. 6 Distributions in the BOB monsoon onset period (5—11 May) of 1989 of (a) the mean surface sensible heat flux (interval in 25 W m^{-2}), and (b) the mean 200 hPa temperature (unit in 0.1 ℃, interval in 0.5 ℃). Stippling indicates where the elevation of the Tibetan Plateau is higher than 3 km

Now let us turn to consider the mechanical forcing of the plateau. The westerly flows at 850 or 700 hPa are much below the crest of the Tibetan Plateau. When they impinge upon the plateau at its southwestern corner before the monsoon onset, according to the research of Wu (1984), the deflecting effect of the orography must prevail over the climbing effect, and airflow should go around the plateau. An anticyclonic pattern over the Arabian Sea and the India subcontinent and cyclonic pattern over the northern part of the BOB will be formed. The latter in weather practice is termed the India-Burma trough, which appears long before the Indian monsoon onset (Yin, 1949). In a set of numerical experiments based on a dynamical circulation model and designed by Zhu (refer to Wu et al., 1996), it was shown that this India-Burma trough exists only when the Tibetan Plateau is presented. This supports the aforementioned postulation. Due to such mechanical forcing, the lower-tropospheric southerly inflow must be intensified in the northeastern coast of the Bay of Bengal. Figure 7 is the Hovmöller diagram of both the pentad mean wind vector at 850 hPa and the OLR averaged over $10°-20°$N. After the middle of April, persistent southerly inflow is observed between 100° and 110°E. The India-Burma trough is already discernible at the beginning of May and develops until late May (Fig. 8). During this period, northerlies and northeasterlies in the eastern front of an anticyclone that is centered over the southern Arabian Peninsula (Fig. 8a) develop over the Arabian Sea and Indian subcontinent. By using daily observations we found that by 10 May, convergence is intensified in the eastern front of the India-Burma trough. From Fig. 7 we see that deep convection with OLR$<$220 W m^{-2} occurs first along the western coast of Burma. About one week later, enhanced southerlies and deep convection start to appear in the SCS, and the SCS monsoon onset occurs. As to the Indian region, it is only when the aforementioned northerlies and northeasterlies are replaced by strong southerlies that the onset of the South Asian monsoon occurs, which is already in early June. In Fig. 7, the top branch of the "cactus" shape of the 220 W m^{-2} OLR isoline represents the deep convection over the eastern BOB region. The second branch is to its east and represents the onset of the SCS monsoon, while the third is to its west and represents the onset of the Indian monsoon. These mean that the earliest development of systematic deep convection and southerlies is over the eastern coast of the Bay of Bengal and before the onset of the SCS monsoon. It can therefore be defined as the BOB monsoon onset.

Our further concern is, what is the role played in the SCS monsoon onset of the deep convection over Burma after the BOB monsoon onset. In Figs. 8 and 9, the fields of streamlines at 850 and 200 hPa before and after the SCS monsoon onset are presented. During the BOB onset period (5—11 May) at 850 hPa (Fig. 8a) a deep trough extends from Burma to the Bay of Bengal. A prominent anticyclone is centered over Saudi Arabia and controls the Arabian Sea and the Indian subcontinent. From the Indian Ocean to the western Pacific Ocean, the cross-equator flow is very weak, limited only to the region from 90° to 120°E. Southerlies are confined to the longitude domain just next to the east of the Plateau. At 200 hPa (Fig. 9a), the pattern of streamline curvature above the North Indian Ocean is opposite to that at 850 hPa: a deep trough is over the Arabian Sea and Indian subcontinent, while an anticyclone is located to the east, over the Bay of Bengal and the Indochina Peninsula, and centered near Rangoon. The southerlies at 850 hPa and the anticyclone at 200 hPa appearing in the BOB monsoon onset period are in correspondence with the strong development of deep convection identified by OLR$<$220 W m^{-2} as shown in Fig. 7 by the top branch of the OLR cactus. The development of the 850 hPa Burma trough and 200-hPa anticyclone is significant in triggering the SCS monsoon onset. First, it causes strong upper-layer divergence just over the SCS region (Fig. 9a). Second, by pumping the intensified crossing-equator northeasterlies in upper layers and southwesterlies in lower layers in the longitude domain from 90° to

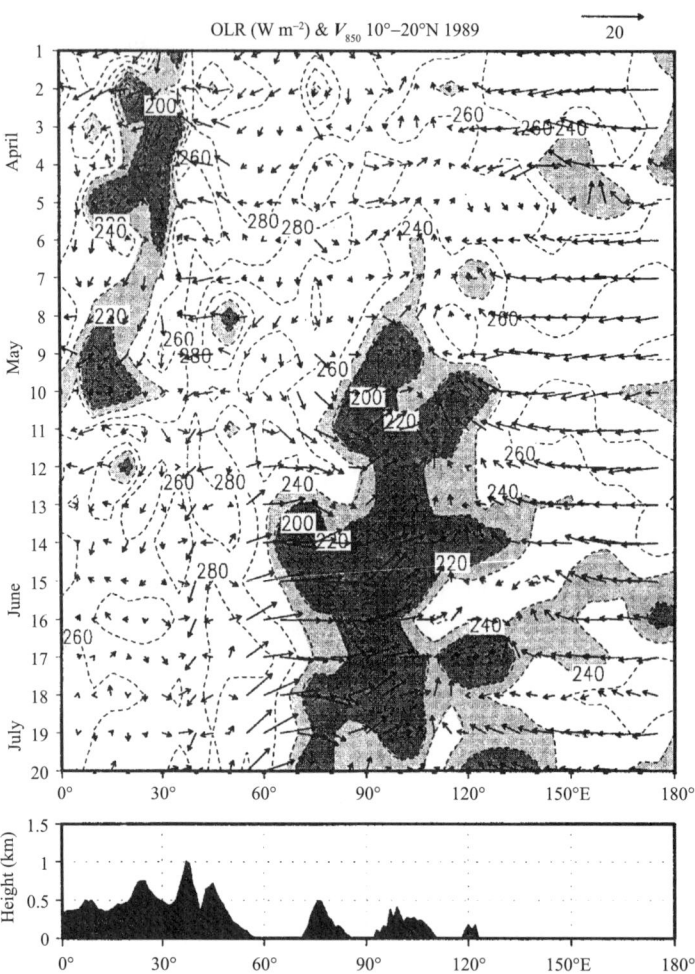

Fig. 7 The evolutions during April—mid-July 1989 of the longitudinal distribution of 850 hPa wind and outgoing longwave radiation (OLR, in contour) meridionally averaged from 10° to 20°N. Contour interval is 20 W m^{-2}. The numbers along the vertical coordinate denote the pentad order counted from 1 April. The bottom panel shows the meridional mean orographic height (km) averaged between 10° and 20°N

120°E, it switches on a forced secondary meridional circulation with rising branch in the SCS region. Such structure develops further (figures not shown) after the BOB monsoon onset. By about 20 May the 850 hPa crossing-equator southwesterly flow has become very strong and ranged from the Indian Ocean to the SCS (Fig. 8b). Much intensified convergence is located just over the SCS region, and the onset of the SCS monsoon occurs. At the same time, a small-size 850 hPa cyclone and a similar size 200 hPa anticyclone (Fig. 9b) appear over the north SCS, in correspondence with the deep convection shown in Fig. 7 by the second branch of the small value ($<$ 220 W m^{-2}) OLR cactus.

Furthermore, during and after the onset of the SCS monsoon the circulation patterns over the monsoon region shifts westward. By late May at 850 hPa (Fig. 8b), the Burma trough has moved to Bihar, and the anticyclone over the Arabian Sea and India has been weakened. At 200 hPa (Fig. 9b), the trough originally over the Arabian Sea has moved to North Africa, while the anticyclone that was over Burma has now settled over the south of the Tibetan Plateau. To its south, upper-layer easterlies are intensified, and divergence prevails over the northern Indian Ocean. All these are then mature for the onset of the Indian monsoon, which occurred in early June of 1989 as signified by the third branch of the small-value cactus shown in Fig. 7, about 20 days later than the monsoon onset in the SCS region. The general

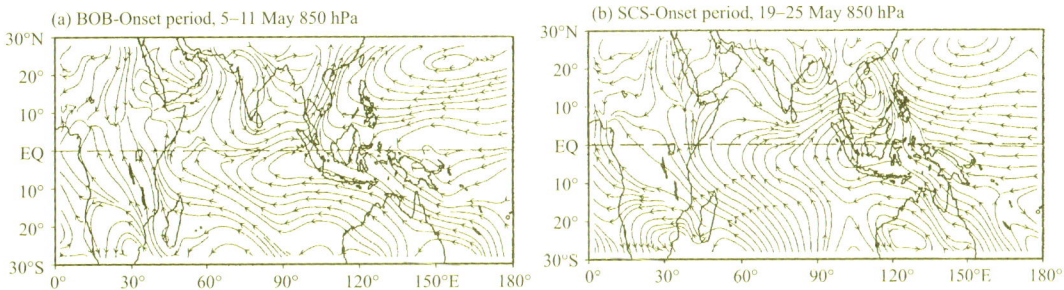

Fig. 8 Distributions of the 850 hPa streamline in 1989 during the BOB monsoon onset period 5—11 May (a), and the SCS monsoon onset period 19—25 May (b)

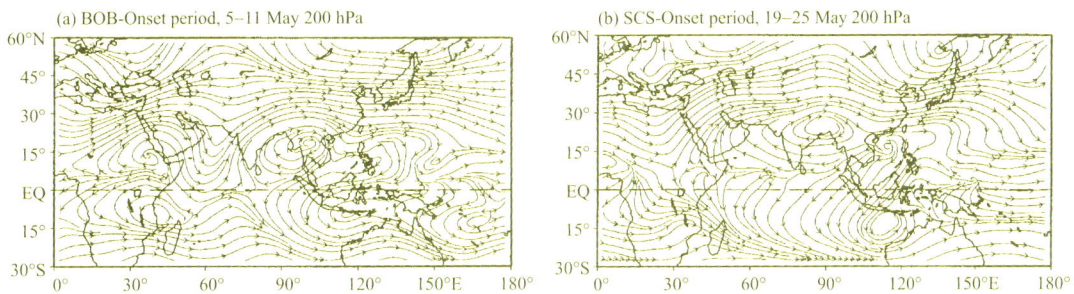

Fig. 9 The same as Fig. 8 except for the 200 hPa streamline

features shown in Figs. 7—9 can be seen from the evolutions of the 10-yr mean 850 hPa flow for the period 1980—1989 calculated by Nakazawa (1992, Fig. 2a). In which the strong impinging 850 hPa flow toward the Tibetan Plateau in May is observed to the southwestern corner of the plateau. The early development of southerlies along the western coast of the Indochina Peninsula and the existence of anticyclonic flow over the Arabian Sea in May, and the replacement of northerlies by southerlies along the western coast of India in early June are all in agreement with those shown in Figs. 7—9.

The three-stage characteristics of the Asian monsoon outbreak revealed in the study is also in good agreement with the annual march of the rainy season in the Asian region as identified from OLR diagnosis. Using the 12yr OLR data from 1975 to 1987 (except for 1978), Nakazawa (1992) analyzed the climate mean marching of OLR distribution from 1 May to 20 June. The results (refer to his Fig. 2b) show that deep convection signified by OLR being less than 240 W m^{-2} appears first over the western coast of Burma, then extend to the SCS region by 21 May and southwestern India in early June. Using the pentad mean OLR data of the Climate Analysis Center (CAC) NMC, which are on a 2.5°×2.5° grid and over the period from June 1974 to December 1989, Wang (1994) found that over southern Asia the rainy season, which is defined as the pentad mean OLR being lower than or equal to 230 W m^{-2}, begins at the northern tip of Sumatra in early April and reaches the southeastern Bay of Bengal in early May, followed by northeastward and northwestward propagation. He also reported that the high reflective clouds (HRC) atlas of Garcia (1985) shows that, "from April to May an HRC maximum in the equatorial eastern Indian Ocean (2°N, 95°E) shifts to Northern Sumatra and the southern Andaman Sea (10°N, 95°E). Therefore, the HRC maximum moves steadily northwestward along the west coast of Burma until August when it reaches its northern most position in the head of the Bay of Bengal (90°E, 18°N)". In addition, the marching of such a low OLR center is accompanied by the intensification of either southwesterlies or southerlies, as shown in Fig. 7. All these support our postulation that the BOB monsoon onset is the earliest monsoon onset in Asia during the spring seasonal transition.

5 Phase locking of low-frequency oscillations and the timing of the Asian monsoon onset

Undoubtedly, the thermal and mechanical forcings of the Tibetan Plateau have important impacts on the Asian monsoon climate, including its onset, lull, revival, and withdrawal. However, significant sensible heating of the plateau to the atmosphere usually begins before the spring equinox, much earlier than the monsoon onset. The intrinsic mechanism linking the Plateau heating directly to the timing of monsoon onset does not seem to be obvious. To find the mechanism that is associated with the timing of the Asian monsoon onset, in conducting this study we have analyzed different frequency oscillations in different latitude regions. Some significant oscillations associated with monsoon onset have been obtained and are presented below.

a. Two-to three-week oscillation of the extratropical temperature

As discussed earlier, the elevated heating of the plateau warms the air column above and sucks lower-tropospheric air from surroundings. This becomes one of the key topics of monsoon dynamics. However, the temperature rise in the troposphere is by no means uniform. From Fig. 4 we see that there are three spells of rapid increase: from the end of April to the beginning of May, 14−21 May, and around 10 June. These three spells correspond to the onsets of the BOB, SCS, and South Asian monsoons, respectively. The period between two successive spells is about two to three weeks, with 20 days on average. Take a 15−25-day bandpass filter and act on the 200 hPa temperature evolution. The filtered curve is shown in Fig. 10, which is generally in phase with the original evolution. The letters A—H indicate the key phases of the evolution. To see the features of temperature evolution at each of the key phases, we employ the same bandpass filter to the original ECMWF 200 hPa temperature series for each grid point and for the period January to July 1989, and extract the filtered temperature fields at the times corresponding to the eight key phases as shown in Fig. 10, then present these fields in Fig. 11. From this figure the following becomes apparent.

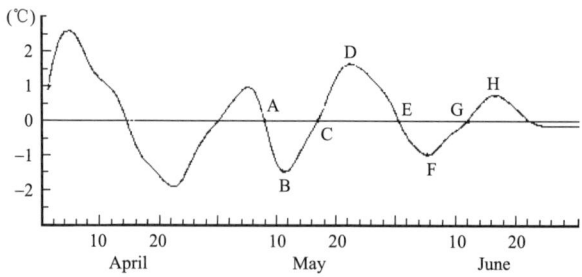

Fig. 10 The evolutions in the seasonal transition period (1 April—30 June) 1989 of the temperature departure from the corresponding period mean at 200 hPa and averaged over the Tibetan Plateau area (27.5°—37.5°N, 80°—100°E). The data used have been treated by using a 15−25-day bandpass filter. The ordinate denotes the departure of temperature (°C). The letters A—H denote different phases of the low-frequency oscillation

1) Significant TTO of the 200 hPa temperature with amplitude greater than 1.0 °C occurs mainly in the extratropics. It can be detected in the Tropics only over the western Pacific (at phase C) and over the SCS region when the East Asian monsoon onset appears (at phase D). Whether the former is the ex-

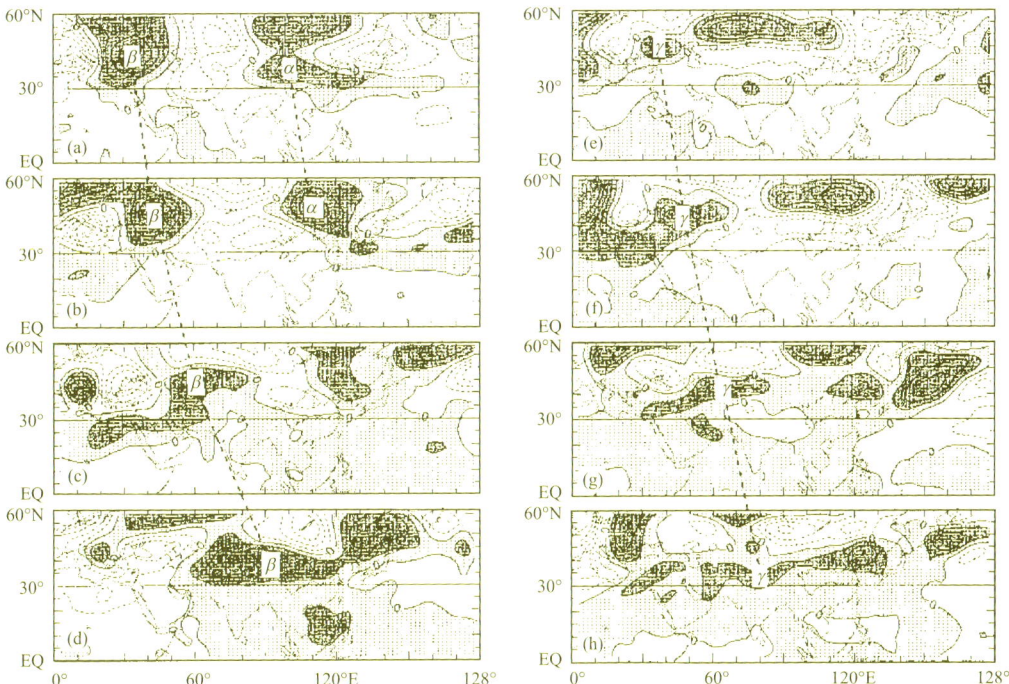

Fig. 11 Distributions of the 200 hPa filtered temperature in 1989 at different phases of the low-frequency oscillation as shown in Fig. 10. The data used have been processed by a 15—25-day bandpass filter. The interval is 1 ℃. Stippling denotes warm anomaly; and heavy stippling, warmer than 1 ℃: (a) 7 May, (b) 14 May, (c) 17 May, (d) 22 May, (e) 30 May, (f) 4 June, (g) 11 June, and (h) 15 June

tension of the propagation of the warm TTO α requires further work and will not be discussed here. The latter may result partly from the arrival of the warm phase of the TTO, and partly from the large latent heat release due to condensation since it appears at the time typhoon 8903 intruded into the SCS and the SCS monsoon onset occurred, and since such kind of δ-function-type warming may be decomposed into different harmonics.

2) Before the East Asian monsoon onset, the temperature variation in the Tropics east of the meridian 90°E is in phase with that over the Tibetan Plateau. When the TTO of the 200 hPa (or 300 hPa, or 500 hPa, figures not shown) temperature over the plateau evolves from its valley (B) to its crest (D), the upper-tropospheric temperature over the tropical region east of 90°E also increases.

3) The warm ridge that controls the area over the Tibetan Plateau and northern China during the monsoon onset (phase D) can be traced backward to the European origin. At phase A, while the previous warm TTO α departs the Plateau, another warm center, β is just over the Mediterranean Sea. This warm ridge moves gradually east—southeastward with a speed of about 4° of longitude per day. From phase B onward, the air over the plateau and the western Pacific is getting warmer. By phase D when this warm ridge starts to invade the plateau area, the onset of East Asian monsoon occurs.

4) The warm ridge γ over the plateau at phase H, which is associated with the Indian monsoon onset, can also be traced backward to the European origin. The warm ridge γ over the Mediterranean Sea at phase E follows system β and propagates in a similar manner and at a similar speed.

In summary, during the spring transition period the low-frequency oscillation TTO of the upper-tropospheric temperature dominates the extratropical regions. In early May, when the warm ridge of the TTO mode develops over Europe and moves east-southeastward into the Tibetan Plateau area, favorable backgrounds for East Asian monsoon onset is then created.

b. 30- to 60-day oscillations of the tropical divergence

To see how the low-frequency oscillations affect the onset of the Asian monsoon, we choose a rectangular block bounded by the meridians 85° and 120°E and the latitude circles 10° and 40°N as shown in Fig. 12. The eastern part of the Tibetan Plateau, and the main parts of the SCS and the Bay of Bengal are all within the block. The onsets of the BOB monsoon and SCS monsoon both occur within this block as well. It can be regarded as the most sensitive region for the East Asian monsoon onset, and defined as the "East Asian monsoon area" (EAMA). The cross sections AB, CD, and EF are made to present how the low-frequency oscillations affect the region from its north, south, and east directions, respectively. No obvious low-frequency oscillations from the west were observed to enter the region. This is possibly because the northern part of the western boundary is occupied by the plateau, and its southern part is on the downwind side of the surface anticyclone that exists over the Arabian Sea before the onset of the Southeast Asian monsoon.

In the lower panel of Fig. 13, we present the Hovmöller diagrams of the horizontal divergence at 200 hPa that is averaged over the longitude range from 85° to 95°E and along the C (20°S, 90°E) to D (30°N, 90°E) cross section shown in Fig. 12. From April to June, there are three main divergence centers that develop in the southern Tropics, respectively, in early April, May, and June, with a period of about one month. The latter two centers are coupled with lower-tropospheric convergence (figures not shown). These centers propagate northward toward the Northern Hemisphere at a speed of about 1.2°—1.5° of latitude per day. Since the MJO is characterized by lower-layer convergence and upper-layer divergence, and since their arrival provides favorable conditions for the development of strong convection and heavy rainfall, it can be defined as the rising phase of the MJO. The arrival of the main rising phase of the MJOb in the Andaman Sea in early May is in correspondence with the BOB monsoon onset. As discussed in the previous sections, the BOB monsoon onset then generates a favorable environment for the SCS monsoon onset. As the MJOb reaches the southern boundary at 10°N of the East Asian monsoon area, the SCS monsoon onset then occurs. From Fig. 13 it becomes prominent that the onset of the East Asian monsoon in the middle of May and the onset of the South Asian monsoon in early June are both associated with the arrival of the MJO at the latitude belt 10°—20°N.

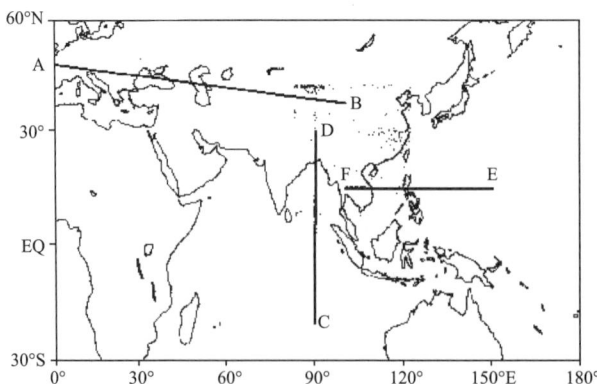

Fig. 12 The locations of the East Asian monsoon area (EAMA) as defined by the block, and the cross sections AB, CD, and EF for presenting the propagation of different low-frequency oscillations as shown in Figs. 13 and 14

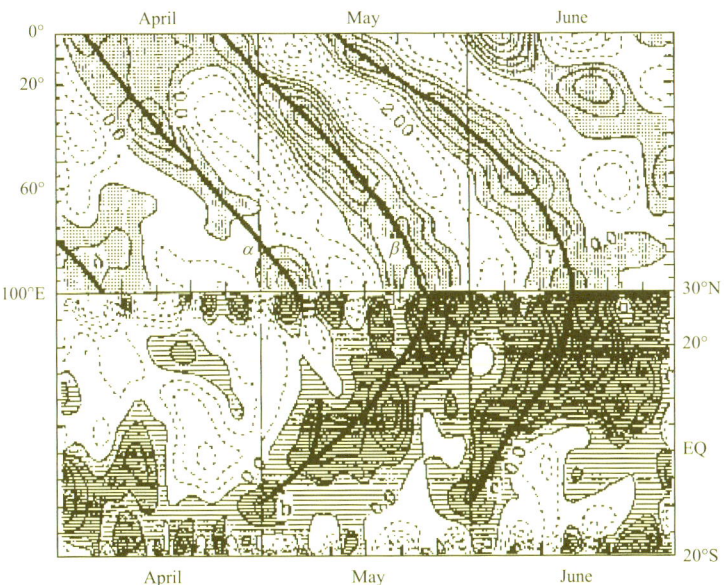

Fig. 13 The evolutions in the seasonal transition period of 1989 and along the AB cross section shown in Fig. 12 of the 200 hPa temperature, which has been proceeded by a 15 – 25-day bandpass filter (upper panel, interval in 1.0 ℃), and along the CD cross section of the 200 hPa divergence (lower panel, interval in 2.0×10^{-6} s^{-1})

c. Two-to three-week oscillations of the tropical divergence

In the right panel of Fig. 14, is shown the evolution of the divergence at 200 hPa averaged between 10° and 20°N and along the E(160°E) to F(100°E) cross section as indicated in Fig. 12. The corresponding evolution at 850 hPa is similar but with opposite signs, and not shown here. From May to June, there are four strong divergence perturbations (a – d) that appear over the western Pacific and then propagate westward, appearing as another kind of TTO. Since these strong upper-layer divergence centers are coupled with strong lower-layer convergence, the passage of such TTO always causes vigorous upward motion and torrential rain. It has been identified from weather charts that TTOb and TTOc correspond respectively to typhoon 8903 (T3) and 8905 (T5). TTOb reaches the SCS region and becomes intensified there by 20 May, the time of the East Asian monsoon onset. TTOc propagates at a speed of about 3.5° of longitude per day. It reaches the eastern coast of the Bay of Bengal by 10 June when the South Asian monsoon onset occurs.

d. Phase locking of low-frequency oscillations and monsoon onset

The upper panel of Fig. 13 and the left panel of Fig. 14 present the eastward propagation of the TTO mode of the upper-tropospheric temperature along the A (45°N, 0°) to B (37.5°N, 100°E) cross section as shown in Fig. 12. They summarize the propagation feature of the TTO shown in Fig. 11. From April to June, there are four warm surges of the TTO and each propagates from Europe and invades the plateau area. The first warm surge that reaches the plateau region in the first half of April does not match the northward propagating MJO and the westward propagating TTO of the tropical divergence, and monsoon onset does not appear. However, the second, third, and fourth warm surges over the plateau are in phase with the arrivals of the northward propagating MJO at the southern boundary (10°N) of the EAMA (Fig. 13, lower panel) and the westward propagating TTO at the eastern boundary of the EAMA (Fig. 14, right panel). The MJO and TTO have different oscillation periods by nature. To understand why they can be phase locked in the EAMA, consider the divergence at 200 hPa of more than 2×10^{-6} s^{-1} as a presentation of strong rising phase. Then from Fig. 13 we see that the lifetime of the strong rising phase of this MJO is short in the Southern Hemisphere but rather sustained while propagating toward the EAMA. At its southern border, the strong rising phase of the MJOb persists from early May to 22 May, and that of the MJOc persists for the whole of June. The persistency of the strong rising phase of this MJO thus provides the possibility for the phase—locking of it with the two TTOs. In other words, all the favorable phases of these low-frequency oscillations arrive in the EAMA at about the same time. They act together in the region and cause vigorous development of strong convection and torrential rain over large areas. The BOB, SCS, and South Asian monsoon onsets thus occur correspondingly. It can therefore be concluded that the timing of monsoon onset is determined by the phase locking within the EAMA of different kinds of low-frequency oscillations from different directions.

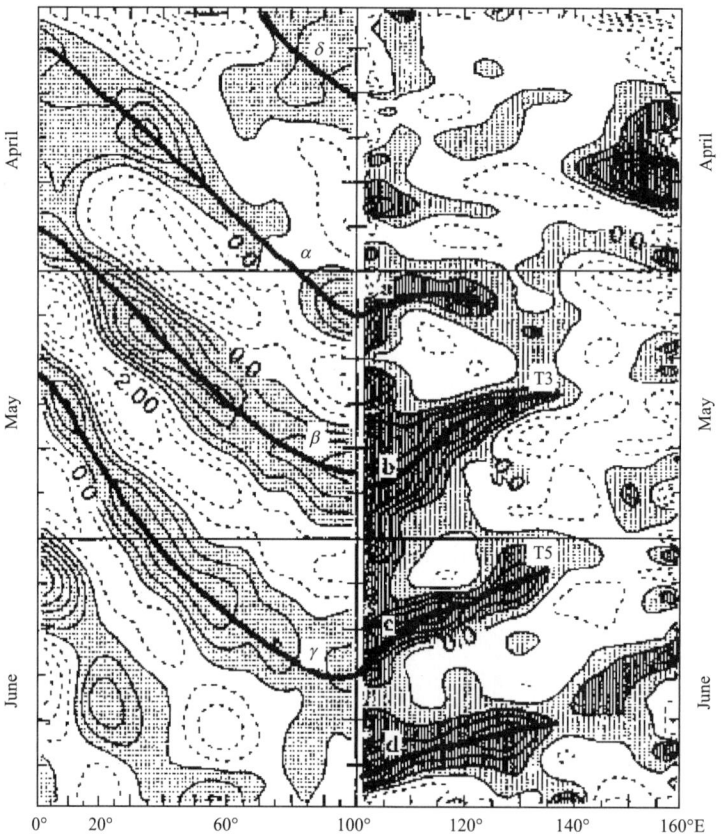

Fig. 14 (Left) Same as the upper panel of Fig. 18. (Right) The evolution in the seasonal transition period of 1989 and along the EF cross section shown in Fig. 12 of the 200 hPa divergence averaged meridionally from 10° to 20°N. Interval is 2.0×10^{-6} s^{-1}. Stippling indicated divergence. T3 and T5 denote typhoon 8903 and typhoon 8905, respectively

6 Conclusions and discussion

It has been confirmed in this case study that during the seasonal transition of 1989 there exists complete reversal in thermal contrast between land and sea surface in the Asian subtropical region. Large surface latent and sensible heat fluxes are observed over the western Pacific during winter months but over the Tibetan Plateau in summer months. The changes in latent heat flux are more or less accompanied with the onset of the Asian monsoon. However, the maximum center of surface sensible heat flux over the western and central Tibetan Plateau is established in early spring. It warms up the atmospheric column aloft by several degrees Celsius per day. Lower-layer air in the surrounding area of the plateau are then sucked toward the plateau. The north to south gradient of the upper-tropospheric temperature to the south of the plateau is gradually reversed, in favor of the development of tropical easterlies. All these provide suitable conditions for the seasonal transition of the circulation in East Asia and the onset of a monsoon.

The onset of the SCS monsoon in 1989 was by 20 May, and the onset of the South Asian monsoon was about 20 days later. The reasons that caused the monsoon onset in East Asia leading that in South Asia were explored. Both the thermal forcing and mechanical forcing were proven to be important factors in causing the earlier occurrence of the SCS monsoon. Thermally, the strong surface sensible heating of western and central Tibet in collaboration with the westerly advection of temperature makes the air temperature over the eastern plateau warmer than that over the western plateau. Strong southerly inflow toward the plateau in the spring season thus occurs first in the eastern part of the Bay of Bengal. Mechanically, the orographically deflected westerly forms low-layer anticyclonic flow over the Arabian Sea and South Asia subcontinent and cyclonic flow over the Bay of Bengal. Deep convection and strong southerlies thus develop vigorously in the eastern front of the Burma trough, and a lower-layer low and an upper-layer high are observed near Rangoon. A low value OLR center appears first in this region in the latitude belt from $10°$ to $20°N$. This is defined as the onset of the BOB monsoon. After the BOB monsoon onset, the upper-layer anticyclone, which is centered near Rangoon, drives strong northerlies toward the Southern Hemisphere over the SCS region. Strong lower-layer return flows of southerlies and southwesterlies then develop in the region. The onset of the SCS monsoon thus occurs. After this onset, the warm South Asian high develops over the Tibetan Plateau, and the monsoonal meridional circulation with southerlies in the lower layer and northerlies aloft shifts westward to the Arabian Sea. The surface anticyclone circulation that occupied the Arabian Sea during the onset of the East Asian monsoon has retreated to the north of the Arabian Peninsula. Cyclonic circulation then prevails over the Arabian Sea, and the onset of the South Asian monsoon occurs. The overall structure discussed in this paragraph has been confirmed by examining the weather data of the last 14 yr and will be published in a parallel paper.

Based upon these results, we suggest that the whole procedure of the outbreak of the Asian monsoon is better divided into three stages. The first is the BOB monsoon onset, followed by the SCS monsoon onset, and the consequent Indian monsoon onset. Such division provides linkage of the Asian monsoon onset to the Tibetan Plateau forcing more closely. It is due to the thermal as well as mechanical forcing of the Tibetan Plateau that the BOB monsoon onset occurs, and the SCS monsoon onset is then induced.

Another important point revealed in this study is about the timing of the monsoon onset. It is proved that monsoon onset occurs when different kinds of the low-frequency oscillations are in phase in the EAMA region. When the warm phase of the TTO of midlatitude upper-layer temperature, which is originally over Europe, arrives in the EAMA concurrently with the arrival of the rising phase of the northward propagating MJO of the tropical divergence, which originates in the Southern Hemisphere, favorable background for the development of atmospheric overturning over a large area is then created. When the tropical TTO of strong convection that originates over the western tropical Pacific and propagates westward reaches this EAMA region, vigorous convection and torrential rain are triggered, and the onsets of the Asian monsoon occur. In other words, the onsets of the Asian monsoon occur when the favorable phases of different kinds of the low-frequency oscillations are locked over the EAMA region.

All the calculations performed in this study have been repeated by using the data for 1988 rather than 1989 and very similar results have been obtained. These will be published in a separate paper. However, since the results obtained here are based on a case study, they should be considered preliminary, and further research is required. Nevertheless, it is still not clear how the MJO of the divergence develop in the Southern Hemisphere. Is it merely a tropical phenomenon, or generated in the southern middle latitudes? Tao et al. (1983) analyzed the low-frequency characteristics of the upper-layer zonal wind component at middle latitudes in the Southern Hemisphere and the difference in tropospheric temperature between 25° and 45°S in the period from May to July of 1979. Besides the most developed TTO, the MJO was identified in the locations of 100° and 120°E. However, how this Southern Hemispheric midlatitude MJO is related to the Asian monsoon is unclear. Another unsolved question is associated with the mechanism that drives the MJO propagating northward to the monsoon region. Is it the lower-tropospheric return flow from the winter hemisphere to the summer hemisphere, or the thermal contrast between land and sea? In fact, there must be other problems we have not thought of. In any circumstances, to further understand the monsoon dynamics, we need to consider the monsoon a spectacular phenomenon that occurs in a system composed of the atmosphere, hydrosphere, and lithosphere, and under the interaction among these subsystems.

References

CHANG C P, CHEN G T J, 1995. Tropical circulations associated with southwest monsoon onset and westerly surges over the South China Sea[J]. Monthly Weather Review, 123(11): 3254-3267.

CHEN G T J, CHANG C P, 1980. The structure and vorticity budget of an early summer monsoon trough (Mei-yu) over southern China and Japan[J]. Monthly Weather Review, 108(7): 942-953.

CHEN L X, REITER E R, FENG Z Q, 1985. The atmospheric heat source over the Tibetan Plateau: May—August 1979 [J]. Monthly Weather Review, 113(10): 1771-1790.

CHEN L X, ZHU Q G, LUO H B, et al, 1991. The East Asian Monsoon[M]. Beijing: China Meteor Press: 362.

CHEN L X, SONG Y, MURAKAMI M, 1996. The characteristics of large scale convective system variation during the onset and prevailing periods of summer monsoon over the South China Sea and its relation to the air-sea interaction. Part Ⅰ: The characteristics of convective system change during the onset period of summer monsoon. Atmospheric Circulation to Global Change[M]. Beijing: China Meteor Press: 314-328.

CHEN T C, 1987. 30—50 day oscillation of 200 hPa temperature and 850 hPa height during the 1979 northern summer[J]. Monthly Weather Review, 115(8): 1589-1605.

CHEN T C, CHEN J M, 1995. An observational study of the South China Sea monsoon during the 1979 summer: Onset and life cycle[J]. Monthly Weather Review, 123(8): 2295-2318.

FLOHN H, 1957. Large-scale aspects of the "summer monsoon" in South and East Asia[J]. Journal of the Meteorological Society of Japan, 35A: 180-186.

FLOHN H, REITER E R, 1969. Contributions to a Meteorology of the Tibetan Highlands[M]. Colorado State University, Fort Collins, CO: 120.

GARCIA O, 1985. Atlas of Highly Reflective Clouds for the Global Tropics[M]. NOAA, Environmental Research Lab, 1971—1983 U S Dept of Commerce: 365.

HAHN D G, MANABE S, 1975. The role of mountains in the South Asian monsoon circulation[J]. Journal of Atmospheric Sciences, 32(8): 1515-1541.

HE H Y, MCGINNIS J W, SONG Z S, et al, 1987. Onset of the Asian summer monsoon in 1979 and the effect of the Tibetan Plateau[J]. Monthly Weather Review, 115(9): 1966-1995.

HIRASAWA N, KATO K, TAKEDA T, 1995. Abrupt change in the characteristics of the cloud zone in subtropical East Asia around the middle of May[J]. Journal of the Meteorological Society of Japan, 73(2): 221-239.

JOHNSON D R, YANAI M, SCHOAK T K, 1987. Global and Regional Distributions of Atmospheric Heat Sources and

Sinks During the GWE[M]. Oxford: Oxford University Press: 271-297.

KRISHNAMURTI T N, BHALME H N, 1976. Oscillation of the monsoon system. Part I: Observational aspects[J]. Journal of Atmospheric Sciences, 33(10): 1937-1954.

KRISHNAMURTI T N, ARDANUY P, RAMANATHAN Y, et al, 1981. On the onset vortex of the summer monsoon [J]. Monthly Weather Review, 109(2): 344-363.

KRISHNAMURTI T N, SUBRAHMANYAM D, 1982. The 30-50 day mode at 850 hPa during MONEX[J]. Journal of Atmospheric Sciences, 39(9): 2088-2095.

KRISHNAMURTI T N, JAYAKUMAR P K, SHENG J, et al, 1985. Divergent circulation on the 30 to 50 day time scale [J]. Journal of Atmospheric Sciences, 42(4): 364-375.

LI C F, YANAI M, 1996. The onset and interannual variability of the Asian summer monsoon in relation to land-sea thermal contrast[J]. Journal of Climate, 9(2): 358-375.

LORENC A C, 1984. The evolution of the planetary-scale 200 hPa divergent flow during the FGGE year[J]. Quarterly Journal of the Royal Meteorological Society, 110(464): 427-441.

LUO H B, YANAI M, 1983. The large-scale circulation and heat sources over the Tibetan Plateau and surrounding areas during the early summer of 1979. Part I: Precipitation and kinematic analyses[J]. Monthly Weather Review, 111(5): 922-944.

LUO H B, YANAI M, 1984. The large-scale circulation and heat sources over the Tibetan Plateau and surrounding areas during the early summer of 1979. Part II: Heat and moisture budgets[J]. Monthly Weather Review, 112(5): 966-989.

MADDEN R A, JULIAN P R, 1971. Detection of a 40-50 day oscillation in the zonal wind in the tropical Pacific[J]. Journal of Atmospheric Sciences, 28(5): 702-708.

MADDEN R A, JULIAN P R, 1972. Description of global scale circulation cells in the tropics with a 40-50 day period[J]. Journal of Atmospheric Sciences, 29(6): 1109-1123.

MATSUMOTO J, 1992. The seasonal changes in Asian and Australian monsoon regions[J]. Journal of the Meteorological Society of Japan, 70(1B): 257-273.

MURAKAMI M, 1976. Analysis of summer monsoon fluctuations over India[J]. Journal of the Meteorological Society of Japan, 54(1): 15-31.

MURAKAMI T, DING Y H, 1982. Wind and temperature changes over Eurasia during the early summer of 1979[J]. Journal of the Meteorological Society of Japan, 60(1): 183-196.

MURAKAMI T, CHEN L X, XIE A, 1986. Relationship among seasonal cycles, low frequency oscillations, and transient disturbances as revealed from outgoing longwave radiation data[J]. Monthly Weather Review, 114(8): 1456-1465.

MURAKAMI T, MATSUMOTO J, 1994. Summer monsoon over the Asian continent and Western North Pacific[J]. Journal of the Meteorological Society of Japan, 72(5): 719-745.

NAKAZAWA T, 1992. Seasonal phase lock of intraseasonal variation during the Asian summer monsoon[J]. Journal of the Meteorological Society of Japan, 70(1B): 597-611.

Staff Members of the Academia Sinica, 1957. On the general circulation over eastern Asia (I)[J]. Tellus, 9(4): 432-446.

TAO S Y, ZHU F K, WU T Q, 1963. Synoptic diagnoses of the summer subtropical high over China and surrounding ocean area. On the Summer Weather Systems in the Subtropics in China[M]. Beijing: Science Press: 106-123.

TAO S Y, HE S X, YANG Z F, 1983. Observational study on the onset of summer monsoon over the east Asia during the 1979 monsoon experiment period[J]. Chinese Journal of Atmospheric Sciences, 7(4): 347-355.

TAO S Y, CHEN L X, 1987. A Review of Recent Research on the East Asian Summer Monsoon in China[M]. Oxford: Oxford University Press: 60-92.

WANG B, 1994. Climatic regimes of tropical convection and rainfall[J]. Journal of Climate, 7(7): 1109-1118.

WU G X, 1984. The nonlinear response of the atmosphere to large-scale mechanical and thermal forcing[J]. Journal of the Atmospheric Sciences, 41(16): 2456-2476.

WU G X, ZHU B Z, GAO D Y, 1996. The impact of Tibetan Plateau on local and regional climate. Atmospheric Circulation to Global Change[M]. Beijing: China Meteor Press: 425-440.

YASUNARI T, 1979. Cloudiness fluctuations associated with the Northern Hemisphere summer monsoon[J]. Journal of the Meteorological Society of Japan, 57(3): 227-242.

YASUNARI T, 1980. A quasi-stationary appearance of 30—40 day period in the cloudiness fluctuations during the summer monsoon over India[J]. Journal of the Meteorological Society of Japan, 58(3): 225-229.

YE D Z, 1981. Some characteristics of the summer circulation over the Qinghai-Xizang (Tibet) Plateau and its neighborhood [J]. Bulletin of the American Meteorological Society, 62(1): 14-19.

YE D Z, 1982. Some aspects of the thermal influences of the Qinghai-Tibetan Plateau on the atmospheric circulation[J]. Arch Meteor Geophys Bioklim, 31(3): 205-220.

YE D Z, TAO S Y, LI M C, 1959. The abrupt change of circulation over the Northern Hemisphere during June and October. The Atmosphere and the Sea in Motion[M]. New York: The Rockefeller Institute Press and Oxford University Press: 249-267.

YE D Z, GAO Y X, 1979. The Meteorology of the Qinghai-Xizang (Tibet) Plateau[M]. Beijing: Science Press: 278.

YIN M T, 1949. A synoptic-aerologic study of the onset of the summer monsoon over India and Burma[J]. Journal of Atmospheric Sciences, 6(6): 393-400.

The Influence of the Mechanical and Thermal Forcing of the Tibetan Plateau on the Asian Climate

WU Guoxiong[1], LIU Yimin[1], WANG Tongmei[1,2], WAN Rijin[1,2], LIU Xin[3], LI Weiping[4]
WANG Zaizhi[1], ZHANG Qiong[1], DUAN Anmin[1], LIANG Xiaoyun[4]

(1 State Key Laboratory of Atmospheric Sciences and Geophysical Fluid Dynamics (LASG), Institute of Atmospheric Physics (IAP)Chinese Academy of Sciences, Beijing 100029; 2 Graduate University of Chinese Academy of Sciences, Beijing 100039; 3 Institute of Tibetan Plateau Research, Chinese Academy of Science, Beijng 100085; 4 National Climate Center, China Meteorological Administration, Beijing 100086)

Abstract: We attempt to provide some new understanding of the mechanical as well as thermal effects of the Tibetan Plateau (TP) on the circulation and climate in Asia through diagnosis and numerical experiments. The air column over the TP descends in winter and ascends in summer and regulates the surface Asian monsoon flow. Sensible heating on the sloping lateral surfaces appears from our experiments to be the major driving source. The retarding and deflecting effects of the TP in winter generate an asymmetric dipole zonal-deviation circulation with a large anticyclone gyre to the north and a cyclonic gyre to the south. Such a dipole deviation circulation enhances the cold outbreaks from the north over East Asia, results in a dry climate in South Asia and a moist climate over the Indochina Peninsula and South China, and forms the persistent rainfall in early spring (PRES) in South China. In summer the TP heating generates a cyclonic spiral zonal-deviation circulation in the lower troposphere, which converges towards and rises over the TP. It is shown that because the TP is located in the east of the Eurasian Continent, in summertime the meridional winds and vertical motions forced by the Eurasian continental-scale heating and the TP local heating are in phase over the eastern and central parts of the continent. The monsoon in East Asia and the dry climate in the Middle Asia are therefore intensified.

Key words: Sensible Heat Driving Air-Pump over the TP (TP-SHAP), dipole circulation pattern, cyclonic spiral circulation pattern, persistent rainfall in early spring (PRES)

1 Introduction

The Tibetan Plateau (TP, also Qinghai-Xizang Plateau in China) extends over the area of 27°—45°N, 70°—105°E, covering a region about one quarter of the size of the Chinese territory. Its mean elevation is more than 4000 m above sea level with the 8844 m (near 300 hPa) peak of Mount Everest standing on its southern fringe. The mean TP altitude lies above 40% of the atmosphere. Because of the lower air pressure, various radiation processes over the plateau, particularly in the boundary layer, are quite distinct from those over lower-elevated regions (Liou and Zhou, 1987; Smith and Shi, 1992; Shi and Smith, 1992). While the TP receives strong solar radiation at the surface, the other parts of Asia at such a level are already in the cold middle troposphere. Geographically the TP is located in the subtropics with westerly winds to the north and easterlies to the south in summer, but provides a barrier to the subtropical westerly jet in winter. Perturbations forced by the TP can generate Rossby waves in the westerlies, which can propagate downstream and influence the circulation anomaly elsewhere. It is also located over the central and eastern parts of the Eurasian continent, facing the Indian Ocean to the south and the Pacific Ocean to its east. Therefore, the Tibetan Plateau can exert profound thermal and dynamical influences on the circulation, energy and water cycles of the climate system.

Before the 1950s, most of the studies concerning the influence of large-scale topography upon atmospheric circulation and climate focused on its mechanical aspects (Queney, 1948; Charney and Eliassen, 1949; Bolin, 1950; Yeh, 1950). In 1957, Yeh et al. (1957) and Flohn (1957), respectively found that the Tibetan Plateau is a heat source for the atmosphere in summer. Using the then available observations, Yeh et al. calculated each term of the heat budget equation and reported that the TP in winter is a weak heat sink, but a large heat source in summer. Since then, the temporal and spatial distributions of the heating field over the Tibetan Plateau and their impacts on weather and climate have become an active area of research and many significant results have been achieved.

Recently, Yanai and Wu (2006) gave a thorough review of the past studies about the effects of the TP. The review starts from the research in the 1950s on the jet stream, the warm South Asian high, and the early progress of TP research in China. The review then goes over studies concerning the mechanical effects of the TP on large-scale motion, the winter cold surge, and the summer negative vorticity source over the TP. The review also covers the importance of the thermal influences of the TP on the seasonal circulation transition and Asian monsoon onset based on different data sets and numerical experiments (Ye and Gao, 1979; Tao and Chen, 1987; Wu et al., 1997; Liu X et al., 2001; Liu Y M et al., 2001; Wu et al., 2002; Wu, 2004; Wu et al., 2004). Great efforts are made in reviewing the evaluation of the heating source on the TP through the analyses of spatial and temporal distributions of $<Q_1>$ and $<Q_2>$ in the notation used by Yanai et al. (1973) and based on observations from the First GARP Global Experiment (FGGE) (December 1978—November 1979) and the Qinghai-Xizang Plateau Meteorology Experiment (QXPMEX) (conducted from May to August 1979 by Chinese meteorologists (Zhang et al., 1988)).

Scientists and visiting scholars at the University of California at Los Angeles (UCLA) headed by Professor M. Yanai evaluated the TP heating source (Yanai et al., 1973, 1992; Yanai and Li, 1994; Shen, et al., 1984; Liou and Zhou, 1987; Nitta, 1983; Luo and Yanai, 1983, 1984; He et al., 1987) and studied the seasonal changes in the large-scale circulation, thermal structure and heat sources and moisture sinks over the Tibetan Plateau and the surrounding areas (Yanai et al., 1992). They also employed the global analyses prepared by the European Centre for Medium-Range Forecasts (ECMWF) to study the onset and interannual variability of the Asian summer monsoon in relation to land-sea thermal contrast (Li and Yanai, 1996), the Australian summer monsoon (Hung et al., 2004b) and its relationship with the Asian summer monsoons (Hung and Yanai, 2004a).

In China, in the last decade, a series of research programs have been organized to support the activities related to the World Climate Research Program (WCRP). They were funded by the Chinese Academy of Sciences (CAS), the National Natural Science Foundation of China (NSFC) and the Ministry of Sciences and Technology of China (MOST). The Second Atmospheric Tibetan-Plateau Field Experiment (TIPEX) was organized in 1998 to provide valuable extra in situ observations (Zhou et al., 2000). The publication of the ECMWF and the National Center of Environment Prediction and National Center for Atmospheric Research (NCEP/NCAR) (Kalnay et al., 1996) reanalysis datasets provided other important data sources for the relevant study. In addition, the progress in the development of the Global Ocean-Atmosphere-Land System climate model (GOALS) (Wu et al., 1997b; Zhang et al., 2000) at the National Key Laboratory of Atmospheric Sciences and Geophysical Fluid Dynamics (LASG) made numerical experiments available.

Great efforts have been made to understand the mechanism concerning how the TP forcing, either mechanical or thermodynamic, can affect the regional as well as global climate. Some of the results are summarized as one part of this study, and can be considered as a complement to the review of Yanai and Wu (2006). In another part of this study, diagnosis and numerical experiments are used to get new insights on our understanding.

The NCEP/NCAR data have an overall intrinsic limit of reliability (Trenberth and Guillemot, 1998; Yanai and Tomita, 1998; Annamalai et al., 1999; Hung et al., 2004b) and should be used with caution. From 1995 to 1999 there was a China-Japan cooperation Experiment of Surface Energy and Water Cycle over the TP. Six automatic meteorology stations (AMS) were installed at Lhasa ($91°08'E$, $29°40'N$), Rikeze ($88°53'E$, $29°15'N$), Lingzi ($94°28'E$, $29°34'N$), Nagqu ($92°04'E$, $31°29'N$), Gaize ($84°25'E$, $32°09'N$), and Shiquanhe ($80°05'E$, $32°30'N$). Based on these observations, the surface fluxes were calculated by Li et al. (2000, 2001). Duan and Wu (2005) compared the surface sensible heat flux at these six stations with those of NCEP/NCAR at the adjacent grid points. They found that in most cases, the temporal variations in the surface sensible heat flux of the NCEP/NCAR and AMS-based results are similar on the monthly mean scale at all six stations. The main difference exists in the magnitude. Apart from at Rikeze, the surface sensible heat flux of NCEP/NCAR is always weaker than that based on AMS, which may lead to an underestimation of the impacts of the sensible heat flux on the TP on climate. For the latent heating, by comparing the July mean precipitation fields of the NCEP/NCAR and Xie-Arkin's data (Xie and Arkin, 1996, 1998) in the same period from 1978 to 1998, Duan and Wu found remarkable differences in East Asia between the two data sets. However, over the TP area, the two data sets possess a similar spatial distribution with decreasing rainfall from the southeastern to northwestern TP, and a maximum rainfall center of more than 14 mm per day over the north of the Bay of Bengal (BOB). For assessing the upper-level routine data such as geopotential height, temperature and wind, Duan (2003) compared the NCEP/NCAR with another reanalysis dataset GEWEX Asian Monsoon Experiment in Intensified Observation Period (GAME-IOP) (Tanaka et al., 2001) with the same horizontal resolution ($2.5° \times 2.5°$) and vertical levels during April to October 1998. This latter dataset assimilates the observations from the field experiments TIPEX, Huihe Basin Experiment (HUBEX), South China Sea Monsoon Experiments (SCSMEX) and GAME-Tibet, and some radiosonde as well as wind profile observations from Thailand, India, and Vietnam, etc. The results show that the discrepancy between the two datasets is small for the monthly mean time-scale. Therefore, in the following diagnoses when NCEP/NCAR reanalysis is employed, we confine ourselves to the investigations of large-scale and long-term monthly mean diabatic heating over the TP and its impacts on large-scale circulations.

This study composes two main parts. The fist part presented in Section 2 is more theoretical. The air column over the TP descends in winter and ascends in summer each year, working as an air-pump and regulating the atmospheric circulation. After discussing the thermal features of the TP, the significance of the surface sensible heating, particularly on its sloping surface, in the effective working of such an air-pump is discussed, and the concept of the TP sensible heat driving air-pump (TP-SHAP) is further stressed. The seasonal variation of heating over the TP and its impact on the winter and summer circulations in the lower troposphere are also discussed in this part. By analyzing the zonal-deviation wind field at 850 hPa and comparing the results with those from numerical experiments, a wintertime "asymmetric dipole" pattern and a summertime "convergent spiral" pattern of the zonal deviation flows are obtained. It is shown that the former is mainly due to the mechanical forcing of the TP, whereas the latter is mainly forced by the TP-SHAP. The second part of this study is presented in Section 3. Here it is demonstrated how the wintertime dipole-type forcing and the summertime spiral-type forcing of the TP influence the Asian climate. Subsection 3.1 focus on how the mechanical forcing of the TP in late winter and early spring contributes to the occurrence of the persistent rainfall (PRES) over Southern China. Subsection 3.2 describes how the summertime continental forcing over Eurasia and the local forcing due to the TP work together to intensify the monsoon in East Asia and the dry climate in the Middle Asia. A Summary and outlook are given in Section 4.

2 Theory

2.1 The Tibetan Plateau Sensible Heat Driving Air-Pump (TP-SHAP)

2.1.1 Observation Evidence

Fig. 1a shows the climate-mean seasonal evolutions of the surface sensible heating and the column integrated total heating over the TP area (80°—100°E, 27.5°—37.5°N) as shown by the heavy rectangle in Fig. 3. The calculation is based on the 18-year (1980—1997) monthly mean NCEP/NCAR Reanalysis. It demonstrates that the TP is a heat sink in the winter months from November to February, but a heat source in the summer months from March to October, in agreement with the earliest finding of Yeh et al. (1957) and Ye and Gao (1979). Fig. 1b shows the seasonal variation of the longitudinal distribution of the surface sensible heat flux averaged from 27.5° to 37.5°N where the TP is located. In winter, while the downward surface sensible heat flux goes from atmosphere to land over the TP region, strong upward sensible heat fluxes are observed over the western Pacific. This is because the frequent wintertime cold outbreaks moving towards the near-sea from the Asian continent uptake huge energy from the underlying ocean surface. In summer the longitudinal thermal contrast between land and sea is reversed: strong upward surface heating occurs over land, while the heating over ocean becomes tranquil or even downward. This is because in summertime the subtropical western Pacific Ocean is covered by a stable and warm anticyclone which suppresses effectively the in-situ air-sea interactions. It is so strong that the west—east land-sea thermal contrast across the eastern coast of China (~120°E) is greatly enhanced, and the reversal of the thermal contrast between winter and summer becomes remarkable. These are in favor of the intensification of northerly winds in winter and southerlies in summer over East Asia, contributing to the formation of the strong East Asian monsoon. Furthermore, the Asian monsoon onset is associated with the reversal of the meridional temperature gradient in the upper troposphere along the subtropics (Flohn, 1968; Li and Yanai, 1996). The earliest reversal occurs over the eastern BOB because the strong heating over the TP in spring can increase the temperature in the northern subtropics earlier than other longitude domain (Wu and Zhang, 1998; Mao et al., 2002a, 2002b). The evolution of the 18-year mean column integrated total diabatic heating rate averaged between 80° and 100°E (Fig. 1c) shows that before March the heating in this longitude domain is negative everywhere except near the equator. From March onwards, the heating over the main part of the TP becomes positive (see also the dashed curve with open circles in Fig. 1a) while the heating over the BOB to its south and over the continental area to its north keeps negative till the early May when the BOB monsoon onset occurs (Wu and Zhang, 1998). This demonstrates again that the TP heating in spring is important for the Asian monsoon onset.

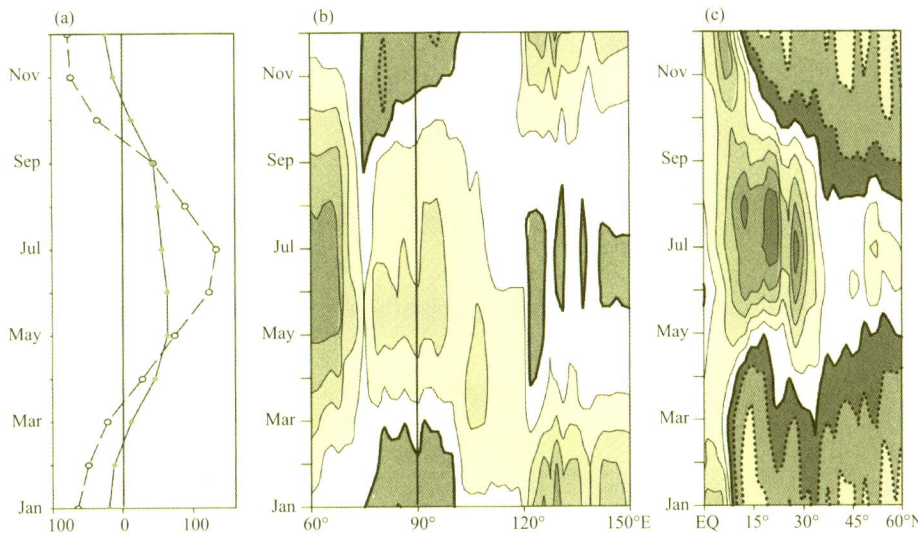

Fig. 1 Seasonal evolutions of (a) column integrated total heating (dashed curve with open circles) and surface sensible heating (solid curve with solid circles) averaged over TP (80°—100°E, 27.5°—37.5°N); (b) surface sensible heating averaged over 27.5°—37.5°N; and (c) column integrated total heating averaged over 80°—100°E calculated from the monthly climatology of NCEP/NCAR reanalysis for 1980—1997. Unit is W m^{-2}. Contour interval is 30 W m^{-2} in (b) and 50 W m^{-2} in (c). Values of more than 30 W m^{-2} in (b) and 50 W m^{-2} in (c) are lightly shaded and in thin-solid contour, while those of negative in (b) and (c) are bounded by heavy solid curves and indicated by heavy shading and dotted contours

Figure 2 shows the cross-sections along 30°N and 90°E of the January- (a and c) and July- (b and d) mean potential temperature and wind vector projected on the corresponding cross-sections in which the vertical velocity w has been multiplied by 10^3. In January along 30°N (Fig. 2a), cold temperatures over continents and warm temperatures over oceans are prominent. Atmospheric cooling indicated by air descent ($\mathbf{V} \cdot \nabla\theta < 0$) prevails in the free troposphere. Along 90°E (Fig. 2c), the strongest cooling is over the TP and its southern slope in association with the strong heating and ascent over the equator. During summer, the warmest center of potential temperature is just over the TP. This is in agreement with the existence of the warm temperature center in July over the plateau (Yanai et al., 1992). Strong ascent prevails in the area to its east from the TP to eastern China (Fig. 2b) and to its south from the TP to the BOB (Fig. 2d), penetrating the isentropic surfaces almost perpendicularly and indicating the existence of a strong heating source over the area in summer (Kuo and Qian, 1982; Liu X et al., 2002; Wu, 2004).

Fig. 3 shows the deviations from the annual mean of the January-mean and July-mean streamlines 10

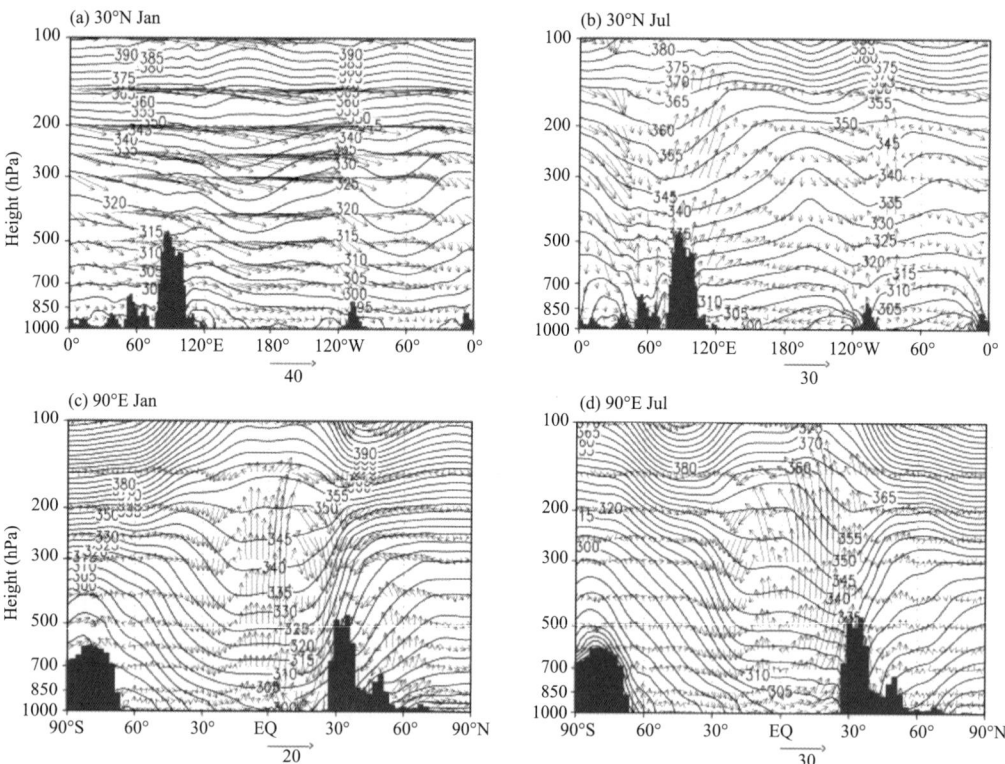

Fig. 2 January (left) and July (right) mean cross-sections of potential temperature (contours, interval is 5 K) and vertical circulation (vectors) along 30°N (a and c) and along 90°E (b and d) from NCEP/NCAR reanalysis (1986—1995). The shaded area is the in situ vertical profile of orography. From Yanai and Wu (2006)

m above the surface calculated from the NCEP II reanalysis for the period 1979—1998. It is evident that the deviation air currents flow from the winter hemisphere into the summer hemisphere, and diverge out of continents and converge towards oceans in the mid-latitude areas in the winter hemisphere, but converge towards the continents and diverge out of oceans in the summer hemisphere. A remarkable circulation reversal appears over the Asian-Australia (A-A) monsoon area. In January the surface air is diverged out of the TP region towards South Africa and Australia in the southern hemisphere, whereas in July the surface air is converged into the TP region from South Africa and Australia, characterizing the seasonal reversal of the A-A monsoon flows.

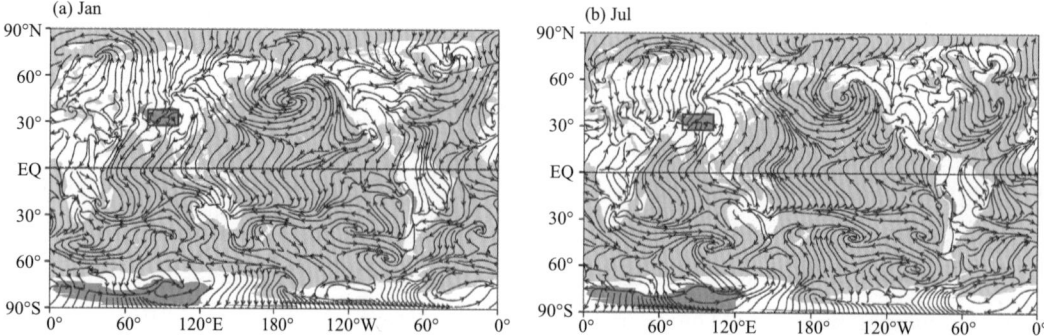

Fig. 3 Distributions of the monthly mean stream field composed of horizontal wind deviations at 10m from the corresponding annual means of 1979—1998 NCEP II reanalysis. (a) January; (b) July. Dark shadings highlight elevations higher than 3000 m. Rectangle indicates the Tibetan Plateau area as defined in the text

2.1.2 Sensitivity Numerical Experiments: Summer

From Figs. 2 and 3, it looks as if the cooling and descent of the air column above the Tibetan Plateau in winter could account for the surface divergence flow, whereas the heating and ascent of the air column above the TP in summer could explain the surface flow convergence towards the TP area, though the huge heating and ascent south of the TP at the equator (Fig. 2c) can weaken the effect of decent over the TP in winter. In other words, the TP air-pump above the TP may play significant roles in regulating the A-A monsoon. However, were there no surface sensible heating especially on the sloping surfaces, such a TP air-pump as identified from Fig. 2 would not be able to converge/diverge the surface air flows from/into the surrounding areas in the lower layers.

To illustrate the significance of the sensible heating over the sloping surfaces in the operation of the TP air-pump, a series of aqua-planet experiments have been conducted based on an atmospheric general circulation model. The model system used in this paper is a new version of the GOALS Spectral Atmospheric Model developed at LASG, Institute of Atmospheric Physics (IAP) (GOALS-SAMIL, version R42L9) (Wu et al., 2003; Wang et al., 2004). It has 42 rhomboidally truncated spectral waves in the horizontal with resolution equivalent to 2.81° (lon) ×1.66° (lat). In the vertical, it has 9 levels in a σ-coordinate system. The model uses a unique dynamical framework by introducing a reference atmosphere and using a semi-implicit time integration scheme (Wu et al., 1997b). The physical processes of the model are quite self-contained, including a k-distribution radiation scheme (Shi, 1981), Slingo's cloud diagnosis scheme (Slingo, 1987), and the simplified simple biosphere model (SSiB) (Xue et al., 1991; Liu and Wu, 1997). For the present purpose, aqua-planet experiments are designed with an idealized topography

$$h(\lambda,\varphi) = \min\left[3 \text{ km}, h_0 \cos\left(\frac{\pi}{2}\frac{\lambda-\lambda_0}{\lambda_d}\right)\cos\left(\frac{\pi}{2}\frac{\varphi-\varphi_0}{\varphi_d}\right)\right] \quad \begin{array}{l} -\lambda_d \leqslant \lambda \leqslant \lambda_d \\ -\varphi_d \leqslant \varphi \leqslant \varphi_d \end{array} \quad (2.1)$$

being introduced into the "aqua-planet" in which the earth's surface in the model is covered merely by ocean. In (2.1), h is elevation, λ is longitude, φ is latitude, and λ_d and φ_d are the half-width in longitude and latitude, respectively. To mimic the TP, the following parameters are set:

$$\lambda_0 = 90°\text{E}, \varphi_0 = 30°\text{N}, \lambda_d = 40° \text{ longitude}, \varphi_d = 16.5° \text{ latitude, and } h_0 = 5 \text{ km} \quad (2.2)$$

In addition, a value of 3 km is set for the elevation limit so that a platform over the plateau can be formed. The ground surface is covered by grass alone. Such a treatment on ground type (soil type is 7) can minimize the impact of land-sea thermal contrast and let us concentrate on the influence of the elevated heating over the topography on the circulations. By using such a design, the surface of the topography is divided into two parts: the sloping surface and the platform top surface as demonstrated on the left panels of Fig. 4. In all experiments, the cloud amounts are prescribed by using the climate zonal means (Wu et al., 2004). The solar angle is set on July 15th so that perpetual July experiments are assumed, and the initial fields are taken from the July means of the original model integration. All the integrations are performed for 540 days, and the results averaged from the last 360 days are taken for the following diagnosis.

There are four experiments, i. e., NOSH, ALLSH, SLPSH, and TOPSH:

(a) NOSH: None of the "TP" surfaces possess surface sensible heating;

(b) ALLSH: All of the "TP" surfaces possess surface sensible heating;

(c) SLPSH: Only the sloping "TP" surface possesses surface sensible heating; and

(d) TOPSH: Only the top "TP" surface possesses surface sensible heating.

Then the differences between the three pairs, i.e., ALLSH-NOSH, SLPSH-NOSH and TOPSH-NOSH, can be considered: the impacts on the circulation of the TP due to all surface sensible heating; sloping-surface sensible heating alone; and top-surface sensible heating alone. Their differences in horizontal wind and vertical motion at the lower model level $\sigma=0.991$ are presented in Fig. 4. Coastlines are added into the figure only for reference and comparison. In the all surface sensible heating case (Fig. 4a), surface air from the Arabian Sea, Indian subcontinent, BOB, Indochina Peninsula and the other neighborhood areas are converged into the TP area and ascend over the TP. The enhanced upward movement of more than -0.2 Pa s^{-1} is on its southern and eastern slopes where the release of latent heating further intensifies the ascent. The sloping-surface sensible heating produces similar circulations (Fig. 4b) as in the case with the surface sensible heating impose on all the TP surfaces (Fig. 4a) except over the platform: due to the absence of the in-situ surface sensible heating, air descends over its top in association with radiation cooling. On the other hand, the mere top-surface sensible heating produces convergence flow only above the platform where the elevation is already higher than 3 km (Fig. 4c). There is no convergence at the lower elevations, and the main ascent appears over the eastern TP as a stationary Rossby-wave response to the top surface heating. The results shown above can be explained by using the right panels in Fig. 4. In the presence of surface sensible heating on the sloping surfaces (Figs. 4a and b), the heated air particles at the sloping surface penetrate the isentropic surfaces and slide upward. The air in the lower elevation layers in the surrounding areas is therefore pulled into the plateau region, forming strong rising motion and even heavy rainfall over the TP. On the contrary, in the top-heating-only case (Fig. 4c), the platform heating can result in air convergence above the plateau and cannot pull air from below. This is because when an air particle is traveling in the lower layer and impinging on the TP, it has to stay at the same isentropic surface due to the absence of diabatic heating from the sloping lateral surface of the TP. Therefore, the air particle just goes around the TP at a rather horizontally located θ-surface and no apparent ascent occurs. Therefore, there are no significant impacts on monsoon rainfall.

2.1.3 Sensitivity Numerical Experiments: Winter

Similar design is also applied to the perpetual January experiments, in which the solar angle is fixed on January 15th. In January the surface sensible heat flux on the TP is weak and negative in NCEP/NCAR reanalysis (Fig. 1a) as well as in the GOALS model. To highlight the effects of the TP's surface cooling on circulation, instead of diagnosing the model outputs, a unified negative surface sensible heat flux of (-30 W m^{-2}) is imposed on the relevant TP surfaces in the experiments ALLSH, SLPSH and TOPSH. By following the same procedures as in the perpetual July experiments, the corresponding results at the $\sigma=0.991$ surface for such perpetual January experiments are obtained and shown in Figs. 4d, 4e and 4f. When all the TP surfaces have a negative sensible heat flux (ALLSH), an anticyclone circulation appears to the northwest and a cyclone circulation appears to the northeast of the TP in the difference flow field between ALLSH and the experiment without surface heating (NOSH) (Fig. 4d). Easterly difference flow then slips down the western slope of the TP, while westerly difference flow slips down its eastern slope, and the surface air on the TP is diverged towards the surroundings. The difference in the wind field and vertical motion on the $\sigma=0.991$ surface are shown in Fig. 4e for the difference between the experiment in which the negative sensible heat flux is only imposed on the sloping lateral surface (SLPSH) and the experiment NOSH. The down sliding flow on both the western and eastern sloping surfaces and the diverging flow in the neighborhood of the TP are similar to those presented in Fig. 4d,

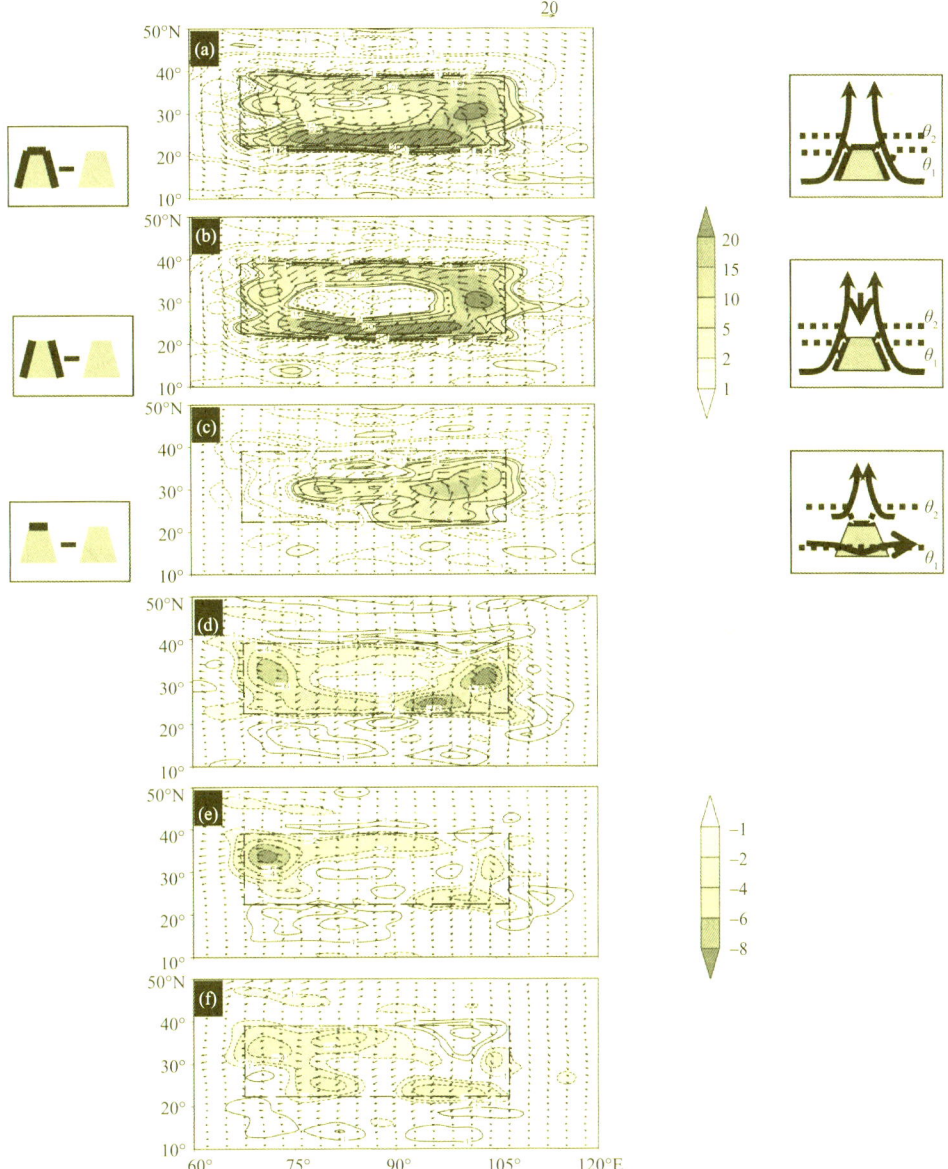

Fig. 4 Distributions of difference wind (vector) and vertical velocity ($-\omega$, shading, 10^{-2} Pa s^{-1}) at the $\sigma=$ 0.991 surface from the perpetual July experiments (a) ALLSH-NOSH, (b) SLPSH-NOSH, and (c) TOPSH-NOSH; and from the perpetual January experiments (d) ALLSH-NOSH; (e) SLPSH-NOSH; and (f) TOPSH-NOSH. Left panels indicate the experiment designs, and right panels, mechanism interpretations. See text for details

though the intensity becomes weaker. When the negative sensible heat flux is only imposed on the top surface (TOPSH), the corresponding difference fields between TOPSH and NOSH are shown in Fig. 4f. In this case, the outward flow to the east and west of the TP is much weaker. Results from the perpetual January experiments then imply that the surface cooling of the TP on its sloping lateral boundary in wintertime also plays an important role in diverging cold air to the surrounding areas. We can now come to an important conclusion that the air-pump over the TP can regulate the A-A monsoon only because the cooling/heating of its sloping surface can diverge/converge the air in the lower layer into/from the surrounding areas. In other words, the Air-Pump of the TP as defined by Wu et al. (1997a, 2004) is a Sensible-Heat driving Air-Pump, and can be expressed as TP-SHAP. Comparing Figs. 4d—f with Figs.

4a—c, we see that in January the diverging impact of the TP-SHAP on the surrounding circulation is not as strong as its converging impact in summertime. This is because in summer the converging impact of the TP-SHAP is intensified by the condensation heating over the TP. In other words, there exists positive feedback between the small-scale convection over the TP and the large-scale convergent spiral circulation in the lower troposphere in the surroundings, manifesting as a kind of the Conditional Instability of the Second Kind (CISK, Charney and Eliassen, 1964). Besides, in the following we will see that in the winter season the mechanical forcing of the TP is more important.

The above finding is of practical significance both in numerical model development and field experiment design. For years we have tried to get observational data over the TP. In 1979 and 1998, China organized two large-scale field experiments in association with the FGGE and GEWEX Programs (Zhang et al., 1988; Zhou et al., 2000), respectively. The main observation networks for the two experiments cover much of the TP platform. Although many important features concerning the thermal state and the energy balance on the TP have been revealed, not much is related to the circulations and monsoon. Results from this study show that in order to better understand the impacts of the energy state over the TP on the water cycle associated with the A-A monsoon, we need to pay more attention to the energy state on the sloping surface of the TP, particularly on its southern and eastern slopes.

2.2 TP Forcing and Seasonal Variation of the General Circulation over Asia

2.2.1 TP-SHAP and the Abrupt Seasonal Transition of the Monsoon Circulation

The operation of the TP-SHAP has profound impacts on the Asian climate. It anchors the earliest Asian monsoon onset venue over the region from the eastern BOB to the western part of the Indochina Peninsula (Wu and Zhang, 1998; Liang et al., 2005). Its forcing also contributes to the abrupt seasonal transition of the nearby circulation as first reported by Yeh et al. (1959). To demonstrate this, the low-resolution version R15L9 of the GOALS-SAMIL model is employed to initiate a set of sensitivity experiments. In the first experiment, for the calculation of radiation the cloud distributions are prescribed by using satellite remote sensing data. The observed distributions of sea surface temperature and sea ice from 1979 to 1988, which were set for the AMIP experiments, are introduced as the prescribed lower boundary conditions to integrate the model for 10 years. This is defined as the CON run. The second experiment is the same as CON except that the sensible heating over the Tibetan Plateau region where the elevation is higher than 3 km is not allowed to heat the atmosphere aloft. This is achieved by switching off the sensible heating term in the thermal dynamic equation at the grid points over the plateau region. This experiment is then defined as the NSH run. Since the cloud amounts for the calculation of radiation are prescribed, there is no cloud-radiation feedback in the model; and since the ground surface temperature in the model is calculated based on the thermal equilibrium assumption, the energy budgets at the ground surface in the two experiments are kept unchanged. Therefore, the differences between these two experiments can be considered as resulting purely from the sensible heating over the plateau.

Fig. 5 shows the mean seasonal evolution of the simulated zonal wind component and the ridgeline of the subtropical anticyclone as identified by the $u=0$ contour at 300 hPa and along 90°E. In the CON run, the westerly jet is strong in winter with its center located near 35°N, but is weak in summer with its center north of 40°N. The ridgeline of the subtropical anticyclone is located near 12°N in the winter months from November to May, and near 30°N in the summer months from June to October. An abrupt northward jump of the ridgeline from its winter location to its summer location occurs near the end of May. In one pentad it jumps north by more than 15°, mimicking the abrupt seasonal transition reported by Yeh et al. (1959). However, in the NSH run, such an abrupt seasonal transition disappears, and the northward migration of the ridgeline of the subtropical anticyclone becomes rather smooth. This is because in spring, as the surface sensible heating over the TP increases, the anticyclone in the upper troposphere intensifies gradually. Soon after the Asian monsoon onset occurs over the BOB and the Southern China Sea, due to the large latent heat release associated with monsoon rainfall, the strong northerly winds over East Asia and the strong South Asian high over the south of the TP are forced (Wu et al., 1999, Wu and Liu, 2000, Liu Y M et al., 1999, 2001). The South Asian high is so strong that the wintertime westerly winds to the south are reversed to easterly winds, and the westerlies to the north are intensified. The northward "jump" of the westerly winds and of the subtropical anticyclone is then completed. In the absence of the elevated sensible heating over the TP, the Asian monsoon and the summertime South Asian high are rather weak, and no apparent abrupt changes in circulation are observed.

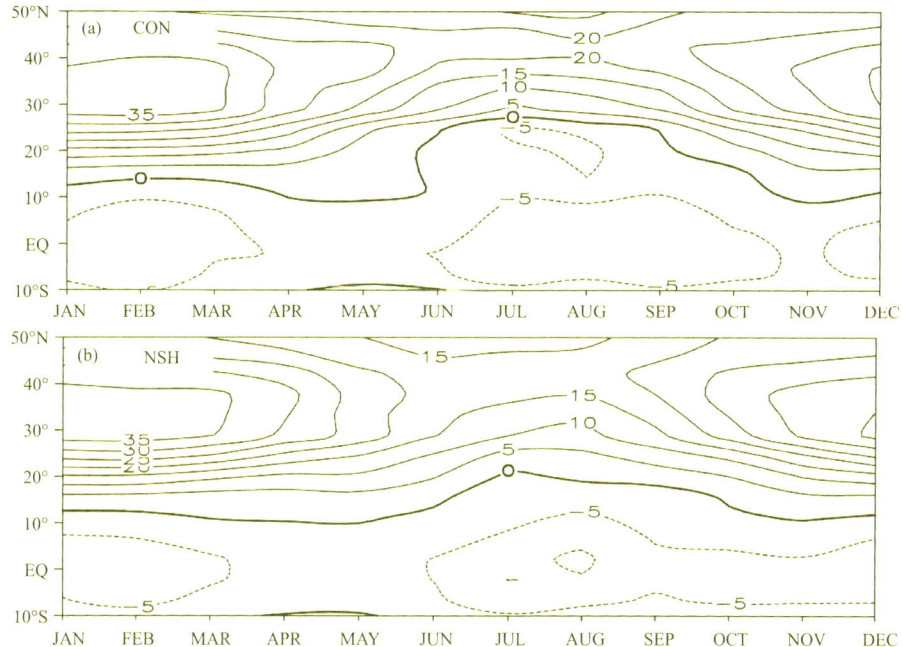

Fig. 5 Seasonal evolution of the zonal wind component (unit is m s^{-1}) and the ridgeline of the subtropical anticyclone (represented by $u=0$) along 90°E at 300 hPa in the model experiment (a) CON, and (b) NSH

2.2.2 TP Forcing and the Associated Winter and Summer Circulation Patterns

The seasonal evolution of the thermal state averaged over the TP area can be calculated by employing the monthly mean NCEP/NCAR Reanalysis. Since the surface elevation varies greatly in different parts of the area, and the atmospheric diffusive heating associated with surface sensible heat flux vanishes rapidly with increasing height, the σ-vertical coordinate is adopted for plotting different heating profiles averaged over the area in different seasons. As shown in Fig. 6, radiation always cools the atmosphere over the TP, with a maximum radiation cooling rate of 2—3 K d^{-1} at $\sigma=0.85$. Condensation heating due to latent heat release is significant during the summer. It warms the atmosphere by more than 3 K d^{-1} over most of the troposphere, with a maximum of more than 5 K d^{-1} at $\sigma\approx 0.95$. Condensation heating is also important in spring, which compensates for the radiation cooling in the free atmosphere. It becomes much weaker in autumn and winter. The most striking feature appears in the vertical profiles of sensible heating. Firstly, it is a lower layer heating that approaches zero at $\sigma=0.85$ in January and at $\sigma=0.75$ in April before the monsoon onset. Secondly, the surface heating is negative in January (~-2 K d^{-1}) but strongly positive in July (~12 K d^{-1}), possessing the largest seasonal cycle and accounting for the diverging effect in winter and converging effect in summer of the aforementioned TP-SHAP on the surrounding atmosphere. Because the near-surface sensible heating is very strong, the shape of the total heating profile follows that of sensible heating quite well in all seasons. At the surface the total surface heating flux is downward in winter but strongly upward in summer (see also Fig. 1a). The whole total heating profile is negative in January but positive in July. It then demonstrates that the whole atmospheric column over the TP is a heat sink in winter and a heat source in summer.

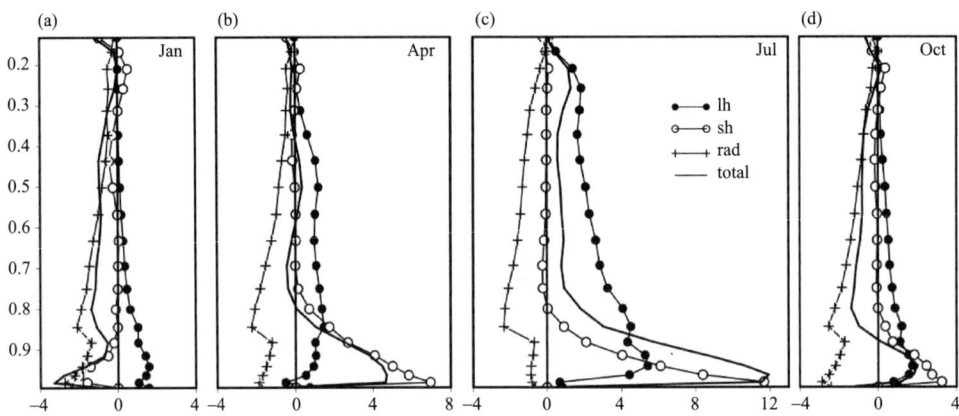

Fig. 6 Vertical heating profiles in difference seasons over the Tibetan Plateau area (80°—100°E, 27.5°—37.5°N) from NCEP/NCAR for 1980—1997. lh: Latent heating; sh: sensible heating; rad: radiation cooling; and total: total heating. Unit is K d^{-1}. (a) January; (b) April; (c) July; (d) October

Because Fig. 6 is based on the NCEP/NCAR reanalysis, and because reanalysis products are model dependent, caution is needed when these results are employed for research. Table 1 shows the surface sensible heat flux and the column integrated total heating averaged over the TP area in different months and calculated for the period 1980 to 1999 from different sources. The nature of model dependence of the reanalysis data can be seen clearly in Table 1. The surface sensible heat flux over the TP calculated from NCEP/NCAR and NCEP II are quite similar, possessing negative flux in January and similar annual cycles, although the fluxes in April and July obtained from NCEP II are about 8 W m^{-2} larger than from NCEP/NCAR, and closer to observations (Duan and Wu, 2005). However, these results are different from their counterparts calculated from ERA-40. In ERA-40, the flux in January is no longer negative, and its maximum appears in spring rather than summer. For the volume integrated total heating the two data sets, NCEP/NCAR and ERA-40, possess similar seasonal changes; but the latter always exceeds the former by more than 30 W m^{-2}, with the minimum 34.28 W m^{-2} in July and maximum of 82.38 W m^{-2} in April. Therefore, although the general features concerning the TP heating (Fig. 6) are in agreement with other reanalysis and in situ observations (Duan and Wu, 2005), the magnitudes presented here are different and the results obtained below should be considered qualitative in nature.

The seasonal variation of the thermal state over the TP has profound influence on the general circulation of the atmosphere. Fig. 7 depicts the winter (a) and summer (b) mean distributions of potential temperatures at 850-hPa and the stream field of the wind deviations from their corresponding zonal means. The distributions of the stream field presented in this figure are similar to the corresponding stream fields of Wu et al. (their Figs. 5a and 5d, 2005), which are composed of the wind difference between two numerical experiments with and without the topography over the Eurasian continent. Therefore, the features of Figure 7 could be considered to be largely due to the impacts of the topography on circulation. In winter months the topography retards the impinging mid-latitude westerly winds and deflects the flow from zonal to circumcolumnar (Fig. 7a). The deviation streamlines depict an "asymmetric dipole shape" with an anticyclone to the north of the TP and a cyclone to the south. The streamlines then converge over Eastern China and go into its "Eastern pole", but come out of its "Western pole" and diverge over the Middle Asia. The anticyclonic deviation circulation gyre in high latitudes advects warm air to the north to its west but cold air to the south to its east. As a result, the isotherms in the high latitudes of Asia tilt from northwest to southeast, and the temperature at 130°E is colder than that at 50°E by 10 K at 40°N and 14 K at 50°N. On the other hand, the cyclonic deviation circulation gyre in low latitudes advects dry air southward to the South Asia Subcontinent, but moist air northward to the Indochina Peninsula and South China. As a result, a prolonged dry season in South Asia and persistent rainy season in Southeast Asia and South China are observed before the Asian monsoon onset. In the summer months, the strong TP heating excites a huge cyclonic deviation circulation over East Asia, and the strong pulling of the TP-SHAP makes the surrounding flows converge into the TP area. Therefore, the summer pattern of the deviation stream field at 850 hPa resembles a huge "cyclonic spiral", and the TP-SHAP looks like a spiral pump. In fact, the summer TP is an important genesis location of vortices that can propagate eastward and result in torrential rain along the Yangtze River (Tao, 1980).

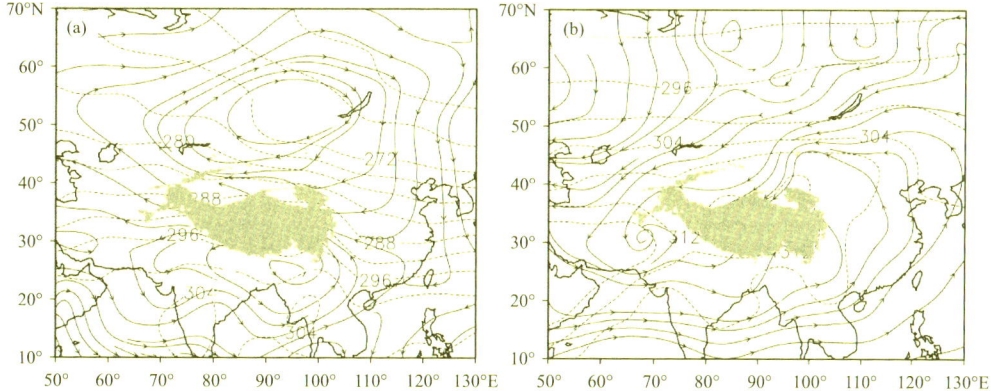

Fig. 7 Distributions at 850 hPa of the potential temperature (unit is K) and of the stream fields composed of wind deviations from the corresponding zonal means based on the NCEP/NCAR Reanalysis for 1968—1997. (a) Winter average between December and February; (b) Summer average between June and August

3 Influence of the TP-SHAP on Asian Climate

3.1 Persistent Rains in Early Spring (PRES) over South China

Besides the well-known Asian monsoon, another less-known unique feature of Asian rainfall is the occurrence of Persistent Rains in Early Spring (PRES) over Southern China. From late February to early May (Pentad 12—26) before the onset of the Asian monsoon, persistent rains occur over a large area south of the Yangtze River (Fig. 8). Southern China is an important agricultural area with abundant rice fields. Heavy PRES are often accompanied by cold temperatures, which can rot rice seedlings and cause agricultural catastrophe. However, the formation mechanism is still unclear. Tian and Yasunari (1998) noticed that the PRES occurs because there is ample water vapor transported into the region by southerly winds from the South China Sea (SCS). They then proposed a mechanism of time-lag in the spring warming between land and sea to explain the formation of the PRES. In early spring the Indochina Peninsula west of the SCS warms more rapidly than the Western North Pacific to the east of SCS. The surface pressure gradient is then oriented eastward across the SCS, forming southerlies and contributing to the occurrence of PRES. However, a recent study of Wan and Wu (2007) shows that such a heating time-lag mechanism also exists between Mexico and the Western North Atlantic in the same period, but there is no PRES in North America. It seems that the time-lag heating mechanism could be a necessary but not a sufficient condition for the formation of PRES.

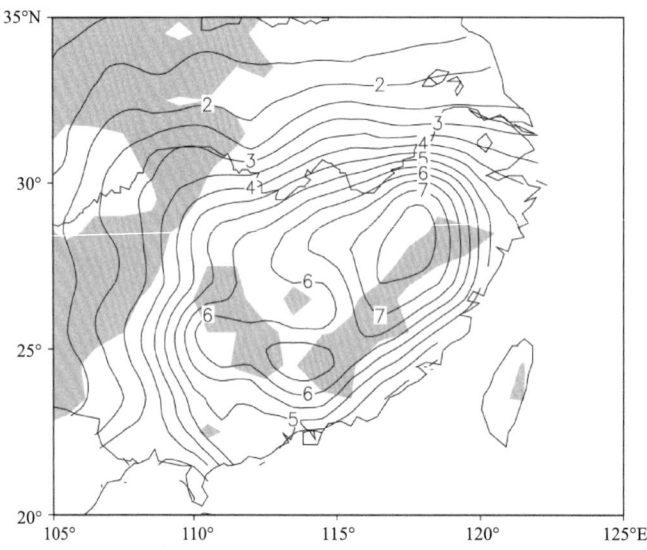

Fig. 8 Distribution of mean rainfall in the period of PRES (Pentad 12—26) over China averaged between 1951 and 2000. Unit is mm d^{-1}. Shading indicates where the elevation exceeds 600 meters. From Wan and Wu (2007)

As demonstrated in Fig. 7a, before the monsoon onset the TP can generate a dipole type circulation pattern. The deflected flow brings cold air from the north and moist air from south which then converge over eastern China. This may contribute to the formation of PRES. To verify this hypothesis, a series of numerical experiments were conducted by Wan and Wu (2007). For our purpose, the R42L9 GOALS-SAMIL is employed again. Results of its long-term integration show the ability of the model in simulating the observed mean climate. Figure 9 displays the distributions of the simulated wind vector at 850 hPa (a) and rainfall (b) in April. Strong southwesterly winds on the southeastern flank of the TP and strong northwesterly winds to the far northeast agree well with observations, reflecting the deflection impacts of the TP on the basic flows as shown in Fig. 7a. The general features of PRES as shown in Fig. 8 are also well captured (Fig. 9b). There is an obvious rain area to the south of the Yangtze River, with a maximum of about 8 mm d^{-1}, consistent with observations (Fig. 8).

To focus on physical mechanisms, perpetual April integrations are designed in the following experiments. The solar angle is set to 10 April. Each experiment is integrated for 36 months, and the results from the last 30 months are retrieved for the following diagnosis. Experiments start with a case of "no mountains in Eurasia" in which the Eurasia Continent is under cut and leveled to 0 km while the topography in other continents are kept unchanged. The other experiments are the same as the first except the Eurasian topography has increasing heights at an interval of 1 km. Fig. 10 shows the wind vector at 850

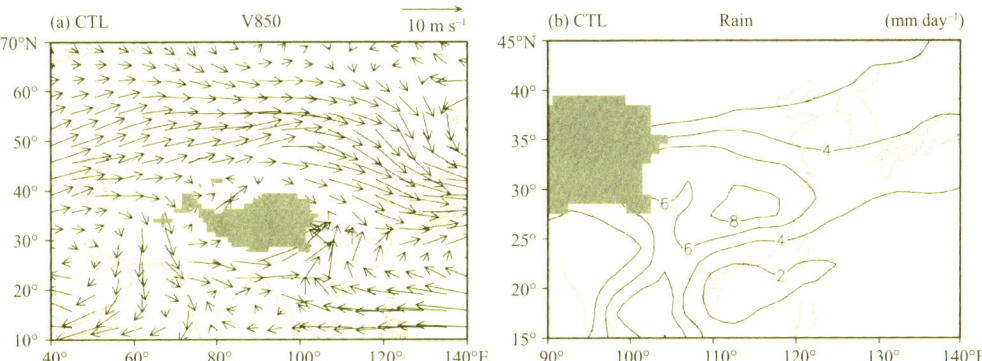

Fig. 9 Distributions of (a) wind vector at 850 hPa and (b) rainfall (unit: mm d^{-1}) in spring (March and April means) averaged over 10 years as simulated in the model GOALS-SAMIL. The shaded areas are where elevation exceeds 2500 m. Based on Wan and Wu (2007)

hPa and rainfall in the experiments when the elevation of the TP is leveled at 0, 2, 4 and 6 km, respectively.

In the "no TP" experiment, the westerly jet in Eurasia does not split, and a zonal-oriented subtropical anticyclone belt dominates in the middle-low latitudes (Fig. 10a). Significant rainfall appears only in the southwest of the Indochina Peninsula, in association with the easterly perturbation forced by land-sea thermal contrast (Fig. 10b). After the elevation of the TP reaches 2 km, the westerly jet starts to split into its northern and southern branches (Fig. 10c). They meet downstream of the TP, forming a strong Asian jet. The belt of subtropical anticyclone breaks, and a prominent anticyclonic center appears over the western Pacific. Therefore, the distribution of rainfall takes the shape of PRES (Fig. 10d). In the 4 km TP experiment, the split westerly jets on both north and south sides of the TP are strengthened, the circulation pattern (Fig. 10e) is already close to the simulation counterpart as shown in Fig. 9a, and the PRES appears (Fig. 10f). In the 6 km TP experiment, although the basic circulation pattern does not change very much, the two split westerly jets are much stronger (Fig. 10g), and the winter time TP Dipole circulation also becomes stronger compared to the control simulation (CON) in which the seasonal variation is included (Fig. 9a). As a result, the PRES are weakened, and the main rainfall center moves to the eastern TP (Fig. 10h) where the convergent entrance pole of the TP dipole in the wintertime deviation streamline pattern is located.

The above experiments prove that PRES are formed mainly due to the deflecting effects of the TP. When the deflected flow—cold air from north and warm, moist air from south—meet downstream of the TP PRES are formed. The experiments further demonstrate that the intensity and location of the PRES are to some extent influenced by the elevation of the TP.

3.2 Continental-Scale Heating, Tibetan Plateau Heating and East Asian Climate

The atmospheric thermal response to diabatic heating is to generate a cyclonic circulation near the surface and anticyclonic circulation in the upper layer, with a rising in the east and a sinking in the west (Gill, 1980; Hoskins, 1991; Wu and Liu, 2000). Wu and Liu (2003) and Liu et al. (2004) found that in the summer months in the subtropics, the dominant heating over each continent and its adjacent oceans are organized in the following order from west to east: Longwave radiation cooling (LO) over the western off-shore region; surface sensible heating (SE) over the west; condensation heating (CO) over the east; and double dominant heating (D) over the eastern off-shore region. In association with such a

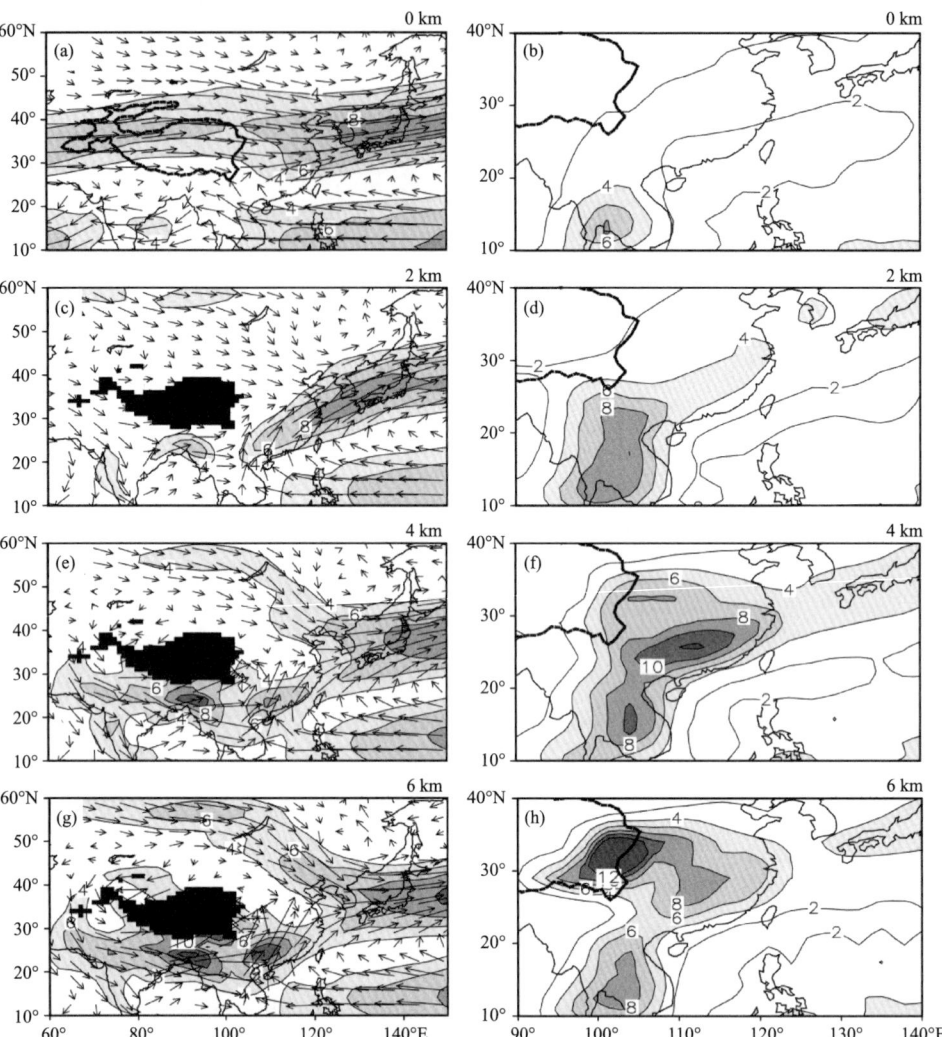

Fig. 10 Distributions of wind vector and isotach at 850 hPa (left panels, unit is m s^{-1}, shaded indicate exceeding 4 m s^{-1}) and rain (right panels, unit is mm d^{-1}) in the perpetual spring sensitivity experiments with different TP elevations and averaged over 30 months. The black shaded in left panels and bold solid curve in right panels are the main part of TP. The TP maximum elevation is 0 km in (a) and (b), 2 km in (c) and (d), 4 km in (e) and (f), and 6 km in (g) and (h). From Wan and Wu (2007)

LOSECOD quadruplet heating, a cyclonic circulation at the surface and anticyclonic circulation in the upper troposphere are forced over land, whereas a surface anticyclonic circulation and upper tropospheric cyclonic circulation are forced over the ocean. Under the continental-scale heating, ascending motion and a moist climate appear over the eastern continent, whereas descending motion and dry climate prevail over its west.

As shown in Fig. 6c, the TP is a strong heat source in summer months with the strongest heating of about 12 K d^{-1} near the surface. Such a forcing should produce a shallow cyclonic circulation near the surface and anticyclonic circulation over the deep layers aloft, with a rising in the east and a sinking in the west (Duan and Wu, 2005). This is observed in the summer not only over the TP, but also over the Iran Plateau (Fig. 11). The meridional wind cross-section (a) demonstrates that over each mountain there exist a shallow cyclonic circulation near the surface and anticyclonic circulation in the deep upper layers. The cross-section of vertical motion (b) shows that over each mountain range, rising air predominates over the central and eastern parts whereas sinking is observed over the west. It is important to note that the rising motion over East Asia excited by the continental heating and the rising excited by the local TP heating link up into a single stretch (Fig. 11b), and become very strong (~ -0.1 Pa s^{-1}). Conversely, over Middle Asia, the descending motions forced by both continental scale heating and topography heating are in phase. The water cycle in this region is suppressed, and the local climate becomes rather dry.

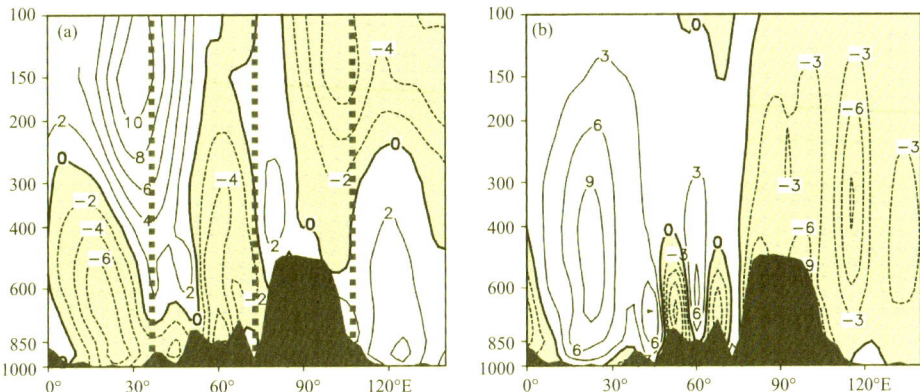

Fig. 11 Longitude-pressure cross-sections along 32.5°N of the July mean meridional wind (a, unit is m s^{-1}) and vertical velocity in p-coordinates (b, unit is 10^{-2} Pa s^{-1}) from NCEP/NCAR for 1981—1999. From Duan and Wu (2005)

The above diagnosis implies that because the TP and Iran Plateau are located in the middle and eastern parts of the Eurasian continent, the water cycle over East Asia associated with the Asian monsoon is intensified. To prove this reasoning, three numerical experiments were designed by using the GOALS-SAMIL. This time an idealized land distribution mimicking the distribution of Eurasian and Africa Continents and the India and Indochina Subcontinents (as marked in Fig. 12) is added to the aqua-planet. Each experiment is integrated for 10 years, and the results averaged from the last 8 years are retrieved for diagnosis. The first experiment does not have any mountains. In the second experiment an idealized ellipsoidal topography with a maximum height of 5 km is embedded in the land area centered at 87.5°E, 32.5°N to mimic the TP. In the third experiment, this idealized ellipsoidal topography is moved westward to 60°E, 32.5°N, just to the north of the "Arabian Sea". The results presented in Fig. 12 show that in the absence of topography (Fig. 12a), the Eurasian continent in summer months forces a huge continental scale cyclone at 850 hPa with southerly flow of 4 m s^{-1} off its eastern coast and northerly flow of 6 m s^{-1} off its western coast. A weak East Asian monsoon of 8 mm d^{-1} is forced over its southeastern corner, with the rain band tilting from southwest to northeast. The northeast tilt of the summer rain band in the absence of the TP contributes to the observed east—west contrast between monsoonal climate in East Asia and dryness in Sahara. Based on a highly simplified experiment with idealized coastline, Xie and Saiki (1999) discussed this tilted rain band due to the northward moisture advection on the east coast and southward advection on the west coast of a major continent. Chou et al. (2001) used an intermediate-complexity atmospheric model coupled with a simple land-surface model and a mixed-layer ocean model to investigate the processes involved in an idealized monsoon occurring on a single rectangular continent. They found that the inclusion of an upward ocean heat transport favors continental convection. The monsoon circulation then produces a moisture transport from the ocean regions that allows substantial progression of convection into the subtropics over the eastern portion of the continent. Chou et al. consider this east—west asymmetry to be partly due to the interactive Rodwell-Hoskins mechanism (Rodwell and Hoskins, 1996).

When the topography is centered near 90°E (Fig. 12b), two cyclonic systems in the 850 hPa wind field can be clearly identified. One is the continental scale cyclone that is similar to the one presented in Fig. 12a, with the intensity of the northerly flow of 6 m s^{-1} off its western coast almost unchanged. The other is just encircling the TP with the northerly flow on its western flank and the southerly flow on its eastern flank that overlaps with the southerly flow forced by the continental heating and is enhanced to 10 m s^{-1}. The northerly flow on the west is quite strong (greater than 8 m s^{-1}). It advects cold and dry air from the Asian inland area and contributes to the formation of the desert climate in the Middle Asia. He et al. (1987) showed that the dry descent over Middle Asia after the onset of the Indian monsoon is induced by the upper layer divergence over the monsoon rainfall region (their Fig. 12c). Yanai et al. (1992) described this dry descent over Iran-Afghanistan region in more detail. Rodwell and Hoskins (1996) interpret the formation of the desert climate in the Middle Asia in terms of the westward Rossby wave forcing that is triggered by the latent heating of the Asian monsoon and interacts with the westerly flow over the Middle Asia. Here we show that the TP forcing can also contribute to the formation of the desert climate in the Middle Asia. However, when the topography is moved westward to 60°E (Fig. 12c), while the cyclonic circulation forced by continental heating does not change very much compared with its counterpart in the no-topography experiment as shown in Fig. 12a, the cyclonic circulation that encircles the TP in Fig. 12b has moved westward. The southerly flow associated with the continental heating is about 8 m s^{-1}, and that associated with the topography forcing also about 8 m s^{-1}. However, these two southerly bands are quite separated, and the strong "East Asia monsoon" rain belt that exists in the previous experiment (Fig. 12b) now splits into two belts, one over the southeastern corner of the continent that is rather similar to its counterpart in the no-topography case (Fig. 12a), and the other on the eastern flank of the TP. On the other hand, the northerly flow on the western flank of the TP over-

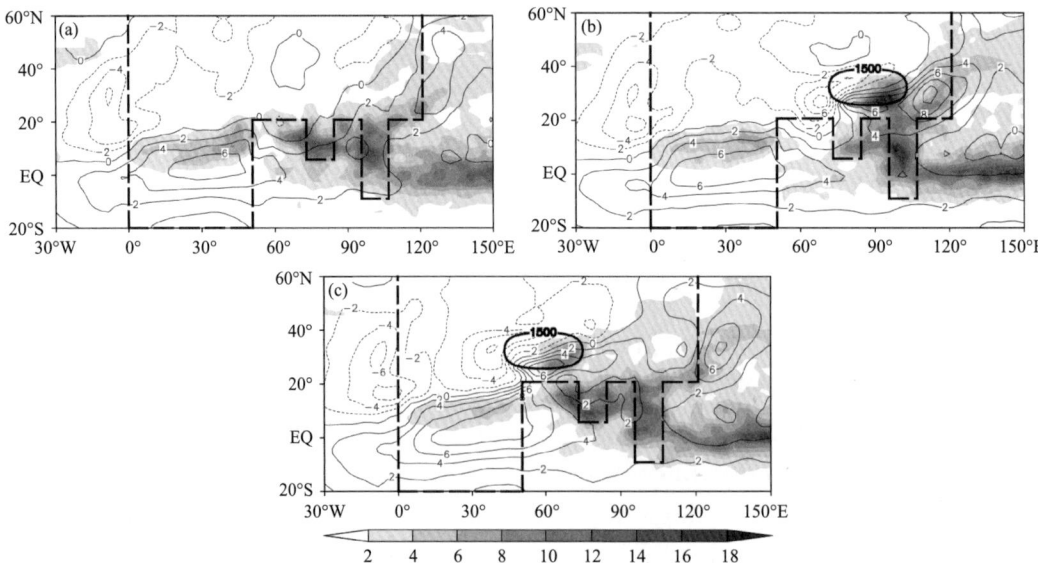

Fig. 12 Distributions of the 8-year July-mean meridional wind at 850 hPa (isoline, interval in 2 m s^{-1}) and rainfall (shading, mm d^{-1}) in the idealized experiment (a) No Mountain; (b) Mountain centered at (87.5°E, 32.5°N); (c) Mountain centered at (60.0°E, 32.5°N). The heavy dashed line denotes the idealized coastlines; and the heavy ellipse in (b) and (c) indicates the location of orography at the elevation of 1500 m

laps with that forced by continental heating thus the area with northerly flow stronger than -2 m s^{-1} is expanded. We can then reach the conclusion that the strong East Asian monsoon and the associated water cycle occur because the continental forcing and TP forcing act to enhance the rising motion and water cycle over the East Asian area.

4 Summary and Outlook

The huge TP is one of the earth's most impressive topographical features. Its mechanical forcing and unique elevated thermal forcing have a large influence on global as well as regional climate. The TP is a weak heat sink in winter and a strong heat source in summer. Mainly driven by the sensible heating at its sloping surfaces, the wintertime cold air particles above the TP descend along the cooling sloping surfaces, being pumped into the surrounding areas; whereas the summertime warm ascending air particles above the TP pull the air from below, and the lower tropospheric air from the surrounding areas are converged towards the TP area and climb up the heating sloping surfaces. It is shown that such a TP-air-pump is mainly driven by the surface sensible heating on TP particularly on its sloping lateral surfaces. It is defined as the Sensible Heat driving Air-Pump of the Tibetan Plateau. It contributes significantly to the seasonal reversal of the airflows in the Asian-Australian monsoon areas. The impact of the pulling on the atmospheric circulation in summer is stronger than its expelling impact in winter because in the summer months there is a CISK-like positive feedback between the small-scale convection over the TP and the large-scale convergent spiral of the lower tropospheric circulation in the surrounding area. The absence of the sensible heating on the sloping lateral surface of the TP can shut off such a CISK mechanism and significantly weakens the TP-Air-Pump.

In winter as a large obstacle, the TP retards the westerly jet flow and deflects it into northern and southern branches. The deviation stream flow then appears as an asymmetric dipole with the convergent entrance on its eastern flank and the divergent outgoing on its western flank. The huge anticyclonic deflected flow in the north has an important impact on atmospheric temperature distributions due to its horizontal advection, making East Asia much colder than the Middle Asia. Its cyclonic deflected flow in the south has an important impact on the dry climate in South Asia and the moist climate in Southeast Asia and South China. In late winter and early spring over South China, the southward moving cold and dry air that flows along the northern anticyclonic circulation of the TP-Dipole meets with the northward moving warm and moist air that flows along the southern cyclonic circulation of the TP-Dipole. Persistent rainfall in early spring (PRES) therefore occurs over South China till the Asian monsoon onset. In late spring, the TP heating also contributes to the establishment and intensification of the South Asian high and the abrupt seasonal transition of the surrounding circulations. In summer, the TP heating produces a large-scale cyclonic circulation in the lower troposphere. Such a forcing in conjunction with the TP-Air-Pump causes the deviation stream field to resemble a cyclonic spiral, converging towards and rising over the TP.

The TP together with Iran Plateau are located in the central and eastern parts of the Eurasian Continent. The meridional and vertical motions generated by the Eurasian continental-scale heating are in phase with those generated by the TP local-scale heating over Asia. Southerly flow rises over East Asia and the northerly flow sinks over the Middle Asia and therefore becomes very strong. The Asian monsoon climate and the Middle Asia dry climate in summer are intensified by the TP heating.

The influence of the TP on the Asian monsoon onset is another important issue. This will be reported in a separate paper. The monsoon system is a complex system (Webster et al., 1998). Despite the great research efforts made on this issue, many aspects are still unclear. Unresolved questions include how radiation processes affect the thermal state of the TP, how the aerosol-cloud-radiation-monsoon circulation feedback occurs in the TP area, how the air-sea exchange processes interact with such a feedback, and how the TP-Air-Pump in conjunction with the thermal state over Indian Ocean and Australia affect the Asia-Australia monsoon? These are topics that need to be studied with the cooperation between GEWEX and other programs under the WCRP umbrella.

References

ANNAMALAI H, SLINGO J M, SPERBER K R, et al, 1999. The mean evolution and variability of the Asian summer monsoon: Comparison of ECMWF and NCEP-NCAR reanalyis[J]. Monthly Weather Review, 127(6): 1157-1186.

BOLIN B, 1950. On the influence of the earth's orography on the general character of the westerlies[J]. Tellus, 2(3): 184-195.

CHARNEY J G, ELIASSEN A, 1949. A numerical method for predicting the perturbations of the middle latitude westerlies [J]. Tellus, 1(2): 38-54.

CHARNEY J G, ELIASSEN A, 1964. On the growth of the hurricane depression[J]. Journal of the Atmospheric Sciences, 21(1): 68-75.

CHOU C, NEELIN J D, SU H, 2001. Ocean-atmosphere-land feedbacks in an idealized monsoon[J]. Quarterly Journal of the Royal Meteorological Society, 127(576): 1869-1891.

DUAN A, 2003. The influence of thermal and mechanical forcing of Tibetan Plateau upon the climate patterns in East Asia [D]. Beijing: Institute of Atmospheric Physics, Chinese Academy of Sciences.

DUAN A, WU G, 2005. Role of the Tibetan Plateau thermal forcing in the summer climate patterns over subtropical Asia [J]. Climate Dynamics, 24(7): 793-807.

FLOHN H, 1957. Large-scale aspects of the "summer monsoon" in South and East Asia[J]. Journal of the Meteorological Society of Japan, 35A: 180-186.

FLOHN H, 1968. Contributions to a meteorology of the Tibetan highlands[J]. Department of Atmospheric Science, 130.

GILL, ADRIAN E, 1980. Some simple solutions for heat-induced tropical circulation[J]. Quarterly Journal of the Royal Meteorological Society, 106(449): 447-462.

HE H, MCGINNS JOHN W, SONG Z S, et al, 1987. Onset of the Asian summer monsoon in 1979 and the effect of the Tibetan Plateau[J]. Monthly Weather Review, 115(9): 1966-1995.

HOSKINS B J, 1991. Towards a PV-θ view of the general circulation[J]. Tellus A: Dynamic Meteorology and Oceanography, 43(4): 27-35.

HUNG C W, YANAI M, 2004a. Factors contributing to the onset of the Australian summer monsoon[J]. Quarterly Journal of the Royal Meteorological Society, 130(597): 739-758.

HUNG C W, LIU X D, YANAI M, et al, 2004b. Symmetry and asymmetry of the Asian and Australian summer monsoons [J]. Journal of Climate, 17(12): 2413-2426.

KALNAY E, KANAMITSU M, KISTLER R, et al, 1996. The NCEP/NCAR 40-year reanalysis project[J]. Bull Amer Meteor Soc, 77: 437-471.

KUO H, QIAN Y F, 1982. Numerical simulation of the development of mean monsoon circulation in July[J]. Monthly Weather Review, 110(12): 1879-1897.

LI C, YANAI M, 1996. The onset and interannual variability of the Asian summer monsoon in relation to land-sea thermal contrast[J]. Journal of Climate, 9(2): 358-375.

LI G P, DUAN T Y, GONG Y F, et al, 2000. The bulk transfer coefficients and surface fluxes on the western Tibetan Plateau[J]. Chinese Science Bulletin, 45(13): 1221-1226.

LI G P, DUAN T Y, HAGIONYA SHIGENORI, et al, 2001. Estimates of the bulk transfer coefficients and surface fluxes over the Tibetan Plateau using AWS data[J]. Journal of the Meteorological Society of Japan, 79(2): 625-635.

LIANG X Y, LIU Y M, WU G X, et al, 2005. Effect of the Tibetan Plateau on the site of onset and intensity of Asian summer monsoon[J]. Acta Meteorological Sinica, 63(5): 799-805.

LIOU K N, ZHOU X J, 1987. Progress and prospects proceedings of the Beijing international radiation symposium Beijing, China, August 26-30, 1986[R]. Beijing: Science Press and American Meteorological Society: 699.

LIU X, WU G X, LI W P, et al, 2001. Thermal adaptation of the large-scale circulation to the summer heating over the Tibetan Plateau[J]. Progress in Natural Science, 11(2): 207-214.

LIU X, LI W P, WU G X, 2002. Interannual variations of the diabatic heating over the Tibetan Plateau and the Northern Hemispheric circulation in summer[J]. Acta Meteor Sin, 60: 267-277.

LIU H, WU G X, 1997. Impacts of land surface on climate of July and onset of summer monsoon: A study with an AGCM, plus SSiB[J]. Adv Atmos Sci, 14: 289-308.

LIU Y M, WU G X, LIU H, et al, 1999. Impacts of spatial differential heating on the formation and variation of the subtropical anticyclone, Ⅲ. Condensation latent heating and South Asian high and the subtropical anticyclone over western Pacific[J]. Acta Meteorological Sinica, 57: 525-538.

LIU Y M, WU G X, LIU H, et al, 2001. Condensation heating of the Asian summer monsoon and the subtropical anticyclone in the eastern hemisphere[J]. Climate Dynamics, 17(4): 327-338.

LIU Y M, WU G X, REN R C, et al, 2004. Relationship between the subtropical anticyclone and diabatic heating[J]. Journal of Climate, 17(4): 682-698.

LUO H, YANAI M, 1983. The large-scale circulation and heat sources over the Tibetan Plateau and surrounding areas during the early summer of 1979. Part Ⅰ: Precipitation and kinematic analyses[J]. Monthly Weather Review, 111(5): 922-944.

LUO H, YANAI M, 1984. The large-scale circulation and heat sources over the Tibetan Plateau and surrounding areas during the early summer of 1979. Part Ⅱ: Heat and moisture budgets[J]. Monthly Weather Review, 112(5): 966-989.

MAO J, LIU Y, WU G X, 2002a. Study on modal variation of the subtropical high and its mechanism during seasonal transition. Part Ⅰ: Climatological features of subtropical high structure[J]. Acta Meteorological Sinica, 60(4): 400-408.

MAO J, LIU Y, WU G X, 2002b. Study on modal variation of subtropical high and its mechanism during seasonal transition part Ⅱ: Seasonal transition index over Asian Monsoon Region[J]. Acta Meteorological Sinica, 60(4): 409-420.

NITTA T, 1983. Observational study of heat sources over the eastern Tibetan Plateau during the summer monsoon[J]. Journal of the Meteorological Society of Japan, 61(4): 590-605.

QUENEY P, 1948. The problem of air flow over mountains: A summary of theoretical studies[J]. Bulletin of the American Meteorological Society, 29(1): 16-26.

RODWELL M J, HOSKINS B J, 1996. Monsoons and the dynamics of deserts[J]. Quarterly Journal of the Royal Meteorological Society, 122(534): 1385-1404.

SHEN Z, DEN W, PAN S, 1984. An outline of the Qinghai-Xizang Plateau heat source observation experiment[C]// Qinghai-Xizang Plateau Meteorological Experiment (I): 1-9.

SHI G Y, 1981. An accurate calculation and representation of the infrared transmission function of the atmospheric constituents[D]. Sendai: Tohoku University of Japan: 191.

SHI L, SMITH E A, 1992. Surface forcing of the infrared cooling profile over the Tibetan Plateau. Part Ⅱ: Cooling-rate variation over large-scale plateau domain during summer monsoon transition[J]. Journal of the Atmospheric Sciences, 49(10): 823-844.

SLINGO J, 1987. The development and verification of a cloud prediction scheme for the ECMWF model[J]. Quarterly Journal of the Royal Meteorological Society, 113(477): 899-927.

SMITH E A, SHI L, 1992. Surface forcing of the infrared cooling profile over the Tibetan Plateau. Part Ⅰ: Influence of relative longwave radiative heating at high altitude[J]. Journal of the Atmospheric Sciences, 49(10): 805-822.

TAO S Y, 1980. Torrential Rain in China[M]. Beijing: China Meteorological Press: 225.

TAO S Y, CHEN L, 1987. A Review of Recent Research on the East Asian Summer Monsoon in China. Monsoon Meteorology[M]. Oxford: Oxford University Press: 60-92.

TIAN S F, YASUNARI T, 1998. Climatological aspects and mechanism of spring persistent rains over central China[J].

Journal of the Meteorological Society of Japan, 76(1): 57-71.

TRENBERTH K E, GUILLEMOT C J, 1998. Evaluation of the atmospheric moisture and hydrological cycle in the NCEP/NCAR reanalysis[J]. Climate Dynamics, 14(3): 213-231.

WAN R, WU G, 2007. Mechanism of the spring persistent rains over southeastern China[J]. Science in China Series D: Earth Sciences, 50(1): 130-144.

WANG Z Z, WU G X, WU T W, et al, 2004. Simulation of Asian monsoon seasonal variations with climate model R42L9/LASG[J]. Advances in Atmospheric Sciences, 21(6): 879-899.

WEBSTER P J, MAGANA V O, PALMER T N, et al, 1998. Monsoons: Processes, predictability, and the prospects for prediction[J]. Journal of Geophysical Research: Oceans, 103(C7): 14451-14510.

WU G, 2004. Recent progress in the study of the Qinghai-Xizang Plateau climate dynamics in China[J]. Quaternary Sciences, 24(1): 1-9.

WU G X, LI W, GUO H, et al, 1997a. Sensible heat driven air-pump over the Tibetan Plateau and its impacts on the Asian summer monsoon[C]//Collections on the Memory of Zhao Jiuzhang. Beijing: Science Press.

WU G, ZHANG X H, LIU H, et al, 1997b. The LASG global ocean-atmosphere-land system model of LASG (GOALS/LASG) and its performance in simulation study[J]. Quart J Appl Meteor, 8: 15-28.

WU G, ZHANG Y, 1998. Tibetan Plateau forcing and the timing of the monsoon onset over South Asia and the South China Sea[J]. Monthly Weather Review, 126(4): 913-927.

WU G, LIU Y M, LIU P, 1999. Impacts of spatial differential heating on the formation and variation of the subtropical anticyclone. Ⅰ. Scale analysis[J]. Acta Meteorological Sinica, 57: 257-263.

WU G X, LIU Y M, 2000. Thermal adaptation, overshooting, dispersion and subtropical anticyclone. Ⅰ. Thermal adaptation and overshooting[J]. Chinese Journal of Atmospheric Sciences, 24: 433-446.

WU G X, SUN L, LIU Y M, et al, 2002. Impacts of land surface processes on summer climate[C]. Selected papers of the Fourth Conference on East Asian and Western Pacific Meteorology and Climate. CHANG C P, et al. World Scientific: 64-76.

WU G, LIU Y, 2003. Summertime quadruplet heating pattern in the subtropics and the associated atmospheric circulation[J]. Geophysical Research Letters, 30(5): 1201.

WU G, LIU Y M, LIU P, 2004. Adaptation of the atmospheric circulation to thermal forcing over the Tibetan Plateau[C]. Observation, Theory and Modeling of Atmospheric Variability, Selected Papers of Nanjing Institute of Meteorology Alumni in Commemoration of Professor Jijia Zhang, 92-114.

WU G, WANG J, LIU X, et al, 2005. Numerical modeling of the influence of Eurasian orography on the atmospheric circulation in different seasons[J]. Acta Meteorological Sinica, 63(5): 603-612.

WU T M, LIU P, WU Z Z, 2003. The performance of atmospheric component model R42L9 of GOALS/LASG[J]. Advances in Atmospheric Sciences, 20(5): 726-742.

XIE P, ARKIN P A, 1996. Analyses of global monthly precipitation using gauge observations, satellite estimates, and numerical model predictions[J]. Journal of Climate, 9(4): 840-858.

XIE P, ARKIN P A, 1998. Global monthly precipitation estimates from satellite-observed outgoing longwave radiation[J]. Journal of Climate, 11(2): 137-164.

XIE S P, SAIKI N, 1999. Abrupt onset and slow seasonal evolution of summer monsoon in an idealized GCM simulation[J]. Journal of the Meteorological Society of Japan, 77(4): 949-968.

XUE Y, SELLERS P J, KINTER J L, et al, 1991. A simplified biosphere model for global climate studies[J]. Journal of Climate, 4(3): 345-364.

YANAI M, ESBENSEN S, CHU J H, et al, 1973. Determination of bulk properties of tropical cloud clusters from large-scale heat and moisture budgets[J]. Journal of Atmospheric Sciences, 30(4): 611-627.

YANAI M H, LI C F, SONG Z S, 1992. Seasonal heating of the Tibetan Plateau and its effects on the evolution of the Asian summer monsoon[J]. Journal of the Meteorological Society of Japan, 70(1B): 319-351.

YANAI M, LI C, 1994. Mechanism of heating and the boundary layer over the Tibetan Plateau[J]. Monthly Weather Review, 122(2): 305-323.

YANAI M, TOMITA T, 1998. Seasonal and interannual variability of atmospheric heat sources and moisture sinks as determined from NCEP-NCAR reanalysis[J]. Journal of Climate, 11(3): 463-482.

YANAI M, WU G X, 2006. Effects of the Tibetan Plateau. The Asian Monsoon[M]. Cambridge: Cambridge University Press: 513-549.

YE D Z, GAO Y X, 1979. Meteorology of the Qinghai-Xizang Plateau[M]. Beijing: Science Press: 278.

YEH T C, 1950. The circulation of the high troposphere over China in the winter of 1945—46[J]. Tellus, 2(3): 173-183.

YEH T C, LO S W, CHU P C, 1957. The wind structure and heat balance in the lower troposphere over the Tibetan Plateau and its surroundings[J]. Acta Meteorological Sinica, 28: 108-121.

YEH T, LUO S W, 1959. The abrupt change of circulation over Northern Hemisphere during June and October[J]. Acta Meteorological Sinica, 29: 249-263.

ZHANG J J, ZHU B Z, ZHU F K, et al, 1988. Advances in the Qinghai-Xizang Plateau Meteorology-the Qinghai-Xizang Plateau Meteorological Experiment (1979) and Research[M]. Beijing: Science Press: 268.

ZHANG X H, SHI G Y, LIU H, et al, 2000. IAP Global Ocean-Atmosphere-Land System Model[M]. Beijing, New York: Science Press: 252.

ZHOU M Y, XU X D, BIAN L G, et al, 2000. Observational, Analytical, and Dynamic Study of the Atmospheric Boundary Layer of the Tibetan Plateau[M]. Beijing: China Meteorological Press: 125.

Astronomical and Hydrological Perspective of Mountain Impacts on the Asian Summer Monsoon

HE Bian, WU Guoxiong, LIU Yimin, BAO Qing

(State Key Laboratory of Numerical Modeling for Atmospheric Sciences and Geophysical Fluid Dynamics, Institute of Atmospheric Physics, Chinese Academy of Sciences, Beijing 100029)

Abstract: The Asian summer monsoon has great socioeconomic impacts. Understanding how the huge Tibetan and Iranian Plateaus affect the Asian summer monsoon is of great scientific value and has far-reaching significance for sustainable global development. One hypothesis considers the plateaus to be a shield for monsoon development in India by blocking cold-dry northerly intrusion into the tropics. Based on astronomical radiation analysis and numerical modeling, here we show that in winter the plateaus cannot block such a northerly intrusion; while in summer the daily solar radiation at the top of the atmosphere and at the surface, and the surface potential temperature to the north of the Tibetan Plateau, are higher than their counterparts to its south, and such plateau shielding is not needed. By virtue of hydrological analysis, we show that the high energy near the surface required for continental monsoon development is maintained mainly by high water vapor content. Results based on potential vorticity-potential temperature diagnosis further demonstrate that it is the pumping of water vapor from sea to land due to the thermal effects of the plateaus that breeds the Asian continental monsoon.

1 Introduction

The precipitation associated with the Asian summer monsoon (ASM) covers a vast area (Wu et al., 1998; Wang and LinHo, 2002) and is important to the regional economy and culture. However, its anomalies usually bring drought or flood and can cause serious damage. Revealing the driving mechanism of its formation and variation has long been the aim of both paleoclimate and modern climate studies because such knowledge can contribute to the improvement of weather forecasts and climate prediction and projection, and to sustainable social development. The onset and maintenance of the ASM are highly related to the Tibetan Plateau (TP) (Yeh et al., 1957; Flohn, 1957; Hahn and Manabe, 1975; Yanai et al., 1992, 2006; Romatschke and Houze, 2011) and Iranian Plateau (IP) topography (TIP) (Wu and Zhang, 1998; Wu et al., 2012a, 2012b), and their thermal forcing and mechanical forcing are considered the main driving mechanism of the ASM besides the land-sea thermal contrast. The thermal forcing hypothesis suggests that (Yanai et al., 1992; Wu et al., 1997; Wu et al., 2007; Wu et al., 2012a) since more than 85% of atmospheric water vapor content is confined to a boundary layer below 3 km, a mechanism is required to transport this water vapor from the surface to the free atmosphere to form monsoonal clouds. The surface sensible heating of the TIP can pump the low-layer moist air upward to higher layers much like a sensible-heat-driven air-pump (SHAP), forming the precipitation of the East ASM (EASM) and the northern branch of the South ASM (SASM) (Wu et al., 2012b). Although this shows that the TIP-SHAP can pump water vapor vertically from the surface to the free atmosphere, how the TIP-SHAP can transport water vapor horizontally from sea to land to support the continental monsoon is still unrevealed. The mechanical forcing hypothesis focuses on the formation of the SASM (Boos and Kuang, 2010, 2013): the Himalayas, the southern border of the TP, prevent the intrusion of cold and dry extra tropical air into North India, where the local subcloud moist entropy or equivalent potential temperature θ_e is large and monsoon convection can develop from a local instability. This hypothesis is facing challenges from both observations and numerical experiments: in winter there is cold/dry northerly invasion into India despite the existence of the TIP, and in summer there is no cold/dry equatorward flow in other global monsoon regions where no huge mountain ranges exist (Wu et al., 2003; Wu et al., 2009). In addition, although both hypotheses emphasize the importance of high surface energy in producing monsoon convection, the relative contributions of water vapor and temperature to this surface energy and the roles played by the TIP have not been discussed. Most of these studies have been mainly confined to a meteorological framework (Rajagopalan and Molnar, 2013; Abe et al., 2013; Chen et al., 2014), and they have left several critical unsolved problems: whether such a TIP mechanical blocking mechanism exists for local monsoon development over North India? How the high surface energy in the continental monsoon area is produced, is it due to high surface temperature, or high surface water vapor content? and how and why remote TIP thermal forcing can help the development of the ASM over continental regions. This study tries to shed new light on these fundamental issues by employing astronomical, hydrological, and dynamical approaches.

Monsoon convection usually develops over a region where θ_e is high (Emanuel et al., 1994). Because θ_e is a function of temperature T and moisture q, convection can develop over the tropical ocean away from the equator where surface temperature T_{sur} is high (Prive and Plumb, 2007a). Tropical continents usually possess high T_{sur}, which is necessary but not sufficient for convection to develop. T_{sur} over the Saudi Arabian Peninsula is very high, but there is no monsoon there. Thus, the fundamental question concerning the formation of the continental monsoon is how a high θ_e over land is generated (Prive and Plumb, 2007b). By analyzing the solar radiation at the top of the atmosphere (TOA) and the downward shortwave radiation at the surface, as well as by using numerical modeling, we demonstrate here that the TIP cannot shield the tropics from cold and dry northerly intrusion whenever such cold/dry northerly advection exists; and in summer, when the daily solar radiation at the TOA and at the surface, and the surface potential temperature to the north of the TIP, are higher than to its south, there is no cold/dry northerly advection into India and the plateau shielding is not needed. With hydrological diagnosis we show that near the surface, moisture q_{sur} is more significant than potential temperature θ_{sur} in producing high θ_e; thus, water vapor transport from the sea becomes crucial for continental monsoon formation.

Furthermore, from the potential vorticity-potential temperature (PV-θ) perspective, we show that the high θ_{sur} over the TIP, which is produced by local strong surface sensible heating, can generate a huge near-surface cyclonic circulation, and it is because of this unique cyclonic circulation that water vapor can be transported from sea to land and gives rise to the continental monsoon over North India and East Asia.

2 Astronomical perspective: no northerly intrusion into Indian monsoon area in summer

At the boreal summer solstice, the solar zenith angle at noon is zero along latitude $\phi = 23.5°$N. At this time, 6.5 degrees away from the Tropic of Cancer and at the top of the atmosphere, the solar radiation (SR) at latitude A (30°N, where Shanghai is located) in the subtropics is the same as that at latitude B (17°N, where Hyderabad is located) in the tropics (Fig. 1a). However, due to the tilting of the Earth's axis, the length of day (LOD) in boreal summer increases with latitude, and the LOD at A ($AA_1/AA_2 \times 24$ hours) is about one hour longer than at B ($BB_1/BB_2 \times 24$ hours). Thus, the daily SR (DSR) at 30°N is 476 Wm^{-2}, about 6% more than that (449 Wm^{-2}) at 17°N (Methods). Climate mean data from the Clouds and the Earth's Radiant Energy System (CERES) (Kato et al., 2013) also show that in June, July, and August the DSR is, respectively, 474, 466, and 436 Wm^{-2} at 30°N and 444, 443, and 438 Wm^{-2} at 17°N (Fig. 1b); and the JJA mean is 458 Wm^{-2} at 30°N but 442 Wm^{-2} at 17°N. These data indicate that in boreal summer the subtropical region receives more solar radiation than the tropical region. Diagnosis of the first year of CERES daily data shows that in winter the DSR is high in the tropics and low in the high latitudes, whereas in summer it is higher in the extra tropics than in the tropics, with a maximum of more than 480 Wm^{-2} located near 45°N (Fig. 1c). Analysis of the daily occurrence of a northerly at the near-surface level ($\sigma = 0.99$) in the same year and based on ERA-interim data (Dee et al., 2011) demonstrates that across the Indian monsoon domain of 75°—100°E, cold/dry northerly advection events exist from the subtropics to the tropics in the winter half-year despite the TIP mountain barrier (Fig. 1d). On the contrary, in summer there is no northerly occurrence in this SASM region. This may be due to the fact that the summertime DSR is higher in the extra tropics than in the tropics, as shown in Fig. 1c.

To verify the above inference, numerical experiments are performed based on the Spectral Atmospheric Model of LASG/IAP (SAMIL) (Methods). The observed seasonally varying SST is used to prescribe the lower boundary condition. The DSR in these experiments is the same as that shown in Fig. 1c. The near-surface moist entropy, measured in terms of equivalent potential temperature

$$\theta_e = \theta \exp\left(\frac{Lq}{C_p T}\right) \tag{1}$$

and precipitation, are diagnosed from the experiment outputs and compared to their counterparts in ERA40 (Uppala et al., 2005), where L, C_p, and q denote, respectively, latent heat of vaporization, specific heat at constant atmospheric pressure, and specific humidity. The first experiment is simply the SAMIL climate integration, defined as the CON run. The second experiment, named NoTIP, is the same as CON except the TIP is removed (Methods). The daily outputs from these experiments show similar features to those observed: Although in winter there is strong northerly advection into the tropics, in summer there is hardly any northerly intrusion from the subtropics into the tropical monsoon region even when the TIP is removed in the NoTIP run (Figs. 1e and 1f).

In winter, cold air flow certainly cannot climb over the massive, high TIP to reach India. How, then, can the cold/dry northerly advection invade India from the subtropics to the tropics in the winter half-year despite the TIP mountain barrier, as shown in Fig. 1d? By diagnosing the ERA40 reanalysis we found that this is because the frequent cold/dry north westerlies or northeasterlies in the lower tropo-

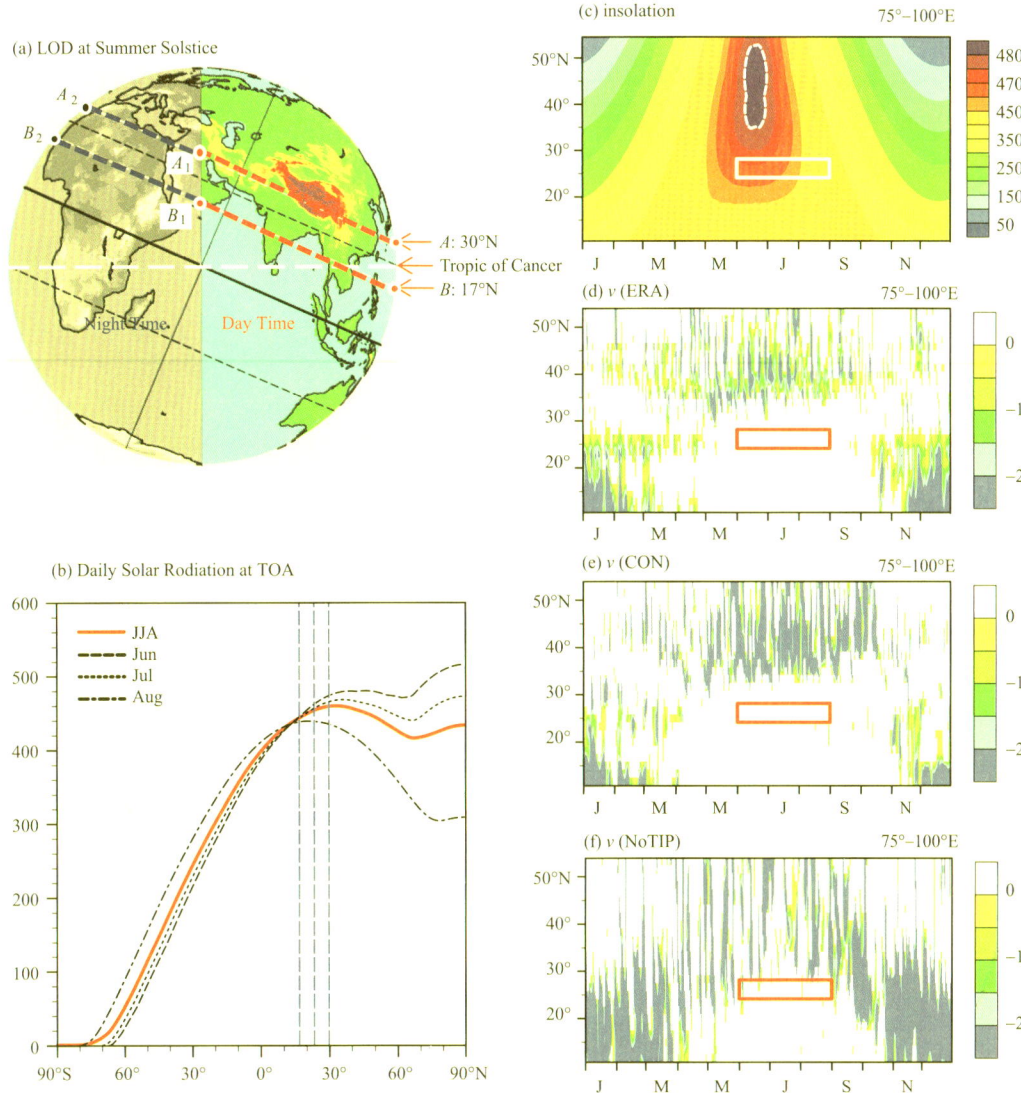

Fig. 1 (a) Intensity of solar radiation (SR) and the length of day LOD at the TOA and at summer solstice; (b) latitude distributions of DSR in summer months; and annual evolution of (c) DSR (white-dashed denotes 480-contour), and of surface northerly occurrence along 75°—100°E from (d) ERA-interim in 2001, (e) CON, and (f) NoTIP. The square in (c—f) indicates the SASM region (24°—28°N) during JJA. Unit is (Wm^{-2}) in (b) and (c), and (ms^{-1}) in d—f. The three dashed lines in (b) indicate the latitudes of 30°N, 23.5°N, and 17°N, respectively

sphere can go around the southern rim of the giant TIP and invade India (Fig. 2a), implying that the TIP cannot protect India from northerly intrusion in winter. This feature is well simulated in the CON run, as shown in Fig. 2b.

The seasonal variation of the DSR at the TOA makes the subtropical continent become the atmospheric heat sink in winter and the source in summer. The continental heating in summer produces a continental-scale cyclonic circulation near the surface and forms a moist/monsoon-type climate over the eastern continent and a dry/desert-type over its west (Wu and Liu, 2003; Wu et al., 2009). The DSR at the TOA can also affect the surface wind in the monsoon region. This is because the surface wind is basically determined by the surface pressure gradient, which is closely related to the surface temperature gradient. Since the surface temperature is significantly influenced by the solar radiation reaching the earth's surface and the surface characteristics, and since the surface downward shortwave radiation is determined by the DSR and local cloudiness, both the DSR and local cloudiness can significantly affect the surface wind distribution. Figure 2c and 2d presents respectively the distributions of cloudiness and downward shortwave radiation at the surface in July 2001. It shows that in the SASM sector of 75°—100°E, there is more cloud (>80%) to the south of the TIP and less (<50%) to its north (Fig. 2c). Consequently, the downward shortwave radiation at the surface is low (<210 Wm^{-2}) to the south of the TIP and high (>270 Wm^{-2}) to its north (Fig. 2d). This will then influence the distributions of surface potential temperature and wind in the ASM area.

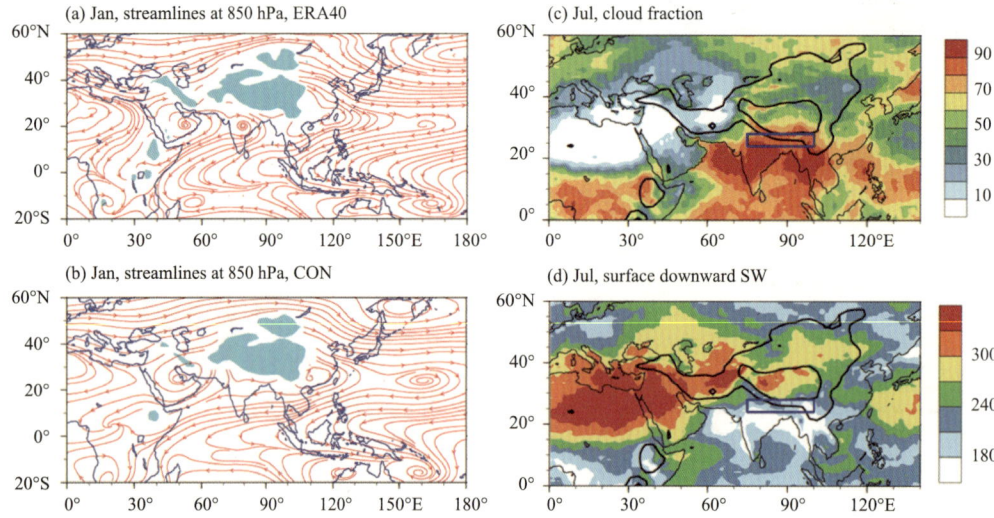

Fig. 2 Climatological mean January streamfield at 850 hPa produced from (a) ERA40 and (b) the CON experiment; and the monthly mean for July 2001 of (c) cloud fraction (%) and (d) downward shortwave radiation at the surface (Wm^{-2}) produced from CERES

The JJA mean potential temperature and streamline at the near-surface level $\sigma = 0.99$ in ERA40, CON, and NoTIP are presented in the top row of Fig. 3. The maximum surface potential temperature θ_{sur} is over the TIP in ERA40 and CON (Figs. 3a and 3e). In NoTIP such a center over the TIP region disappears due to the lowered elevation in the region, and the maximum surface potential temperature θ_{sur} is located over the subtropics to midlatitudes (Fig. 3i), where both the DSR and downward shortwave radiation are high (Figs. 1c and 2d). In all cases the surface flows converge toward the warm θ_{sur} area. Consequently, a mean surface southerly prevails over the North Indian monsoon region, which can explain why there is hardly any transient northerly intrusion into India in summer. It is important to note that in other subtropical continental monsoon regions where no huge mountain ranges exist, the observed summer monsoon-type climate presents over the eastern continent where poleward flow dominates (Wu and Liu, 2003; Wu et al., 2009). This then justifies the NoTIP results that even when there is NoTIP, the southerly can dominate the South Asian monsoon region, and there should be no frequent cold/dry northerly advection there.

In boreal summer (JJA), the observed climate-mean state exhibits a high θ_e and precipitation over the Indian Ocean, Indian Peninsula, and along the southern slopes of the TIP, and minimum precipitation over the central Arabian Peninsula (Fig. 3c). This basic pattern in the observations is reasonably simulated in the CON experiment (Fig. 3g). However, in the NoTIP run over North India, the surface θ_e weakens remarkably and the monsoon rainfall disappears (Fig. 3k). Is this due to the removal of the TIP such that the cold/dry advection can destroy the north branch of the SASM? If this were true, then re-

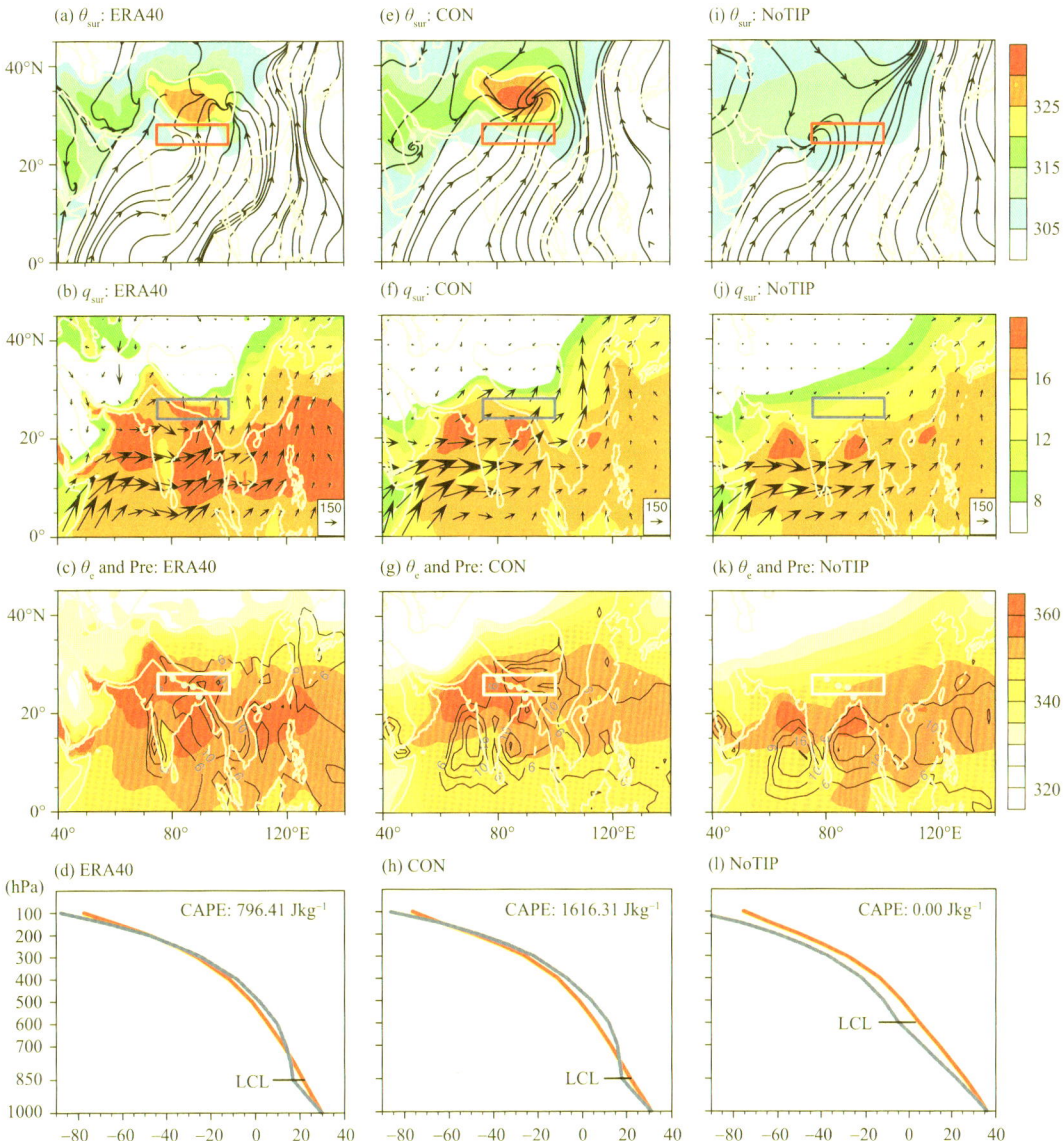

Fig. 3　JJA-means of θ_{sur} (K) and stream field (top row), q_{sur} (gkg^{-1}) and 850 hPa water vapor fluxes (vectors, kgm^{-1} s^{-1}) (second row), and θ_e (K) and precipitation (contours, mmd^{-1}) (third row); and of profiles of environmental temperature (K, red) and rising air-parcel temperature (K, blue) (bottom row) averaged over the four pale blue grid-points over North India shown in the third row, calculated from ERA-40 (a—d), CON (e—h), and NoTIP (i—l). The square indicates the SASM region of (24°—28°N, 75°—100°E)

placing the TIP barrier would recover the observed mean state. To test this, a third experiment, TIP_NS, is designed, which is also the same as CON except that the surface sensible heating of the TIP region is not allowed to heat the atmosphere, so that the TIP merely acts as a barrier to block the tropics from the subtropics. The results shown in Fig. 4c are similar to those in the NoTIP run (Fig. 3k): over North India the simulated θ_e is still low and the monsoonal precipitation does not recover. In the subtropical continental area, the surface potential temperature in the TIP_NS experiment is even weaker compared to the NoTIP experiment because in the NoTIP run the surface sensible heating over the TIP platform, which is 500 m above sea level, remains, indicating the importance of TIP thermal impacts. Actually, the mean surface air temperatures averaged from the four grid points as denoted in Figs. 3c, 3g, and 3k, which are located over northeastern India and associated with the maximum monsoon precipitation in the reanalysis (Fig. 3c), are all around 30 ℃ in ERA40 and in all experiments (bottom row of Fig. 3), indicating again that cold advection does not occur in the region even if the TIP is removed. The above results thus imply that a significant decrease of θ_e in this region in either NoTIP or TIP_NS (Figs. 3k and 4c) must be due mainly to the reduced water vapor content (Figs. 3j and 4b).

3 Hydrological perspective: water vapor transport from the sea is crucial for the continental monsoon

Figs. 3 and 4 present the relative contributions of θ_{sur} (top row) and q_{sur} (second row) to θ_e (third row) in ERA40 and all experiments. The distribution of θ_{sur} is relatively uniform except over mountainous regions, and it is warmer over land than over the ocean, while the distribution of q_{sur} is resembles that of θ_e in all cases. In ERA40 and CON (Figs. 3a—b, 3e—f) except over the TIP, the high θ_{sur} is located over the Arabian Peninsula, while the q_{sur} maxima are located over the oceans and Northeast India. In the NoTIP run, the distribution of θ_{sur} is more uniform in the subtropics (Fig. 3i), while q_{sur} decreases remarkably over Northeast India (Fig. 3j), similar to the decrease of θ_e in the same region (Fig. 3k). The distributions of θ_{sur}, q_{sur}, and θ_e (Figs. 4a—c) in the TIP_NS experiment resemble, respectively, those in the NoTIP experiment (Figs. 3i—k).

The lifting condensation level (LCL) is a measure of the water vapor content in the lower troposphere: the lower the LCL, the higher the water vapor content. In the bottom row of Fig. 3 are shown the LCL distributions over Northeast India. The LCL is low (at 850 hPa) in ERA40 and the CON run, but high (above 700 hPa) in NoTIP, indicating that the water vapor content is much lower in NoTIP than in ERA40 and CON. The Convective Available Potential Energy (CAPE), which is measured by calculating the temperature difference between a rising air parcel (blue dashed) and the environment (red solid), is also shown in the figure. ERA40 shows active convection in the North India region, with a CAPE of 796.41 Jkg^{-1} (Fig. 3d). This feature is well simulated in CON, with a stronger CAPE of 1616.31 Jkg^{-1} (Fig. 3h). In NoTIP, however, the potential convection diminishes because the temperature of the air parcel is colder than the environment at all levels, and the CAPE disappears (Fig. 3l). We then infer that in NoTIP, although θ_{sur} over North India does not change compared to θ_{sur} in ERA40 and CON, q_{sur} becomes much lower, causing the reduced surface energy and suppressed rainfall there (Fig. 3k).

The difference between CON and TIP_NS can to a certain extent present the thermal impacts of the TIP-SHAP. The θ_e difference (Fig. 4f) is significantly positive over North India and East Asia, accompanied by a remarkable increase in precipitation. The difference in θ_{sur} is mainly limited to over the two mountainous regions (Fig. 4d) and is insignificant over North India, while the difference in q_{sur} is remarkably positive over North India and East Asia (Fig. 4e), as is the θ_e difference (Fig. 4f).

It is important to note the similarity in the distributions of q_{sur} and θ_e (Figs. 4e and 4f). Actually, from (1) the dependence of the relative change in surface θ_e on the changes in θ_{sur} and q_{sur} can be evaluated as:

$$\frac{\Delta \theta_e}{\theta_e} = \frac{L}{C_p} \frac{\Delta q_{sur}}{T_{sur}} + \frac{\Delta \theta_{sur}}{\theta_{sur}} \qquad (2)$$

or,

$$\left|\frac{\Delta \theta_e}{\theta_e}\right| = 2.5 \times 10^3 \left|\frac{\Delta q_{sur}}{T_{sur}}\right| + \left|\frac{\Delta \theta_{sur}}{\theta_{sur}}\right| \qquad (3)$$

E M T

Where "|a|" indicates a nondimensional variable "a". The distributions of the three terms (T, M, and E) in (3) for the difference between CON and TIP_NS are presented, respectively, in Figs. 4g—i. It becomes evident that except over the mountainous regions, the q_{sur} term (Fig. 4h) prevails over the θ_{sur} term (Fig. 4g) in determining the change in surface entropy (Fig. 4i) over North India and East Asia.

Local water vapor content is influenced by the convergence of water vapor transport. In ERA40 and the CON run, the 850 hPa water-vapor conveyer belt in the tropics deviates northward toward North India and the southern slope of the TP (Figs. 3b, 3f), resulting in the high q_{sur} there. In the NoTIP and TIP_NS runs, such northward water vapor transport is remarkably reduced, resulting in the low q_{sur} over North India (Figs. 3j and 4b). Consequently, continental rainfall is suppressed (Figs. 3k and 4c), and the northern branch of the SASM disappears. In the CON—TIP_NS case, the increased q_{sur} over the continent is also accompanied by much stronger water vapor transport from the Indian Ocean to the tropical and subtropical Asian continent (Fig. 4e).

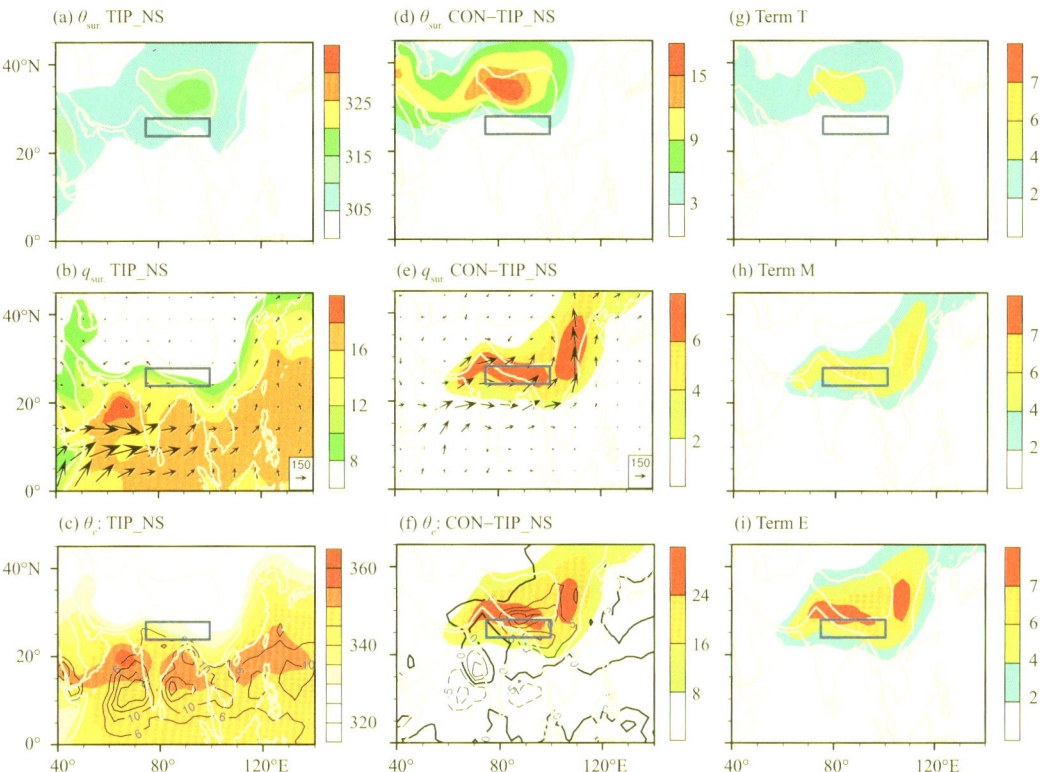

Fig. 4 (a—c) and (d—f) are the same as Fig. 3 (a—c) but, respectively, for TIP_NS and (CON—TIP_NS); and (g—i) the JJA-mean distributions of the T, M, and E terms in Formula (3). The square indicates the SASM region of (24°—28°N, 75°—100°E)

Local surface evaporation is another source of atmospheric water vapor. To evaluate its relative contribution to the monsoon rainfall, we calculated the regional means of evaporation and precipitation over the boxed region (24°—28°N, 75°—100°E) shown in Figs. 3 and 4, which cover North India and the southern slope of the Himalayas. Results demonstrate that local evaporation is much less than precipitation (Fig. 5) both in ERA40 (2.9 versus 9.5 mmd^{-1}) and in CON (3.8 versus 11.4 mmd^{-1}). This means that even if all the surface water vapor that evaporated from the ground were condensed and recycled locally, it would account for only about one-third of the local precipitation, not to mention that the precipitation recycling ratio is far below 0.5 in the Indian area (Pathak et al., 2014). In both the NoTIP and TIP_NS runs, the local evaporation does not decrease much, but the precipitation decreases substantially, from 11.4 mmd^{-1} in CON to 3.5 mmd^{-1} in NoTIP and 3.6 mmd^{-1} in TIP_NS. The results of CON minus TIP_NS show that the thermal impact of the TIP can increase evaporation by merely 2.0 mmd^{-1}, but it can increase precipitation by 7.7 mmd^{-1}, which accounts for more than two-thirds of the total precipitation in CON. All these results indicate that local evaporation does not contribute significantly to the successive strong precipitation of the ASM, while the water vapor transport from outside does. It can be concluded that water vapor advection from the Indian Ocean induced by the TIP-SHAP can make a substantial contribution to producing high entropy and convection over North India and East Asia and forming the continental monsoon.

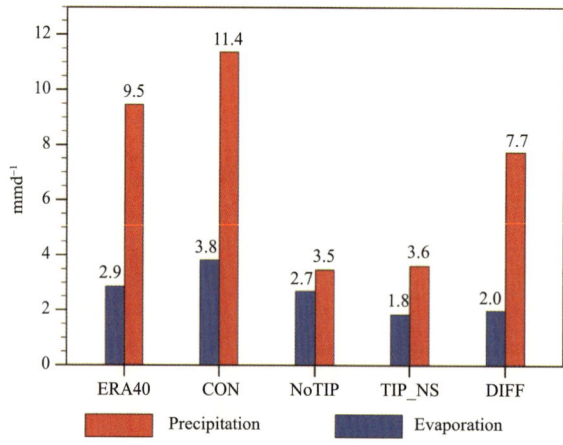

Fig. 5 Hydrological budgets (mmd^{-1}) in the SASM region of (24°—28°N, 75°—100°E) for ERA-40 and experiments CON, NoTIP, and TIP_NS. DIFF denotes the difference (CON−TIP_NS)

4 Dynamics of horizontal water vapor transport: thermal pumping of the TIP

Why can the TIP-SHAP transport water vapor horizontally from sea to land? According to the PV-θ theory (Thorpe, 1985; Hoskins et al., 1985; Hoskins, 1991), a warm surface potential temperature (θ_{sur}) anomaly can generate a positive potential vorticity (PV) and cyclonic circulation near the surface. The difference between the CON and TIP_NS runs (Fig. 6a) shows that the surface sensible heating over the TIP can result in a positive θ_{sur} anomaly, with a maximum of 8 K over the TIP. Accordingly, two surface cyclonic circulations are generated surrounding the IP and TP, respectively, forming an enhanced strong southwesterly over South Asia and a southerly over East Asia and leading to the effective water vapor transport from sea to land. These results can be well interpreted from the above PV-θ perspective. It is this fundamental response of θ_{sur} to the surface sensible heating over the TIP (TIP-SHAP) and the resultant atmospheric circulation that generate the high surface entropy and rain belt over North India, the southern slope of the TP, and East Asia, forming the northern branch of the SASM as well as the East Asian monsoon.

To evaluate the dependence of the TIP-SHAP on moist processes, we carry out two more experiments, CON_dry and TIP_NS_dry, under dry conditions, which are reproduced respectively from the CON and TIP_NS experiments except that the water vapor in the atmosphere is set to zero so that moist convection is excluded (Methods). The differences in θ_{sur} and the surface circulation between CON_dry and TIP_NS_dry are shown in Fig. 6b. The θ_{sur} difference is positive, with a similar intensity of 8 K, but it is more confined to the plateau area; and the surface cyclonic circulation and the onshore wind induced by the TIP-SHAP in dry conditions are, although weaker, similar to those under moist conditions (Fig. 6a), demonstrating that the initiation of the TIP-SHAP does not rely on moist processes.

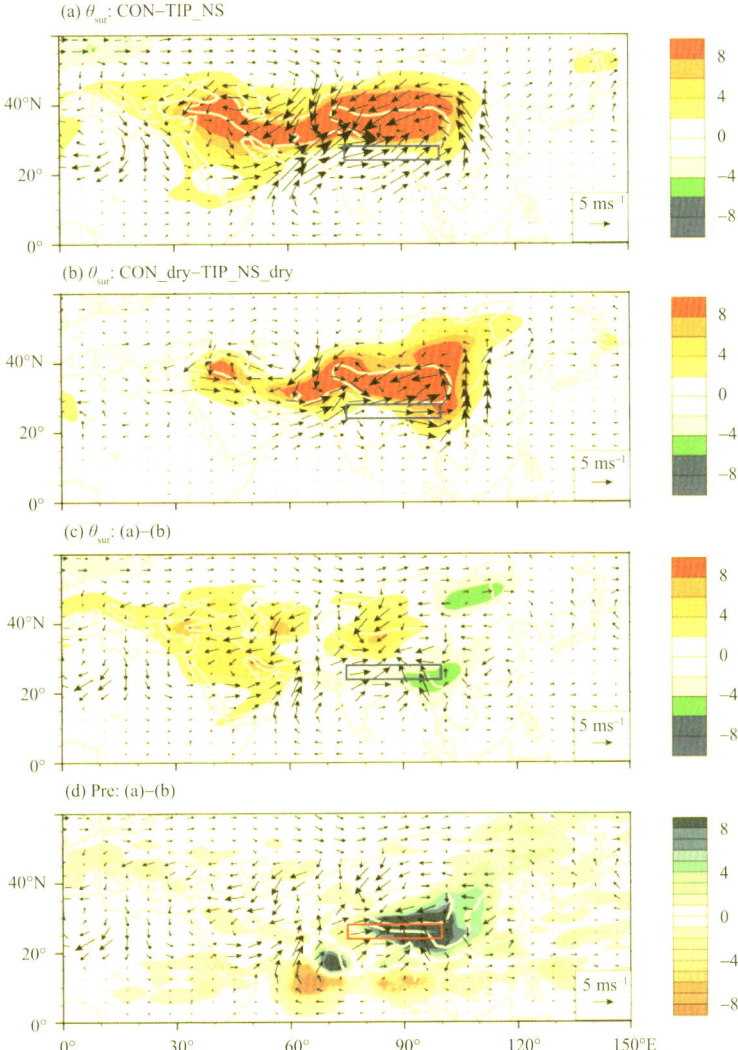

Fig. 6 JJA-mean differences of θ_{sur} (shading, K) and surface circulation (vectors, ms^{-1}) between (a) CON and TIP_NS, (b) CON_dry and TIP_NS_dry, (c) difference between (a) and (b); (d) same as (c) but for precipitation (mmd^{-1}). The square indicates the SASM region of (24°–28°N, 75°–100°E)

However, moist processes have strong positive feedback on the general circulation of the atmosphere. The depth of the positive θ difference between the runs with and without surface sensible heating over the TIP extends to above 200 hPa in the moist experiment (Fig. 7a) but is quite shallow in the dry experiment (Fig. 7b). The differences in θ_{sur} and surface circulation between Figs. 6a and 6b as shown in Fig. 6c present the impact of moist processes on the pumping effect of the TIP-SHAP: they intensify the cyclonic circulation surrounding the TIP with a remarkable convergence over its southeast and produce increased θ_{sur} to the west of 90°E and decreased θ_{sur} to the east and over North India. This θ_{sur} difference coincides well with the precipitation difference (Fig. 6d): the area with increased rainfall corresponds to reduced surface θ_{sur} and vice versa, presenting a negative feedback of monsoon rainfall on surface energy, though it is secondary.

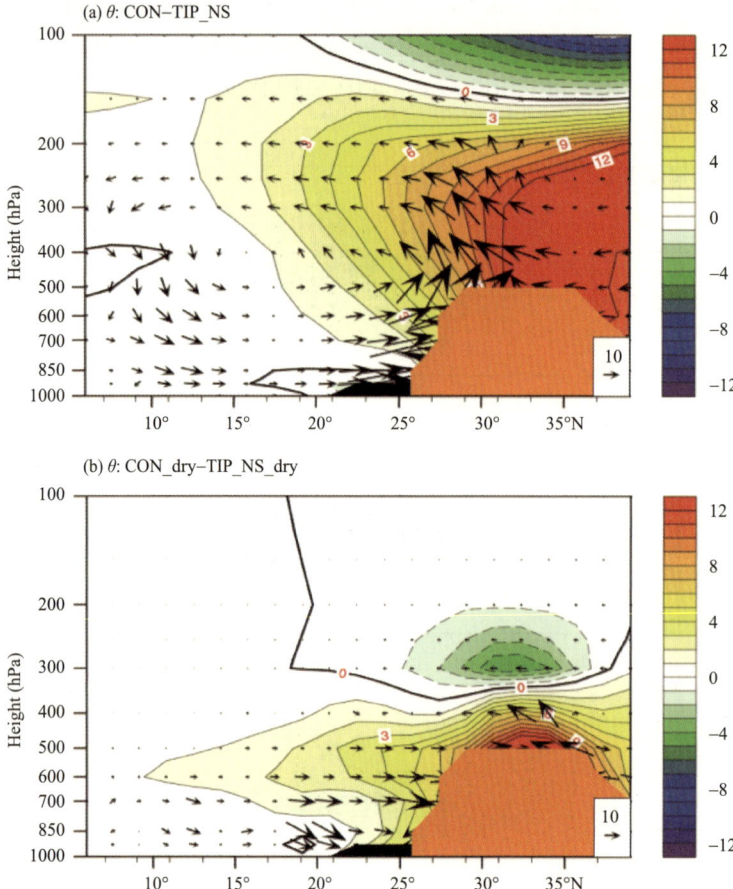

Fig. 7　JJA-mean differences along 90°E of θ (K) and circulation (vectors, v in ms^{-1}, $-\omega$ has been amplified by 300 and is in hPas^{-1}) of (a) CON−TIP_NS, and (b) CON_dry−TIP_NS_dry

5　Summary and discussion

Based on astronomical radiation analysis, reanalysis data diagnosis, and numerical experiments, we have demonstrated that in winter the Tibetan-Iranian Plateau cannot shield India from cold/dry northerly intrusion since the frequent cold/dry north westerlies or northeasterlies can go around the giant TIP by surrounding its southern rim intruding into India. In summer, on the other hand, in the South Asian summer monsoon sector the midlatitudes and subtropics receive more solar energy and possess higher surface potential temperature than North India, and there is no cold/dry advection from the north into the SASM area so that TIP shielding does not exist. Results from the hydrological analysis demonstrate that over North India local evaporation can account for only a small portion of precipitation, and the change in surface entropy is essentially attributed to the change in water vapor. The PV-θ diagnosis reveals that the cyclonic circulation surrounding the TIP in the lower layers and the remarkable onshore wind along 20°N are the atmospheric response to the thermal forcing of the Tibetan-Iranian-Plateau sensible-heat-driven air-pump (TIP-SHAP); and its generation does not rely on moist processes, although these processes have significant positive feedback on the cyclonic circulation. The TIP-SHAP-induced cyclonic circulation transports enormous quantities of water vapor from the oceans to the far north Indian Peninsula, the foothills of the Himalayas, and East Asia, leading to high surface entropy and strong and persistent precipitation there. Therefore, the TIP-SHAP has a dominant influence on moist convection and the continental summer monsoon over North India and East Asia.

Results from this study imply that protecting the ecosystems of the TIP and its thermal status can not only improve the local environment, but can also influence the global climate, particularly the Asian monsoon.

6 Methods

6.1 Data Sources

The climate incoming Daily Solar Radiation (DSR) used in Fig. 1b is from Clouds and the Earth's Radiant Energy System (CERES) (Kato et al., 2013) Level3B datasets at http://ceres.larc.nasa.gov/index.php.

The DSR for 2001 used in Fig. 1c is from CERES level3 datasets.

The daily near-surface wind and air temperature for 2001 used in Fig. 1d are from the European Centre for Medium-Range Weather Forecasts (ECMWF) ERA-interim (Dee et al., 2011) dataset at http://apps.ecmwf.int/datasets/data/interim-full-daily.

The monthly mean reanalysis dataset used in Figs. 2—5 is from ERA-40 between 1958 and 2002 at http://apps.ecmwf.int/datasets.

All of the variables are interpolated from pressure coordinates to sigma coordinates at the $s=0.99$ level near the surface.

The precipitation dataset used in Fig. 3c is from the Global Precipitation Climatology Project (GPCP) (Adler et al., 2003) monthly mean data from 1979 to the present at http://www.esrl.noaa.gov/psd/data/gridded/data.gpcp.html.

6.2 Calculation of the Daily Solar Radiation (DSR) at Summer Solstice

At the summer solstice, the solar declination angle $\delta=23.5°N$. The calculations of the daily sunshine duration at latitude A ($\varphi_A=30°N$, 6.5 degrees north of δ) and B ($\varphi_B=17°N$, 6.5 degrees south of δ) can be calculated as shown below.

The relation between the solar zenith angle Z and the hour angle h satisfies:

$$\cos Z = \sin\varphi\sin\delta + \cos\varphi\cos\delta\cos h \qquad (4)$$

At sunrise t_1 and sunset t_2, Z equals 90°. Therefore, the hour angles h_A and h_B at sunrise and sunset for latitudes A and B are estimated by: $h_A = \arccos(-\tan30°\tan23.5°)$

$$\sin30°\sin23.5° + \cos30°\cos23.5°\cos h_A = 0 \qquad (5)$$

$$\sin17°\sin23.5° + \cos17°\cos23.5°\cos h_B = 0 \qquad (6)$$

$$h_A = \arccos(-\tan30°\tan23.5°) \qquad (7)$$

$$h_B = \arccos(-\tan17°\tan23.5°) \qquad (8)$$

Consequently, the sunshine duration (daytime) is $2h_A$ (unit: radians) at ϕ_A and $2h_B$ at ϕ_B for the units in hours, the sunshine duration becomes:

$$\frac{24}{2\pi} \times 2h_A = \arccos(-\tan30°\tan23.5°) \times \frac{24}{\pi} \approx 13.9 \qquad (9)$$

$$\frac{24}{2\pi} \times 2h_B = \arccos(-\tan17°\tan23.5°) \times \frac{24}{\pi} \approx 13.0 \qquad (10)$$

(9) and (10) suggest that the sunshine duration at 30°N is about one hour longer than at 17°N. Meanwhile, the total shortwave radiation at the top-of-atmosphere (TOA) in an integral for daytime is:

$$Q = \int_{t_1}^{t_2} S \left(\frac{d_m}{d}\right)^2 (\sin\varphi\sin\delta + \cos\varphi\cos\delta\cos h) dt \tag{11}$$

Where $S=1365 \text{ Wm}^{-2}$ and denotes the solar constant, $d = 1.521 \times 10^8$ m and denotes the distance between the sun and the earth, $d_m = 1.496 \times 10^8$ m and denotes its mean value, and $dt = \frac{T}{2\pi}dt$ with T being 86400 seconds. Then

$$Q = \frac{T}{2\pi} S \left(\frac{d_m}{d}\right)^2 \int_{-h}^{h} (\sin\varphi\sin\delta + \cos\varphi\cos\delta\cos h) dh \tag{12}$$

Therefore, the total solar insolation at latitudes A and B is:

$$Q_A = 86400/2\pi \times 0.96 \times 1365 \times (2h_A\sin30°\sin23.5° + 2\cos30°\cos23.5°\sin h_A) \tag{13}$$

$$Q_B = 86400/2\pi \times 0.96 \times 1365 \times (2h_B\sin17°\sin23.5° + 2\cos17°\cos23.5°\sin h_B) \tag{14}$$

From (13) and (14), $Q_A \approx 41 \times 10^6 \text{ Jm}^{-2}$ and $Q_B \approx 38.8 \times 10^6 \text{ Jm}^{-2}$, respectively, which is identical to 476 Wm^{-2} at 30°N and 448.9 Wm^{-2} at 17°N for the daily solar radiation (DSR) at the TOA.

6.3 Model introduction and experiment design

The model used in this study is the Spectral Atmospheric Model of LASG/IAP (known as SAMIL) (Wu et al., 1996; Bao et al., 2010, 2013), which has the horizontal resolution R42 (2.81° longitude × 1.66° latitude) with 26 vertical layers in σ-p hybrid coordinates, extending from the surface to 2.19 hPa. The mass flux cumulus parameterization of Tiedtke (1989) is used to calculate convective precipitation. The cloud scheme is a diagnostic method parameterized by low-layer static stability and relative humidity (Slingo, 1980, 1989). A stratocumulus scheme is also employed, based on a statistical cloud scheme (Dai et al., 2004). A nonlocal scheme is employed to calculate the eddy-diffusivity profile and turbulent velocity scale, and the model incorporates nonlocal transport effects for heat and moisture (Holtslag and Boville, 1993). The radiation scheme employed is the Edwards-Slingo scheme (Edwards et al., 1996), but with some improvement by Sun(2011, 2005). SAMIL is coupled with the land model NCAR CLM3 (Oleson et al., 2004) in this study.

Five experiments are carried out in this study. The detailed experimental design is shown in Table 1. The CON run is forced by the observed monthly mean sea surface temperature (SST) (Hurrell et al., 2008) averaged between 1990 and 1999 and the realistic topography. The NoTIP run is the same as the CON except the Tibetan-Iranian Plateau (TIP) is removed where the topography is above 500 m. The TIP_NS run is also the same as the CON but the vertical diffusion heating in the atmosphere is set to zero over the TIP region where the elevation is above 500 m during the integration. CON_dry and TIP_NS_dry are, respectively, the same as CON and TIP_NS except the water vapor in the atmosphere is set to zero during the integration in both experiments. All of the experiments are integrated for 10 years, and the mean of the last 5 years is calculated as the equilibrium state.

Table 1 Experiment design

Experiments	TIP Topography	TIP Sensible heating	Moist process
CON	Yes	Yes	Yes
NoTIP	No	Yes	Yes
TIP_NS	Yes	No	Yes
CON_dry	Yes	Yes	No
TIP_NS_dry	Yes	No	No

References

ABE M, HORI M, YASUNARI T, et al, 2013. Effects of the Tibetan Plateau on the onset of the summer monsoon in South Asia: The role of the air-sea interaction[J]. Journal of Geophysical Research: Atmospheres, 118(4): 1760-1776.

ADLER R F, HUFFMAN G J, CHANG A, et al, 2003. The version-2 global precipitation climatology project (GPCP) monthly precipitation analysis (1979—present)[J]. Journal of hydrometeorology, 4(6): 1147-1167.

BAO Q, WU G, LIU Y, et al, 2010. An introduction to the coupled model FGOALS1.1-s and its performance in East Asia [J]. Advances in Atmospheric Sciences, 27(5): 1131-1142.

BAO Q, LIN P, ZHOU T, et al, 2013. The flexible global ocean-atmosphere-land system model, spectral version 2: FGOALS-s2[J]. Advances in Atmospheric Sciences, 30(3): 561-576.

BOOS W R, KUANG Z, 2010. Dominant control of the South Asian monsoon by orographic insulation versus plateau heating[J]. Nature, 463(7278): 218-222.

BOOS W R, KUANG Z, 2013. Sensitivity of the South Asian monsoon to elevated and non-elevated heating[J]. Scientific Reports, 3(1): 1-4.

CHEN G S, LIU Z, KUTZBACH J E, 2014. Reexamining the barrier effect of the Tibetan Plateau on the South Asian summer monsoon[J]. Climate of the Past, 10(3): 1269-1275.

DAI F, YU R, ZHANG X, et al, 2004. A statistically-based low-level cloud scheme and its tentative application in a general circulation model[J]. Acta Meteorological Sinica, 3: 12.

DEE D P, UPPALA S M, SIMMONS A J, et al, 2011. The ERA-Interim reanalysis: Configuration and performance of the data assimilation system[J]. Quarterly Journal of the Royal Meteorological Society, 137(656): 553-597.

EDWARDS J, SLINGO A, 1996. Studies with a flexible new radiation code. I: Choosing a configuration for a large-scale model[J]. Quarterly Journal of the Royal Meteorological Society, 122(531): 689-719.

EMANUEL K A, NEELIN D J, BRETHERTON C S, 1994. On large-scale circulations in convecting atmospheres[J]. Quarterly Journal of the Royal Meteorological Society, 120(519): 1111-1143.

FLOHN H, 1957. Large-scale aspects of the "summer monsoon" in South and East Asia[J]. Journal of the Meteorological Society of Japan, 35A: 180-186.

HAHN D G, MANABE S, 1975. The role of mountains in the south Asian monsoon circulation[J]. Journal of the Atmospheric Sciences, 32(8): 1515-1541.

HOLTSLAG A, BOVILLE B, 1993. Local versus nonlocal boundary-layer diffusion in a global climate model[J]. Journal of Climate, 6(10): 1825-1842.

HOSKINS B J, 1991. Towards a PV-θ view of the general circulation[J]. Tellus A: Dynamic Meteorology and Oceanography, 43(4): 27-35.

HOSKINS B J, MCINTYRE M E, ROBERTSON A W, 1985. On the use and significance of isentropic potential vorticity maps[J]. Quarterly Journal of the Royal Meteorological Society, 111(470): 877-946.

HURRELL J W, HACK J J, SHEA D, et al, 2008. A new sea surface temperature and sea ice boundary dataset for the community atmosphere model[J]. Journal of Climate, 21(19): 5145-5153.

KATO S, LOEB N G, ROSE F G, et al, 2013. Surface irradiances consistent with CERES-derived top-of-atmosphere shortwave and longwave irradiances[J]. Journal of Climate, 26(9): 2719-2740.

OLESON K W, DAI Y, BONAN G, et al, 2004. Technical description of the community land model (CLM)[R]. Tech. Note NCAR/TN-461+STR.

PATHAK A, GHOSH S, KUMAR P, 2014. Precipitation recycling in the Indian subcontinent during summer monsoon [J]. Journal of Hydrometeorology, 15(5): 2050-2066.

PRIVE N C, PLUMB R A, 2007a. Monsoon dynamics with interactive forcing. Part I: Axisymmetric studies[J]. Journal of the Atmospheric Sciences, 64(5): 1417-1430.

PRIVE N C, PLUMB R A, 2007b. Monsoon dynamics with interactive forcing. Part II: Impact of eddies and asymmetric geometries[J]. Journal of the Atmospheric Sciences, 64(5): 1431-1442.

RAJAGOPALAN B, MOLNAR P, 2013. Signatures of Tibetan Plateau heating on Indian summer monsoon rainfall variability[J]. Journal of Geophysical Research: Atmospheres, 118(3): 1170-1178.

ROMATSCHKE U, HOUZE JR R A, 2011. Characteristics of precipitating convective systems in the South Asian monsoon [J]. Journal of Hydrometeorology, 12(1): 3-26.

SLINGO A, 1989. A GCM parameterization for the shortwave radiative properties of water clouds[J]. Journal of Atmospheric Sciences, 46(10): 1419-1427.

SLINGO J M, 1980. A cloud parametrization scheme derived from GATE data for use with a numerical model[J]. Quarterly Journal of the Royal Meteorological Society, 106(450): 747-770.

SUN Z, 2005. Parameterizations of radiation and cloud optical properties[R]. BMRC Research Rep: 1-6.

SUN Z, 2011. Improving transmission calculations for the Edwards-Slingo radiation scheme using a correlated-k distribution method[J]. Quarterly Journal of the Royal Meteorological Society, 137(661): 2138-2148.

THORPE A J, 1985. Diagnosis of balanced vortex structure using potential vorticity[J]. Journal of the Atmospheric Sciences, 42(4): 397-406.

TIEDTKE M, 1989. A comprehensive mass flux scheme for cumulus parameterization in large-scale models[J]. Monthly Weather Review, 117(8): 1779-1800.

UPPALA S M, KALLBERG P W, SIMMON A J, et al, 2005. The ERA-40 re-analysis[J]. Quarterly Journal of the Royal Meteorological Society, 131(612): 2961-3012.

WANG B, LINHO, 2002. Rainy season of the Asian-Pacific summer monsoon[J]. Journal of Climate, 15(4): 386-398.

WU G, LIU H, ZHAO Y, et al, 1996. A nine-layer atmospheric general circulation model and its performance[J]. Advances in Atmospheric Sciences, 13(1): 1-18.

WU G, ZHANG Y, 1998. Tibetan Plateau forcing and the timing of the monsoon onset over South Asia and the South China Sea[J]. Monthly Weather Review, 126(4): 913-927.

WU G, LIU Y, 2003. Summertime quadruplet heating pattern in the subtropics and the associated atmospheric circulation [J]. Geophysical Research Letters, 30(5): 1201.

WU G, LIU Y, ZHANG Q, et al, 2007. The influence of mechanical and thermal forcing by the Tibetan Plateau on Asian climate[J]. Journal of Hydrometeorology, 8(4): 770-789.

WU G, LIU Y, ZHU X, et al, 2009. Multi-scale forcing and the formation of subtropical desert and monsoon[J]. Annales Geophysicae, 27(9): 3631-3644.

WU G, LIU Y, DONG B, et al, 2012a. Revisiting Asian monsoon formation and change associated with Tibetan Plateau forcing: I. Formation[J]. Climate Dynamics, 39(5): 1169-1181.

WU G, LIU Y, HE B, et al, 2012b. Thermal controls on the Asian summer monsoon[J]. Scientific Reports, 2(1): 1-7.

YANAI M, LI C, SONG Z, 1992. Seasonal heating of the Tibetan Plateau and its effects on the evolution of the Asian summer monsoon[J]. Journal of the Meteorological Society of Japan, 70(1B): 319-351.

YANAI M, WU G, 2006. Effects of the Tibetan Plateau. The Asian Monsoon[M]. Cambridge: Cambridge University Press: 513-549.

YEH T C, LO S W, CHU P C, 1957. The wind structure and heat balance in the lower troposphere over the Tibetan Plateau and its surroundings[J]. Acta Meteorological Sinica, 28: 108-121.

Weakening Trend in the Atmospheric Heat Source over the Tibetan Plateau during Recent Decades. Part I: Observations

DUAN Anmin, WU Guoxiong

(State Key Laboratory of Numerical Modelling for Atmospheric Sciences and Geophysical Fluid Dynamics (LASG), Institute of Atmospheric Physics (IAP), Chinese Academy of Sciences (CAS), Beijing 100029, China)

Abstract: The trend in the atmospheric heat source over the Tibetan Plateau (TP) during the last four decades is evaluated using historical observations at 74 meteorological stations in the period of 1961—2003 and satellite radiation data from 1983 to 2004. It is shown that, in contrast to the strong surface and troposphere warming, the sensible heat (SH) flux over the TP exhibits a significant decreasing trend since the mid-1980s. The largest trend occurs in spring, a season of the highest SH over the TP. The subdued surface wind speed contributes most to the decreasing trend. At the same time, the radiative cooling effect in the air column enhances persistently. Despite of the fact that the in situ latent heating presents a weak increasing trend, the springtime atmospheric heat source over the TP losses its strength during recent two decades. Further investigation suggests that the weakened SH over the TP may be part of the global circulation shift.

1 Introduction

The Tibetan Plateau (TP) is located in the subtropical central and eastern Eurasian continent, acting as a huge, intense, and elevated heat source with strong sensible heating (SH) in the surface layers. The mechanical and thermal forcing of the TP plays a crucial role not only in the onset and maintenance of the Asian summer monsoon (e.g., Yeh et al., 1957; Flohn, 1957, 1960; Yanai et al., 1992; Li and Yanai, 1996; Li et al., 2001; Zhao and Chen, 2001; Kitoh, 2004; Yanai and Wu, 2006; Wu et al., 2007), but also in the development of weather systems over East China (Tao and Ding, 1981) and even the boreal summer climate pattern (Hahn and Manabe, 1975; Broccoli et al., 1992; Duan et al., 2005a, hereafter DW2005).

Growing evidence has shown that a striking climate warming occurred over the TP during the second half of the 20th century (e.g., Liu and Chen, 2000; Zhu et al., 2001; Niu et al., 2004). Most recent analysis indicates that the increased surface air temperature was accompanied by a diminished diurnal range in surface air temperature, warmed troposphere, and cooled lower stratosphere (Duan et al., 2006a). The decrease of the in situ cloud amount and the accompanied change of radiation process are important for the increase of surface air temperature and the diminished diurnal range (Duan and Wu, 2006b; hereafter DW2006). Moreover, numerical simulation results given by both regional and global climate models further suggest that the recent climate warming over the TP is likely to be induced by the enhanced anthropogenic greenhouse gases effect (Chen et al., 2003; Duan et al., 2006b)

The climate warming over the TP should be accompanied by a change in the atmospheric heat source, not only its intensity, but also its diurnal variation and seasonal evolution. Zhang et al. (2004) found that the snow depth over the TP exhibits a sharp increase during spring (March and April) after the 1970s, which implies excessive precipitation and land surface cooling. Similar results have also been obtained by a more recent study (Zhu and Ding, personal communication), which hypothesized that there be a suppressed atmospheric heat source over the TP in spring. So far, however, there has not been any investigation concerning the decadal change of the heat source over the TP and its possible connection with the global circulation shift. The main objectives of this study, therefore, are to determine quantitatively the long-term trend in the atmospheric heat source over the TP and to identify its possible connection with the global circulation shift.

The structure of this paper is as follows. The data and analysis procedures used in this study are described briefly in Section 2. Section 3 introduces the climatology of the atmospheric heat source/sink over the TP, and Section 4 describes the change of SH over the central and eastern TP (CE-TP) and over the western TP (W-TP). In Section 5, decadal trends of the latent heating and radiative cooling are investigated to give an overall picture of the change in the atmospheric heating status over the TP. The relationship between the diabatic heating over the TP and the global climate change in decadal timescale is assessed in Section 6, followed by conclusions and discussions in Section 7.

2 Data and Methodology

a. Data

The data used in this study include the following sources:

1) The regular surface meteorological observations with an initial quality control for the TP region provided by China Meteorological Administration (CMA). Variables are collected four times daily (0000 LST, 0600 LST, 1200 LST, and 1800 LST at 90°E Lhasa time, 6 h earlier than UTC) and include surface air temperature (T_a), ground surface temperature (T_s), wind speed at 10 m above the surface (V_0), station surface pressure (P_s), and daily accumulative precipitation (P_r). We only analyze the data from 1961 to 2003 because most of the plateau meteorological observatories do not have continuous data until the 1960s.

2) Monthly records of the 12 radiosonde stations in the TP from 1980 to 2004 archived by CMA. They are monthly-mean air temperature, wind speed, and geopotential height at 16 standard pressure levels (1000, 925, 850, 700, 500, 400, 300, 250, 200, 150, 100, 70, 50, 30, 20, and 10 hPa, respectively). Since the height of the TP is almost at 600 hPa and balloons frequently burst over 10 hPa, the levels selected for the present study are limited to those from 500 to 20 hPa. Previous studies have shown that the quality of the radiosonde data in China is generally quite good (Wang et al., 2005; Zhou et al., 2005).

3) The International Satellite Cloud Climatology Project (ISCCP, http://isccp.giss.nasa.gov/projects/flux.html; Rossow et al., 1991, 1999) radiation data are also utilized. Radiation fluxes include the downward and upward shortwave and longwave radiation fluxes at the top of the atmosphere and on the ground. The data period is from July 1983 to June 2005, with a horizontal resolution of 2.5° by 2.5°.

The locations and elevations of the stations are given in Fig. 1a, with more detailed information presented in appendix. The highest and lowest stations are Bange (31.3°N, 90.02°E; 4700 m above sea level) and Zhaojue (28°N, 102.85°E; 2132 m), respectively. Minxian (34.43°N, 104.01°E; 2315 m), Dingri (28.63°N, 87.08°E; 4300 m), Huapin (26.63°N, 101.27°E; 2245 m), and Tuole (38.80°N, 98.42°E; 3367 m) outline the east, west, south, and north borders of the CE-TP area. Most of these stations are located in Qinghai and Xizang (Tibet) areas in China, and a few from the adjacent areas in Gansu, Sichuan, and Yunnan Provinces of China.

The data coverage is adequate to depict the trend in the domain except for the western plateau where there are only three stations: Shiquanhe (80.08°E, 32.50°N; 4278 m) founded in 1960, Gaize (84.42°E, 32.15°N; 4415 m) founded in 1973, and Pulan (81.25°E, 30.28°N, 3900 m) founded in 1973). Hence this study focuses mainly on the change of the atmospheric heat source in the CE-TP. The area averages over the CE-TP (26°—39°N, 85°—105°E) and W-TP (30°—33°N, 80°—85°E) are obtained using 54 and 6 grid points when the ISCCP data is used, as indicated by the two rectangular boxes in Fig. 1a.

As the first step of the data analysis, the missing data are filled by averaging the values of the previous year and the following year at the same time. This operation is done only once for every missing value. When there is still insufficient data after this, the missing value is replaced by its corresponding mean value at the same station. The missing T_a, V_0, and P_r account for less than 0.5% of the total records, hence the quality of T_a, V_0, and P_r is reasonably good. Missing T_s is close to 8% of the total record, and most occurs before 1980s. To ensure a reliable outcome, the observations that suffer from such missing T_s records are not included in the statistics before 1980, and only 37 observatories in the period 1961—2003 and 71 observatories in 1980—2003 are chosen for this study.

b. Analysis procedures

Atmospheric heat source/sink (E) is a physical quantity used to discuss the heat budget in an air column. For a given location, an atmospheric heat source (sink) is defined when there is a net heat gain (loss) within a given period. The expression of E is therefore defined as

$$E = SH + LH + RC \tag{1}$$

where SH denotes the local surface sensible heat transfer, LH is the latent heat released to the atmosphere by precipitation, and RC the net radiation flux of the air column. We can calculate E term by term or estimate it by using the apparent heat source Q_1 derived from the thermodynamic budget equation (Yanai, 1961).

In this study, we calculate SH by the bulk aerodynamic method

$$SH = C_p \rho C_{DH} V_0 (T_s - T_a) \tag{2}$$

where $C_p = 1005$ Jkg^{-1}K^{-1} is the specific heat of dry air at constant pressure, ρ is the air density that decreases exponentially with increasing elevation, C_{DH} is the drag coefficient for heat, and V_0 is the mean wind speed measured at 10 m above ground. This procedure is widely used in the TP-related studies (e.g., Yeh and Gao, 1979; Chen et al., 1985; Li et al., 2001). The bulk aerodynamic method obviously depends on the choice of C_{DH}, which varies from location to location (turbulence regime, surface roughness, and so on) and differs widely among different studies (Ye and Wu, 1998). For a fixed location and during our studied period from 1961 to 2003, the changes in ρ and C_{DH} should be small, and their impacts on the change of SH are negligible. Therefore, the wind speed V_0 and ground-air temperature difference ($T_s - T_a$) are the two key factors influencing the trend in SH. Here we assume $\rho = 0.8$ kg m^{-3} (Yeh and Gao, 1979) and $C_{DH} = 4 \times 10^{-3}$ (Li et al., 1996) for the CE-TP and $C_{DH} = 4.75 \times 10^{-3}$ for the W-TP (Li et al., 2000). These simplifications will not affect our final results.

Cautions are needed concerning the applicability of the bulk aerodynamic Eq. (2) to the TP studies. On the sloping lateral surfaces of the TP, because of the pumping effects of the TP SH driving air-pump (TP-SHAP; Wu et al., 1997, 2007), ascending air flows penetrate the isentropic surfaces in sharp angles and compensate the surface SH. In such circumstances the heat flux is carried by thermal convection, and the time-mean surface SH can be formulated approximately as ($V_0 \cdot \nabla \theta$), where θ is potential temperature. However, over the vast TP platform, the near-ground θ surfaces become relatively flat, and the near-surface thermal advection becomes smaller as well. By employing four automatic weather stations over the TP, and using the gradient method, Li et al. (2000) calculated the SH based on Eq. (2) and found that the results are comparable with other data outputs. In 1998, China launched the second TP field observation campaign with several observation sites located on the platform. By employing the corresponding observation data and comparing the results produced from different methods, Zhou et al. (2000) also indicated that the SH calculated based on Eq. (2) is comparable with those from other methods. These suggest that the bulk aerodynamic method be applicable over the TP platform. Nevertheless, to ensure that our analysis is reliable, the associated changes in wind speed V_0 and the land-air temperature difference ($T_s - T_a$) are also considered when investigating the change of SH.

LH can be calculated by precipitation via the following formula:

$$\text{LH} = P_r \times L_w \times \rho \tag{3}$$

where $L_w = 2.5 \times 10^{-6}$ Jkg^{-1} is condensation heat coefficient

$$\text{RC} = R_\infty - R_0 = (S_\infty^\downarrow - S_\infty^\uparrow) - (S_0^\downarrow - S_0^\uparrow) - (F_0^\downarrow - F_0^\uparrow) - F_\infty \tag{4}$$

where R_∞ and R_0 are the net radiation values at the top of the atmosphere and at the earth's surface, respectively. Variables S and F denote shortwave and longwave radiation fluxes, the subscripts ∞ and 0 denote the top of the atmosphere and the ground surface, and the superscripts $\downarrow$ and $\uparrow$ represent downward and upward transports.

Simple linear regression is used to calculate the trend, and the sliding t test is adopted to check the abrupt change. Using x_i to indicate a climatic variable with a sample size n, and using t_i to indicate the time corresponding to x_i, a linear equation is assumed to show the relationship between x_i and t_i:

$$x_i = a + b t_i (i = 1, 2, 3, \ldots, n)$$

where a and b are the regression constant and linear regression coefficient (i.e., linear variation rate (LVR)), respectively. They can be estimated by using the least squares method:

$$b = \frac{\sum_{i=1}^n x_i t_i - \frac{1}{n} (\sum_{i=1}^n x_i)(\sum_{i=1}^n t_i)}{\sum_{i=1}^n t_i^2 - \frac{1}{n} (\sum_{i=1}^n t_i)^2}, a = \bar{x} - b \bar{t} \tag{5}$$

where

$$\bar{x} = \frac{1}{n} \sum_{i=1}^n x_i, \bar{t} = \frac{1}{n} \sum_{i=1}^n t_i$$

After calculating b, a statistical test is employed to detect the significance of the trend. Here the correlation coefficient r between t and x is examined: $|r| > 0.29$ and $|r| > 0.37$ denote that the LVR passes 95% and 99% confidence test for a 43-year (1961—2003) sample size, and $|r| > 0.38$ and $|r| > 0.49$, for a 24-year (1980—2003) sample size.

To compare the change in amplitudes of different variables with different units, the Relative Change Rate (RCR) is also employed in this work, which is defined as,

$$RCB = (B_e - B_b)/B_b \tag{6}$$

where B_e and B_b are respectively the last/end and first/beginning values of the LVR of a time series. Unless stated otherwise, all significant changes reported in this study pass 95% confidence level.

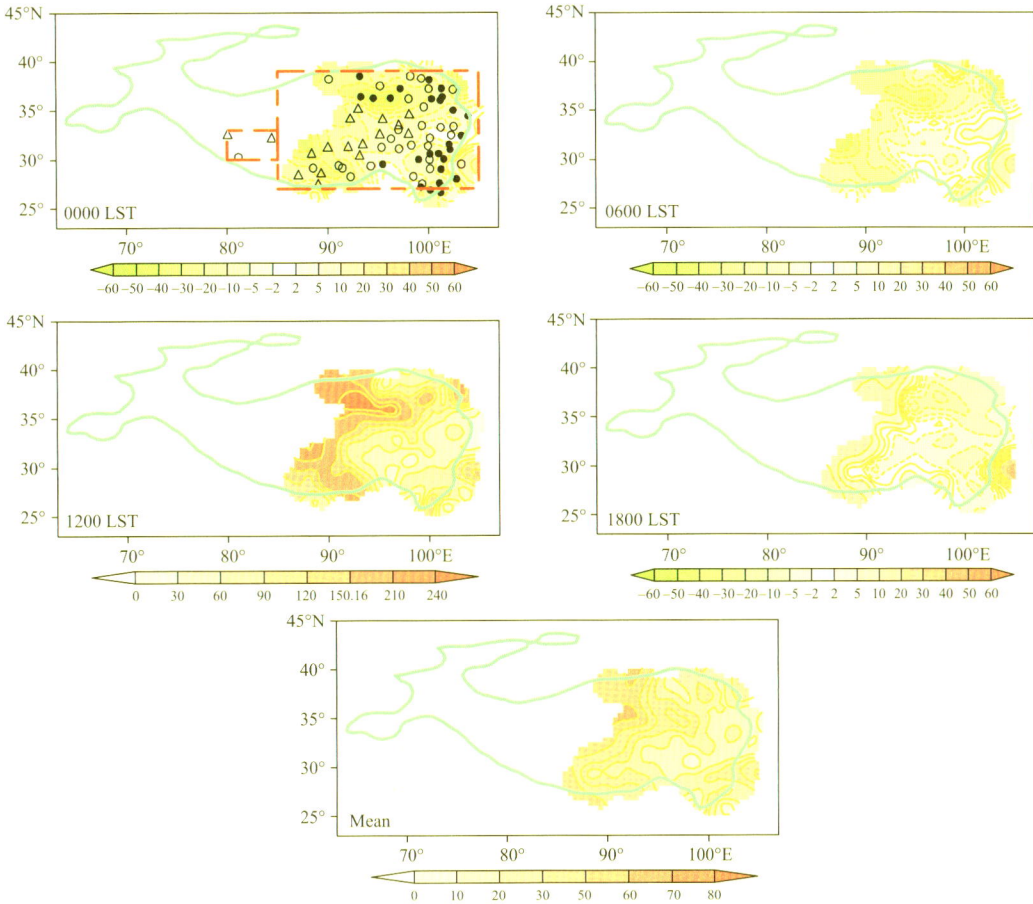

Fig. 1 Climatologically annual mean sensible heat flux (SH) on the central and eastern Tibetan Plateau (CE-TP) during 1980—2003 in units of Wm^{-2}. Triangles, open circles, and solid circles in (a) denote stations equal to or higher than 4000, 3000, and 2000 m, respectively. The thick solid line outlines the Tibetan Plateau area with the average altitude higher than 2500 m. Two boxes in Fig. 1a denote W-TP (left) and CE-TP (right)

3 Climatology of the atmospheric heat source/sink over the TP

Before discussing the trend of the heating status over TP, it is necessary to examine the climatology, in terms of spatial distribution and seasonal evolution.

a. CE-TP

Fig. 1 exhibits the annual-mean SH over the CE-TP during 1980—2003 at 0000 LST, 0600 LST, 1200 LST, 1800 LST, and their daily mean, respectively. In general, 0000 LST and 0600 LST represent the night because it is still dark at 0600 LST most time of the year except for the mid-summer. Most stations have negative SH at night, which results from a negative value of $(T_s - T_a)$, thanks to warmer T_a than T_s, and indicates the downward SH transfers from air to land surface. The minimum SH at 0000 LST exceeds -60 Wm^{-2} within Caidam Basin (near $36°-37°N$, $90°-97°E$) which is located in the northern CE-TP. At noon (1200 LST), intensely positive SH spreads over the whole CE-TP region, with a maximum of above 240 Wm^{-2}, again in Caidam Basin. The intensity in the central and northern TP is notably larger than that in the southeast TP. The noticeable difference in SH across the TP is directly related to the complex orography, underlying surface characteristics, and local climate. As is well known, the eastern TP is characterized by relatively lower elevation, exuberant vegetation, as well as warm and humid climate, while the W-TP is characterized by higher elevation, sparse vegetation and semiarid climate. Hence the magnitude of SH over the CE-TP is reasonably smaller than that over the W-TP, as will be shown later. The average SH at local mid-day for the CE-TP region is around 100 Wm^{-2}, approximately equivalent to 10 $Kday^{-1}$ heating rate in the surface air layer. This is the reason why the TP is called a gigantic SH air pump (Wu et al., 1997), which generates the deep and well-mixed layer of potential temperature and tropospheric heating in the afternoon (Luo and Yanai, 1983; Yanai et al., 1994). At 1800 LST, which is before sunset in the central TP but after sunset in the eastern TP for most parts of the year, SH remains positive to the west of 90°E but becomes negative to the east. For the annual mean, the 71-station-averaged SH over the CE-TP is $\sim$40 Wm^{-2}, which is about half of that given by Yeh and Gao (1979) who set the C_{DH} as high as 8×10^{-3}, but agrees well with Chen et al. (1985) and Zhao (1999). It is worth noting that the strongest SH source or sink is located in the northern TP instead of the central TP.

The annual cycle of SH averaged over the 71 stations and for the period 1980—2003 is displayed in Fig. 2. The SH at midnight (0000 LST) is always negative throughout the year. It increases from the minimum of -22.8 Wm^{-2} in February to the maximum of -5.8 Wm^{-2} in July, and then gradually decreases to the minimum. The SH at dawn (0600 LST) is transferred from air to land surface year round except for June and July, reaching the minimum in February and the maximum (1.4 Wm^{-2}) in June. The perennial strong surface SH at local mid-day (1200 LST) ranges from 138 Wm^{-2} (in September) to 287 Wm^{-2} (in April), and exceeds 200 Wm^{-2} from February to May. The SH at 1800 LST is positive from April to September and negative during the rest months, with the maximum of 23 Wm^{-2} in June and the minimum of -30 Wm^{-2} in January. Since the SH at noon is one order larger, the daily-mean SH exhibits a similar change with its maximum of 67 Wm^{-2} in April and minimum of 21 Wm^{-2} in December.

Due to the lack of 6-hourly precipitation data, the climatology of the seasonally averaged LH over the CE-TP is calculated instead (Fig. 3). Large LH occurs over the southeast and south-central parts of the plateau, with maximum values along the valley of the Yarlung Zangbo Jiang (Brahmaputra River). LH decreases gradually toward northwest. One prominent feature of the TP climate is the vigorous summer monsoon, and the magnitude of LH in JJA is much larger than other times of the year. During the rainy season, LH over most areas usually exceeds 80 Wm^{-2} (about 3 $mmday^{-1}$ rainfall), overwhelming SH or RC over the TP.

Domain-averaged ISCCP data is used to estimate the net RC over the TP. The data at the 54 grids in the CE-TP and at the 6 grids in the W-TP are extracted, and averaged to obtain RC for each region. The RC averaged from the 54 grids together with the 71-station-averaged SH and LH is shown in Table 1 to show the magnitudes of the individual components of E in the CE-TP. Clearly, the TP is an atmospheric heat source in spring (March—May, MAM) and summer (June—August, JJA), but becomes a heat

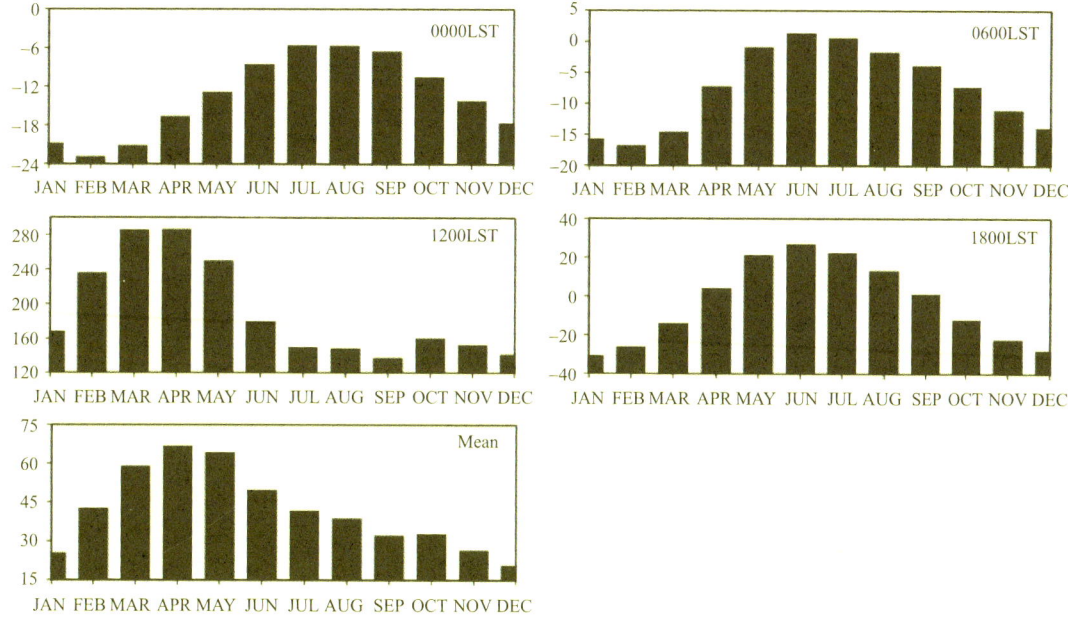

Fig. 2 Climatologically (1980—2003) annual cycle of the 71-station-averaged SH on CE-TP in units of Wm^{-2}

sink in autumn (September—November, SON) and winter (December—February, DJF). Most contribution to the total diabatic heating comes from SH in MAM and from LH in JJA, which is usually more than twice the SH in summer. On the other hand, RC ranges from -60 to -90 Wm^{-2}, always tending to make the air column a heat sink. Despite many differences in details, the results demonstrated here agree qualitatively with those given by Yeh and Gao (1979), Chen et al. (1985), Zhao and Chen (2001), and DW2005.

b. W-TP

Both the diurnal and annual ranges of SH in the W-TP are noticeably larger than those in the CE-TP. Fig. 4 and Tables 1 and 2 exhibit the climatology of SH in W-TP (calculated with aforementioned three stations). The strongest SH with the intensity of 290 Wm^{-2} at 1200 LST and the largest diurnal range with the value of 332.8 Wm^{-2} are detected at Gaize, but the daily-mean SH at this station (63.7 Wm^{-2}) is smaller than that at Shiquanhe (76.2 Wm^{-2}). The diurnal range of SH in the W-TP even exceeds 400 W m^{-2} in May, and the annual range of daily-mean SH is more than 100 Wm^{-2}. Furthermore, the SH maximum in the W-TP lags that in the CE-TP by one or two months except for 1800 LST. This phenomenon is directly related to the fact that the rainy season starts first in the southeastern TP from late April to early May and then propagates westward until it reaches the W-TP in late June or early July (Yeh et al., 1979).

As shown in Table 1, the seasonal evolution of E in the W-TP is very different from that in the CE-TP. The LH in the W-TP is always a small term in the energy budget Eq. (1). The spring heat source over the W-TP is two times larger than that over the CE-TP. The SH in MAM and JJA exceeds 100 Wm^{-2}, about double that in the CE-TP. Nevertheless, the atmospheric heat sink in SON and DJF is weaker than that in the CE-TP mainly due to the less RC.

In general, the TP acts as an especially strong sensible heat source in spring and summer with the daily maximum in the local afternoon. A distinct feature of the SH over the TP is the large diurnal range but much weaker annual range. Both the diurnal and annual ranges decease gradually from northwest toward southeast, while the annual cycle over the semiarid W-TP lags that over the humid CE-TP by about one month.

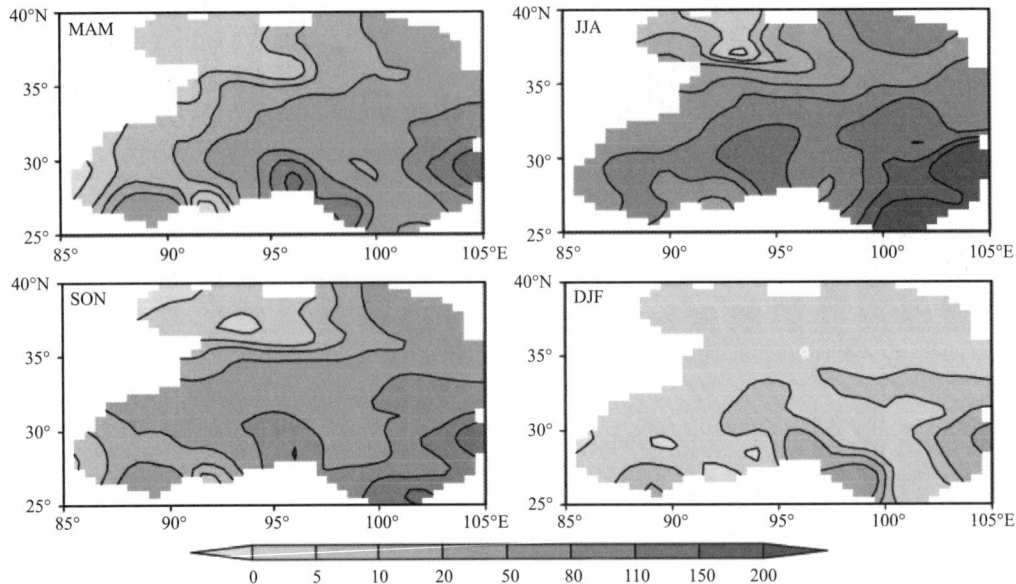

Fig. 3 Climatologically latent heat flux (LH) over CE-TP during 1980—2003 in units of Wm^{-2}

Table 1 Climatology(1984—2003) of the atmospheric heat source/sink and each component over TP in units of Wm^{-2}

Region	Component	MAM	JJA	SON	DJF	Annual
CE-TP	SH	62	42	29	27	40
	LH	30	101	37	4	43
	RC	−61	−80	−90	−81	−78
	E	31	63	−24	−50	5
W-TP	SH	102	101	47	15	66
	LH	6	25	6	4	10
	RC	−45	−63	−64	−63	−59
	E	63	63	−11	−44	18

4 Trend in SH over the TP

a. Trends in SH at different times of the day

Because SH reaches its daily minimum and maximum, respectively, at the night and the afternoon, in Fig. 5 we show the time sequences of T_a, T_s, V_0, and SH at 0000 LST and 1200 LST for the CE-TP to estimate their long-term trends and to find out their differences between day and night.

At 0000 LST, T_a increases rapidly after the late 1960s and reaches the peak at 1999. The LVR during 1961—2003 is of 0.29 ℃ decade^{-1} and the trend is above 99.9% confidence level. The most significant (above 99.9% confidence level) abrupt warming is detected in 1986. The increase in T_a is more evident during 1980—2003, and its LVR is 0.4 ℃ decade^{-1} for the 71-station-average. Note that both the interannual variation and long-term trend of the 71-station-averaged T_a are very similar to those of 37-station-averaged T_a during 1980—2003. This gives us confidence in using the 37 stations for the CE-TP during the whole period 1961—2003.

At 0000 LST, T_s and T_a have similar interannual variations and decadal trend. The LVR of T_s during 1961—2003 is 0.2 ℃ decade^{-1}, and the LVR for 71-station-averaged T_s during 1980—2003 is 0.37 ℃ decade^{-1}. Both the trend in T_s and its abrupt warming in 1986 are statistically significant. Smaller increase in T_s than in T_a leads to a reduction in ($T_s - T_a$). On the other hand, V_0 decreases continuously since early1970s with a LVR of -0.09 ms^{-1} decade^{-1}. The decrease in V_0, together with the reduction in ($T_s - T_a$) due to the larger increase in T_a than in T_s, leads to a persistent increase of nocturnal SH since 1973. This implies decreased SH transfer from air to land surface. The LVR of SH at 0000 LST during 1961—2003 is 0.3 Wm^{-2} decade^{-1}, and the LVR during 1980—2003 for the 71-station average is 1.2 Wm^{-2} decade^{-1}. The increasing trend at 0000 LST is significant only during 1980—2003.

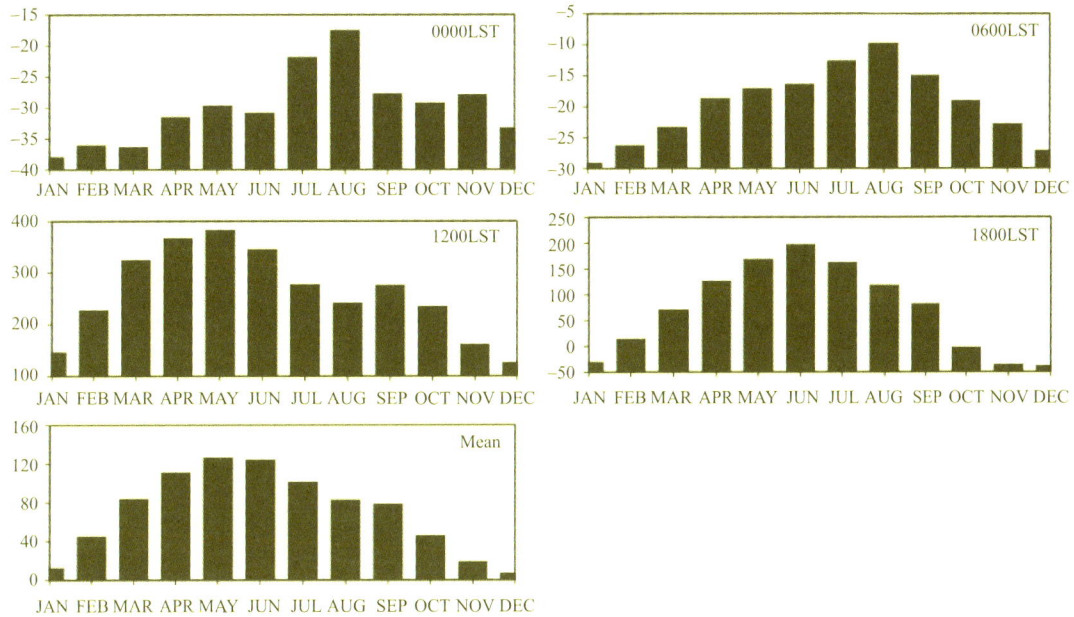

Fig. 4 Climatologically (1980—2003) annual cycle of the 3-station-averaged SH on W-TP in units Wm^{-2}

The trends in T_a and V_0 at 1200 LST are similar to those at 0000 LST, except for small changes in amplitude. During 1961—2003, the LVRs of T_a and V_0 are 0.2 ℃ decade^{-1} and -0.06 ms^{-1} decade^{-1}, respectively. However, the trends in T_s during 1961—2003 for the 37-station average and that during 1980—2003 for the 71-station average are opposite, with the LVRs of -0.19 ℃ decade^{-1} and 0.47 ℃ decade^{-1}, respectively. The combined effect of the changes in T_a and T_s lead to a decreasing trend of ($T_s - T_a$) during 1961—2003 but a slightly increasing trend during 1980—2003 (figures not shown here). Similar to 0000 LST, the surface wind speed V_0 at 1200 LST decreases significantly since 1970s. The weakened V_0 induces a significant weakening trend of SH at noon with the LVR of -16.3 Wm^{-2} decade^{-1} during 1980—2003.

Because the daily maximum SH usually occurs at noon, and because the SH trend at 1200 LST declines remarkably, a significant decreasing trend of the daily mean SH then occurs during 1980—2003 over the CE-TP. During this period, a increasing trend of 0.68 Wm^{-2} decade^{-1} exists at 0600 LST, while a decreasing trend of -0.03 Wm^{-2} decade^{-1} exists at 1800 LST. Therefore, the long-term trend in SH over the the CE-TP can be summarized as a weaker increasing trend at night and a stronger decreasing trend during day.

Table 2 Annual-mean SH and its diurnal range at three stations in the W-TP during 1980−2003 in units of Wm^{-2}

Station	0000 LST	0600 LST	1200 LST	1800 LST	Daily mean	Diurnal range
Shiquanhe	−25.9	−13.5	255.7	88.4	76.2	281.6
Caze	−42.8	−24.7	290.0	28.6	63.7	332.8
Pulan	−21.0	−21.0	234.1	94.0	71.5	255.1

b. Trends in SH during winter and spring

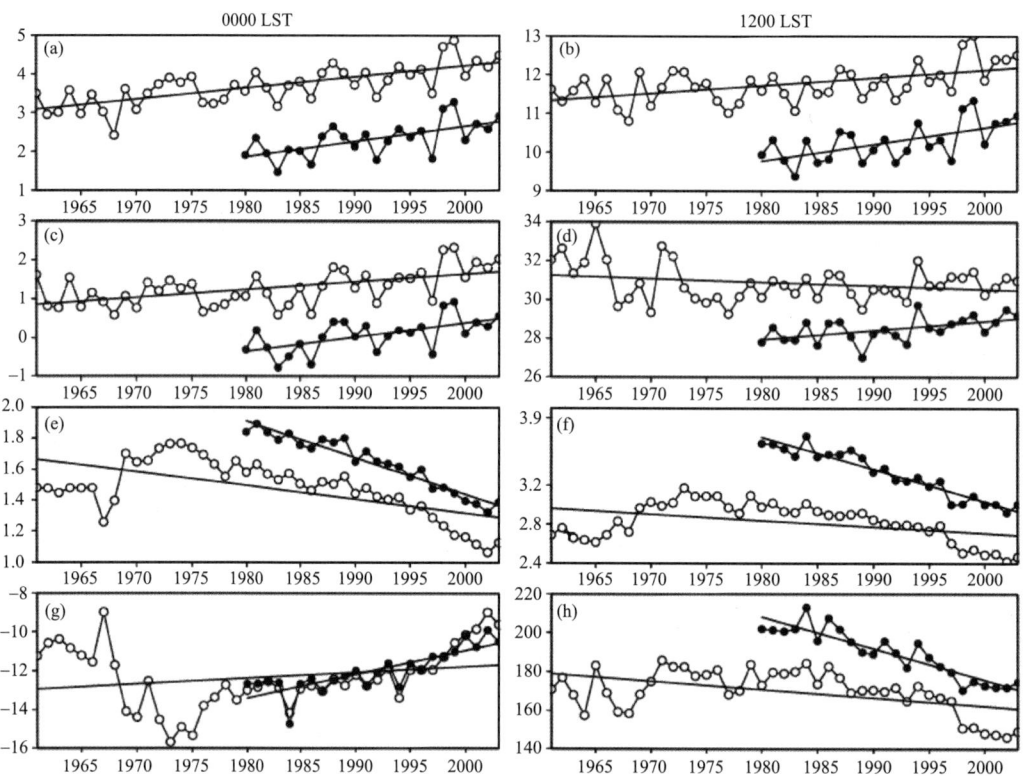

Fig. 5 Temporal evolution of the T_a (a and b), T_s, (c and d), V_0 (e and f), and SH (g and h,) on the CE-TP at 0000 LST (left panels) and 1200 LST (right panels). The units of T_a and T_s are ℃, V_0 ms^{-1}, and SH Wm^{-2}. Curves with open circles are 71-station-averaged and closed circles are 37-station-averaged

As shown in Section 3, the annual minimum and maximum SH over the CE-TP occur in DJF and MAM, respectively. Fig. 6 shows the long-term trends in SH during these two seasons. Both T_a and T_s increase significantly during 1980−2003 regardless of seasons. However the increase in DJF is more rapid in T_a (0.49 ℃ decade^{-1}) than in T_s (0.42 ℃ decade^{-1}), which lead to a slightly decreasing trend of ($T_s - T_a$) of −0.07 ℃ decade^{-1}. The opposite happened in MAM, the increase in T_s (0.4 ℃ decade^{-1}) is more rapid than that in T_a (0.36 ℃ decade^{-1}), which lead to a slightly increasing trend of ($T_s - T_a$) of 0.04 ℃ decade^{-1}. On the other hand, V_0 decreases sharply since 1970s. This causes a decreasing trend in DJF (−2.3 Wm^{-2} decade^{-1}) and a larger decreasing trend in MAM (−5.4 Wm^{-2} decade^{-1}). As a result, a diminished annual cycle of SH over the CE-TP is observed. For summer and autumn, decreasing SH trends are also significant with the LVRs of −3.1 and −2.6 Wm^{-2} decade^{-1}, respectively (figures not shown).

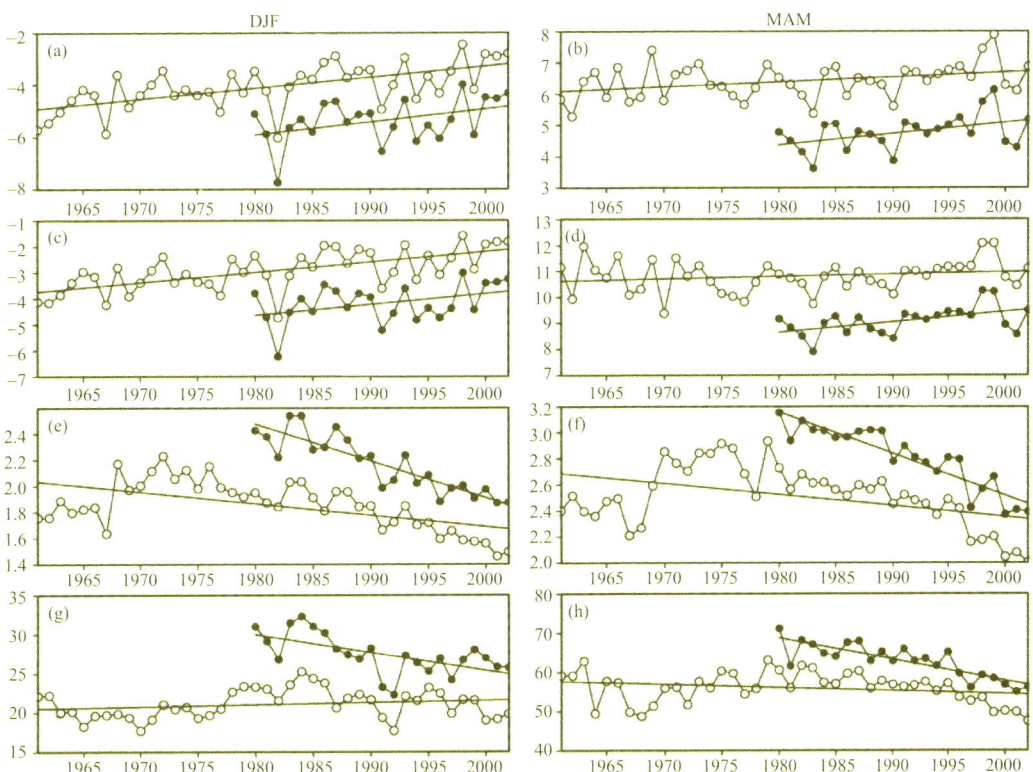

Fig. 6 Temporal evolution of the T_a (a and b), T_s (c and d), V_0 (e and f), and SH (g and h) on the CE-TP in DJF (left panels) and MAM (right panels). The units of T_a and T_s are ℃, V_0 is ms^{-1}, and SH is Wm^{-2}. Curves with open circles are 71-station-averaged and closed circles are 37-station-averaged

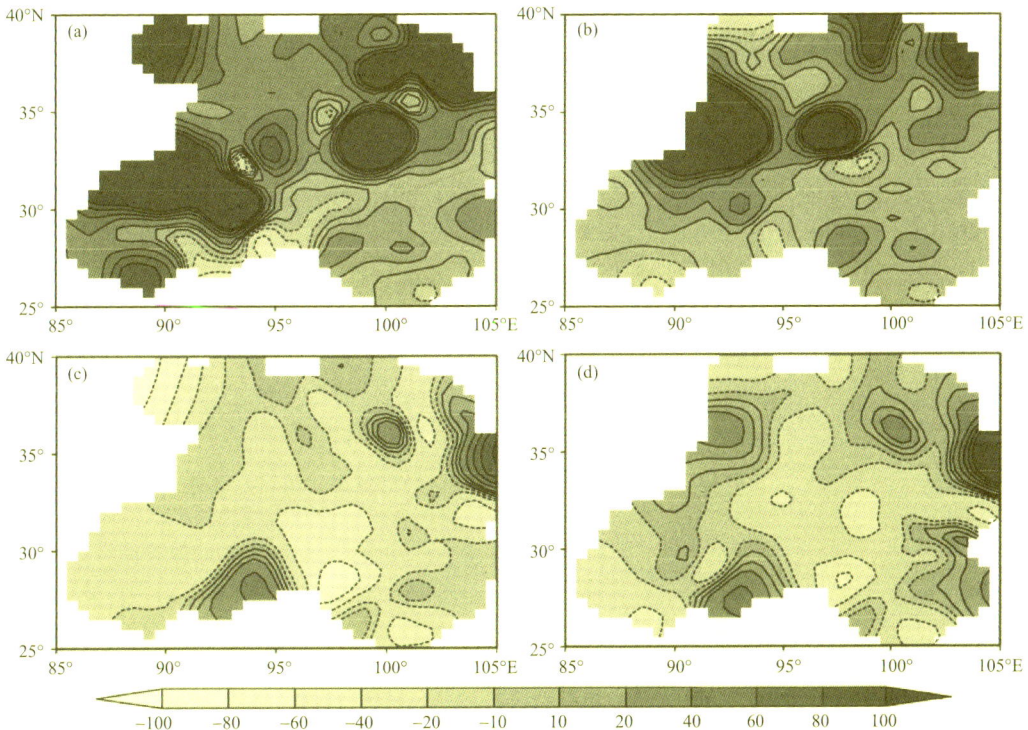

Fig. 7 Spatial distribution of relative change rates (RCR) in unit of percent in CE-TP during 1980–2003
(a): T_a, (b): T_s, (c): V_0, and (d): SH

c. Spatial distribution of the SH trend

Due to the vast domain with complicated topography, climate change over the TP may vary with locations. To compare the change in terms of amplitude among various meteorological variables over the CE-TP, Fig. 7 shows the spatial distribution of RCR given by Eq. (6) for T_a, T_s, V_0, and SH over the CE-TP during 1981—2003. In contrast to the almost coherent increases in T_a and T_s across the CE-TP, V_0 and SH experience obvious decreasing trend for most parts of the CE-TP. The 71-station-averaged RCR of T_a, T_s, V_0) and SH are 57%, 32%, −22%, and −14%, respectively. There are only 11 of 71 stations that show an increasing trend in SH. It is necessary to note that the change in T_a, T_s, V_0, and SH is not linearly correlated with elevation.

Over the W-TP (Fig. 8), positive trends of T_a, T_s, and negative trend of V_0 at both 0000 LST and 1200 LST are as the same as their counterparts over the CE-TP. Due to varying V_0 and $(T_s - T_a)$ at the three stations, the LVRs of SH are also be different throughout the whole period. An abrupt change point in SH at 1200 LST can be detected. The SH shows an increasing trend before 1989 and a decreasing trend afterwards. Therefore, there is a decreasing trend in SH over the W-TP, with the largest amplitude in spring (Table 3).

d. Relationship with the declined East Asian summer monsoon

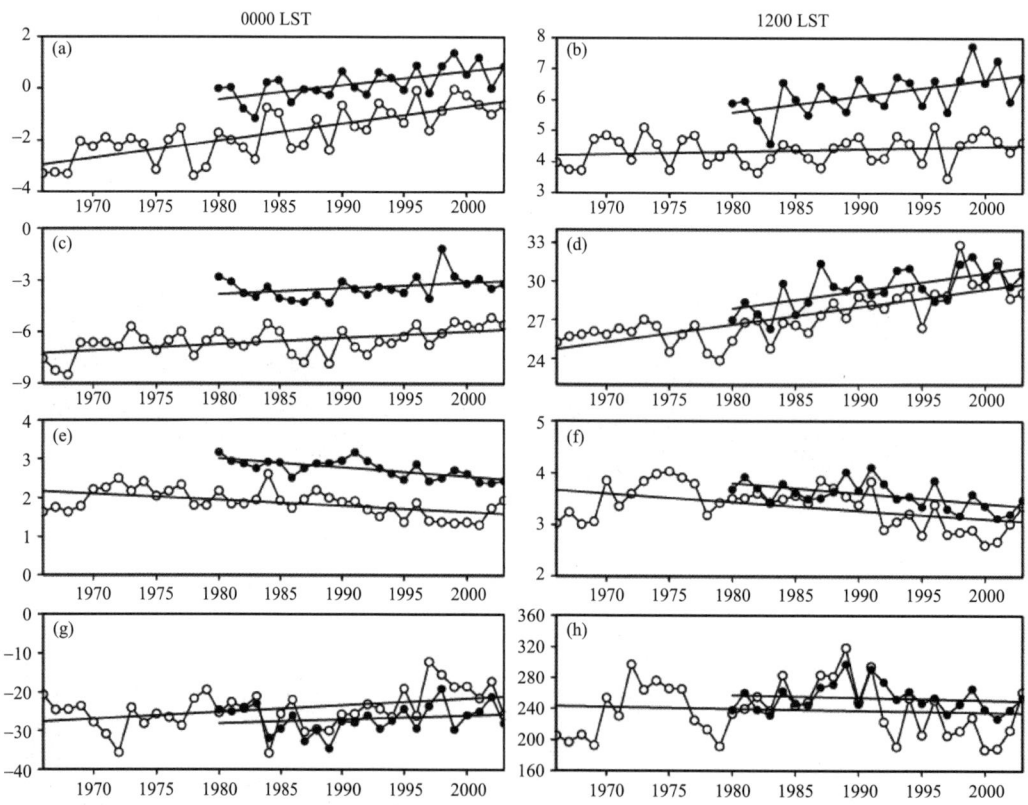

Fig. 8 Temporal evolution of the T_a (a and b), T_s, (c and d), V_0 (e and f,), and SH (g and h,) on the W-TP at 0000 LST (left panels) and 1200 LST (right panels). The units of T_a and T_s are ℃, V_0 is ms^{-1}, and SH is Wm^{-2}. Curves with open circles denote the case at Shiquanhe and closed circles denote the average of three stations in W-TP (i.e., Shiquanhe, Gaze, and Pulan)

As the primary heat source before the rainy season (e.g., Yeh et al., 1979; Yanai et al., 1992), SH over the TP exerts a significant influence on the East Asian summer monsoon in terms of both the timing of the onset (Wu et al., 1998) and the subsequent rainfall pattern and intensity (Zhao et al.,

2001; Duan et al., 2005b). The above-normal SH over the TP before July may lead to positive rainfall anomalies over the TP, along the reaches of the Yangtze River and Huaihe River, in Sichuan Basin and over the Yunnan-Guizhou Plateau, but negative rainfall anomalies in the regions over the north, northeast, and west of the TP. On decadal timescales, the significant weakening trend of the spring SH over the TP may modulate the East Asian summer monsoon to some degree.

As a matter of fact, the surface wind speed over most parts of China shows a steady declining trend during 1969—2000, accompanying by the decreased winter and summer monsoons in the East Asia (Xu et al., 2006). This results in increased droughts in the northern China and flood along the Yangtze River Valley (Yu et al., 2004). Xu et al. (2006) argued that the decline of the summer monsoon is linked to the summer cooling in central South China, which is likely induced by the air pollution, as well as by the warming over the South China Sea and the western North Pacific Ocean. However, the mechanism for the decadal trend in the East Asian summer monsoon may be more complicated. As shown earlier, the steady decline of V_0 over the TP starts in early 1970s as well, implying that a similar mechanism may exist. And this mechanism is more likely related to the global circulation shift. The connection between the trend in SH over the TP and the global circulation shift will be discussed in Section 6. It is not yet clear how the suppressed summer monsoon is linked to the SH trends over the TP, a puzzle that needs future investigation.

Table 3 Trend in the atmospheric heat source/sink E and its components over the TP in units of $Wm^{-2} decade^{-1}$. Analysis period for SH and LH is 1980—2003 and for RC is 1984—2004

Region	Component	MAM	JJA	SON	DJF	Annual
CE-TP	SH	−5.4	−3.1	−2.6	−2.3	−3.4
	LH	1.5	0.5	0.4	0.3	0.7
	RC	−8.1	−9.7	−14.4	−12.7	−11.2
	E	−12.0	−12.3	−16.6	−14.7	−13.9
W-TP	SH	−3.0	−6.1	0.2	1.1	−2.0
	LH	−1.4	1.3	−1.6	0.4	0.3
	RC	4.5	−3.6	−7.0	−1.4	−1.8
	E	0.1	−8.4	−8.4	0.1	−4.2

5 Trend in LH and RC over the TP

The trends in LH and RC over the TP are further investigated here. In contrast to the weakening trend of the surface heat source, an evident strengthening trend of LH over the CE-TP exists in MAM and DJF (Fig. 9). Spatial distribution of the trend of LH varies with seasons. A widespread increasing trend can be detected in MAM except at few stations located in the northeast and southeast TP. However, the JJA is featured by a pattern with an increasing trend in the north and south TP but a decreasing trend in the middle. The pattern in SON is somewhat similar to that in MAM but with a larger area of weakening trend in the eastern plateau. Precipitation in DJF is much less than the rest of the year, but it does exhibit a significant growing trend, especially in the midst of the CE-TP. This induces a positive trend in LH. These results are consistent with earlier studies that showed snow depth over the TP increases persistently during the recent decades (Zhang et al., 2004)

Fig. 10 shows the temporal evolution of the domain-averaged net shortwave, longwave radiation fluxes, and their combination term (RC) over the CE-TP and W-TP during 1984—2004. The heating effect from solar radiation and cooling effect from longwave radiation are enhanced simultaneously over the whole TP area. However, the decrease in longwave radiative cooling is stronger, which lead to an intensified RC especially over the CE-TP. The change of radiation effect in atmosphere is closely related to the change of the in situ cloud. Based on the observations from the same stations as used in this study, DW2006 showed that the total cloud amount over the TP area decreases persistently since the mid-1970s. Therefore the increase trends in shortwave heating as well as in longwave cooling as shown in Fig. 10 may be related to the reduced cloud amount over the TP. However, the physical and chemical processes responsible for the change of cloud amount are complicated and beyond the scope of this study.

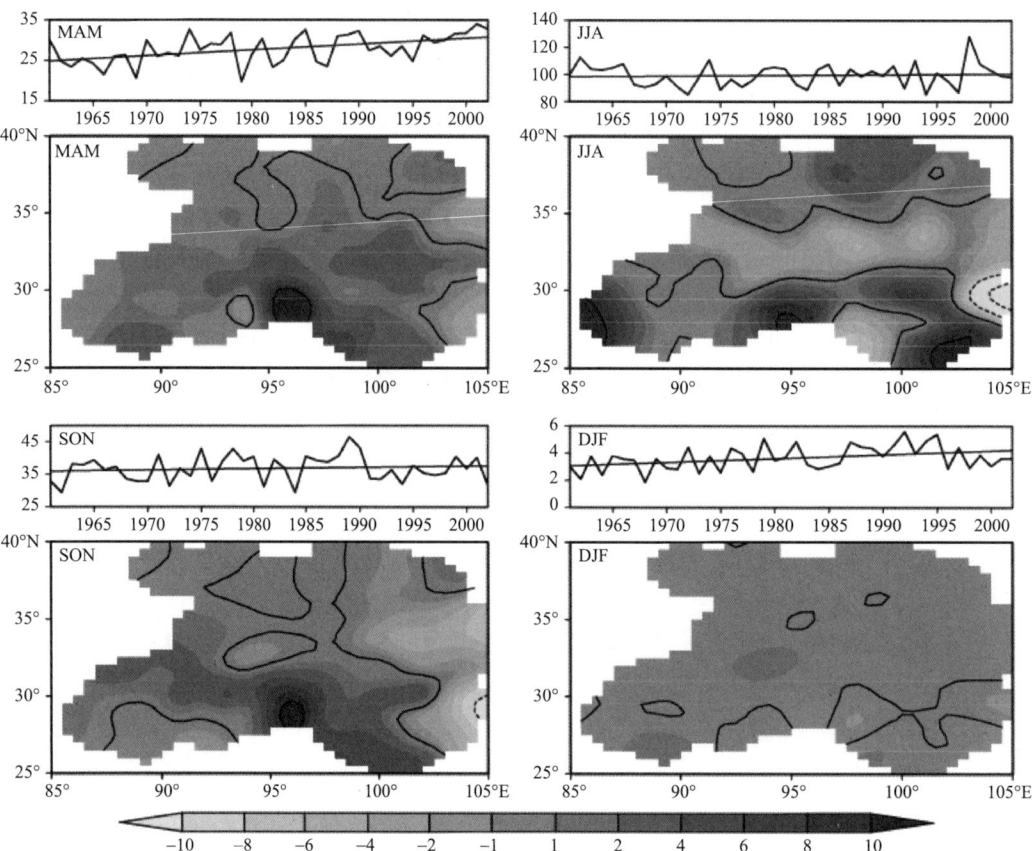

Fig. 9 Temporal evolution (upper panel, in units of Wm^{-2}) and spatial distribution of the LVRs over CE-TP during 1961—2003 (lower panel, in units of $Wm^{-2} decade^{-1}$) for seasonal LH over CE-TP during 1961—2003

The trends of E and its individual components are summarized in Table 3. Due to the difference in data length between the station observations and ISCCP satellite observations and the difference between using station average and domain average, some inconsistency may occur when using these data sets. But the resulted errors should be small due to the huge sample size of the data. Over the CE-TP, both SH and RC exhibit a decreasing trend in all seasons, though the amplitude of the latter is clearly larger than the former. Although the LH presents a reversed increasing trend throughout the whole year, it is too small to compensate the weakening effect induced by the decreasing trend in either SH or RC. The trend in annual-mean SH and LH during 1980—2003 are −3.4 and 0.7 $Wm^{-2} decade^{-1}$, respectively, and the trend in annual-mean RC during 1984—2004 is −11.2 $Wm^{-2} decade^{-1}$. Therefore, the annual-mean atmospheric heat source decreases at a rate of about −13.9 $Wm^{-2} decade^{-1}$ during the last two decades.

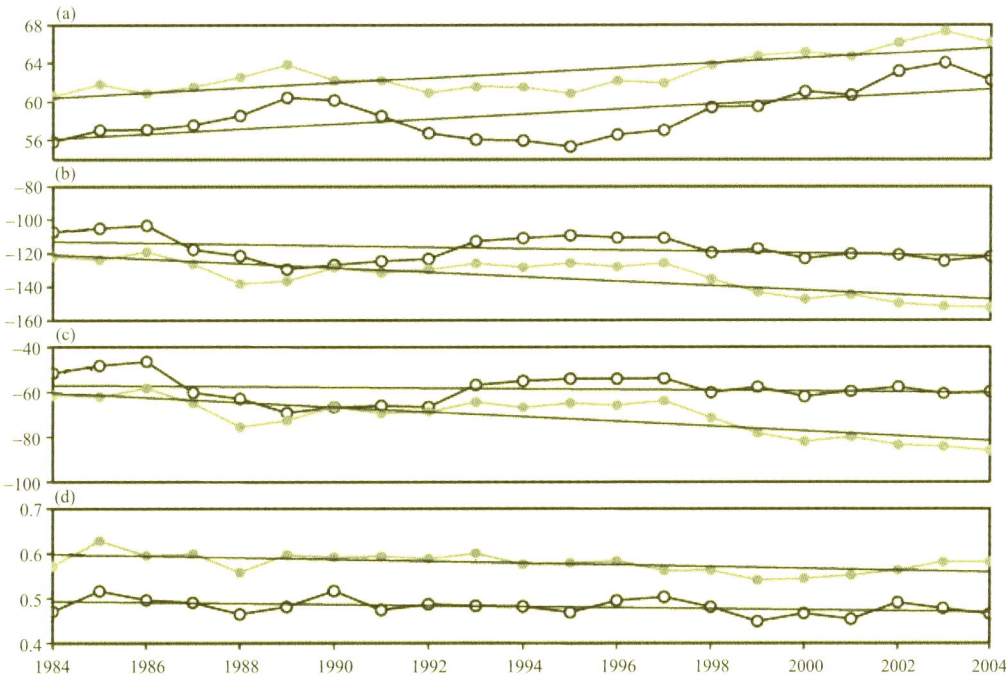

Fig. 10 Temporal evolution of the annual mean (a): net shortwave radiation flux, (b): net longwave radiation flux, (c): net total radiation flux of the air column, and (d): mean cloud fraction over CE-TP (solid circle) and W-TP (open circle) in 1984—2004. The units of radiation flux is Wm^{-2}. Cloud fraction varies from 0 to 1 tenths of sky cover

The air column integrated diabatic heating over the W-TP is also featured by a negative trend, though some differences between the W-TP and the CE-TP exist. For instance, the largest weakening for SH occurs in summer ($-6.1\ Wm^{-2}\ decade^{-1}$) instead of in spring ($-3.0\ Wm^{-2}\ decade^{-1}$), and small positive trends exist in autumn ($0.2\ Wm^{-2}\ decade^{-1}$) and winter ($1.1\ Wm^{-2}\ decade^{-1}$). The trend in LH is characterized by a weakening in spring ($-1.4\ Wm^{-2}\ decade^{-1}$) and autumn ($-1.6\ Wm^{-2}\ decade^{-1}$), but a strengthening in summer ($1.3\ Wm^{-2}\ decade^{-1}$) and winter ($0.4\ Wm^{-2}\ decade^{-1}$). Hence its annual change is negligible ($0.3\ Wm^{-2}\ decade^{-1}$). RC intensifies for much of the year except in spring ($4.5\ Wm^{-2}\ decade^{-1}$), and it contributes to tamper the heat source over the W-TP with an annual trend of $-1.8\ Wm^{-2}\ decade^{-1}$. All these changes generate an annual trend for the entire air column of $-4.2\ Wm^{-2}\ decade^{-1}$, which is particularly significant in summer and autumn ($-8.4\ Wm^{-2}\ decade^{-1}$ in both).

In summary, due to the overwhelming role of the SH before the summer monsoon onset and the much larger decreasing trend in SH and RC than the increasing trend in LH, the atmosphere heat source over the TP losses its strength in all seasons. In addition, the weakening trend in SH is more evident over the CE-TP than over the W-TP, and more evident in spring and summer than in autumn and winter.

6 Relationship between changes in SH over the TP and the general circulation

We have shown that during the past two decades the surface wind speed and sensible heat flux over the TP have weakened. The question is whether the weakening SH over the TP is only a localized phenomenon or it is a result of the global climate change. We hope to address at least some aspects of this issue here.

Surface sensible heating is a shallow one in the atmosphere. Its impact decreases upward exponentially and vanishes by $\sigma=0.8$ ($\sigma=$ pressure/surface pressure). The atmospheric thermal adaptation to the surface sensible heating is characterized by generating cyclonic vorticity in a shallow layer near the surface and anticyclonic vorticity in a deep layer aloft (e. g., Wu and Liu, 2000 and 2004; Yanai and Wu, 2006). Yanai and Li (1994) confirmed that over the TP before the monsoon onset when SH contributes significantly to the total heating, there is a large ($T_s - T_a$) and strong SH during the day, which generates a thin layer of superadiabaitc lapse rates near the surface, a low-level warm low at around ~ 600 hPa, as well as the strong dry convection over the TP. Accompanying the weakening SH, an increased surface pressure and thus a weakened warm low in the surface layers appears above the TP (figure not shown). This indicates that the diminished ($T_s - T_a$) and SH over the TP during the last two decades have contributed to the weakened the surface cyclonic circulation.

The stratosphere is one of the key players in determining the memory of the climate system (Baldwin et al., 2003). Figure11 presents the LVRs of air temperature, geopotential height, and wind speed during 1980—2004, which are averaged from the 12 radiosonde stations in the TP. There exist a clear cooling trend in the upper troposphere and lower stratosphere (UTLS) and a warming trend in the mid and lower troposphere (MLT) over the TP. Cooling in the UTLS exists in February, May, and August, as three centres of above -1.2 ℃ decade^{-1}, while warming in the MLT is much weaker with two centres of about 0.3 ℃ decade^{-1} in March and May. Similar results can be seen in Fig. 2 of Zhou and Zhang (2005). They found that ozone depletion in the UTLS is particularly obvious in winter and spring over the TP, which induces less absorbing solar radiation in the UTLS but more in the MLT and caused cooler UTLS and warmer MLT.

As a result of the temperature change, the geopenital height displays a decreasing trend in the UTLS and an increasing trend in the MLT (Fig. 11b). Consistent with the seasonally-dependent change of temperature, the largest geopotential height increase occurs in late winter and early spring with the average LVR of more than 1 dagpm decade^{-1}.

The change of wind speed seems more complicated (Fig. 11c). Although for the most part of the year (except for May and November) there exist a weakening trend in the troposphere and an intensifying trend in the stratosphere. The largest weakening trend (-1.2 ms^{-1} decade^{-1}) in tropospheric wind speed occurs during summer season (June to September) rather than in spring. However, a peculiar strengthening trend happens in May and November (0.3 to 1.2 ms^{-1} decade^{-1}). These two month are normally transition seasons between winter and summer, when the subtropical westerly jets retreats northward or marches southward across the TP. The trend then indicates an increasing westerly over the TP during the transition seasons, which might influence the climate downstream.

The changes in temperature and geopotential height (Figs. 11a and 11b) show that the variations in the UTLS and the MLT have opposite signs. In spring, the altitudes of the positive trends in temperature and geopotential height even reach the tropopause. This certainly cannot be explained as a local response to the decreasing trend in SH over the TP since the latter would generate a positive geopotential height trend and a negative temperature trend only near the surface according to the thermal adaptation theory. On the other hand, the weakening and strengthening trends in the wind speed over the TP occur throughout the whole troposphere. They are related to the changes in advection, Coriolis force, and pressure gradient terms in the momentum equation. We may infer that the weakening in surface wind speed and SH over the TP during the last two decades is associated with the global climate change. The confirmation of this inference, however, requires further investigation.

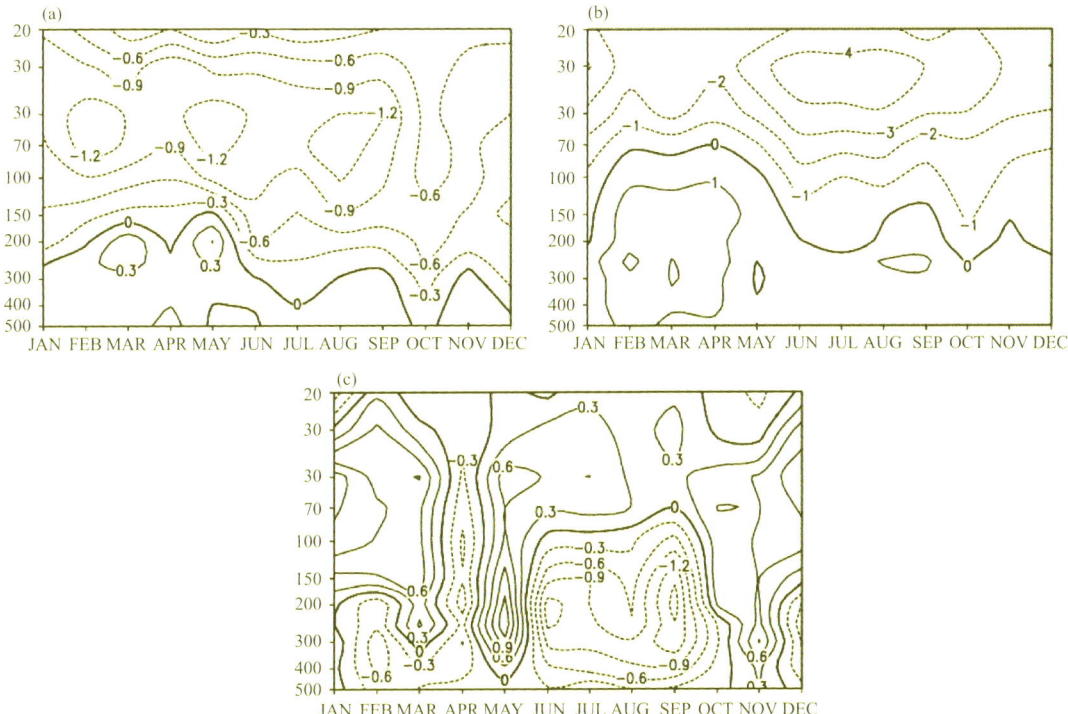

Fig. 11 Time-altitude sections of the trends for (a) air temperature (in unit of ℃ decade^{-1}), (b) geopotential height (in unit of dagpm decade^{-1}), and (c) wind speed (in unit of ms^{-1} decade^{-1}) of 12 radiosonde stations means over the TP during the period from 1980—2003

7 Summary and discussions

Using historical records, we have demonstrated that both SH and RC over the plateau underwent pronounced changes when a striking climate warming occurred there during the last two decades. The main findings are summarized as follows:

1) A statistically-significant decreasing trend in SH during the last two decades has been detected in the CE-TP. The linear tendency of 71-station-averaged SH during 1980—2003 is -3.8 Wm^{-2} decade^{-1} with a relatively decreasing rate of -14%. The largest decreasing trend occurs at local noon and in spring when SH reaches the daily and annual maxima. It in turn leads to the reduction in both the diurnal and annual ranges of SH.

2) Meanwhile, a relatively weaker increasing trend in LH (0.7 Wm^{-2} decade^{-1}) during 1980—2003 and a stronger decreasing trend in RC (-11.2 Wm^{-2} decade^{-1}) during 1984—2003 are also detected. The combination of these changes lead to a weakened atmospheric heating in spring and summer and an enhanced cooling in autumn and winter over the CE-TP.

3) The annual mean atmospheric heat source over the W-TP also declines (-4.2 Wm^{-2} decade^{-1}) mainly due to the weakened SH in spring and summer and the enhanced RC in summer and autumn.

4) The decreasing trend in SH over the TP is induced mainly by the reduced surface wind speed, and also influenced to a certain degree by the diminished ground-air temperature difference. Furthermore, the reduced surface wind speed happens when there are changes of temperature, geopotential

height and wind speed in the troposphere as well as in the lower stratosphere. All these changes cannot be explained as a local response to the decreasing trend in SH. Most likely, they are linked to the global climate changes.

Some key issues still remain unresolved. For example, why dose the T_s have opposite trends during 1960—2003 and 1980—2003 while T_a remains in an increasing trend? How can we distinguish the contribution by the change in local greenhouse gases from that by global atmospheric circulation shift due to the enhanced greenhouse effect globally? More importantly, the surface sensible heating over the TP can generate a strong air-pump and regulate the Asian monsoon. Recent studies have reported that the Asian monsoon circulation has been weakened during the past decades (Wang et al., 2004; Xu et al., 2006). To what extent the weakening of the SH over the TP may affect the suppressed Asian monsoon remains unknown. To answer these questions, in particular to understand the influence of the changes in heat source over the TP on the Asian summer monsoon at decadal time scales, a numerical modeling study is imperative. The numerical results will be presented in a separated paper.

References

BALDWIN M P, THOMPSON D W J, SHUCKBURGH E F, et al, 2003. Weather from the stratosphere[J]. Science, 301(5631): 317-319.

BROCCOLI A J, MANABE S, 1992. The effects of orography on midlatitude Northern Hemisphere dry climates[J]. Journal of Climate, 5(11): 1181-1201.

CHEN B, CHAO W C, LIU X, 2003. Enhanced climatic warming in the Tibetan Plateau due to doubling CO_2: A model study[J]. Climate Dynamics, 20(4): 401-413.

CHEN L, REITER E R, FENG Z, 1985. The atmospheric heat source over the Tibetan Plateau: May—August 1979[J]. Monthly Weather Review, 113(10): 1771-1790.

DUAN A, LIU Y, WU G, 2005a. Heating status of the Tibetan Plateau from April to June and rainfall and atmospheric circulation anomaly over East Asia in midsummer[J]. Science in China Series D: Earth Sciences, 48(2): 250-257.

DUAN A, WU G, 2005b. Role of the Tibetan Plateau thermal forcing in the summer climate patterns over subtropical Asia [J]. Climate Dynamics, 24(7): 793-807.

DUAN A, WU G, ZHANG Q, et al, 2006a. New proofs of the recent climate warming over the Tibetan Plateau as a result of the increasing greenhouse gases emissions[J]. Chinese Science Bulletin, 51(11): 1396-1400.

DUAN A, WU G, 2006b. Change of cloud amount and the climate warming on the Tibetan Plateau[J]. Geophysical Research Letters, 33: L22704.

FLOHN H, 1957. Large-scale aspects of the "summer monsoon" in South and East Asia[J]. J Meteor Soc Japan, 35A: 180-186.

FLOHN H, 1960. Recent investigation of the mechanism of the "Summer monsoon" of Southern and eastern Asia[C]. Proc Symp Monsoon of the World. New Delhi: Hind vnion press: 75-88.

HAHN D G, MANABE S, 1975. The role of mountains in the South Asian monsoon circulation[J]. Journal of the Atmospheric Sciences, 32(8): 1515-1541.

KITOH A, 2004. Effects of mountain uplift on East Asian summer climate investigated by a coupled atmosphere-ocean GCM[J]. Journal of Climate, 17(4): 783-802.

LI C, YANAI M, 1996. The onset and interannual variability of the Asian summer monsoon in relation to land-sea thermal contrast[J]. Journal of Climate, 9(2): 358-375.

LI G, DUAN T, GONG Y, 2000. The bulk transfer coefficients and surface fluxes on the western Tibetan Plateau[J]. Chinese Science Bulletin, 45(13): 1221-1226.

LI W, WU G, LIU Y, et al, 2001. How the surface processes over the Tibetan Plateau affect the summertime Tibetan Anticyclone-numerical experiments[J]. Chinese Journal of Atmospheric Sciences Chinese Edition, 25(6): 809-816.

LIU Y M, 2007. Review and new insights in the sensible heat driving air-pump over the Tibetan Plateau and its influence on the climate in Asia[J]. Hydro, 8: 770-789.

LIU X, CHEN B, 2000. Climatic warming in the Tibetan Plateau during recent decades[J]. International Journal of Climatology: A Journal of the Royal Meteorological Society, 20(14): 1729-1742.

LUO H, YANAI M, 1983. The large-scale circulation and heat sources over the Tibetan Plateau and surrounding areas during the early summer of 1979. Part II: Heat and moisture budgets[J]. Monthly Weather Review, 112(5): 966-989.

MANABE S, BROCCOLI A J, 1990. Mountains and Arid climates of middle latitudes[J]. Science, 247(4939): 192-195.

NIU T, CHEN L, ZHOU Z, 2004. The characteristics of climate change over the Tibetan Plateau in the last 40 years and the detection of climatic jumps[J]. Advances in Atmospheric Sciences, 21(2): 193-203.

ROSSOW W B, SCHIFFER R A, 1991. ISCCP cloud data products[J]. Bulletin of the American Meteorological Society, 72(1): 2-20.

ROSSOW W B, SCHIFFER R A, 1999. Advances in understanding clouds from ISCCP[J]. Bulletin of the American Meteorological Society, 80(11): 2261-2288.

TAO S Y, DING Y H, 1981. Observational evidence of the influence of the Qinghai-Xizang (Tibet) Plateau on the occurrence of heavy rain and severe convective storms in China[J]. Bulletin of the American Meteorological Society, 62(1): 23-30.

WANG Z Y, DING Y H, HE J H, 2004. An updating analysis of the climate changes in China in recent 50 years [J]. Acta Meteor Sin, 62: 228-236.

WANG Y, REN G, 2005. Change in free atmospheric temperature over China during 1961—2004[J]. Climate and Environment Research, 10(4S): 780-790.

WU G, 1997. Sensible heat driven air-pump over the Tibetan Plateau and its impacts on the Asian summer monsoon[C]// YE D Z. Essays in Honor Zhao Jiuzhang. Beijing: Science Press: 116-126.

WU G, ZHANG Y, 1998. Tibetan Plateau forcing and the timing of the monsoon onset over South Asia and the South China Sea[J]. Monthly Weather Review, 126(4): 913-927.

WU G X, ZHOU M Y, XU X D, et al, 2007. Review and new insights in the sensible heat driving air-pump over the Tibetan Plateau and its influence on the climate in Asia[J]. Hydro, 8: 770-789.

XU M, CHANG C P, FU C, et al, 2006. Steady decline of East Asian monsoon winds, 1969—2000: Evidence from direct ground measurements of wind speed[J]. Journal of Geophysical Research: Atmospheres, 111(D24).

YANAI M, 1961. A detailed analysis of typhoon formation[J]. Journal of the Meteorological Society of Japan, 39(4): 187-214.

YANAI M, LI C, SONG Z, 1992. Seasonal heating of the Tibetan Plateau and its effects on the evolution of the Asian summer monsoon[J]. Journal of the Meteorological Society of Japan, 70(1B): 319-351.

YANAI M, LI C, 1994. Mechanism of heating and the boundary layer over the Tibetan Plateau[J]. Monthly Weather Review, 122(2): 305-323.

YANAI M, WU G X, 2006. Role of the Tibetan Plateau on Asia monsoon[J]. The Asian Monsoon: 513-629.

YE D Z, WU G X, 1998. The role of the heat source of the Tibetan Plateau in the general circulation[J]. Meteorology and Atmospheric Physics, 67(1): 181-198.

YEH T C, LO S W, CHU P C, 1957. The wind structure and heat balance in the lower troposphere over Tibetan Plateau and its surroundings[J]. Acta Meteorological Sinica, 28: 108-121.

YEH T, GAO Y, 1979. Meteorology of the Qinghai-Xizang (Tibet) Plateau[M]. Beijing: Science Press: 278.

YU R, WANG B, ZHOU T, 2004. Tropospheric cooling and summer monsoon weakening trend over East Asia[J]. Geophysical Research Letters, 31(22):212.

ZHANG Y, LI T, WANG B, 2004. Decadal change of the spring snow depth over the Tibetan Plateau: The associated circulation and influence on the East Asian summer monsoon[J]. Journal of Climate, 17(14): 2780-2793.

ZHAO P, 1999. The heating status over the Tibetan Plateau and its relationship with the air-sea interaction[D]. Beijing: Chinese Academy of Meteorological Sciences: 229.

ZHAO P, CHEN L, 2001. Climatic features of atmospheric heat source/sink over the Qinghai-Xizang Plateau in 35 years and its relation to rainfall in China[J]. Science in China Series D: Earth Sciences, 44(9): 858-864.

ZHOU M, XU X D, BIAN L G, et al, 2000. Observation in the atmospheric boundary layer and investigation of dynamic meteorology over the Tibetan Plateau[M]. Beijing: China Meteorological Press: 125.

ZHOU S, ZHANG R, 2005. Decadal variations of temperature and geopotential height over the Tibetan Plateau and their relations with Tibet ozone depletion[J]. Geophysical Research Letters, 32(18): 109-127.

ZHU W, CHEN L, ZHOU Z, 2001. Several characteristics of contemporary climate change in the Tibetan Plateau[J]. Science in China Series D, 44 (supp): 410-420.

The Role of Air-Sea Interactions in Regulating the Thermal Effect of the Tibetan-Iranian Plateau on the Asian Summer Monsoon

HE Bian[1,2,3], LIU Yimin[1,3], WANG Ziqian[1,4], BAO Qing[1]

(1 State Key Laboratory of Numerical Modeling for Atmospheric Sciences and Geophysical Fluid Dynamics (LASG), Institute of Atmospheric Physics (IAP), Chinese Academy of Sciences, Beijing 100029, China; 2 Key Laboratory of Meteorological Disaster of Ministry of Education, Nanjing University of Information Science and Technology, Nanjing 210044, China; 3 University of Chinese Academy of Sciences, Beijing 100029, China; 4 School of Atmospheric Sciences, and Guangdong Province Key Laboratory for Climate Change and Natural Disaster Studies, Sun Yat-sen University, Guangzhou 510275, China)

Abstract: The thermal effect of the Tibetan-Iranian Plateau (TIP) on the Asian summer monsoon and the role of air-sea interactions over the Indian Ocean in regulating the effects of the TIP are explored. The results demonstrate that the direct thermal effect of the TIP produces a lower troposphere cyclonic circulation in the area surrounding the TIP and increases the continental precipitation over South and East Asia. It also decreases the precipitation over the tropical Indian Ocean and increases the sea-surface temperature (SST) of the tropical Indian Ocean with a large gradient zone located along 10°N but decreases SST of the western coast of Indonesia. In the lower troposphere, the air-sea interaction induced by the TIP thermal forcing produces an anticyclonic circulation surrounding the TIP and a stronger westerly flow to the south of the anticyclone. A circulation dipole thus forms to the south of the TIP. Together with this horizontal dipole, a meridional circulation dipole is generated to the south of the TIP, which is characterized by strong air ascent from 10 to 15°N where a strong westerly flow occurs, and the descent of air over the southern slope of the TIP and south Hemisphere. These results demonstrate that the indirect effect of the air-sea interaction over the Indian Ocean induced by the TIP thermal forcing is to counteract its direct effect on the Asian summer monsoon. The uncertainty of this indirect effect is also discussed.

Keywords: air-sea interactions, Asian summer monsoon, Tibetan-Iranian Plateau, topographical heating

1 Introduction

The Tibetan Plateau (TP), the largest highland in the world, is located in the central and eastern parts of the Eurasian continent and has a strong effect on the climate of Asia (Manabe and Terpstra, 1974; Tao and Ding, 1981; Huang, 1985; Zhao and Chen, 2001; Wu et al., 2007, 2012a; Wang et al., 2008; Zhou et al., 2009; Liu et al., 2012; Boos and Kuang, 2010, 2013). During the boreal winter, atmospheric circulations are mainly modulated by the mechanical forcing of the TP (Queney, 1948; Bolin, 1950; Yeh, 1950), which can excite gravity waves and Rossby waves and is important in the development of stationary waves during the northern winter (Charney and Eliassen, 1949; Held et al., 2002). During the boreal summer, the TP has been found to be a heat source (Flohn, 1957; Yeh et al., 1957) closely linked to the onset, formation and evolution of the Asian summer monsoon (ASM) (Wu and Zhang, 1998; Hsu and Liu, 2003; Duan and Wu, 2005; Liu et al., 2007; Wu et al., 2012b). Based on station records, Ye and Gao (1979) found that the increase in atmospheric heating over the TP during the boreal summer is dominated by vertical diffusive heating from the surface of the TP. Subsequently, Wu et al. (1997, 2007) suggested that the sensible heating on the slope surface of the TP is the major driving force in the transport of abundant water vapor from the ocean surface to the land, leading to monsoonal precipitation over the Asian continent, known as the sensible heat-driven air pump (SHAP).

In recent decades, the effects of the TP on atmospheric circulations have been investigated through simulation with atmospheric general circulation models (AGCMs) and coupled CGCMs (CGCMs). By comparing a simulation with mountains to one without mountains in an AGCM with a prescribed sea-surface temperature (SST), Hahn and Manabe (1975) found that the presence of the TP is instrumental in maintaining the South Asian low-pressure system, leading to high temperatures in the middle and upper troposphere over the TP, causing the monsoon climate to extend farther north, with similar results found using different models (Xu et al., 2009, 2010). More detailed topographic experiments have been carried out by Kitoh et al. (Kitoh, 2004; Kitoh et al., 2010), who assessed the effects of different mountain heights in a CGCM and found that the ASM precipitation moves gradually inland when forced by progressive mountain uplift. In this scenario, the Pacific subtropical anticyclone and the associated trade winds also become stronger. These studies emphasized the importance of the high elevation of the TP on maintaining the ASM. Notably, Wu et al. (2007) documented that the thermal forcing of the TP is more important than the mechanical forcing in maintaining the ASM. Through a series of sensitivity experiments involving changes in topographic and thermal conditions, they determined that sensible heating on sloping lateral surfaces appears to be the major source of the forcing. The warm ascending air that develops over the TP in the summer pulls air from below, and the lower tropospheric air from the surrounding areas converges on the TP region before climbing the heated sloping surfaces. Moreover, Wu et al. (2012b) showed that, in addition to the thermal effect of the TP, the thermal effect of the Iranian Plateau (IP) is also important in maintaining the ASM. Sensible heating of the IP causes a cyclonic anomaly in the lower troposphere and transports a large amount of water vapor from the Arabian Sea (AS) to northern India, which favors intense monsoonal precipitation. Therefore, the thermal effects of the TP and the IP should be considered as an integral part of the large-scale topographic forcing affecting the dynamics of the Asian monsoon (Wu et al., 2012b).

Despite these efforts, our understanding of the thermal effect of the TP on the ASM remains insufficient, since most of the sensitivity experiments used a prescribed SST, and the Tibetan-Iranian Plateau (TIP) heating effect on the SST was not considered. Meanwhile, using CGCMs, several studies (Kitoh et al., 2004, 2010; Okajima and Xie, 2007; Koseki et al., 2008) have shown that the TP significantly influences the SST and the associated changes in the Asian climate. Abe et al. (2003) revealed that the SST in the equatorial Indian Ocean increases with mountain uplift, resulting in increase of local precipitation. This increase is caused by the ocean surface dynamics due to the enhanced monsoonal circulation. Kitoh (2004) demonstrated that mountain uplift results in an increased SST within the western tropical Pacific and the Maritime Continent, and a decreased SST within the western Indian Ocean and the central subtropical Pacific in the summertime. Abe et al. (2013) examined the effects of the presence of the TP on the onset of the ASM using both CGCMs and AGCMs, and found that the onset of the South Asian summer monsoon in the presence of the TP is strongly related to the air-sea interactions during the pre-monsoon season. The lower SST seen in the AS in the presence of the TP is related to the later onset of summer monsoonal precipitation in South Asia, while the heat transport in the mixed layer is responsible for changes in the SST in the absence of the TP. Using a regional model, Wang et al. (2018) found TP heating could cool the SST over northern Indian Ocean by enhancing south westly winds. Despite these studies, the responses of the SST and the associated changes in the mixed layer to the uplifted thermal effects of the TP remain unclear. Therefore, it is important to analyze changes in the mixed layer to investigate the impacts of the thermal forcing associated with the TIP on the subsurface Indian Ocean. To address these issues, it is necessary to carry out thermal sensitivity experiments with CGCMs.

The thermal effects of the TIP on the formation of the ASM are re-evaluated here, and the associated air-sea interactions are analyzed through comparison of sensitivity experiments performed with a CGCM and an AGCM. Section 2 introduces the datasets and models, the experiment design, and the model evaluation. Section 3 presents the direct influence of the TIP heating on the ASM based on the CGCM simulations. Section 4 presents the direct influence of the thermal effect of the TIP on the circulation and thermal status of the Indian Ocean. Section 5 discusses the role of air-sea interactions in modulating the thermal effect of the TIP on the ASM, which is the so-called indirect influence of the thermal effect of the TIP on the ASM. The uncertainty of such an indirect influence is evaluated in Section 6. Finally, the conclusion and discussion are presented in Section 7.

2 Datasets, model, experiment design, and model evaluation

2.1 Datasets

The precipitation data were derived from the Global Precipitation Climatology Project (GPCP) monthly mean dataset (Adler et al., 2003), which is constructed on a 2.5°×2.5° grid over the globe and covers the period from 1979 to the present. This dataset is available at http://www.esrl.noaa.gov/psd/data/gridded/data.gpcp.html.

The multilevel air temperature, specific humidity, and wind field from the European Centre for Medium-range Weather Forecasts (ECMWF) ERA-Interim (Dee et al., 2011) were used. The dataset includes a 1.5°×1.5° grid resolution and covers the period from 1979 to the present and is available at http://apps.ecmwf.int/datasets/.

2.2 Model

The Flexible Global Ocean-Atmosphere-Land System Model spectral version 2 (FGOALS-s2) was used here as the climate system model, and was composed of four individual components: version 2 of the spectral atmospheric model (SAMIL2) developed at the State Key Laboratory of Numerical Modeling for Atmospheric Sciences and Geophysical Fluid Dynamics, Institute of Atmospheric Physics (LASG/IAP) (Wu et al., 1996; Bao et al., 2010), version 2 of the LASG/IAP Climate system Ocean Model

(LICOM2) (Liu et al., 2013), version 3 of the Community Land Model (CLM3) (Oleson et al., 2004), and version 5 of the Community Sea Ice Model (CSIM5) (Briegleb et al., 2004). The fluxes were exchanged between these components using version 6 of the coupler module from the National Center for Atmospheric Research (Collins et al., 2006). The basic performance of FGOALS-s2 is described in Bao et al. (2013).

2.3 Experiment design

To investigate the direct and indirect influence of the thermal effect of the TIP on the ASM and the associated air-sea interactions, we conducted a series of experiments as summarized in Table 1. The first experiment (CON) was performed using FGOALS-s2. The CGCM model was integrated over 90 years, with results for the last 50 years analyzed. The second experiment (TIPNS) was the same as the CON experiment, but the TIP surface was not allowed to heat the atmosphere. That is, in integrating the atmospheric thermodynamic equation, the vertical diffusion heating at the surface was set to zero during the summer months (June—July—August) for grid points within the region extending from 30° to 110°E and from 10° to 45°N with elevations above 500 m. This method is the same as that used in Wu et al. (2012b) and He et al. (2015). In addition, two AGCM experiments (CON_OBSST and TIPNS_OBSST) were carried out using the SAMIL2 component, which used the same setting as the CON and TIPNS runs, except that the SST and sea ice were forced by the observed climatological monthly mean. The AGCM was integrated over 30 years, and the mean values for the last 20 years were analyzed.

Table 1 Experiment design

Name	Description
CON	Climate Model Intercomparison Project experiment using the external forcings prescribed as their climatological values. The model is integrated for 90 years, and the mean values from the last 50 years are analyzed.
TIPNS	The same as for the CON experiment, but the surface sensible heating is not allowed to heat the atmosphere during June—July—August over the domain (30°—110°E/10°—45°N) where the TIP topography is above 500 m. The change of total sensible heat flux is 178 TW.
CON_OBSST	Atmospheric Model Intercomparison Project run using the observed climatological SST and sea ice at the lower boundary. The other forcing fields are prescribed as their climatological values. The model is integrated for 30 years, and the mean values from the last 20 years are analyzed.
TIPNS_OBSST	As for the CON_OBSST experiment, but the surface sensible heating is not allowed to heat the atmosphere during June—July—August over the domain (30°—110°E/10°—45°N), where the TIP topography is above 500 m. The change of the total sensible heat flux is 172 TW.
CON_CPSST	Same as the CON_OBSST experiment, but the SST is prescribed by outputs from the CON experiment.
TIPNS_CPSST	Same as the TIPNS_OBSST experiment, but the SST is prescribed by outputs from the CON experiment.
CON_NSSST	Same as the CON_OBSST experiment, but the SST is prescribed by outputs from the TIPNS experiment.
TIPNS_NSSST	Same as the TIPNS_OBSST experiment, but the SST is prescribed by outputs from the TIPNS experiment.

The direct influence of the TIP thermal forcing was estimated by calculating the differences between the CON and TIPNS experiments and between the CON_OBSST and TIPNS_OBSST experiments. We calculated the changes in the total sensible heat flux for the AGCM and CGCM using the method described in Ma et al. (2014), with the resulting values of 172 and 178 TW, respectively (Table 1). Student's t-test was also applied to assess the statistical significance of the model results.

To investigate the indirect influence of the TIP thermal forcing, two additional pairs of AGCM experiments CON_CPSST, TIPNS_CPSST, CON_NSSST, and TIPNS_NSSST were carried out (Table 1). In the CON_CPSST and TIPNS_CPSST experiments, the SST outputs from the CON experiments were archived and employed to replace the observed SST in CON_OBSST and TIPNS_OBSST as the lower boundary condition. Similarity, in CON_NSSST and TIPNS_NSSST, the SST outputs from TIPNS were prescribed. In other words, the AGCM experiments are driven by the SST generated from the corresponding CGCM, and not by the observed SST. Thus, the indirect effect of TIP thermal forcing was evaluated by comparing the difference between a pair of CGCM experiments with and without the surface sensible heating over the TIP and a pair of AGCM experiments with the same experimental settings, which can be expressed as

$$DD\_CPSST = (CON - TIPNS) - (CON\_CPSST - TIPNS\_CPSST) \quad (1)$$

or

$$DD\_NSSST = (CON - TIPNS) - (CON\_NSSST - TIPNS\_NSSST) \quad (2)$$

It is noteworthy that this method is a little different to the methods presented in previous literature (Kitoh, 2004; Okajima and Xie, 2007), which used the observed SST to force the AGCM model to estimate the indirect effect of the mountain. The different approach was taken because the indirect effect can be largely isolated from the influence of the SST difference between the CGCM and the observations by using the SST from the CGCM. However, it is widely known that CGCMs suffer from SST simulation bias. To evaluate the influence of this kind of bias on the indirect effect, we also estimated the indirect effect of TIP thermal forcing by using the traditional method:

$$DD\_OBS = (CON - TIPNS) - (CON\_OBSST - TIPNS\_OBSST) \quad (3)$$

The common features derived from these results are confirmed as the TIP indirect effect on the ASM based on the modeling efforts, and their differences and associated SST biases are also discussed in detail in Section 6.

2.4 Model evaluation

The performance of the model in capturing the basic pattern of the ASM was evaluated first, with the simulated precipitation and lower tropospheric wind speed produced by the CON_OBSST run shown in Fig. 1b. Compared with the observational data (Fig. 1a), the CON_OBSST run captures the main features of the ASM precipitation pattern, with more than 4 mm day^{-1} of rainfall covering the area over the Indian Ocean, the Asian continent, and the western Pacific Ocean, while also simulating the maximum precipitation center over the eastern AS, the Bay of Bengal (BOB), and the southern slope of the TP. The Somali Jet, the monsoonal trough over the BOB, and the anticyclone over the western Pacific are also well simulated. However, the model underestimates the precipitation over northern New Guinea and overestimates the precipitation over the AS, the BOB, and the southern slope of the TP. Moreover, the simulated rain belt over the western Pacific is located to the north of that seen in the observational data. The overall monsoonal features simulated in the CON experiment (Fig. 1c) are consistent with those simulated in the CON_OBSST run. There were a series of improvements from the use of the CGCM. First, the simulated precipitation pattern over the western Pacific in the CON run is more realistic and closer to the observations. Moreover, the centers of maximum precipitation in the AS, the BOB, and the southern slope of the TP are weaker in the CON run than in the CON_OBSST run and are also more consistent with the observations. The above evaluation demonstrates that FGOALS-s2 and its atmospheric component SAMIL2 are capable of capturing the major characteristics of the ASM and can be used for the following sensitivity experiments.

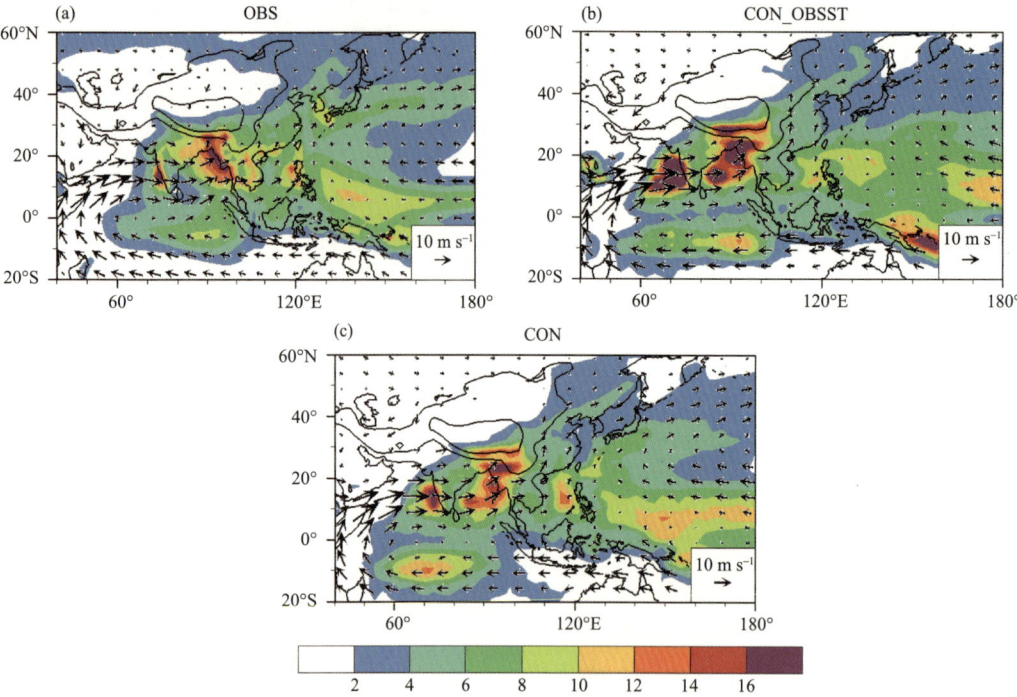

Fig. 1 Climate mean June—July—August precipitation (shaded, mm day^{-1}) and 850 hPa wind speed (vector, m s^{-1}) from (a) reanalysis data (1979—2008), with precipitation from the GPCP dataset and wind speed from ERA-Interim Reanalysis, and from the numerical experiments (b) CON_OBSST and (c) CON. The black contours denote elevations of 1000 and 3000 m, respectively

3 Direct influence of the thermal effect of the Tibetan-Iranian Plateau on the Asian summer monsoon

We first investigate the response of the ASM to topographic heating in the CGCM. For this purpose, the surface sensible heating over the TIP is removed and the result (TIPNS) is shown in Fig. 2a, and then compared with its counterparts in the control experiment. Compared with Fig. 1c, Fig. 2a shows a similar pattern of the ASM over the ocean even if the thermal effect of the TIP is removed, which is consistent with previous studies (Hahn and Manabe, 1975; Xu et al., 2009; Wu et al., 2012b) and shows that the large-scale, land-sea thermal contrast is of fundamental importance for the formation of the tropical ASM. However, in the sensitivity experiment, the monsoonal rain belt barely extends over the Asian continent, and the East Asian monsoon is significantly weakened.

To examine the direct thermal impacts of the TIP on changes in the ASM in the lower troposphere, we show the precipitation and 850 hPa wind differences between the CON and TIPNS runs (Fig. 2b). The thermal effect of the TIP causes increased precipitation over the southern slope of the TP and the East Asian continent, and decreased precipitation over central Asia, and almost the entire Indian Ocean. The increase in monsoonal precipitation over the Asian continent is also accompanied by cyclonic circulation around the TIP in the lower troposphere. The direct effects on precipitation, i.e. changes in precipitation are similar over the Asian continent, but different over the Indian Ocean and the western Pacific between the CGCM (Fig. 2b) and AGCM results (Fig. S2c in OSM), implying that air-sea interaction may play significant roles in modulating the TIP impacts on the oceanic ASM.

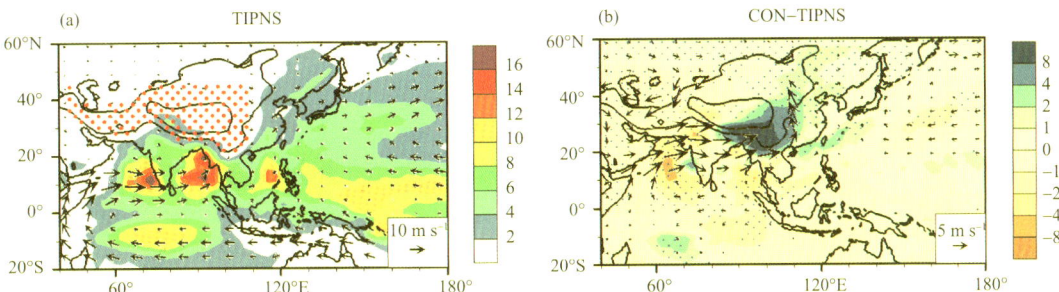

Fig. 2 June—July—August mean precipitation (shaded, mm day^{-1}) and 850 hPa wind speed (vector, m s^{-1}) produced by the numerical experiments (a) TIPNS, and the differences (b) between the CON and TIPNS experiments. The red dots in (a) denote the regions where the sensible heating was modified. The vectors and blue dots in (b) denote the winds and precipitation difference passing the 95% significance in the t-test respectively. The black contours denote elevations of 1000 and 3000 m, respectively

The vertical structure of the ASM also varies in response to the thermal effect of the TIP. As many studies have demonstrated (Schneider and Lindzen, 1977; Schneider, 1977, 1987; Held and Hou, 1980), the atmospheric meridional circulation adopts two distinct regimes in response to axisymmetric diabatic heating, specifically the thermal equilibrium regime in the extra-tropics and the angular momentum conservation regime in the tropics. Under the regime of angular momentum conservation in the ASM region, a strong ascending motion is generated by the easterly vertical wind shear (Wu et al., 2015). In the tropics, the horizontal temperature advection is weak, and the diabatic heating is mainly balanced by the adiabatic ascent and cooling. Figure 3 presents the June—July—August (JJA) mean meridional circulation zonal averaged over the longitudinal range of 75°—110°E in different experiments and the differences between them. As shown in Fig. 3a, the vertical velocity component shows strong ascending motion over the South Asian Ocean region and the southern slope of the TIP in the CON run. The ridge line of the subtropical anticyclone is located above the southern slope of the TIP. When the sensible heating of the TIP is removed, the major ascending branch over the southern slope of the TIP disappears, and the South Asian high ridge line shifts southward to nearly 20°N in the TIPNS (Fig. 3b) in the troposphere. However, the vertical wind speeds over the Indian Ocean in the TIPNS run (Fig. 3b) are stronger than those in the CON run (Fig. 3a). Their difference (CON-TIPNS) demonstrates that the thermal effect of the TIP generates a monsoonal-type meridional circulation, with an ascending branch located over the southern slope of the TIP and a descending one over the tropical Indian Ocean (Fig. 3c). In summary, the ascending branch is responsible for the northern branch of the southern ASM.

4 Direct influence of the thermal effect of the TIP on the physical processes in the mixed layer of the Indian Ocean

4.1 Sea-surface temperature, mixed-layer temperature, and ocean current

As changes in SST are crucial to the surface diffusive heat flux and moisture flux from the ocean into the atmosphere, such temperature changes can significantly influence the ASM (Yeh et al., 1957; Duan and Wu, 2005). We show the changes in the skin-surface temperatures over land and SST over ocean due to the thermal effect of the TIP in Fig. 4. The differences in skin-surface temperatures between the CON and TIPNS runs over the Asian continent show a decrease over the Indian subcontinent

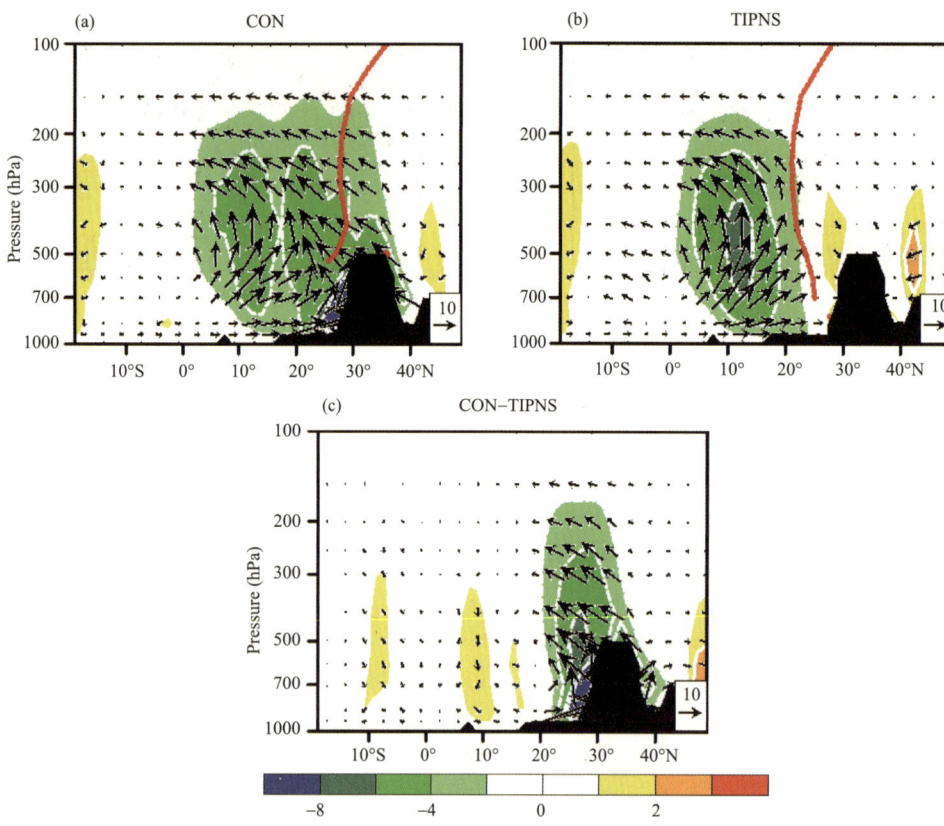

Fig. 3 Zonal averaged (75°—110°E) June—July—August mean vertical velocity (shaded, 10^{-2} Pa s^{-1}) and the meridional circulation (vector, v in m s^{-1}, -3ω in 10^{-2} Pa s^{-1}) from the (a) CON, (b) TIPNS, and (c) CON−TIPNS experiments. The red line denotes the subtropical ridge line that satisfies $u = 0$ and $\partial u/\partial y > 0$. The black shading indicates the topography

and central-eastern China, as well as an increase over the whole TIP region. In the ocean domain, the SST decreases over the AS, the BOB, the eastern equatorial Indian Ocean, and the subtropical western Pacific, and increases over the tropical Indian Ocean and the mid-latitudes of the northwestern Pacific. These decreases resemble those noted by Abe et al. (2013), who found that the SST decreased in the AS when the TP was removed in a CGCM and attributed these differences to the changes in heat flux driven by anomalous ocean currents.

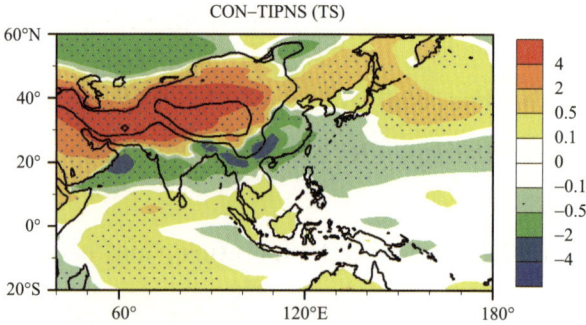

Fig. 4 June—July—August mean skin-surface temperature differences (K) between the CON and TIPNS experiments. The blue dots denote the values that pass a t-test with a significance level of 95%. The black contours indicate elevations of 1000 and 3000 m, respectively

In general, the depth of the mixed layer in the Indian Ocean is approximately 40—50 m during the boreal summer, where the mixed layer in the ocean is often defined as the layer between the sea surface and the lowest depth with a vertical temperature difference less than 0.5 ℃ (Montégut et al., 2004; Lin et al., 2011; He and Wu, 2013). To obtain the three-dimensional features of ocean temperature and current, in Fig. 5 we show the spatial characteristics of the ocean temperature and current differences between the CON and TIPNS runs at depths of 5, 25, 45, and 85 m, representing the near surface, the middle of the mixed layer, the intersection between the mixed layer and the thermocline, and the thermocline, respectively.

The seawater at the depth of 5 m (Fig. 5a) cools within the northern AS, the northern BOB, and the western coast of Indonesia, and warms within most of the tropical Indian Ocean. This pattern is very similar to the changes in SST shown in Fig. 4, with both panels showing a low SST off Sumatra and a high SST in the western Indian Ocean, accompanied by 850 hPa wind and precipitation anomalies as shown in Fig. 2b, which resemble the Indian Ocean Dipole pattern along the equator (Saji et al., 1999). Forced by changes in the surface wind stress (red vectors), water flows to the west along the equator (black vectors) and turns northward along the eastern coast of Africa and the Arabian Peninsula into the northern AS. At 25 m depth (Fig. 5b), in the middle of the mixed layer, the changes in temperatures and currents are similar to those at 5 m, except that the current in the northern AS is slightly weaker. At 45 m depth (Fig. 5c), close to the bottom of the mixed layer, the cooling of seawater within the AS is significantly weakened, whereas the cooling over the BOB almost disappears. At this depth, the western tropical Indian Ocean is warmer than at 5 m, and the cooling along the western coast of Indonesia is stronger. At greater depths below the thermocline, such as at 85 m (Fig. 5d), the seawater warms over the entire Indian Ocean. Moreover, the ocean currents flow from the northern AS to the tropics, which

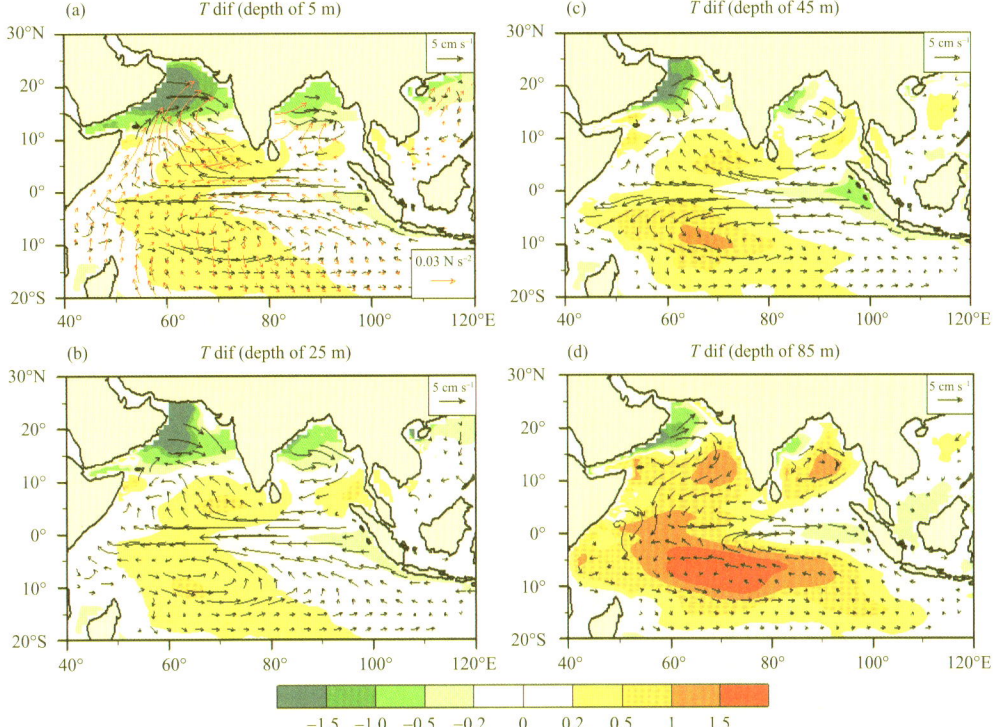

Fig. 5 June—July—August mean differences between the CON and TIPNS runs for the seawater temperatures (shaded, ℃) and ocean currents (black vectors, cm s^{-1}) at the depths of (a) 5 m, (b) 25 m, (c) 45 m, and (d) 85 m. The red arrows in (a) denote the surface wind stress (10^{-2} N s^{-2})

then turn eastward along the equator to the western coast of Indonesia, before reversing direction relative to those above the thermocline.

4.2 Surface radiation and heat fluxes

Changes in SST are strongly affected by the fluxes of radiation and heat at the sea surface. To understand the possible causes of the changes in SST, we first calculate the surface radiation budget in the Indian Ocean. The surface net radiation budget can be expressed as

$$F_{net}^{sfc} = F_{sw}^{D}(1 - A_{sfc}) + F_{lw}^{D} - F_{lw}^{U} \tag{4}$$

where F_{net}^{sfc} represents the surface net downward radiation; F_{sw}^{D} represents the surface downward shortwave radiation; A_{sfc} represents the surface albedo; and F_{lw}^{D} and F_{lw}^{U} represent the surface downward and upward longwave radiation, respectively. Changes in the skin-surface temperature are influenced by changes in downward radiation, which are in turn affected by changes in the cloud distribution. Therefore, we show the changes in the total cloud cover resulting from the thermal forcing of the TIP (CON-TIPNS) in Fig. 6a, which shows that the total cloud cover increases over the western tropical Indian Ocean and the South Asian continent, with a maximum centered over the southeastern slope of the TP. The cloud cover decreases prominently over central Asia and slightly over the Maritime Continent in the tropics. The pattern of change in total cloud cover results in a strong decrease in F_{sw}^{D} (Fig. 6b) over the South Asian continent, a weak decrease over the western Indian Ocean, and an increase over the Maritime Continent and the eastern tropical Indian Ocean. The net surface shortwave radiation $F_{sw}^{D}(1 - A_{sfc})$ (Fig. 6c) shows a similar pattern to F_{sw}^{D}, indicating a relatively small impact of the change in surface albedo, with almost all the change in net downward solar energy resulting from the change in insolation. The surface downward longwave radiation F_{lw}^{D} increases primarily over the South Asian continent and the Indian Ocean (figure not shown) consistent with the cloud-cover distribution. Consequently, F_{net}^{sfc} decreases significantly over the AS, the BOB, the northwestern Pacific Ocean, and southern South Asia (Fig. 6d). Over the Indian Ocean, the distribution is similar to the distribution of changes in downward shortwave radiation, F_{sw}^{D} (Fig. 6b). By comparing this distribution to the observed distribution of changes in

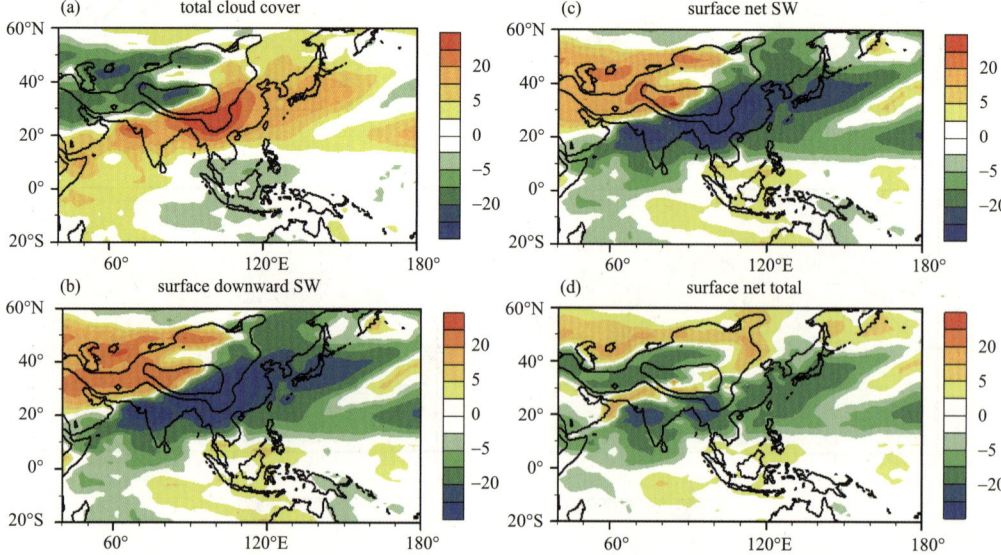

Fig. 6 June−July−August mean differences between the CON and TIPNS runs of (a) total cloud cover (%), (b) surface downward shortwave radiation (W m^{-2}), (c) surface downward net shortwave radiation (W m^{-2}), and (d) surface total net downward radiation (W m^{-2}). The black contours indicate elevations of 1000 and 3000 m, respectively

SST (Fig. 4), we find that the warming in the southern Indian Ocean is quite consistent with the local increase in F_{net}^{sfc}, and the cooling over the AS, the BOB, and the monsoon region in the northern Indian Ocean is consistent with the local decrease in F_{net}^{sfc}. However, the SST warming in the central tropical Indian Ocean cannot be explained by the change in F_{net}^{sfc} alone. Further analysis is provided in the next section.

Figure 7 shows the influence of the thermal effect of the TIP on changes in the surface heat fluxes and the associated water-vapor transport (CON−TIPNS). The surface sensible heat flux (Fig. 7a) primarily increases over the whole TIP due to the experiment design, and decreases over the ASM region, which is connected to the increase in precipitation over this area (Fig. 2b). The surface sensible heat flux also decreases over the northern AS and the BOB in good agreement with the decrease in the SST change (Fig. 4). However, the changes in the surface sensible heat flux are quite weak over the whole Indian Ocean. The surface latent heat flux increases strongly over East Asia and the southeastern TP (Fig. 7b) in response to the thermal effect of the TIP. Over the ocean, the surface latent heat flux decreases over the northwestern Pacific, the central AS, and the tropical Indian Ocean, but increases over the western Indian Ocean. The surface latent heat flux decreases by more than 10 W m^{-2} in the eastern AS, and this change coincides with the maximum in situ SST warming, which exceeds 0.5 K (Figs. 4 and 5a). The decrease in the surface latent heat flux in the area from the central AS to the north of the equator, which is connected with a decrease in evaporation (Fig. 7c), also contributes to the local SST increases, as shown in Fig. 4. This result implies that the thermal forcing of the TIP reduces the surface latent heat flux and contributes to the SST increase over the central and eastern tropical Indian Ocean through reducing the surface wind speed in this region (Fig. 5a). In correspondence with enhanced circulation at 850 hPa surrounding the TIP (Fig. 2b), the thermal effect of the TIP induces a large amount of water vapor to the south of the TIP and to the north of approximately 10°N, contributing to the remarkable increase in the vertically integrated water vapor over subtropical South and East Asia, whereas the vertically integrated water vapor decreases prominently over the area to the northwest of the TIP (Fig. 7d).

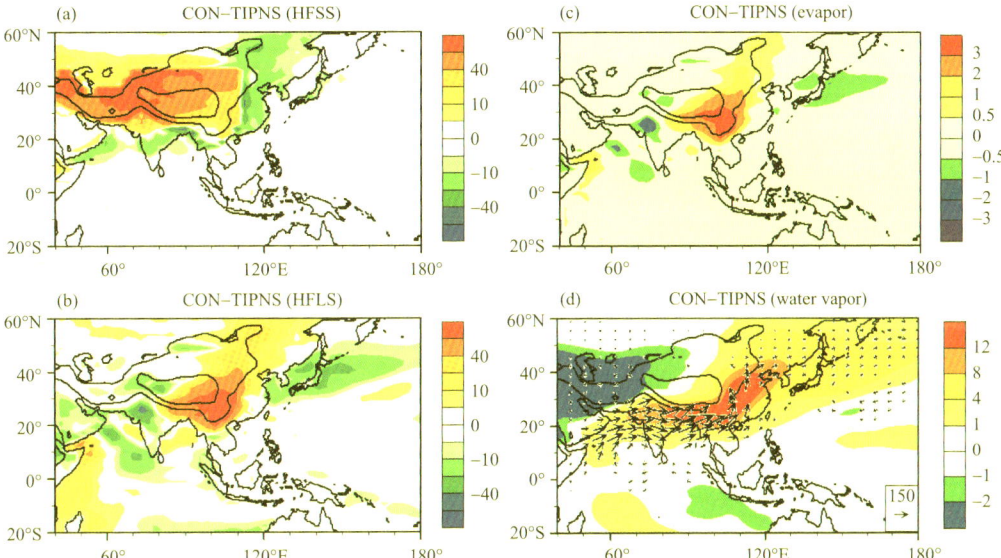

Fig. 7 June−July−August mean differences between the CON and TIPNS runs of (a) sensible heat fluxes (W m^{-2}), (b) latent heat fluxes (W m^{-2}), (c) surface evaporation (mm day^{-1}), and (d) vertically integrated water vapor (shaded, kg m^{-2}) and the 850 hPa water vapor fluxes (vector, kg m^{-1} s^{-1}). The black contours indicate elevations of 1000 and 3000 m, respectively

4.3 Energy budget in the mixed layer of the Indian Ocean

The changes in SST may also be attributed to changes in the physical processes occurring in the ocean mixed layer. Here, we analyze the causes of these SST changes in the mixed layer using the heat budget equation

$$\underbrace{\frac{\partial T}{\partial t}}_{A} = \underbrace{\frac{Q_{net} - Q_{pen}}{\rho_0 c_p H}}_{B} - \underbrace{u\frac{\partial T}{\partial x} - v\frac{\partial T}{\partial y}}_{C\quad\quad D} - \underbrace{w\frac{\partial T}{\partial z}}_{E} + \varepsilon \tag{5}$$

where T represents the temporally averaged mixed-layer temperature, $H=45$ m represents the mixed-layer depth, Q_{net} represents the net surface heat flux, Q_{pen} represents the shortwave penetration through depth H, u represents the zonal current, v represents the meridional current, and w represents the current vertical velocity component. The vertical means of the oceanic terms are calculated as

$$\overline{(\ )} = \frac{1}{H}\int_{-H}^{0} \mathrm{d}z \tag{6}$$

The seawater has a density (ρ_0) of 1.029×10^3 kg m^{-3} and a specific heat (c_p) of 3996 J kg^{-1} K^{-1}. The calculation of Q_{pen} follows the solar radiation penetration parameterization scheme introduced by Paulson and Simpson (1977) as

$$Q_{pen} = \mathrm{SW} \times [R \times e^{-H/L_1} + (1-R)e^{-H/L_2}] \tag{7}$$

where SW is the downward shortwave radiation at the sea surface, $R = 0.58$, $L_1 = 0.35$ m and $L_2 = 23$ m. In Eq. (5), term A denotes the local temperature change; terms B to E are the net heat fluxes, the horizontal advection, and the vertical mixing, respectively; and the last term, ε, represents the sum of the horizontal and vertical diffusion of temperature, which is relatively small (Lin et al., 2007).

To further understand the temperature changes in the mixed layer of the Indian Ocean, we calculate each term of the budget in Eq. (5) and analyze their relative contributions to the temperature changes within this layer. Figure 8 shows the differences between the CON and TIPNS runs for each term in Eq. (5). The difference for term A (Fig. 8a), which is the difference of the local change in vertical mean ocean temperature in the mixed layer, shows a similar distribution to that of the mean changes in SST (Fig. 4), with significant cooling within the northern AS, the BOB, and off the western coast of Indonesia, and warming within the middle tropical Indian Ocean. The sum of the B, C, D, and E terms on the right-hand side of Eq. (5) is shown in Fig. 8g, indicating a similar pattern to that of the vertically integrated mixed-layer temperature change (Fig. 8a), implying the calculations on both sides of Eq. (5) are generally balanced.

Term B is related to the surface net heat flux, which decreases by 0.1—0.5 °C month^{-1} in the western Indian Ocean, the northeastern AS, and the northern BOB, but increases primarily over the northwestern AS (Fig. 8b). Importantly, the pattern of change in the heat fluxes is consistent with the changes in the surface latent heat flux (Fig. 7b) but with opposite signs, indicating the importance of ocean surface evaporation in the surface energy budget in these areas. Within the tropical Indian Ocean, the differences in term B over the western Indian Ocean have an opposite sign to that over the central and eastern Indian Ocean. The zonal advection term C (Fig. 8c) generates cooling in the northern AS and warming over the western Indian Ocean, but also a weak cooling within the BOB and off the western coast of Indonesia. The meridional advection term D (Fig. 8d) also shows a strong cooling over the AS, a weak cooling over the BOB and along 10°S in the Southern Hemisphere, and a weak warming near 5°N in the Northern Hemisphere. The vertical mixing term E (Fig. 8e) shows prominent warming in the

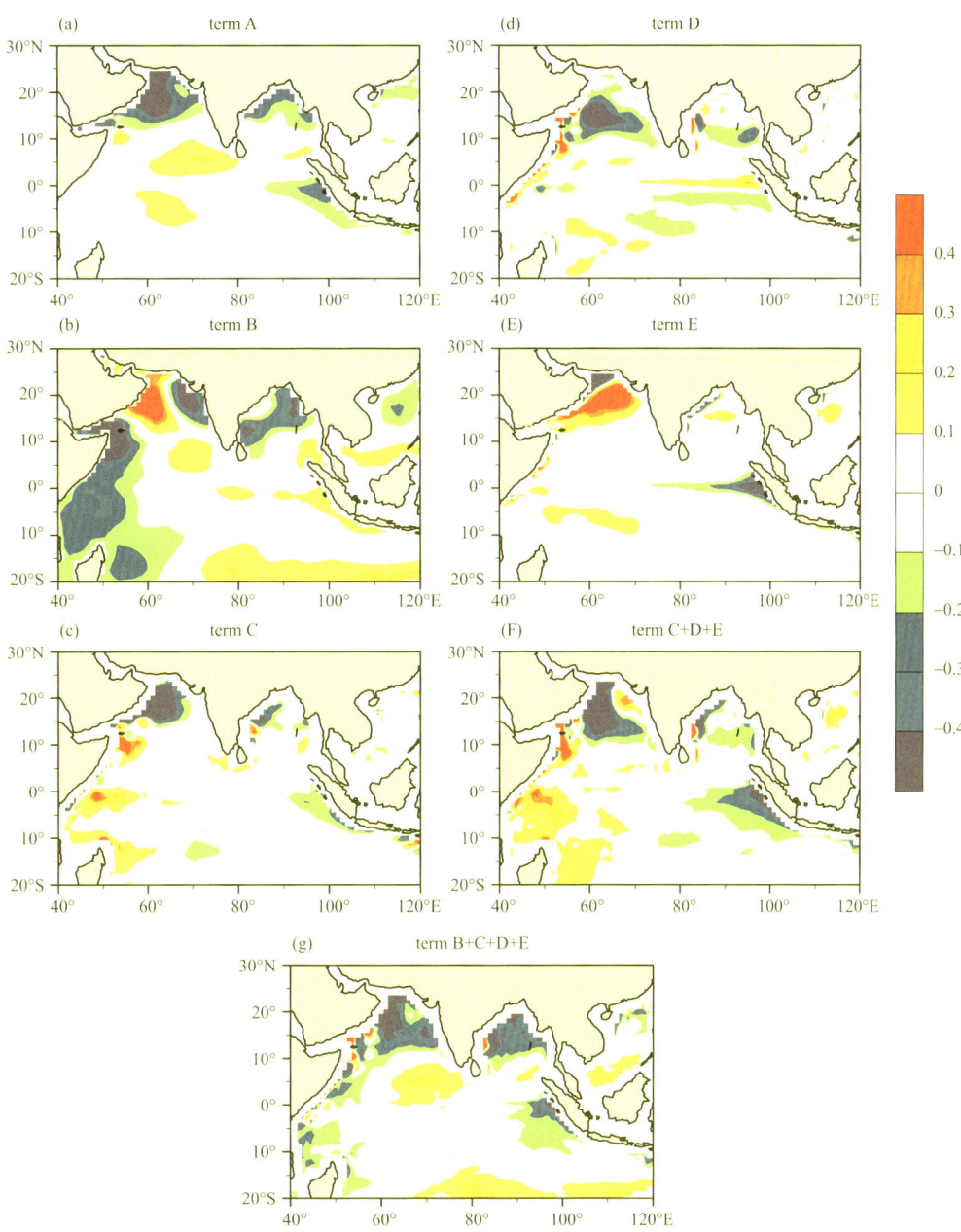

Fig. 8 June—July—August mean differences between the CON and TIPNS runs of the mixed layer (0—45 m) heat budget (℃ month^{-1}) calculated using Eq. (5) for the terms (a) A, (b) B, (c) C, (d) D, and (e) E; (f) the sum of terms C, D, and E; and (g) the sum of terms B, C, D, and E

northern AS and in the Seychelles, implying strong downwelling in these ocean regions. This term also shows strong cooling along the western coast of Indonesia extending to the equator, which implies strong upwelling in the mixed layer. To compare the contributions of changes in heat flux and advection, terms C, D, and E are summed, with the results presented in Fig. 8f, showing that the warming over the western Indian Ocean resulting from advection is mostly counteracted by the cooling from the heat flux (Fig. 8b). Moreover, the cooling within the northern AS, the BOB, and off the western coast of Indonesia, is similar to the changes in term A, showing the importance of horizontal advection and upwelling/downwelling in determining the temperature change in these regions. However, the cooling in the northern AS and the BOB as shown in Fig. 8b is closely connected to the changes in the radiation flux

(Fig. 6d).

Based on the above analysis of the changes in the temperature, circulation, and heat budget of the mixed layer in the South Asian oceans, the major effects of the thermal forcing of the TIP on the changes in subsurface temperatures and ocean dynamics can be summarized as follows. The thermal forcing of the TIP enhances the westerlies over the northern AS and the BOB (Fig. 2b) in the lower troposphere, thereby intensifying the near sea-surface wind stress (Fig. 5a). The decrease in the net surface heat flux, together with the horizontal advection, contributes to the cooling of the mixed layer within the northern AS and the BOB. The cooling off the western coast of Indonesia is mainly caused by local upwelling. The warming within the tropical Indian Ocean is a consequence of the contributions from multiple causes, including the increased heat fluxes, horizontal advection, and downwelling. An exchange of seawater between the mixed layer and the thermocline is also observed. The seawater in the mixed layer flows from the tropics into the northern AS, where the seawater sinks into the deep ocean (Fig. 8e).

The results shown in Figs. 5 and 8e and f clearly demonstrate that the thermal forcing exerted by the TIP can generate remarkable westward tropical ocean currents in the mixed layer of the Indian Ocean (Figs. 5a—c). These currents sink and penetrate the thermocline into the deep ocean in the western tropical Indian Ocean, then flow eastward towards the eastern Indian Ocean in the thermocline (Fig. 5d). Finally, these ocean currents ascend in the eastern tropical Indian Ocean, bringing cold water from the deep ocean upward, penetrating the thermocline, and reaching the near-surface layers (Fig. 8e). These processes produce a trans-thermocline longitudinal circulation in the tropical Indian Ocean that includes downwelling currents in the west and upwelling currents in the east. An SST anomaly pattern similar to an Indian Ocean Dipole is therefore produced, with a reduced SST located off the western coast of Indonesia, and an increased SST located in the western Indian Ocean.

5 Indirect influence of the thermal effect of the TIP on the Asian summer monsoon

5.1 Indirect effects on precipitation and circulation

As defined in Section 2, the indirect effect of TIP thermal forcing can be expressed as Eq. (1) or (2). For the right-hand side of these two equations, the first term (CON−TIPNS) was presented in Fig. 2b, and the last terms (CON_CPSST−TIPNS_CPSST) and (CON_NSSST−TIPNS_NSSST) can be found in Figs. S2a, and S2b in OSM. Comparison of the differences of the monsoonal precipitation between the coupled (Fig. 2b) and uncoupled (Figs. S2a and b in OSM) runs shows that the largest differences occur over the Indian Ocean, which indicates the important role of the indirect effect in regulating the thermal impacts of the TIP on the ASM. For the DD_CPSST case, the precipitation anomaly (Fig. 9a) mainly decreases over the southern slope of the TP, the western AS, and the tropical Indian Ocean, and is accompanied by an anomalous anticyclonic circulation around the TIP at the 850 hPa level. Meanwhile, the precipitation difference increases mainly in the northeastern AS, south Iranian Plateau, most parts of the BOB, and the Indian mainland, which is accompanied by anomalous westerly surface winds over the AS. In general, the distribution pattern of the differences in rainfall shown in Fig. 9a is the opposite pattern to the rainfall differences between the CON_CPSST and TIPNS_CPSST runs shown in Fig. S2a in OSM. This result suggests that the indirect effect on the ASM tends to counteract the TIP-

SHAP direct forcing, which reduces the amount of precipitation falling over the southern Himalayas but intensifies rainfall from the northeastern AS to the western Pacific.

The changes in precipitation are closely related to the changes in the meridional circulation in ASM region. The vertical circulation of DD_CPSST is shown in Fig. 9c. The figure indicates that a strong ascent of air over the Indian mainland (10°—20°N), accompanied by a strong descent over south slope of the TP. The above features correspond to the in situ difference in surface easterly winds and reduced precipitation (Fig. 9a). These results indicate that the indirect effect of the thermal forcing of the TIP resulting from the air-sea interactions tends to counteract its direct effect by enhancing the monsoonal meridional circulation around 10°—20°N and generating an anti-monsoonal meridional circulation further north.

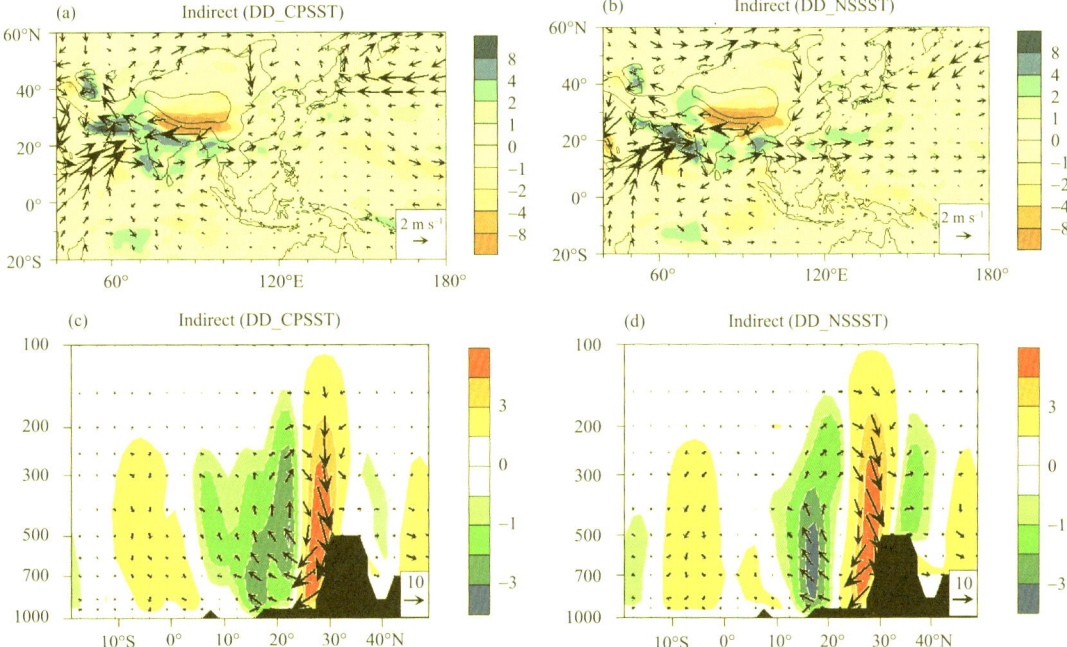

Fig. 9 Indirect effect of the TIP thermal forcing on the JJA mean ASM generated from (a and c) [(CON−TIPNS)−(CON_CPSST−TIPNS_CPSST)] and (b and d) [(CON−TIPNS)−(CON_NSSST−TIPNS_NSSST)]. (a and b) Precipitation (shading, mm day^{-1}) and 850 hPa wind speed (vectors, m s^{-1}); and (c and d) the 75°—110°E mean vertical velocity component (shaded, 10^{-2} Pa s^{-1}) and meridional circulation (vectors, v in m s^{-1}, −3 ω in 10^{-2} Pa s^{-1}). The black contours indicate elevations of 1000 and 3000 m, respectively. The black shading in (c and d) indicates the topography

For the DD_NSSST case, the difference patterns of precipitation and circulation shown in Fig. 9b and meridional circulation shown in Fig. 9d are similar to their counterparts in the DD_CPSST case as shown in Figs. 9a and c, respectively. These results imply that the indirect impacts demonstrated in Fig. 9 are well preserved and not very sensitive to the imposed SST. Fig. 10 shows the indirect effect on the skin-surface temperature. Because the SST difference between the CON_CPSST and TIPNS_CPSST runs, or between the CON_NSSST and TIPNS_NSSST runs, is zero, the indirect impact on the skin-surface temperature over oceans is exactly equal to the CON−TIPNS result (Fig. 4). The temperature difference mainly appears as a positive-negative-positive pattern in the meridional direction over the 70°—120°E section. Corresponding to this sandwich-like pattern of skin-surface temperature difference, strong anomalous westerly winds occur at the 850 hPa level along 15°N, and easterlies are produced near the equator (Figs. 9a and b). In combination with the anticyclonic circulation surrounding the TIP, a prominent surface circulation dipole develops to include a cyclone to the north of 15°N and an anticyclone to its south. This circulation dipole is accompanied by higher precipitation along the zonal belt between 10°N and 15°N and lower precipitation to its south and north. This rainfall pattern accompanies a stronger air ascent along 10°—20°N and air descent over the southern slope of the TIP and over the southern tropics (Figs. 9c and d). Thus, the presence of air-sea coupling in the CGCM increases the strength of the southern branch of the South Asian summer monsoon compared with that for the AGCM, whereas its northern branch is weakened. Thus, the similar responses to the TIP indirect forcing of the precipitation, skin-surface temperature, and circulation between the DD_NSST and DD_CPSST (Figs. 9 and 10) confirms the indirect effect of TIP thermal forcing is not sensitive to the different SST, which was adopted from CGCM experiments.

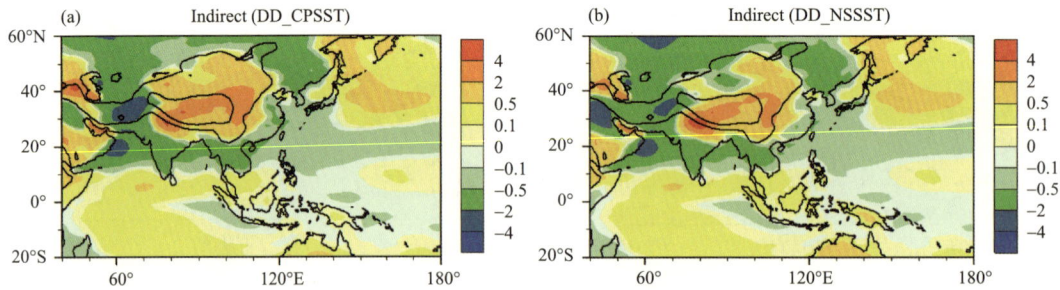

Fig. 10 Indirect effect of the TIP thermal forcing on the JJA mean skin-surface temperature (K) generated from (a) [(CON−TIPNS)−(CON_CPSST−TIPNS_CPSST)] and (b) [(CON−TIPNS)−(CON_NSSST−TIPNS_NSSST)]. The black contours indicate elevations of 1000 and 3000 m, respectively

5.2 Hydrological cycle and surface fluxes related to the indirect effects

We present here the responses of the hydrological cycle, the surface net heat fluxes, and the radiation budget to the TIP indirect forcing as demonstrated in the DD_CPSST. In terms of the hydrological cycle, the indirect effects of the TIP heating on the changes in water-vapor transport and evaporation are shown in Figs. 11a and e, respectively. The transport of water vapor decreases over TP in association with the anticyclone anomaly, which contributes to the decrease in precipitation. The water vapor increases mainly from Indian mainland, and northern BOB to East Asia, associated with the intensified westerlies. The changes of the evaporation anomaly (Fig. 11e) are weak compared with the changes in water vapor in the atmosphere (Fig. 11a). Therefore, the increase of precipitation in DD_CPSST is mainly attributed to the increase of water vapor over Asian mainland. However, in the northern Indian Ocean, the decreased SST over the northern AS (Fig. 10a) leads to local decreases in evaporation (Fig. 11e), which is the main cause of the decrease in precipitation over the western AS (Fig. 9a). For sensible heat flux (Fig. 11b), there is typically a positive anomaly over the TP and negative anomaly over the Indian mainland and the Indo-China peninsula, which is closely connected to the changes in the precipitation and radiation budget.

In terms of changes in surface radiation budgets, the cloud cover (Fig. 11c) increases over South Asia and northern AS but decreases over the TP, which leads to the increased F_{sw}^D in southern edge of the TP and the decreased F_{sw}^D in the Iranian Plateau, northern AS, and South Asian sub-continent (Fig. 11g). The responses of F_{lw}^D are weak over the Indian Ocean, but an increase of more than 5 W m^{-2} occurs over the Indian mainland and East Asia (Fig. 11d). The spatial pattern of F_{net}^{sfc} (Fig. 11h) is similar to that of F_{sw}^D

(Fig. 11g), indicating the dominant role of cloud cover and shortwave radiation in the radiation changes induced by the indirect effect. The radiation effect leads to a reduced SST in the northern AS and BOB, and an increased SST mainly in the western and tropical Indian Ocean.

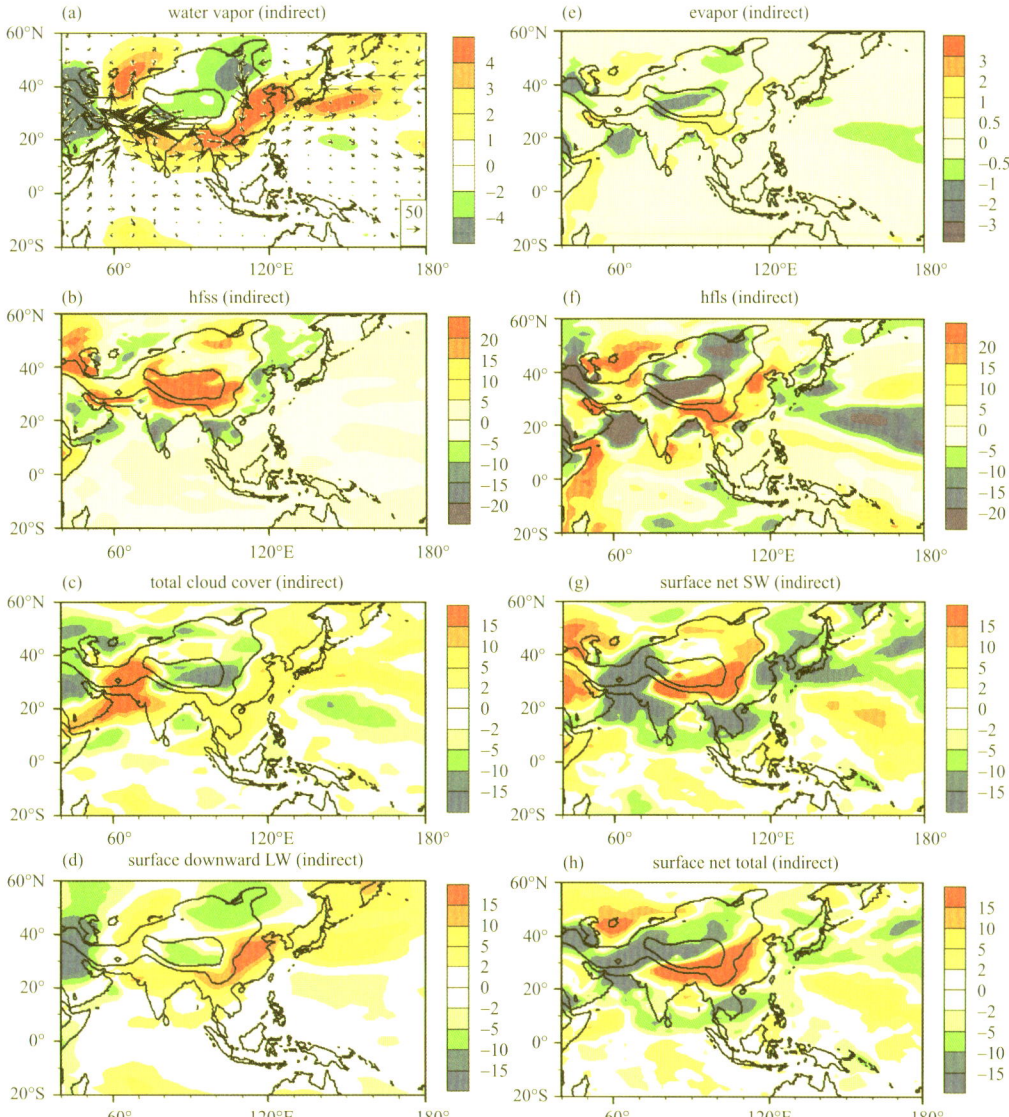

Fig. 11 Indirect effect of the thermal forcing associated with the TIP on the JJA mean ASM [(CON−TIPNS) −(CON_CPSST−TIPNS_CPSST)]. (a) Vertically integrated water vapor (shaded, kg m^{-2}) and 850 hPa water vapor fluxes (vectors, kg m^{-1} s^{-1}), (b) surface sensible heat fluxes (W m^{-2}), (c) total cloud cover, (d) surface downward longwave radiation fluxes (W m^{-2}), (e) surface evaporation (mm day^{-1}), (f) surface latent heat fluxes (W m^{-2}), (g) surface net shortwave radiation fluxes (W m^{-2}), and (h) surface net radiation fluxes (W m^{-2}). The black contours indicate elevations of 1000 and 3000 m, respectively

In short, the indirect thermal effect of the TIP suggests that oceanic feedbacks to the atmosphere have an important role. Specifically, these feedbacks generate sandwich-type SST and circulation anomaly over Indian Ocean and South Asian continent, associated with cloud, radiation, and water vapor changes. Together, these processes reflect a complete physical representation of the response of the ASM to the indirect thermal effect of the TIP through changes in the local hydrological processes, which counteracts the direct thermal effect of the TIP on the ASM and reduces its strength.

6 Uncertainties in evaluating the indirect effects

Because of model bias, the SST generated from the CGCM (i. e., CPSST) is different from the observed SST. Thus, the indirect effects obtained from the DD_CPSST run may be also different from those obtained from the experiments based on the observed SST. To improve our understanding of the TIP thermal effects on ASM, it is important to analyze the influence of the SST bias, and to assess the uncertainty of the indirect effects of the TIP thermal forcing.

6.1 Impacts of the sea-surface temperature bias

Figure 12a shows the bias of the SST generated from the CGCM. The bias is calculated from the difference of the mean SST during JJA between the coupled control experiment (CON) and the observed SST, which is used to drive the AGCM experiment (CON_OBSST). In the Indian Ocean, the cold bias occurs over the AS, the BOB, and off the western coast of Sumatra, while the warm bias occurs in the tropical and southern Indian Ocean. In the western Pacific, the warm bias mainly occurs in mid-high latitudes and along the equator, while the cold bias appears in both the northern and southern tropics. Notice that the influence of SST bias on the control and sensitivity runs can be expressed as

$$D\_SH = CON - CON\_OBSST \qquad (8)$$
$$D\_NS = TIPNS - TIPNS\_OBSST \qquad (9)$$

For D_SH = (CON−CON_OBSST), in both the CGCM experiment CON and the AGCM experiment CON_OBSST, the surface sensible heating exists over the TIP, where the difference inherent in D_SH exhibits the main impacts of the SST bias (Fig. 12a). Similarly, for D_NS = (TIPNS−TIPNS_OBSST), in both the CGCM experiment TIPNS and the AGCM experiment TIPNS_OBSST, the surface sensible heating over the TIP is excluded, and the difference contained in the D_NS run also exhibits the main impacts of the SST bias (Fig. 12c).

The distributions of JJA mean precipitation and the wind speed at 850 hPa calculated from the D_SH and D_NS runs are presented in Figs. 12b and d, respectively. Interestingly, the distributions of the difference in wind speed and precipitation are similar. In particular, the distribution of the difference in the precipitation over the ocean resembles that of the SST bias, respectively (Figs. 12a and c), where a positive (negative) precipitation difference appears where the SST bias is positive (negative). The differences in the SST bias are mainly located over the AS and BOB, where D_NS (Fig. 12c) shows weaker SST negative bias over the south AS and BOB but positive SST bias over the north AS. Weak precipitation occurs over northern Indian Ocean and extremely strong precipitation over the south slope of TP in the CON_CPSST and CON_NSSST because of the SST bias, as shown in Figs. S1b and c in the OSM. These results demonstrate that the SST bias is an important issue that should be considered when conducting numerical experiments. The uncertainties are identified in the following sub-section.

6.2 Uncertainty of the indirect effects

To further evaluate the degree of the impacts of SST bias on the simulation, and the uncertainty of the indirect effects of the TIP thermal forcing, we estimate the indirect effects of TIP following the approach used by Kitoh (2004) and Okajima and Xie (2007), who defined the indirect effect of the TIP thermal forcing (DD_OBS) as shown in Eq. (3) using the observed SST-driven AGCM. Since Eq. (3) can be rewritten as

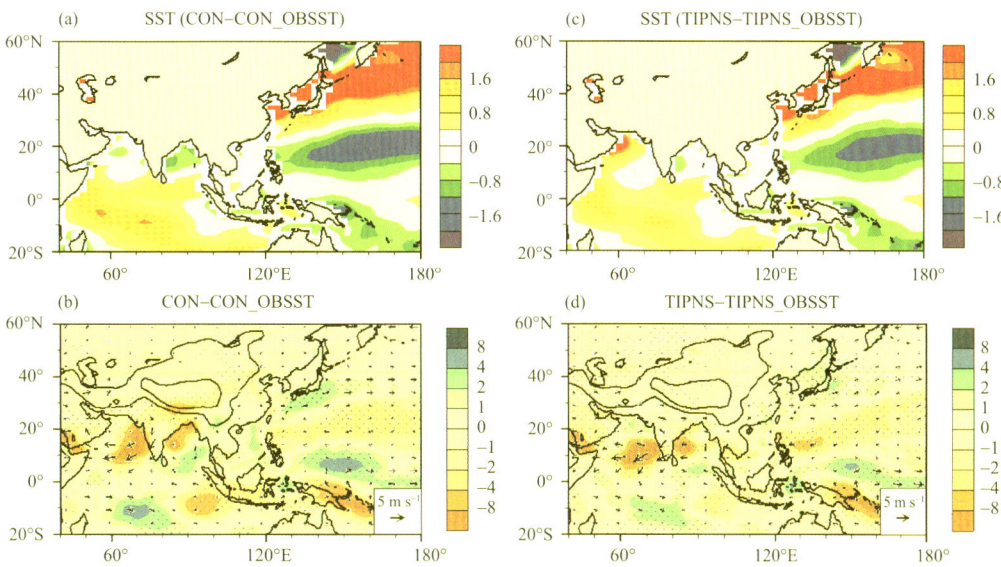

Fig. 12 June—July—August mean difference of (a) mean SST (K) between the CON and CON_OBSST runs, and (c) the TIPNS and TIPNS_OBSST, and of precipitation (shaded, mm day^{-1}) and 850 hPa wind speed (vectors, m s^{-1}) between (b) the CON and CON_OBSST runs, and (d) the TIPNS and TIPNS_OBSST runs. The blue dots in (b) and (c) denote the precipitation difference passing the 95% significance t-test, and the black contours denote elevations of 1000 and 3000 m, respectively

$$DD\_OBS = (CON - CON\_OBSST) - (TIPNS - TIPNS\_OBSST) \quad (10)$$

and since the first and second terms on the right-hand side of Eq. (10) possess similar distributions (Figs. 12b and d) and suffered from similar SST biases (Figs. 12a and c), the experiment design in Eq. (3) could mitigate the impacts of the SST bias to some extent.

The distributions of the precipitation, wind speed at 850 hPa, skin-surface temperature, and meridional circulations produced from the DD_OBS are presented in Fig. 13. In comparison with Figs. 9 and 10, the indirect effects obtained from these three sets of experiments (DD_OBS, DD_CPSST, and DD_NSSST) possess the following common features:

a) Near the surface, a remarkable anticyclone circulation surrounds the TIP; to its south is a cyclonic circulation located over the northern Indian Ocean and another anticyclone circulation located further south over the tropical Indian Ocean, exhibiting a triple horizontal circulation pattern. An intensified tropical westerly flow is located over the northern Indian Ocean between the cyclone circulation in the north, and the anticyclone circulation in the south (Figs. 9a and b and 13a).

b) The precipitation triple pattern in Figs. 9a and b is also reproduced over the South Asian summer monsoon area in this experiment, with increased precipitation appearing along with the intensified tropical westerly flow, and reduced precipitation over the southern slope of the TIP to the north, and over the tropical Indian Ocean to the south.

c) A dipole meridional circulation develops over the Indian Ocean, with the ascending arm located over the tropical Indian Ocean in coordination with the increased precipitation and two descending arms located over the southern slope of the TIP and the southern tropical Indian Ocean (Figs. 9c and d and 13c).

Results from the above analysis imply that indirect effects on the ASM of the TIP thermal forcing presented in Section 5 have captured certain basic features in the changes of circulation and precipitation.

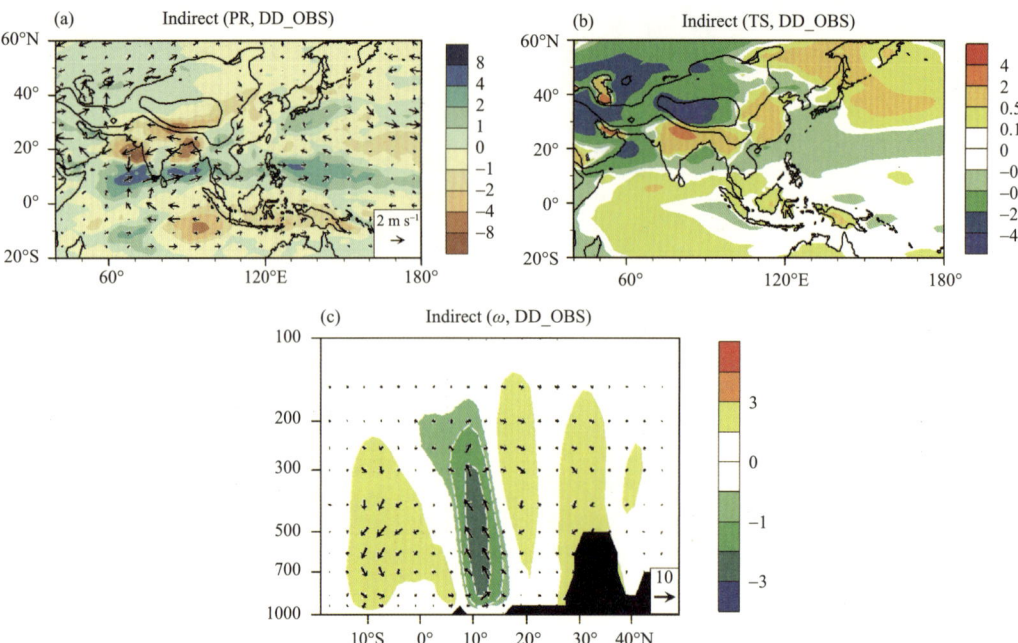

Fig. 13 Indirect effect (DD_OBS) of the thermal forcing associated with the TIP on the JJA mean ASM [(CON_TIPNS)−(CON_OBSST−TIPNS_OBSST)]. (a) Precipitation (shaded, mm day^{-1}) and 850 hPa wind speed (vectors, m s^{-1}); (b) surface temperatures (shaded, K); and (c) the 75°–110°E mean vertical velocity (shaded, 10^{-2} Pa s^{-1}) and meridional circulation (vector, v in m s^{-1}, 3 ω in 10^{-2} Pa s^{-1}). The black contours indicate elevations of 1000 and 3000 m, respectively. The black shading in (c) indicates the topography

Compared with the DD_OBS run, however, several noticeable differences can be identified from the DD_CPSST and DD_NSSST runs:

d) The cyclonic circulation over the northern Indian Ocean of the "triple horizontal circulation pattern" shrinks and shifts northward, and, accordingly, the intensified westerly flow shifts northward. At the same time, the anticyclone circulation in the tropical Indian Ocean extends westward, contributing to the enhanced southwesterly winds over the northwestern AS.

e) The ascending arm of the "dipole meridional circulation" over the tropical Indian Ocean also shifts northward and is almost three times stronger in DD_CPSST than in DD_OBS.

f) Consequently, remarkable precipitation differences between the DD_OBS experiment and the DD_CPSST and DD_NSSST experiments occur over the northern AS and the BOB: The precipitation decreases over these regions in the DD_OBS run (Fig. 13a) but increases in the DD_CPSST and DD_NSSST runs (Figs. 9a and b), demonstrating the largest uncertainty in the evaluation.

At present, it is difficult to explore what causes this uncertainty since biases may excite further feedbacks in the model simulations. However, the existence of these differences reminds us that there is some uncertainty when the CGCMs and AGCMs are employed to study the indirect impact of the TIP thermal forcing, and thus caution is required. For example, the differences of precipitation and circulation between CON_CPSST and CON_OBSST (Fig. S3 in OSM) suggest that SST bias could cause excessive precipitation on the south slope of the TP and reduced precipitation over the AS and BOB, accompanied by enhanced westerly flow over the south TP and easterly flow over the Indian Ocean. This result suggests that the indirect effect revealed by DD_CPSST is overestimated over the south slope of the TP and underestimated over the Indian Ocean.

7 Discussion and Conclusion

Based on a series of experiments using a CGCM and one of its components, an AGCM, the thermal effects of the TIP on the changes in the ASM during the boreal summer were investigated. The influences of the direct TIP thermal forcing and indirect forcing induced by the air-sea interaction on the changes in monsoonal precipitation, circulation, and SST were analyzed.

The direct thermal effects of large-scale elevated topography TIP cause the precipitation to increase over the southern slope of the TP and East Asia, whereas the precipitation decreases over the northwestern Indian subcontinent. This behavior is also accompanied by cyclonic circulation around the TIP in the lower troposphere in both the AGCMs and CGCMs. The thermal forcing of the TIP also exerts a direct effect on the thermal status and circulation over the Indian Ocean. The changes in the SST and the temperatures of the mixed layer in the Indian Ocean are mainly caused by changes in atmospheric radiative forcing, surface latent heat fluxes, and the associated ocean dynamics. Meanwhile, the increased shortwave radiation over the southern Indian Ocean contributes to the increased SST. The cooling of the mixed layer within the northern AS and the BOB is mainly induced by decreases in the surface heat flux and horizontal advection, whereas the cooling over western Indonesia is mainly caused by local upwelling.

Finally, the major roles of the indirect effect of the TIP thermal forcing in affecting the ASM are summarized in the schematic diagram shown in Fig. 14.

(1) The surface heating of the TIP drives cyclonic circulation in the lower troposphere around the mountains (a) and transports a large amount of water vapor towards the TP, shifting the main rain belt into the Asian inland region. In contrast, precipitation is reduced over almost the entire Indian Ocean in the Northern Hemisphere.

(2) In response to the TIP thermal forcing, the SST in the tropical Indian Ocean is warmer in the case with the TIP surface heating, and a difference in zonal ocean circulation along the equator is produced between the mixed layer and the deep ocean through the thermocline, as shown by the blue arrows in (c). The downwelling in the western Indian Ocean takes heat from the near surface to the deep ocean, and increases the ocean temperatures, whereas the upwelling in the eastern Indian Ocean leads to a reduced SST off the western coast of Indonesia.

(3) The heating of the TIP-induced air-sea interactions over the Indian Ocean as indicated by the dashed double-sided arrow between (b) and (c) generates non-uniform spatial SST anomalies (Figs. 10 and 13b). The atmosphere influences the ocean mainly through radiation (downward-pointing red arrow) and the surface wind stress. In contrast, the sea surface feeds back into the atmosphere (upward-pointing purple arrow) mainly through changes in surface latent heat fluxes and evaporation. Thus, the ocean triggers upward vertical motion over the tropical Indian Ocean (upward-pointing heavy red arrow between (a) and (b)) and the associated descent air over the southern tropical Indian Ocean and the southern slope of the TP regions (downward-pointing heavy red arrow between (a) and (b)), forming a meridional circulation dipole over the tropical Indian Ocean.

(4) For the indirect effect (b) near the surface, the northern meridional circulation corresponds to the difference in easterly flow located to the south of the TP, whereas the cross-equator southern meridional circulation corresponds to the easterly flow anomaly in the southern tropics and westerly flow anomaly in the northern tropics where the meridional gradient of the SST is large. Consequently, a tri-

pole horizontal difference circulation pattern composed of anticyclonic-cyclonic-anticyclonic (A-C-A) motion is formed over the TIP and the Indian Ocean (fine red curves in (b)). Therefore, we conclude that the indirect effect of the TIP thermal forcing plays the role of counteracting its direct effect on the ASM.

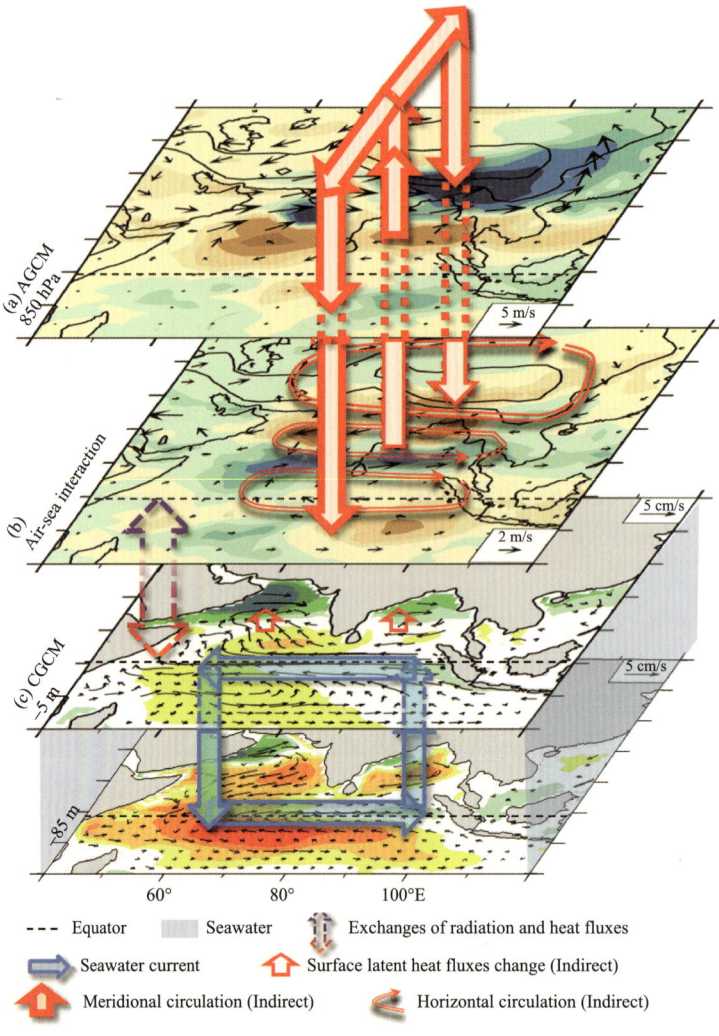

Fig. 14 Schematic diagram presenting the thermal effect of the TIP on the ASM. The direct effect of the TIP thermal forcing generates a cyclonic circulation in the lower troposphere around the TIP, while the indirect effect of the TIP thermal forcing generates an anticyclonic flow surrounding the TIP, and a cyclone-anticyclone circulation dipole to its south in the lower troposphere (fine red arrows), coupled with a pair of meridional circulations (bold red arrows), counteracting the TIP direct impact on the ASM. See text for details

It is important to note that comparison of the model outputs from CGCM and AGCM to study the indirect effects of the TIP thermal forcing on the ASM involves uncertainty. The SST generated in the CGCM is usually different from the observed SST employed to drive the AGCM. Thus, the difference in model output between the experimental pairs of the CGCM and the corresponding experimental pairs of the AGCM not only indicates the indirect effect of the TIP thermal forcing, but also involves some unwanted noise signals induced by the bias of the SST of the CGCM simulated from the observed SST. Sorting out the noise signals helps reduce the uncertainty in determining the indirect effect of the TIP forcing. Improving the CGCM performance and using multi-model ensemble can reduce the model bias as well as the noise signal, so as to mitigate the uncertainty.

We have investigated the influence of the surface sensible heating of the TIP on the climate mean ASM in the CGCM and compared it with that of the AGCM to reveal its indirect effect on the ASM. However, we did not address how this indirect effect associated with the TIP thermal forcing affects the ASM variability. It is interesting to note that, along the equator, the SST response to TIP heating resembles the pattern of the Indian Ocean Dipole. Whether or not they have connections on an interannual scale needs further investigation. Moreover, the ASM and the associated SST variation also have close relations with the activities of the El Nino and the Southern Oscillation (Kawamura, 1998). It is also unclear whether the thermal effect of the TIP is also influenced by these climate system factors, or by other physical processes, such as a reduction in cloud cover, responsible for the increase in the incoming solar radiation, leading to an enhancement in the heating of the TIP. Further studies on the interactions between the thermal effect of the TIP and SST variability and their relative contributions to the variability of the ASM will help in further understanding the dynamics of the ASM and in improving climate predictions.

References

ABE M, KITOH A, YASUNARI T, 2003. An evolution of the Asian summer monsoon associated with mountain uplift-simulation with the MRI atmosphere-ocean coupled GCM[J]. Journal of the Meteorological Society of Japan, 81(5): 909-933.

ABE M, HORI M, YASUNARI T, et al, 2013. Effects of the Tibetan Plateau on the onset of the summer monsoon in South Asia: The role of the air-sea interaction[J]. Journal of Geophysical Research: Atmospheres, 118(4): 1760-1776.

ADLER R F, HUFFMAN G J, CHANG A, et al, 2003. The version-2 global precipitation climatology project (GPCP) monthly precipitation analysis (1979—present)[J]. Journal of Hydrometeorology, 4(6): 1147-1167.

BAO Q, WU G X, LIU Y, et al, 2010. An introduction to the coupled model FGOALS1.1-s and its performance in East Asia[J]. Advances in Atmospheric Sciences, 27(5): 1131-1142.

BAO Q, LIN P, ZHOU T, et al, 2013. The flexible global ocean-atmosphere-land system model, spectral version 2: FGOALS-s2[J]. Advances in Atmospheric Sciences, 30(3): 561-576.

BOLIN B, 1950. On the influence of the earth's orography on the general character of the westerlies[J]. Tellus, 2(3): 184-195.

BOOS W R, KUANG Z, 2010. Dominant control of the South Asian monsoon by orographic insulation versus plateau heating[J]. Nature, 463(7278): 218-222.

BOOS W R, KUANG Z, 2013. Sensitivity of the South Asian monsoon to elevated and non-elevated heating[J]. Scientific Reports, 3(1): 1-4.

BRIEGLEB B P, BITZ C M, HUNKE E C, et al, 2004. Scientific description of the sea ice component in the community climate system model, version three[J]. NCAR Tech, 70.

CHARNEY J G, ELIASSEN A, 1949. A numerical method for predicting the perturbations of the middle latitude westerlies [J]. Tellus, 1(2): 38-54.

COLLINS W D, BITZ C M, BLACKMON M L, et al, 2006. The community climate system model version 3 (CCSM3)[J]. Journal of Climate, 19(11): 2122-2143.

DEE D P, UPPALA S M, SIMMONS A J, et al, 2011. The ERA-interim reanalysis: Configuration and performance of the data assimilation system[J]. Quarterly Journal of the Royal Meteorological Society, 137(656): 553-597.

DUAN A, WU G, 2005. Role of the Tibetan Plateau thermal forcing in the summer climate patterns over subtropical Asia [J]. Climate Dynamics, 24(7): 793-807.

FLOHN H, 1957. Large-scale aspects of the "summer monsoon" in South and East Asia[J]. Journal of the Meteorological Society of Japan, 35A: 180-186.

HUAHN D G, MANABE S, 1985. The role of mountains in the South Asian monsoon circulation[J]. Journal of the Atmospheric Sciences, 32(8): 1515-1541.

HE B, WU G, LIU Y, et al, 2015. Astronomical and hydrological perspective of mountain impacts on the Asian summer monsoon[J]. Scientific Reports, 5(1): 1-12.

HE Z, WU R, 2013. Coupled seasonal variability in the South China Sea[J]. Journal of Oceanography, 69(1): 57-69.

HELD I M, HOU A Y, 1980. Nonlinear axially symmetric circulations in a nearly inviscid atmosphere[J]. Journal of the Atmospheric Sciences, 37(3): 515-533.

HELD I M, TING M, WANG H, 2002. Northern winter stationary waves: Theory and modeling[J]. Journal of Climate, 15(16): 2125-2144.

HSU H H, LIU X, 2003. Relationship between the Tibetan Plateau heating and east Asian summer monsoon rainfall[J]. Geophysical Research Letters, 30(20): 2066.

KAWAMURA R, 1998. A possible mechanism of the Asian summer monsoon-ENSO coupling[J]. Journal of the Meteorological Society of Japan, 76(6): 1009-1027.

KITOH A, 1997. Mountain uplift and surface temperature changes[J]. Geophysical Research Letters, 24(2): 185-188.

KITOH A, 2004. Effects of mountain uplift on East Asian summer climate investigated by a coupled atmosphere - ocean GCM[J]. J Clim, 17(4):783-802.

KITOH A, MOTOI T, ARAKAWA O, et al, 2010. Climate modelling study on mountain uplift and Asian monsoon evolution[J]. Geological Society, London, Special Publications, 342(1): 293-301.

KOSEKI S, WATANABE M, KIMOTO M, 2008. Role of the midlatitude air-sea interaction in orographically forced climate[J]. Journal of the Meteorological Society of Japan, 86(2): 335-351.

LIN P, LIU H, ZHANG X, 2007. Sensitivity of the upper ocean temperature and circulation in the equatorial Pacific to solar radiation penetration due to phytoplankton[J]. Advances in Atmospheric Sciences, 24(5): 765-780.

LIN P, LIU H, YU Y, et al, 2011. Response of sea surface temperature to chlorophyll-a concentration in the tropical Pacific: Annual mean, seasonal cycle, and interannual variability[J]. Advances in Atmospheric Sciences, 28(3): 492-510.

LIU H, LIN P, YU Y, et al, 2013. The baseline evaluation of LASG/IAP climate system ocean model (LICOM) version 2 [J]. Acta Meteorological Sinica, 26(3): 318-329.

LIU Y, HOSKINS B, BLACKBURN M, 2007. Impact of Tibetan orography and heating on the summer flow over Asia[J]. Journal of the Meteorological Society of Japan, 85B: 1-19.

LIU Y, WU G, HONG J, et al, 2012. Revisiting Asian monsoon formation and change associated with Tibetan Plateau forcing: II. Change[J]. Climate Dynamics, 39(5): 1183-1195.

MA D, BOOS W, KUANG Z, 2014. Effects of orography and surface heat fluxes on the South Asian summer monsoon[J]. Journal of Climate, 27(17): 6647-6659.

MANABE S, TERPSTRA T B, 1974. The effects of mountains on the general circulation of the atmosphere as identified by numerical experiments[J]. Journal of Atmospheric Sciences, 31(1): 3-42.

MONTÉGUT C D B, MADEC G, FISCHER A S, et al, 2004. Mixed layer depth over the global ocean: An examination of profile data and a profile-based climatology[J]. Journal of Geophysical Research, 109: C12003.

OKAJIMA H, XIE S P, 2007. Orographic effects on the northwestern Pacific monsoon: Role of air-sea interaction[J]. Geophysical Research Letters, 34: L21708.

OLESON K W, DAI Y, BONAN B, et al, 2004. Technical description of the community land model (CLM)[R]. Tech Note NCAR/TN-461+STR, 173.

PAULSON C A, SIMPSON J J, 1977. Irradiance measurements in the upper ocean[J]. Journal of Physical Oceanography, 7(6): 952-956.

QUENEY P, 1948. The problem of air flow over mountains: A summary of theoretical studies[J]. Bulletin of the American Meteorological Society, 29(1): 16-26.

SAJI N, GOSWAMI B N, VINAYACHANDRAN P N, et al, 1999. A dipole mode in the tropical Indian Ocean[J]. Nature, 401(6751): 360-363.

SCHNEIDER E K, 1977. Axially symmetric steady-state models of the basic state for instability and climate studies. Part II. Nonlinear calculations[J]. Journal of Atmospheric Sciences, 34(2): 280-296.

SCHNEIDER E K, 1987. A simplified model of the modified Hadley circulation[J]. Journal of Atmospheric Sciences, 44

(22): 3311-3328.

SCHNEIDER E K, LINDZEN R S, 1977. Axially symmetric steady-state models of the basic state for instability and climate studies. Part I: Linearized calculations[J]. J Atmos Sci, 34(2): 263-279.

TAO S Y, DING Y H, 1981. Observational evidence of the influence of the Qinghai-Xizang (Tibet) Plateau on the occurrence of heavy rain and severe convective storms in China[J]. Bulletin of the American Meteorological Society, 62(1): 23-30.

WANG Z, DUAN A, YANG S, 2018. Potential regulation on the climatic effect of Tibetan Plateau heating by tropical air-sea coupling in regional models[J]. Climate Dynamics, 52(3): 1685-1694.

WANG B, BAO Q, HOSKIN W B, et al, 2008. Tibetan Plateau warming and precipitation changes in east Asia[J]. Geophysical Research Letters, 35(14): L14702.

WU G, LIU H, ZHAO Y C, 1996. A nine-layer atmospheric general circulation model and its performance[J]. Advances in Atmospheric Sciences, 13(1): 1-18.

WU G X, LI W, GUO H, et al, 1997. Sensible heat driven air-pump over the Tibetan Plateau and its impacts on the Asian summer monsoon[C]//YE D Z. Essays in Honor Zhao Jiuzhang. Beijing: Science Press: 116-126.

WU G, ZHANG Y, 1998. Tibetan Plateau forcing and the timing of the monsoon onset over South Asia and the South China Sea[J]. Monthly Weather Review, 126(4): 913-927.

WU G, LIU Y, ZHANG Q, et al, 2007. The influence of mechanical and thermal forcing by the Tibetan Plateau on Asian climate[J]. Journal of Hydrometeorology, 8(4): 770-789.

WU G, LIU Y, DONG B, et al, 2012a. Revisiting Asian monsoon formation and change associated with Tibetan Plateau forcing: I. Formation[J]. Climate dynamics, 39(5): 1169-1181.

WU G, LIU Y, HE B, et al, 2012b. Thermal controls on the Asian summer monsoon[J]. Scientific Reports, 2(1): 1-7.

WU G, HE B, LIU Y, et al, 2015. Location and variation of the summertime upper-troposphere temperature maximum over South Asia[J]. Climate Dynamics, 45(9): 2757-2774.

XU Z, FU C, QIAN Y, et al, 2009. Relative roles of land-sea distribution and orography in Asian monsoon intensity[J]. Journal of the Atmospheric Sciences, 66(9): 2714-2729.

XU Z, QIAN Y, FU C, 2010. The role of land-sea distribution and orography in the Asian monsoon. Part II: Orography [J]. Advances in Atmospheric Sciences, 27(3): 528-542.

YE D Z, GAO Y X, 1979. Meteorology of the Qinghai-Xizang Plateau[M]. Beijing: Science Press: 278.

YEH T C, LUO S W, CHU P C, 1957. The wind structure and heat balance in the lower troposphere over the Tibetan Plateau and its surroundings[J]. Acta Meteorological Sinica, 28: 108-121.

YEH T C, 1950. The circulation of the high troposphere over China in the winter of 1945—46[J]. Tellus, 2(3): 173-183.

ZHAO P, CHEN L X, 2001. Interannual variability of atmospheric heat source/sink over the Qinghai-Xizang (Tibetan) Plateau and its relation to circulation[J]. Adv Atmos Sci, 18: 106-116.

ZHOU X J, ZHAO P, CHEN J M, et al, 2009. Impacts of thermodynamic processes over the Tibetan Plateau on the Northern Hemispheric climate[J]. Science in China Series D, 39: 1473-1486.

第4部分

亚洲季风的若干基本问题

综述 8

亚洲季风的气候特征和本质

刘伯奇[1], 毛江玉[2]

(1 中国气象科学研究院,北京 100081;2 中国科学院大气物理研究所,北京 100029)

摘要:季风泛指冬夏季节盛行风向相反的气候现象。季风形成的本质是太阳年循环下海陆热力差异的季节性变化。亚洲季风是全球季风系统中最明显、范围最大的区域季风系统。冬季盛行冷、干的东北季风,而夏季盛行暖、湿的西南季风,冬、夏季风环流系统的季节转换具有爆发性的突变特征。亚洲季风包含南亚季风和东亚季风,两者的气候特征和维持机制差异明显。其中,南亚季风降水的形成和索马里跨赤道气流的水汽输送以及赤道辐合带的季节性经向移动有关。在此基础上,吴国雄院士团队揭示了南亚次大陆纬向海陆分布对赤道辐合带附近热带对流的强迫作用,提出了印度夏季风强迫对流发展的新理论。同时,该团队还通过数值试验证明东亚季风的形成不仅依赖于行星尺度的海陆分布,还受青藏高原大地形动力和热力强迫调制,两者的共同作用在北半球副热带地区形成独特且强盛的东亚季风气候。

1 引言

季风(monsoon),是一个古老而又有新意的气候学概念。阿拉伯人很早已发现了季风,并称之为"Mausim",意思为季节。17世纪后期英国哈莱首先提出季风是由于海陆热力性质的不同和太阳辐射的季节变化而产生的以一年为周期的大型海陆直接环流(Halley,1686)。在中国古代对季风有各种不同的名称,如信风、黄雀风、落梅风、舶风,所谓舶风即夏季从东南洋面吹至我国的东南季风。北宋苏东坡《船舶风》诗中有"三时已断黄梅雨,万里初来船舶风"之句。20世纪30年代,竺可桢(1934)系统地阐述了东南季风与中国雨量的关系。本质上,季风是海陆热力差异导致的在大陆和海洋之间大范围风向随季节变化的气候现象(Halley,1686;Hadley,1735)。现代人们对季风现象的认识包含以下三个基本观点:季风是大范围地区的盛行风向随季节改变的现象;冬季风寒冷干燥少雨,夏季风温暖潮湿多雨;随着盛行风向的变换,冬季和夏季的天气气候有显著不同。

季风包括了多个空间尺度,即区域季风、洲际季风和全球季风等。其中,全球季风系统的典型特征是热带和副热带地区全球尺度大气环流持续的季节性反转(Qian et al.,1998;Trenberth et al.,2000;Lau et al.,2007;Wang and Ding,2008;An et al.,2015)。全球季风系统包括热带和副热带全球季风系统。其中热带全球季风系统主要分布在亚洲和澳大利亚热带地区、热带非洲、南美洲和热带东太平洋地区。其中热带亚洲-澳大利亚季风系统又包含南亚季风(包括印度和孟加拉湾季风)、南海季风-西北太平洋季风、热带澳大利亚季风和海洋大陆季风。而副热带全球季风则以东亚副热带季风为最显著,同时还有北美季风、北非季风、青藏高原季风、副热带北大西洋和北太平洋季风,以及南澳大利亚季风、南非季风和副热带南太平洋季风。

通讯作者:刘伯奇,liubq@cma.gov.cn
资助项目:国家自然科学基金重点项目 41830969

早在17世纪末期,科学家就发现了海洋和陆地热力差异对季风形成的重要作用,并提出行星尺度的"海-陆"风模型(Halley,1686),形成了季风动力学雏形。随着观测网络的逐渐完善,人们对季风的认识也不断加深。20世纪,大气环流与大尺度动力学的快速发展带动了季风动力学的发展,现代季风动力学变成以大气长波理论、大气环流形成机理和热带大气动力学为核心,以季风的形成演变和发展变率为研究对象,以理论分析和数值模拟为研究手段的可试验的现代科学。

南亚季风和东亚副热带季风作为亚洲季风系统最重要的成员,影响着全球超过50%人口的生产生活,其形成和维持机制是季风动力学研究的长期主题之一。本文旨在通过回顾亚洲季风的气候特征和本质,总结亚洲季风动力学近30年来的主要进展,特别是吴国雄院士团队所取得的理论和数值试验方面的进展(例如,Liu and Wu,1997;吴国雄和张永生,1998;Xu and Wu,1999;Liu et al.,2002;Liang et al.,2005;毛江玉和吴国雄,2006;Wang et al.,2004;王同美和吴国雄,2008;Wu et al.,2009;吴国雄 等,2010,2013,2016;Wu et al.,2012;Liu et al.,2012;Liu et al.,2013;吴国雄和刘伯奇,2014),并在此基础上展望季风动力学的未来发展方向。

2 亚洲季风的气候特征

亚洲季风特指盛行于亚洲区域的季风,是最为活跃的气候系统,表现为冬季和夏季盛行风向相反的气候特征,其典型代表是南亚季风和东亚季风。亚洲季风年际干湿变率大,对社会经济、农业生产和交通运输等国民经济的发展具有重要影响。南亚季风和东亚季风两者既紧密联系、又相对独立(Tao and Chen,1987),其主要差异在于印度季风属于热带季风,而东亚季风兼有热带和副热带季风属性。东亚副热带季风受到中、高纬度系统的影响比较大,因而东亚季风系统更加复杂。现在人们已逐步认识到,西北太平洋地区季风也是亚洲季风系统的重要组成部分。

亚洲季风系统在夏季和冬季的主要环流成员完全不同(图1),夏季主要季风环流系统包括:马斯克林高压、澳大利亚高压、西太平洋高压、印度北部和南部季风槽、东非越赤道低空急流、南海低空急流、副热带西南低空急流、印度北部及南海地区和江淮流域降水与云覆盖、对流层上层青藏高压、热带东风急流等;冬季主要季风环流系统包括:西伯利亚高压、印度尼西亚季风槽、对流层低层季风涌、马来西亚南部和印度尼西亚的降水与云覆盖、对流层上层的西太平洋高压、高空副热带西风急流等。季风环流系统由若干个成员组成,而且夏季风环流系统要比冬季风环流系统复杂得多。因此,亚洲夏季风的气候特征受到气象学界更广泛的关注。

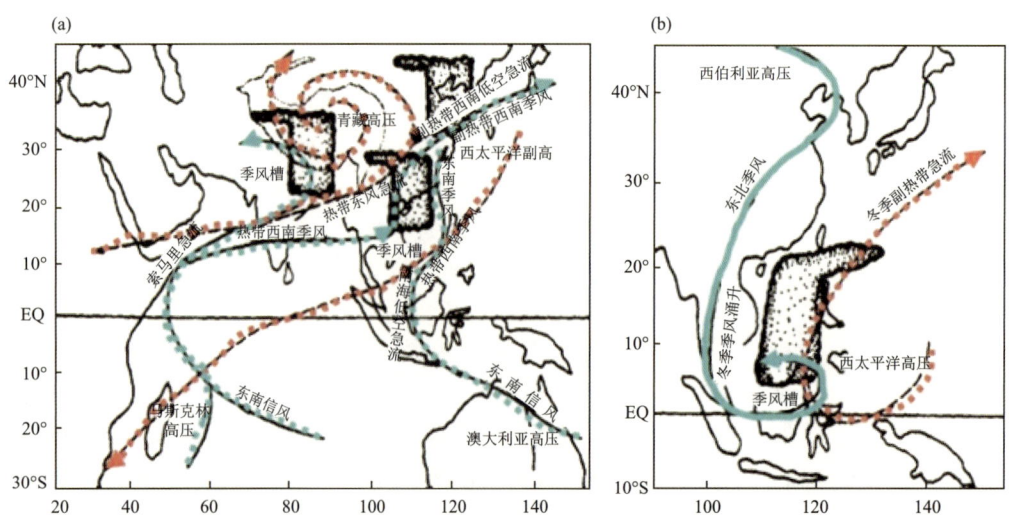

图1 亚洲夏季风系统(a)和冬季风系统(b)示意图
(实线表示低层系统,虚线表示高层系统)(摘自伍荣生,1999)

2.1 南亚夏季风

南亚夏季风又包含孟加拉湾夏季风和印度夏季风两个重要成员,属于热带季风。其中孟加拉湾季风是指位于孟加拉湾东北部(90°~100°E,10°~20°N)的季风现象,而印度季风是指位于印度半岛和阿拉伯海地区(60°~80°E,10°~20°N)的季风现象。

2.1.1 孟加拉湾夏季风

在气候学上,孟加拉湾夏季风是亚洲区最早爆发的子系统,其爆发时间平均是4月底5月初(Mao and Wu,2007)。孟加拉湾夏季风爆发过程的基本特征包括:高空南亚高压形成(Liu et al.,2013),副高脊面向北倾斜,低空副热带高压带断裂,对流层中上部经向温度梯度反转(Mao and Wu,2007),对流层低层季风爆发性涡旋形成发展(吴国雄 等,2010),印缅槽加强,同时孟加拉湾南部的热带对流从赤道(苏门答腊)附近自南向北推进。此外,孟加拉湾夏季风爆发过程与青藏高原的强迫作用关系紧密(毛江玉和吴国雄,2006),高空南亚高压季节变化与低空季风爆发性涡旋的垂直耦合是孟加拉湾夏季风爆发的关键物理过程(Liu et al.,2013,2014),而来自中纬度的偏东气流所造成的低层正压不稳定流场也有利于孟加拉湾夏季风的爆发。但是,目前对孟加拉湾冬季风的研究仍较少。

孟加拉湾夏季风的爆发时间和强度都受ENSO事件及其引起的印度洋-西太平洋暖池海温异常影响(Mao and Wu,2007),表现出明显的年际变化。孟加拉湾夏季风爆发后,对流层低层的热带西风和副热带西风在孟加拉湾地区打通,大量水汽随着西南风暖湿气流输送到我国南方地区,进而加强东亚副热带夏季风。另一方面,孟加拉湾夏季风爆发后,季风对流向东传播,并于5月中下旬到达南海地区,引起南海夏季风爆发。因此,孟加拉湾夏季风的爆发标志着亚洲热带夏季风建立进程的开始。

2.1.2 印度夏季风

印度夏季风通常于5月底6月初建立,标志着印度地区进入雨季。印度夏季风建立的特征包含:高空副热带西风急流北跳至青藏高原北部,热带东风急流到达印度南部地区;低空索马里越赤道气流加强,阿拉伯海上空的低空急流加速、西南风明显增大,以及印度季风槽建立和热带对流自南向北推进至印度地区(Yin,1949;Koteswaram,1958;Findlater,1969;Joseph and Sijikumar,2004;Rao,1976;Joseph et al.,2006)。

印度夏季风的建立时间和强度都具有明显的多尺度变化特征。在季节内尺度上,印度夏季风的建立、活跃和中断都与赤道印度洋上空明显的30~60天季节内振荡北传有关(Goswami et al.,2006)。在年际尺度上,ENSO事件、印度洋海温异常和欧亚大陆冬春积雪是影响印度夏季风环流和降水的重要因素(Liu et al.,2014)。在年代际尺度上,印度夏季风降水具有55~60年的准周期年代际振荡,这种年代际变化既与太平洋海温的类ENSO型年代际变率有关(Krishnan and Sugi,2003),也受人类活动排放气溶胶引起的大气辐射平衡变化的影响(Ueda et al.,2006)。印度夏季风作为全球最强最显著的子季风系统之一,能够影响慢变的边界层条件,进而影响ENSO事件的发生和演变(Yasunari,1990)。一方面,印度夏季风降水能够改变印度洋的低空东传异常西风,从而引起滞后1~2季节的成熟ENSO事件。另一方面,印度夏季风能够改变Walker环流造成太平洋的异常低空信风,进而影响ENSO事件。此外,印度夏季风降水能够激发出向下游传播的大气定常波列,来影响东亚夏季风的环流和降水异常(Liu and Ding,2008)。

2.2 东亚夏季风

东亚夏季风包含南海-西北太平洋夏季风和东亚副热带夏季风两个主要成员,前者具有典型的热带季风属性,而后者则属于副热带季风系统(朱乾根和何金海,1985)。中国南海-西北太平洋季风又包含南海季风和西北太平洋季风,是指位于中国南海-西北太平洋地区(5°~22.5°N,105°~150°E)的季风现象(Wang and LinHo,2002)。东亚副热带季风是出现在东亚副热带地区(20°N以北)的季风现象,受东亚地区海陆热力性质对比影响,东亚副热带季风表现为冬、夏季节东亚副热带地区盛行风向和降水主雨带的季节性变化，其形成和演变与东亚地区的天气气候异常联系十分紧密。

2.2.1 中国南海-西北太平洋夏季风

中国南海-西北太平洋夏季风始于每年5月第4候（第28候）前后南海夏季风的建立，约5天后，季风对流到达西北太平洋，西北太平洋夏季风随之爆发。因此，南海夏季风也被视为东亚夏季风爆发的前兆。中国南海-西北太平洋夏季风的爆发十分迅速激烈，主要特征有：高层南亚高压在中南半岛南部建立并迅速加强北移；低层印缅槽加强，赤道印度洋西风加强并向东向北迅速扩展和传播，以及相伴随的中低纬度相互作用和西太平洋副高连续东撤。而在北半球秋季，中国南海-西北太平洋夏季风的撤退过程则相对平缓（朱乾根和何金海，1985；李崇银和屈昕，2000；He et al.，2006）。

中国南海-西北太平洋季风环流和降水的变化既受热带环流系统影响，又受中高纬冷空气活动调控，表现出十分明显的多尺度变率（Wang et al.，2009）。在季节内尺度上，热带的30～60天和准双周季节内振荡是造成中国南海-西北太平洋季风的建立、活跃和中断的重要因素。在年际尺度上，ENSO事件、印度洋海温、中国南海-西北太平洋局地海洋热容量和青藏高原的冬春季节积雪都会影响中国南海-西北太平洋季风降水和环流。在年代际尺度上，中国南海-西北太平洋季风在20世纪90年代初发生了显著变化，表现为1993/1994年之后，夏季南海北部季风降水明显增多，秋季南海南部降水减少，冬、春季南海中部降水增多，并且中国南海-西北太平洋季风与东亚副热带季风的关系也发生了年代际调整。此外，中国南海-西北太平洋季风和ENSO事件的联系在20世纪70年代后明显加强，同时中国南海-西北太平洋冬季风在过去的60年内显著增强。

中国南海-西北太平洋季风的降水和环流异常还具有明显的全球气候效应。一方面，中国南海-西北太平洋夏季风降水异常能够激发出向东北方向传播的"太平洋-日本"（亦称东亚）遥相关型波列，从而影响东亚副热带夏季风。另一方面，中国南海-西北太平洋夏季风能够通过引起西太平洋的赤道异常东风，进而影响ENSO事件的消亡过程，而中国南海-西北太平洋冬季风也能够造成次年冬季赤道中太平洋的海温异常，并触发ENSO事件（穆明权和李崇银，1999）。

2.2.2 东亚副热带夏季风

东亚副热带夏季风的气候特征包括：低空盛行偏南风，伴随着华南前汛期、江淮梅雨和华北-东北雨季的先后建立。东亚副热带夏季风有时特指北纬30°N附近的江淮梅雨（Ding and Chan，2005）。与亚洲热带夏季风相比，东亚副热带夏季风除了受热带环流系统（如，西太副高、南亚高压、季风槽等）影响外，还受中纬度环流系统（如，阻塞高压、东亚大槽、东北冷涡等）调制，从而导致梅雨锋面系统的时空变化（Ding and Chan，2005；Ding et al.，2020）。

东亚副热带季风环流和降水存在多时间尺度变率，不同尺度的影响因子也各不相同（He and Liu，2016）。(1) 就其季节内变率而言，存在明显的时空差异，这与青藏高原春季感热的季节内变化、中纬度急流中的涡度异常移动、西太平洋副高的短期振荡和热带季节内振荡（Madden-Julian oscillation，简写MJO）的经向传播等因素有关；(2) 就其年际变率而言，主要外强迫因子是ENSO事件、印度洋海温异常（包括海盆一致模和印度洋偶极子模（Indian Ocean dipole，简写IOD））、欧亚大陆及青藏高原冬春积雪，以及东亚地区土壤湿度；(3) 就其年代际变率而言，东亚副热带夏季风自20世纪70年代以来明显减弱，而自20世纪90年代以来又开始加强，这与印度洋-海洋大陆热带对流的加强以及青藏高原热状况的年代际变率都有联系，而这既与气候系统自然变率（如太平洋年代际振荡（Pacific decadal oscillation，简写PDO））有关，也受人类活动排放的温室气体和气溶胶影响。

3 亚洲季风的本质

一般而言，全球季风的基本驱动力是大气层顶接收的太阳短波辐射的季节循环，而经、纬向海陆热力对比,环流-降水反馈和高大地形则显著加强了区域季风。具体而言,在太阳辐射年循环的强迫作用下,海-陆-气系统通过产生或改变跨赤道气压梯度的季节变化,进而形成了季风,其中主要的物理过程包括:陆地和海水热容量差异导致的海陆热力对比,海洋和陆地上热量存储和垂直输送方式的差异,湿过程对不同加热方式的调控作用,不同类型加热引起的跨赤道气压梯度力,以及海洋中动力过程造成的经向热量输送(Webster et al.,1998;Trenberth et al.,2000)。然而,对亚洲夏季风而言,南亚夏季风和东亚夏季风,尤其是东亚副热带夏季风的维持机制明显不同。

3.1 南亚季风的维持机制

作为热带季风系统,南亚季风的本质是赤道辐合带(ITCZ)的季节性经向摆动(Saha S and Saha K,1980;Gadgil,2003)。ITCZ的季节性移动和赤道附近的惯性不稳定机制有关(Tomas and Webster,1997;Tomas et al.,1999)。吴国雄和刘伯奇(2014)则提出了"强迫对流发展"的南亚夏季风维持机制,具体而言:北半球由冬到夏,随着太阳直射点越过赤道向北回归线移动,北印度洋海表温度快速升高,热带印度洋出现"北暖南冷"的跨赤道海温经向梯度,这时阿拉伯海热带地区的表面气压呈南高北低分布,对应着近赤道的纬向地转风。由于近赤道的惯性力很小,在气压梯度力的驱动下,近地层大气跨赤道从南半球进入北半球(图2)。该偏南气流首先穿越惯性不稳定区,产生非地转东风,使西风减速;接着穿越绝对涡度零线进入惯性稳定区,产生非地转西风,使西风加速。在惯性力作用下,上述非地转纬向风在惯性不稳定区使穿越等压线的南风加强,加强了局地经圈环流;在惯性稳定区使穿越等压线的南风减弱。总的结果是在惯性不稳定区,穿越等压线的南风加强,地转西风减弱;相反在惯性稳定区,穿越等压线的南风减弱,地转西风加强(图2a)。为了维持大气的准平衡态,在绝对涡度0线以南,大气辐散;而在绝对涡度0线以北,大气辐合(图2b);南风在绝对涡度0线附近最大。于是,低空辐合和对流降水在绝对涡度0线北侧被激发产生(图2c)。相应地,ITCZ偏离赤道向北半球移动。

但是,上述机制只能解释热带海洋上ITCZ的季节性位移和南亚夏季风的洋面对流,却无法解释印度半岛上空的季风降水。吴国雄和刘伯奇(2014)的研究证明,跨赤道经向气压梯度在阿拉伯海南部对应着低空西风急流,而南、北海陆热力差异在阿拉伯海东南部和印度西南海岸的向东增强导致该处形成明显的纬向地转动量平流。该平流作用诱发的强迫对流发展使低空辐合中心出现在最大西风轴线北侧,相应地季风对流降水出现在印度半岛西南海岸,印度夏季风建立(图2d)。之后,在青藏高原热力强迫作用下,该

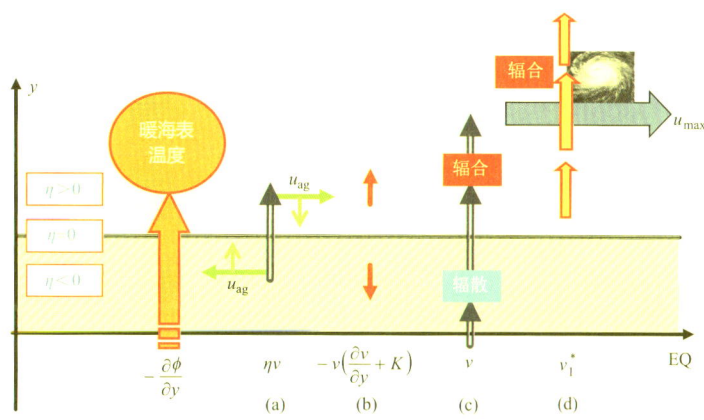

图2 强迫和惯性不稳定对流发展示意图。(a)在跨赤道表面气压梯度力驱动下,向北的气块在惯性不稳定区($\eta<0$)/稳定区($\eta>0$)纬向风减弱/增强(ηv,绿色粗实箭头),穿越等压线的南风加强/减弱($-\beta y u_{ag}$,绿色细实箭头);(b)为平衡纬向地转偏差风在南北向产生的惯性加/减速,摩擦和平流作用($-v\left(\frac{\partial v}{\partial y}+K\right)$)必须有反向的减/加速;(c)导致在绝对涡度0线以南/北产生辐散/辐合;(d)强迫对流发展:经向气压梯度的纬向变化还强迫产生附加的经向风,在位于更北的最大纬向风(u_{max})附近及其以北区域形成辐合,导致强迫对流发展和印度夏季风爆发(摘自吴国雄和刘伯奇,2014)

低空辐合区继续北抬,季风雨带随之北上,印度夏季风逐步向北推进(Wu et al.,2012)。因此,"强迫对流发展"理论是对传统 ITCZ 动力学的创新发展,解释了印度夏季风形成维持的根本原因。

3.2 东亚季风的维持机制

与印度季风不同,东亚季风区位于 10°N 以北,并且季风降水可以深入东亚大陆内陆地区,这无法用 ITCZ 的季节性移动解释。数值试验表明,东亚季风形成的本质是"行星尺度海陆分布-青藏高原大地形强迫"共同作用的结果(Liang et al.,2005;梁潇云 等,2006)。在无陆地分布的水球试验中,虽然 ITCZ 由冬至夏的季节性移动能够引起赤道附近降水的季节差异,但无法在赤道外地区产生季风现象(图3)。当加入行星尺度亚洲大陆理想陆地以后,冬、夏季盛行风向角转变出现在 20°N 的东亚地区,但局地夏季风降水依然很弱,降水大值中心仍位于赤道附近(图4)。在此基础上,进一步考虑亚非大陆的空间差异后,由冬到夏的跨赤道气流明显加强,印度次大陆和中南半岛的季风降水变得更接近观测,但 20°N 以北的东亚季风降水依然很弱(图5)。研究表明,在增加青藏高原大地形以后,模式模拟的东亚副热带夏季风降水明显加强(图6)。因此,亚洲行星尺度海陆分布和青藏高原大地形强迫是东亚季风,尤其是东亚副热带季风形成维持的根本原因。

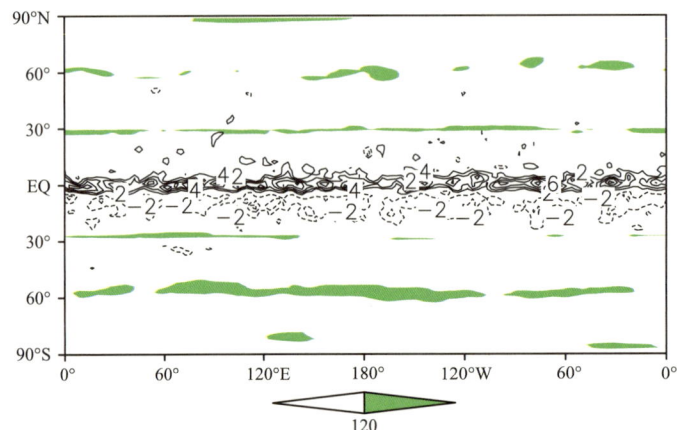

图 3 水汽试验中 7 月和 1 月地面风向的偏转(阴影区为偏转角度大于 120°的区域)和降水差异(7 月减去 1 月)(等值线间隔 2 mm·d^{-1})(摘自梁潇云 等,2006)

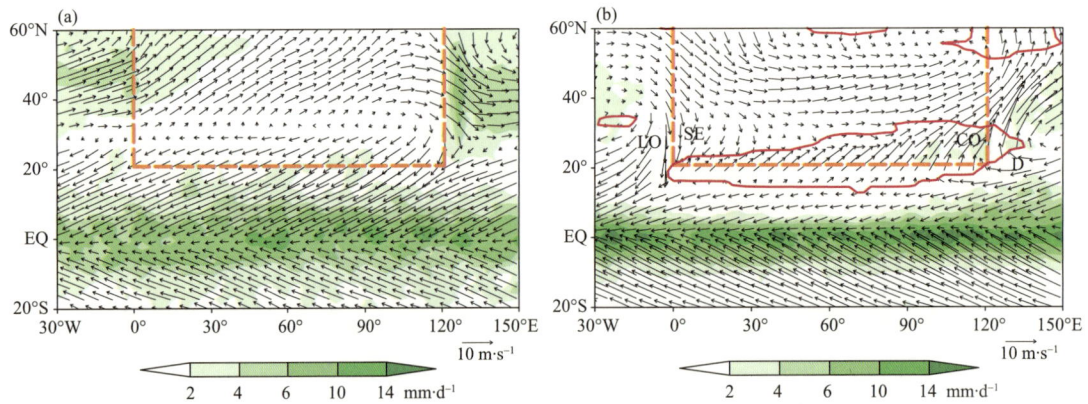

图 4 亚洲大陆理想地形试验的地面风场(矢量,m·s^{-1})和降水场(阴影,mm·d^{-1})。(a)1月,(b)7月。(b)中的红色曲线所包围区域为 7 月与 1 月风向偏转角大于 120°的区域,橘黄色粗虚线为海陆分界线(摘自梁潇云 等,2006)

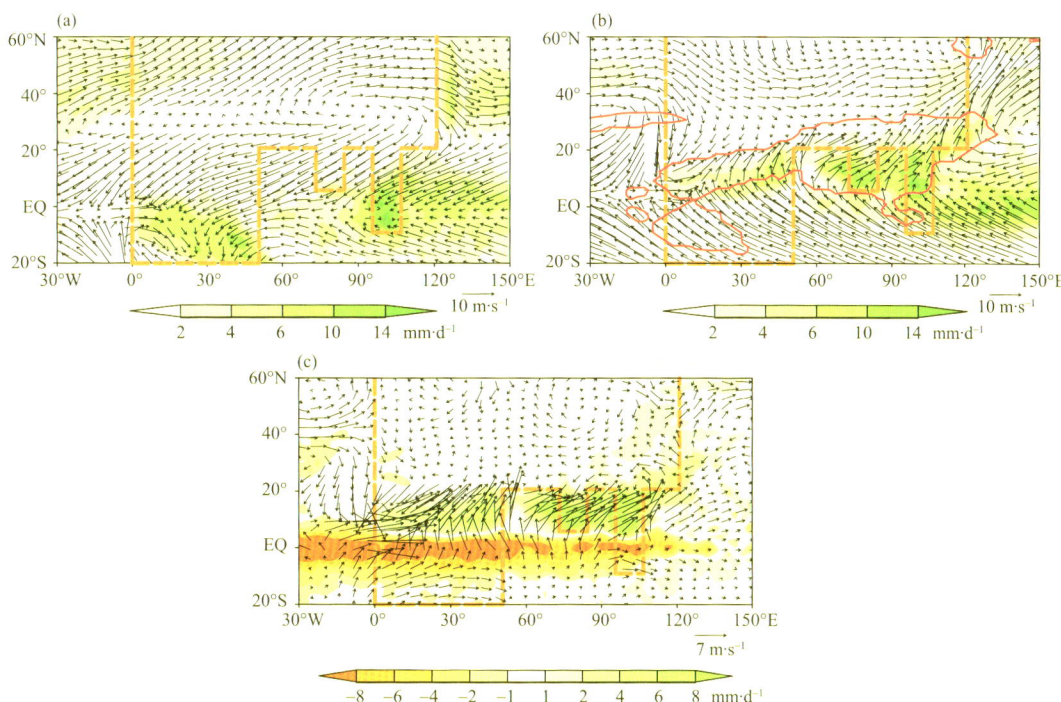

图5 亚非大陆理想地形试验的地面风场(矢量,m·s⁻¹)和降水场(阴影,mm·d⁻¹)。(a)1月,(b)7月,(c)和亚洲大陆理想地形试验之差。(b)中的红色曲线所包围区域为7月与1月风向偏转角大于120°的区域,橘黄色粗虚线为海陆分界线(摘自梁潇云 等,2006)

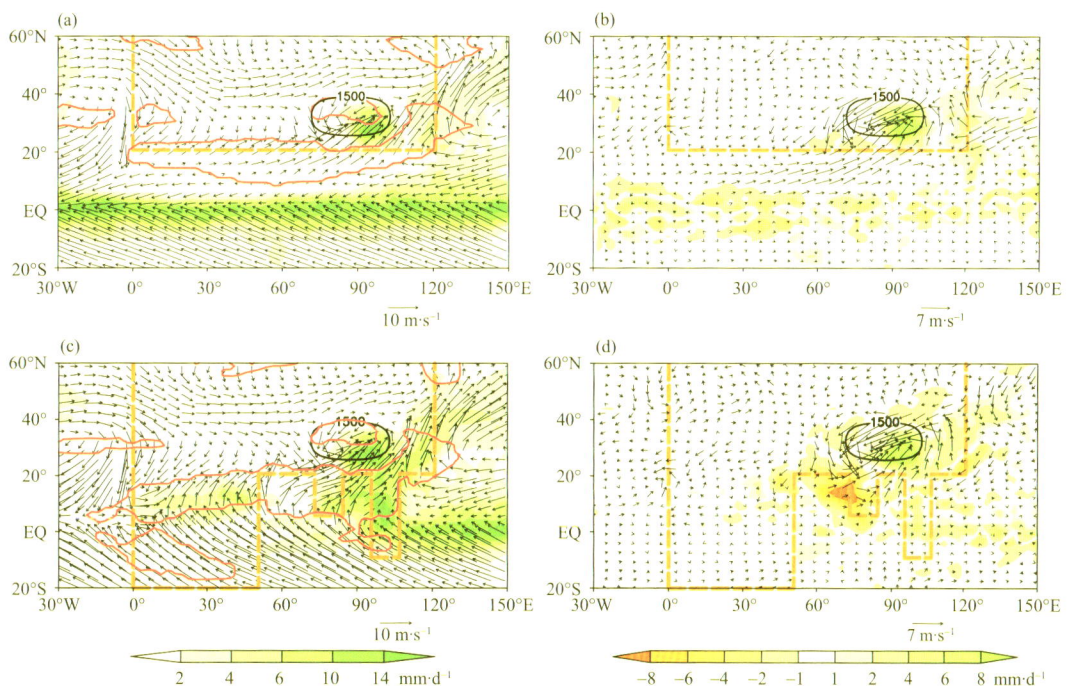

图6 青藏高原大地形强迫试验结果。(左列)7月地面风场(矢量,m·s⁻¹)和降水(阴影,mm·d⁻¹)及其(右列)与无高原试验的差异(摘自梁潇云 等,2006)

4 总结和讨论

文中简要回顾了亚洲季风气候的基本特征,重点总结了南亚季风与东亚季风形成维持机理的差异,特别是基于数值模拟试验揭示出青藏高原热力和动力强迫在东亚副热带季风气候形成和变异中的关键作用,反映了吴国雄院士团队在季风动力学和季风形成机理方面的重要进展。

经典季风动力学研究以天气和气候动力学理论为基础,首先根据季风环流的特点,对控制大气运动的方程组进行简化推演,使其能够准确反映季风环流的变化特征,然后利用数学解析方法,分析运动方程中不同变量之间的数学关系,接着结合季风环流的演变规律,将数学关系和实际物理过程相结合,从而解释季风环流及其变化规律的动力学机制。随着数值模式的发展,利用数值模拟研究与季风有关的科学问题成为季风动力学研究的新方向,使我们可以考虑季风区更加复杂的物理过程。近年来,随着气候变化问题越来越受到关注,利用动力学理论和复杂气候系统、甚至地球系统模式分析全球气候变化背景下多圈层相互作用对季风的影响将成为季风动力学的新方向。

参考文献

李崇银,屈昕,2000.伴随南海夏季风爆发的大尺度大气环流演变[J].大气科学,24(1):1-14.
梁潇云,刘屹岷,吴国雄,2006.热带海陆分布和青藏高原在亚洲夏季风形成中的作用[J].地球物理学报,49(4):983-992.
毛江玉,吴国雄,2006.青藏高原热状况和海温异常对亚洲季风季节转换年际变化的影响[J].地球物理学报,49(5):1279-1289.
穆明权,李崇银,1999.东亚冬季风年际变化的ENSO信息Ⅱ.模拟资料分析[J].气候与环境研究,4:176-184.
王同美,吴国雄,2008.南亚海陆热力差异及其对热带季风区环流的影响[J].热带气象学报,27(1):37-43.
吴国雄,张永生,1998.青藏高原的热力和机械强迫作用以及亚洲季风的爆发 I.爆发地点[J].大气科学,22(6):825-838.
吴国雄,关月,王同美,等,2010.春季孟加拉湾涡旋形成及其对亚洲夏季风爆发的激发作用[J].中国科学:地球科学,40(11):1459-1467.
吴国雄,段安民,刘屹岷,等,2013.关于亚洲夏季风爆发的动力学研究的若干近期进展[J].大气科学,37(2):211-228.
吴国雄,刘伯奇,2014.强迫和惯性不稳定对流发展在印度夏季风爆发过程中的作用[J].中国科学:地球科学,44:783-796.
吴国雄,何编,刘屹岷,等,2016.青藏高原和亚洲夏季风动力学研究的新进展[J].大气科学,40(1):22-32.
伍荣生,1999.现代天气学原理[M].北京:高等教育出版社:316.
朱乾根,何金海,1985.亚洲季风建立及其中期振荡的高空环流特征[J].热带气象,1:9-18.
竺可桢,1934.东南季风与中国之雨量[J].地理学报,1:1-27.
AN Z S, WU G X, LI J P, et al, 2015. Global monsoon dynamics and climate change[J]. Annu Rev Earth Planet Sci, 43:29-77.
DING Y H, CHAN J C L, 2005. The East Asian summer monsoon: An overview[J]. Meteor Atmos Phys, 89:117-142.
DING Y H, LIANG P, LIU Y J, et al, 2020. Multiscale variability of Meiyu and its prediction: A new review[J]. J Geophys Res Atmos, 125. doi:10.1029/2019JD031496.
FINDLATER J, 1969. A major low-level air current near the Indian Ocean during the northern summer[J]. Quart J Roy Meteor Soc, 195:362-380.
GADGIL S, 2003. The Indian monsoon and its variability[J]. Annu Rev Earth Planet Sci, 31:429-467.
GOSWAMI B N, WU G X, YASUNARI T, 2006. The annual cycle, intraseasonal oscillations, and roadblock to seasonal predictability of the Asian summer monsoon[J]. J Climate, 19:5078-5099.
HADLEY G, 1735. Concerning the cause of the general trade winds[J]. Philos Trans Roy Soc A, 39:58-62.
HALLEY E, 1686. An historical account of the trade winds and monsoons observable in the seas between and near the tropics with an attempt to assign the physical cause of the said wind[J]. Philos Trans Roy Soc A, 16:153-168.
HE J H, WEN M, WANG L J, et al, 2006. Characteristics of the onset of the Asian summer monsoon and the importance of Asian-Australian "Land Bridge"[J]. Adv Atmos Sci, 23:951-963.
HE J H, LIU B Q, 2016. The East Asian subtropical summer monsoon: Recent progress[J]. J Meteor Res, 30:135-155.

JOSEPH P V, SIJIKUMAR S, 2004. Intraseasonal Variability of the Low-Level Jet Stream of the Asian Summer Monsoon [J]. J Clim, 17: 1449-1458.

JOSEPH P V, SOORAJ K P, RAJAN C K, 2006. The summer monsoon onset process over South Asia and an objective method for the date of monsoon onset over Kerala[J]. Int J Climatol, 26: 1871-1893.

KOTESWARAM P, 1958. The easterly jet stream in the tropics[J]. Tellus,10: 43-57.

KRISHNAN R, SUGI M, 2003. Pacific decadal oscillation and variability of the Indian summer monsoon rainfall[J]. Clim Dyn,21: 233-242.

LAU W K M, KIM K M, LEE M I, 2007. Characteristics of diurnal and seasonal cycles in global monsoon systems[J]. J Meteor Soc Japan,85A: 403-416.

LIANG X Y, LIU Y M,WU G X, 2005. The role of land-sea distribution in the formation of the Asian summer monsoon [J]. Geophys Res Lett, 32, doi:10.1029/2004GL021587.

LIU B Q, WU G X, MAO J Y, et al, 2013. Genesis of the South Asian high and its impact on the Asian summer monsoon onset[J]. J Climate, 26: 2976-2991.

LIU B Q, WU G X, REN R C, 2014. Influences of ENSO on the vertical coupling of atmospheric circulation during the onset of South Asian summer monsoon[J]. Clim Dyn,45: 1859-1875.

LIU H, WU G X, 1997. Impacts of land surface on climate of July and onset of summer monsoon: A study with an AGCM plus SSiB[J]. Adv Atmos Sci,14: 273-282.

LIU Y M, CHAN J C L, MAO J Y, et al, 2002. The role of Bay of Bengal convection in the onset of the 1998 South China Sea summer monsoon[J]. Mon Wea Rev,130: 2731-2744.

LIU Y M, WU G X, HONG J L, et al, 2012. Revisiting Asian monsoon formation and change associated with Tibetan Plateau forcing: II. Change[J]. Clim Dyn,39: 1183-1195.

LIU Y Y, DING Y H, 2008. Teleconnection between the Indian summer monsoon onset and the Meiyu over the Yangtze River Valley[J]. Science in China Series D: Earth Sciences,51:1021-1035.

MAO J Y, WU G X, 2007. Interannual variability in the onset of the summer monsoon over the eastern Bay of Bengal[J]. Theor Appl Climatol,89: 155-170.

QIAN W H, ZHU Y F, XIE A, et al, 1998. Seasonal and interannual variations of upper tropospheric water vapor band brightness temperature over the global monsoon regions[J]. Adv Atmos Sci, 15:337-345.

RAO Y P, 1976. Southwest Monsoon[M]. India Meteorological Department;Met. Monograph:1-376.

SAHA S, SAHA K,1980. A hypothesis on onset, advance and withdrawal of the Indian summer monsoon[J]. Pure Appl Geophys, 118: 1066-1075.

TAO S Y, CHEN L X, 1987. A Review of Recent Research on East Asian Summer Monsoon in China. Monsoon Meteorology[M]. Oxford:Oxford University Press:60-92.

TOMAS R A, WEBSTER P J, 1997. The role of inertial instability in determining the location and strength of near-equatorial convection[J]. Q J R Meteor Soc, 123: 1445-1482.

TOMAS R A, HOLTON J R, WEBSTER P J, 1999. The influence of cross-equatorial pressure gradients on the location of near-equatorial convection[J]. Q J R Meteor Soc, 125: 1107-1127.

TRENBERTH K E, STEPANIAK D P, CARON J M, 2000. The global monsoon as seen through the divergent atmospheric circulation[J]. J Climate,13: 3969-3993.

UEDA H, IWAI A, KUWAKO K, 2006. Impact of anthropogenic forcing on the Asian summer monsoon as simulated by eight GCMs[J]. Geophys Res Lett,33, doi:10.1029/ 2005GL025336.

WANG B, LINHO, 2002. Rainy season of the Asian-Pacific summer monsoon[J]. J Climate,15: 386-398.

WANG B, DING Q, 2008. Global monsoon: Dominant mode of annual variation in the tropics[J]. Dyn Atmos Oceans, 44: 165-183.

WANG B, HUANG F, WU Z W, et al, 2009. Multi-scale climate variability of the South China Sea monsoon: A review [J]. Dyn Atmos Oceans, 47: 15-37.

WANG Z Z, WU G X, WU T W, et al, 2004. Simulation of the seasonal variations of Asian monsoon with the climate model R42L9/LASG[J]. Adv Atmos Sci,21: 879-889.

WEBSTER P J, MAGANA V O, PALMER T N, et al, 1998. Monsoons: Processes, predictability, and the prospects for prediction[J]. J Geophys Res,103: 14451-14510.

WU G X, LIU Y M, ZHU X Y, et al, 2009. Multi-scale forcing and the formation of subtropical desert and monsoon[J]. Annals of Geophysics, 27: 3631-3644.

WU G X, LIU Y M, DONG B W, et al, 2012. Revisiting Asian monsoon formation and change associated with Tibetan Plateau forcing: Ⅰ. Formation[J]. Clim Dyn,39: 1169-1181.

XU J J, WU G X, 1999. Dynamic features and maintenance mechanism of the Asian summer monsoon subsystem[J]. Adv Atmos Sci,16: 523-536.

YASUNARI T,1990. Impact of Indian monsoon on the coupled atmosphere/ocean system in the tropical Pacific[J]. Meteor Atmos Phys,44: 29-41.

YIN M T, 1949. A synoptic-aerologic study of the onset of the summer monsoon over India and Burma[J]. J Atmos Sci,6: 393-400.

综述 9

亚洲夏季风爆发的阶段性及其物理机制

毛江玉[1], 刘伯奇[2], 颜京辉[3]

(1 中国科学院大气物理研究所,北京 100029;2 中国气象科学研究院,北京 100081;
3 中国气象局地球系统数值预报中心,北京 100081)

摘要:亚洲夏季风爆发标志着亚洲地区大气环流系统由冬季型向夏季型的转换,表现为盛行风向的反转和降水的迅速增加。文中总结了吴国雄院士团队在亚洲热带夏季风爆发基本规律和动力学方面的主要进展及创新性成果。发现亚洲夏季风爆发首先始于孟加拉湾东部、随之中国南海、最后印度的"三阶段"特征;基于副热带高压脊面倾斜与热成风理论,从副热带高压带形态变异的角度证实了亚洲夏季风爆发的阶段性,由此提出以脊面附近对流层中上层经向温度梯度作为表征夏季风爆发的新指数。青藏高原的动力和热力强迫对亚洲夏季风爆发地点具有锚定作用。一方面,高原的机械阻挡作用维持了孟加拉湾春季暖池,通过局地海-气相互作用激发局地爆发性涡旋而使得低空副高带断裂;另一方面,高原的热力作用通过调制亚洲季风区北侧的经向温度梯度,引导了季风爆发的"三阶段"。在此期间,亚洲南部局地海-陆-气相互作用在印度半岛东侧形成独特的"季风爆发屏障"现象,而高空南亚高压的形成与变异是亚洲夏季风爆发"三阶段"物理联系的纽带。因此,历时近一个月的亚洲夏季风阶段性爆发是发生在特定的地理环境下、受特定的动力-热力学规律驱动的接续过程。

1 引言

亚洲季风是全球最强、范围最广的大气环流系统,它的演变影响着全球半数以上的人口生存问题(Webster et al.,1998;吴国雄 等,2004,2013)。亚洲季风一般分为南亚季风和东亚季风两大子系统(Tao and Chen,1987)。亚洲季风又可分为印度、孟加拉湾、中国南海和西北太平洋四个热带子季风系统和以长江流域梅雨为代表的东亚副热带子季风系统(Wu and Zhang,1998;Wu and Wang,2000;毛江玉 等,2002a;吴国雄 等,2013;Wu and Liu,2014)。所以,亚洲夏季风是因地域而分阶段爆发的(Wu and Zhang,1998;毛江玉 等,2002a)。尽管夏季风爆发表现为子季风区低空盛行风向的反转和降雨量的迅速增加,但本质是副热带高压(简称副高)脊面附近经向温度梯度的反转(毛江玉 等,2002a,2002b),也就是亚洲区大气环流由冬季型向夏季型的转换。

气候学上,亚洲夏季风最早于 5 月初在孟加拉湾东部爆发、5 月中旬末在中国南海爆发、6 月初在印度爆发(Wu and Zhang,1998;毛江玉 等,2002a,2002b;吴国雄 等,2013),最后于 7 月初在西北太平洋爆发(Wu and Wang,2000)。随着印度夏季风的建立,东亚副热带季风雨带自 6 月初由华南向北推进,6 月中旬到达江淮流域形成梅雨;7 月中旬继续北推至华北-东北地区,而与此同时江淮流域"出梅",华北-东北雨季随之开始。因此,亚洲热带季风爆发是东亚夏季风全面建立和中国进入主汛期的前兆信号。由于亚洲夏季风在各子季风区的爆发具有突变性,且涉及复杂的多时间尺度海-陆-气相互作用,所以亚洲夏季风爆发的阶段性及不同进程的相关物理机制是季风动力学研究的主要问题之一。本文重点回顾近 30 年来吴国雄院士团队在亚洲热带夏季风爆发机理方面的研究进展(Wu and Zhang,1998;Mao and Wu,2007;吴国雄 等,2013;Wu and Liu,2014)。

风和降水是定义季风爆发的两个基本物理量。与 Ramage(1971)用 1 月和 7 月稳定持续的盛行风反向定义季风区域不同,关于亚洲夏季风(ASM)爆发和演变的近期研究大都用 850 hPa(或 700 hPa)风场的反转作为判据(Webster and Yang,1992;Wu and Zhang,1998;张永生和吴国雄,1998;Wang and Fan,1999),或用降水的急剧增加(或向外长波辐射 OLR 的突然减少)到某个阈值作为判据(Yoshino,1966;Wang,2006)。由于资料和判据的差异,出现了不同的亚洲夏季风爆发等时线图(Tao and Chen,1987;Tanaka,1992;Lau and Yang,1997;Wang and LinHo,2002)。这些等时线图在印度子季风区大致一致,但在其他子季风区则存在明显差异。仅使用降水作为爆发判据不能把季风性降水与其他类型降水区分开来,因此,需要综合使用风向变化和降水指标,才能准确描述亚洲夏季风的爆发进程。

文中将首先回顾吴国雄院士团队基于大气经向热力梯度、地面风向角转变和降水的亚洲夏季风爆发综合定义方法,然后总结团队在亚洲夏季风爆发动力学方面的研究进展,包括:低空副高带断裂、高空南亚高压生成与亚洲夏季风爆发的联系,特别是副高脊面倾斜方向的改变与大气经向温度梯度对夏季风爆发的指示意义;局地海-气相互作用对孟加拉湾爆发性涡旋生成和高空南亚高压发展的影响机理;青藏高原对亚洲夏季风爆发地点的锚定作用以及海-陆-气相互作用对亚洲夏季风爆发进程的调制。在此基础上,本文将展望季风爆发动力学的未来研究方向。

2 亚洲夏季风爆发日期的定义和季风爆发"三阶段"

亚洲夏季风爆发前后最显著的变化是经向海陆热力对比反转、盛行风向角转向和降水激增。早在 20 世纪 90 年代末,吴国雄和张永生(1998,1999)在观测和再分析资料非常有限的情况下分析了 1989 年春季亚洲南部表面感热和潜热通量、850 hPa 风场以及 OLR 的时空演变。基于该个例分析发现,亚洲夏季风的爆发由三个接续的阶段组成,即 5 月上旬在孟加拉湾东岸爆发的第一阶段,5 月 20 日在中国南海爆发的第二阶段,6 月 10 日在印度爆发的第三阶段。随后,毛江玉等(2002a,2002b)将亚洲季风爆发与副高带形态变异联系起来,利用多年的 NCEP 再分析资料,通过追踪气候平均逐日、逐候副高脊面和脊线的变化,发现了副高带低空断裂现象、对流层中上层副高脊面倾斜方向转变现象,由此提出以对流层中上层(200~500 hPa)平均气温的经向温度梯度(MTG)表征夏季风爆发的新指数。基于"季节转换轴"的生成和演变,毛江玉等(2002a,2002b)从气候学上证实了亚洲夏季风爆发的"三阶段"。

实际上,MTG 能够客观地描述亚洲夏季风区经向海陆热力对比的季节转变(毛江玉等,2002a,2002b)。根据热成风关系,季风区 MTG 由负转正对应着局地垂直东风切变的建立,这是热带季风环流建立的基本特征,因此,MTG 转向时间和垂直东风切变、低空西南季风和季风对流建立时间之间具有很好的对应关系。吴国雄等(2013)指出,传统基于西南风或者 850 hPa 风向角变化的定义方式具有局限性,例如,在印度地区,偏东风被偏西风替代出现在 4 月下旬,远早于强降水出现的时间;而在中南半岛北部—中国华南地区,终年盛行偏西风或西南风,因此这些区域很难用 850 hPa 风向的反转去定义季风爆发。相比而言,10 m 盛行风向角的季节转换能够有效克服上述局限性,从而更准确地描述季风爆发时的风向转变。同时,应用表面风场去研究季风,也真实表达了阿拉伯语"mausim"的本义。此外,这种显著风向角的改变的北移通常与夏季强降水的出现是同步的。由此可见,用地面风的变化去表征夏季风的爆发能更好地与强降水的出现相匹配。

因此,亚洲夏季风的爆发日期可以用降水和环流变化的综合指标来定义(Liu et al.,2015):
(1)对流层中上部(500~200 hPa)平均 MTG 由负转正并维持 10 天以上;
(2)10 m 风向角改变大于 100°;
(3)海面降水稳定地大于 5 mm·d,陆地降水稳定地大于 3 mm·d。

根据上述定义计算的亚洲热带夏季风爆发等时线分布(图1)清楚地呈现了"三阶段"特征:

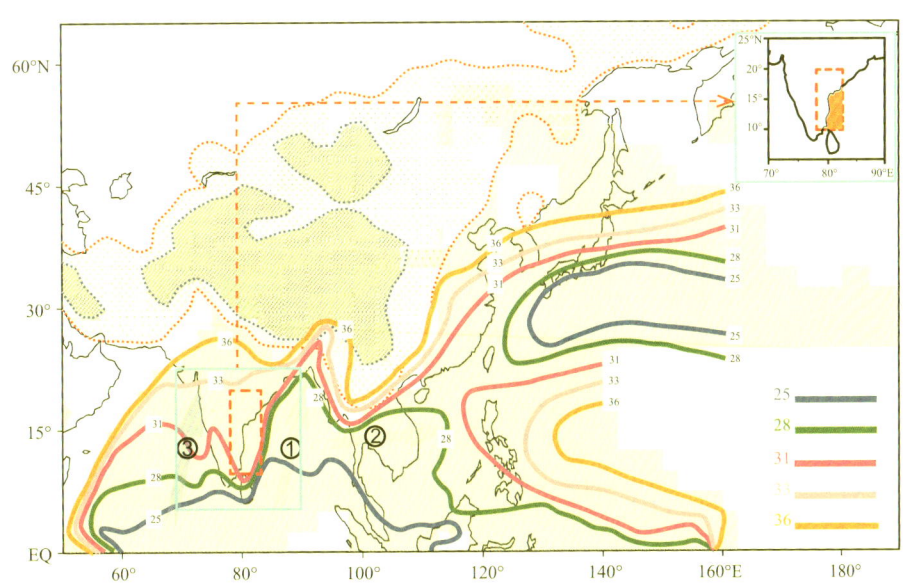

图 1 亚洲夏季风爆发的等时线(候,阴影表示季风区,①、②、③及相关箭头分别表示孟加拉湾、南海和印度夏季风的爆发过程,红色和蓝色点线包围的区域分别表示地形高度超过 500 m 和 1500 m 的地区,"季风爆发屏障"位于印度半岛以东的沿海地区(橙色阴影),灰色阴影表示亚洲季风区)(摘自 Liu et al.,2015)

(1) 孟加拉湾(BOB)夏季风爆发(5 月 1—5 日,25 候):在 25 候,ASM 爆发等时线出现在 BOB 东南部 10°N 附近。这时,海陆热力差异出现夏季型(毛江玉 等,2002a),BOB 夏季风爆发(Wu and Zhang,1998)。然后,等时线规则地北移,5 月底(30 候)到达 BOB 北端。

(2) 中国南海(SCS)夏季风爆发(5 月 15—20 日,28 候):在季风爆发等时线向北推进的同时,也迅速向东传播。26 候到达中印半岛西部;约 10 天后,于 28 候夏季风爆发区域东伸至南海中部,SCS 夏季风爆发;6 月初(31 候),亚洲热带夏季风区域继续东伸至菲律宾东部,热带西太平洋夏季风爆发。

(3) 印度夏季风爆发(6 月 1—5 日,31 候):印度夏季风的爆发源自 25 候时出现在阿拉伯海近赤道的对流降水。它随后逐渐北进,于 31 候抵达印度西南部的喀拉拉邦,印度夏季风爆发。值得注意的是,印度夏季风的爆发过程相对独立。这是因为孟加拉湾夏季风爆发后,无法直接向西传播到印度半岛和阿拉伯海,在印度半岛以东沿海地区形成"季风爆发屏障"(Wu and Zhang,1998;Liu et al.,2015)。该现象是对传统亚洲夏季风爆发研究的重要补充,其形成机制将在第 4 节详述。

3 亚洲夏季风爆发"三阶段"的环流成因

虽然亚洲夏季风爆发"三阶段"各具特点,但都与低空副热带反气旋和高空南亚高压的形态变异相联系。研究表明,在海陆热力对比季节性反转的背景下,季风区高、低空环流的垂直耦合是亚洲夏季风爆发的关键物理过程。

3.1 副高脊面倾斜方向改变、低空副高脊线断裂与季风爆发

亚洲夏季风爆发的"三阶段"对应着亚洲南部季风区对流层副高脊面(纬向风为 0 的东、西风交界面,简称 WEB)的季节性反转(毛江玉 等,2002a,2002b,2002c)。受地转风关系约束,WEB 总是随高度增加偏向暖区:冬季 WEB 随高度向南倾斜,而夏季 WEB 随高度向北倾斜(图 2)。WEB 倾斜方向的演变可以描述季节转换期间副高带的时空结构和大气热力结构,进而揭示季风爆发前后导致亚洲环流变化的机制。

在亚、非季风区,冬季副热带高压带是相对对称的,具有脊线连续的带状结构,脊面随高度增加向南倾

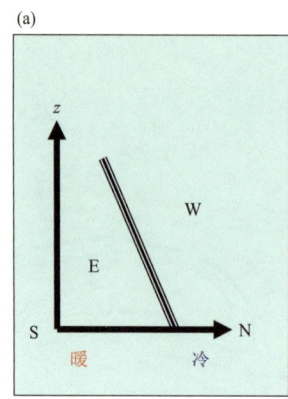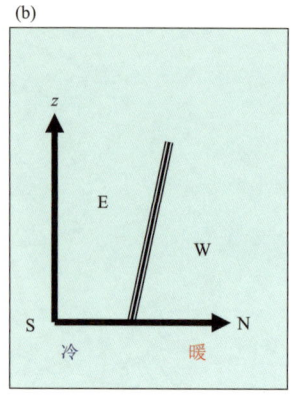

图 2 北半球冬、夏季副高带脊面季节转换示意图。(a)冬季；(b)夏季
（根据 Mao et al.，2004 重新绘制）

斜；夏季副热带高压带中低层是间断的，高层是连续的，脊面随高度增加向北倾斜。夏季型副热带高压建立的三个阶段与亚洲夏季风爆发"三阶段"一一对应（毛江玉 等，2002a，2002b）；5 月副热带高压形态变异最显著，不同地域副热带高压的结构、性质存在较大差异。夏季型副热带高压于 5 月初首先出现在孟加拉湾东部，对应孟加拉湾夏季风爆发；随后，夏季型副高于 5 月第 3 候稳定建立在孟加拉湾东部、中南半岛及南海西部地区，并于 5 月第 4—5 候在南海建立，对应南海夏季风爆发；6 月第 1—2 候，夏季型副高在印度中部建立，对应印度夏季风爆发。

热力学方程诊断结果表明，亚洲夏季风爆发"三阶段"WEB 转变的热力机制不同（毛江玉 等，2002c）。导致孟加拉湾夏季型副高建立的主要因素在季风爆发初期是经向暖平流，爆发以后是下沉运动；引起中国南海地区经向温度梯度反转的因素包括经向暖平流、纬向暖平流和江南地区的非绝热加热，其中经向暖平流的贡献更大；造成印度季风区经向温度梯度逆转的主要原因是下沉增温。

3.2 高空南亚高压的生成变异与季风爆发

WEB 的季节转换不仅能够描述对流层低层副热带反气旋的断裂北抬，还能够反映季风爆发过程中对流层高层南亚高压的变化特征。研究表明，春末夏初南亚高压的形成和形态变异对亚洲夏季风爆发"三阶段"具有重要的指示意义。

在孟加拉湾夏季风爆发前，南亚高压的形成与热带菲律宾群岛南部对流的加强发展有关（Liu et al.，2013），而青藏高原的强迫作用能够通过改变局地背景环流来增强季风爆发前中南半岛的局地对流，令南亚高压西伸发展，从而加强了孟加拉湾南部的高空辐散抽吸作用，触发了孟加拉湾季风爆发性涡旋（MOV），随后高空南亚高压和低空 MOV 在孟加拉湾发生耦合，相应地孟加拉湾夏季风爆发（图 3）（Wu et al.，2015）。此外，与来自中纬度的偏东气流有关的低层正压不稳定流场也有利于孟加拉湾 MOV 的生成和夏季风的爆发（Mao and Wu，2011）。

南亚高压在孟加拉湾季风对流的作用下，向东伸展至南海上空。随着南亚高压的东伸加强，南海上空出现正位涡平流，相应地局地高空上升运动加强，其导致的抽吸作用令低空西太副高开始东撤出南海，季风槽和季风对流逐步建立（图 4a）；随着南海季风对流的不断加强，南亚高压持续东伸发展，与之对应的高空暖中心和近海面暖中心的空间位相叠加，中国南海地区的对流层温度垂直层结被破坏，从而满足了环流的角动量守恒条件，令季风环流的下沉支能够穿越赤道到达南半球，最终引起了跨赤道的季风经向环流，将南、北半球季风系统相联，南海夏季风最终于第 29 候完全爆发（图 4b）（Liu and Zhu，2016）。

在印度夏季风爆发前夕，随着孟加拉湾和南海季风对流不断加强，降水释放的凝结潜热造成高空南亚高压东伸发展，将中纬度高位涡输送到阿拉伯海上空，形成局地"喇叭口"状流场，产生明显的高空抽

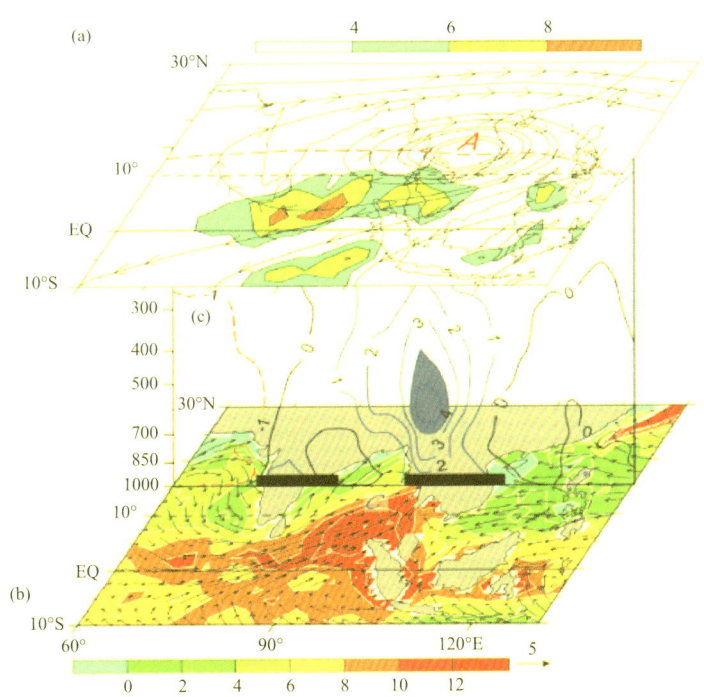

图 3 孟加拉湾夏季风爆发期间高空南亚高压和低空季风爆发性涡旋耦合过程示意图。(a)在春季青藏高原强迫作用下,高空南亚高压加强并向西北方向移动,在孟加拉湾东南部形成高空抽吸环流;(b)孟加拉湾春季暖池有助于季风爆发性涡旋形成北上和季风对流生成;(c)高、低空环流的垂直耦合造成斜压型环流发展和孟加拉湾季风爆发(摘自 Wu et al.,2015)

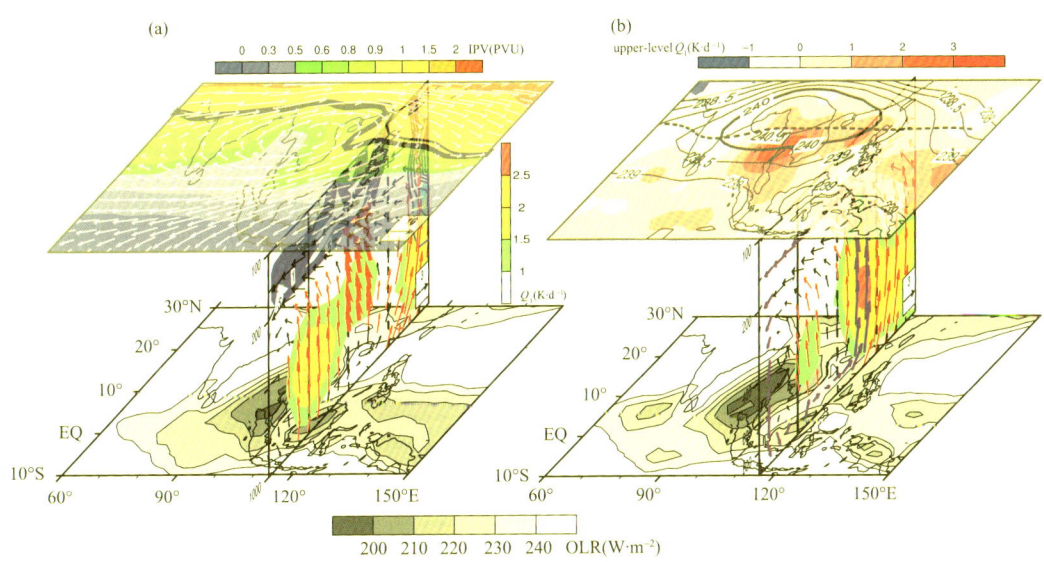

图 4 南海夏季风爆发进程示意图。(a)第 27—28 候(上:360 K 等熵位涡(阴影,PVU)和风场(矢量,m·s^{-1}),中:非绝热加热(阴影,K·d^{-1})、正位涡平流(等值线,10^{-5} PVU·s^{-1})和垂直运动(矢量,加粗箭头表示高空上升运动,10^{-2}Pa·s^{-1}),下:OLR(阴影,W·m^{-2}));(b)第 28—29 候(上:对流层上部非绝热加热(阴影,K·d^{-1})和气温(K);中:非绝热加热(阴影,K·d^{-1})和季风经圈环流(矢量,紫色箭头表示跨赤道季风环流圈,m·s^{-1});下:OLR(阴影,W·m^{-2}))(根据 Liu and Zhu,2016 重新绘制)

吸作用,为夏季风的爆发推进提供了有利的高空背景条件(Zhang et al.,2014)。当其与南北海陆热力对比的纬向差异所强迫的低空辐合中心在印度大陆西南海岸附近垂直耦合引起大气斜压不稳定发展时,激发了印度夏季风爆发(图5)(Wu and Liu,2014)。

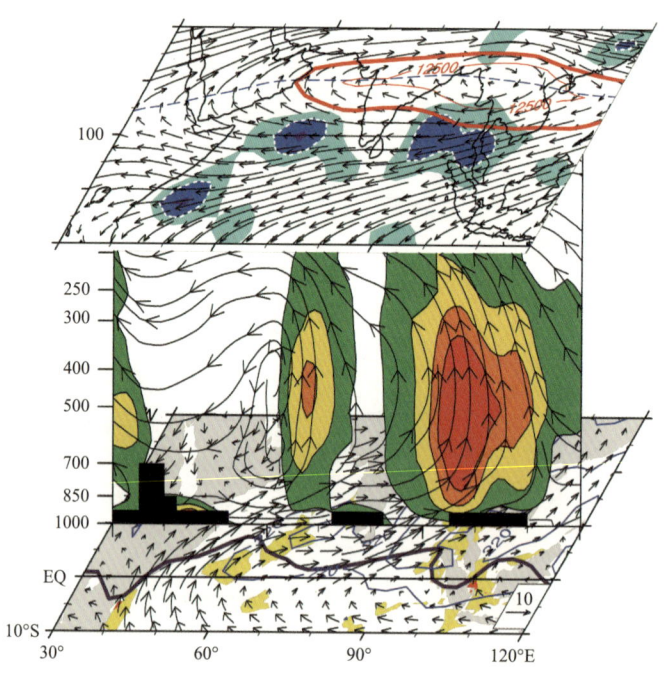

图5 印度夏季风爆发过程示意图。上:南亚高压形态变异引起的高空抽吸,为印度夏季风的爆发提供了有利背景;中:高低空环流耦合是印度夏季风对流爆发的关键;下:强迫对流发展触发了印度季风对流(根据 Wu and Liu,2014重新绘制)

4 亚洲夏季风爆发屏障的形成机制和季风爆发"三阶段"的联系

由于资料和判据的差异,出现了不同的亚洲夏季风爆发等时线图。Tao 和 Chen(1987)以及 Tanaka(1992)提出季风爆发等时线最早出现在中国南海,然后逐渐向西北偏西方向传播。有的学者认为亚洲夏季风首先在孟加拉湾建立后,季风降水于5月底东传至中国南海,而于6月初自孟加拉湾西传至印度(Lau and Yang,1997;Webster et al.,1998;Wang and LinHo,2002)。而 Wu 和 Zhang(1998)通过分析1989年亚洲夏季风的爆发过程,认为亚洲夏季风首先在孟加拉湾地区爆发,之后季风对流向东扩展到达中国南海,然而季风爆发无法从孟加拉湾直接向西传播到达印度地区。Liu 等(2015)通过更全面的资料分析和季风爆发综合定义,将这种季风爆发无法从孟加拉湾地区直接西传至印度和阿拉伯海的现象,称为"季风爆发屏障"。

4.1 季风爆发屏障的形成机制

孟加拉湾夏季风对流引起的环流变化和印度半岛附近的局地海-陆-气相互作用是造成季风"爆发屏障"的主要原因(图6)(Liu et al.,2015)。孟加拉湾夏季风爆发后,季风对流释放出大量凝结潜热,根据Gill 型响应,在孟加拉湾东侧形成上升运动,而下沉运动位于孟加拉湾西侧(图6a)。位于孟加拉湾西部的下沉气流到达印度半岛后,陆面感热明显加强,在表面气温纬向平流的作用下,表面气温上升中心出现在印度东海岸至孟加拉湾西部(图6b)。

孟加拉湾季风对流使孟加拉湾海表温度快速降低,而与季风对流相伴随的表面西南风在孟加拉湾西部造成离岸流,形成局地海温冷却中心(图6c)。在该海温冷却中心与叠加在其上方的表面气温升高中心

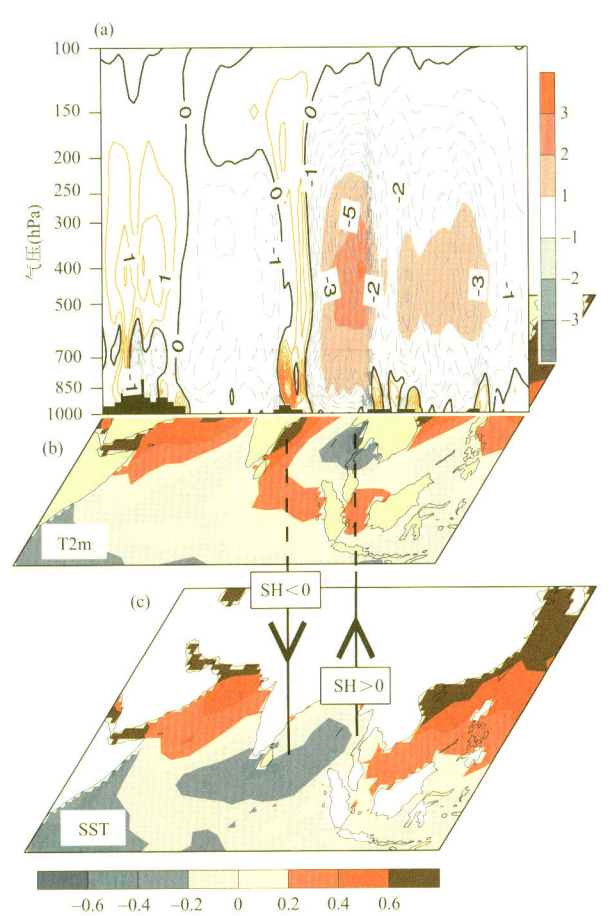

图6 亚洲夏季风"爆发屏障"动力机制示意图((a)孟加拉湾夏季风对流释放的非绝热加热(阴影,K·d^{-1})及对应的垂直运动(实(虚)线表示下沉(上升)运动,10^{-2} Pa·s^{-1});(b)孟加拉湾夏季风爆发前后 2 m 气温的变化(K);(c)孟加拉湾夏季风爆发前后海表温度的变化(K)。(b)与(c)之间的箭头表示海气热量交换的方向)(根据 Liu et al.,2015 重新绘制)

的共同作用下,造成了向下表面感热通量,一方面不利于湿气块到达自由对流高度,另一方面不利于大气有效位能的产生,这时局地大气柱从下表面失去热量,进一步加强了下沉运动,并在 700 hPa 以下的对流层底部形成下沉中心。总之,位于孟加拉湾西部至印度半岛东海岸的局地下沉运动和大气柱明显失热,抑制了季风对流的向西传播,形成了"季风爆发屏障"。

4.2 季风爆发"三阶段"的内在联系

虽然亚洲夏季风爆发"三阶段"各具特色,但三者又紧密联系。孟加拉湾夏季风爆发后,季风对流释放的凝结潜热将改变亚洲南部大气热力结构,进而影响南海和印度夏季风爆发。以 1998 年亚洲夏季风爆发过程为例,当年孟加拉湾夏季风爆发日期为 5 月 15—16 日(Mao et al.,2004)。随着季节的推进,高空反气旋脊线各处都向北推进,但孟加拉湾以西的脊线在 5 月 10—20 日期间从 10°N 到 20°N 只推进 10 个纬距,而东段从赤道向华南推进了约 22 个纬距,尤其是在 16 日孟加拉湾季风爆发后的 4 天里,脊线从南海南部约 7°N 向北跃进了 15 个纬距(图7a)。其原因是孟加拉湾季风爆发后,中印半岛潜热释放显著增加,使南亚高压加速向北推进。在对流层低层,从 10 日—20 日,印度区域的反气旋脊线从 15°N 至 25°N 北推了 10 个纬距,与高空相同。而在 16 日孟加拉湾夏季风爆发后,脊线在孟加拉湾东部断裂,东段南撤至 5°N,至 20 日则南撤到赤道附近,与高层的移动方向正好相反(图7b)。在高层,扣除季节推进的位移 y_1,脊线从初始时刻至终了时刻绕孟加拉湾作反时针旋转。注意脊线南面为盛行东风,北面为盛行西风,上述高层脊线的反时针转动使孟加拉湾东面成为东风加速区,西边为西风加速区(图7c)。而在700 hPa等压面上,西端副高脊线的北移和东端副高脊线的南撤使孟加拉湾东面产生西风加速,西面产生东风加速(图7d)。于是在孟加拉湾东面出现东风垂直切变加强,有利于上升运动发展,引发夏季风爆发向东传播,最终导致南海夏季风爆发;而西面出现西风切变加强,有利于下沉运动发展,抑制了印度夏季风爆发。

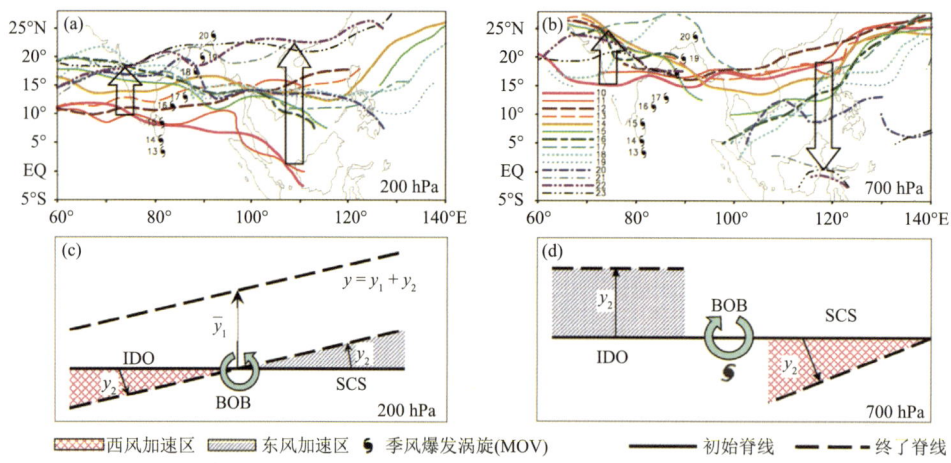

图7 1998年5月10—22日200 hPa(a)和700 hPa(b)等压面上副高脊线的隔日变化,以及孟加拉湾季风爆发期间对流层高(c)、低(d)层副高脊线位移变化示意图。y_1 表示季节位移,y_2 为孟加拉湾季风爆发激发的位移(摘自吴国雄 等,2013)

5 青藏高原对亚洲夏季风爆发地点的锚定作用

虽然亚洲夏季风爆发具有明显的年际差异,但每年总是始于孟加拉湾夏季风爆发。研究表明,青藏高原动力和热力强迫对亚洲夏季风爆发地点具有锚定作用(梁潇云 等,2005;Wu et al.,2012)。梁潇云等(2005)的数值试验证明,当青藏高原位于目前位置(中心约90°E)时,亚洲夏季风首先在孟加拉湾地区爆发(图8a);当高原被西移30个纬距至60°E时,亚洲夏季风则首先在阿拉伯海爆发(图8b),表明青藏高原大地形决定了亚洲夏季风的爆发地点。

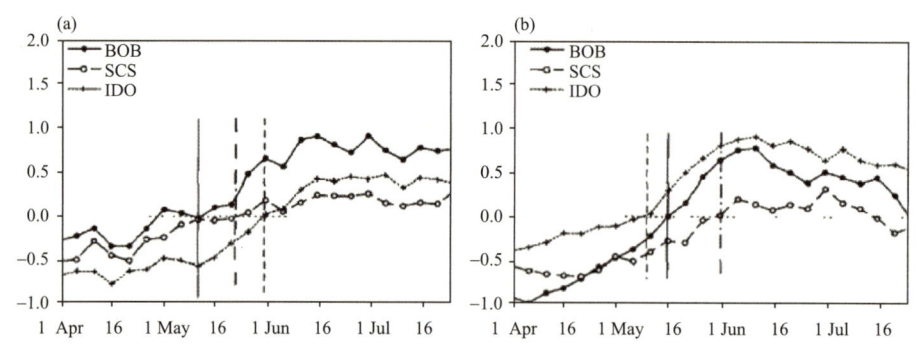

图8 理想地形和海陆分布试验中亚洲季风区MTG的演变(10^{-6} K·m^{-1})
(a)青藏高原位于90°E;(b)青藏高原位于60°E(摘自梁潇云 等,2005)

Wu等(2012)的研究表明,青藏高原通过调制孟加拉湾春季暖池和局地海气相互作用锚定了亚洲夏季风的爆发地点。冬季,青藏高原激发的偶极型定常波(王同美 等,2008)在印度中低空形成干冷的西北气流,使地气温差加大,春季的印度大陆成为强的表面感热源(>150 W·m^{-2})。它强迫出强大的陆面低压,在孟加拉湾西北沿海产生强大的低层西南气流,与近赤道西风一起形成了孟加拉湾中北部大范围的反气旋环流和东南隅的气旋环流。水汽于是从孟加拉湾北部、阿拉伯海和南印度洋向孟加拉湾东南部辐合(图9a)。BOB 西部的西南气流在其沿岸激发出 Sverdrup 离岸海流,表面暖海水向东堆积,下层冷海水上翻,使西部的海表温度(SST)变冷(图9b)。而在反气旋控制下的孟加拉湾中北部,天气晴好风小,海表面能量收入高达 240 W·m^{-2} 以上,而因感热和潜热失去的能量却小于 100 W·m^{-2}。加之海洋混合层的厚度一般只有 20 m,强大的能量盈余用于加热浅薄的混合层使 SST 迅速升高,孟加拉湾春季暖池由此形成(图9b)。暖池南部气温较高,表面感热加热较大,加之那里有水汽辐合上升(图9a)释放潜热,于是气温和加热场的异常呈正相关,大气获得大量的有效位能,为季风爆发涡旋的形成提供了有利条件(图9c)。

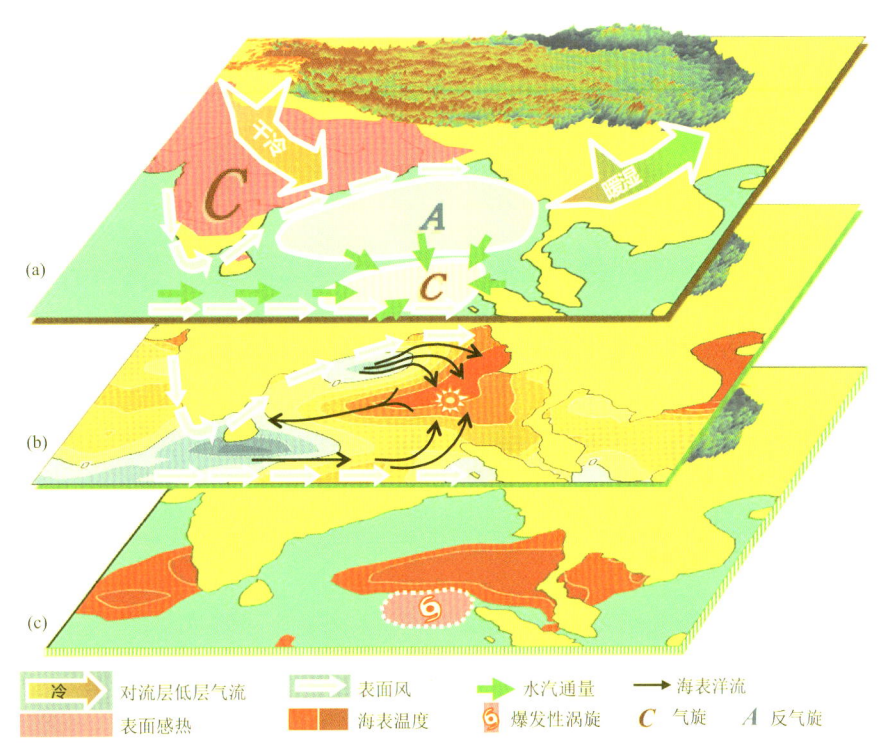

图 9 春季在青藏高原强迫和南亚海陆热力对比共同作用下(a)、孟加拉湾暖池形成(b)和季风爆发涡旋激发(c)的示意图(摘自 Wu et al.,2012)

6 春季孟加拉湾涡旋形成及其对亚洲夏季风爆发的激发作用

在孟加拉湾春季 SST 暖池的影响下,孟加拉湾夏季风爆发呈现以季风爆发性涡旋北上的天气特征。吴国雄等(2010)指出北印度洋和亚洲热带区域春季强烈的海-陆-气相互作用是激发孟加拉湾季风爆发涡旋发生及亚洲夏季风爆发的根本原因。在孟加拉湾春季暖池形成后,热带印度洋的经向海温梯度和非洲大陆的加热在 5 月初形成了索马里跨赤道气流,它在沿赤道惯性振荡的过程中与位于北部阿拉伯海上空的反气旋同相叠加,惯性槽中的西风激增,海表向大气释放大量感热。其所强迫的气旋涡度叠加在惯性槽上使其急剧增长,大气的惯性运动被破坏。气流以气旋式弯曲偏离惯性路径而转向北部的孟加拉湾暖池,与孟加拉湾北部反气旋南侧的偏东气流在斯里兰卡以东海面辐合。由于该处海洋对大气加热,最后形成了孟加拉湾季风爆发涡旋。该涡旋向北部暖海水区移动的过程中不断发展,最终使冬季纬向分布的副热带高压带在孟加拉湾东部断裂,涡旋东部和南部的热带西南风与原高压带北部的副热带西风打通联接,激烈对流在孟加拉湾东部和中南半岛西部发生,亚洲夏季风于是在孟加拉湾地区爆发(图10)。

近期研究表明,孟加拉湾季风爆发性涡旋北上后,还能够通过和青藏高原大气热源的相互作用,触发南海夏季风爆发(Liu and Zhu,2020)。以 2019 年为例,当年南海夏季风爆发极端偏早,这与 4 月下旬孟

图 10 在南亚海气相互作用背景下孟加拉湾季风爆发涡旋形成示意图(引自吴国雄 等,2010)

加拉湾的北上超强台风"法尼"联系紧密。4月底,孟加拉湾超强台风"法尼"生成并北移,孟加拉湾夏季风随即爆发。在台风登陆后,大量水汽被输送到青藏高原东部地区,加强了局地大气热源。随后,尽管台风快速耗散消亡,但高原热源异常一方面加强了海陆热力对比,另一方面加强了南亚高压和南海上空的高空辐散抽吸效应,这种"升尺度"作用最终令当年南海夏季风爆发异常偏早(图11)。

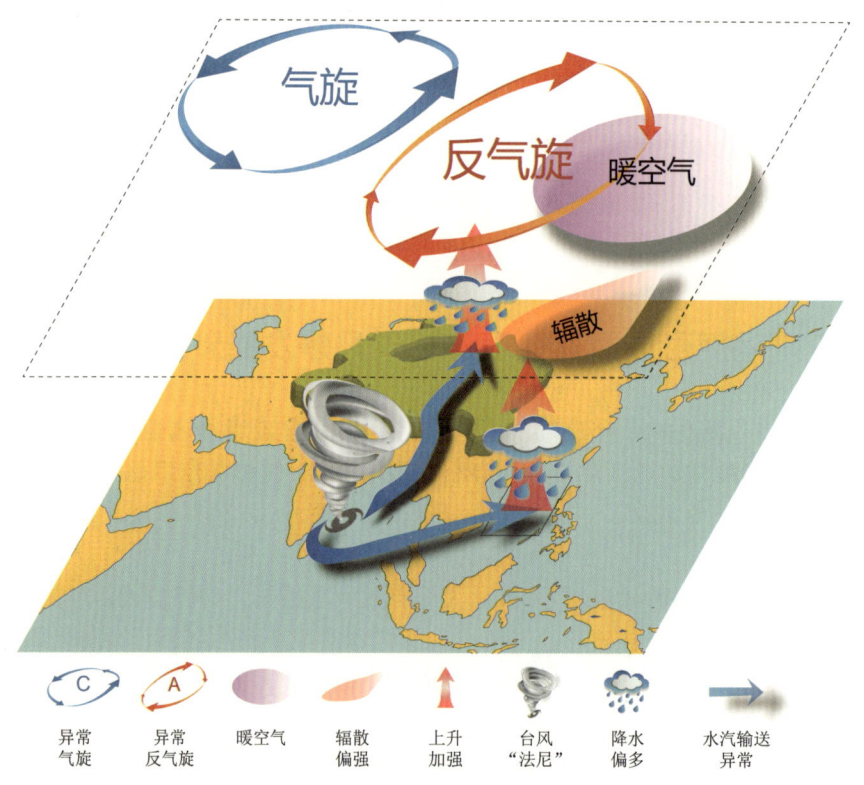

图 11 孟加拉湾季风爆发性涡旋通过青藏高原大气热源持续性异常触发 2019 年南海夏季风提早爆发的示意图(摘自 Liu and Zhu,2020)

7 总结与讨论

亚洲夏季风的爆发推进既与大气内部动力过程有关,又受大气外强迫因子影响。文中总结了吴国雄院士团队在亚洲夏季风爆发定义、季风环流垂直耦合机制、海-陆-气相互作用对季风爆发的影响机理和青藏高原对亚洲夏季风爆发的锚定作用等方面的一些研究进展,得到的关于季风爆发的新认识总结为以下三点:

(1)亚洲夏季风爆发过程存在明显的"三阶段"特征。季风区海陆热力对比季节转换背景下,高、低空季风环流的垂直耦合是亚洲夏季风爆发的关键物理过程。

(2)亚洲夏季风爆发的"三阶段"既相对独立,又紧密联系。春末夏初对流层副热带高压带结构变异特别是高空南亚高压的生成及形态变异是联系亚洲夏季风爆发各阶段的环流纽带。

(3)青藏高原动力强迫对亚洲夏季风的爆发地点具有锚定作用。亚洲夏季风爆发"三阶段"是青藏高原动力和热力强迫下季风区海-陆-气相互作用的产物。

需要指出的是,亚洲夏季风爆发过程十分复杂,现有气候系统模式依然无法准确模拟亚洲夏季风爆发的气候特征,说明季风爆发期间大气海洋瞬变过程向季节进程的跨时间尺度相互作用仍不清楚。同时,亚洲夏季风爆发还具有明显的年际和年代际变率(Mao and Wu,2007;Liu and Zhu,2016;Wu and Mao,2019),不同年份亚洲夏季风各阶段爆发日期之间的关系尚不明确,而动力模式对季风爆发日期的预测时效仍未突破天气预报上限(Yan et al.,2021),说明对大气准定常外强迫(如 ENSO、高原冬春积雪等)如何影响亚洲夏季风爆发瞬变过程的机理也有待继续探索。

参考文献

梁潇云,刘屹岷,吴国雄,2005.青藏高原对亚洲夏季风爆发位置及强度的影响[J].气象学报,63(5):799-805.

毛江玉,吴国雄,刘屹岷,2002a.季节转换期间副热带高压带形态变异及其机制的研究Ⅰ:副热带高压结构的气候学特征[J].气象学报,60(4):400-408.

毛江玉,吴国雄,刘屹岷,2002b.季节转换期间副热带高压带形态变异及其机制的研究Ⅱ:亚洲季风区季节转换指数[J].气象学报,60(4):409-420.

毛江玉,吴国雄,刘屹岷,2002c.季节转换期间副热带高压带形态变异及其机制的研究Ⅲ:热力学诊断[J].气象学报,60(6):647-659.

王同美,吴国雄,万日金,2008.青藏高原的热力和动力作用对亚洲季风环流的影响[J].高原气象,27:1-9.

吴国雄,张永生,1998.青藏高原的热力和机械强迫作用以及亚洲季风的爆发Ⅰ.爆发地点[J].大气科学,22(6):825-838.

吴国雄,张永生,1999.青藏高原的热力和机械强迫作用以及亚洲季风的爆发Ⅱ.爆发时间[J].大气科学,23(1):51-61.

吴国雄,毛江玉,段安民,等,2004.青藏高原影响亚洲夏季风气候研究的最新进展[J].气象学报,62(5):528-540.

吴国雄,关月,王同美,等,2010.春季孟加拉湾涡旋形成及其对亚洲夏季风爆发的激发作用[J].中国科学:地球科学,40(11):1459-1467.

吴国雄,段安民,刘屹岷,等,2013.关于亚洲夏季风爆发的动力学研究的若干近期进展[J].大气科学,37(2):211-228.

张永生,吴国雄,1998.关于亚洲夏季风爆发及北半球季节突变的物理机理的诊断分析Ⅰ:季风爆发的阶段性特征[J].气象学报,56(5):513-528.

LAU K M, YANG S, 1997. Climatology and interannual variability of the Southeast Asian summer monsoon[J]. Adv Atmos Sci,14:141-162.

LIU B Q, WU G X, MAO J Y, et al, 2013. Genesis of the South Asian high and its impact on the Asian summer monsoon onset[J]. J Climate,26: 2976-2991.

LIU B Q, LIU Y M, WU G X, et al, 2015. Asian summer monsoon onset barrier and its formation mechanism[J]. Clim Dyn,45:711-726.

LIU B Q, ZHU C W, 2016. A possible precursor of the South China Sea summer monsoon onset: Effect of the South Asian high[J]. Geophys Res Lett,43. doi:10.1002/2016GL071083.

LIU B Q, ZHU C W, 2020. Boosting effect of tropical cyclone "Fani" on the onset of the South China Sea summer monsoon in 2019[J]. J Geophys Res Atmos,125: e2019JD031891.

MAO J Y, CHAN J C L, WU G X, 2004. Relationship between the onset of the South China Sea summer monsoon and the structure of the Asian subtropical anticyclone[J]. J Meteor Soc Japan,82: 845-859.

MAO J Y, WU G X, 2007. Interannual variability in the onset of the summer monsoon over the eastern Bay of Bengal[J]. Theor Appl Climatol,89: 155-170.

MAO J Y, WU G X, 2011. Barotropic process contributing to the formation and growth of tropical cyclone Nargis[J]. Adv Atmos Sci,28: 483-491.

RAMAGE C, 1971. Monsoon Meteorology[M]. New York:Academic Press: 296.

TANAKA M, 1992. Intraseasonal oscillation and the onset and retreat dates of the summer monsoon over East, Southeast Asia and the Western Pacific region using GMS high cloud amount data[J]. J Meteor Soc Japan,70: 613-629.

TAO S Y, CHEN L X, 1987. A Review of Recent Research on the East Asian Summer Monsoon in China[M]. Oxford:Oxford University Press: 60-92.

WANG B, 2006. The Asian Monsoon[M]. New York:Springer: 612.

WANG B, FAN Z, 1999. Choice of South Asian summer monsoon indices[J]. Bull Amer Meteor Soc,80:629-638.

WANG B, LINHO, 2002. Rainy season of the Asian-Pacific summer monsoon[J]. J Climate,15: 386-398.

WEBSTER P J, YANG S, 1992. Monsoon and ENSO, selectively interactive systems[J]. Quart J Roy Meteor Soc,118: 877-926.

WEBSTER P J, MAGANA V O, PALMER T N, et al, 1998. Monsoons: Processes, predictability, and the prospects for prediction[J]. J Geophys Res,103: 14451-14510.

WU G X, ZHANG Y S, 1998. Tibetan Plateau forcing and timing of the monsoon onset over South Asian and the South China Sea[J]. Mon Wea Rev,126: 913-927.

WU G X, GUAN Y, LIU Y M, et al, 2012. Air-sea interaction and formation of the Asian summer monsoon onset vortex over the Bay of Bengal[J]. Clim Dyn,38: 261-279.

WU G X, LIU B Q, 2014. Roles of forced and inertially unstable convection development in the onset process of Indian summer monsoon[J]. Science China: Earth Sciences,57: 1438-1451.

WU G X, DUAN A M, LIU Y M, et al, 2015. Tibetan Plateau climate dynamics: Recent research progress and outlook[J]. Nat Sci Rev,2: 100-116.

WU R G,WANG B, 2000. Interannual variability of summer monsoon onset over the Western North Pacific and the underlying processes[J]. J Climate,13: 2483-2501.

WU X F, MAO J Y, 2019. Decadal changes in interannual dependence of the Bay of Bengal summer monsoon onset on ENSO modulated by the Pacific decadal oscillation[J]. Adv Atmos Sci,36: 1404-1416.

YAN Y H, LIU B Q, ZHU C W, 2021. Subseasonal predictability of South China Sea summer monsoon onset with the ECMWF S2S forecasting system[J]. Geophys Res Lett,48. doi:10.1029/2021GL095943.

YOSHINO M M, 1966. Four stages of the rainy season in early summer over East Asia (Part Ⅱ) [J]. J Meteor Soc Japan, 44: 209-217.

ZHANG Y N, WU G X, LIU Y M,et al, 2014. The effects of asymmetric potential vorticity forcing on the instability of South Asia high and Indian summer monsoon onset[J]. Science China: Earth Sciences,57: 337-350.

综述 10

华南春雨的气候成因和季节内振荡

万日金[1]，潘蔚娟[2]，毛江玉[3]

(1 中国气象局上海台风研究所，上海 200030；2 广东省气象局，广州 510080；
3 中国科学院大气物理研究所，北京 100029)

摘要："华南春雨"是东亚地区冬、夏季节转换过程中存在的独特气候现象。文中总结了吴国雄院士团队在华南春雨的时空特征、形成机制和季节内变率方面的重要进展。首次提出"江南春雨"的气候学概念，通过东亚与北美同纬度带气候平均的春季环流和降水分布的对比分析，指出了国际上关于华南春季持续降水的"季节增暖迟滞效应"的局限性。重新界定了华南春雨的起讫时间和空间分布范围——即华南春雨开始于第 13 候，对应高原主体感热加热由冷源转为热源，终止于第 27 候，随着副高脊面在中国南海区域由南倾转为北倾，南海季风爆发。华南春雨空间分布受南岭、武夷山脉地形的影响显著，其主雨带与山脉地形重合。资料诊断分析发现，春季高原东南侧出现西南风风速中心导致下游华南地区强烈的风速和水汽辐合是华南春雨气候形成的直接原因，而该西南风正是高原机械和热力强迫的结果；基于敏感性数值模拟试验，证明青藏高原的机械和热力强迫是华南春雨气候形成的根本原因。揭示出 10～20 天准双周振荡是华南春雨期间最显著的季节内变率，青藏高原表面感热异常的准双周振荡显著地调控环绕高原的低空"偶极型"环流的演变及其与高空低频波列的垂直耦合。

1 引言

中国位于欧亚大陆东南端，大部分地区处在副热带，西部盘踞着"世界屋脊"青藏高原，东部面临世界最大的太平洋(吴国雄 等，2002)。因而，冬季受西伯利亚冷高压南伸的副热带分支所控制；夏季则是热带西南季风盛行而产生的主雨季。然而，在冬、夏季节转换的春季，中国长江流域及其以南的广大地区通常阴雨连绵、甚至发生强对流天气(李麦村 等，1977；LinHo et al.，2008)。对于广东和福建两省的一些地区，仅 3—4 月的降水量就能占到全年总降水量的 25%～35%(LinHo et al.，2008；Wu and Mao，2016)。可见，春季是江南地区梅雨季之前的一个季节性多雨时段(毛江玉 等，2002c)，因而，万日金和吴国雄(2006，2008)称之为"江南春雨"。毛江玉等(2002c)指出江南地区春季降水所形成的非绝热加热源非常显著，该热源对后期亚洲季风的季节转换有重要影响。因而，"江南春雨"或"华南春雨"是东亚地区冬、夏季节转换过程中存在的独特气候现象(Pan et al.，2013)。在 20 世纪 90 年代后期，Tian 和 Yasunari(1998)(以下简称 T&Y)提出"春季持续降水(英文简称 SPR)"的概念。该 SPR 主雨区分布在 110°E 以东、长江中下游以南，这与万日金和吴国雄(2006)定义的"江南春雨"范围一致，所以本文统一称为"华南春雨"(South China spring rain，简称 SCSR)。T&Y 认为，春季东亚陆地增暖快于海洋，即春季季增暖的时滞效应，导致陆地低压发展和海洋高压增强以及南海出现偏南暖湿气流，并因初春从华南到日本南部的雨量同时快速增长而推断 SCSR 为海陆分布影响的结果。然而，吴国雄院士很早就意识到青藏高原的感热气泵效应可能对中国东部春季降水产生重要影响(吴国雄 等，1997a)，并且注意到，与亚洲东南部处在同一纬度带的北美东南部,春季也同样出现季节增暖的时滞现象,却没有出现类似 SCSR 的显著雨带,从而对 SCSR 气候成因是海陆分布增暖时滞效应的观点提出了质疑(万日金和吴国雄,2006)。文中主要总结吴国雄院士团队在华南春雨气候特征、形成机制以及季节内变率方面的进展及创新性成果(Wan and Wu,2006;万日金和吴国雄,2008;Wan et al.,2009;Pan et al.,2013)。

2 华南春雨的时空分布

2.1 华南春雨的气候特征及海陆分布成因说的疑问

华南春雨一般始于第 12 候而止于第 26 候(Tian and Yasunari,1998),多年(1951—2000 年)平均的 SCSR 降水量分布(图 1 左)表明,SCSR 雨带位于长江中下游以南,中心区域大致与南岭、武夷山脉重合,位于 24°~30°N,110°~120°E,中心强度达 6~7 mm·d^{-1}。从逐候降水的时间演变(图 1 右)可见,除了显著降水集中在春季(3—5 月上半月)以外,沿着 28°N 的降水大值轴线,在春季的每个月里都有一个极大降水中心,这清楚地反映 SCSR 存在季节内振荡现象(Pan et al., 2013)。从 SCSR 期间全球降水分布图(图略)可见,在赤道以外,北半球主雨带一般都出现在大洋西部与大陆相邻的海洋上,它们都是由来自极地的冷空气与来自热带洋面的暖空气交绥而形成的极锋锋区所致。值得注意的是,在亚洲东南部的陆地上却出现了一个中心区雨量大于 6 mm·d^{-1} 的显著雨带 SCSR,但在北美东南部陆地上相应区域并没有出现类似的显著雨带。

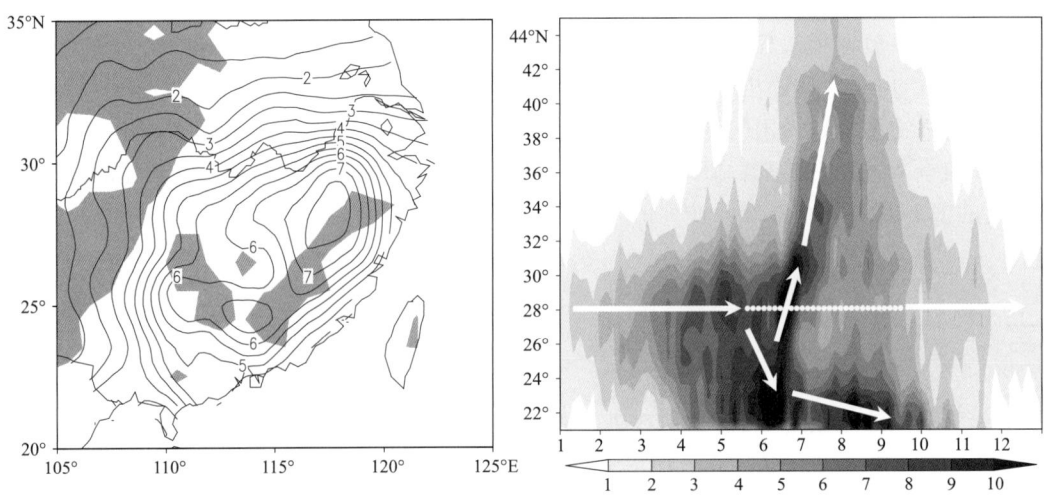

图 1 左:1951—2000 年 12—26 候候平均降水图(单位:mm·d^{-1}),阴影区为地形高度平滑后超过 600 m 的区域。右:1951—2000 年 110°~120°E 经度平均纬度时间(全年共 73 候)剖面图,横坐标值为各候相应月份(摘自万日金和吴国雄,2006)

在中低纬度海陆分布方面,东亚和北美存在诸多可比性。中低纬度东亚和北美都处于北半球的副热带,西北侧是大陆,东南侧是广阔的海洋,它们都受同性质的极锋锋面雨带影响。图 2 是气候平均的 SCSR 时期东亚与北美地区 850 hPa 温度、高度、风场、水汽辐合空间分布图,图中 A 区(12.5°~22.5°N,95°~105°E)、B 区(10°~20°N,130°~140°E)分别为 T&Y 选取的东亚陆海对比区域;此外,万日金和吴国雄(2006)在北美陆海对比区域又选取了 C 区(15°~25°N,95°~105°W)和 D 区(12.5°~22.5°N,60°~70°W)与东亚的情形相比较。

在温度场上(图 2a,b)明显可见,在两大陆的南部,西部的陆地(A,C)的温度分别高于东部的海洋(B,D)的温度,区域平均北美 C、D 的温差为 4.1 ℃,而东亚 A、B 的温差为 3.2 ℃,可见北美的东西温度梯度甚至比东亚的还大,如季节增暖的时滞效应理论成立,则北美应出现比东亚更强的类似 SCSR 的雨带。

在850 hPa高度场上(图2c,d)上,在东亚,青藏高原的东南部西南暖低涡发展强盛,其与南海副高之间的气压梯度显著,南海副高明显要比北美的墨西哥高压弱;而在北美,墨西哥高原北部只出现了很浅的地形槽,其与墨西哥高压之间的气压梯度反而较下游小。

对应于高度场,东亚与北美的低层850 hPa风场具有显著不同(图2e,f)。由于块状高原大地形的分流作用,东亚的西风急流带明显分为南北两支,大致在(30°N,120°E)处再度会合;在高原的东南侧出现一个西南急流风速中心,最大风速达到7 m·s^{-1},使得其下游华南地区西南风速辐合明显,从而直接导致了水汽的辐合明显(图2g)。该水汽通量散度辐合区控制了华南大部。而在北美只有一支显著的西风急流带,其风速中心位于大西洋西部(图2f),在北美东南部上游没有出现西南风风速中心,只有风速和风向很强的辐散和微弱的水汽辐合(图2h)。可见,正是高原东南侧的西南急流风速中心的出现导致了华南春雨

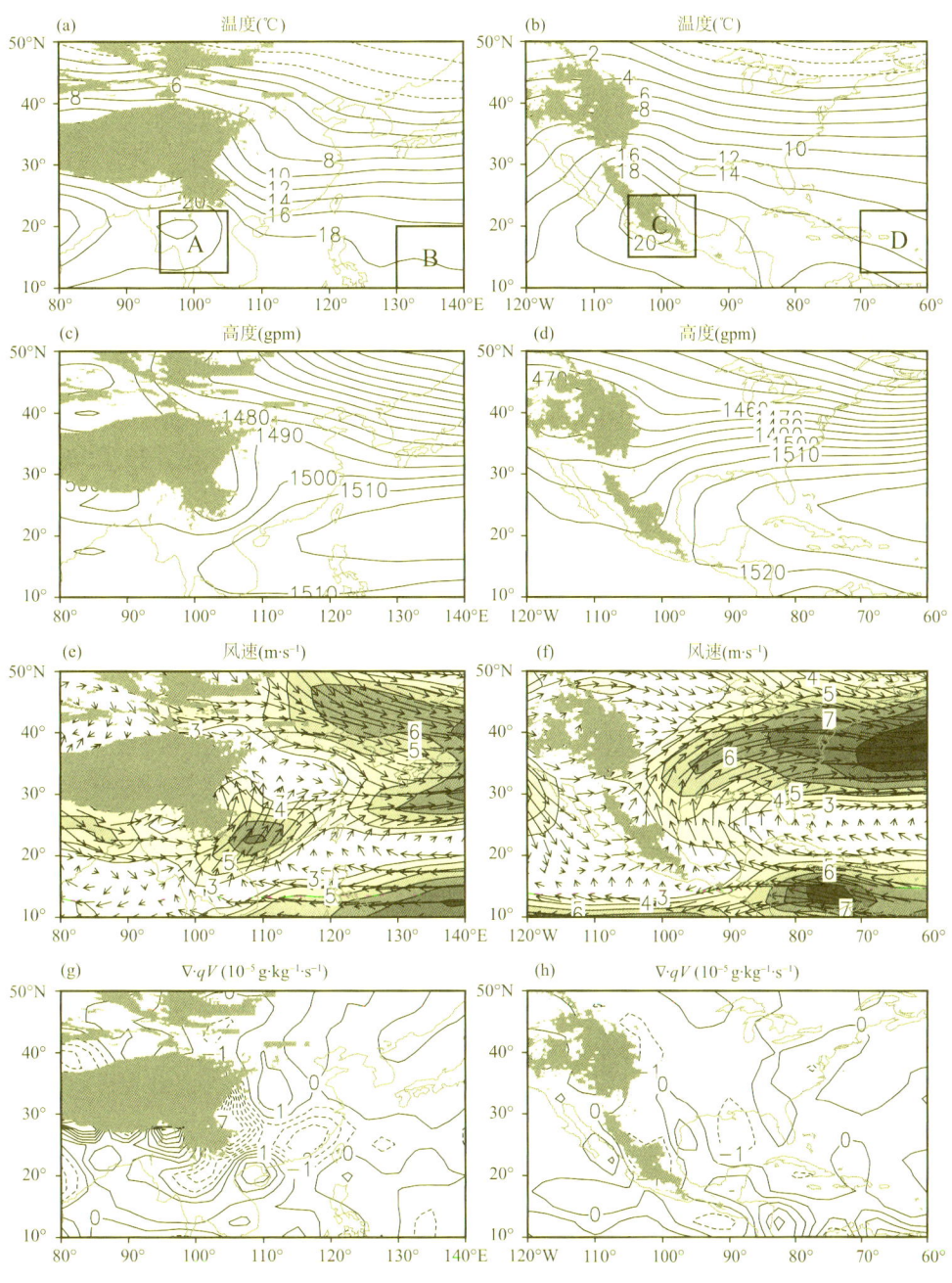

图2 SCSR期(12—26候)东亚(左列)与北美(右列)850 hPa温度、高度、风场、水汽辐合对比。
图中地形阴影区高度达到1500 m(摘自万日金和吴国雄,2006)

SCSR 的形成。

从东亚与北美 850 hPa 风场纬向平均时间演变图可以看得更清楚,见图 3a(东亚:110°～120°E)和图 3b(北美:80°～90°W)。图 3a 中阴影区风速超过 3 m·s^{-1},虚线框表示 SCSR 的时空分布大致区域。可见,西南风大值带南起 20°N,北至 27.5°N,维持近 2.5 个月,直至南移,恰好与 SCSR 的建立、维持和减弱相对应。紧接着,该副热带西南风与南海夏季风爆发的热带西南风合并,从而华南进入前汛期降水盛期。6 月中期,华南的西南风再次加强并向北突进,长江中下游地区进入梅雨季节;7 月中期,强劲的西南风扩展至 35°N,中国北方进入主汛期,而此时的中国南海西南风速进一步加大,华南进入多台风季节。西南风的季节演变与中国东部降水的时间演变(图 1 右)是完全一致的。在北美中低纬地区(图 3b),全年主要为东南风控制,在初春 30°N 附近也出现了西南气流,但该西南气流不能稳定维持,而是随季节向北退缩。由此可见,华南独特的西南风对于 SCSR 的形成至关重要。

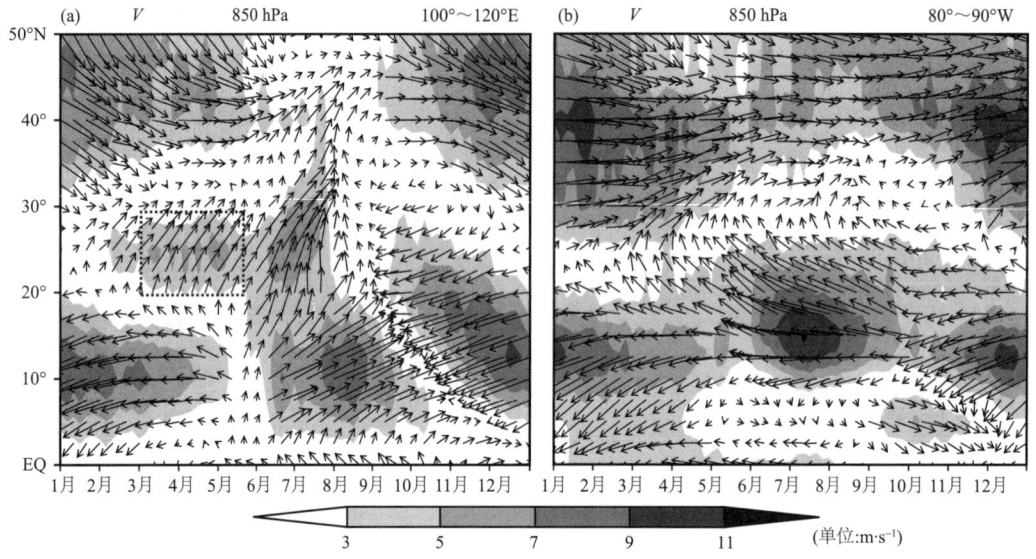

图 3 东亚与北美 850 hPa 上纬向平均风场纬度-时间演变图,横坐标为月份,纬向平均区域分别为(a)东亚(110°～120°E),(b)北美(90°～80°W),图中阴影区风速大于 3 m·s^{-1}(摘自万日金和吴国雄,2006)

2.2 华南春雨建立与终结

SCSR 建立与结束前后纬向平均(110°～120°E)的经圈环流合成剖面图(图略)显示,在 SCSR 建立前后,SCSR 初期的环流基本上仍是比较稳定的冬季环流形势。而由东亚与北美 850 hPa 风场纬向平均时间演变图(图 3a)可以看到,在 3 月第 1 候,高原东南侧西南风风速大于 3 m·s^{-1} 区迅速同时向南北方向扩展,并维持到 5 月中旬(第 27 候),其后在南北方向同时收缩。

在北半球的 10°～20°N 之间的中国南海上空,在对流层中低层的春季,东西风分界面(副高脊面)随高度升高南倾,表明此时温度场分布是南高北低(毛江玉 等,2002a,2002b)。而在 SCSR 结束前后,第 26、27 候低层是下沉气流,但到了第 28 候都转变为上升气流;此外,随高度升高时副高脊面在第 27 候及之前一直向南倾,但第 28 候及之后都变为北倾,表明此时南北温度梯度分布出现反转,环流由冬季型转变为夏季型,同时南海季风爆发。

图 4 是 SCSR 主雨区 A 区降水 R_{SPR}、江南上游 B 区 850 hPa 西南风速 V_{sw}、高原东南部 C 区地表感热通量 S_{SE}、高原主体 D 区地表感热通量 S_{MN} 所在的区域(图 4 左)及各物理量区域平均季节演变(图 4 右)。由图可见,A 区降水 R_{SPR} 从 1 月到 5 月中几乎是直线上升的,在 3 月第 1 候达到 4 mm·d^{-1},到 5 月第 4 候出现明显的下降;在 3 月第 1 候,V_{sw} 超过 5 m·s^{-1},S_{SE} 达到 43 W·m^{-2};在 5 月第 4 候,V_{sw} 降至 5 m·s^{-1} 以下,S_{SE} 降至 43 W·m^{-2} 以下。仔细对比可以发现,这两条曲线几乎都以 4 月第 2 候为中心呈前后对称分

布,说明春季高原东南部感热加热对西南暖低涡气旋性环流西南风紧密相关,同时也为 SCSR 时域的划分提供了有力的依据。

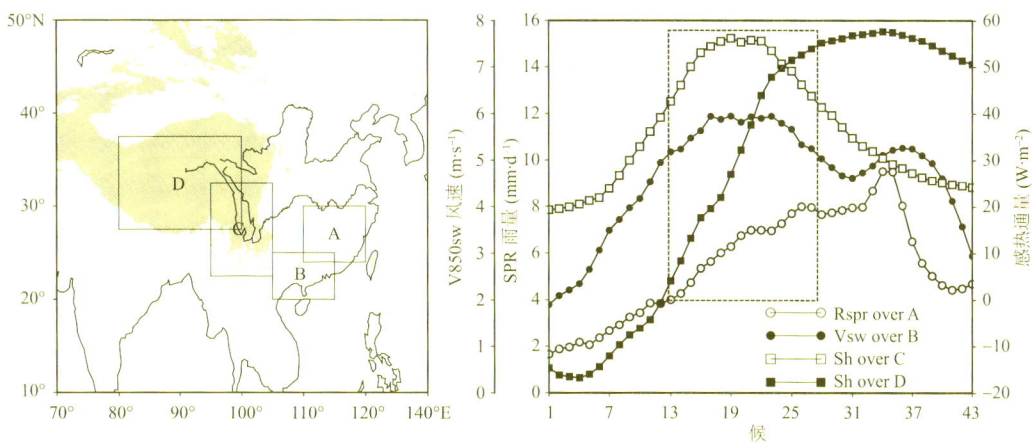

图 4 物理量相关区域(左)及其物理量区域平均逐候演变(右):SCSR 区(A:24°~30°N,110°~120°E)降水 RSPR、850 hPa 西南风区(B:20°~25°N,105°~115°E)西南风速 V_{sw}、高原东南部(C:22.5°~32.5°N,95°~105°E)和高原主体(D:27.5°~37.5°N,80°~100°E)地表感热通量 S_{SE}、S_{MN}(摘自万日金和吴国雄,2008)

值得注意的是,青藏高原主体(D 区)平均的地表感热通量 S_{MN} 从 2 月第 6 候的 -1 W·m^{-2} 升为 3 月第 1 候的 $+4$ W·m^{-2},标志着高原在感热加热方面从热汇变为热源,其在低层产生的热力环流分量不再是高压反气旋性环流,而是低压气旋性环流,从而加强了其东南侧的西南风,对于 SCSR 的建立具有重要的促进作用。

可见,在 3 月第 1 候(全年第 13 候),高原主体和高原东南部的感热加热、高原东南侧西南风速、SCSR 区的西南风速和雨量都提升到一个新的水平,4 mm·d^{-1} 的雨强在 28°N 附近建立,标志着 SCSR 的建立;在 5 月第 3 候(全年第 27 候)以后,上述物理量都迅速减小,中国东南部雨带中心南移至南海(图略),南海副高脊线北倾,南海季风爆发,标志着 SCSR 期的结束。因此以 3 月第 1 候(全年第 13 候)和 5 月第 3 候(全年第 27 候)作为 SCSR 的建立和终结时间是比较适当的。

2.3 华南春雨的空间分布

南岭、武夷山脉是横亘在中国南方的大型山系,武夷山脉呈东北—西南走向,南岭山脉呈东西走向,两支山脉组合成一个反"L"型结构,SCSR 的雨带中心轴线也呈反"L"型分布,与南岭和武夷山脉的走向完全一致,这充分说明了 SCSR 雨带的分布与山脉地形有密切关系。使用数值模式敏感性试验可以进一步说明山脉地形对于 SCSR 雨带的影响。模式选自中国科学院大气物理研究所(IAP)大气科学和地球流体力学数值模拟国家重点实验室(LASG)发展的全球大气环流谱模式 SAMIL-R42L9(吴国雄 等,1997b)。该模式能很好地再现观测气候基本模态。模式试验结果表明(图略),在无华南山脉地形的情形下,华南 SCSR 期降水将减少 1~3 mm·d^{-1},而华南沿海和长江以北降水将略有增加,即中国东部总降水将减少,南北分布将更为均匀,雨带会北移到长江流域 30°N 附近;这说明在 SCSR 期,华南山脉对南方暖湿气流的阻挡作用更为明显。在加高山脉高度 300 m 后,长江以南的降水明显增强 1~2 mm·d^{-1},而长江以北降水有所减少,即降水南北分布将更不均匀,雨带南移到山脉附近,与实况位置更为接近。由此可见,山脉地形对于冷暖空气的阻挡、强迫抬升和增雨的作用。

总之,SCSR 雨带的位置和强度明显受南岭、武夷山脉地形的影响,山脉地形能阻挡、抬升冷暖空气,加强锋生,增强降水,使雨带位置与山脉分布重合。SCSR 主雨区覆盖整个南岭、武夷山脉地区,特别是在南岭山脉南侧的迎风坡也出现一个次大降水中心,如果以平均 6 mm·d^{-1} 的雨强为标准,华南春雨的中心区范围取为长江(30°N)以南、110°E 以东的中国东南部地区。此区域比传统所称的长江以南、南岭以北的"江南"的范围更为宽广,称之为"华南"更为准确。因此,将之前"江南春雨"(万日金和吴国雄,2006;万日金和吴国雄,2008)的名称修改为"华南春雨"(Pan et al.,2013)是合适的。

3 华南春雨的气候成因机制

3.1 青藏高原的机械和热力强迫与华南春雨的关系

首先考察高原的机械强迫效应。很明显,由于青藏高原,甚至是其东南部的云贵高原,其高度远远超过爬坡与绕流的临界高度。结果在 SCSR 期,中纬度的西风气流被高原分流为南北两支西风带,其南支沿喜马拉雅山脉南麓继续东进,又遇云贵高原,根据涡度守恒原理,在云贵高原东南侧的背风坡,气流的气旋性加大,转向东北,成为西南风(图2e)。所以西南绕流是高原东南侧西南风的主要部分之一。

再考察高原的热力强迫效应。图5是欧亚大陆选定区域(左)及其区域平均物理量气候平均季节演变图(右)。由图可见,在 SCSR 期,欧亚大陆南部(G 区)与北部(E 区)之间 850 hPa 上的温差 T_{G-E}(长短虚线)由 30 ℃ 迅速减小到 18 ℃ 以下,欧亚中纬度地区(F 区)850 hPa 上的平均西风风速 U_F(点线)也随之由 3 m·s^{-1} 以上减小到 2 m·s^{-1} 以下,无疑,这是由于热成风效应减退引起的。在 SCSR 期,中纬度平均西风气流的减弱意味着高原的入流减弱,其绕流也应相应减弱,而令人惊讶的是,高原东南侧(B 区)850 hPa 上西南风风速 V_{sw}(实线)从冬到春却一直处于上升过程之中,直至在 3 月中期升至峰值(6 m·s^{-1} 以上)。如果该西南风仅是高原入流的绕流,那为什么在春季入流减小的同时其绕流却增强呢?

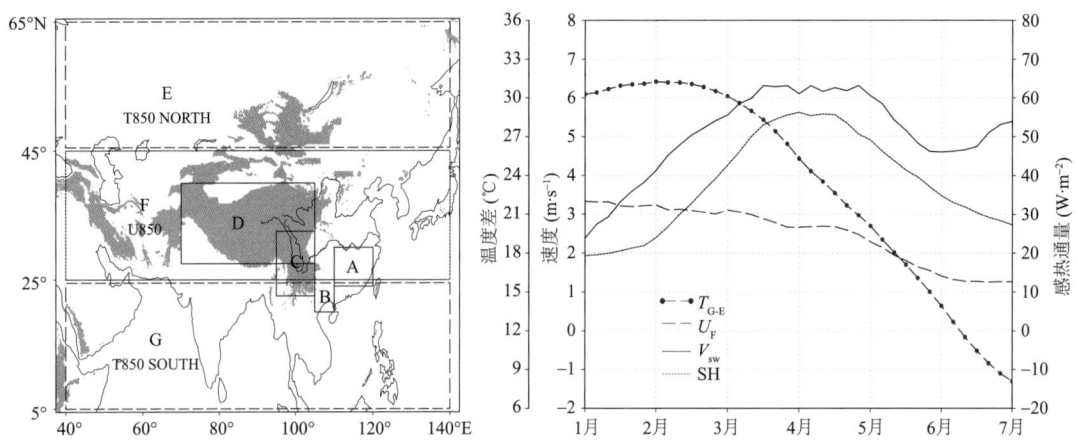

图5 区域及相应变量候平均上半年季节演变曲线,阴影区为地形超过 1500 m 区域。T_{G-E}:850 hPa 低纬南部 F(5°～25°N,40°～140°E)与高纬北部 D(45°～65°N,40°～140°E)的温度差。U_F:850 hPa 中纬度 E(25°～45°N,40°～140°E)平均西风风速。V_{sw}:850 hPa 高原东南侧 B(20°～25°N,105°～110°E)平均西南风速。SH:高原东南部 C(22.5°～32.5°N,95°～105°E)平均地面感热通量。另外,A(24°～30°N,110°～120°E)为 SCSR 区,D(27.5°～40°N,70°～105°E)为高原主体区域(摘自万日金和吴国雄,2006)

再考察高原东南部(C 区)的地面感热通量 SH(点线)。由于地处低纬,SH 在上半年一致为正,1 月增长缓慢,2 月增长加快,直至 3 月中期增至全年峰值,此后维持大约 1 个月,由于云覆盖的增加在 4 月后期开始减弱。总体来看,SH 与 V_{sw} 的变化趋势几乎是一致的。这种高度一致性表明高原东南部的感热与高原东南侧的西南风风速密切相关。根据热力适应理论(吴国雄 等,1999),地表非绝热加热会产生正的位涡源维持气旋性环流。地面感热加热的增强,除引起表层内能增加,还引起气压的下降,等熵面下凹,表面有暖性气旋式涡旋生成。所以不难理解在高原东南部出现了温度脊和低涡(图2a,c),该低涡称作西南暖低涡(SWWL),在 SWWL 的东南侧是西南风,可见热力强迫的西南风是高原东南侧西南风的又一主要部分。随着季节的推进,暖湿的上升气流逐渐增强,不稳定能量的释放和云覆盖的增加,4 月后期感热开始下降,但由于潜热加热的增强,SWWL 还能够继续维持。到 5 月初,SWWL 与南海副热带高压之间的气压梯度迅速减小,V_{sw}也随之减小。

综上所述,正是青藏高原的存在催生了西南风风速中心,高原的机械强迫和热力强迫在 SCSR 的气候形成中起了根本作用。因而,利用数值模式敏感性试验进一步验证以上论点。

3.2 高原隆升敏感性数值试验

再次使用上述 SAMIL-R42L9 模式进行逐步抬升高原地形高度的敏感性试验(以下简称 LFTP),具体方法如下。将太阳高度角固定在 4 月 10 日(此时 SCSR 显著增强),先将包括高原的欧亚大陆削平(地形高度降至 0 km),然后再削至 1 km,如此以 1 km 为间隔逐步增加削除高度并积分直至 5 km。当超过 5 km 时,仅增加青藏高原主体范围(D 区)内高度大于 2500 m 的区域的高度,直到 9 km(图略)。这样共有 10 个试验,模式共积分 30 年。

图 6 是高原在隆升过程中处于各个高度时所对应的 850 hPa 风场(左列)和降水量场(右列),选代表性的高度 0 km、2 km、4 km、6 km。很明显,当没有高原地形时(图 6a),欧亚低层西风带没有分支,中低纬为副高环流所控制,东亚降水量很小,无 SCSR 雨带(图 6e);2 km 时西风带出现明显南北分支,南亚次大陆完全由波状西风气流控制,高原东南侧西南绕流明显,华南地区的西南风明显增强(图 6b),SCSR 雨带初具雏形(图 6f);4 km 时高原南北两支西风和高原绕流增强形成急流,在高原东南侧出现西南风速中心(图 6c),SCSR 雨带中心形成(图 6g)。6 km 时,环流形势没有明显改变,但高原南北两侧西风急流进一步增强,高原绕流西南风速进一步增大,气旋性弯曲加大,华南反气旋性涡度发展(图 6d),华南降水明显减少,东亚雨带推进到长江及其以北地区,同时雨带中心转移至高原东南部(图 6h)。可见,高原东南侧西南风速中心的存在对于 SCSR 的形成至关重要,而它只有在高原大地形隆起到一定高度后才能出现。

再考察高原东南侧西南风速与高原隆升过程中的加热之间的关系。在春季高原隆升过程中,总加热逐渐增大,它对大气的加热作用不断加强,在 3 km 以下主要来自感热加热缓慢增加,在 3 km 以上主要来自潜热加热作用迅速增强。在最初的高原机械强迫绕流使 V_{sw} 迅速增加后,V_{sw} 的增长又与高原的总加热几乎线性一致(图略)。这充分说明,在春季,高原的隆升不仅使西风带分流形成西南绕流,还引起高原总非绝热加热的迅速增加,并造成了低层正涡源,使低层气旋式环流加强,使高原东南侧的西南风更加强大,进而导致了高原东南侧西南风风速中心的出现和华南春雨的形成。

4 华南春雨的季节内振荡

由于华南春雨期间的西南风取决于青藏高原的机械和热力强迫(万日金和吴国雄,2006),所以华南地区每年春季降水的主要特点是阶段性的连阴雨,即降水变化具有显著的季节内振荡(ISO)特征。Pan 等(2013)基于多年的春季降水资料,利用小波和功率谱分析发现 10~20 天准双周振荡是华南春雨最主要的 ISO 分量,并揭示了影响华南春雨期间 10~20 天 ISO 的大气低频环流结构、演变及动力学机制。

4.1 华南春雨 10~20 天低频振荡的环流结构和演变

合成分析表明,华南春雨 10~20 天 ISO 干/湿转换,对应着对流层低层环绕青藏高原的反气旋式/气旋式环流的交替(Pan et al., 2013)。在低层 850 hPa(图略),当华南春雨 ISO 为干位相时,高原北侧的气旋式环流和环绕高原的反气旋环流组成一个非对称的"偶极型"结构,控制着亚洲地区,其辐合中心在高原西北侧,辐散中心在高原东南侧;高原东侧上空是异常反气旋环流中心,华南受东北风异常控制、盛行辐散气流,不利于降水发生。其后,伴随着北部的气旋式环流及其相连的辐合区向东南方向移动,高原北侧转为反气旋环流异常;环绕高原的巨大反气旋式环流逐步减弱,高原南侧的西风和南海地区的偏南风逐步加强,形成围绕高原的气旋式环流异常,华南地区受西南风异常影响,盛行辐合气流,转入湿位相,此时分布形态与干位相相反。之后,当半封闭气旋离开华南并向东传播时,华南地区的西南风异常以及上游的高原南侧的西风异常逐步减弱,并转为盛行东北风,受辐散气流影响,进入干位相,完成一个振荡过程。

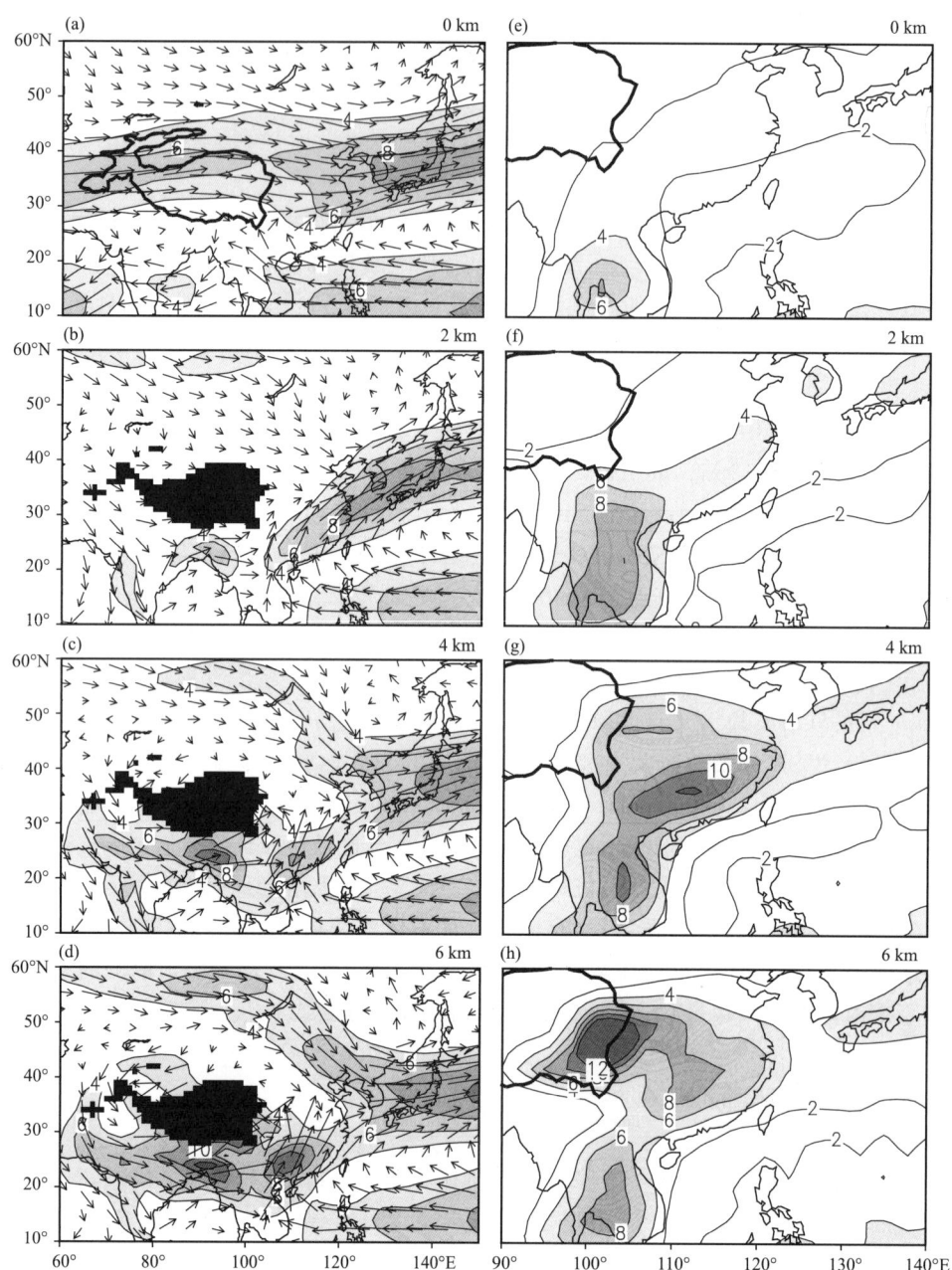

图 6 高原隆升试验中各高度时 850 hPa 上风场(左图,阴影区为风速大于 4 m·s^{-1} 区域)和地面降水量场(右图,单位:mm·d^{-1}),图中黑色阴影区和粗虚线为高原主体位置(摘自万日金和吴国雄,2006)

实际上,青藏高原绕流的强度很大程度上取决于高原表面感热异常引起的热力强迫,南支西风气流随着高原热力作用加强,为华南地区输送更多的暖湿空气,有利于华南春雨的形成(吴国雄 等,1997a,Wu et al.,2007;梁潇云 等,2005;万日金和吴国雄,2006;王同美 等,2008;Wan et al.,2009)。这种高原表面感热异常对南支西风的调制作用,在季节内时间尺度也同样得到体现。10~20 天滤波 10 m 流场和表面感热通量在 ISO 演变合成图(图7)显示:在华南降水干位相时,高原上为感热加热负异常,对应着辐散流场,气流沿斜坡向外辐散,伴随着 850 hPa 上环绕高原的反气旋。此时高原以北地区存在显著加热正异常区域。在华南降水湿位相时,青藏高原表面为感热加热正异常,高原的"感热气泵"效应使得周围气流向高原辐合,在 850 hPa 等压面层观察到高原南侧西风绕流加强,此时华南地区辐合最强,转入湿位相。这种演变过程说明华南春雨 10~20 天振荡的"偶极型"环流强烈地受到春季高原感热异常的调控。

在高层 200 hPa 上,对流层上部存在一个显著的西北—东南向的波列(图略)。在干位相时,高原西北部是东北—西南向的带状的强辐散异常区域,伴随着闭合的反气旋环流,气旋、反气旋和气旋中心交替位于乌拉尔山脉、巴基斯坦和印度东海岸;另一个气旋环流中心在日本及其附近地区,与南海上空的反气旋有关。最显著的辐合在中国东部,与对流层低层的强辐散相对应。随后,波列向东南方向传播,中国东部转受逐步加强的辐散中心控制,至湿位相时到达极值,华南地区高层强辐散与低层强辐合相协调,上升运动发展。

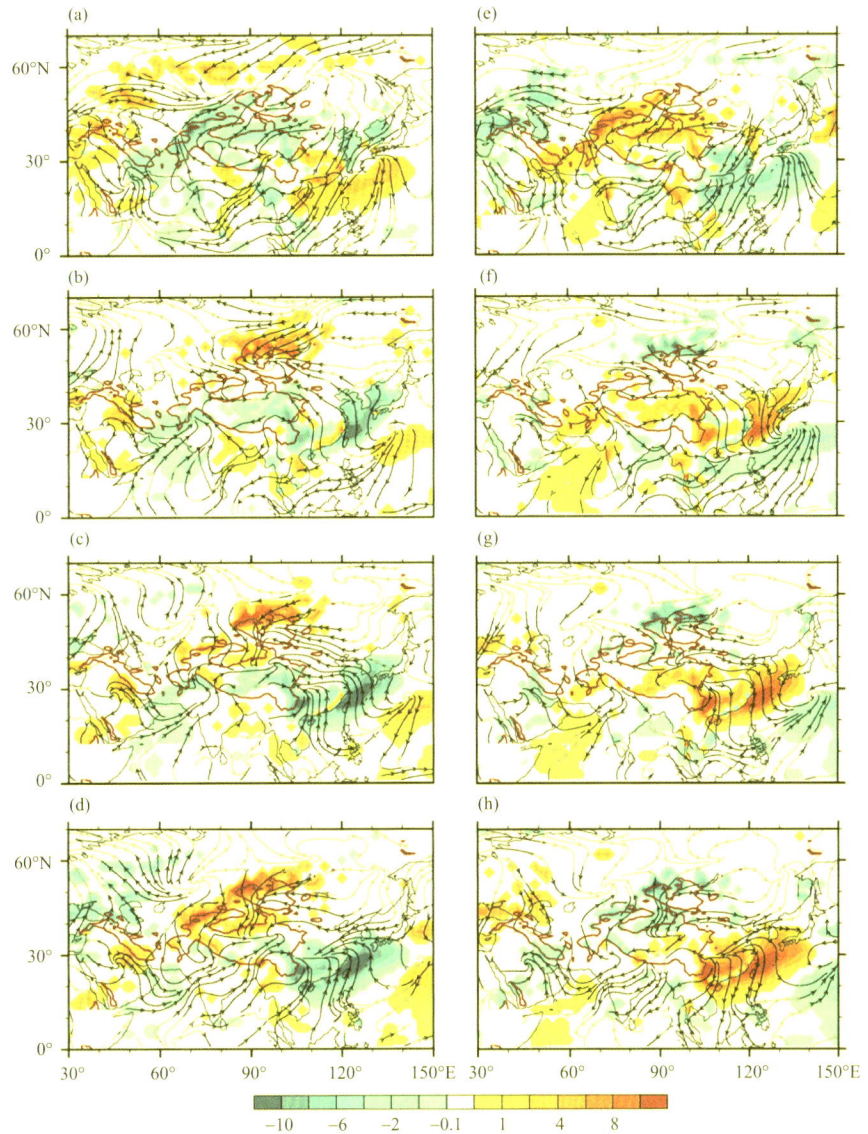

图 7　10～20 天滤波 10 m 流场(流线图,黑色线代表经向风或者纬向风分量通过 5% 显著性水平的统计 t 检验)和地表感热通量(阴影,单位:W·m^{-2},通过 5% 显著性水平统计检验)合成图,(a)—(h)分别对应 ISO 不同位相。实线包围的是 1500 m 以上的地形(摘自 Pan et al.,2013)

4.2　华南春雨 10～20 天振荡的斜压性

10～20 天滤波的 30°～40°N 纬向垂直剖面风场、高度场和气温场和 105°～120°E 经向垂直剖面风场、散度和气温场(图略)显示,对流层中高层显现出明显的斜压特征。干位相时,气温负异常中心和下沉运动区位于位势高度负异常中心的西侧,高原东侧强下沉运动区从 700 hPa 左右延伸至对流层高层,对流层高层的冷空气这时已入侵至中国东部;高原西侧上空为槽前脊后形势,盛行西风且有弱的上升运动,受暖异常控制。之后温度正异常越过高原,高原表面感热增加,气流向高原上空汇合;中国东部地区转为槽前脊后形势,上升运动发展,伴随的冷空气下沉有助于暖空气抬升,有效位能释放;气温正异常区域同时扩展,在 200 hPa 附近高度出现温度正异常大值中心时,低层的温度异常中心以及上升运动强度也达到最强,华南处于湿位相。

在干位相时,随高度向北倾斜的气温负异常带从对流层低层延伸至 200 hPa,并伴随沿假相当位温面的强劲北风和下沉运动,华南地区受冷空气影响、温度锋区位于中国南海北部地区。在 60°N 以北对流层中下层为冷异常控制。两个冷异常区域间为显著的暖异常区域。之后北部的冷空气南下,华南的暖空气受其抬升作用,有利于斜压能量从有效位能转化为动能。当冷空气前锋到达华南地区,上升运动达到最强,华南转为湿位相。因此,伴随着相关能量转化的斜压不稳定是华南区域春季降水 10~20 天准双周振荡的形成和传播的动力学机制。

5 总结与讨论

文中总结了吴国雄院士团队近年来在华南春雨(江南春雨)气候成因和季节内变率方面的研究进展。利用多源资料诊断,并通过华南同纬度带的东亚与北美地区春季环流结构和降水演变的对比分析,论证了关于"华南春季持续降水"的"季节增暖时滞效应"认识的局限,由此结合数值模拟试验证实了青藏高原的机械和热力强迫是华南春雨气候形成的根本原因。提出了"江南春雨"的气候学概念,明确其时空分布范围和季节内振荡机制。主要结论归纳如下:

华南春雨是东亚地区除南海季风降水、梅雨、台风降水等夏季风降水之外独特的季节性降水,是东亚地区冬、夏季节转换过程中存在的独特气候现象。华南春雨开始于第 13 候,对应于青藏高原主体感热加热由冷源转为热源,高原东南部感热加热迅速增强,华南地区西南风迅速加大,SCSR 雨强迅速增长;华南春雨终止于第 27 候,对应高原东南部感热加热、华南西南风、SCSR 雨强同时迅速回落,其后南海副高脊面由南倾转为北倾,标志着亚洲季风环流由冬季型转入夏季型,南海季风爆发。华南春雨空间分布受南岭和武夷山脉地形的影响显著,其主雨带与山脉地形重合,位于 30°N 以南,110°E 以东的中国广大南方地区。山脉地形还显著增强了华南春雨的雨量。

春季高原东南侧西南风风速中心的出现是 SCSR 气候形成的直接原因,它在下游引起强烈的风速辐合,是华南出现强烈的水汽输送和辐合的关键。高原东南侧西南风速中心的出现是高原机械和热力强迫的结果,该西南风不仅是高原的绕流,更是高原热力强迫环流的一部分。高原的隆升不仅使西风带分流、绕流,还引起高原总非绝热加热的迅速增加,产生气旋性涡源,增强了高原东南部的西南绕流,并导致高原东南侧西南风速中心的出现。高大青藏高原的机械强迫和热力效应是华南春雨气候形成的根本原因。

华南春雨的季节内变化以 10~20 天准双周振荡为主,准双周振荡的干/湿转换对应着对流层低层环绕青藏高原的反气旋式/气旋式环流的交替;青藏高原表面感热异常的准双周振荡显著地调控环绕高原的低空"偶极型"环流的演变及其与高空低频波列的垂直耦合。斜压不稳定是华南春雨 10~20 天准双周振荡形成和传播的动力学机制。

在实际业务预报中,华南沿海各省仍然按着传统将汛期分为 4—6 月的前汛期和 7—9 月的后汛期,但根据本文的回顾,其前汛期既包括华南春雨时期的 4—5 月上半月共 1.5 个月,又包括 5 月下半月南海季风爆发后的 1.5 个月,分别为两个不同性质、具有不同气候背景的降水时段,似乎将华南的前汛期分为华南春雨时期和南海季风爆发期两个时段更为合理。季风是由海陆热力对比的反转变化引起的,它不仅与大陆的面积大小、纬度位置有关,还与大陆上的大地形有关。亚洲季风之所以如此显著,青藏高原的热力作用是重要原因之一;同样,华南春雨之所以如此显著,高原的动力和热力作用是根本原因。由初春华南至日本南部雨量同时快速增长就推论 SCSR 是季节增暖时滞效应即海陆分布影响的观点,只看到海陆分布影响的普遍性,而没有认识到大地形影响的特殊性。而正是这样的特殊性,在全球各大陆地和海洋上塑造了各具特色的气候类型。

参考文献

梁潇云,刘屹岷,吴国雄,2005.青藏高原隆升对春、夏亚洲大气环流的影响[J].高原气象,24(6):837-845.

李麦村,潘菊芳,田生春,等,1977.春季连续低温阴雨天气的预报方法[M].北京:科学出版社:3-6.

毛江玉,吴国雄,刘屹岷,2002a.季节转换期间副热带高压带形态变异及其机制的研究Ⅰ:副热带高压结构的气候学特征[J].气象学报:60(4):400-408.

毛江玉,吴国雄,刘屹岷,2002b.季节转换期间副热带高压带形态变异及其机制的研究Ⅱ:亚洲季风区季节转换指数[J].气象学报,60(4):409-420.

毛江玉,吴国雄,刘屹岷,2002c.季节转换期间副热带高压带形态变异及其机制的研究Ⅲ:热力学诊断[J].气象学报,60(6):647-659.

万日金,吴国雄,2006.江南春雨的气候成因机制研究[J].中国科学D辑:地球科学,36(10):936-950.

万日金,吴国雄,2008.江南春雨的时空分布[J].气象学报,66(3):310-319.

王同美,吴国雄,万日金,2008.青藏高原的热力和动力作用对亚洲季风区环流的影响[J].高原气象,27(1):1-9.

吴国雄,李伟平,郭华,等,1997a.青藏高原感热气泵和亚洲夏季风[C]//赵九章纪念文集.北京:科学出版社:116-126.

吴国雄,张学洪,刘辉,等,1997b.LASG全球海洋-大气-陆面模式(GOALS/LASG)及其模拟研究[J].应用气象学报,8(增刊):15-28.

吴国雄,刘屹岷,刘平,1999.空间非均匀加热对副热带高压带形成和变异的影响Ⅰ:尺度分析[J].气象学报,57(3):257-263.

吴国雄,丑纪范,刘屹岷,等,2002.副热带高压形成和变异的动力学问题[M].北京:科学出版社:314.

LINHO, HUANG X L, LAU N C, 2008. Winter-to-spring transition in East asia: A planetary-scale perspective of the South China spring rain onset[J]. J Climate, 21: 3081-3096.

PAN W J, MAO J Y, WU G X, 2013. Characteristics and mechanism of the 10—20-day oscillation of spring rainfall over Southern China[J]. J Climate, 26: 5072-5087.

TIAN S F, YASUNARI T,1998. Climatological aspects and mechanism of spring persistent rains over central China[J]. J Meteor Soc Japan,76: 57-71.

WAN R J,WU G X,2006. Mechanism of the spring persistent rains over Southeastern China[J]. Science in China Ser D,36(10):936-950.

WAN R J,ZHAO B K,WU G X,2009. New evidences on the climatic causes of the formation of the spring persistent rains over Southeastern China[J]. Adv Atmos Sci, 26: 1081-1087.

WU G X,LIU Y M,WANG T M,et al, 2007. The influence of mechanical and thermal forcing by the Tibetan Plateau on Asian climate[J]. J Hydrometeor, 8: 770-789.

WU X F,MAO J Y,2016. Interdecadal modulation of ENSO-related spring rainfall over South China by the Pacific decadal oscillation[J]. Clim Dyn, 47: 3203-3220.

Air-Sea Interaction and Formation of the Asian Summer Monsoon Onset Vortex over the Bay of Bengal

WU Guoxiong[1], GUAN Yue[1,2], LIU Yimin[1], YAN Jinghui[3], MAO Jiangyu[1]

(1 LASG, Institute of Atmospheric Physics, CAS, Beijing 100029, China; 2 Graduate University of Chinese Academy of Sciences, Beijing 100039, China; 3 National Climate Center, China Meteorological Administration, Beijing 100080, China)

Abstract: In spring over the southern Bay of Bengal (BOB), a vortex commonly develops, followed by the Asian summer monsoon onset. An analysis of relevant data and a case study reveals that the BOB monsoon onset vortex is formed as a consequence of air-sea interaction over BOB, which is modulated by Tibetan Plateau forcing and the land-sea thermal contrast over the South Asian area during the spring season.

Tibetan Plateau forcing in spring generates a prevailing cold northwesterly over India in the lower troposphere. Strong surface sensible heating is then released, forming a prominent surface cyclone with a strong southwesterly along the coastal ocean in northwestern BOB. This southwesterly induces an local offshore current and upwelling, resulting in cold sea surface temperatures (SSTs). The southwesterly, together with the near-equatorial westerly, also results in a surface anticyclone with descending air over most of BOB and a cyclone with ascending air over the southern part of BOB.

In the eastern part of central BOB, where sky is clear, surface wind is weak, and ocean mixed layer is shallow, intense solar radiation and low energy loss due to weak surface latent and sensible heat fluxes act onto a thin ocean layer, resulting in the development of a unique BOB warm pool in spring. Near the surface, water vapor is transferred from northern BOB and other regions to southeastern BOB, where surface sensible heating is relatively high. The atmospheric available potential energy is generated and converted to kinetic energy, thereby resulting in vortex formation. The vortex then intensifies and moves northward, where SST is higher and surface sensible heating is stronger. Meanwhile, the zonal-mean kinetic energy is converted to eddy kinetic energy in the area east of the vortex, and the vortex turns eastward. Eventually, southwesterly sweeps over eastern BOB and merges with the subtropical westerly, leading to the onset of the Asian summer monsoon.

Keywords: air-sea interaction, upwelling, BOB warm pool, vortex, monsoon onset

1 Introduction

The Asian monsoon seasonal transition from winter to summer in East and South Asia is characterized by abrupt changes in the general circulation. Ananthakrishnan et al. (1981) pointed out that the summer monsoon onset over the southeast Bay of Bengal (BOB) starts toward the end of April, earlier than any land monsoon onset. Recent studies have also proposed that the earliest onset of the Asian summer monsoon occurs in BOB, following by the onset over South China Sea (SCS), and lastly over India (e.g., Lau et al., 1997; Wu et al., 1998; Mao et al., 2002a, 2007). Associated with the Indian summer monsoon onset is a type of low-level vortex known as the onset vortex (Krishnamurti et al., 1981), which is formed at around 10°N in the Bay of Bengal (BOB) and east of the Arabian Sea (Ananthakrishnan et al., 1968). The vortex moves northward and generally develops rapidly into a tropical storm, resulting in a region of convergence and subsequent strengthening of the westerly to its south. In turn, the enhanced westerly leads to the Indian monsoon onset and advances northward with the onset vortex (Vinayachandran et al., 2007). Many previous studies have described features of the vortex, including the distribution of precipitation and the vortex structure (e.g., Pisharoty et al., 1957; Rao et al., 1970; Sharma et al., 1971). Subsequent studies suggested that barotropic instability associated with horizontal wind shear leads to the formation of the vortex, and that diabatic heat supplies the energy to the vortex required for its development (Krishnamurti, 1981, 1985; Krishnamurti et al., 1981, 1982). Mak et al. (1982) found that vertical wind-shear and baroclinic instability play an important role in the development of the vortex.

The onset vortex is also influenced by the boundary condition connected with the local sea surface temperature (SST) (Joseph, 1990). As is well known, atmospheric convection in the tropics is highly sensitive to SST and to fluctuations in surface flux of the underlying ocean. SST above 28 ℃, in combination with other factors, is necessary to produce organized deep convection in the tropical atmosphere (Gray, 1968, 1975, 1998; Harr et al., 2004). In the Indian Ocean, warm SST is generally found in three locations (Vinayachandran et al., 2007): the western equatorial Indian Ocean, the southeastern Arabian Sea (SEAS), and eastern BOB, with the latter being warmest during April. At about one week before the Indian monsoon onset, a warm pool (SST>30.5 ℃) exists over the Arabian Sea (Seetaramayya et al., 1984). The warm SST in the SEAS leads to the formation of a vortex that affects the Indian monsoon onset over Kerala in early June (Joseph, 1990; Shenoi et al., 1999, 2005). Furthermore, Rao et al. (1999) reported that the onset vortex generally forms over this region because it is related to the location of the maximum heat content in the upper layer of the tropical ocean, as this is a potential energy source for the vortex. Furthermore, by use of numerical experiments, Kershaw (1985, 1988) demonstrated that the use of anomalies or accurate SST data results in more accurate predictions of summer monsoon onset than the use of climatological SST.

Most of the above studies focused on the Arabian Sea onset vortex. Actually the BOB monsoon onset in early May is also influenced by the development of the onset vortex in BOB (Lau et al., 1998; He et al., 2002; Liu et al., 2002; Yan, 2005). This onset vortex moves northward over the BOB area, splitting the ridge of the subtropical high and leading to the monsoon onset. Yan (2005) proposed a physical mechanism for interpreting the eastward advance of the Asian monsoon onset from BOB to the South China Sea (SCS). Vinayachandran et al. (2007) discussed the structure of the onset vortex, reporting that the warmest SST is formed earliest over eastern BOB, which is then followed by the BOB monsoon onset.

Despite these advances, many important and basic questions remain unanswered concerning the BOB monsoon onset, including how the warm SST is formed over BOB in spring, how it influences and interacts with the BOB monsoon onset vortex, and where the energy required for the vortex genesis comes from since either barotropic or baroclinic instability merely explains the energy conversion within the atmosphere. In this context, the present study analyzes observational data related to air-sea interaction in BOB in the spring of 2003, and investigates how this interaction contributed to the formation of a spring warm pool in the eastern part of central BOB and to the development of a vortex that occurred from May 8 to 19, which led to the Asian summer monsoon onset.

The remainder of the manuscript is organized as follows. Section 2 descripts the data used in this study. Section 3 outlines the general behavior of the BOB monsoon onset vortex and its connection with the seasonal transition from winter to summer. Section 4 considers the formation and development of the vortex under a special circulation pattern over BOB for the year 2003. Also examined is the oceanic influence on vortex development and the energy transformation associated with evolution of the vortex. Section 5 examines how the spring BOB warm pool is formed, and how SST in BOB evolves before and after formation of the vortex. For this purpose, we calculate the energy budget in the mixed layer, including surface net radiation, heat flux, and cold entrainment due to mixing. In Section 6, we investigate the influence of the vortex on the Asian summer monsoon onset. Finally, a summary and discussion are provided in Section 7.

2 Data description

This study analyzed the following datasets: (1) daily global sea-surface pressure with a resolution of 1.125° latitude×1.1.25° longitude, from the reanalysis of the European Centre for Medium-Range Forecasts (ECMWF) (ERA-40, Uppala, 2005); (2) daily global sea-surface winds with a high horizontal resolution of 0.25° latitude × 0.25° longitude, from the National Oceanic and Atmospheric Administration (NOAA) (Zhang et al., 2006) and National Climatic Data Center (NCDC) (Smith and Reynolds, 2004), for which wind speeds are generated by blending observations from six satellites and for which wind directions from the National Center for Environment Prediction (NCEP) Reanalysis 2 (NRA-2) (Kanamitsu et al., 2002) are interpolated into the blended speed grids; (3) daily optimum interpolation SST (OISST) analysis data (AVHRR+AMSR-E products) from the NOAA Satellite and Information Service with the same resolution as sea surface winds and weekly OISST with horizontal resolution of 1° latitude×1° longitude; (4) daily SST, ocean latent heat flux, sensible heat flux, net radiation flux, longwave and shortwave radiation at the sea surface with a resolution of 1° latitude × 1° longitude obtained from the Objectively Analyzed Air-Sea Fluxes (OAFlux) for the Global Oceans, sponsored by the Woods Hole Oceanographic Institution (WHOI) Cooperative Institute for Climate and Ocean Research (CICOR); and (5) pentad total downward heat flux at the surface, ocean current, salinity, mixed-layer depth, and potential temperature profiles (1/3° latitude × 1° longitude, 40 levels) taken from the NCEP Global Ocean Data Assimilation System (GODAS) (Ji et al., 1995) for accurate air-sea processes. Because OAFlux only provides heat flux data over the sea, we also employed daily diabatic heating and wind data at 10 m height and at various pressure levels from NCEP/NCAR Reanalysis 1 to analyze the temporal variations in land heating.

The vortex records are retrieved from the tropical cyclone Best Track Data from 1979 to 2009, as archived at the Joint Typhoon Warning Center (JTWC). This dataset provides data on cyclone category for the period after 2000, but not before. To overcome the period limitation on cyclone category in the Best Track Data, we use the "maximum sustained wind speed in knots" provided by JTWC to determine the vortex category, with reference to the Saffir-Simpson Hurricane Scale (see Table 1). Using this approach, the vortex categories prior to 2000 are determined on a daily basis.

Table 1 Categories of monsoon onset vortex based on the Saffir-Simpson Hurricane Scale and using the "maximum sustained wind speed in knots" provided by JTWC

Category	Super typhoon	Typhoon	Tropical storm	Tropical depression	Tropical cyclone	Disturbance
Abbreviation	ST	TY	TS	TD	TC	DB
Maximum wind speed V in knots	$\geqslant 136$	$\geqslant 64$	$64 > V \geqslant 34$	$34 > V$	$34 > V$	$34 > V$

3 Climatology: BOB vortex and seasonal transition from winter to summer

In spring of most years, a vortex develops over southern BOB, subsequently moving northward and followed by the Asian summer monsoon onset over eastern BOB. This vortex is generally referred to as the BOB monsoon onset vortex (Lau et al., 1998). Figure 1 shows the tracks of the BOB monsoon onset vortex from 1979 to 2009. The monsoon onset date is defined as the time when the meridional temperature gradient ($\partial T/\partial y$) in the upper troposphere (between 500 and 200 hPa) changes from negative to positive, as proposed by Mao et al. (2002). The monsoon onset vortex and its track are defined according to the following principle: if and only if the cyclone is formed before the BOB monsoon onset and if its life cycle covers the onset time, then the tropical cyclone is defined as the BOB monsoon onset vortex, and the corresponding tropical cyclone track is defined as the vortex track. According to this principle, the BOB monsoon onset vortex is sought for each year, and the dates of the BOB Asian summer monsoon onset for years in which a monsoon onset vortex developed is listed in Table 2. Only two monsoon onset vortexes were identified for the period before the 1990s, mainly due to a lack of tropical cyclone data. For the period after 1990, 13 onset vortexes are identified from the 20-year record. In other words, the strong BOB monsoon onset vortex occurs in most years, at least for the period after 1990. The corresponding tracks of the onset vortexes are shown in Fig. 1, revealing the following common features:

(1) All the onset vortexes originate in southern BOB, south of 10°N.

(2) All the vortexes move northward before turning northeastward in northern BOB (north of 15°N), except in 1979 and 1990 when the onset vortex rapidly developed into a typhoon after crossing 10°N and then moved northwestward to make landfall over the east coast of India.

(3) All the onset vortexes became tropical storms or typhoons upon landfall.

(4) Most of the vortexes occurred in late April and early May, in accordance with the BOB monsoon onset.

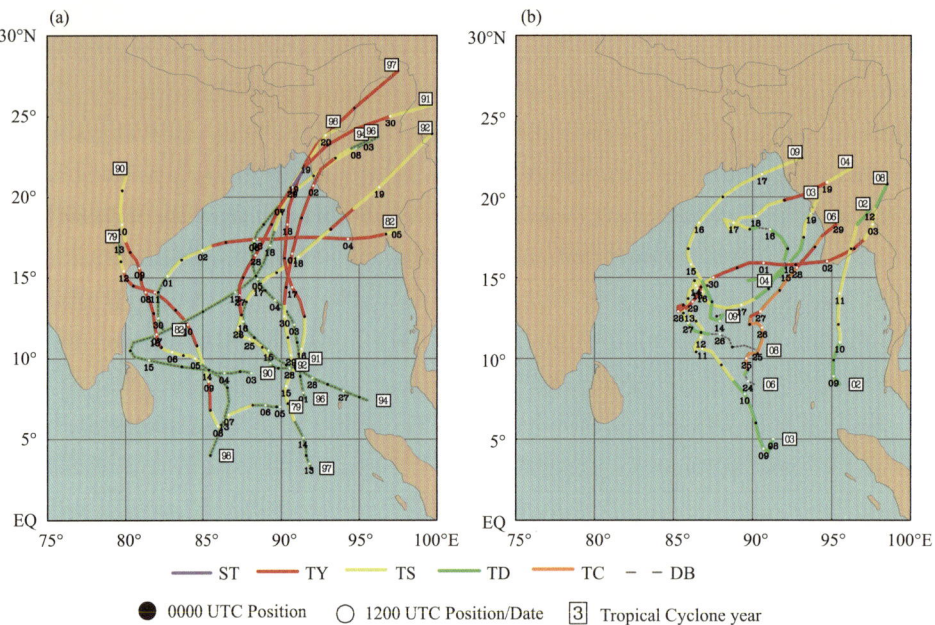

Figure 1 Tracks of the BOB summer monsoon onset vortex from JTWC (a) before 2000 and (b) after 2000. For each track, different categories of vortex (i.e., ST, TY, TS, TD, TC, and DB; see Table 1) are indicated by line color. The year of each onset vortex (listed in Table 2) is marked by the number at each end of each track. Numbers along each track denote the dates of vortex activity; black and white dots along each track denote 0000 and 1200 UTC, respectively

Table 2 Dates of the BOB Asian summer monsoon onset for years in which a monsoon onset vortex developed

Year	1979	1982	1990	1991	1992	1994	1996	1997	1998	2002	2003	2004	2006	2008	2009
Date	May 11	May 4	May 15	Apr 26	May 14	May 01	May 03	May 18	May 18	May 11	May 12	May 10	Apr 29	Apr 27	Apr 16

By searching the low geopotential height centers at 850 and 700 hPa surfaces before the monsoon onset, the monsoon onset vortex can also be identified from NCEP/NCAR reanalysis data, yielding similar features to those listed above (Yan, 2005).

The monsoon is considered a response of the land-sea thermal contrast to the seasonal march of the solar angle (e.g., Webster et al., 1998). In tropical Asia during winter, cold land is located to the north and the warm Indian Ocean to the south; whereas in summer, warm land is located to the north and colder ocean to the south. Therefore, the meridional temperature gradient is reversed during the change in season.

Li et al. (1996) used the upper tropospheric temperature difference between 30°N and 5°N as an index with which to measure the seasonal transition in the Asian area, and Webster et al. (1998) used the zonal wind difference between 850 and 200 hPa as an index with which to study the monsoon onset. These two indexes are similar in terms of describing the meridional thermal contrast, due to the constraint of the following thermal wind relation:

$$\frac{\partial u}{\partial z} \propto -\frac{\partial T}{\partial y}$$

Note that in south Asia (5°—20°N) during winter, there exists a zonally orientated ridgeline of high pressure at every level. The tropical easterlies exist to the south of this ridgeline, whereas the subtropical westerlies prevail to the north. Vertical tilting of the ridgelines at different levels is consistent with the local vertical shear of zonal wind, and can be used as a measure of the meridional thermal contrast (Mao et al., 2002, 2004). Before the Asian monsoon onset, $\partial T/\partial y < 0$, $\partial u/\partial z > 0$, and the ridgelines tilt southward with increasing height. After the monsoon onset, $\partial T/\partial y > 0$, $\partial u/\partial z < 0$, and the ridgelines tilt northward with height. In other words, the ridgelines always tilt toward the warmer region with increasing height, and the moment when the ridgelines become vertical (indicating $\partial T/\partial y = 0$) is the timing of monsoon onset in the region.

Figure 2 shows the evolution, from 30 April to 14 May, of the climate-mean distributions of the ridgelines at 700, 500, and 200 hPa (Figs. 2a—2c, respectively), and of the relative vorticity at 700 hPa (Figs. 2d—2f). The data used for the analysis are from the NCEP/NCAR reanalysis averaged from 1979 to 2003. By the end of April, the ridgelines over the Asian sector generally show a southward tilt with height (Fig. 2a), representing a typical winter pattern. In early May (Fig. 2b), while the ridgelines still tilt southward with height over most areas, they tilt northward over eastern BOB, representing a summer pattern (Fig. 2b) and indicating that the summer monsoon onset happened in the region. Such a rapid seasonal transition occurs with breaking of the ridgeline at 700 hPa. By mid-May (Fig. 2c), the ridgeline at 500 hPa also breaks throughout the region. While tilting of the ridgeline maintains a winter pattern over Saudi Arabia, the Arabian Sea, India, and the Western Pacific, it takes a summer pattern over eastern BOB and the Indochina Peninsula, where the summer monsoon onset has already occurred.

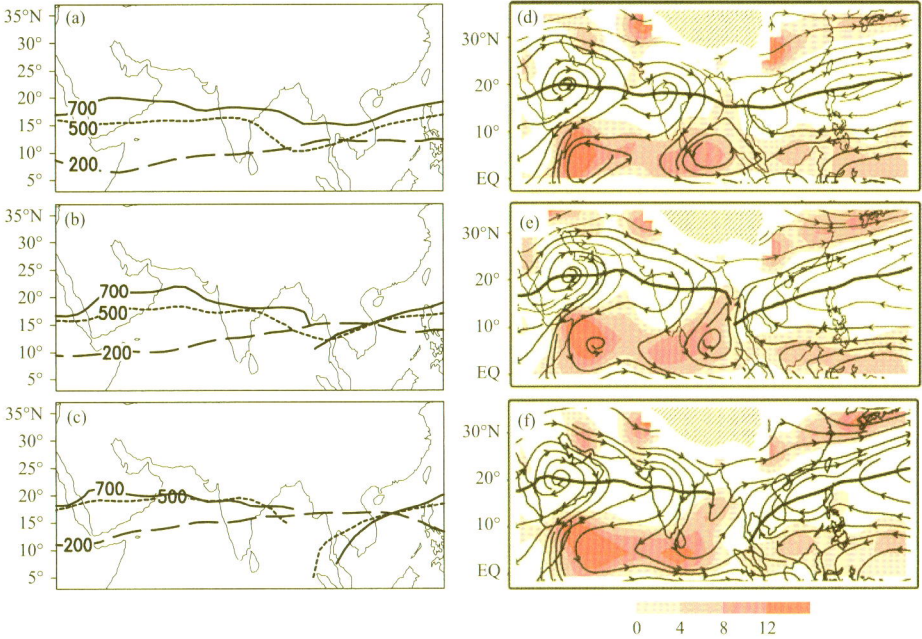

Figure 2 Climatological ridgelines (thick curve; numbers indicate the isobaric surface) of the subtropical anticyclone at 700, 500, and 200 hPa for (a) 30 April, (b) 6 May, and (c) 14 May. (d)—(f) are the same as (a)—(c) except for streamlines and vorticity (shading, 10^{-5} s^{-1}) at 700 hPa, with the ridgeline (thick curve) superimposed

To reveal the mechanism responsible for breaking of the ridgeline over eastern BOB in early May, Figs. 2d—2f shows the stream field and relative vorticity at 700 hPa. Before the BOB monsoon onset, a cyclonic flow exists south of the Tibatan Plateau, anticyclone centers are observed over Saudi Arabia, India, and the Western Pacific, while geopotential height is relatively low over eastern BOB (Fig. 2d). A continuous ridgeline is located at 15°N. The two centers over land can be attributed to strong regional surface sensible heating, whereas the center over the Western Pacific may be due to intensified rainfall over the Indochina Peninsula (Liu et al., 2004). A strong vortex, with vorticity of about 1×10^{-5} s^{-1}, exists over southern BOB. By the time of BOB monsoon onset (Fig. 2e), the BOB vortex has moved northeastward, the anticyclone over India has weakened, the anticyclone over the Western Pacific has extended southwestward and the western end of its ridgeline has shifted southward from 15°N to 10°N. Consequently, the tropical southwesterly located east of the vortex merges with the subtropical westerly located north of the ridgeline, resulting in breaking of the ridgeline. One week later (Fig. 2f), the BOB vortex has disappeared and the strong southwesterly sweeps over a large area ranging from eastern BOB to the Indochina Peninsula, South China, and the Western Pacific. The earliest Asian summer monsoon onset has arrived in eastern BOB and over the Indochina Peninsula. Therefore, the climate data indicate that development of the BOB vortex plays an important role in leading the Asian summer monsoon onset. Below, we examine a case study with the aim of understanding how the monsoon onset vortex develops over the BOB area.

4 Oceanic influences on the development of the onset vortex

4.1 Evolution of SST with development of the onset vortex

In 2003, a vortex developed during early May, leading to the BOB monsoon onset. This vortex is choosen for current study because it was formed in the southeastern BOB, took a track first northwestward then northeastward, and had a longer lifetime. Its life cycle is typical of the season. Fig. 3 demonstrates the track and intensity evolution in terms of sea surface pressure of the vortex. Its daily location and intensity are determined by searching the low-pressure center at sea-level in the region based on the ERA-40 reanalysis. The resultant track (Fig. 3a) is similar to that provided by JTWC, although not exact. We can therefore diagnose its formation processes, its interaction with the underlying ocean, and its contribution to the Asian summer monsoon onset, focusing on the BOB area of 80°—97°E in longitude and 0°—20°N in latitude.

The BOB monsoon onset in 2003 occurred on 12 May, as measured by the monsoon onset index defined by Mao et al. (2002, 2004). On 8 May, before the BOB monsoon onset, a vortex developed over southern BOB and moved northward until 19 May (Fig. 3a). The vortex attained its lowest pressure at sea level on 13 May before weakening and disappearing on 19 May (Fig. 3b). To understand the evolution of the vortex, we recognize three stages: Stage I corresponds to the vortex gestational phase, prior to 8 May; Stage II represents the vortex developing phase, from 8 to 13 May; and Stage III corresponds to the vortex decaying phase, from 14 to 19 May. Unless otherwise indicated, 5 and 8 May are chosen as being representative of Stage I, while 11 and 14 May are representative of Stages II and III, respectively (see Fig. 3b).

Figure 4 shows the evolution of SST before and after formation of the BOB monsoon vortex. In Stage I, a warm pool (SST>30.5 ℃) occurs in the eastern part of central BOB (Figs. 4a and b), immediately north of the location where the vortex formed. During Stage II, the vortex propagates northwestwards toward the warm pool (Fig. 4c) and cold SST is generated behind the vortex in association with precipitation and ocean mixing, as discussed below. During Stage III, after the vortex has matured on 13 May (Fig. 4d), the warm pool decreases in size and the vortex begins to weaken. The vortex track is confined to the area of SST>30 ℃, and the evolution of the vortex is influenced by the warm SST pool.

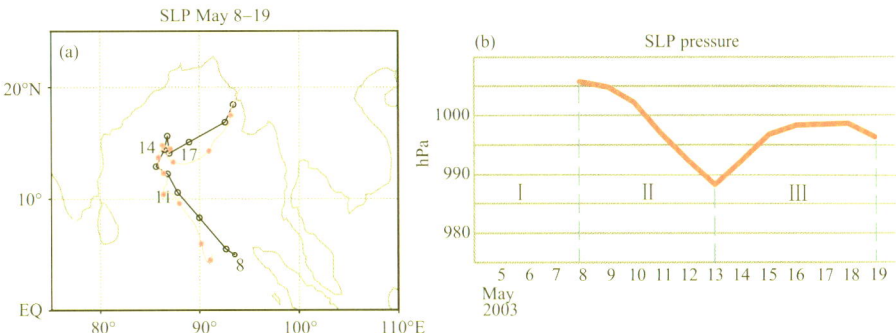

Figure 3 Evolution of the vortex from 8 to 19 May 2003, showing (a) the track of the vortex center (identified by sea level pressure), with numbers indicating the date on which the center was at that location, and (b) the intensity of the vortex, in sea level pressure from ERA-40. The pink curve in (a) denotes the track produced from the JTWC data

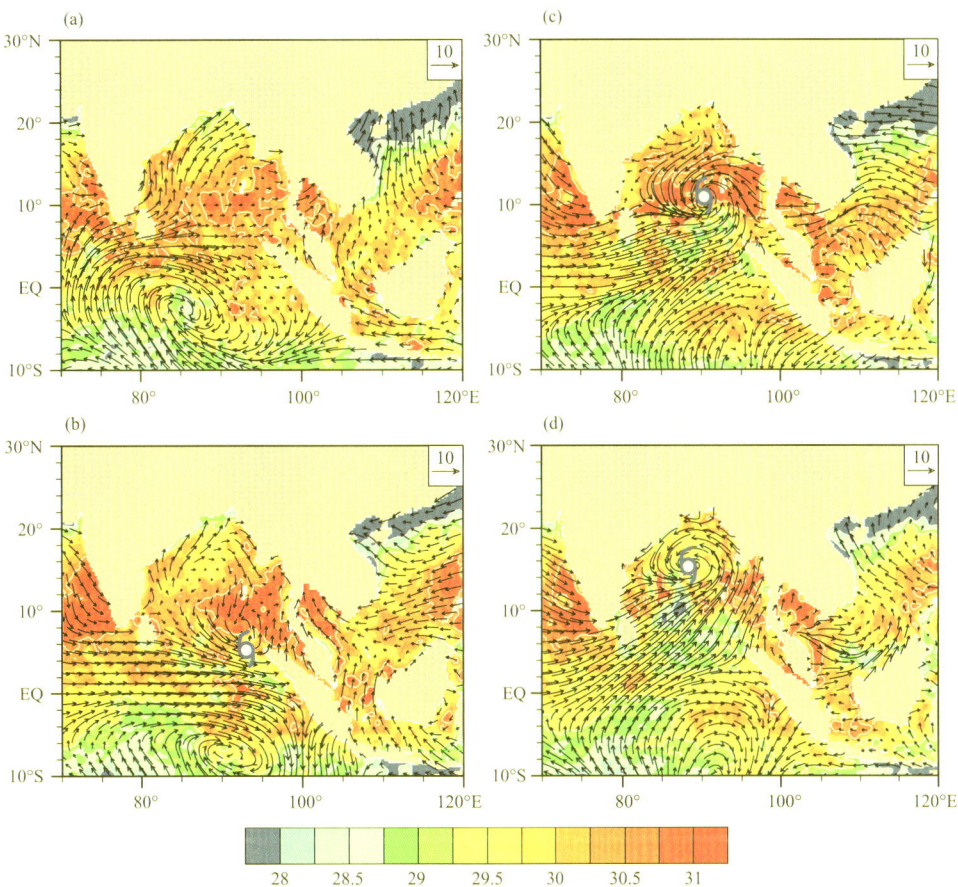

Figure 4 SST (shading, ℃) from OISST and sea surface wind from NCDC (vectors, m·s^{-1}) on (a) 5 May, (b) 8 May, (c) 11 May, and (d) 14 May 2003. The vortex temporal position is marked by the typhoon symbol

4.2 Evolution of sensible heating

As a boundary condition, SST clearly changed with movement of the vortex in 2003, although it did not directly influence the atmosphere. The air-sea temperature difference determines the occurrence of heating from the sea to the atmosphere, or vice versa. According to the concept of thermal adaption (Wu et al., 2000), whenever sensible heating occurs from the underlying surface to the atmosphere, cyclonic circulation is generated in the lower layers of the atmosphere.

Figure 5 shows the surface sensible heating (upward is positive) during the three stages of the vortex life cycle. In the northern BOB area, a common feature is observed throughout the whole period: the ocean heats the atmosphere in eastern BOB but cools the atmosphere in western BOB. This occurs because SST is warmer in the east but colder in the west (Fig. 4) due to local upwelling, as discussed below. Therefore, the eastern part of northern BOB becomes a heat source for the atmosphere, whereas the western part becomes a heat sink. This finding explains why the vortex track, as shown in Fig. 3a, is maintained over eastern BOB.

During the gestational stage on 5 May (Fig. 5a), weak sensible heating appears throughout the entire area of the BOB warm pool. In the southeastern BOB region on 8 May (Fig. 5b), sensible heat is enhanced with a maximum exceeding 15 W·m^{-2}, centered at (95°E, 5°N), and the regional surface wind shows cyclonic shear in response to surface forcing induced by this sensible heating. Consequently, the onset vortex is formed in the region. Surface sensible heating around the center of the vortex is intensified during its developing phase. On 11 May (Fig. 5c), the sensible heat flux near the vortex center exceeds 35 W·m^{-2}, and the surface cyclonic circulation already occupies the entire northern BOB. During the decaying phase, the surface sensible heating associated with the vortex is severely weakened (Fig. 5d); consequently, the onset vortex is weakened. To south of the vortex, as precipitation and ocean mixing result in a rapid decline in SST (Fig. 4d), surface sensible heating becomes negative.

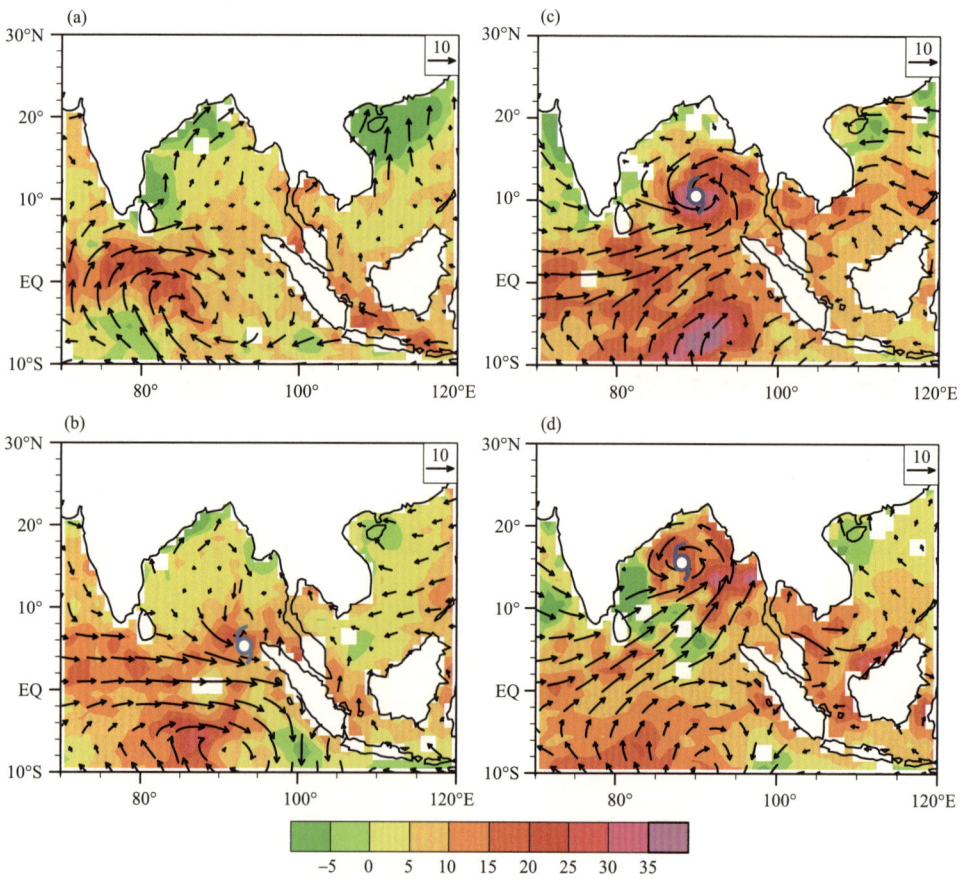

Figure 5 Daily sensible heat flux from OAFlux (shading, W·m^{-2}) and sea surface wind (vectors, m·s^{-1}) from NCDC on (a) 5 May, (b) 8 May, (c) 11 May, and (d) 14 May 2003. The vortex temporal position is marked by the typhoon symbol

4.3 Water vapor transport and precipitation over BOB

The above analysis reveals that the location of the SST warm pool in the eastern Central BOB, along with the remarkable surface sensible heating around the vortex center, plays an important role in the development and evolution of the vortex. It is unclear why this vortex is formed over the southeastern corner of BOB or from where the energy is sourced for its development.

Figure 6 shows the surface vorticity and divergence wind, calculated based on 10 m wind data of the NCEP/NCAR reanalysis. During the gestational phase (Figs. 6a and 6b), anticyclonic vorticity dominates the northern BOB area (north of 10°N), while cyclonic vorticity dominates the southeastern area. This pattern is in good correspondence with the distribution of surface wind (Figs. 5a and 5b): because strong surface wind develops along the equator and over the northwestern coastal ocean region of BOB, the horizontal wind shear must be anticyclonic north of 10°N, but cyclonic to the south. Therefore, surface air is divergent in the northern BOB region and convergent in the southeastern region. Accordingly, the southeastern corner of BOB, where the monsoon onset vortex developed on 8 May, is characterized by positive vorticity and wind convergence (Fig. 6b). After the vortex has formed, it is accompanied by remarkable positive vorticity of more than 4×10^{-5} s^{-1} and strong surface wind convergence from the eastern Arabian Sea, India, and the Southern Hemisphere toward the vortex center (Figs. 6c and 6d).

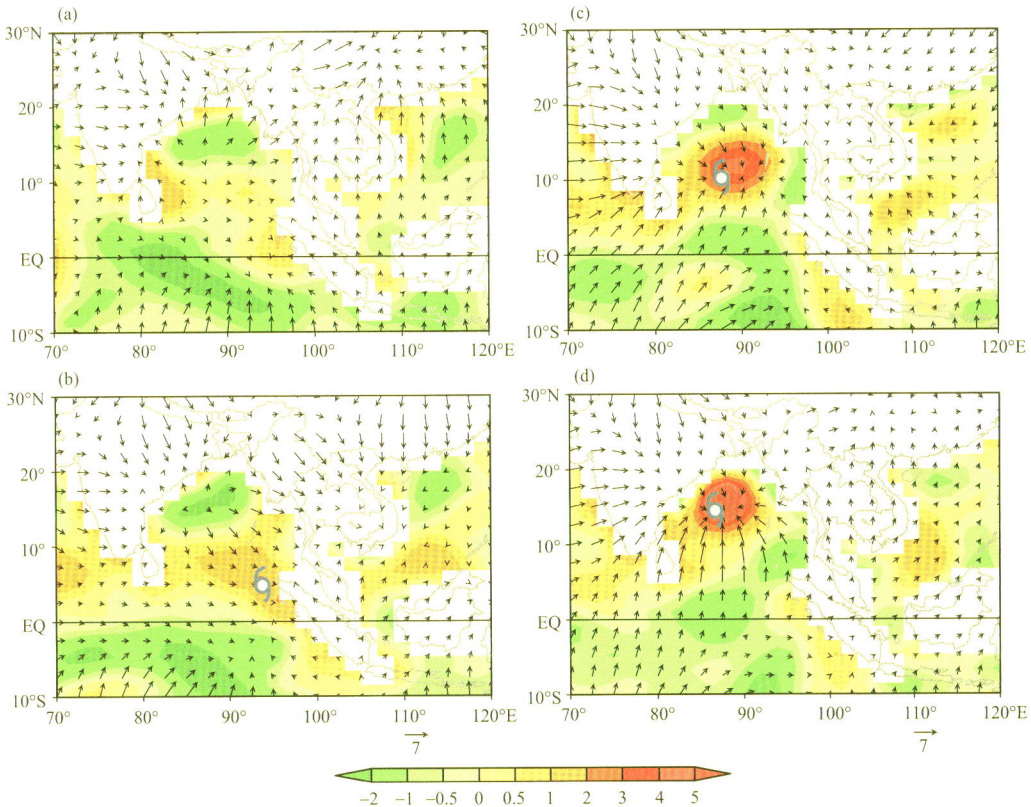

Figure 6 Relative vorticity (shading, 10^{-5} s^{-1}) and divergent wind (vectors) at 10 m above sea level from NCEP-1 on (a) 5 May, (b) 8 May, (c) 11 May, and (d) 14 May 2003. The vortex temporal position is marked by the typhoon symbol

The ocean supplies not only surface sensible heat, but also abundant water vapor to the atmosphere, as is evident by comparing the surface sensible heat flux (Fig. 5) with the sum of surface sensible and latent heat fluxes (Fig. 11), since surface evaporation is proportional to the surface latent heat flux. Figure 7 shows daily precipitation rate, vertically integrated water vapor flux Vq, and its divergence $\nabla \cdot (Vq)$. Because

$$\nabla \cdot (Vq) = \nabla \cdot (V_\psi q) + \nabla \cdot (V_\chi q) = \nabla \cdot (V_\chi q) + V_\psi \cdot \nabla q$$

where V_ψ and V_χ are, respectively, the rotational part and divergent part of V, to illustrate the contribution due to divergent wind, only $V_\chi q$ (not Vq) and its divergence are plotted in Fig. 7. During the gestational phase, water vapor flux is divergent over most of northern BOB but convergent in its southeastern corner, resulting in enhanced precipitation in this latter region (Figs. 7a and 7b). Latent heat release warms the atmosphere and favors the generation of available potential energy and development of the vortex in the lower atmosphere. During the developing phase, the convergence of water vapor flux in the region surrounding the vortex is greatly intensified and is accompanied by heavy precipitation (Fig. 7c), providing sufficient energy for vortex growth. In the decaying phase (Fig. 7d), the convergence of water vapor flux weakens, resulting in a reduction in the size of the area of intense precipitation over BOB.

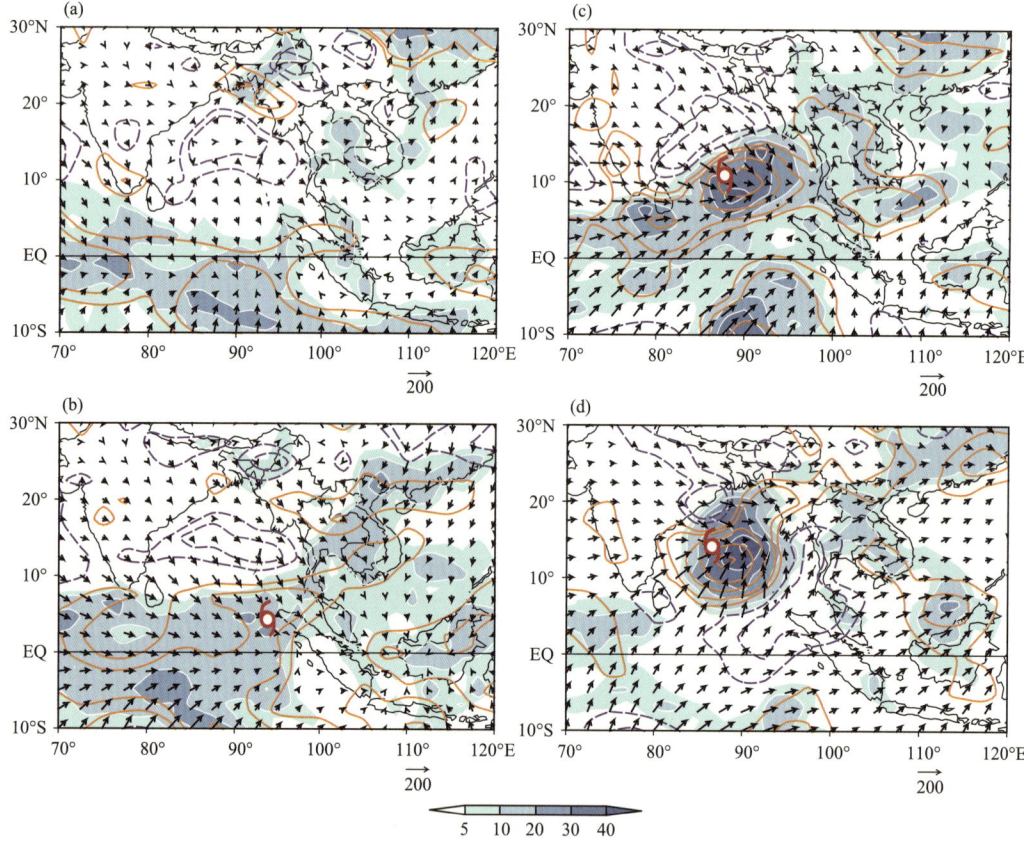

Figure 7 Precipitation (shading, mm·day^{-1}), vertically integrated water vapor transport $v \times q$ (vectors, kg·m^{-1}·s^{-1}) due to divergent wind, and water vapor transport divergence $\nabla \cdot (v \times q)$ (contours, 10^{-4} kg·m^{-2}·s^{-1}) from NCEP-1 on (a) 5 May, (b) 8 May, (c) 11 May, and (d) 14 May 2003. The dashed line indicates divergence and the solid line indicates convergence. The position of the vortex center is marked by a red typhoon symbol. The contour interval is 0.5 in the case that the absolute value of divergence is less than 1×10^{-4} kg·m^{-2}·s^{-1} (otherwise 1.0)

4.4 Energetics and development of the BOB vortex

Barotropic instability is commonly used to interpret the formation of the monsoon onset vortex in the tropical atmosphere (e.g., Krishnamurti, 1981). Because the growth of eddy kinetic energy is proportional to $(-\overline{u'v'}\frac{d\overline{u}}{dy})$, where an over-bar denotes quantities of mean flow and the prime indicates quantities of eddies (as departures from the mean flow), then $(-\overline{u'v'}\frac{d\overline{u}}{dy}>0)$ implies the kinetic energy conversion from zonal flow to eddies, favoring vortex development. However, the barotropic kinetic energy conversion from zonal mean to eddy is negligible during the gestational phase of the vortex examined in the present case study (not shown). In fact, our analysis reveals that the baroclinic instability associated with the conversion from available potential energy to kinetic energy contributes to the triggering of vortex formation.

Figures 8a and 8b shows the distribution of $(-\omega T)$ on 7 and 8 May, respectively, at the time of vortex formation. $-\omega T$ is generally negative over northern BOB, but positive over the south. This finding indicates that over northern BOB, where surface anticyclonic circulation is dominant, the total energy conversion is from kinetic to available potential; in contrast, over southern BOB, where surface cyclonic circulation prevails, the conversion is from available potential energy to kinetic energy. Over the southeastern corner of BOB on 8 May, the positive value of $-\omega T$ (i. e., the growth rate of kinetic energy) increases and reaches 10 Pa·K·s^{-1} at the time of vortex formation in the region.

During the gestational phase, the surface air temperature T and SST in the region are warm (Figs. 4a and 4b), and surface sensible heating Q is enhanced (>15 W·m^{-2}; Fig. 5b); consequently, available potential energy is generated ($QT>0$) and stored in the region. In the case that the irrotational flow brings water vapor into the region from northern BOB, the Arabian Sea, and the Southern Hemisphere (Figs. 6a, 6b, 7a, and 7b), the air rises and becomes warmer due to the latent heat release associated with rainfall, leading to $-\omega T>0$. The stored available potential energy is then released to enhance the kinetic energy, triggering vortex formation. Nevertheless, barotropic instability still plays an important role during the developing and decaying phases of the vortex. The large positive value of $(-\overline{u'v'}\frac{d\overline{u}}{dy})$ in the region east of the vortex on 11 and 14 May (Figs. 8c and 8d, respectively, the zonal mean is averaged over 60°—120°E) indicates that during these phases, zonal kinetic energy in the vortex domain is converted to eddy kinetic energy over the area east of the vortex, and the vortex develops and moves eastward.

5 Influence of the atmosphere and ocean mixing on SST evolution in BOB

5.1 Formation of the BOB SST warm pool in spring

An analysis of weekly OISST data demonstrates that in 2003 over the BOB area, SST increased gradually from below 27.5 ℃ in mid-January to a maximum of 30.3 ℃ in early May, then decreased sharply afterwards (Fig. 9a). The spatial distribution of SST increase from the week centered on 6 April to the week centered on 27 April (herein referred to as Week 6 April and Week 27 April, respectively) reveals that about 1 month before vortex formation, SST in the Indian Ocean became cooler south of the equator and became warmer north of the equator, over most areas (Fig. 9b). Over the Arabian Sea and South China Sea, SST warming was greater in the west; whereas over BOB, warming was greater in the east, with remarkable cooling in the northwest and southwest. SST was cold over the northwestern Arabian Sea and SCS in early April (Fig. 9c); consequently, a warm pool (SST>30.75 ℃) formed in the central and eastern BOB by the end of April (Fig. 9d).

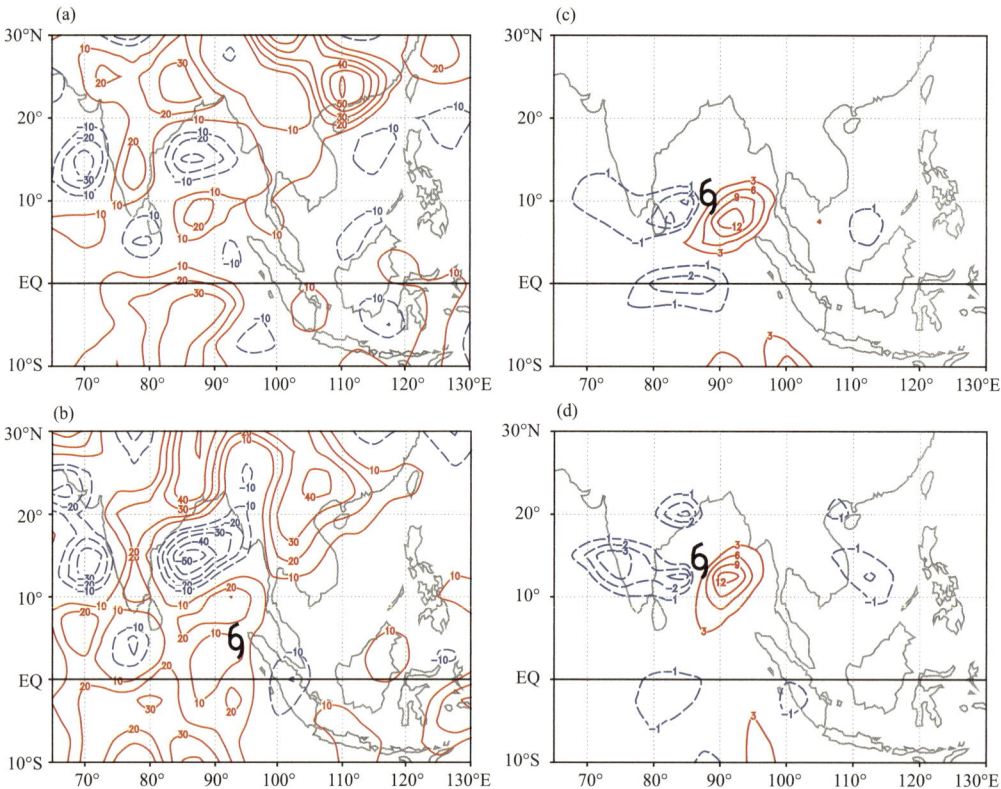

Figure 8 Distribution at 850 hPa of $-\omega T$ (Pa·K·s^{-1}) on (a) 7 May and (b) 8 May 2003, and distribution of $-\overline{u'v'}\dfrac{d\overline{u}}{dy}$ (m^2·s^{-3}) from NCEP-1 on (c) 11 May and (d) 14 May 2003. The position of the vortex center is marked by a black typhoon symbol

In the tropics, SST and ocean motion are strongly influenced by surface wind. Figure 4 shows the evolution of surface wind, in addition to SST. During the gestational stage, two strong surface-wind belts developed in the BOB area (Figs. 4a and 4b). South of BOB, a near-equator cyclone centered at (85°E, 5°S) developed in the Southern Hemisphere (Fig. 4a), associated with a strong cross-equator flow from the Southern Hemisphere between 70° and 80°E, which was accompanied in turn by the development of a near-equator westerly immediately south of Sri Lanka. As the cyclone propagated eastward, the westerly belt intensified and extended eastward to approach Java Island by 8 May (Fig. 4b). Another strong wind belt is observed over the western coastal ocean of BOB. This belt induced a pronounced surface anticyclone over northern BOB (Figs. 4a and 4b) that provided clear sky and sunny conditions in the atmosphere, driving a convergent surface ocean current in the area with down-welling (Section 5.2 or Figure 12). These two strong wind belts occurred either side of the northern BOB; while in the central and eastern parts of BOB the surface winds were weak, where the warm pool formed. During the developing phase, the warm pool remained north of the vortex (Fig. 4c), whereas during its decaying phase the warm pool was weakened due to the intensification of surface wind and precipitation (Fig. 4d).

The appearance of the BOB warm pool in spring is at least partly attributed to the weak local surface wind, because SST is determined externally by the net surface heat flux and because the sea-surface sensible heat flux and latent heat flux are affected by surface wind. At the sea surface, the net heat flux Q_N, which is determined by the surface net downward radiation Q_R, and upward latent heat release Q_L and sensible heat release Q_S, can be expressed as

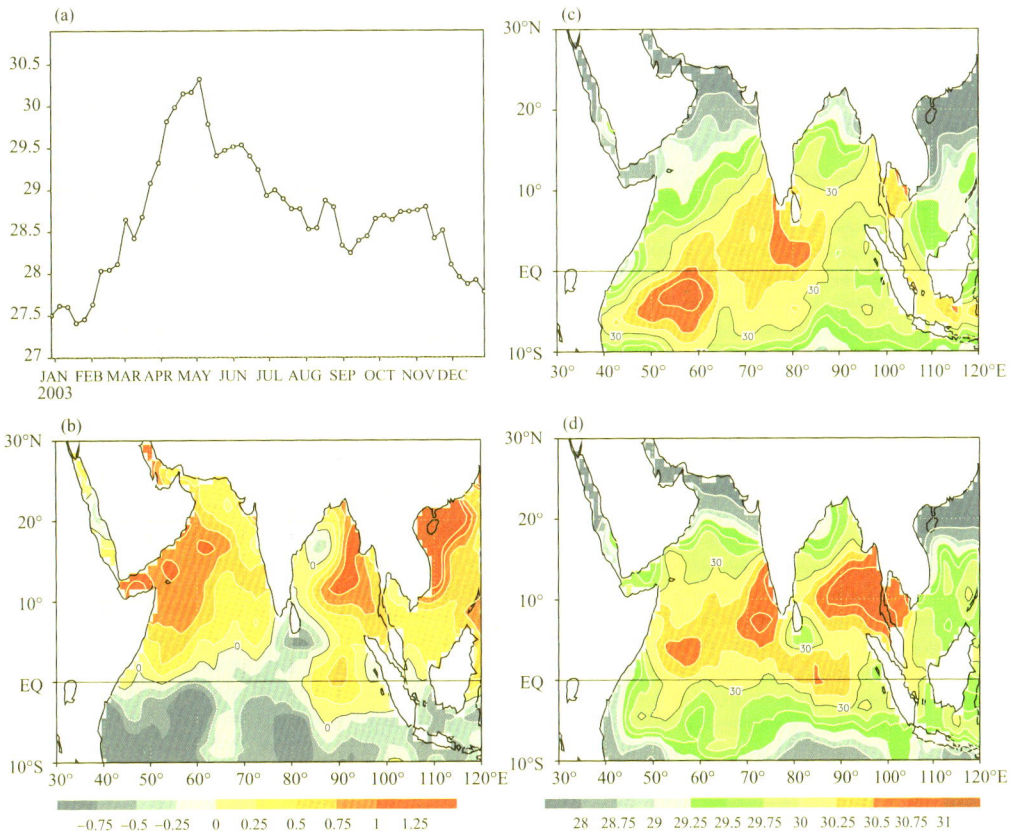

Figure 9 (a) Time series of area-averaged weekly SST (℃) from OISST over the BOB area (80°—97°E, 0°—20°N). (b) SST increase from the Week 6 April to the Week 27 April. (c) Weekly SST distribution for Week 6 April. (d) As in (c), except for Week 27 April

$$Q_N = Q_R - (Q_L + Q_S)$$

In general, Q_R is a heat source for the sea and $(Q_L + Q_S)$ is a dissipation sink. The net radiation Q_R is determined by the full-sky downward shortwave radiation subtracted from the upward longwave radiation at the surface.

Over the central and northern BOB, Q_R was strongly positive ($>$ 220 W · m^{-2}) during the gestational phase (Fig. 10a), supplying plentiful heat to the sea over almost the entire BOB area, resulting in enhanced SST. Consequently, SST across most of BOB was warmer than that in other areas (see Figs. 4a and 4b). With formation of the monsoon onset vortex on 8 May (Fig. 10b), cloud and convection developed around the vortex, and Q_R in the region became weaker ($<$ 80 W · m^{-2}). During the developing and decaying phases of the vortex, convection and precipitation were further intensified in the region surrounding the monsoon onset vortex; consequently, Q_R became negative (Figs. 10c and 10d) and SST became cooler (Figs. 4c and 4d).

Figure 11 shows the sum of upward latent heat flux and sensible heat flux $(Q_L + Q_S)$. Compared with Figure 5, Q_S is much smaller than $(Q_L + Q_S)$ in Fig. 11, and the release of latent heat flux prevails over the surface sensible heat flux in terms of cooling the ocean. During the gestational phase, $(Q_L + Q_S)$ was very large near the equator, immediately beneath the strong equatorial surface westerly, but was relatively small in the eastern part of central BOB (Figs. 11a and 11b), where the surface wind was weak.

Taken together, Figs. 11a and 10a reveal that prior to formation of the Asian monsoon onset vor-

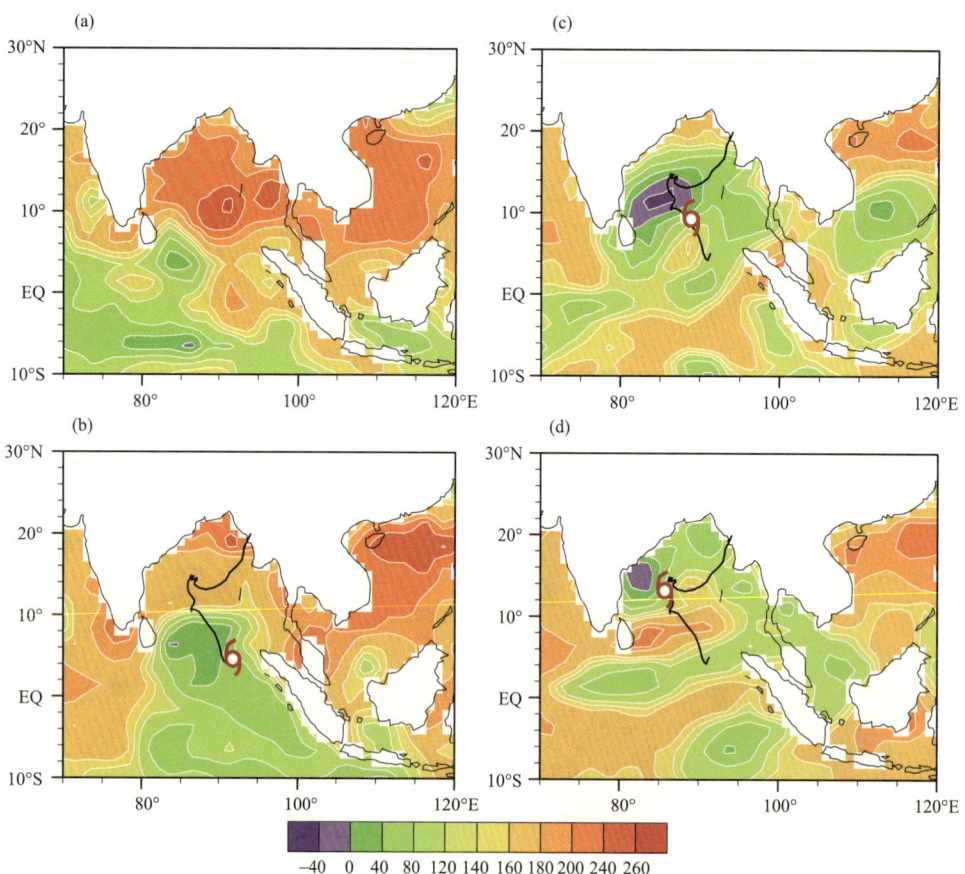

Figure 10 Net surface radiation flux (W · m^{-2}, downward full-sky shortwave radiation minus upward longwave radiation, with downward being positive) derived from OAFlux data on (a) 5 May, (b) 8 May, (c) 11 May, and (d) 14 May 2003. Thick black curve denotes the vortex track, with its position marked by the typhoon symbol

tex, the eastern part of central BOB experienced relatively little heat loss because of the weak surface wind and strong insolation because of the clearer sky under the controlling surface anticyclonic circulation. As shown in Fig. 12, the ocean mixed layer depth in the northern BOB is only 10—20 m in the gestational period. Strong solar radiation (Fig. 10) together with weak energy loss (Fig. 11) act onto a very thin (10—20 m) mixed layer (Fig. 12), the SST in this region presents a much more rapid increase than that in other parts of BOB (Fig. 9b). Consequently, a BOB warm pool formed in the region during early spring (Figs. 4a and 9d). Note that the controlling surface anticyclonic circulation over northern BOB was related to strong southwesterly forcing over the northwestern coastal ocean; therefore, we can attribute the formation of the spring BOB warm pool to the heterogeneous nature of surface wind forcing over the BOB area. During the developing and decaying phases, as strong cyclonic circulation developed over the entire BOB and moved northward, a strong surface southwesterly sweeps over most of BOB (Figs. 11c and 11d) and ($Q_L + Q_S$) increases northward, with a center exceeding 300 W · m^{-2}. SST over BOB decreases and the BOB warm pool shrinks to the area within the Bay of Siam (Fig. 4d).

5.2 Influence of vertical mixing on SST

SST (equivalent to mixed-layer temperature, MLT) in the tropical ocean is known to be determined by the net heat budget Q of the mixed layer, which in turn is primarily governed by the net surface heat

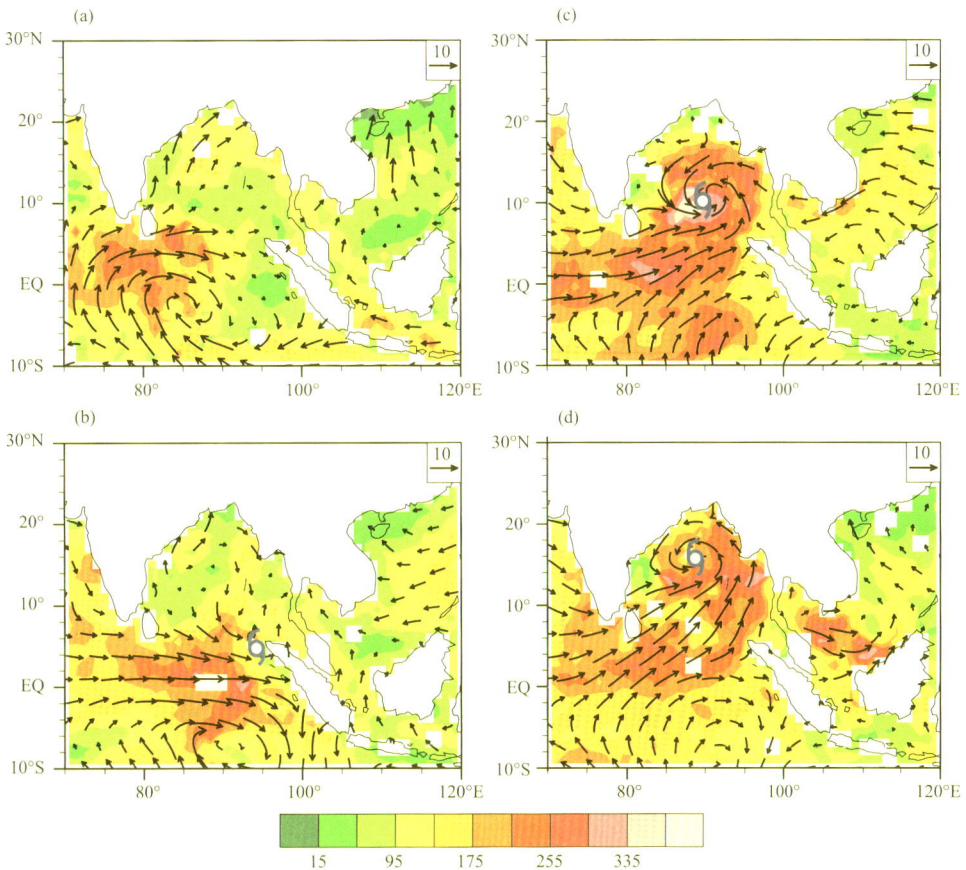

Figure 11　Surface upward latent heat flux plus sensible heat flux from OAFlux (shading, with upward positive; W·m^{-2}), and sea surface wind (vectors, m·s^{-1}) on (a) 5 May, (b) 8 May, (c) 11 May, and (d) 14 May 2003. The vortex position is marked by the typhoon symbol

fluxes Q_N, the lateral advection of heat Q_v, and vertical mixing Q_e. Q_e measures the entrainment of heat across the bottom of the mixed layer (McPhaden and Hayes, 1991); i.e.,

$$Q = Q_N - Q_v - Q_e$$

where

$$Q_v = \rho_0 C_p h \mathbf{V} \cdot \nabla T$$

and

$$Q_e = \rho_0 C_p h W_e \Delta T_h / h$$

where $\mathbf{V}$ is the horizontal current velocity; ∇T is the horizontal temperature gradient in the mixed layer; W_e is the entrainment velocity at the base of the mixed layer; ΔT_h is the temperature drop below the mixed layer, which is calculated as the difference between the temperature at the mixed-layer depth (MLD) and at 5 m below the MLD; ρ_0 is the density of sea water; C_p is the specific heat of sea water; and h is MLD. In the current study, NCEP GODAS pentad data (e.g., $\mathbf{V}$, W_e, and temperature profile) are used to calculate the terms in the above equations. Since the vertical velocity W_e was obtained based on the horizontal divergence/convergence of the Ekman transport (Gill, 1982), it represents an entrainment rate from below at the base of the mixed layer.

Before evaluating the net heat budget in the mixed layer, we diagnosed the evolution of the mixed layer depth in 2003, averaged over the BOB area. The evolution (figure not shown) is similar to that of SST shown in Fig. 9a. The MLD remains as deep as 50 m in the winter months under the wintertime northeasterly monsoon circulation. From mid-March, as the winter monsoon ceases, the MLD gradually becomes thinner and attains a minimum of 30 m by the end of April. As the BOB vortex develops in early May and with the ensuing onset of the Asian summer monsoon, MLD in BOB shows a marked deepening. Because the monsoon onset vortex develops during 8—19 May, the second and third pentads in May are chosen for analysis to illustrate the spatial distribution of the MLD during the gestational and developing/decaying phases of the vortex.

During the gestational phase, the MLD is less than 20 m over the Central BOB in the area ahead of the trajectory of the vortex (Fig. 12a). In the third pentad of May (May 11—15), during the developing and decaying phases of the vortex, the depth of the mixed layer increases, exceeding 50 m near the vortex center by 15 May (Fig. 12b). This finding indicates that during early May, MLD is small. When the BOB vortex develops, the surface wind is so strong that the local vertical mixing is expected to be important in influencing the net heat budget of the mixed layer and SST.

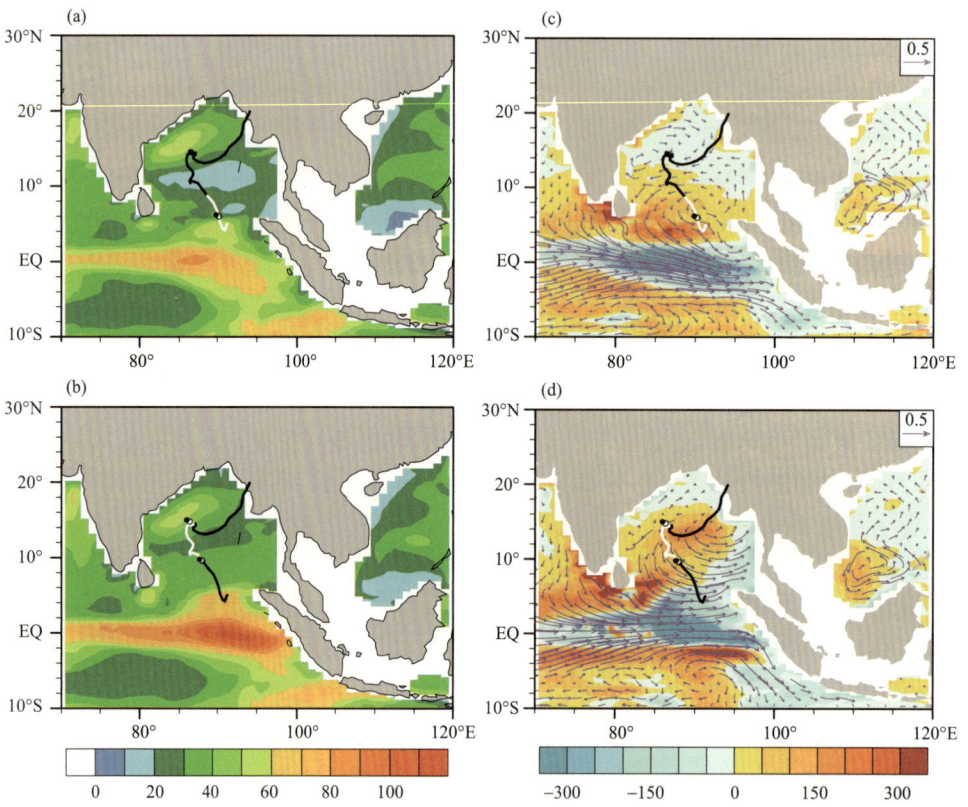

Figure 12　Mixed-layer depth (unit: m) from GODAS in (a) the second pentad and (b) the third pentad of May 2003. (c)—(d) As in (a)—(b) except for vertical mixing heat flux Q_e (W·m^{-2}) across the bottom of the mixed-layer and sea current at 5 m depth in the ocean (vectors). Thick black curve denotes the vortex track, with white segments indicating the track during the pentad of interest

In the net heat budget equation, Q_V is an order of magnitude smaller than Q_e or Q_N in the tropics (data not shown), meaning that it can be ignored. The characteristics of Q_N and its evolution during the life cycle of the BOB vortex have been described in previous sections (Figs. 10 and 11). Here, we seek to assess the importance of vertical mixing in governing the evolution of SST. In fact, vertical mixing Q_e represents the upward Ekman pumping forced by sea surface wind.

Figures 12c and 12d demonstrate that strong downwelling occurs along the equator and to the south of the maximum axis of the surface westerly. Downwelling also appears in northern BOB, where a surface anticyclone occurs. Strong upwelling occurs in southern BOB, particularly north of the maximum axis of the surface westerly. Upwelling also appears offshore from India, associated with an offshore current related to the development of a strong local surface southwesterly east of the surface cyclonic circulation that occurs over India. During the second pentad of May, the sea surface wind is weak over most of northern BOB; consequently, Q_e (Fig. 12c) is small in this region (< 50 W·m^{-2}) during the gestational stage of the vortex. It is much weaker than the net surface radiation Q_R (> 240 W·m^{-2}; Figs. 10a and 10b) or net surface flux (55—135 W·m^{-2}; Figs. 11a and 11b).

In the third pentad of May (11—15 May), the vortex intensifies over northern BOB, where the pre-existing downwelling is replaced by strong upwelling. MLD near the vortex center becomes deeper (exceeding 50 m by 15 May; Fig. 12b), and Q_e around the track of the vortex over northern BOB exceeds 150 W·m^{-2} (Fig. 12d), which is comparable with or even stronger than the net surface radiation Q_R ($<$ 100 W·m^{-2}; Figs. 10c and 10d). Because the mixed layer remains shallow in front of the vortex track, and because mixing induced by the vortex is strong when the vortex arrives, the cold, deep ocean water is stirred into the mixed layer. This mixing process, together with the strong surface latent and sensible flux release ($>$150 W·m^{-2}; Figs. 11c and 11d), contributes to a rapid decrease in SST (Fig. 4d) along the passage of the vortex, and to the decay of the vortex.

6 External modulation of air-sea interaction over BOB and Asian monsoon onset

The above analyses reveal the critical importance of the surface wind distribution in modulating air-sea interaction in BOB and in generating the spring BOB warm pool and the BOB onset vortex of the Asian summer monsoon. Therefore, it is necessary to investigate the origin of this surface wind pattern.

Figures 13a and 13b show the evolution from 1 March to 30 June of the surface latent heat flux and surface sensible heat flux, respectively, averaged between 10° and 20°N across the Arabian Sea, Indian Subcontinent, BOB, Indochina Peninsula, and the SCS. In the spring, prior to formation of the BOB vortex, despite the occurrence of mild variations over the ocean, there exists a remarkable difference in surface thermal features between the Indian Subcontinent and the Indochina Peninsula. At this time, the Indian Subcontinent is experiencing the dry season, when surface evaporation is weak and the surface sensible heat flux exceeds 120 W·m^{-2}. In contrast, the climate of the Indochina Peninsula at this time is warm and moist, where the surface latent heat flux exceeds 125 W·m^{-2} and surface sensible heating is generally less than 90 W·m^{-2}. By 8 May, at which time the monsoon vortex is formed, the averaged sensible heat flux over the Indochina Peninsula is already less than 30 W·m^{-2}.

This thermal contrast is seen more clearly from the spatial distributions of surface latent heat flux and wind at 850 hPa on 8 May (Fig. 13c). Evaporation is weak over the dry surface of India, whereas it is stronger over the wet Indochina Peninsula, where rainfall is abundant (Fig. 7b). Of note, the prevailing wind at 850 hPa is northwesterly over India but southerly over the Indochina Peninsula. According to Wu et al. (2007) and Wang et al. (2008), this pattern of prevailing wind is typical of the season because mechanical and thermal forcing of the Tibetan Plateau generates a flow dipole with anticyclonic circulation to the north and cyclonic circulation to the south. The prevailing northerly over India brings cold and dry air from the north, thereby intensifying the land-air temperature difference, leading in turn to strong surface sensible heat release (see Fig. 13d, which shows the distributions of surface sensible heat flux and 10 m wind on 8 May). Over most of India, surface sensible heating exceeds 150 W·m^{-2}.

Because the sensible heating Q over India decreases with increasing height, negative vorticity is generated in the lower troposphere (Wu et al., 2000). In the absence of zonal vorticity advection and following the Sverdrup vorticity balance, we have

$$\beta v = (f + \zeta)\theta_z^{-1} \frac{\partial Q}{\partial z}, \quad \theta_z \neq 0$$

indicating that positive planetary vorticity should be transferred from the north to maintain the local vorticity balance, thus leading to the development of a northerly in return. As this process continues, the surface sensible heating and generation of negative vorticity form a positive feedback loop via the development of the northerly wind, and intense surface sensible heat flux is maintained, similar to the mechanism involved in the formation of desert areas (see Wu et al., 2009). According to the atmospheric thermal adaptation theory (Hoskins, 1991; Wu et al., 2000), remarkable surface cyclonic circulation must form in response to such strong surface thermal forcing. Consequently, a strong surface northwesterly forms along the west coast of India and a southwesterly forms along the east coast (Fig. 13d).

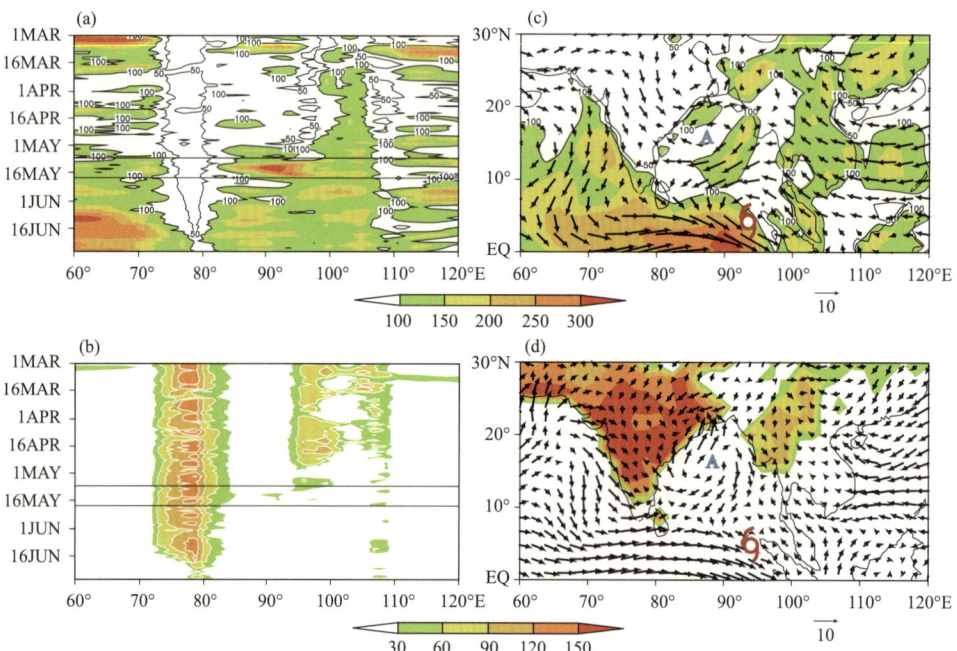

Figure 13 Time-longitude cross-sections (averaged between 10° and 20°N) of (a) surface latent heat flux and (b) surface sensible heat flux from March to June, 2003. (c) surface latent heat flux (shading) and 850 hPa wind (vectors), and (d) surface sensible heat flux (shading) and 10 m surface wind (vectors) on 8 May 2003 from NCEP-1. Units are $W \cdot m^{-2}$ for flux and $m \cdot s^{-1}$ for wind. The two horizontal lines in (a) and (b) denote, respectively, the times of the genesis and disappearing of the monsoon onset vortex

The development of a strong southwesterly over northwestern BOB, combined with the development of the near-equator westerly over southern BOB, induces a surface anticyclone in northern BOB and a cyclone in the south. This pattern of surface circulation remains over BOB at 850 hPa (Fig. 13c), and is even discernable at 700 hPa prior to formation of the monsoon vortex (Figs. 14a and 14b). However, unlike in the lower layers, anticyclonic circulation over northern BOB at 700 hPa occurs as part of a continuous "ridgeline" of the subtropical anticyclone belt (thick line along 18°—20°N in Fig. 14b). A zonal belt of easterlies prevails to the south of the ridgeline, whereas a zonal westerly prevails to the north, representing a typical pattern of the wintertime circulation. After the BOB vortex has formed on 8 May, as the vortex develops and moves northward, part of the ridgeline west of 95°E is also shifted northwards to about 20°N (Fig. 14c). With further development and northward movement of the vortex, the originally continuous easterly belt in the tropics is strongly interrupted: The southerly and southwesterly over the eastern portion of the vortex merge with the westerly at higher latitudes, the ridgeline splits over eastern BOB and western Indochina Peninsula in the longitude domain between 93° and 105°E on 12 May (data not shown), and the Asian summer monsoon commences.

Before the monsoon onset (Fig. 14c), all the ridgelines tilt southward from the lower to higher troposphere, corresponding to the general winter feature of warm in the south and cold in the north. After the monsoon onset (Fig. 14d), the ridgelines at 850 and 700 hPa show strong northward movement in the longitude domain west of 93°E, but show southward movement in the longitude domain east of 105°E. Consequently, the ridgelines between 93° and 105°E are broken in the lower troposphere, and warm and moist air is transported from eastern BOB to the southeastern Tibetan Plateau and southern China. By May 14th, although the ridgelines still tilt southward with increasing height west of 93° and east of 110°E, they tilt northward in the region between 93° and 105°E, indicating the establishment of the general summer feature of warm air in the north and cold air in the south. It is evident that the development of the BOB vortex in 2003 played an important role in leading the earliest Asian summer monsoon onset and the seasonal transition from winter to summer in eastern BOB and over the western Indochina Peninsula.

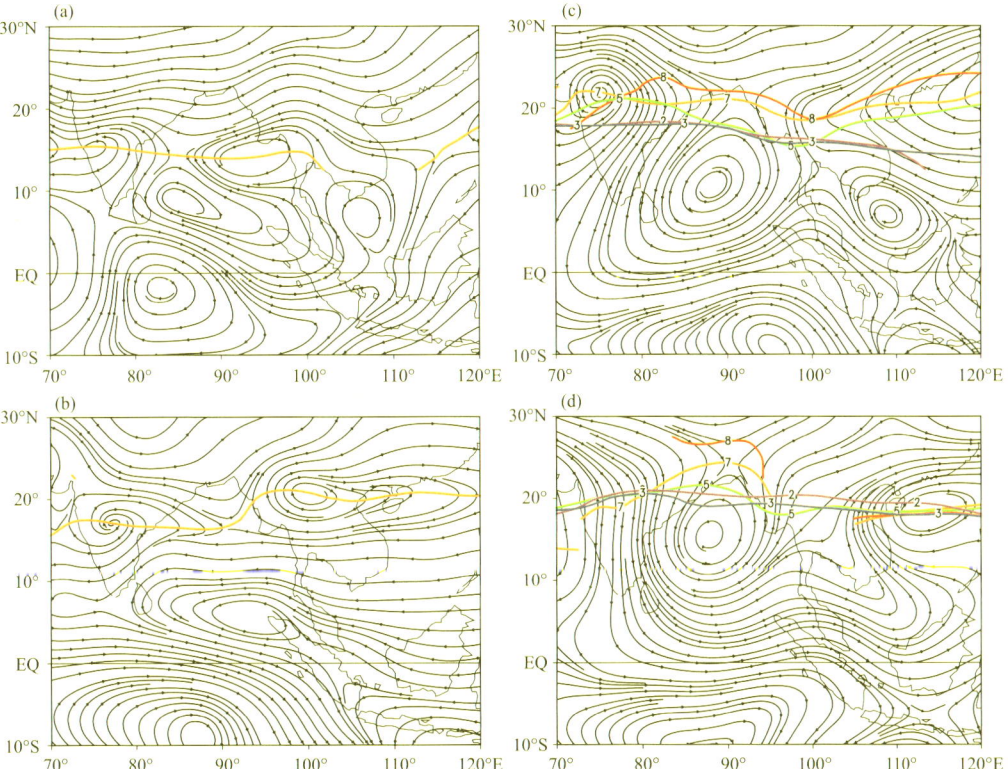

Figure 14 Streamline and ridgeline (thick curve) of the subtropical high at 700 hPa from NCEP-1 on (a) 5 May, (b) 8 May, (c) 11 May, and (d) 14 May 2003. In (c) and (d), the ridgelines at five isobaric surfaces (850, 700, 500, 300, and 200 hPa) are shown with different colors and are marked by 8, 7, 5, 3, and 2, respectively

7 Summary and discussion

In 2003, the monsoon onset vortex formed over the southeastern corner of BOB on 8 May before propagating northward and northeastward, attaining maximum intensity on 13 May before disappearing on 19 May. The wintertime continuous ridgeline of the subtropical anticyclone at 700 hPa was broken by 12 May, followed by the onset of the BOB summer monsoon.

Prior to formation of the vortex, BOB was surrounded by strong surface wind on its southern and western boundaries, with weak wind over the main area of BOB. An offshore ocean current and upwelling were confined to the western coastal BOB, resulting in regional SST cooling. Surface ocean currents are converged from west to east BOB. On the other side, surface anticyclonic circulation and descending air developed over northern BOB, whereas surface cyclonic circulation and ascending air were confined to the south. Because ocean mixing is weak and ocean mixed layer is shallow during this season, intense solar radiation and low surface energy loss acting onto a thin ocean mixed layer in northern BOB resulted in rapid surface warming, and warm SST formed in the eastern part of central BOB, producing a remarkable springtime BOB warm pool.

By early May of 2003, the background conditions of the atmosphere over BOB were favorable for vortex formation. Near the equator, the strong tropical westerly wind brought large amounts of water vapor from Southern Hemisphere and the Arabian Sea eastward to southeastern BOB, from early May. At the same time, water vapor was diverged from the anticyclone over northern BOB and converged to southeastern BOB. The southeastern BOB is located immediately south of the BOB warm pool, which is an area of stronger surface sensible heating. The convergence of water vapor and the apparent surface sensible heating created favorable conditions for the development of convection in this region, and potential energy was generated and transformed to the kinetic energy required for vortex formation. The spring BOB warm pool and stronger surface sensible heating were located immediately north of the vortex in favor of the generation of available potential energy ($\int_S QT \mathrm{d}s > 0$), meaning that the vortex propagated northward thereafter. During the developing and decaying phases of the vortex over eastern BOB, zonal-mean kinetic energy was transformed to eddy kinetic energy due to the local barotropic instability on the eastern side of the vortex, thereby supplying energy to sustain the vortex and resulting in its eastward movement.

It is clear that the development of a strong southwesterly over the northwestern offshore ocean in BOB is a critical prerequisite for the development of the BOB warm pool and for formation of the monsoon onset vortex. The development of such a strong southwesterly flow in spring season is a consequence of the strong surface thermal forcing of the Indian Subcontinent, induced in turn by Tibetan Plateau forcing. During spring, the Tibetan Plateau, together with its thermal forcing, generates a cyclonic circulation to its south in the lower troposphere (Wu et al., 2007; Wang et al., 2008), with a northwesterly prevailing over India and a southeasterly over the Indochina Peninsula. At the end of April, the surface sensible heat flux is weak over the Indochina Peninsula, whereas cold advection over India induces an enhanced land-air temperature difference, resulting in turn in intensification of surface sensible heat release. The strong sensible heat flux ($> 150 \text{ W} \cdot \text{m}^{-2}$) over India, which is potentially maintained until the end of May, generates strong surface cyclonic circulation. In turn, this results in a strong surface northwesterly along the west coast of India and a strong southwesterly along the east. Given these observations, the generation of the BOB warm pool and BOB summer monsoon onset vortex can be considered as a consequence of air-sea interaction across BOB, modulated by Tibetan Plateau forcing and the land-sea thermal contrast in South Asia during spring, as shown schematically by Fig. 15, which reveals the following features.

(a) In late spring, the atmospheric circulation and climate over South Asia is strongly modulated by the mechanical as well as thermal forcing of the Tibetan Plateau and the land-sea thermal contrast across the area.

(b) The atmospheric forcing thus creates an environment with clear sky, weak surface wind and shallow mixed layer over the central BOB. There strong solar radiation and weak energy loss act onto a very thin mixed layer, leading to the generation of a short-life SST warm pool in the eastern part of central BOB.

(c) The BOB warm pool and the associated strong surface sensible heat lease from the ocean in return produce available potential energy in the atmosphere and generate a vortex which leads to the occurrence of earliest Asian monsoon onset over the eastern BOB.

In 2003, the onset vortex moved northward, with intensified surface sensible flux near its center. During its developing and decaying phases, eddy kinetic energy was converted from zonal-mean kinetic energy over the eastern portion of the vortex, and its track turned northeastward. Development of the vortex produced abundant cloud and strong surface wind, reduced the solar radiation on the sea surface, and resulted in enhanced sea surface latent and sensible heat release from the ocean to the atmosphere in the region surrounding the vortex, along with strong mixing and upwelling of cold water in the underlying ocean. These phenomena resulted in a rapid drop in SST within BOB, resulting in turn in gradual decay of the vortex, which by this time was located over northern BOB. The strong southwesterly on the southern and eastern sides of the vortex swept over eastern BOB, merging with the subtropical westerly over southern China. Consequently, the wintertime zonally aligned anticyclone belt became broken over eastern BOB, the regional meridional temperature gradient was reversed, and the Asian summer monsoon onset occurred.

(a) Modulation on the atmospheric circulation of the Tibetan Plateau and land-sea thermal contrast in South Asia

The cold and dry northwesterly over India, which is induced by Tibetan Plateau forcing, generates strong surface sensible heating and cyclone circulation. The strong southwesterly along the western offshore region of BOB, combined with the equatorial westerly, force surface anticyclonic circulation with descending air over northern BOB and cyclonic circulation with ascending air over southern BOB. Large amounts of water vapor are transported from northern BOB and other areas toward southeastern BOB.

(b) Surface ocean current and variation of SST in BOB

The strong southwesterly along the western offshore region of BOB results in an offshore current and upwelling, and cold SST is induced along the western BOB. Warm surface water is converged to the eastern part of central BOB, which is under the influence of an anticyclone, with clear sky, weak surface wind and shallow mixed layer. Strong solar radiation together with weak energy loss act onto a very thin mixed layer. Consequently, SST decreases in the west BOB but increases rapidly in the east BOB prior to the formation of monsoon onset vortex.

(c) Formation of the BOB warm pool and Asian summer monsoon onset vortex

A remarkable springtime warm pool is formed in the eastern part of central BOB just to the north of the surface cyclone, which is marked by strong surface sensible heating. Convection is triggered, and

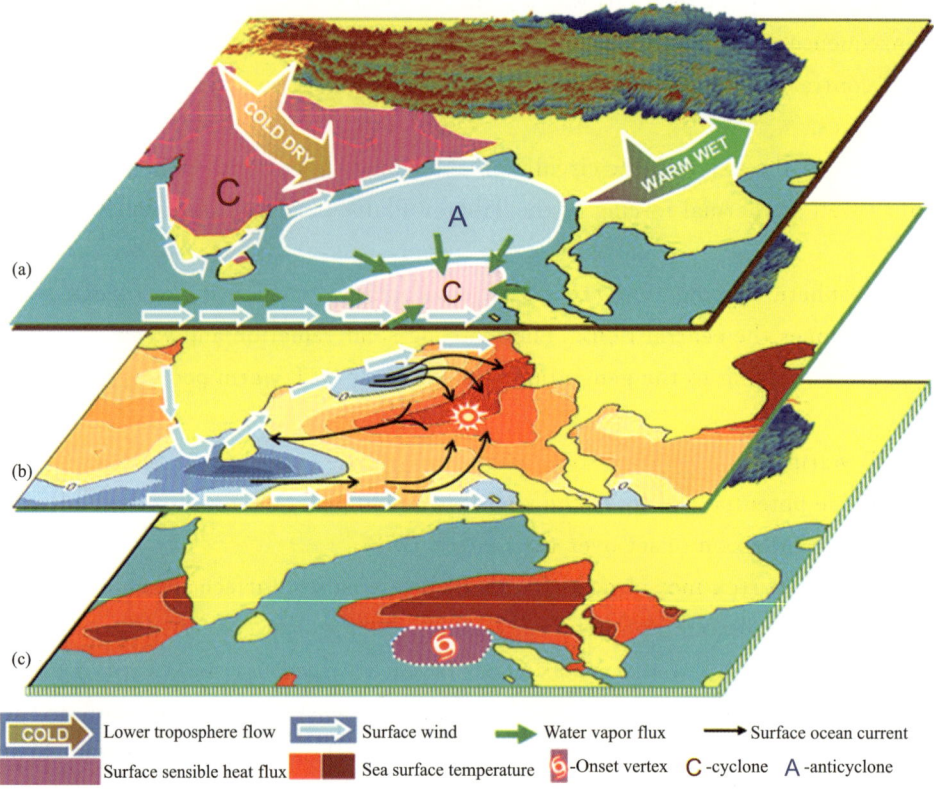

Figure 15 Schematic diagram showing the formation of the BOB monsoon onset vortex as a consequence of regional air-sea interaction modulated by the land-sea thermal contrast in South Asia and Tibetan Plateau forcing in spring

available potential energy is generated and transformed to kinetic energy, thereby generating the BOB monsoon onset vortex.

The results demonstrated in Fig. 13 present the modulations on the South Asian circulation and climate of the Tibetan Plateau forcing and the land-sea thermal contrast across the tropical Asia. By diagnosing the 1998 case in which the MOV appeared over the southwestern BOB near Sri Lanka, a similar configuration was obtained (Wu et al., 2011). Because this feature is common in spring, therefore the schematic diagram Fig. 15 should be considered as representative. However, the equatorial westerly flow varies from year to year. Although the flow is rather zonal in 2003, it fluctuated as a disturbed inertial flow along the equator in 1998. Thus, further understanding the variation of the equatorial flow may help in improving the Asian monsoon onset prediction. In addition, air-sea interaction involves many complicated coupling and feedback processes. Given that the present results are based on a case study, they should be considered as preliminary. Nevertheless, the BOB summer monsoon onset vortex occurs in most years, and has a strong influence on the evolution of the Asian monsoon. Furthermore, the track and life cycle of the onset vortex vary among different years, indicating that its influence on monsoon evolution also varies. Therefore, the dynamic mechanism related to the formation and movement of the BOB monsoon onset vortex represents an important topic that requires further investigation. In this regard, a high-resolution and coupled modeling study would be very useful to reveal more details of the vortex development.

References

ANANTHAKRISHNAN R, RAMAKRISHNAN A R, JAMBUNATHAN R,1968. Synoptic features associated with onset of southwest monsoon over Kerala[R]. Pune, India: Forecasting Manual Report No. IV-18. 2.

ANANTHAKRISHNAN R, PATHAN J M, ARALIKATTI S S,1981. On the northward advance of the ITCZ and the onset of the southwest monsoon rains over the southeast Bay of Bengal[J]. Int J Climatol, 1: 153-165.

GILL A E,1982. Atmosphere-Ocean Dynamics[M]. San Diego, CA: Academic Press: 30: 662.

GRAY W M,1968. Global view of the origin of tropical disturbances and storms[J]. Mon Wea Rev, 96: 669-700.

GRAY W M,1975. Tropical Cyclone Genesis[M]. Fort Collins USA: Dept of Atmospheric Science, Colorado State University: 121.

GRAY W M,1998. The formation of tropical cyclones[J]. Meteor Atmos Phys, 67:37-69.

HARR P A, CHAN J C L,2004. Monsoon impacts on tropical cyclone variability[C]. Hangzhou: The Third International Workshop on Monsoons (IWM-Ⅲ): 512-542.

HE J H, SHI X H,2002. Splitting and eastward withdrawal of the subtropical high belt during the onset of the South China Sea summer monsoon and their possible mechanism[J]. J Nanjing University (Natural Science), 38: 318-330.

HOSKINS B J,1991. Towards a PV-θ view of the general circulation[J]. Tellus, 43AB: 27-35.

JI M,LEETMAA A,DEBER J,1995. An ocean analysis system for seasonal to interannual climate studies[J]. Mon Wea Rev, 123(2): 460-481.

JOSEPH P V,1990. Warm pool over the Indian Ocean and monsoon onset[J]. Tropical Ocean Atmosphere Newsletter, 53: 1-5.

KANAMITSU M, EBISUZAKI W, WOOLLEN J, et al.2002. NCEP-DOE AMIP-Ⅱ reanalysis (R-2)[J]. Bull Amer Meteor Soc, 83: 1631-1643.

KERSHAW R,1985. Onset of the south-west monsoon and sea-surface temperature anomalies in the Arabian Sea[J]. Nature, 315: 561-563.

KERSHAW R,1988. Effect of a sea surface temperature anomaly on a prediction of the onset of the southwest monsoon over India[J]. Quart J Roy Meteor Soc, 114: 325-345.

KRISHNAMURTI T N,1981. Cooling of the Arabian Sea and the onset-vortex during 1979[R]. Venice, Italy: Recent Progress in Equatorial Oceanography: A Report of the Final Meeting of SCOR WORKING GROUP 47: 1-12.

KRISHNAMURTI T N,1985. Summer monsoon experiment—A review[J]. Mon Wea Rev, 113: 1590-1626.

KRISHNAMURTI T N, ARDANUY P, RAMANATHAN Y, et al.1981. On the onset vortex of the summer monsoon [J]. Mon Wea Rev, 109: 344-363.

KRISHNAMURTI T N, RAMANATHAN Y,1982. Sensitivity of the monsoon onset to differential heating[J]. J Atmos Sci, 39: 1290-1306.

LAU K M, YANG S,1997. Climatology and interannual variability of the Southeast Asian summer monsoon[J]. Adv Atmos Sci, 14: 141-162.

LAU K M, WU H T, YANG S,1998. Hydrologic processes associated with the first transition of the Asian summer monsoon: A pilot satellite study[J]. Bull Am Meteor Soc, 79: 1871-1882.

LI C, YANAI M,1996. The onset and interannual variability of the Asian summer monsoon in relation to land-sea thermal contrast[J]. J Climate, 9: 358-375.

LIU Y, CHAN J, MAO J, et al.2002. The Role of Bay of Bengal convection in the onset of the 1998 South China Sea summer monsoon[J]. Mon Wea Rev, 130: 2731-2744.

LIU Y, WU G, REN R,2004. Relationship between the subtropical anticyclone and diabatic heating[J]. J Climate, 17: 682-698.

MAK M, KAO C Y J,1982. An instability study of the onset-vortex of the southwest monsoon, 1979[J]. Tellus, 34: 358-368.

MAO J,WU G,LIU Y,2002a. Study on modal variation of subtropical high and its mechanism during seasonal transition Part Ⅰ: Climatological features of subtropical high structure[J]. Acta Meteor Sinica, 60(4): 400-408.

MAO J, WU G,2002b. Study on modal variation of subtropical high and its mechanism during seasonal transition Part Ⅱ: seasonal transition index over Asian monsoon region[J]. Acta Meteor Sinica, 60(4): 409-420.

MAO J, CHAN J, WU G,2004. Relationship between the onset of the South China Sea summer monsoon and the structure of the Asian subtropical anticyclone[J]. J Meteor Soc Japan, 82: 845-859.

MAO J, WU G,2007. Interannual variability in the onset of the summer monsoon over the eastern Bay of Bengal[J]. Theor Appl Climatol, 89: 155-170.

MCPHADEN M, HAYES S,1991. On the variability of winds, sea surface temperature, and surface layer heat content in the western equatorial Pacific[J]. J Geophys Res, 96: 3331-3342.

PISHAROTY P R, ASNANI G C,1957. Rainfall around monsoon depressions over India[J]. Indian J Meteor Geophys, 8: 1-6.

RAO K V, RAJAMANI S,1970. Diagnostic study of a monsoon depression by geostrophic baroclinic model[J]. Indian J Meteor Geophys, 21: 187-194.

RAO R R, SIVAKUMAR R,1999. On the possible mechanisms of the evolution of a mini-warm pool during the pre-summer monsoon season and the genesis of onset vortex in the south-eastern Arabian Sea[J]. Quart J Roy Meteor Soc, 125: 787-809.

SEETARAMAYYA P, MASTER A,1984. Observed air-sea interface conditions and a monsoon depression during MONEX-79[J]. Meteor Atmos Phys, 33: 61-67.

SHARMA M C, SRINIVASAN V,1971. Centres of monsoon depressions as seen in satellite pictures[J]. Indian J Meteor Geophys, 22: 357-358.

SHENOI S S C, SHANKAR D, SHETYE S R,1999. On the sea surface temperature high in the Lakshadweep Sea before the onset of the southwest monsoon[J]. J Geophys Res, 104: 15703-15712.

SHENOI S, SHANKAR D, GOPALAKRISHNA V, et al,2005. Role of ocean in the genesis and annihilation of the core of the warm pool in the southeastern Arabian Sea[J]. Mausam, 56: 147-160.

SMITH T M, REYNOLDS R W,2004. Improved extended reconstruction of SST (1854-1997)[J]. J Climate, 17(12): 2466-2477.

UPPALA,2005. The ERA-40 reanalysis[J]. Quart J Roy Meteor Soc, 131: 2961-3012.

VINAYACHANDRAN P N, SHANKAR D, KURIAN J, et al,2007. Arabian Sea mini warm pool and the monsoon onset vortex[J]. Curr Sci, 93: 203-214.

WANG T M, WU G,2008. Land-sea thermal contrast over South Asia and its influences on tropical monsoon circulation [J]. J Tropic Meteor, 24: 37-43.

WEBSTER P J, MAGAÑA V O, PALMER T N, et al,1998. Monsoons: Processes, predictability, and the prospects for prediction[J]. J Geophys Res, 103: 14451-14510.

WU G, ZHANG Y,1998. Tibetan Plateau forcing and the timing of the monsoon onset over South Asia and the South China Sea[J]. Mon Wea Rev, 126: 913-927.

WU G, LIU Y,2000. Thermal adaptation, overshooting, dispersion, and subtropical anticyclone Part Ⅰ: Thermal adaptation and overshooting[J]. Chin J Atmos Sci, 24(4): 433-446.

WU G, ZHANG Q, DUAN A M, et al,2007. The influence of the mechanical and thermal forcing of the Tibetan Plateau on the Asian climate[J]. J Hydrometeor, 8: 770-789.

WU G, LIU Y, ZHU X, et al,2009. Multi-scale forcing and the formation of subtropical desert and monsoon[J]. An Geophys, 27: 3631-3644.

WU G X, GUAN Y, WANG T M, et al,2011. Vortex genesis over the Bay of Bengal in spring and its role in the onset of the Asian summer monsoon[J]. Sci China Earth Sci, 54(1): 1-9.

YAN J H,2005. Asian summer monsoon onset and advancing process and the variation of the Subtropical High[D]. Beijing: Dissertation, Graduate University of Chinese Academy of Sciences.

ZHANG H M, BATES J J, REYNOLDS R W,2006. Assessment of composite global sampling: Sea surface wind speed[J]. Geophys Res Lett, 33(17): L17714.

Interannual Variability in the Onset of the Summer Monsoon over the Eastern Bay of Bengal

MAO Jiangyu, WU Guoxiong

(State Key Laboratory of Numerical Modeling for Atmospheric Sciences and Geophysical Fluid Dynamics (LASG), Institute of Atmospheric Physics, Chinese Academy of Sciences, Beijing 100029, China)

Abstract: Climatological characteristics associated with the summer monsoon onset over the eastern Bay of Bengal (BOB) are examined in terms of the westerly-easterly boundary surface (WEB). The vertical tilt of the WEB depends on the horizontal meridional temperature gradient (MTG) near the WEB under the constraint of the thermal wind balance. The switch of the WEB tilt firstly occurs between 90° and 100°E during the first pentad of May, with the 850 hPa ridgeline splitting over the BOB and heavy rainfall commencing over the eastern BOB, indicating the BOB summer monsoon (BOBSM) onset. The area-averaged MTG (200−500 hPa) is proposed as an index to define the BOBSM onset. A comparison of the onset determined by the MTG, 850 hPa zonal wind, and OLR shows that the MTG index is more effective in characterizing the interannual variability of the BOBSM onset.

Strong precursor signals are found prior to an anomalous BOBSM onset. Composite results show that early (late) BOBSM onset follows excessive (deficient) rainfall over the western Pacific and anomalous lower tropospheric cyclonic circulation zonally extending from the northern Indian Ocean into the western Pacific, with the strong (weak) equatorial westerly anomalies in the preceding winter and spring. Prior to an early (late) BOBSM onset, significant positive (negative) thickness anomalies exist around the Tibetan Plateau, accompanied by anomalous upper tropospheric anticyclonic (cyclonic) circulation. The interannual variations of the BOBSM onset are significantly correlated with anomalous sea surface temperature relevant to ENSO through changing the walker circulation and local Hadley circulation, leading to the middle and upper tropospheric temperature anomalies over the Asian sector. The strong precursor signals around the Tibetan Plateau may be partly caused by local snow cover anomalies, and an early (late) BOBSM onset is preceded by less (more) snow accumulation over the Tibetan Plateau during the preceding winter.

1 Introduction

The onset of the summer monsoon is one of the most important sub-seasonal phenomena within the annual cycle. A late or early onset of the monsoon may have devastating effects on agriculture even if the mean annual rainfall is normal. Therefore forecasting the timing of the onset is critical as it defines the ploughing and planting times in agrarian societies in the monsoon regions (Webster et al., 1998).

The monsoon may be thought of as the circulation responding to the annual cycle of solar heating in an interactive ocean-atmosphere-land system (Webster et al., 1998), and it is controlled by the thermal contrast between land and ocean. Thus the monsoon onset occurs along with a transition from the winter season to the summer season. During monsoon onset some dramatic changes occur in the large-scale atmospheric structure over the monsoon region. Among them are a rapid increase of the daily rainfall, a change of the low-level wind direction, a sudden establishment of the "monsoon vortex" and so on (e. g., Krishnamurti et al., 1981; Krishnamurti, 1985; Chan et al., 2000; Ding et al., 2001). Based on these characteristics around monsoon onset, a great variety of local indices with objective or subjective criteria have been proposed to define the onset dates over different parts in the Asian monsoon region. Therefore, the monsoon onset appears to be defined in various ways. For example, Ananthakrishnan et al. (1988) used observational station rainfall with objective criteria to determine the southern Indian monsoon onset dates over Kerala, and Xie et al. (1998) used a combined zonal wind (850 hPa) and outgoing longwave radiation (OLR) index to define the South China Sea (SCS) summer monsoon onset dates for individual years. Based on rain gauge observations and proxy rainfall data derived from satellites, many previous studies have investigated the climatological onset dates of the Asian summer monsoon (e. g., Tao et al., 1987; Tanaka, 1992; Lau et al., 1997; Webster et al., 1998). However, the disagreements exist significantly among these results due to the use of different datasets and different definitions of the onset. Since there are large differences between regional monsoon sub-systems in terms of large-scale circulation, and since the monsoon onset of each sub-system has considerable interannual variability, it is difficult to seek a universal definition for monsoon onset. Therefore the characteristics associated with monsoon onset should be further explored.

Wu and Zhang (1998) found that in 1989, the Asian summer monsoon onset consists of three consequential stages. The first is over the eastern BOB, the second over the SCS, and the third over South Asia. Similar situation took place in 1998 (Xu et al., 2001). Recently, Wang et al. (2002) attempted to build up a universal rainfall-based definition for the quantitative description of monsoon climatology in the entire Asian-Pacific domain, and they proposed the relative climatological pentad mean (CPM) rainfall rate as a parameter to define the monsoon onset, peak, and withdrawal of the rainy season. The relative CPM rainfall rate is derived from the difference between the pentad mean and the January mean precipitation rates, thus the monsoon onset date is defined as the pentad when the relative CPM rainfall rate exceeds 5 mm day^{-1}. It is shown that the earliest summer monsoon onset occurs over the eastern Bay of Bengal (5°—15°N, 90°—100°E) around the end of April through early May in the Asian-Pacific monsoon regime (see Fig. 1 of Wang et al., 2002). Following this initial stage, the monsoon onset extends northeastward and northwestward. LinHo et al. (2002) also designed the Asian-Pacific summer monsoon calendar at various key locations, in which the southern Bay of Bengal and southern Burma undergo the earliest onset and the longest rainy season. In an investigating of the relationship between the onset of the Asian summer monsoon and the structure of the Asian subtropical anticyclone, Mao et al. (2004) found that the switch in the tilt of the ridge surface of the subtropical anticyclone belt is closely associated with the Asian monsoon onset. The ridge surface of the subtropical anticyclone is defined by the boundary between the westerly to the north and the easterly to the south (or the westerly-easterly boundary surface (WEB) in brief). It is obvious that the tilt of the WEB firstly changes from southward to northward between 90° and 100°E during the first pentad of May, with the 850 hPa ridgeline splitting over the Bay of Bengal (BOB), indicating the summer monsoon onset over the eastern BOB (see Fig. 3 of Mao et al. 2004). These results suggest that the BOB summer monsoon (BOBSM) onset may be considered as a precursor to the subsequent establishment of the monsoon rainy season. In other words, study of the BOBSM onset has important significance for understanding the Asian monsoon variability.

Geographically, the BOB and the Tibetan Plateau are roughly located along the same longitudinal band. It is intuitively obvious that the Tibetan Plateau should have an important impact on an early/late monsoon onset. The roles played by the Tibetan Plateau in the evolution of the Asian monsoon have been discussed in many studies (e.g., Flohn, 1957; Yanai et al., 1992; Wu and Zhang, 1998). As suggested by Flohn (1957), Li et al. (1996) found that the reversal of the meridional temperature gradient (MTG), indicated by the difference of the upper tropospheric (200—500 hPa) temperature between 30° and 5°N, first occurs on the south side of the Tibetan Plateau, and they suggested that the onset of the Asian summer monsoon is concurrent with the reversal of the MTG in the upper troposphere. However, as shown by Wang et al. (2002), the Asian summer monsoon is not established simultaneously over different parts of Asia. In which latitudinal extent and layers is the MTG capable of indicating the monsoon onset for a particular region? Mao et al. (2004) gave some preliminary analyses about the relationship between the MTG and the WEB, and how the MTG determines the SCS summer monsoon onset. The potential applications of the MTG in measuring the monsoon onset over the BOB and other regions deserve further study.

It is well known that the Asian monsoon is significantly linked with ENSO (e.g., Meehl, 1987; Ju et al., 1995; Kawamura, 1998; Zhang et al., 2002). Webster et al. (1992) studied the interannual variability of the broad-scale Asian summer monsoon and identified strong signals of the monsoon variability in the upper tropospheric westerlies over subtropical Asia during the winter and spring seasons, and also suggested that the monsoon and ENSO are selectively interactive systems. Joseph et al. (1994) found that most delayed southern Indian monsoon onsets are associated with warm sea surface temperature (SST) anomalies at and south of the equator in the Indian and Pacific Oceans and cold SST anomalies in the tropical and subtropical oceans to the north during the season prior to the monsoon onset, indicating that ENSO is one of the important factors influencing the interannual variability of the Asian monsoon onset. Some studies show that snow cover, as an external forcing, can influence the interannual variability of the atmospheric circulation due to changes in the surface energy balance and land surface hydrological processes (e.g., Walsh et al., 1985; Barnett et al., 1989; Yang et al., 1996). Hahn et al. (1976) found the Indian summer monsoon rainfall is negatively related to the Eurasian snow cover in the preceding spring. Dey et al. (1986) showed the southern Indian monsoon onset is also associated with the Himalayan snow cover area. Heavy snowfall during winter leads to a weakened summer monsoon circulation due to a weak heat contrast between the Eurasian continent and the Indian Ocean (Yasunari et al., 1991). Based on the observational snow depth dataset over the Tibetan Plateau, Wu et al. (2003) re-examined the relationship of the Tibet winter snow anomaly with the subsequent summer rainfall over the Asian monsoon region. How and to what extent the BOBSM onset is significantly linked with these forcings on the interannual timescale need to be answered.

In view of the priority of the BOBSM onset in the Asian monsoon regime, the objectives of this paper are to identify the BOBSM onset date based on the MTG and to examine the factors during the preceding winter and spring that influence the interannual variability of the BOBSM onset. The slowly varying boundary conditions such as SST and snow cover associated with an anomalous monsoon onset are investigated in terms of the contributions to the land-sea thermal contrast.

Section 2 describes the data used in this study. Some characteristics of an abnormal BOBSM are examined in section 3. The precursory signals during the preceding winter and spring and factors that significantly influence the interannual variability of the BOBSM onset are investigated in section 4. Summary and discussion are given in section 5.

2 Data

The primary data are the global monthly and daily means of the National Centers for Environmental Prediction-National Center for Atmospheric Research (NCEP-NCAR) reanalysis products (Kalnay et al., 1996), which include wind, geopotential height, temperature, vertical velocity, and relative humidity at 17 standard pressure levels, with a horizontal resolution of 2.5°×2.5° from 1958 to 2001. The daily mean climatology is based on the period 1968—1996. They are obtained directly from the website of the National Oceanic and Atmospheric Administration-Cooperative Institute for Research in Environmental Sciences (NOAA-CIRES) Climate Diagnostics Center.

The Climate Prediction Center (CPC) Merged Analysis of Precipitation (CMAP; Xie et al., 1997) from 1979 to 2001 is derived from merging rain gauge observations, five different satellite estimates, and numerical-model outputs. It is used to describe the monsoonal rainfall over the entire Asian land and oceans. Since the CMAP dataset merges multi-source estimates, the uncertainties contained in each individual estimate are significantly reduced. The CMAP is available in pentads, with a resolution of 2.5°×2.5°. Daily outgoing longwave radiation (OLR) from NOAA, spanning from June 1974 to December 2001 (except for eight months in 1978), is also a commonly used proxy of tropical convective activity.

Monthly SST for the period 1958—2001 from NOAA extended reconstructed SST datasets is used to investigate the impacts of ENSO on the interannual variability of the BOBSM onset. Also used is the monthly snow depth (SD) over the Tibetan Plateau from 60 stations for the period 1960—1998. The SD is used to examine qualitatively the influences of the Tibetan Plateau snow cover on the atmospheric thickness and wind fields.

3 Characteristics of the anomalous BOB monsoon onset

3.1 Relationship between the switch of the WEB tilt and the BOBSM onset

Mao et al. (2004) utilized the WEB to understand the Asian summer monsoon onset. Because the summer monsoon onset occurs as a transition from cold-dry to warm-moist season, along with the lower tropospheric wind reversal, it is inevitably related to the changes of temperature, pressure, and wind fields in the entire troposphere through geostrophic and hydrostatic balance constraints. The subtropical anticyclone is a key system that connects the tropical and mid-latitude atmospheric circulations, with its ridgeline representing the maximum in the geopotential height (Li et al., 1998). Since the WEB is also the vertical shear zone between easterlies and westerlies, it represents the three-dimensional structure of the subtropical anticyclone. The WEB can thus be used as a proxy for the essential features of both the pressure and wind fields (Mao et al., 2004). On the other hand, the thermal wind balance suggests that the tilt of the WEB depends on the horizontal MTG in the vicinity of the WEB. Therefore, the WEB always tilts vertically towards the warmer region when the geostrophic relation is valid.

Using climatological pentad mean datasets, Mao et al. (2004) surveyed in detail the switch of the WEB tilt and its association with the Asian summer monsoon onset during the transitional season (see their Fig. 3). It is shown that the "seasonal transition axis" (STA) first establishes over the eastern BOB and Indochina Peninsula during the first pentad of May. The STA represents a critical state at which the ridge axis in the troposphere is "vertical", and it can be identified by the crossing points between the 500 and 200 hPa ridgelines. When the WEB tilts northward over eastern BOB and Indochina Peninsula from the second pentad of May, deep convection develops, which represents the onset of the Asian summer monsoon. This change of the WEB tilt indicates that a switch from the winter to the summer monsoon starts here, because when the tilt of the WEB changes from southward to northward, the MTG in the vicinity of the WEB changes from negative to positive. The MTG is thus a good indicator for measuring the BOBSM onset.

It should be mentioned that in practice, the STA is not convenient to be used as an index for the onset definition, but it can be highlighted as the critical state at which the MTG is equal to zero. Further, the location of the STA helps to determine a domain to define the onset based on the MTG.

3.2 Identifying the onset date based on the MTG

Seasonal variations of rainfall, 850 hPa winds, and MTG over the eastern BOB are shown in Fig. 1. The heavy precipitation exceeding 6 mm day^{-1} is found over the equatorial region almost throughout the year (Fig. 1a). From the end of April to early May, the rapid northward extension of the heavy rainbelt takes place from 5°N to 15°N, accompanied by the commencement of the southwesterlies at low levels (Fig. 1b), indicating a BOBSM onset. Subsequently, the strong southwesterlies maintains until the end of September. Note also that the reversal of the zonal wind direction is primarily confined to the latitudinal band 5°—15°N, and such a reversal is distinct around early May, suggesting that the 850 hPa zonal winds can reflect the BOBSM onset. Moreover, the mean position of the WEB is located within this latitudinal band (see Fig. 3 of Mao et al., 2004). Considering these features along with the result shown by Wang et al. (2002), we choose the eastern BOB domain (5°—15°N, 90°—100°E) in this study. When 850 hPa winds change from easterlies to westerlies, the OLR value rapidly drops to near or below the critical value of 230 W m^{-2} (Fig. 1b).

According to Wang et al. (2002), the BOB is a unique region where the South Asian monsoon links with the East Asian monsoon. The former is a typical tropical monsoon system, whereas the latter is a combined tropical-midlatitude system. The properties of the lower tropospheric southwesterlies over the BOB during monsoon onset are thus more complex. As suggested by Zhang et al. (2002), there are at least three different branches that contribute to the circulation over the BOB and Indochina Peninsula, in which the southwesterlies leading to BOBSM onset consist of the subtropical westerlies from the Arabian anticyclone and tropical westerlies from the Somali cross-equatorial flow. Therefore, one may argue whether the 850 hPa zonal wind index can be constructed to describe the BOB circulation change before, during, and after the monsoon onset. Use of the MTG as an index can avoid this doubt. It can be seen that an abrupt reversal of the sign in the MTG occurs in the middle and upper troposphere between 200 and 600 hPa around 2 May (Fig. 1c), which corresponds to the time of the establishment of the STA over the eastern BOB. Note that the reversal of the MTG in the lower layer (700—1000 hPa) is much earlier than that in the upper layers, which may be caused by stronger sensible heating over Burma. Although Li et al. (1996) pointed out the significance of the MTG (200—500 hPa) to the Asian summer monsoon onset, the mean reversal time by using their calculation (temperature difference between 30°

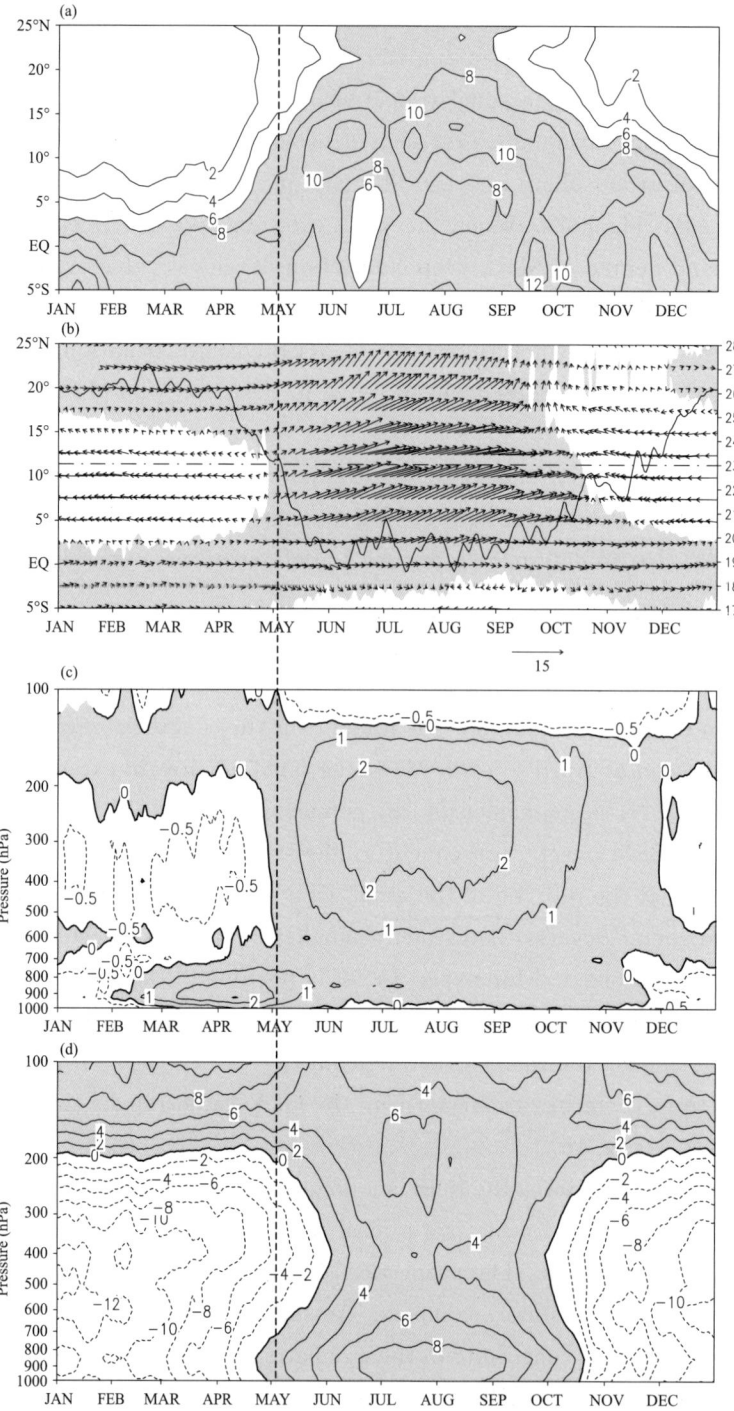

Fig. 1 Seasonal evolutions of rainfall, 850 hPa winds, and atmospheric meridional temperature gradient over the eastern BOB (5°—15°N, 90°—100°E). (a) Latitude-time cross section (90°—100°E) of the climatological pentad mean Merged Analysis Precipitation from Climate Prediction Center (mm day^{-1}), with shading indicating the precipitation rate exceeding 6 mm day^{-1}; (b) Latitude-time cross section (90°—100°E) of the daily mean 850 hPa winds (m s^{-1}), with shading denoting the zonal wind greater than zero. Thick solid line denotes the time series of the area-averaged daily mean OLR (W m^{-2}) over the eastern BOB; (c) Pressure-time cross section of the area-averaged daily mean meridional temperature gradient (10^{-6} K km^{-1}) over the eastern BOB; (d) Pressure-time cross section of the meridional temperature gradient (10^{-6} K km^{-1}) indicated by the difference between 30° and 5°N and averaged between 90° and 100°E. The shading in (c) and (d) denotes the meridional temperature gradient greater than zero

and 5°N) is around the end of May (Fig. 1d). This date is nearly one month later than those shown in Figs. 1a to 1c, suggesting that the upper tropospheric temperature difference between 30° and 5°N is not suitable as an index for defining the BOBSM onset. Above discussion indicates that in determining the monsoon onset date, the use of the MTG reversal in the vicinity of the WEB is more appropriate since the onset of the low-level monsoon southwesterlies occurs almost simultaneously with the switch of the WEB tilt in the upper troposphere or a break of the lower tropospheric ridgelines.

It is obvious that the BOBSM onset dates (first pentad of May or 2 May) as determined by the above four meteorological quantities (Figs. 1a to 1c) are fairly consistent, and are also accordant with the results given by Wang et al. (2002). These four elements from several different aspects reflect the main climatological characteristics of the BOBSM onset. Because the wind patterns and origins over the BOB are very complicated as stated above, and the longer daily CMAP and OLR records are unavailable, in this paper, we use merely the area-averaged upper tropospheric (200—500 hPa) MTG over the eastern BOB as a primary index to identify the BOBSM onset date.

Following Mao et al. (2004), the onset date for each individual year is defined as the day when the following criteria are first satisfied: (1) the area-averaged upper tropospheric (200—500 hPa) MTG over the eastern BOB (5°—15°N, 90°—100°E) changes from negative to positive; (2) the MTG remains positive for more than 10 days. The duration threshold is now selected as 10 days to assure that a true change in the large-scale circulation has occurred because the period of synoptic-scale disturbances is about one week based on previous studies (e.g. Chan et al., 2000). For a comparison, an identical definition is applied to the 850 hPa zonal wind (U_{850}) and OLR, with the threshold of OLR chosen as 230 W m^{-2}.

The BOBSM onset dates for each individual year based on the above definition are displayed in Fig. 2. It is found that in a great number of individual years, the dates determined by these three indices are very similar, but there are large differences in a few cases, e.g., in 1967, 1982, 1989, 1990. In those inconsistent years, the occurrence of convection was usually later than the reversals of the MTG and U_{850}. This inconsistency need to be further examined. These three onset time series are well correlated with each other, passing the 99% confidence level. The correlation coefficients of MTG with U_{850} and with OLR are 0.919 and 0.807, respectively, and the correlation coefficient between U_{850} and OLR is 0.797. In contrast, the MTG exhibits a better correlation with U_{850} than OLR does. These indicate that the MTG index is more effective in characterizing the interannual variation of the BOBSM onset.

3.3 Interannual variability of the BOBSM onset

The long-term mean onset dates determined by the MTG, U_{850} and OLR are 2 May, 2 May, and 1 May, with standard deviations of 10.2, 10.2, and 12.2 days respectively (Fig. 2). The BOBSM onset dates differ significantly from one year to another, with the range between the earliest and the latest onset dates exceeding one month. Now we examine the interannual variability of the BOBSM onset on the basis of the time series of onset dates derived from the MTG index.

Using one standard deviation as the threshold, we classify an early (late) BOBSM onset as one with its anomaly less (greater) than one standard deviation. Thus, the early onset years selected are 1959, 1974, 1975, 1982, 1984, 1985, 1991, 1999, 2000, while the late onset years are 1963, 1978, 1983, 1990, 1992, 1993, 1997, 1998. Since the abnormal early and late BOBSM onset years are classified, the precursory signals appearing in the preceding winter and spring can be examined in terms of the atmospheric circulation, SST, etc.

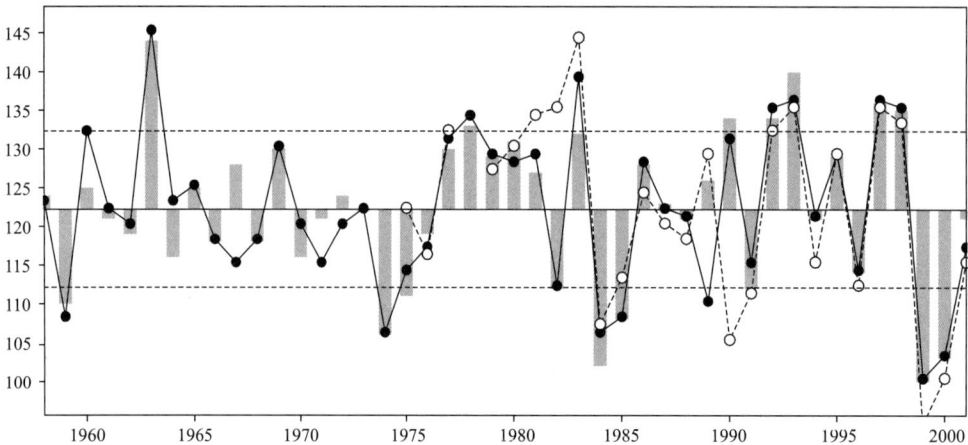

Fig. 2 Time series of the BOBSM onset dates defined by the MTG (bars), 850 hPa zonal wind (solid circles), and OLR (open circles) for the period 1958−2001. All dates are Julian dates in the calendar year (e. g., day 121 = 1 May, etc.). The long-term mean onset date for each of three time series is indicated by a solid line. One standard deviation of the time series derived from the MTG index is indicated by the two dashed lines

4 Precursory signals associated with the interannual variability of the BOBSM onset

4.1 Atmospheric circulation anomalies

To examine the remote influences on the interannual variability of the BOBSM onset during the preceding winter and spring, composite analyses are made for early and late onset categories. Composite differences in the 850 hPa winds and precipitation for late winter to spring between early and late onset categories show that an anomalous cyclone is present over the BOB, SCS, and western Pacific, with significant westerlies prevailing in the equatorial region and easterlies in the northwestern BOB (Fig. 3a). Significant easterly anomalies exist over the equatorial central-eastern Pacific. Strong southerly anomalies are also found over east of the Philippine Sea, due to the convergence between the anomalous equatorial westerlies and easterlies. As a result, positive rainfall anomalies occur in the vicinity of the Philippines (Fig. 3d). The entire flow pattern together with anomalous rainfall is similar to the analytical solution given by Gill (1980). The cyclonic circulation around the Philippines, accompanied by enhanced precipitation, becomes more intense in March, when the equatorial westerlies south of the BOB strengthen and extend northward (Figs. 3b and 3e). The twin-cyclone or double-low structure suggested by Chen et al. (1993) in the Indian Ocean appears in April, in which the cyclone to north of the equator is centered over the west coast of the BOB so that the eastern BOB is dominated by significant southwesterlies (Fig. 3c). Importantly, the equatorial westerlies between these two cyclones are strengthened and extend northward further, leading to more moisture transport from the tropical Indian Ocean to the eastern BOB at the low levels. The heavy rainfall thus occurs over the eastern BOB and western Indochina Peninsula (Fig. 3f), implying that the BOBSM bursts earlier than normal. The strengthening of the BOB cyclonic circulation and the equatorial westerlies may be attributed to a Rossby wave response to the enhanced convection over the SCS and western Pacific (Gill, 1980), since the significant rainfall domain continuously extends westward from the western Pacific to the Indian Ocean. The westward extension of the significant rainfall anomalies around the equator demonstrates a westward Rossby wave propagation.

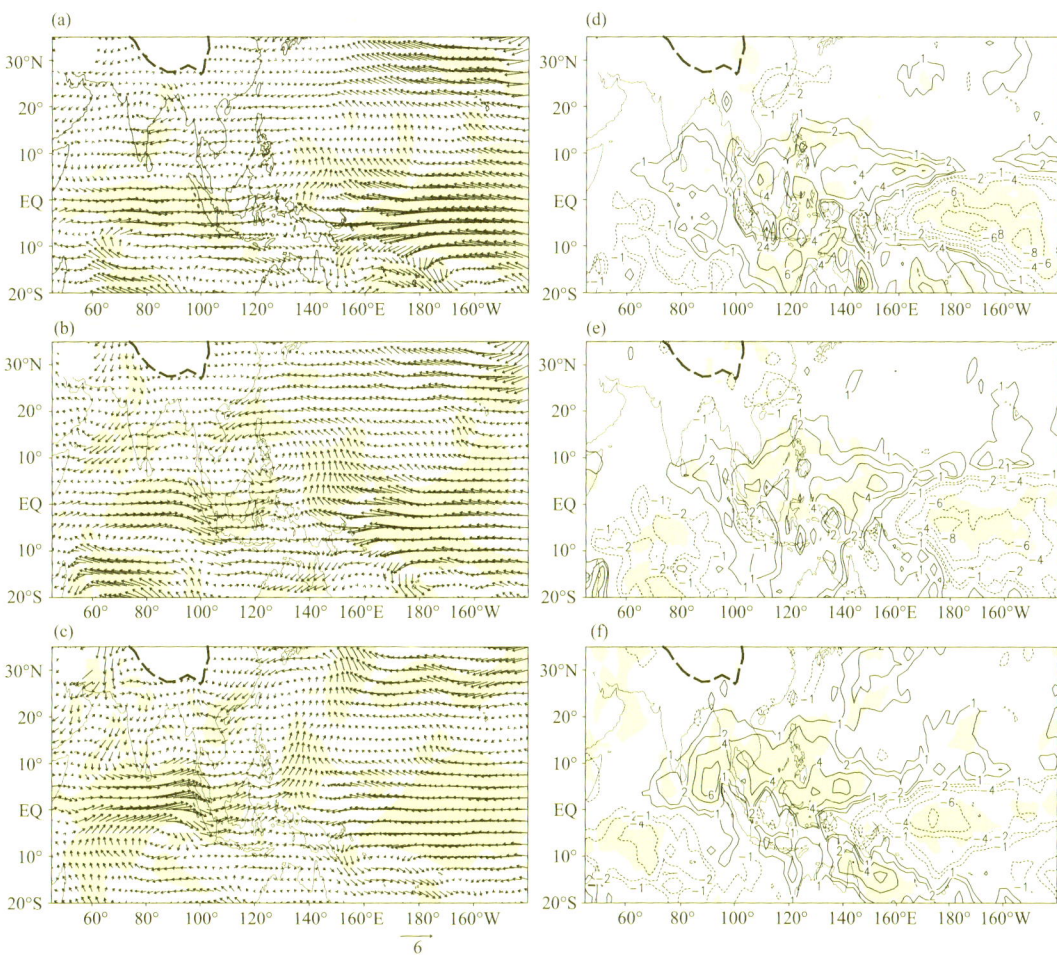

Fig. 3 The monthly early-minus-late composite patterns of the differences in 850 hPa wind field from (a) to (c) and CMAP rainfall from (d) to (f) for February to April in the preceding late winter and spring. Shading indicates that the *t*-test passes the 95% significance level. In the left panel, shading denotes that at least one of the wind components (zonal or meridional) passes the 95% significance level. Thick dashed line shows the Tibetan Plateau above 3000 m

Composite patterns of the 200 hPa wind differences illustrate that prior to the BOBSM onset, an anomalous anticyclone exists over the vicinity of the Tibetan Plateau for the early onset category. The center of the anticyclone is located over the east of the Tibetan Plateau in February, then moves southwestward so that anomalous easterlies prevail over the northern Indian Ocean from 5°S to 25°N (figures not shown), implying that the upper tropospheric westerly jet south of the Tibetan Plateau is weaker in the winter and spring seasons. In summary, an early (late) BOBSM onset follows an excessive (deficient) rainfall over the western Pacific, an anomalous lower tropospheric cyclone (anticyclone) extending from the northern Indian Ocean to the western Pacific, strong (weak) equatorial westerly anomalies, and an anomalous upper tropospheric anticyclone (cyclone) around the Tibetan Plateau with easterly (westerly) anomalies in the preceding winter and spring.

Because the configurable variation of the subtropical anticyclone is closely related to the Asian monsoon onset, the differences in atmospheric circulation between early and late BOBSM onset should be reflected by the WEB. Composite patterns of the WEB for early and late onset are presented in Fig. 4. The remarkable differences are found within the BOB and Indochina Peninsula longitudes. In March, the ridgelines of the early onset category are closer between 90° and 100°E than those of the late onset category, indicating that the southward tilt of the WEB or the MTG in the region in early onset years is usually weaker than that in late onset years (Figs. 4a and 4d). By April, the WEB above 500 hPa of the early onset category becomes vertical between 90° and 100°E, while the WEB of the late onset category still exhibits a significantly southward tilt (Figs. 4b and 4e). In May, although the WEB tilt in both categories exhibits a northward tilt over the BOB and Indochina Peninsula, some substantial differences exist in the tilting extent and the location of the STA (Figs. 4c and 4f). Here the STA can be identified by the crossing point stacked up with the ridgelines at most levels. In the early category, the 200 and 300 hPa ridgelines are located to the north of 15°N, while the STA is situates over the northeastern SCS. Consistently, the upper tropospheric anticyclone is located farther north than normal (see also Fig. 2 of Mao et al., 2004), with the lower tropospheric southwesterlies prevailing in broader longitudes. For the late category, the STA is only located at the western coast of the SCS, and the northward tilt of the WEB is smaller.

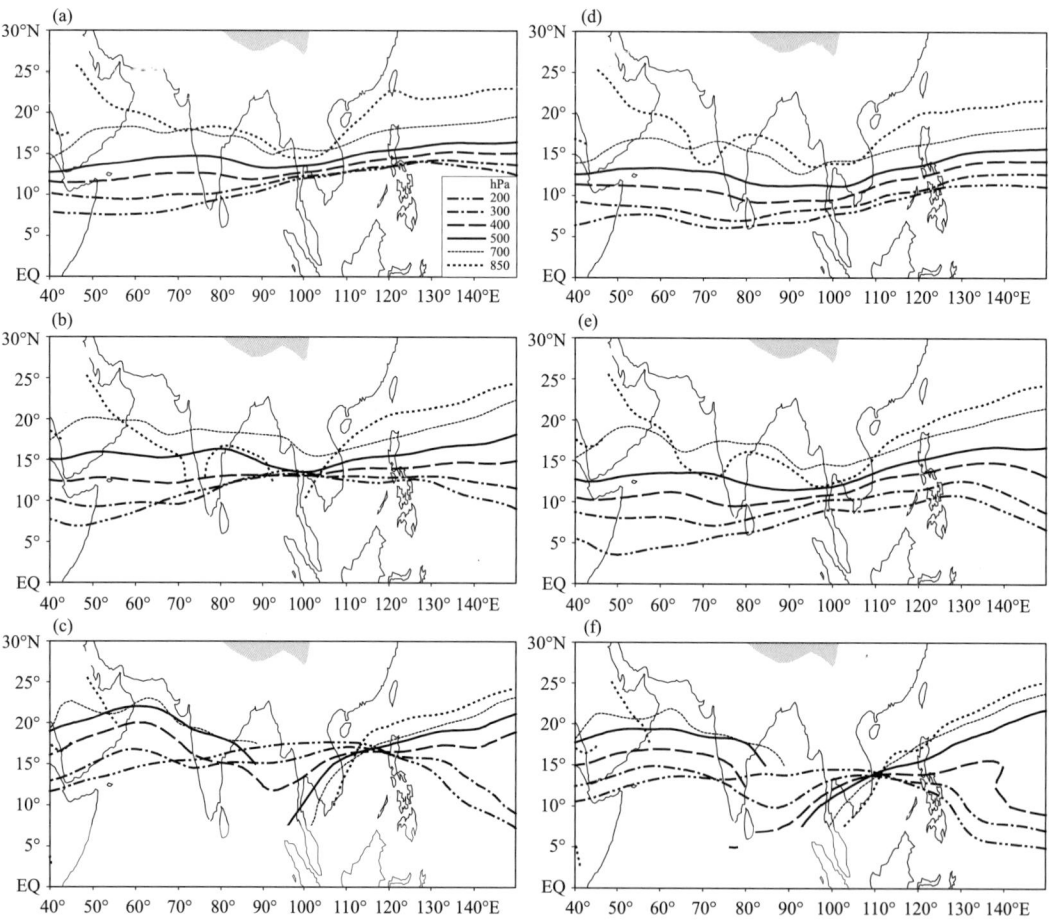

Fig. 4 Composite projections of the WEB for early (left panel) and late (right panel) BOBSM onset categories. Projections for March to May are displayed from (a) to (c) and from (d) to (f) respectively. Thick curves denote the subtropical anticyclone ridgelines on various isobaric surfaces (indicated by the numbers in legend). Shading denotes terrain above 3000 m

4.2 Temperature and thickness anomalies

Lagged correlations between the BOBSM onset date and the upper tropospheric (200—500 hPa) thickness for January to April show that the significantly positive correlation areas appear over the eastern Pacific, indicating remote influences of the Pacific Ocean on the BOBSM onset (not shown). Pronouncedly positive correlation areas exist over the Lake Baikal in February and March. It is stressed here that the negative correlation is particularly strong around the Tibetan Plateau. A small negative correlation area is observed over East China in January, then becomes larger and stronger, and moves steadily westward, forming a large significant negative area over the Tibetan Plateau. Such a significant correlation area reflects the thermal effect of the Tibetan Plateau. That is, the thermal anomaly of the Tibetan Plateau in the winter and spring may have an important impact on the temperature anomalies to the north side of the WEB, leading to the change of the land-sea thermal contrast across the WEB.

To examine the detailed features over the Tibetan Plateau, composite patterns of upper tropospheric (200—500 hPa) thickness and 600 hPa temperature differences between the early and late BOBSM onset categories are shown in Fig. 5. Corresponding to the correlation pattern, an area of significantly positive thickness anomalies is observed over central China in February (Fig. 5a), and then extends westward, with its center just over the southeastern Tibetan Plateau (Fig. 5b), forming an anomalously strong meridional contrast of thickness characterized by positive anomalies over the Tibetan Plateau north of 15°N and negative anomalies over the tropical Indian Ocean. In April (Fig. 5c), the thickness anomaly area is centered on the south side of the Tibetan Plateau (along 90°E), resulting in a pattern similar to that shown in Li et al. (1996). Because the Tibetan Plateau surface is close to the 600 hPa isobaric surface, the 600 hPa temperature anomalies are chosen to highlight the thermal effects of the Tibetan Plateau (Figs. 5d to 5f). Similar to upper tropospheric thickness anomalies, significant differences in 600 hPa temperature are found over the Tibetan Plateau and Indian Ocean, especially in April, indicating a contribution to the thermal contrast between land and ocean in the middle troposphere. It turns out that early (late) BOBSM onset is preceded by positive (negative) temperature anomalies over the Tibetan Plateau in the middle and upper troposphere. Notice that significant thickness anomalies with a butterfly pattern exist over the Pacific sector, which indicates the atmospheric response to ENSO heating over the equatorial eastern Pacific region. Such a thickness anomaly pattern may be associated with the temperature anomalies over the Tibetan Plateau (Miyakoda et al., 2003).

Our correlation and composite patterns are similar to those discussed by Miyakoda et al. (2003), who investigated pre-monsoon signals of the South Asian monsoon over the Tibetan Plateau and their association with ENSO (see their Figs. 1 and 2). They defined an index that is also called MTG, indicating the Asian summer monsoon intensity, and they calculated the lagged correlation between this host index and the upper tropospheric (200—500 hPa) thickness for different seasons. As in Miyakoda et al. (2003), in addition to the high correlation over the equatorial region, the westward migration of high correlation around the Tibetan Plateau is very significant from winter to summer. These results suggest that the upper tropospheric temperature anomaly over the Tibetan Plateau can foreshows not only the South Asian summer monsoon intensity but also the BOBSM onset, and it may provide a precursory background for anomalous BOBSM onset and anomalous intensity of the South Asian summer monsoon.

4.3 SST and Snow anomalies

Since the atmospheric anomalies cannot persist for several months without any external forcing, the aforementioned persistent signals associated with abnormal BOBSM onset are difficult to explain by atmospheric processes themselves. As suggested by Miyakoda et al. (2003), the presence of these precursory signals may be related to anomalous SST forcing. We thus investigate remote influences involving snow cover and SST.

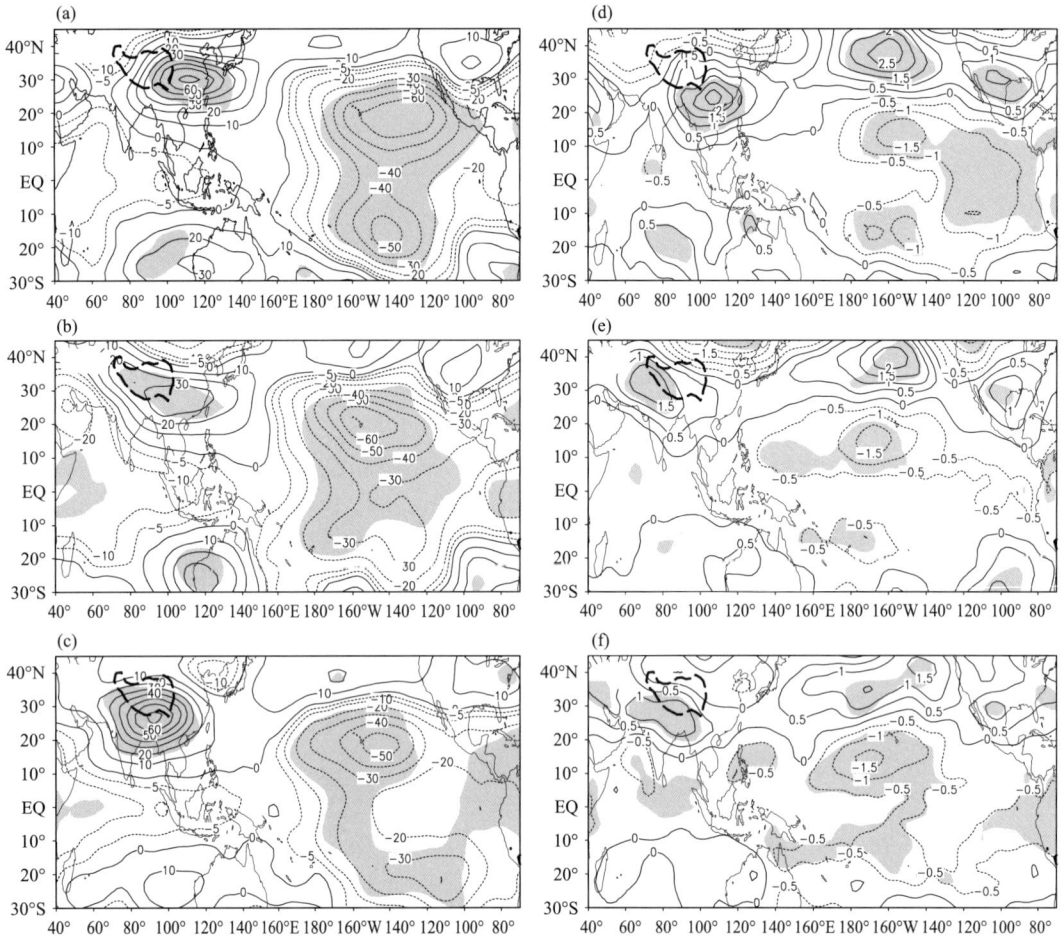

Fig. 5 The monthly early−minus−late composite patterns of the differences in upper tropospheric (200−500 hPa) thickness from (a) to (c) and 600 hPa air temperature from (d) to (f) for February to April in the preceding late winter and spring. Shading indicates that the t-test passes the 95% significance level. Thick dashed line shows the Tibetan Plateau above 3000 m

Figure 6 shows lagged correlations between the BOBSM onset date and SST during the preceding winter and spring. Significant positive correlations are observed over the equatorial central and eastern Pacific, with negative correlations extending from the equatorial western Pacific to the North Pacific of middle latitude. The positive correlations are also found over the equatorial Indian Ocean from March to May, reflecting that the SST anomalies over the Indian Ocean are usually later than those over the equatorial Pacific during an ENSO event (Trenberth, 1990; Sewell et al., 2001). The entire pattern of correlations strongly resembles the SST anomaly in an El Niño event from December to April. To further validate the relationship between ENSO and the abnormal BOBSM onset, we also calculate the correlation coefficient between the normalized time series of the BOBSM onset date and Nino 3.4 SST index for the period 1958−2001. The correlation coefficient of the BOBSM onset date with Nino 3.4 SST index of February is 0.44, passing the 99% confidence level. The BOBSM-ENSO relationship seems to have undergone a decadal change. Before 1967 the magnitude of the onset date anomalies was weak and its correlation with the eastern Pacific SST anomalies was weakly negative, while the correlation has been stronger since 1968, with a correlation coefficient value of 0.58 for the period 1968−2001. Actually, the correlation coefficient of the BOBSM onset date with Nino 4 SST index is higher than with Nino 3.4 SST index for December to March. The above facts suggest that El Niño (La Niña) event is followed by late (early) BOBSM onset.

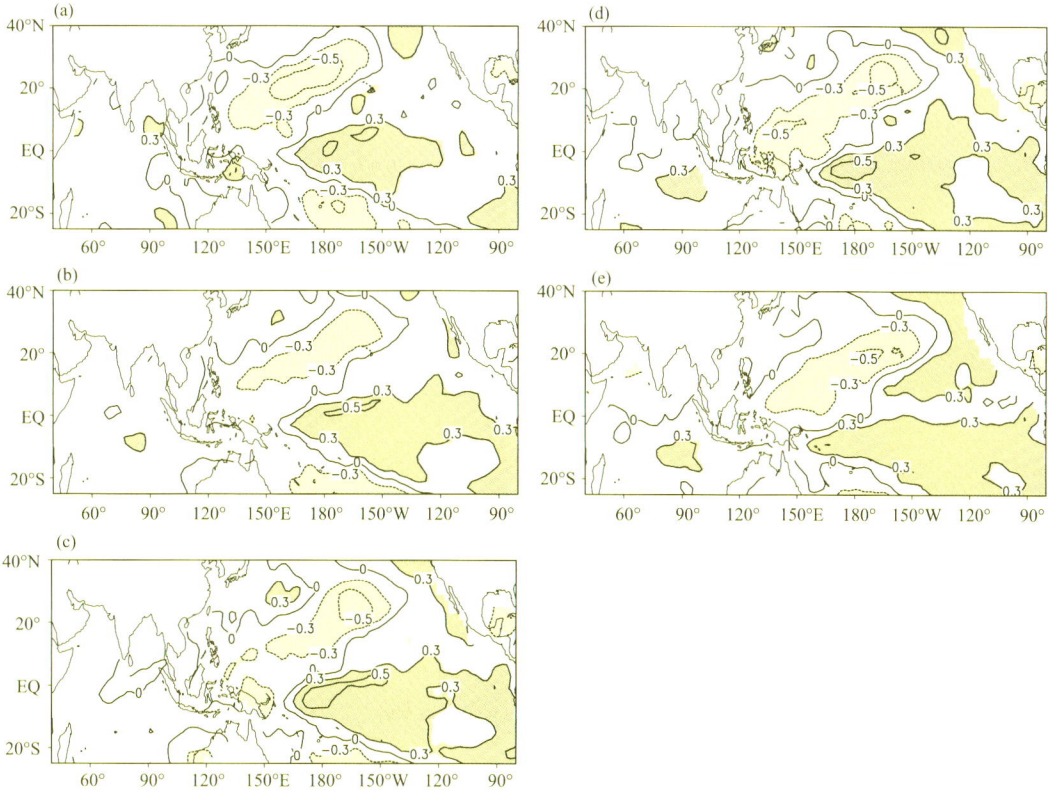

Fig. 6 Lagged correlations between time series of the BOBSM onset date and SST anomalies from December (previous year) to April for (a) to (e). Critical positive (negative) values of the correlation at the 95% confidence level are shaded

Zhang et al. (2002) found that the interannual variation of the Asian summer monsoon onset over the Indochina Peninsula is closely related to El Niño/La Niña during boreal spring. They suggested that the changes of the Walker circulation and the local Hadley circulation as a direct response to ENSO lead to anomalous convective activities over the equatorial Pacific. Similar situations can be seen from Fig. 3. From February to April, significantly positive rainfall anomalies over the western Pacific are concurrent with westerly anomalies over the Indian Ocean-western Pacific, while negative rainfall anomalies correspond to easterly anomalies over the central-eastern Pacific. These suggest that the changes of the Walker circulation and the local Hadley circulation forced by ENSO may be responsible for an early/late BOBSM onset.

Namias et al. (1988) pointed out that persistent SST anomalies in the North Pacific are usually associated with the atmospheric teleconnection patterns. The impact of ENSO on the BOBSM onset can be further understood by the study of Miyakoda et al. (2003), which suggested that the air temperature anomaly between 200 and 500 hPa and its westward movement over the Asian sector along $20°-35°N$ are associated with the butterfly pattern over the Pacific sector. In our study, composite patterns of thickness anomalies (Fig. 5) are similar to the result from Miyakoda et al. (2003) for strong and weak monsoons (see their Fig. 8). The butterfly pattern over the Pacific sector is essentially a Gill-type response to the tropical heat source (Gill, 1980). Corresponding to the butterfly pattern, the symmetrical circulations associated with Hadley cells are thus produced in both hemispheres when positive heat sources of a short longitudinal length appear in the equator. Such a flow pattern has been verified by some numerical simulations (e. g., Ting et al., 1990). Wu and Newell (1998) used the Gill-type model to examine the tropical atmospheric response to an El Niño event. A local heat source can warm the entire tropical troposphere when the heat source appears in the equatorial eastern Pacific. The released latent heat and its forced adiabatic subsidence elsewhere in the Tropics can warm the atmosphere. The tropospheric atmosphere thus becomes warm over the equatorial Indian Ocean, affecting the land-sea thermal contrast over the Asian sector.

In addition to association with the teleconnection forced by ENSO, the signals over the Tibetan Plateau may be partly caused by local snow cover anomalies. Some previous studies (e. g., Namias, 1985; Leathers et al., 1993) pointed out that the variability of surface temperature can be regulated, to some extent, by the change in snow cover. Figure 7 shows the anomalies of accumulated snow depth (ASD) for the early and late BOBSM onset categories. Because the Tibetan Plateau snow cover begins to occur in mid-September and develops very quickly after mid-October or early November, the ASD from November (of the previous year) to March can be used as a representative of snow mass variation for each year (Li, 1993; Wu et al., 2003). At most stations over the Tibetan Plateau, ASD anomalies are negative in the early onset category (Fig. 7a), corresponding to positive temperature (600 hPa) and thickness (200—500 hPa) anomalies (Fig. 5). In contrast, most stations exhibit positive anomalies in the late onset category (Fig. 7b). These indicate that an early (late) BOBSM onset is associated with less (more) Tibetan Plateau snow accumulation during the preceding winter.

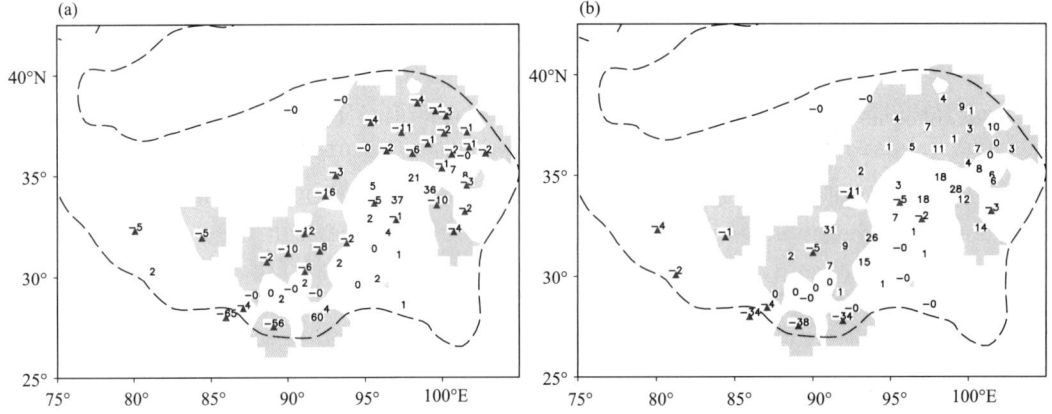

Fig. 7　Composite distributions of Nov−Mar accumulated snow depth anomalies (cm month^{-1}) over the Tibetan Plateau for (a) early and (b) late BOBSM onset categories. Negative anomalies are marked with triangles. Shading denotes the regions where the t-test is significant at 90% confidence level. Dashed line shows the Tibetan Plateau above 2500 m

5　Conclusions and discussions

Following Mao et al. (2004), we examine the climatological characteristics associated with the summer monsoon onset over the eastern Bay of Bengal (BOB) by considering the variation of the westerly-easterly boundary surface (WEB). During the transitional season, the WEB tilts from southward to northward, suggesting a replacement between the winter and the summer monsoon. The switch of the WEB tilt firstly occurs between 90° and 100°E during the first pentad of May and coincides with the 850 hPa ridgeline splitting and heavy rainfall commencing over the eastern BOB, indicating the onset of the

BOB summer monsoon. The climatological onset date determined by the upper tropospheric (200—500 hPa) MTG is similar to those defined by zonal wind at 850 hPa, CMAP precipitation, and OLR. Therefore, the area-averaged MTG (200—500 hPa) near the WEB is used to define the BOBSM onset. Our study shows that the reversal of the MTG can capture the essential features of the BOBSM onset.

The interannual variability of the BOBSM onset is investigated based on the time series of onset dates derived from the MTG. The long-term mean onset date is 2 May, with a standard deviation of 10.1 days. Composite results show that early (late) BOBSM onset follows excessive (deficient) rainfall over the western Pacific and anomalous lower tropospheric cyclonic (anticyclonic) circulation in the northern Indian Ocean-western Pacific region, coincident with strong (weak) equatorial westerly anomalies in the preceding winter and spring. Changes of atmospheric temperature and thickness are observed to be dynamically consistent with those in the upper tropospheric wind field. Prior to an early (late) BOBSM onset, significantly positive (negative) thickness anomalies appear over the Tibetan Plateau and migrate westward, accompanied by an anomalous upper tropospheric anticyclonic (cyclonic) circulation. These strong signals manifest that the Tibetan Plateau has an important impact on the middle and upper tropospheric temperature anomalies, leading to a change in the land-sea thermal contrast.

The interannual variations of the BOBSM onset date are significantly correlated to the SST anomalies during the preceding winter and spring, suggesting a remote impact of the El Niño (La Nina) event on BOBSM onset. An El Niño (La Niña) event is followed by late (early) BOBSM onset. The changes of the Walker circulation and the local Hadley circulation forced by ENSO may be one of the mechanisms responsible for regulating early/late BOBSM onset. The SST anomalies in the Indian Ocean and western Pacific may cause a change in the land-sea thermal contrast over the Asian sector. On the other hand, the heat source in the equatorial eastern Pacific can warm the entire tropical tropospheric atmosphere, and, thus, can also affect the land-sea thermal contrast. Our study suggests that the precursor signals around the Tibetan Plateau are partly caused by local snow cover anomalies. An early (late) BOBSM onset is usually preceded by a less (more) snow accumulation over the Tibetan Plateau during the preceding winter.

It is well known that the summer monsoon circulation can be explained by the diabatic forcing due to condensation heating, sensible heating, and radiation cooling (e. g., Rodwell et al., 1996; Ose, 1998). On the one hand, the diabatic heating contributes as an external forcing to the increase of local temperature based on the thermodynamic equation. On the other hand, the monsoon circulation responds to differential heating between the regions of cooling and heating. These features thus indicate that there exists a nonlinear feedback between heating and monsoon circulation dynamics. Krishnamurti (1985) suggested that the Indian monsoon can largely be viewed as a response of circulation to the differential heating, and the mode of migration of the heating region from northern Burma in the month of May to the foothills of the Himalayas (eastern Tibet) in the month of June is a critical factor in determining the onset of the Indian Monsoon. These imply that the temporal and spatial variation of the heat source should play some role in the seasonal transition associated with the BOBSM onset. However, there are no directly observed heating data available. The NCEP-NCAR heating fluxes as model output products are the six-hour averages starting at the reference time, thus the heating distribution as derived from these flux data can only help us understand qualitatively the role of differential heat in the BOBSM onset. The distributions of diabatic heating anomalies can be roughly found from Fig. 3, because released latent heating is dominant component that composes the total diabatic heating, especially over ocean areas. As suggested by Krishnamurti et al. (1982), such a west-east heating difference between Indian Ocean and western Pacific is favorable for an acceleration of low-level divergent westerly winds, further enhancing the rotational winds over the BOB, via energetic transform among the rotational, divergent and available potential energies.

Wu and Zhang (1998) emphasized the importance of sensible heating over the Tibetan Plateau in the increase of air column temperature. Climatologically, sensible heating flux greater than 50 W m^{-2} occurs over the entire Tibetan Plateau in April to May. Such diabatic heating flux can warm the middle and upper tropospheric atmosphere. In this sense, sensible heating flux anomalies may partly contribute to positive thickness anomalies around the plateau, because horizontal and vertical temperature advections are also responsible for the increase of air column. Based on snow depth anomalies shown in Fig. 7, it may be inferred that the sensible heating flux anomalies to a certain extent influence the BOBSM onset. Of course, zonal and meridional variations of the differential heating may lead cooperatively to an abnormal BOBSM onset.

References

ANANTHAKKRISHNAN R, SOMAN M K,1988. The onset of the south-west monsoon over Kerala: 1901—1980[J]. J Climatol, 8: 283-296.

BARNETT T P, DÜMENIL L, SCHLESE U, et al,1989. The effect of Eurasian snow cover on regional and global climate variations[J]. J Atmos Sci, 46: 661-685.

CHAN J, WANG Y, XU J,2000. Dynamic and thermodynamic characteristics associated with the onset of the 1998 South China Sea summer monsoon[J]. J Meteor Soc Japan, 78: 367-380.

CHEN T C, CHEN J M,1993. The 10—20-day mode of the 1979 Indian monsoon: Its relation with the time variation of monsoon rainfall[J]. Mon Wea Rev, 121: 2465-2482.

DEY B, KATHURIA S N,1986. Himalayan Snow Cover Area and Onset of Summer Monsoon over Kerla, India[J]. India: Mausam, 37: 193-196.

DING Y, LIU Y,2001. Onset and the evolution of the summer monsoon over the South China Sea during SCSMEX field experiment in 1998[J]. J Meteor Soc Japan, 79: 255-276.

FLOHN H,1957. Large-scale aspects of the "summer monsoon" in South and East Asia[J]. J Meteor Soc Japan, 35A: 180-186.

GILL A E,1980. Some simple solutions for heat-induced tropical circulation[J]. Quart J Roy Meteor Soc, 106: 447-462.

HAHN D G, SHUKLA J,1976. An apparent relationship between Eurasian snow cover and Indian monsoon rainfall[J]. J Atmos Sci, 33: 2461-2462.

JOSEPH P V, EISCHEID J K, PYLE R J, et al,1994. Interannual variability of the onset of the Indian summer monsoon and its association with atmospheric features, El Niño, and sea surface temperature anomalies[J]. J Climate, 7: 81-105.

JU J, SLINGO J M,1995. The Asian summer monsoon and ENSO[J]. Quart J Roy Meteor Soc, 121: 1133-1162.

KALNAY E, COAUTHORS,1996. The NCEP/NCAR 40-year reanalysis project[J]. Bull Amer Meteor Soc, 77: 437-471.

KAWAMURA R,1998. A possible mechanism of the Asian summer monsoon-ENSO coupling[J]. J Meteor Soc Japan, 76: 1009-1027.

KRISHNAMURTI T N,1985. Summer monsoon experiment—A review[J]. Mon Wea Rev, 113: 344-363.

KRISHNAMURTI T N, ARDANUY P, RAMANATHAN Y, et al,1981. On the onset vortex of the summer monsoons [J]. Mon Wea Rev, 109: 344-363.

KRISHNAMURTI T N, RAMANATHAN Y,1982. Sensitivity of the monsoon onset to differential heating[J]. J Atmos Sci, 39: 1290-1306.

LAU K M, YANG S,1997. Climatology and interannual variability of the Southeast Asian summer monsoon[J]. Adv Atmos Sci, 14: 141-162.

LEATHERS D J, ROBINSON D A,1993. The association between extremes in north American snow cover extent and United States temperatures[J]. J Climate, 6: 1345-1355.

LI C, YANAI M.1996. The onset and interannual variability of the Asian summer monsoon in relation to land-sea thermal contrast[J]. J Climate, 9: 358-375.

LI J, CHOU J.1998. Dynamical analysis on splitting of subtropical high-pressure zone—Geostrophic effect[J]. Chinese Sci Bull, 43: 1285-1288.

LI P J.1993. Characteristics of snow cover in western China[J]. Acta Geograph Sin, 48: 505-514.

LINHO, WANG B.2002. The time-space structure of Asian summer monsoon—A fast annual cycle view[J]. J Climate, 15: 2001-2019.

MAO J, CHAN J, WU G.2004. Relationship between the onset of the South China Sea summer monsoon and the structure of the Asian subtropical anticyclone[J]. J Meteor Soc Japan, 82: 845-859.

MEEHL G A.1987. The annual cycle and interannual variability in the tropical Pacific and Indian Ocean regions[J]. Mon Wea Rev, 115: 27-50.

MIYAKODA K, KINTER III J L, YANG S.2003. The role of ENSO in the South Asian monsoon and pre-monsoon signals over the Tibetan Plateau[J]. J Meteor Soc Japan, 81: 1015-1039.

NAMIAS J.1985. Some empirical evidence for the influence of snow cover on temperature and precipitation[J]. Mon Wea Rev, 113: 1542-1553.

NAMIAS J, YUAN X, CAYAN D R.1988. Persistence of North Pacific sea surface temperature and atmospheric flow patterns[J]. J Climate, 1: 682-703.

OSE T.1998. Seasonal change of Asian summer monsoon circulation and its heat source[J]. J Meteor Soc Japan, 76: 1045-1063.

RODWELL M J, HOSKINS B J.1996. Monsoons and the dynamics of deserts[J]. Quart J Roy Meteor Soc, 122: 1385-1404.

SEWELL R D, LANDMAN W A.2001. Indo-Pacific relationships in terms of sea-surface temperature variations[J]. Int J Climatol, 21: 1515-1528.

TANAKA M.1992. Intraseasonal oscillation and the onset and retreat dates of the summer monsoon over the east, southeast and Western North Pacific region using GMS high cloud amount data[J]. J Meteor Soc Japan, 70: 613-629.

TAO S, CHEN L.1987. A review of recent research on the East Asian summer monsoon in China[C]//CHANG C P, KRISHNAMURTI T N. Monsoon Meteorology. Oxford: Oxford University Press: 60-92.

TING M, HELD I M.1990. The stationary wave response to a tropical SST anomaly in an idealized GCM[J]. J Atmos Sci, 47: 2546-2566.

TRENBERTH K E.1990. Recent observed interdecadal climate changes in the Northern Hemisphere[J]. Bull Amer Meteor Soc, 71: 988-993.

WALSH J E, JASPERSON W H, ROSS B.1985. Influences of snow cover and soil moisture on monthly air temperature[J]. Mon Wea Rev, 113: 756-768.

WANG B, LINHO, 2002. Rainy season of the Asian-Pacific summer monsoon[J]. J Climate, 15: 386-398.

WEBSTER P J, YANG S.1992. Monsoon and ENSO: Selectively interactive systems[J]. Quart J Roy Meteor Soc, 118: 877-926.

WEBSTER P J, MAGAÑA V O, PALMER T N, et al.1998. Monsoons: Processes, predictability, and the prospects for prediction[J]. J Geophys Res, 103: 14451-14510.

WU G, ZHANG Y.1998. Tibetan Plateau forcing and the timing of the monsoon onset over South Asia and the South China Sea[J]. Mon Wea Rev, 126(4): 913-927.

WU T, QIAN Z.2003. The relation between the Tibetan winter snow and the Asian summer monsoon and rainfall: An observational investigation[J]. J Climate, 16: 2038-2051.

WU Z X, NEWELL R E.1998. Influence of sea surface temperatures on air temperatures in the tropics[J]. Clim Dyn, 14: 275-290.

XIE A, CHUNG Y S, LIU X, et al.1998. The interannual variations of the summer monsoon onset over the South China Sea[J]. Theor Appl Climatol, 59: 201-213.

XIE P, ARKIN P A.1997. Global precipitation: A 17-year monthly analysis based on gauge observations, satellite estimates and numerical model outputs[J]. Bull Amer Meteor Soc, 78: 2539-2558.

XU J, CHAN J C L, 2001. First transition of the Asian summer monsoon in 1998 and the effect of the Tibet-tropical ocean thermal contrast[J]. J Meteor Soc Japan, 79: 241-253.

YANAI M, LI C, SONG Z, 1992. Seasonal heating of the Tibetan Plateau and its effects on the evolution of the Asian summer monsoon[J]. J Meteor Soc Japan, 70: 319-351.

YANG S, LAU K M, 1996. Precursory signals associated with the interannual variability of the Asian summer monsoon[J]. J Climate, 9: 949-964.

YASUNARI T, KITOH A, TOKIOKA T, et al, 1991. Local and remote responses to excessive snow mass over Eursasia appearing in the northern spring and summer climate—A study with the MRI-GCM[J]. J Meteor Soc Japan, 69: 473-487.

ZHANG Y, LI T, WANG B, et al, 2002. Onset of the summer monsoon over the Indochina Peninsula: Climatology and interannual variations[J]. J Climate, 15: 3206-3221.

Thermal Controls on the Asian Summer Monsoon

WU Guoxiong[1], LIU Yimin[1], HE Bian[1,2], BAO Qing[1], DUAN Anmin[1], JIN Feifei[3]

(1 State Key Laboratory of Numerical Modeling for Atmospheric Sciences and Geophysical Fluid Dynamics, Institute of Atmospheric Physics, Chinese Academy of Sciences, Beijing 100029; 2 Nanjing University of Information Science & Technology Nanjing 210044; 3 Department of Meteorology, University of Hawaii, Honolulu, Hawaii, USA, Hawaii 96822)

Abstract: The Asian summer monsoon affects more than sixty percent of the world's population; understanding its controlling factors is becoming increasingly important due to the expanding human influence on the environment and climate and the need to adapt to global climate change. Various mechanisms have been suggested; however, an overarching paradigm delineating the dominant factors for its generation and strength remains debated. Here we use observation data and numerical experiments to demonstrates that the Asian summer monsoon systems are controlled mainly by thermal forcing whereas large-scale orographically mechanical forcing is not essential: the South Asian monsoon south of 20°N by land-sea thermal contrast, its northern part by the thermal forcing of the Iranian Plateau, and the East Asian monsoon and the eastern part of the South Asian monsoon by the thermal forcing of the Tibetan Plateau.

1 Introduction

The monsoon is generally considered an atmospheric response to seasonal changes in land-sea thermal contrast, induced by the annual cycle of the solar zenith angle (Webster, et al., 1998; Meehl, 1994). The Asian summer monsoon (ASM) is the strongest element of the global monsoon system (Flohn, 1957; Chang, 2004; Wang, 2006). In addition to land-sea contrast, it is affected by large-scale mountain-ranges such as the Tibetan Plateau (TP) (Yeh et al., 1957; Ye et al., 1979; Yanai et al., 1992, 2006), which serve, in winter, as a giant wall across almost the whole Eurasian continent that blocks cold outbreaks from the north and confines the winter monsoon to the eastern and southern Asia (Chang et al., 2006). The air over the TP descends in winter and ascends in summer, driving the surrounding surface air that diverges from the TP in winter and converges toward it in summer, much like a sensible-heat-driven air-pump (SHAP) (Wu,1997;Wu et al., 2007). Because in summer the impinging flow toward the TP is weak, TP-SHAP has been suggested to be dominant over mechanical forcing in controlling the ASM (Yanai et al., 2006; Wu et al., 2007; Held, 1983; Chen, 2001), including its subsystems: the South Asian summer monsoon (SASM) and the East Asian summer monsoon (EASM).

本文原载于 *Scientific Reports*, 2012, 2: 404

Variations in the ASM and its controlling factors are known to be influenced by natural fluctuations (Lau et al., 2006; Li, 2010; Wu et al., 2009) along with anthropogenic greenhouse-gas emissions and environmental pollution (Allan, 2011; Ramanathan et al., 2008). The TP surface sensible heat flux has weakened in recent decades, due mainly to global warming (Duan et al., 2006, 2008). The interception of solar radiation by atmospheric brown clouds leads to surface dimming (Ramanathan et al., 2008). In summer, the greater dimming over land than over adjacent oceans suggests a weakening in land-sea contrast; the dimming trend over the northern Indian Ocean leads to a decrease in local evaporation and less moisture being fed to the monsoonal inflow(Ramanathan et al., 2005), as well as a decrease in the meridional sea-surface-temperature gradient (Ramanathan et al., 2005; Chung et al., 2007). These factors may contribute to the weakening of monsoonal rainfall (Ramanathan et al., 2008). In addition, the darkening of snow and ice due to the deposition of soot reduces surface albedo and enhances solar absorption (Clarke et al., 1985; Warren et al., 1985), and heating of the troposphere over the TP due to increasing amounts of dust and black carbon aerosols can lead to enhanced land-atmosphere warming, which in turn accelerates snow melt and glacier retreat (Lau et al., 2010; Ramana et al., 2004).

Global warming and widespread glacier-mass loss and snow-cover reduction are projected to accelerate throughout the 21st century (Solomon et al., 2007). These changes are expected to have a strong influence on certain regions of the ASM in which TP thermal forcing plays an important role. Many studies have examined the separate influences of individual factors on the ASM (Flohn, 1957; Yeh et al., 1957; Ye et al., 1979; Yanai et al., 1992, 2006; Wu, 1997; Wu et al., 2007); however, their relative and synthetic contributions to the ASM remain unclear, also it remains debated whether the "Himalayas wall" can block dry air from the north and contribute to the formation of the SASM (Boos et al., 2010). In addition the impacts of another Plateau, the Iranian Plateau (IP), lower than the TP but with the same size, has never been paid attention. With the aim of resolving these issues, in the present study we performed numerical experiments to investigate the various factors that control different aspects of the ASM.

2 Results

2.1 Influence of land-sea thermal contrast on the ASM

The influence of land-sea thermal contrast and plateau forcing on the ASM is investigated by employing a general circulation model (GCM), FGOALS/SAMIL (Methods). The model is integrated with prescribed, seasonally varying sea surface temperature (SST) and sea ice. The controlled climate integration is referred to as the CON experiment. The modelled precipitation (Fig. 1a) shows some bias compared with observations (Adler et al., 2003; Kanamitsu et al., 2002) (Fig. 1b), with a stronger Somali jet and enhanced rainfall over most of the ASM domain, but it captures the main features of the ASM, performing reasonably well in simulating the maximum centers over the western coast of India, the Bay of Bengal (BOB), and the southeastern slopes of the TP.

Since the monsoon is traditionally considered an atmospheric response to the seasonal land-sea thermal contrast (Webster, et al., 1998; Meehl, 1994), it is reasonable to infer that the precipitation forced not by orography but by the land-sea distribution alone could be considered as a monsoon prototype. A no-mountain experiment NMT is thus designed, being the same as the CON run except that all the mountains worldwide are removed (Methods Table 1). The modelled precipitation (Fig. 1c) is confined to south of 20°N, with the maximum centers (>18 mmd^{-1}) located between 10°N and 15°N, as in the control run. Remarkable changes compared with the control run are seen in the subtropical area: the SASM north of 20°N and the EASM are substantially reduced. Because our main concern is how the extensive Asian mountains of the IP and TP (IPTP) influence the ASM, in the next experiment only the IPTP is removed. The simulated precipitation pattern (Fig. 1d) is similar to that in NMT (Fig. 1c) and is considered the component of the ASM that is induced by land-sea thermal contrast alone. The experiment is thus termed the L_S experiment (See Methods Experiment Design).

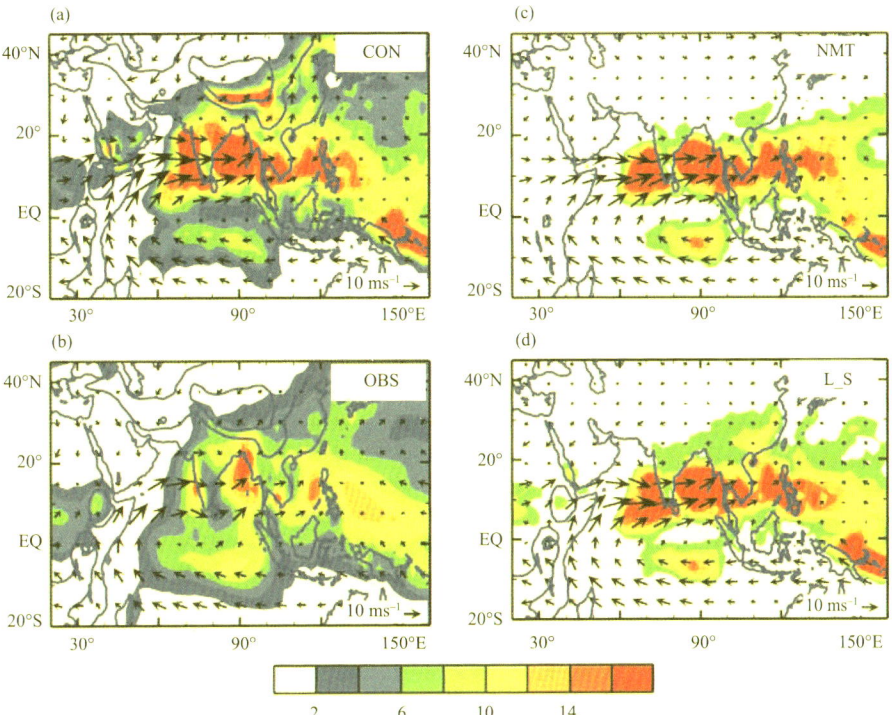

Figure 1　Impacts of land-sea thermal contrast on the Asian summer monsoon, showing the summer precipitation rate (color shading, unit mmd^{-1}) and 850 hPa winds (vectors) for (a) the control experiment CON; (b) observations averaged over the period 1979—2009 from Global Precipitation Climatology Project (GPCP) for precipitation and from NCEP-DEO AMIP-II Reanalysis (R−2) for winds; (c) experiment NMT in which the global surface elevations are set to zero; and (d) experiment L_S in which only the elevations of the Iranian Plateau (IP) and the Tibetan Plateau (TP) are set to zero. Thick contours indicate elevations higher than 1500 m and 3000 m

2.2　Influence of IPTP mechanical insulation on the ASM

The differences (DIFF) in circulation and precipitation between CON (Fig. 1a) and L_S (Fig. 1d), as shown in Fig. 2a, are forced by mechanisms other than land-sea thermal forcing. Such mechanisms are required to (1) produce a cyclonic circulation at 850 hPa over the subtropical continent between 20° and 40°N, circumambulating the IPTP; (2) reduce precipitation over tropical oceans and the northwestern Pacific; and (3) increase precipitation mainly over the Asian continent, with maximum centers over India, the northern BOB, the southern slopes of the TP, and eastern Asia.

The absence of precipitation over northern India in the L_S experiment might reflect the removal of the "IPTP insulator", which results in the southward advection of dry, cold air from the subtropics and a lack of tropical convective instability and rainfall. Were this the case, merely adding the IPTP (but not allowing its surface-sensible-heating to heat the atmosphere) into the L_S experiment, which is defined as the IPTP_M experiment (See Methods Experiment Design), would be sufficient to produce the monsoon rainfall in the northern South Asia. However, the results in Fig. 2b indicate that this is not the case. In the IPTP_M experiment, the patterns of both precipitation and circulation at 850 hPa are similar to those in the L_S experiment. Similarly, if we merely add IP and TP separately into the L_S experiment (i.e., the IP_M and TP_M experiments, respectively), the resultant precipitation and circulation distributions (Figs. 2c and 2d, respectively) are also similar to those in the L_S experiment. These results demonstrate that in summer, mechanical insulation of the IP and TP has a minor influence on the generation of the ASM, as it cannot produce the required compensating rainfall and precipitation patterns (Fig. 2a).

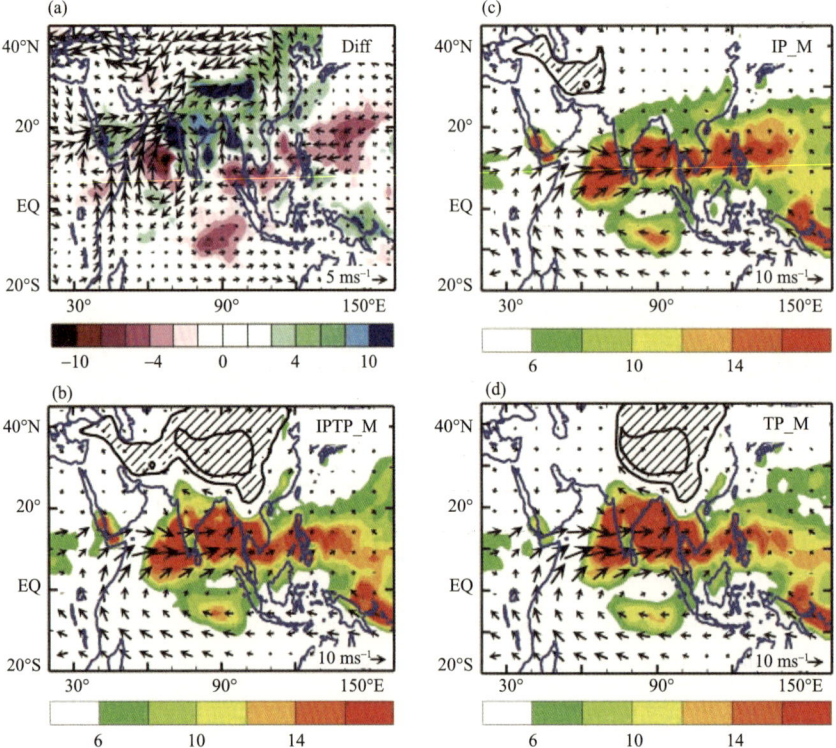

Figure 2 Impacts of mountain mechanical forcing on the Asian summer monsoon, showing the summer precipitation rate (color shading, unit mmd^{-1}) and 850 hPa winds (vectors) for (a) the difference (DIFF) between the CON and L_S experiments, indicating the compensating rainfall and circulation required to make up the total monsoon; (b) experiment IPTP_M in which the IP and TP mechanical forcing exists; (c) experiment IP_M in which the IP's mechanical forcing exists; and (d) experiment TP_M in which the TP's mechanical forcing exists. Thick black contours surrounding grey-hatched regions indicate elevations higher than 1500 m and 3000 m

2.3 Influence of IPTP thermal forcing on the ASM

Three sets of experiments were designed with surface sensible heating on the IP (IP_SH), TP (TP_SH), and IPTP (IPTP_SH) (Methods Experiment Design), in order to study the influence of orographically elevated thermal forcing on the ASM. In IP_SH (Fig. 3a), the IP thermal forcing generates a cyclonic circulation encircling the IP, similar to the western parts of the compensating circulation in Fig. 2a. The forcing also results in reduced precipitation, mainly over the tropical Indian Ocean and the northwest Pacific, and increased precipitation over the Asian continent west of 100°E (especially over Pakistan, northern India, and the southwestern slopes of the TP), a pattern similar to that of the compensating precipitation west of 100°E, indicating the important role of the IP in generating the northern SASM.

In the TP_SH experiment (Fig. 3b), TP thermal forcing also generates a cyclonic circulation encircling the TP. Correspondingly, reduced precipitation occurs west of 80°E; in contrast, increased precipitation occurs east of 80°E, especially over the BOB, the southern slopes of the TP, and East Asia. They are similar to the compensating precipitation and circulation patterns in the region east of 80°E (Fig. 2a), indicating that TP thermal forcing plays a dominant role in the generation of the EASM and the eastern part of the SASM.

In the IPTP_SH experiment (Fig. 3c), the elevated IPTP heating results in reduced precipitation in tropical oceans, and increased precipitation over the Asian continent to the north. The heating also generates a cyclonic circulation at 850 hPa over the Asian subtropical continental areas, with relatively isolated centers over the IP and TP. The results shown in Fig. 3c are basically equivalent to the linear addition of the results in Figs. 3a and 3b, indicating the important but contrasting roles of IP and TP thermal forcing in different parts of the ASM. More significantly, the precipitation and circulation patterns generated by IPTP thermal forcing (Fig. 3c) are close to those required to compensate the ASM (Fig. 2a). This result demonstrates that in addition to land-sea thermal contrast, the thermal forcing of large mountain ranges in Asia is an important factor in producing the ASM, especially over continental areas.

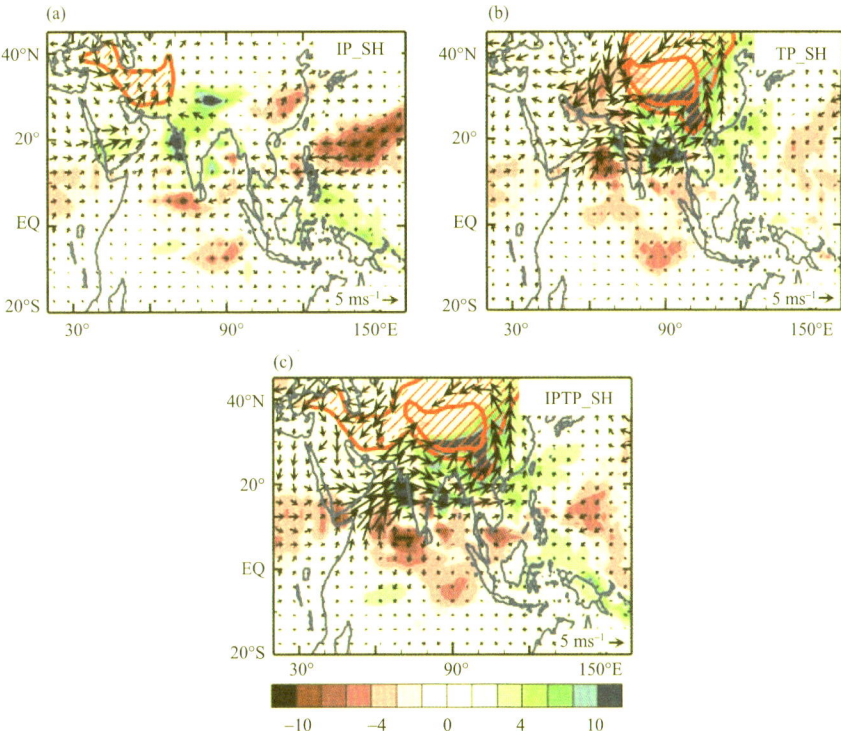

Figure 3 Impacts of mountain thermal forcing on the Asian summer monsoon, showing the summer precipitation rate (color shading, unit mmd^{-1}) and 850 hPa winds (vectors) generated due to the elevated surface sensible heating of (a) the Iranian Plateau (IP_SH); (b) the Tibetan Plateau (TP_SH); and (c) the IP and TP (IPTP_SH). Thick red contours surrounding red-hatched regions indicate elevations higher than 1500 m and 3000 m

2.4 Influence of climbing versus deflecting orographic effects on the ASM

More than 85% of the total atmospheric water vapor, as measured by specific humidity, generally resides in a layer below 3 km above sea level. In order for monsoon clouds and precipitation to form, lower-tropospheric water vapor must be lifted by vertical motions forced either internally or externally; consequently, high near-surface moist entropy and warm upper-level temperature are coupled (Emanuel et al., 1994). One of the internal forcing is the type of cold/warm fronts (Holton, 2004). This mechanism is important at middle and high latitudes, especially in winter, but is not important in the tropics in summer because the air temperature in the tropics is relatively uniform.

The mechanical forcing of mountains is an important external forcing: air flow impinging upon mountains is either deflected to produce encircling flow or lifted to produce climbing flow (Queney, 1948; Wu, 2004). Consequently, clouds and precipitation are generated around mountains. However, if a mountain is higher than several hundred meters, the conservation constraint of angular momentum and energy means that the airflow passes around the mountain rather than rising over it (Wu, 1984).

Thermal forcing can also generate atmospheric ascent, because large-scale atmospheric potential temperature (θ) increases with height. According to the steady-state thermodynamic equation

$$\mathbf{V} \cdot \nabla \theta = Q \tag{1}$$

where $\mathbf{V}$ is air velocity, in regions of heating ($Q>0$), air should penetrate isentropic surfaces upward. There are several types of atmospheric heating. Shortwave radiation is weakly absorbed directly by the atmosphere. In the absence of cloud, longwave radiation can easily escape into space. Condensation heating normally occurs above the cloud base. Whereas surface sensible heating can increase the near-surface entropy, result in the development of convective instability and trigger atmospheric ascent; and is effective in generating atmospheric ascent in the lower troposphere.

If surface sensible heating occurs on a mountain slope, and if the mountain is high enough, large amounts of moisture in lower layers are readily transported to the free atmosphere (Wu et al., 2007). The TP in summer is a heat source for the atmosphere and has a strong influence on weather and climate (Yanai et al., 2006; Wu, 1997; Wu et al., 2007; Yanai et al., 1994). When a moist and warm southwesterly approaches the TP, the air becomes heated, starts to penetrate isentropic surfaces, and slides upward along its sloping surface.

Figure 4 shows the distribution of precipitation and streamlines at the $\sigma = 0.89$ surface, which is about 1 km from the surface. In the CON experiment (Fig. 4a), when the water conveyer belt originating from the Southern Hemisphere meanders eastward through the South Asian subcontinent, the effects of land-sea thermal forcing mean that severe precipitation centers are formed along 15°N. The rest of the water vapor is transported to sustain the East Asian Monsoon, although some swerves northward over northern India and the BOB. The pumping effect of TP-SHAP results in the convergence of air toward the TP. The upward streamlines are subperpendicular to the TP contours, eventually forming a cyclonic circulation at the southeastern corner of the TP. Consequently, heavy monsoon rainfall occurs over northern India and western China, with a maximum center (>18 mmd^{-1}) appearing over the southeastern slopes of the TP. The condensation heating of this rainfall center generates cyclonic circulation in the lower layer and further intensifies the EASM, implying a positive feedback between precipitation and circulation (Eady, 1950).

In the IPTP_M experiment (Fig. 4b), in contrast, when the water vapor flux from the main water conveyer belt approaches the TP, it is not heated and the airflow remains at the same isentropic surface

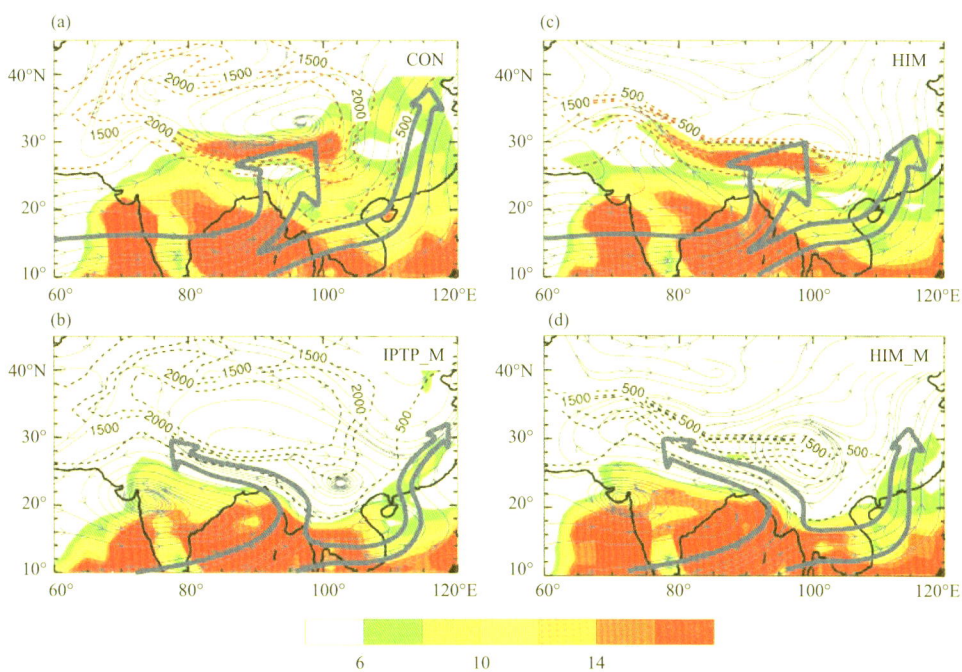

Figure 4 Mechanism Relative contributions of the climbing and deflecting effects of mountains, showing the summer precipitation rate (color shading, unit: mmd^{-1}) and streamlines at the $\sigma = 0.89$ level for (a) the CON experiment; (b) the IPTP_M experiment; (c) the HIM experiment; and (d) the HIM_M experiment. Dashed contours surround elevations higher than 1500 m and 3000 m, with red and black colors respectively indicating with and without surface sensible heating of the mountains. Dark blue open arrows denote the main atmospheric flows impinging on the TP, either climbing up the plateau (a and c) or moving around the plateau, parallel to orographic contours (b and d)

(Equation (1)). Consequently, the streamlines do not climb up the TP; instead, they move around the mountains, parallel to orographic contours. Thus, no monsoon develops over northern India and the TP, and the EASM is substantially weakened. These results indicate that the thermal forcing of large-scale mountains plays a dominant role in the generation of the northern and eastern parts of the SASM and the EASM.

A recent study based on numerical sensitivity experiments (Boos et al., 2010) emphasized the importance of the insulation effect of the Himalayas in producing the SASM. This finding is questionable because the experiment design took into account both the mechanical effect and surface sensible heating of the Himalayas and adjacent mountains (Fig. 4c); thus, the TP-SHAP mechanism still operates for the SASM. To demonstrate this point, we employed the current GCM and designed a sensitivity experiment (HIM) in which we followed Boos and Kuang's approach exactly in considering the effect of the Himalayas on the SASM (Methods Experiment Design), as indicated by the polygon in Fig. 4c. In addition to this tracking experiment, another experiment (HIM_M) was performed, which was the same as the HIM experiment except that the Himalayas' surface-sensible-heating was not allowed to heat the atmosphere (Fig. 4d). The results show that the HIM experiment (Fig. 4c) yields a similar SASM to that in the CON experiment (Fig. 4a), the same consequence as in Boos and Kuang. However, in the HIM_M case (Fig. 4d), the impinging tropical flow cannot climb up the TP; instead, it splits into eastern and western branches that each move around the TP, subparallel to orographic contours. The western branch forms a return flow, causing a southward shift in the Indian monsoon trough from its usual location over northeastern India to over central India. In this situation, the part of the SASM over northeastern India disappears, and the EASM is markedly weakened. These results demonstrate that the insulation effect of the TP is insignificant in terms of the ASM. Instead, it is the thermal forcing of the Tibetan/Iranian Plateau that plays a dominant role in the generation of the northern part of the SASM and the EASM.

2.5 Structure of the SASM

A striking feature of the present experiments is the insensitivity of the southern part of the SASM to IPTP forcing: in all experiments (Figs. 1 and 2), the intensity and spatial distribution of precipitation south of 20°N show little change compared with the control while the configuration or thermal status of the Tibetan/Iranian Plateau shows a marked change. Figures 5a and 5b show 80°–90°E longitudinally averaged latitude-height cross sections from the CON and IPTP_M experiments, demonstrating that the vertical velocity is divided into a southern branch and a northern branch at about 25°N. In the CON run (Fig. 5a), strong rising associated with the southern SASM is located over the northern Indian Ocean. Ascending air is also dominant above the TP, with maxima located near the surface, indicating the importance of surface sensible heating in generating orographic ascent. In the IPTP_M experiment (Fig. 5b), the lack of surface heating on the TP results in two remarkable sinking centers over its slopes; thus, the northern branch of the SASM disappears over northern India. However, the intensity and location of the southern branch is largely unchanged. In fact, in all the experiments the southern SASM branch remains steady, with a center ($>18\times10^{-2}$ Pa s^{-1}) at about 400 hPa, locked to the south of the coastline. The insensitivity of the southern branch of the SASM to orographic change indicates that the land-sea thermal contrast plays a dominant role in its generation and variation.

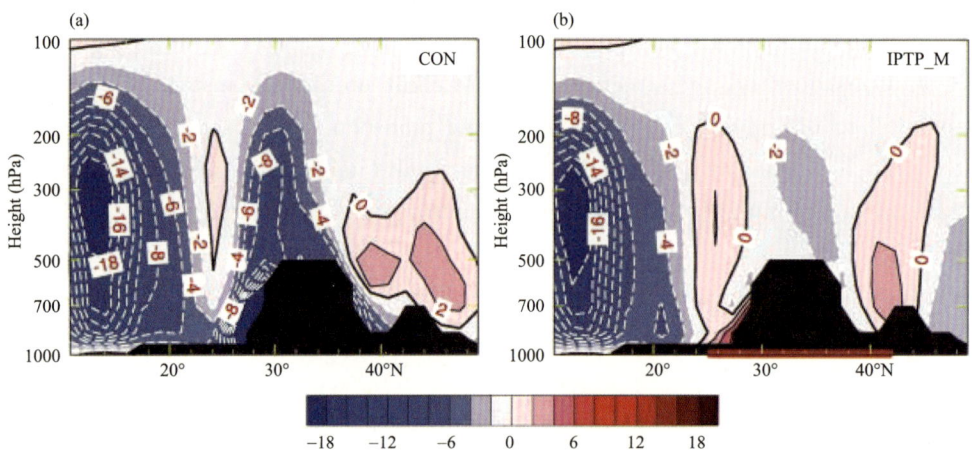

Figure 5 Structure of the South Asian summer monsoon, showing 80°–90°E longitudinally averaged vertical-meridional cross-sections of pressure vertical velocity (contour interval, 2×10^{-2} Pa s^{-1}) for experiments (a) CON; and (b) IPTP_M

The above discussion is summarized schematically in Figure 6. The meridional circulation of the SASM can be divided into southern and northern branches. Its southern branch is located in the tropics: water vapor that originates from the Southern Hemisphere and is transported along the zonally oriented "water-vapor conveyer belt" is lifted upward due to the land-sea thermal contrast, forming monsoon precipitation there. The northern branch occurs along the southern margin of the IPTP in the subtropics. When the conveyer belt approaches the TP, part of its water vapor is hauled away and turned northward, then lifted upward by the IPTP-SHAP, resulting in heavy precipitation in the monsoon trough over northern India and along the foothills and slopes of the TP. The rest of the water vapor along the water conveyer belt is transported northeastward to sustain the EASM, which is controlled by the land-sea thermal contrast and thermal forcing of the TP. These results highlight the dominant roles of the land-sea thermal contrast and IPTP thermal forcing in influencing the ASM.

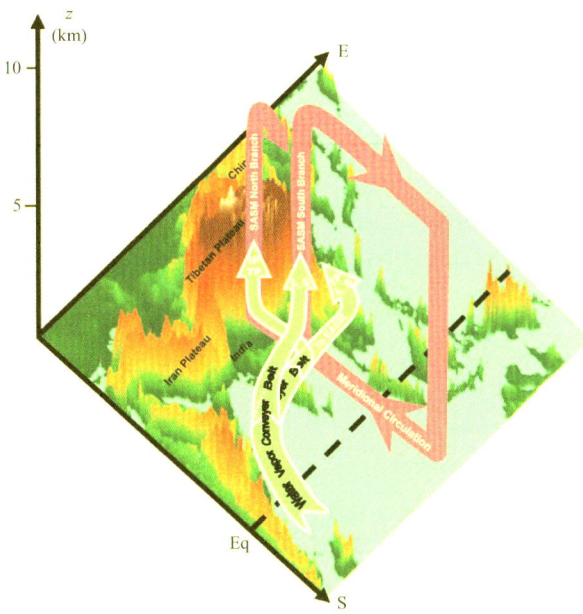

Figure 6 Schematic diagram showing the gross structure of the South Asian summer monsoon. For the southern branch, water vapor along the conveyer belt is lifted up due mainly to land-sea (L_S) thermal contrast in the tropics; for the northern branch the water vapor is drawn away from the conveyer belt northward toward the foothills and slopes of the TP, and is uplifted to produce heavy precipitation that is controlled mainly by IPTP-SHAP; the rest of the water vapor is transported northeastward to sustain the East Asian summer monsoon, which is controlled by the land-sea thermal contrast as well as thermal forcing of the TP

3 Discussion

In nature, the influences of the IPTP orography and its surface sensible heating cannot be separated. The significance of the dominance of IPTP thermal forcing in influencing the ASM lies in the fact that, over the modern-day orography, the thermal status of the IPTP varies due to natural and anthropogenic factors, as reviewed above. By focusing on changes in the thermal status of the IPTP, the dominance of thermal controls on the ASM may provide us with a tangible way of identifying climate trends in the Asian summer monsoon in a warming world, and of improving weather forecasts, climate predictions, and projections in areas affected by the Asian monsoon.

4 Methods

4.1 General Circulation Model Description

The atmospheric general circulation model (GCM) employed in the present study is SAMIL (Spectral Atmospheric Model of IAP/LASG), as developed at the State Key Laboratory of Numerical Modeling for Atmospheric Sciences and Geophysical Fluid Dynamics/Institute of Atmospheric Physics (LASG/IAP), Beijing, China. SAMIL is a spectral model with rhomboidal truncation at wave-number 42. It has 26 vertical layers with a top at 2.1941 hPa (Wu et al., 2003). As an atmospheric component of the climate system model FGOALS-s developed at LASG/IAP, SAMIL is driven by the sixth generation of NCAR Coupler using NCAR CLM3 as a land surface component (Bao et al., 2010). Its physical parameterizations include SES2 (Sun, 2005), the updated radiation parametric scheme developed by Edwards et al. (1996), diagnostic cloud parameterization, the mass flux cumulus parameterization of Tiedtke (1989) for deep, shallow, and mid-level convection, and a nonlocal PBL parameterization scheme (Holtslag et al., 1993). Observed mean monthly sea surface temperature (SST) and sea ice with a seasonal cycle, as obtained from AMIP II (Taylor et al., 2000), were prescribed. All experiments were integrated for 7 years, and the monsoon climate was calculated from the last five June—August periods.

4.2 Experiment Design

The model was integrated with prescribed, seasonally varying SST and sea ice to form a controlled-climate integration first, which is referred to as the CON experiment.

The orographic setting in each experiment was achieved by changing only the prescribed surface elevations. The no-surface-sensible-heating experiments were designed as follows: While the surface energy balance was kept unchanged, the surface sensible heat released at the elevation above 500 m was not allowed to heat the atmosphere; i.e., the vertical diffusive heating term in the atmospheric thermodynamic equation was set to zero. Details of the design for each experiment are given in Table 1 and explained below.

Table 1 Experiment design for examination of the dominant control of land-sea thermal contrast, as well as orographically mechanical insulation and elevated thermal impacts on the various parts of the Asian summer monsoon. The sign ($\sqrt{}$) indicates that the corresponding element has been included in the related experiment

Experiment Abbreviation	Land-sea thermal contrast	Orographically mechanical insulation		Orographically elevated thermal control	
		Iran Plateau (IP)	Tibet Plateau (TP)	Iran Plateau (IP)	Tibet Plateau (TP)
CON	√	√	√	√	√
NMT	√				
L_S	√				
IPTP_M	√	√	√		
IP_M	√	√			
TP_M	√		√		
IPTP_SH				√	√
IP_SH				√	
TP_SH					√
HIM	√	√	Southern slope only	√	√
HIM_M	√	√	Southern slope only		

• Influence of Land-Sea Thermal Contrast on the ASM

For experiments NMT and L_S (see Fig. 1): the elevation over the whole globe is set to zero in NMT. Since the main purpose of this study is to compare the relative contributions of land-sea thermal contrast and large-scale Asian mountain ranges on different parts of the ASM, in the L_S experiment only the elevations of the Iranian Plateau and Tibetan Plateau are set to zero.

• Influence of IPTP Mechanical Insulation on ASM

For the mechanical forcing experiments IPTP_M, IP_M, and TP_M (see Fig. 2), the Iranian Plateau and Tibetan Plateau together, the Iranian Plateau alone, and the Tibetan Plateau alone are retained from the CON experiment, respectively, but their surface sensible heating is not allowed to warm the atmosphere.

• Influence of IPTP Thermal Forcing on the ASM

In designing the thermal forcing experiments IP_SH, TP_SH, and IPTP_SH (see Fig. 3), three no-surface-sensible-heating experiments (IP_NS, TP_NS, and IPTP_NS) were first designed and integrated, in which the global orographic distribution is the same as in the CON experiment while the surface sensible heating on the atmosphere over the IP, TP, and IPTP, respectively, was removed from the CON experiment. Because the only difference between the CON and IP-NS (TP_NS, IPTP_NS) experiments is whether the elevated surface sensible heating on the atmosphere occurs over the IP (TP, IPTP), their integration differences in Fig. 3a (Figs. 3b, 3c) are qualitatively considered as the thermal impact of the IP (TP, IPTP) on the ASM, and are assigned as the IP_SH (TP_SH, IPTP_SH) experiment.

In designing the HIM and HIM_NS experiments(see Fig. 4), we followed exactly the original work of Boos and Kuang. The HIM experiment is the same as CON except for "removing the bulk of the Tibetan Plateau while preserving a zonal elongated band of mountains to the south, east and west of the plateau," as denoted by the polygon in Figs. 4c and 4d. The HIM_NS experiment is the same as the HIM, except that the surface sensible heat flux on the HIM mountain surfaces is not allowed to heat the atmosphere.

References

ADLER R F, HUFFMAN G J, CHANG A, et al.2003. The version 2 global precipitation climatology project (GPCP) monthly precipitation analysis (1979—present) [J]. J Hydrometeor, 4: 1147.

ALLAN R P.2011. Human influence on rainfall[J]. Nature, 470: 345.

BAO Q, WU G, LIU Y, et al.2010. An introduction to the coupled model FGOALS1.1-s and its performance in East Asia [J]. Adv Atmos Sci, 27: 1131-1142.

BOOS W R, KUANG Z M.2010. Dominant control of the South Asian monsoon by orographic insulation versus plateau heating[J]. Nature, 463: 218-223.

CHANG C P.2004. East Asian Monsoon[M]. New Jersey: World Scientific.

CHANG C P,WANG Z, HENDON H.2006. The Asian winter monsoon[M]// WANG B. The Asian Monsoon. Chichester: Springer: Chapter 3.

CHEN P. 2001. Thermally forced stationary waves in a quasigeostrophic system[J]. J Atmos Sci, 58: 1585-1594.

CHUNG C, RAMANATHAN V.2007. Relationship between trends in land precipitation and tropical SST gradient[J]. Geophys Res Lett, 34(16):16809.

CLARKE A, NOONE K.1985. Soot in the Arctic: A cause for perturbation in radiative transfer[J]. J Geophys Res, 19: 2045-2053.

DUAN A M, WU G X, ZHANG Q.2006. New proofs of the recent climate warming over the Tibetan Plateau as a result of the increasing greenhouse gases emissions[J]. Chinese Sci Bulletin, 51 (11): 1396-1400.

DUAN A M, WU G X.2008. Weakening trend in the atmospheric heat source over the Tibetan Plateau during recent decades. Part Ⅰ: Observations[J]. J Climate, 21: 3149-3164.

EADY E T,1950. The cause of the general circulation of the atmosphere[J]. Centen Proc Roy Meteor Soc: 156-172.

EDWARDS J M, SLINGO A.1996. Studies with a flexible new radiation code. Ⅰ: Choosing a configuration for a large-scale model[J]. Quart J Roy Meteor Soc, 122: 689-720.

EMANUEL K A, NEELIN J D, BRETHERTON C S, 1994. On large-scale circulations in convecting atmospheres[J]. Quart J Roy Meteor Soc, 120: 1111-1143.

FLOHN H, 1957. Large-scale aspects of the "summer monsoon" in South and East Asia[J]. J Meteor Soc Japan Ser II, 35A: 180-186.

HELD I M, 1983. Stationary and quasi-stationary eddies in the extra-tropical troposphere: Theory[M]// HOSKINS B J, PEARCE R P. Large-Scale Dynamical Processes in the Atmosphere. London: Academic Press.

HOLTON J R, 2004. An Introduction to Dynamic Meteorology[M]. Amsterdam : Elsevier Academic Press.

HOLTSLAG A A M, BOVILLE B A, 1993. Local versus nonlocal boundary-layer diffusion in a global climate model[J]. J Climate, 6: 1825-1842.

KANAMITSU M, EBISUZAKI W, WOOLLEN J, et al, 2002. NCEP-DOE AMIP-II reanalysis (R-2)[J]. Bull Amer Meteor Soc, 83: 1631-1643.

LAU N C, WANG B, 2006. Interactions between the Asian monsoon and the El Nino/Southern Oscillation[M]// WANG B. The Asian Monsoon. Chichester: Springer: Chapter 12.

LAU W K M, KIM M K, KIM K M, et al, 2010. Enhanced surface warming and accelerated snow melt in the Himalayas and Tibetan Plateau induced by absorbing aerosols[J]. Environ Res Lett, 5: 025204.

LI T, 2010. Monsoon climate variabilities[M]// SUN D Z, FRANK B. Climate Dynamics: Why Does Climate Vary. Geophys Monogr Ser.

MEEHL G A, 1994. Coupled ocean-atmosphere-land processes and South Asian monsoon variability[J]. Science, 265: 263-267.

QUENEY P, 1948. The problem of air flow over mountain: a summary of the theoretical studies[J]. Bull Amer Meteor Soc, 29: 16-26.

RAMANA M V, RAMANATHAN V, PODGORNY I A, et al, 2004. The direct observations of large aerosol radiative forcing in the Himalayan region[J]. Geophys Res Lett, 31: L05111.

RAMANATHAN V, CHUNG C, KIM D, et al, 2005. Atmospheric brown clouds: Impacts on South Asian climate and hydrologic cycle[J]. Proc Natl Acad Sci USA, 102: 5326-5333.

RAMANATHAN V, CARMICHAEL G, 2008. Global and regional climate changes due to black carbon[J]. Nature Geosci, 1: 221-227. https://doi.org/10.1038/ngeo156.

SOLOMON S D, QIN D, MANNING M, et al, 2007. Contribution of working group I to the fourth assessment report of the intergovernmental panel on climate change [M]// Climate Change 2007: The Physical Science Basis. Cambridge: Cambridge Univ Press.

SUN Z, 2005. Parameterizations of radiation and cloud optical properties[R]. BMRC Research Report: 107-112.

TAYLOR K E, WILLIAMSON D, ZWIERS F, 2000. The sea surface temperature and sea-ice concentration boundary conditions for AMIP II simulations [R]. Congress of the European Pain Federation: PCMDI Report No. 60, 25.

TIEDTKE M A, 1989. Comprehensive mass flux scheme for cumulus parameterization in large-scale models[J]. Mon Wea Rev, 117: 1779-1800.

WANG B, 2006. The Asian Monsoon[M]. Berlin Heidelberg: Springer.

WARREN S, WISCOMBE W, 1985. Dirty snow after nuclear war[J]. Nature, 313: 467-470.

WEBSTER P J, MAGAÑA V O, PALMER T N, et al, 1998. Monsoons: Processes, predictability, and the prospects for prediction[J]. J Geophys Res, 103: 14451-14510.

WU G X, 1984. The nonlinear response of the atmosphere to large-scale mechanical and thermal forcing[J]. J Atmos Sci, 41: 2456-2476.

WU G X, 1997. Sensible heat driven air-pump over the Tibetan Plateau and its impacts on the Asian summer monsoon[C]// YE D Z. Collections on the Memory of Zhao Jiuzhang. Beijing: Science Press.

WU G X, 2004. Recent progress in the study of the Qinghai-Xizang plateau climate dynamics in China[J]. Quarternary Sci, 24: 1-9.

WU G, LIU Y, DUAN A, et al, 2007. The Influence of the mechanical and thermal forcing of the Tibetan Plateau on the Asian climate[J]. J Hydrometeor, 8: 770-789.

WU G X, LIU Y, ZHU X, et al. 2009. Multi-scale forcing and the formation of subtropical desert and monsoon[J]. Ann Geophys, 27: 3631-3644.

WU T W, LIU P, WANG Z Z, et al. 2003. The performance of atmospheric component model R42L9 of GOALS/LASG[J]. Adv Atmos Sci, 20: 726-742.

YANAI M, LI C, SONG Z. 1992. Seasonal heating of the Tibetan Plateau and its effects on the evolution of the Asian summer monsoon[J]. J Meteor Soc Japan, 70: 319-351.

YANAI M, LI C. 1994. Mechanism of heating and the boundary layer over the Tibetan Plateau[J]. Mon Wea Rev, 122: 305-323.

YANAI M, WU G X. 2006. Effects of the Tibetan Plateau[M]. Berlin Heidelberg: Springer.

YEH T C, LO S W, CHU P C. 1957. The wind structure and heat balance in the lower troposphere over Tibetan Plateau and its surroundings[J]. Acta Meteor Sinica, 28: 108-121.

YE D Z, GAO Y X. 1979. Meteorology of the Qinghai-Xizang Plateau[M]. Beijing: Science Press.

Roles of Forced and Inertially Unstable Convection Development in the Onset Process of Indian Summer Monsoon

WU Guoxiong[1], LIU Boqi[1,2]

(1 State Key Laboratory of Numerical Modeling for Atmospheric Sciences and Geophysical Fluid Dynamics (LASG), Institute of Atmospheric Physics, Chinese Academy of Sciences, Beijing 100029, China; 2 Key Laboratory of Meteorological Disaster of Ministry of Education (KLME), Nanjing University of Information Science and Technology, Nanjing 210044 China)

Abstract: The NCEP/NCAR R1 reanalysis data are employed to investigate the impact of forced and inertial instability in the lower troposphere over the Arabian Sea on the onset process of Indian summer monsoon (ISM), and to reveal the important role of zonal advection of zonal geostrophic momentum played in the forced unstable convection. Results show that during the ISM onset the zero absolute vorticity contour ($\eta = 0$) shifts northward due to the strong cross-equatorial pressure gradient in the lower troposphere over southern Arabian Sea. Thus a region with negative absolute vorticity is generated near the equator in the Northern Hemisphere, manifesting the evident free inertial instability. When a southerly passes through this region, under the influence of friction a lower convergence that facilitates the convection flourishing at the lower latitudes appears to the north of zero absolute vorticity contour. However, owing to such a traditional inertial instability, the convection is confined near the equator which does not have direct influence on the ISM onset. On the contrary in the region to the north of the zero absolute vorticity contour and to the south of the low pressure center near the surface, although the atmosphere there is inertially stable, the lower westerly jet can develop and bring on the apparent zonal advection of zonal geostrophic momentum. Both theoretical study and diagnosing analysis present that such a zonal advection of geostrophic momentum is closely associated with the zonal asymmetric distribution of meridional land-sea thermal contrast, which induces a convergence center near and further north of the westerly jet in the lower troposphere over the southwestern coast of the Indian Peninsula, providing a favorable lower circulation for the ISM onset. It illustrates that the development of convection over the Arabian Sea in late spring and early summer is not only due to the frictional inertial instability but also strongly affected by the zonal asymmetric distribution of land-sea thermal contrast. Moreover, before the ISM onset due to the eastward development of the South Asian high (SAH) in the upper troposphere, high potential vorticity is transported to the region over the Arabian Sea.

本文原载于 Science China Earth Sciences, 2014, 57: 1438-1451

Then a local trumpet-shaped stream field is generated to cause the evident upper divergence-pumping effect which favors the ISM onset. When the upper divergence is vertically coupled with the lower convergence resulted from the aforementioned forced unstable convection development near the southwestern coast of Indian Peninsula, the atmospheric baroclinic unstable development is stimulated and the ISM onset is triggered.

Key words: forced convection development, Indian summer monsoon onset, zonal advection of zonal geostrophic momentum, South Asian high

The Asian monsoon, including the South Asian monsoon and East Asian monsoon, is the strongest monsoon system in the world. The Asian monsoon can be divided into the tropical monsoon, subtropical monsoon, and temperate monsoon according to the climate belt. The tropical monsoon is composed of the Indian monsoon, the Bay of Bengal (BOB) monsoon, the South China Sea (SCS) monsoon, and the northwestern Pacific monsoon (Lau et al., 2009). The onset and evolution of Asian summer monsoon is complicated and spectacular. The tropical summer monsoon builds up from early May over the BOB to early June over India, lasting for about 40 days. But the East Asian summer monsoon takes about 50 days, starting from South China in early June and subsequently propagating to the North China in middle July when the Meiyu along the Yangtze River basin ends. The Indian summer monsoon (ISM) is one of the most important members in the Asian summer monsoon region. Its onset and intra-seasonal oscillation are significant for the variability and forecast of both Meiyu over the Yangtze River valley and precipitation over North China in boreal summer (Liu et al., 2008a, 2008b). Therefore, it is significant to understand the ISM onset process and the relevant mechanism correctly for improving climate prediction and disaster prevention and mitigation in China.

The climatological onset time of ISM is around the end of May and the beginning of June. The ISM onset dates are similar, determined by different criteria with different meteorological elements, such as rainfall, wind and moisture, etc (Ananthakrishnan et al., 1968; Prasad et al., 2005; Taniguchi et al., 2006; Wang et al., 2009). Recently Mao et al. (2004, 2007) used the seasonal overturns of the meridional temperature gradient (MTG) in the upper troposphere to define the onset dates of Asian summer monsoon. This definition is useful for studying the interaction among different subsystems of the Asian summer monsoon. It was reported that the seasonal northward migration of the intertropical convergence zone (ITCZ) is associated with the ISM onset featured by the advancing of monsoon depression and rainfall from the equator to the north (Saha et al., 1980; Gadgil, 2003). In detail, the precipitation in the monsoon region (70°—110°E, 10°—30°N) increases with the strengthened cross-equatorial Somali flow, which accelerates the lower-level jet and enhances the southwesterly over the Arabian Sea. Simultaneously the ITCZ is moving northward, and the ISM trough is established (Saha, 2010). Thus the seasonal northward shifting of ITCZ is of great importance for the ISM onset and advancement. The numerical simulation performed by Krishnamurti et al. (1983) showed that in the boundary layer of the ISM region, the geostrophic balance and advection balance (i.e., air parcel acceleration) existed to the north and the south of ITCZ, respectively. Such acceleration of cross-equatorial Asian summer monsoon circulation over the southwestern Indian Peninsula may be linked with the inertial instability. Tomas et al. (1997) declared that the northward propagation of ITCZ over the ISM region is intimately correlated with the seasonal northward migration of near-equatorial zero absolute vorticity contour ($\eta = 0$) in the lower troposphere. It was presented that the negative absolute vorticity belt ($\eta < 0$) near the equator in the North Hemisphere (NH) reflects the inertial instability in the planet boundary layer (PBL). They also suggested that the criterion of inertial instability is $f \cdot \eta < 0$, in which f is the Coriolis parameter. If an air parcel passes northward through the inertially unstable ($f \cdot \eta < 0$) region, it will be accelerated under the inertial effect. When the air parcel passes through the inertial stable ($f \cdot \eta > 0$) area, it will be decelerated, forming a lower level convergence to the northern neighbor of the $\eta = 0$ contour. The northward migration of the zero absolute vorticity contour is caused by the increased cross-equatorial pressure gradient force associated with the intensified meridional gradient of sea surface pressure (SLP). Furthermore, Tomas et al. (1999) adopted an analytic method to demonstrate the possible dynamical mechanism for the near-equatorial inertial instability in boreal summer. They stressed the important influence of PBL drag coefficient (frictional effect) on the inertial instability, and successfully simulated the inertial instability and relative elements in the lower troposphere over the eastern Pacific and Atlantic Ocean in July by a simple one-dimensional PBL model. However, the simulation over the Indian Ocean is distinctly different from the observation. Then what's the reason for such a discrepancy? Dose the distribution of inertial instability in summer resemble that in spring? Given the close association between inertial instability and seasonal northward displacement of ITCZ in spring, is the theory of inertial instability applicable to the ISM onset from late spring to early summer in the NH? This paper aims to solve these scientific questions.

1 Data and methodology

1.1 Data description

The NCEP/NCAR R1 daily reanalysis dataset (1979—2009) with horizontal resolution of 2.5°× 2.5° is utilized in this paper, including the wind, air temperature, geopotential height, specific humidity, and SLP fields (Kalnay et al., 1996). The wind, air temperature, and geopotential height fields are distributed on the 12 standard isobars from 1000 hPa to 100 hPa, whereas the specific humidity field is on the 8 standard isobars from 1000 hPa to 300 hPa. The daily outgoing longwave radiation (OLR) from 1979 to 2009 is supplied by NOAA (Liebmann and Smith, 1996). The daily precipitation observed by satellites (1997—2009) is from the Tropical Rainfall Measuring Mission (TRMM) offered by NASA (Huffman et al., 1995). The heat flux on the sea surface is attained from OAFlux datasets supported by the Woods Hole Oceanographic Institution (WHOI, Yu et al., 2007).

1.2 Definition of ISM onset date

Following Mao et al. (2004), the ISM onset date is identified as the day when the following criteria are satisfied: (1) the area-averaged upper-tropospheric (500—200 hPa) MTG over the ISM region (60° —85°E, 10°—20°N) changes from negative to positive and (2) the MTG remains positive for more than 10 days. The ISM onset date defined by MTG criteria (Table 1) are similar to that by other methods (Prasad et al., 2005; Taniguchi et al., 2006; Xavier et al., 2007; Wang et al., 2009). To investigate the climatic features of inertial instability during the ISM onset accurately, the composite techniques based on the ISM onset date are adopted to remove the marked interannual variability of the ISM onset. For each individual year the onset date is defined as the zero date (D_0), with the dates before (after) the onset date being labeled as negative (positive) days. All the variables are chosen from -30 days (D_{-30}) to $+30$ days (D_{+30}) relative to the ISM onset date each year. The following analysis is all based on the composite results.

Table 1 ISM onset date

Year	Onset date	Year	Onset date
1979	Jun-05	1995	Jun-01
1980	Jun-01	1996	May-25
1981	May-27	1997	Jun-06
1982	May-30	1998	Jun-06
1983	Jun-13	1999	May-19
1984	May-24	2000	May-14
1985	May-23	2001	May-22
1986	Jun-01	2002	Jun-02
1987	Jun-03	2003	Jun-01
1988	May-19	2004	May-09
1989	May-19	2005	May-30
1990	May-11	2006	May-20
1991	May-05	2007	Jun-02
1992	Jun-10	2008	Jun-03
1993	Jun-06	2009	May-19
1994	Jun-01	2010	May-23
		average	May-28

2 Dynamical characteristics of forced and inertially unstable convection development

Tomas et al. (1999) have examined the streamline field related with inertial instability by using a simple PBL model only forced by cross-equatorial pressure gradient. They adopted the homogeneous Helmhotz equation with frictional effect in PBL, but ignored other forcing in the inhomogeneous term. However, the cross-equatorial pressure gradient could lead to the lower jet and strong zonal advection over the tropical Arabian Sea in spring. As an air parcel moving eastward along the lower jet under the influence of pressure gradient force, its meridional speed is changed by the forcing of background pressure, affecting the spatial distribution of horizontal divergence and convection. Thus it is essential to consider the forcing of background pressure when studying the inertial instability issue over the ISM region. Here we start with deducing the control equation based on Tomas et al. (1999)'s framework. The 2-dimension control equation in PBL is:

$$\begin{cases} \dfrac{\mathrm{D}u}{\mathrm{D}t} = fv - fv_g - Ku = f(v - v_g) - Ku & (1) \\ \dfrac{\mathrm{D}v}{\mathrm{D}t} = -fu + fu_g - Kv = -f(u - u_g) - Kv & (2) \end{cases}$$

in which $u_g = -\dfrac{1}{f}\dfrac{\partial \phi}{\partial y}$ and $v_g = \dfrac{1}{f}\dfrac{\partial \phi}{\partial x}$ are the zonal and meridional component of geostrophic wind, respectively; total derivative $\dfrac{\mathrm{D}}{\mathrm{D}t} = \dfrac{\partial}{\partial t} + u\dfrac{\partial}{\partial x} + v\dfrac{\partial}{\partial y}$; and u, v, f, ϕ and K are respectively for the zonal wind, meridional wind, Coriolis parameter, geopotential, and the frictional coefficient. Before the ISM onset the strong cross-equatorial meridional gradient of sea surface temperature (SST) and sea level pressure (SLP) render the zonal geostrophic wind much larger than the meridional one over the Arabian Sea. We can thus assume $v_g = \dfrac{1}{f}\dfrac{\partial \phi}{\partial x} \approx 0$. Then Eqs. (1) and (2) can be written as:

$$\frac{D^2 v}{Dt^2} + \lambda^2 v = f\left(u\frac{\partial u_g}{\partial x} + \frac{\partial u_g}{\partial t}\right) + Kf(2u - u_g) \tag{3}$$

in which

$$\lambda^2 = f\eta - K^2$$

and absolute vorticity

$$\eta = f - \frac{\partial u_g}{\partial y}$$

Eq. (3) is an atmospheric motion equation controlling the forced inertial movement. The RHS of Eq. (3) is the inhomogeneous terms representing the external forcing. If the inhomogeneous terms are zero, Eq. (3) will degenerate to an equation depicting the atmospheric free inertial oscillation with the following homogeneous solution:

$$v_0 = V e^{i\lambda t} \tag{4}$$

The inertial instability criteria are:

$$\lambda^2 = f\left(f - \frac{\partial u_g}{\partial y}\right) - K^2 \begin{cases} < 0, & \text{unstable} \\ = 0, & \text{neutral} \\ > 0, & \text{stable} \end{cases} \tag{5}$$

When the variations of u_g and u are much slower than that of v, the particular solution of Eq. (3) can be obtained as:

$$v^* = v_1^* + v_2^* = \lambda^{-2}\left[f\left(\frac{\partial u_g}{\partial t} + u\frac{\partial u_g}{\partial x}\right) + Kf(2u - u_g)\right] \tag{6}$$

in which,

$$v_1^* = \lambda^{-2} f\left(\frac{\partial u_g}{\partial t} + u\frac{\partial u_g}{\partial x}\right) = -\lambda^{-2}\left(\frac{\partial}{\partial t} + u\frac{\partial}{\partial x}\right)\left(\frac{\partial \phi}{\partial y}\right) \tag{7}$$

when $u_g = u_g(y)$,

$$v^* = v_2^* = \lambda^{-2} Kf(2u - u_g) \tag{8}$$

Eq. (7) indicates the influence on the meridional current passing through the isobars due to the zonal change in zonal geostrophic flow. Since the zonal geostrophic flow is proportional to the north−south pressure gradient, which is closely associated with the meridional land-sea thermal contrast, the particular solution v_1^* in Eq. (7) thus implies that the zonal asymmetric distribution of north−south land-sea thermal contrast can affect the meridional motion of air parcel. In other words, the atmospheric inertial motion is also influenced by the external forcing of background streamline field. The solution v_2^* shown in Eq. (8) indicates the effect of drag coefficient (friction) on the cross-isobaric meridional flow and inertial instability. Its dynamical connotation has been stressed in detail by Tomas et al. (1999). In this paper we concentrate on how the zonal variation of zonal geostrophic flow influences the meridional motion passing through the isobars, i.e., the particular solution v_1^*.

Traditionally inertial instability is applied to understand the development of convection in the ITCZ (Tomas and Webster, 1997; Tomas et al., 1999). To examine its effect on the ISM onset, a brief analysis is carried out on the near-equatorial stationary atmospheric motion by referring to Tomas et al. (1999)'s work. Then Eqs. (1) and (2) can be predigested to:

$$\begin{cases} 0 = v\eta - Ku \tag{9a} \\ 0 = -v\left(\frac{\partial v}{\partial y} + K\right) - \beta y u_{ag} \tag{9b} \end{cases}$$

Note that the β-plane approximation ($f = \beta y$) is adopted here and u_{ag} is the zonal ageostrophic wind. From late May to early June over the eastern Arabian Sea, a lower westerly jet is situated between 10°

and 15°N. The relative vorticity is negative to the south of the jet. Then the zero absolute vorticity contour is near 4°N. An inertially unstable ($\eta < 0$) region is located between the south of zero absolute vorticity contour and the equator, while the inertially stable ($\eta > 0$) area is located to the north of the zero absolute vorticity contour (Fig. 1). Meanwhile the northward cross-equatorial gradient of SST leads to a high SLP in the south and a low one in the north of the Arabian Sea, corresponding to the zonal geostrophic flow near the equator. Near the equator the inertial force is small, and the pressure gradient force thus drives the air in the PBL to move northward crossing the equator from the Southern Hemisphere. The southerly passing through the inertial instability region forms an ageostrophic easterly to decelerate the westerly (Eq. (9a)). Subsequently the southerly enters the inertially stable region and generates an ageostrophic westerly to accelerate the zonal eastward flow (Eq. (9a)). Under the influence of inertial force, the zonal ageostrophic wind enhances the southerly crossing the isobars and the local meridional circulation gets intensified in the inertially unstable region, but weakens the southerly in the inertially stable area (Eq. (9b)). In summary, in the inertial unstable region the southerly crossing the isobar is strengthened and the geostrophic westerly is weakened, whereas in the inertial stable region the southerly crossing the isobar gets damped and the geostrophic westerly is reinforced (Fig. 1a). The quasi-equilibrium state of atmosphere requires the occurrence of divergence ($\partial v/\partial y > 0$) to the south but convergence ($\partial v/\partial y < 0$) to the north of the zero absolute vorticity contour in the lower troposphere (Fig. 1b, Eq. (9b)). Therefore, the maximum of southerly occurs near the zero absolute vorticity contour and lower level convergence and convectional precipitation are produced to its north (Fig. 1c).

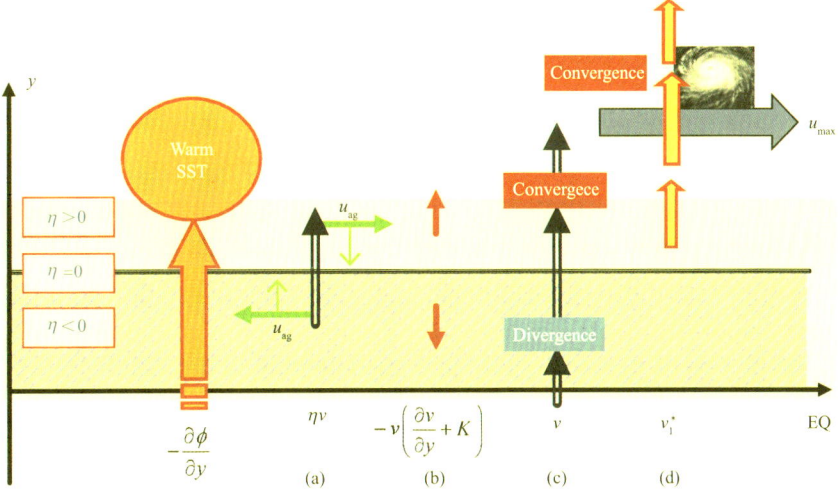

Fig. 1 Schematic diagram of the forced and inertially unstable convection development. (a) Driven by the cross-equatorial sea surface pressure gradient force, when an air parcel passes through the inertially unstable region ($\eta < 0$)/ inertially stable region ($\eta > 0$), the zonal wind is weakened/ strengthened (ηv, bold solid green arrows) and the southerly crossing the isobar gets increased/decreased ($-\beta y u_{ag}$, thin solid green arrows); (b) in order to balance the meridional inertial acceleration/deceleration induced by the zonal ageostrophic wind, the frictional and advection effects ($-v\left(\frac{\partial v}{\partial y} + K\right)$) have to decelerate/accelerate the southerly. (c) Thus the divergence/ convergence is established to the south/ north of the zero absolute vorticity contour. (d) Forced convection development: the zonal variation of meridional pressure gradient can generate strong additional meridional wind, lower-layer convergence is produced near and to the north of the zonal wind maximum (u_{max}) which is located to the north of zero absolute vorticity contour, causing the forced convection development and leading to the ISM onset

The traditional theory of inertial instability described above can be applied to explain the frequent occurrence of convection along the ITCZ over the near-equatorial Arabian Sea before the ISM onset. But such convection is too far away from the Indian Peninsula to induce the ISM onset. Notice that the particular solution (v_1^*) is associated with both the zonal speed of air parcel and the zonal distribution of meridional pressure gradient, i. e., the influence of external forcing on the meridional motion. If the air parcel takes sufficient time to mix fully with the surrounding atmosphere as it moves eastward along the westerly, the individual variation of zonal geostrophic wind then will be proportional to the zonal gradient of zonal geostrophic westerly. Under such circumstances and based on Eq. (7), over the area where the pressure gradient increases eastward, the northerly is accelerated in the inertial unstable region, while enhanced southerly appears to the north of zero absolute vorticity contour with its maximum occurring near the westerly jet axis. Thereby an additional forced convergence forms near or to the north of the zonal westerly maximum, and air ascent and convection are stimulated (Fig. 1d). Since the lower westerly jet axis closing to the southwestern coast of Indian Peninsula is located near 10°—15°N over the Arabian Sea, such a forced convection development could affect the ISM onset more directly. In the next section the NCEP/NCAR R1 reanalysis data will be used to verify the impact of the forced convection development on the ISM onset process.

3 Convection development during the ISM onset and relevant dynamical analysis

Figure 2 describes the evolution of precipitation and low level divergence during the ISM onset. Prior to the ISM onset the zero absolute vorticity contour stays to the south of 4°N, accompanied by the low-level convergence and rainfall center situated mainly between 10°N and the equator (Fig. 2a to c). When the ISM onsets, the zero absolute vorticity contour shifts northward rapidly, so does the convergence in the lower troposphere. Then the strong lower convergence dominates the northern Arabian Sea, consistent with the evident rainfall migrating northward to the southwestern coast of Indian Peninsula (Fig. 2d). After that the zero absolute vorticity contour maintains near 6°N, and the rainfall belt keeps advancing northward towards the Indian subcontinent (Figs. 2e and 2f). During the whole onset period, there exists conspicuous divergence and negative absolute vorticity in the lower troposphere between the zero absolute vorticity contour and the equator in the NH. Generally, the precipitation is moving northward in the ISM onset process with the distinctly northward displacement of the zero absolute vorticity contour and convergence in the lower troposphere over the Arabian Sea. The northward shifting of zero absolute vorticity contour represents the strengthening of inertial instability in situ. According to the traditional theory of inertial instability (Tomas et al., 1999), the lower convergence should be located on the north side close to the zero absolute vorticity contour. However, the lower level convergence shown in Figure 2 is far away from the north edge of absolute vorticity zero contour, but close to the southwestern Indian Peninsula and the southeastern Arabian Sea, so that the traditional theory of inertial instability cannot explain such a phenomenon.

3.1 Effect of traditional inertial instability on the ISM onset

Figure 2 shows that the zero absolute vorticity contour moves northward most evidently in the region from 55°—65°E over the southern Arabian Sea. From the 55°—65°E averaged latitude-pressure cross section of the zero absolute vorticity contour (Fig. 3a), it can be seen that the northward migration of the zero absolute vorticity contour over the Arabian Sea is restricted in the PBL under 700 hPa, re-

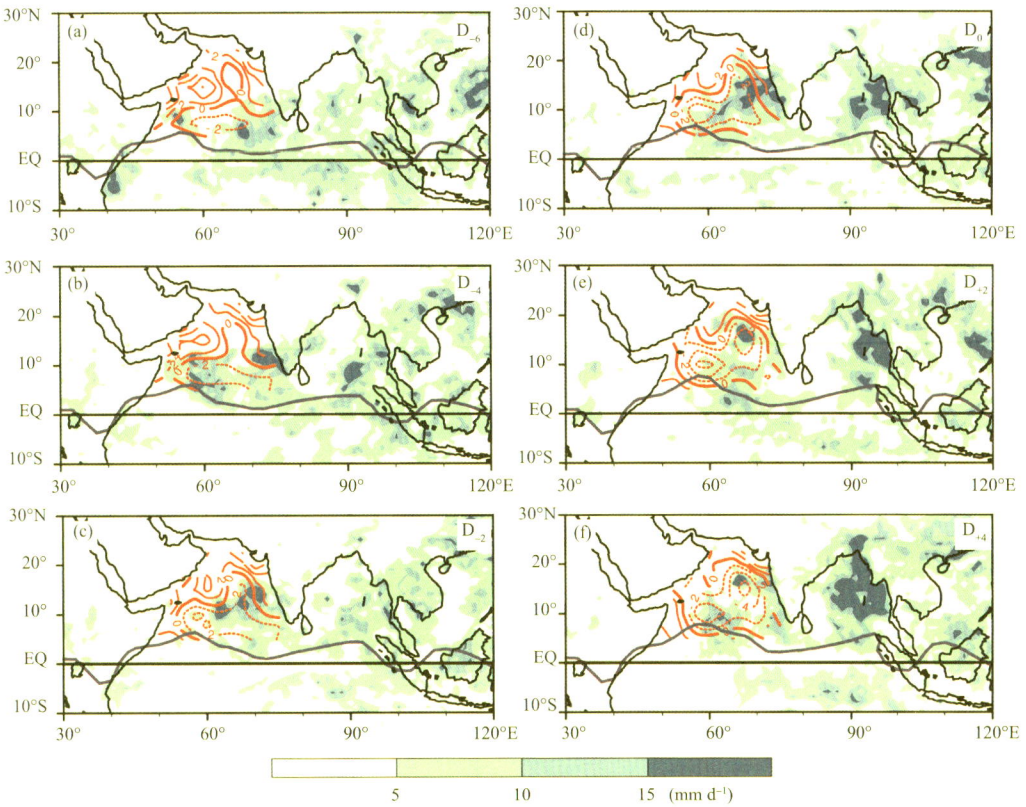

Fig. 2 Evolution during the ISM onset of the TRMM precipitation (shading, mm d^{-1}) and 925 hPa divergence (contours, dotted lines are for the convergence, interval is 2×10^{-6} s^{-1})

sembling the results obtained by Tomas and Webster (1997). The reason for the northward movement of the zero absolute vorticity contour is the strong cross-equatorial gradient of SLP in the ISM region (Fig. 3b). Before the ISM onset the enhanced poleward SLP gradient favors the intensifying and northward shifting of westerly jet, and pushes the zero absolute vorticity contour northward. Simultaneously, the pressure on the north of zero absolute vorticity contour decreases and the poleward gradient of SLP is enhanced, leading the zero contour to move towards the higher latitudes. It is due to this positive feedback that the zero absolute vorticity contour can move northward during the ISM onset. As latitude increases gradually the Coriolis parameter f is increasing, so that northward migration of the zero absolute vorticity contour stops at 6°N where $f>\zeta$ and the absolute vorticity becomes positive. Furthermore, the local SST over the Arabian Sea is high before the ISM onsets. After the ISM onsets, the summer monsoon engenders more cloud and rainfall to reduce the shortwave radiation reaching the sea surface and the enhanced sea surface wind stimulates the cold water upwelling and latent and sensible heat release. Thus the local SST is cooled quickly (Fig. 3c). Consequently, the primary influence of traditional inertial instability on the ISM onset is manifested in the seasonal northward migration of ITCZ (Tomas et al., 1997; Gadgil, 2003), resulting in the reinforcement and northward movement of lower convergence near the equator only.

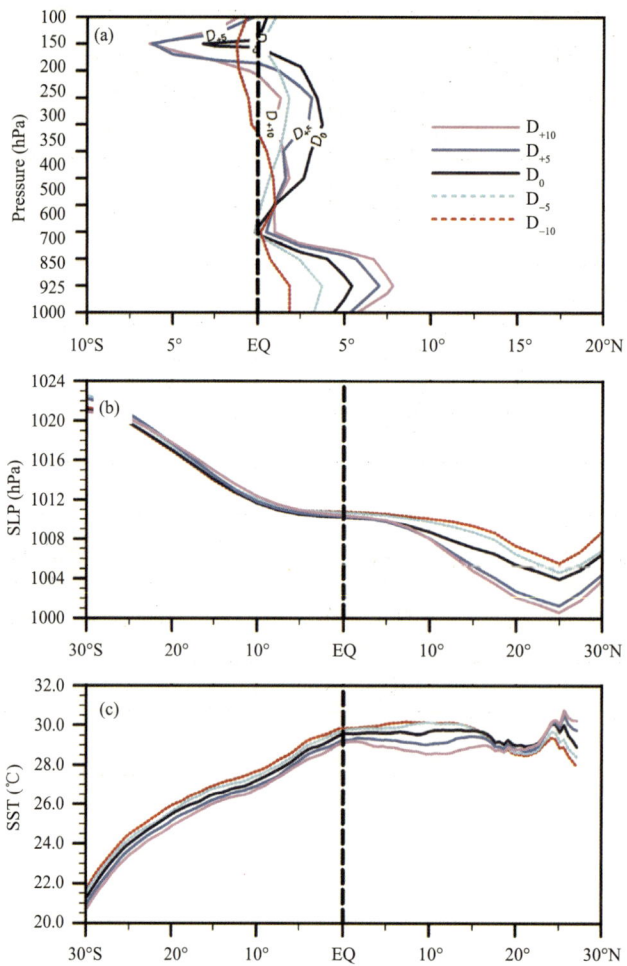

Fig. 3 Evolution of 55°—65°E averaged vertical profile of zero absolute vorticity contour (a, unit is 10^{-6} s^{-1}) and longitudinal profile of SLP (b, unit is hPa) and SST (c, unit is ℃)

3.2 Influence on the ISM onset of forced convection development induced by zonal variation of meridional SLP gradient

The lower convergence caused by traditional inertial instability is too close to the equator to affect the ISM onset directly. Before the ISM onsets, as the cross-equatorial gradient of SLP gets strengthened, the zonal flow in the tropics is accelerated, generating a westerly jet in the lower troposphere over the southern Arabian Sea (Figs. 4a to c). Then the enhanced lower level jet could increase the sensible heat flux on the sea surface, which facilitates the convection flourishing. After the ISM builds up, the convection cools the SST and decreases the sensible heating flux on the sea surface (Figs. 4d to f). The convection and lower level convergence is intensified most rapidly over the southwestern Indian Peninsula, rather than the southwestern Arabian Sea where the zero absolute vorticity contour shifts northward most evidently. Given that the strong zonal momentum advection resulted from the lower westerly jet on the north of the zero absolute vorticity contour, it can be speculated that the external forcing should play an important role in the northward movement of convergence due to the forced convection development. In a steady state, the external forcing solution in Eq. 7 can be written as:

$$v_1^* \approx \lambda^{-2} f\left(u \frac{\partial u_g}{\partial x}\right) = -\lambda^{-2} \left(u \frac{\partial}{\partial x}\right)\left(\frac{\partial \phi}{\partial y}\right) \qquad (10)$$

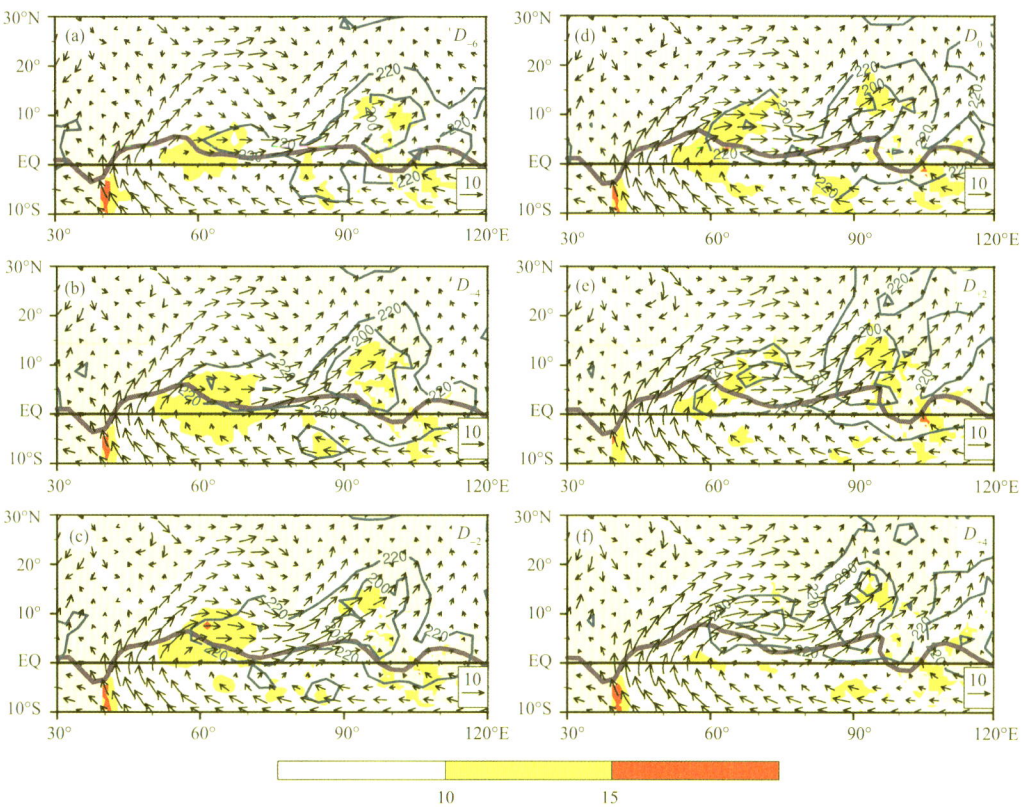

Fig. 1 Evolution of sea surface sensible heating flux (shading, unit is W m^{-2}), OLR (blue contours, unit is W m^{-2}) and 925 hPa wind (vectors, unit is m s^{-1}) during the ISM onset. The bold lines are for the zero absolute vorticity contours at 925 hPa

Figure 5 portrays the evolution of geopotential height at 925 hPa during the ISM onset. The zonal difference of low-level geopotential height is inconspicuous near the equator. But to the north of 5°N, there is a high pressure over the Arabian Sea with a low one over the Indian Peninsula, making the meridional pressure gradient increase eastward over the eastern Arabian Sea from 5°N to 15°N. Based on Eq. 10, the air parcel moving eastward along such a background pressure field should be driven to turn to the north, thereby producing an additional lower layer convergence to the north of the basic flow.

The evolvement of v_1' at 925 hPa during the ISM onset is shown in Figure 6. The v_1' over the eastern Arabian Sea is concentrated near the equator to the south of 6°N before the ISM onsets (Figs. 6a to c). During the ISM onset process, however, the low-level westerly jet over the southern Arabian Sea accelerates and shifts northward because of the increasing cross-equatorial gradient of SLP. The westerly jet in the lower troposphere produces the zonal advection of zonal geostrophic momentum, which leads the most evident response of meridional wind to the east of 70°E, corresponding to the convergence near the southwestern coast of Indian Peninsula (Figs. 6d to f). This strengthened low layer convergence in the NCEP/NCAR R1 reanalysis is in accordance with the intensified monsoon convection obtained from the TRMM satellite data as shown in Figure 2. It becomes evident that the zonal advection of zonal geostrophic momentum can force meridional wind to develop further downstream at higher latitudes, resulting in the low layer convergence and convection development over the eastern Arabian Sea further away from the north edge of the zero absolute vorticity contour.

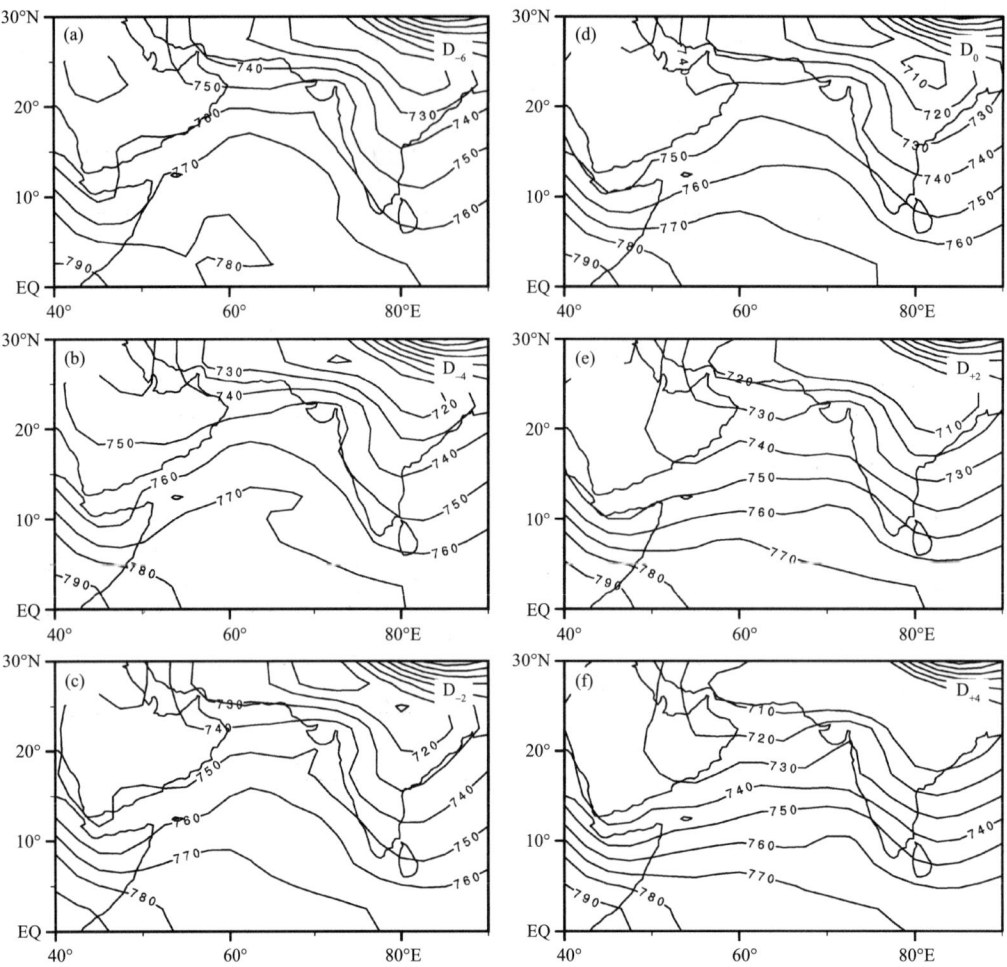

Fig. 5 Evolution of 925 hPa geopotential height field (gpm) during the ISM onset

Briefly, the strengthened inertial instability during the ISM onset can affect the evolution of streamline field in the lower troposphere. The inertial instability over the Arabian Sea is characterized by the northward migration of the zero absolute vorticity contour due to the intensified cross-equatorial pressure gradient. Negative absolute vorticity appears near the equator in the NH. As the meridional current passing through this region, a dipole of southern divergence and northern convergence establishes straddling the zero absolute vorticity contour in the lower troposphere. Moreover, the cross-equatorial southerly enhances the zonal wind to the north of zero absolute vorticity contour, forming a low layer westerly jet over the tropical Arabian Sea. Because over the eastern Arabian Sea the north—south land-sea thermal contrast increases eastward, air parcel moving through this region along the lower westerly jet leads to strong zonal advection of zonal geostrophic momentum and deviates northward, causing the centers of convergence and convection formed to the north of jet axis in the lower troposphere over the southeastern Arabian Sea and the southwestern coast of Indian Peninsula to the east of 70°E. This forced convection development creates favorable conditions for the ISM onset.

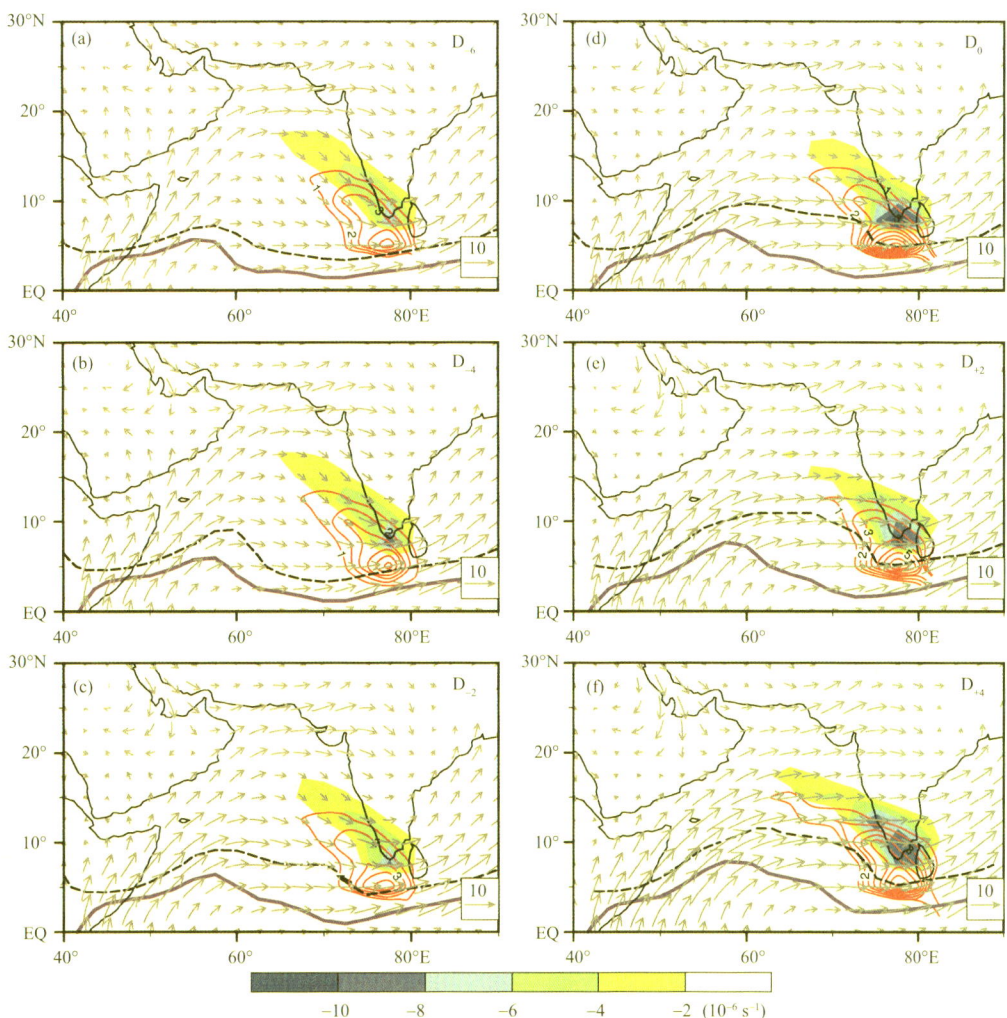

Fig. 6 Evolution of the meridional wind $v_1^* \approx -\lambda^{-2}\left(u\frac{\partial}{\partial x}\right)\left(\frac{\partial \phi}{\partial y}\right)$ (contours, unit is m s^{-1}) induced by forced convection development and its meridional convergence ($\partial v_1^*/\partial y$) at 925 hPa (shading, unit is 10^{-6} s^{-1}). Vectors, purple solid lines and black dashed lines are for the winds (m s^{-1}), the zero absolute vorticity contour and the maximum axis of westerly at 925 hPa, respectively

4 Vertical coupling between upper and lower circulations during the ISM onset

4.1 characteristics of the upper circulation

Except for the lower circulation, the contribution of the upper circulation to the ISM onset is also significant. Specifically, the SAH experiences evident variation during this period (Zhang et al., 2013; Liu et al., 2013). Prior to the ISM onset, the SAH situated over the Indochina Peninsula gets strengthened and expends eastward (Figs. 7a to c) due to the monsoon convection over the BOB and SCS (Figs. 4a to c). Then a conspicuous trumpet-shaped streamline field is established on its southwest just over the Arabian Sea, consistent with the strengthening of local divergence in the upper troposphere. During the ISM onset, the SAH expands further eastward and the upper divergence keeps intensifying over the Arabian Sea. According to Zhang et al. (2013), the eastward expansion of the SAH enhances the northerly on its east, bringing high potential vorticity (PV) from mid-latitude southward to form a high PV belt. The high PV is then transported to the southwest of the SAH over the southern Arabian Sea by the easterly jet in the upper troposphere. Subsequently a local cyclonic curvature appears as a response to the high PV. The southeasterly of the cyclonic curvature, together with the northeasterly to its north, thus forms the trumpet-shaped streamline field on the southwest of the SAH, producing the upper divergence center, which supplies a favorite background of upper pumping to the ISM onset. Meanwhile the cross-equatorial pressure gradient pushes the zero absolute vorticity contour northward to accelerate the lower westerly jet, resulting in the lower convergence and convection to the east of 70°E near the southwestern coast of Indian Peninsula. Then how does the upper circulation couple with the lower system in vertical to influence the ISM onset?

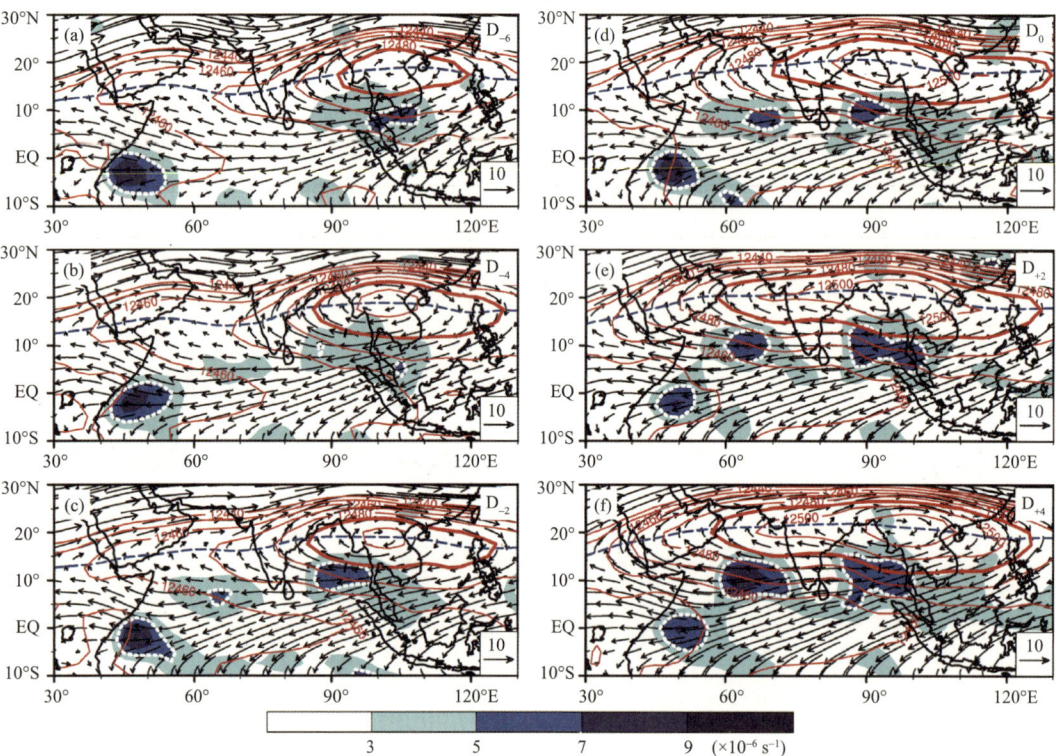

Fig. 7 Evolution of wind (vectors, unit is m s^{-1}), geopotential height (contours, interval is 20 gpm) and divergence (shading, interval is 2×10^{-6} s^{-1}, white dotted lines are for the maximum of divergence) at 200 hPa during the ISM onset

4.2 Vertical coupling between the upper and lower circulations

According to the vorticity equation:

$$\frac{D(f+\zeta)}{Dt} = \left(\frac{\partial}{\partial t} + u\frac{\partial}{\partial x} + v\frac{\partial}{\partial y}\right) \cdot (f+\zeta) = f\frac{\partial w}{\partial z} \quad (11)$$

at a steady state, $\frac{\partial}{\partial t} = 0$, assuming that w has normal mode solution in vertical, and taking partial derivative with respect to z on the both sides of Eq. (11) leads to:

$$w \approx \frac{1}{f} \cdot \frac{\partial}{\partial z}[-\boldsymbol{V} \cdot \nabla(f+\zeta)] \quad (12)$$

Eq. (12) delineates the effect of vertical shear of absolute vorticity advection on the vertical motion. That is, the ascending (descending) is accompanied by the advection of absolute vorticity increasing (decreasing) with height.

Figure 8 is the latitude-pressure cross section of vertical shear of absolute vorticity advection along 75°E during the ISM onset, indicating the favorite dynamical background for the ascending. As analyzed above, both the lower convergence and convection develop most rapidly along this longitude. Prior to the ISM onset, two near-equatorial ascending centers are located to the south of 5°N, including an upper center near 250 hPa and a lower one near 850 hPa (Figs. 8a and 8b). On the D_{-2} the upper center stretches to about 10°N, but the lower center stays in place (Fig. 8c). On the D_0, the upper center goes on enhancing with the development of lower center, so the vertical coupled ascending establishes from 5°N to 10°N. The collaboration of upper and lower centers promotes the ISM onset (Fig. 8d). After that the upper center keeps expanding to the north of 10°N, while the lower center is still situated near 10°N.

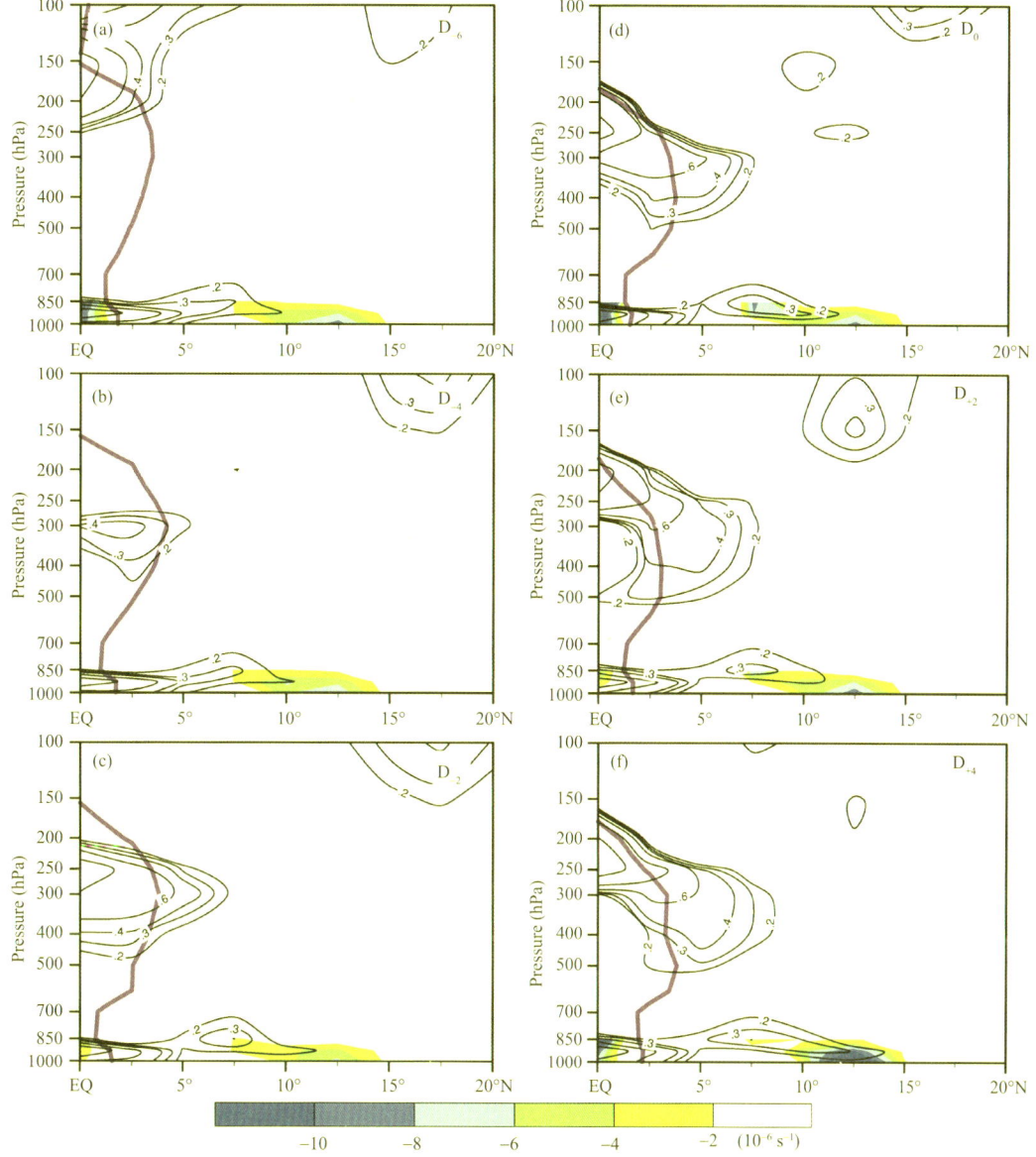

Fig. 8 Evolution of pressure-latitude cross section along 75°E of the vertical shear of absolute vorticity advection $(-\frac{1}{f_0+\beta y} \cdot \frac{\partial}{\partial p}[-u\frac{\partial \zeta}{\partial x} - v\frac{\partial}{\partial y}(f+\zeta)], f_0 = f|_{lat=2.5°})$ (contour, unit is 10^{-7} Pa^{-1} s^{-1}), and the meridional gradient of v_1' ($\partial v_1'/\partial y$) (shading, unit is 10^{-6} s^{-1}). Bold solid lines are for the zero absolute vorticity contours along this longitude

The divergence-pumping effect owing to the upper center facilitates the northward propagation of monsoon rainfall (Figs. 7e and 7f). In fact, the upper center corresponds to the positive advection of absolute vorticity related with the high PV transport on the south of SAH, whereas the lower center is consistent with the northward migration of the zero absolute vorticity contour and the intensification of the forced convection development. In detail, when the SAH in the upper troposphere strengthens and elongates eastward, the high PV is transported to the southern Arabian Sea on D_{-2} (Fig. 7c), the local upper divergence is enhanced, and the upper ascending center stretches northward (Fig. 8c). Later, the northward shifting of the intensified divergence and ascending center aloft provide a background of upper pumping, stimulating the ISM onset and the northward advancing of the monsoon rainfall belt. In the lower troposphere, as the convergence associated with forced convection development and westerly jet begins to grow up near the southwestern coast of Indian Peninsula on D_0 (Fig. 6d), the lower ascending center develops and migrates northward, and the ISM builds up (Fig. 8d). Finally, the zero absolute vorticity contour moves further northward, and the lower convergence and monsoon rainfall also move to higher latitudes (Figs. 8e and 8f), favoring the northward propagation of ISM.

The observed vertical motion exhibits similar features in both upper and lower troposphere (Fig. 9) as those calculated and presented in Figure 8. Before the ISM onset, the two weak ascending centers are to the south of 5°N near the equator. The upper center is located at 250 hPa and and the lower center at 850 hPa. The descending is dominant to the north of 10°N (Fig. 9a). Then the upper divergence-pumping effect is strengthened to enhance the upper ascending center above the southern Arabian Sea. Meanwhile the zero absolute vorticity contour and the lower ascending center both move northward, and the ascending center along the southwestern coast of Indian Peninsula starts to stretch to the north of 10°N (Figs. 9b and 9c). When the ISM onsets, the strong pumping maintains in the upper troposphere, accompanied by the northward shifting of lower ascending center and monsoon rainfall (Fig. 9d). The preliminary establishment of summer monsoon over the India is marked by the lower ascending center arriving at 10°N (Fig. 9e and 9f). It is indicated that the observed ascending generally resembles the diagnosed one in the upper and lower troposphere by comparing Figure 8 with Figure 9. Therefore, the ascending stimulated by vertical shear of absolute vorticity advection is very important for the ISM onset. There exist some discrepancies between observation and diagnosis. First, due to the β-plane approximation in the diagnosis, the ascending is damped away from the equator, so the northward movement of low-level ascending in the diagnosis is weaker than that resulting from the observation. Besides, the apparent ascending of observation in the middle troposphere is not represented by the diagnosis. The reason is that only the ascending forced by internal dynamical process is contained in this diagnosis, while the ascending attributed to external thermal forcing (e.g., the condensation heating) is excluded.

Accordingly, before the ISM onset in the upper troposphere the SAH is enhanced and elongating eastward, and the high PV in the mid-latitude is transported to the southern Arabian Sea. Thus the local cyclonic curvature occurs on the southwest of SAH, and an upper trumpet-shaped streamline field with strong divergence is formed, providing pumping effect aloft and a favorable background for the ISM onset. In the lower troposphere, however, the convergence due to the forced convection development moves northward along the southwestern coast of Indian Peninsula. When the lower layer convergence is coupled with the upper layer divergence in vertical, the ascending in the ISM region is explosively intensified and heavy precipitation appears. Then the ISM builds up. Afterwards the ISM migrates northward gradually, manifested in the northward advancing of monsoon rainfall belt.

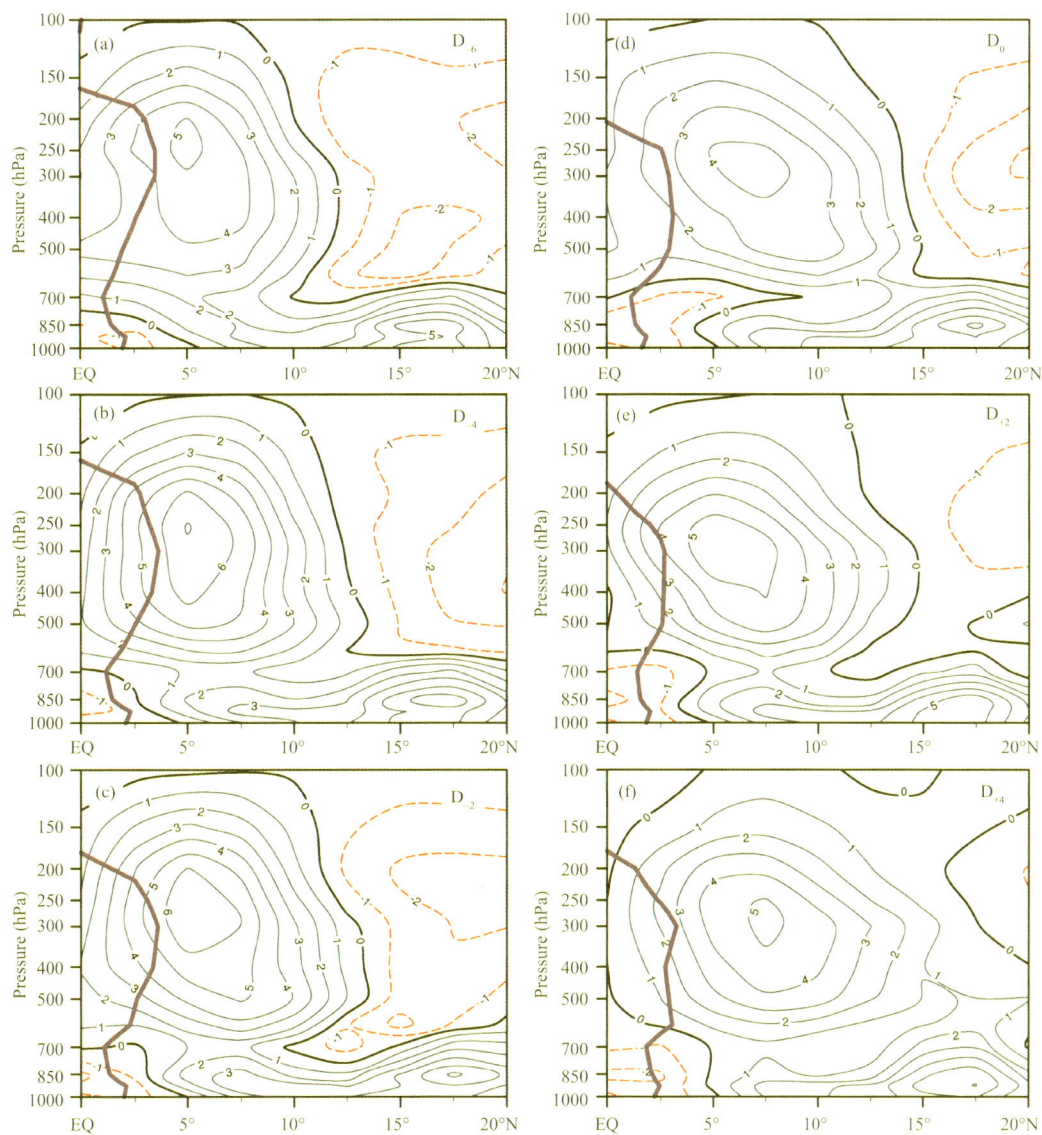

Fig. 9 The same as Fig. 8, but is about the vertical motion (10^{-2} Pa s^{-1}) obtained from the NCEP/NCAR R1 reanalysis

5 Summary and discussion

The ISM is an important member of the Asian monsoon system, and its onset process reflects the abrupt circulation transition from winter to summer over South Asia. This paper uses the NCEP/NCAR R1 reanalysis datasets and the composite technology to investigate the climatological characteristics of ISM onset. The impact on the ISM onset of the frictional inertial instability in the lower troposphere is analyzed. It is highlighted that the zonal differential land-sea thermal contrast can induce low-level zonal advection of zonal geostrophic momentum and the resultant forced convection development is significant for the ISM onset. The importance of vertical coupling between upper and lower circulation during the ISM onset is also studied. The primary results are summarized as follows.

(1) The evident cross-equatorial pressure gradient exists in the PBL of Arabian Sea from late spring to early summer. It causes the zero absolute vorticity contour to move to the north from the equator,

establishing a near-equatorial region with negative absolute vorticity in the NH. As the southerly passing through this region, the maximum of meridional wind takes place near the zero absolute vorticity contour, producing a dipole of southern divergence and northern convergence straddling the zero contour. Usually such a frictional instability occurs near the equator, and does not exert a direct impact on the ISM onset. On the other hand, the cross-equatorial pressure gradient generates a lower westerly jet over the southern Arabian Sea, and the zonal eastward increasing land-sea thermal contrast from the southeastern Arabian Sea to the southwestern Indian Peninsula produces a conspicuous local zonal advection of zonal geostrophic momentum. The advection induces a forced convection development with a low-level convergence located to the north of maximum westerly jet. Consequently the monsoon convection and heavy rainfall appear over the southeastern Arabian Sea and the southwestern coast of Indian Peninsula, marking the onset of ISM. Afterwards, as the forced convection development moving northward, the ISM precipitation advances northward over the Indian inland area.

(2) In addition to the forced convection development in the lower troposphere, the SAH evolution in the upper troposphere also plays a significant role in the ISM onset. Prior to the ISM onset, due to the latent heating released by the BOB and SCS monsoon, the SAH is strengthened and extends eastward. As a result the enhanced northerly on the east of the SAH brings the high PV from mid-latitude to the tropics, where the tropical easterly transports the high PV to the southern Arabian Sea in the upper troposphere. A local cyclonic curvature appears as a response to the high PV advection and forms, together with the cross equatorial northerly to the south, a trumpet-shaped streamline field on the southwest of the SAH, leading to stronger upper divergence-pumping. This provides another favorable background for the ISM onset. When the lower convergence moves northward below the upper divergence, the upper and lower circulation is fully coupled in vertical, the atmosphere becomes baroclinically unstable (Hoskins et al., 1985), and convection and severe rainfall develop near the southwestern coast of Indian Peninsula. Finally the ISM builds up.

This study demonstrates the climate mean features of the ISM onset, and sheds new light on the dynamic mechanism for further understanding the physical connotation of seasonal northward migration of monsoon rainfall belt over the ISM region. However, the interannual variability of ISM onset is so strong that the standard deviation of ISM onset date is about 9 days, implying the distinct difference of onset process from year to year (Joseph et al., 2006; Gadgil, 2011). The ISM onset time can be influenced by the ENSO events (Joseph et al., 1994; Goswami, 2005a), the condition of underlying surface and surrounding snow cover (Yang et al., 1996; Robock et al., 2003), the intraseasonal oscillation (Wu et al., 1998; Goswami, 2005b), and the extra-tropical circulation anomalies (Chang et al., 2001). In particular, the effect of ENSO events is most prominent since it is a global scale forcing which persists from previous winter to concurrent summer. Joseph et al. (1994) proposes that the ISM onset is postponed after the occurrence of strong El Niño event in previous winter, which may induce the SST anomalies over the Indian Ocean and the Pacific. Fasullo et al. (2003) use the vertically integrated water vapor transport to define the ISM onset date, and propose that there is a significant positive correlation between the ISM onset date and the Niño-3 SST anomalies averaged from June to September. Xavier et al. (2007) identify the ISM onset date based on the meridional difference of air temperature averaged in the upper troposphere to confirm its significant and steady positive correlation with the ENSO events. Nevertheless, previous studies concentrate mainly on the statistic relationship between ENSO events and ISM onset date, little investigation was done about the influence of ENSO on the ISM onset process and relevant mechanism, especially the effect on the circulation in the middle and upper troposphere. This paper highlights the important role of both the forced convection development in the lower troposphere and the SAH variation in the upper troposphere played in the ISM onset. To further understand the ISM onset process, more studies are required to examine the ENSO effect on the ISM onset via its influence on the dynamical processes revealed in this study.

References

ANANTHAKRISHNAN R, SRINIVASAN V, RAMAKRISHNA A R, et al. 1968. Synoptic features associated with onset of southwest monsoon over Kerala[R]. Pune, India: Forecasting Manual Report No. IV-18. 2.

CHANG C P, HARR P, JU J. 2001. Possible roles of Atlantic circulations on the weakening Indian monsoon rainfall-ENSO relationship[J]. J Climate, 14: 2376-2380.

FASULLO J, WEBSTER P J. 2003. A hydrological definition of Indian monsoon onset and withdrawal[J]. J Climate, 16: 3200-3211.

GADGIL S. 2003. The Indian monsoon and its variability[J]. Ann Rev Earth Planet Sci, 31: 429-467.

GADGIL S. 2011. South Asian monsoon: Interannual variation[C]//CHANG C P, DING Y H, LAU N C, et al. The Global Monsoon System: Research and Forecast (2nd). New York: World Scientific: 25-42.

GOSWAMI B N. 2005a. South Asian summer monsoon: An overview[C]//CHANG C P, WANG B, LAU N C, et al. The Global Monsoon System: Research and forecast (2nd). New York: World Scientific: 47-71.

GOSWAMI B N. 2005b. South Asian monsoon[C]//LAU K-M, WALISER D E. Intraseasonal Variability in the Atmosphere-Ocean Climate System. London: Springer: 19-61.

HOSKINS B J, MCINTYRE M E, ROBERTSON A W. 1985. On the use and significance of isentropic potential vorticity maps[J]. Quart J Roy Meteor Soc, 111: 877-946.

HUFFMAN G J, ADLER R F, RUDOLF B, et al. 1995. Global precipitation estimates based on a technique for combining satellite-based estimates, rain gauge analysis, and NWP model precipitation information[J]. J Climate, 8: 1284-1295.

JOSEPH P V, EISCHEID J K, PYLE R J, et al. 1994. Interannual variability of the onset of the Indian summer monsoon and its association with atmospheric features, El Niño, and sea surface temperature anomalies[J]. J Climate, 7: 81-105.

JOSEPH P V, SOORAJ K P, RAJAN C K. 2006. The summer monsoon onset process over South Asia and an objective method for the date of monsoon onset over Kerala[J]. Int J Climatol, 26: 1871-1893.

KALNAY E, KANAMITSU M, KISTLER R, et al. 1996. The NCEP/NCAR 40-year reanalysis project[J]. Bull Amer Meteor Soc, 77: 437-471.

KRISHNAMURTI T N, RAMANATHAN Y, PAN H L, et al. 1983. Cumulus parameterization and rainfall rates II[J]. Mon Wea Rev, 111: 816-828.

LAU N C, PLOSHAY J J. 2009. Simulation of synoptic and sub-synoptic scale phenomena associated with the East Asian summer monsoon using a high-resolution GCM[J]. Mon Wea Rev, 137: 137-160.

LIEBMANN B, SMITH C A. 1996. Description of a complete (interpolated) outgoing longwave radiation dataset[J]. Bull Amer Meteor Soc, 77: 1275-1277.

LIU B Q, WU G X, MAO J Y, et al. 2013. Genesis of the South Asian high and its impact on the Asian summer monsoon onset[J]. J Climate, 26: 2976-2991.

LIU Y Y, DING Y H. 2008a. Teleconnection between the Indian summer monsoon onset and the Meiyu over the Yangtze River valley[J]. Sci China Ser D-Earth Sci, 51: 1021-1035.

LIU Y Y, DING Y H. 2008b. Analysis and numerical simulations of the teleconnection between Indian summer monsoon and precipitation in North China[J]. Acta Meteor Sinica, 22: 489-501.

MAO J, CHAN J, WU G. 2004. Relationship between the onset of the South China Sea summer monsoon and the structure of the Asian subtropical anticyclone[J]. J Meteor Soc Japan, 82: 845-859.

MAO J Y, WU G X. 2007. Interannual variability in the onset of the summer monsoon over the eastern Bay of Bengal[J]. Theor Appl Climatol, 89: 155-170.

PRASAD V S, HAYASHI T. 2005. Onset and withdrawal of Indian summer monsoon[J]. Geophys Res Lett, 32: L20715. doi:10.1029/2005GL023269.

ROBOCK A, MU M, VINNIKOV K, et al, 2003. Land surface conditions over Eurasia and Indian summer monsoon rainfall [J]. J Geophys Res, 108: 4131. doi: 10.1029/2002JD002286.

SAHA K, 2010. Chapter 4: Monsoon over Southern Asia (comprising Pakistan, India, Bangladesh, Myanmar and Countries of Southeastern Asia) and adjoining Indian Ocean (Region-I)[C]//SAHA K. Tropical Circulation Systems and Monsoons. London: Springer: 89-121.

SAHA S, SAHA K, 1980. A hypothesis on onset, advance and withdrawal of the Indian summer monsoon[J]. Pure Appl Geophys, 118: 1066-1075.

TANIGUCHI K, KOIKE T, 2006. Comparision of definitions of Indian summer monsoon onset: Better representation of rapid transitions of atmospheric conditions[J]. Geophys Res Lett, 33: L02709. doi:10.1029/2005GL024526.

TOMAS R A, WEBSTER P J, 1997. The role of inertial instability in determining the location and strength of near-equatorial convection[J]. Quart J Roy Meteor Soc, 123: 1445-1482.

TOMAS R A, HOLTON J R, WEBSTER P J, 1999. The influence of cross-equatorial pressure gradients on the location of near-equatorial convection[J]. Quart J Roy Meteor Soc, 125: 1107-1127.

WANG B, DING Q, JOSEPH P V, 2009. Objective definition of the Indian summer monsoon onset[J]. J Climate, 22: 3303-3316.

WU G X, ZHANG Y S, 1998. Tibetan Plateau forcing and the timing of the monsoon onset over South Asia and the South China Sea[J]. Mon Wea Rev, 126: 913-927.

XAVIER P K, MARZIN C, GOSWAMI B N, 2007. An objective definition of the Indian summer monsoon season and a new perspective on the ENSO-monsoon relationship[J]. Quart J Roy Meteor Soc, 133: 749-764.

YANG S, LAU K M, 1996. Precursory signals associated with the interannual variability of the Asian summer monsoon[J]. J Climate, 9: 949-964.

YU L, JIN X, WELLER R A, 2007. Annual, seasonal, and interannual variability of air-sea heat fluxes in the Indian Ocean [J]. J Climate, 20: 3190-3209.

ZHANG Y N, WU G, LIU Y, et al, 2013. Effect of zonal asymmetric instability of potential vorticity on the South Asian high and the India summer monsoon onset[J]. Sci China Earth Sci, 57: 337-350. doi: 10.1007/s11430-013-4664-8.

Asian Summer Monsoon Onset Barrier and Its Formation Mechanism

LIU Boqi[1,2], LIU Yimin[1], WU Guoxiong[1], YAN Jinghui[3], HE Jinhai[1,2], REN Suling[1,3]

(1 State Key Lab of Atmospheric Sciences and Geophysical Fluid Dynamics (LASG), Institute of Atmospheric Physics, Chinese Academy of Sciences, Beijing 100029, China; 2 Key Laboratory of Meteorological Disaster (Nanjing University of Information Science and Technology), Ministry of Education, Nanjing 210044, China; 3 China Meteorological Administration, Beijing 100081, China)

Abstract: The onset process of Asian summer monsoon (ASM) is investigated based on diagnostic analysis of observations of precipitation and synoptic circulation. Results show that after the ASM commences over the eastern Bay of Bengal (BOB) around early May, the onset can propagate eastwards towards the South China Sea and western Pacific but is blocked on its westward propagation along the eastern coast of India. This blocking, termed the "monsoon onset barrier (MOB)", presents a Gill-type circulation response to the latent heating released by BOB monsoon convection. This convective condensation heating generates summertime (wintertime) vertical easterly (westerly) shear to its east (west) and facilitates air ascent (descent). The convection then propagates eastward but gets trapped on its westward path. To the east of the central BOB, the surface air temperature (SAT) cools faster than the underlying sea surface temperature (SST) due to monsoon onset. Thus more sensible heat flux supports the onset convection to propagate eastward. To the west of the central BOB, however, the land surface sensible heating over the Indian Peninsula is strengthened by the enhanced anticyclone circulation and air descent induced by the BOB monsoon heating. The strengthened upstream warm horizontal advection then produces a warm SAT center over the MOB region, which together with the in situ cooled SST reduces the surface sensible heating and atmospheric available potential energy to prevent the occurrence of free convection. Therefore, it is the change in both large-scale circulation and air-sea interaction due to BOB summer monsoon onset that contributes to the MOB formation.

Key words: monsoon onset barrier; Bay of Bengal summer monsoon; Gill-type response; air-sea interaction

1 Introduction

Monsoon occurs as a consequence of the seasonal reversal of circulation induced by an atmospheric response to the seasonal transition of land-sea thermal contrast. Monsoon onset is characterized by rapid changes in the prevailing wind direction and rainfall intensity. However, rainfall is both the result of, and a driving force for, the atmospheric circulation (Eady, 1950). The atmospheric response and feedback to the huge latent heating of monsoon convection makes the monsoon onset process highly complex. The evolution of Asian summer monsoon (ASM) onset, which is accompanied by an abrupt enhancement of atmospheric energy and the water cycle, takes approximately one month (Wu et al., 2013) and has great impact on society and economic growth in Asian countries.

The ASM onset process is characterized by three successive phases: onset begins over the southeastern Bay of Bengal (BOB), followed by onset over the South China Sea (SCS), and finally over India (Wu et al., 1998; Wang et al., 2002; Mao et al., 2003; Mao et al., 2007; Yang et al., 2012). The BOB summer monsoon onset is associated with an overturning of the meridional air temperature gradient (MTG) in the mid-upper troposphere (Mao et al., 2007). In the lower troposphere, the development of monsoon onset vortex (MOV) in early May over the BOB plays an important role in the onset (Lau et al., 1998; Liu et al., 2002; Vinayachandran et al., 2007). It has been demonstrated that MOV formation and development are dominated by the effects of local air-sea interaction (Wu et al., 2011, 2012a), barotropic and baroclinic instability (e.g., Krishnamurti et al., 1981; Mak et al., 1982; Mao et al., 2011), and the pumping effect in the upper troposphere (Liu et al., 2013). All the above factors are strongly influenced by the Tibetan Plateau forcing (Wu et al., 2011, 2012a).

The SCS summer monsoon onset is characterized by multi-scale activities (Ding et al., 2005; Wang et al., 2009a). The mechanism for the establishment of SCS summer monsoon can be classified into two categories: one is atmospheric internal variability, including the local intraseasonal oscillation (Chen et al., 1995; Zhou et al., 2005; Wu, 2010) and effect of neighboring weather systems (Chan et al., 2000; Xu et al., 2001; Liu et al., 2002; Tong et al., 2009); the other is external forcing, such as the thermal and mechanical effects of local orography (Qian et al., 2001; Xie et al., 2003; Xu et al., 2008) and air-sea interaction (Zhou et al., 2007).

For the Indian summer monsoon onset process, there are traditionally two different viewpoints. One considers Indian summer monsoon onset as the seasonal northward movement of the Intertropical Convergence Zone (ITCZ) (e.g., Saha et al., 1980; Srinivasan et al., 1993; Gadgil et al., 1998; Gadgil, 2003). The other treats the onset as a westward propagation of the ASM onset process, which commences over the eastern BOB (e.g., Tao et al., 1987; Tanaka, 1992). It has also been proposed that Indian summer monsoon onset is determined by the arrival of the active phase of 30—60-day intraseasonal oscillation (ISO) at the Indian Peninsula (Goswami et al., 2001; Goswami, 2005). In addition, the mechanical and thermal effects of the Tibetan Plateau on ASM onset have been highlighted by many researchers (Tao et al., 1987; Li et al., 1996; Hung et al., 2004a, 2004b; Abe et al., 2013). In particular, the significant impact of the surface sensible heat driven air-pump (SHAP) of the Iranian and Tibetan plateaus on ASM onset and evolution has also been emphasized (Wu et al., 2007, 2012b).

Despite remarkable accomplishments achieved in the past decades concerning ASM dynamics, many questions remain unanswered. One example of an area yielding many such questions is the study of the propagation of ASM onset. It is still unclear how ASM onset evolves and what is the controlling mechanism. By analyzing ASM onset isochrones based on scarce data, Tao et al. (1987) suggested that the earliest ASM onset is over the SCS in mid May. It then propagates westward gradually and eventually reaches the Indian subcontinent. Other earlier studies proposed that after the commencement of ASM over the BOB, ASM rainfall can propagate eastward and westward concurrently (Lau et al., 1997;

Webster et al., 1998; Wang et al., 2002). In contrast, based on an analysis of ECMWF reanalysis data, Wu et al. (1998) proposed that ASM onset begins over the BOB in early May, then extends eastward to the SCS in mid May, but it cannot propagate westward directly to India. Based on the NCEP/NCAR reanalysis products (Kalnay et al., 1996) and CMAP rainfall data (Xie et al., 1997), Yan (2005) further revealed that the westward propagation of ASM onset gets trapped over the western coast of the BOB, which is termed the "monsoon onset barrier" (MOB) in this study. However, the mechanism responsible for the formation of the MOB remains unclear.

The objective of this study is to investigate the general characteristics of the MOB based on rainfall and reanalysis datasets, and explore the possible mechanism for its formation. The remainder of the paper is organized as follows. Section 2 describes the data and method applied. The propagation of ASM onset based on different precipitation data products is investigated in section 3. Section 4 explores the possible mechanism for the MOB formation. Finally, a summary and discussion of the key findings are presented in section 5.

2 Data and method

2.1 Data description

To reduce the uncertainty in any individual dataset, three types of precipitation data, i.e., the CPC Merged Analysis of Precipitation (CMAP) (Xie et al., 1997) from 1979 to 2010 provided by the National Oceanic and Atmospheric Administration (NOAA), version 1.2 of the Global Precipitation Climatology Project (GPCP) one-degree daily precipitation records (Huffman et al., 2001) from 1997 to 2010, and the Tropical Rainfall Measurement Mission Project (TRMM) daily rainfall production (3B42 Version7) from 1998 to 2010 provided by the National Aeronautics and Space Administration (NASA), are used to depict the MOB. The horizontal resolutions of the CMAP, GPCP and TRMM datasets are 2.5°×2.5°, 1.0°×1.0°, and 0.25°×0.25°, respectively. The 3-D wind field, air temperature on standard isobars and at 2 m height, total cloud cover and surface heat flux, with a horizontal resolution of 1.5°×1.5°, are extracted from the ERA-Interim reanalysis data from 1979 to 2010 (Dee et al., 2011) to describe the ASM onset characteristics. The Outgoing Longwave Radiation (OLR, Liebmann et al., 1996) from 1979 to 2010 is provided by NOAA with a horizontal resolution of 2.5°×2.5°. Over the ocean surface, daily sensible heat flux, sea surface temperature (SST) and surface air temperature (SAT) with a horizontal resolution of 1°×1° from 1985 to 2010 are obtained from the Objectively Analyzed air-sea Fluxes (OAFlux), which is archived by the Woods Hole Oceanographic Institution (WHOI) Cooperative Institute for Climate and Ocean Research (CICOR).

2.2 Method

In this study, both the changes in circulation and precipitation are used to define ASM onset. As proposed by Mao et al. (2003, 2007), ASM onset is identified as the time when the ridge surface of the subtropical anticyclone starts to tilt northward towards the warm land area from winter to summer. Based on the thermal wind relationship, this tilting process can be expressed by the reversal of the MTG in the upper troposphere (500—200 hPa), implying the change in sign of $\partial u/\partial z$, or $-\partial T/\partial y$, along the ridge surface. This criterion is actually equivalent to the monsoon onset indices proposed by Webster et al. (1992) based on the change in vertical shear of zonal wind ($\partial u/\partial z$) and by Li et al. (1996) according to the change in meridional temperature difference between 5° and 30°N ($-\partial T/\partial y$). The MTG definition reflects the essence of the onset process of the BOB, SCS, and Indian summer monsoons.

Precipitation is an important element related to ASM onset. ASM onset can also be determined when a dramatic increase in rainfall occurs (Yoshino, 1966; Wang, 2006), with a threshold of greater than 5 mm d^{-1} over ocean and 3 mm d^{-1} over land. Moreover, since the heavy precipitation is usually accompanied by an evident change in wind direction, it is necessary to take the wind direction change into account when identifying ASM onset. Here, the wind direction (θ) is as usual defined as

$$\theta = \begin{cases} 0° & \text{if } u = 0, v < 0 \\ 270° - \arctan(v/u) & \text{if } u > 0 \\ 180° & \text{if } u = 0, v > 0 \\ 90° - \arctan(v/u) & \text{if } u < 0 \end{cases} \quad (1)$$

in which 0°, 90°, 180° and 270° represent northerly, easterly, southerly and westerly wind, respectively. And the deviation of wind direction on each day (θ_t) from the mean value in January ($\bar{\theta}_1$) is treated as the wind direction change ($\Delta\theta$), i.e.,

$$\Delta\theta = \begin{cases} (\theta_t - \bar{\theta}_1) - 360°, & \text{if } \theta_t - \bar{\theta}_1 > 180° \\ |(\theta_t - \bar{\theta}_1) + 360°|, & \text{if } \theta_t - \bar{\theta}_1 < -180° \end{cases} \quad (2)$$

The summer monsoon onset is marked by a greater than 100° change of surface wind direction at 10 m height (Wu et al., 2013). The thresholds of 100°, 120° and 140° have been tested to determine ASM onset, and the results show no significant difference. One reason for choosing the threshold of 100° is that the wind direction change during ASM onset in spring is smaller than that in July (usually 120°). The other reason is that the wind direction change is less than 120° over the northeastern BOB (15°—23°N) due to the influence of the Indo-Burma trough.

Therefore, the ASM onset date on each grid can be defined as the day when the following ASM onset criteria (ASMOC) are satisfied:

(1) The upper tropospheric (500—200 hPa) MTG changes from negative to positive and remains positive for more than 10 days;

(2) The wind direction change at 10 m height is greater than 100°;

(3) The rainfall is steadily greater than 5 mm d^{-1} over the ocean and 3 mm d^{-1} over the land.

If the area-averaged MTG, wind direction change and rainfall over the BOB (5°—15°N, 85°—100°E), the SCS (5°—20°N, 110°—120°E) and India (10°—20°N, 60°—75°E) meet the ASMOC, the summer monsoon is considered to commence in those specific regions. The onset dates of summer monsoon over the BOB, the SCS and India defined by the ASMOC are significantly correlated with those defined by other criteria (Table 1), including the zonal wind at 850 hPa (Wang et al., 2004, 2009b; Tian et al., 2010), the OLR or precipitation (Ananthakrishnan et al., 1988; Wang et al., 1997), the MTG in the middle and upper troposphere (Webster et al., 1998; Mao et al., 2007), and the vertically integrated diabatic heating (Xavier et al., 2007) or water vapor (Taniguchi et al., 2006; Prasad et al., 2005). The climatology obtained by arithmetically averaging the ERA-Interim reanalysis is utilized to describe the general features of ASM onset. Since ASM onset begins over the eastern BOB accompanied by a large release of latent heat to the atmosphere, the BOB monsoon convection plays an important role in the subsequent progress of ASM onset (Liu et al., 2002). To highlight the climatological effect of BOB monsoon convection, a composite analysis technique introduced below is also employed in this study. For each individual year, the onset date is defined as day zero (D_0), with the dates before (after) the onset date of BOB summer monsoon being labeled as negative (positive) days. All the variables are chosen from 30 days before (D_{-30}) to 30 days after (D_{+30}) the BOB summer monsoon onset date each year. The climatology of a particular variable for each of the 61 days (D_{-30} to D_{+30}) during BOB summer monsoon onset is thus calculated by arithmetically averaging the data over each corresponding day.

Table 1 Linear correlation coefficients of onset dates defined by the ASMOC in this study with others over the BOB, SCS, and India. All values have passed the 99% confidence level

BOB	
U_{850} (Mao and Wu, 2007)	0.892
OLR (Mao and Wu, 2007)	0.635
MTG (Mao and Wu, 2007)	0.804
SCS	
Wang et al. (2004)	0.504
Wang and Wu (1997)	0.699
Tian and Wang (2010)	0.557
India	
Ananthakrishnan and Soman (1988)	0.665
Wang and Wu (1997)	0.699
Taniguchi and Koike (2006)	0.739
Xavier et al. (2007)	0.678
Wang et al. (2009b)	0.734
Prasad and Hayashi (2005)	0.739

3 Propagation of Asian summer monsoon onset

Evolutions of daily mean precipitation derived from CMAP, GPCP, TRMM and ERA-Interim data averaged over 10°—20°N are shown in Figure 1 for the period 1 April to 30 June. Despite some dissimilarity in the rainfall evolutions in different datasets, they generally depict the ASM onset process well. Along the latitudinal zone between 10° and 20°N, rainfall intensification first occurs over the BOB from late April to early May, and then expands eastward with its maximum reaching the SCS around mid May. Over the Arabian Sea and Indian Peninsula, however, no westward propagation of monsoon rainfall is found after the BOB summer monsoon onset. It is not until the end of May and early June that the rainfall over the Arabian Sea and Indian subcontinent enhances to greater than 5 mm day^{-1} or 3 mm day^{-1}, representing the Indian summer monsoon onset. The three-phase feature of ASM onset, as first revealed in a case study by Wu and Zhang (1998), can be identified clearly from the climatological means based on different data sets. Since the spatial and temporal distributions of CMAP rainfall are similar to their counterparts from satellite observations and reanalysis data, and since the CMAP data set is high quality and has a long history, it is appropriate to use CMAP precipitation to study ASM onset. Hereafter, we adopt CMAP precipitation to define the rainfall criterion for ASM onset.

According to the MTG summer monsoon onset criteria, climatologically the summer monsoon onset times over the BOB, the SCS, and the Indian subcontinent are determined as early May, mid May, and late May, respectively (Fig. 2a). These are also plotted in Figure 1 for comparison purposes. The three-phase monsoon onset identified by the precipitation is comparative to that depicted by the MTG, and is prominent based on both the MTG analysis and the precipitation.

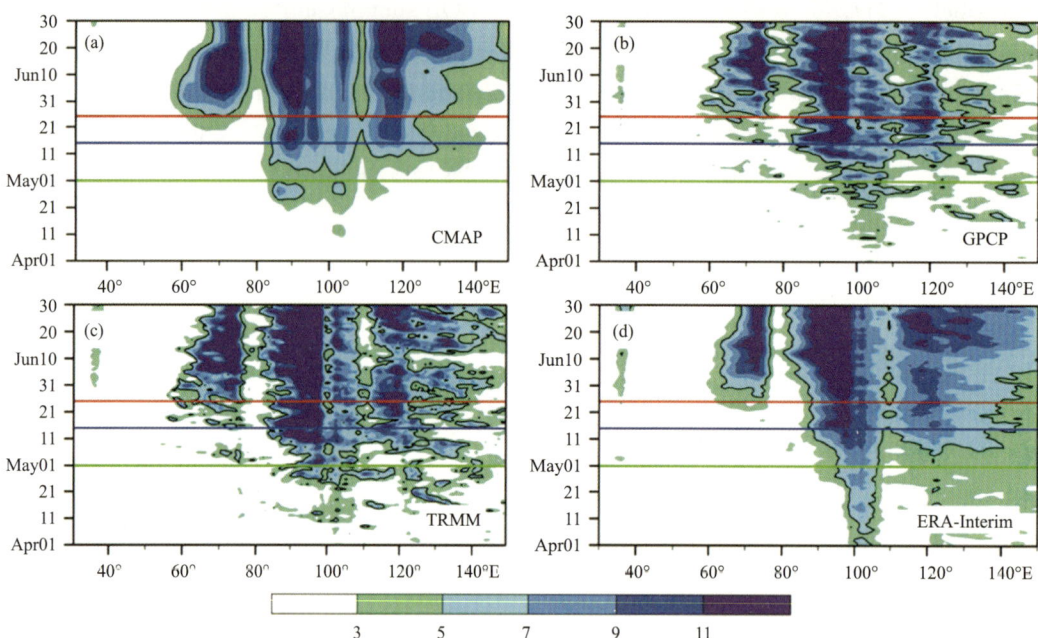

Fig. 1 Evolutions of daily mean precipitation derived from (a) CMAP, (b) GPCP, (c) TRMM and (d) ERA-Interim averaged over 10°—20°N for the period 1 April to 30 June. The contour interval is 2 mm d^{-1}, and solid black curves indicate the 5 mm d^{-1} contour. Red, blue and green bold lines denote, respectively, the onset dates of summer monsoon over the BOB (5°—15°N, 85°—100°E), SCS (10°—20°N, 110°—120°E), and India (10°—20°N, 60°—75°E) regions defined by the MTG index, as shown in Fig. 2a

The ASM onset process can be depicted more clearly by analyzing its climatological daily or pentad isochrones (Tao et al., 1987; Lau et al., 1997; Webster et al., 1998; Wang et al., 2002). Due to the limitation of available data, as well as the different onset criteria and diversity of datasets used in various studies, remarkable discrepancies exist in previous work with respect to the isochrones of ASM onset. Most of these isochrones of summer monsoon onset are based on the seasonal rainfall variation (Tao et al., 1987; Lau et al., 1997; Wang et al., 2002). However, since ASM onset is accompanied by a seasonal reversal of the prevailing wind direction, it is insufficient to determine ASM onset based solely on the rainfall criterion without considering the change in wind direction. Accordingly, it is more appropriate to identify ASM onset in light of the seasonal transition of the MTG, the wind direction change, and the rainfall variation.

Here, the ASM onset isochrones are calculated over the ASM region based on the ASMOC described in section 2. The ASM region is identified as the area where all three of the onset criteria are satisfied for at least one pentad in boreal summer. The results shown in Figure 2b demonstrate that ASM starts with BOB summer monsoon onset at Pentad 25 (1—5 May). It then expands eastwards, crosses the Indochina Peninsula, and reaches the SCS at Pentad 28 (16—20 May), corresponding to the SCS summer monsoon onset. However, after ASM onset is established over the BOB around early May, it cannot propagate westward to reach India and the Arabian Sea directly. In other words, the westward propagation of ASM onset is blocked near the western coast of the BOB, forming a MOB in situ (Fig. 2b). Here the MOB region is defined as the ocean portion in the definition region which is located in the western coast of the BOB and the eastern coastal waters of Indian Peninsula bounded between 78° and 82.5°E and between 10° and 20°N. In this region the ASMOC is not satisfied and the tropical rainfall greater than 5 mm d^{-1} is limited to the south of 10°N from April to early June (Fig. 2c). Because of the existence of MOB, the summer monsoon onset process over the Arabian Sea/India appears as a rainy belt advancing northward from the equator to the north during Pentads 25—31. The rainy belt reaches the southwestern coast of Indian Peninsula at Pentad 30 (26—30 May), indicating the Indian summer monsoon onset, and by the time the one-month process of tropical ASM onset has completed.

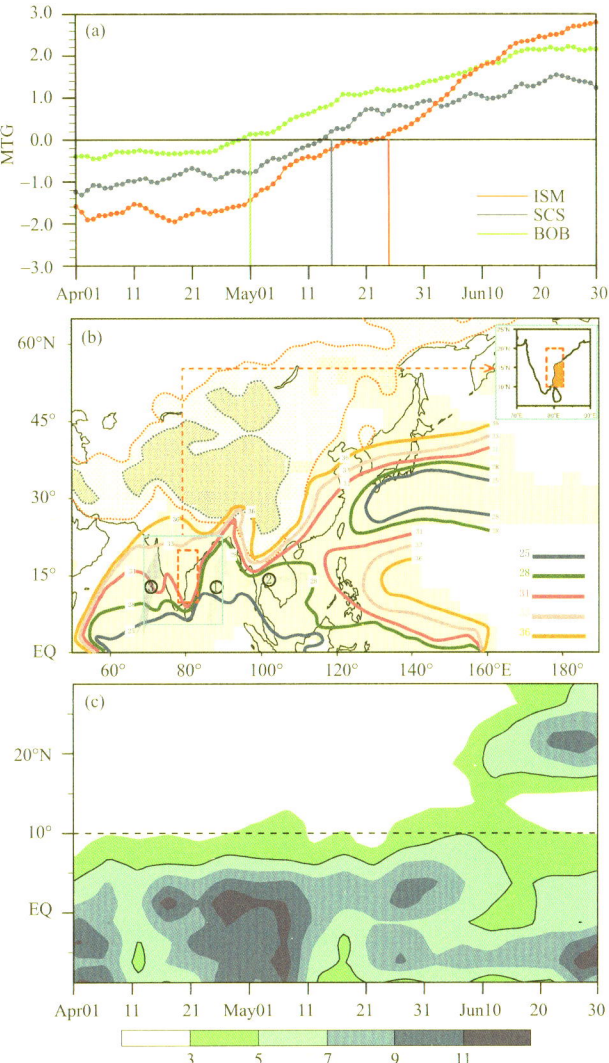

Fig. 2 (a) Time series of the MTG (10^{-6} K m^{-1}) averaged over the BOB (5°—15°N, 85°—100°E), SCS (10°—20°N, 110°—120°E), and India (10°—20°N, 60°—75°E). Green, blue and red bold lines denote the onset date of summer monsoon over the BOB, SCS, and India, respectively. (b) The climatological pentad-isochrones indicating the onset process of Asian summer monsoon. The arrows with the circled numbers 1, 2 and 3 represent the onset phases over the BOB, SCS and India. The area encircled by red and blue dot lines represents the topography greater than 500 m and 1500 m, respectively. Orange shading within the definition region (box bounded by red dashed lines between 78° and 82.5°E and between 10° and 20°N) over the eastern coastal waters of Indian Peninsula in the inserted map denotes the Monsoon Onset Barrier (MOB) region. The gray-filled area represents the Asian summer monsoon region where the ASMOC are satisfied (see text for details). (c) Time-latitude crossing section of CMAP rainfall (mm d^{-1}) from April to June and averaged between 78° and 82.5°E in which the MOB is located, solid black curves indicate the 5 mm d^{-1} contours

4 Possible mechanism for the MOB formation

Figures 2b and 2c indicate that the MOB is generated after the BOB summer monsoon onset. To investigate the effect of BOB summer monsoon convection on the MOB formation, we define the pre-onset and post-onset stage as the mean state from D_{-10} to D_{-1} and from D_{+1} to D_{+10}, respectively. To highlight the effect of BOB monsoon convection, the onset change is defined as the difference between the post-onset and pre-onset stages. This 21-day period centered on the onset day is long enough for Rossby waves to develop but short enough to isolate the onset signal from the more general seasonal evolution. Based on a diagnostic analysis of various elements in different onset stages and their onset changes, the impact of BOB monsoon onset on the local circulation and air-sea interaction can be revealed.

4.1 Impact of BOB convection on the large-scale atmospheric circulation

The distribution of diabatic heating in the middle troposphere and the relevant atmospheric circulation at different levels associated with BOB summer monsoon onset are shown in Figure 3. In the pre-onset stage of BOB summer monsoon, the diabatic heating at 400 hPa released by tropical convection is situated over the southern BOB along the equator (Figs. 3a—c). The only rainfall located from 10° to 20°N is the pre-monsoon convection over the Indochina Peninsula, which is generated under the mechanical and thermal forcing of the Tibetan Plateau prior to the BOB summer monsoon onset (Liu et al., 2013). Meanwhile, in the upper troposphere the South Asian high (SAH) settles at 10°N with its center over the southern SCS (Fig. 3a), and in the middle and lower troposphere a continuous subtropical anticyclone belt is located along 10° to 20°N from Arabian Peninsula to western Pacific (Figs. 3b and 3c). After the BOB summer monsoon onset in early May, as shown in Figure 2b, massive diabatic heat release (Q_1) takes place over the northeastern BOB due to the monsoon convection (Figs. 3d—f); the SAH at 200 hPa has migrated northwestward over the Indochina Peninsula (Fig. 3d); and a deep trough in the middle and lower troposphere, named the Indo-Burma trough, has established over the northern BOB (Figs. 3e and 3f). The evolution of circulation during the BOB summer monsoon onset can be further depicted by the onset changes in Q_1 at 400 hPa and the wind field on different isobaric surfaces (Figs. 3g—i). A meridional dipole pattern of Q_1 at 400 hPa, presenting a positive center over the northern and central east BOB and a negative one near the equator, implicates the northward migration of monsoon convection over the BOB. As the SAH elongates northwestwards, the onset change of 200 hPa circulation manifests itself as an anticyclone over the Tibetan Plateau to the north of the Q_1 center over the northeastern BOB (Fig. 3g). In the middle and lower troposphere, a closed cyclone is located over the BOB at 500 hPa and 700 hPa (Figs. 3h and 3i), consistent with the formation of the Indo-Burma trough and the splitting of the subtropical anticyclone belt. Apparently, the whole picture presents a Gill-type circulation response to the BOB monsoon heating (Gill, 1980).

The temporal evolutions of vertical motion over different regions are depicted in Figure 4. As a Gill-type Rossby-wave response, a strengthening descent is situated over the MOB region to the west of the BOB monsoon convection (Fig. 4a), especially in the lower troposphere below 500 hPa. After the BOB summer monsoon onset, the monsoon convection over the BOB releases a mass of condensation heating after D_0. Then the descent over the MOB region, which settles below 500 hPa before D_0, is enhanced with the strengthening downward airflow from 500 to 200 hPa after the BOB summer monsoon onset

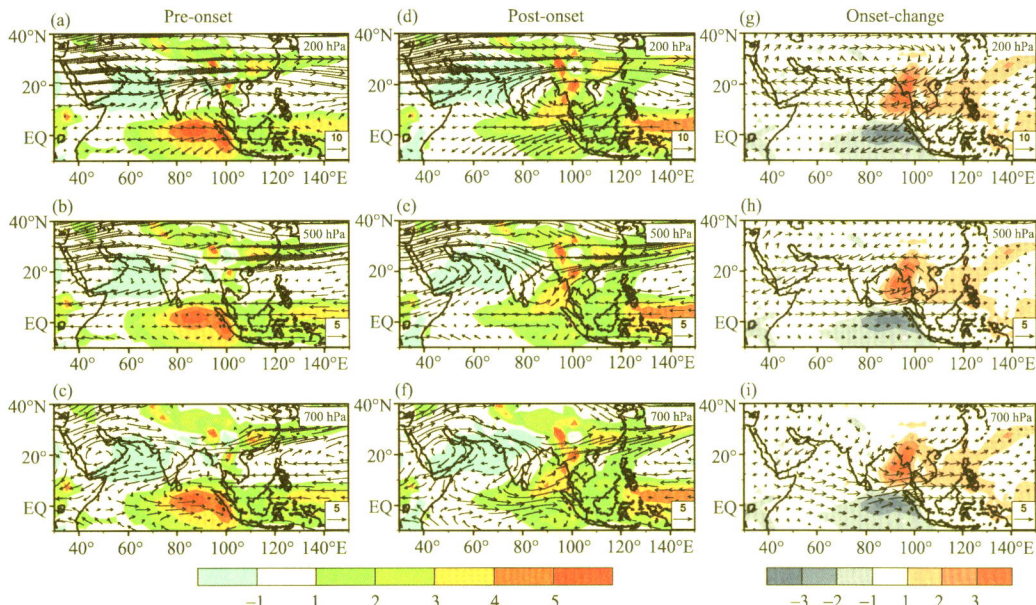

Fig. 3 Horizontal distributions of wind field (vectors) at (a, d) 200 hPa, (b, e) 500 hPa, and (c, f) 700 hPa, and diabatic heating Q_1 at 400 hPa (shading) in the pre-onset (a—c) and post-onset (d—f) stages of BOB summer monsoon. The onset change of Q_1 at 400 hPa (shading) and the wind field (vectors) at 200 hPa, 500 hPa and 700 hPa are presented in (g—i), respectively. Units are m s^{-1} for wind and K day^{-1} for Q_1

(Fig. 4a). Consequently, the condensation heating released by the BOB summer monsoon convection is conducive to the formation of the MOB by enhancing the descent to the west of the BOB. In the meantime, a transition from local descent to ascent occurs over the BOB and to its east, accompanied by the remarkable intensified ascent with its maximum near 500 hPa (Figs. 4b and 4c). Furthermore, the ascent over the SCS develops in the whole troposphere after D_{+4}, implying the incubation of the SCS summer monsoon onset (Fig. 4c).

Figure 4 also shows that the vertical shear of zonal wind is different over the MOB region, BOB and SCS regions during ASM onset. Before D_0, the vertical westerly shear below 200 hPa controls the above three regions. While the vertical westerly shear persists over the MOB region after the BOB summer monsoon onset, it weakens and even reverses to vertical easterly shear over the BOB and SCS after D_0 and D_{+4}, respectively. Actually, the vertical shear of zonal wind in the latitudinal band between 10° and 20°N is associated with the location of the subtropical anticyclone ridgeline ($u=0$) at different levels. In the upper troposphere, as the ridgeline migrates gradually northward during the BOB monsoon onset, the easterly is strengthened over the whole ASM region, as shown in Figures 3g and 5a. Note that after the BOB summer monsoon onset, the ridgeline at 200 hPa moves northward slowly to the west of 90°E, but faster to its east. This is because the SAH is evidently expanding northeastward due to the latent heat released by monsoon convection over the BOB and SCS (Fig. 3g). In the lower troposphere, however, the ridgeline migration is quite zonal asymmetric (Fig. 5b). Before and after the BOB monsoon onset, the ridgeline remains stable to the west of 90°E, but migrates southward to its east. The southward withdrawal of the ridgeline over the east is mainly due to the development of cyclonic circulation from the BOB to the SCS, as shown in Figures 3h and 3i. Thus, the vertical tilting of the ridgeline between 500 hPa and 200 hPa over the Indian Peninsula remains southward, maintaining the wintertime pattern (Fig. 5c); whereas, it is northward over the Indochina Peninsula (from 100° to 110°E) and has changed

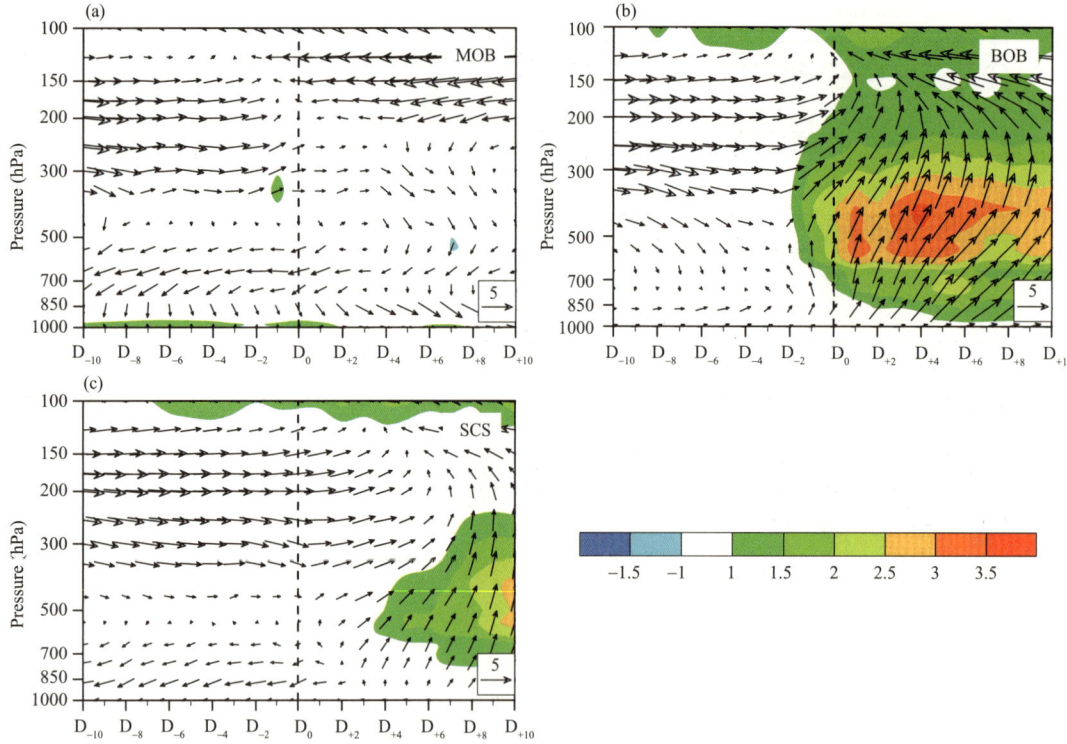

Fig. 4 Pressure-time cross section of local zonal circulation (vectors; m s^{-1}) and Q_1 (shading; K day^{-1}) over the (a) MOB region (orange shading shown in Fig. 2b), (b) BOB (10°—20°N, 90°—100°E), and (c) SCS (10°—20°N, 110°—120°E)

to the summertime pattern (Fig. 5d). These changes suggest that, after the BOB summer monsoon onset, vertical westerly shear still maintains over the Indian Peninsula, corresponding to the warm-in-the-south and cold-in-the-north type pattern associated with winter (Fig. 5e); whereas, strong easterly shear has established to the east of the BOB, implying the development of the warm-in-the-north and cold-in-the-south type pattern, typical of summer, over the eastern ASM area (Fig. 5f). Since a vertical easterly shear strengthens air ascent, the monsoon rainfall expands eastward to the Indochina Peninsula and the SCS. Moreover, based on the thermal wind relationship, the occurrence of vertical easterly shear prompts the MTG reversal from negative to positive and favors the establishment of summer circulation. Consequently the SCS summer monsoon builds up by Pentad 28, following the BOB summer monsoon onset (Fig. 2b). In contrast, the stable vertical westerly shear accompanied by air descent over the MOB region and the Indian Peninsula inhibits the development of monsoon convection. Hence, the westward extension of ASM onset is blocked.

4.2 Onset changes of thermal structure in the lower troposphere

As shown in Figures 1 and 2, the MOB manifests itself as suppressed monsoon convection over east coastal waters of the Indian Peninsula. Hence, it is necessary to first analyze the thermal status over the MOB region to check the atmospheric convective instability. Generally, the spatial distribution of convective available potential energy (CAPE) is used to identify the regions where the atmosphere is convectively unstable, and it determines the potential of deep convection.

However, a high value of CAPE does not necessarily result in strong convection (e.g., Thompson et al., 1979; McBride et al., 1999; Sobel et al., 2004; Yano et al., 1991), as the simulated air parcel

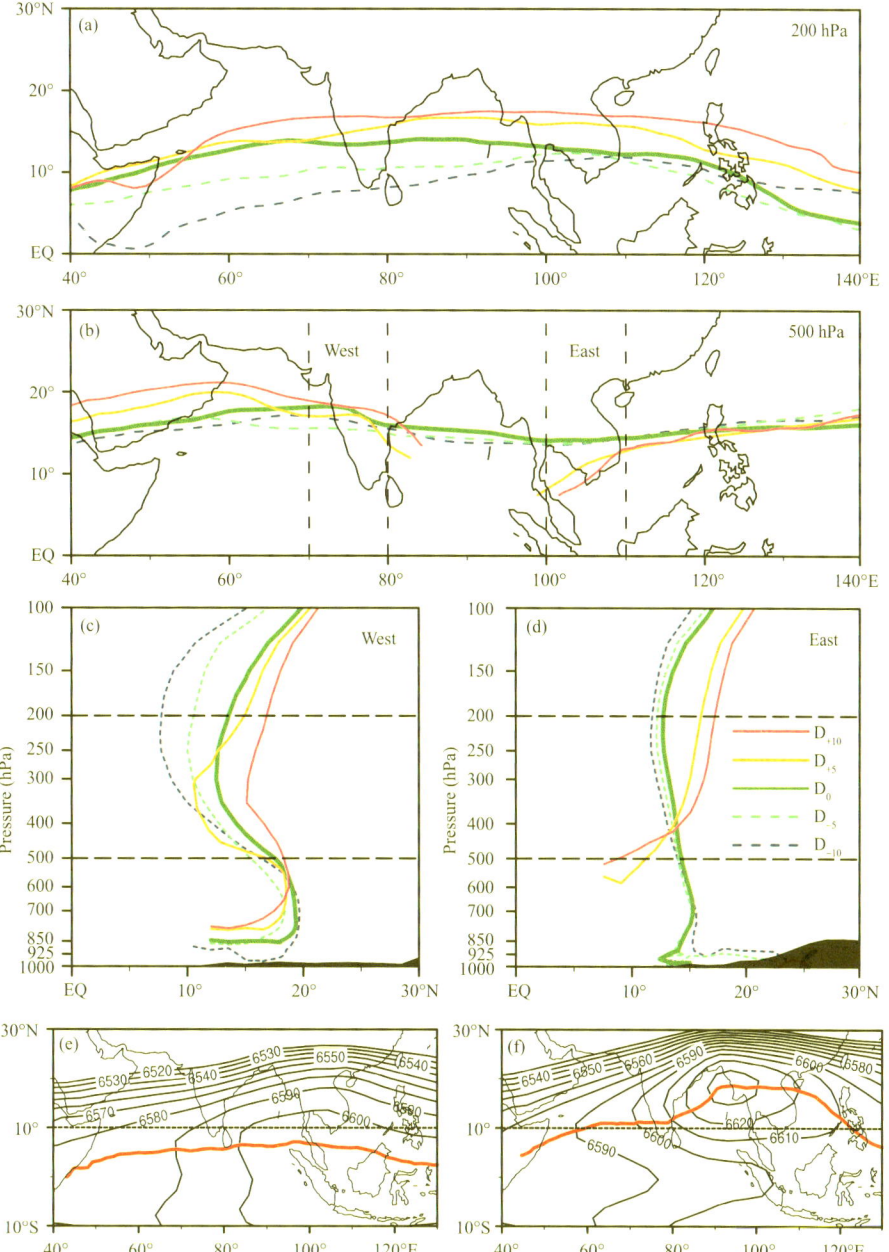

Fig. 5 Evolution of the ridgeline from D_{-10} to D_{+10} with respect to the BOB summer monsoon onset date at (a) 200 hPa and (b) 500 hPa. Panels (c) and (d) respectively present vertical cross sections of the ridgeline of subtropical anticyclone averaged over the "west" and the "east" region, as indicated in (b). Shaded areas in (c) and (d) denote topography. Panels (e) and (f) are the horizontal distribution of the thickness between 500 and 200 hPa (gpm) in the pre-onset and post-onset stage, respectively. The red solid line in (e) and (f) represents the maximum axis of air column thickness

needs to overcome a usually stable layer between the surface and the level of free convection (LFC). The intensity of this stable layer is identified by convective inhabitation energy (CINE), which is expressed as

$$\mathrm{CINE} = -\int_{P_{\mathrm{SFC}}}^{P_{\mathrm{LFC}}} R_d (T_{ve} - T_{vp}) \mathrm{d}(\ln P) \tag{3}$$

Where P_{SFC} is the surface pressure level (1000 hPa in this study); P_{LFC} is the pressure at LFC for a parcel that has risen from 1000 hPa; T_{ve} is the virtual temperature of the environment at pressure level P through which the parcel rises; T_{vp} is the virtual temperature of the parcel at pressure level P through which the parcel rises; and R_d is the dry gas constant.

As CINE defines the energy of a parcel needed to reach the LFC and therefore to be able to develop convection, CINE describes the limiting factor, which can prevent convection even though very high values of CAPE may exist. Thus, the occurrence of convection is closely associated with the CINE intensity. The larger the CINE, the more difficult it is for the convection to take place.

The horizontal distribution of CINE during the BOB summer monsoon onset is presented in Figure 6. Before the BOB summer monsoon onset, the CINE is larger than 100 J kg^{-1} over the Indochina Peninsula, and it is greater than 200 J kg^{-1} over the Indian Peninsula. The relatively small CINE over the Indochina Peninsula is related with the local pre-monsoon rainfall. Furthermore, the CINE is very small over the southern BOB and southern Arabian Sea (Fig. 6a), suggesting active convection in the tropics. Subsequently, monsoon convection develops over the Indochina Peninsula after the BOB summer monsoon onset (Figs. 2b and 4b), accompanied by an obvious in situ decrease in CINE (Fig. 6b). In addition, the CINE starts to diminish over the central and northern Arabian Sea, predicting the independent development of Indian summer monsoon convection (Fig. 2b). Note that the CINE becomes more organized and even larger to the west of the BOB monsoon convection, especially over the MOB region (Fig. 6b). Such evolution of CINE can be presented more evidently in its onset change. Figure 6c shows that

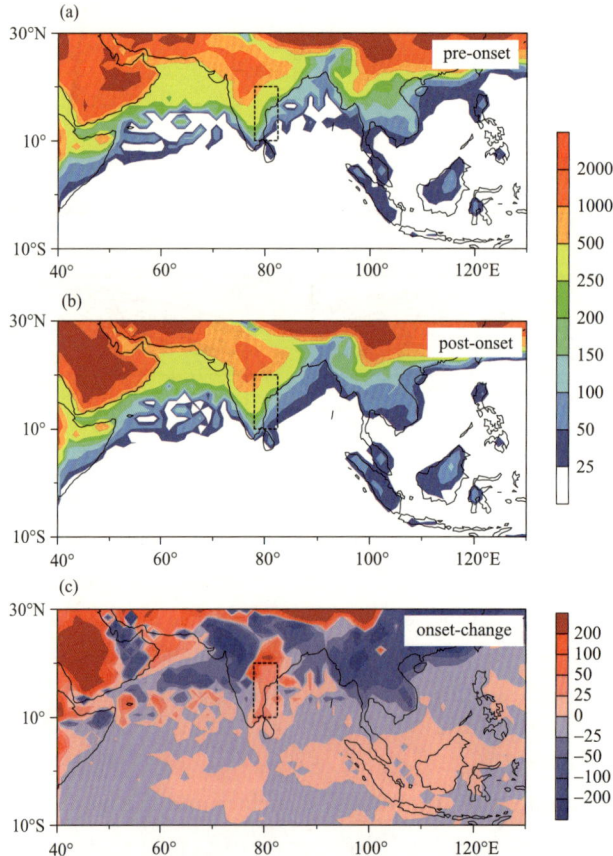

Fig. 6 Horizontal distributions of CINE (J kg^{-1}) in the pre-onset and post-onset stage. The onset change of CINE is presented in (c). The ocean portion in the box surrounded by dashed lines is the defined MOB region

the CINE is increased over the MOB region, but decreased over the eastern BOB after the BOB summer monsoon onset. Over the MOB region, the descent is enhanced (Fig. 4a), leading to a drier and more stable layer and increased CINE in the lower troposphere; whereas, the developed ascent over the eastern BOB is consistent with the less-stable lower troposphere and the decreased CINE. Moreover, the onset change of CINE is determined by variation of the pressure at LFC and the virtual temperature difference between environmental and air parcel ($T_{ve} - T_{vp}$) according to Eq. (3). Figure 7a shows that after the BOB summer monsoon onset, the temperature difference between environment and air parcel is increased evidently over the MOB region, but decreased over the eastern BOB. While the pressure at LFC is decreased over the northern BOB with its minimum near 90°E, but increased over the eastern BOB (Fig. 7b). The overlapping in the MOB region between the enhanced CINE and the increased temperature difference (Figs. 6c and 7a) implies that the onset change in temperature difference between environment and air parcel is more important for the increase of CINE over the MOB region after the BOB summer monsoon onset.

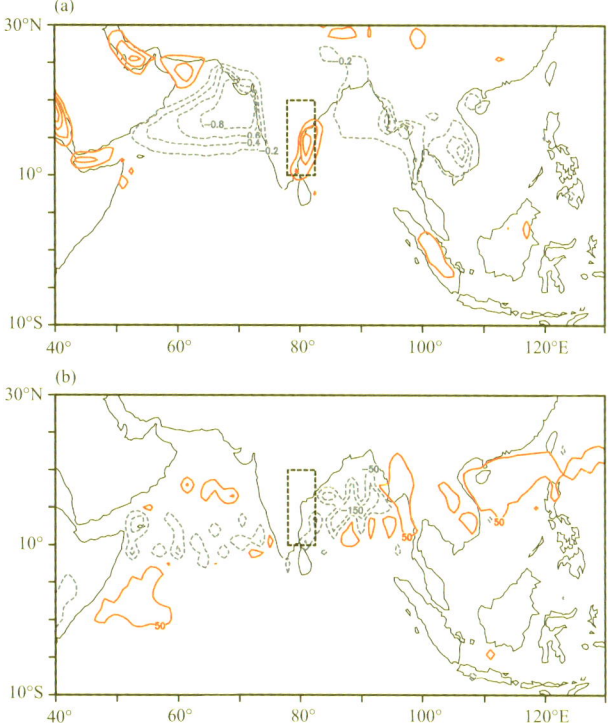

Fig. 7 Onset changes of (a) the difference between environmental and air parcel virtual temperature ($T_{ve} - T_{va}$, interval is 0.2 K) averaged from 1000 hPa to 900 hPa and (b) the pressure of lifting free convection (interval is 100 hPa). The ocean portion in the box surrounded by dashed lines is the defined MOB region

The increased onset change in CINE in the MOB region suggests that it requires more energy to uplift the air parcel to the LFC from the surface over the MOB region and the development of monsoon convection becomes more difficult in this region.

4.3 Impact of BOB convection on air-sea interaction

Air-sea heat flux is composed of solar radiation, longwave radiation, latent heat flux, and sensible heat flux. Their onset changes are portrayed in Figure 8. After the BOB summer monsoon onset, due to the abrupt increase in cloud amount over the BOB region, the downward solar radiation is decreased obviously while the downward longwave radiation is increased slightly. Thus the total downward radiation heating received by the sea surface of the BOB is reduced (Figs. 8a and 8b). Simultaneously, the upward latent heat flux increases over the BOB (Fig. 8c). Furthermore, the upward sensible heat flux is enhanced over the eastern BOB but weakened over the western BOB, with its minimum over the MOB region (Fig. 8d).

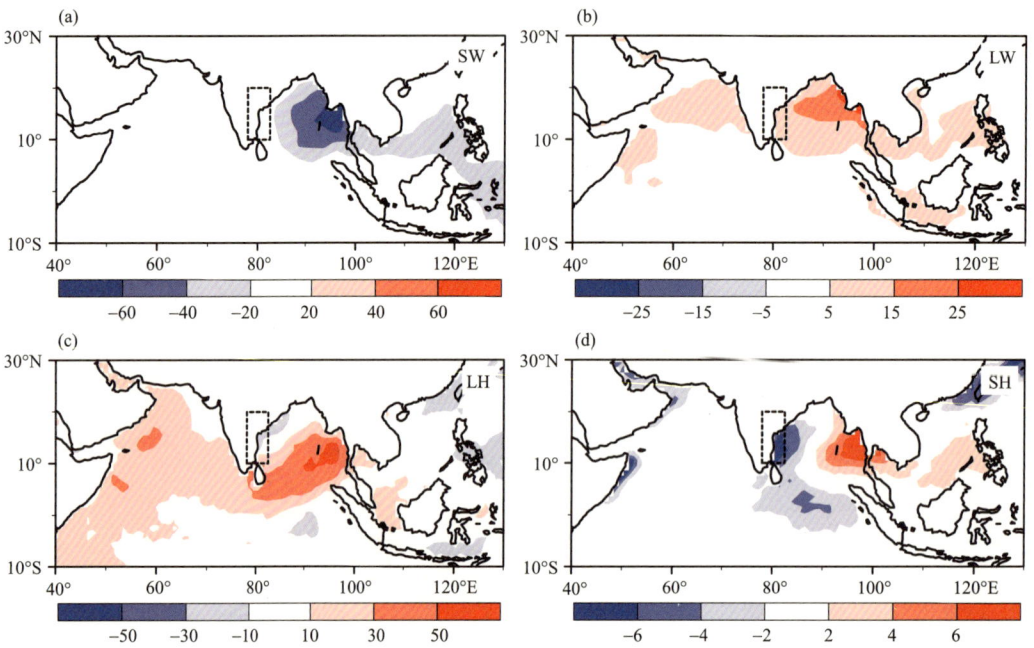

Fig. 8　Onset changes of heating flux (W m^{-2}) at the sea surface: (a) net downward solar radiation; (b) net downward longwave radiation; (c) upward latent heating flux; and (d) upward sensible heating flux. The radiation and flux are extracted from the ISCCP and OAFlux datasets, respectively. Downward is positive in (a) and (b), while upward is positive in (c) and (d). The ocean portion in the box surrounded by dashed lines is the defined MOB region

Because of the smaller scale of sensible heat flux, the SST variation is primarily controlled by the onset changes in solar radiation, longwave radiation and latent heat flux. When the BOB summer monsoon builds up, a local cyclone is formed in the lower troposphere over the BOB (Fig. 3i). On one hand, the monsoon convection increases the cloud amount (Figs. 9a and 9b) to reduce the shortwave solar radiation arriving at the sea surface (Fig. 8a); on the other hand, it strengthens the sea-surface southwesterly, which remarkably increases the latent heat release from the sea surface (Fig. 8c). As a result, the monsoon onset convection over the BOB decreases the SST over most of the BOB by diminishing the shortwave solar radiation and strengthening the sea-surface heat release (Fig. 9c).

Although the sensible heat flux has little influence on SST variation, it is very important for the atmospheric circulation. First, a cloud-free atmosphere does not absorb radiation heating effectively, while the sensible heating from the bottom can generate available potential energy (APE) efficiently in a region S where the heating Q and the temperature T are positively correlated, i.e., $\int_S (TQ) \mathrm{d}s > 0$. Second, it can act as a trigger to release positive CAPE in the atmospheric column by elevating a moist air parcel to the LFC from the underlying surface in the boundary layer. Therefore, as shown in Figure 8d, the positive sensible heat flux then increases over the eastern BOB where the ASM onset propagates, whereas

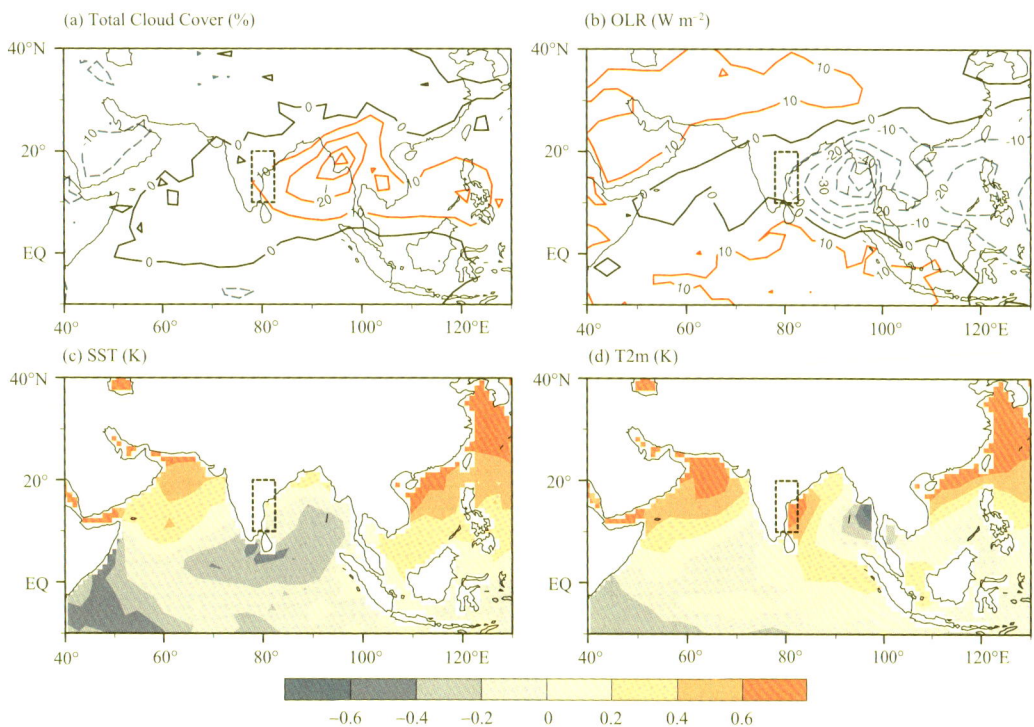

Fig. 9 Onset changes in (a) total cloud cover (%), (b) OLR (W m^{-2}), (c) SST (K), and (d) air temperature at 2 m height (K). The ocean portion in the box surrounded by dashed lines is the defined MOB region

the negative change of sensible heat develops over the MOB region to inhibit any uplifting of moist parcels, and then the monsoon convection is suppressed. Consequently, the strong negative sensible heat over eastern coastal waters of the Indian Peninsula becomes another factor contributing to the MOB formation after the BOB summer monsoon onset.

More importantly, the polarity of sensible heating over the BOB is ascribed to the difference between SST and surface air temperature (SAT; represented by air temperature at 2 m height). Figures 9c and 9d implicate that although the SST is cooling over most of the BOB, the SAT variation manifest itself as a very zonal asymmetric distribution, say that the SAT is cooling over the eastern BOB but warming over the western BOB after the BOB summer monsoon onset. Also the warming SAT center is over the MOB region. Actually, after the BOB summer monsoon onset over the eastern BOB, the cooling of SST is weaker than that of SAT, and so positive sensible heating is enhanced. However, over the western BOB, especially over the MOB region, the SST cooling and SAT warming engender negative sensible heating (Figs. 9c and 9d). Thus the zonal asymmetry of onset change in sea-surface sensible heat flux is primarily ascribed to the SAT variation over the BOB after the BOB summer monsoon onset. In addition to the cooling of SST, the warming of SAT over the MOB region also contributes to the local strong negative sensible heat flux, which supports the MOB genesis.

To investigate the possible reason for the SAT variation before and after the BOB summer monsoon onset, we treat the variables at 925 hPa as proxies for those near the surface because of the similarity of wind and air temperature distributions between 925 hPa and the surface (figure not shown). Their evolution during monsoon onset is shown in Figure 10. For the 10-day average state the time-dependent terms are weak enough to be ignored and the thermodynamic equation can be written in its stationary form:

$$0 \approx \frac{Q_1}{c_p} - \boldsymbol{V} \cdot \nabla T - \omega \frac{\partial T}{\partial p} \tag{4}$$

where $\boldsymbol{V}$, T and Q_1 are horizontal wind, air temperature, and diabatic heating at 925 hPa, respectively; and c_p is 1004 J K^{-1} kg^{-1}, representing the specific heat of dry air at constant pressure. The balance relationship of Eq. (4) varies from place to place in different episodes of BOB summer monsoon onset. In the pre-onset stage (Figs. 10a—c), over the Arabian Sea and the BOB, the diabatic cooling is primarily balanced by adiabatic warming associated with local subsidence. Meanwhile, the diabatic heating over

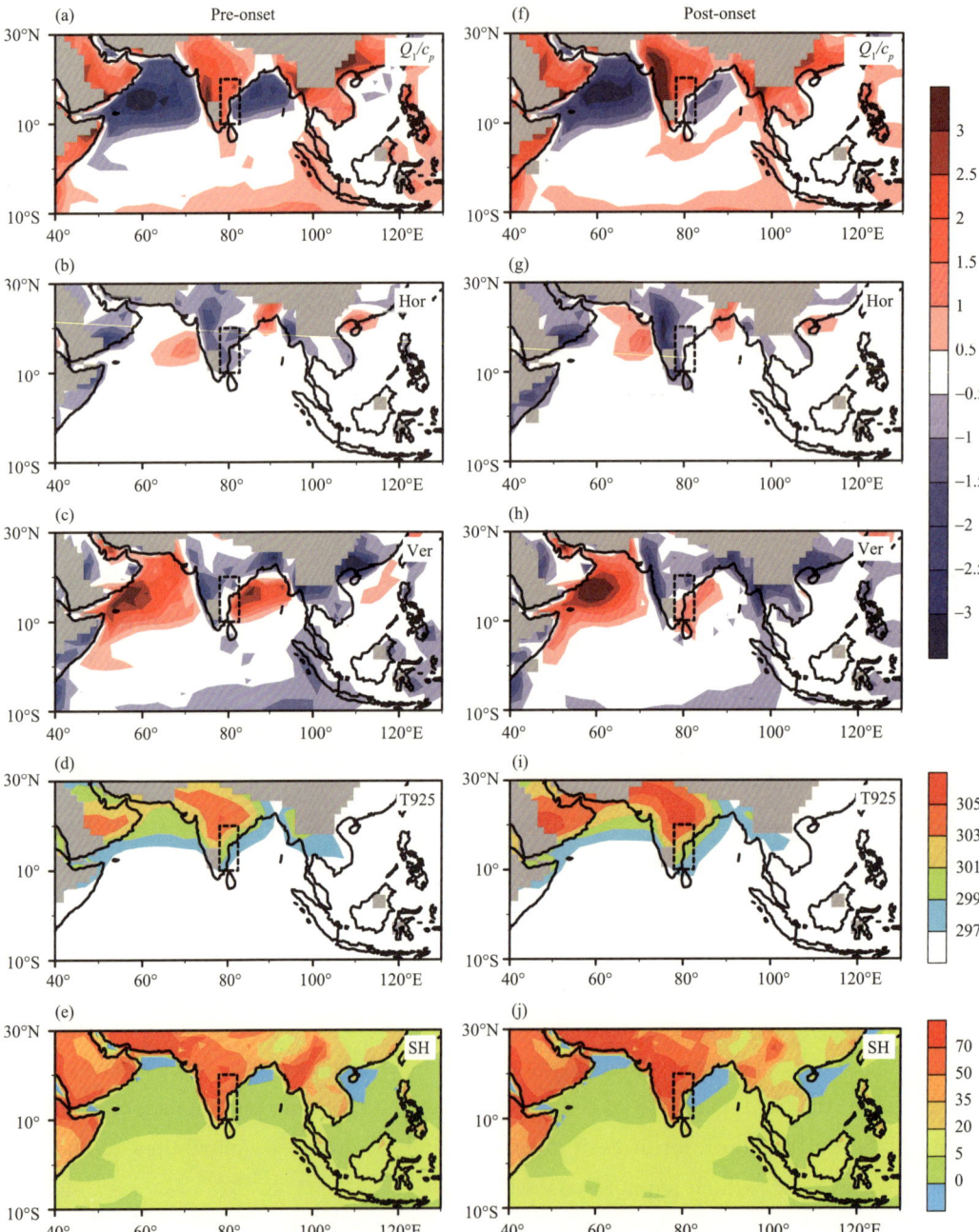

Fig. 10 Horizontal distributions of (a, f) diabatic heating, (b, g) horizontal temperature advection, (c, h) adiabatic heating, and (d, i) air temperature (units: K) at 925 hPa; and (e, j) upward surface sensible heating flux (units: W m^{-2}) in the pre-onset (a—e) and post-onset (f—j) stages. Units are K day^{-1} in (a—c) and (f—h). All calculations are based on the ERA-Interim dataset. Gray shading denotes topography. The ocean portion in the box surrounded by dashed lines is the defined MOB region

the Indian Peninsula and Indochina Peninsula is compensated by both cold temperature advection and adiabatic cooling, with the latter prevailing over the former. The 925 hPa air temperature is warm over the Indian Peninsula but cold over Indochina Peninsula where the pre-monsoon rainfall takes place (Fig. 10d). The surface sensible heat flux is strong over land but weak or even negative over ocean in this period (Fig. 10e).

However, such equilibrium is changed in the post-onset period (Figs. 10f—h). The descent is still located over the Arabian Sea, where the adiabatic heating, together with relatively weak warm temperature advection, balances the local diabatic cooling. Over the MOB region, the diabatic cooling near the surface is due mainly to the enhanced negative sensible heating (Figs. 10f and 10j), and is balanced by the adiabatic warming due to the local descent after the BOB summer monsoon onset (Figs. 4a and 10h). To the west of MOB region there is stronger large-scale sinking over the western Indian Peninsula with little convection. Notice that the land surface sensible heat flux is increased to strengthen the near-surface diabatic heating (Figs. 10f and 10j), which is balanced by strong cold temperature advection (Fig. 10g) and relatively weak adiabatic cooling (Fig. 10h) in the lower troposphere. The air temperature then increases with the enhanced surface sensible heating over the Indian Peninsula (Fig. 10i). Moreover, to the east of MOB region a deep monsoon convection and strong ascent controls the Indochina Peninsula, where the diabatic heating induced by convective condensation is in equilibrium with adiabatic cooling, even at the near-surface level (Figs. 10f and 10h). Therefore, during the BOB onset process, the diabatic effect at 925 hPa is balanced by the adiabatic process over the ocean, including the MOB region, the eastern BOB and the Arabian Sea, but is compensated by both horizontal temperature advection and the adiabatic effect over the Indian Peninsula to the west of MOB region.

The reason for the strengthened SAT warming and negative sensible heat flux over the MOB region after the BOB summer monsoon onset can be investigated by the onset change in each term of Eq. (4) at 925 hPa, as displayed in Figure 11. After the BOB summer monsoon onset, the diabatic heating due to monsoon convection is strengthened over the northeastern BOB to the east of MOB region, and balanced by adiabatic cooling associated with the enhanced ascent (Figs. 11a and 11d). Meanwhile, the strengthened anticyclone over the western Indian Peninsula to the west of MOB region (Fig. 11b) enhances the local surface sensible heating (Fig. 11f) to increase the near-surface diabatic heating (Fig. 11a). Subsequently, the enhanced westerly over the Indian Peninsula (Fig. 11b) brings the heated air downstream to the MOB region. Hence, cold and warm temperature advection is strengthened over the Indian Peninsula and the MOB region, respectively (Fig. 11c). In other words, the intensified cold advection tends to balance the stronger diabatic heating over land (Figs. 11a and 11c); while the reinforced warm advection increases the air temperature over the MOB region, forming the local warm air (Fig. 11e) and producing negative sensible heating and diabatic cooling (Figs. 11f and 11a) which is in equilibrium with the adiabatic warming due to local descent (Figs. 4a and 11d). The maintenance of negative surface sensible heating over the MOB region after the BOB summer monsoon onset reduces the local APE since heating and temperature there are negatively correlated, and prevents the moist air parcel from uplifting to the LFC, further suppressing monsoon convection over the MOB region. Both the descent in the free atmosphere and the negative sensible heating in the boundary layer contribute to the MOB formation over eastern coastal waters of the Indian Peninsula, where the ASM onset is blocked. However, over the east of the BOB, the enhanced sensible heat flux, with a maximum of 6 W m^{-2} (Figs. 8d and 11f), provides a favorable condition for the uplifting of air parcel, and then facilitates the eastward propagation of monsoon onset. Simultaneously, the upper divergence-pumping associated with the SAH variation is also conducive to the occurrence of monsoon rainfall over the northeastern BOB; and the vertical coupling between the upper and lower circulation is vital for the formation and maintenance of cyclonic circulation in the lower troposphere over the eastern BOB (Liu et al., 2013). As a result, the monsoon precipitation advances eastward to the Indochina Peninsula and eventually reaches the SCS.

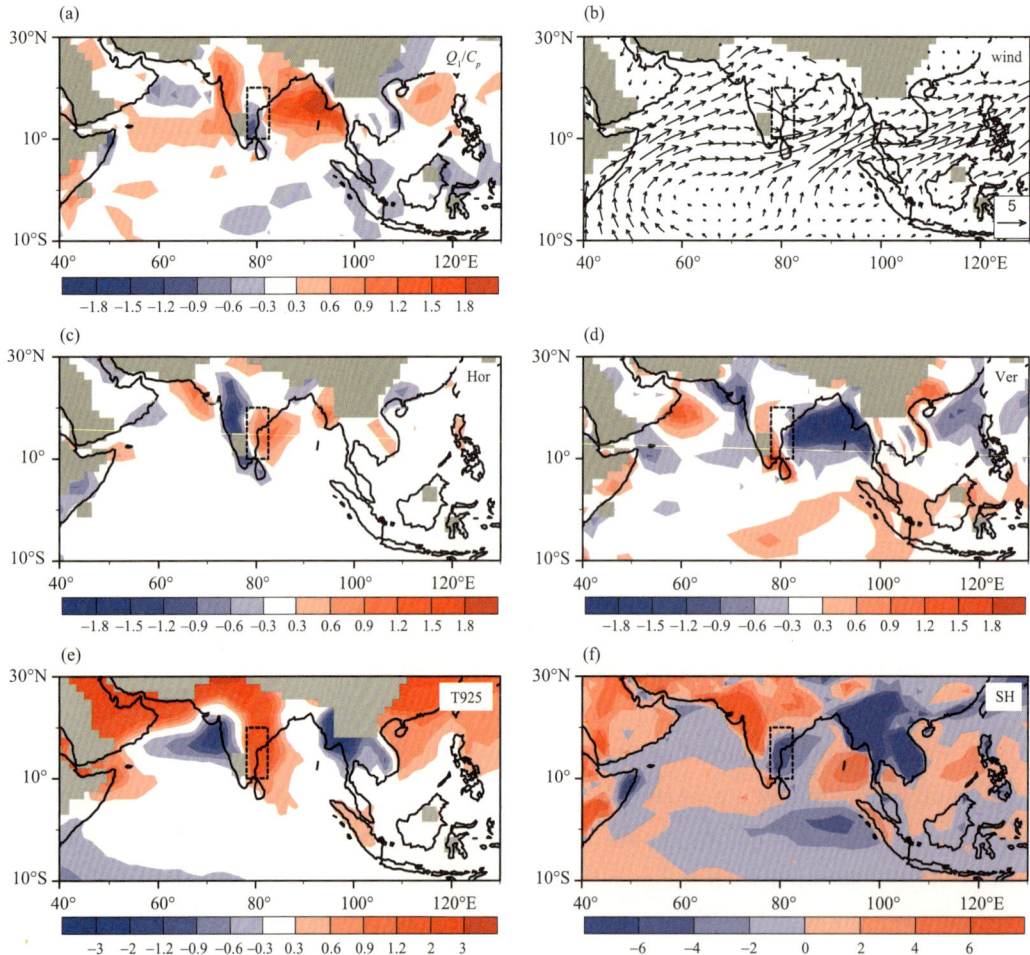

Fig. 11 Onset changes in (a) diabatic heating, (b) wind field (units: m s^{-1}), (c) horizontal temperature advection, (d) adiabatic heating, and (e) air temperature (units: K) at 925 hPa; and (f) upward surface sensible heating flux (units: W m^{-2}). Units in (a), (c) and (d) are K day^{-1}. Gray shading denotes topography. The ocean portion in the box surrounded by dashed lines is the defined MOB region

5 Summary and discussion

The ERA-Interim reanalysis dataset and latest rainfall products derived from satellite observation were used to investigate the ASM onset process. The results showed that the evolution of monsoon precipitation over India and the Arabian Sea is distinct from that over the BOB and the SCS during ASM onset, i. e., there is a discontinuity between the monsoon onset process over the Indian subcontinent to the Arabian Sea region and that over the BOB and SCS. After ASM onset occurs over the eastern BOB, it continues to advance eastward but cannot propagate westward. Actually, ASM onset propagation is blocked over the eastern coastal waters of Indian Peninsula to form a "monsoon onset barrier" (MOB) in situ, which is a conspicuous feature during the ASM onset process.

We also discussed the relevant mechanism responsible for the MOB formation. It was elucidated that the atmospheric feedback from the monsoon convective heating during BOB monsoon onset causes the remarkable changes of both the circulation in the free atmosphere and the local air-sea interaction in the boundary layer of the ASM region, and contributes to the MOB formation and ASM onset sequence.

(1) After summer monsoon is established over the BOB in early May, the condensation heating released by monsoon convection forces a large-scale circulation change. In detail, vertical westerly shear and air descent occurs over the MOB region and the Indian Peninsula, but vertical easterly shear and air ascent appears over the Indochina Peninsula and SCS. In the upper troposphere, the strong condensation heating released by the BOB monsoon convection leads to the strengthening and northwestward migration of the SAH. In the middle and lower troposphere, the Indo-Burma trough is generated over the northeastern BOB, which results in the southeastward retreat of the ridgeline of the subtropical anticyclone from 17.5° to 5°N. The entire pattern of circulation change in the troposphere presents a distinct Gill-type circulation response to the condensation heating over the eastern BOB. Consequently, in the free atmosphere vertical easterly shear develops to the east of the BOB, whereas the wintertime vertical westerly shear is maintained over the MOB region and the India. This large-scale circulation pattern is conducive to the development of air ascent to the east of the BOB and descent to its west, which inhibits the ASM onset from propagating westward directly but allows its eastward propagation.

(2) In the boundary layer, the BOB monsoon convection reduces the basin-wide SST through cloud-radiation and wind-evaporation-SST feedback (Xie, 1996). Meanwhile, the surface air temperature experiences strong cooling over the eastern BOB due to the monsoon rainfall development, but undergoes remarkable warming over the MOB region ascribed to the strong warm temperature advection from the Indian subcontinent. Consequently, the positive sensible heating over the area east of the central BOB is increased. This helps maintain the surface cyclonic circulation and air ascent in conjunction with the upper-pumping associated with the SAH evolution, and is thus favorable for the eastward expansion of monsoon rainfall to the SCS. Meanwhile, the negative sensible heating develops over the MOB region, which reduces the local APE, prevents the moist air parcel from uplifting to the LFC, and then suppresses the development of monsoon convection. Consequently, monsoon convection cannot be triggered, and summer monsoon rainfall is suppressed to generate the MOB to the west of the BOB monsoon convection.

Although ASM onset cannot advance westward from the BOB to India and the Arabian Sea directly, monsoon convection over the BOB and SCS can facilitate Indian summer monsoon onset indirectly via generating zonal asymmetric instability of the SAH (Liu et al., 2007) and producing divergence-pumping over the Arabian Sea in the upper troposphere (Wu et al., 2013; Zhang et al., 2014). From this viewpoint, the summer monsoon onsets over the BOB, SCS and India should be considered as an open-dissipative integral process controlled by complex land-air-sea interaction, including the influence of large-scale mountain ranges.

References

ABE M, HORI M, YASUNARI T, et al, 2013. Effects of the Tibetan Plateau on the onset of the summer monsoon in South Asia: The role of the air-sea interaction[J]. J Geophys Res Atmos, 118: 1760-1776.

ANANTHAKKRISHNAN R, SOMAN M K, 1988. The onset of the south-west monsoon over Kerala: 1901−1980[J]. J Climatol, 8: 283-296.

CHAN J C L, WANG Y, XU J, 2000. Dynamic and thermodynamic characteristics associated with the onset of the 1998

South China Sea summer monsoon[J]. J Meteor Soc Japan, 78: 367-380.

CHEN T C, CHEN J M,1995. An observational study of the South China Sea monsoon during the 1979 summer: Onset and life cycle[J]. Mon Wea Rev, 123: 2295-2318.

DEE D P, DE R P, UPPALA S M, et al,2011. The ERA-Interim reanalysis: Configuration and performance of the data assimilation system[J]. Quart J Roy Meteor Soc, 137: 553-597.

DING Y H, CHAN J C L,2005. The East Asian summer monsoon: An overview[J]. Meteorol Atmos Phys, 89: 117-142.

EADY E T,1950. The cause of general circulation of atmosphere[J]. Cent Proc Roy Meteor Soc: 156-172.

GADGIL S,2003. The Indian monsoon and its variability[J]. Annu Rev Earth Planet Sci, 31: 429-467.

GADGIL S, SAJANI S,1998. Monsoon precipitation in the AMIP runs[J]. Clim Dyn, 14: 659-689.

GILL A E,1980. Some simple solutions for heat-induced tropical circulation[J]. Quart J Roy Meteor Soc, 106: 447-462.

GOSWAMI B N,2005. Intraseasonal Variability in the Atmosphere-Ocean Climate System[C]//LAU WKM, WALISER D E. South Asian Monsoon. London: Springer-Praxis: 19-61.

GOSWAMI B N, MOHAN R S A,2001. Intra-seasonal oscillations and inter-annual variability of the Indian summer monsoon[J]. J Climate, 14: 1180-1198.

HUFFMAN G J, ADLER R F, MORRISSEY M M, et al,2001. Global precipitation at one-degree daily resolution from multi-satellite observations[J]. J Hydrometeor, 2: 36-50.

HUNG C W, YANAI M,2004a. Factors contributing to the onset of the Australian summer monsoon[J]. Quart J Roy Meteor Soc, 130: 739-758.

HUNG C W, LIU X D, YANAI M, et al,2004b. Symmetry and asymmetry of the Asian and Australian summer monsoons[J]. J Climate, 17: 2413-2426.

KALNAY E, KANAMITSU M, KISTLER R, et al,1996. The NCEP/NCAR 40-year reanalysis project[J]. Bull Amer Meteor Soc, 77: 437-471.

KRISHNAMURTI T N, ARDANUY P, RAMANATHAN Y, et al,1981. On the onset vortex of the summer monsoon[J]. Mon Wea Rev, 109:344-363.

LAU K M, YANG S,1997. Climatology and interannual variability of the southeast Asian summer monsoon[J]. Adv Atmos Sci, 14: 141-162.

LAU K M, WU H T, YANG S,1998. Hydrologic processes associated with the first transition of the Asian summer monsoon: A pilot satellite study[J]. Bull Am Meteor Soc, 79: 1871-1882.

LI C F, YANAI M,1996. The onset and interannual variability of the Asian summer monsoon in relation to land-sea thermal contrast[J]. J Climate, 9: 358-375.

LIEBMANN B, SMITH C A,1996. Description of a complete (interpolated) outgoing longwave radiation dataset[J]. Bull Amer Meteor Soc, 77: 1275-1277.

LIU B Q, WU G X, MAO J Y, et al,2013. Genesis of the South Asian high and its impact on the Asian summer monsoon onset[J]. J Climate, 26: 2976-2991.

LIU Y, CHAN J, MAO J, et al,2002. The role of Bay of Bengal convection in the onset of the 1998 South China Sea summer monsoon[J]. Mon Wea Rev, 130: 2731-2744.

LIU Y M, BRIAN H, MICHAEL B,2007. Impact of Tibetan orography and heating on the summer flow over Asia[J]. J Meteor Soc Japan, 85B: 1-19.

MAK M, KAO C Y J,1982. An instability study of the onset-vortex of the southwest monsoon, 1979[J]. Tellus, 34:358-368.

MAO J Y, WU G, LIU Y,2003. Study on the variation in the configuration of subtropical anticyclone and its mechanism during seasonal transition-Part I: Climatological features of subtropical high structure[J]. Acta Meteor Sin, 17: 274-286.

MAO J Y, WU G X,2007. Interannual variability in the onset of summer monsoon over the eastern Bay of Bengal[J]. Theor Appl Climatol, 89: 155-170.

MAO J Y, WU G X,2011. Barotropic process contributing to the formation and growth of tropical cyclone Nargis[J]. Adv Atmos Sci, 28: 483-491.

MCBRIDE J L, FRANK W M,1999. Relationships between stability and monsoon convection[J]. J Atmos Sci, 56: 24-36.

PRASAD V S, HAYASHI T,2005. Onset and withdrawal of Indian summer monsoon[J]. Geophys Res Lett. doi: 10.1029/2005GL023269.

QIAN Y F, WANG S Y, SHAO H,2001. A possible mechanism effecting the earlier onset of southwesterly monsoon in the South China Sea compared to the Indian monsoon[J]. Meteor Atmos Phys, 76: 237-249.

SAHA S, SAHA K,1980. A hypothesis on onset, advance and withdrawal of the Indian summer monsoon[J]. Pageoph, 118: 1066-1075.

SOBEL A H, YUTER S E, BRETHERTON C S, et al,2004. Large-scale meteorology and deep convection during TRMM KWAJEX[J]. Mon Wea Rev, 132: 422-444.

SRINIVASAN J, GADGIL S, WEBSTER P J,1993. Meridional propagation of large-scale monsoon convective zones[J]. Meteor Atmos Phys, 52: 15-35.

TANAKA M,1992. Intraseasonal oscillation and the onset and retreat dates of the summer monsoon over the east, southeast and Western North Pacific region using GMS high cloud amount data[J]. J Meteor Soc Japan, 70:613-629.

TANIGUCHI K, KOIKE T,2006. Comparison of definitions of Indian summer monsoon onset: Better representation of rapid transitions of atmospheric conditions[J]. Geophys Res Lett. doi: 10.1029/2005GL024526.

TAO S Y, CHEN L X,1987. A review of recent research on the East Asian summer monsoon in China[C]//CHANG C P, KRISHNAMURTI T N. Monsoon Meteorology. Oxford: Oxford University Press: 60-92.

THOMPSON R M, PAYNE S W, RECKER E E, et al,1979. Structure and properties of synoptic-scale wave disturbances in the intertropical convergence zone of the eastern Atlantic[J]. J Atmos Sci, 36: 53-72.

TIAN Y, WANG Q,2010. Definition of the South China Sea summer monsoon onset[J]. Chinese J Oceanol Limnol, 28: 1281-1289.

TONG H W, CHAN J C L, ZHOU W,2009. The role of MJO and mid-latitude fronts in the South China Sea summer monsoon onset[J]. Clim Dyn, 33: 827-841.

VINAYACHANDRAN P, SHANKAR D, KURIAN J, et al,2007. Arabian Sea mini warm pool and monsoon onset vortex [J]. Curr Sci, 93: 203-214.

WANG B,2006. The Asian Monsoon[M]. London: Springer-Praxis: 788.

WANG B, WU R G,1997. Peculiar temporal structure of the South China Sea summer monsoon[J]. Adv Atmos Sci, 14: 177-194.

WANG B, LINHO,2002. Rainy season of the Asian-Pacific summer monsoon[J]. J Climate, 15: 386-398.

WANG B, LINHO, ZHANG Y, et al,2004. Definition of South China Sea monsoon onset and commencement of the east Asia summer monsoon[J]. J Climate, 17: 699-710.

WANG B, HUANG F, WU Z, et al,2009a. Multi-scale climate variability of the South China Sea monsoon: A review[J]. Dyn Atmos Ocean, 47: 15-37.

WANG B, DING Q, JOSEPH P V, et al,2009b. Objective definition of the Indian summer monsoon onset[J]. J Climate, 22: 3303-3316.

WEBSTER P J, YANG S,1992. Monsoon and ENSO: Selectively interactive systems[J]. Quart J Roy Meteor Soc, 118: 877-926.

WEBSTER P J, MAGANA V O, PALMER T N, et al,1998. Monsoons: Processes, predictability, and the prospects for prediction[J]. J Geophys Res, 103: 14451-14510.

WU G X, ZHANG Y S,1998. Tibetan Plateau forcing and the timing of the monsoon onset over South Asian and the South China Sea[J]. Mon Wea Rev,126:913-927.

WU G X,LIU Y M,ZHANG Q,et al,2007. The influence of the mechanical and thermal forcing of the Tibetan Plateau on the Asian climate[J]. J Hydrometeor, 8: 770-789.

WU G X, GUAN Y, WAMG T, et al,2011. Vortex genesis over the Bay of Bengal in spring and its role in the onset of the Asian summer monsoon[J]. Sci China Earth Sci, 54: 1-9.

WU G X, GUAN Y, LIU Y, et al,2012a. Air-sea interaction and formation of the Asian summer monsoon onset vortex over the Bay of Bengal[J]. Clim Dyn, 38: 261-279.

WU G X, LIU Y, HE B, et al,2012b. Thermal controls on the Asian summer monsoon[J]. Sci Rep, 2:404. doi:10.1038/srep00404.

WU G X, DUAN A, LIU Y, et al,2013. Recent advances in the study on the dynamics of the Asian summer monsoon onset[J]. Chinese J Atmos Sci, 37: 211-228.

WU R G,2010. Subseasonal variability during the South China Sea summer monsoon onset[J]. Clim Dyn, 34: 629-642.

XAVIER P K, MARZIN C, GOSWAMI B N, et al,2007. An objective definition of the Indian summer monsoon season and a new perspective on the ENSO-monsoon relationship[J]. Q J R Meteorol Soc, 133: 749-764.

XIE S P,1996. Westward propagation of latitudinal asymmetry in a coupled ocean-atmosphere model[J]. J Atmos Sci, 53: 3236-3250.

XIE S P, ARKIN P,1997. Global Precipitation: A 17-year monthly analysis based on gauge observations, satellite estimates, and numerical model outputs[J]. Bull Amer Meteor Soc, 78: 2539-2558.

XIE S P, PING S,2003. Summer upwelling in the South China Sea and its role in regional climate variations[J]. J Geophys Res, 108: 3261. doi:10.1029/2003JC001867.

XU H, XIE S P, WANG Y, et al,2008. Orographic effects on South China Sea summer climate[J]. Meteor Atmos Phys, 100: 275-289.

XU J, CHAN J C L,2001. First transition of the Asian summer monsoon in 1998 and the effect of the Tibet tropical ocean thermal contrast[J]. J Meteor Soc Jpn, 79: 241-253.

YAN J H,2005. Asian summer monsoon onset and advancing process and the variation of the subtropical high[D]. Beijing: Graduate University of Chinese Academy of Sciences.

YANG X, YAO T, YANG W, et al,2012. Isotopic signal of earlier summer monsoon onset in the Bay of Bengal[J]. J Climate, 25: 2509-2516.

YANO J, EMANUEL K A,1991. An improved model of the equatorial troposphere and its coupling with the stratosphere[J]. J Atmos Sci, 48: 377-389.

YOSHINO M M,1966. Four stages of the rainy season in early summer over East Asia (Part II)[J]. J Meteor Soc Jpn, 44:209-217.

ZHANG Y N, WU G, LIU Y, et al,2014. Effect of zonal asymmetric instability of potential vorticity on the South Asian high and the India summer monsoon onset[J]. Sci China Earth Sci, 57: 337-350.

ZHOU W, CHAN J C L,2005. Intraseasonal oscillations and the South China Sea summer monsoon onset[J]. Int J Climatol, 25: 1585-1609.

ZHOU W, CHAN J C L,2007. ENSO and the South China Sea summer monsoon onset[J]. Int J Climatol, 27: 157-167.

The Nature of the Thermal Forcing of the Asian Summer Monsoon

WU Guoxiong, LIU Yimin, HE Bian

(State Key Laboratory of Numerical Modeling for Atmospheric Sciences and Geophysical Fluid Dynamics,
Institute of Atmospheric Physics, Chinese Academy of Sciences, Beijing 100029)

Abstract: Monsoon is characterized by the reversal of prevailing surface wind between winter and summer accompanied by the occurrence of heavy precipitation. Traditionally, monsoon is considered as a consequence of the response of atmospheric circulation to the seasonal change of land-sea thermal contrast induced by the annual cycle of solar radiation. Here it is demonstrated that land-sea thermal contrast is only one part of the monsoon story. The idealized experiments presented in this study reveal that the seasonal change of land-sea thermal contrast alone is not a sufficient condition for the existence of the monsoon. The seasonal change of thermal forcing of the Tibetan-Iranian Plateau plays a significant role in generating the continental Asian summer monsoon. The nature of the Asian monsoon is indeed the response of atmospheric circulation to the seasonal changes of land-sea thermal contrast and thermal status of large-scale topography induced by the annual cycle of solar radiation.

Key words: Asian summer monsoon, land-sea thermal contrast, TIP thermal pumping

In this review, the results based on numerical experiments are used to reveal the nature of the Asian summer monsoon (ASM). It starts with a set of aqua-plane experiments to understand the roles of large-scale mountains in the formation of the ASM in contrast to those of land-sea distribution. Then, a series of experiments based on an atmospheric general circulation model are employed to demonstrate the significance of the thermal forcing of the Tibetan-Iranian Plateau (TIP) in forcing the vertical transport of water vapor, large-scale air ascent, and horizontal transport of water vapor from ocean to inland in the ASM area. Discussions on different hypothesis are also provided.

1 Aqua-planet experiment

Wu et al. (2012) developed a series of idealized numerical experiments based on an atmospheric general circulation model SAMIL-R42L9 (Wu et al., 2003) to understand how the land-sea distribution and the large-scale mountain ranges, the Tibetan-Iranian Plateau (TIP) affect the formation of the ASM. To conduct sensitivity experiments, the entire surface of the model Earth is first covered with water to form an aqua-planet experiment (Exp AQU). The sea surface temperature (SST) used is the zonally averaged climatological SST provided by the Second Atmospheric Model Intercomparison Program (AMIP-II, Fiorino, 2000), which has seasonal variation. The zonal mean of the model climatology of wind, temperature, humidity, and surface pressure from Exp AQU are taken as initial values for other idealized experiments. Four types of land distributions with different geometries are embedded in the aqua-planet for different experiments (Table 1). In Exp MID, a continent is located over 0°—120°E and 30°—90°N is investigate the influence of a continent at middle and high latitudes on circulations. In Exp SUB, the southern boundary of the continent in Exp MID is extended 10° southward into the subtropics, mimicking the main part of the Eurasian continent. In Exp TRO, three square-shaped tropical lands over (0°—50°E, 35°S—20°N), (75°—85°E, 5°—20°N), and (95°—105°E, 9°S—20°N) are introduced to represent the tropical African, Indian, and Indochina Peninsula subcontinents, respectively, which are then integrated into the main Eurasian continent as in Exp SUB to form the "Afro-Eurasian continent". Exp TIP uses the same continent as that in Exp TRO, but adds an idealized TIP to investigate its influence on the monsoon. All of these experiments were integrated for 10 years, and the means of the last 8 years were used for the analyses.

Table 1 Experiment design of the idealized aqua-planet experiments

Experiment	Distribution of land
MID	Higher-latitude continent (0°—120°E, 30°—90°N)
SUB	Subtropical continent (0°—120°E, 20°—90°N)
TRO	Tropical continent (SUB land and 0°—50°E, 35°S—20°N; 75°—85°E, 5°—20°N; and 95°—105°E, 9°S—20°N)
TIP	TRO and Tibetan-Iranian Plateau (elliptic topography TP with maximum altitude of 5000 m, centered at (87.5°E, 32.5°N), and elliptic topography IR with maximum altitude of 3000 m, centered at (53.4°E, 32.5°N)

In Exp AQU, the July mean precipitation largely occurs near the equator along the "Inter-tropical convergent zone (ITCZ)", as obtained in Liang et al. (2006). For a continent located at extratropical latitudes, the experiments (Exps MID and SUB) produce insignificant monsoon rainfall. A very weak monsoon rainband along the southern continental boundary coexists with the ITCZ along the equator, and tropical water vapor substantially converges to the ITCZ as in Exp AQU, which effectively prohibits the interaction of circulation systems between the two hemispheres. The inclusion of a tropical land (Fig. 1c) produces cross-equatorial flow, which destroys the local ITCZ, pushes it to about 10°—15°N across the Africa-Indian Ocean sector, becoming a tropical convergent zone (TCZ, Fig. 1c), and forms the tropical summer monsoon there termed as the southern branch of the South Asian summer monsoon (SASM). It also enhances the monsoon over the southern and southeastern boundary of the extratropical continent as shown in Fig. 1b.

However, the land-sea distribution alone does not produce significant precipitation over continent in all the above experiments. The uplifted orographic thermal forcing in summer generates a cyclonic circulation in the lower troposphere, with a southerly to its east and a northerly to its west. Consequently, TIP forcing strengthens the circulation coupling between the subtropics and tropics, and between the lower and upper troposphere. Abundant water vapor is transported from the tropical ocean to the subtropical and extratropical continent, where it condenses during the course of northward travel (Fig. 1d). The difference between the experiments with and without the TIP as shown in Fig. 1e reveals that the TIP forcing in summer is to produce the northern branch of the SASM and the Eastern ASM (EASM), and reduce the precipitation of the southern branch of the SASM west of 80°E. This implies that only when the large scale mountain, the TIP, is presented in the experiment, can the northern branch of the SASM and the EASM be generated (Fig. 1d).

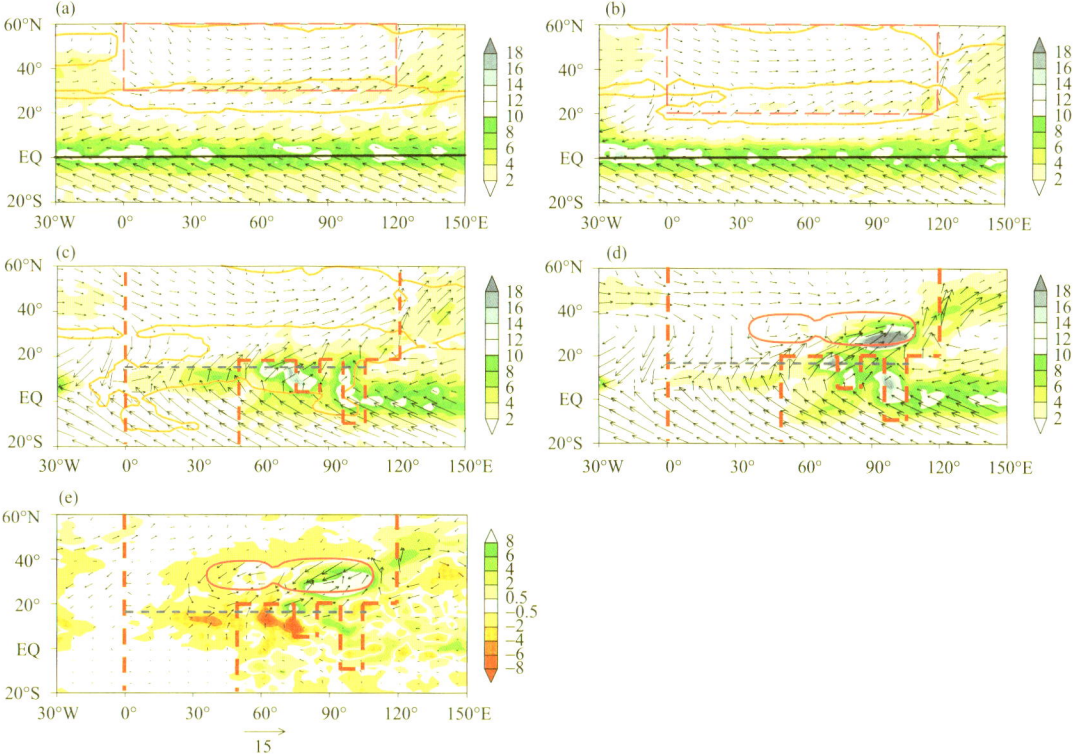

Figure 1 July mean wind vectors at $\sigma=0.991$ (arrows; unit in m·s^{-1}) and precipitation (shading; unit in mm·day^{-1}) in Exp of (a) MID, (b) SUB, (c) TRO, (d) TIP, and (e) the difference (TIP−TRO). The orange curve in (a—c) indicates the area where the surface wind reversal between Jan and Jul is larger than 120°; the heavy curve denotes the orographic contour at 700 m, the heavy red dashed line denotes the continent boundary, and the heavy blue dashed line indicates the location of the southern branch of the SASM

2 Sensible heat-driven air-pump(shap) of large-scale mountains

The TIP is located in the northern subtropics across most of the Eurasian continent, characterized by prominent atmospheric elevated heating in summer and cooling in winter. The resultant air descent in winter and ascent in summer above the TIP occur remarkably over the vast subtropical areas in Eurasia continent year after year, working as a huge air-pump standing in situ, and regulating the seasonal change of the Asian circulation.

Wu et al. (2007) demonstrated that the surface sensible heating (SH) on the sloping lateral surfaces of the TP is the major driver of the air pump. According to the thermodynamic equation, in the presence of surface SH on the sloping surface of a mountain, an air parcel moving along an isentropic surface that impinges on the topography is heated as it encounters the SH along the sloping surface of the mountain, and is able to penetrate the isentropic surface and ascend. In the area surrounding the topography, air from lower levels is lifted to the plateau region, producing strong ascent and heavy rainfall over the TP. In the case of no SH on the slope surface of the TP, air parcels travelling in the lower layer and impinging on the TP must remain on the same isentropic surface. The air parcel moves around the

TP on an approximately horizontal isentropic surface without ascending and has no significant impact on rainfall. It appears that air pumping is driven by the surface SH on the TP slopes, and is therefore referred to as TP-SHAP (Wu, 1997). Results from the numerical experiments of Wu et al. (2012) demonstrate that such a thermal pumping also exists in the Iranian Plateau forcing, and the pumping effect of Tibetan and Iranian Plateaus (TIP) forms a TIP-SHAP. Since more than 85% of the atmospheric water vapor is concentrated in a surface layer below 3 km, such a vertical pumping of water vapor is important for the development of cloud and precipitation over the slopes of the plateau, which forms a significant atmospheric heating source in the subtropics.

The pumping effect of the surface sensible heating of the TIP in the control experiment (CON) causes the surface moist air flow climb up the TIP on its southeast (Fig. 2a) where the streamlines of the surface current are almost perpendicular to the plateau contours. On the contrary the experiment TIP_NS in which although the surface sensible heating still exists over the TIP's surface, it is not allowed to heat the in situ atmosphere (Fig. 2b), the surface air currents cannot climb up the TIP, and just go around the TIP horizontally with their streamlines parallel to the TIP's contours. All the above experiment results indicate that the pumping of the TIP SHAP is significant for the formation of the northern branch of the SASM and the EASM.

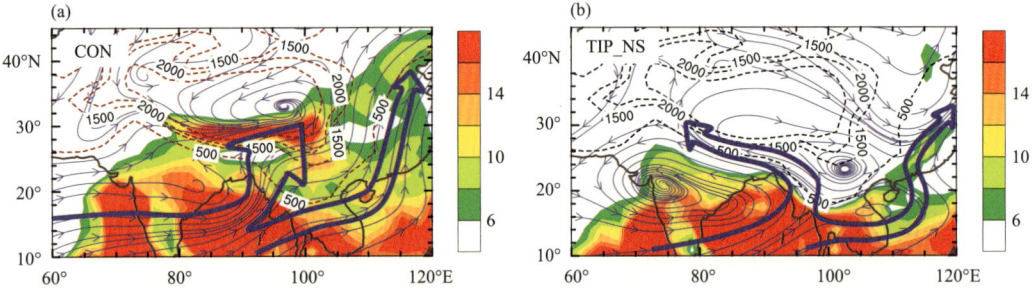

Figure 2 JJA mean precipitation rate (color shading, unit: mm • d^{-1}) and streamlines at the $\sigma=0.89$ level for (a) the CON experiment, and (b) TIP_NS experiment. Dashed contours indicate elevations above sea level, with red and black colors respectively indicating with and without surface sensible heating of the mountains. Dark blue open arrows denote the main atmospheric flows impinging on the TP, either climbing up the plateau (a) or moving around the plateau, parallel to orographic contours (b) (Reproduced from Wu et al., 2012)

3 Large-scale air ascent

In the tropics, the planetary vorticity is small and the Rossby deformation radius is large. The atmospheric response to an axisymmetric forcing usually takes a regime of angular momentum conservation (AMC, Plumb et al., 1992). This AMC regime can be divided into two types (Wu et al., 2016b): one is ITCZ-induced Hadley-type meridional circulation (H-AMC) with its rising/sinking arm located in the equatorial/subtropical region (Fig. 3a); the other is the monsoonal-type meridional circulation (M-AMC) with its rising arm located in the subtropical region and sinking arm in the Southern Hemisphere (Fig. 3b). Since the atmospheric vertical velocity w is proportional to the vertical differential potential vorticity (PV=q) advection ($w \propto f\frac{\partial}{\partial z}(-\mathbf{V} \cdot \nabla q)$) (Hoskins et al., 2003), and since PV advection decreases with height along the H-AMC meridional circulation and increases with height along the M-AMC meridional circulation, the H-AMC is thus accompanied by descending air and a vertical westerly shear;

whereas the M-AMC is accompanied by ascending air and a vertical easterly shear. Therefore, the large-scale background of ascending motion in the ASM area must be related to the monsoonal-type meridional circulation, M-AMC.

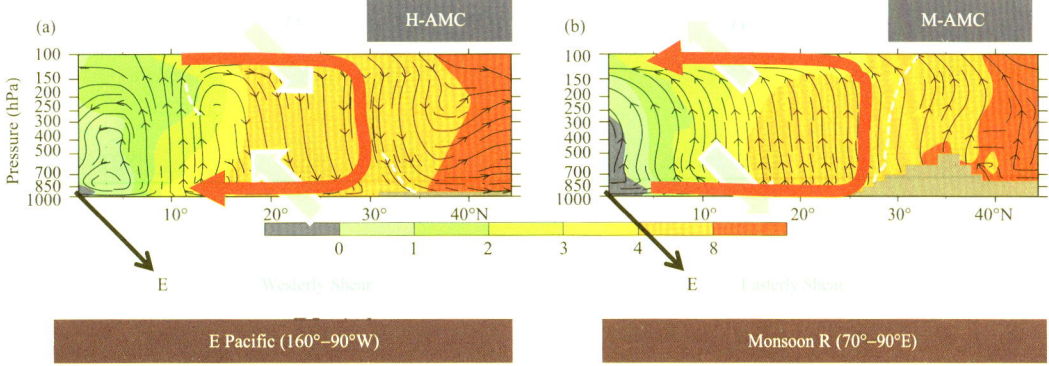

Figure 3 Cross sections of climate July mean (1979—1998) meridional circulation (lines with arrows), absolute vorticity (color shading; units: 10^{-5} s^{-1}), and the zonal wind (u) line where $u=0$ (white dashed line) for the (a) East Pacific (160°—90°W) regional mean, and (b) Asian monsoon area (70°—90°E) regional mean. Gray shading denotes topography, heavy red curve denotes the main meridional circulation, and the blue arrow indicates the zonal wind component fv

To reveal the cause by which the M-AMC type meridional circulation is generated, the WRF model version 3.4.1 (Skamarock et al., 2008) with 45 km horizontal resolution and 35 vertical layers was used to execute a pair of numerical experiments, each incorporated six summers (2003—2008) with the initial conditions set at 0000 UTC 1 May (Wu et al., 2015). The integration time was 4 months, i.e., every experiment ended at 1800 UTC 31 August, and output from the final 3 months (June—August, JJA) were analyzed. The normal experiment with the WRF model was named the control run (CTL). The other experiment is the sensitivity experiment (TP_NS), which has the same setting as the CTL run except the surface SH over the TP area where terrain is higher than 2 km over the main TP (23°—40°N, 70°—105°E) was not allowed to heat the atmosphere. The results obtained from the difference between CTL and TP_NS can be considered as the influence of the surface sensible heating over the main part of the TP. The TP surface SH generates remarkable difference anticyclone circulation and warm center over the TP at 300 hPa (Fig. 4b), and much stronger anticyclone circulation but cold temperature center at 100 hPa (Fig. 4a). Thus the static stability at the center of the difference anticyclone is the minimum (Fig. 4c). It was proved (Wu et al., 2016a) that there exists a critical pressure level p_c, where the horizontal temperature gradient vanishes ($\nabla_{p_c} T(x,y) \equiv 0$) and the circulation possesses the strongest anticyclone vorticity. Therefore, at this critical level, both the absolute vorticity and potential vorticity in the anticyclone area are minimum compared to its upper and lower levels. These imply that the TIP heating changes the atmospheric thermal structure aloft, forming minimum absolute vorticity and potential vorticity in the subtropics near the tropopause. Subject to the constraint of AMC, a monsoonal type of meridional circulation is produced in the ASM area with its ascending arm located in the subtropics (Fig. 4d). Since the PV advection is increased with height within this monsoonal type of meridional circulation, large scale air ascent thus prevails over the vast Asian monsoon area, providing a favorable background for monsoon development.

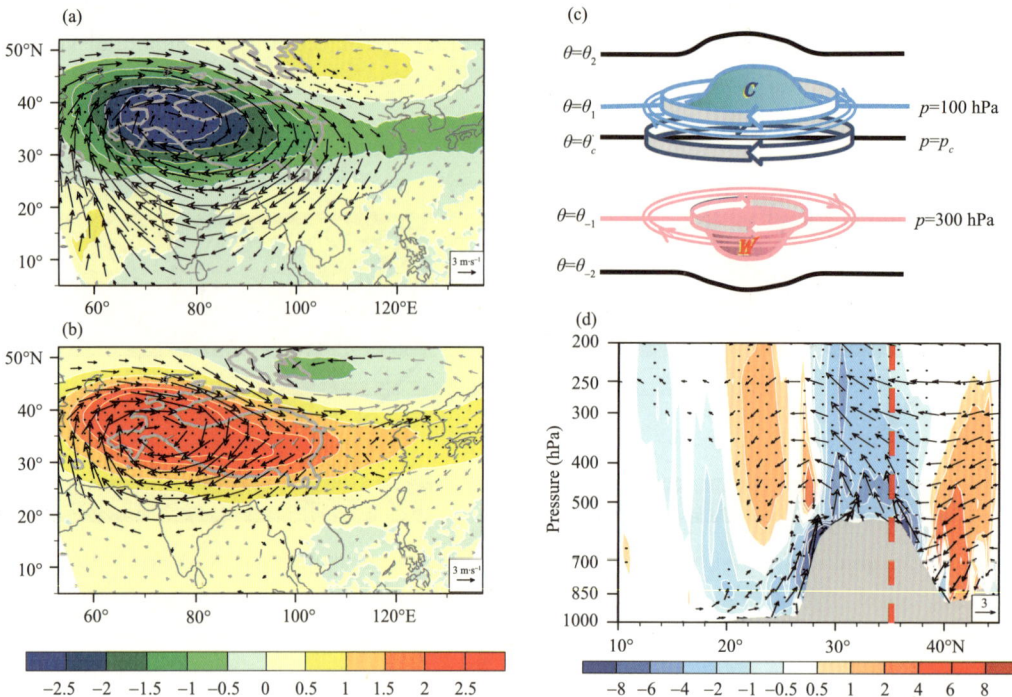

Figure 4 Summer mean differences in air temperature (shading, unit: K) and wind field (vectors, unit: m·s^{-1}) between CTL and TP_NS at (a) 100 hPa and (b) 300 hPa; (c) Schematic diagram indicating the formation of the area of minimum PV forcing near the tropopause due to thermal forcing over the main TP, Vector indicates anticyclone circulation; "C" and blue color denotes cold temperature, and "W" and pink color denotes warm temperature; and (d) Pressure-latitude cross sections of the vertical circulation (vectors, v and $-50\times\omega$) and vertical velocity (shading, unit: 0.02 Pa·s^{-1}) averaged from 85°E to 95°E. Dotted regions denote statistical significance of the difference above the 95% level. Only vertical circulation difference with statistical significance above the 95% level is plotted (Reproduced from Wu et al., 2016)

4 Water vapor transport from ocean to inland

Continental monsoon development requires tremendous water vapor supply which cannot be fulfilled by local evaporation from land surface. Instead, it depends significantly on the in situ convergence of water vapor flux. Thus water vapor transport from the source area is critical for the development of continental monsoon. The key problem here is how the circulation which is responsible for the water vapor transport is formed.

Due to the PV conservation constraint, a 10 ℃ warm anomaly on the surface can initiate a cyclonic circulation near the surface (Hoskins et al., 1985). Based on the AGCM SAMIL, He et al. (2015) conducted a pair of numerical experiments in which the control run is the normal climate integration, whereas the sensitivity run is the same except the sensible heating over the TIP is not allowed to heat the atmosphere. The differences between the two experiments demonstrate that the surface SH over the TIP can generate a strong surface warm cyclonic circulation surrounding the TIP of more than 8 ℃ (Fig. 5a). In another parallel set of experiments in which the specific humidity was set to zero, similar results were obtained (Fig. 5b), implying the essential role of the strong TIP surface sensible heating in initiating the warm surface cyclonic circulation. It is due to this TIP-SHAP induced strong cyclonic circulation that abundant water vapor is transported from the ocean surface to the inland area to feed the Asian continental monsoon rainfall.

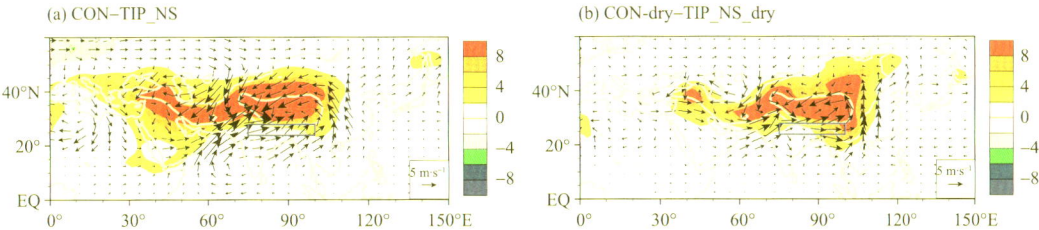

Figure 5　JJA-mean differences of near-surface ($\sigma=0.991$) potential temperature θ (shading, K) and circulation (vectors, m·s^{-1}) between (a) CON and TIP_NS, (b) As for (a) but for a dry atmosphere. The square indicates the SASM region of (24°—28°N, 75°—100°E) (Reproduced from He et al., 2015)

Some recent studies (e.g. Boos et al., 2010) hypothesized that the Himalayas acts as a thermal insulator that prohibits the cold and dry northerly air intrusion from middle latitudes so that the lower layer atmosphere with high level "moist static energy" over India can couple with the warm center in the upper troposphere through monsoonal convection, and the "physical" insulation influence of the Himalayas is equal to or more important than the thermal influence of the TP for the SASM maintenance. However, in summer there does not exist such a northerly air intrusion either in observation or in modeling and the "thermal insulator" is not required; whereas in winter the Himalayas cannot protect India from cold and dry northerly intrusion (He et al., 2015). This is because in winter the northerly flow can go around the TIP, becoming northwesterly or northeasterly flows and intruding India. In summer compared to in India, the solar radiation at the top of the atmosphere is stronger and the cloud cover is diminished to the north of the TP. There the surface shortwave radiation is much stronger, resulting in the development of a hot and low-pressure surface to the north of the TP. More importantly, in summer under the continental-scale thermal forcing, a continental-scale surface cyclonic circulation dominates the Eurasian continent (Wu et al., 2009). Thus the prevailing wind in the lower troposphere is southerly across the TP. Therefore, the development of the SASM does not require a thermal insulator to its north. As to the high "moist static energy" over the Indian Subcontinent, it is basically due to the high moisture content there (He et al., 2015). As discussed above, it is due to the TIP thermal forcing that the water vapor can be transported from ocean area to inland area to create high "moist static energy" near the surface over India. It does not mean that the TIP's mechanical forcing can be ignored. As shown in Fig. 2a, when the moist southerly flow converges toward the TIP, in addition to the flow climbing up the mountain, there exists a branch of air flow deflected by the mountain, contributing to the formation of the monsoon trough over the Indian Subcontinent and affecting the detailed distribution and variation of the ASM.

5　Conclusion

Results based on numerical experiments demonstrate that the Asian summer monsoon is thermally controlled (Fig. 6). While land-sea thermal contrast significantly determines the southern branch of the South Asian summer monsoon in the tropics, the large scale mountain TIP's thermal pumping is fundamental for the formation of the northern branch of the South Asian summer monsoon and the East Asian summer monsoon. It is therefore concluded that the nature of the Asian summer monsoon is a consequence of the response of atmospheric circulation to the seasonal changes in land-sea thermal contrast and in the thermal status of the large-scale plateau TIP which is induced by the annual cycle of solar radiation.

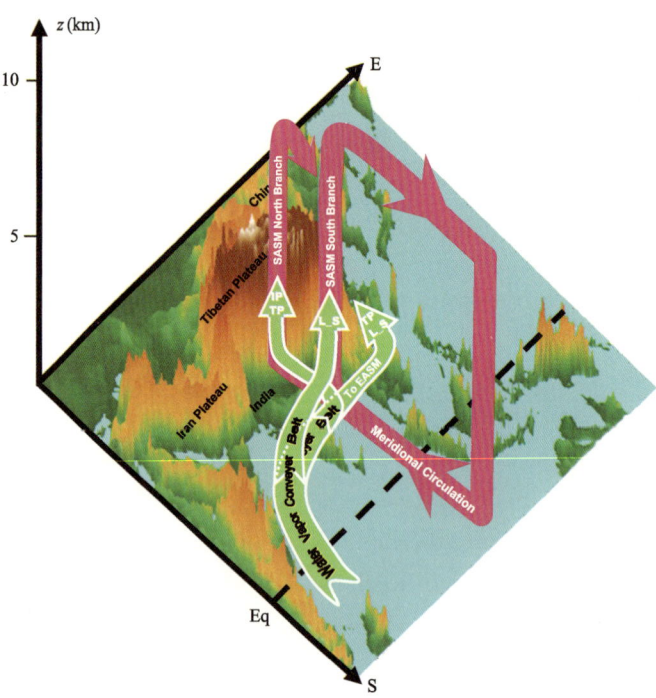

Figure 6 Schematic diagram showing the gross structure of the Asian summer monsoon. For the southern branch, water vapor along the conveyer belt is lifted up due mainly to land-sea thermal contrast in the tropics; for the northern branch the water vapor is drawn away from the conveyer belt northward toward the foothills and slopes of the TP, and is uplifted to produce heavy precipitation that is controlled mainly by TIP-SHAP; the rest of the water vapor is transported northeastward to sustain the East Asian summer monsoon, which is controlled by the land-sea thermal contrast as well as thermal forcing of the TIP (Wu et al., 2012)

References

BOOS W R, KUANG Z M, 2010. Dominant control of the South Asian monsoon by orographic insulation versus plateau heating[J]. Nature, 463. doi:10.1038/nature0870.

FIORINO M, 2000. AMIP Ⅱ Sea Surface Temperature and Sea Ice Concentration Observations[R]. Lawrence Livermore National Laboratory: PCMDI Report.

HE B, WU G, LIU Y, et al, 2015. Astronomical and hydrological perspective of mountain impacts on the Asian summer monsoon[J]. Sci Rep, 5: 17586. doi: 10.1038/srep17586.

HOSKINS B J, MCINTYRE M E, ROBERTSON A W, 1985. On the use and significance of isentropic potential vorticity maps[J]. Quart J Roy Meteor Soc, 111: 877-946.

HOSKINS B J, PEDDER M, JONES D W, 2003. The omega equation and potential vorticity[J]. Q J R Met Soc, 129: 3277-3303.

LIANG X Y, LIU Y M, WU G X, et al, 2006. Roles of tropical and subtropical land-sea distribution and the Qinghai-Xizang Plateau in the formation of the Asian summer monsoon[J]. Chin J Geophys Ch, 49(4): 983-992.

PLUMB R A, HOU A Y, 1992. The response of a zonally symmetric atmosphere to subtropical thermal forcing: Threshold behavior[J]. J Atmos Sci, 49: 1790-1799.

SKAMAROCK W C, KLEMP J B, DUDHIA J, et al, 2008. A description of the advanced research WRF version 3[R].

NCAR Technical Note NCAR/TN-475+STR. doi: 10.5065/D68S4MVH.

WU G X,1997. Sensible heat driven air-pump over the Tibetan Plateau and its impacts on the Asian summer monsoon[C]// YE D Z. Collections on the Memory of Zhao Jiuzhang. Beijing: Science Press: 116-126.

WU G X, LIU Y M, ZHANG Q, et al,2007. The influence of mechanical and thermal forcing by the Tibetan Plateau on Asian climate [J]. J Hydrometeor, 8: 770-789.

WU G X, LIU Y, ZHU X, et al,2009. Multi-scale forcing and the formation of subtropical desert and monsoon[J]. Ann Geophys, 27: 3631-3644.

WU G X, DONG B, LIANG X, et al,2012. Revisiting Asian monsoon formation and change associated with Tibetan Plateau forcing: I. Formation[J]. Climate Dyn, 39(5): 1169-1181.

WU G X, DUAN A, LIU Y, et al,2015. Tibetan Plateau climate dynamics: Recent progress and outlook[J]. National Science Review, 2a: 100-116.

WU G X, HE B, LIU Y, et al,2016a. Recent progresses on dynamics of the Tibetan Plateau and Asian summer monsoon [J]. Chinese J Atmos Sci, 40(1): 1-11.

WU G X, ZHUO H F, WANG Z Q, et al,2016b. Two types of summertime heating over the Asian large-scale orography and excitation of potential-vorticity forcing I. Over Tibetan Plateau[J]. Science China Earth Sciences, 59 (10): 1996-2008.

WU T W, LIU P,2003. The performance of atmospheric component model R42L9 of GOALS/LASG[J]. Adv Atmos Sci, 20: 726-742.

第5部分

位涡及位涡源的若干基本问题

综述 11

倾斜涡度发展理论和全型涡度方程及应用

崔晓鹏[1,2]，姚秀萍[3]，郑永骏[4]

(1 中国科学院大气物理研究所，北京 100029；2 中国科学院大学，北京 100049；
3 中国气象局气象干部培训学院，北京 100081；4 南京信息工程大学大气科学学院，南京 210044)

摘要：回顾了倾斜涡度发展(SVD)理论和全型垂直涡度方程，并简要梳理了上述理论和方程在天气学和气候学研究中的应用。SVD 理论和全型垂直涡度方程从经典位涡理论出发，通过引入湿过程和细化考虑斜压性特征，突破了经典位涡理论在对流层中低层大气过程研究中的局限，为天气系统形成机理和气候变化机制研究提供了重要理论工具和定量化诊断工具，大幅度拓展了经典位涡理论及其应用范畴。SVD 理论和全型垂直涡度方程已经在我国不同地区暴雨与强对流成因、热带气旋强度变化与登陆热带气旋暴雨机理、温带气旋发展机制、副热带高压变异与南亚高压活动机理、持续性高温成因、城市群发展对季风爆发影响机制等诸多天气和气候学领域得到广泛应用。天气-气候一体化研究、多时空尺度相互作用研究是当前大气科学研究的热点和未来研究趋势，在这些热点研究领域，倾斜涡度发展理论和全型垂直涡度方程及其未来的不断拓展必将持续发挥重要作用。

1 引言

涡度作为气象学和海洋学中的经典物理量，其相关研究最早可以追溯到 19 世纪末和 20 世纪初 Bjerknes 的开创性工作。Rossby(1939)指出，绝对涡度的垂直分量在大尺度大气运动分析研究中至关重要，进而形成了正压大气中涡度的"可反演性"思想，并构建了位涡概念的雏形(Rossby，1940)；Ertel (1942)正式提出了位涡的概念，即 Ertel 位涡(Ertel potential vorticity，简称 Ertel PV)，并指出在绝热、无摩擦的干空气中，Ertel PV 严格守恒。位涡作为一个可综合表征大气动力学和热力学特性的物理量，成为继大气位温和大气比湿之后，空气质块追踪识别的第三个重要物理量(Starr and Neiburger，1940)；位涡不仅具有守恒特性(Ertel，1942)，同时具有可反演性(Davis，1992a，1992b)，对于全面描述大气运动的热、动力学特性和规律极为重要，基于上述位涡特性的等熵位涡(isentropic potential vorticity，简称 IPV)分析在大气科学研究中逐渐展开，并受到高度重视(Starr and Neiburger，1940；Platzman，1949；Kleinschmidt，1950a，1950b，1951，1955，1957；Reed and Sanders，1953；Reed，1955；Obukhov，1964；Danielsen，1967，1968；McIntyre and Palmer，1983，1984；Uccellini et al.，1985 等)。例如，基于 IPV 分析方法，Reed (1955)研究指出，平流层高 PV 空气可以沿着等熵面下滑至对流层(Reed and Danielsen，1959；Danielsen，1968；Shapiro，1976，1978)，从而对对流层中天气系统的发展和演变造成重要影响。Hoskins 等(1985)对 Ertel PV 和 IPV 分析方法在大气科学研究中的应用及其重要价值做出了系统性的回顾和总结。

通讯作者：崔晓鹏，xpcui@mail.iap.ac.cn

位涡(Ertel PV)及相应的 IPV 分析方法对于中高纬度和大尺度大气环流运动研究具有极为重要的意义和实用价值,但对流层内(尤其是对流层中低层)中小尺度天气系统活动显著的湿过程特性、强烈致灾性天气系统的斜压性特征,以及灾害天气形成过程中的多尺度作用等诸多因素,使得 Ertel PV 经典概念和 IPV 分析方法在上述天气系统机理分析中的具体应用遇到一定限制,水汽条件和(湿)等熵面倾斜等特性的引入成为必然。Bennetts and Hoskins(1979)从布西内斯克(Boussinesq)近似出发,引入相变潜热的影响,推导得到湿球位涡变化方程;吴国雄等(1995)进一步从完整的大气原始运动方程出发,通过引入饱和大气中水汽凝结潜热的作用,推导得到湿位涡(moist potential vorticity,简称 MPV)变化方程,并严格证明了在绝热、无摩擦的饱和湿空气中,湿位涡守恒;进一步,基于这一守恒特性,研究了对流层大气湿斜压过程中垂直涡度的发展问题(吴国雄和蔡雅萍,1997),指出在湿等熵坐标系中,涡旋垂直涡度的发展与大气对流稳定度的减小、等熵面上的辐合和潜热释放等过程有关。由于传统的 IPV 分析方法在对流层天气系统研究应用中受到等熵面倾斜特性的限制,吴国雄等(吴国雄 等,1995;吴国雄和蔡雅萍,1997;Wu and Liu,1997)继而发展了 z 坐标系和 p 坐标系中的倾斜涡度发展理论(slantwise vorticity development theory,简称 SVD 理论),指出除大气对流稳定度之外,环境垂直风切变的增大或水平湿斜压性的增长等,均可以在湿等熵面(等相当位温面)倾斜的背景下,引起大气局地涡度垂直分量(垂直涡度)的快速增长;且湿等熵面越倾斜,这种由于大气斜压性加强所引起的涡旋垂直涡度的发展越剧烈;而在对流不稳定的饱和大气中,倾斜涡度的发展必然伴有低空急流的存在。从 SVD 理论出发,吴国雄和刘还珠(Wu and Liu,1997;吴国雄和刘还珠,1999a)推导得到了包含不稳定、斜压性、环境风垂直切变、摩擦效应和非绝热加热等完整热、动力学因素的全型垂直涡度方程,并详细阐明了全型垂直涡度方程和经典垂直涡度方程的异同,以及前者的独特优势和物理内涵(吴国雄和刘还珠,1999a;吴国雄,2001)。

涡度和位涡同为大气科学研究领域不可或缺且广泛应用的经典物理量,两者在定义上存在着无法分割的密切联系,但在大气科学领域的具体应用过程中往往单独进行诊断应用。而倾斜涡度发展理论明确构建了上述两个经典物理量之间的理论联系,且全型垂直涡度方程进一步构建了两者之间的量化关系,这为天气系统发生、发展机理中的多尺度作用和全物理过程耦合研究奠定了坚实的理论基础,同时提供了强有力的定量化诊断工具。倾斜涡度发展理论的提出和全型垂直涡度方程的构建,大幅拓展了经典位涡理论及其应用范围,上述理论和方程无论对于梅雨锋暴雨天气系统、西南涡、高原涡、中纬度爆发性气旋、热带气旋和热带扰动等相关的天气学研究(吴国雄 等,1995;吴国雄和蔡雅萍,1997;Wu and Liu,1997;吴国雄和刘还珠,1999a;余晖和吴国雄,2001;崔晓鹏 等,2002;Cui et al.,2003;Yao et al.,2007,2009,2020b;郑永骏 等,2013;吴国雄 等,2013;Wu et al.,2020),还是气候学研究(吴国雄 等,1999b;刘屹岷 等,1999a,1999b),均具有极为重要的指导意义和实际应用价值。尤其值得一提的是,在当今地球系统集成研究和天气气候一体化研究的大趋势下,学界对于多圈层相互作用和多尺度系统相互作用机理的深入认识愈加重视,倾斜涡度发展理论和全型垂直涡度方程在上述相互作用机理的深入揭示方面展现出极大的应用潜力和重要的实用价值(吴国雄 等,1999b;刘屹岷 等,1999a,1999b;Yao and Sun,2016;Yao et al.,2020a)。

本小节尝试回顾和梳理倾斜涡度发展理论和全型垂直涡度方程及其应用研究的相关学术进展,并对未来相关研究方向做出适当展望,不妥和不足之处敬请斧正。

2 倾斜涡度发展理论

吴国雄等基于系统的开创性研究(吴国雄 等,1995;吴国雄和蔡雅萍,1997;Wu and Liu,1997),创新性提出并建立了倾斜涡度发展(SVD)理论,对大气中(湿)等熵面倾斜结构(锋面结构)附近常见的涡旋系统(垂直涡度)的发生、发展给出了合理且清晰的理论解释,为相关天气系统发生、发展机理的深入揭示提供了强有力的理论基础。下面简要回顾 SVD 理论的相关研究工作。

2.1 湿位涡方程和湿位涡的守恒性

在绝热、无摩擦的干大气中,Ertel PV 严格守恒(Ertel,1942),在此基础上的等熵位涡(IPV)分析在大气科学相关研究领域中得到了极为广泛的应用(Hoskins et al.,1985)。IPV 分析对于研究对流层中、上层和中、高纬度天气系统的发展和移动具有重要意义和实用价值,但在对流层中低层以及中低纬度地区,其应用存在一定局限性,其中,Ertel PV 对对流层中低层和中低纬度地区水汽影响考虑的不足是关键之一。吴国雄等(1995)推导出包含水汽影响的湿位涡方程,并证明在绝热、无摩擦的饱和湿空气中,湿位涡严格守恒,这为位涡理论和相关分析方法的拓展应用,以及 SVD 理论的提出和全型涡度方程的构建奠定了基础。

欧拉形式的大气运动方程可写为,

$$\frac{\partial \mathbf{V}}{\partial t} + \nabla\left(\frac{V^2}{2} + \varphi\right) - \mathbf{V} \times (\nabla \times \mathbf{V} + 2\mathbf{\Omega}) = -\alpha \nabla p + \mathbf{F} \tag{1}$$

对其求旋度,可得大气三维涡度变化方程,

$$\frac{\partial \boldsymbol{\zeta}_a}{\partial t} - \nabla \times (\mathbf{V} \times \boldsymbol{\zeta}_a) = \nabla p \times \nabla \alpha + \nabla \times \mathbf{F} \tag{2}$$

式中,$\boldsymbol{\zeta}_a = \nabla \times \mathbf{V} + 2\mathbf{\Omega}$ 为大气绝对涡度。

大气热力学方程可写为,

$$C_p \frac{T}{\theta} \frac{\mathrm{d}\theta}{\mathrm{d}t} = -L \frac{\mathrm{d}q}{\mathrm{d}t} + Q_d \tag{3}$$

式中,Q_d 为除去凝结潜热以外的其他形式的非绝热加热。

对包含了水汽效应的相当位温($\theta_e = \theta \exp\left(\frac{Lq}{C_p T}\right)$)取自然对数,并对其求时间导数,进而代入式(3)中,略去高阶小量,得到,

$$\frac{\mathrm{d}\theta_e}{\mathrm{d}t} = \frac{\theta_e}{C_p T} Q_d = Q \tag{4}$$

利用 $\nabla \theta_e$ 点乘大气三维涡度方程式(2),并利用矢量计算规则以及式(4),可得,

$$\frac{\mathrm{d}(\boldsymbol{\zeta}_a \cdot \nabla \theta_e)}{\mathrm{d}t} = -(\boldsymbol{\zeta}_a \cdot \nabla \theta_e) \nabla \cdot \mathbf{V} + (\nabla p \times \nabla \alpha) \cdot \nabla \theta_e + \boldsymbol{\zeta}_a \cdot \nabla Q + \nabla \theta_e \cdot (\nabla \times \mathbf{F}) \tag{5}$$

对式(5)两边同时乘以大气比容($\alpha = \rho^{-1}$,ρ 为空气密度),并利用大气连续性方程,经过简单推导可得湿位涡变化方程,

$$\frac{\mathrm{d}(\alpha \boldsymbol{\zeta}_a \cdot \nabla \theta_e)}{\mathrm{d}t} = \alpha(\nabla p \times \nabla \alpha) \cdot \nabla \theta_e + \alpha \boldsymbol{\zeta}_a \cdot \nabla Q + \alpha \nabla \theta_e \cdot (\nabla \times \mathbf{F}) \tag{6}$$

式中,$Q = \frac{\theta_e}{C_p T} Q_d$。设 $\mathbf{F}_\xi = \nabla \times \mathbf{F}$(涡度摩擦耗散)以及 $P_m = \alpha \boldsymbol{\zeta}_a \cdot \nabla \theta_e$(湿位涡),进一步得到如下形式的湿位涡方程,

$$\frac{\mathrm{d}P_m}{\mathrm{d}t} = \alpha(\nabla p \times \nabla \alpha) \cdot \nabla \theta_e + \alpha \boldsymbol{\zeta}_a \cdot \nabla Q + \alpha \nabla \theta_e \cdot \mathbf{F}_\xi \tag{7}$$

在绝热、无摩擦($\mathbf{F}_\xi = Q = 0$)大气条件下,上述湿位涡方程(7)变为,

$$\frac{\mathrm{d}P_m}{\mathrm{d}t} = \alpha(\nabla p \times \nabla \alpha) \cdot \nabla \theta_e \tag{8}$$

当初始不饱和空气沿干绝热上升至抬升凝结高度,达到饱和时,等 θ_e 面同等 P 面和等 α 面的相交轴平行,进而得到,

$$\frac{\mathrm{d}P_m}{\mathrm{d}t} = \alpha(\nabla p \times \nabla \alpha) \cdot \nabla \theta_e = 0 \tag{9}$$

即,绝热、无摩擦饱和湿大气中湿位涡守恒。

2.2 倾斜涡度发展(SVD)理论

借助上述(湿)位涡守恒特性,吴国雄等(吴国雄 等,1995;吴国雄和蔡雅萍,1997;Wu and Liu,1997)重点研究了大气(湿)斜压过程中,涡旋系统垂直涡度的发展问题,提出了倾斜涡度发展理论,指出,大气中的涡旋系统易于在等熵面较陡立的地方发生、发展。下面以干大气为例,从位涡守恒特性出发,对 SVD 理论做简要介绍。

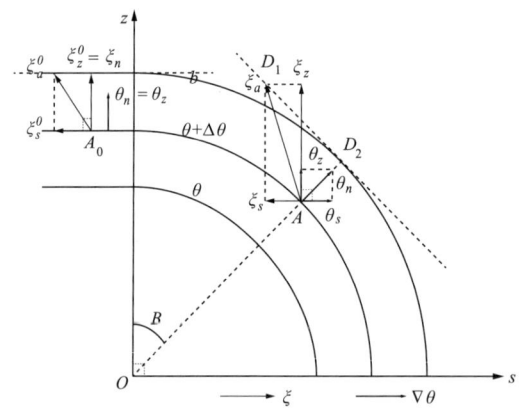

图 1 倾斜涡度发展理论示意图(参考(吴国雄和蔡雅萍,1997)中图 3 绘制)

参照吴国雄等(1995)、吴国雄和蔡雅萍(1997)和 Wu 和 Liu(1997),如图 1 所示,假定垂直坐标轴(z)左侧的等熵面为平行平面,而右侧的等熵面弯曲成圆,进一步,为简化说明,设等熵面的梯度 $\Delta\theta = \theta_n$ 为常数,同心圆"b"与 $\theta + \Delta\theta$ 等熵面相距为 $|\xi_n|$,则 Ertel 位涡可写为,

$$P_E = \xi_n \theta_n = \xi_s \theta_s + \xi_z \theta_z \tag{10}$$

进而有,

$$\xi_z = \frac{\xi_n \theta_n - \xi_s \theta_s}{\theta_z} = \frac{P_E - \xi_s \theta_s}{\theta_z}, \quad (\theta_z \neq 0) \tag{11}$$

进一步写为,

$$\xi_z = \frac{\xi_n}{\cos B} - \xi_s \tan B, \quad \left(|B| \neq \frac{\pi}{2}\right) \tag{12}$$

式中,$\xi_z = \alpha\zeta_z$ 和 $\xi_s = \alpha\zeta_s$ 分别表示比绝对涡度矢量($\boldsymbol{\xi}_a$)的垂直和水平分量,而 $\xi_n = \alpha\zeta_n$ 为 $\boldsymbol{\xi}_a$ 在位温梯度方向上的分量;B 为偏离 z 轴的角度($-\frac{\pi}{2} < B < \frac{\pi}{2}$),且设顺时针转向水平 s 坐标轴方向时,B 角度为正;θ_s/θ_z 表示等熵面的倾斜程度,$\tan B = \frac{\theta_s}{\theta_z}$,$\cos B = \frac{\theta_z}{\theta_n}$,$\theta_n = |\nabla\theta|$,而 θ_s 和 θ_z 分别为水平及垂直方向上的位温梯度。

由式(11)可见,垂直涡度的变化与 θ_z(大气对流稳定度)、θ_s(大气水平斜压性)和 ξ_s(比绝对涡度水平分量,主要与环境水平风速的垂直切变有关)有关。在位温和位涡守恒前提下,当空气质点 A_0 沿 $\theta + \Delta\theta$ 等熵面自左向右移动时,在 z 轴左侧,满足"盒子定律"(Wu and Liu,1997),即,不管环境水平风的垂直切变如何变化,比绝对涡度 $\boldsymbol{\xi}_a$ 的垂直分量 $\xi_z (=\xi_n)$ 不变(垂直涡度不发展);而但当 A_0 移至 z 轴右侧,沿 $\theta + \Delta\theta$ 等熵面下滑角度 B 至点 A 时,根据"外切平面定律"(Wu and Liu,1997),比绝对涡度 $\boldsymbol{\xi}_a$ 的矢量末端 D_1 必位于通过圆"b"上点 D_2 的外切平面 D_1D_2 上。在北半球天气系统生成的地区,一般有 $\xi_n > 0$,同时,由于 A 点处有 $\boldsymbol{n} \cdot \boldsymbol{k} > 0$,故 $\cos B = \frac{\theta_z}{\theta_n} > 0$;因此,依据式(12),当 $C_D = \frac{\xi_s \theta_s}{\theta_z} < 0$ 得到满足时(C_D 可称为 SVD 指数,注意,由图 1 所示,因起始空气质点 A_0 位于 z 轴左侧时,有 $\theta_s = 0$,初始时刻 $C_D = 0$,因此,当空气质点沿等熵面下滑过程中 C_D 的变化($\dot{C}_D = \frac{\mathrm{D}C_D}{\mathrm{D}t}$)小于零,即上述 $C_D < 0$ 条件等价于 $\dot{C}_D < 0$),式(12)

可写成,

$$\xi_z = \frac{\xi_n}{\cos B} + |\xi_s \tan B|, \quad \left(|B| \neq \frac{\pi}{2}\right) \tag{13}$$

此时,ξ_z 成为 B 的递增函数,意味着沿等熵面下滑的大气质块的垂直涡度将增大,而当等熵面十分陡立时,ξ_z 将迅速增大。

上述理论描述了大气中等熵面倾斜区域附近涡旋系统垂直涡度的发展,故而称为倾斜涡度发展理论。需要指出的是,尽管 SVD 理论是从位涡守恒框架下提出的,但在受非绝热加热、斜压性和摩擦等因素影响,天气系统位涡随时间不断变化的实际大气中,其呈现的涡度发展的理论内涵依然稳定成立,因此,SVD 理论的重要意义和实用价值并不因位涡守恒与否而改变。此外,尽管这里仅仅给出干大气中的情形,但对于湿空气情形,上述理论推演过程依然成立,具体参见吴国雄和蔡雅萍(1997)和 Wu 和 Liu(1997)。

3 全型垂直涡度方程

从 SVD 理论出发,吴国雄和刘还珠(Wu and Liu,1997;吴国雄和刘还珠,1999a)推导得到了包含不稳定、斜压性、环境风垂直切变、摩擦效应和非绝热加热等完整热、动力学因素的全型垂直涡度方程,从方程角度严格证明了倾斜涡度发展理论,并详细阐明了全型垂直涡度方程和经典垂直涡度方程的异同,以及前者的物理内涵和独特优势(吴国雄和刘还珠,1999a;吴国雄,2001)。

对于干空气,对式(11)两边同时求时间导数,经过一系列推导,即可得到干空气情形下的全型垂直涡度方程,

$$\frac{D\zeta_z}{Dt} + \beta v + (f + \zeta_z)\nabla \cdot \boldsymbol{V} = \frac{1}{\alpha}\frac{D}{Dt}\left(\frac{P_E}{\theta_z} - C_D\right), \theta_z \neq 0 \tag{14}$$

而对于湿空气,上述 Ertel 位涡(P_E)由湿位涡(P_M)替代,大气静力稳定度指数(θ_z)变为对流稳定度指数 θ_{ez}(相当位温垂直梯度),而 SVD 指数(C_D)亦变成 C_M(SVD 湿指数),则湿空气情形下的全型垂直涡度方程写为,

$$\frac{D\zeta_z}{Dt} + \beta v + (f + \zeta_z)\nabla \cdot \boldsymbol{V} = \frac{1}{\alpha}\frac{D}{Dt}\left(\frac{P_M}{\theta_{ez}} - C_M\right), \theta_{ez} \neq 0 \tag{15}$$

式中,$C_M = \frac{\xi_s \theta_{es}}{\theta_{ez}}$ 为包含环境水平风垂直切变、水平湿斜压性和湿大气对流稳定度等热动力学信息的 SVD 湿指数。

由式(14)和(15)可以很容易地证明(吴国雄和刘还珠,1999a),在大气中等熵面倾斜的地区,倾斜涡度发展理论无论对于干空气还是湿空气均成立,同时,无论位涡(湿位涡)守恒与否,SVD 理论描述的垂直涡度发展过程也均存在。

经典垂直涡度方程是通过对动量方程求旋度得到的,其形式如下,

$$\frac{d\zeta_z}{dt} + \beta v + (f + \zeta_z)\nabla \cdot \boldsymbol{V} = (f + \zeta_z)\frac{\partial w}{\partial z} + \left(\frac{\partial u}{\partial z}\frac{\partial w}{\partial y} - \frac{\partial v}{\partial z}\frac{\partial w}{\partial x}\right) + N_z + \boldsymbol{k} \cdot \boldsymbol{F}_\xi \tag{16}$$

式中,N_z 为力管项。与如上经典垂直涡度方程(16)相比,全型垂直涡度方程(15)显示包含了大气稳定度、斜压性、环境风垂直切变、摩擦效应和非绝热加热等影响垂直涡度发展的热、动力学因素,十分有利于各类天气系统发生、发展机理的诊断分析研究,在天气学和气候学相关领域研究中均有着广阔的应用前景。

全型垂直涡度方程强调了在等熵面出现倾斜时,涡旋系统涡度的发生、发展。而当等熵面平坦时,全型垂直涡度方程与经典垂直涡度方程完全一致(吴国雄,2001;Yan et al.,2005)。

4 倾斜涡度发展理论和全型垂直涡度方程的应用

倾斜涡度发展理论和全型垂直涡度方程提出和建立之后,便在江淮暴雨成因(吴国雄 等,1995;吴国

雄和蔡雅萍,1997;Cui et al.,2003;姜勇强 等,2004;张滨 等,2005;雷正翠 等,2008;王炳坤 等,2009;李银娥 等,2015;Shen et al.,2020)、西南区域暴雨和强对流形成机理(Wu and Liu,1997;吴国雄和刘还珠,1999a;李英和段旭,1999,2000;段旭和李英,2000;邱朝桂 等,2007;郑永骏 等,2013;吴国雄 等,2013;Wu et al.,2022)华北和华中暴雨与强对流成因(赵宇 等,2004b;龚佃利 等,2005,2019;杨晓霞 等,2006,2007;吴君 等,2007;王淑云 等,2009;范学峰 等,2012;李娜 等,2013)、华南暴雨和强对流机理(蒙伟光 等,2004;覃昌柳 等,2008;Wu et al.,2010;何编 等,2012)、西北地区暴雨和强对流成因(井宇 等,2010;李安泰 等,2012;白涛 等,2013;黄艳,2013;黄艳和裴江文,2014;王伏村 等,2016)、热带气旋强度变化机理和登陆热带气旋暴雨成因(王淑静 等,1997;余晖和吴国雄,2001;赵宇和吴增茂,2004a;端义宏 等,2005;赵宇 等,2006;范学峰,2007;赖绍钧 等,2007;徐文慧和倪允琪,2009;朱健和罗律,2009;黄亿 等,2009;徐文慧 等,2010;刘峰 等,2011;丁治英 等,2012;Zhou et al.,2013;孙力 等,2015)、台前飑线机理(Lin et al.,2021)、海洋爆发性气旋与江淮气旋入海发展机制(崔晓鹏 等,2002;马雷鸣 等,2002)、暴雪成因(马新荣 等,2008;王子谦 等,2010;于碧馨 等,2016;孙俊 等,2018)等天气学研究领域,以及副热带高压变异机理(吴国雄 等,1999b;刘屹岷 等,1999a,1999b;林建 等,2005;康志明 等,2013;沈阳 等,2019;钟中 等,2020)、南亚高压活动机制(刘屹岷 等,1999b;周兵 等,2004;葛静 等,2015)、持续性高温成因(吴振鹏 等,2016)和城市群发展对南海夏季风爆发影响机制(余荣 等,2016)等气候学研究领域得到了广泛的应用,对相关领域理论研究的深化起到了极为重要的推动作用。

　　灾害性天气系统的活动具有显著的斜压性和湿过程特性,且多尺度作用对热带气旋、暴雨和强对流等致灾性天气系统的形成以及灾害性天气的发生至关重要。倾斜涡度发展理论基于经典位涡理论,聚焦灾害性天气最易出现的倾斜等熵面区域,借助湿过程和斜压性特征的引入,透过现象看本质,创新性理论解析了倾斜等熵面地区灾害性天气系统的发展过程物理图像和机理,而在SVD理论架构下提出的全型垂直涡度方程则给出了具体和强有力的诊断分析工具,因此得到了天气学领域广泛的关注和应用。例如,我国暴雨多发、频发、分布极为广泛,影响极大,但不同区域多尺度环流和水汽特征、地形和下垫面特点等影响因素存在很大差异,且具有显著和不同的多时间尺度变化规律,造成不同地区暴雨形成机理明显不同。但绝大多数暴雨系统均与不同尺度的倾斜等熵面密切关联;从这一共性特征出发,利用倾斜涡度发展理论(吴国雄 等,1995;吴国雄和蔡雅萍,1997;Wu and Liu,1997),即可揭示和理解不同区域、不同季节暴雨天气系统的发生、发展以及移动过程机理,而利用全型垂直涡度方程可进一步定量化诊断分析上述天气系统的发展和移动以及暴雨的成因(吴国雄 等,1995;吴国雄和蔡雅萍,1997;Wu and Liu,1997;吴国雄和刘还珠,1999a;Cui et al.,2003;赵宇 等,2004b;Wu et al.,2010;何编 等,2012;郑永骏 等,2013;吴国雄 等,2013;李娜 等,2013;Wu et al.,2022)。短时强降水、冰雹等强对流天气常由雷暴和飑线等中小尺度系统所引发,强对流天气分布广、区域差异大、局地性强、时间变化块、形成机理复杂、预报难度大。但借助等熵面倾斜这一常见特征,利用倾斜涡度发展理论和全型垂直涡度方程,同样可以深入揭示其成因(龚佃利 等,2005,2019;覃昌柳 等,2008;黄艳,2013;黄艳和裴江文,2014;李英和段旭,1999,2000;段旭和李英,2000;Lin et al.,2021)。未来利用上述理论和方程,借助时空精细化数据对等熵面倾斜程度变化的细致识别与分析,亦有可能对暴雨和强对流天气发生时间和具体位置展开分析,进而给出预报线索。热带气旋一般生成和发展于正压环境中,斜压性往往不利于其发生、发展,强盛斜压能量的侵入则会造成其变性为温带气旋,尽管如此,热带气旋环流内部依然存在斜压结构,例如,眼壁区域的斜压结构变化伴随的涡度发展可能引起热带气旋强度突然变化(余晖和吴国雄,2001;端义宏 等,2005),而热带气旋环流内由小股冷空气、地形、非绝热加热等因素造成的局部等熵面倾斜结构,则可能与热带气旋环流内显著非均匀分布的强降水有关(王淑静 等,1997;徐文慧和倪允琪,2009;黄亿 等,2009;徐文慧 等,2010;刘峰 等,2011;Zhou et al.,2013;孙力 等,2015)。

　　倾斜涡度发展理论和全型垂直涡度方程在大尺度和气候学研究领域同样可以发挥重要作用。例如,利用全型垂直涡度方程和SVD理论,借助尺度分析和适度约化,便可深入揭示非绝热加热的空间非均匀分布对副热带高压形态变异的影响机理、暴雨过程反馈对副热带高压变动的调控机制(吴国雄 等,1999b;

刘屹岷 等,1999a,1999b;周兵 等,2004;林建 等,2005;康志明 等,2013;沈阳 等,2019;钟中 等,2020),以及空间非均匀分布非绝热加热和暴雨反馈效应对南亚高压活动的影响机理(刘屹岷 等,1999b;周兵 等,2004;葛静 等,2015)等。此外,基于倾斜涡度发展理论,利用全型垂直涡度方程尺度分析和适度约化,还可深入认识大尺度环流异常维持特征和机制,揭示持续性高温天气的成因(吴振鹏 等,2016),以及大规模城市群发展导致的区域性陆面增温和温差变化,通过引发环流异常,对南海夏季风提前爆发的可能影响机制(余荣 等,2016)等。

5　结论与展望

　　文中回顾了倾斜涡度发展(SVD)理论和全型垂直涡度方程,并简要梳理了天气学和气候学研究领域中,应用上述理论和方程开展的研究工作进展。

　　涡度和位涡是大气科学研究领域广泛应用的经典物理量,两者有着不可分割的密切联系,SVD理论和全型垂直涡度方程从经典位涡理论出发,通过引入湿过程和细化考虑斜压性特征,突破了经典位涡理论在对流层中低层大气过程研究中的局限,阐明了上述两个经典物理量之间的理论联系,构建了两者之间的量化关系;为天气系统形成机理和气候变化机制研究奠定了坚实理论基础,并提供了定量化诊断工具,大幅拓展了经典位涡理论及其应用范畴。SVD理论和全型垂直涡度方程在江淮暴雨机理、西南、华南、西北和华北等地区暴雨与强对流成因、热带气旋强度变化机理和登陆热带气旋暴雨成因、海洋爆发性气旋与江淮气旋发展机制、副热带高压变异机理和南亚高压活动机制,以及持续性高温成因和城市群发展对南海夏季风爆发影响机制等诸多天气和气候学研究领域得到了广泛应用,显著推动了相关领域理论研究的进展。

　　值得指出的是,大气环流是一个整体,天气和气候研究、大尺度和中小尺度过程研究等并非是完全独立和平行的研究范畴。气候特征为具体的天气系统发展演变提供了背景条件,而天气过程又会对气候背景造成反馈影响,大尺度环流及其演变对中小尺度天气系统的发生发展起到重要的调控作用,同样,中小尺度系统的发展和演变也会在细节尺度上给大尺度环流背景带来改变。天气-气候一体化研究、多时空尺度相互作用研究是当前的研究热点,亦是未来大气科学研究的大趋势。可以相信,在这些热点研究领域,倾斜涡度发展理论和全型垂直涡度方程及其未来的不断拓展必将持续发挥重要作用。

参考文献

白涛,李崇银,王铁,等,2013. 干侵入对陕西"2008·07·21"暴雨过程的影响分析[J]. 高原气象,32(2):345-356.
崔晓鹏,吴国雄,高守亭,2002. 西大西洋锋面气旋过程的数值模拟和等熵分析[J]. 气象学报,60(4):385-399.
丁治英,邢蕊,徐海明,等,2012. 南亚高压南部环境位涡对台风加强的影响分析[J]. 热带气象学报,28(5):675-686.
端义宏,余晖,伍荣生,2005. 热带气旋强度变化研究进展[J]. 气象学报,63(5):636-645.
段旭,李英,2000. 滇中暴雨的湿位涡诊断分析[J]. 高原气象,19(2):253-259.
范学峰,2007. "海棠"台风远距离强降水分析[J]. 气象与环境科学,30(4):30-33.
范学峰,王新敏,胡燕平,2012. 一次缓慢东移的黄河气旋暴雨的诊断分析[J]. 气象与环境科学,35(3):10-16.
葛静,王黎娟,张良瑜,2015. 春末夏初南亚高压活动与青藏高原及周边热力强迫的关系[J]. 大气科学学报,38(5):611-619.
龚佃利,吴增茂,傅刚,2005. 2001年8月23日华北强风暴动力机制的数值研究[J]. 气象学报,63(4):504-515.
龚佃利,庞华基,王俊,等,2019. 2006年4月28日山东强飑线过程中尺度结构和动力机制分析[J]. 海洋气象学报,39(3):64-73.
何编,孙照渤,李忠贤,2012. 一次华南持续性暴雨的动力诊断分析和数值模拟[J]. 大气科学学报,35(4):466-476.
黄艳,裴江文,胡素琴,等,2013. 新疆喀什两次超级单体致雹风暴特征对比分析[J]. 气象科学,33(6):693-700.
黄艳,裴江文,2014. 2012年新疆喀什一次罕见冰雹天气的中尺度特征[J]. 干旱气象,32(6):989.
黄亿,寿绍文,傅灵艳,2009. 对一次台风暴雨的位涡与湿位涡诊断分析[J]. 气象,35(1):65-73.
姜勇强,陈中一,周祖刚,等,2004. 倾斜涡度发展与β中尺度低涡[J]. 解放军理工大学学报:自然科学版,5(6):81-87.

井宇,井喜,屠妮妮,等,2010. 黄土高原低值对流有效位能区β中尺度大暴雨综合分析[J]. 高原气象,29(1):78-89.
康志明,桂海林,王小光,2013. 2009年夏季西太平洋副热带高压北抬原因初探[J]. 气象,39(1):46-56.
赖绍钧,何芬,赵汝汀,等,2007. "龙王"(LONGWANG)台风过程湿位涡的诊断分析[J]. 气象科学,27(3):266-271.
雷正翠,夏文梅,武金岗,等,2008. 一次暴雨的湿位涡分析及EVAD技术应用[J]. 气象科学,28(3):283-288.
李安泰,何宏让,张云,2012. 引起舟曲特大泥石流灾害的"8·8"暴雨过程中尺度特征分析[J]. 气象科学,32(2):169-176.
李娜,冉令坤,周玉淑,等,2013. 北京"7·21"暴雨过程中变形场引起的锋生与倾斜涡度发展诊断分析[J]. 气象学报,71(4):593-605.
李银娥,王艳杰,李武阶,等,2015. 低层锋生型暴雨特征合成分析[J]. 气象科学,35(2):223-229.
李英,段旭,1999. 倾斜涡度的发展与云南冬季强降水[J]. 南京气象学院学报,22(4):705-710.
李英,段旭,2000. 湿位涡在云南冰雹天气分析中的应用[J]. 应用气象学报,11(2):242-248.
林建,毕宝贵,何金海,2005. 2003年7月西太平洋副热带高压变异及中国南方高温形成机理研究[J]. 大气科学,29(4):594-599.
刘峰,丁治英,梁艳,等,2011. "莫拉克"台风暴雨过程中湿位涡场的演变特征[J]. 暴雨灾害,30(2):161-166.
刘屹岷,刘辉,刘平,等,1999a. 空间非均匀加热对副热带高压形成和变异的影响Ⅱ:陆面感热与东太平洋副高[J]. 气象学报,57(4):385-396.
刘屹岷,吴国雄,刘辉,等,1999b. 空间非均匀加热对副热带高压形成和变异的影响Ⅲ:凝结潜热加热与南亚高压及西太平洋副高[J]. 气象学报,57(5):525-538.
马雷鸣,秦曾灏,端义宏,等,2002. 大气斜压性与入海江淮气旋发展的个例研究[J]. 海洋学报,24(S1):95-104.
马新荣,任余龙,丁治英,2008. 青藏高原东北侧一次暴雪过程的湿位涡分析[J]. 干旱气象,26(1):57.
蒙伟光,王安宇,李江南,等,2004. 华南暴雨中尺度对流系统的形成及湿位涡分析[J]. 大气科学,28(3):330-341.
邱朝桂,张云谨,刘开宇,等,2007. 贵阳地区暴雨和冰雹湿位涡对比诊断分析[J]. 云南大学学报:自然科学版,29(S1):177-182.
沈阳,孙燕,吴海英,等,2019. 2018年5月江苏极端降水事件发生前副热带高压异常及原因分析[J]. 气象科学,39(2):214-225.
孙俊,邓国卫,夏炳江,2018. 川西高原中部一次极端暴雪成因分析[J]. 气象科技,46(3):584-593.
孙力,董伟,药明,等,2015. 1215号"布拉万"台风暴雨及降水非对称性分布的成因分析[J]. 气象学报,73(1):36-49.
覃昌柳,黎惠金,董良淼,2008. 湿位涡在广西冰雹大风天气过程诊断分析中的应用[J]. 气象科技,36(5):541-546.
王炳坤,莫静华,张丽,2009. 一次梅雨暴雨过程的双峰结构分析[J]. 气象研究与应用,30(2):21-25.
王伏村,许东蓓,姚延锋,等,2016. 一次陇东大暴雨的锋生过程及倾斜涡度发展[J]. 高原气象,35(2):419-431.
王淑静,周黎明,陈高峰,1997. 解释台风暴雨落区判据的探讨[J]. 应用气象学报,8(2):167-174.
王淑云,寿绍文,周连科,2009. 干侵入对"0310"暴雨形成过程的影响[J]. 自然灾害学报,18(6):129-134.
王子谦,朱伟军,段安民,2010. 孟湾风暴影响高原暴雪的个例分析:基于倾斜涡度发展的研究[J]. 高原气象,29(3):703-711.
吴国雄,2001. 全型涡度方程和经典涡度方程比较[J]. 气象学报,59(4):385-392.
吴国雄,蔡雅萍,唐晓菁,1995. 湿位涡和倾斜涡度发展[J]. 气象学报,53(4):387-405.
吴国雄,蔡雅萍,1997. 风垂直切变和下滑倾斜涡度发展[J]. 大气科学,21(3):273-282.
吴国雄,刘还珠,1999a. 全型垂直涡度倾向方程和倾斜涡度发展[J]. 气象学报,57(1):1-15.
吴国雄,刘屹岷,刘平,1999b. 空间非均匀加热对副热带高压带形成和变异的影响Ⅰ:尺度分析[J]. 气象学报,57(3):257-263.
吴国雄,郑永骏,刘屹岷,2013. 涡旋发展和移动的动力和热力问题Ⅱ:广义倾斜涡度发展[J]. 气象学报,71(2):198-208.
吴君,汤剑平,邱庆国,等,2007. 切变线暴雨过程中湿位涡的中尺度时空特征[J]. 气象,33(10):45-51.
吴振鹏,武媛,李乃杰,等,2016. 1999年北京持续高温天气过程的诊断分析[J]. 气象与环境科学,39(1):82-88.
徐文慧,倪允琪,2009. 登陆台风环流内的一次中尺度强对流过程[J]. 应用气象学报,20(3):267-275.
徐文慧,倪允琪,汪小康,等,2010. 登陆台风内中尺度强对流系统演变机制的湿位涡分析[J]. 气象学报,68(1):88-101.

杨晓霞,万丰,刘还珠,等,2006. 山东省春秋季暴雨天气的环流特征和形成机制初探[J]. 应用气象学报,17(2):183-191.

杨晓霞,李春虎,杨成芳,等,2007. 山东省2006年4月28日飑线天气过程分析[J]. 气象,33(1):74-80.

于碧馨,张云惠,宋雅婷,2016. 2012年前冬伊犁河谷持续性大暴雪成因分析[J]. 沙漠与绿洲气象,10(5):44-51.

余晖,吴国雄,2001. 湿斜压性与热带气旋强度突变[J]. 气象学报,59(4):440-449.

余荣,江志红,马红云,2016. 中国东部城市群发展对南海夏季风爆发影响的模拟研究[J]. 大气科学,40(3):504-514.

张滨,周林,关皓,2005. 1998年夏季江淮地区强暴雨过程的湿位涡诊断分析[J]. 解放军理工大学学报:自然科学版,6(4):399-403.

赵宇,吴增茂,2004a. 9711号北上台风演变及暴雨过程的位涡诊断分析[J]. 中国海洋大学学报:自然科学版,34(1):13-21.

赵宇,张兴强,杨晓霞,2004b. 山东春季一次罕见暴雨天气的湿位涡分析[J]. 南京气象学院学报,27(6):836-843.

赵宇,龚佃利,刘诗军,等,2006. "99·8"山东特大暴雨形成机制的数值模拟分析[J]. 高原气象,25(1):95-104.

郑永骏,吴国雄,刘屹岷,2013. 涡旋发展和移动的动力和热力问题 I:PV-Q观点[J]. 气象学报,71(2):185-197.

钟中,王天驹,胡轶佳,2020. 2019年6月西太平洋副热带高压脊线位置异常的机制分析[J]. 气象科学,40(5):639-648.

周兵,何金海,徐海明,2004. 暴雨过程对副热带高压变动的影响[J]. 应用气象学报,15(4):394-406.

朱健,罗律,2009. 超强台风"韦帕"的暴雨机制及湿位涡分析[J]. 气象科学,29(6):742-748.

BENNETTS D A, HOSKINS B J, 1979. Conditional symmetric instability-a possible explanation for frontal rainbands[J]. Quarterly Journal of the Royal Meteorological Society, 105(446): 945-962.

CUI X P, GAO S T, WU G X, 2003. Up-sliding slantwise vorticity development and the complete vorticity equation with mass forcing[J]. Advances in Atmospheric Sciences, 20(5): 825-836.

DANIELSEN E F, 1967. Transport and diffusion of stratospheric radioactivity based on synoptic hemispheric analyses of potential vorticity[R]. Dept of Met Penn State Univ, Report NYO-3317-3.

DANIELSEN E F, 1968. Stratospheric-tropospheric exchange based on radioactivity ozone and potential vorticity [J]. J Atmos Sci, 25: 502-518.

DAVIS C A, 1992a. Piecewise potential vorticity inversion[J]. Journal of the Atmospheric Sciences, 49(16): 1397-1411.

DAVIS C A, 1992b. A potential-vorticity diagnosis of the importance of initial structure and condensational heating in observed extratropical cyclogenesis[J]. Monthly Weather Review, 120(11): 2409-2428.

ERTEL H, 1942. Ein neuer hydrodynamischer wirbelsatz[J]. Meteor Zeitschrift Braunschweig, 6:277-281.

HOSKINS B J, MCINTYRE M E, ROBERTSON A W, 1985. On the use and significance of isentropic potential vorticity maps[J]. Quarterly Journal of the Royal Meteorological Society, 111(470): 877-946.

KLEINSCHMIDT E, 1950a. Über aufbau und entstehung von zyklonen (1. Teil)[J]. Met Runds, 3: 1-6.

KLEINSCHMIDT E, 1950b. Über aufbau und entstehung von zyklonen (2. Teil)[J]. Met Runds, 3: 54-61.

KLEINSCHMIDT E, 1951. Über aufbau und entstehung von zyklonen (3. Teil)[J]. Met Runds, 4: 89-96.

KLEINSCHMIDT E, 1955. Die entstehung einer höhenzyklone über nordamerika[J]. Tellus, 7: 96-110.

KLEINSCHMIDT E, 1957. Dynamic meteorology[J]. Handbuch der Physik, 48: 112-129.

LIN X, YIN S, CAI Y, et al, 2021. Comparative analysis of dry intrusion in the different position of pre-TC squall line on typhoon Lekima (1909) and Matsa (0509)[J]. Tropical Cyclone Research and Review, 10(1): 43-53.

MCINTYRE M E, PALMER T N, 1983. Breaking planetary waves in the stratosphere[J]. Nature, 305(5935): 593-600.

MCINTYRE M E, PALMER T N, 1984. The "surf zone" in the stratosphere[J]. Journal of Atmospheric and Terrestrial Physics, 46(9): 825-849.

OBUKHOV A M, 1964. Adiabatic invariants of atmospheric processes[J]. Meteor Gidrol, 2: 3-9.

PLATZMAN G W, 1949. The motion of barotropic disturbances in the upper troposphere[J]. Tellus, 1(3): 53-64.

REED R J, 1955. A study of a characteristic type of upper-level frontogenesis[J]. Journal of Atmospheric Sciences, 12(3): 226-237.

REED R J, SANDERS F, 1953. An investigation of the development of a mid-tropospheric frontal zone and its associated vorticity field[J]. Journal of Atmospheric Sciences, 10(5): 338-349.

REED R J, DANIELSEN E F, 1959. Fronts in the vicinity of the tropopause[J]. Archiv für Meteorologie, Geophysik und

Bioklimatologie, Ser. A, Meteorologie und Geophysik, 11(1): 1-17.

ROSSBY C G, 1939. Relation between variations in the intensity of the zonal circulation of the atmosphere and the displacements of the semi-permanent centers of action[J]. Journal of Marine Research, 2(1): 38-55.

ROSSBY C G, 1940. Planetary flow patterns in the atmosphere[J]. Quarterly Journal of the Royal Meteorological Society, 66: 68-87.

SHAPIRO M A, 1976. The role of turbulent heat flux in the generation of potential vorticity in the vicinity of upper-level jet stream systems[J]. Monthly Weather Review, 104(7): 892-906.

SHAPIRO M A, 1978. Further evidence of the mesoscale and turbulent structure of upper level jet stream-frontal zone systems[J]. Monthly Weather Review, 106(8): 1100-1111.

SHEN Y, SUN Y, LIU D Y, 2020. Effect of condensation latent heat release on the relative vorticity tendency in extratropical cyclones: A case study[J]. Atmospheric and Oceanic Science Letters, 13(4): 275-285.

STARR V P, NEIBURGER M, 1940. Potential vorticity as a conservative property[J]. J Marine Res, 3: 202-210.

UCCELLINI L W, KEYSER D, BRILL K F, et al, 1985. The Presidents' Day cyclone of 18-19 February 1979: Influence of upstream trough amplification and associated tropopause folding on rapid cyclogenesis[J]. Monthly Weather Review, 113(6): 962-988.

WU G X, LIU H Z, 1997. Vertical vorticity development owing to down-sliding at slantwise isentropic surface[J]. Dynamics of atmospheres and oceans, 27(1-4): 715-743.

WU G X, MA T T, LIU Y M, et al, 2020. PV-Q perspective of cyclogenesis and vertical velocity development downstream of the Tibetan Plateau[J]. Journal of Geophysical Research: Atmospheres, 125(16): e2019JD030912.

WU G, TANG Y, HE B, et al, 2022. Potential vorticity perspective of the genesis of a Tibetan Plateau vortex in June 2016 [J]. Climate Dynamics: 1-17.

WU L, HUANG R, HE H, et al, 2010. Synoptic characteristics of heavy rainfall events in pre-monsoon season in South China[J]. Advances in Atmospheric Sciences, 27(2): 315-327.

YAN J H, WU G X, CUI X P, 2005. Comparison of two kinds of atmospheric vorticity equations[J]. Chinese Physics Letters, 22(3): 769-772.

YAO X P, WU G X, ZHAO B K, et al, 2007. Research on the dry intrusion accompanying the low vortex precipitation [J]. Sci China Ser D-Earth Sci, 50(9): 1396-1408.

YAO X P, WU G X, LIU Y M, 2009. Case study on the impact of the vortex in the easterlies over the tropical upper troposphere on the subtropical anticyclone over Western Pacific [J]. Acta Meteorologica Sinica, 23(3): 363-373.

YAO X P, SUN J Y, 2016. Thermal forcing impacts of the easterly vortex on the east-west shift of the subtropical anticyclone over the western Pacific Ocean [J]. J Trop Meteorol, 22(1): 51-56.

YAO X P, GAO Y, MA J L, 2020a. MPV-Q^* view of vorticity development in a saturated atmosphere [J]. Atmospheric Research, 244: 105058.

YAO X P, ZHANG Q, ZHANG X, 2020b. Potential vorticity diagnostic analysis on the impact of the easterlies vortex on the short-term movement of the subtropical anticyclone over the western Pacific in the Mei-yu period [J]. Adv Atmos Sci, 37(9): 1019-1031.

ZHOU L, DU H, ZHAI G, et al, 2013. Numerical simulation of the sudden rainstorm associated with the remnants of Typhoon Meranti (2010) [J]. Advances in Atmospheric Sciences, 30(5): 1353-1372.

综述 12

地表位涡制造、位涡环流特征及在气候动力学中的应用

何编[1]，生宸[1,2]，谢永坤[3]

(1 中国科学院大气物理研究所，北京 100029；2 中国科学院大学，北京 100049；
3 兰州大学西部生态安全协同创新中心，兰州 730000)

摘要：位涡是大气动力学中一个重要物理量，由于在绝热无摩擦条件下具有天然的守恒性质，它被广泛用于天气和气候问题的动力学机理分析研究。文中主要回顾了吴国雄院士团队近年来在位涡气候动力学方面的研究新进展。从原始的位涡方程出发，研究团队给出了广义坐标系下的位涡方程，利用位涡和位涡收支方程研究了全球和区域的气候异常现象。研究结果表明了全球变暖背景下高层位涡平流是北极增暖影响欧亚大陆变冷的关键因素；揭示了春季跨赤道位涡输送对北半球欧亚大陆上空大气位涡三维结构的形成及欧亚大陆近地面气温冷异常有重要影响。在已有研究基础上，研究结果进一步证明青藏高原地表位涡制造是亚洲夏季风在大陆上形成的关键因素，是调节东亚地区夏季降水年际变率的重要强迫源。以上研究为理解气候变化中的动力学问题提供了新视角。

关键词：位涡；青藏高原；全球变暖；跨赤道位涡输送；亚洲季风

1 引言

气候变化对人类社会的经济生活等方面有非常重要的影响。地球气候系统包括多圈层的相互作用，因此气候变化的原因是非常复杂的，这是相关研究的一个难点问题，同时也是国际前沿的热点科学问题。气候系统由大气圈、水圈、岩石圈、冰雪圈和生物圈五大圈层组成，圈层之间的相互作用是气候系统中的基本物理过程。其中，发生在海-陆-气界面上的能量、动量和水分交换是导致气候系统异常的基本原因。值得注意的是，大气环流反过来对海-陆-气相互作用的调控非常重要，这是因为发生在海面和陆面上的能量、动量和物质交换过程与大气底层的环流和热状况紧密相关。例如，地球表面的两种非常重要的非绝热加热，感热和潜热通量就受到地球表层大气的环流、温度、湿度和稳定度的影响。大气对非绝热加热的响应非常快速，热带对流的扰动在大气中经 7 天左右的传播就可以影响全球。换言之，研究大气的动力和热力状况以及发生在大气中的动力和热力过程是气候系统动力学的重要内容。

气候系统是一个开放耗散系统，外部加热和摩擦耗散对于气候变化非常重要。因此，研究气候变化的原因就是研究大气环流对外部加热和摩擦耗散的响应。位涡（PV）是一个研究气候变化非常理想的物理量，它综合了大气动力和热力信息，其最重要的一个特性是绝热无摩擦条件下的守恒性。在下部大气，等熵面和地面相交，出现边界线。在中部大气中，高纬度处于平流层，热带处于对流层，沿等熵面的物质能够

通讯作者：何编，heb@lasg.iap.ac.cn

跨越对流层顶,在平流层和对流层之间自由交换,因而这里是对流层和平流层相互作用的重要区域。上部大气则完全处于热带平流层以上(图1)。在上部大气和中部大气中等熵面内的 PV 守恒(Haynes and McIntyre,1987,1990),PV 既不能被制造也不能被消耗,只能从一个地区输送到另一个地区;而在下部大气中由于等熵面与地表相交而存在边界线,这里存在着 PV 的制造或耗散。因此,从大气动力学角度来说,影响大气环流的外部涡度源汇位于地球表面(Aebischer and Schär,1996;Schär et al.,2003;Flamant et al.,2004)。分析大气环流位涡制造的变化不仅可以为我们提供一个研究全球变暖对气候变化影响的新途径和新方法,还有希望揭示全球变暖影响区域和全球气候变化的新机制。文中回顾了吴国雄院士带领他的研究团队近几年来在位涡动力学方面的研究新进展,包括广义坐标系下的位涡方程及地表位涡的气候态分布特征;全球变暖背景下北极增暖和欧亚大陆变冷的联系;南、北半球之间的位涡跨赤道输送过程;以及全球最显著的地表位涡制造区的青藏高原地表位涡对亚洲夏季风系统的可能影响。上述工作为全球和亚洲气候动力学研究提供了新视角和新证据。

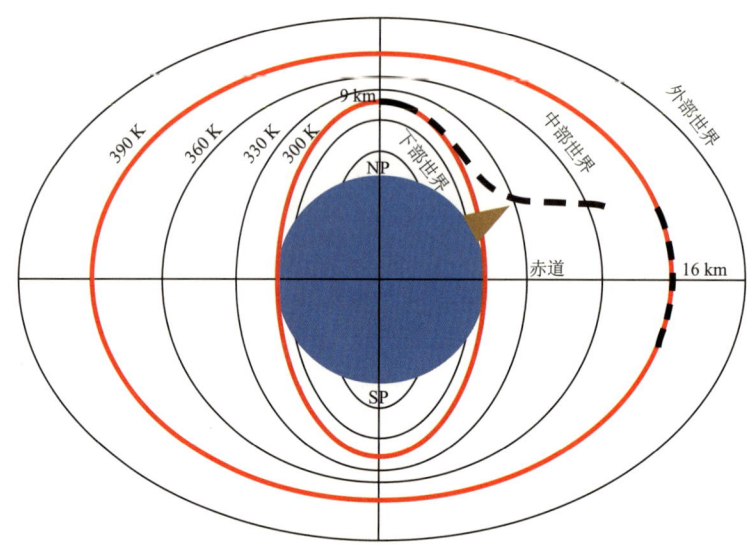

图 1 三层世界分布示意图。黑实线表示等熵面,黑断线表示大气对流层顶。蓝色圆表示地球。棕色的三角表示青藏高原。NP 和 SP 分别表示南极和北极(摘自 Sheng et al.,2021)

2 位涡理论及广义坐标系下的地表位涡

2.1 广义坐标系下的位涡方程

位涡 P 是指单位质量气块的三维绝对涡度矢量($\boldsymbol{\xi}_a$)与三维位温(θ)梯度的数量积,即:

$$P = \alpha \boldsymbol{\xi}_a \cdot \nabla \theta \tag{1}$$

式中,α 为比容,它是密度 ρ 的倒数。位涡的变化($\frac{\mathrm{d}P}{\mathrm{d}t}$)满足如下位涡变化方程(Ertel,1942;Hoskins et al.,1985):

$$\frac{\mathrm{d}P}{\mathrm{d}t} = \frac{\partial P}{\partial t} + \boldsymbol{V} \cdot \nabla P = \alpha \nabla \times \boldsymbol{F} \cdot \nabla \theta + \alpha \boldsymbol{\xi}_a \cdot \nabla \dot{\theta} \tag{2}$$

式中,$\boldsymbol{F}$ 是摩擦力,$\dot{\theta}$ 是非绝热加热率。

引入单位体积位涡 W(位涡密度,或位涡物质)的概念,位涡及位涡变化方程还可以写成如下形式:

$$W = \rho P = \boldsymbol{\xi}_a \cdot \nabla \theta \tag{3}$$

$$\frac{\mathrm{d}W}{\mathrm{d}t} = \frac{\partial W}{\partial t} + \nabla \cdot \boldsymbol{V} W = \nabla \times \boldsymbol{F} \cdot \nabla \theta + \boldsymbol{\xi}_a \cdot \nabla \dot{\theta} \tag{4}$$

传统的 PV 及其收支分析是在等压或者等熵坐标系下进行的。然而，在近地面层，由于高大地形如青藏高原的存在，等压或者等熵面与高原相交，因此传统等压或等熵 PV 理论不适用于研究青藏高原地形表面的热力、动力作用对大气环流和位涡制造的影响研究。在全球数值模式中，目前大多采用 σ-p 混合坐标。这样在近地面层采用地形坐标可以较好地解析大地形附近的热力和动力过程，而在对流层中高层慢慢过渡到气压坐标有利于解析自由大气中的平流过程。因此，为了理解位涡在全球的三维传输和分布状况，有必要引入适用于 σ-p 混合坐标系下的位涡方程（Sheng et al., 2021）。对于大尺度而言，垂直速度的水平变化量级较小，可以忽略，这样大大简化了位涡方程在不同坐标系下的形式。对于任意变量 F，引入一个广义的垂直坐标系 (x,y,e,t)，有 $F(x,y,e,t)=F(x,y,z(x,y,e,t),t)$，其中 e 是 z 的单调函数，并且是 e 坐标系统的垂直轴（通常情况下，e 等于 p、z 或 θ）。在忽略垂直速度的水平变化的情况下，水平坐标转换项相互抵消，有如下关系成立：

$$P_e = \alpha_e\, \boldsymbol{\xi}_{ae}\cdot \nabla_e\theta = \alpha_z\, \boldsymbol{\xi}_{az}\cdot \nabla_z\theta = P \tag{5}$$

上式意味着 PV 具有垂直坐标不依赖性，即 PV 在任意坐标下均具有相同的计算格式。对于 σ-p 坐标系的近地表而言，位涡及位涡变化方程可以进一步具体写为如下形式：

$$P_h = g\left[\frac{\partial v}{\partial p}\left(\frac{\partial \theta}{\partial x}\right)_h - \frac{\partial u}{\partial p}\left(\frac{\partial \theta}{\partial y}\right)_h\right] - g\left[f + \left(\frac{\partial v}{\partial x}\right)_h - \left(\frac{\partial u}{\partial y}\right)_h\right]\frac{\partial \theta}{\partial p} \tag{6}$$

$$\underbrace{\phantom{g\left[\frac{\partial v}{\partial p}\left(\frac{\partial \theta}{\partial x}\right)_h - \frac{\partial u}{\partial p}\left(\frac{\partial \theta}{\partial y}\right)_h\right]}}_{\text{PV 水平项}} \quad \underbrace{\phantom{g\left[f + \left(\frac{\partial v}{\partial x}\right)_h - \left(\frac{\partial u}{\partial y}\right)_h\right]\frac{\partial \theta}{\partial p}}}_{\text{PV 垂直项}}$$

$$\frac{\partial P_h}{\partial t} = \underbrace{-\boldsymbol{V}\cdot\nabla_h P_h}_{\text{局地变化}\;\;\text{平流}} + \underbrace{\alpha_h\,\boldsymbol{\xi}_{ah}\cdot\nabla_h\dot{\theta}}_{\text{加热}} + \underbrace{\alpha_h\nabla_h\times\boldsymbol{F}\cdot\nabla_h\theta}_{\text{摩擦}}$$

$$= -u\frac{\partial P_h}{\partial x} - v\frac{\partial P_h}{\partial y} - \omega_h\frac{\partial P_h}{\partial p} - \tag{7}$$

$$g\left\{-\frac{\partial v}{\partial p}\left(\frac{\partial \dot\theta}{\partial x}\right)_h + \frac{\partial u}{\partial p}\left(\frac{\partial \dot\theta}{\partial y}\right)_h + \left[f + \left(\frac{\partial v}{\partial x}\right)_h - \left(\frac{\partial u}{\partial y}\right)_h\right]\frac{\partial \dot\theta}{\partial p}\right\} -$$

$$g\left\{-\frac{\partial F_y}{\partial p}\left(\frac{\partial \theta}{\partial x}\right)_h + \frac{\partial F_x}{\partial p}\left(\frac{\partial \theta}{\partial y}\right)_h + \left[f + \left(\frac{\partial F_y}{\partial x}\right)_h - \left(\frac{\partial F_x}{\partial y}\right)_h\right]\frac{\partial \theta}{\partial p}\right\}$$

式中，

$$\omega_h = \left[\omega - \frac{\partial p}{\partial t} - u\left(\frac{\partial p}{\partial x}\right)_h - v\left(\frac{\partial p}{\partial y}\right)_h\right] \tag{8}$$

式(6)—(7)是诊断再分析资料中地表 PV 和 PV 收支的两个主要公式。在下一小节，将给出青藏高原地表位涡的气候态空间分布特征。

2.2 青藏高原地表位涡的气候态分布特征

一般而言，PV 的水平项较小，在以往研究中，常常用 PV 的垂直项来代表 PV 本身。图 2 给出了基于 MERRA 再分析资料的夏季（6—8 月平均）气候态地表 PV 及其垂直和水平项分布。整体来看，由于科里奥利效应，地表 PV 随纬度增加。在冬季（图 2a），最大的地表 PV 中心位于青藏高原北部，其量级约为高原东部的 2 倍。与之相比，在夏季（图 2b），地表 PV 较小，最大中心移向伊朗高原，青藏高原上的大值中心位于青藏高原平台中北部。地表 PV 的垂直项（图 2c、d）分布与地表 PV 总量（图 2a、b）非常相似，这意味着气候态 PV 的垂直项是主要贡献项。

从欧亚大陆大范围来看，地表 PV 水平项很小。然而仔细分析发现，在青藏高原陡峭地形处却并非如此，尤其是在青藏高原南北侧斜坡附近（图 2e、f）。图 2e 给出了冬季地表 PV 的水平项，可以看到水平项在高原平台几乎均为负值，尤其是在天山、昆仑山附近。正值出现在塔里木盆地，喜马拉雅山以及高原东南侧，其中大值中心为青藏高原南坡。夏季（图 2f）地表 PV 的水平项大值均出现在高原边界处，最大值位于伊朗高原东部，喜马拉雅山，昆仑山和青藏高原东南角。由此可见，尽管地表 PV 垂直项主导了地表 PV 的大尺度分布，但地表 PV 水平项在陡峭的高原南北侧斜坡处十分重要。

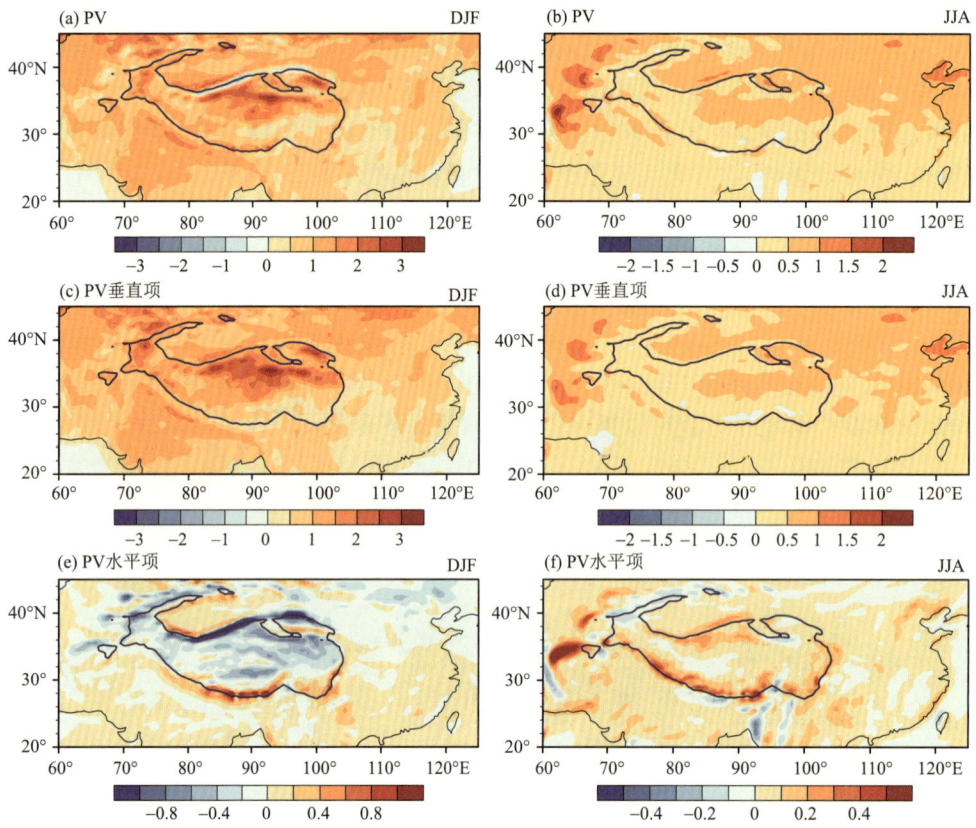

图 2　MERRA2 资料的气候态冬季(a)PV 本身；(c)PV 垂直项；(e)PV 水平项的分布(单位：PVU，1 PVU = 10^{-6} K·m^2·kg^{-1}·s^{-1})。(b)，(d)，(f)分别同图(a)，(c)，(e)，但为夏季。蓝色实线表示青藏高原 3000 m 等高线(摘自 Sheng et al.，2021)

为了定量描述地表 PV 水平项的重要性，图 3 进一步给出了冬夏季沿 85°~95°E 平均的地表位涡水平分量与垂直分量之比绝对值的经向分布。图 3a 为 MERRA2 结果，可以看到曲线呈双峰结构，分别出现在青藏高原南北侧斜坡处，其中南侧的峰值远大于北侧。无论是冬季还是夏季(图 3a)，高原南坡的水平项均为垂直项的 2 倍以上，北坡约为 0.5 倍。图 3b 为大气模式 FAMIL2(图 3b)的结果。图中可以看出，FAMIL2 与 MERRA2 的结果非常相似，但是 FAMIL2 量值偏小。这主要是由于 FAMIL2 水平分辨率较 MERRA2 低，因此地形较为平滑，相应的梯度计算会偏弱一些。这一结果进一步说明，无论是在再分析资料还是模式资料中，在陡峭的青藏高原南北边界上，地表 PV 水平项确实十分重要，在青藏高原南坡水平项甚至可达垂直项的 2 倍以上。

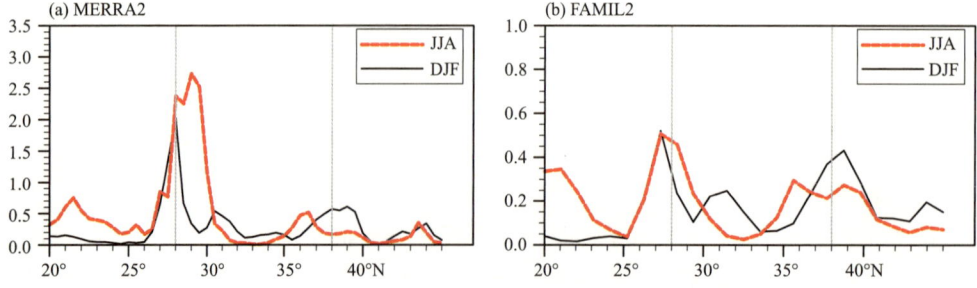

图 3　冬夏季沿 85°~95°E 平均的地表位涡水平项与垂直项之比的绝对值。(a)MERRA2 结果；(b)FAMIL2 结果。灰线(28°E，38°E)表示高原的南北边界(摘自 Sheng et al.，2021)

3 地表位涡制造与位涡三维输送在全球气候研究中的应用

3.1 位涡观点下北极增暖和欧亚大陆变冷的联系

过去几十年观测的气温变化显示,冬季北极地区的显著增暖与欧亚大陆中高纬的变冷同时存在,形成了暖北极-冷欧亚的偶极型分布。这种暖-冷偶极的两极之间是否存在因果关系是学术界研究且存在争论的热点问题。本质上讲,对该问题的研究属于全球变暖背景下区域增温差异的研究范畴。针对该问题,吴国雄院士带领研究团队利用位涡反演结合位涡收支分析的动力学方法进行了研究(Xie et al.,2020)。该研究是利用位涡理论研究气候变化问题的范例。研究表明,冬季纬向非均匀的北极增暖能够通过局地的位涡制造以及向中高纬地区的位涡平流,对欧亚大陆冬季气候变化产生影响。其中,北极增暖的空间型和垂直结构,以及气候态的经向位涡梯度等要素是北极增暖影响中高纬的动力过程中的关键因素,具体阐述如下。

在垂直方向,暖北极表现为底部(近地表)的增暖最强,冷欧亚表现为对流层中部(500 hPa 左右)和近地表两个强变冷中心。其中,暖北极底部最强的垂直结构对应于从低到高垂直递减的非绝热加热异常。基于位涡收支方程(式(2)),垂直递减的非绝热加热制造负位涡,即非绝热制造项中 $\frac{\partial \dot{\theta}}{\partial z} < 0$。因此,暖北极的强地表加热直接引起巴伦支喀拉海区域的负位涡制造。非绝热加热制造的负位涡进一步引起局地的反气旋环流异常(如图 4a 和 b)。仅有局地的非绝热位涡制造和反气旋环流发展时(即正压情形),北极增暖无法远程影响到欧亚大陆。然而,巴伦支喀拉海区域的北极增暖的南侧正好处于中高纬大气斜压性极大的位置。由于强斜压性区即是位涡梯度很大的区域,因此当地表反气旋环流发展时必然引起位涡的平流,从而激发位涡的再分布过程,以及进一步对环流的影响,直至达到新的平衡。

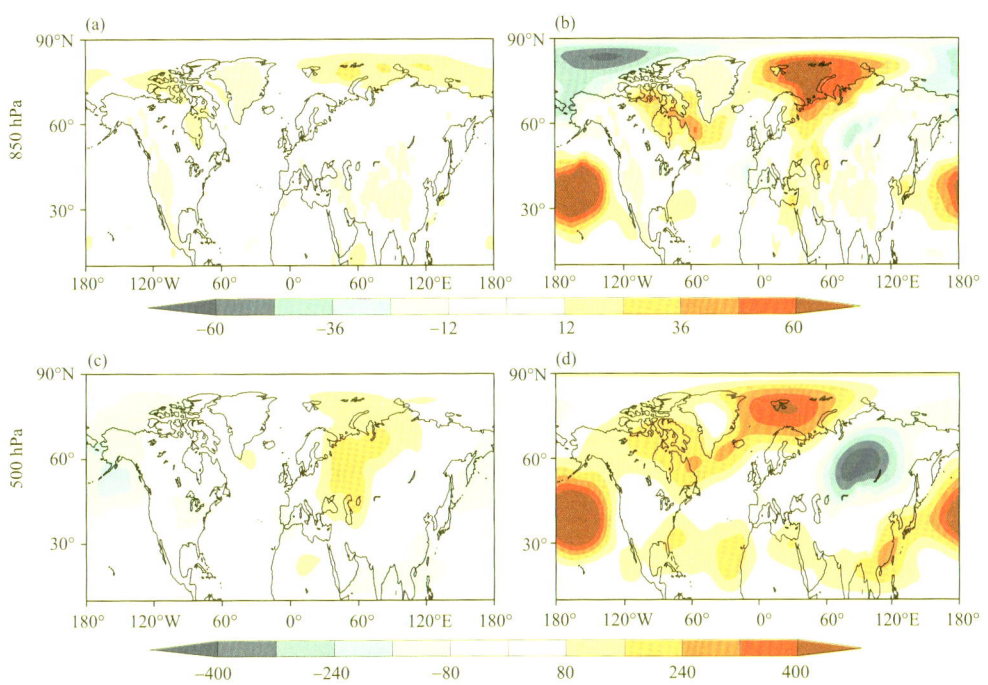

图 4 分片准地转位涡反演得到的不同垂直层次位涡异常对 850 hPa(a、b)和 500 hPa(c、d)位势异常(1979—2017 的线性变化)的贡献。(a)、(b)分别为 825 hPa 以下和 800~250 hPa 层次的位涡异常对 850 hPa 位势异常的贡献。(c)、(d)分别为 450 hPa 以下和 400~250 hPa 层次的位涡异常对 500 hPa 位势异常的贡献(摘自 Xie et al.,2020)

通过分片位涡反演方法,量化了不同垂直层次位涡异常对对流层低层(850 hPa)和中层(500 hPa)位势异常的贡献(图4)。由于环流符合地转关系,这里位势异常直接代表环流异常。结果表明,无论在对流层低层还是中层、无论暖北极对应的反气旋环流还是冷欧亚对应的气旋环流均由更高层次的位涡异常所主导。然而,较高层次的非绝热加热及其垂直梯度相对低层小很多,根据位涡收支方程可知较高层次非绝热加热造成的位涡异常很弱。因此,高层次的位涡异常是由位涡平流造成的再分配主导。研究表明,暖北极对应的近地面的强非绝热加热是激发后续位涡平流再分配的因子。

非绝热加热在低层制造的位涡异常的量值虽大,但这种大值位涡主要由垂直稳定度,即热力变化主导。因此,如图4a所示,低层的位涡异常对环流的直接影响较小。然而,由于背景经向位涡梯度的存在(图5),非绝热加热制造的位涡引起的环流异常通过平流进一步造成了不同区域的位涡异常。具体过程及最终达到的平衡态利用图5所示的理想情形为例进行说明。这里的理想情形为非绝热加热和摩擦制造的位涡为零,且垂直运动造成的位涡输送可忽略,即平衡态的位涡收支方程(式(7))中仅有水平的位涡平流项。此时,局地位涡的异常仅由位涡平流引起的再分配所决定,即图5中线性化的平衡关系$\overline{V} \cdot \nabla q' = -V' \cdot \nabla \overline{q}$。对流层中、高层均满足这种关系,例如图5中的500 hPa示意图与观测非常一致。

暖北极对应反气旋环流可理想化为图5中的由西面的南风到东面的北风的分布。在背景位涡分布从南向北递增的情形下,反气旋环流异常造成西面的负位涡平流和东面的正位涡平流,即$-V' \cdot \nabla \overline{q}$的西负东正型分布。求解平衡关系$\overline{V} \cdot \nabla q' = -V' \cdot \nabla \overline{q}$,即可得到平衡态的位涡异常分布。结果表明,平衡态的位涡异常与环流异常对背景位涡的平流存在相位差,即最终的正(负)位涡中心位于异常环流引起的正(负)位涡平流的下游。由于巴伦支喀拉海区域的北极增暖的南侧正好处于位涡梯度极大的区域,因此暖北极反气旋南侧的位涡平流能够得到不断发展,直到形成位涡异常和环流异常的重合。在此过程中,暖北极区域最初的非绝热加热制造的负位涡也由于从南向北的负位涡平流得到进一步发展,从而形成强大的反气旋环流异常(图4、5)。同时,暖北极下游(东南方位)由于位涡平流形成了正位涡中心及其对应的气旋性环流,并且通过自上而下的影响引起欧亚大陆的变冷(图4、5)。由于位涡平流受背景的位涡分布影响,因此位涡平流造成的异常主要在背景的位涡梯度大的区域,即欧亚大陆变冷的区域正好是位涡梯度大值区。

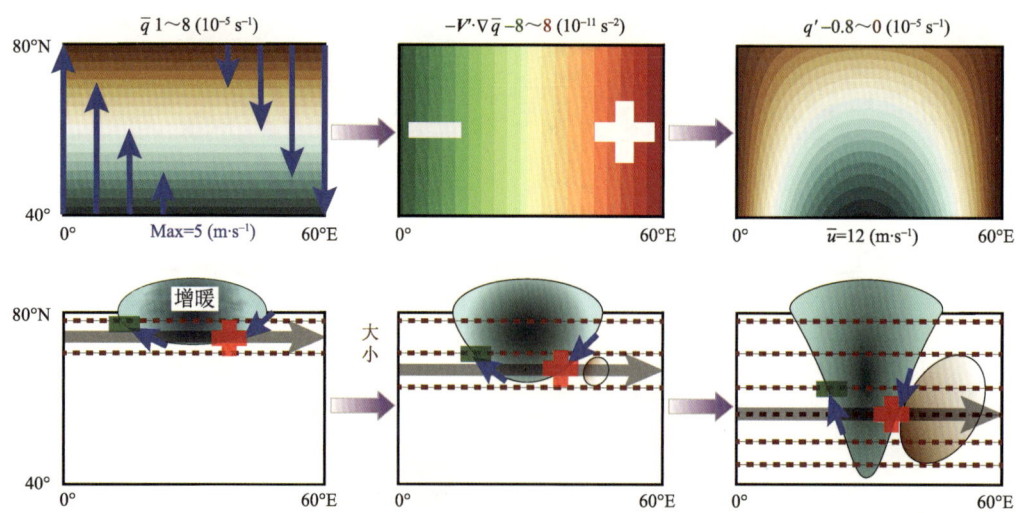

图5 位涡平流引起位涡再分布的示意图(上)以及基于位涡观点的暖北极引起冷欧亚的动力机制示意图(下)。上部第一幅图中给定反气旋性切变的理想环流异常以及由南向北递增的背景位涡分布,其中风速最大值为5 m·s^{-1}。上部第二幅图中为异常环流对背景位涡的平流,即$-V' \cdot \nabla \overline{q}$的分布。上部第三幅图为由平衡关系$\overline{V} \cdot \nabla q' = -V' \cdot \nabla \overline{q}$求解得到的平衡态位涡异常,其中$\overline{V}$给定为12 m·s^{-1}的纬向风。下部为从初始的局地暖北极到最终平衡的暖北极-冷欧亚型对应的位涡平流再分布的过程。其中,正负号表征位涡平流的符号,棕色虚线表征由南到北递增的背景位涡分布(摘自Xie et al.,2020)

3.2 大气位涡的跨赤道输送对春季欧亚大陆地面气温的影响

产生自地球表面的 PV 可以在大气内部自由传输,甚至沿等熵面从一个半球传输到另一个半球。如果考虑半球大气被无限高的等熵面、地面和赤道剖面包裹,由于 PV 通量不能穿越等熵面,那么半球的地表和赤道剖面就成为决定半球 PV 总量变化的两个边界。在长期平均下,地表的 PV 通量应该与跨赤道 PV 通量(cross-equatorial PV flux,CEPVF)相平衡。由此来看,CEPVF 即可以视为一个"窗口",通过这个"窗口",即可检测地表 PV 通量的行为,地表通量的一切行为亦可反映在这个"窗口"上。由于地表 PV 通量与地面气温(SAT)密切相关,因此只需要诊断热带的 CEPVF 的变化就能够获得整个半球 SAT 变化的信息。本小节利用 CEPVF 研究了春季跨赤道 PV 输送对欧亚大陆地面气温的影响(Sheng et al.,2022a)。

通量形式的 PV 收支方程可写为如下形式(Haynes and McIntyre,1987,1990):

$$\frac{\partial W}{\partial t} = -\nabla \cdot \boldsymbol{J} \tag{9}$$

式中,PV 通量矢量为:

$$\boldsymbol{J} = \boldsymbol{V}W - \boldsymbol{\xi}_a \dot{\theta} - \boldsymbol{F}_\xi \theta \tag{10}$$

式中,$\boldsymbol{V}$ 为三维风速矢量,$\dot{\theta} = \frac{\mathrm{d}\theta}{\mathrm{d}t}$ 为非绝热加热率,$\boldsymbol{F} = (F_x, F_y)$ 为摩擦力,$\boldsymbol{F}_\xi$ 为摩擦涡度。

由于散度的旋度为 0,将式(3)代入式(9)中可得:

$$\frac{\partial \boldsymbol{\xi}_a \theta}{\partial t} = -\boldsymbol{J} \tag{11}$$

将方程(11)在 Δt 时段内积分,并取如下定义:

$$\boldsymbol{J}_s = -\boldsymbol{\xi}_a \theta \tag{12}$$

可以得到:

$$\boldsymbol{J}_s = \int_{t_0}^{t_0+\Delta t} \boldsymbol{J} \mathrm{d}t + C \tag{13}$$

式中,C 是与时间独立的定常场。由式(13)可见,通量 $\boldsymbol{J}_s$ 可被视为在给定的 Δt 时段内 PV 通量矢量 $\boldsymbol{J}$ 的累积效应。式(13)也表明,在给定时段内,$\boldsymbol{J}_s$ 的变化与 $\boldsymbol{J}$ 的变化相平行。

由式(12),我们可以进一步得到:

$$W = -\nabla \cdot (\boldsymbol{J}_s) \tag{14}$$

基于式(14),仿照大气的辐合 $C = -\nabla \cdot \boldsymbol{V}$,其中 $\boldsymbol{V}$ 为大气环流,定义 $\boldsymbol{J}_s$ 为 PV 环流(PV circulation,PVC),其中 PVC 的辐合等于 PV 本身。

由于跨赤道 PV 通量主要由平流通量主导(图略),在下文中计算自由大气中的 CEPVF 时,仅考虑平流通量,即 $\boldsymbol{J}_a = \boldsymbol{V}W$。CEPVF 计算如下:

$$\text{CEPVF} = \boldsymbol{J}_a \cdot \boldsymbol{j} = J^y = vW \tag{15}$$

赤道剖面上,垂直和纬向积分的 CEPVF 可以写为:

$$\{\text{CEPVF}\} = \int_{EQ} \int_{p_t}^{p_s} \text{CEPVF} \, \mathrm{d}p \mathrm{d}x = \int_{EQ} \int_{p_t}^{p_s} vW \mathrm{d}p \mathrm{d}x \tag{16}$$

由式(16),定义 CEPVF 指数(CEPVFI)为 {CEPVF} 的标准化时间序列:

$$\text{CEPVFI} = \{\text{CEPVF}\}'/\sigma \tag{17}$$

式中,σ 为 {CEPVF} 的标准差,{CEPVF}$'$ 为 {CEPVF} 的时间异常。正 CEPVFI 表示异常向北的 CEPVF,负 CEPVFI 表示异常向南的 CEPVF。

由于 380 K 等熵面几乎和热带对流层顶(Wilcox et al.,2012)重合,又由于在赤道 100 hPa 等压面几乎与 380 K 等熵面重合,因此可将 100 hPa 等压面视为热带对流层顶。我们计算了积分上边界取为 380 K 和 100 hPa 时的 CEPVFI,二者相关高达 0.91,通过显著性水平为 0.01 的检验。由于原始数据位于等压面上,因此直接利用等压面数据可以避免插值误差。由于我们重点关注对流层中的信号,因此式(16)中的积分上边界(p_t)取为 100 hPa。

图 6a 给出了春季 CEPVFI 的时间序列,CEPVFI 正位相代表存在向北的 PV 输送异常,CEPVFI 负位相代表存在向南的 PV 输送异常。为了探究 CEPVF 相关的气候效应,图 6b 展示了欧亚大陆中高纬 CEPVFI 回归的 SAT 场。可以看到,对应于 CEPVFI 正位相,整个欧亚大陆中高纬呈现大范围显著冷异常,其中存在三个显著的冷中心,分别位于地中海区域,贝加尔湖西南部以及俄罗斯远东地区。这一异常的 SAT 分布型与欧亚大陆中高纬 SAT 的 EOF1(Chen et al.,2016)十分相似,CEPVFI 与欧亚大陆中高纬 SAT 年际变率 EOF 分解的 PC1 的相关系数为 0.36,通过了显著性水平为 0.05 的检验。这进一步说明 CEPVF 与欧亚大陆中高纬度 SAT 紧密相关。

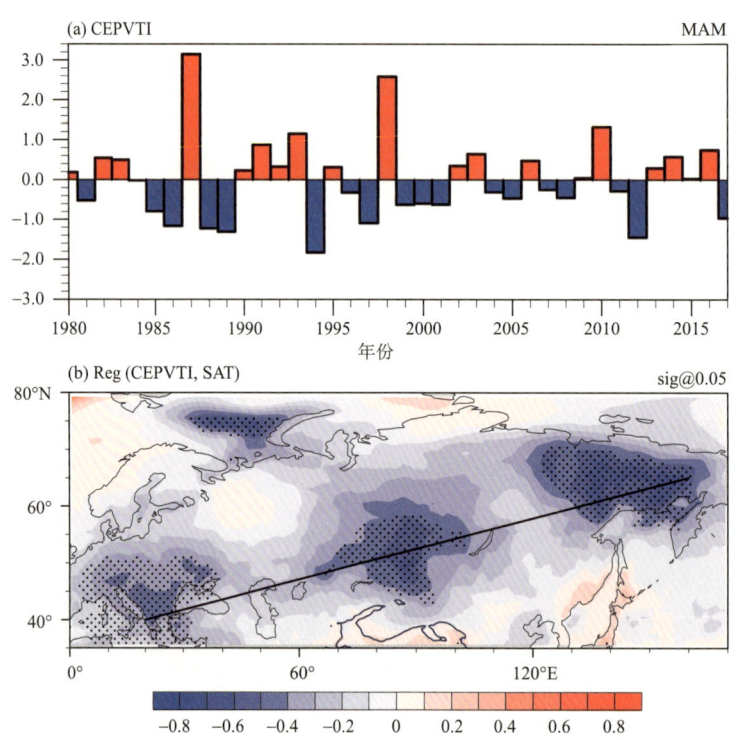

图 6 (a)CEPVFI 标准化时间序列;(b)北半球春季,CEPVFI 回归的欧亚大陆中高纬 SAT(阴影,℃)。打点区域表示超过显著性水平为 0.05 的检验。黑色倾斜实线为三个温度异常中心所在连线(摘自 Sheng et al.,2022a)

我们进一步分析了 CEPVFI 和北半球春季大气内部 PV 的相关系数分布的空间及垂直特征。结果表明,CEPVFI 和北半球大气 PV 异常的三维结构有紧密联系,并且和地表的三个冷中心结构有很好的对应关系(图略)。进一步研究发现,由 ENSO 和热带南大西洋造成的赤道纬向垂直环流的改变是驱动跨赤道 PV 输送,进而造成欧亚大陆温度异常的原因。"由 ENSO 和热带南大西洋驱动跨赤道 PV 输送影响欧亚大陆地面气温"的物理过程示意图如图 7 所示。一旦热带地区出现非绝热加热,赤道纬向环流将会发生改变(Bjerknes,1969)。异常的赤道纬向环流将会引发异常的 CEPVF。由于 PV 为 PVC 的辐合,发生在北半球赤道边界上的 CEPVF 变化将会激发北半球内部 PVC 的调整,进而导致北半球内部 PV 的改变。最终,在 PV-θ 等机制的作用下,欧亚大陆中高纬的地表气温冷异常得以出现和维持。

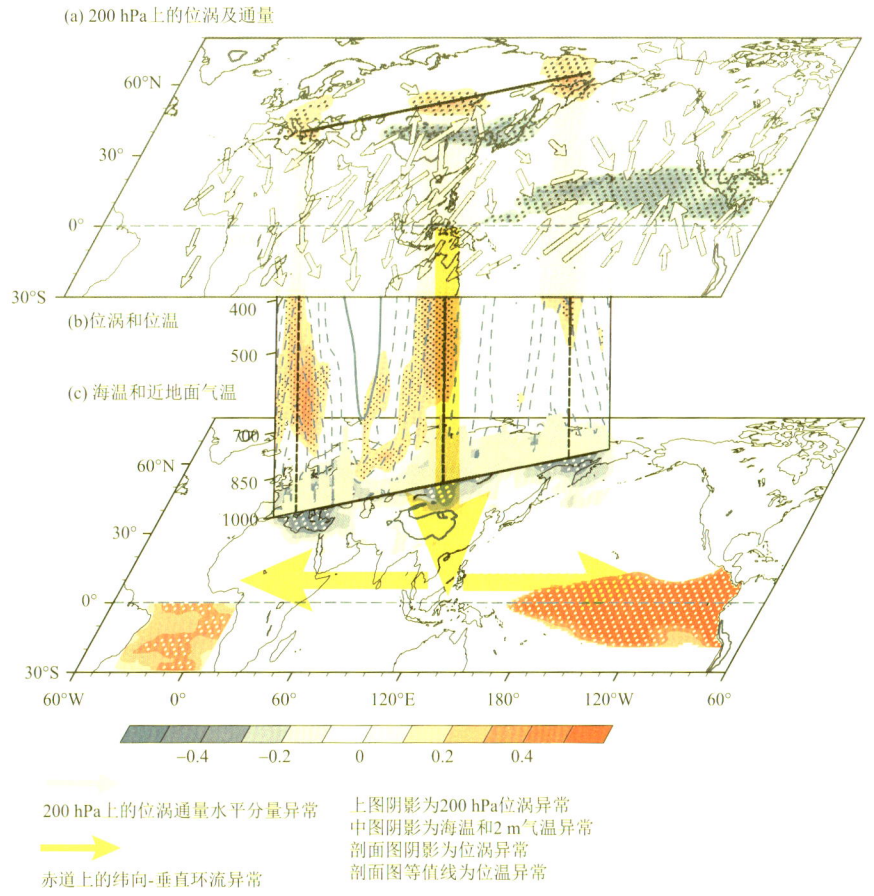

图 7 CEPVF 影响欧亚大陆中高纬地面气温的机制示意图。(a)200 hPa PV(阴影)和水平 PV 通量(箭头);(b)PV(阴影)和位温(线条)的垂直剖面;(c)海洋 SST 和陆地气温。黄色箭头表示赤道纬向-垂直环流(摘自 Sheng et al.,2022a)

4 青藏高原地表位涡制造及其对亚洲夏季风的影响

4.1 青藏高原地表位涡季节演变及与亚洲夏季风的联系

青藏高原热力强迫对亚洲夏季风形成具有重要影响。青藏高原表面的感热加热过程是青藏高原热力强迫的重要驱动力。已有研究指出,青藏高原表面感热是将周围水汽抽吸到高原附近并且抬升形成季风降水,从而进一步驱动季风环流的重要原因(叶笃正 等,1957;叶笃正和高由禧,1979;吴国雄 等,1997;Wu et al.,2007)。然而,利用青藏高原地表感热通量(SSHF)开展计算,用于诊断青藏高原表面热力强迫作用对大气环流和气候的影响不够准确。一方面是由于感热通量不能完全表征青藏高原的抬升加热效应;另一方面是在夏季,青藏高原感热加热还会引起局地的潜热释放和对流发展,使得感热通量降低。因此,不少研究计算高原的感热通量和亚洲季风降水的关系都不是很好(Rajagopalan and Molnar,2013;Zhang et al.,2019)。针对上述问题,最新的研究基于位涡理论框架,使用了青藏高原地表位涡的变化来衡量青藏高原热动力强迫以及研究其和亚洲夏季风之间的关系(He et al.,2022)。

在气候态和季节演变尺度,青藏高原地表热力强迫也可以通过位涡来表达。基于式(1),我们进一步简化了在地形坐标系下,近地面位涡(SPV)的表达式:

$$\text{SPV} = -\left[-\frac{g}{p_s}(f + \xi_{\sigma_1}) \frac{\theta_s - \theta_{\sigma_1}}{1 - \sigma_1} \right] \tag{18}$$

式中，g 为重力加速度，p_s 为地表气压，f 为科氏参数，ξ 为相对对涡度，θ 为位温，σ 为地形坐标系数。该表达式包含了地形效应、近地表绝对涡度和地气位温差，能够较好地表征青藏高原的抬升加热效应。我们进一步计算了青藏高原附近地区（25°～40°N，并且地形高度大于 500 m）气候态平均的 SPV 如图 8 所示，比较了感热通量和亚洲夏季风降水的年循环特征。结果表明，青藏高原在 4 月由冷源向热源转变，SPV 在 6—8 月达到最大值（图 8c），这与亚洲夏季风降水的季节演变过程高度一致（图 8a）。相比较而言，青藏高原附近的感热通量（图 8b）在 3—5 月达到最大值，在盛夏季节反而有所降低。因此，利用 SPV 来表征青藏高原夏季的热力强迫作用并研究其与季风的关系具有显著优势。

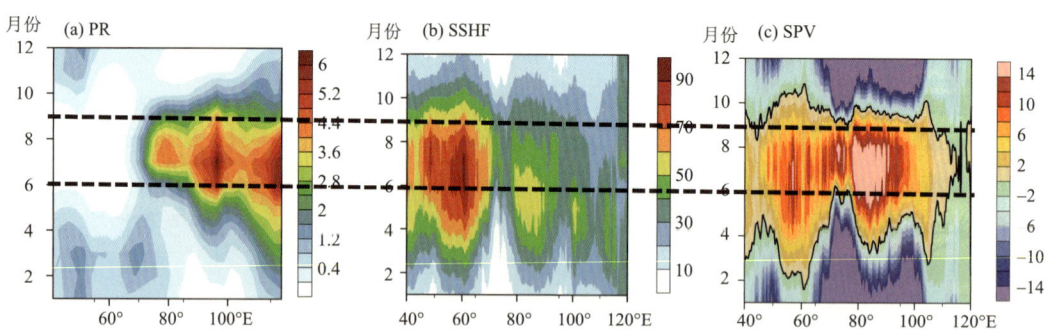

图 8 气候平均（1979—2014）的各要素年循环的时间-经度分布图（25°～40°N 平均并且地形高度大于 500 m）。(a) GPCP 观测降水（mm·d^{-1}）；(b) ERA5 再分析的表面感热通量（W·m^{-2}）；(c) ERA5 再分析地表位涡（PVU）。黑色点线表明北半球 6—8 月（摘自 He et al.，2022）

图 9 为合成分析结果。基于区域 SPV 总和计算了 1979—2014 年的夏季平均高原 SPV 指数（TP-SP-VI），并以 1 倍标准差为临界点，将数据分为高/低 TP-SPVI 年份，其合成的亚洲季风降水和环流的结果如图 9c 所示。结果表明，高原地表位涡偏强时，高原附近对流层低层存在一个气旋式环流异常，该异常与青藏高原东南坡和华南降水增加和印度洋降水减少有密切联系。这与之前众多数值试验中，高原感热增强导致的环流和降水异常的结果高度一致。进一步定量计算了 TP-SPVI 与降水的线性关系发现其线性回归系数可达 0.47（图 9b）。因此，高原 SPV 指数可以很好地定量表征青藏高原热动力强迫作用。在未来的研究中，该变量可作为模式评估和气候表征的指标，以量化青藏高原在全球气候变化中的作用（Sheng et al.，2021；Yu et al.，2021，2022）。

4.2 青藏高原地表位涡的年际变率对东亚夏季降水的影响

在年际尺度，我们还分析了青藏高原地表位涡对东亚夏季降水的影响（Sheng et al.，2022b）。图 10a 给出了夏季青藏高原地表 PV（TPSPV）的 EOF1，如图所示，除了青藏高原南坡和东南侧有小范围的正值外，整个高原平台几乎为一致负值，呈 PV 负一致模（PVNUM）结构，该模态对整个 TPSPV 变率的解释方差为 36%。PVNUM 所对应的时间序列为 PVNUMI，如图 10d 中红线所示。我们进一步给出了东亚夏季降水的 EOF1（图 10b），及其对应的时间序列记为 PreI（图 10d 蓝色虚线）。由图 10b 可见，东亚夏季降水的最大变率中心出现在长江流域，朝鲜，日本（YKJ，Yangtze River-South Korea-Japan）和华南地区。YKJ 区域的降水异常与华南地区相反，呈现南北偶极子结构。同时，华北地区也出现了较弱的降水偶极子结构。整体来看，降水的主模态与 Zuo 等（2011）的结果一致。

利用多变量经验正交分解（MVEOF）分解，图 10c 给出了夏季东亚中纬度 200 hPa 风场的主模态，即 MAS（midlatitude Asian summer）环流模态。MAS 模态最早由 Wu（2002）利用 NCEP 资料对 1948—1998 年 6—9 月东亚区域进行 MVEOF 分解得到。MAS 正位相的主要特征为：青藏高原西侧和辽东半岛附近为气旋式环流，青藏高原东南侧为反气旋式环流，由高纬度到低纬度呈现"气旋-气旋-反气旋（CCA）"型环流结构。反之，MAS 负位相呈现"反气旋-反气旋-气旋（AAC）"型环流结构。MAS 环流模态的时间序列记为 MASI，如图 10d 黑色虚线所示。

第 5 部分　位涡及位涡源的若干基本问题

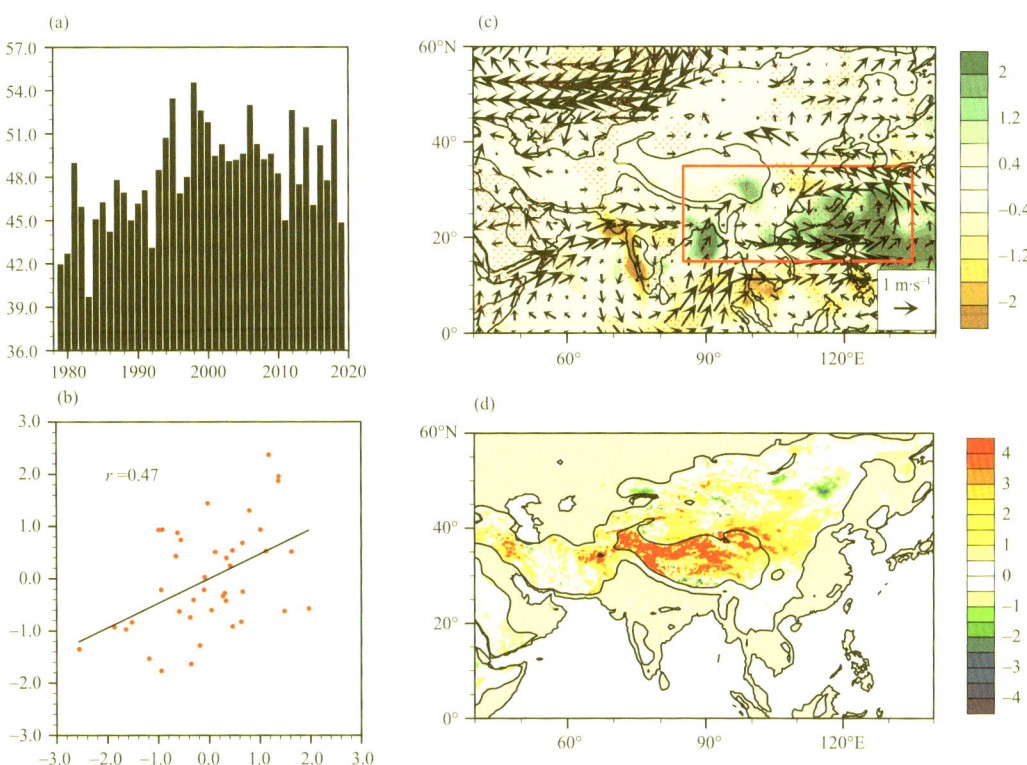

图 9　(a)1979—2019 年北半球夏季(JJA)青藏高原地表位涡指数(TP-SPVI)的时间序列柱状图(PVU·km^{-2});(b)标准化的 TP-SPVI 和降水的线性回归,其中水平坐标为 TP-SPVI,垂直坐标为区域平均降水(15°～35°N,85°～135°E)。线性回归系数为 0.47;(c)合成分析的夏季降水(mm·d^{-1})和 850 hPa 风场(m·s^{-1})异常。其中红点表明通过了 0.01 显著性 t 检验的区域。红色矩形区域范围为(15°～35°N,85°～135°E);(d)合成分析的夏季地表位涡(PVU)空间分布(摘自 He et al.,2022)

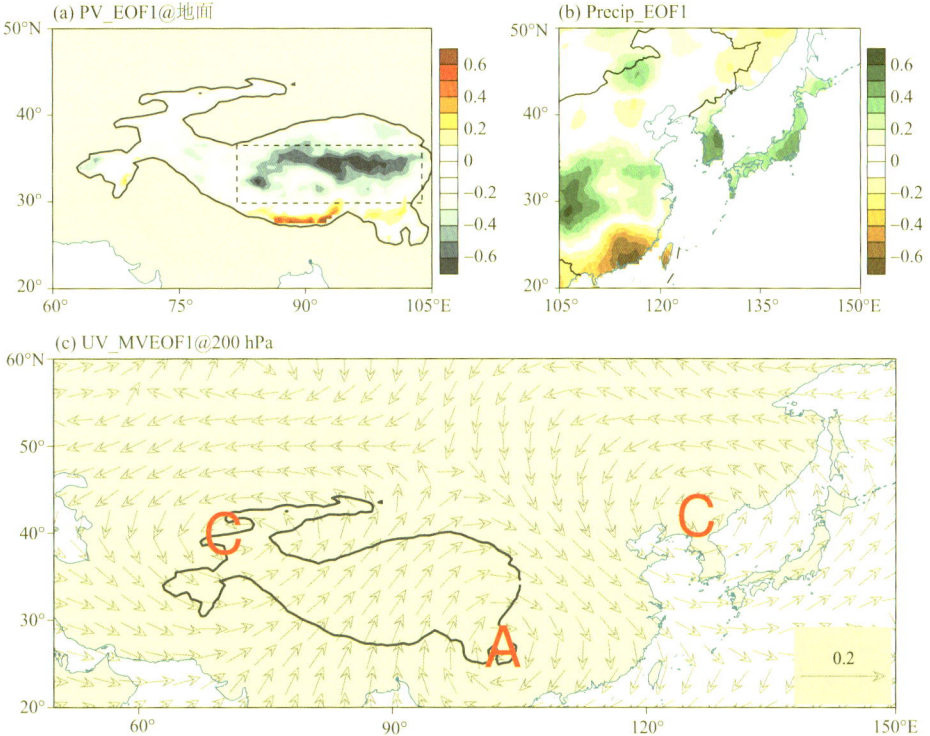

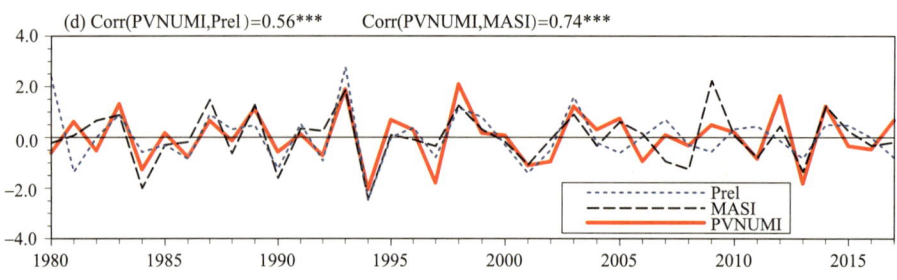

图10 北半球夏季各变量的EOF1。(a)TPSPV,虚线框表示TPSPV变率最大区域;(b)东亚降水;(c)200 hPa风场;(d)(a—c)图的PC1时间序列。"***"表示通过显著性水平为0.01的检验,蓝色实线表示青藏高原地形2000 m等高线(摘自Sheng et al.,2022b)

为了进一步分析TPSPV强迫与上层环流和东亚夏季降水的关系,图10d给出了PVNUMI(红实线)、PreI(蓝虚线)和MASI(黑虚线)的时间序列。由图可见,PVNUMI与PreI和MASI的相关系数分别高达0.56和0.74,均通过了显著性水平为0.05的检验。这意味着东亚夏季降水,东亚夏季环流与TPSPV强迫具有密切相关。我们综合了统计分析和数值试验的结果,最后揭示了TPSPV影响东亚夏季降水的机制,如图11所示。PVNUM能够通过减弱地表静力稳定度激发出地面气旋式环流(图11b)。由地面气旋式环流所引发的纬向偶极子加热模态可以导致青藏高原上空出现典型的MAS异常环流(图11a)。最终,嵌入在环流异常中的水汽输送异常引起了东亚夏季降水异常(图11c)。

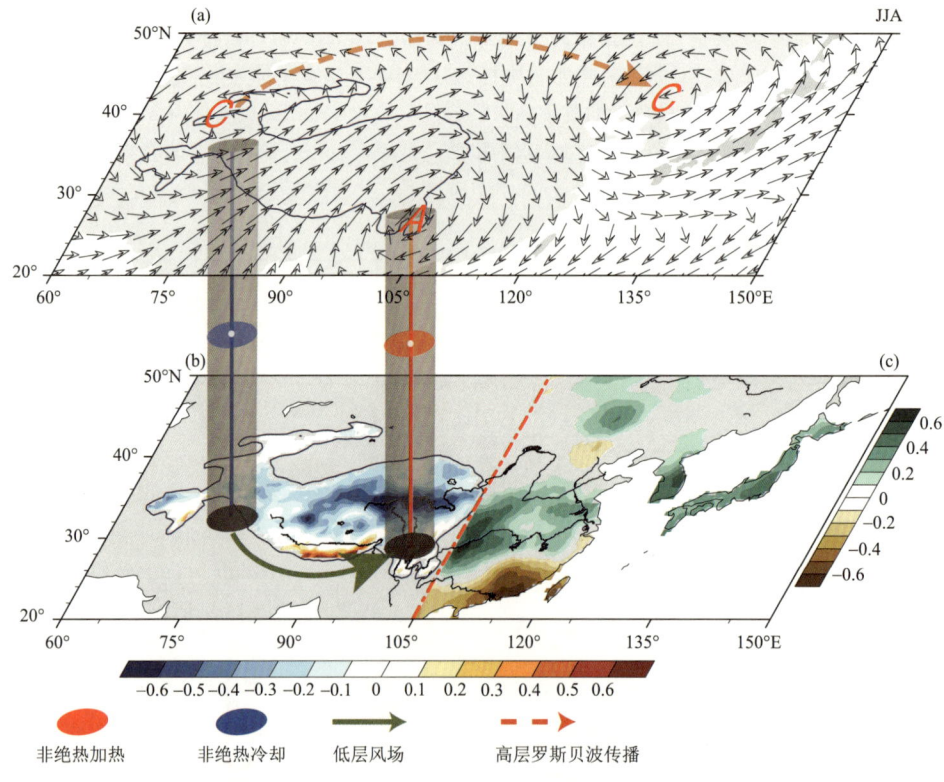

图11 TPSPV强迫对东亚夏季降水的年际影响示意图。(a)200 hPa MAS环流模态;(b)TPSPV异常主模态;(c)东亚夏季降水主模态。红色(蓝色)圆圈表示非绝热加热(冷却)。绿色和红色箭头分别表示低层风场、高层罗斯贝波传播(摘自Sheng et al.,2022b)

5 总结和展望

文中主要回顾了吴国雄院士团队近年来在位涡气候动力学方面的新进展。研究团队从原始的位涡方程出发,推导了适用于混合坐标系的位涡方程,对全球变暖背景下,地表位涡的制造以及位涡在大气中的传输和影响开展了深入研究。从理论上,基于尺度分析,表明位涡中垂直速度的水平变化项可忽略,证明了位涡及位涡方程在形式上的坐标不依赖性,并导出了广义坐标系下的位涡及位涡方程。基于上述理论,团队研究了北极增暖和欧亚大陆变冷之间的联系,发现高层位涡平流是北极增暖影响欧亚大陆变冷的关键因素,揭示了春季跨赤道位涡输送对北半球欧亚大陆上空大气位涡三维结构的形成及欧亚大陆近地面气温冷异常具有重要影响。在区域尺度上,揭示了青藏高原地表位涡的季节变化和年际变率特征,并进一步指出青藏高原地表位涡强迫的增强有利于亚洲夏季风的形成,与季风降水的季节演变一致。高原地表位涡强迫还可以激发高低层环流异常,调节东亚地区夏季降水的年际变率。以上研究是位涡理论在气候动力学中和气候变化研究中的创新性前沿应用。未来研究团队将进一步深化位涡理论研究,对影响位涡收支的摩擦耗散以及非绝热加热等过程开展详细分析,为理解全球变暖下的气候动力学问题提供新的前沿视角。

参考文献

吴国雄,李伟平,郭华,等,1997. 青藏高原感热气泵和亚洲夏季风[C]//叶笃正. 赵九章纪念文集. 北京:科学出版社: 116-126.

叶笃正,罗四维,朱抱真,1957. 西藏高原及其附近的流场结构和对流层大气的热量平衡[J]. 气象学报,28(2):108-121.

叶笃正,高由禧,1979. 青藏高原气象学[M]. 北京:科学出版社:278.

AEBISCHER U, SCHÄR C, 1996. Low-level potential vorticity and cyclogenesis to the lee of the Alps[J]. Journal of the Atmospheric Sciences, 55(2): 186-207.

BJERKNES J, 1969. Atmospheric teleconnections from equatorial pacific[J]. Monthly Weather Review, 97(3): 163-172.

CHEN S, WU R, LIU Y, 2016. Dominant modes of interannual variability in eurasian surface air temperature during boreal spring[J]. Journal of Climate, 29(3): 1109-1125.

ERTEL H, 1942. Ein neuer hydrodynamische wirbelsatz[J]. Meteorologische Zeitschrift Braunschweig, 59: 33-49.

FLAMANT C, RICHARD E, SCHAR C, et al, 2004. The wake south of the Alps: Dynamics and structure of the lee-side flow and secondary potential vorticity banners[J]. Quarterly Journal of the Royal Meteorological Society, 130(599): 1275-1303.

HAYNES P H, MCINTYRE M E, 1987. On the evolution of vorticity and potential vorticity in the presence of diabatic heating and frictional or other forces[J]. Journal of the Atmospheric Sciences, 44(5): 828-841.

HAYNES P H, MCINTYRE M E, 1990. On the conservation and impermeability theorems for potential vorticity[J]. Journal of the Atmospheric Sciences, 47(16): 2021-2031.

HE B, SHENG C, WU G X, et al, 2022. Quantification of seasonal and interannual variations of the Tibetan Plateau surface thermodynamic forcing based on the potential vorticity[J]. Geophysical Research Letters, 49(5).

HOSKINS B, MCINTYRE M E, ROBERTSON A W, 1985. On the use and significance of isentropic potential vorticity maps[J]. Quarterly Journal of the Royal Meteorological Society, 111(470): 877-946.

RAJAGOPALAN B, MOLNAR P, 2013. Signatures of Tibetan Plateau heating on Indian summer monsoon rainfall variability[J]. Journal of Geophysical Research Atmospheres, 118(3): 1170-1178.

SCHÄR C, SPRENGER M, LÜTHI D, et al, 2003. Structure and dynamics of an alpine potential-vorticity banner[J]. Quarterly Journal of the Royal Meteorological Society, 129(588): 825-855.

SHENG C, WU G X, TANG Y Q, et al, 2021. Characteristics of the potential vorticity and its budget in the surface layer over the Tibetan Plateau[J]. International Journal of Climatology, 41: 439-455.

SHENG C, WU G X, HE B, et al, 2022a. Linkage between cross-equatorial potential vorticity flux and surface air temperature over the mid-high latitudes of Eurasia during boreal spring[J]. Climate Dynamics.

SHENG C, HE B, WU G X, et al, 2022b. Interannual influences of the surface potential vorticity forcing over the Tibetan Plateau on east Asian summer rainfall[J]. Advances in Atmospheric Sciences, 39(7): 1050-1061.

WILCOX L J, HOSKINS B J, SHINE K P, 2012. A global blended tropopause based on era data. Part Ⅰ: Climatology [J]. Quarterly Journal of the Royal Meteorological Society, 138(664): 561-575.

WU G X, LIU Y M, WANG T M, et al, 2007. The influence of mechanical and thermal forcing by the Tibetan Plateau on Asian climate[J]. Journal of Hydrometeorology, 8(4): 770-789.

WU R, 2002. A mid-latitude Asian circulation pattern in boreal summer and its connection with the Indian and East Asian summer monsoons[J]. International Journal of Climatology, 22(15): 1879-1895.

XIE Y, WU G, LIU Y, et al, 2020. Eurasian cooling linked with arctic warming: Insights from PV dynamics[J]. Journal of Climate, 33(7): 2627-2644.

YU W, LIU Y, YANG X Q, et al, 2021. Impact of north atlantic SST and Tibetan Plateau forcing on seasonal transition of springtime South Asian monsoon circulation[J]. Climate Dynamics, 56(1-2): 559-579.

YU W, LIU Y M, XU L L, et al, 2022. Potential impact of spring thermal forcing over the Tibetan Plateau on the following winter El Nino-Southern Oscillation[J]. Geophysical Research Letters, 49(6).

ZHANG H, LI W, LI W, 2019. Influence of late springtime surface sensible heat flux anomalies over the Tibetan and Iranian Plateaus on the location of the South Asian high in early summer[J]. Advances in Atmospheric Sciences, 36(1): 93-103.

ZUO Z Y, ZHANG R H, ZHAO P, 2011. The relation of vegetation over the Tibetan Plateau to rainfall in China during the boreal summer[J]. Climate Dynamics, 36(5-6): 1207-1219.

> 综述 13

垂直运动方程以及青藏高原位涡强迫的天气效应

马婷婷[1], 刘屹岷[1,2], 毛江玉[1], 张冠舜[3], 汤艺琼[4]

(1 中国科学院大气物理研究所,北京 100029;2 中国科学院大学,北京 100049;
3 中国气象局广州热带海洋气象研究所,广州 510640;4 无锡学院,无锡 214105)

摘要:位涡理论和垂直运动方程是 20 世纪中纬度天气气候动力学中最重要的两个成就。文中回顾了大气垂直运动方程的研究进展,详细地介绍了新型准地转垂直运动方程。该方程将垂直速度的发展与非绝热加热和准地转位涡平流联系在一起,在理解由位涡异常引起天气系统发生、发展的机制中具有较好的应用前景。在此基础上,回顾了吴国雄院士团队近年来在高原上空高位涡系统的形成机制,以及高原高位涡系统在东移的过程中引起长江中下游地区极端降水的发生等方面的最新研究进展。以上研究为青藏高原大地形影响下游激烈天气过程的发展提供了新思路。

1 引言

在大气科学领域取得的众多成果中,垂直运动方程和位涡理论是中纬度天气、气候动力学中十分重要的两个成就,二者被广泛应用于天气气候的诊断分析和研究工作中。垂直运动的计算与诊断早在 20 世纪 50 年代就开始受到关注,如何准确计算垂直速度是一代又一代气象学者不懈的追求。位涡理论的提出可以向前追溯到 20 世纪 40 年代 Rossby(1940) 和 Ertel(1942),以及 20 世纪 50 年代 Kleinschmidt(1955) 的工作。但直到 20 世纪 80 年代,Hoskins 等(1985)对位涡理论进行了全面系统的回顾之后,位涡理论的应用才开始蓬勃发展。在相当长的一段时间内,二者独自发展,并无交集。

1.1 垂直运动方程

垂直运动方程的发展大多起源于 Sutcliffe(1947) 的"发展理论"。基于准地转近似(风速和气压梯度保持近似平衡),Sutcliffe(1947)提出了一个"发展方程",将对地面气旋的发展有指示意义的大气散度的垂直变化与"热成风涡度"(由热成风计算得到的涡度)平流结合在一起。在此基础上,Bushby(1952)提出了准地转垂直运动方程的近似形式,严格形式的方程由 Thompson 于 1961 年提出(Thompson,1961)。然而,由于忽略了热力过程,这些方程仅在描述中纬度大尺度天气系统的移动和发展时有较好的应用效果(Hoskins et al.,1978,1980)。Petterssen(1956) 和 Djuric(1969)进一步将垂直速度的发展与对流层高层的涡度平流和温度平流相联系,拓展了垂直运动方程的适用性。但 Petterssen(1956)同时也意识到在大多数情况下,这两个因子之间存在相互抵消的情形,单独考虑其中任何一个因子可能会得到错误的结论。Holton(1972)在准地转理论的基础上提出了绝热条件下垂直运动的诊断方程,指出涡度平流

通讯作者:马婷婷,matingting@lasg.iap.ac.cn
资助项目:国家自然科学基金项目 41905068

随高度的变化和温度平流的拉普拉斯是决定垂直速度的两个因子。后来,他又将非绝热加热考虑在内,提出了经典的准地转垂直运动方程(Holton,2004)。该方程的提出为理解中纬度天气系统的发展提供了有利的诊断工具,但影响因子之间相互抵消的缺点依然存在。

为了克服影响因子之间相互抵消的问题,Trenberth(1978)采用类似 Sutcliffe(1947)的方法,将垂直运动方程强迫项中的涡度平流改为由热成风引起的涡度平流,克服了对流层中层存在明显抵消现象的缺点。Hoskins 等(1978)从准静力平衡、准地转、绝热无摩擦、f 平面近似的等压坐标系原始方程组出发,对准地转垂直运动方程进行了形式上的简化,将两个强迫项合并成一个矢量 Q 的散度的形式,由此提出了准地转 Q 矢量的概念。该方法不仅解决了以往垂直运动方程中两个强迫因子相互抵消的问题,而且其物理意义清楚,计算简单,一经提出很快便被广泛应用于天气诊断分析中。但在当时的条件下,该方法只能利用观测数据进行诊断,很难应用于图形化的模式预报产品中,且过多的假设和简化限制了准地转 Q 矢量在中尺度天气过程中的应用。随后,广大气象学者对准地转 Q 矢量进行了一系列的改进,如 Sanders 等(1990)提出了一种简单的方法来估计天气图中 Q 矢量的方向和相对大小。中国气象学者在这一方面也取得了相当多的研究成果,先后提出了湿 Q 矢量(张兴旺,1998;岳彩军 等,2003)、非地转 Q 矢量(张兴旺,1999;Yue et al.,2008;Yue,2009)、非地转湿 Q 矢量(姚秀萍 等,2000,2001;Yao et al.,2004;Yang et al.,2007)等概念。

Hoskins 等(2003)在绝热条件下,利用准地转热力学能量方程将垂直速度分解为伴随着系统发展的随等熵面位移的垂直速度分量($w_{ID} = -N^{-2}(\partial b/\partial t)|_C$,其中,$b = g\theta/\theta_0$,为浮力;$\theta_0 = 300\ K$;$N = \left[\frac{g}{\theta_0}\frac{d\Theta(z)}{dz}\right]^{1/2}$,为基准态的浮力频率;$\Theta(z)$ 为特定水平区域和时间平均的基准态位温分布,仅为高度(z)的函数)和沿等熵面滑动的垂直速度分量($w_{IG} = -N^{-2}(\mathbf{V}_g - \mathbf{C})\cdot\nabla_h b$,其中,$\mathbf{C}$ 为参考系移动的水平速度)。从定义式可以看出,w_{ID} 与系统热力结构的变化有关,而 w_{IG} 则是在系统热力结构确定的情况下产生的垂直速度。他们进一步提出了一个准地转垂直运动方程(见 Hoskins et al.,2003,方程(11))将 w_{ID} 的发展与大气内部准地转位涡平流随高度的变化紧密联系在一起。由此,不仅解决了影响因子之间相互抵消的问题,还将垂直速度与位涡理论有机结合在一起。

1.2 地形位涡强迫

一般而言,位涡的分布呈现出由低纬向高纬、由低层向高层增大的特征,边界层内尤其是近地表附近位涡值很小。这使得早期对位涡强迫的研究多集中在中高纬度、对流层中高层。近年来,随着位涡理论的不断应用和发展,尤其是在 Hoskins(1991)首次明确了与地表相交的等熵面边界处存在位涡的制造之后,越来越多的研究(Held et al.,1999;Schneider et al.,2003;Egger et al.,2015)指出,自由大气中的位涡与地表位涡具有密切联系,地表制造的位涡可以输送到自由大气中,影响自由大气中位涡的收支。

目前,与地表位涡强迫及其影响有关的研究工作主要集中在山脉等高大地形处,这一方面与大地形地表非绝热加热和摩擦作用较强有关,另一方面,大地形地表附近的环流有利于将地表制造的位涡向自由大气内部输送,造成下游地区大气位涡异常,从而影响下游地区的天气。Thorpe 等(1993)通过 1987 年德国锋面野外试验期间的观测资料和数值模拟,探讨了低空气流经过阿尔卑斯山脉过程中,山脉对其附近大气位涡异常的影响。观测资料显示在阿尔卑斯山脉下游北侧存在 200 km 宽的位涡负异常区,而南侧是位涡正异常区。这种山脉下游的正负位涡异常区不仅局限在近地层内,而且向上伸展到对流层中层。数值试验的结果表明,阿尔卑斯山斜坡的边界层摩擦效应是其下游南、北两侧正、负位涡异常产生的主要原因。Aebischer 等(1998)指出,山脉产生的位涡异常会以平流的方式输送到地形以外的下游地区,是中、小尺度涡旋发展的源泉。Schneider 等(2003)定义"广义位涡"为传统位涡和地表制造的位涡之和,并在此基础上模拟了气流过山时,在山脉背风坡一侧出现的地表"涡旋对"的演变,发现"涡旋对"在发展的过程中可以引起高层大气位涡的异常。之后,很多学者研究了阿尔卑斯山位涡强迫作用对欧洲极端降水事件的影响(Martius et al.,2006;McTaggart-Cowan et al.,2010;Awan et al.,2017)。Rudari 等(2004)发现,在秋季对流层中层大西洋东北部至中部地区之间常有低压槽活动。当这些低压槽经过阿尔卑斯山时,次级的中尺度气旋会频繁地在阿尔卑斯山背风坡一侧的热亚那湾发展起来,导致地中海北部地区极端强降水事件的发生。

青藏高原作为北半球覆盖面积最广、平均海拔最高的高原,其巨大的动力和热力作用对我国乃至全球天气气候的影响已经取得了相当重要和丰富的研究进展(Tao et al.,1981;Wu G X et al.,1998,2006,2012;Wang et al.,2008;Lv et al.,2018;Sun et al.,2019;Wu Z W et al.,2012)。尽管早在20世纪90年代,Hoskins(1991)就已经指出,夏季青藏高原地表感热制造的正位涡是高原近地层气旋式环流得以维持的主要原因,但是目前关于高原地表位涡强迫及其影响的研究仍然相对较少。

基于以上回顾,文中全面总结了吴国雄院士团队近年来在垂直运动方程的发展以及青藏高原位涡强迫方面的研究进展。同时,基于新型准地转垂直运动方程分析了高原位涡强迫对极端天气过程的影响。这些研究成果为青藏高原影响中国东部极端天气发生、发展机制的研究提供了新视角。

2 新型准地转垂直运动方程

基于Hoskins等(2003)的工作,Wu等(2020)在准地转理论框架下将该垂直运动方程推广到非绝热大气中。等压坐标系下的准地转热力学能量方程:

$$\frac{\partial \theta}{\partial t} + (\boldsymbol{V}_g - \boldsymbol{C}) \cdot \nabla \theta = -\frac{\partial \Theta(p)}{\partial p}\omega + \dot{\theta} \tag{1}$$

式中,$\boldsymbol{V}_g = \boldsymbol{k} \times f^{-1}\nabla \Phi$,为地转风;$\Theta(p)$同$\Theta(z)$,仅为高度($p$)的函数,其垂直梯度 $\frac{\partial \Theta(p)}{\partial p}$ 在后文简写为 Θ_p;$\dot{\theta}$ 为非绝热加热率。由方程(1)可以将垂直速度分解为随等熵面位移的垂直速度分量(ω_{ID})、沿等熵面滑行的垂直速度分量(ω_{IG})和非绝热加热引起的垂直速度分量(ω_Q):

$$\omega = \omega_{\mathrm{ID}} + \omega_{\mathrm{IG}} + \omega_Q \tag{2}$$

式中,

$$\omega_{\mathrm{ID}} = -\Theta_p^{-1}\left(\frac{\partial \theta}{\partial t}\right)\bigg|_C \tag{3}$$

$$\omega_{\mathrm{IG}} = -\Theta_p^{-1}(\boldsymbol{V}_g - \boldsymbol{C}) \cdot \nabla \theta \tag{4}$$

$$\omega_Q = -\Theta_p^{-1}\dot{\theta}\big|_C \tag{5}$$

为了得到非绝热大气中垂直速度与准地转位涡平流之间的联系,首先导出准地转位涡方程。具体过程如下:利用关系式 $\frac{\partial \Phi}{\partial p} = -\frac{R}{p}\left(\frac{p}{p_0}\right)^\kappa \cdot \theta = -\Pi(p) \cdot \theta$ 改写热力学能量方程(1),得到如下形式的方程:

$$\frac{\partial}{\partial t}\left(\frac{1}{\Sigma^2}\frac{\partial \Phi}{\partial p}\right) + (\boldsymbol{V}_g - \boldsymbol{C}) \cdot \nabla\left(\frac{1}{\Sigma^2}\frac{\partial \Phi}{\partial p}\right) = -\omega + \frac{\dot{\theta}}{\Theta_p} \tag{6}$$

式中,Φ为位势;$\Pi(p) = \frac{R}{p}\left(\frac{p}{p_0}\right)^\kappa$;$\Sigma^2 = \Sigma^2(p) = -\Pi(p)\Theta_p = \left(\frac{\theta_0}{g}\right)^2 \Pi^2(p) N^2$;$p_0 = 1000$ hPa。

引入准地转涡度方程:

$$\frac{\partial \eta_g}{\partial t} + (\boldsymbol{V}_g - \boldsymbol{C}) \cdot \nabla \eta_g = f\frac{\partial \omega}{\partial p} \tag{7}$$

式中,$\eta_g = f + f^{-1}\nabla_h^2\Phi$,为准地转绝对涡度。将方程(6)对$p$偏微商并乘以$f$后,与方程(7)相加消去 $f\frac{\partial \omega}{\partial p}$ 项,再利用热成风关系化简方程可得:

$$\frac{\partial}{\partial t}\left[f + f^{-1}\nabla_h^2\Phi + f\frac{\partial}{\partial p}\left(\frac{1}{\Sigma^2}\frac{\partial \Phi}{\partial p}\right)\right] + (\boldsymbol{V}_g - \boldsymbol{C}) \cdot \nabla\left[f + f^{-1}\nabla_h^2\Phi + f\frac{\partial}{\partial p}\left(\frac{1}{\Sigma^2}\frac{\partial \Phi}{\partial p}\right)\right] = f\frac{\partial}{\partial p}\left(\frac{\dot{\theta}}{\Theta_p}\right) \tag{8}$$

式中,定义 $q_g = f + f^{-1}\nabla_h^2 \Phi + f\dfrac{\partial}{\partial p}\left(\dfrac{1}{\Sigma^2}\dfrac{\partial \Phi}{\partial p}\right)$,为准地转位涡,方程(8)即为自由大气中非绝热条件下的准地转位涡方程。

将准地转位涡 q_g 对 p 偏微商并乘以 f 可得:

$$f\frac{\partial q_g}{\partial p} = \nabla_h^2 \frac{\partial \Phi}{\partial p} + f^2 \frac{\partial^2}{\partial p^2}\left(\frac{1}{\Sigma^2}\frac{\partial \Phi}{\partial p}\right) \tag{9}$$

将 $\dfrac{\partial \Phi}{\partial p} = -\Pi(p)\cdot\theta = \dfrac{\Sigma^2}{\Theta_p}\cdot\theta$ 代入方程(9),并将方程对时间 t 偏微商可得:

$$f\frac{\partial}{\partial p}\left(\frac{\partial q_g}{\partial t}\right) = \Sigma^2 \nabla_h^2\left(\Theta_p^{-1}\frac{\partial \theta}{\partial t}\right) + f^2 \frac{\partial^2}{\partial p^2}\left(\Theta_p^{-1}\frac{\partial \theta}{\partial t}\right) \tag{10}$$

将方程(3)和方程(8)代入方程(10)的左右两端即可得到非绝热条件下新型准地转垂直运动方程:

$$\begin{cases} \left(\Sigma^2 \nabla_h^2 + f^2 \dfrac{\partial^2}{\partial p^2}\right)\omega_{\mathrm{ID}} = f\dfrac{\partial}{\partial p}[(\boldsymbol{V}_g - \boldsymbol{C})\cdot\nabla q_g] - f^2 \dfrac{\partial^2}{\partial p^2}\left(\dfrac{\dot\theta}{\Theta_p}\right) \\ \omega_{\mathrm{ID}} = \Theta_p^{-1}(\boldsymbol{V}_g - \boldsymbol{C})\cdot\nabla\theta - \Theta_p^{-1}\dot\theta,\text{水平边界处} \end{cases} \tag{11}$$

即随等熵面位移的垂直速度分量(ω_{ID})受大气内部准地转位涡平流随高度的变化和非绝热加热的影响,当准地转位涡平流随高度增加或大气非绝热加热为正值时,ω_{ID} 均为上升运动。其中,边界条件满足水平边界上的垂直速度为零。因此,水平暖平流或水平边界上的非绝热加热会导致局地等熵面向下的位移,反之亦然。在绝热条件下,方程(11)化简为 Hoskins 等(2003)中方程(11)的形式。

虽然新型准地转垂直运动方程仅给出垂直速度一个分量的诊断关系式,但是它在诊断由位涡平流引起的天气过程发展中十分便利。此外,垂直速度三个分量之间存在相互影响的过程,通过分析这三个分量之间相互作用的过程,能够直观地再现天气系统发生、发展到消亡的机制。对新型准地转垂直运动方程的应用将在 3.2 节中详细介绍。

3 青藏高原位涡强迫

根据位涡方程,青藏高原巨大的热力作用和表面摩擦必然使得高原成为重要的位涡源汇区。刘新等(2001)的研究指出高原上空的加热作用在高原近地层制造正位涡、在高空制造负位涡。这些正、负位涡源是高原低层气旋式环流、高层反气旋式环流得以稳定维持的重要原因。他们的研究还进一步指出高原上空气柱有净负位涡通过东边界和北边界向周边大气输出。由此可见,青藏高原的位涡强迫效应不仅局限在高原上,也影响着周边的大气环流。

3.1 高原位涡强迫对高原涡等高位涡系统生成的影响

夏季,高原上空对流活动非常活跃,这些对流活动常常伴随着东移的高位涡系统,如高原涡、切变线等。这不仅能够为高原局地带来降水,东移出高原的系统还能够造成我国东部地区的灾害性天气(Tao et al., 1981;张顺利 等,2002;傅慎明 等,2011;郑永骏 等,2013;Fu et al., 2019)。以往的观点认为这些系统的生成与大尺度大气环流条件和高原热力条件息息相关。Li 等(2011,2014)认为低空 500 hPa 的辐合场配合高空与南亚高压、西风急流有关的 200 hPa 辐散场有利于高原涡的生成。同时,来自印度洋和阿拉伯海充足的水汽供应也是有利于高原涡生成的重要因素(郁淑华 等,2001)。另外,一些研究认为高原大气潜热释放对高原涡的生成起至关重要的作用(Dell'Osso et al., 1986;Wang, 1987;Li et al., 2011),但也有研究强调了地表感热的作用(Shen et al., 1986;罗四维 等,1991;李国平 等,2002)。

近来的研究表明这些系统的生成与高原地表和低空之间的加热差异有关(马婷 等,2020;Ma et al., 2022)。图1给出了 2020 年 6 月 1 日—8 月 2 日青藏高原位涡、位涡收支及与这一过程相关的大气非绝热加热的日变化特征。由图可见,高原地表和低空之间的加热差异表现出显著的日变化特征。具体来讲,日出后,高原地表感热的增强造成地表和低空之间非绝热加热随高度递减,使近地层产生位涡耗散,不利于高位涡系统的生成;而随着地表温度的升高,蒸发的加强削弱了地表的非绝热加热,同时 400 hPa 附近云的形成释放潜热。两者共同造成了地表和低空之间非绝热加热随高度增加,有利于正位涡的制造;入夜后,400 hPa 附近的潜热释放略有减弱,而地表由于辐射冷却使得近地层非绝热加热随高度增加的垂直梯度仍然维持。当夜间的位涡制造异常强时,便有利于高位涡系统的生成。这也解释了观测中高原涡的生成频率在 18—24 时(当地时间)最高、06—12 时最低(Li et al.,2018;Lin et al.,2020)的特征。

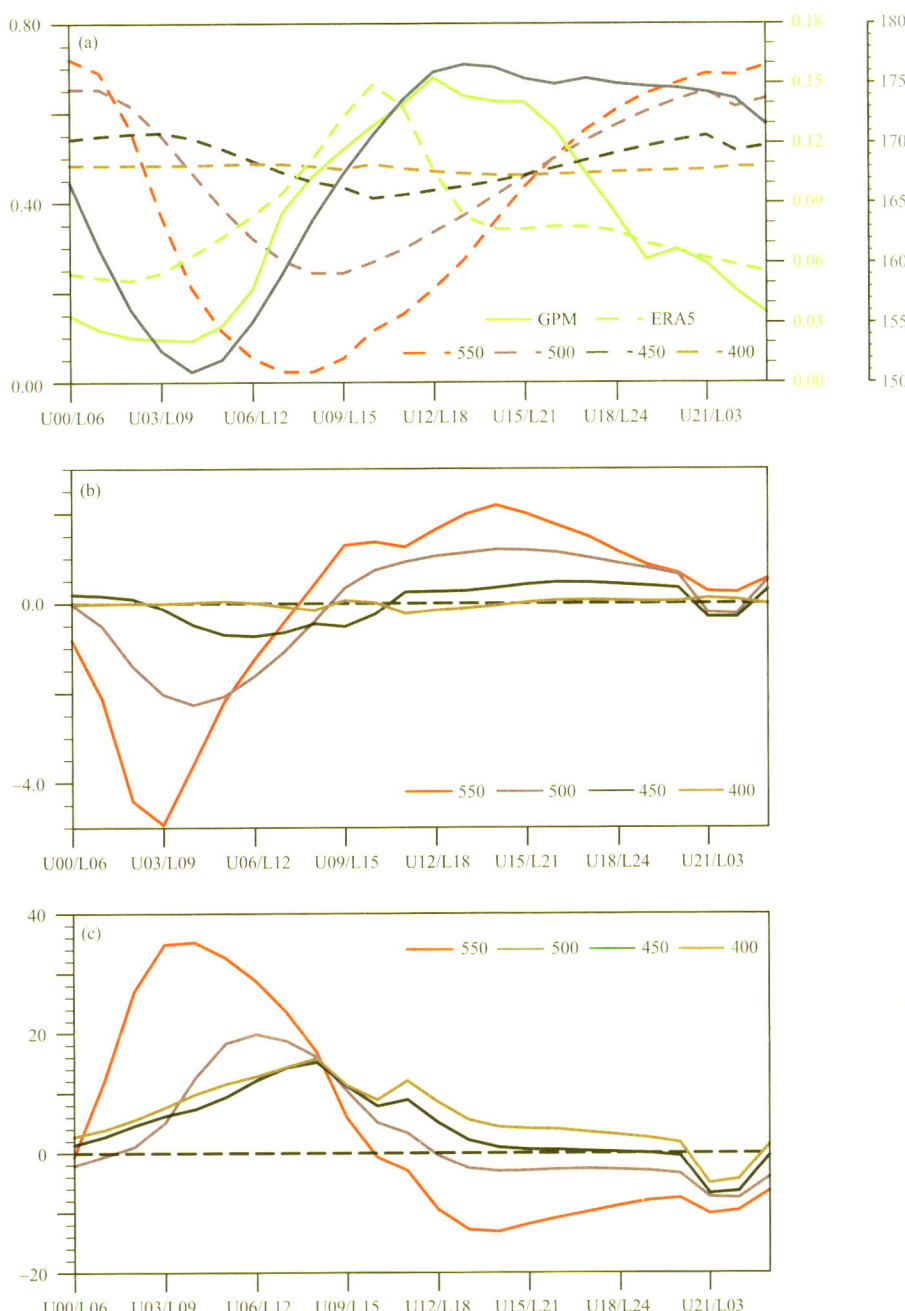

图 1 2020 年 6 月 1 日—8 月 2 日高原中部(31°~37°N,85°~96°E)550~400 hPa (a)位涡(PVU)、(b)位涡倾向(10^{-5} PVU·s^{-1})和(c)非绝热加热(10^{-5} K·s^{-1})的日变化特征。(a)中绿色实线和点线分别为 GMP 和 ERA5 资料提供的降水数据,蓝色实线为中国静止气象卫星 FY-2G 提供的红外 1 通道数据(摘自 Ma et al.,2022)

Wu 等(2020,2022)指出非绝热过程对高原涡等高位涡系统的生成也起着至关重要的作用。Wu 等(2020)提出位涡重构理论,指出即使在绝热无摩擦的大气中,通过位涡内部其他分量(如大气密度、水平涡度、斜压性及静力稳定度)的变化仍可以导致系统垂直涡度的显著发展。如 2016 年 6 月 28 日 00 时一次高原涡生成的前期,气块从 A_0 点绝热下滑到 A 点时(图 2),其将经历静力稳定度减小、斜压性增强的过程。根据位涡重构理论和倾斜涡度发展理论,垂直涡度都将得到剧烈发展,有利于低涡生成。利用位涡重构理论,他们探究了 2008 年 1 月 17—18 日青藏高原东部近地层位涡显著增强的原因(Wu et al.,2020),揭示了这些位涡之后的东移导致了 2008 年初南方第二次雨雪冰冻灾害过程(陶诗言 等,2008)的机制。

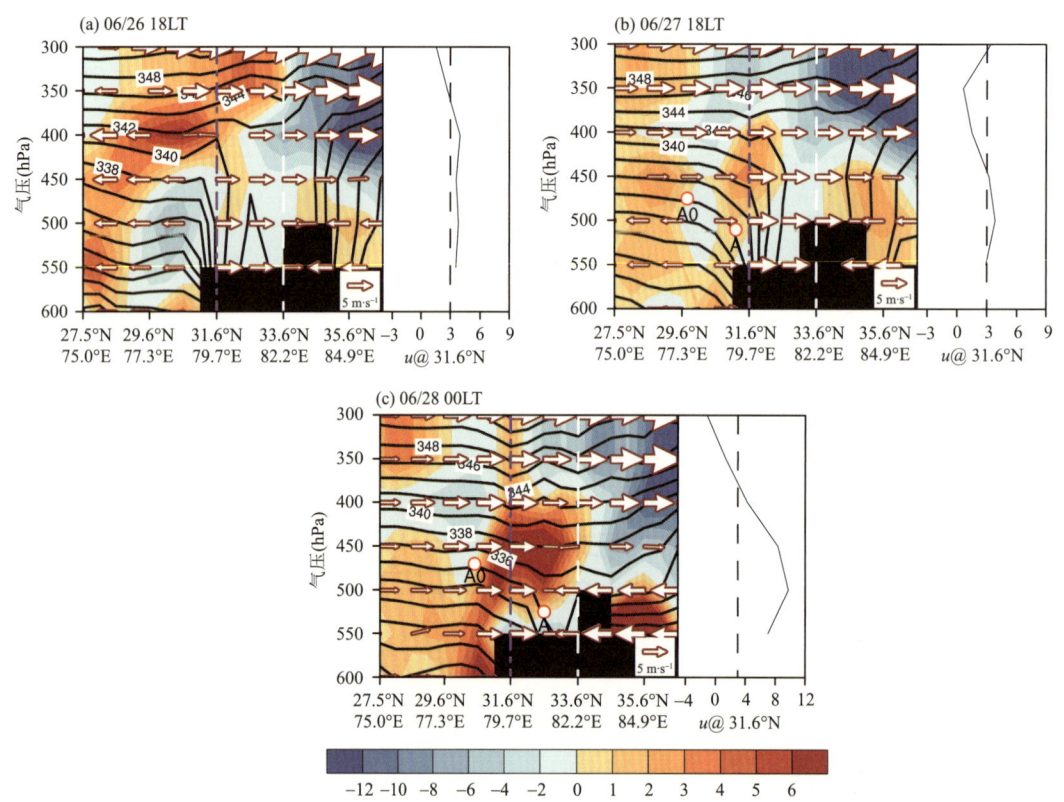

图 2 沿低涡上游气流方向且过低涡中心的水平风场(矢量,m·s^{-1})、位温(黑色等值线,K)和水平涡度(阴影,m·s^{-1}·Pa^{-1})在(a)6 月 26 日 18 时、(b)6 月 27 日 18 时和 (c) 6 月 28 日 00 时的剖面图。右侧附图为低涡上游紫色虚线标识位置的水平风(垂直于剖面)速度的垂直廓线;白色虚线为高原涡的生成位置(摘自 Wu et al.,2022)

3.2 高原位涡强迫对下游天气过程的影响

高原上生成的高位涡系统往往在低层风场的作用下向东移动,其中大部分系统在高原东部局地消散,只有小部分系统能够移出高原。这些移出高原的系统往往与下游地区的灾害性天气密切相关。研究表明,东移出高原的系统既可以单独引起下游地区降水异常(师春香 等,2000;Li et al.,2008;Wu et al.,2020),又可以通过诱发西南涡或与其他系统相互作用为下游地区带来强降水(傅慎明 等,2011;Zhang et al.,2021)。本节将回顾新型准地转垂直运动方程在高原高位涡系统东移影响下游地区极端降水中的应用。

Wu 等(2020)利用新型准地转垂直运动方程探究了 2008 年 1 月 17—21 日高原高位涡系统东移引起长江中下游地区极端雨雪事件发生发展的过程。在初始阶段(图 3 ST Ⅰ),高原东部的高位涡系统在对流层中层西风带中向下游平流,造成对流层中下层气旋式环流和偏离等熵面位移的垂直速度(ω_{ID})的发展。上升的 ω_{ID} 激发了低空辐合和南风的发展。一方面,发展增强的南风携带着水汽沿自南向北倾斜的等熵面向上运动,造成沿等熵面上滑的垂直速度(ω_{IG})的发展,加强了上升运动;另一方面,南风造成对流层低层经向负位涡平流增强,加强了对流层中下层准地转位涡平流随高度增加的环流背景,反过来进一步加强了 ω_{ID}。随着上升运动的不断发展,水汽凝结释放潜热,导致与非绝热加热有关的垂直速度(ω_Q)的发展增强(图3 ST II)。凝结潜热的释放一方面有利于大气位温增加,导致 ω_{ID} 减弱,还可以通过减弱降水区以南位温的经向梯度从而导致 ω_{IG} 减弱;另一方面,非绝热加热在高空制造负位涡,使得降水区上空位涡减小,由此加强了降水区上游位涡大值中心与降水区上空位涡低值中心之间的水平位涡梯度,导致水平位涡平流增强,使得降水形势得以维持。在这一过程中,南风的不断发展使得低空的经向负位涡平流不断增强且向上移动,当负位涡平流中心恰好位于纬向正位涡平流中心下方时,水平位涡平流随高度的增加达到最大,垂直运动发展到最强(图3 ST III)。之后经向负位涡平流中心与纬向正位涡平流中心重合,使对流层中下层水平位涡平流随高度增加的环流背景减弱,上升运动和降水也相应减弱(图3 ST IV)。由此可见,在降水发生发展的过程中,自高原向东移动的高位涡系统对下游地区垂直上升运动的发展具有激发作用,垂直速度的三个分量之间存在着相互影响、相互作用的过程,非绝热加热对垂直速度和位涡平流具有非常强的反馈作用,也正是位涡平流及其与非绝热加热之间的反馈过程主导了降水发生发展的过程。

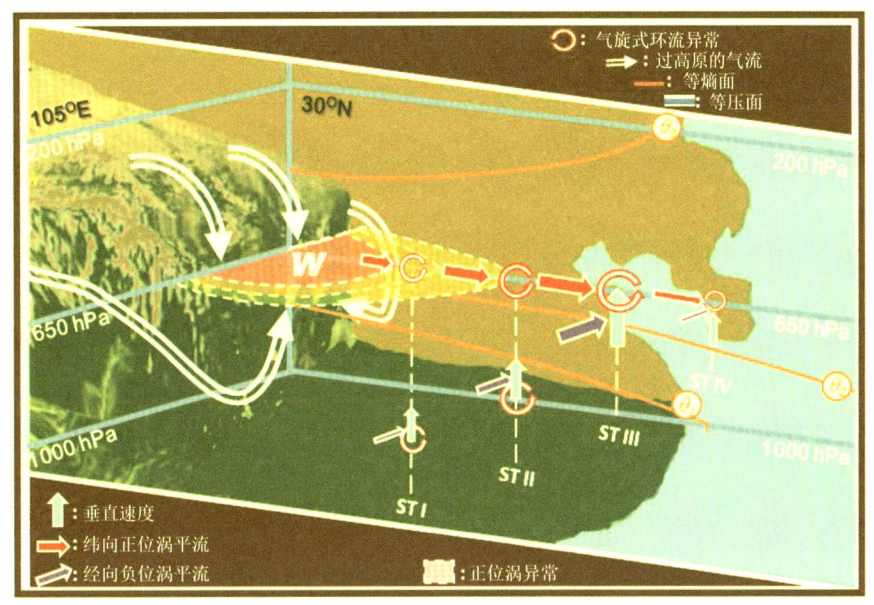

图3 高原东部位涡平流引起下游降水发生发展的示意图(摘自 Wu et al., 2020)

Charney 等(1962)的工作证实等压坐标系下准地转位涡的局地变化及平流与等熵坐标系下的 Ertel 位涡的局地变化及平流近似成正比。Hoskins 等(2014)进一步证明在非绝热大气中,当罗斯贝数(Ro)足够小、理查森数(Ri)足够大,且二者满足 $Ri^{-1} \ll Ro$ 时,Ertel 位涡(P)和准地转位涡(q_g)满足 $\left(\dfrac{\partial}{\partial t}+\boldsymbol{V}_g\cdot\nabla_h\right)_p\left[-g\dfrac{\partial\Theta(p)}{\partial p}q_g\right]\cong\left(\dfrac{\partial}{\partial t}+\boldsymbol{V}_h\cdot\nabla_h\right)_\theta P$。在此基础上,Zhang 等(2021)将新型准地转垂直运动方程中准地转位涡平流拓展到 Ertel 位涡平流的形式,并利用该拓展方程诊断了2016年6月30日—7月6日期间高原涡东移诱发西南涡、协同影响长江中下游极端暴雨的过程。他们的研究发现,高原涡生成后持续东移至高原东部,发展为一个垂直方向上向东倾斜的高位涡系统,与高原涡相关的水平位涡平流激发了对流层中下层上升的 ω_{ID},造成局地降水和潜热释放。垂直运动的发展和非绝热加热的产生也激发了四川盆地东部西南涡的发展(图4a)。之后,与西南涡相关的非绝热加热的垂直不均匀性导致低空出现较强的正位涡制造,造成低空西南涡增强的同时配合着对流层中层和高原涡有关的正的位涡平流,有利于高原涡和西南涡在垂直方向上发生合并(图4b)。合并后的高原涡-西南涡作为一个高位涡系统,引起对流层中层等熵面变得更加密集,从而导致西风气流在长江中下游地区上空沿着等熵面上滑,产生较强的上升 ω_{IG}。另一方面,加强的西南涡增大了与副高之间的气压梯度,引起低空南风气流的加强,从而导致长江中下游地区水汽输送增强以及降水增强。之后,高原涡和西南涡发生分离(图4c),分离后的高原涡继续向东移动,并通过影响局地等熵面的结构维持其东部上升的 ω_{IG},导致降水进一步加强,使得 ω_Q 也随之加强。在此期间,西南涡基本维持在原地,为下游降水输送水汽。随着高原涡继续向东移动(图4d),高原涡后方触发的下沉 ω_{IG} 依次出现在四川盆地东部和长江中下游地区,显著削弱了上升运动和西南涡的强度,使得向长江中下游地区输送的水汽减少,强降水过程逐渐减弱。在整个高原涡移出高原并协同西南涡引起下游极端降水的过程中,垂直速度三个分量之间也存在相互作用的过程,ω_{IG} 的发展对总的垂直速度的发展具有显著的超前效应。

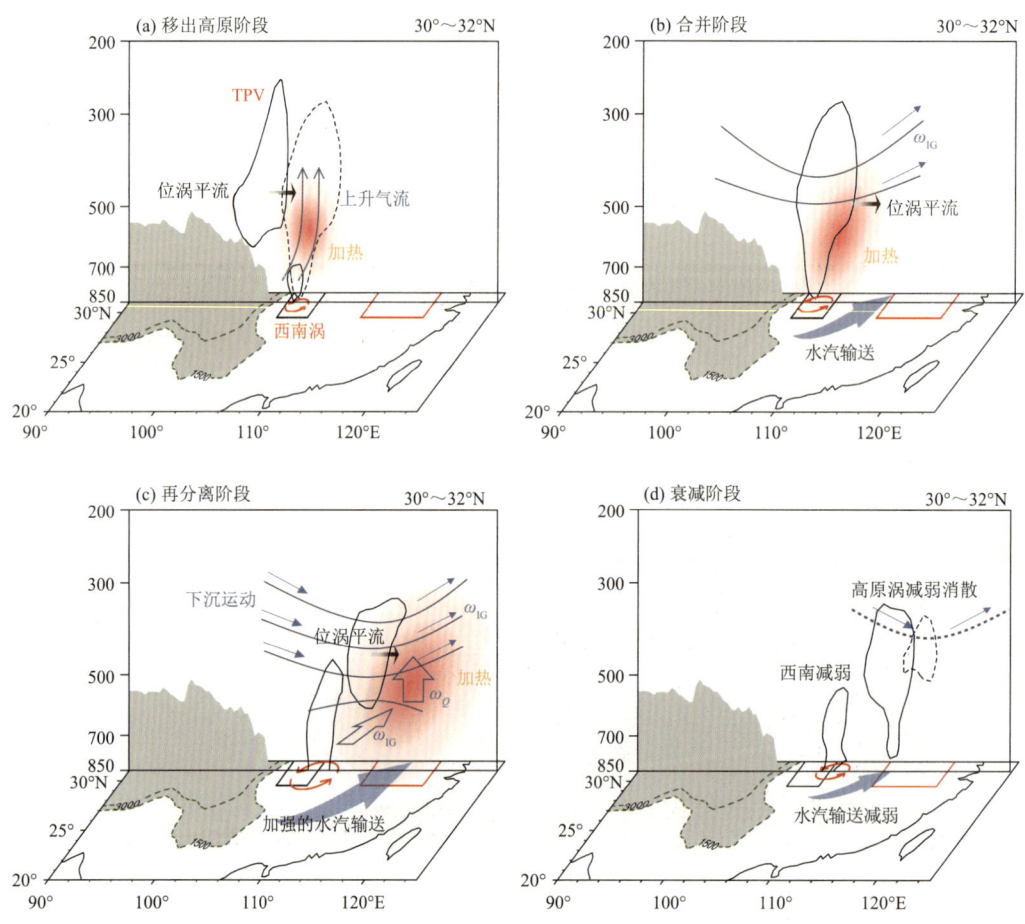

图4　2016年6月30日—7月6日高原涡和西南涡协同影响长江中下游地区极端暴雨过程的示意图
(摘自 Zhang et al,2021)

由以上回顾可以看出,在这两个个例中,与高原高位涡系统东移相伴随的准地转位涡平流均对下游上升运动的发展有激发作用,在降水发生发展的过程中垂直速度的三个分量之间均存在相互作用的过程,正是这三个分量之间的相互作用主导了降水发生、发展和消亡的过程。但不同个例之间也存在明显的差异,如冬季个例中 ω_{ID} 的发展对垂直速度的影响更大,而夏季个例中则是 ω_{IG} 更为明显。那么,这两个个例中所表现出来的共性在其他个例中是否也会存在?它们之间的差异是个例之间的差异还是冬、夏季天气系统发展的不同特征?垂直速度三个分量之间相互作用的过程是否还有其他特征?对这些问题的回答还有待下一步研究工作的深入开展。

4　总结与讨论

文中主要回顾了吴国雄院士团队近年来在准地转垂直运动方程的发展和青藏高原位涡强迫方面的新进展。首先,从准地转热力学能量方程和涡度方程出发,理论导出了新型准地转垂直运动方程,证明随等熵面位移的垂直速度(ω_θ)的发展受准地转位涡平流的垂直梯度和大气非绝热加热的影响。在简要回顾了前人对高原涡等高位涡系统生成机制的研究基础上，基于位涡重构理论，进而重点阐述了非绝热过程（高原地表和低空之间加热差异的日变化）和绝热过程在高原高位涡系统生成中的作用。最后，利用新型准地转垂直运动方程，详细回顾了高原高位涡系统东移过程中，准地转位涡平流激发上升运动及垂直速度三个分量之间的相互作用导致下游地区降水发生、发展及消亡的过程。这些研究的开展为深入理解青藏高原影响下游极端天气机制提供了新思路。

新型准地转垂直运动方程尚属新理论，它的应用价值还需要从更多的实践中进行揭示。期望本文的回顾将有助于扩大该方程的应用，对深入理解极端天气过程发生、发展和消亡的机制有所裨益。

参考文献

傅慎明，孙建华，赵思雄，等，2011. 梅雨期青藏高原东移对流系统影响江淮流域降水的研究[J]. 气象学报，69：581-600.

李国平，赵邦杰，杨锦青，2002. 地面感热对青藏高原低涡流场结构及发展的作用[J]. 大气科学，26：519-525.

刘新，吴国雄，李伟平，等，2001. 夏季青藏高原加热和大尺度流场的热力适应[J]. 自然科学进展，11：33-39.

罗四维，杨洋，吕世华，1991. 一次青藏高原夏季低涡的诊断分析研究[J]. 高原气象，10：1-12.

马婷，刘屹岷，吴国雄，等，2020. 青藏高原低涡形成、发展和东移影响下游暴雨天气个例的位涡分析[J]. 大气科学，44：472-486.

师春香，江吉喜，方宗义，2000. 1998长江大水期间对流云团活动特征研究[J]. 气候与环境研究，5：279-286.

陶诗言，卫捷，2008. 2008年1月我国南方严重冰雪灾害过程分析[J]. 气候与环境研究，13(4)：337-350.

姚秀萍，于玉斌，2000. 非地转湿Q矢量及其在华北特大台风暴雨中的应用[J]. 气象学报，58：436-446.

姚秀萍，于玉斌，2001. 完全Q矢量的引入及其诊断分析[J]. 高原气象，20：208-213.

郁淑华，何光碧，2001. 对流层中上部水汽对高原低涡形成影响的数值试验[J]. 南京气象学院学报，24：553-559.

岳彩军，寿亦萱，寿绍文，等，2003. Q矢量的改进与完善[J]. 热带气象学报，19：308-316.

张顺利，陶诗言，2002. 青藏高原对1998年长江流域天气异常的影响[J]. 气象学报，60：442-452.

张兴旺，1998. 湿Q矢量表达式及其应用[J]. 气象，24：3-7.

张兴旺，1999. 修改的Q矢量表达式及其应用[J]. 热带气象学报，15：162-167.

郑永骏，吴国雄，刘屹岷，2013. 涡旋发展和移动的动力和热力问题 I PV-Q观点[J]. 气象学报，71：185-197.

AEBISCHER U, SCHAR C, 1998. Low-level potential vorticity and cyclogenesis to the lee of the Alps[J]. J Atmos Sci, 55：186-207.

AWAN N K, FORMAYER H, 2017. Cutoff low systems and their relevance to large-scale extreme precipitation in the European Alps [J]. Theor Appl Climatol, 129：149-158.

BUSHBY F H, 1952. The evaluation of vertical velocity and thickness tendency from Sutcliffe's theory [J]. Quart J Roy Meteor Soc, 78：354-362.

CHARNEY J G, STERN M E, 1962. On the stability of internal baroclinic jets in a rotating atmosphere [J]. J Atmos Sci, 19：159-172.

DELL'OSSO L, CHEN S J, 1986. Numerical experiments on the genesis of vortices over the Qinghai-Tibet Plateau [J]. Tellus A, 38A：236-250.

DJURIC D, 1969. Note on estimation of vertical motion by the omega equation[J]. Mon Wea Rew, 97：902-904.

EGGER J, HOINKA K P, SPENGLER T, 2015. Aspects of potential vorticity fluxes: Climatology and impermeability [J]. J Atmos Sci, 72：3257-3267.

ERTEL H, 1942. Ein neuer hydrodynamische wirbelsatz [J]. Meteorology Z Braunschweig, 59：33-49.

FU S M, MAI Z, SUN J H, et al, 2019. Impacts of convective activity over the Tibetan Plateau on plateau vortex, southwest vortex and downstream precipitation [J]. J Atmos Sci, 76：3803-3830.

HELD I M, SCHNEIDER T, 1999. The surface branch of the zonally averaged mass transport in the troposphere [J]. J Atmos Sci, 56：1688-1697.

HOLTON J R, 1972. An Introduction to Dynamic Meteorology [M]. New York and London: Academic Press：319.

HOLTON J R, 2004. An Introduction to Dynamic Meteorology [M]. London: Elsevier Academic Press: 164-168.

HOSKINS B J, 1991. Towards a PV-θ view of the general circulation [J]. Tellus, 43AB: 27-35.

HOSKINS B J, DRAGHICI I, DAVIES H C, 1978. A new look at the omega-equation [J]. Quart J Roy Meteor Soc, 104: 31-38.

HOSKINS B J, PEDDER M A, 1980. The diagnosis of middle latitude synoptic development [J]. Quart J Roy Meteor Soc, 106: 707-719.

HOSKINS B J, MCITNYRE M E, ROBERTSON A W, 1985. On the use and significance of isentropic potential vorticity maps [J]. Quart J Roy Meteor Soc, 111: 877-946.

HOSKINS B J, PEDDER M, JONES D W, 2003. The omega equation and potential vorticity [J]. Quart J Roy Meteor Soc, 129: 3277-3303.

HOSKINS B J, JAMES I N, 2014. Fluid Dynamics of the Midlatitude Atmosphere [M]. John Wiley & Sons, Ltd, 405.

KLEINSCHMIDT E, 1955. Die entstehung einer hohenzyklone uber nordamerika [J]. Tellus, 7: 96-110.

LI L, ZHANG R H, WEN M, 2011. Diagnostic analysis of the evolution mechanism for a vortex over the Tibetan Plateau in June 2008 [J]. Adv Atmos Sci, 28: 797-808.

LI L, ZHANG R H, WEN M, 2014. Diurnal variation in the occurrence frequency of the Tibetan Plateau vortices [J]. Meteorology and Atmospheric Physics, 125: 135-144.

LI L, ZHANG R H, WEN M, 2018. Diurnal variation in the intensity of nascent Tibetan Plateau vortices [J]. Quart J Roy Meteor Soc, 144: 2524-2536.

LI Y D, WANG Y, YANG S, et al, 2008. Characteristics of summer convective systems initiated over the Tibetan Plateau. Part I: Origin, track, development, and precipitation [J]. J Appl Meteor Climatol, 47: 2679-2695.

LIN Z Q, GUO W D, JIA L, et al, 2020. Climatology of Tibetan Plateau vortices derived from multiple reanalysis datasets [J]. Clim Dyn, 55: 2237-2252.

LV M M, YANG S, LI Z N, et al, 2018. Possible effect of the Tibetan Plateau on the "upstream" climate over West Asia, North Africa, South Europe and the North Atlantic [J]. Clim Dyn, 51: 1485-1498.

MA T T, WU G X, LIU Y M, et al, 2022. Abnormal warm sea-surface temperature in the Indian Ocean, active potential vorticity over the Tibetan Plateau, and severe flooding along the Yangtze River in summer 2020 [J]. Quart J Roy Meteor Soc, 148: 1001-1019.

MARTIUS O, ZENKLUSEN E, SCHWIERZ C, et al, 2006. Episodes of Apline heavy precipitation with an overlying elongated stratospheric intrusion: A climatoloty [J]. Int J Climatol, 26: 1149-1164.

MCTAGGART-COWAN R, GALARNEAU T J, BOSART L F, et al, 2010. Development and tropical transition of an Alpine lee cyclone. Part I: Case analysis and evaluation of numerical guidance [J]. Mon Mea Rev, 138: 2281-2307.

PETTERSSEN S, 1956. Weather Analysis and Forecasting [M]. New York: The McGraw-Hill Co:428.

ROSSBY C G, 1940. Planetary flow patterns in the atmosphere [J]. Quart J Roy Meteor Soc, 66(suppl): 68-87.

RUDARI R, ENTEKHABI A, ROTH G, 2004. Terrain and multiple-scale interactions as factors in generating extreme precipitation events [J]. J Hydrometeor, 5: 390-404.

SANDERS F, HOSKINS B J, 1990. An easy method for estimation of Q-vectors from weather maps [J]. Wea Forecasting, 5: 346-353.

SCHNEIDER T, HELD I M, GARNER S T, 2003. Boundary effects in potential vorticity dynamics [J]. J Atmos Sci, 60: 1024-1040.

SHEN R, REITER E R, BRESCH J F, 1986. Some aspects of the effects of sensible heating on the development of summer weather systems over the Tibetan Plateau [J]. J Atmos Sci, 43: 2241-2260.

SUN R Z, DUAN A M, CHEN L L, et al, 2019. Interannual variability of the North Pacific mixed layer associated with the spring Tibetan Plateau thermal forcing [J]. J Climate, 32: 3109-3130.

SUTCLIFFE R C, 1947. A contribution to the problem of development [J]. Quart J Roy Meteor Soc, 73: 370-383.

TAO S Y, DING Y H, 1981. Observational evidence of the influence of the Qinghai-Xizang (Tibet) Plateau on the occurrence of heavy rain and severe convective storms in China [J]. Bulletin American Meteorological Society, 62: 23-30.

THOMPSON P D, 1961. Numerical Weather Analysis and Prediction [M]. New York: Macmillan: 170.

THORPE A, VOLKERT H, HEIMANN D, 1993. Potential vorticity of flow along the Alps [J]. J Atmos Sci, 50(11): 1573-1590.

TRENBERTH K E, 1978. On the interpretation of the diagnostic quasigeostrophic omega equation[J]. Mon Wea Rev, 107: 682-703.

WANG B, 1987. The Development Mechanism for Tibetan Plateau Warm Vortices [J]. J Atmos Sci, 44: 2978-2994.

WANG B, BAO Q, HOSKINS B J, et al, 2008. Tibetan Plateau warming and precipitation changes in East Asia [J]. Geophys Res Lett, 35: L14702.

WU G X, ZHANG Y S, 1998. Tibetan Plateau forcing and the timing of the monsoon onset over South Asia and the Southern China Sea [J]. Mon Wea Rev, 126: 913-927.

WU G X, LIU Y M, WANG T M, et al, 2006. The influence of mechanical and thermal forcing by the Tibetan Plateau on Asian climate [J]. J Hydrometeorology, 8: 770-789.

WU G X, LIU Y M, HE B, et al, 2012. Thermal Controls on the Asian Summer Monsoon [J]. Nature Sci Rep, 2: 404.

WU G X, MA T T, LIU Y M, et al, 2020. PV-Q perspective of cyclogenesis and vertical velocity development downstream of the Tibetan Plateau [J]. J Geophys Res, 125: e2019JD030912.

WU G X, TANG Y Q, HE B, et al, 2022. Potential vorticity perspective of the genesis of a Tibetan Plateau vortex in June 2016 [J]. Clim Dyn, 58: 3351-3367.

WU Z W, LI J P, JIANG Z H, et al, 2012. Modulation of the Tibetan Plateau snow cover on the ENSO teleconnections: From the East Asian summer monsoon perspective [J]. J Climate, 25: 2481-2489.

YANG S, GAO S T, WANG D H, 2007. Diagnostic analyses of the ageostrophic Q vector in the non-uniformly saturated, frictionless, and moist adiabatic flow [J]. J Geophys Res, 112: 1-9.

YAO X P, YU Y Y, SHOU S W, 2004. Diagnostic analyses and application of the moist ageostrophic vector Q [J]. Adv Atmos Sci, 21: 96-102.

YUE C J, 2009. The Q vector analysis of the heavy rainfall from Meiyu front cyclone: A case study [J]. Acta Meteorologica Sinica, 66: 3-7.

YUE C J, SHOU S W, 2008. A modified moist ageostrophic Q vector [J]. Adv Atmos Sci, 25: 1053-1061.

ZHANG G S, MAO J Y, LIU Y M, et al, 2021. PV perspective of impacts on downstream extreme rainfall event of a Tibetan Plateau vortex collaborating with a southwest China vortex [J]. Adv Atmos Sci, 38: 1835-1851.

湿位涡和倾斜涡度发展

吴国雄,蔡雅萍,唐晓菁

(中国科学院大气物理研究所,大气科学和地球流体力学数据模拟国家重点实验室(LASG),北京 100029)

摘要:从完整的原始方程出发,在导出精确形式的湿位涡方程的基础上,证得绝热无摩擦的饱和湿空气具有湿位涡守恒的特性。并由此去研究湿斜压过程中涡旋垂直涡度的发展。结果表明,在湿等熵坐标中,涡旋的发展与对流稳定度的减少,等熵面上的辐合和潜热的释放有关。由于等熵位涡分析的应用受等熵面倾斜的限制,又进而发展了 z 坐标及 p 坐标中的倾斜涡度发展理论。指出无论是湿对称不稳定或对流不稳定大气,还是湿对称稳定或对流稳定大气,除对流稳定度的影响外,风的垂直切变的增加或水平湿斜压的增加均能因湿等熵面的倾斜而引起垂直涡度的增长。湿等熵面的倾斜越大,这种由干湿斜压性加强所引起的涡旋发展更激烈。在梅雨锋附近及其南侧暖湿区的北端,湿等熵面十分陡立,是涡旋发展及暴雨发生的重要地区。

对 1991 年 6 月 12—15 日江淮流域暴雨过程的湿位涡分析表明,湿位涡分析,尤其是等压面上湿位涡量 MPV1 和 MPV2 的分析不仅在中高纬有效,在低纬度及低对流层也十分有效,是暴雨诊断和预报的有力工具。

关键词:湿位涡;倾斜涡度发展;湿对称不稳定

1 引言

大气中大暴雨的发生发展与低空流场的辐合、垂直运动的急速发展有关,并常伴有气旋性垂直涡度的急剧增大(陶诗言,1986)。因而研究气旋性涡度的发展机制是研究暴雨发生发展的一个重要内容。人们或用线性化小扰动方程,通过对运动场特征模进行分析,去研究背景场对特征模变化的影响;或利用大气运动的某种守恒特性,通过分析该特性在不同尺度运动间的转化,去研究特定系统的发展。在绝热无摩擦的干空气中存在具有严格守恒特性的 Ertel(1942)位势涡度

$$P_E = \alpha \boldsymbol{\zeta}_a \cdot \nabla \theta$$

式中,α 和 θ 为比容和位温,$\boldsymbol{\zeta}_a$ 为三维绝对涡度。Hoskins 等(1985)曾对 Ertel 位涡在大气运动诊断中的应用进行系统的分析。一般而言,P_E 对许多中高纬度天气系统的移动和发展具有指示性。但在对流层低层,尤其在低纬度,P_E 变得很弱。再者,P_E 不包括水汽分布的影响,因此在降水过程的应用中存在局限性。

Bennetts 和 Hoskins(1979)曾从布西内斯克(Boussinesq)近似出发,引入潜热作用,导得湿球位涡变化方程。本文从严格的原始运动方程出发,把饱和大气中水汽凝结潜热的作用引进位涡的分析中,以研究与暴雨相联系的垂直涡度的变化。第 2 节首先从原始方程系出发导出湿位势涡度方程。第 3 节证明了绝热无摩擦的饱和大气中湿位涡的守恒性。利用湿位涡的这一守恒特征,在第 4 至第 6 节分别研究了湿等

本文原载于《气象学报》,1995,53(4):387-405

熵坐标、等高坐标及等压坐标中湿位涡的特征及相应的发展理论；证明在等高坐标和等压坐标中，大气的干湿斜压性及风垂直切变的增长可以导致系统气旋性垂直涡度的激烈发展。第 7 节通过对 1991 年 6 月 12—15 日发生在江淮流域下游的持续大暴雨进行等压面湿位涡分析，研究倾斜涡度发展理论在暴雨分析和预报中的应用。若干讨论和结论由第 8 节给出。

2 湿位涡方程

对欧拉形式的动量方程

$$\frac{\partial \boldsymbol{v}}{\partial t} + \nabla\left(\frac{v^2}{2} + \varphi\right) - \boldsymbol{v} \times \boldsymbol{\zeta}_a = -\alpha \nabla p + \boldsymbol{F}_v \tag{1}$$

求涡度，可得涡度方程

$$\frac{\partial \boldsymbol{\zeta}_a}{\partial t} - \nabla \times (\boldsymbol{v} \times \boldsymbol{\zeta}_a) = \nabla p \times \nabla \alpha + \boldsymbol{F}_\zeta \tag{2}$$

式中，$\boldsymbol{v}$ 为 3 维风矢，φ 为外作用力的势函数，$\boldsymbol{F}_v$ 为表面摩擦力，$\boldsymbol{F}_\zeta = \nabla \times \boldsymbol{F}_v$ 为涡度摩擦耗散，$\boldsymbol{\zeta}_a$ 为绝对涡度。设 $\boldsymbol{\Omega}$ 为地球自转角速度，则有

$$\boldsymbol{\zeta}_a = \nabla \times \boldsymbol{v} + 2\boldsymbol{\Omega} \tag{3}$$

考虑了潜热释放及其他形式非绝热加热 Q_d 的热力方程可表示为

$$c_p \frac{T}{\theta} \frac{d\theta}{dt} = -L \frac{dq}{dt} + Q_d \tag{4}$$

式中，各量为气象常用量，$\frac{d}{dt} = \frac{\partial}{\partial t} + \boldsymbol{v} \cdot \nabla$。由相当位温定义

$$\theta_e = \theta \cdot e^{\frac{Lq}{c_p T}} \tag{5}$$

代入式(4)，并略去高阶小量，得

$$\frac{d\theta_e}{dt} = \left(\frac{\partial}{\partial t} + \boldsymbol{v} \cdot \nabla\right)\theta_e = \frac{\theta_e}{c_p T} Q_d \equiv Q \tag{6}$$

用 $\nabla \theta_e$ 点乘式(2)，并利用矢量关系

$$\nabla \theta_e \cdot \nabla \times (\boldsymbol{v} \times \boldsymbol{\zeta}_a) = -\nabla \cdot [\nabla \theta_e \times (\boldsymbol{v} \times \boldsymbol{\zeta}_a)]$$

$$\nabla \theta_e \times (\boldsymbol{v} \times \boldsymbol{\zeta}_a) = \boldsymbol{v}(\boldsymbol{\zeta}_a \cdot \nabla \theta_e) + \boldsymbol{\zeta}_a\left(\frac{\partial \theta_e}{\partial t} - Q\right)$$

得

$$\left(\frac{\partial}{\partial t} + \boldsymbol{v} \cdot \nabla\right)(\boldsymbol{\zeta}_a \cdot \nabla \theta_e) + (\boldsymbol{\zeta}_a \cdot \nabla \theta_e) \nabla \cdot \boldsymbol{v}$$
$$= (\nabla p \times \nabla \alpha) \cdot \nabla \theta_e + \nabla \theta_e \cdot \boldsymbol{F}_\zeta + \boldsymbol{\zeta}_a \cdot \nabla Q \tag{7}$$

其次，用比容 α 乘上式两边并利用连续方程

$$\frac{d\alpha}{dt} - \alpha \nabla \cdot \boldsymbol{v} = 0 \tag{8}$$

便可得到方程

$$\frac{dP_m}{dt} = \alpha(\nabla p \times \nabla \alpha) \cdot \nabla \theta_e + \alpha \nabla \theta_e \cdot \boldsymbol{F}_\zeta + \alpha \boldsymbol{\zeta}_a \cdot \nabla Q \tag{9}$$

式中，

$$P_m = \alpha \boldsymbol{\zeta}_a \cdot \nabla \theta_e \tag{10}$$

为湿空气位势涡度，或简称湿位涡(MPV)。它等于单位质量气块的绝对涡度在 $\nabla \theta_e$ 方向的投影与 $|\nabla \theta_e|$ 的积。式(9)则为精确形式的湿位涡方程。对干空气而言，$q = 0$。这时式(9)蜕化为干空气的位涡方程，而湿位涡式(10)则蜕化为 Ertel 位涡(Ertel, 1942)。

3 湿位涡的守恒性

无摩擦、湿绝热大气满足

$$\boldsymbol{F}_\zeta = 0, Q = 0$$

这时湿位涡方程(9)成为

$$\frac{\mathrm{d}P_m}{\mathrm{d}t} = \alpha(\nabla p \times \nabla \alpha) \cdot \nabla \theta_e \quad (\boldsymbol{F}_\zeta = 0, Q = 0) \tag{11}$$

由于压容力管项 $(\nabla p \times \nabla \alpha)$ 对单位横截面的积分等于气压梯度力沿其周界所做的功,根据式(3),该做功量引起横截面上绝对涡度的变化。因此式(11)的物理意义就易于理解:湿位涡的个别变化等于单位质量气块因气压梯度力所引起的绝对涡度个别变化在 $\nabla\theta_e$ 方向上的投影与 $|\nabla\theta_e|$ 的积。尽管式(11)右端在形式上与 Bennetts 和 Hoskins 在 Bousinesq 近似下所导得的形式略有差别,但易于证明,它们在本义上是一致的。

设状态为 (p_0, T_0, q) 的未饱和空气受扰动沿干绝热上升至抬升凝结高度(LCL)达到饱和态 (p, T, q)。此后 $(p \leqslant p_c)$ 上升气块便处于饱和状态,因此有

$$\theta_e(p_0, T_0, q) = \theta_e[p, T, q(T)] = \theta_e(p, \alpha) \quad \text{当}(p \leqslant p_c) \tag{12}$$

这就是说,当 $(p \leqslant p_c)$ 时,等 θ_e 面与等 p 面和等 α 面的相交轴平行,于是 $\alpha(\nabla p \times \nabla \alpha) \cdot \nabla \theta_e \equiv 0$。因此,对饱和空气(这时 $p = p_c, T = T_d$)或未饱和空气受抬升至凝结高度以上,有

$$\frac{\mathrm{d}P_m}{\mathrm{d}t} \equiv 0 \quad (p \leqslant p_c) \tag{13}$$

或

$$P_m = \alpha \boldsymbol{\zeta}_a \cdot \nabla \theta_e \equiv 常数 \quad (p \leqslant p_c) \tag{14}$$

换言之,在绝热无摩擦的饱和大气中,湿位涡是守恒的。这时

$$\theta_e = \theta(p_c, T_c) \cdot \mathrm{e}^{\frac{Lq}{c_p T_c}}$$

另由 θ 的守恒性知 $\theta(p_c, T_c) = \theta(p_0, T_0)$,所以有

$$\theta_e = \theta(p_0, T_0) \cdot \mathrm{e}^{\frac{Lq}{c_p T_c}} \tag{15}$$

上式中 T_c 的计算较为繁琐。由于在温度-对数压力图上等 q 线的斜率很大, T_c 与露点温度 T_d 接近。因此在分析中可近似地用 T_d 代替 T_c,也就是说可以由马格努斯公式从 q 和 p 直接计算 T_c。

与干空气的动力特征类似,$P_m > 0$ 的大气是湿对称稳定的;$P_m < 0$ 的大气是湿对称不稳定的(Bennett 和 Hoskins,1979)。

4 湿等熵面上的湿位涡及发展理论

在绝热无摩擦的饱和大气中,当 θ_e 面与等压面的交角很小时,如取 θ_e 为垂直坐标,这时由于沿湿等熵面 θ_e 的梯度为零,则湿位涡守恒取如下简单的形式

$$P_m = -g\zeta_\theta \frac{\partial \theta_e}{\partial p} \approx 常数 \tag{16}$$

式中,ζ_θ 为 ζ_a 在垂直方向的投影。定义对流稳定度

$$N_m = \frac{g}{\theta_0} \frac{\partial \theta_e}{\partial z} = \frac{\rho g^2}{\theta_0} \cdot \frac{\partial \theta_e}{\partial p} \tag{17}$$

则由式(16)可得

$$\alpha \zeta_\theta N_m \approx 常数 \tag{18}$$

因此在湿等熵坐标中,当气块从对流稳定度大的区域向小的区域移动时,或从等 θ_e 面密集区向扇开区移

动时,气旋性涡度将增长。由式(16)还可得

$$\frac{1}{\zeta_\theta}\frac{\mathrm{D}}{\mathrm{D}t}(\zeta_\theta) = -\frac{1}{\theta_{\mathrm{ep}}}\frac{\mathrm{D}}{\mathrm{D}t}\theta_{\mathrm{ep}} \tag{19}$$

式中,$\frac{\mathrm{D}}{\mathrm{D}t}$ 为拉格朗日坐标系的全微分。由 θ_e 定义(5)及 θ 坐标中的连续方程

$$\frac{\partial}{\partial t}\left(\frac{\partial p}{\partial \theta}\right) + \nabla_\theta \cdot \left(\frac{\partial p}{\partial \theta}\boldsymbol{v}\right) + \frac{\partial}{\partial \theta}\left(\dot{\theta}\frac{\partial p}{\partial \theta}\right) = 0 \tag{20}$$

易证得对流稳定度的变化方程

$$\frac{1}{\theta_{\mathrm{ep}}}\frac{\mathrm{D}}{\mathrm{D}t}\theta_{\mathrm{ep}} = r\left[\nabla_\theta \cdot \boldsymbol{v} + \frac{\mathrm{D}}{\mathrm{D}t}\left(\frac{Lq}{c_pT}\right) + \frac{\theta}{\theta_p}\frac{\mathrm{D}}{\mathrm{D}t}\frac{\partial}{\partial p}\left(\frac{Lq}{c_pT}\right)\right] \tag{21}$$

式中,$r = \frac{\theta_p}{\theta_{\mathrm{ep}}}\mathrm{e}^{\frac{Lq}{c_pT}}$,及绝对涡度的变化方程为

$$\frac{\mathrm{D}}{\mathrm{D}t}(\zeta_\theta) = -\zeta_\theta r\left[\nabla_\theta \cdot \boldsymbol{v} + \frac{\mathrm{D}}{\mathrm{D}t}\left(\frac{Lq}{c_pT}\right) + \frac{\theta}{\theta_p}\frac{\mathrm{D}}{\mathrm{D}t}\frac{\partial}{\partial p}\left(\frac{Lq}{c_pT}\right)\right] \tag{22}$$

上述结果指出,θ 面上的辐合及水汽的凝结均能引起对流稳定度的减少,从而导致系统气旋性涡度的增加。与一般的涡度方程比较,式(22)的一个显著特点是引进了凝结加热的效应。

图1为典型的横跨梅雨锋南北剖面。在副热带地区有自北向南的冷空气与自热带副热带向北的暖湿空气在锋面 A 附近汇合。根据式(22),气旋性涡度将增长。在 I,II 区,当气块向南移动时,N_m 的减小将导致气旋性涡度增加。其次,在锋面 A 南面对流层的中部存在中性稳定层($\frac{\partial \theta_e}{\partial p} = 0$)。在该层以下的高湿区为对流不稳定层。当气块在 III、IV 区向北运动时,θ_{ep} 减少。根据式(19),气旋性涡度也将增加。而且,一旦梅雨锋冷空气侵入该区,对流不稳定能将释放从而导致对流发展。最后,考虑近地表的感热加热,由湿位涡方程

$$\frac{\mathrm{d}}{\mathrm{d}t}\left(\alpha\zeta_\theta\frac{\partial\theta_e}{\partial z}\right) = \alpha\zeta_\theta\frac{\partial Q}{\partial z}$$

知,地表加热时 $\frac{\partial Q}{\partial z} < 0$,在 II 区导致近地面气柱气旋性涡度减少,在 III 区则使气柱的气旋性涡度增加。概言之,梅雨锋北面气旋性涡度的消长取决于对流稳定度变化的绝热效应及地表非绝热过程的相对大小。而梅雨锋南面的 III 区中,各因子均促使正涡度增长,是气旋性系统最易发展的区域。一旦气块在运动中释出降水,由式(22)知,其气旋性涡度将进一步增长。因此强降水易发生在梅雨锋附近及其南侧。

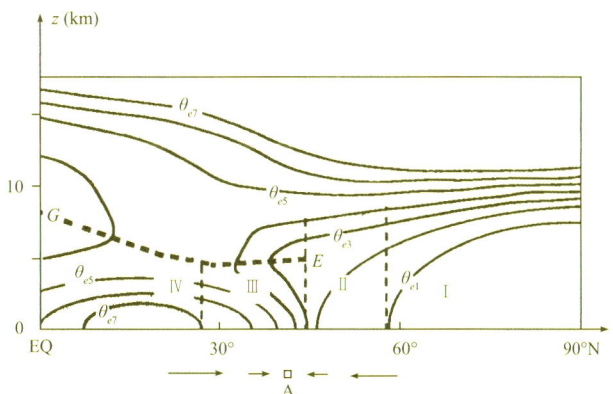

图1 在典型的横跨梅雨锋 A 的南北剖面中湿等熵面 θ_e(细实线)的示意分布

(粗断线为中性对流稳定层的位置。阴影区示出近地层的高 θ_e 区,箭矢代表冷暖空气流的运动形态)

1991年6月,东亚地区的高空形势为西太平洋副热带高压稳定偏强;孟加拉湾低槽偏强偏东;中纬多短波槽活动。中旬,自西北地区向南的冷空气与来自孟加拉湾的低空 SW 气流在江淮流域交绥,11—15

日该地区出现持续暴雨(董立清,1991)。在日雨量分布图(图2)上,雨强中心在12—15日的强度均超过或接近150 mm,造成严重的洪涝。图3给出1991年6月9—16日每天00时沿115°E的南北向剖面的演变。起始时,31°N以南的暖空气为对流不稳定大气($\theta_{ez}<0$)。高纬干冷空气前锋的θ_e密集带自50°N逐渐南移。11日该θ_e锋面侵入南方的湿对流不稳定(MCI<0)区,暴雨开始出现在θ_e陡峭和密集区中,即图1的Ⅲ区中。此后南北气流在江淮地区汇合对峙,该地暴雨持续不断。14日以后,高纬有一强的副极锋生成并南移。15日该新鲜冷空气并入江淮梅雨锋系中;暖湿空气撤至29°N以南;一次梅雨锋暴雨过程遂告结束。仔细分析图3发现,这次降水过程其实可分为两个阶段。11—12日降水主要发生在锋面南部。天气图上那里存在低空冷性切变及地表高温区(董立清,1991)。因此暴雨的发生应与地表加热,风场辐合及对流不稳定能的释放有关。13—15日,降水发生在锋区内,那里的对流不稳定已消失($\theta_{ez}>0$)。但由于这时江淮地区对流层底层有低空西南气流发展及暖切变性维持(董立清,1991),根据式(22),降水涡旋系统的发展应与风场辐合及凝结潜热的释放有关。

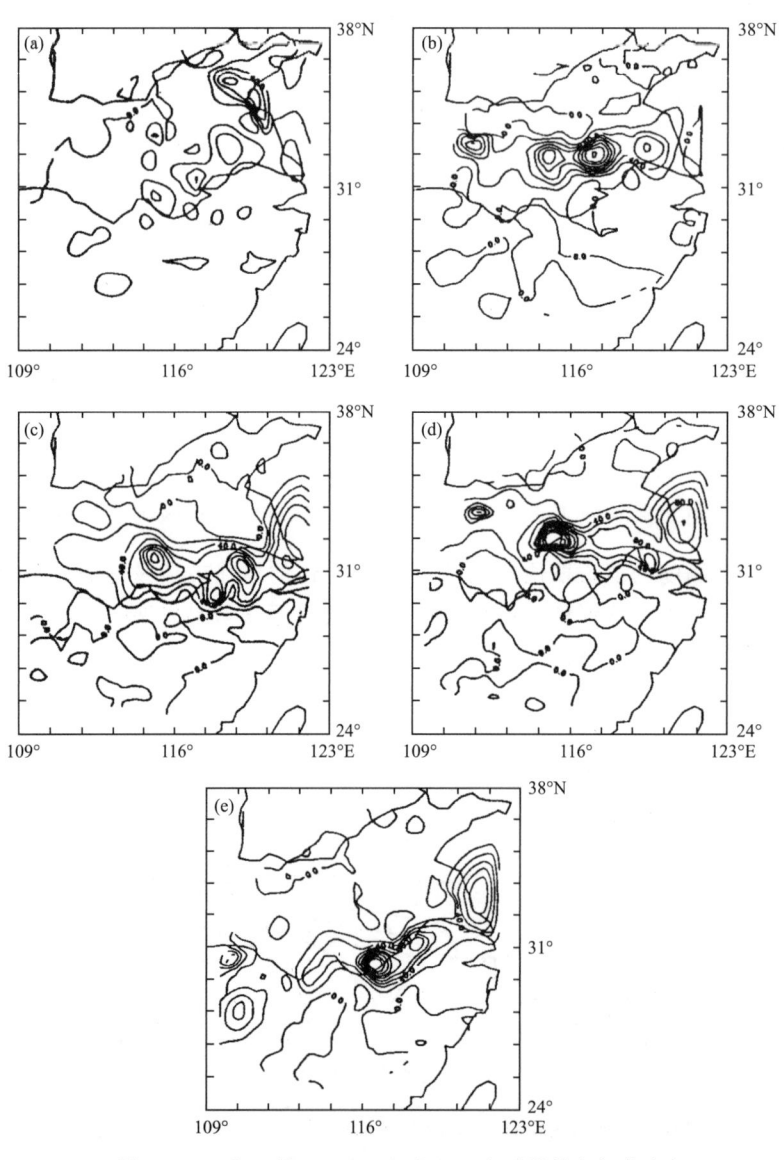

图2　1991年6月11—15日逐日24 h雨量的空间分布

(等值线间隔为20 mm)

(a)11日;(b)12日;(c)13日;(d)14日;(e)15日

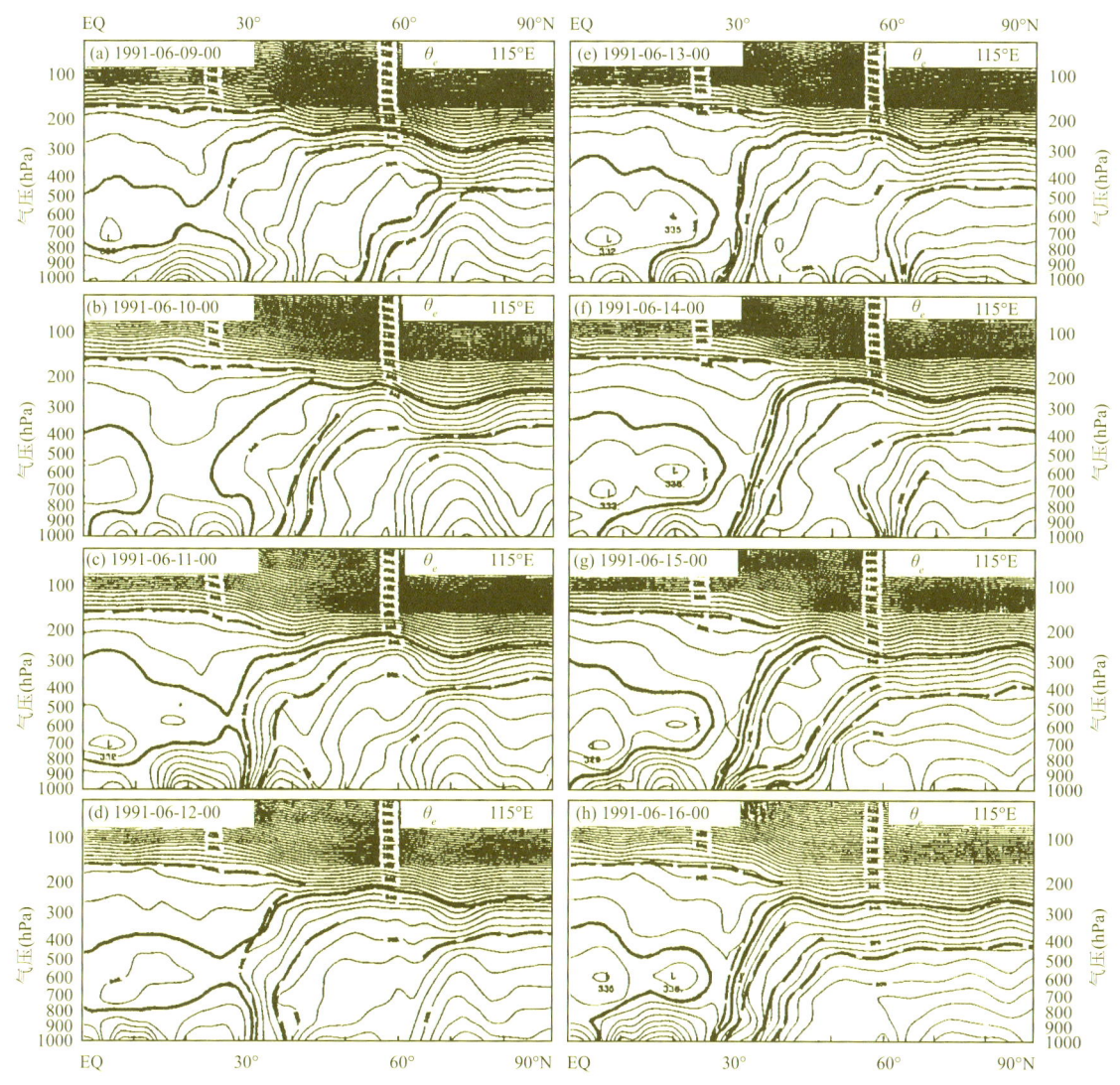

图 3 1991 年 6 月 9—16 日逐日 00 时沿 115°E 剖面上 θ_e 的分布
(粗线为 $\theta_e = 345$ K 等值线,粗断线标出 θ_e 锋区及对流层顶所在的位置)

图 4 为此期间 $\theta_e = 345$ K 湿等熵面上湿位涡(MPV)的演变。在制图时,对每一格点采用自上面下确定 θ_e 高度的办法。因此,在 θ_e 探空出现折叠的格点上,θ_e 面至少应有双页结构。由于下层 $\frac{\partial \theta_e}{\partial z} < 0$,表明该处为对流不稳定。因一般 ζ_0 为正,故下层一般为湿对称不稳定区。图 4 表明,从 6 月 9 日起中国江淮流域以南低空存在对流不稳定。同期沿对流层顶附近的高值 P_m 的天气尺度扰动自西向东移动时发展加强。11 日该扰动前锋入侵对流不稳定区,江淮流域开始出现持续性大暴雨。14 日贝加尔湖以西有另一高值 P_m 系统发展东移,江淮地区的湿对流不稳定区也逐渐南撤。至 16 日该系统已侵入华北;对流不稳定区也已撤至华南沿海。江淮持续暴雨过程结束。

图 5 为该期间上页 $\theta_e = 345$ K 湿等熵面的气压高度的演变。高值中心表明该 θ_e 面呈伸向低空之"漏斗状"。当气压值超过 1000 hPa 时,即该漏斗已及地面时,表明地表的高湿高温区已与高空的高 θ_e 打通。因此该区基本为图 1 中之Ⅲ和Ⅳ区。其北部的等高线密集带对应着图 3 中的锋区及图 1 中的Ⅱ区。最强涡旋发展区于是应位于该密集带南侧附近。从图 2 及图 5 可以看出,此次暴雨期间,暴雨基本上发生在 θ_e 锋面附近之第Ⅲ区及第Ⅱ区中,与上面关于涡度发展理论一致。

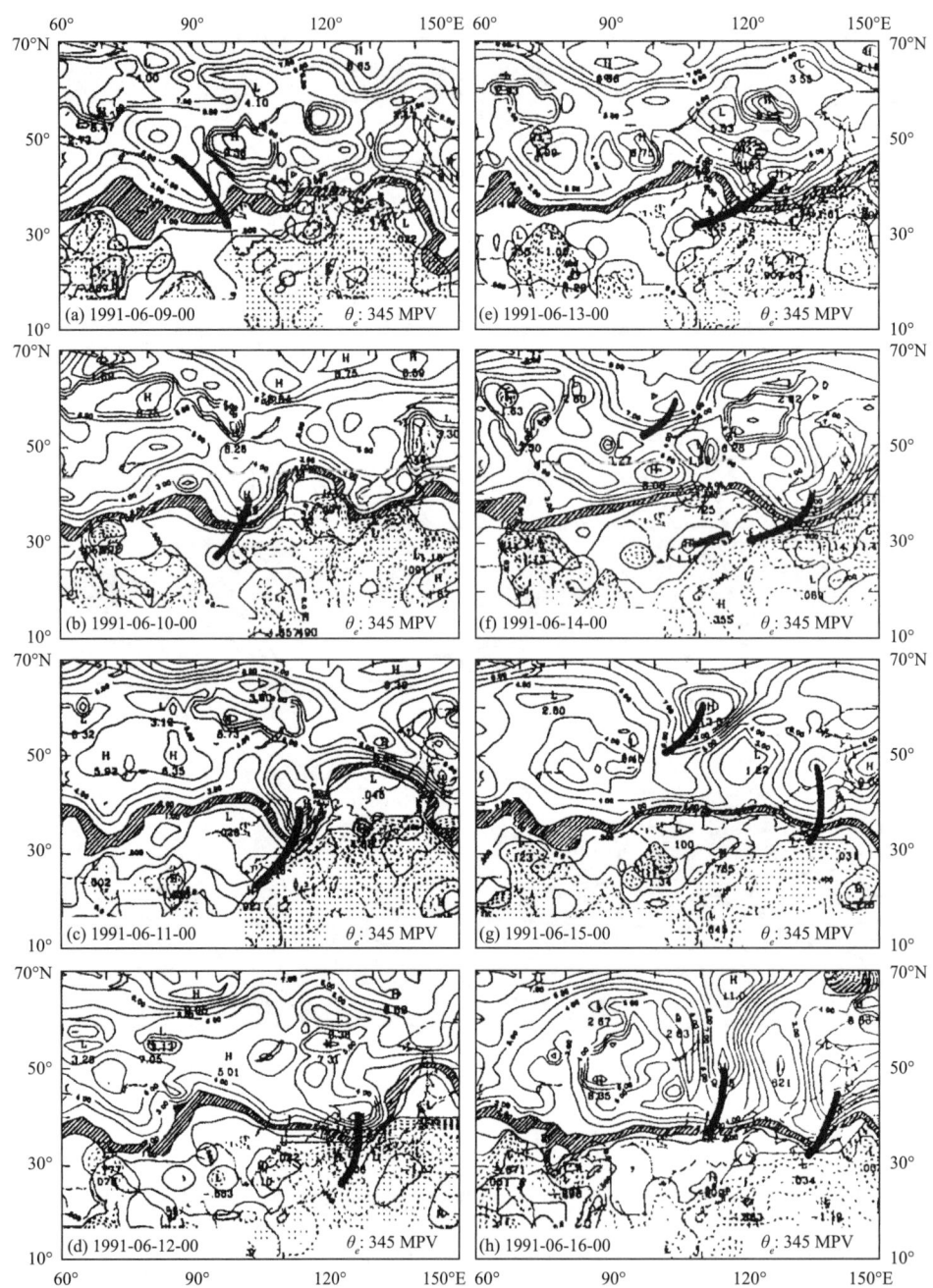

图 4　1991 年 6 月 9—16 日逐日 00 时湿等熵面 $\theta_e = 345$ K 上湿位涡(MPV)分布的连续演变

(点区示出该 θ_e 面的下页面,虚线表示 MPV 为负值。粗线示出引起江淮暴雨开始和结束的中高纬高值 MPV 系统的位置。等值线间隔当 MPV<0 时为 0.2 PVU(1 PVU = 1.0×10^{-6} m²·s⁻¹·K·kg⁻¹);当 MPV<1.0 PVU 时为 0.5 PVU;否则为 1.0 PVU。图中 1.0~2.0 PVU 区域加斜线,它近似表示对流层顶断裂带所处的纬带)

5　倾斜涡度发展理论

注意到只有当 θ_e 水平分布时,式(16)才是准确成立的。当 θ_e 面倾斜时,精确的湿位涡守恒应表示为

$$\alpha \zeta_\theta |\nabla \theta_e| = 常数$$

这里的 ζ_θ 为 ζ_a 在 θ_e 梯度方向上的投影。当 $\nabla \theta_e$ 不变时,湿位涡的守恒于是表现为 $\alpha \zeta_a$ 的矢量端迹必须位

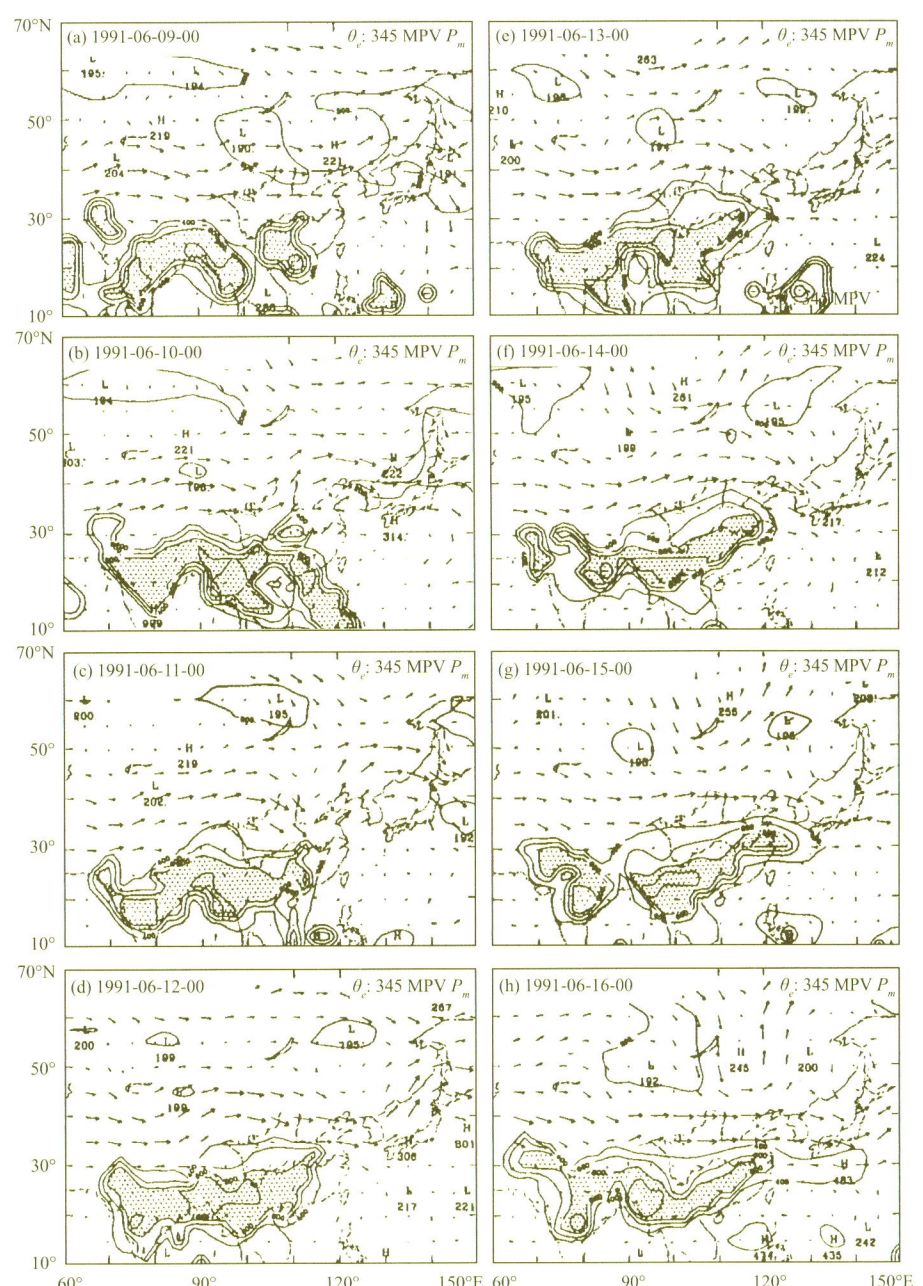

图 5 1991 年 6 月 9—16 日逐日 00 时湿等熵面 $\theta_e=345$ K 之上页面气压高度分布的连续演变
(等值线间隔为 200 hPa。点区表示该面低于 800 hPa 的区域)

于同一 θ_e 等值线上(参见图 6)。如采用 z 坐标,则有

$$P_m = \alpha\zeta_z \frac{\partial \theta_e}{\partial z} + \alpha\zeta_s \frac{\partial \theta_e}{\partial s}$$

式中,ζ_z 和 ζ_s 分别为 ζ_a 的垂直和水平分量,$|\zeta_s| = \left|\dfrac{\partial v_s}{\partial z}\right|$。这时湿位涡守恒便表述为

$$P_m = \alpha\zeta_z \frac{\partial \theta_e}{\partial z} + \alpha\zeta_s \frac{\partial \theta_e}{\partial s} = \alpha\zeta_\theta |\nabla\theta_e| = 常数 \tag{23}$$

上式表明,单位质量垂直涡度的变化与对流稳定度、风的垂直切变及湿斜压度($\dfrac{\partial \theta_e}{\partial s}$)有关。下面将按对流

稳定($\frac{\partial \theta_e}{\partial z} > 0$)和对流不稳定($\frac{\partial \theta_e}{\partial z} < 0$)两种情形来分析风的垂直切变和湿斜压度的变化对$\zeta_a$变化的影响。为简单起见，我们假定$\frac{\partial}{\partial x} \equiv 0$。

(1) 设大气是对流稳定的，$\frac{\partial \theta_e}{\partial z} > 0$；且 $\nabla \theta_e = $ 常数。设其单位质量涡度$\alpha \zeta_a$在$\nabla \theta_e$方向上的投影(图6a)为$\overline{AB} = \alpha \zeta_\theta = \alpha \zeta_a \cdot \frac{\nabla \theta_e}{|\nabla \theta_e|}$，则位涡守恒表现为$\overline{AB}$不变。亦就是说，在尔后运动中，质块的绝对涡度$\alpha \zeta_a$的矢端必位于过$B$点的等$\theta_e$线$\theta_e(B)$上。另一方面，在图6a中，由式(23)给出的$P_m$在$z$坐标中的分解成为

$$P_m = \alpha \zeta_z \frac{\partial \theta_e}{\partial z} + \alpha \zeta_y \frac{\partial \theta_e}{\partial y}, \quad \zeta_y = \frac{\partial u}{\partial z} \tag{24}$$

设由于某种原因，水平风的垂直切变从$\alpha \zeta_y$增大至$\alpha \zeta_y'$，由于新的合成矢量$\alpha \zeta_a'$之端点必须位于$\theta_e(B)$上，因此对应的垂直涡度也从$\alpha \zeta_z$增加为$\alpha \zeta_z'$。

(2) 仍设大气为对流稳定，且$\frac{\partial \theta_e}{\partial z} = $ 常数> 0。另设由于某种原因，其湿斜压度从$\left|\frac{\partial \theta_e}{\partial y}\right|$增至$\left|\frac{\partial \theta_e'}{\partial y}\right|$，如图6b所示。这意味着$\nabla \theta_e$及湿等熵面均逆时钟转过$\beta$角。如变化后质块的涡度$\alpha \zeta_a'$在$\nabla \theta_e'$上的投影为$\overline{AB'}$，则位涡守恒表现为(a) $\overline{AB'} = \overline{AB} |\nabla \theta_e| / |\nabla \theta_e'|$，以及(b) $\alpha \zeta_a'$的矢端必位于等熵线$\theta_e'(B')$上。既然$\alpha \zeta_a'$的水平投影$\alpha \zeta_y$不变，由图6b知垂直涡度必从$\alpha \zeta_z$增至$\alpha \zeta_z'$。

上述的分析表明，在湿位涡守恒的约束下，即使湿对称稳定的大气($P_m > 0$)具有对流稳定性($\frac{\partial \theta_e}{\partial z} > 0$)，由于湿等熵面的倾斜，气块的垂直涡度仍可由于水平风垂直切变的增加或湿斜压性的增加而增加。

(3) 设大气为对流不稳定，$\frac{\partial \theta_e}{\partial z} < 0$；并且$\nabla \theta_e = $ 常数。如图6c所示。这时，$P_m = \alpha \zeta_a' \cdot \nabla \theta_e = -\overline{AB} |\nabla \theta_e| < 0$，大气为湿对称不稳定。又设质块$A$位于低空急流轴上空，因此有$\alpha \zeta_y = \alpha \frac{\partial u}{\partial z} < 0$。假定由于某种原因，风的垂直切变从$|\alpha \zeta_y|$增长至$|\alpha \zeta_y'|$，与(a)的情况类似，垂直涡度也必须由$\alpha \zeta_z$增长至$\alpha \zeta_z'$。

(4) 设大气为对流不稳定，$\frac{\partial \theta_e}{\partial z} = $ 常数< 0，这时大气为湿对称不稳定(图6d)。仍设质块A位于低空急流轴上空，$\alpha \zeta_y = \alpha \frac{\partial u}{\partial z} < 0$。假定由于某种原因，湿斜压度从$\left|\frac{\partial \theta_e}{\partial y}\right|$增加至$\left|\frac{\partial \theta_e'}{\partial y}\right|$，则$\nabla \theta_e$顺时针方向转过$\beta$角至$\nabla \theta_e'$。$\theta_e$也同向转过$\beta$角至$\theta_e'$。与(2)的情况类似，$\theta_e$坐标中湿位涡的守恒要求$\overline{AB'} = \overline{AB} |\nabla \theta_e| / |\nabla \theta_e'|$；$z$坐标中则表现为新的$\alpha \zeta_a'$必在等熵线$\theta_e'(B')$上使其水平投影恰为$\alpha \zeta_y$的点。根据图6d，这时垂直涡度便从$\alpha \zeta_z$增至$\alpha \zeta_z'$。

上面的讨论表明，在湿位涡守恒的制约下，无论大气是湿对称稳定，还是湿对称不稳定；是对流稳定，还是对流不稳定，由于θ_e面的倾斜，大气水平风垂直切变或湿斜压性的增加能够导致垂直涡度的显著发展。湿等熵面θ_e的倾斜越大，这种发展越激烈。在图1所示的低纬对流不稳定区，当介于θ_e和$\theta_e + \Delta \theta_e$之间的气柱从Ⅳ区移向Ⅲ区时，等熵面的演变与图6d相似，且偏转角β可以很大。当有低空气流配合时，气旋性涡度发展可以很急剧。因此，图1中的Ⅲ区可以是暴雨的常发地区。当中纬度大气从Ⅰ区移向Ⅱ区时，系统气旋式涡度的增长则可用图5b予以解释。

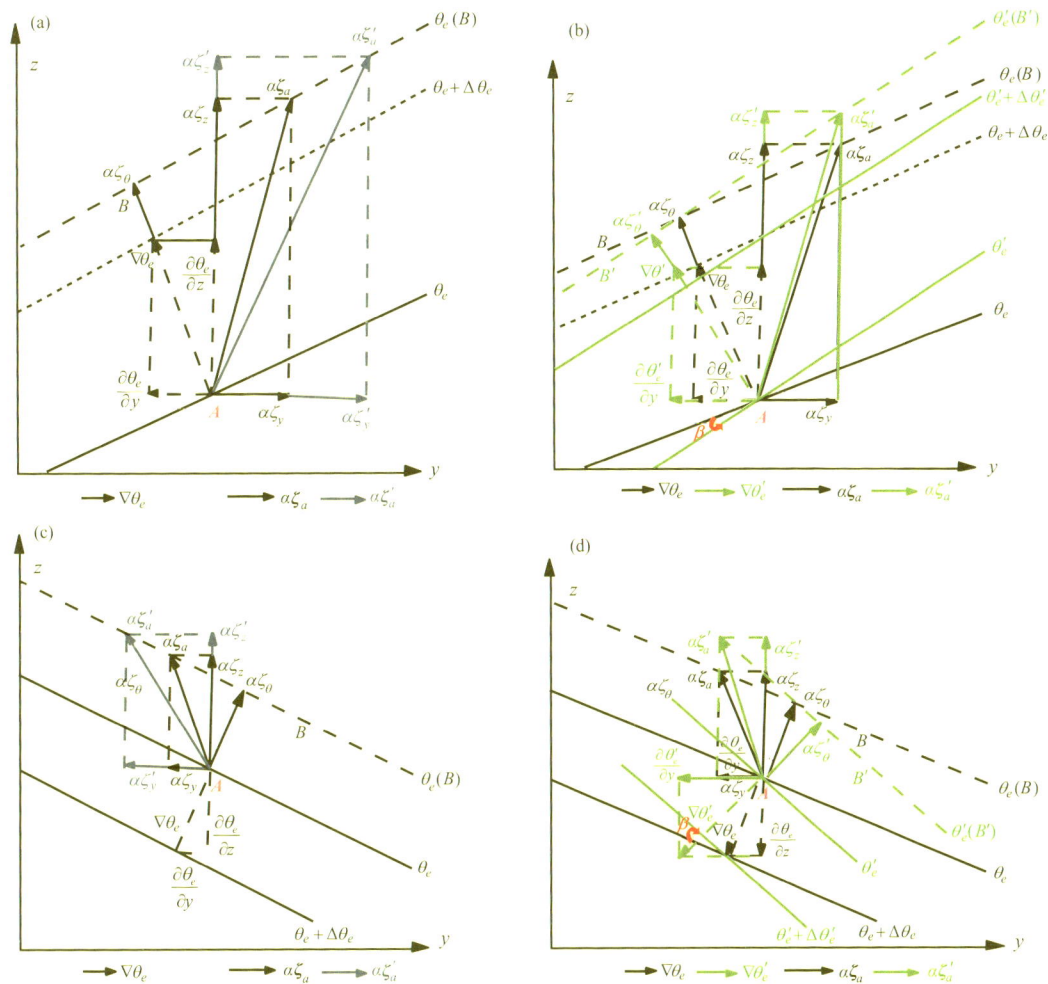

图 6 在湿位涡守恒约束下,对流稳定(图 a 和 b)及对流不稳定(图(c)和(d))大气中,由于水平风垂直切变的增长(图(a)和(c))或湿斜压性 $\left|\dfrac{\partial \theta_e}{\partial y}\right|$ 的增长(图(b)和(d)),导致垂直涡度增长的示意图

(细实线为等 θ_e 线,细虚线为变化后的 θ_e' 线,粗虚线为单位质量绝对涡度的矢量端迹(详见正文)

(a) $\nabla\theta_e =$ 常数,$\dfrac{\partial \theta_e}{\partial z} > 0$;(b) $\dfrac{\partial \theta_e}{\partial z} =$ 常数 > 0;(c) $\nabla\theta_e =$ 常数,$\dfrac{\partial \theta_e}{\partial z} < 0$;(d) $\dfrac{\partial \theta_e}{\partial z} =$ 常数 < 0

6 等压面上的湿位涡和倾斜涡度发展理论

在式(13)和(14)中引进静力近似,取 p 为垂直坐标,并假定垂直速度的水平变化比水平速度的垂直切变小得多,此可得湿位涡的表达式

$$P_m = -g(f\mathbf{k} + \nabla_p \times \mathbf{v}) \cdot \nabla_p \theta_e = \text{常数} \tag{25}$$

如定义湿位涡的第一分量为垂直分量,第二分量为等压面上的水平分量,即

$$\begin{cases} \text{MPV1} = -g\zeta_p \dfrac{\partial \theta_e}{\partial p} \\ \text{MPV2} = -g\mathbf{k} \times \dfrac{\partial \mathbf{v}}{\partial p} \cdot \nabla_p \theta_e \end{cases} \tag{26}$$

式中,$\zeta_p = f + \left(\dfrac{\partial v}{\partial x} - \dfrac{\partial u}{\partial y}\right)_p$,则等压坐标中湿位涡守恒可表达为

$$P_m = \text{MPV1} + \text{MPV2} = \alpha\zeta_\theta |\nabla\theta_e| = \text{常数} \tag{27}$$

由于天气尺度的运动基本满足静力近似,式(27)与 z 坐标中的式(23)基本是等价的。因此上面关于倾斜涡度发展的理论(参见图 6)在 p 坐标中也基本成立。

为进一步理解式(26)及(27)的物理含义,暂设与梅雨锋相联系的风场满足地转关系,即

$$\boldsymbol{v} = f^{-1} \boldsymbol{k} \times \nabla_p \varphi$$

由此得

$$\frac{\partial \boldsymbol{v}}{\partial p} = -f^{-1} R^* \boldsymbol{k} \times \nabla_p \theta \tag{28}$$

式中,

$$R^* = c_p^{-1} p^{-1} R\pi \tag{29}$$

式中,埃克斯纳(Exner)函数

$$\pi = c_p (p p_0^{-1})^\kappa \tag{30}$$

把式(28)代入式(25),得

$$P_m = -g \left(\zeta_p \frac{\partial \theta_e}{\partial p} + f^{-1} R^* \nabla_p \theta \cdot \nabla_p \theta_e \right) = 常数 \tag{31}$$

用 $fg/(\alpha \theta_e)$ 乘上式两边,并定义

$$\begin{cases} 惯性稳定度 & F^2 = f\zeta_p \\ 对流稳定度 & N_m^2 = \dfrac{g}{\theta_0} \dfrac{\partial \theta_e}{\partial z} \\ 干斜压度 & S^2 = \dfrac{g}{\theta_0} |\nabla_p \theta| \\ 湿斜压度 & S_m^2 = \dfrac{g}{\theta_0} |\nabla_p \theta_e| \end{cases} \tag{32}$$

则可得静力近似和地转近似下湿位涡守恒的另一种表述

$$P_m = \frac{\alpha \theta_0}{fg} (F^2 N_m^2 - S^2 S_m^2 \cos\alpha) = 常数 \quad \alpha = (\nabla_p \theta, \nabla_p \theta_e) \tag{33}$$

从上式及式(31)可导得地转平衡下绝对涡度个别变化的方程

$$\frac{\mathrm{D}}{\mathrm{D}t} \zeta_p = -\zeta_p \frac{\mathrm{D}}{\mathrm{D}t} \ln\left(\frac{\partial \theta_e}{\partial p}\right) - \left(g \frac{\partial \theta_e}{\partial p}\right)^{-1} \frac{\mathrm{D}}{\mathrm{D}t} \left(S^2 S_m^2 \cos\alpha \frac{\alpha \theta_0}{fg} \right) \tag{34}$$

一般地 $|\alpha| < \dfrac{\pi}{2}$,尤其当大气为饱和时,$\alpha = 0$。于是从式(33)和(34),我们可得如下结论:

(1)在地转近似约束下,湿位涡 P_m 的水平分量必为负值,即式(27)中的 MPV2<0;

(2)若湿对称稳定的大气起始时刻为惯性稳定和对流稳定($F^2 > 0, N_m^2 > 0$),则在尔后的运动中也应是惯性稳定和对流稳定,换言之,在地转近似约束下,起始稳定的系统不可能通过平流过程变为不稳定;

(3)在干、湿斜压度不改变的情况下,对流稳定度绝对值的减少将导致绝对涡度的增大;

(4)在湿对称稳定的大气中,当对流稳定度不变时,系统干斜压度或湿斜压度的增加将导致绝对涡度增加(参见图 6a 和 6b);

(5)在湿对称不稳定的大气中,当对流稳定度不变时,系统干或湿斜压度的增加将导致绝对涡度减少。

由于在地转场合,图 6c 和 6d 中的 $\alpha \zeta_y$ 和 $\dfrac{\partial \theta_e}{\partial y}$ 具有不同符号,于是本结论就变得十分明白。

实际大气中大暴雨过程常发生在对流不稳定场合,常有干、湿斜压度和垂直涡度同时增长。在低空常有 MPV2>0。这些均与上述地转结论相悖。因此,在对流不稳定场和用地转近似作梅雨锋面持续暴雨分析是不甚妥当的。

7　等压面上的湿位涡分析

本节将通过分析 1991 年 6 月 11—15 日暴雨过程中等压面上的 P_m 的演变去检验湿位涡在梅雨锋暴

雨分析中的指示性。图 7 为 200 hPa 上 MPV1 的演变。图中 2.0 和 3.0 PVU 等值线之间的区域用阴影示出，它近似地表示了对流层顶附近扰动的演变情况。从该图可以看出，9 日位于新疆上空的高值 MPV1 扰动在东移中不断加强并向南发展，11 日扫过黄河和长江中游地区，江淮流域开始出现暴雨。以后在 14 日有新的高值 MPV1 系统在西伯利亚发展移向东南方，至 16 日已扫过华北上空，江淮暴雨也告结束。把图 7 与图 4 比较，发现两者在中高纬度具有十分相似的特征，对流层顶附近天气系统的演变也十分一致。这是因为在中高纬地区 θ_e 为 345 K 的湿等熵面大致位于 200 hPa 附近（参见图 3）。还因为在 200 hPa 面上 MPV2≪MPV1，因此 P_m 与 MPV1 相当。在中低纬地区，由于 $\theta_e=345$ K 面的水平倾斜较大，用水平风场 (u,v) 内插得到的 ζ_θ 已不与 θ_e 面垂直。这时图 4 所给出的湿位涡分析已存在较大的误差。

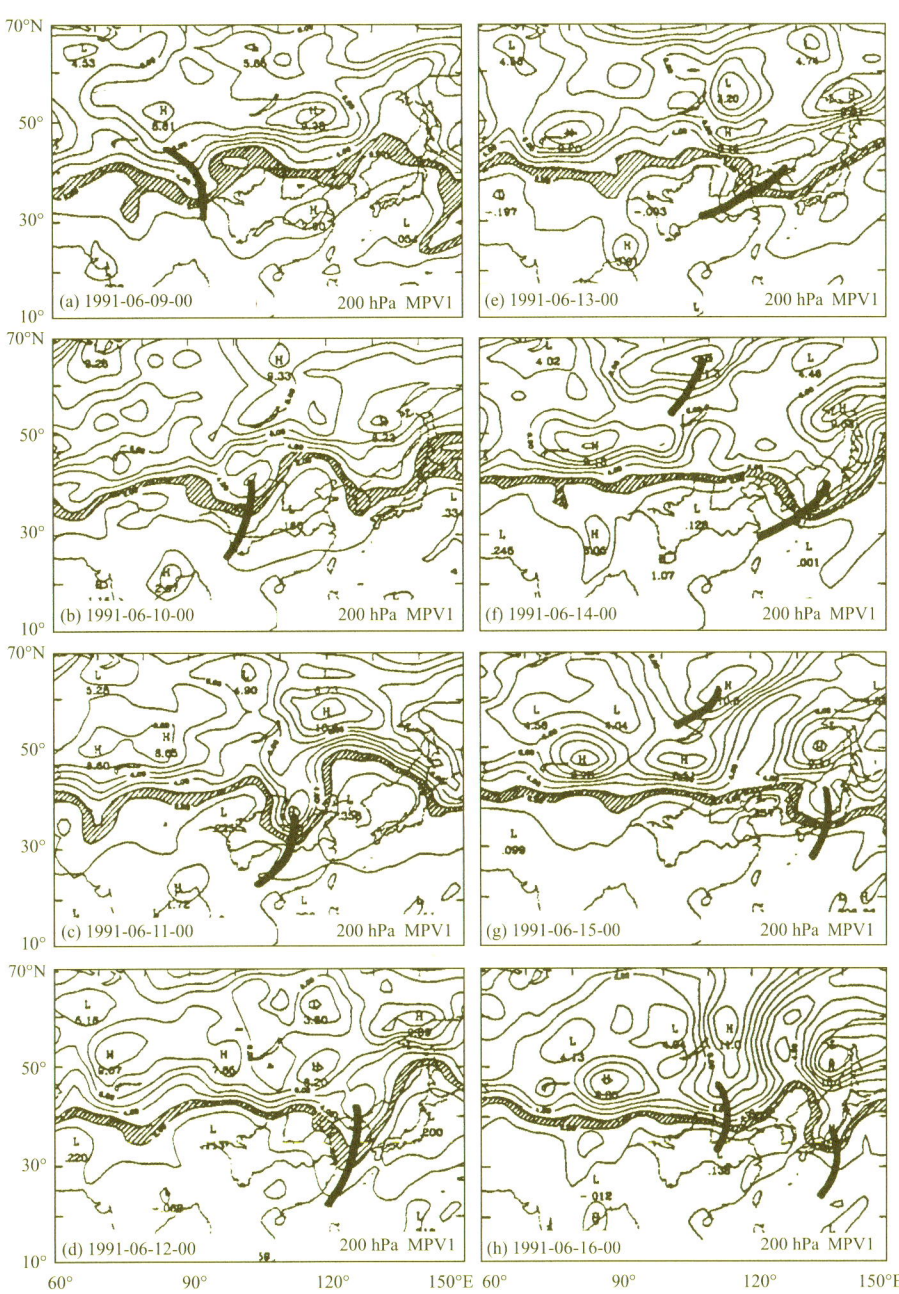

图 7　1991 年 6 月 9—16 日逐日 00 时 200 hPa 面上湿位涡垂直分量 MPV1 的连续演变（粗线示出引起江淮持续暴雨的开始和结束的中高纬高值 MPV 系统的位置。等值线间隔为 1.0 PVU。图中 2.0 和 3.0 PVU 之间的区域加斜线表示）

图 8 为同期 700 hPa 上 MPV1 的时空演变。由图可见低空的 MPV1 分布与高空的 MPV1 有良好的对应关系。例如,9 日位于中国西北的高值 MPV1 系统有规则地向东南方向移动。11 日起,华北对流稳定的高值 MPV1 空气与华南对流不稳定的空气在江淮中下游地区对峙。持续暴雨即出现在高值 MPV1 前方的对流不稳定区中。16 日随着对流稳定区向江南推进,江淮地区的大暴雨结束。

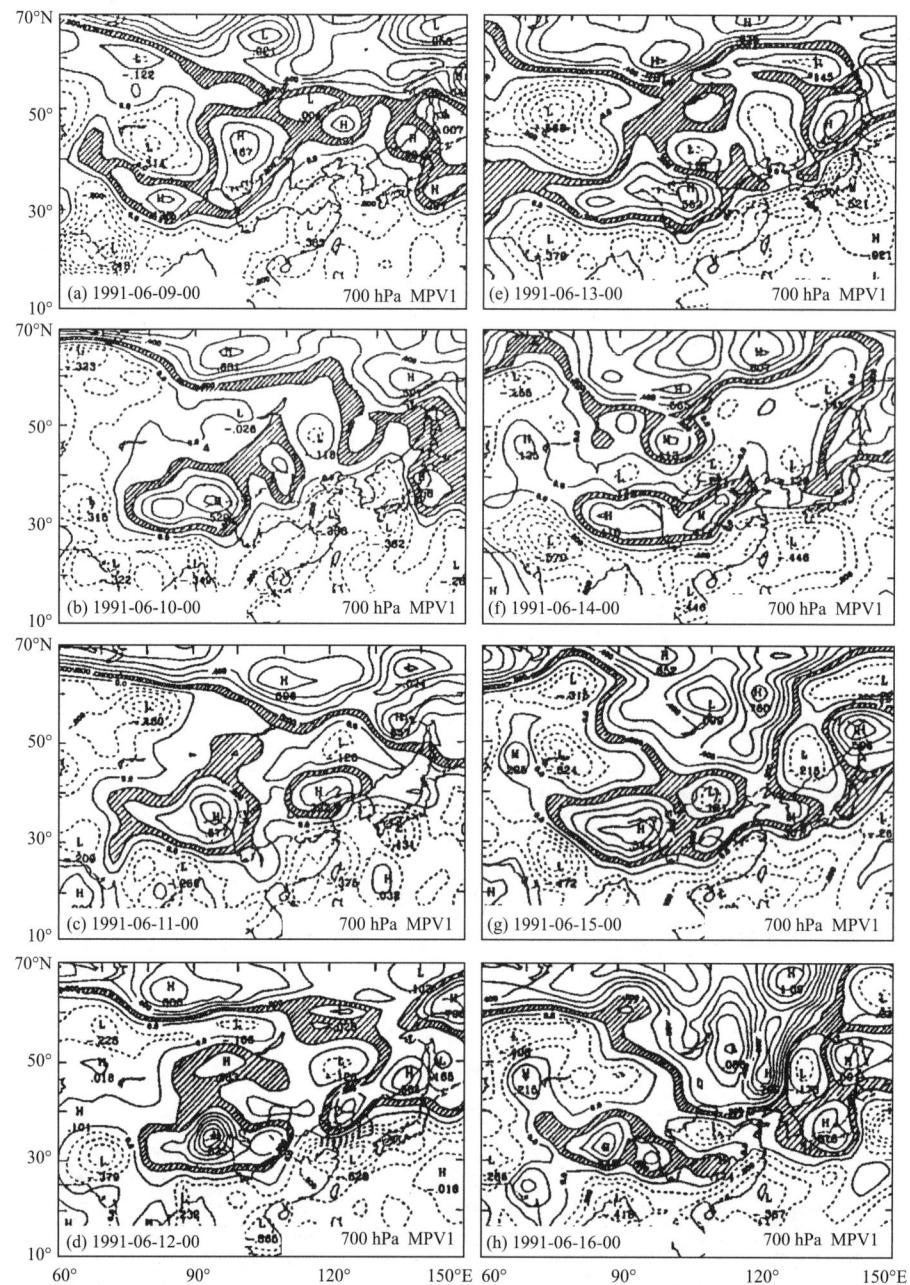

图 8　1991 年 6 月 9—16 日逐日 00 时 700 hPa 面上湿位涡垂直分量 MPV1 的连续演变
(等值线间隔为 0.1 PVU。负值区用虚线表示。图中 0.1 和 0.2 PVU 之间的区域加斜线表示)

图 9 为同期 700 hPa 上湿位涡水平分量 MPV2 的时空演变。在此期间沿长江一带一直有一狭长的正值 MPV2 区。这是因为在此期间,700 hPa 以下存在低空急流(董立清,1991),700 hPa 附近的暖高湿区又位于江南,因此长江一带有 $\frac{\partial u}{\partial z}<0$,$\frac{\partial \theta_e}{\partial y}<0$,使 MPV2 出现正值。从式(33)知,在地转关系成立时,

该项应为负值。江淮地区在暴雨期间 700 hPa MPV2 为正表明,该区域风压场之间不存在地转关系。根据图 6c 和 6d,低空急流的加强(即 $\left|\dfrac{\partial u}{\partial z}\right|$ 增大)或暖湿气流的加强(即 $\left|\dfrac{\partial \theta_e}{\partial y}\right|$ 增大)均可导致 MPV2 增加及 ζ_z 发展,有利于降水加剧。因此,低对流层大的正值 MPV2 的移动可作为低空急流和暖湿气流活动或涡旋活动的示踪。在图 9 中,9 日位于青藏高原东南部的高 MPV2 区沿长江流域东伸。11 日从该源区中分裂出的中心移至江淮流域的对流不稳定区(MPV1<0),大暴雨开始出现。此后该高值 MPV2 区在该地维持,江淮大暴雨持续。在此期间,位于高原东南侧的高值 MPV2 区原地减弱消失,意味着暖湿气流不断衰亡。16 日随着 700 hPa 对流不稳定区退出江淮,及正值 MPV2 区移进东海,江淮流域的暴雨过程结束。

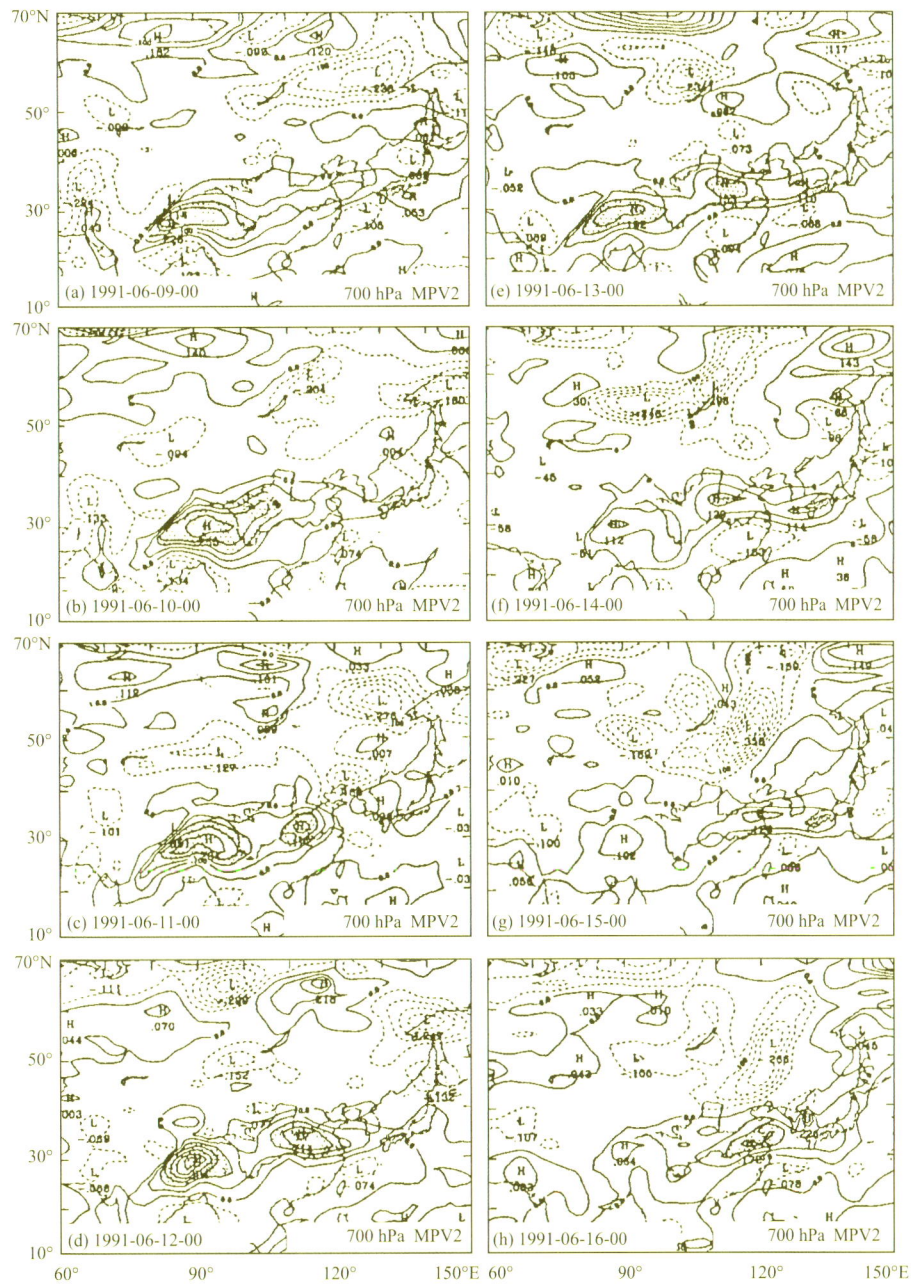

图 9 1991 年 6 月 9—16 日逐日 00 时 700 hPa 面上湿位涡水平分量 MPV2 的连续演变
(等值线间隔为 0.05 PVU。负值区用虚线表示。加点区表示 MPV2>0.1 PVU 的区域)

8 讨论和结论

本文从原始方程系统出发，导得精确形式的湿位涡变化方程，并证明绝热无耗散的饱和大气之湿位涡是守恒的。

在导出 θ_e 坐标、z 坐标及 p 坐标下湿位涡的表达式后，根据湿位涡的守恒特征，讨论了倾斜涡度发展理论。在 θ_e 坐标中，对流稳定度的减少，等熵面上的辐合及潜热释放均可导致法向涡度的增加。在梅雨锋南侧的暖湿区中，上述诸因子及地表的感热加热均有利于系统气旋性涡度的增加，是暴雨最易发展的区域。但当 θ_e 面不水平时，ζ_θ 包含有水平分量，使湿等熵位涡分析的应用受到限制。在等高坐标或等压坐标中，除了上述因子外，水平风垂直切变或湿斜压度 $\nabla\theta_e$ 的增长都可引起气旋性涡度的增加。这种特征主要是由于 θ_e 面的水平倾斜造成的。倾斜越大，气旋性涡度增长越激烈。在梅雨锋附近及其南侧的暖湿区（图 1 的 Ⅲ 区以及 Ⅱ 区），θ_e 面十分陡立。当偏南气流移向该区时，便有气旋性涡度激烈发展，导致对流性暴雨发生。

本文所讨论的等高面或等压面上的倾斜涡度发展理论与等熵面上的发展理论相比较有显著优点。在锋面附近 θ_e 面存在水平倾斜，θ_e 坐标中的法向涡度一般并不垂直于地表。因此 ζ_θ 的改变并不等价于垂直涡度的改变。当选用 z 坐标或 p 坐标时，P_m 的水平分量和垂直分量的总和守恒。这时 $\zeta_z(\zeta_p)$ 的改变由对流稳定度，风的垂直切变及湿斜压度的变化共同决定。因此风垂直切变的增强或湿斜压度的增强能引起垂直气旋性涡度发展。当取地转近似时，湿位涡的守恒表述为对单位质量气柱而言，其惯性稳定度和对流稳定度的乘积与干、湿斜压度乘积之差为常数。本研究表明，地转近似并不适用于梅雨锋中对流不稳定性的持续暴雨的分析。

最后值得指出的是，本文所讨论的倾斜涡度发展理论与 p 坐标中水平涡度向垂直涡度转化的抬举发展理论是不一样的。后者伴随着涡度的增加应有水平涡度的减少。这里所讨论的理论基于湿位涡守恒特征，它需要湿等熵面的倾斜。在本理论中，垂直涡度的发展与水平涡度或湿斜压度的发展同时发生，且强度比抬举发展要强得多。风垂直切变、湿斜压度及垂直涡度同时发展的现象在梅雨锋暴雨及其他锋生过程中是经常发生的。因此倾斜涡度发展过程在大气中具有更普遍的特征。在夏季，暖湿气流十分活跃，其前锋地区 θ_e 面十分陡立，湿斜压涡度发展变得十分激烈，大暴雨容易发生。因此，倾斜涡度发展应是暴雨形成的一种重要机制。

由于历史的原因，一些有关持续暴雨的分析和预报等气候动力学研究非常强调 500 hPa 的流型演变。毫无疑问，500 hPa 的环流能为暴雨预报提供重要的背景。但在本文的分析中，中低纬度 500～600 hPa 恰好处在湿空气中性对流稳定 $\dfrac{\partial \theta_e}{\partial z}=0$ 的层次（参见图 3）。从热力学角度来说，中对流层似乎不是进行暴雨分析的理想层次。相反，在暴雨发展过程中，低空倾斜涡度发展过程十分活跃。当湿等熵面较陡立时，与低空急流和暖湿气流相联系的湿位涡水平分量急速发展，并触发垂直涡度急剧增长，导致暴雨发生。因此，低空湿位涡分析和倾斜涡度发展应当是持续暴雨动力天气和动力气候研究的一个重要内容。

致谢：国家气象中心董立清先生为本次暴雨分析提供了基本素材。

参考文献

董立清, 1991. 苏皖连降大到暴雨, 全国大部气温正常[J]. 气象, 17(9)：58-61.

陶诗言, 1980. 中国之暴雨[M]. 北京：气象出版社：225.

BENNETTS D A, HOSKINS B J, 1979. Conditional symmetric instability-a possible explanation for frontal rainbands[J]. Quart J Roy Meteor Soc, 105：945-962.

ERTEL H, 1942. Ein neuer hydrodynamischer wirbelsatz[J]. Meteorolog Zeitschr Braunschweig, 6：277-281.

HOSKINS B J, MCINTYRE M E, ROBERTSON A W, 1985. On the use and significance of isentropic potential vorticity maps[J]. Quart J Roy Meteor Soc, 111：877-946.

代表性论文 25

影响大气涡度发展的若干热力过程
——Ⅱ. 全型涡度方程

吴国雄[1],刘还珠[2],刘屹岷[1],刘平[1]

(1 中国科学院大气物理研究所,北京 100029;2 国家气象中心,100081)

在第Ⅰ部分的研究中,我们已经阐述了在位涡普适原理的支配下,大气运动如何向外部加热适应的问题;揭示了位涡方程在理论研究中的优越性。但在应用研究及日常业务中,使用垂直相对涡度方程(以下简称涡度方程)更为方便。这是因为垂直涡度与天气系统有很好的对应关系,其急剧发展常伴有剧烈的灾害天气。就副热带高压而言,用垂直涡度可直接表征其变异。本章的目的在于从位涡方程出发,导出更为精确的全型垂直涡度倾向方程。在此基础上以研究大气内部热力结构所诱发的涡度变化,称之为涡度变化的内部热力强迫。进而讨论与外部加热相关的"涡度变化的外部热力强迫"对副热带高压变异的影响。

涡度是流体动力学的一个基本概念(见 Holton,1992),它描述一个质块在空间中如何旋转,并与各种流体现象紧密联系。从行星尺度波动到天气尺度的气旋,其垂直涡度分量 ζ 尤为重要。气旋、低涡、台风、冷暖锋面具有大的正 ζ,而阻塞和副高则以大范围的负涡度为特征,它们的生成、发展和减弱均可用垂直涡度的变化加以表述(叶笃正 等,1988)。

垂直涡度方程(下面简称涡度方程,或 VE)描述流体质块在运动中的涡度变化,因此是大气科学和流体力学中的一个基本方程,被广泛地用于天气和气候的分析、预报和理论研究。

经典的涡度方程在过去的世纪中得到广泛应用,在推动长波理论、能量频散和数值天气预报的发展中起着重要作用。由于它是直接从动量方程导出的,具有明显的动力特征。天气实践表明(陶诗言,1980),激烈的天气过程和气候异常往往与大气的稳定度和斜压性的变异紧密相关,但这些热力因子均不显式出现在该方程中,因此经典涡度方程的应用存在局限性。

基于 Ertel(1942)位涡理论,吴国雄等(吴国雄 等,1999a;Wu et al.,1998)导得全型涡度方程,该方程不仅克服了经典涡度方程的上述局限性,还包含了外界动力和热力强迫对涡度变化的影响,因此更具普遍性。本章在他们的研究基础上,首先介绍全型涡度方程的推导,接着对上述两种涡度方程进行比较,证明了全型涡度方程新的物理内涵及其天气气候意义;最后阐明利用全型涡度方程研究副热带高压形态变异的原理。

1 涡度变化的内部热力强迫——倾斜涡度发展(SVD)

1.1 全型垂直涡度方程

副热带高压以负涡度分布为特征,其变化规律可以用涡度方程加以描述。传统的涡度方程

本文原载于《副热带高压形成和变异的动力学问题》,北京:科学出版社,2002:125-142

$$\frac{D}{Dt}(f+\zeta_z) + (f+\zeta_z)\nabla_h \cdot \boldsymbol{V} = N_z + \left(\frac{\partial u}{\partial z}\frac{\partial w}{\partial y} - \frac{\partial v}{\partial z}\frac{\partial w}{\partial x}\right) + \boldsymbol{k} \cdot \boldsymbol{F}_\zeta \tag{1}$$

是通过对动量方程求旋度再点乘垂直单位矢量得到的，具有明显的平面特征和动力特征。大量的天气实践表明，稳定度的变化，斜压性的改变，以及与风的垂直切变相关的高、低空急流的发生、发展常与涡度变化相联系。但这些热力项并不直接出现在该涡度方程中，使应用涡度方程进行涡度分析受到局限。为克服这一缺陷，吴国雄等（吴国雄 等，1999a；Wu et al.，1998）从位涡方程

$$\frac{DP}{Dt} = \alpha \nabla\theta \cdot \boldsymbol{F}_\zeta + \alpha \boldsymbol{\zeta}_a \cdot \nabla Q \tag{2}$$

出发，导得全型垂直涡度倾向方程，并由此得出倾斜涡度发展（SVD）理论以研究涡度变化的内部热力强迫机制。为简化讨论，引进比绝对涡度

$$\boldsymbol{\xi}_a = \alpha \boldsymbol{\zeta}_a$$

根据定义

$$P = \alpha \boldsymbol{\zeta}_a \cdot \nabla\theta \tag{3}$$

采用 z-坐标，把 P 分解为垂直分量（下标 z 表示）和水平分量（下标 s 表示，s 为沿着 θ 的水平梯度方向的单位矢量）

$$P = \boldsymbol{\xi}_a \cdot \nabla\theta = \alpha(f+\zeta_z)\theta_z + \alpha\nabla\times\boldsymbol{V} \cdot \nabla_h\theta = \xi_z\theta_z + \xi_s\theta_s \tag{4}$$

式中，

$$\xi_s\theta_s = \alpha\nabla\times\boldsymbol{V} \cdot \nabla_h\theta = \xi_x\theta_x + \xi_y\theta_y = -\alpha\frac{\partial v}{\partial z}\frac{\partial \theta}{\partial x} + \alpha\frac{\partial u}{\partial z}\frac{\partial \theta}{\partial y} = \alpha\frac{\partial V_m}{\partial z}\frac{\partial \theta}{\partial S}$$

式中，θ 的下标表示求偏微商的自变量，m 为平行于水平面上的 θ 线并指向 s 右侧的水平单位矢量。

由(4)式可得"比垂直绝对涡度"

$$\xi_z = (P - \xi_s\theta_s)/\theta_z = \frac{P}{\theta_z} - C_D, \quad \theta_z \neq 0 \tag{5}$$

这里热力学参数

$$C_D = \alpha\nabla\times\boldsymbol{V} \cdot \nabla_h\theta/\theta_z = \xi_s\theta_s/\theta_z, \quad \theta_z \neq 0 \tag{6}$$

为位涡水平分量与静力稳定度参数 θ_z 之比，它正比于风垂直切变与斜压性之积除以静力稳定性。将式(5)对时间求全微商，代入式(1)，并利用下述关系式

$$\frac{D\zeta_z}{Dt} = \alpha\left[\frac{D\zeta_z}{Dt} + \beta v + (f+\zeta_z)\nabla\cdot\boldsymbol{V}\right] \tag{7}$$

可得

$$\begin{aligned}\frac{D\zeta_z}{Dt} &+ \beta v + (f+\zeta_z)\nabla\cdot\boldsymbol{V} \\ &= -\frac{1}{\alpha\theta_z^2}\left[(P-\xi_s\theta_s)\frac{D\theta_z}{Dt} + \xi_s\theta_z\frac{D\theta_z}{Dt} + \theta_z\theta_s\frac{D\xi_s}{Dt}\right] + \frac{1}{\theta_z}\nabla\theta\cdot\boldsymbol{F}_\zeta + \\ &\quad \frac{1}{\theta_z}\boldsymbol{\zeta}_a\cdot\nabla Q\end{aligned} \tag{8}$$

式(8)即为"全型垂直涡度倾向方程"。它不仅包含动力项（左端），还包含有静力稳定度的变化 $\frac{D\theta_z}{Dt}$，斜压性的改变 $\frac{D\theta_s}{Dt}$，以及风垂直切变的发展 $\frac{D\xi_s}{Dt}$ 等热力因子对垂直涡度发展的贡献；还包含有摩擦耗散 $\boldsymbol{F}_\zeta$ 和非绝热加热 Q 对涡度发展的贡献。当等熵面呈水平分布时，$\theta_s \equiv 0$。则在绝热无摩擦的情况下有

$$\frac{D}{Dt}[\alpha(f+\zeta_z)\theta_z] = \frac{D}{Dt}[\text{IPV}] \equiv 0 \tag{9}$$

此即为 Rossby 位涡守恒方程，或等熵位涡（IPV $= \alpha(f+\zeta_z)\theta_z$）守恒方程（Hoskins et al.，1985）。因此，Rossby 位涡或等熵位涡守恒只是等熵面为水平面分布时的特例。

1.2 位涡守恒的盒子定律和外切平面定律

在无摩擦绝热的大气中,由式(2)知位涡守恒,由定义式(3),可得

$$P = \alpha \zeta_\theta |\nabla \theta| = \text{const} \tag{10}$$

式中,ζ_θ 是绝对涡度 ζ_a 在 $\nabla\theta$ 上的投影。基于此,我们先研究位涡守恒关系中的两个基本特性(吴国雄 等,1999a)。

(1) 盒子定律

当 θ 面为平行平面时,$|\nabla\theta|$ 为常数。设想一个盒子的底面和顶面各为等熵面 θ_1 和 θ_2,其无量纲厚度 $|\Delta\theta| = |\alpha\zeta_\theta|$。$A$ 为底面 θ_1 上的一个质点,也是比绝对涡度矢量 $\alpha\zeta_a$ 的原点。则其运动过程中位温守恒和位涡守恒的特征可用图 1a 表示,并陈述为如下的盒子定律:

无论质点 A 在盒子的一个面 θ_1 上如何运动和旋转,其比绝对涡度矢量 $\alpha\zeta_a$ 的端点必定始终位于盒子的另一个对面 $\theta_2(=\theta_1+\Delta\theta)$ 上。

根据盒子定律,当 θ 面为水平面时,绝对涡度水平分量(即风的垂直切变)的变化是与其垂直分量的变化无关的。这即为 Rossby 位涡和等熵位涡守恒的情况。

(2) 外切平面定律

为简明起见,设 θ 面为同心球面(图 1b),这时 $|\nabla\theta|$ 为常数。假设内球面 θ_1 和外球面 θ_2 之间的无量纲厚度为 $|\Delta\theta| = |\alpha\zeta_\theta|$,$A$ 为内球面 θ_1 上的一个点,也是其比绝对涡度矢量 $\alpha\zeta_a$ 的原点。令矢径 OA 延伸交 θ_2 面于点 B,平面 P 为球面 θ_2 过点 B 的外切平面。则其运动过程中位温守恒和位涡守恒的特征可用图 1b 表示,并陈述为如下的外切平面定律:

无论质点 A 在内球面 θ_1 上如何运动和旋转,其比绝对涡度矢量 $\alpha\zeta_a$ 的端点必定始终位于球面 $\theta_2(=\theta_1+\Delta\theta)$ 过点 B 的外切平面 P 上。

由于在这种场合,$\alpha\zeta_a$ 在 $\nabla\theta$ 上的投影 $\overline{AB} \equiv |\alpha\zeta_\theta|$,根据位涡守恒式(10),上述定律不证自明。外切平面定律和盒子定律不同点在于,当 $\alpha\zeta_a$ 的一个分量改变时,另一个分量也必须有相应的改变。换言之,当 $(\alpha\zeta_a)$ 的水平分量已知时,其垂直分量可根据外切平面 P 唯一确定。这就为下面讨论的倾斜涡度发展(slantwise vorticity development,简写为 SVD)理论提供依据。在建立盒子定律和外切平面定律时,我们假设 $|\nabla\theta|$ 为常数,θ 面平行或同心。这一假定并不影响下面的讨论。在下一节中我们将看到,由于位涡守恒式(10),对垂直涡度的变化而言,$|\nabla\theta|$ 的变化总是被 $\alpha\zeta_\theta$ 的相应变化所补偿。

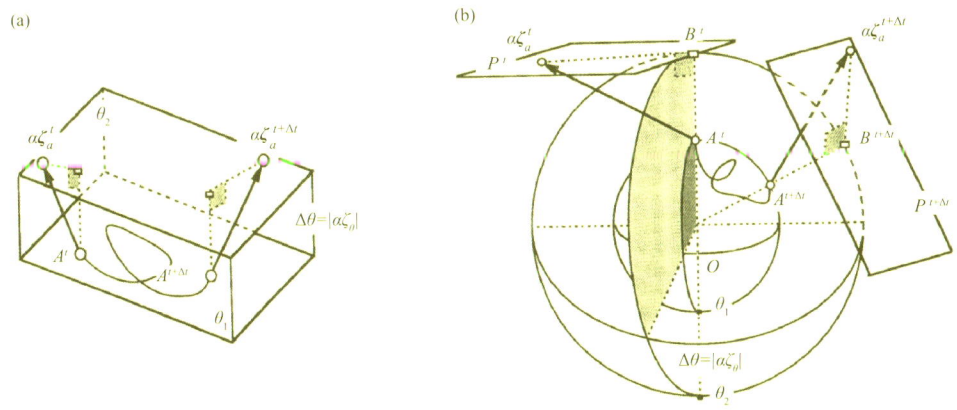

图 1 (a)盒子定律:当 θ 面为平面时,不管质块在盒子的一个面 θ_1 上如何移动和转动,其比绝对涡度矢量($\alpha\zeta_a$)的矢端必定总位于盒子的另一个面 $\theta_2(=\theta_1+\Delta\theta)$ 上。(b)外切平面定律:当 θ 面为曲面时,不管质块在 θ_1 面上如何移动和转动,其比绝对涡度($\alpha\zeta_a$)的矢端必定总位于另一个面 $\theta_2(=\theta_1+\Delta\theta)$ 过点 B 的外切平面 P 上

1.3 倾斜涡度发展(SVD)

采用 θ 坐标时,令 $\boldsymbol{n}$ 表示沿 $\nabla\theta$ 方向的单位矢量,则式(3)可写为

$$P = \xi_n \theta_n$$

式中,$\xi_n = \alpha \xi_\theta, \theta_n = |\nabla \theta|$。上式与(4)式联立可得

$$\xi_z \theta_z + \xi_s \theta_s = \xi_n \theta_n \tag{11}$$

于是

$$\xi_z = (\xi_n \theta_n - \xi_s \theta_s)/\theta_z, \quad \theta_z \neq 0 \tag{12}$$

上式表明,当位涡守恒时,θ_n 的改变并不会引起垂直涡度的变化,因为它的变化总必须由 ξ_n 的变化所补偿($\xi_n \theta_n = \text{const}$)。这为下面应用外切平面定律提供了根据。

为研究等熵面的倾斜如何影响涡度发展,在图 2 中我们假定垂直 z-坐标的右端等熵面为平行平面,在其左端弯曲成圆。为简明起见,设等熵面梯度 $\nabla\theta = \theta_n$ 为常数,同心圆"b"与 $\theta + \Delta\theta$ 面相距 $\overline{AB} = |\xi_n|$。当起始质点 A_0 沿 $\theta + \Delta\theta$ 面左行时,在 z-轴右侧满足盒子定律,不管风垂直切变如何变化,其比绝对涡度的垂直分量 $\xi_z (= \xi_n)$ 不发展。当 A_0 移至 z-轴左侧,沿 $\theta + \Delta\theta$ 面下滑一个角度 β(β 偏离 z-轴顺时针向转向 $\boldsymbol{S}$ 方向为正)至点 A 时,根据外切平面定律,其比绝对涡度 $\boldsymbol{\xi}_a$ 的端点 D 必须位于通过圆 b 上点 B 的外切平面 DB 上。由于

$$\tan\beta = \theta_s/\theta_z, \quad (-\pi/2 < \beta < \pi/2)$$

还由于 $\boldsymbol{n} \cdot \boldsymbol{k} > 0$,于是

$$\cos\beta = \theta_z/\theta_n > 0$$

把上述表达式代入式(12)有

$$\xi_z = \frac{\xi_n}{\cos\beta} - \xi_s \tan\beta, \quad |\beta| \neq \pi/2 \tag{13}$$

当北半球气旋生成时,$\xi_n > 0$。一般而言,ξ_s、θ_s 和 θ_z 各项可正可负。但重要的是,在式(12)中当

$$C_D = \xi_s \theta_s / \theta_z < 0 \tag{14}$$

时,式(13)可写成

$$\xi_z = \frac{\xi_n}{\cos\beta} + |\xi_s \tan\beta|, \quad |\beta \neq \pi/2| \tag{15}$$

式中,ξ_z 是 β 的递增函数。这意味着,当式(12)满足时,沿等熵面下滑的质块其垂直涡度将发展。用上置短线表示线段,则表达式(15)中下滑角 β 与 ξ_n 的关系还可用图 2 的几何图像表示为

$$\xi_z = \overline{CD} = \overline{DE} + \overline{EC} = \overline{AB}/\cos\beta + \overline{AC}|\tan\beta|$$

为满足条件式(14),在制作该图时,我们已假定风垂直切变不变,于是 $\xi_s = \xi_s^0 > 0$。否则,当 ξ_s 为负值,条件式(14)不满足。根据图 2,ξ_z 随 $|\beta|$ 的增加而增加。当等熵面十分陡立时,ξ_z 可变得很大,亦即

$$\xi_z \to \infty \quad |\beta| \to \pi/2 \tag{16}$$

由于这时涡度的发展是由质块沿倾斜等熵面下滑所致,故可谓之下滑"倾斜涡度发展"。另一方面,当等熵面接近水平时

$$\xi_z \to \xi_n, \quad |\beta| \to 0$$

只有在这种情形,Rossby 位涡和 IPV 才守恒。

根据式(5)和(7)还可得到垂直涡度变化和 SVD 相联系的方程如下

$$\frac{\mathrm{D}\zeta_z}{\mathrm{D}t} + \beta v + (f + \zeta_z) \nabla \cdot \boldsymbol{V} = \frac{1}{\alpha} \frac{\mathrm{D}}{\mathrm{D}t}\left(\frac{P}{\theta_z} - C_D\right), \quad \theta_z \neq 0 \tag{17}$$

或

$$\frac{\mathrm{D}\zeta_z}{\mathrm{D}t} + \beta v + (f + \zeta_z) \nabla \cdot \boldsymbol{V}$$

$$= \frac{1}{\alpha}\left[P\frac{\mathrm{D}}{\mathrm{D}t}\left(\frac{1}{\theta_z}\right) - \frac{\mathrm{D}}{\mathrm{D}t}C_D\right] + \frac{1}{\theta_z}\nabla\theta \cdot \boldsymbol{F}_\xi + \frac{1}{\theta_z}\boldsymbol{\zeta}_a \cdot \nabla Q, \quad \theta_z \neq 0 \tag{18}$$

式(17)实际上是"全型垂直涡度方程"式(8)的另一种表述形态。由此得到在绝热无摩擦的情况下,SVD 的充分必要条件为

$$C_D(t+\Delta t) - C_D(t) < P\left[\frac{1}{\theta_z(t+\Delta t)} - \frac{1}{\theta_z(t)}\right] \tag{19}$$

在图 2 所示的例子中,式(19)右端为正。由于起始时刻在 A_0 处 $C_D=0$(因为 $\theta_s=0$),因此在 A 处 C_D 为负,将有涡度发展。换言之,在这种情况下式(14)是 SVD 的充分条件。还由于当 θ-面十分陡立时式(17)右端趋于无穷,因此涡度发展将变得十分剧烈。

中纬度地区,对流层中低层的 θ-面呈倾斜分布。那里应是 SVD 的常发区,对应着温带锋面气旋的发展。夏季青藏高原东侧 θ-面常常陡立倾斜。SVD 的发展可以非常迅猛,导致低涡的剧烈发展。为检验 SVD 对涡度发展的贡献,在下一节中我们将利用一个绝热等 θ 坐标模式,分析沿等熵面下滑的质点在满足充要条件式(14)时对一次西南低涡形成的贡献。

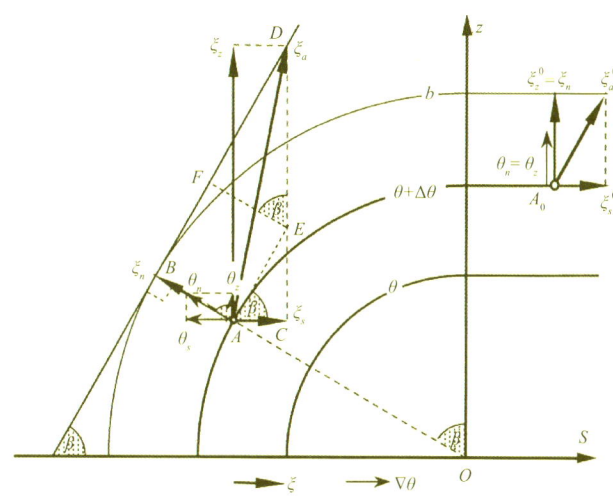

图 2 在干大气中,当质点沿着倾斜等熵面下滑且满足 $C_D \equiv \xi_s \theta_s/\theta_z < 0$ 时垂直涡度发展图示。当质块 A_0 在水平 θ 面上移动时,垂直分量 $\xi_z = \xi_n$ 与水平分量 ξ_s 无关。当其沿倾斜等熵面下滑一个角 $\beta(<0)$ 到点 A 时,$\xi_z = \dfrac{\xi_n}{\cos\beta} - \xi_s \tan\beta$ 可表示为 $\xi_z = \overline{CD} = \overline{DE} + \overline{EC}$,或 $\xi_z = \dfrac{\overline{AB}}{\cos\beta} + \overline{AC}|\tan\beta|$。于是 ξ_s 或 $|\beta|$ 的增加均会导致 ξ_z 发展

1.4 倾斜涡度发展的数值模拟

中、低纬度环流的相互作用可以通过倾斜涡度发展这一涡度内强迫机制去影响副热带高压的变异。为理解倾斜涡度发展这一物理过程,本节通过一次低涡发展过程的数值模拟以说明沿倾斜等熵面移动的质块所诱发的涡度变化。1981 年 7 月 11—15 日,四川省境内发生特大暴雨。雨量集中在 12—14 日,降水中心总雨量超过 366 mm,损失严重,引起大气科学界的关注。许多科学家(Zhou et al.,1984;Hovermale,1984;Anthes et al.,1984;Chen et al.,1984;Kuo et al.,1986)对此次过程进行了分析和模拟试验。共同的结论是暴雨是由一个 α-尺度的西南低涡迅猛发展引起的。在 1981 年 7 月 11 日 12Z 的天气图上,青藏高原东南部出现气旋式环流,它在 7 月 12 日移到四川盆地成为低涡,中心位于(104°E,28°N)。与此同时,高原东北侧有一中纬槽东移发展。12 日在高原东部形成"北槽南涡"的形势,四川地区的大暴雨开始出现。

尽管西南低涡的发展与水汽凝结潜热的释放有关,但 Wu 等(1985)的研究表明,其形成主要是由高原的机械强迫作用所致。这使我们可以用 Bleck(1984)的绝热模式来模拟该低涡的形成。该模式用 θ-面作

为垂直坐标,在垂直方向从 305 K 到 365 K,以 5 K 为间隔共有 13 层。水平格距为 80 km。时间步长为 60 s。其基本方程由动量方程

$$\left(\frac{\partial u}{\partial t}\right)_\theta + \mathbf{V} \cdot \nabla_\theta u - fv = -\left(\frac{\partial M}{\partial x}\right)_\theta + F_x$$

$$\left(\frac{\partial v}{\partial t}\right)_\theta + \mathbf{V} \cdot \nabla_\theta v + fu = -\left(\frac{\partial M}{\partial y}\right)_\theta + F_y$$

连续方程

$$\frac{\partial}{\partial t}\left(\frac{\partial p}{\partial \theta}\right) + \nabla_\theta \cdot \left(\mathbf{V}\frac{\partial p}{\partial \theta}\right) = 0$$

热力方程

$$\frac{D\theta}{Dt} = 0$$

以及静力方程

$$\frac{\partial M}{\partial \theta} = \pi$$

构成。式中 F_x 和 F_y 分别为 x 和 y 方向的摩擦力;$M = \Phi + C_p T$ 为蒙哥马利函数;$\pi = C_p \left(\frac{p}{p_0}\right)^{R/C}$ 为 Exner 函数。每一变量在每一时间步长的边界条件是由初始时刻和终了时刻该量的观测值线性内插得到的。模式的细节,尤其是对低边界的处理可参见 Bleck(1984)。赵宇澄等(1996)对该模式引进中国的地形分布,选择 1981 年 7 月 11 日 00Z 的气象场作为初始场,共积分 60 小时,其前 24 小时的积分即准确地模拟出低涡的形成。这里我们将利用这一积分结果来说明 SVD 是如何导致"北槽南涡"形势的形成。由于四川盆地在该低涡形成前后,$\theta = 315$ K 接近 750 hPa(见图 4 和图 5),刚好通过该低涡的上部。下面将对这个 θ 面上各种变量的分布进行分析。

图 3 是 1981 年 7 月 11 日 00Z $\theta = 315$ K 面上起始场的分布。最为显著的特征是该面在高原上空高高抬起,最高点达 493 hPa 的高度(图 3a);而在高原周围迅速下垂,在四川盆地和戈壁沙漠低达 817 hPa 和 863 hPa。由于模式是绝热的,起始时位于该面上的气块在其后的运动中也必将位于该面上。由于该面在高原周围陡立下倾,该面上由低气压区指向高压而横越高原边缘的风速将沿着该陡立 θ 面下滑。在图 3—图 5 中,这些下滑风被加粗的、带有箭头和风矢的矢量表示。在起始场中,沿高原东侧 $\xi_n = \alpha(f + \xi_n)$ 为正(见图 3b),θ_n 也为正,那里的大气是对称稳定的,即 $P > 0$。根据 θ 的分布,θ 的水平梯度(θ_s)的方向 $\mathbf{S}$ 从高原指向外部。由于高原东侧的偏南风在近地层随高度增加而增加(Zhou and Hu,1984),那里风的垂直切变 $\xi_s\left(= \frac{\partial V_m}{\partial z}\right)$ 与 θ_s 反向,因此热力因子 $C_D = \xi_s \theta_s / \theta_z$ 为负值。这意味着高原东侧的起始场满足 SVD 的充分条件式(11),有利于该处垂直涡度发展。图 3b 为该面上起始场相对涡度的分布。由于相邻 θ 面上 ζ_n 的分布与图 3b 相似(图略),把 ξ_ε 内插到其间的等压面将不影响涡度分布的总体特征。为简单起见,我们将用该 θ 面上的涡度分布去透视相邻等压面上或等高面上的涡度分布。

图 3b 上出现两个正涡度中心。一个在高原东北侧,强度为 3.15×10^{-5} s^{-1};一个在高原东南侧,强度为 4.33×10^{-5} s^{-1}。前者与东移的中纬西风槽相联系;后者与即将生成的气旋性涡旋有关。两者均出现在风沿陡立 θ-面下滑的区域(图 3a)。在该图上,华东、渤海、西太平洋、中印半岛和孟加拉湾也有正涡度中心,它们与特定的天气系统相联系。由于这些中心出现在 θ-面相对平坦的区域,与本节主题无关,这里不分析其演变,只集中考察高原东侧上述两涡度中心的发展。

12 小时后(图 4),$\theta = 315$ K 面上的气压分布(图 4a)变化不大。但四川盆地上空的"暖漏斗"从 817 hPa 下沉至 842 hPa。与起始场比较发现,高原东北侧和东南侧的下滑运动显著加强。下滑区扩展,风速更增,风矢更垂直于等压线。高原东南侧的下滑区一分为二:西面一支在 98°E 附近;东面一支下滑前锋已达四川盆地。与迅速加强的下滑运动相对应,垂直涡度在此期间急速发展(图 4b)。高原东北侧的正涡度区扩大东移,强度达 6.46×10^{-5} s^{-1},比起始场强一倍以上。高原东南侧涡度发展更为迅猛。它分为两

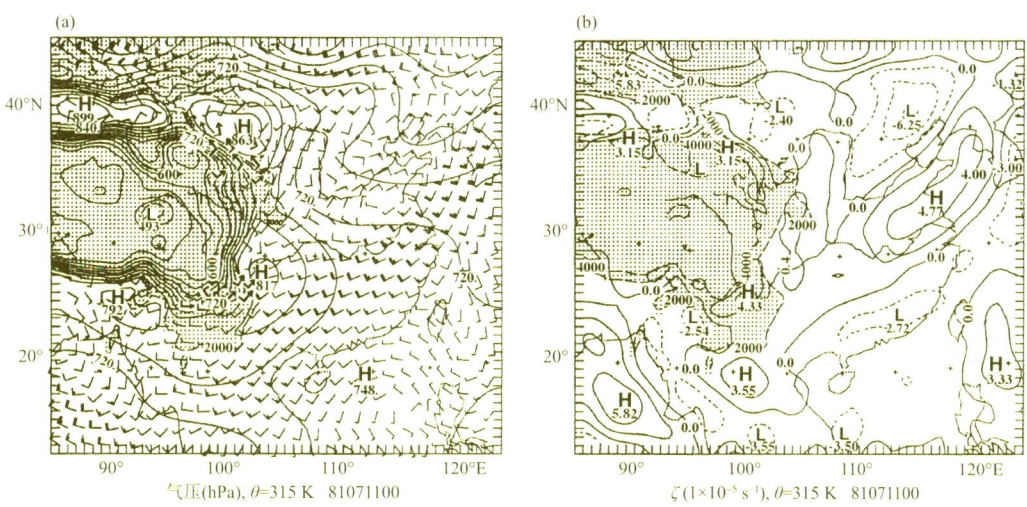

图3　1981年7月11日00Z $\theta=315$ K面上等压线(间隔30 hPa)和风速(a),以及垂直涡度(b,间隔2.00×10^{-5} s^{-1})的分布。(a)中风矢的每一横线代表4 m·s^{-1},带有箭头的加粗风矢表明那里风沿着陡斜的θ面下滑

个中心:西侧中心强度达8.42×10^{-5} s^{-1},近于加倍;东侧的中心也达4×10^{-5} s^{-1}以上,并开始进入四川盆地。在此时刻,西南低涡实际上已生成,只不过只有一部分中心出现在四川盆地而已。

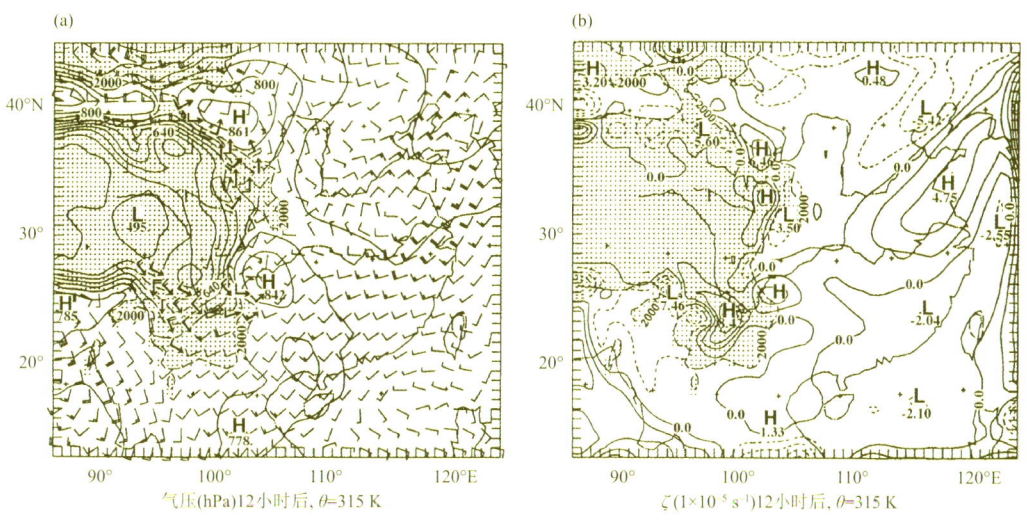

图4　同图3,但为1981年7月11日12Z模式模拟得到的分布。等值线间隔在(a)中为40 hPa,在(b)中为2.00×10^{-5} s^{-1}

到7月12日00Z,$\theta=315$ K面上环绕高原的气压场分布仍大致相似(图5a)。四川上空的暖漏斗缓慢东移,继续向下伸展达857 hPa。高原东侧的下滑区也加强东移。北面增强的下滑气流使中纬槽进一步加深,垂直涡度强达7.19×10^{-5} s^{-1}(图5b)。在南面,西部98°E附近的下滑气流减弱,使100°E附近的涡度中心也减弱。在东部,虽然θ面的倾斜在南面减弱,但那里从横断山脉下滑的西南风大大加强,这使西南低涡的强度变化不大。再者,暖漏斗的东移及下滑气流的加强导致西南低涡也东移,低涡的整个中心已出现在四川盆地。至此,已可观测到完整的西南低涡。新生的低涡中心位于(104°E,28°N),正好与观测到的低涡中心位置重合。上述分析表明,Bleck模式能很好地模拟该西南低涡的形成和中纬西风槽的加深;该西南低涡的形成则可用下滑SVD理论加以解释。质块在高原东北侧和东南侧沿着等θ面绝热下滑而诱发垂直涡度发展的过程与图2所描述的一致,使"北槽"和"南涡"的垂直涡度在12小时内的增加分别达到3.11和4.09×10^{-5} s^{-1}。

以往,人们常用过山气流沿着山脉下沉引起气柱垂直伸展来解释低涡的生成(如Wu et al.,1985;赵

宇澄 等,1996)。对此,我们利用式(8)和连续方程将涡度方程中的垂直延展项准确地写成

$$\frac{D\zeta_z}{Dt} \propto -(f+\zeta_z)\frac{1}{\alpha}\frac{D\alpha}{Dt} = (1-\kappa)(f+\zeta_z)\frac{\omega}{p}$$

取 $p=700$ hPa, $\omega=1\times10^{-3}$ hPa·s^{-1}, $\varphi=30°$N,以及 $\Delta t=0.5$ d,那么下沉气柱由于密度压缩所诱发的涡度增长的值应是 $\Delta\zeta_z=0.5\times10^{-5}$ s^{-1}。这对所研究的西南低涡的发展($\sim 5\times10^{-5}$ s^{-1})来说是太弱了。由此看来,SVD 应当是该西南低涡形成的主要机制。

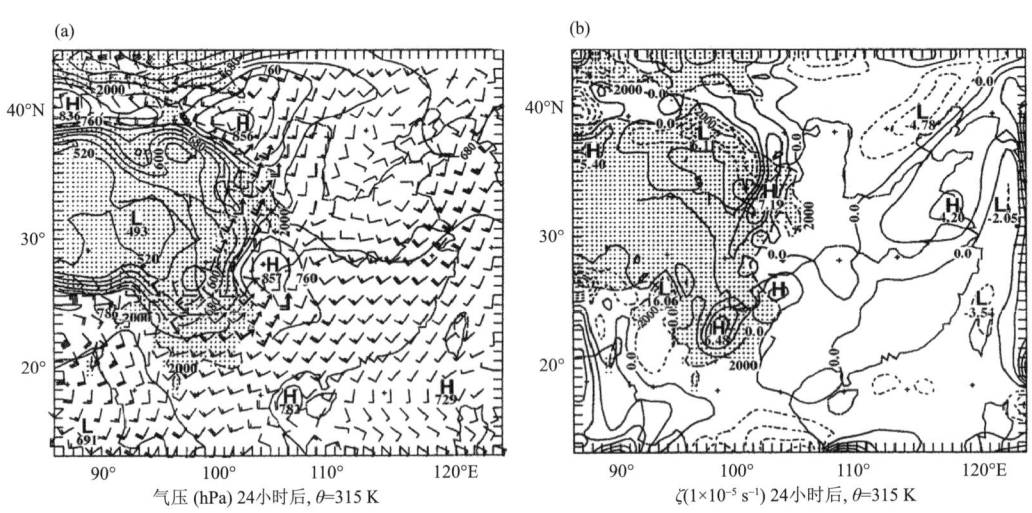

图 5 同图 4,但为 1981 年 7 月 12 日 00Z 模式模拟得到的分布

2 全型涡度方程和传统涡度方程的比较

从三维位涡方程出发导得的全型垂直涡度倾向方程(8)不仅包括动力因子的影响,还包括热力因子的作用。容易看到这一方程与传统的涡度倾向方程(1)明显不同,全型涡度方程中许多与水平量有关的项(如水平摩擦力项、水平涡度变化项和水平温度梯度变化项)、温度层结变化项和非均匀加热项等均不出现在式(1)中。这是由于式(1)一开始就只取三维涡度方程垂直分量的缘故。本节将进一步就上述两种涡度方程的差异作深入分析,并阐明在全型涡度方程中新出现各项的物理意义。

如定义

$$F(\zeta) = D\zeta/Dt + \beta v + (f+\zeta)\nabla\cdot\mathbf{V} \tag{20}$$

则传统涡度方程(1)可写为

$$F_{\text{TVE}}(\zeta) = (\boldsymbol{\zeta}_a\cdot\nabla)w + N_z + \mathbf{k}\cdot\mathbf{F} \tag{21}$$

如把三维绝对涡度 $\boldsymbol{\zeta}_a$ 分解为垂直分量和水平分量,并采用通常的气象近似,式(21)可展开为

$$F_{\text{TVE}}(\zeta) = \underbrace{(f+\zeta)\partial w/\partial z}_{(a)} + \underbrace{(\partial u/\partial z\cdot\partial w/\partial y - \partial v/\partial z\cdot\partial w/\partial x)}_{(c)} + N_z \tag{22}$$

式中,(a)和(c)各为(垂直)伸展项和扭曲项。式(22)即为教科书上常用的涡度方程。

另一方面,全型涡度方程(17)可写为

$$F_{\text{CVE}}(\zeta) = \rho D(P/\theta_z - C_D)/Dt, \theta_z \neq 0 \tag{23}$$

注意到式(14)中的 $|\nabla_h\theta|/\theta_z$ 表示 θ 面的倾斜,所以参数 C_D 代表着单位质量气块的水平涡度向垂直涡度的映射。而 $\dot{C}_D$ 代表着这种映射率对垂直涡度变化的影响。把式(14)代入式(23),全型涡度方程还可写成

$$F_{\text{CVE}}(\zeta) = -\frac{1}{\alpha\theta_z}(\xi_z\dot{\theta}_z + \xi_s\dot{\theta}_s + \theta_s\dot{\xi}_s), \quad \theta_z \neq 0 \tag{24}$$

式中,$\theta_s = |\nabla_h\theta|$ 为 $\nabla\theta$ 的水平分量的强度,而 ξ_z 和 ξ_s 分别为比绝对涡度的垂直和水平分量,即

$$\boldsymbol{\xi} = \alpha \boldsymbol{\zeta}_a = \xi_s \boldsymbol{s} + \xi_z \boldsymbol{k}$$

$$\xi_z = \boldsymbol{k} \cdot \alpha \boldsymbol{\zeta}_a = \alpha(f + \zeta)$$

$$\xi_s = \boldsymbol{s} \cdot \alpha \boldsymbol{\zeta}_a = \alpha \nabla \times \boldsymbol{V} \cdot \nabla_h \theta / |\nabla_h \theta|$$

式中, s 为沿着 θ 的水平梯度方向的单位矢量,即 $\boldsymbol{s} = \nabla_h \theta / |\nabla_h \theta|$. 利用如下关系

$$\frac{\mathrm{D}}{\mathrm{D}t}\theta_z = \frac{\partial}{\partial z}\frac{\mathrm{D}\theta}{\mathrm{D}t} - \frac{\partial \boldsymbol{V}}{\partial z} \cdot \nabla \theta$$

$$\frac{\mathrm{D}}{\mathrm{D}t}\theta_s = \frac{\partial}{\partial s}\frac{\mathrm{D}\theta}{\mathrm{D}t} - \frac{\partial \boldsymbol{V}}{\partial s} \cdot \nabla \theta$$

式(24)可写成

$$F_{\mathrm{CVE}}(\zeta) = \underbrace{(\boldsymbol{\zeta}_a \cdot \nabla)w}_{\text{(a\&c)}} + \underbrace{(\boldsymbol{\zeta}_a \cdot \nabla)\boldsymbol{V}_h \cdot \nabla_h \theta / \theta_z}_{\text{(b\&d)}} - \underbrace{\alpha^{-1}\dot{\xi}_s |\nabla_h \theta|/\theta_z}_{\text{(e)}}, \quad \theta_z \neq 0 \tag{25}$$

式(23)、(24)和(25)即为全型涡度方程的不同表达形式。

2.1 两方程的一致性

式(25)中的(e)项描述了水平涡度变化对垂直涡度发展的贡献;下面将证明,(b & d)项描述了稳定度和斜压性的改变对垂直涡度发展的贡献,它们均不存在于经典涡度方程中。注意到只有当 θ 面为倾斜时这些项才存在。当 θ 面水平分布时, $\nabla_h \theta \equiv 0$, (e) 和 (b & d) 项也为零。另一方面经典涡度方程(21)中的力管数项 N_z 也不显式出现在全型涡度方程(25)中。但由于力管的方向与 θ 面平行,当 θ 面为水平时力管也呈水平分布,这时力管数 $N_z \equiv 0$。于是,在无摩擦时(有摩擦不影响结论),两方程均只保留了(a & c)项,即

$$F_{\mathrm{CVE}}(\zeta) = F_{\mathrm{TVE}}(\zeta) = \boldsymbol{\zeta}_a \cdot \nabla w$$
$$= (f + \zeta)\partial w/\partial z + (\partial u/\partial z \cdot \partial w/\partial y - \partial v/\partial z \cdot \partial w/\partial x), \text{ 如果} \nabla_h \theta \equiv 0 \tag{26}$$

这时两个方程趋于一致。这就是说,当 θ-面为水平面时,除了 β-效应外,质块运动中的垂直伸展和扭曲作用控制着其涡度的变化。

2.2 两方程的差异

方程(25)与(21)的显著区别是增加了与 θ-面倾斜相联系的"斜压项"(e 和 b & d)。暂不考虑水平涡度变化项(e),先集中分析新增项(b & d)与经典项(a & c)的相对重要性。为此我们引入一个比率 $\Re$ 以表征两项的相对大小,即

$$\Re = |(\text{b\&d}):(\text{a\&c})| = |(\boldsymbol{\zeta}_a \cdot \nabla)\boldsymbol{V}_h \cdot \nabla_h \theta / \theta_z : (\boldsymbol{\zeta}_a \cdot \nabla)w| = \left|\frac{\boldsymbol{V}_h}{w}\right| \tan\delta \tag{27}$$

式中, $\tan\delta = |\nabla_h \theta / \theta_z|$ 代表了 θ 面的倾斜程度,由式(27)的定义可见, $\Re > 1$ 意味着倾斜项(b & d)比经典项(a & c)更重要,它意味着当 θ 面的倾斜程度

$$\tan\delta = |\nabla_h \theta / \theta_z| \geqslant |w/V_h| \tag{28}$$

时,倾斜项对涡度发展的贡献要远超过延伸项和扭曲项而不能被忽略。对于大尺度运动来说,式(28)的阈值为 $10^{-3} \sim 10^{-2}$,相当于一个典型冷锋的倾斜程度,表明倾斜项对锋面气旋的发展是重要的。当这一倾斜度变得很大时,由于 θ 面倾斜所诱发的涡度发展可以远远超过延伸和扭曲效应而变得非常激烈,导致严重天气事件的发生(吴国雄 等,1995)。

2.3 全型涡度方程新的物理内涵

把式(25)中的三维绝对涡度分解为垂直和水平分量,可得如下全型涡度方程的展开式

$$F_{\text{CVE}}(\zeta) = (f+\zeta)\partial w/\partial z + (f+\zeta)\frac{\partial \boldsymbol{V}_h}{\partial z}\cdot\nabla_h\theta/\theta_z +$$
$$\text{(a)} \qquad\qquad\text{(b)}$$

$$(\partial u/\partial z\cdot\partial w/\partial y - \partial v/\partial z\cdot\partial w/\partial x) + \zeta_s\frac{\partial \boldsymbol{V}_h}{\partial s}\cdot\nabla_h\theta/\theta_z - \alpha^{-1}\dot{\xi}_s\theta_s/\theta_z, \quad \theta_z\neq 0 \qquad (29)$$
$$\text{(c)} \qquad\qquad\qquad\qquad\text{(d)} \qquad\qquad\text{(e)}$$

式中，垂直延伸项(a)和扭曲项(c)与经典涡度方程(1)一致，其物理意义已在教科书中讨论过，这里不再赘述。关于水平涡度变化的贡献项(e)涉及其他的动力过程，将另文予以讨论。这里集中分析式(29)中项(b)和(d)的物理意义。为简单起见，假设大气是无摩擦和绝热的，满足地转和静力平衡，并采用 p-坐标。这时 Ertel 位涡成为(参见 Wu et al., 1998)

$$p = -g(f+\zeta)\theta_p - \lambda|\nabla_p\theta|^2 = \text{const} \qquad (30)$$

式中，$\nabla_p\theta$ 为 p-坐标面上 θ 的水平梯度，$\lambda = \alpha g/f_0$，而 θ_0 为 θ 在 $p_0 = 1000$ hPa 上的值。易于证明，这时全型涡度方程(24)变为

$$F_{\text{CVE}}(\zeta) = \frac{1}{g\theta_p}(\xi_p\dot{\theta}_p + \xi_s\dot{\theta}_s + \theta_s\dot{\xi}_s), \quad \theta_p\neq 0 \qquad (31)$$

式中，

$$P = \xi_p\theta_p + \xi_s\theta_s, \nabla\theta = \nabla_p\theta - \rho g\theta_p\boldsymbol{k}, \boldsymbol{\xi} = \alpha\boldsymbol{\zeta}_a = -g\boldsymbol{k}\times\frac{\partial \boldsymbol{V}_h}{\partial p} + \alpha(f+\zeta_p)\boldsymbol{k},$$

$$\xi_s = \boldsymbol{s}\cdot\alpha\boldsymbol{\zeta}_a = \alpha\nabla\times\boldsymbol{V}\cdot\nabla_h\theta/|\nabla_h\theta|$$

按照推导式(29)的途径，可得与 p-坐标下相对应的涡度方程，这时，与式(29)中对应的(b)项和(d)项变为

$$\text{(b)} = \left[(f+\zeta_p)\frac{\partial \boldsymbol{V}_h}{\partial p}\cdot\nabla_p\theta/\theta_p\right], \text{(d)} = \left[-\alpha g^{-1}\zeta_s\frac{\partial \boldsymbol{V}_h}{\partial s}\cdot\nabla_p\theta/\theta_p\right] \qquad (32)$$

式中，ζ_p 为 p-坐标中的垂直涡度分量。为方便讨论，这里暂假定水平位温的梯度方向指向东，即 $\boldsymbol{s} = \boldsymbol{i}$。

(1) 静力稳定度的影响

把式(32)中的项(b)写为标量形式

$$\text{(b)} = (f+\zeta_p)u_p\theta_x\theta_p^{-1}$$

设大气是惯性稳定 $(f+\zeta_p > 0)$ 和静力稳定 $(\theta_p < 0)$；其位温随 x 线性增加 $(\theta_x > 0)$；其风速 u 在 $p = p_0$ 处为 0，在 $p = p_1 < p_0$ 处增加为 $u_1 > 0$ (如图 6 所示)。这时 (b) > 0。注意到这样的风垂直切变将使 θ-面的倾斜增加，它虽然不影响斜压性 θ_x，但使静力稳定度 $(-\theta_p)$ 减小了。根据式(30)，垂直涡度 ζ_p 增加了。因此项(b)表达了与风垂直切变相联系的静力稳定度变化所导致的垂直涡度发展。

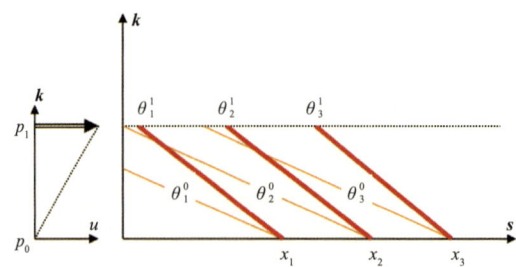

图 6 在惯性和静力稳定大气中，风的垂直切变改变了静力稳定度，导致垂直涡度发生变化。图中 θ_i^0 表示起始时刻 θ 随 x 的分布，θ_i^1 表示由于风的垂直切变引起的终了时刻 θ 随 x 的分布

(2) 斜压性的影响

把式(32)中的项(d)写成标量形式

$$\text{(d)} = -\alpha g^{-1}\zeta_x u_x\theta_x\theta_p^{-1}$$

同样，设大气是惯性稳定和静力稳定的，位温 θ 随 x 增加，从热成风关系可推知 $\zeta_x = \rho g v_p = -g(f\theta_0)\theta_x <$

0。于是项(d)的符号取决于 u_x。

首先,考虑水平风为辐合的情况,即 $u_x < 0$。这时项(d)>0,气旋性涡度将发展。这是因为依据图7a,风的辐合将导致起始温度梯度 $(\theta_3^0 - \theta_1^0)/\Delta x^0$ 增大至 $(\theta_3^1 - \theta_1^1)/\Delta x^1$ 而出现锋生。根据式(30),垂直涡度将增加。

反之,考虑水平风为辐散的情况,即 $u_x > 0$。这时项(d)<0,气旋性涡度将减小。这是因为依据图7b,风的辐散将导致温度梯度减小而出现锋消。根据式(30),垂直涡度将减小。

简言之,项(d)代表着与锋生、锋消相伴随的斜压性的改变所诱发的气旋涡度的生消。由此可与经典的锋面气旋理论联系起来。

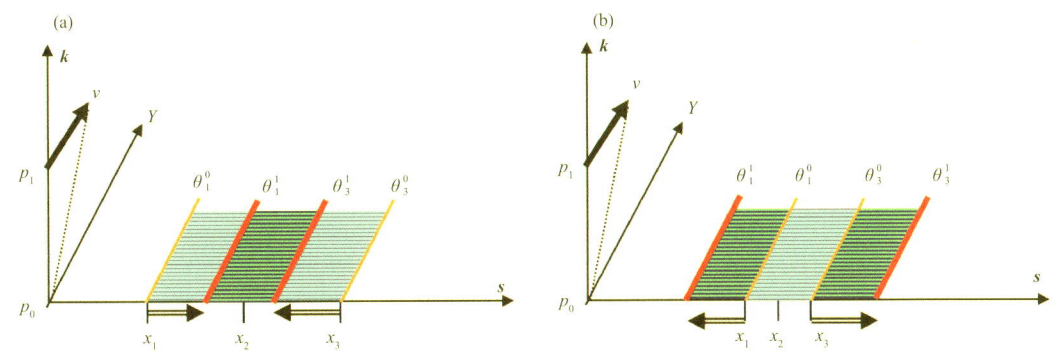

图 7 由于沿 θ 的梯度方向的水平风速出现辐合(a)或辐散(b),大气的斜压性改变,出现锋生(a)或锋消(b),从而导致垂直涡度增长(a)或消弱(b)

(3)涡度发展的总体观念

根据上述讨论,全型涡度方程(29)中右边各项可分别定义为垂直延伸项、稳定度项、扭曲项、斜压项或锋生项,以及垂直风切变项。尽管每项的物理涵义不同,然而其对涡度发展的效应是不可分的,这是因为它们与大气的热力结构紧密相联。举例说,当位温面的倾斜发生变化时,大气的斜压性和稳定度将改变,项(b)和(d)对涡度发展开始起作用,风的垂直切变也起了变化。于是项(a)、(c)和(e)也开始发生变化而影响涡度发展。因此,在诊断涡度变化的过程中,仅强调单一因素而不考虑其他因素将是片面与不适宜的。这就是说,应该从总体观点去分析涡度变化的原因。注意到由式(6)所定义的热力参数 C_D 由斜压性、静力稳定度和风的垂直切变所组成,而且式(29)右端(a)至(e)5项在式(23)中被结合成一个单一的涡度强迫项。于是,全型涡度方程(23)提供了从总体观点诊断涡度发展的手段,而对应的分解形式的方程(29)则提供了从不同角度确定涡度发展机制的工具。在本章上一节提出的"倾斜涡度发展"(SVD)的概念(吴国雄 等,1995,1999a;Wu et al.,1998)即是根据涡度发展的全局观念导得的。

注意到当采用 θ-坐标时,$\nabla_h \theta \equiv 0$,这时 $C_D = 0$,于是全型涡度方程(23)退化为所谓的等熵位涡方程(Hoskins et al.,1985)

$$D[\alpha(f+\zeta_\theta)\theta_z]/Dt = D(IPV)/Dt = 0$$

副热带至中纬度大气的位温面呈倾斜分布。沿着这一纬带,SVD 在引发短期气旋和反气旋涡度发展中应是非常重要的,也可能是中高纬度环流影响副热带高压短期变异的一种机制。有关这方面的研究刚刚开始,可望在不久的将来取得重要进展。

必须指出的重要一点是,本节所讨论的内强迫与下节讨论的外强迫是不能截然分开的。因为大气的热力结构本身(即 C_D 值)就与大气所受到的外部加热密切相关。其实,在上节讨论热力适应中,外部加热场就引起了大气热力结构的变化。沿着加热区的边界,θ 面出现倾斜,与斜压位涡 $W_h < 0$ 相对应,$C_D < 0$。当负位涡 W 在下层输入的同时,穿越边界的质点经历了倾斜涡度发展。从这一意义上说,大气的热力结构是大气向外部加热场适应的过程中所形成的。因此,从下一节开始,将直接研究非绝热加热对副高形成的影响。

3 涡度变化的外部热力强迫——副热带高压的形态变异

本节利用尺度分析研究空间非均匀加热对副热带高压形成和变异的影响(吴国雄等,1999b)。为此,在全型涡度方程(18)中暂不考虑大气热力结构所致的涡度内部强迫及表面摩擦的外强迫。

仅考虑外热源作用的情形,方程(18)简化为

$$\frac{\partial \zeta_z}{\partial t} + \boldsymbol{V} \cdot \nabla \zeta_z + \beta v = -(f + \zeta_z)\nabla \cdot \boldsymbol{V} + \frac{1}{\theta_z}\boldsymbol{\zeta}_a \cdot \nabla Q, \theta_z \neq 0 \tag{33}$$

式中,Q 为热力学方程中的非绝热加热率。当只考虑垂直方向非均匀分布的表面感热加热率(Q^{SH})和凝结潜热释放 Q^{CO} 时,热力学方程可写为

$$\frac{\mathrm{D}\theta}{\mathrm{D}t} = Q = Q^{SH} + Q^{CO} \tag{34}$$

式中,

$$\begin{cases} Q^{SH} = -\dfrac{\theta}{\rho c_p T}\dfrac{\partial F_{SH}}{\partial z} \approx \dfrac{1}{c_p}\left(\dfrac{\theta}{\rho T}\right)^* \dfrac{F_{SH}}{H} \\ Q^{CO} = \dfrac{L}{c_p}\left(\dfrac{\theta}{\rho T}\right)^* \dfrac{\rho_w P}{H} \end{cases} \tag{35}$$

式中,F_{SH} 为表面感热通量,ρ_w 和 ρ 分别为水和大气的密度,P 为单位面积上的降水率,即雨强($m \cdot s^{-1}$),H 为加热的垂直特征尺度,($*$)表示 H 层中的平均值。L 为水汽潜热,$L\rho_w P$ 表示单位气柱中因降水 P 所致的凝结潜热释放量。其他为气象常用符号。如取 F_{SH} 为 100 W·m^{-2},P 为每天 10 mm 降水,H 为 10^4 m,$|(\theta/\rho T)^*|$ 的量级为 10^0,则 Q^{SH} 和 Q^{CO} 量级为 10^{-5} K·s^{-1}。

这里针对惯性和静力稳定的大气,分别讨论副热带地区表面感热加热(Q^{SH})、深对流凝结加热(Q^{CO})以及水平加热差异等因子对涡度强迫的贡献,由此以研究外部热源强迫对北半球副热带高压带形态变异的影响。

3.1 感热加热(SH)对副热带反气旋形态的影响

首先考虑垂直向非均匀受热。由于 SH 以近地层加热为主,这时,除了近地面层外,式(33)中的非绝热加热项

$$\left(\frac{\partial \zeta_z}{\partial t}\right)_z^{SH} \sim \frac{(f+\zeta_z)}{\theta_z}\frac{\partial}{\partial z}(Q^{SH}) < 0 \tag{36}$$

这就是说,SH 加热产生了反气旋涡源。在时间尺度很短时,SH 加热区上空将出现深厚的反气旋。对时间尺度长的 SH 加热,局地变化项可略,大气可取为准定常态。取感热的特征垂直厚度,$\Delta z \sim 1$ km,$\Delta \theta \sim 10$ K,三维散度 $\nabla \cdot \boldsymbol{V} \sim 10^{-7}$ s^{-1},则式(33)右端两项的量级分别为 10^{-11} 和 10^{-10}。由方程(36)知,这时有

$$\boldsymbol{V} \cdot \nabla \zeta_z + \beta v < 0 \tag{37}$$

由于副热带地区低空为东西风交界处,$u \approx 0$,水平平流作用很小。根据式(37)中,

$$\beta v < 0 \tag{38}$$

在 SH 的作用下,纬向均匀的副热带反气旋带便出现断裂:在 β 效应作用下,SH 上空出现强的北风($v<0$),其西侧为反气旋,东侧为气旋(图8)。又由于在南暖北冷的背景温度场中,西风随高度增加,使副热带高空受西风控制,$u>0$。根据式(37)

$$u\frac{\partial \zeta}{\partial x} < 0 \tag{39}$$

那里的平流作用使均匀风带出现波动,在 SH 区的下游出现反气旋式环流(图 8 上部)。根据热力适应原理,在近地层则有气旋式涡度生成(图 8 下部)。

盛夏副热带大陆西岸有很强的感热加热($F_{SH}>100\ W\cdot m^{-2}$)。副热带低层出现北太平洋和北大西洋副高，陆面上为低气压；高层的反气旋则出现在陆地上，大洋上空为低槽，与图8的情况相似。降冬强烈的感热加热出现在大陆东岸沿海地区。使低空大洋上出现气旋，大陆上出现反气旋；对流层高层大洋上为脊，陆地上为槽，也与图8的情况吻合。由此可推论，表面感热加热对冬、夏气候基本态的形成十分重要。

由于表面感热通量F_{SH}为$100\ W\cdot m^{-2}$所产生的涡源$\left(\dfrac{\partial \zeta_z}{\partial t}\right)_z^{SH}$的量级为$(-10^{-10}\ s^{-2})$，它在24小时内即可强迫出强大的副热带反气旋系统。足见感热加热对大气环流的形成非常重要。

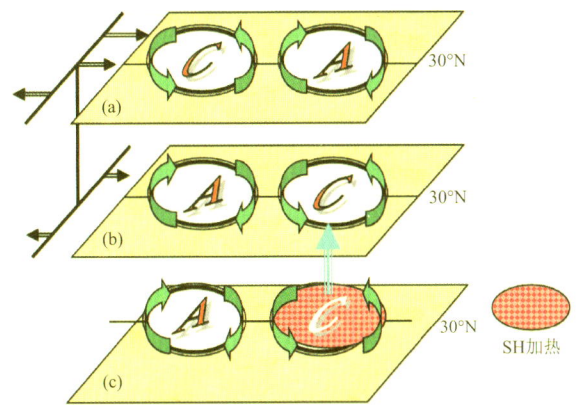

图8 沿副热带(30°N)表面感热加热(Q^{SH})的垂直分布对副热带高压带形态变异的影响。阴影区为加热中心，"A"和"C"分别表示"Q^{SH}"所激发的反气旋和气旋偏差环流。图中左侧箭头表示纬向平均流场空间分布。(a) 高层；(b) 低层；(c) 近地层

3.2 深对流降水对副热带反气旋形态的影响

深对流降水引起高层凝结潜热释放(Q^{CO})。其特征尺度H可取7 km左右。在其下方直至近地层产生了正涡源，即

$$\left(\frac{\partial \zeta_z}{\partial t}\right)_z^{CO} \sim \frac{(f+\zeta_z)}{\theta_z}\frac{\partial}{\partial z}(Q^{CO}) > 0 \tag{40}$$

在时间尺度很短时，它将加强气旋的发展。在时间尺度很长时，沿副热带东西风的分界处这时有

$$u \approx 0, \beta v > 0 \tag{41}$$

β效应将使南风发展($v>0$)，在其西侧为气旋式环流，其东侧为反气旋式环流(图9中)。

在最大加热中心以上的高层，由于

$$\left(\frac{\partial \zeta_z}{\partial t}\right)_z^{CO} \sim \frac{(f+\zeta_z)}{\theta_z}\frac{\partial}{\partial t}(Q^{CO}) < 0 \tag{42}$$

类似地有

$$\beta v < 0 \tag{43}$$

于是在对流降水区西侧为反气旋式环流，东侧为气旋式环流(图9上)。在近地面层，热力适应的结果形成气旋式环流(图9下)。

传统认为，当西太平洋副高向大陆西伸时，利于降水在华东、华南发生。上述分析(图9)则表明，华东华南降水的出现本身也能诱发中低空($u=0$处)偏南气流的发展及西太副高西伸；在高空有偏北气流发展及南亚高压的加强。看来两者有互为因果的相互作用。

3.3 水平非均匀加热对副热带反气旋形态的影响

当只考虑水平非均匀加热时，方程(33)可写成

$$\dot{\zeta}_z + \beta v + (f+\zeta_z)\nabla\cdot\boldsymbol{V} = -\theta_z^{-1}(v_z Q_x - u_z Q_y),\quad \theta_z \neq 0 \tag{44}$$

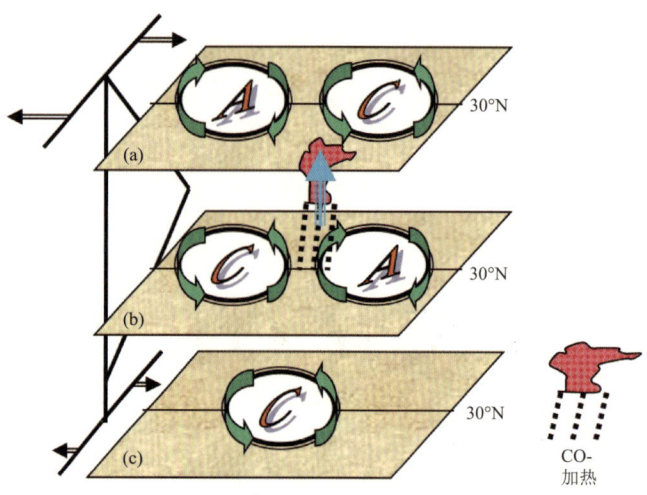

图 9 同图 8,但为凝结潜热加热(Q^{CO})垂直分布的影响,阴影区为加热中心
(a) 高层;(b) 中层;(c) 近地层

应用 p-坐标并利用热成风关系,得到

$$\dot{\zeta} + \beta v + (f+\zeta)\nabla\cdot\mathbf{V} = R(fP\theta_p)^{-1}(T_xQ_x + T_yQ_y),\quad \theta_p \neq 0 \qquad (45)$$

式中, $-\theta_p = \theta_z/\rho g$ 为气压坐标中的静力稳定度。上式表明,当水平非均匀加热与水平温度同位相配置,则有反气旋涡度在加热区发展,反之则有气旋涡度发展(图 10)。取如下的变量尺度

$$\begin{cases}(\Delta u,\Delta v) \sim 10\text{ m}\cdot\text{s}^{-1}, \theta_z \sim 10^{-2}\text{ K}\cdot\text{m}^{-1}, \Delta T \sim 10\text{ K},\\ (\Delta x,\Delta y) \sim 10^6\text{ m}, Q \sim 10^{-5}\text{ K}\cdot\text{s}^{-1}\end{cases}$$

得到方程(45)右端水平非均匀加热的量级为 10^{-11}s^{-2},比垂直非均匀加热项小一个量级以上。这意味着在加热区内,方程(33)中的垂直非均匀加热比水平非均匀加热更为重要。但在加热区外部,$\partial Q/\partial z \equiv 0$,于是水平非均匀加热成为主要的外部加热强迫,从而支配着大气的涡度变化。在实际大气或模式大气中,水平扩散过程使热源区中的加热向外界扩散。因此在热源区外存在着一个因水平非均匀加热所产生的涡度强迫场。当空气质块带着其自身的涡度进入该强迫场时,其轨迹将改变,好似带电粒子进入磁场一般。与水平非均匀加热相联系的这种涡度强迫场的存在对形成大气环流的形态有深远的影响。由于水平非均匀加热效应,给定的副热带热源能够在中纬度西风带中激发出新的涡度强迫源,并影响全球的环流异常。

图 10 水平非均匀加热场和温度场的配置对副热带高压带形态变异的影响示意图。
当两者同位相配置时有反气旋涡度制造;当两者反位相配置时有气旋涡度制造

上述讨论表明,空间非均匀加热对副热带反气旋形态变异有非常重要的影响,尺度分析表明,一般说来垂直方向非均匀的加热分布与水平非均匀的加热分布比较,前者对副热带反气旋变异的影响比后者要大一个量级以上。从气流平均状态看,夏季近地面大洋东部和大陆西部副热带反气旋位置和加热的配置与图 8 较为相近;而 500 hPa 西太平洋副热带反气旋及 200 hPa 南亚高压位置和加热的配置与图 9 较为接近。由此推测,对北半球夏季气候平均副热带反气旋的形成来说,感热加热对东北太平洋和大西洋的副高形成十分重要,而季风降水的潜热释放对西北太平洋副高及南亚高压的形成应是不可忽视的。刘屹岷等(1999a,1999b)已通过一系列的数值试验和模拟检验了感热和深对流潜热加热对北半球夏季各个副高单体的不同贡献。

参考文献

刘屹岷,刘辉,刘平,等,1999a.空间非均匀加热对副热带高压形成和变异的影响.Ⅱ.陆面感热与东太平洋副高[J].气象学报,57(4):385-396.

刘屹岷,吴国雄,刘辉,等,1999b.空间非均匀加热对副热带高压形成和变异的影响.Ⅲ.凝结潜热加热与南亚高压及西太平洋副高[J].气象学报,57(5):525-538.

陶诗言,1980.中国之暴雨[M].北京:气象出版社.

吴国雄,蔡雅萍,唐晓菁,1995.湿位涡和倾斜涡度发展[J].气象学报,53:387-405.

吴国雄,刘还珠,1999a.全型垂直涡度倾向方程和倾斜涡度发展[J].气象学报,57(1):1-15.

吴国雄,刘屹岷,刘平,1999b.空间非均匀加热对副热带高压带形成和变异的影响.Ⅰ.尺度分析[J].气象学报,57(3):257-263.

叶笃正,李崇银,王必魁,1988.动力气象学[M].北京:科学出版社:340.

赵宇澄,吴国雄,纪立人,等,1996.等熵模式对一次西南低涡过程的模拟和等熵位涡分析[M]//台风,暴雨数值预报新技术的研究.北京:气象出版社:94-102.

ANTHES R A, HEAGERSON P L,1984. A comparative numerical simulation of the Sichuan flooding catastrophe. Proc[C]//The Chinese Academy of Sciences and the U.S. National Academy of Sciences. First Sino American Workshop on Mountain Meteorology. Beijing:Ame Meteor Soc:519-524.

BLECK R,1984. An isentropic model suitable for lee cyclogenesis simulation[J]. Rivista Meteor Acronaut,43:189-194.

CHEN S J, DELL'OSSO L,1984. Numerical prediction of the heavy rainfall vortex over the eastern Asian monsoon region[J]. J Meteor Soc Japan,62:730-747.

ERTEL H,1942. Ein neuer hydrodynamischer wirbelsatz[J]. Meteorology Z Braunschweig,59:271-281.

HOLTON J R,1992. An Introduction to Dynamic Meteorology[M]. San Diego, California:Academic Press, INC:3:511.

HOSKINS B J, MCINTYRE M E, ROBERTSON A W,1985. On the use and significance of isentropic potential vorticity maps[J]. Quart J Roy Meteor Soc,111:877-946.

HOVERMALE J B,1984. Numerical experiments with the Sichuan flooding catastrophe[C]//The Chinese Academy of Sciences and the U.S. National Academy of Sciences. First Sino American Workshop on Mountain Meteorology. Beijing:Ame Meteor Soc:243-264.

KUO Y H, CHENG L S, ANTHES R A,1986. Mesoscale analysis of Sichuan flood catastrophe[J]. Mon Wea Rev,114:1984-2003.

WU G X, CHEN S J,1985. The effect of mechanical forcing on the formation of a mesoscale vortex[J]. Quart J Roy Meteor Soc,111:1049-1070.

WU G X, LIU H Z,1998. Vertical vorticity development owing to down-sliding at slantwise isentropic surface[J]. Dyn Atmos Oce,27:715-743.

ZHOU X P, HU X F,1984. A brief analysis and numerical simulation of the Sichuan extra ordinary heavy rainfall event[C]//The Chinese Academy of Sciences and the U.S. National Academy of Sciences. First Sino American Workshop on Mountain Meteorology. Beijing:Ame Meteor Soc:555-565.

Eurasian Cooling Linked with Arctic Warming: Insights from PV Dynamics

XIE Yongkun[1], WU Guoxiong[1,2], LIU Yimin[1,2], HUANG Jianping[3]

(1 LASG, Institute of Atmospheric Physics, Chinese Academy of Sciences, Beijing 100029, China; 2 College of Earth and Planetary Sciences, University of Chinese Academy of Sciences, Beijing 100049, China; 3 Key Laboratory for Semi-Arid Climate Change of the Ministry of Education, College of Atmospheric Sciences, Lanzhou University, Lanzhou 730000, China)

Abstract: The three-dimensional connections between Eurasian cooling and Arctic warming since 1979 were investigated using potential vorticity (PV) dynamics. We found that Eurasian cooling can be regulated by Arctic warming through PV adaptation and PV advection. Here, PV adaptation refers to the adaptation of PV to forcing and coherent dynamic-thermodynamic adaptation to PV change. In a PV perspective, first, the anticyclonic circulation change over the Arctic is produced by a negative PV change through PV adaptation, in which the change means the linear trend from 1979 to 2017. The negative PV change is directly regulated by Arctic warming because the vertical structure of Arctic warming is stronger at lower levels, which generates a negative PV change through the diabatic heating effect. Second, the circulation change produces a change in horizontal PV advection due to the existence of climatological PV gradients. Thus, as a balanced result, both the circulation change and PV change extend to the midlatitudes through horizontal PV advection and PV adaptation. Eventually, Eurasian cooling at the surface and in the lower troposphere is dominated by PV changes at the surface through PV adaptation. Meanwhile, enhanced Eurasian cooling in the middle troposphere is dominated by top-down influences of upper-level PV change through PV adaptation. Nevertheless, the upper-level PV changes are still contributed to by horizontal PV advection associated with Arctic warming. Overall, the general dynamics connecting Eurasian cooling with Arctic warming are demonstrated in a PV view.

1 Introduction

During the past several decades, amplified Arctic warming is one of the most remarkable phenomena regarding global temperature change (Stouffer and Manabe, 2017; Xie et al., 2019). Meanwhile, remarkable regional cooling on decadal or multidecadal scales occurs in the context of overall global warming, such as the Eurasian cooling (Easterling and Wehner, 2009; Cohen et al., 2014; Shepherd,

2016). This kind of decadal variability is generally blamed for internal climate variability (Trenberth, 2015; Steinmann et al., 2015). On the one hand, the dominant role of internal climate variability associated with Atlantic and Pacific Oceans on multidecadal variability of temperature has been proposed by previous studies (Held, 2013; Kosaka and Xie, 2014; Steinman et al., 2015; Dai et al., 2015; Huang et al., 2017; Luo et al., 2017a). On the other hand, as directly observed in Fig. 1, Arctic warming is one of the strongest signatures during the past several decades (1979−2017), which overwhelms the signals of internal climate variability directly reflected by the temperature change over the Atlantic and Pacific Oceans. Recently, more and more studies have put their efforts into investigating the relative or joint effect of the warm Arctic and internal climate variability on the cold Eurasian pattern. Specific attention was paid to the internal climate variability of the upstream region(i.e., North Atlantic) (e.g., Luo et al., 2016, 2017a; Sung et al., 2018; Yang et al., 2018; Luo et al., 2019a). An exact attribution relies on the model simulation such as the fingerprint method (Ding et al., 2019), while intermodel discrepancies make a robust attribution far from realizing (Cohen et al., 2018; Smith et al., 2019).

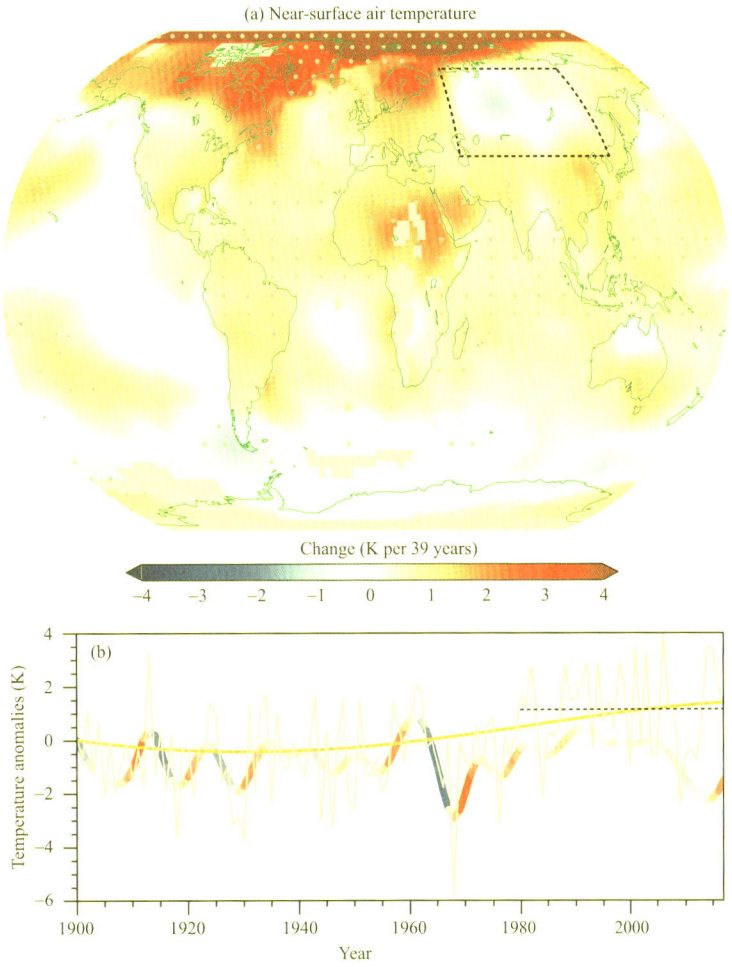

Fig. 1 (a) Near-surface air temperature change (K) (linear trend multiplied by the time span) for the period of 1979−2017 during boreal winter, in units of K. The results are based on GISTEMP. The gray dot indicates that the linear trend is significant at the 99% confidence level based on a two-tailed Student's t-test. (b) Time series of near-surface air temperature anomalies (relative to 1951−1980) averaged over region of (40°−70°N, 50°−130°E), outlined by rectangle in (a). The raw series is the gray line, and the orange and blue-red lines indicate the long-term trend (IMF6 from EEMD) and decadal/multidecadal variability (sum of IMFs 3−5), respectively. The dashed black line indicates the trend line of the raw series from 1979−2017

Apart from a full attribution, this paper focuses on the general dynamics that could link the Arctic warming with Eurasian cooling. Regarding dynamics, there are extensive works published. Here, a detailed review those of all the previous works is not necessary because some excellent reviews are already available, such as those of Cohen et al. (2014, 2018), Overland et al. (2016), Coumou et al. (2018), and Screen et al. (2018). Previous works suggest many possible pathways linking the Arctic and the midlatitude. The main pathways include Rossby waves (Francis and Vavrus, 2012; Sung et al., 2018; Li et al., 2019), meandering of jet stream (Screen and Simmonds, 2013, 2014; Francis and Vavrus, 2015; Di Capua and Coumou, 2016), the Arctic oscillation (Yang et al., 2016), the North Atlantic oscillation (Yang et al., 2018; Luo et al., 2019a), polar vortexes (Cohen et al., 2014), the Siberian High (Ye at al., 2018; Sung et al., 2018), atmospheric blocking (Yao et al., 2017; Luo et al., 2017b, 2018, 2019a, 2019b; Wegmann et al., 2018), stratospheric anomalies (Zhang et al., 2018), and eddy energy propagation (Gu et al., 2018).

Overall, more and more light are being shed on the dynamics of the Arctic and midlatitude linkages. As seen above, the complexity of the connections increased at the same time. And some pathways are tightly connected, such as the North Atlantic Oscillation and atmospheric blocking (Yang et al., 2018; Luo et al., 2019a). Thus, this study examines the general dynamic and thermodynamic features of atmospheric circulation without preference as to the aforementioned specific pathways. To achieve this objective, we do the investigation using potential vorticity (PV) dynamics. We use PV dynamics because PV inherently depicts seamless coupling between dynamic and thermodynamic aspects (Ertel, 1942; Yeh and Chu, 1958; Charney and Stern, 1962). In other words, any observed circulation and temperature change can, on a fundamental level, be attributed to PV. In this paper, piecewise PV inversion and the PV equation are used (Davis and Emanuel, 1991; Hartley et al., 1998; Zhao et al., 2007; Egger, 2008; Spengler and Egger, 2012; Egger et al., 2017). Piecewise PV inversion (PPVI) quantitatively tells the effects of arbitrary PV change on the entire atmosphere, while the PV equation tells how the PV change occurred. Importantly, we show that PV inversion can also be used in the climate change realm, although it is commonly used in synoptic cases.

This paper focuses directly on the persistent change of the mean state from 1979—2017 rather than weather or interannual variability. The remainder of this paper is arranged as six parts. Section 2 describes all the data and methods used in this paper. Essential characteristics (e.g., spatial pattern and vertical structure) of Eurasian cooling and Arctic warming are presented in section 3. Section 4 addresses the concept of PV adaptation based on the ideal case. Section 5 and 6 address the vertical coupling among different levels and horizontal coupling among different locations, respectively. Concluding remarks are provided in section 7.

2 Data and method

a. Reanalysis data

ERA-interim data are used in this study. The original model used to produce ERA-interim data is an approximately 0.75° grid (Dee et al., 2011), but the European Centre for Medium-Range Weather Forecasts (ECMWF) provides many choices of the resolution. Here, we choose a coarse gridded output of 1.5°×1.5° at 37 pressure levels because the analysis of high-resolution six-hourly data spanning thirty-eight years (1979—2017) would be too cumbersome. A coarse resolution output will not affect our results because the original model resolution did not change for even a coarse output and we do not investigate small scale features. The climatology is the mean from 1979—2017, and the change from 1979—2017 is calculated as the linear trend for each variable multiplied by 39 years. In the following, overbar "$\overline{A}$" represents the climatology, while the prime "A'" indicates the change without specification. The boreal winter is defined as December to the following February.

b. Piecewise PV inversion (PPVI)

The essential idea of PPVI is analogous to the electric charge in the electric field (Hoskins et al., 1985; Hartley et al., 1998; Schneider et al., 2003; Egger and Spengler, 2018). Here, PV corresponds to the electric charge, while temperature or circulation corresponds to properties of the electric field. The minimalist description of PV inversion is $\Phi = L^{-1}(q)$, where Φ and q are the geopotential and pseudo- (or quasigeostrophic) PV (Charney and Stern, 1962), respectively. The operator L is a three-dimensional Laplacian-like operator, while L^{-1} is the inverse operator. Notably, L is an operator rather than a coefficient, which ensures the three-dimensional coupling among varied vertical levels and horizontal locations can be described by PV inversion more than just correspondence between Φ and q. The superposition principle of a linear operator L allows piecewise in PV inversion, namely, one-to-one correspondence of Φ_n and q_n in $\sum_1^n \Phi_n = \sum_1^n L^{-1}(q_n)$ (Hartley et al., 1998).

In pressure (p) coordinates, pseudo-PV is in the form

$$q = f + \frac{1}{f_0} \nabla^2 \Phi + \frac{\partial}{\partial p}\left[\frac{f_0}{S}\frac{\partial \Phi}{\partial p}\right] \quad (1)$$

where f is the Coriolis parameter and ∇^2 is a two-dimensional Laplacian operator (Holton and Hakim, 2013). $\overline{S} = -\overline{\alpha}\frac{\partial \ln \overline{\theta}}{\partial p}$ represents the static stability of the basic state and is only a function of pressure. Namely, the specific volume $\overline{\alpha}$ and potential temperature $\overline{\theta}$ are averaged over a large domain (northward of 10°N in this study). The f-plane assumption is adopted for relative vorticity. Namely, f in the denominator of $\frac{1}{f}\nabla^2 \Phi$ is replaced by $f_0 = 2\Omega \sin(\varphi_0)$, in which $\varphi_0 = 45°N$ and $\Omega = 7.292 \times 10^{-5}$ s^{-1} are earth's angular speed of rotation. In this paper, PPVI is used to investigate change rather than the mean state. Here, the PV change is in the form

$$q' = \frac{1}{f_0}\nabla^2 \Phi' + \frac{\partial}{\partial p}\left[\frac{f_0}{S}\frac{\partial \Phi'}{\partial p}\right] = \frac{1}{f_0}\nabla^2 \Phi' + \frac{\partial}{\partial p}\left[\frac{f_0}{S}\right]\frac{\partial \Phi'}{\partial p} + \frac{f_0}{S}\frac{\partial^2 \Phi'}{\partial p^2} \quad (2)$$

The technical details, such as step-by-step producer and boundary conditions, and the accuracy of PPVI producer are described in APPENDIX. Despite pseudo-PV, Ertel PV can also be inverted (Davis and Emanuel, 1991). Nevertheless, pseudo-PV is adopted in this research because Ertel PV inversion is much more computationally intensive for a hemispheric domain and quasi-geostrophic approximation is excellent for mesoscale systems. Hereafter, PV directly refers to pseudo-PV without specification.

c. PV equation

The PV equation is in the form

$$\frac{dq}{dt} = -f_0 \frac{\partial Q}{\partial p} + \boldsymbol{k} \cdot \nabla \times \boldsymbol{F} \quad (3)$$

where $Q = \frac{\overline{\alpha}\dot{\theta}}{\overline{S}\theta}$ indicates the diabatic heating ($\dot{\theta}$ represents the material change rate of potential temperature), and $\boldsymbol{F}$ indicates the horizontal frictional force. The effect of friction is difficult to specifically evaluate from reanalysis data (Egger et al., 2015). Physically, friction tends to reduce the relative motion between two objects at an interface. Following Smagorinsky (1953) and Yeh and Chu (1958), the effect of boundary layer friction is approximated to be directly proportional to the vorticity perturbation. Specifically, the frictional term in the PV equation is expressed as $\boldsymbol{k} \cdot \nabla \times \boldsymbol{F} = -\mu \cdot \frac{1}{f_0} \nabla^2 \Phi'$, where μ is the positive frictional coefficient. Hence, it is clear that friction always tends to offset the change in PV. However, friction cannot alter the sign of PV change. Simply put, friction makes the eventual observed change more moderate than it would otherwise be. We do not further investigate friction in the following.

The Eulerian form PV equation without friction is

$$\frac{\partial q}{\partial t} = -\boldsymbol{V}_g \cdot \nabla q - f_0 \frac{\partial Q}{\partial p} \tag{4}$$

where $\boldsymbol{V}_g$ is geostrophic wind. In a climate change sense, the equilibrium form should be adopted, namely, $-\boldsymbol{V}_g \cdot \nabla q - f_0 \frac{\partial Q}{\partial p} = 0$, by neglecting the local tendency term. Here, the linear perturbation form of the equilibrium PV equation is $-(\boldsymbol{V}_g \cdot \nabla q)' - f_0 \frac{\partial Q'}{\partial p} = 0$. By neglecting the high-order term, the PV equation in the perturbation form is finally

$$-\boldsymbol{V}'_g \cdot \nabla \overline{q} - \overline{\boldsymbol{V}}_g \cdot \nabla q' - f_0 \frac{\partial Q'}{\partial p} = 0 \tag{5}$$

The perturbation expansion was applied to $(\boldsymbol{V}_g \cdot \nabla q)'$ in Eq. (5). Namely, there is $(AB)' = AB - \overline{AB} = A'\overline{B} + \overline{A}B' + A'B' \approx A'\overline{B} + \overline{A}B'$ for $A = \overline{A} + A'$ and $B = \overline{B} + B'$ because $A'B'$ is negligible for the large-scale features of PV advection. Note that the overbar and prime have a different meaning with the traditional mean-eddy interaction formulations. Nevertheless, the effect of the eddy is inherently included in our results because all the results are calculated directly from six-hourly data.

d. Ensemble empirical mode decomposition (EEMD)

Ensemble empirical mode decompositcon (EEMD) is used to split the time series of a variable into different oscillatory components with intrinsic time scales (Wu et al., 2011; Ji et al., 2014). Following Huang et al. (2017), we obtain six intrinsic mode functions (IMFs). As shown in Fig. A2, IMFs 1 and 2 are annual to interannual variabilities, IMFs 3 to 5 are decadal to multidecadal variabilities, and IMF 6 is a long-term trend. Here, the detailed parameters and steps to perform EEMD follow Huang et al. (2017).

e. Plumb flux

Plumb flux is used to identify the wave train of the stationary Rossby wave. The method to calculate Plumb flux is the same as described in Plumb (1985), namely the expression (7.1) in that paper. Here, the domain used to calculate the static stability, expression (7.2) therein, is northward of 10°N.

3 Appearance and structure of Eurasian cooling and Arctic warming

As shown in Fig. 1, the most apparent feature of wintertime temperature change from 1979−2017 is the amplified Arctic warming, also known as Arctic amplification (Stouffer and Manabe, 2017). Meanwhile, the contrasting characteristic is cooling to various degrees over many regions globally, especially Eurasian cooling in the midlatitude regions. To better understand the Eurasian cooling, a series of temperature evolution averaged over the area (40°−70°N, 50°−130°E) is shown in Fig. 1b. EEMD (see section 2d) is used to identify the variability of temperature on various time scales. First, the overall change is a rise in temperature, namely, long-term warming. Besides, large interannual variability is involved, which is common in the high latitudes due to large internal atmospheric circulation variability (Wallace et al. 2012). Meanwhile, the decadal variability is indeed remarkable. Such a large magnitude of decadal variability will accelerate or decelerate warming by favoring or offsetting long-term warming trends. Taken together, the Eurasian cooling is mainly associated with the observed downward-trending decadal variability. However, the underlying mechanisms of this downward-trending decadal variability in temperature from 1979—2017 need to be explored further.

To get a global view, the vertical structure of Eurasian cooling and Arctic warming is examined further. Fig. 2 shows a vertical-latitudinal cross section of the changes in temperature and geopotential height averaged over (60°—120°E). Eurasian cooling occurs not only at the surface but also throughout almost the entire troposphere (Fig. 2a). The strongest cooling in the troposphere presents at about 500 hPa. Meanwhile, the Arctic warming also occurs throughout the whole troposphere. The strongest Arctic warming occurs at the surface, which is also suggested by Screen and Simmonds (2010) and Cohen et al. (2014). There is also remarkable cooling or warming occurring in the stratosphere, and we will examine the stratospheric influences on troposphere in section 5.

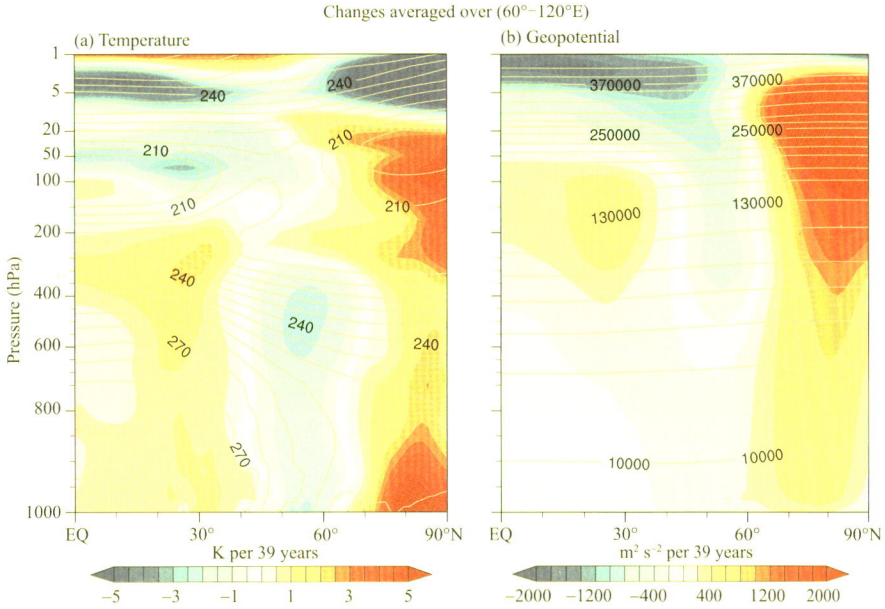

Fig. 2 (a) Vertical-latitudinal cross section of temperature changes (linear trend multiply the time span) for the period of 1979—2017 averaged over region of (60°—120°E) during boreal winter. The gray contour line indicates the corresponding climatology for the period of 1979—2017. (b) Same as (a) but for geopotential

Under geostrophic approximation, circulation could be represented by geopotential height in pressure coordinates. In pressure coordinates, the hydrostatic relation is $\frac{\partial \Phi}{\partial p} = -\frac{R}{p}T$, and the relation is applicable for both the mean state and change in the pressure coordinates. Thus, the temperature change should be proportional to the vertical gradient of geopotential change at the arbitrary pressure level. The relation is confirmed by the results in Fig. 2. The results show that the absolute changes in the vertical gradient of geopotential are larger in areas where the absolute value of temperature changes is larger for the same pressure level. Thus, the tight connections between atmospheric circulation change and regional temperature change are evident.

4 Concept of PV adaptation based on the ideal case

In the PV view, the seamless dynamic and thermodynamic coupling are straightforwardly indicated by Eq. (2), in which any change in geopotential Φ' could induce the changes in both circulation (represented by vorticity term $\frac{1}{f_0}\nabla^2\Phi'$) and temperature (represented by static stability term $\frac{\partial}{\partial p}\left(\frac{f_0}{S}\frac{\partial \Phi'}{\partial p}\right)$ and supported by results shown in Fig. 2). The classic Ertel PV conservation picture concisely and vividly depicts the inherent dynamic and thermodynamic coupling as shown in Fig. 3a. The corresponding pseudo-PV conservation picture is shown in Fig. 3b, which is similar to Fig. 3a but replacing the isentropic surface with the isobaric surface. PV conservation picture is shown in a Lagrangian view rather than an Eulerian view, which is generally adopted in climate change. In Eulerian view, as indicated by Eq. (4), local PV conservation is very hard to satisfy because advection of PV widely exists and will change local PV even for the absence of diabatic heating and friction.

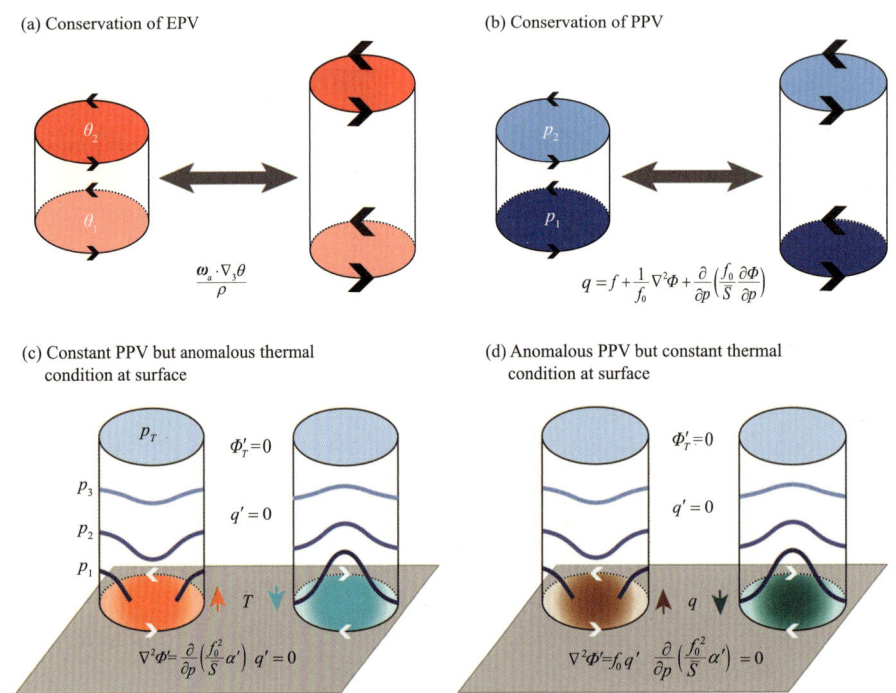

Fig. 3 Schematic views of (top) PV conservation (a)–(b) and PV adaptation (c)–(d). (a)–(b) show PV conservation for both EPV (Ertel PV) and pseudo-PV (PPV). (c)–(d) show two ideal cases of PV adaptation. PV adaptation means the adaptation of PV (PV change) to forcing and coherent dynamic/thermodynamic adaptation to PV change. (c) PV adaptation under constant PPV but anomalous surface thermal conditions. (d) PV adaptation under anomalous PPV but constant surface thermal conditions. All the variables have the same meaning as described in section 2. Subscript "T" indicates the top of the atmosphere. The gray parallelogram indicates the earth's surface. The visual illustration in Figs. 3c and d is similar to the traditional PV conservation illustration (Figs. 3b) but with a fixed top and bottom

However, the picture of PV conservation still helps to get insights. Here, two ideal cases extended from the PV conservation picture are used for some instructive discussions. Two specific questions emerged when applying PV dynamic to climate change: one is what is the effect of PV change; the other is why the PV changed. As already addressed in the introduction section, PV inversion and the PV equation can answer the two questions. However, a more fundamental question that has no direct help to solve the practical question but that helpful to conceptual understanding is "why should the PV change?".

Here take the effect of diabatic heating in generating PV as an example, which has a simple mathematical expression as $\frac{\partial q}{\partial t} \propto -f_0 \frac{\partial Q}{\partial p}$. Thus, diabatic heating generates positive, negative, and zero PV change when Q is increased, decreased, or unchanged with height, respectively. Here, an ideal case extended from the PV conservation is shown in Fig. 3c. In this ideal case, at first, there is temperature change present at the surface. Then, the atmosphere strongly adapts to the surface temperature change so that the diabatic heating of atmosphere (induced by the surface temperature change and corresponding vertical motion of atmosphere) is even in the vertical direction (i.e., $\frac{\partial Q}{\partial p} = 0$). In such a case, the surface temperature change induced no PV change, but geopotential greatly changes, and the balance relation is $\nabla^2 \Phi' = \frac{\partial}{\partial p}\left[\frac{f_0^2}{S}\alpha'\right]$ according to Eq. (2), where the hydrostatic relation $\frac{\partial \Phi'}{\partial p} = -\alpha'$ is introduced.

In practice, the ideal case in Fig. 3c can be quantified by using PPVI according to the superposition principle, which needs to keep only the change in surface thermal condition in the PPVI, as shown in Fig. 4b. The ideal case (Fig. 4b) with surface heating change but keeping a constant PV will make the geopotential change of atmosphere much stronger than the real case (Fig. 4a). In other words, changed PV in response to surface diabatic heating makes the change in motion of atmosphere more moderate. This conclusion also holds for the friction as demonstrated in section 2c. Therefore, PV change works as a stabilizer of the atmosphere in a fundamental understanding. This is an answer to "why should the PV change?".

Opposite to the ideal case in Fig. 3c, the other ideal case is shown in Fig. 3d, which prescribes a PV change at the surface but without surface thermal condition change ($\frac{\partial}{\partial p}\left(\frac{f_0^2}{S}\alpha'\right) = 0$ rather than a zero temperature change). In Fig. 3d, the balance relation is $\nabla^2 \Phi' = f_0 q'$. Thus, geopotential must have a negative change in response to a positive PV change (according to $\Phi' \propto -\nabla^2 \Phi'$), and vice versa. Regarding the ideal case in Fig. 3c, for surface warming stronger at the surface than above, $\frac{\partial}{\partial p}\left(\frac{f_0^2}{S}\alpha'\right)$ is positive under the current climatological $\overline{S}$. Therefore, geopotential change must be negative according to $\nabla^2 \Phi' = \frac{\partial}{\partial p}\left[\frac{f_0^2}{S}\alpha'\right]$, and vice versa.

The real case is the sum of Figs. 3c and 3d. Namely, for surface warming stronger at the surface than above, the balance in Fig. 3c tends to generate negative geopotential, while the stronger diabatic heating at the surface than above also induces negative PV, which tends to generate positive geopotential. Thus, the eventually observed geopotential change depends on the net effects after summing Figs. 3c and 3d together. In other words, the reality is the sum of one rule "warming is cyclonic and cooling is anticyclonic" and another rule "a positive PV change is cyclonic and a negative PV change is anticyclonic". In short, from a "forcing" (i.e., diabatic heating and friction in Eq. (3)) to final observed changes of atmosphere, two key adaptation processes are involved. One is the PV change as a direct adaptation to forcing (Eq. (3)). The other is the commonly observed change of atmosphere (e.g., temperature and circulation) as an adaptation to PV change. Therefore, in brief, the adaptation of PV to forcing and

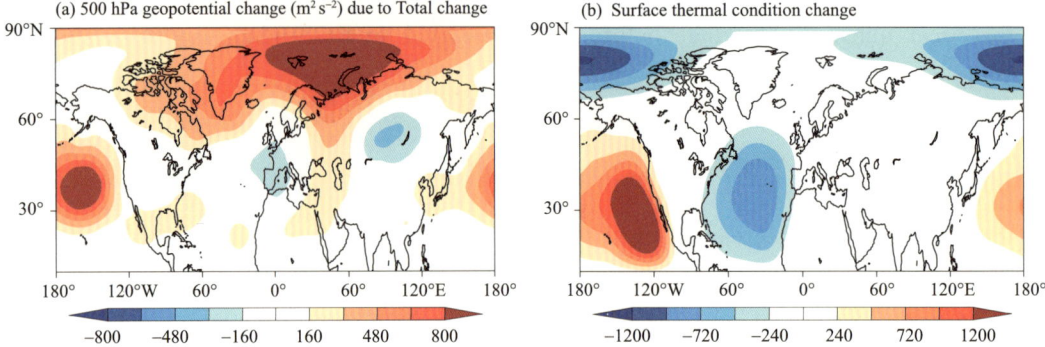

Fig. 4　Inverted geopotential changes (linear trend multiply the time span) from PPVI with three pieces, at the 500 hPa level, during boreal winter from 1979—2017, in units of m²s⁻². The three pieces are surface thermal condition change (b), surface and tropospheric PV change (not shown), and stratospheric total change (shown in Fig. 5f). (a) The sum of all the three pieces. Note that the ranges of the two color bars are different

coherent dynamic-thermodynamic adaptation to PV change is referred to as "PV adaptation". The nature of PV adaptation is the seamless dynamic-thermodynamic coupling.

5　Vertical coupling among different levels

The previous section discusses the conceptual understanding of the "stabilizer" effect of PV change and introduces the concept of PV adaptation based on two ideal cases. In practice, the concept of PV adaptation can be quantified by using PPVI and further used to understand the vertical coupling between different levels. In a real case, the net influence of the surface on the atmosphere is determined by the surface total change (i. e., changes in both surface thermal condition and PV). Thus, a new three-piece PPVI scheme was performed based on this consideration. The three pieces are the surface total change, tropospheric PV change, and stratospheric total change.

Fig. 5 shows the results of the new PPVI scheme. At 850 hPa, the geopotential change is dominated by surface total change (Fig. 5 left). The tropospheric PV change shows a tiny contribution to the geopotential change at 850 hPa (Fig. 5c), and even the 850 hPa level itself is located in the troposphere. The dominant surface contribution means that the change at 850 hPa is determined by the surface PV change via PV adaptation.

In sharp contrast, both surface total change and tropospheric PV change have great influences on the geopotential change at 500 hPa (Fig. 5 right). The positive geopotential change at 500 hPa corresponding to Arctic warming is determined by both surface total change and the tropospheric PV change. However, the negative geopotential change corresponding to Eurasian cooling (Figs. 1 and 2) is dominated by the tropospheric PV change (Fig. 5d), while surface total change shows negligible influence (Fig. 5b). The dominant contribution of tropospheric PV change means that enhanced Eurasian cooling at 500 hPa is mainly determined by tropospheric PV change via PV adaptation.

In the new piecewise scheme, the tropospheric PV change piece covers a wide vertical range, which includes the 850 and 500 hPa target levels themselves. In this sense, we cannot determine the vertical coupling among different levels within the troposphere. To make the vertical coupling clearer, two additional schemes of PPVI with five pieces are performed. In the new piecewise schemes, the surface and stratospheric pieces are the same as those in Fig. 5, while the tropospheric PV change piece is divided in-

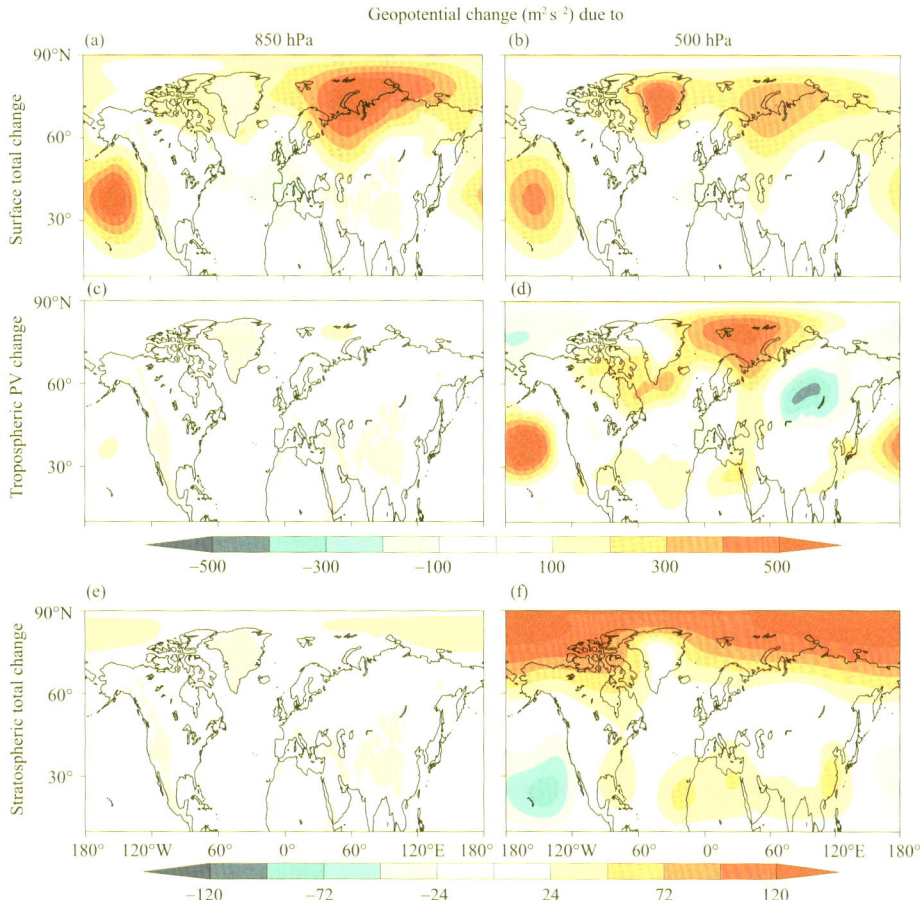

Fig. 5 Inverted geopotential changes (linear trend multiply by the time span) from PPVI with three pieces, at the 850 (Left) and 500 (Right) hPa levels, during the boreal winter from 1979—2017, in units of $m^2 s^{-2}$. The three pieces are surface total change (Top), tropospheric PV change (Middle), and stratospheric total change (Bottom). Note that the ranges of the two color bars are different

to three pieces in two ways. Regarding the 850 hPa level, the three pieces of tropospheric PV change are 900 hPa and below, 875—825 hPa, and 800—250 hPa. Meanwhile, for 500 hPa level, the three pieces are 600 hPa and below, 550—450 hPa, and 400—250 hPa.

Fig. 6 shows the results of the two new PPVI schemes. In Fig. 6, the two pieces 900 hPa and below and 875—825 hPa are merged for the 850 hPa level (Fig. 6a). Similarly, the two pieces 600 hPa and below and 550—450 hPa are merged together for 500 hPa (Fig. 6c). Because the contribution of the 900 hPa and below piece (not shown) is very tiny relative to the other two pieces, so does the 600 hPa and below piece (not shown). Thus, Figs. 6a and 6c represent the effects of a PV change within the vertical levels of 850 hPa and 500 hPa, respectively. Meanwhile, the 800—250 hPa and 400—250 hPa pieces represent the influences from upper levels (relative to 850 hPa and 500 hPa levels, respectively). As shown in Fig. 6, the influences from the upper-level PV change are larger than those of local (vertical levels) PV change on geopotential change at both the 850 and 500 hPa levels. Simply put, in the vertical direction, the influences of tropospheric PV change on both 850 hPa and 500 hPa levels are mainly top-down.

Notably, the top-down influence of PV change on 850 hPa (Fig. 6b) is negligible (relative to the bottom-up influence shown in Fig. 5a). In contrast, the top-down influence of PV change on the 500 hPa

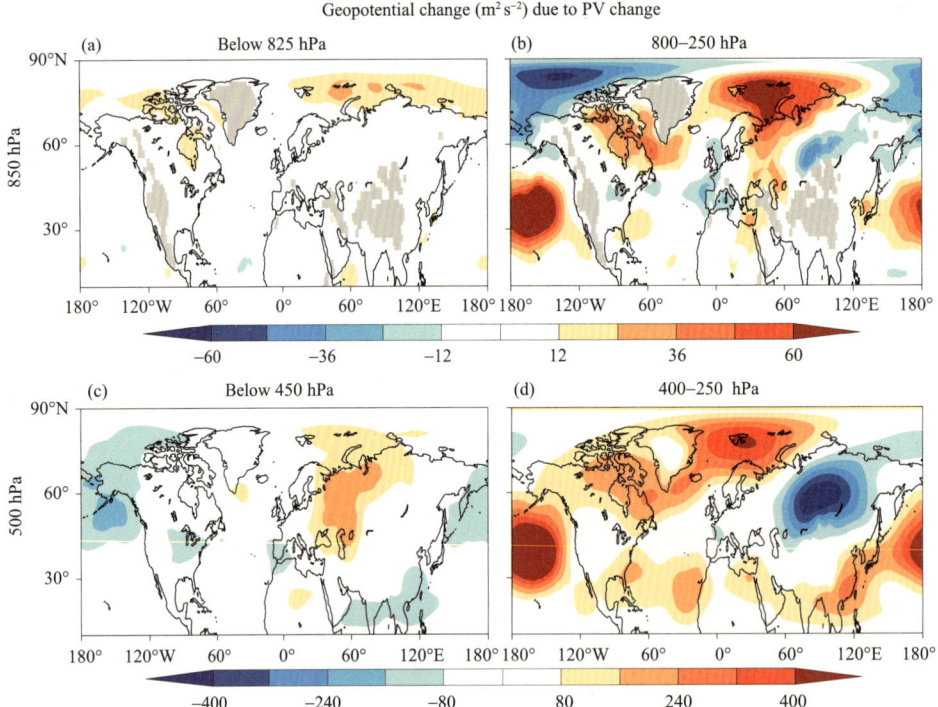

Fig. 6 Inverted geopotential changes (linear trend multiplied by the time span) from PPVI with five pieces, at the 850 (Top) and 500 (Bottom) hPa levels, during the boreal winter from 1979—2017, in units of $m^2 s^{-2}$. Two of the five pieces are surface and stratospheric total change. For 850 hPa, the remaining three pieces are tropospheric PV change from the surface to 900 hPa, 875—825 hPa, and 800—250 hPa (b). For 500 hPa, the remaining three pieces are tropospheric PV change from the surface to 600 hPa, 550—450 hPa, and 400—250 hPa (d). The sum of two pieces (surface to 900 hPa and 875—825 hPa) is shown, because the surface to 900 hPa piece has a very small value. For similar reasons, (c) is also the sum of two pieces (surface to 600 hPa and 550—450 hPa). Again, please notice the range of the color bar

geopotential change is as large as the influence of the surface total change (Figs. 5c, 6d). In particular, the negative geopotential change at 500 hPa over the Eurasian cooling region is almost completely generated by the top-down influence of PV change. Thus, enhanced Eurasian cooling at 500 hPa is mainly determined by the top-down influence of PV change via PV adaptation. According to Figs. 5 e and f, stratospheric total change show tiny direct influence on both 850 and 500 hPa levels. Nevertheless, the stratosphere may indirectly affect 850 and 500 hPa via a top-down mechanism, such as a gradual influence through the upper troposphere to 500 and 850 hPa. Therefore, the effect of stratospheric total change on troposphere should be reexamined.

To do so, the stratospheric influence on the 250 hPa level was examined, where the stratosphere is simply chosen as 225—1 hPa. Fig. 7 shows the geopotential change at 250 hPa from three-piece PPVI same as that in Fig. 5. As shown in Fig. 7, stratospheric total change show very little influence on the 250 hPa geopotential change relative to that of the tropospheric PV change. In addition, stratospheric influence mainly occurred over the Arctic rather than over the Eurasian cooling region (Fig. 7b). The stratospheric influence on tropospheric Eurasian cooling is negligible. Nevertheless, on the synoptic scale, the stratospheric influence on the troposphere may be crucial (Hartley et al., 1998; Cohen et al., 2014; Zhang et al., 2018). In summary, through PV adaptation, the bottom-up surface influence dominates the 850 hPa level, while both bottom-up surface and top-down upper-level (troposphere) influences control 500 hPa level. However, enhanced Eurasian cooling at 500 hPa is solely determined by the top-down influence.

6 Horizontal coupling due to PV advection and adaptation

In the last section, we explored the vertical coupling among different levels. However, the vertical coupling is not enough to tell the whole story because the linkage between Eurasian cooling and Arctic warming must have a horizontal pathway. On the basis of vertical coupling, the horizontal coupling is explored. To access the horizontal coupling, we just need to determine whether the local PV change is contributed by horizontal PV advection.

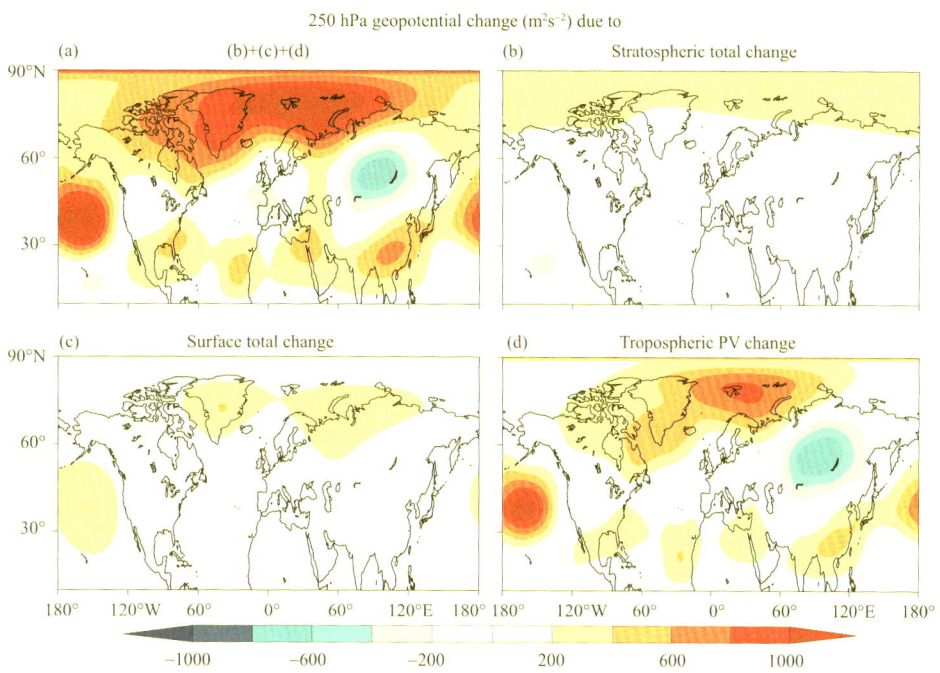

Fig. 7 Same as Fig. 5 but for results at the 250 hPa level

Here, only horizontal PV advection at the levels 850 and 500 hPa is examined because the changes show equivalent barotropic structure (Fig. 2). Regarding climate change, the horizontal PV advection term is linearized as $-V'_g \cdot \nabla \bar{q} - \bar{V}_g \cdot \nabla q'$ according to Eq. (5). The PV change itself in the equilibrium PV equation is directly reflected in the horizontal advection term. This is very different from the synoptic case, in which the PV change is reflected in the local tendency term ($\frac{\partial q}{\partial t}$ in Eq. (4)). In terms of physical understanding, the nature of the difference between climate and the synoptic case is a complete or ongoing PV adaptation.

Fig. 8 shows the PV climatology and change overlaid with the wind change and climatology for both 850 hPa and 500 hPa. The combinations of wind change/PV climatology and wind climatology/PV change represent $-\bar{V}'_g \nabla \bar{q}$ and $-\bar{V}'_g \cdot \nabla q'$, respectively. As shown in Fig. 8a, over the Eurasian cooling-Arctic warming region, the wind change flows across the isoline of climatological PV (with a large crossing angle). Thus, PV advection via wind change/PV climatology is vital. The situation at 500 hPa is similar (Fig. 8c).

Now, physical understanding of the relationship between horizontal PV advection change and PV

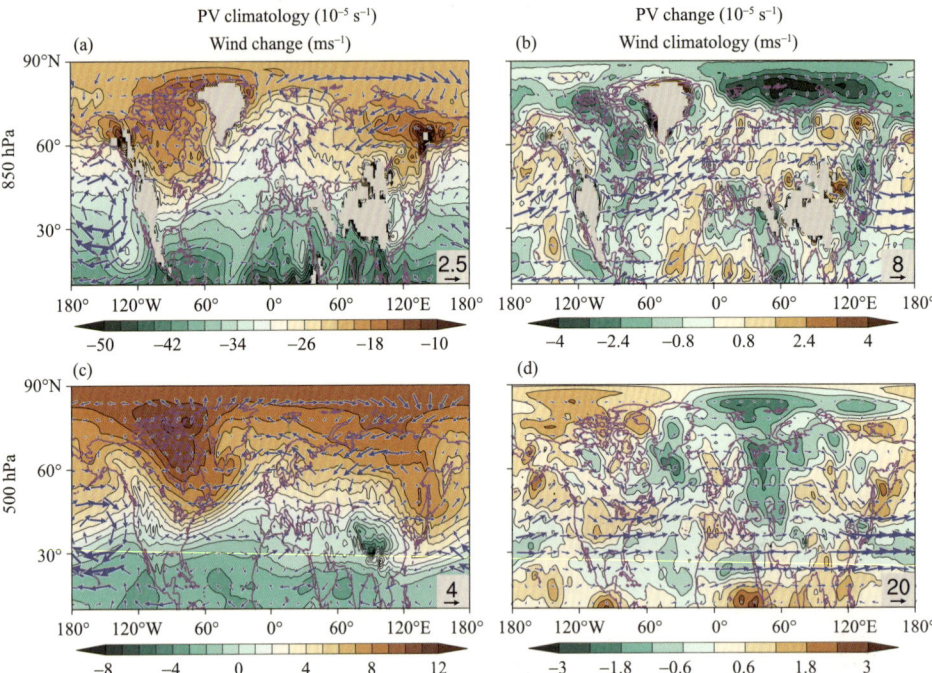

Fig. 8 (Left) PV climatology (filled, in units of $10^{-5}\,s^{-1}$) and horizontal wind change (vector, linear trend multiply the time span, in units of ms^{-1}) during the boreal winter from 1979—2017. (Right) Same as (Left) but for PV change and horizontal wind climatology. (Top) and (Bottom) are for the 850 and 500 hPa levels, respectively

change itself is a key to move forward. To make things clearer, we first examine an ideal case. According to Eq. (5), the balance relation is $-\boldsymbol{V}'_g \cdot \nabla \bar{q} + \bar{\boldsymbol{V}}_g \cdot \nabla q' = 0$ when diabatic heating and friction are neglected. The situation at 500 hPa is closer to this ideal case than that at 850 hPa. Therefore, we use the featured values of wind and PV at 500 hPa. Symbolically, the domain is chosen as (0°—60°E, 40°—80°N), and the magnitudes of the variables in both climatology and change are roughly set according to the featured values in Fig. 8 (bottom). For simplicity, the climatological wind is pure zonal (uniformly 12 ms^{-1}), the wind change is pure meridional (with anticyclonic zonal shear of 5 to -5 ms^{-1} from west to east, and northward is positive), and PV is zonally uniform but with a meridional gradient (1—8 $10^{-5}\,s^{-1}$ from south to north). This ideal case is shown in Fig. 9 (top), in which the overall patterns do not change with, although the magnitudes will change with, different values of selected parameters.

Under the PV climatology in Fig. 9 top-left, the southward wind change has positive PV advection, whereas the opposite is true for the northward wind change. The PV advection change induced by $-\boldsymbol{V}'_g \cdot \nabla \bar{q}$ for a given zonal shear value because of the wind change is shown in Fig. 9 top-middle. Then, the term $-\bar{\boldsymbol{V}}_g \cdot \nabla q'$ is balanced with $-\boldsymbol{V}'_g \cdot \nabla \bar{q}$. After the balance relation is solved, the PV change is shown in Fig. 9 top-right, where the PV change at the western boundary is set as zero. The balance between $-\boldsymbol{V}'_g \cdot \nabla \bar{q}$ and $-\bar{\boldsymbol{V}}_g \cdot \nabla q'$ will eventually make the distribution of PV change induced by $-\boldsymbol{V}'_g \cdot \nabla \bar{q}$ move downstream. Thus, the eventual PV change does not directly overlap with the PV advection change implied by the circulation change ($-\boldsymbol{V}'_g \cdot \nabla \bar{q}$).

With the help of the ideal case (Fig. 9 top), a real change at the 500 hPa level is not difficult to understand (Figs. 8c, d). The phenomenon that occurred around the Eurasian cooling region (Figs. 8c, d) is similar to the ideal case (Fig. 9 top). PV advection induced by the wind change upstream of the Eurasian cooling region (Fig. 8c) was advected by the climatological wind downstream (Fig. 8d). Eventually, the horizontal PV advection change generated a positive PV change over the Eurasian cooling region and

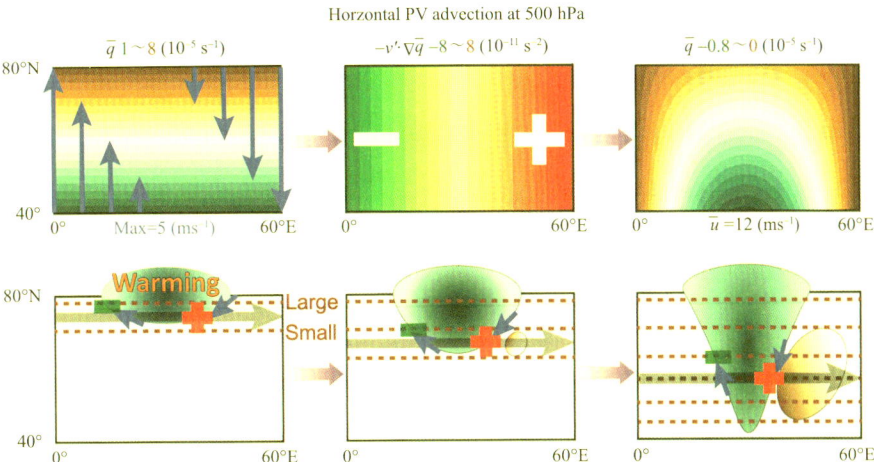

Fig. 9 (top) An ideal case of zero net horizontal PV advection (i.e., $\overline{\mathbf{V}}' \cdot \nabla\overline{q} + \overline{\mathbf{V}} \cdot \nabla q' = 0$), with prescribed climatology of PV ($\overline{q}$) and wind ($\overline{\mathbf{V}}$) and change of wind ($\mathbf{V}'$). The prescribed $\overline{q}$ and $\mathbf{V}'$ are shown as filled and blue arrows in (top-left). $\overline{q}$ is zonally uniform, and $\mathbf{V}'$ is purely meridional (with a maximum value of 5 ms^{-1}). $\overline{\mathbf{V}}$ is a uniform zonal wind ($\overline{u}$) of 12 ms^{-1}. These prescribed values approximately correspond to the situation at 500 hPa (Fig. 8 bottom, within the domain of approximately (0°—60°E, 40°—80°N)). (Top-middle) Distribution of $-\mathbf{V}' \cdot \nabla\overline{q}$ under prescribed values, in units of 10^{-11} s^{-2}. (top-right) Distribution of q' calculated from $\mathbf{V}' \cdot \nabla\overline{q} + \overline{\mathbf{V}} \cdot \nabla q' = 0$, in which q' at 0° longitude is prescribed as zero. (Bottom) Schematic view of the evolution of horizontal PV advection under a situation with enhanced regional warming over the Arctic, a meridional PV gradient (larger to the north, brown dashed lines indicate isoline of PV), and purely westerly wind (gray arrow). (Bottom) is based on results shown in (Top). Similar to (Top), the blue arrow indicates wind anomaly, "small" and "large" indicate the relative value of $\overline{q}$, and the minus and plus symbols indicate the sign of $-\mathbf{V}'_g \cdot \nabla\overline{q}$. The filled symbol indicates a change in PV (q')

a negative PV change over the upstream regions (Fig. 8d). In contrast, the situation at 850 hPa (Figs. 8a, b) seems different than that at 500 hPa and in the ideal case (Figs. 8c, 8d, 9 top). This discrepancy is expected because the diabatic heating effect is essential in the lower troposphere. Nevertheless, the general concept that PV advection induced by the upstream wind change will be advected by the climatological wind to downstream, as that is the compensation between $-\mathbf{V}' \cdot \nabla\overline{q}$ and $-\overline{\mathbf{V}}_g \cdot \nabla q'$, exists independent of whether force exists.

In extratropical regions, it is inevitable for an anticyclonic (or cyclonic) circulation change to cross the isoline of climatological PV because large PV gradients are widespread over the extratropical regions (e.g., Figs. 8a, c). Therefore, the anticyclonic circulation change induced by Arctic warming will extend southward via horizontal PV advection and PV adaptation. The schematic view in Fig. 9 (bottom) shows the extension of the Arctic warming-induced anticyclonic circulation change. As shown in Fig. 9 (bottom-right), first the local anticyclonic circulation change induced by Arctic warming advects positive PV to the adjacent area to the southeast and negative PV to the adjacent area to the southwest. No PV advection is shown on the northern side of the anticyclonic circulation anomaly (Fig. 9, bottom) because the PV gradient is very small in very northern regions (Figs. 8a, c). Second, the PV change induced by meridional wind advection is advected by climatological wind downstream. As shown in Fig. 9 (top), after a temporal balance between meridional and zonal PV advection is achieved, the negative PV change on the southern side of the anticyclonic anomaly is enhanced. In turn, the enhanced PV change induces a northward extension of the anticyclonic circulation anomaly via PV adaptation. At the same time, PV advection by climatological wind induces positive PV change in the downstream regions (east of the anticyclonic circulation anomaly). The final balanced PV change is schematically shown in Fig. 9 (bottom-right).

The illustration of the horizontal PV advection at 500 hPa (Fig. 8 bottom) is identical to the schematic view in Fig. 9 bottom. In Fig. 8 (bottom), the negative PV change in the Arctic Ocean induced by Arctic warming, the northward extension of this negative PV change, and the positive PV change over the downstream Eurasian cooling region are all observed. As demonstrated in the last section, Eurasian cooling at 500 hPa is dominated by top-down influence. The situation of horizontal PV advection at levels higher than 500 hPa is similar to that at 500 hPa. Simply put, enhanced Eurasian cooling at 500 hPa is regulated by both horizontal coupling (with Arctic warming) at higher levels and top-down vertical coupling.

Based on the PV advection perspective, the horizontal coupling is further examined using PPVI. Fig. 10 shows the results of a new PPVI scheme based on geographical piecewise. In the new scheme, there are three horizontal pieces, including 70°—90°N, 30°—70°N, and 10°—30°N. For each piece, the total change (changes in the thermal condition at the bottom and top boundaries and the PV for the entire atmosphere) is set to the observed values within the piece and is set to zero outside of each of the pieces. As shown in Fig. 10, the total change within each horizontal piece present evident influences on the circulation out of the piece. As shown in Fig. 10a, Arctic warming-related total change generates a clear influence on the circulation change over the adjacent area to the south (Europe). Similarly, total change over the midlatitude domain 30°—70°N also has a clear influence on adjacent regions (Fig. 10b). The extended influences shown in Fig. 10 confirm the extension process of the PV change schematically shown in Fig. 9 (bottom). Specifically, Fig. 10 suggests that the local PV change can further advect PV to distant regions through its extended influences on circulation.

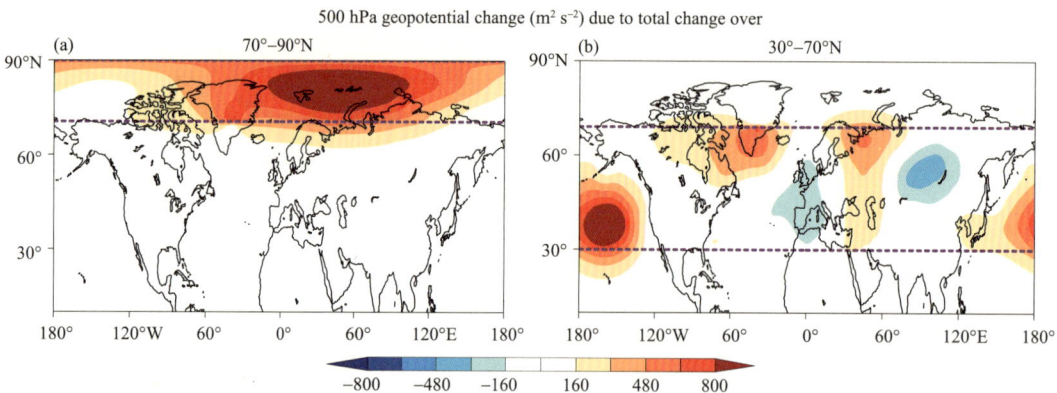

Fig. 10 Inverted 500 hPa geopotential changes (linear trend multiply the time span) from PPVI with three pieces, during the boreal winter from 1979—2017, in units of m² s⁻². The three pieces are total change over regions of 70°—90°N (Top), 30°—70°N (Bottom), and 10°—30°N (not shown). The purple dashed lines indicate the boundary of each piece. The lateral boundary conditions (geopotential change) at 90° and 10°N are observed and zero for 70°—90°N piece, and both zero for 30°—70°N piece, zero and observed for 10°—30°N piece

Actually, from a traditional wave view, the 500 hPa geopotential change associated with Eurasian cooling can be connected to the anomalous stationary wave that originated from upstream regions (e.g., Sung et al., 2018; Li et al., 2019). As shown in Fig. 11a, two wave trains pointed to the Eurasian cooling region. One is the western branch from Europe, and the other is the southeastern branch from the eastern coast of the Eurasian continent. Given the background of climatological eastward oriented stationary wave train (Fig. 11b), the upstream wave train means enhancement of the upstream influences the area from Europe to the Eurasian cooling region (Fig. 11a), while the downstream wave train means a reduction in the downstream influences from the Eurasian cooling region. Thus, the upstream wave train from Europe may contribute to the Eurasian cooling. However, the point is that the eddy streamfunction change over European regions connects with Arctic warming as demonstrated above. Such a wave view is a particular aspect of the PV view we proposed in Figs. 9—10 because anomalous wave depends on the existence of PV change (Hoskins et al., 1985; Plumb, 1985, 1986).

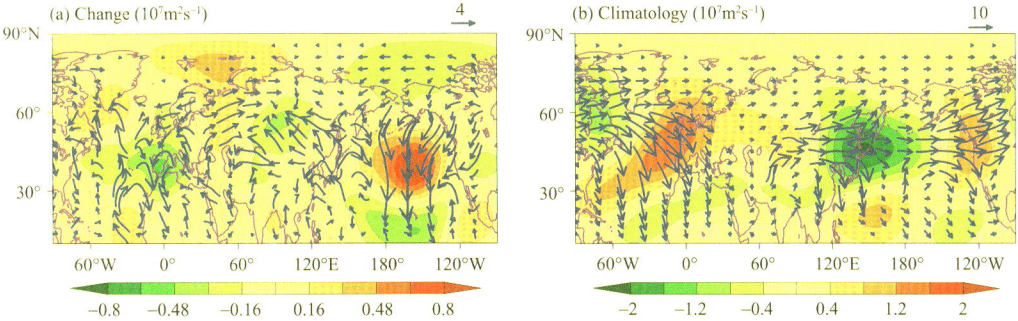

Fig. 11 (a) Changes (linear trend multiply the time span) in eddy (with zonal mean value removed) streamfunction (filled, in units of 10^7 m^2 s^{-1}) and Plumb flux (vector, in units of m^2 s^{-2}) at 500 hPa for the period of 1979—2017 during boreal winter. (b) Climatology for the period of 1979—2017 corresponding to (a)

7 Concluding remarks

By combining PV inversion with the PV equation, this paper explored the general dynamics that could connect the Eurasian cooling with Arctic warming from the PV view. We drew a schematic view that synthetically summarized the major conclusions from this paper in Fig. 12. The major results of this paper are summarized as follows:

1) Both Arctic warming and Eurasian cooling from 1979—2017 occurred throughout the entire troposphere. Eurasian cooling is large at both the surface and middle troposphere (approximately 500 hPa), while Arctic warming is most significant in lower layers.

2) Through PV adaptation, anticyclonic circulation change is generated by negative PV change, and the negative PV change is generated by the diabatic heating effect of Arctic warming that is stronger at lower levels. Here, PV adaptation refers to the adaptation of PV to forcing and coherent dynamic-thermodynamic adaptation to PV change.

3) Enhanced Eurasian cooling in the middle troposphere (around 500 hPa) is dominated by the top-down influence. Stratospheric influence on the troposphere change from 1979—2017 is negligible, especially for middle and low latitudes.

4) Change in horizontal PV advection is the cause of the horizontal coupling between the Arctic and the midlatitudes. The anticyclonic circulation change associated with Arctic warming and the climatological PV gradients are essential to horizontal PV advection change.

5) Taking the vertical and horizontal couplings together, enhanced Eurasian cooling at both the surface and in the middle troposphere is affected by the Arctic warming through PV advection and PV adaptation. Eurasian cooling at the surface and in the lower troposphere is dominated by PV advection and

PV adaptation at the surface. Meanwhile, enhanced Eurasian cooling in the middle troposphere is dominated by upper-level horizontal PV advection and corresponding PV adaptation.

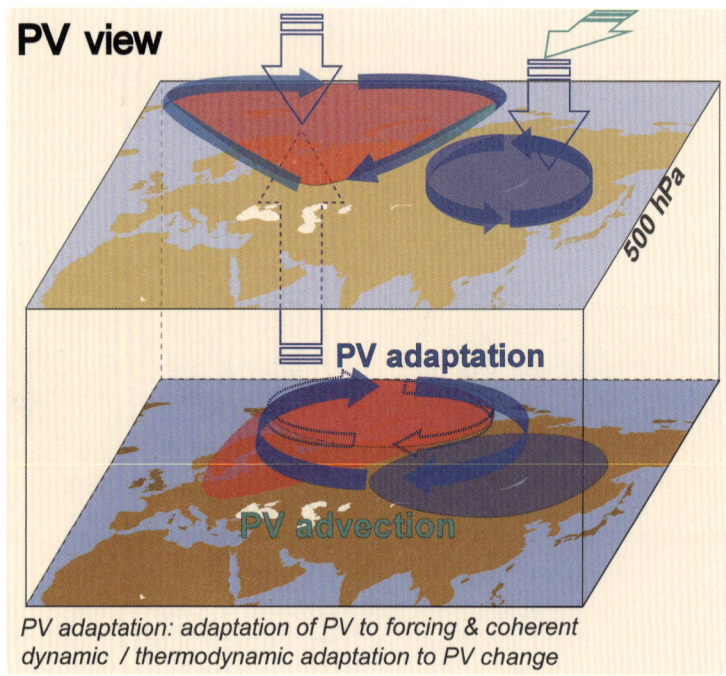

Fig. 12 Schematic view of the whole situation described in this paper. The red and blue shading indicate warming and cooling, respectively. The dashed blue arrows indicate circulation anomalies with PV adaptation but without horizontal PV advection. The mixing blue and turquoise filled arrows indicate the final circulation anomalies with both PV adaptation and advection. The topmost map indicates the 500 hPa level. The blue and turquoise arrows with tails indicate the direction of influences from PV adaptation and advection, respectively

In summary, we demonstrated a PV view picture of the three-dimensional linkages between wintertime Eurasian cooling and Arctic warming from 1979—2017. The whole situation is described by using only the concept of PV adaptation and PV advection. Technically, this paper presents an example of the use of PV dynamics in the climate change realm.

In plain language, the spatial pattern and vertical structure of Arctic warming and climatological PV distribution are crucial factors that determine the linkage between mid-latitude and Arctic. First, the zonal asymmetry and uneven vertical warming are crucial for a large circulation change directly corresponding to the Arctic warming. Second, the large climatological PV gradient is crucial to the southward extension of the anticyclonic circulation change associated with Arctic warming and corresponding downstream influences on Eurasia.

Overland et al. (2016) summarized the complexity of mid-latitude and Arctic connections. According to our findings, the complexity is partly due to the complexity of PV adaptation and PV advection. As we concluded above, the complexity is associated with the spatial pattern and vertical structure of Arctic warming and climatological PV distribution (Kug et al., 2015; Koenigk et al., 2016; Luo et al., 2018). Recently, Luo et al. (2019b) further suggested the crucial role of climatological PV gradient in controlling the linkage between Arctic and midlatitude weather extremes through regulating atmospheric blocking.

Overall, our findings could be beneficial for both theoretical understanding and practical modeling.

For example, attention should be paid to climatological PV distribution when performing emergent constraints on the climate model. However, it should keep in mind that this paper only depicts the general dynamical linkages between Arctic warming and Eurasian cooling, while the Arctic warming itself can be contributed to by both anthropogenic forcing and internal climate variability (Ding et al., 2019; Mori et al., 2019). Furthermore, regarding the direct connections between sea ice and the midlatitudes, recent model-based studies suggest that sea ice has minimal influence on Eurasian cooling on both long-term and interannual timescales (Ogawa et al., 2018; Blackport et al., 2019).

APPENDIX

Step-by-step procedure of piecewise PPVI:

1) To perform numerical integration, the equation must be nondimensionalized. The nondimensional form of Eq. (2) is $q^* = C_1 \nabla^2 \Phi^* + C_2 \frac{\partial \Phi^*}{\partial p^*} + C_3 \frac{\partial^2 \Phi^*}{\partial p^{*2}}$, where $\Phi = \Phi_0 \Phi^*$, $q = f_0 q^*$, $p = p_0 p^*$, $C_1 = \Phi_0 / f_0^2 a^2$, $C_2 = \frac{\partial}{\partial p}\left(\frac{1}{S}\right)$, $C_3 = \frac{1}{p_0 S}$, $\Phi_0 = 10^5 \text{ m}^2 \text{s}^{-2}$, and $p_0 = 10^5 \text{Pa}$. Note that p in C_2 is dimensional. For simplicity, the asterisk ($*$) is omitted hereafter in this section.

2) To solve the second-order linear partial differential equations for Φ', we need to apply the finite-difference method. Here, we use the central differentiation method to discretize the equation. Specifically, we have $\nabla^2 A = \frac{1}{\cos^2 \varphi_j} \frac{A_{i+1,j,k} + A_{i-1,j,k} - 2A_{i,j,k}}{\delta \lambda^2} + \frac{A_{i,j+1,k} + A_{i,j-1,k} - 2A_{i,j,k}}{\delta \varphi^2} - \tan\varphi_j \frac{A_{i,j+1,k} - A_{i,j-1,k}}{2\delta\varphi}$, $\frac{\partial A}{\partial p} = \frac{A_{i,j,k+1} - A_{i,j,k-1}}{p_{k+1} - p_{k-1}}$, and $\frac{\partial^2 A}{\partial p^2} = \frac{A_{i,j,k+1} + A_{i,j,k-1} - 2A_{i,j,k}}{(p_{k+1} - p_k)(p_k - p_{k-1})}$ for arbitrary variable A, in which λ and φ are longitude and latitude, and subscripts i, j, and k indicate the meridional, zonal, and vertical direction indexes of the three-dimensional spatial grid, respectively. Thus, six neighboring points are needed for any grid in the iteration process. The expressions used for the iteration are as follows: $\Phi'_{i,j,k} = [-q' + C_4(\Phi'_{i+1,j,k} + \Phi'_{i-1,j,k}) + C_5 \Phi'_{i,j+1,k} + C_6 \Phi'_{i,j-1,k} + C_7 \Phi'_{i,j,k+1} + C_8 \Phi'_{i,j,k-1}]/C_9$, where $C_4 = \frac{C_1}{\cos^2 \varphi_j \delta \lambda^2}$, $C_5 = C_1\left(\frac{1}{\delta\varphi^2} - \frac{\tan\varphi_j}{2\delta\varphi}\right)$, $C_6 = C_1\left(\frac{1}{\delta\varphi^2} + \frac{\tan\varphi_j}{2\delta\varphi}\right)$, $C_7 = \frac{C_2}{p_{k+1} - p_{k-1}} + \frac{C_3}{(p_{k+1} - p_k)(p_k - p_{k-1})}$, $C_8 = \frac{C_3}{(p_{k+1} - p_k)(p_k - p_{k-1})} - \frac{C_2}{p_{k+1} - p_{k-1}}$, $C_9 = \frac{2C_1}{\cos^2 \varphi_j \delta \lambda^2} + \frac{2C_1}{\delta \varphi^2} + \frac{2C_3}{(p_{k+1} - p_k)(p_k - p_{k-1})}$ and $\delta\lambda$ and $\delta\varphi$ are longitude and latitude intervals, respectively, for an equally divided horizontal rectangular grid. The successive overrelaxation method is adopted for the final iteration:

$$\Phi'^{m+1}_{i,j,k} = (1-\omega)\Phi'^{m}_{i,j,k} + \omega[-q' + C_4(\Phi'_{i-1,j,k} + \Phi'_{i-1,j,k}) + C_5 \Phi'_{i,j+1,k} + C_6 \Phi'_{i,j-1,k} + C_7 \Phi'_{i,j,k+1} + C_8 \Phi'_{i,j,k-1}]/C_9 \quad (A1)$$

where m indicates the round of integration and ω is the weight for successive overrelaxation, with a value of 1.6 in this study.

3) Boundary conditions for the geopotential are needed for numerical solution. The lateral and horizontal boundary conditions are the Dirichlet and Neumann conditions, respectively. For simplicity, the target area within a relatively large rectangular domain is commonly confined and outlined by latitude and longitude lines. Here, we chose the domain northward of 10°N, so that there are no boundaries on the east and west sides. Thus, the only lateral boundaries are at the North Pole and 10°N. We chose the geopotential at the southern boundary to be zero because it is far from our target zone (Hartley et al., 1998; Davis, 1992). However, we may not make the same choice for the northern boundary because we

are exploring extratropical aspects. In this sense, we first obtain the inverted geopotential from the piecewise PPVI using the zero boundary condition at the North Pole. Second, the boundary condition is given by portioning the total geopotential from reanalysis at the North Pole to each piece based on the percentage. This percentage is calculated as the mean geopotential north of 80°N, expect at the North Pole, in each piece relative to the total geopotential perturbation. Finally, the piecewise inversion is performed again by using the new boundary conditions at the North Pole. In fact, all our conclusions are not influenced even using simple zero north boundary. The horizontal boundary conditions are given by using the hydrostatic balance $\frac{\partial \Phi}{\partial p} = -\alpha$ at both the surface and the top.

4) Eventually, the iteration is performed according to Eq. (A1) for the interior layers. Meanwhile, the iteration at the surface and top boundaries should be performed according to Eq. (A2) with the given boundary conditions.

$$\Phi'^{m+1}_{i,j,k} = (1-\omega)\Phi'^{m}_{i,j,k} + \omega[-q' + C_4(\Phi'_{i+1,j,k} + \Phi'_{i-1,j,k}) + C_5\Phi'_{i,j+1,k} + C_6\Phi'_{i,j-1,k} + C_{10}]/C_{11} \quad (A2)$$

where $C_{10} = -C_2\alpha' - C_3\frac{\partial \alpha'}{\partial p}$, and $C_{11} = \frac{2C_1}{\cos^2\varphi_j \delta\lambda^2} + \frac{2C_1}{\delta\varphi^2}$. The iteration is terminated when the absolute value of geopotential differences between two iterations is less than 1 m² s⁻² in all three-dimensional grids.

The performance of PPVI must be assessed before exploring the results from PPVI. To test the accuracy of PPVI, we perform a simple inversion regarding the total PV change during the boreal winter from 1979—2017 over the domain northward of 10°N. Fig. A1 shows the original and inverted geopotential changes both at 850 hPa and 500 hPa levels. Geopotential changes at both 850 and 500 hPa from PPVI are identical to the original results directly from reanalysis. Thus, our PPVI procedure is accurate. Figure 2 shows the six IMFS obtained by EEMD that was performed on the gray line in Fig. 1b.

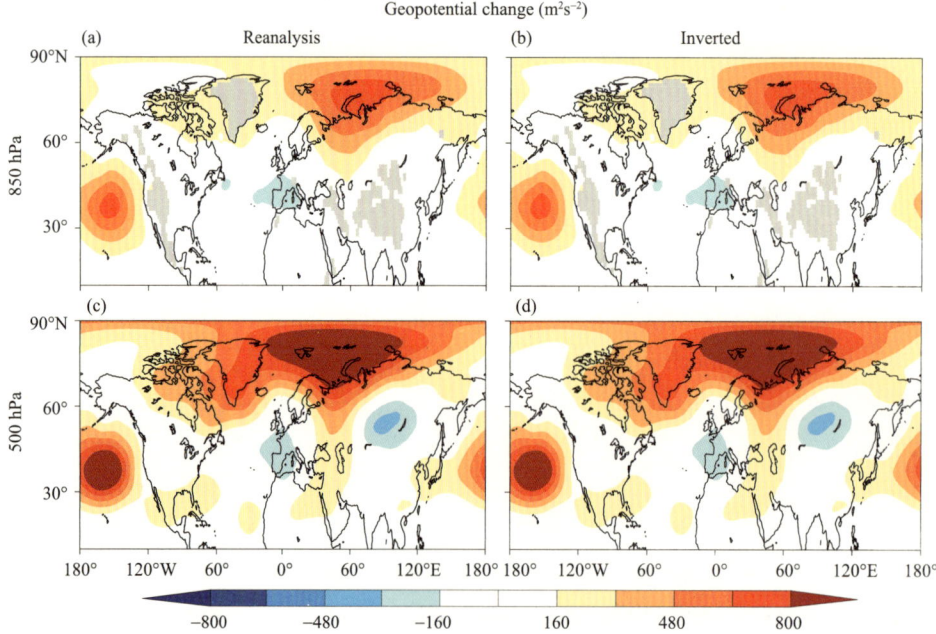

Fig. A1 Geopotential changes (linear trend multiply the time span) during boreal winter of 1979—2017, in units of m² s⁻². (Left) Results directly from reanalysis data. (Right) Results from PV inversion. (Top) and (Bottom) Results at 850 hPa and 500 hPa levels, respectively. The domain of performing inversion is northward of 10°N. The absolute value of max discrepancies between (a) and (b), (c) and (d) are 6.6, 2.4 m² s⁻², respectively

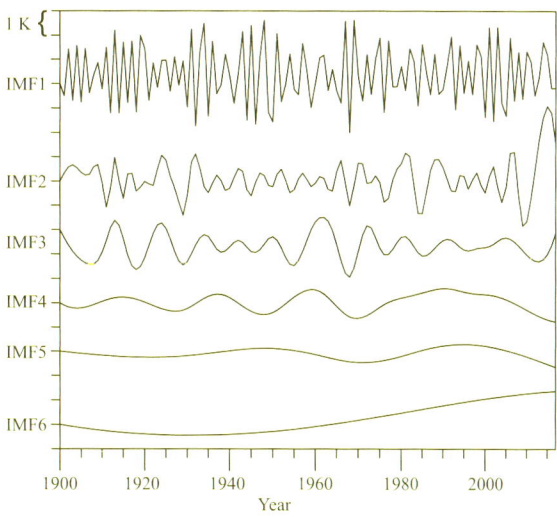

Fig. A2   IMFs 1-6 of EEMD corresponding to Fig. 1b

References

BLACKPORT R, SCREEN J A, VAN DER WIEL K, et al. 2019. Minimal influence of reduced Arctic sea ice on coincident cold winters in mid-latitudes[J]. Nat Clim Chang, 9: 697-704.

CHARNEY J G, STERN M E. 1962. On the stability of internal baroclinic jets in a rotating atmosphere[J]. J Atmos Sci, 19(2): 159-172.

COHEN J, SCREEN J A, FURTADO J C, et al. 2014. Recent Arctic amplification and extreme mid-latitude weather[J]. Nature Geoscience, 7: 627-637.

COHEN J, ZHANG X, FRANCIS J,et al. 2018. Arctic change and possible influence on mid-latitude climate and weather. A US CLIVAR White Paper[J]. Arctic Change & Possible Influence on Mid-latitude Climate & Weather: 41.

COUMOU D, DI CAPUA G, VAVRUS S, et al. 2018. The influence of Arctic amplification on mid-latitude summer circulation[J]. Nat Commun, 9: 2959.

DAI A, FYFE J, XIE S P, et al. 2015. Decadal modulation of global surface temperature by internal climate variability[J]. Nature Clim Change, 5: 555-559.

DAVIS C A. 1992. Piecewise potential vorticity inversion[J]. J Atmos Sci, 49:1397-1411.

DAVIS C A, EMANUEL K A. 1991. Potential vorticity diagnostics of cyclogenesis [J]. Mon Weather Rev, 119: 1929-1953.

DEE D P, AND UPPALA S M, SIMMON A J, et al. 2011. The ERA-Interim reanalysis: Configuration and performance of the data assimilation system[J]. Q J Roy Meteor Soc, 137:553-597.

DI CAPUA G D, COUMOU D. 2016. Changes in meandering of the Northern Hemisphere circulation [J]. Environmental Research Letters, 11: 094028.

DING Q, SCHWEIGER A, L'HEUREUX M, et al. 2019. Fingerprints of internal drivers of Arctic sea ice loss in observations and model simulations[J]. Nature Geoscience, 12: 28-33.

EASTERLING D R,WEHNER M F. 2009. Is the climate warming or cooling? [J]. Geophys Res Lett, 36: L08706.

EGGER J. 2008. Piecewise potential vorticity inversion: Elementary tests[J]. J Atmos Sci, 65:2015-2024.

EGGER J, HOINKA K P, SPENGLER T. 2015. Aspects of potential vorticity fluxes: Climatology and impermeability[J]. J Atmos Sci, 72: 3257-3267.

EGGER J,HOINKA K P, SPENGLER T. 2017. Inversion of potential vorticity density[J]. J Atmos Sci, 74: 801-807.

EGGER J, SPENGLER T. 2018. Nonuniqueness of attribution in piecewise potential vorticity inversion[J]. J Atmos Sci, 75: 875-883.

ERTEL H. 1942. Ein neuer hydrodynamischer wirbelsatz[J]. Met Z, 59: 277-281.

FRANCIS J A,VAVRUS S J, 2012. Evidence linking Arctic amplification to extreme weather in mid-latitudes[J]. Geophys Res Lett, 39, L06801.

FRANCIS J A,VAVRUS S J, 2015. Evidence for a wavier jet stream in response to rapid Arctic warming[J]. Environmental Research Letters, 10: 014005.

GU S,ZHANG Y,WU Q, et al, 2018. The linkage between Arctic sea ice and midlatitude weather: In the perspective of energy[J]. J Geophys Res, 123: 11536-11550.

HARTLEY D E,VILLARIN J T,BLACK R X, et al, 1998. A new perspective on the dynamical link between the stratosphere and troposphere[J]. Nature, 391: 471-474.

HELD I M, 2013. Climate science: The cause of the pause[J]. Nature, 501: 318-319.

HOLTON J R,HAKIM G J, 2013. An Introduction to Dynamic Meteorology[M]. Elsevier,Boston,USA: Academic press, 532.

HOSKINS B J,MCINTYRE M E, W ROBERTSON A, 1985. On the use and significance of isentropic potential vorticity maps[J]. Q J Roy Meteor Soc, 111: 877-946.

HUANG J,XIE Y,GUAN X, et al, 2017. The dynamics of the warming hiatus over the Northern Hemisphere[J]. Clim Dyn, 48: 429-446.

JI F,WU Z,HUANG J, et al, 2014. Evolution of land surface air temperature trend[J]. Nature Clim Change, 4: 462-466.

KOENIGK T, CAIAN M, NIKULIN G, et al, 2016. Regional Arctic sea ice variations as predictor for winter climate conditions[J]. Clim Dyn, 46: 317-337.

KOSAKA Y, XIE S P, 2014. Recent global-warming hiatus tied to equatorial Pacific surface cooling[J]. Nature, 501: 403-407.

KUG J S,JEONG J H , JANG Y S, et al, 2015. Two distinct influences of Arctic warming on cold winters over north America and East Asia[J]. Nature Geoscience, 8: 759-762.

LI X, WU Z, LI Y, 2019. A link of China warming hiatus with the winter sea ice loss in Barents-Kara Seas[J]. Clim Dyn, 53: 2625-2642.

LUO D, XIAO Y, DIAO Y, et al, 2016. Impact of Ural blocking on winter warm Arctic-cold Eurasian anomalies. Part II: The link to the North Atlantic Oscillation[J]. J Climate, 29: 3949-3971.

LUO D, CHEN Y, DAI A, et al, 2017a. Winter Eurasian cooling linked with the Atlantic multidecadal oscillation[J]. Environmental Research Letters, 12: 125002.

LUO D, YAO Y, DAI A, et al, 2017b. Increased quasi stationarity and persistence of winter Ural blocking and Eurasian extreme cold events in response to Arctic warming. Part II: A theoretical explanation[J]. J Climate, 30: 3569-3587.

LUO D, CHEN X, DAI A, et al, 2018. Changes in atmospheric blocking circulations linked with winter arctic warming: A new perspective[J]. J Climate, 31: 7661-7678.

LUO B, WU L, LUO D, et al, 2019a. The winter midlatitude-Arctic interaction: Effects of North Atlantic SST and high-latitude blocking on Arctic sea ice and Eurasian cooling[J]. Clim Dyn, 52: 2981-3004.

LUO D, CHEN X,OVERLAND J E, et al, 2019b. Weakened potential vorticity barrier linked to recent winter Arctic sea ice loss and midlatitude cold extremes[J]. J Climate, 32: 4235-4261.

MORI M, KOSAKA Y, WATANABE M, et al, 2019. A reconciled estimate of the influence of Arctic sea-ice loss on recent Eurasian cooling[J]. Nat Clim Change, 9: 123-129.

OGAWA F, KEENLYSIDE N, GAO Y Q, et al, 2018. Evaluating impacts of recent Arctic sea ice loss on the Northern Hemisphere winter climate change[J]. Geophys Res Lett, 45: 3255-3263.

OVERLAND J E, DETHLOFF K, FRANCIS J A, et al, 2016. Nonlinear response of mid-latitude weather to the changing Arctic[J]. Nat Clim Change, 6: 992-999.

PLUMB R A, 1985. On the three-dimensional propagation of stationary waves[J]. J Atmos Sci, 4: 217-229.

PLUMB R A, 1986. Three-dimensional propagation of transient quasi-geostrophic eddies and its relationship with the eddy forcing of the time-mean flow[J]. J Atmos Sci, 43: 1657-1678.

SCHNEIDER T, M HELD I, GARNER S T, 2003. Boundary effects in potential vorticity dynamics[J]. J Atmos Sci, 60: 1024-1040.

SCREEN J A, SIMMONDS I, 2010. The central role of diminishing sea ice in recent Arctic temperature amplification[J]. Nature, 464: 1334-1337.

SCREEN J A, SIMMONDS I, 2013. Caution needed when linking weather extremes to amplified planetary waves[J]. Proc Natl Acad Sci, 110: E2327-E2327.

SCREEN J A, SIMMONDS I, 2014. Amplified mid-latitude planetary waves favour particular regional weather extremes[J]. Nat Clim Change, 4: 704-709.

SCREEN J A, BRACEGIRDLE T J, SIMMONDS I, 2018. Polar climate change as manifest in atmospheric circulation[J]. Current Climate Change Reports, 4: 383-395.

SHEPHERD T G, 2016. Effects of a warming Arctic[J]. Science, 353: 989-990.

SMAGORINSKY J, 1953. The dynamical influence of large-scale heat sources and sinks on the quasi-stationary mean motions of the atmosphere[J]. Q J Roy Meteor Soc, 79: 342-366.

SMITH D, SCREEN J A, DESER C, et al, 2019. The Polar amplification model intercomparison project (PAMIP) contribution to CMIP6: Investigating the causes and consequences of polar amplification[J]. Geoscientific Model Development, 12: 1139-1164.

SPENGLER T, EGGER J, 2012. Potential vorticity attribution and causality[J]. J Atmos Sci, 69: 2600-2607.

STEINMAN B A, MANN M E, MILLER S K, 2015. Atlantic and Pacific multidecadal oscillations and Northern Hemisphere temperatures[J]. Science, 347: 988-991.

STOUFFER R J, MANABE S, 2017. Assessing temperature pattern projections made in 1989[J]. Nat Clim Change, 7: 163-165.

SUNG M, KIM S, KIM B, et al, 2018. Interdecadal variability of the warm Arctic and cold Eurasia pattern and its north Atlantic origin[J]. J Climate, 31: 5793-5810.

TRENBERTH K E, 2015. Has there been a hiatus? [J]. Science, 349: 691-692.

WALLACE J M, FU Q, SMOLIAK B V, et al, 2012. Simulated versus observed patterns of warming over the extratropical Northern Hemisphere continents during the cold season[J]. Proc Natl Acad Sci, 109: 14337-14342.

WEGMANN M, ORSOLINI Y J, ZOLINA O, 2018. Warm Arctic-cold Siberia: comparing the recent and the early 20th-century Arctic warmings[J]. Environmental Research Letters, 13: 025009.

WU Z, HUANG N E, WALLACE J M, et al, 2011. On the time-varying trend in global-mean surface temperature[J]. Clim Dyn, 37: 759-773.

XIE Y, HUANG J, MING Y, 2019. Robust regional warming amplifications directly following the anthropogenic emission[J]. Earth's Future, 7: 363-369.

YANG X Y, YUAN X, TING M, 2016. Dynamical link between the Barents-Kara sea ice and the Arctic oscillation[J]. J Climate, 29: 5103-5122.

YANG Z, HUANG W, WANG B, et al, 2018. Possible mechanisms for four regimes associated with cold events over East Asia[J]. Clim Dyn, 51: 35-56.

YAO Y, LUO D, DAI A, et al, 2017. Increased quasi-stationarity and persistence of winter Ural blocking and Eurasian extreme cold events in response to Arctic warming. Part Ⅰ: Insights from observational analyses[J]. J Climate, 30: 3549-3568.

YE K, JUNG T, SEMMLER T, 2018. The influences of the Arctic troposphere on the midlatitude climate variability and the recent Eurasian cooling[J]. J Geophys Res, 123: 10162-10184.

YEH T C, CHU P C, 1958. Some Fundamental Problems of the General Circulation of the Atmosphere[M]. Beijing: Science Press: 159.

ZHAO B, WU G, YAO X, 2007. Potential vorticity structure and inversion of the cyclogenesis over the Yangtze River and Huaihe River valleys[J]. Adv Atmos Sci, 24: 44-54.

ZHANG P, WU Y, SIMPSON I R, et al, 2018. A stratospheric pathway linking a colder Siberia to Barents-Kara Sea sea ice loss[J]. Science advances, 4: EAAT6025.

PV-*Q* Perspective of Cyclogenesis and Vertical Velocity Development Downstream of the Tibetan Plateau

WU Guoxiong[1,2], MA Tingting[1,3], LIU Yimin[1,2,4], JIANG Zhihong[3]

(1 State Key Laboratory of Numerical Modeling for Atmospheric Sciences and Geophysical Fluid Dynamics (LASG), Institute of Atmospheric Physics (IAP), Chinese Academy of Sciences (CAS), Beijing 100029, China; 2 College of Earth Science, University of Chinese Academy of Sciences, Beijing 100049, China; 3 Key Laboratory of Meteorological Disaster of Ministry of Education (KLME)/ Collaborative Innovation Center on Forecast and Evaluation of Meteorological Disasters (CIC-FEMD), Nanjing University of Information Science and Technology, Nanjing 210044, China; 4 CAS Center for Excellence in Tibetan Plateau Earth Sciences, Beijing 100101, China)

Abstract: This study reviews the development of the omega equation and the relation between omega and potential vorticity (PV) advection, and considers the application of PV theory to investigate the impact of large-scale mountains on downstream weather development. A diabatic quasi-geostrophic omega equation is introduced to reveal the feedback of diabatic heating (*Q*) on PV advection and vertical velocity. A challenge therein concerning the use of the diagnostic omega equation to interpret weather system development is considered from the PV-*Q* perspective based on a severe weather event that occurred downstream of the Tibetan Plateau (TP) on 17–21 January 2008. Results demonstrate that owing to PV restructuring, positive PV was generated over the eastern flank of the TP, and its eastward advection triggered development of isentropic displacement vertical velocity and cyclogenesis in the lower troposphere. A converging southeasterly wind accompanied with ascending isentropic gliding vertical velocity in the lower troposphere was induced to the east of the cyclone center, which transported not only warm moist air but also negative PV from the south, contributing to developments of both local diabatic ascent and precipitation and eastward migration of the cyclone. During the cyclone life cycle, three omega components induced by different processes interacted with each other, and diabatic heating associated with precipitation exerted considerable feedback to vertical motion as well as PV advection. It is the vertical differential PV advection and feedback from diabatic heating that control the evolution of the circulation and the associated precipitation downstream of the TP.

Keywords: PV restructuring and PV advection, omega equation, feedback of diabatic heating *Q*, Tibetan Plateau, cyclogenesis

本文原载于 *Journal of Geophysical Research: Atmospheres*, 2020, 125(16)

Plain Language Summary

The development of the omega equation and potential vorticity (PV) theory represents two fundamental accomplishments in relation to extratropical weather and climate dynamics in the 20th century. This study reviews the development of the omega equation. The omega equation constructed by Hoskins et al. (2003) is highlighted as this equation establishes the direct link between vertical displacement velocity from isentropic surface ω_{ID} and vertical variation of horizontal PV advection. The application of omega equation and PV theory to the impact of large-scale mountains on downstream weather development is also considered. Because the vertically differential PV advection is closely related with cyclogenesis, this ω_{ID}- equation is supposed to be applicable to interpreting weather system development. However, ω_{ID} is only one small part of the total vertical velocity, and the corresponding omega equation is a diagnostic equation rather than a prognostic one. How best to use this diagnostic equation to understand the development of full vertical velocity due to the triggering of ω_{ID} and the evolution of weather system has become a challenge.

This challenge has been tackled by applying an established diabatic quasi- geostrophic omega equation and the PV theory to a severe weather event associated with cyclogenesis that occurred downstream of the Tibetan Plateau (TP) on 17—21 January 2008. Results demonstrate that the development of the cyclone system is strongly tied both to the restructuring of PV near the surface of the eastern TP due to airflow convergence in the region and to the subsequent downstream eastward positive PV advection. Because the TP is located in western China with topography extending more than 5 km above sea level, the eastward advection of the generated positive PV over the TP's eastern flank which is in the midtroposphere can trigger the ascending development of various vertical velocity components below and air convergence and cyclogenesis in the lower troposphere. The converging southeasterly wind sliding upward along the northward sloped isentropic surfaces transports not only abundant water vapor from the ocean to South China to form precipitation and the resultant diabatic air ascent, but also negative PV advection which strengthens cyclogenesis further. The life cycle of cyclogenesis is controlled by the vertical coupling between the zonal positive PV advection in the mid-troposphere and the meridional negative PV advection in the lower troposphere. Furthermore, the released latent heat associated with precipitation can feedback on both PV advection and vertical velocity development.

It is the vertical variation of PV advection and interaction with diabatic heating that control the development of vertical velocity, the evolution of the cyclogenesis, and the associated precipitation downstream of the TP.

1 Introduction

Among various significant achievements in atmospheric sciences in the previous century, the developments of the omega equation and potential vorticity (PV) theory represent the two fundamental accomplishments in relation to extratropical weather and climate dynamics. Applying these established theories to interpret the impacts of orography on weather and climate has significantly advanced our understanding of the effects of mountains, which has contributed to the improvement of both weather forecasts and climate prediction.

1.1 Omega equation

Accurate diagnosis of vertical velocity has long been one of the challenging issues for meteorologists. With the application of quasi-geostrophic theory in meteorology, various forms of the omega equation have been used to interpret the vertical motion of systems (e.g., Sutcliffe, 1947; Bushby, 1952). Progressive studies linked vertical velocity with the advection of vorticity in the upper troposphere and the Laplacian of thermal advection (Petterssen, 1956; Djuric, 1969; Holton, 1972). For a diabatic atmosphere, the conventional omega equation in p-coordinates can be expressed as (Holton, 2004):

$$L\{\omega\} = \left(\Sigma^2 \nabla_h^2 + f^2 \frac{\partial^2}{\partial p^2}\right)\omega = f\frac{\partial}{\partial p}(\boldsymbol{V}_g \cdot \nabla \eta_g) + \Pi(p)\nabla_h^2(\boldsymbol{V}_g \cdot \nabla \theta) - \Pi(p)\nabla_h^2 \dot{\theta} \tag{1}$$

where the parameters are:

$$\begin{cases} \Sigma^2 = \Sigma^2(p) = -\Pi(p)\Theta_p \\ \Pi(p) = \frac{R}{p}\left(\frac{p}{p_0}\right)^\kappa \end{cases} \tag{2}$$

In the above, $\dot{\theta}$ is diabatic heating rate ($\dot{\theta} = \frac{d\theta}{dt} = Q$), $\boldsymbol{V}_g = \boldsymbol{k} \times f^{-1}\nabla\Phi$ is geostrophic wind, $\eta_g = f + f^{-1}\nabla^2\Phi$ is absolute geostrophic vorticity, $\Theta(p)$ is a standard potential temperature distribution averaged over a horizontal domain and a period of interest, θ_p is its vertical gradient, and the other parameters are used following their conventional meteorological notation. This equation provides a basis for understanding the dynamics of midlatitude weather systems. Each term on the right-hand side of Eq. (1) has clear interpretation as a separate physical process, i.e., increase with height of absolute vorticity advection, or warm advection, or diabatic heating will lead to ascent of air. However, in practice there is a drawback when diagnosing the individual forcing terms in Eq. (1) because considerable cancelation often exists between them. Thus, as highlighted by Hoskins et al. (1978), the effect of each term in isolation could be misleading when attempting to diagnose the magnitude and even the sign of the vertical velocity.

To overcome this drawback, Trenberth (1978) proposed an alternative but complementary approach, suggesting that upward motion is present in the midtroposphere where there is advection of cyclonic vorticity by the thermal wind. Hoskins et al. (1978) developed a $\boldsymbol{Q}$-vector approach for operational vertical motion estimation, in which the first two terms on the right-hand side of Eq. (1) are combined and the quasi-geostrophic forcing of vertical motion is written in a single term (the so-called divergence of the $\boldsymbol{Q}$-vector), which means the cancelation problem is negated. This improved method, which is more exact than the approach proposed by Trenberth (Dunn, 1991), was used widely during subsequent years. However, it was found inconvenient in operational application because of the difficulty in determining the $\boldsymbol{Q}$-vector directly from weather charts (Dunn, 1991). For this, Sanders and Hoskins (1990) proposed an easy method for estimation of the directions and relative magnitudes of $\boldsymbol{Q}$-vectors from weather maps.

Considerable effort has also been concentrated on linking the upper-level PV anomaly or PV advection to vertical velocity in the lower troposphere (Hoskins et al., 1985; Thorpe, 1985; Waugh and Polvani, 2000; Funatsu and Waugh, 2008). Hoskins et al. (2003, HPJ03 hereafter) made a breakthrough by formulating the link of vertical velocity with PV advection in z-coordinates. Based on the adiabatic thermodynamic equation, the vertical velocity of the adiabatic flow was partitioned into two components: the isentropic gliding (IG, hereafter) vertical velocity (w_{IG}) and the isentropic displacement (ID, hereafter) vertical velocity (w_{ID}). The w_{IG} ($w_{IG} = -N^{-2}(\boldsymbol{V}_g - \boldsymbol{C}) \cdot \nabla_h b$, $b = g\theta/\theta_0$, where θ_0 is a constant potential temperature, $N = (g\theta_0^{-1}\Theta_z)^{1/2}$ is the basic state Brunt-Väisälä buoyancy frequency, $\Theta(z)$ and Θ_z are the same as $\Theta(p)$ and Θ_p, respectively, except for p-coordinates, $\boldsymbol{C}$ is the horizontal velocity of a moving frame of reference, and ∇_h is the horizontal gradient operator) is associated with the horizontal motion of a particle moving along a sloping isentropic surface, whereas the w_{ID} ($w_{ID} = -N^{-2}(\partial b/\partial t)|_C$) measures the rate of vertical displacement of a particle that is associated with the development of the thermal field as well as the "development" of the system's circulation. It was further shown that the component of w_{ID} satisfies a conventional omega equation but with forcing determined solely by the vertical gradient of quasi-geostrophic PV (q_g, QGPV hereafter) advection (Eq. 11 in HPJ03):

$$\begin{cases} \left[N^2 \nabla_h^2 w_{ID} + f^2 \frac{\partial}{\partial z}\left(\frac{1}{\rho_r}\frac{\partial \rho_r w_{ID}}{\partial z}\right)\right] = f \frac{\partial}{\partial z}[(\mathbf{V}_g - \mathbf{C}) \cdot \nabla_h q_g] \\ q_g = f + \nabla_h^2 \psi + \frac{f^2}{\rho_r}\frac{\partial}{\partial z}\left(\frac{\rho_r}{N^2}\frac{\partial \psi}{\partial z}\right) \\ \text{Boundary condition: } w_{ID} = N^{-2}(\mathbf{V}_g - \mathbf{C}) \cdot \nabla_h b \text{ on } z = \text{constant} \end{cases} \quad (3a)$$

Where ψ is the geostrophic stream function, ρ_r the basic-state density. Therefore, the ID vertical velocity becomes a solely response to the interior redistribution of QGPV, i.e., increase of QGPV advection with increasing height will result in vertical air ascent ($w_{ID}>0$), and vice versa. The cancellation problem is negated, and the mechanism is concise and lucid. The solution of w_{ID} can be considered the sum of two parts: one associated with interior QGPV advection and the other associated with boundary temperature advection. In the presence of diabatic heating H ($H(x, y, z) = db/dt$), a diabatic ID vertical velocity w_{DID} that is identical to w_{ID} and a diabatic IG vertical velocity w_{DIG} ($w_{DIG} = -N^{-2}(\mathbf{V}_g - \mathbf{C}) \cdot \nabla_h b + N^{-2}H$) were defined. It was shown that the omega equation for w_{DID} is the same as for w_{ID} (Eq. 37 in HPJ03), and that the effect of heating on ID vertical velocity is present only implicitly in its impact on the tendencies of QGPV (Eq. 38a in HPJ03) and boundary temperature (Eq. 38b in HPJ03):

$$\begin{cases} \left[N^2 \nabla_h^2 w_{ID} + f^2 \frac{\partial}{\partial z}\left(\frac{1}{\rho_r}\frac{\partial \rho_r w_{ID}}{\partial z}\right)\right] = -f \frac{\partial}{\partial z}\frac{\partial q_g}{\partial t}\bigg|_C \\ \frac{\partial q_g}{\partial t}\bigg|_C = -(\mathbf{V}_g - \mathbf{C}) \cdot \nabla_h q_g + \frac{gf}{\theta_0 \rho_r}\frac{\partial}{\partial z}\left(\frac{\rho_r}{N^2}H\right) \\ \frac{\partial b}{\partial t}\bigg|_C = -(\mathbf{V}_g - \mathbf{C}) \cdot \nabla_h b + H \text{ on horizontal boundary} \end{cases} \quad (3b)$$

Once the tendencies are expressed in terms of system-relative advection and heating as pointed out in HPJ03, the effects of heating on w_{ID} will become explicit which will be performed in this study.

This novel omega equation developed by HPJ03 provides a tool for understanding how the interior redistribution of PV and boundary temperature advection can initiate vertical motion associated with system development. However, w_{ID} is only a part of the full vertical velocity, and mere use of w_{ID} to interpret the development of weather system may be misleading. Furthermore, all omega equations are diagnostic by nature, providing only the link between vertical velocity and the dynamic and thermodynamic conditions at a given time $t=t_0$. At t_0, the different components of vertical velocity ($w_{ID}(t_0)$ and $w_{IG}(t_0)$) can be calculated separately, albeit compensation or cancellation may exist as demonstrated in Figure 1 of HPJ03. However, during the development and evolution of a weather system ($t>t_0$), these different components can interact with each other as will be discussed below. Therefore, determining the approach for how best to employ the omega equation of w_{ID} to interpret the evolution of a weather system becomes a challenge.

1.2 Impact of mountains on downstream PV development

PV theory has been used widely in dynamic analysis of weather systems, as reviewed in Hoskins et al. (1985). The celebrated hydrodynamic PV equation developed by Ertel (1942) is:

$$\begin{cases} \dfrac{\mathrm{d}P}{\mathrm{d}t} = \alpha(\boldsymbol{\zeta}_a \cdot \nabla\dot{\theta} + \nabla\times \boldsymbol{F} \cdot \nabla\theta) \\ P = \alpha\boldsymbol{\zeta}_a \cdot \nabla\theta \end{cases} \quad (4)$$

where α is specific volume, $\boldsymbol{\zeta}_a$ is 3D absolute vorticity, $\boldsymbol{F}$ is 3D frictional acceleration in the momentum equation, and ∇ is a 3D gradient operator. The PV equation (4) considers the impact of both diabatic heating and friction; therefore, it is appropriate for application to an open dissipative system and suitable for climate studies.

During previous decades, the effect of near-surface PV generation on atmospheric circulation has attracted much attention. For example, Haynes and McIntyre (1987, 1990) and Schneider et al. (2003) introduced the concept of PV density (PVD, $W = \rho \cdot P = \boldsymbol{\zeta}_a \cdot \nabla\theta$, where $\rho = \alpha^{-1}$ is air density). They demonstrated that the total PV change along an isentropic surface intersecting the ground is determined by the integral of the normal component of PVD flux along the boundary where the isentropic surface intersects the ground surface in the part of the "underworld" defined by Hoskins (1991). The studies of Held and Schneider (1999), Koh and Plumb (2004), Schneider (2005), Egger et al. (2015), and others have also demonstrated that the surface boundary PV can substantially influence interior atmospheric circulation. These studies have indicated that PV in the free atmosphere is linked closely to surface PV generation. However, the questions of how PV is generated at the ground surface and how the generated surface PV might be transferred into the interior of the atmosphere to influence the atmospheric circulation remain unclear and challenging.

One example is the formation of PV anomalies in the lee of high terrain, which has long been the subject of research directed toward understanding the influence of mountains on downstream weather and climate. In a study of PV change of airflow parallel to the Alps, Thorpe et al. (1993) found the Alps a significant source of PV anomalies in the lower troposphere, with negative and positive PV anomalies on the northern and southern sides of the Alps, respectively. Such anomalies, which are advected away from the orography, become an important part of the tropospheric PV budget. Using high-resolution numerical model simulations, Aebischer and Schär (1998) showed that when the direction of the synoptic-scale wind is perpendicular to the Alps, low-level elongated PV banners are generated downstream. Furthermore, individual pairs of banners with anomalously positive and negative PV values can be attributed to flow splitting, either on the scale of the entire Alpine region or on that of individual massifs and peaks of the model topography. These banners evolve and grow when the orographically generated PV anomalies are advected downstream. In short, the PV of air flowing past high mountains can be altered and PV banners generated downstream of a complex mountain barrier can affect the downstream circulation and weather.

The Tibetan Plateau (TP), located in western China, covers an area of 2.5×10^6 km^2, i.e., approximately one quarter of the entire land territory of China. It extends to elevations of >5 km above sea level. Many of the synoptic-scale systems that cause damaging floods in China have their origins in the region of the TP and its surroundings (Tao and Ding, 1981; Chen and Dell'osso, 1984; Yasunari and Miwa, 2006). Previous studies (Shi et al., 2000; Wang et al., 2008; Zheng et al., 2013; Wang et al., 2014; Ma et al., 2020) have focused primarily on the influence of the TP on severe weather events during summer due to its unique thermodynamic forcing (Yeh et al., 1957; Ding et al., 1994; Chen et al., 1996; Wu et al., 1997, 2007, 2017; Li et al., 2002, 2016; Liu et al., 2020). In winter, a strong westerly flow that dominates the subtropical troposphere over Asia is split into northern and southern branches in the lower troposphere when it impinges the western TP (Yeh, 1950). Over the eastern TP, near-surface airflows with different origins converge from the north and the south, resulting in circulation perturbations that are then advected eastward by the westerly winds (Wan and Wu, 2007; Li et al., 2011; Pan et al., 2013; Ma et al., 2019; Yu et al., 2019). Similar to the Alps, circulation change in the region of the TP could act as a significant source of PV anomalies in the lower troposphere and generate downstream PV perturbations. The Yangtze River Basin, which is located downstream of the TP, is one area in China that frequently receives heavy precipitation associated with weather system development irrespective of season. However, the mechanism via which weather system development in winter might be related to PV generation in the lee of the TP remains unclear and constitutes another research challenge.

In this study, we employ a case diagnosis to elucidate how a winter weather system developed downstream of the TP in the diabatic atmosphere from the perspective of PV and the omega equation in p-coordinates. Two datasets are used for the diagnosis: 6 h cumulative precipitation data from 629 rain gauge stations distributed across China, obtained from the China Meteorological Administration (https://data.cma.cn/data/cdcdetail/dataCode/SURF_CLI_CHN_MUL_DAY_V3.0.html), and MERRA-2 data produced by the Global Modeling and Assimilation Office (Gelaro et al., 2017) based on the terrain-following hybrid σ-p coordinates of the GEOS-5 model. Three collections of MERRA-2 data are used: the basic assimilated meteorological fields on both pressure levels and model levels, and atmospheric state variables at the surface level provided on a 0.625°×0.5° (longitude × latitude) grid and sampled at 3 h intervals.

The remainder of the paper is arranged as follows. Section 2 introduces the circulation background of the selected weather event and analyzes the PV restructuring of the airflow passing the TP as well as the evolution of the related circulation. Section 3 presents descriptions of both the evolution of PV advection downstream of the TP and the triggering of cyclogenesis and vertical velocity by focusing on adiabatic processes. The interaction between the ID vertical velocity and the IG vertical velocity is also explored. Section 4 provides a PV-Q perspective on vertical velocity development, including the interaction between the IG vertical velocity and the vertical velocity induced by diabatic heating Q, as well as the feedback of heating on PV forcing and the ID vertical velocity. The evolution of cyclogenesis associated with meridional and zonal PV advection is analyzed in Section 5. Conclusions are provided in Section 6.

2 Background of the weather event and PV evolution downstream of the TP

2.1 Circulation background of the weather event

In January and early February 2008, southern China, including the Yangtze River Basin, experienced extreme and persistent low temperatures and severe freezing precipitation just before the Chinese Spring Festival. Detailed analysis of the synoptic-scale process (Tao and Wei, 2008) indicates that the freezing precipitation occurred because of the existence of an inversion layer in the lower troposphere in Southern China as is commonly observed in the season; and that this event resulted from the combined effects of multiple factors, which included the western Pacific subtropical high in the lower troposphere, a quasi-stationary front over southern China, and the eastward propagation of a cold trough from central Asia across the TP in the mid-troposphere. However, the mechanism via which such an extreme weather event might be linked to the forcing of the TP remains unclear. Thus, a synoptic episode that occurred during 17—21 January 2008 is selected for diagnosis from the perspective of PV and the omega equation.

During this period, the observed maximum daily precipitation amount, including freezing rain and snow, exceeded 30 mm in some areas (Shi et al., 2008, also refer to Fig. 1i), causing severe damage to local transportation, electrical power lines, and agriculture.

Figure 1 presents the evolution of the daily tropospheric circulation and precipitation over central and eastern China during 17−19 January 2008. At 500 hPa, westerly flow swept over the TP and East Asia. No remarkable perturbation was observed in the subtropics on 17 January (Fig. 1a). However, on the following day, a center of weak positive vorticity appeared over southwestern China (Fig. 1b), which became enhanced and expanded eastward (Fig. 1c). At 650 hPa, a center of high relative vorticity appeared first over the eastern flank of the TP on 17 January (Fig. 1d). It intensified gradually and then spread eastward during the following two days (Figs. 1e and 1f). Coincident with the intensification of the center of relative vorticity, a trough developed to the east of the TP on 18 January, which became strengthened on 19 January. The center of relative vorticity at 850 hPa was located along the Yangtze River Basin (~30°N) and to the east of the corresponding relative vorticity center at 650 hPa (Figs. 1g−1i), which enhanced gradually during 18−19 January over southern China. The distribution of the atmospheric circulation in the mid- and lower troposphere favored development of ascending air and precipitation. Organized large-scale precipitation started to develop on 18 January over South China (Fig. 1h), which increased considerably as it moved eastward during the following day (Fig. 1i). The observed

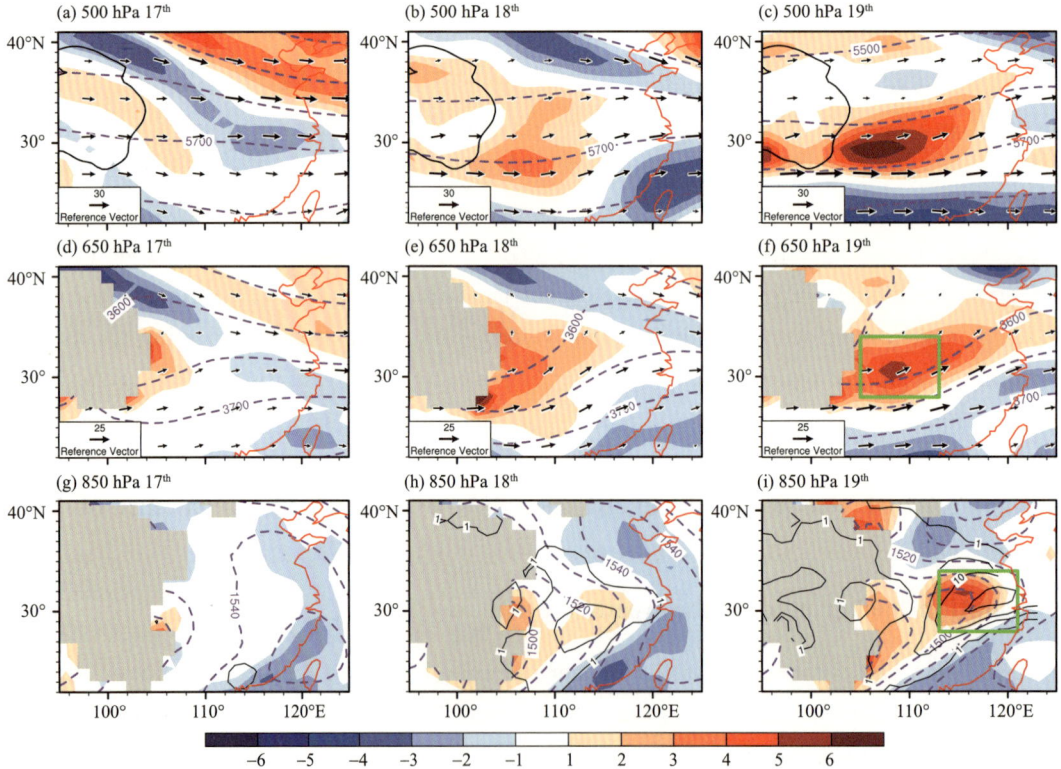

Figure 1 Distributions of daily mean relative vorticity (shading, 10^{-5} s^{-1}), geopotential height (purple dashed line, m), and wind (vector, m s^{-1}) at (a)−(c) 500 hPa, (d)−(f) 650 hPa, and (g)−(i) 850 hPa on 17 January (left column), 18 January (middle column), and 19 January 2008 (right column). Box area V (28°−34°N, 105°−113°E) in (f) denotes the key area of relative vorticity growth, and box area R (28°−34°N, 113°−121°E) in (i) denotes the key area of precipitation and vertical velocity. Black contour in (a)−(c) indicates the elevation of 3000 m. Gray area in (d)−(i) denotes the Tibetan Plateau. Black contours in (g)−(i) indicate daily precipitation of 1, 5, 10, and 20 mm

maximum daily precipitation over the Yangtze River Basin (28°—34°N, 113°—121°E; box area R in Fig. 1i) on 19 January exceeded 10 mm and it was as high as 20 mm in some areas.

Figure 2 presents the evolution during 17—18 January 2008 and along 33°N of the longitude-height distributions of Ertel PV, circulation (u, $-\omega$), and potential temperature (θ). A striking feature in the configuration of the isentropic surfaces is their eastwardly dispersed distribution, i. e. , from west to east, the isentropic surfaces tilt upward in the middle and upper troposphere but downward in the lower troposphere. This is because a strong cold trough known as the Northeast Asia Grand Trough exists in the upper troposphere over Northeast Asia in winter. Severe cold air is transported steadily southward by the northerly wind associated with the cold trough to East China, where the temperature in the free atmosphere is very low. The isentropic surfaces in the upper troposphere thus slope upward. Conversely, in the lower troposphere, the cold land surface to the west and the warm ocean surface to the east cause the isentropic surfaces to slope downward. Consequently, static stability decreases eastward downstream of the TP. This climatic feature is common in winter over East China. When adiabatic airflow propagates eastward downstream of the TP under the climatic background of decreasing static stability, its vertical vorticity is easily increased. This result at least partly explains the increase of the center of relative vorticity at 650 hPa during its eastward movement from the eastern flank of the TP on 17 January to eastern China on 19 January (Figs. 1d—f).

In Fig. 2, it is shown that the atmosphere over the region east of 110°E was characterized by dominant air descent and strong cooling on early 17 January, but that this changed to dominant air ascent and strong heating on 18 January. This feature is associated with the generation and eastward propagation of high PV over the eastern flank of the TP. At 06 UTC (Fig. 2b) of 17 January, a center of high PV ($>$ 1.5 PVU) appeared on the surface over the eastern flank of the TP with weak air ascent nearby to its east. Subsequently, the center of high PV extended eastward gradually; by 00 UTC on 18 January (Fig. 2e), the forefront of the 1.5 PVU contour had already passed 110°E and air ascent developed over eastern China, implying a potential link of vertical velocity development to PV advection.

2.2 PV restructuring in the region of the TP

The PV equation (Eq. (4)) indicates that external forcing, i. e. , either diabatic heating or frictional dissipation, can change the atmospheric circulation. Even in adiabatic and frictionless circumstances in which PV is conserved, the circulation can still change via internal conversion among, or "restructuring" of, the different PV components, i. e. , the 3D absolute vorticity (ζ_a), static stability ($\frac{\partial \theta}{\partial z}$), baroclinicity ($\frac{\partial \theta}{\partial x}$, $\frac{\partial \theta}{\partial y}$), and specific volume ($\alpha$) associated with convergence. The highly elevated TP intersects the isobaric and isentropic surfaces in the mid- and lower troposphere. The airflow in the lower troposphere that is split by the TP converges near the surface in the lee of the TP, resulting in a circulation anomaly due to changes in the PV structure that accompanies surface convergence. The downstream westerly flow might experience further change in PV structure associated with static stability variation. Such a process of PV restructuring induced by TP forcing could have considerable effects on the ambient atmospheric circulation, its thermal structure, and downstream weather development. Here via diagnosing the above selected cyclogenesis event, we analyze how the PV structure of the airflow changes in the region of the TP and how subsequent advection of PV downstream of the TP affects the development of the weather system and its vertical motion.

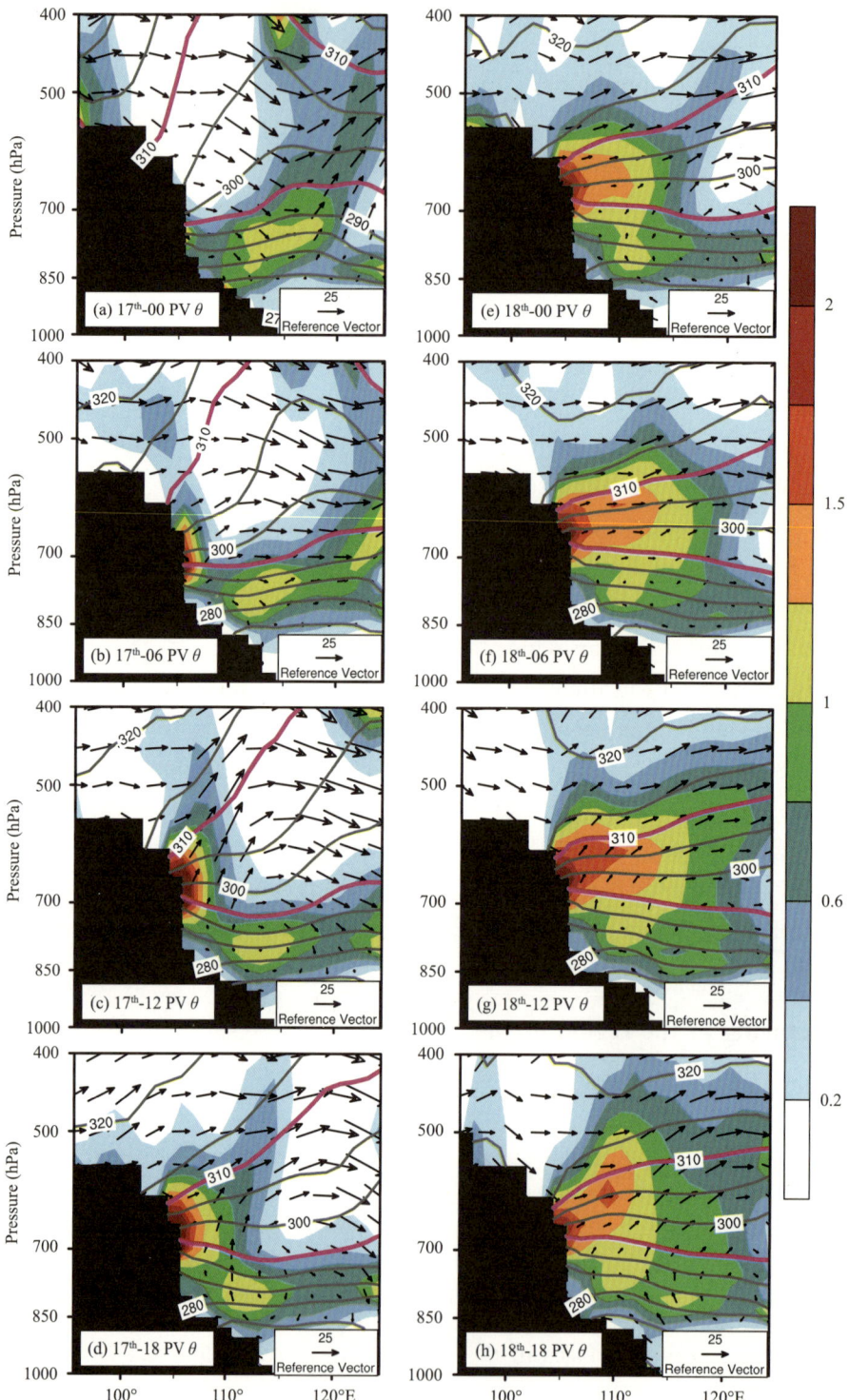

Figure 2 Vertical cross section along 33°N of PV (shading, PVU; 1 PVU = 10^{-6} K m² s⁻¹ kg⁻¹), potential temperature (contour, K), and reanalysis wind ($u\boldsymbol{i} + \omega_{OB}\boldsymbol{k}$, vector, units: u in m s⁻¹ and ω_{OB} in Pa s⁻¹ (values multiplied by a factor of −50 J) from (a) 00 UTC on 17 January to (h) 18 UTC on 18 January 2008 at 6 h intervals). Heavy magenta contours denote isentropic temperature of 295 and 310 K

The system of interest is the relative vorticity center initially located over the eastern TP at 650 hPa, which subsequently propagated eastward and intensified over the Yangtze River Basin (Figs. 1d—

f). Unlike the PV banner in the lee of Alps which usually possesses pairs of anomalously positive and negative PV values, the relative vorticity center over the eastern TP presents as a positive sheet. Its formation is related to PV restructuring within the boundary layer in the region of the TP. To demonstrate this, the PV is partitioned into two components:

$$P = \alpha W$$

in which $W = \zeta_a \cdot \nabla\theta$ is the amount of PV per unit volume, i.e., PV density W (Haynes and McIntyre, 1987, 1990; Schneider et al., 2003). From the PV equation (4), the PVD equation can be obtained:

$$\frac{dW}{dt} = \alpha^{-1}\left(\frac{dP}{dt} - W\frac{d\alpha}{dt}\right) = -W\nabla \cdot V + \zeta_a \cdot \nabla\dot\theta + \nabla \times F \cdot \nabla\theta \quad (5)$$

It implies that in addition to diabatic heating and friction, convergence of airflow associated with the change in specific volume can also cause a change in W. In terrain-following hybrid σ-p coordinate

$$\sigma = (p - p_T)/(p_S - p_T)$$

the local change of W_σ becomes

$$\frac{\partial W_\sigma}{\partial t} = -\nabla_\sigma \cdot [VW_\sigma - \zeta_{a\sigma}\dot\theta - \theta\nabla_\sigma \times F_\sigma] \quad (6)$$
$$\quad\quad\quad (\text{I})\quad\quad (\text{II})\quad\quad (\text{III})$$

in which p_S is pressure at the ground surface (depending on elevation), p_T is pressure at the top of the domain ($p_T = 0.01$ hPa in MERRA-2), subscript σ denotes the variable in σ-coordinates, and the gradient operator $\nabla_\sigma = \frac{\partial}{\partial x}\bm{i} + \frac{\partial}{\partial y}\bm{j} + \frac{\partial}{\partial \sigma}\bm{k}$. Using the MERRA-2 terrain-following hybrid σ-p coordinate data, the influence of airflow convergence on PV restructuring in the surface boundary layer can be estimated, the results of which are presented in Fig. 3. The local change (Eq. (6) term Ⅰ) and the contributions from both PVD flux convergence (term Ⅱ) and diabatic heating (term Ⅲ) are shown in the first, second, and third columns, respectively. For attribution purposes, the contribution from air convergence ($-W_\sigma\nabla_\sigma \cdot V$) which is one component of term Ⅱ is also shown in the right column. During this period, two notable centers of positive PVD tendency appear over the eastern TP south of 35°N. The southern center, which is located near 26°N with elevation just above 1.5 km, is associated with a stationary cold front over southern China. The northern center, located around 33°N with elevation of 3—4 km (first column), aligns well with the position of the positive vorticity center at 650 hPa shown in Fig. 1d. The formation of these two centers is largely due to PVD flux convergence (second column), whilst the contribution from diabatic heating is small (third column). Partitioning the W flux convergence into two components shows that the contribution from air convergence ($-W_\sigma\nabla_\sigma \cdot V$) is dominant (right column), whereas the component associated with the advection of W ($-V \cdot \nabla_\sigma W_\sigma$) is small, particularly over the eastern flank of the TP (figures not shown). These results indicate that mainly due to the conversion between specific volume α and PV density W, the near-surface convergence of airflow in the region of the TP contributes to the increase of PVD over the eastern flank of the TP.

2.3 Evolution of circulation and vorticity

Figure 4 presents the evolutions during 17—20 January 2008 of the relative vorticity (purple curve) averaged over box area V (28°—34°N, 105°—113°E, Fig. 1f) at 650 hPa, together with the precipitation (black curve), vertical velocity at 650 hPa (red curve), and vertical integral of convergence of water vapor flux from the surface to 650 hPa (blue curve) averaged over box area R (28°—34°N, 113°—121°E, Fig. 1i) which is located downstream of box area V. The evolution of the relative vorticity can be separated into four stages: Ⅰ. Initial Stage (12 UTC 17 January to 00 UTC 18 January); Ⅱ. Developing

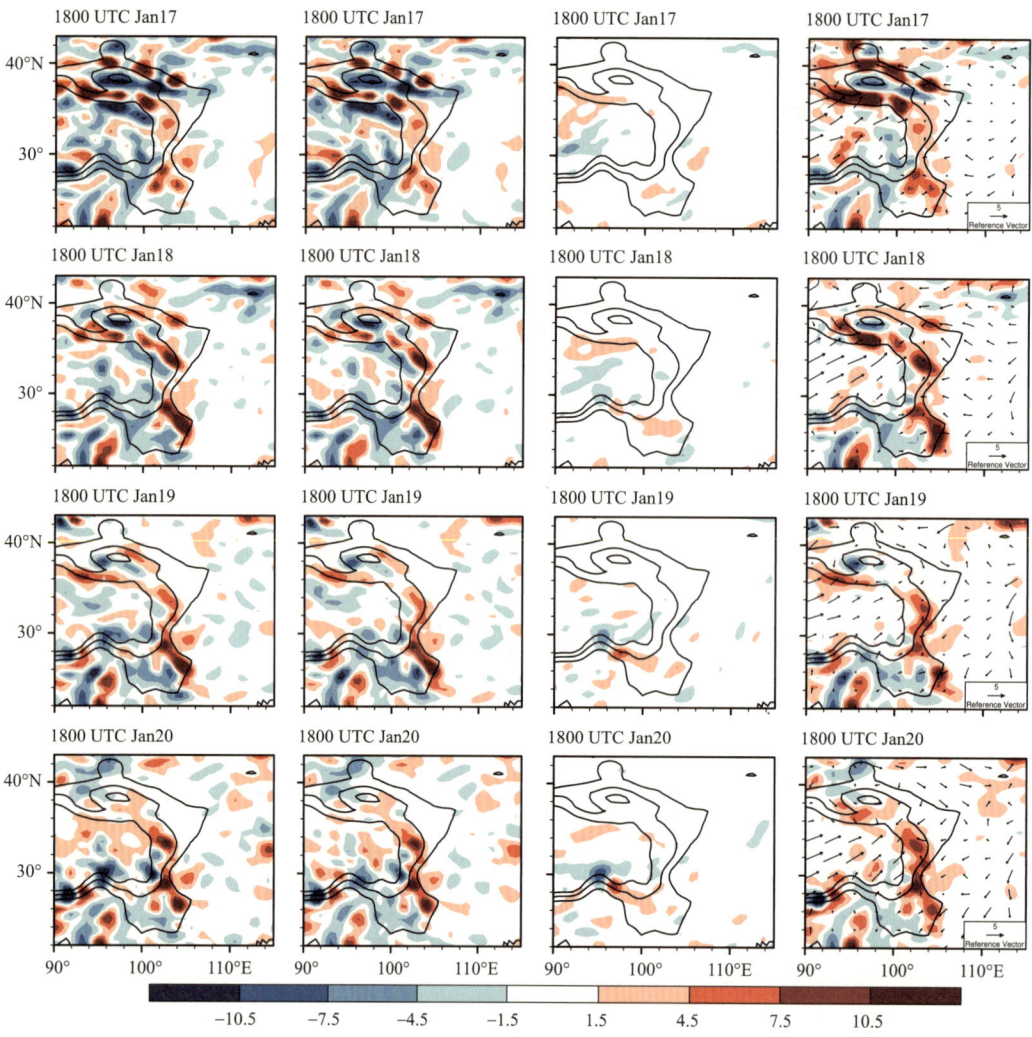

Figure 3 Distributions near the surface at 18 UTC of (left column) the tendency of surface potential vorticity density ($\partial W_\sigma/\partial t$), (second column) W-flux divergence term ($-\nabla_\sigma \cdot (\boldsymbol{V} W_\sigma)$), (third column) diabatic heating term ($\nabla_\sigma \cdot (\boldsymbol{\xi}_{a\sigma} \dot{\theta})$), and (right column) divergence term ($-W_\sigma \nabla_\sigma \cdot \boldsymbol{V}$) and surface wind at 2 m (vector, m s^{-1}) from 17 January (top row) to 20 January (bottom row) 2008. Black contours indicate elevations of 1500, 3000, and 4000 m. Unit in shading is 10^{-7} K s^{-2}

Stage (00 UTC 18 January to 00 UTC 19 January); Ⅲ, Maturation Stage (00 UTC 19 January to 00 UTC 20 January); and Ⅳ, Decaying Stage (00 UTC 20 January to 18 UTC 20 January). In Stage Ⅰ, relative vorticity was triggered to the east of the TP by 12 UTC on 17 January, followed by downstream development of air ascent and lower-layer convergence of water vapor flux, but with no appreciable precipitation (Fig. 1g). During Stage Ⅱ, as rapid development of cyclonic vorticity occurred in area V, air ascent and lower-layer water vapor convergence in the downstream area R intensified quickly with remarkable increase of precipitation (Figs. 1e and 1h). The intensity of the relative vorticity in area V and the precipitation in area R reached their maxima in Stage Ⅲ (Figs. 1f and 1i), before decaying quickly in Stage Ⅳ (Fig. 4). The other elements exhibited similar trends, except the decay of vertical velocity and the lower-layer convergence happened earlier and faster. These results imply that the system development represented by the evolution of relative vorticity at 650 hPa is fundamental in controlling the evolution of the entire weather process, including the vertical motion and the precipitation.

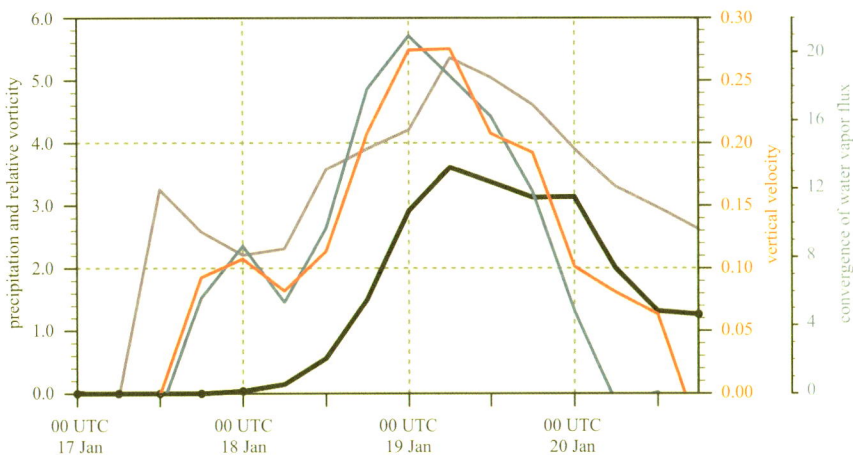

Figure 4 Evolution from 00 UTC 17 January to 18 UTC 20 January of the area-mean relative vorticity (purple curve, 10^{-5} s^{-1}) at 650 hPa averaged over box area V shown in Fig. 1f; and the area-mean 6 h accumulative precipitation (black curve, mm), vertical velocity (red curve, Pa s^{-1}; multiplied by a factor of -1) at 650 hPa, and vertical integral of convergence of water vapor flux from the surface to 650 hPa (blue curve, 10^{-7} kg m^{-2} s^{-1}; multiplied by a factor of -1) averaged over box area R shown in Fig. 1i

3 PV advection and triggering of vertical velocity and cyclogenesis

The seminal work of HPJ03 reveals a dynamic link of ω_{ID} with PV advection. However, ω_{ID} is only a small part of the full vertical velocity, whereas the PV advection or redistribution can lead to the development of full vertical velocity and cyclogenesis (Hoskins et al., 1985). How the triggering of ω_{ID} contributes to the development of full vertical velocity and cyclogenesis is unclear and will be explored here firstly for adiabatic circumstance.

3.1 Triggering of ω_{ID}, ω_{IG} and cyclogenesis

The link between the perspectives of the quasi-geostrophic omega equation and PV in p-coordinates for a diabatic atmosphere can be obtained by following the approach of HPJ03 in z-coordinates for an adiabatic atmosphere. For a frame of reference moving at constant horizontal velocity $\mathbf{C}$, the QGPV and thermodynamic equations can be written, respectively, as follows:

$$\left[\frac{\partial}{\partial t}\bigg|_{C} + (\mathbf{V}_g - \mathbf{C}) \cdot \nabla\right] q_g = f \frac{\partial}{\partial p}\left(\frac{\dot{\theta}}{\Theta_p}\right) \tag{7}$$

and

$$\left[\frac{\partial}{\partial t}\bigg|_{C} + (\mathbf{V}_g - \mathbf{C}) \cdot \nabla\right]\theta = -\Theta_p \omega + \dot{\theta} \tag{8}$$

In the above, the QGPV is defined as

$$q_g = f + f^{-1}\nabla_h^2 \Phi + f\frac{\partial}{\partial p}\left(\frac{1}{\Sigma^2}\frac{\partial \Phi}{\partial p}\right) = f + f^{-1}\nabla_h^2 \Phi + f\frac{\partial}{\partial p}\left(\frac{\theta}{\Theta_p}\right) \tag{9}$$

where the hydrostatic relation is $\partial \Phi/\partial p = -\Pi(p) \cdot \theta$. From Eq. (8), the vertical velocity of diabatic flow (ω) can be divided into three components, i.e., the ω_{ID}, ω_{IG}, and the diabatic heating vertical velocity (ω_Q):

$$\omega = \omega_{ID} + \omega_{IG} + \omega_Q \tag{10}$$

with

$$\omega_{ID} = -\Theta_p^{-1}\left(\frac{\partial \theta}{\partial t}\right)\bigg|_C \tag{11}$$

$$\omega_{IG} = -\Theta_p^{-1}(\boldsymbol{V}_g - \boldsymbol{C}) \cdot \nabla\theta \tag{12}$$

and

$$\omega_Q = \Theta_p^{-1}\dot{\theta} \tag{13}$$

The implication of ω_{ID} and ω_{IG} is the same as what is defined in HPJ03 and interpreted in Section 1.1, whereas the ω_Q term is related to diabatic heating. For a diabatic atmosphere, the ω_{ID} is equivalent to ω_{DID} in HPJ03, whilst the sum of ω_{IG} and ω_Q is equivalent to their w_{DIG}. The separation of ω_Q from ω_{IG} is for the use of following diagnosis. From Eqs. (7) to (11), the omega equation for ω_{ID} in a diabatic atmosphere can be obtained (Appendix):

$$\begin{cases} L\{\omega_{ID}\} = \left(\Sigma^2 \nabla_h^2 + f^2 \frac{\partial^2}{\partial p^2}\right)\omega_{ID} = f\frac{\partial}{\partial p}[(\boldsymbol{V}_g - \boldsymbol{C}) \cdot \nabla q_g] - f^2 \frac{\partial^2}{\partial p^2}\left(\frac{\dot{\theta}}{\Theta_p}\right) \\ \qquad = F_1 + F_2 = F \\ \text{in which } F_1 = f\frac{\partial}{\partial p}[(\boldsymbol{V}_g - \boldsymbol{C}) \cdot \nabla q_g], F_2 = -f^2\frac{\partial^2}{\partial p^2}\left(\frac{\dot{\theta}}{\Theta_p}\right) \\ \omega_{ID} = \Theta_p^{-1}(\boldsymbol{V}_g - \boldsymbol{C}) \cdot \nabla\theta - \Theta_p^{-1}\dot{\theta}, \text{ on horizontal boundary} \end{cases} \tag{14}$$

In a diabatic case ($Q = \dot{\theta} \neq 0$), Eq. (14) is equivalent to Eq. (3b) for w_{DID}. Heating can induce extra local change in potential temperature (Eq. (8)) and it can change ω_{ID} (Eq. (11)). In an adiabatic case ($\dot{\theta} = 0$ and $F_2 = 0$), Eq. (14) is equivalent to Eq. (3a) for w_{ID}. For the weather episode of current case study, the forefront of the 1.5 PVU contour propagates eastward from 107°E on 06 UTC 17 January (Fig. 2b) to 112°E on 18 UTC 18 January (Fig. 2h) with speed $|\boldsymbol{C}|$ (< 4 m s^{-1}) that is much slower than the basic flow $|\boldsymbol{V}_g|$ (which is close to $|\boldsymbol{V}|$ ($\sim 10-20$ m s^{-1}), as shown in Fig. 2), i.e., $\boldsymbol{V}_g - \boldsymbol{C} \approx \boldsymbol{V}_g$. Therefore, according to Eq. (14), ID air ascent ($\omega_{ID} < 0$) is generated when QGPV advection increases with height, and vice versa.

Eq. (14) is only a diagnostic equation for a given time $t = t_0$. A natural extension of the above conclusion is how the initiation of ω_{ID} leads to the triggering of ω_{IG} and cyclogenesis afterwards ($t > t_0$). Let the characteristic horizontal and vertical scales of the system under consideration be L and H ($\approx P_0/\rho_0 g$, where P_0 is the characteristic pressure depth and ρ_0 is the characteristic density), respectively. Then, the ratio between the first and second terms on the left-hand side of Eq. (14) becomes $R = (NH/fL)^2$. For adiabatic large-scale motion in which $L \gg NH/f$, the second term on the left-hand side of Eq. (14) is much larger than the first. Thus, Eq. (14) can be qualitatively approximated to the following:

$$f^2 \frac{\partial^2 \omega_{ID}}{\partial p^2} \approx -f\frac{\partial}{\partial p}\left(\frac{\partial q_g}{\partial t}\bigg|_C\right) = f\frac{\partial}{\partial p}[(\boldsymbol{V}_g - \boldsymbol{C}) \cdot \nabla q_g]$$

Thus,

$$f\frac{\partial \omega_{ID}}{\partial p} \approx -\frac{\partial q_g}{\partial t}\bigg|_C = [(\boldsymbol{V}_g - \boldsymbol{C}) \cdot \nabla q_g]$$

and

$$\frac{d\zeta_g}{dt}\bigg|_{ID} \propto f\frac{\partial \omega_{ID}}{\partial p} = -f\nabla_h \cdot \boldsymbol{V}_{ID} \approx [(\boldsymbol{V}_g - \boldsymbol{C}) \cdot \nabla q_g] \tag{15}$$

For an adiabatic case, Eq. (15) indicates that positive QGPV advection causes horizontal divergence, whereas negative QGPV advection causes horizontal convergence and development of cyclonic relative vorticity. Thus, a configuration with negative PV advection below and positive PV advection aloft, that is an increase of PV advection with increasing height, will lead to cyclogenesis in the lower layer and ascending ω_{ID} in between owing to the atmospheric continuity. Furthermore, the convergence flow (V_{ID}) in the lower layer induced by the negative PV advection should move along the isentropic surfaces, forcing the IG flow ($V_{IG} \equiv V_{ID}$). According to Eq. (12), IG vertical velocity is upward ($\omega_{IG} < 0$) when the flow V_{IG} is along positively sloped isentropic surface and downward when it is along negatively slopped isentropic surface. For a stably stratified atmosphere, isentropic surface always tilts upward towards cold area, adiabatic air flow V_{IG} towards cold area (or warm advection) should ascend while descend if it is towards warm area (or cold advection). For the winter weather event under investigation, as the isentropic surfaces in the lower troposphere tilt upward toward the north, the development of southerly flow to the east of a center of cyclogenesis will result in associated IG air ascent:

$$(\omega_{IG})_v = -v_{IG}(\partial\theta/\partial y)\Theta_p^{-1} < 0 \tag{16}$$

At the same time, the isentropic surfaces in the lower troposphere over East Asia tilt downward toward the east (Fig. 2), the convergence ($\nabla_h \cdot V_{IG} < 0$) in the lower layer will also result in IG ascending easterlies:

$$(\omega_{IG})_u = -u_{IG}(\partial\theta/\partial x)\Theta_p^{-1} < 0. \tag{17}$$

Consequently, ascending IG vertical velocity ($\omega_{IG} < 0$) and cyclogenesis are triggered in the lower layers by the development of ω_{ID} aloft due to the sloping of isentropic surfaces, i. e., the baroclinicity of the atmosphere: The stronger the baroclinicity, The stronger the ascending ω_{IG}.

On the steep eastern flank of the TP, many isobaric surfaces intersect the surface of the TP. The PV generated near the surface at these locations (Figs. 2 and 3) can be advected toward downstream areas and thus influence the in situ circulation. Figure 5 shows the evolutions of PV at 650 hPa, ω_{ID} at 700 hPa, and relative vorticity (ζ), potential temperature (θ), ω_{IG}, and streamlines at 850 hPa. To maintain consistency with Fig. 2, Ertel PV is used for constructing the panels in the left column in Fig. 5. This will not affect evaluation of the impact of QGPV advection because (to the first-order approximation) Ertel PV (P) and QGPV (q_g) satisfy the relation $q_g \approx P/[-g\Theta_p(p)]$ (Nielsen-Gammon and Gold, 2006).

In Stage I (Figs. 5a—c) there was no precipitation. According to Eqs. (14) — (17), PV advection at 650 hPa triggers the ascending ω_{ID} at 700 hPa (Fig. 5a) and ω_{IG} at the lower level of 850 hPa (Fig. 5b). The induced southerly (Fig. 5c) is associated with negative PV advection, which causes air convergence and the formation of cyclonic relative vorticity in situ (Eq. (15)). In a moving frame of reference with its origin located at the forefront of the 1.5 PVU contour of the high PV region over the eastern TP, significant ascending ω_{ID} appears only close to the forefront (Fig. 5a), whereas strong ascending ω_{IG} dominates the vertical motion away from the forefront (Fig. 5b). This is in good correspondence with the analytical result of an idealized case experiment of HPJ03 in which a prescribed constant QGPV is located at a height where the zonal wind speed is a constant westerly (refer to their Eq. (15) and Fig. 2). From Stage II to Stage III (Figs. 5d—i), as eastward PV advection at 650 hPa intensified (Figs. 5d and 5g), the cyclonic relative vorticity at 850 hPa and the associated ascending ω_{IG} (Figs. 5e and 5h), together with the southerly flow to the east of the cyclone center (Figs. 5f and 5i), develop further. The warm advection and negative PV advection to the east of the center become apparent, contributing to the slow eastward movement of the cyclone center (Eq. (15)). The entire process is similar to the schematic of cyclogenesis associated with the arrival of an upper-air PV anomaly over a lower-level baroclinic region (Hoskins et al., 1985). In Stage IV (Figs. 5j—l), as positive PV was advected rapidly to the western

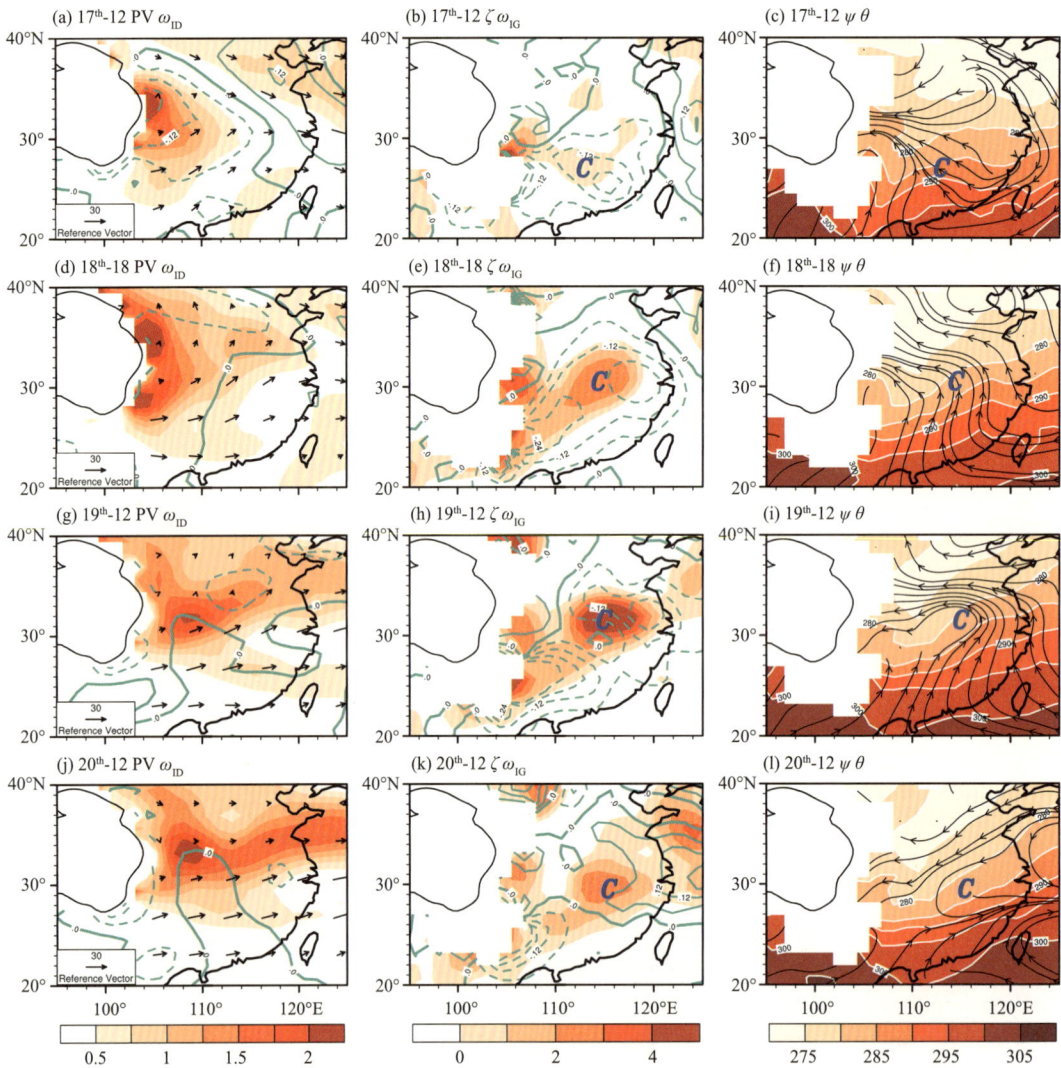

Figure 5 Distributions of (left column) PV (shading, PVU) and wind (vector, m s^{-1}) at 650 hPa, and ω_{ID} (contour, Pa s^{-1}) at 700 hPa; (middle column) relative vorticity (shading, 10^{-5} s^{-1}) and ω_{IG} (contour, Pa s^{-1}) at 850 hPa; and (right column) potential temperature (shading, K) and circulation (streamline) at 850 hPa at (a)—(c) 12 UTC 17 January, (d)—(f) 18 UTC 18 January, (g)—(i) 12 UTC 19 January, and (j)—(l) 12 UTC 20 January. "C" denotes the cyclone circulation center at 850 hPa

Pacific, the main cyclonic center moved from the continent to the Sea of Japan, the prevailing flow at 850 hPa over eastern China changed from southerly to northerly (Fig. 5l), and, following formula (16), the ω_{IG} changed from ascent to descent (Fig. 5k); consequently, the cyclonic vorticity decayed.

3.2 Evolution of different ω components

For a specific location (x, y, p), the ω_{ID}, ω_{IG}, and ω_Q at a given time $(t=t_0)$ can be calculated separately from Eqs. (11—13). They may compensate or cancel to each other and their sum satisfies Eq. (10). However, the results from Fig. 5 demonstrate that during the cyclone evolution $(t>t_0)$, they can influence each other. To reveal how the different omega components interact, the evolutions of ω_{ID}, ω_Q, and total ω at 700 hPa, together with the ω_{IG} and horizontal wind divergence at 850 hPa, are shown in Fig. 6. In Stage I when there was no precipitation, ω_Q was weak and descending. ω_{IG} was weak ascending because of the background southerly wind associated with the depression to the east of the TP (Fig. 1g). Owing to PV forcing, ascending ω_{ID} at 700 hPa was triggered by 06 UTC on 17 January. Consequently, ascending ω_{IG} at 850 hPa increased, leading to increasing vertical stretching and contributing to the development of the cyclonic vorticity and the southerly flow (Figs. 4 and 5c). The negative feedback of the ω_{IG} development on ω_{ID} as will be discussed caused the latter to start to weaken after 18 UTC on 17 January.

In Stage II, ascending ω_{IG} increased continuously, and ascending heating velocity ω_Q started to develop as ω_{ID} changed to descending. It is interesting to see that the intensity of ω_{IG} started to decline at 18 UTC on 18 January before the vorticity and precipitation reached their peaks at 06 UTC on 19 January (Fig. 4). This is because the occurrence of precipitation along the Yangtze River Basin increases the in situ potential temperature and reduces the meridional gradient of potential temperature to its south. According to Eq. (16), the ω_{IG} should be weakened. Despite this, wind convergence at 850 hPa continued to increase and reached its maximum at 00 UTC on 19 January, i.e., at the beginning of Stage III, as did both the convergence of water vapor flux and the air ascent in the precipitation region of area R (Fig. 4). Six hours later, the total vertical air ascent ω at 700 hPa, cyclonic vorticity, and precipitation all reached their maxima (Figs. 4 and 6). As the ω_{IG} continued to weaken and changed from ascent to descent, water vapor supply was reduced and the ascending ω_Q also weakened. The ascending ω_{IG} presents throughout Stage I, II, and most of Stage III, and is much stronger than ω_{ID}. This is similar to the theoretical results of HPJ03 (their Eq. (15)) that the prominent quasi-geostrophic vertical motion can be mainly attributed to isentropic upgliding measured in a frame of reference moving with a speed close to the basic flow. In Stage IV, both ω_{IG} and ω_Q became positive, precipitation was weak, and ω_{ID} returned to ascent. The evolution of different omega components thus demonstrates that different omega components can interact with each other during system evolution and the diabatic heating and the associated ω_Q have strong impacts on both ω_{ID} and ω_{IG}.

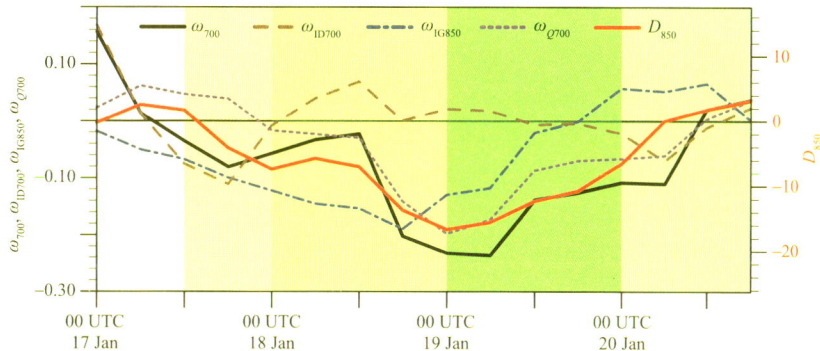

Figure 6 Evolution from 00 UTC 17 January to 18 UTC 20 January at 30°N and averaged over 113°—121°E of vertical velocity (ω_{700}, black curve, Pa s^{-1}), ω_{ID} (ω_{ID700}, brown dashed curve, Pa s^{-1}), and ω_Q (ω_{Q700}, purple dotted curve, Pa s^{-1}) at 700 hPa, and ω_{IG} (ω_{IG850}, blue dashed-dotted curve, Pa s^{-1}) and air divergence (D_{850}, red curve, 10^{-6} s^{-1}) at 850 hPa

3.3 Interaction between ω_{ID} and ω_{IG}

The eastward advection of high PV, which was generated over the eastern flank of the TP, triggered ascending ω_{ID} from below, with air convergence and cyclonic vorticity generation in the lower troposphere (Figs. 7a, 5a, and 5b). For adiabatic atmospheric motion, the converging southeasterly flow in the lower troposphere (Fig. 5c) should move along the isentropic surfaces ($V_{IG} = V_{ID}$) and thus the upward IG flow ($\omega_{IG} < 0$) is generated as shown in Fig. 5b and Fig. 6, and indicated by the solid brown arrow between Fig. 7a and 7b. It is important to note that ω_{ID} is ascending over southern China only in Stage I (Fig. 5a) and that it changes to descending afterwards (Figs. 5d and 5g). From Stage I to Stage II, the ascending southerly ($v_{IG} > 0, \omega_{IG} < 0$, Eq. (16)) develops quickly (Figs. 5e and 5f). It transports warm moist air northward from the ocean to the Yangtze River region (Fig. 5f), causing the increase of local potential temperature that results in descending ω_{ID} (Eq. (11), Fig. 5d). Therefore, ω_{ID} receives a negative feedback from the development of ω_{IG} via warm advection as indicated by the dashed brown arrow ($-\nabla \cdot (V\theta)^-$) between Figs. 7b and 7a, leading to the weakening of ω_{ID} (Figs. 5a and 5d).

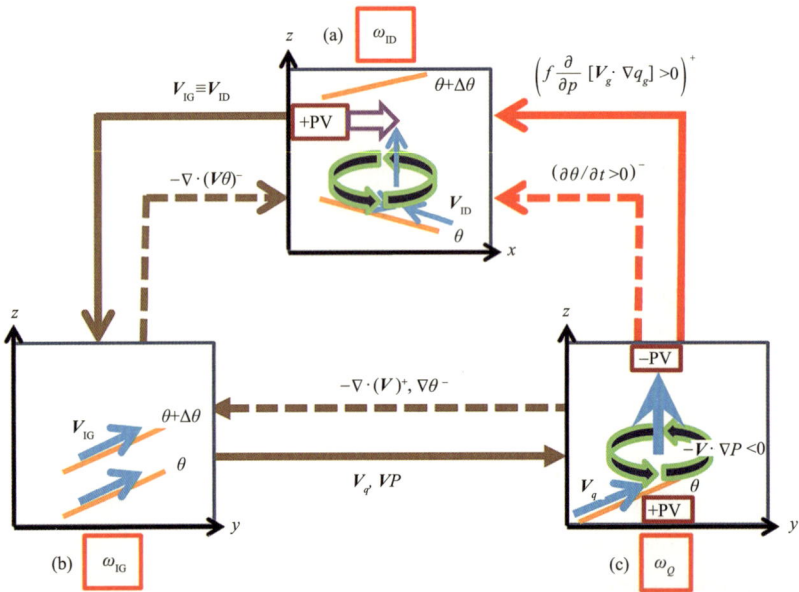

Figure 7 Schematic showing the interaction among different omega components and the feedback of diabatic heating on horizontal PV advection: (a) triggering of cyclogenesis and ascending ω_{ID} by PV advection, (b) triggering of isentropic upgliding wind ($V_{IG} = V_{ID}$) and convergence due to the development of ascending ω_{ID} and generation of ω_{IG} due to the slope of isentropic surfaces, and (c) generation of ω_Q associated with moisture transport (Vq) and intensification of cyclogenesis due to negative PV advection (VP). Dashed brown arrows indicate positive (superscript +) and negative (superscript −) feedback mechanism; and the red solid and dashed arrows denote, respectively, positive (superscript +) and negative (superscript −) feedback of latent heating on ω_{ID}. See text for more details

4 PV-Q perspective on vertical velocity development

The converging southeasterly flow associated with the ascending IG vertical velocity ($\omega_{IG} < 0$) transported abundant water vapor from the ocean to South China and lifted the moist air upward (Figs. 7c and 5f). When the moist air reached the condensation level, precipitation started. The released latent heat (Q) associated with precipitation and condensation will affect the in situ PV (Eq. (7)) as well as vertical velocity (Eq. (14)). The evolution of vertical velocity in the diabatic atmosphere thus needs to be investigated from a PV-Q perspective.

4.1 Interaction between ω_Q and ω_{IG}

Once the latent heat is released ($\dot{\theta} > 0$), following the definition (13) air ascent associated with diabatic heating ($\omega_Q < 0$) is generated. The development of ascending ω_Q intensifies the convergence and cyclonic circulation in the lower layers, and a stronger southerly wind together with enhanced ω_{IG} develops to the east of the cyclone center (Figs. 5e, 5f, 5h, and 5i). Thereby ω_{IG} gets positive feedback from the development of ω_Q, as shown by the part of dashed brown arrow ($-\nabla \cdot (V)^+$) between Figs. 7b and 7c. The southerly wind produces not only warm advection but also negative PV advection in the lower layers (solid brown arrow (Vq, VP) between Figs. 7b and 7c). This is coupled with the arrival of high PV from the west in the upper layer (Figs. 5d and 5g), which leads to baroclinic cyclonic development to the east of the cyclone center and increases the local precipitation and diabatic air ascent ($\omega_Q < 0$), presenting a positive feedback between ω_{IG} and ω_Q. Conversely, latent heating associated with precipitation increases the local potential temperature and reduces the meridional temperature gradient to its south, resulting in weakening of the ascending ω_{IG} (Eq. (16)). As shown in Fig. 6, the ascending ω_{IG} started weakening at 18 UTC on 18 January when the precipitation over the lower reaches of the Yangtze River Basin developed most rapidly (Figs. 1h and 1i). It indicates that ω_{IG} also receives a negative feedback from the development of ω_Q via reducing the upwind gradient of potential temperature (Fig. 7, the part of dashed brown arrow $\nabla\theta^-$ between Figs. 7b and 7c). Subsequently, the local ω_{IG} declined (Fig. 6), and both the water vapor flux convergence and the vertical velocity in this area started to decrease as well (Fig. 4). When ω_{IG} changed to descent late on the 19 January (Figs. 5k and Fig. 6), the intensity of the cyclonic vorticity started to decay (Fig. 6) followed by weakening of the precipitation.

4.2 Feedback of heating on PV forcing and ω_{ID}

It is interesting to see that ω_{ID} changed its sign after the precipitation occurred (Figs. 5 and Fig. 6), whilst the positive zonal PV advection at 650 hPa remained strong (Figs. 5, left column), which implies that diabatic condensation heating associated with precipitation must play a significant role in the evolution of vertical velocity. To reveal how diabatic heating can feedback on PV forcing and ω_{ID}, Equation (14) is employed.

The evolutions during the period 17—20 January of the different forcing terms, together with ω_{ID} at 650 hPa and at 30°N averaged over 113°—121°E, are evaluated using the analysis data and the results are presented in Fig. 8. It demonstrates that the forcing F_1 associated with vertical differential PV advection developed rapidly in Stage I and continued to act as positive forcing on ω_{ID} during the subsequent stages until the end of the Stage IV. The forcing F_2 associated with diabatic heating was weak in Stage I but became strongly negative during Stages II and III after precipitation occurred. Owing to the compensation and interaction between F_1 and F_2, the total forcing ($F = F_1 + F_2$) and ω_{ID} were weak and even changed their signs during these periods. In Stage IV, precipitation was reduced and F_2 became very weak, and the total forcing and ω_{ID} were determined mainly by F_1. All these results demonstrate that the forcing induced by vertically differential PV advection (F_1) plays a significant role in the triggering, development, and decay of the cyclonic circulation, while the initiated precipitation exerts considerable feedback on the ID vertical motion and system development.

Diabatic heating can exert both negative and positive impacts on ω_{ID}, as shown schematically by the red arrows in Fig. 7 between panel (c) and (a). According to Eq. (14), latent heat release can produce

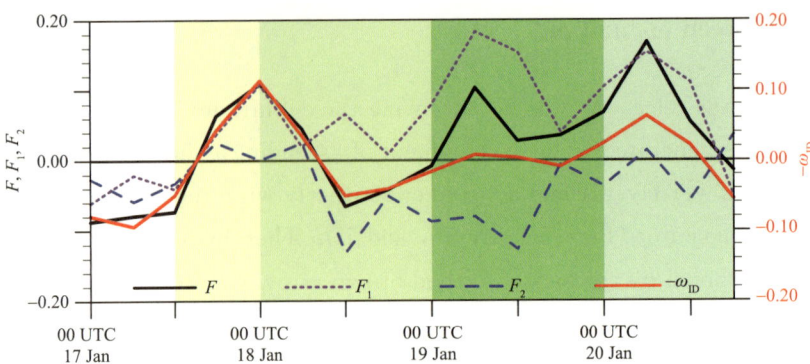

Figure 8 The same as Fig. 6 but for ω_{ID} (red curve, Pa s^{-1}, multiplied by a factor of -1), and the sum of ω_{ID} forcing terms (F, black curve), the vertical differential advection of the quasi-geostrophic PV term (F_1, purple dotted curve), and the diabatic heating term (F_2, blue dashed curve) on the right-hand side of Eq. (14) and at 650 hPa. Unit for ω_{ID} forcing terms is 10^{-18} Pa^{-1} s^{-3}

negative forcing on ω_{ID} ($F_2 < 0$). Actually, by following the definition of Eq. (11), heating increases the local potential temperature and results directly in descending ω_{ID} (Fig. 7, red dashed arrow between Figs. 7c and 7a), as demonstrated in Figs. 5d, 5g, 5j, and 8, and schematically illustrated in Figs. 10d and 10g of HPJ03 for an infinite atmosphere at rest. For an atmosphere at motion, heating can change the PV advection and affect ω_{ID}. This is because, according to the QGPV equation (Eq. (7)), heating generates positive PV below and negative PV above the heating maximum (Fig. 7c). Thus, the zonal PV gradient in the upper layer between the positive PV generated over the eastern TP and the induced negative PV over the heating region is intensified, as is the meridional PV gradient to the south of the heating region in the lower layer. Consequently, the increase of PV advection with height is intensified, forming positive forcing on ω_{ID} ($F_1 > 0$) during the precipitation period as indicated by the red solid arrow between Figs. 7c and 7a. The total forcing for ω_{ID} is determined by the combined effect of F_1 and F_2, and ω_{ID} can change its sign during precipitation. During Stage Ⅳ, diabatic heating and the associated forcing F_2 are reduced, and the feedback of heating on the forcing F_1 becomes weakened as well.

In summary, we can conclude that strong interaction exists among the three omega components, which can significantly influence the circulation development during its life cycle. Although the development of ascending ID omega is important in initiating IG omega and inducing diabatic heating omega, it receives strong feedback from diabatic heating and it can even change its sign, whilst the forcing of the vertical differential PV advection has good correspondence with the evolution of the circulation and associated precipitation.

5 Roles of zonal and meridional PV advection in cyclogenesis

The above results demonstrate that it is the forcing of vertical differential PV advection together with the feedback from diabatic heating that modulates the evolution of the circulation and associated precipitation. This section further discusses how different PV advection coupling between the upper and lower layers influences the evolution of the circulation and associated precipitation.

5.1 Initial Stage (Ⅰ)

The evolution from 12 UTC on 17 January to 00 UTC on 18 January of the vertical cross sections averaged between 113°—121°E of PV, its zonal and meridional advection, as well as wind are shown in Fig. 9. At 12 UTC on 17 January, a center of weak positive zonal PV advection appeared between 700 and 500 hPa at around 30°N (Fig. 9a), in correspondence with the eastward advection of the high PV that originated from the eastern flank of the TP (Fig. 2c). As the TP-induced high PV became enhanced and moved further eastward (Figs. 2d and 2e), positive zonal PV advection increased downstream along the middle and lower reaches of the Yangtze River Basin (Figs. 9c and 9e), and positive relative vorticity and a southerly wind (mainly ω_{IG}) developed in the lower troposphere (Figs. 5a—c). Concurrently, in the lower layer (below 700 hPa), negative meridional PV advection, with its center located above 850 hPa, increased over southern China (Figs. 9a, 9c, and 9e) and precipitation began to develop over southwestern China (Fig. 1).

The geostrophic advection of QGPV and its association with the development of vertical velocity are presented in the right column of Fig. 9. While positive PV advection developed in the mid-troposphere, negative PV advection in the lower tropospheric layer intensified because of the development of the cyclonic circulation and its related southerly winds within that layer. Consequently, PV advection increased with height below 500 hPa at around 30°N, suggesting forcing for ascent (Figs. 9b, 9d, and 9f). This ascent favored the development of convergence within the lower troposphere and the enhancement of southerly wind to the south of 30°N, which enhanced the transport of moist air to the Yangtze River Basin (Fig. 5c).

The above analysis demonstrates that during Stage Ⅰ, forcing for ascent was induced mainly by the positive eastward advection of mid-tropospheric high PV, generated near the surface at the eastern flank of the TP, and by the enhanced negative meridional PV advection in the lower troposphere. Owing to the enhanced vertical gradient of QGPV advection, ascending vertical velocity was diagnosed to the southeast of the TP and in situ precipitation was initiated.

5.2 Developing Stage (Ⅱ)

With the development of precipitation, the influence of diabatic heating associated with latent heat release on both vertical velocity and PV advection should be taken into consideration. Here, the changes in atmospheric circulation in the developing stage are diagnosed using the PV equation (Eq. (7)).

With the development of ascending airflow on the southern side of the center of high PV, the southerly wind was strengthened and negative meridional PV advection in the lower troposphere was enhanced (Fig. 9). Thus, large-scale circulation with positive horizontal PV advection in the mid-troposphere and negative horizontal PV advection in the lower troposphere formed to the south of 30°N (Figs. 5d—f, and 10a), which further reinforced the forcing for quasi-geostrophic ascent (Fig. 10c).

A center of prominent diabatic heating was located in the mid-troposphere near 30°N, which generated a positive (negative) PV tendency below (above) its maximum center (Fig. 10b). The negative PV tendency above the precipitation area enhanced and sustained strong zonal PV advection upstream of the precipitation area. The vertical differential QGPV advection was thus enhanced compared with prior times, leading to further intensification of the vertical velocity between 28°N and 30°N (Fig. 10c).

In summary, during Stage Ⅱ, vertical velocity and its associated precipitation over the middle and lower reaches of the Yangtze River Basin were highly related to the evolution of PV advection in the lower and mid-troposphere. During this period, the feedback of diabatic condensational heating contributed to both the maintenance of the zonal PV gradient, by acting to erode mid-tropospheric PV downstream of the area of high PV generated on the eastern flank of the TP, and the maintenance of the meridional

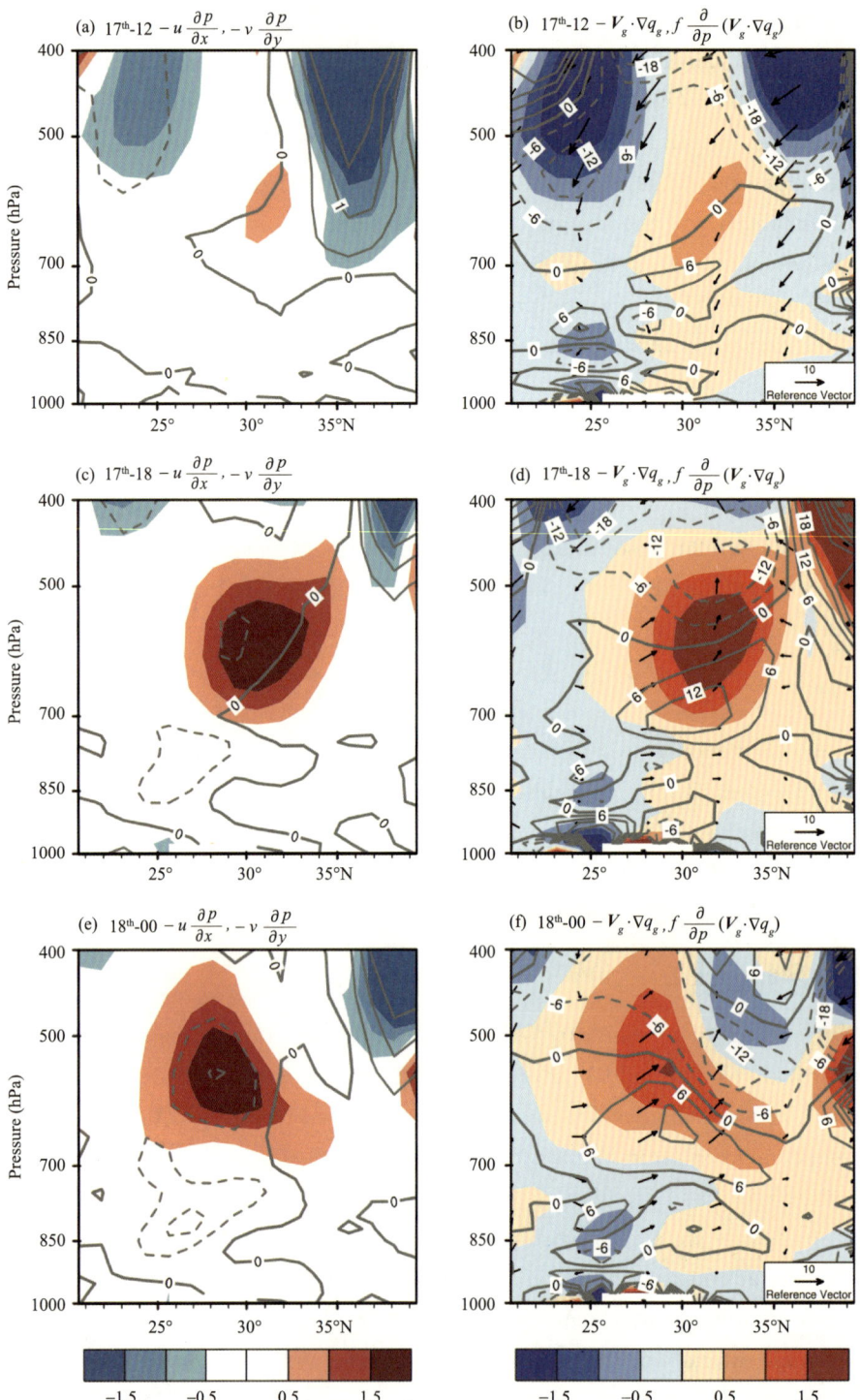

Figure 9 Mean vertical cross section at 113°—121°E of (left column) zonal ($-u\frac{\partial P}{\partial x}$, shading) and meridional ($-v\frac{\partial P}{\partial y}$, contour) PV advection (interval: 0.5×10^{-5} PVU s^{-1}), and (right column) quasi-geostrophic QGPV advection ($-\mathbf{V}_g \cdot \nabla q_g$, shading, interval: 0.5×10^{-9} s^{-2}), ω forcing due to vertical differential quasi-geostrophic QGPV advection ($f\frac{\partial}{\partial p}(\mathbf{V}_g \cdot \nabla q_g)$, contour, interval: 6×10^{-18} s^{-3} Pa^{-1}), and reanalysis wind ($v\mathbf{j} + \omega_{OB}\mathbf{k}$, vector; units: v in m s^{-1}, ω_{OB} in Pa s^{-1} (values multiplied by a factor of -50)) for (a) and (b) 12 UTC 17 January, (c) and (d) 18 UTC 17 January, and (e) and (f) 00 UTC 18 January

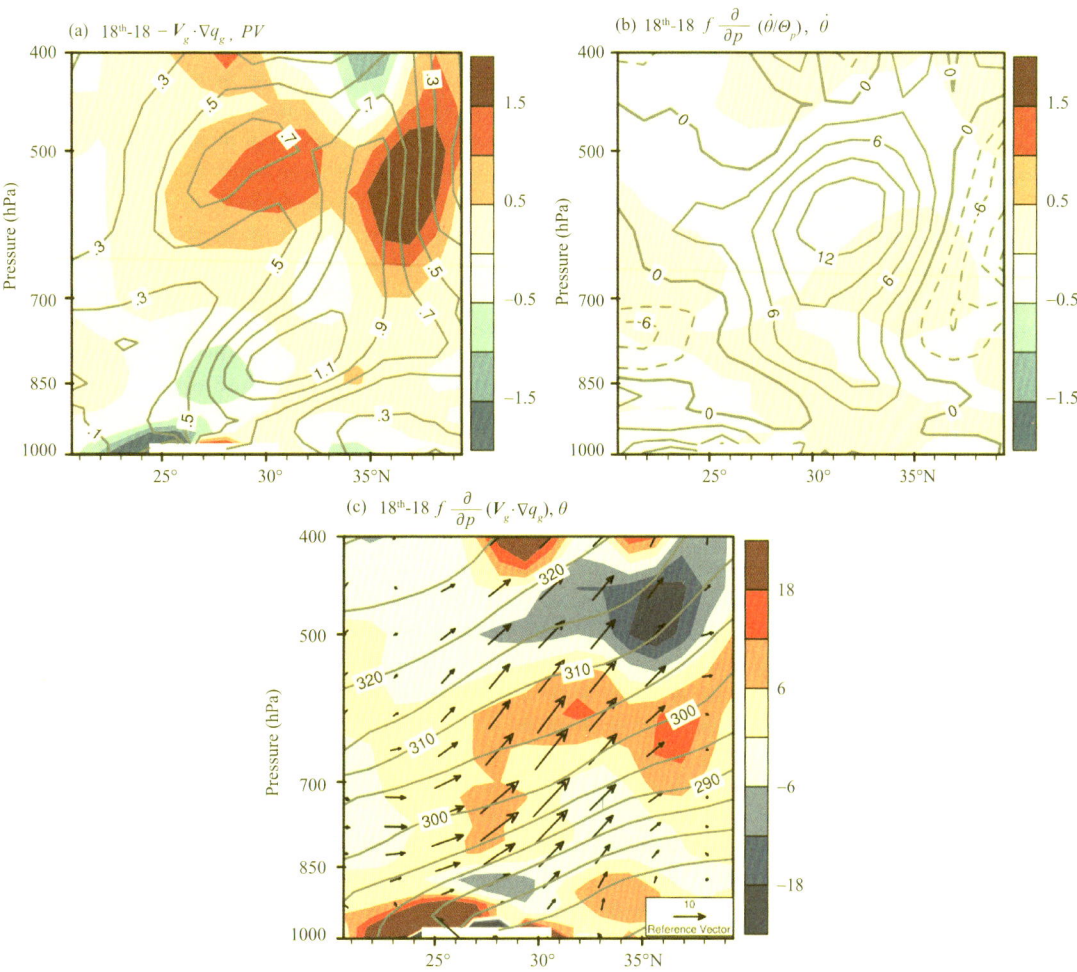

Figure 10 Mean vertical cross section at 113°—121°E of (a) quasi-geostrophic QGPV advection ($-\mathbf{V}_g \cdot \nabla q_g$, shading, interval: 0.5×10^{-9} s^{-2}) and PV (contour, interval: 0.2 PVU), (b) diabatic heating term ($f\frac{\partial}{\partial p}\left(\frac{\dot{\theta}}{\Theta_p}\right)$, shading, interval: 0.5×10^{-9} s^{-2}) and diabatic heating rate ($\dot{\theta}$, contour, interval: 3×10^{-5} K s^{-1}), and (c) ω-forcing due to vertical differential quasi-geostrophic QGPV advection ($f\frac{\partial}{\partial p}(\mathbf{V}_g \cdot \nabla q_g)$, shading, interval: 6×10^{-18} s^{-3} Pa^{-1}), potential temperature (contour, K) and reanalysis wind ($v\mathbf{j} + \omega_{OB}\mathbf{k}$, vector; units: v in m s^{-1}, ω_{OB} in Pa s^{-1} (values multiplied by a factor of -50)) for 18 UTC on 18 January 2008

PV gradient to its south, by acting to create lower tropospheric PV. These processes provide enhanced quasi-geostrophic forcing that enhances the cyclogenesis in the lower troposphere and supports both the development of vertical velocity and the associated precipitation.

5.3 Maturation and Decay Stages (Ⅲ and Ⅳ)

During Stage Ⅲ, as ascent in association with the enhanced southerly flow over the Yangtze River Basin (Figs. 5f and 5i) developed (Fig. 10), the center of negative meridional PV advection was located at a higher altitude and closer to the center of positive zonal PV advection. When the precipitation peaked at 06 UTC on 19 January, a center of slightly stronger negative meridional PV advection was located immediately below the center of positive zonal PV advection, which resulted in the maximum magnitude of the vertical differential QGPV advection, in addition to the lifting of the vertical velocity center compared with earlier times (Figs. 11a and 11b). Subsequently, the center of negative meridional PV advection tended to overlay the positive center of zonal PV advection (Fig. 11c). Owing to prominent cancelation between the positive zonal PV advection and the negative meridional PV advection (Fig. 11e), both the vertical differential PV advection and the vertical velocity weakened (Figs. 11d and 11f). Meanwhile, the southerly wind weakened correspondingly (Fig. 11f), which led to reduced poleward advection of water vapor toward southern China. Consequently, the negative effect of diabatic heating on PV generation in the mid-troposphere decreased (not shown), which reduced the zonal PV gradient between the positive PV center over the eastern flank of the TP and the precipitation center over eastern China. Finally, as the cyclonic circulation decayed, the heavy precipitation diminished rapidly (Fig. 4) and the extreme winter precipitation event ended.

6 Conclusions

This study conducts a brief review of the development of the omega equation and the application of PV theory to the study of the impacts of large-scale mountains on downstream weather development. Several relevant challenges are also identified. The omega equation for ω_{ID} constructed by HPJ03 is highlighted. As this equation establishes the link between ω_{ID} and vertically differential PV advection, and as vertically differential PV advection is closely related with weather system development, this omega equation should be applicable to interpretation of weather system development. However, ω_{ID} is only one part of the total vertical velocity, and the corresponding omega equation is a diagnostic equation. Therefore, how best to use this diagnostic equation to interpret the development of full vertical velocity and the evolution of weather system has become a challenge.

In the current study, this challenge has been tackled by applying the QGPV equation (Eq. (7)), the thermodynamic equation (Eq. (8)), and the established ID omega equation (Eq. (14)) for a diabatic atmosphere to the development of a weather event that occurred downstream of the TP during 17—21 January 2008. During the evolution of this event, positive PV was first generated over the eastern flank of the TP, and its subsequent eastward advection triggered the development of vertical velocity below and the generation of cyclone vorticity in the lower troposphere. The total vertical velocity of diabatic flow ω is partitioned into three components (i.e., ω_{ID}, ω_{IG}, and ω_Q). Consideration of the interactions of these components, together with analysis of the influence of diabatic heating Q on PV forcing, allows the impacts of the TP on PV generation and downstream development of both vorticity and vertical velocity, as well as the feedback of heating on PV advection, to be elucidated. The main results can be summarized as follows.

It has been shown that the development of the cyclone system associated with severe weather downstream of the TP is strongly tied both to the restructuring of PV near the surface due to airflow convergence in the region of the TP and to the subsequent downstream advection of positive PV in the mid-troposphere. The dynamical processes associated with the evolution of the cyclone system are depicted schematically in Fig. 12.

In winter the westerly flow impinging upon the TP in the lower troposphere is split by the TP and it converges over its eastern flank, causing the increase in PV density W and the development of positive relative vorticity in that location. Since in winter the static stability $\partial\theta/\partial z$ near the TP is larger than in areas further downstream, the relative vorticity of the flow is increased easily as the PV advection is eastward. This is because along the westerly jet the vertical wind shear ($\partial u/\partial z, \partial v/\partial z$) and the horizontal

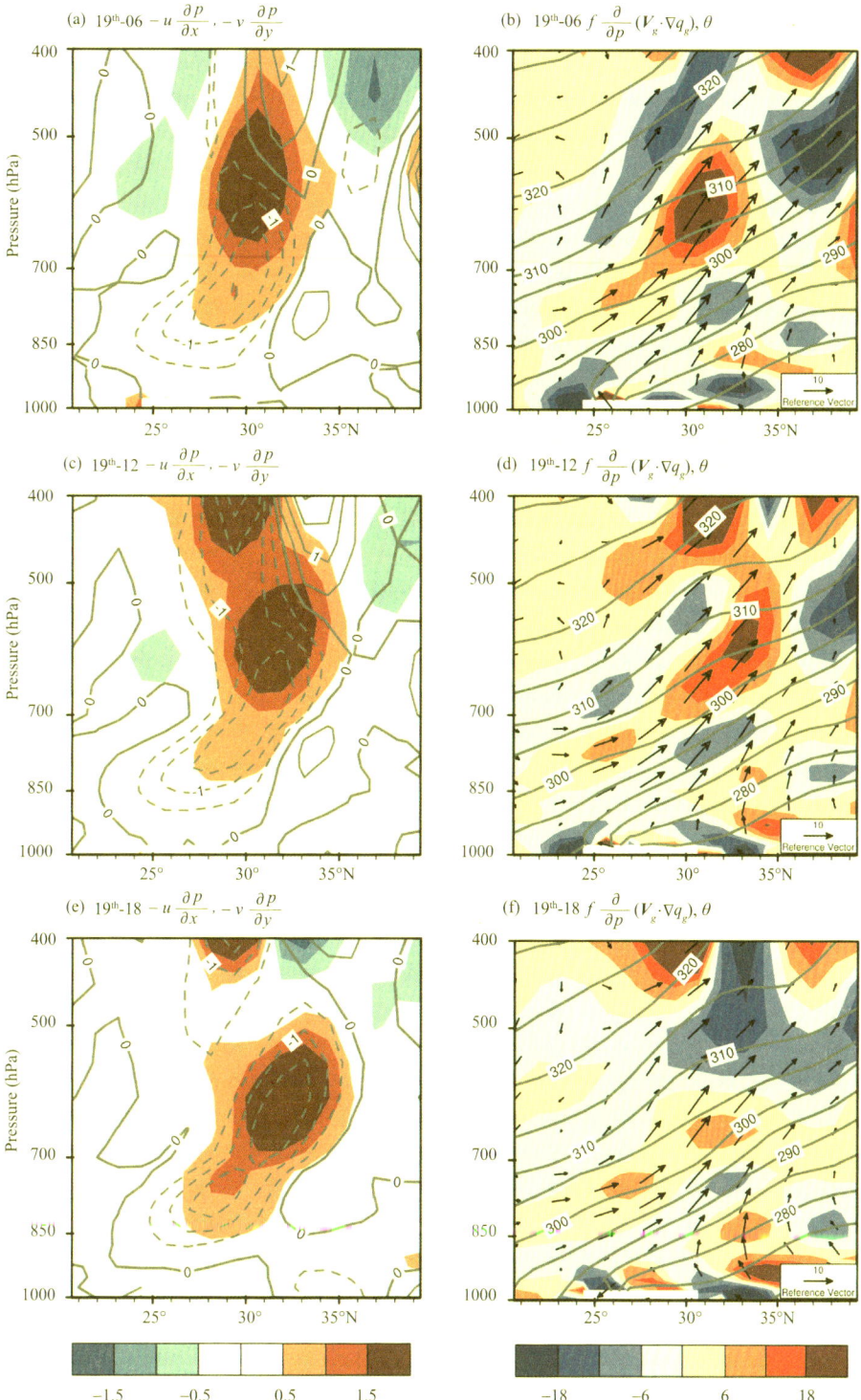

Figure 11 Mean vertical cross section at 113°—121°E of (left column) zonal ($-u\frac{\partial P}{\partial x}$, shading) and meridional ($-v\frac{\partial P}{\partial y}$, contour) PV advection (interval: 0.5×10^{-5} PVU s^{-1}), and (right column) ω forcing due to vertical differential quasi-geostrophic QGPV advection ($f\frac{\partial}{\partial p}(\mathbf{V}_g \cdot \nabla q_g)$, shading, interval: 6×10^{-18} s^{-3} Pa^{-1}), potential temperature (contour, K), and reanalysis wind ($v\mathbf{j} + \omega_{OB}\mathbf{k}$, vector; units: v in m s^{-1}, ω_{OB} in Pa s^{-1} (values multiplied by a factor of -50)) for (a) and (b) 06 UTC on 19 January, (c) and (d) 12 UTC on 19 January, and (e) and (f) 18 UTC on 19 January 2008

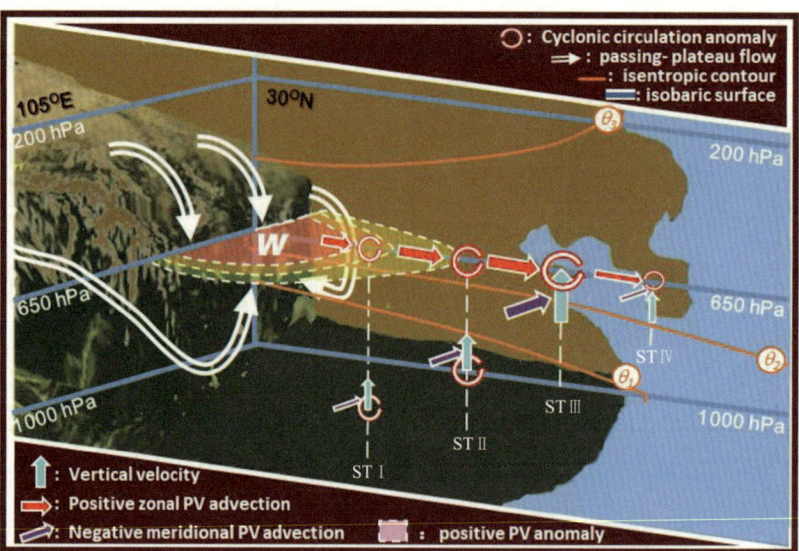

Figure 12 Schematic of PV restructuring in the region of the Tibetan Plateau (TP) and the impact of PV advection on downstream circulation during different stages (ST) of cyclogenesis

Stage Ⅰ: Surface airflow convergence in the lee of the TP increases local PV density W, generating a positive relative vorticity anomaly and initiating light rain near the TP.

Stage Ⅱ: Eastward moving positive vorticity anomaly is intensified owing to reduced static stability. Positive zonal PV advection in the mid-troposphere and increased southerly and negative meridional PV advection below enhances cyclogenesis, air ascent, and precipitation.

Stage Ⅲ: Negative meridional PV advection is located immediately below the center of strong positive zonal PV advection, and cyclonic vorticity, vertical velocity, and precipitation are peaked.

Stage Ⅳ: Negative meridional PV advection tends to overlay the positive center of zonal PV advection. Consequently, the cyclone and air ascent are both weakened and precipitation is diminished.

components of potential vorticity are small so that $P \approx \alpha(f+\zeta) \cdot \partial\theta/\partial z$ is conserved in adiabatic and frictionless circumstances. Thus, the TP plays a role in modulating the PV structure and in amplifying the relative vorticity of the passing airflow.

The downstream growth of PV advection leads to cyclogenesis in the lower troposphere and development of vertical velocity. In the Initial Stage (ST Ⅰ, Fig. 12), positive zonal PV advection increases in the mid-troposphere, resulting in the development of ascending ID vertical velocity ($\omega_{ID}<0$). This is accompanied by wind convergence due to atmospheric continuity and by an increase of in situ relative vorticity in the lower troposphere, where an isentropic southerly wind develops in association with ascending IG vertical velocity ($\omega_{IG}<0$). Consequently, moist air is transported from the south, and negative meridional PV advection is increased, which contribute to quasi-geostrophic forcing for air ascent and cyclogenesis in situ. Thus, weak precipitation is initiated over southwestern China (ST Ⅰ, Fig. 12) and ascending diabatic heating vertical velocity ($\omega_Q<0$) is generated. In the Developing Stage (ST Ⅱ, Fig. 12) over eastern China, the increasing negative meridional PV advection associated with the strengthening southerly wind in the lower troposphere enhances the vertical differential PV advection between the mid- and lower troposphere, leading to enhanced forcing for quasi-geostrophic ascent compared to prior times. Meanwhile, the negative (positive) PV anomaly induced by diabatic condensation heating in the mid- (lower) troposphere weakens (enhances) the local PV over the precipitation region. Consequently, the zonal PV gradient in the mid-troposphere between the positive PV center over the eastern TP and the precipitation center, as well as the meridional PV gradient in the lower troposphere to the south of the precipitation center, are both enhanced. This provides persistent and strong positive zonal PV advection in the mid-troposphere and negative meridional PV advection in the lower troposphere for the development of both cyclonic vorticity and ascending vertical velocity.

During the Maturation Stage (ST Ⅲ, Fig. 12), as the altitude of the ascending southerly flow is increased, the center of negative meridional PV advection is lifted further. When it becomes located immediately below the center of positive zonal PV advection, both cyclonic vorticity and air ascent with precipitation reach their maxima. At subsequent times (Decay Stage; ST Ⅳ, Fig. 12), the negative meridional PV advection substantially cancels the positive zonal PV advection, and both the vertical differential PV advection and the ascending airflow are weakened. The negative (positive) PV generation above (below) the center of convective diabatic heating decreases correspondingly and the zonal (meridional) PV gradient in the mid- (lower) troposphere upstream of the precipitation area is weakened. Consequently, along the middle and lower reaches of the Yangtze River Basin, the vertically differential PV advection decreases, air ascent is weakened, and precipitation diminishes.

It can be summarized that the frequent occurrence of a center of high PV over the eastern flank of the TP in winter is due to the PV restructuring associated with in situ surface convergence, and the eastward advection in the mid-troposphere of positive PV can trigger vertical velocity development and cyclogenesis in the lower troposphere. The evolution of the cyclone life is related closely to the interaction of different components of vertical velocity. It is the vertically differential PV advection and the feedback of diabatic heating that control the development and decay of vertical velocity and the evolution of the lower layer cyclone, as well as the weather downstream of the TP.

Current study only analyzes the impacts of the interior QGPV advection on the development of vertical vorticity. Since the solution of Equation (14) is also determined by boundary conditions, revealing the impacts of boundary temperature advection and diabatic heating on vertical velocity development will help further understanding the underlying physics. In this study, only one winter case is diagnosed to elucidate the proposed PV-Q perspective. Further analyses of other weather events, particularly in summer, are required.

Data Availability Statement

All datasets used in this paper are publicly available. We would like to thank the China Meteorological Administration for releasing the precipitation data (downloaded from https://data.cma.cn/data/cdcdetail/dataCode/SURF_CLI_CHN_MUL_DAY_V3.0.html), and both the Global Modeling and Assimilation Office and the Goddard Earth Sciences Data and Information Services Center for the dissemination of MERRA-2 data (https://gmao.gsfc.nasa.gov/GMAO_products/reanalysis_products.php).

Appendix: Omega equation and PV advection for diabatic motion

Under the quasi-geostrophic framework, the vorticity and temperature equations in p-coordinates can be written as:

$$\left(\frac{\partial}{\partial t}+\boldsymbol{V}_g \cdot \nabla\right)\eta_g = f\frac{\partial \omega}{\partial p} \tag{A1}$$

$$\left(\frac{\partial}{\partial t}+\boldsymbol{V}_g \cdot \nabla\right)\theta = -\frac{\partial \Theta(p)}{\partial p}\omega + \dot{\theta} \tag{A2}$$

From the hydrostatic relation:

$$\frac{\partial \Phi}{\partial p} = -\Pi(p)\theta \tag{A3}$$

Eq. (A2) can be written as

$$\left(\frac{\partial}{\partial t}+\boldsymbol{V}_g\cdot\nabla\right)\frac{\partial\Phi}{\partial p}=-\Sigma^2\omega-\Pi(p)\theta \tag{A4}$$

Eliminating vertical velocity "ω" between the vorticity equation (Eq. (A1)) and temperature equation (Eq. (A4)) yields the frictionless QGPV equation:

$$\left(\frac{\partial}{\partial t}+\boldsymbol{V}_g\cdot\nabla\right)q_g=f\frac{\partial}{\partial p}\left(\frac{\dot\theta}{\Theta_p}\right) \tag{A5}$$

Thus, for a frame of reference moving at constant horizontal velocity $\boldsymbol{C}$, the QGPV and temperature equations can be, respectively, written as

$$\left[\frac{\partial}{\partial t}\bigg|_c+(\boldsymbol{V}_g-\boldsymbol{C})\cdot\nabla\right]q_g=f\frac{\partial}{\partial p}\left(\frac{\dot\theta}{\Theta_p}\right) \tag{A6}$$

and

$$\left[\frac{\partial}{\partial t}\bigg|_c+(\boldsymbol{V}_g-\boldsymbol{C})\cdot\nabla\right]\theta=-\Theta_p\omega+\dot\theta \tag{A7}$$

Applying $f\frac{\partial q_g}{\partial p}$ to Eq. (9) and using the hydrostatic relation (Eq. (A3)) and the definition of ω_{ID} (Eq. (11)) lead to:

$$L\{\omega_{\mathrm{ID}}\}=\Sigma^2\nabla_h^2(\omega_{\mathrm{ID}})+f^2\frac{\partial^2}{\partial p^2}(\omega_{\mathrm{ID}})=-f\frac{\partial}{\partial p}\left(\frac{\partial q_g}{\partial t}\right)\bigg|_C$$

From Eq. (A6), the omega equation for ω_{ID} in a diabatic atmosphere is obtained:

$$L\{\omega_{\mathrm{ID}}\}=\left(\Sigma^2\nabla_h^2+f^2\frac{\partial^2}{\partial p^2}\right)\omega_{\mathrm{ID}}=f\frac{\partial}{\partial p}[(\boldsymbol{V}_g-\boldsymbol{C})\cdot\nabla q_g]-f^2\frac{\partial^2}{\partial p^2}\left(\frac{\dot\theta}{\Theta_p}\right) \tag{A8}$$

References

AEBISCHER U, SCHÄR C, 1998. Low-level potential vorticity and cyclogenesis to the lee of the Alps[J]. Journal of the Atmospheric Sciences, 55(2): 186-207.

BUSHBY F H, 1952. The evaluation of vertical velocity and thickness tendency from Sutcliffe's theory[J]. Quarterly Journal of the Royal Meteorological Society, 78(337): 354-362.

CHEN B M, QIAN Z A, ZHANG L S, 1996. Numerical simulation of the formation and development of vortices over the Qinghai-Xizang Plateau in summer[J]. Chinese Journal of Atmospheric Sciences, 20(4): 491-502.

CHEN S J, DELL'OSSO L, 1984. Numerical prediction of the heavy rainfall vortex over eastern Asia monsoon region[J]. Journal of the Meteorological Society of Japan, 62(5): 730-747.

DING Z Y, LIU J L, LV J N, 1994. The study for the mechanism of forming QXP-vortex on 600 hPa[J]. Plateau Meteorology, 13(4): 411-418.

DJURIC D, 1969. Note on estimation of vertical motion by the omega equation[J]. Monthly Weather Review, 97(12): 902-904.

DUNN L B, 1991. Evaluation of vertical motion: Past, present, and future[J]. Weather and Forecasting, 6(1): 65-75.

EGGER J, HOINKA K P, SPENGLER T, 2015. Aspects of Potential Vorticity Fluxes: Climatology and Impermeability[J]. Journal of the Atmospheric Sciences, 72(8): 3257-3267.

ERTEL H, 1942. Ein neuer hydrodynamischer wirbelsatz[J]. Meteor Z Braunschweig, 59: 33-49.

FUNATSU B M, WAUGH D W, 2008. Connections between potential vorticity intrusions and convection in the eastern tropical Pacific[J]. Journal of the Atmospheric Sciences, 65(3): 987-1002.

GELARO R, MCCARTY W, SUÁREZ M, et al, 2017. The Mordern-Era Retrospective Analysis for Research and Applications, Version 2 (MERRA-2)[J]. Journal of Climate, 30(14): 5419-5454.

HAYNES P H, MCINTYRE M E, 1987. On the evolution of vorticity and potential vorticity in the presence of diabatic heating and frictional or other forces[J]. Journal of the Atmospheric Sciences, 44(5): 828-841.

HAYNES P H, MCINTYRE M E, 1990. On the conservation and impermeability theorems for potential vorticity[J]. Journal of the Atmospheric Sciences, 47(16): 2021-2031.

HELD I M, SCHNEIDER T, 1999. The surface branch of the zonally averaged mass transport in the troposphere[J]. Journal of the Atmospheric Sciences, 56(11): 1688-1697.

HOLTON J R, 1972. An Introduction to Dynamic Meteorology[M]. Salt Lake City, USA: Academic.

HOLTON J R, 2004. An Introduction to Dynamic Meteorology[M]. London, UK: Elsevier Academic.

HOSKINS B J, 1991. Towards a PV-θ view of the general circulation[J]. Tellus, 43(4): 27-35.

HOSKINS B J, DRAGHICI I, DAVIES H C, 1978. A new look at the ω-equation[J]. Quarterly Journal of the Royal Meteorological Society, 104(439): 31-38.

HOSKINS B J, MCINTYRE M E, ROBERTSON A W, 1985. On the use and significance of isentropic potential vorticity maps[J]. Quarterly Journal of the Royal Meteorological Society, 111(470): 877-946.

HOSKINS B J, PEDDER M, JONES D W, 2003. The omega equation and potential vorticity[J]. Quarterly Journal of the Royal Meteorological Society, 129(595): 3277-3303.

KOH T Y, PLUMB R A, 2004. Isentropic zonal average formalism and the near-surface circulation[J]. Quarterly Journal of the Royal Meteorological Society, 130(600): 1631-1653.

LI G P, ZHAO B J, YANG J Q, 2002. A dynamical study of the role of surface sensible heating in the structure and intensification of the Tibetan Plateau vortices[J]. Chinese Journal of Atmospheric Sciences, 26(4): 519-525.

LI G P, LU H G, HUANG C H, et al, 2016. A climatology of the surface heat source on the Tibetan Plateau in summer and its impacts on the formation of the Tibetan Plateau vortex [J]. Chinese Journal of Atmospheric Sciences, 40(1): 131-141.

LI L F, LIU Y M, BO C Y, 2011. Impacts of diabatic heating anomalies on an extreme snow events over South China in January 2008[J]. Climatic and Environmental Research, 16(2): 126-136.

LIU Y M, LU M M, YANG H J, et al, 2020. Land-atmosphere-ocean coupling associated with the Tibetan Plateau and its climate impacts[J]. National Science Review, 7: 534-552.

MA T, LIU Y M, WU G X, et al, 2020. Potential Vorticity diagnosis on the formation, development and eastward movement of a Tibetan Plateau Vortex and its influence on the downstream precipitation[J]. Chinese Journal of Atmospheric Science.

MA T T, WU G X, LIU Y M, et al, 2019. Impact of surface potential vorticity density forcing over the Tibetan Plateau on the South China extreme precipitation in January 2008. Part I: Data analysis[J]. Journal of Meterological Research, 33: 400-415.

NIELSEN-GAMMON J W, GOLD D A, 2006. Dynamical diagnosis: A comparison of quasigeostrophy and Ertel potential vorticity[C]// BOSART L F, BlUESTEIN H B, Synoptic-dynamic meteorology and weather analysis forecasting: A tribute to fred sanders Chapter 9. Boston, MA: American Meteorological Society:183-202.

PAN W J, MAO J Y, WU G X, 2013. Characteristics and mechanism of the 10-20-Day oscillation of spring rainfall over southern China[J]. Journal of Climate, 26(14): 5072-5087.

PETTERSSEN S, 1956. Weather Analysis and Forecasting[J]. New York, USA: McGrawHill.

SANDERS F, HOSKINS B J, 1990. An easy method for estimation of Q-vectors from weather maps[J]. Weather and Forecasting, 5(2): 346-353.

SCHNEIDER T, 2005. Zonal momentum balance, potential vorticity dynamics, and mass fluxes on near-surface isentropes [J]. Journal of the Atmospheric Sciences, 62(6): 1884-1900.

SCHNEIDER T, HELD I M, GARNER S T, 2003. Boundary effects in potential vorticity dynamics[J]. Jouranl of the Atmospheric Sciences, 60(8): 1024-1040.

SHI C X, JIANG J X, FANG Z Y, 2000. A study on the features of severe convection cloud clusters causing serious flooding over Changjiang River Basin in 1998[J]. Climatic Environmental Research, 5(3): 279-286.

SHI N, BUEH C, JI L R, 2008. On the medium-range process of the rainy, snowy and cold weather of South China in early 2008 Part II: Characteristics of the western Pacific subtropical high [J]. Climatic Environmental Research, 13(4): 434-445.

SUTCLIFFE R C, 1947. A contribution to the problem of development[J]. Quarterly Journal of the Royal Meteorological Society, 73: 370-383.

TAO S Y, DING Y H, 1981. Observational evidence of the influence of the Qinghai-Xizang (Tibet) Plateau on the occurrence of heavy rain and severe convective storms in China[J]. Bulletin of the American Meteorological Society, 62(1): 23-30.

TAO S Y, WEI J, 2008. Severe snow and freezing-rain in January 2008 in the southern China [J]. Climatic Environmental Research, 13(4): 337-350.

THORPE A J, 1985. Diagnosis of balanced vortex structure using potential vorticity[J]. Journal of the Atmospheric Sciences, 42(4): 397-406.

THORPE A J, VOLKERT H, HEIMANN D, 1993. Potential vorticity of flow along the Alps[J]. Journal of the Atmospheric Sciences, 50(11): 1573-1590.

TRENBERTH K E, 1978. On the interpretation of the diagnostic quasi-geostrophic omega equation[J]. Monthly Weather Review, 106(1): 131-137.

WAN R J, WU G X, 2007. Mechanism of the Spring Persistent Rains over southeastern China[J]. Science in China D: Earth Sciences, 50(1): 130-144.

WANG B, BAO Q, HOSKINS B, et al, 2008. Tibetan Plateau warming and precipitation changes in east China[J]. Geophysical Research Letters, 35(14): L14702.

WANG Z Q, DUAN A M, WU G X, 2014. Time-lagged impact of spring sensible heat over the Tibetan Plateau on the summer rainfall anomaly in east China: Case studies using the WRF model[J]. Climate Dynamics, 42: 2885-2898.

WAUGH D W, POLVANI L M, 2000. Climatology of intrusions into the tropical upper troposphere[J]. Geophysical Research Lettets, 27(23): 3857-3860.

WU G X, ZHANG X H, LIU H, et al, 1997. Sensible heat driven air-pump over the Tibetan Plateau and its impacts on the Asian summer monsoon[C]// YEH T C. Collection in Memory of Zhao Jiuzhang. Beijing, CHN: Science Press:116-126.

WU G X, LIU Y M, WANG T M, et al, 2007. The influence of mechanical and thermal forcing by the Tibetan Plateau on Asian climate[J]. Journal of Hydrometeorology-special section, 8: 770-789.

WU G X, HE B, DUAN A M, et al, 2017. Formation and variation of the atmospheric heat source over the Tibetan Plateau and its climate effects[J]. Adv Atmos Sci, 34(10): 1169-1184.

YASUNARI T, MIWA T, 2006. Convective cloud systems over the Tibetan Plateau and their impact on meso-scale disturbances in the Meiyu/Baiu frontal zonal. A case study in 1998[J]. Journal of the Meteorological Society of Japan, 84(4): 783-803.

YEH T C, 1950. The circulation of the high troposphere over China in the winter of 1945−1946[J]. Tellus, 2(3): 173-183.

YEH T C, LO S W, CHU P C, 1957. The wind structure and heat balance in the lower troposphere over Tibetan Plateau and its surroundings[J]. Acta Meteorologica Sinica, 28(2):108-121.

YU J H, WU G X, LIU Y M, et al, 2019. Impact of Surface Potential Vorticity Density Forcing over the Tibetan Plateau on the south China extreme precipitation in January 2008. Part II: Numerical simulation[J]. Journal of Meterological Research, 33: 416-432.

ZHENG Y J, WU G X, LIU Y M, 2013. Dynamical and thermal problems in vortex development and movement. Part I: A PV-Q view[J]. Acta Meteorologica Sinica, 27(1):1-14.

Characteristics of the Potential Vorticity and Its Budget in the Surface Layer over the Tibetan Plateau

SHENG Chen[1,2], WU Guoxiong[1,2], TANG Yiqiong[1,3], HE Bian[1,2,4], XIE Yongkun[1], MA Tingting[1], MA Ting[1,3], LI Jinxiao[1], BAO Qing[1,2], LIU Yimin[1,2,4]

(1 State Key Laboratory of Numerical Modeling for Atmospheric Sciences and Geophysical Fluid Dynamics (LASG), Institute of Atmospheric Physics, Chinese Academy of Sciences, Beijing 100029, China; 2 College of Earth and Planetary Sciences, University of Chinese Academy of Sciences, Beijing 100049, China; 3 School of Atmospheric Sciences, Nanjing University of Information Science and Technology, Nanjing 210044, China; 4 CAS Center for Excellence in Tibetan Plateau Earth Sciences, Chinese Academy of Sciences (CAS), Beijing 100101, China)

Abstract: The variability of interior atmospheric potential vorticity (PV) is linked with PV generation at the Earth's surface. The present paper reveals the features of the surface PV and provides a stepping stone to investigate the surface PV budget.

In this study, the formats of the PV and PV budget adopting a generalized vertical coordinate were theoretically examined to facilitate the calculation of the surface PV and its budget. Results show that the formats of the PV and PV budget equations are independent of the vertical coordinate. While the vertical component of the surface PV dominates over the platform of the Tibetan Plateau, the horizontal component plays an important role over the slopes of the Tibetan Plateau, especially the southern slope owing to the strong in-situ meridional gradient of the potential temperature. These results indicate that the employment of complete surface PV not only provides a finer PV structure but also more appropriately reveals its effect on atmospheric circulation.

Diagnosis based on reanalysis and model output demonstrates that the surface PV budget equation is well balanced both in terms of the climate mean and synoptic process, and the surface PV budget in June has a prominent diurnal cycle. The diabatic heating with a minimum in the early morning and a maximum from evening to midnight contributes dominantly to this diurnal cycle. It is further indicated that positive PV generation due to diabatic heating is essential for the formation, development, and movement of the Tibetan Plateau vortex.

Key words: potential vorticity (PV), surface PV, surface PV budget, diabatic heating, Tibetan Plateau

1 Introduction

Potential vorticity (PV; Rossby, 1940; Ertel, 1942) is a physical quantity that inherently couples atmospheric dynamics and thermodynamics. The examination of PV has clarified a wide range of dynamical phenomena in the field of planetary fluids, primarily owing to the conservation and invertibility properties of PV (e.g., Hoskins et al., 1985). Many aspects of the role of PV in atmospheric general circulation, such as the connection between the isentropic PV anomaly and circulation anomaly (Hoskins et al., 1985; Hoskins, 1997, 2015), the relationship between PV and rainstorms (Wu et al., 1995), and the effects of advective PV flux on the upper tropospheric flow (Liu et al., 2007; Ortega et al., 2018), have been documented. In these studies (e.g., Hoskins et al., 1985; Hoskins, 1997, 2015; Wu et al., 1995; Zhao and Ding, 2009; Luo et al., 2018a, 2018b; Ortega et al., 2018), the redistribution of atmospheric interior PV and its weather and climate effects have been of primary interest.

Meanwhile, the studies of Haynes and McIntyre (1987, 1990) and Hoskins (1991) evoked interest in PV at the Earth's surface. The impermeability theorem of PV proposed by Haynes and McIntyre (1987, 1990) suggests that PV can be neither created nor destroyed for a completely closed isentropic surface. A subdivision of the atmosphere into the Overworld, Middleworld, and Underworld (Fig. 1) proposed by Hoskins (1991) showed that there are entirely closed isentropic surfaces in the Overworld and Middleworld, while the isentropic surfaces of the Underworld intersect with the Earth's surface. In terms of the impermeability theorem, Hoskins (1991) demonstrated that total PV is constant in the Overworld and Middleworld but changeable in the Underworld. These studies indicated that the source of global atmospheric PV lies within the Earth's surface. In other words, the globally integrated PV in the atmosphere is determined by its generation at the Earth's surface. Further theoretical studies demonstrated that the variation of PV in the atmospheric interior is closely related to the surface PV, and the PV generated at the surface can be transferred into the atmospheric interior and affects the variation of PV and the associated circulation at a global scale (Held and Schneider, 1999; Schneider, 2005; Egger et al., 2015). However, the potential temperature fluctuates at the Earth's surface and near-surface isentropes may lie above or below the surface, and the investigation of the Earth's surface PV and its budget has thus been difficult.

In addition to the flat surface of the Earth, elevated mountains penetrate isentropic surfaces in the lower troposphere and generate extra internal boundaries of these surfaces as shown in Fig. 1, thereby producing additional PV sources for the atmosphere. Aebischer and Schar (1996), Schar et al. (2003), and Flamant et al. (2004) studied PV around the Alps through observation and numerical simulation and found that the Alps are a source of PV anomalies in the lower troposphere and generate a wake structure with numerous elongated filaments of anomalous PV (i.e., PV banners) that affects downstream windstorms. The gigantic Tibetan Plateau (TP), as the highest and broadest plateau in the world, intensively intersects with many isentropic surfaces in the lower troposphere (Wu et al., 2007, 2018). According to the impermeability theorem, the TP generates a great deal of PV that can be transported into the atmospheric interior (Hoskins, 1991; Ren et al., 2014; Ma et al., 2019; Yu et al., 2019). Given the huge global weather and climate effects of the TP (e.g., Yeh et al., 1957; Flohn, 1957; Ye and Gao, 1979; Yanai, 1992; Wu et al., 2012; Xu et al., 2015; Ren et al., 2019), the associated surface PV and surface PV generation of the TP need to be further explored.

One of the great challenges in studying the surface PV over the TP is that the traditional PV and PV budget are calculated in a pressure or an isentropic coordinate system. As documented above, because the higher pressure level or lower isentropic level is under the mountain surface, there is no data across the wide terrain. Thus, it is almost impossible to calculate the surface PV and surface PV budget directly in the pressure or isentropic coordinate system. The method of interpolating the pressure coordinate system to the terrain-following η coordinate system is widely adopted, but introduces a high interpolation error (Li et al., 2017). Cao and Xu (2011) derived the PV formula in the terrain-following η coordinate system, but many reanalysis model datasets are archived in the hybrid σ-p coordinate system. Although the second Modern-Era Retrospective Analysis for Research and Applications (MERRA2) reanalysis dataset provides PV in the hybrid σ-p coordinate system, the PV product neglects the horizontal component of PV, which could be important in regions with topographic relief. It remains difficult to accurately calculate the PV and PV budget at the Earth's surface, particularly in mountainous areas.

The present paper aims at improving the understanding of the surface PV and providing a stepping stone to investigate the surface PV budget. To this end, we developed an algorithm that calculates the surface PV and surface PV budget and examined the characteristics of surface PV and its budget in the TP area. The remainder of the paper is organized as follows. Section 2 presents the data and model used in the study. Section 3 introduces algorithms for the calculation of surface PV and its budget. Section 4 analyzes the spatial and temporal distributions of the surface PV. Additionally, the vertical and horizontal components of PV in the TP area in reanalysis and model output are compared. Characteristics of the surface PV budget, particularly the diurnal cycle, are investigated in Section 5. The effects of such a diurnal cycle on the formation and development of a Tibetan Plateau vortex (TPV) are briefly presented. Finally, a summary and discussion are provided in Section 6.

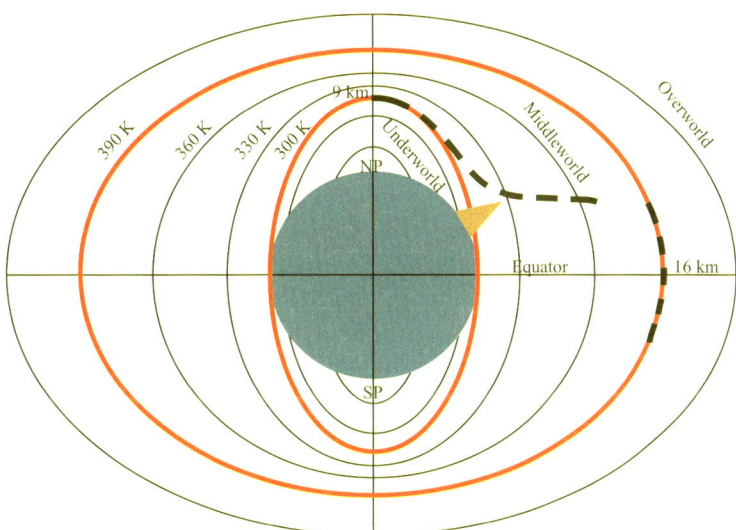

Fig. 1 Schematic distributions of isentropic surfaces and the three-fold world. Solid and dashed black lines respectively indicate isentropic surfaces and the atmospheric tropopause. The blue circle represents the Earth while the brown triangle represents the TP. The North and South Poles are respectively denoted NP and SP

2 Data and model

2.1 Data

We used monthly mean data on the hybrid σ-p model level obtained from MERRA2 for the period 1980—2014 (Rienecker et al., 2011; Lucchesi, 2012). We employed the air temperature, zonal and meridional wind speeds, and pressure in calculating the surface PV for each month. Climate mean values calculated over June, July, and August (JJA) and over December, January, and February (DJF) were respectively used to represent the boreal summer and winter conditions. The horizontal resolution of all MERRA2 data was 0.625°× 0.5° (longitude × latitude).

Three-hourly data of the MERRA2 hybrid σ-p model level, including the air temperature, zonal and meridional wind speeds, vertical velocity, and pressure, for the period 2010—2017 were adopted to analyze the climate mean surface PV budget in June. Data for 27 to 29 June 2016 (universal time) were also used to analyze the surface PV budget during a TPV process.

2.2 Model

The model adopted in this study was the Flexible Global Ocean-Atmosphere-Land System (FGOALS-f2) climate model of the Chinese Academy of Sciences, which was developed at the Institute of Atmospheric Physics/State Key Laboratory of Numerical Modeling for Atmospheric Sciences and Geophysical Fluid Dynamics (Bao et al., 2018). The atmospheric component of FGOALS-f2 is FAMIL2. The FAMIL2 adopts a three-dimensional finite-volume dynamical core (Lin, 2004) over cubed-sphere grids (Putman and Lin, 2007) with six tiles over the globe, which approximates a horizontal resolution of 1°× 1°. The hybrid σ-p coordinates have 32 layers, with the model top being at 2.16 hPa. A detailed physical configuration of the model was presented by He et al. (2019, 2020).

Two AGCM experiments based on FAMIL2 were conducted. The first experiment was an AMIP standard experiment conducted to characterize the climate mean surface PV. Monthly mean data of the FAMIL2 hybrid σ-p model level were generated for the period 1979—2014. The period 1980—2014 was selected to ensure consistency with the time range of the MERRA2 data.

In the second experiment, the generation and early evolution of a TPV from 27 June to 29 June, 2016, were simulated. The forcing sea surface temperature data used to drive FAMIL2 were taken from Optimum Interpolation Sea Surface Temperature version 2 (OISSTv2) (Reynolds et al., 2007; Banzon et al., 2016; Li et al., 2019). Because each term in the PV equation involves a product of two different variables, and because the reanalysis adopted in this study only provides data at three-hourly resolution, the contribution of the transient processes to the total budget may not be presented properly. To address this problem, we produced a dataset with high time resolution of a half hour for the same period by nudging the FAMIL2 to the MERRA2 reanalysis. A nudging method, which involves the addition of terms to the atmospheric equations to relax the predicted state variables toward the reanalysis at 30-min intervals, was adopted to generate a fine-resolution dataset (Hoke and Anthes, 1976). On the grounds of stability, nudging was performed for 1 January to 30 June, 2016. The atmospheric variables used for nudging were the surface pressure, surface geopotential height, three-dimensional atmospheric wind field, and temperature field. The integration time step was set at 30 min. Variables, namely the surface wind, temperature, and pressure, were then used to calculate the surface PV and its budget according to the method introduced in section 3.

3 PV, PV budget equation, and vertical-coordinate independence

3.1 PV and PV budget equation

According to Ertel (1942), the vector form of PV (P) in the z coordinate system and its budget equation are

$$P = \alpha_z \boldsymbol{\xi}_{az} \cdot \nabla_z \theta \tag{1}$$

$$\frac{\mathrm{d}P}{\mathrm{d}t} = \alpha_z \boldsymbol{\xi}_{az} \cdot \nabla_z \dot{\theta} + \alpha_z \nabla_z \times \boldsymbol{F}_f \cdot \nabla_z \theta \tag{2}$$

where $\nabla_z \equiv \left(\frac{\partial}{\partial x}, \frac{\partial}{\partial y}, \frac{\partial}{\partial z}\right)$, $\alpha_z = \frac{1}{\rho}$ is the specific volume, ρ is the density, $\boldsymbol{\xi}_{az} = \nabla_z \times \boldsymbol{V} + f\boldsymbol{k}$ is the three-dimensional absolute vorticity vector, $\boldsymbol{V} = (u, v, w)$ is the three-dimensional wind vector, θ is the potential temperature, $\dot{\theta} = \frac{\mathrm{d}\theta}{\mathrm{d}t}$ is the diabatic heating rate, $\boldsymbol{F}_f = (F_x, F_y)$ is the frictional force, and the subscript z indicates the z coordinate system.

3.2 Vertical-coordinate independence of PV and the PV budget equation

For large-scale motion, $\left|\frac{\partial w}{\partial x}\right|$ and $\left|\frac{\partial w}{\partial y}\right|$ have an order of magnitude of 10^{-8} while $\left|\frac{\partial u}{\partial z}\right|$ and $\left|\frac{\partial v}{\partial z}\right|$ have an order of magnitude of 10^{-3}. The horizontal change of the vertical velocity in the horizontal vorticity can therefore be ignored with high accuracy in the z coordinate system, and the scalar form of PV in the z coordinate system is

$$P = \alpha_z \left\{ -\frac{\partial v}{\partial z}\left(\frac{\partial \theta}{\partial x}\right)_z + \frac{\partial u}{\partial z}\left(\frac{\partial \theta}{\partial y}\right)_z + \left[f + \left(\frac{\partial v}{\partial x}\right)_z - \left(\frac{\partial u}{\partial y}\right)_z\right]\frac{\partial \theta}{\partial z} \right\} \tag{3}$$

Introducing a generalized vertical coordinate system (x, y, e, t) for any variable F, $F(x, y, e, t) = F(x, y, z(x, y, e, t), t)$, where e is a monotonic function of z in the vertical direction and the vertical axis of the e coordinate system (with e usually being equal to p, z or θ). According to the chain rule, we have

$$\left.\begin{array}{l} \left(\dfrac{\partial F}{\partial x}\right)_e = \left(\dfrac{\partial F}{\partial x}\right)_z + \dfrac{\partial F}{\partial z}\left(\dfrac{\partial z}{\partial x}\right)_e \\[2mm] \left(\dfrac{\partial F}{\partial y}\right)_e = \left(\dfrac{\partial F}{\partial y}\right)_z + \dfrac{\partial F}{\partial y_z}\left(\dfrac{\partial z}{\partial y}\right)_e \\[2mm] \dfrac{\partial F}{\partial e} = \dfrac{\partial F}{\partial z}\dfrac{\partial z}{\partial e} \end{array}\right\} \tag{4}$$

Letting α_e indicate the specific volume in the e coordinate system, we have $\delta m = \frac{1}{\alpha_z}\delta x \delta y \delta z = \frac{1}{\alpha_e}\delta x \delta y \delta e$; thus,

$$\alpha_e = \alpha_z \frac{\partial e}{\partial z} \tag{5}$$

The PV in the e coordinate system (P_e) is defined to have the same form as that in the z coordinate system; i.e., the symbols imitate those in formula (3) and we have

$$P_e = \alpha_e \boldsymbol{\xi}_{ae} \cdot \nabla_e \theta$$
$$= \alpha_e \left\{ -\frac{\partial v}{\partial e}\left(\frac{\partial \theta}{\partial x}\right)_e + \frac{\partial u}{\partial e}\left(\frac{\partial \theta}{\partial y}\right)_e + \left[f + \left(\frac{\partial v}{\partial x}\right)_e - \left(\frac{\partial u}{\partial y}\right)_e\right]\frac{\partial \theta}{\partial e} \right\} \quad (6)$$

Letting
$$A = -\frac{\partial v}{\partial e}\left(\frac{\partial \theta}{\partial x}\right)_e$$
$$B = \frac{\partial u}{\partial e}\left(\frac{\partial \theta}{\partial y}\right)_e$$
$$C = \left[f + \left(\frac{\partial v}{\partial x}\right)_e - \left(\frac{\partial u}{\partial y}\right)_e\right]\frac{\partial \theta}{\partial e}$$

we then have $P_e = \alpha_e (A + B + C)$.

Performing vertical coordinate transformation and substituting the conversion relationship (4) into expressions for A, B, and C lead to

$$A = -\frac{\partial v}{\partial e}\left(\frac{\partial \theta}{\partial x}\right)_e = -\frac{\partial v}{\partial z}\frac{\partial z}{\partial e}\left[\left(\frac{\partial \theta}{\partial x}\right)_z + \frac{\partial \theta}{\partial z}\left(\frac{\partial z}{\partial x}\right)_e\right]$$

$$B = \frac{\partial u}{\partial e}\left(\frac{\partial \theta}{\partial y}\right)_e = \frac{\partial u}{\partial z}\frac{\partial z}{\partial e}\left[\left(\frac{\partial \theta}{\partial y}\right)_z + \frac{\partial \theta}{\partial z}\left(\frac{\partial z}{\partial y}\right)_e\right]$$

$$C = \left[f + \left(\frac{\partial v}{\partial x}\right)_e - \left(\frac{\partial u}{\partial y}\right)_e\right]\frac{\partial \theta}{\partial e} = \left\{f + \left[\left(\frac{\partial v}{\partial x}\right)_z + \frac{\partial v}{\partial z}\left(\frac{\partial z}{\partial x}\right)_e\right] - \left[\left(\frac{\partial u}{\partial y}\right)_z + \frac{\partial u}{\partial z}\left(\frac{\partial z}{\partial y}\right)_e\right]\right\}\frac{\partial \theta}{\partial z}\frac{\partial z}{\partial e}$$

The underlined terms correspond to horizontal coordinate transformation items. After adding A, B, and C, the horizontal coordinate transformation items cancel each other out, and we have the relation

$$P_e = \alpha_e \boldsymbol{\xi}_{ae} \cdot \nabla_e \theta = \alpha_z \boldsymbol{\xi}_{az} \cdot \nabla_z \theta = P \quad (7)$$

Equation (7) indicates that for a generalized monotonic e coordinate system, when Eq. (6) holds for P_e, the final result is strictly equal to P. In other words, PV is independent of the vertical coordinate. From Eqs. (5) and (6), we obtain PV in a generalized vertical coordinate system.

The coordinate transformation is carried out for the right-hand side of the PV budget (2). The right-hand side of Eq. (2) is transformed strictly using the conversion relationships Eqs. (4) and (5):

$$\alpha_z \boldsymbol{\xi}_{az} \cdot \nabla_z \dot{\theta} = \alpha_z \left\{ -\frac{\partial v}{\partial z}\left(\frac{\partial \dot{\theta}}{\partial x}\right)_z + \frac{\partial u}{\partial z}\left(\frac{\partial \dot{\theta}}{\partial y}\right)_z + \left[f + \left(\frac{\partial v}{\partial x}\right)_z - \left(\frac{\partial u}{\partial y}\right)_z\right]\frac{\partial \dot{\theta}}{\partial z} \right\}$$
$$= \alpha_e \left\{ -\frac{\partial v}{\partial e}\left(\frac{\partial \dot{\theta}}{\partial x}\right)_e + \frac{\partial u}{\partial e}\left(\frac{\partial \dot{\theta}}{\partial y}\right)_e + \left[f + \left(\frac{\partial v}{\partial x}\right)_e - \left(\frac{\partial u}{\partial y}\right)_e\right]\frac{\partial \dot{\theta}}{\partial e} \right\} \quad (8)$$
$$= \alpha_e \boldsymbol{\xi}_{ae} \cdot \nabla_e \dot{\theta}$$

$$\alpha_z \nabla_z \times \boldsymbol{F}_f \cdot \nabla_z \theta = \alpha_z \left\{ -\frac{\partial F_y}{\partial z}\left(\frac{\partial \theta}{\partial x}\right)_z + \frac{\partial F_x}{\partial z}\left(\frac{\partial \theta}{\partial y}\right)_z + \left[\left(\frac{\partial F_y}{\partial x}\right)_z - \left(\frac{\partial F_x}{\partial y}\right)_z\right]\frac{\partial \theta}{\partial z} \right\}$$
$$= \alpha_e \left\{ -\frac{\partial F_y}{\partial e}\left(\frac{\partial \theta}{\partial x}\right)_e + \frac{\partial F_x}{\partial e}\left(\frac{\partial \theta}{\partial y}\right)_e + \left[\left(\frac{\partial F_y}{\partial x}\right)_e - \left(\frac{\partial F_x}{\partial y}\right)_e\right]\frac{\partial \theta}{\partial e} \right\} \quad (9)$$
$$= \alpha_e \nabla_e \times \boldsymbol{F}_f \cdot \nabla_e \theta$$

where it is seen that the right-hand side of the PV budget equation is also independent of the vertical coordinate.

Substituting Eqs. (7)—(9) into Eq. (2) gives the vertical coordinate independence of the PV budget equation:

$$\frac{\mathrm{d}P_e}{\mathrm{d}t} = \alpha_e \boldsymbol{\xi}_{ae} \cdot \nabla_e \dot{\theta} + \alpha_e \nabla_e \times \boldsymbol{F}_f \cdot \nabla_e \theta \quad (10)$$

Furthermore, in a monotonic e coordinate system, pressure is expressed by $p = p(x,y,e,t)$. We then have

$$\frac{\mathrm{d}p}{\mathrm{d}t} = \omega = \frac{\partial p}{\partial t} + u\left(\frac{\partial p}{\partial x}\right)_e + v\left(\frac{\partial p}{\partial y}\right)_e + \dot{e}\frac{\partial p}{\partial e} \tag{11}$$

and thus

$$\dot{e} = \left[\omega - \frac{\partial p}{\partial t} - u\left(\frac{\partial p}{\partial x}\right)_e - v\left(\frac{\partial p}{\partial y}\right)_e\right]\bigg/\frac{\partial p}{\partial e} \tag{12}$$

Therefore, Eq. (10) can also be written in Eulerian form in a generalized coordinate system as

$$\frac{\partial P_e}{\partial t} = -\mathbf{V} \cdot \nabla_e P_e + \alpha_e \boldsymbol{\xi}_{ae} \cdot \nabla_e \dot{\theta} + \alpha_e \nabla_e \times \mathbf{F}_f \cdot \nabla_e \theta \tag{13}$$

where $\mathbf{V} = (u, v, \dot{e})$ and $\nabla_e \equiv \left(\frac{\partial}{\partial x}, \frac{\partial}{\partial y}, \frac{\partial}{\partial e}\right)$.

3.3 PV and PV equation at the surface layer

The hybrid σ-p system is adopted as vertical coordinate in most current climate models. A surface satisfying the partial difference thus needs to be constructed. Let the hybrid σ-p model levels from top to bottom be $h=1, h=2, \cdots, h=N$. The model level is an isosurface indicated by h, which describes the surface layer and the space above.

According to the vertical-coordinate independence introduced in Section 3.2, the PV and PV budget equations (Eqs. (6) and (13)) at the hybrid σ-p model level are obtained by setting $e=h$:

$$\begin{aligned}P_h &= \alpha_h \boldsymbol{\xi}_{ah} \cdot \nabla_h \theta \\ &= \alpha_h\left\{-\frac{\partial v}{\partial h}\left(\frac{\partial \theta}{\partial x}\right)_h + \frac{\partial u}{\partial h}\left(\frac{\partial \theta}{\partial y}\right)_h + \left[f + \left(\frac{\partial v}{\partial x}\right)_h - \left(\frac{\partial u}{\partial y}\right)_h\right]\frac{\partial \theta}{\partial h}\right\}\end{aligned} \tag{14}$$

$$\frac{\partial P_h}{\partial t} = -\mathbf{V} \cdot \nabla_h P_h + \alpha_h \boldsymbol{\xi}_{ah} \cdot \nabla_h \dot{\theta} + \alpha_h \nabla_h \times \mathbf{F}_f \cdot \nabla_h \theta \tag{15}$$

where $\alpha_h = \alpha_z \frac{\partial h}{\partial z}$, $\mathbf{V} = (u, v, \dot{h})$, $\dot{h} = \left[\omega - \frac{\partial p}{\partial t} - u\left(\frac{\partial p}{\partial x}\right)_h - v\left(\frac{\partial p}{\partial y}\right)_h\right]\bigg/\frac{\partial p}{\partial h}$, and $\nabla_h = \left(\frac{\partial}{\partial x}, \frac{\partial}{\partial y}, \frac{\partial}{\partial h}\right)$.

We substitute $\frac{\partial p}{\partial z} = -\rho g$ and $\alpha_h = \alpha_z \frac{\partial h}{\partial z}$ into Eqs. (14) and (15) and obtain

$$P_h = g\left[\frac{\partial v}{\partial p}\left(\frac{\partial \theta}{\partial x}\right)_h - \frac{\partial u}{\partial p}\left(\frac{\partial \theta}{\partial y}\right)_h\right] - g\left[f + \left(\frac{\partial v}{\partial x}\right)_h - \left(\frac{\partial u}{\partial y}\right)_h\right]\frac{\partial \theta}{\partial p} \tag{16}$$

$$\qquad\qquad\text{PV\_horizontal} \qquad\qquad\qquad \text{PV\_vertical}$$

$$\frac{\partial P_h}{\partial t} = -\mathbf{V} \cdot \nabla_h P_h + \alpha_h \boldsymbol{\xi}_{ah} \cdot \nabla_h \dot{\theta} + \alpha_h \nabla_h \times \mathbf{F} \cdot \nabla_h \theta$$

$$\quad\text{local} \quad\text{advection} \quad\text{heating} \quad\quad\text{friction}$$

$$= -u\frac{\partial P_h}{\partial x} - v\frac{\partial P_h}{\partial y} - \dot{\omega}_h\frac{\partial P_h}{\partial p} - \tag{17}$$

$$g\left\{-\frac{\partial v}{\partial p}\left(\frac{\partial \dot{\theta}}{\partial x}\right)_h + \frac{\partial u}{\partial p}\left(\frac{\partial \dot{\theta}}{\partial y}\right)_h + \left[f + \left(\frac{\partial v}{\partial x}\right)_h - \left(\frac{\partial u}{\partial y}\right)_h\right]\frac{\partial \dot{\theta}}{\partial p}\right\} -$$

$$g\left\{-\frac{\partial F_y}{\partial p}\left(\frac{\partial \theta}{\partial x}\right)_h + \frac{\partial F_x}{\partial p}\left(\frac{\partial \theta}{\partial y}\right)_h + \left[f + \left(\frac{\partial F_y}{\partial x}\right)_h - \left(\frac{\partial F_x}{\partial y}\right)_h\right]\frac{\partial \theta}{\partial p}\right\}$$

where

$$\dot{\omega}_h = \omega - \frac{\partial p}{\partial t} - u\left(\frac{\partial p}{\partial x}\right)_h - v\left(\frac{\partial p}{\partial y}\right)_h \tag{18}$$

Here the horizontal difference is carried out at the hybrid σ-p model level, and the pressure is taken as the vertical difference variable. The terms on the right-hand side of Eq. (16) are referred to as the PV horizontal and vertical components. The terms in Eq. (17) are the local term, advection term, heating term, and friction term. Equations (16) and (17) form a set of equations for PV and the PV budget in the hybrid σ-p coordinate system. The surface PV and surface PV budget are obtained from Eqs. (16)—(18) at the two bottom model levels.

4 Climate features of surface PV

The horizontal component of PV is usually small over the area of a plain, and only the vertical component of PV is used to represent PV. Figure 2 shows the climate mean surface PV and its horizontal and vertical components over the TP in winter and summer. In general, the surface PV increases with latitude, which is mainly due to the f-effect. In winter, the maximum surface PV center is located north of the TP platform (Fig. 2a) and the surface PV on the TP is approximately twice that on the eastern plain. In comparison, in summer, the surface PV decreases and its maximum center shifts to the western TP (Fig. 2b). There is a relatively large center on the north-central TP platform, which is larger than that in western and southern areas. The distribution of the vertical component of the surface PV (Figs. 2c and 2d) is similar to that of the total surface PV in both winter and summer (Figs. 2a and 2b), indicating that the vertical component dominates the surface PV over the main TP platform.

Although the horizontal component of the surface PV is relatively small, it cannot be ignored, espe-

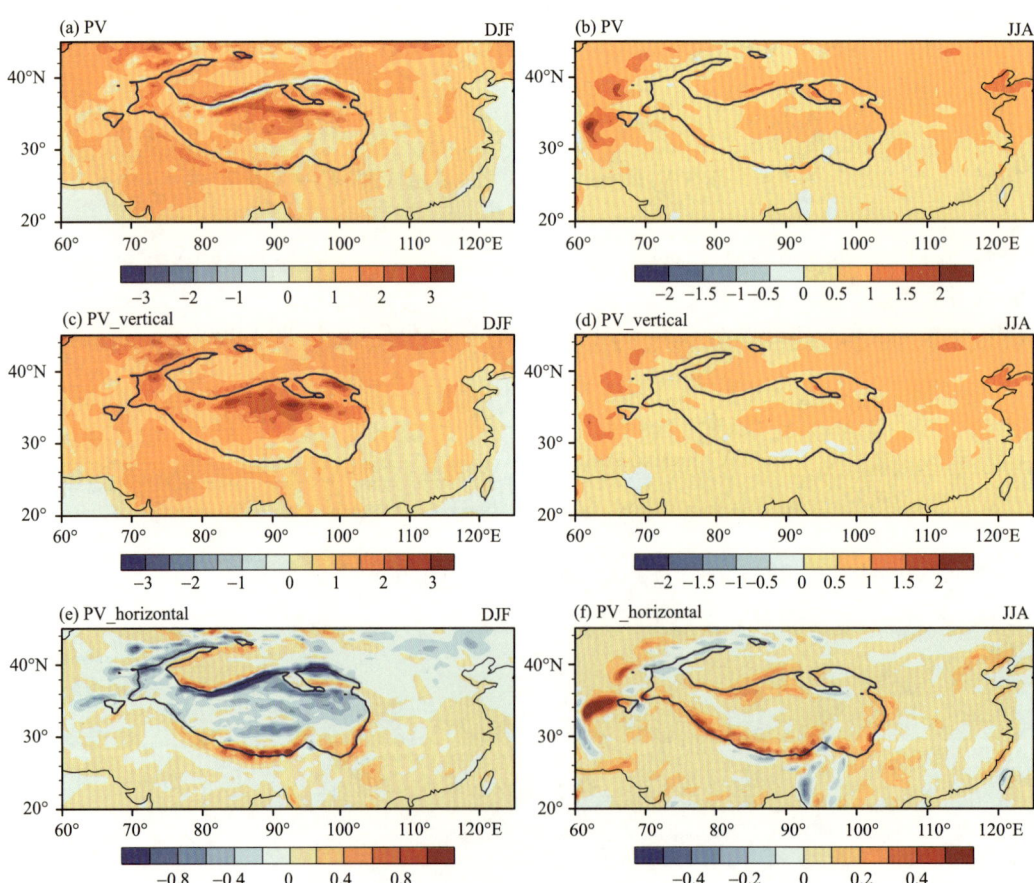

Fig. 2 Climate mean surface distribution in winter calculated from MERRA2 results for (a) the total PV, (c) the vertical PV component, and (e) the horizontal PV component (Units: PVU, 1 PVU = 10^{-6} K m^2 kg^{-1} s^{-1}). (b), (d), (f) are the same as (a), (c), (e), respectively, but for summer. The blue line denotes the TP topographic boundary of 3000 m

cially along the sloping edge of the TP (Figs. 2e and 2f). Figure 2e shows the horizontal component of the surface PV in winter. The component is negative for almost all the TP, especially along the Kunlun and Tianshan mountains. Positive values are found for the Tarim Basin, the Himalayas, and the southeastern slope of the TP, with the maximum center located on the southern slope of the TP. Figure 2f shows the horizontal component of the surface PV in summer. The horizontal component is characterized as being greater along the edge of the TP than on the central TP platform. Its maximum centers are located on the eastern Iranian plateau, the Himalayas, the Kunlun Mountains, and the eastern slope of the TP. One of the most striking features is that the absolute value of the horizontal component of the surface PV along the edge of the TP, especially the northern and southern slopes, is larger than that over the flat areas.

The corresponding results of FAMIL2 are shown in Fig. 3. The magnitudes of the surface PV and its horizontal and vertical components in the model are larger than those in the reanalysis (Fig. 2). This difference in magnitude is mainly due to the difference in the vertical resolution of the surface layer between MERRA2 and FAMIL2. The lowest level in MERRA2 is approximately 60 m above the surface of the TP while that in FAMIL2 is approximately 30 m above the surface. The thickness of the surface layer for the calculation of surface PV is approximately 110 m in MERRA2 and 60 m in FAMIL2. This difference in thickness is amplified in mountain areas because Δp in Eqs. (16) and (17) is proportional to the surface pressure. Consequently, the magnitude calculated from the FAMIL2 output over the TP (Fig. 3) is approximately double that calculated from the MERRA2 data (Fig. 2). Despite this, the distributions are similar. In winter, the maximum surface PV center is located north of the TP platform (Fig. 3a) owing to the vertical component of PV (Fig. 3c). The positive horizontal component (Fig. 3e) is mainly located on the southern slope of the TP and Tarim Basin, and the negative horizontal component covers the TP almost entirely and especially covers the Kunlun and Tianshan mountains near the northern TP. In summer, there is a large positive surface PV center over the western and eastern TP (Fig. 3b). The horizontal component is characterized by the absolute value near the edge of the TP being larger than that on the central TP platform (Fig. 3f). Both the MERRA2 and FAMIL2 results show that the absolute value of the horizontal component of surface PV over the northern and southern slopes of the TP is larger than that over the flat areas.

For further investigation of the importance of the horizontal component of surface PV over the steep northern and southern TP slopes, Fig. 4 shows the ratio of the 85°—95°E mean absolute horizontal component of surface PV to the absolute vertical component. According to MERRA2, there are two peaks near the southern and northern slopes of the TP, where the peak value over the southern slope is much higher than that over the northern slope (Fig. 4a). Both in winter and summer (Fig. 4a), the horizontal component is twice the vertical component near the southern slope and approximately 0.5 times the vertical component near the northern slope of the TP. The ratio calculated from FAMIL2 (Fig. 4b) is close to that calculated from MERRA2 (Fig. 4a) but the magnitudes are smaller for FAMIL2 data, which may be attributed to the weaker terrain gradient and coarser horizontal resolution in FAMIL2. Both the results of the reanalysis and model simulation show that on the steep slopes and especially on the southern TP slope, the horizontal component of the surface PV is comparable to or even far in excess of the vertical component.

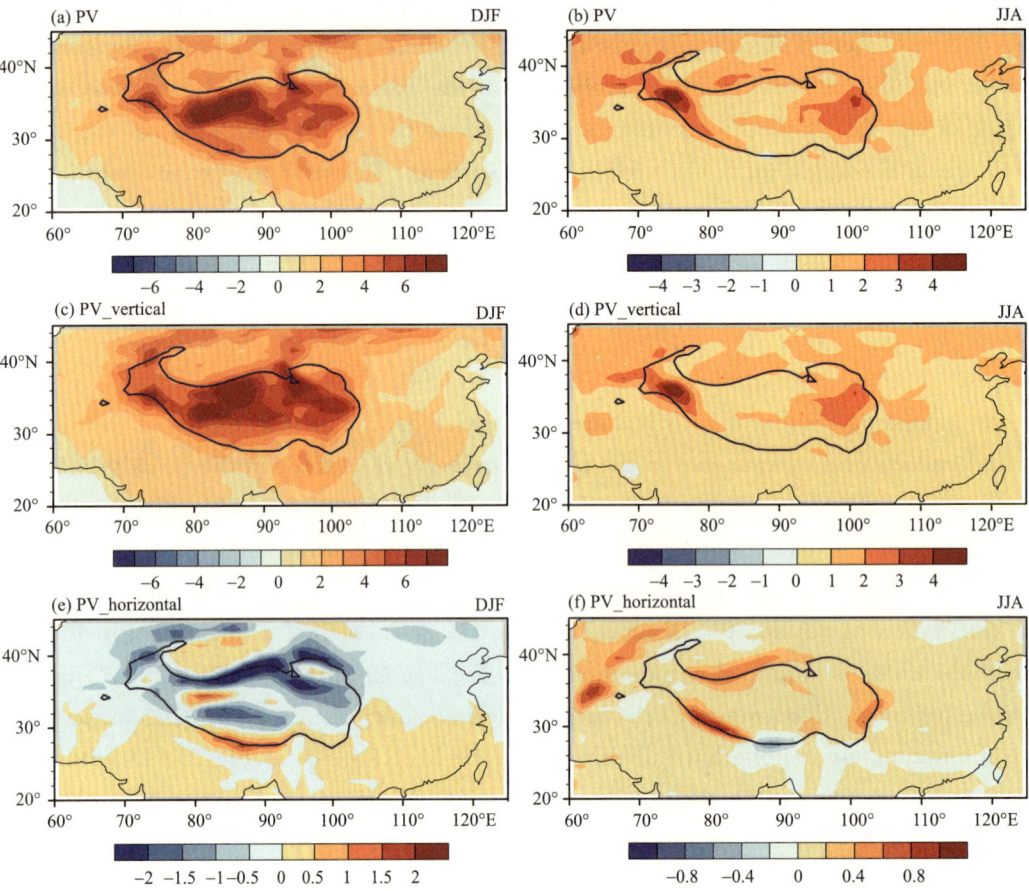

Fig. 3 Same as Fig. 2 but for the results from FAMIL2

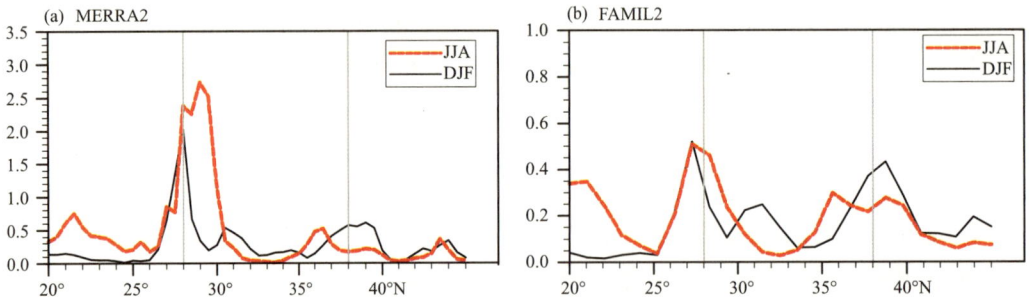

Fig. 4 Ratio of the 85°—95°E mean absolute horizontal component of the surface PV to the absolute vertical component during winter (DJF) and summer (JJA). (a) MERRA2 result, (b) FAMIL2 result. Gray lines (28°E, 38°E) indicate the edges of the southern and northern TP slopes

$$\text{ratio} = \frac{85°-95°\text{E mean } |\text{horizontal component of surface PV}|}{85°-95°\text{E mean } |\text{vertical component of surface PV}|}$$

The horizontal component of the surface PV is further divided into two parts (i. e., $h_1 = g\frac{\partial v}{\partial p}\left(\frac{\partial \theta}{\partial x}\right)_h$ and $h_2 = -g\frac{\partial u}{\partial p}\left(\frac{\partial \theta}{\partial y}\right)_h$) to explore why it is large on steep terrain. We found that the horizontal component of the surface PV is mainly determined by h_2, which is the product of the vertical shear of the zonal wind (Figs. 5a—d) and the meridional gradient of potential temperature (Figs. 5e—h) within the surface layer. In winter, there is a uniform positive mode of $-g\dfrac{\partial u}{\partial p}$ over the whole TP (Figs. 5a and 5e); in summer (Figs. 5b and 5f), $-g\dfrac{\partial u}{\partial p}$ has a dipole pattern that is positive on the southern TP and negative on the northern TP, with no specific signals along the steep slopes. However, there is a strong meridional gradient of potential temperature on the southern and northern slopes both in winter (Figs. 5c and 5g) and summer (Figs. 5d and 5h), and a weak meridional gradient of potential temperature mainly occurs on the central TP and other flat areas. The results document that the strong meridional gradient of potential temperature contributes to the formation of the strong horizontal component of surface PV over the TP slopes.

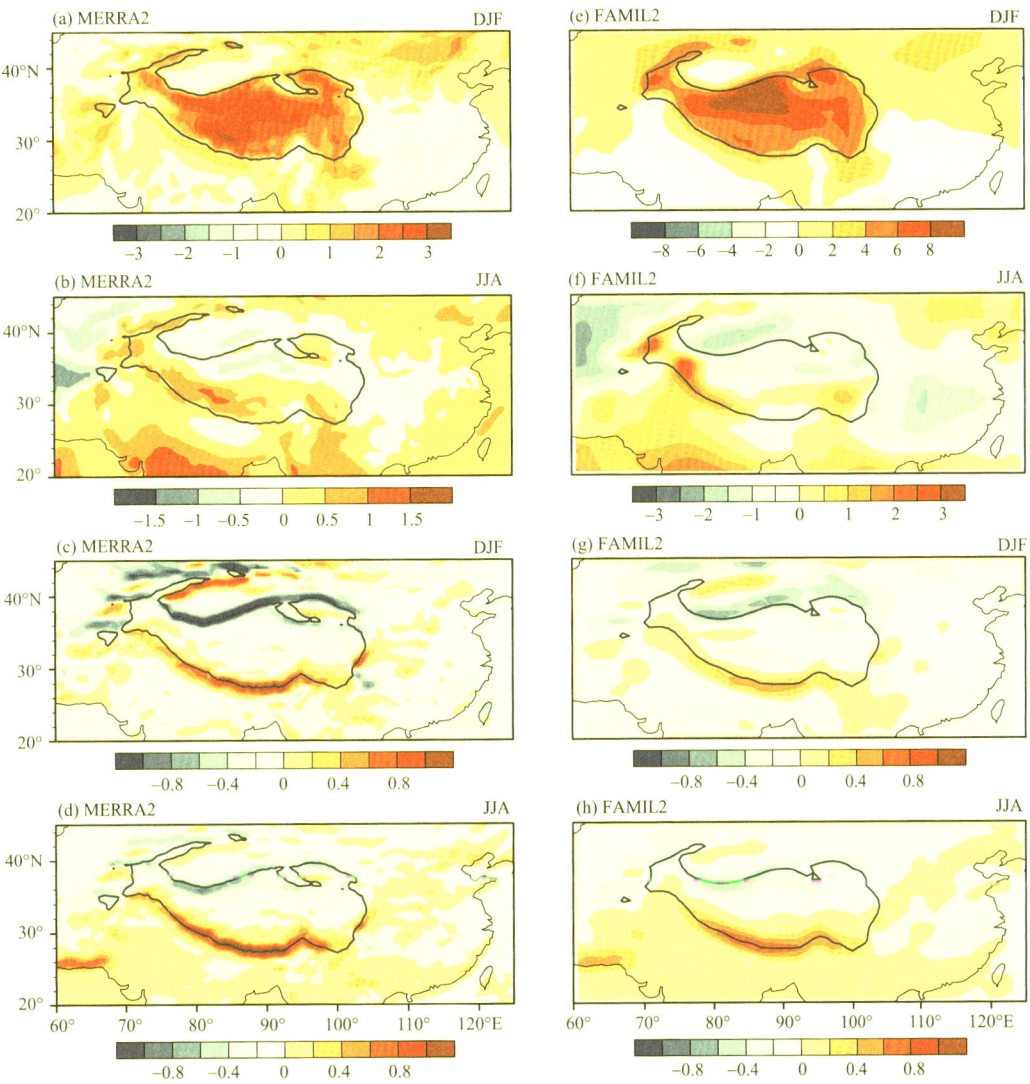

Fig. 5 Climate mean vertical shear of the zonal wind (first two rows, units: 10^{-2} s^{-1} m^3 kg^{-1}) and meridional potential temperature gradient (last two rows, units: 10^{-4} K m^{-1}) calculated from MERRA2 (a—d) and FAMIL2 (e—h) at the surface in boreal winter and summer. The blue line denotes the TP topographic boundary of 3000 m

In this section, a comparison of the surface PV based on data at model levels from FAMIL2 and MERRA2 illustrates that although the surface PV is dominated by its vertical component over flat areas, its horizontal component is important over the slopes of the TP, especially over the southern TP slope.

The prominent horizontal component of the surface PV over the slopes of the TP is mainly due to the strong in-situ meridional gradient of potential temperature. We therefore highlight that the horizontal component of the surface PV should be included in the analysis. In fact, Wu et al. (1995) documented the importance of the PV horizontal component. They showed that when air parcels move along a slantwise isentropic surface, changes in the horizontal component of PV induce vertical vorticity and stimulate precipitation. Moreover, there is feedback between the precipitation and thermal structure of circulation over the southern TP slope (Wu et al., 2016). The thermal forcing of the TP plays a vital role in the Asian summer monsoon (Wu et al., 2007, 2012, 2018). Therefore, to better understand atmospheric phenomena related to the TP, such as the Asian summer monsoon, a more accurate characterization of the vertical and horizontal components of the surface PV (Eq. (16)) is required.

5 Application of the surface PV budget to TPV study

A TPV is a specific synoptic system over the TP in boreal summer, with a typical spatial scale of 400–800 km horizontally and 2–3 km vertically. TPVs not only have an important effect on precipitation over the main body of the TP but also are closely related to the rainfall over southwestern and eastern China (Ye and Gao, 1979; Luo et al., 1994; Li et al., 2014). As a result, the formation and evolution of TPVs have received increasing attention. In general, TPVs form mainly over the central-western plateau in the period from June to August and much less so in May and September; the occurrence frequency of the vortices peaks in June; most of the TPVs die out in-situ while some develop and move eastward and a few move off the TP (Qiao and Zhang, 1994; Chen et al., 1996; Li et al., 2014).

As an example of applying the surface PV budget equation in practice, we selected a TPV event lasting from 27 to 29 June 2016 as a case study. The synoptic analysis of this process at 500 hPa from the PV perspective was carried out by Ma et al. (2019) using MERRA2 data, which provided a weather background for the exploration of the surface PV budget. Instead of examining the synoptic process and evaluating model bias, we mainly focused on the balance of the PV budget based on Eq. (17) and the effect of the heating term.

The occurrence frequency of TPVs peaks in June (Li et al., 2014). Figure 6 shows the climate mean diurnal cycle of the surface PV budget from Lhasa time (LT) 00:00 to LT 21:00 in June. The time interval is 3 hours. In the climate mean state, the contributions of the advection term (Figs. 6b1—b8) and residual term (Figs. 6d1—d8) are small and can be neglected. The local term (Figs. 6a1—a8) is mainly balanced by the heating term (Figs. 6c1—c8). The local change of surface PV (Figs. 6a1—a8) and diabatic heating (Figs. 6c1—c8) show a prominent diurnal cycle with a positive value from afternoon to midnight (LT 15:00—03:00) and negative value from early morning to noon (LT 06:00—12:00). The minimum appears at LT 06:00 over the central-western TP while the maximum appears from LT 21:00 to LT 03:00, implying that the surface PV generated mainly from afternoon to midnight is consumed in the morning. Li et al. (2014) documented that the occurrence frequency of the TPVs processes a robust diurnal variation with a maximum from evening to midnight and a minimum from early morning to noon. Comparing the results of Li et al. (2014) (see their Fig. 2) with the results of the present study (Fig. 6), we infer that the TPVs preferably form at night when the surface PV is generated quickly owing to the strong effect of surface diabatic heating. Additionally, the prominent diurnal cycle of the surface PV budget over the TP is stronger than that in the plain area, indicating the TP is an important area of surface PV variation.

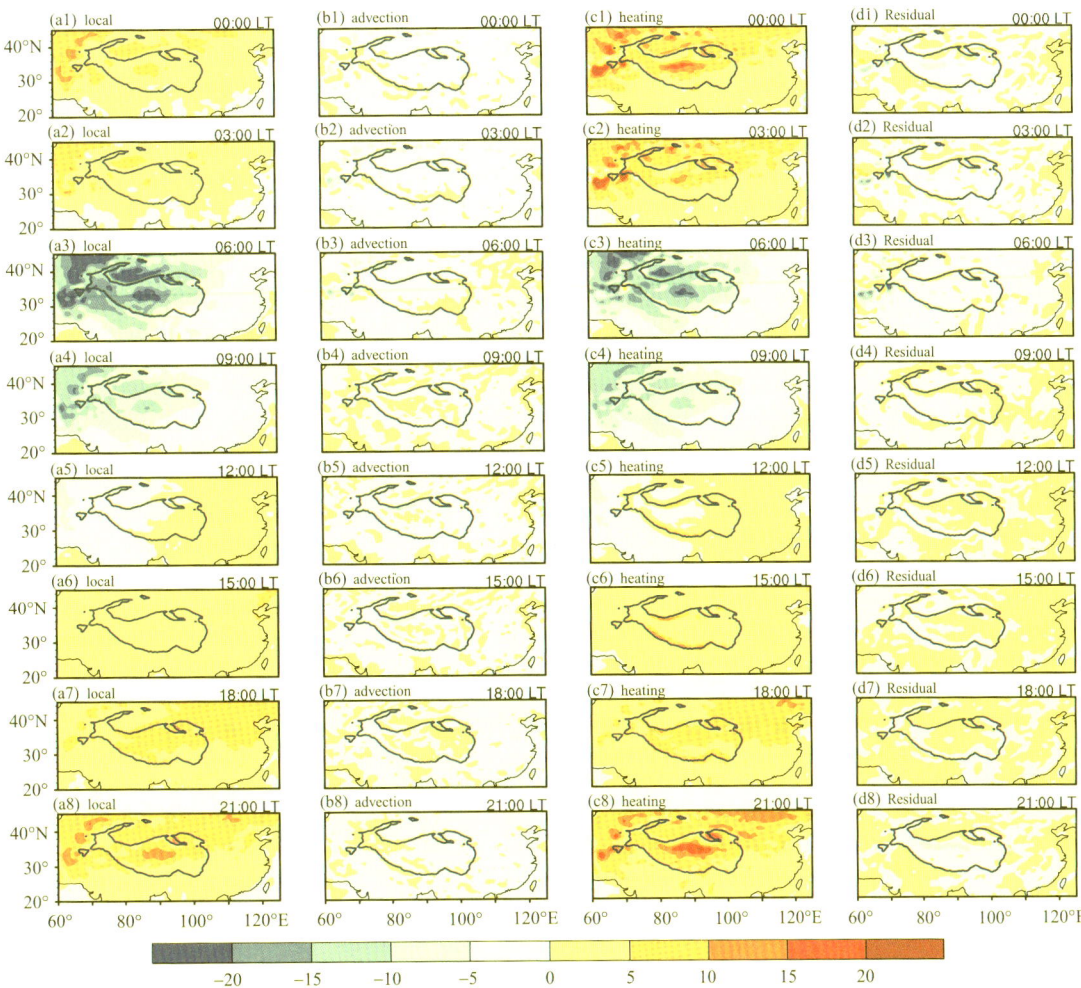

Fig. 6 Climate mean (2010–2017) diurnal cycle from LT 00:00 (top row) to LT 21:00 (bottom row) in 3-hour intervals of each term (units: 10^{-5} PVU s^{-1}) in surface PV budget equation (17) for June: (a1)–(a8) local term, (b1)–(b8) advection term, (c1)–(c8) heating term, and (d1)–(d8) residual term. The blue line denotes the TP topographic boundary of 3000 m

The TPV case considered in this study occurred late at night on 27 June 2016, corresponding to a period of strong PV generation associated with surface diabatic heating in the climate mean state. The TPV developed and moved on the TP on 28 and 29 June and moved off the TP on 30 June (Ma et al., 2019). To examine the balance of the surface PV budget equation during the TPV evolution, the time series of each term in Eq. (17) averaged over the rectangle (20°–45°N, 60°–125°E) was calculated according to MERRA2 and FAMIL2 output at the surface. The time covered the formation and evolution of the TPV from LT 06:00 on 27 June to LT 03:00 on 30 June 2016. Results are shown in Fig. 7. The coefficient of correlation between the local term on the left side of Eq. (17) and the sum2 (advection plus heating) term on the right side of Eq. (17) exceeds 0.95 for both MERRA2 (Fig. 7a) and FAMIL2 (Fig. 7b), surpassing the 99% significance level. Consequently, the residual (local minus sum2) term (Figs. 7a and 7b) is very small. In MERRA2 (Fig. 7a) and FAMIL2 (Fig. 7b), the advection term is smaller than the heating term, which means the heating term dominates the sum2 term and local term. Although the magnitudes in FAMIL2 (Fig. 7b) are larger than their counterparts in MERRA2 (Fig. 7a), as previously mentioned, the diurnal cycle of each term is similar in these two datasets, with the minimum around sunrise and the maximum around sunset, indicating an increase in PV from afternoon to midnight but a decrease from early morning to noon. The small residual and the consistency in the PV budget between MERRA2 and FAMIL2 indicate that the surface PV budget equation (17) is well balanced in the synoptic process.

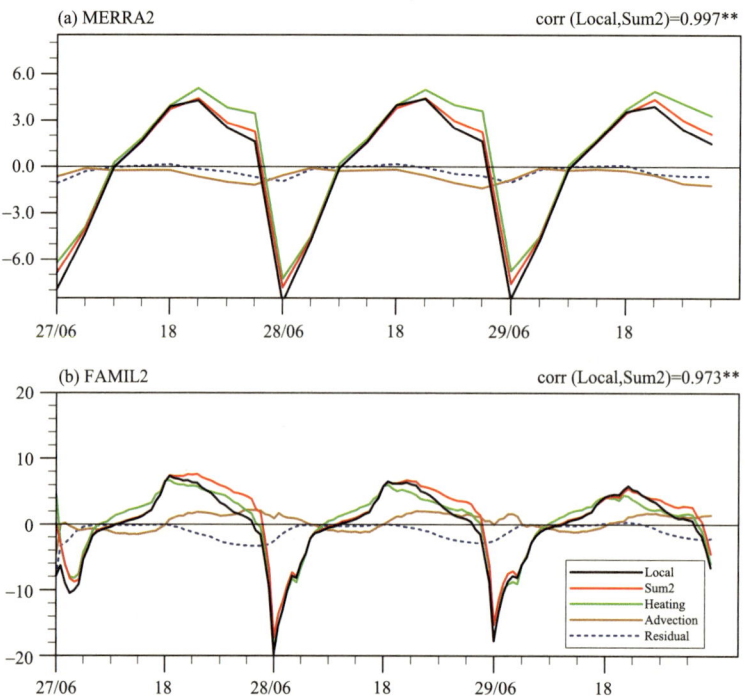

Fig. 7 Evolution from LT 06:00 on 27 June to LT 03:00 on 30 June 2016 and averaged over the rectangle (20°−45°N, 60°−125°E) of the local term (black curve), sum of the advection and heating terms (red curve, sum2), heating term (green curve), advection term (brown curve), and residual term (blue dashed curve) at the surface based on Eq. (17). (a) MERRA2, (b) FAMIL2. Units: 10^{-11} K m^2 kg^{-1} s^{-2}. " * * " indicates surpassing the 99% significance level

The coefficients of correlation between the local term and heating term and between the local term and advection term during the TPV process are given in Table 1. The correlation between the local term and heating term is as high as 0.986 and 0.966 for MERRA2 and FAMIL2, surpassing the 99% significance level in both datasets. However, the correlation between the local term and advection term is weak and does not surpass the significance test. The results show that diabatic heating plays an important role in the local change of surface PV.

Table 1 Coefficient of correlation between the local term and the sum2 (heating plus advection) term (second row), heating term (third row), and advection term (bottom row) for MERRA2 (middle column) and FAMIL2 (rightmost column) during the TPV process from 27 to 29 June 2016. " * * " indicates surpassing the 99% significance level

Correlation (Local term)	MERRA2	FAMIL2
Sum2 term	0.997**	0.973**
Heating term	0.986**	0.966**
Advection term	−0.124	0.127

The lowest hybrid σ-p levels in MERRA2 and FAMIL2 have different altitudes. Therefore, for comparison of the evolutions of the PV budget in the two model outputs, each term in Eq. (17) calculated from MERRA2 and FAMIL2 is interpolated to the same pressure level (500 hPa). The results from MERRA2 reanalysis for the period from 27 to 29 June 2016 are shown in Fig. 8. From the top row to the bottom rows are the local term, sum2 term, advection term, heating term, and residual term at LT 15:00 while the left, middle, and right columns are respectively for 27, 28, and 29 June 2016. The residual term is small except at the northern edge (Figs. 8a5—c5), and the patterns of the local term (Figs. 8a1—c1) and sum2 term (Figs. 8a2—c2) are similar during the TPV generation (left column of Fig. 8), development (middle), and movement (right) stages, indicating a reasonable balance of the PV budget based on Eq. (17), as previously discussed. During the development stage of the TPV, there is positive PV generation to the southeast of the TPV center (Fig. 8b1), which leads to the local PV increase and southeastward movement of the growing TPV. Moreover, the positive PV generation (Fig. 8b1) is primarily due to the heating term (Fig. 8b4), which means that the diabatic heating effect has a noticeable effect on the positive PV generation and the movement of the TPV. However, the advection term (Fig. 8b3) makes the opposite contribution and tends to cancel part of the diabatic heating effect (Fig. 8b4). During the movement stage, although the negative advection term (Fig. 8c3) is stronger than that in the development stage (Fig. 8b3), the positive diabatic heating effect (Fig. 8c4) is more powerful than the negative advection effect, resulting in strong positive PV generation (Fig. 8c1) to the south of the TPV center. Finally, the TPV further strengthens and moves off the TP.

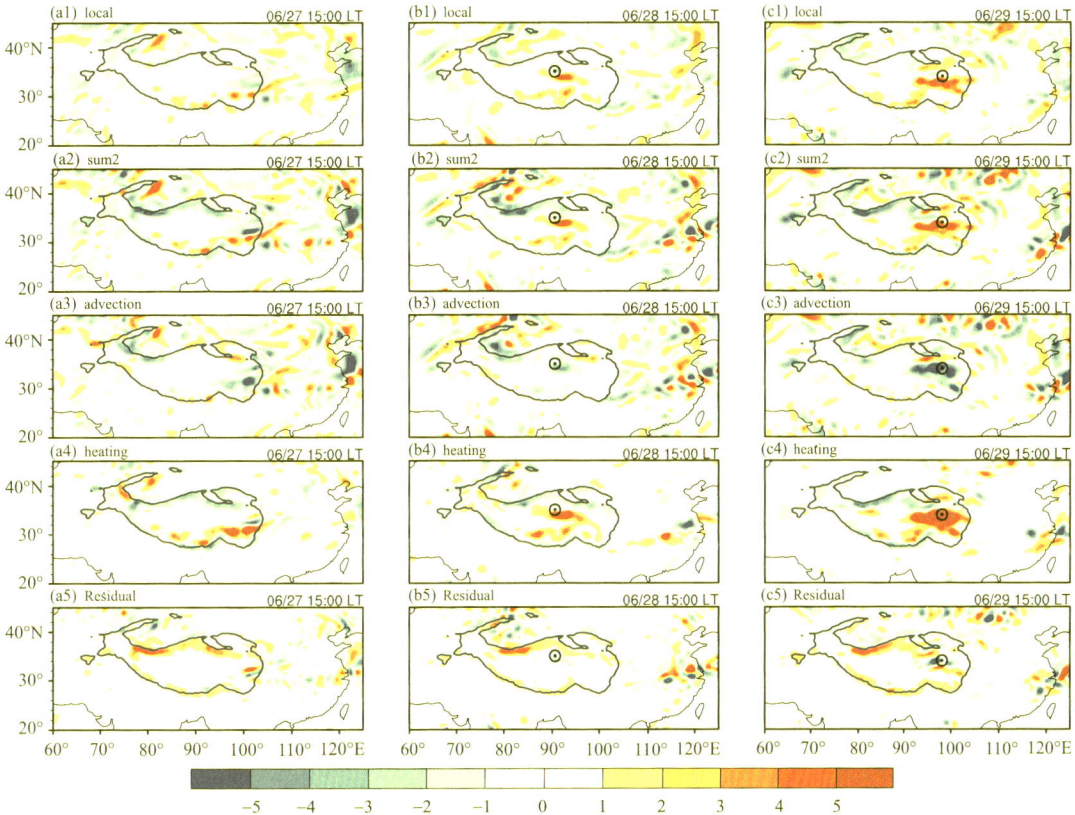

Fig. 8 Distributions at 500 hPa of (a1—c1) the PV local term, (a2—c2) the sum of the advection and heating (sum2) terms, (a3—c3) the advection term, (a4—c4) the heating term, and (a5—c5) the residual term at local time 15:00 in Lhasa on 27 (a1—a5), 28 (b1—b5), and 29 (c1—c5) June 2016. Units: 10^{-11} K m^2 kg^{-1} s^{-2}. The black dot and circle indicate the TPV

The results obtained from FAMIL2 (Fig. 9) are smoother than those obtained from MERRA2 (Fig. 8), which is mainly due to the coarser horizontal resolution and higher time resolution. Despite this, the distributions are similar. As expected, the residual term (Figs. 9a5—c5) is small in FAMIL2, and the patterns of the local term (Figs. 9a1—c1) and the sum2 term (Figs. 9a2—c2) are similar, implying a good balance of the PV budget Eq. (17). The positive PV generation (Figs. 9a1—c1) is dominated by the heating term (Figs. 9a4—c4), especially in the movement stage (Figs. 9c1 and c4). The advection term (Figs. 9b3—c3) mainly contributes to the in-situ PV reduction and tends to partly cancel the strong diabatic heating effect (Figs. 9a4—c4). These results agree with those of MERRA2. The results of FAMIL2 and MERRA2 consistently show that diabatic heating has an important effect on the development and movement of the TPV center associated with positive PV generation. The results of dynamic analyses here are consistent with the results of previous statistical studies (e.g., Luo et al., 1994; Li et al., 2002, 2016), which showed that the occurrence, development, and extinction of the TPV are closely related to the diabatic heating effect.

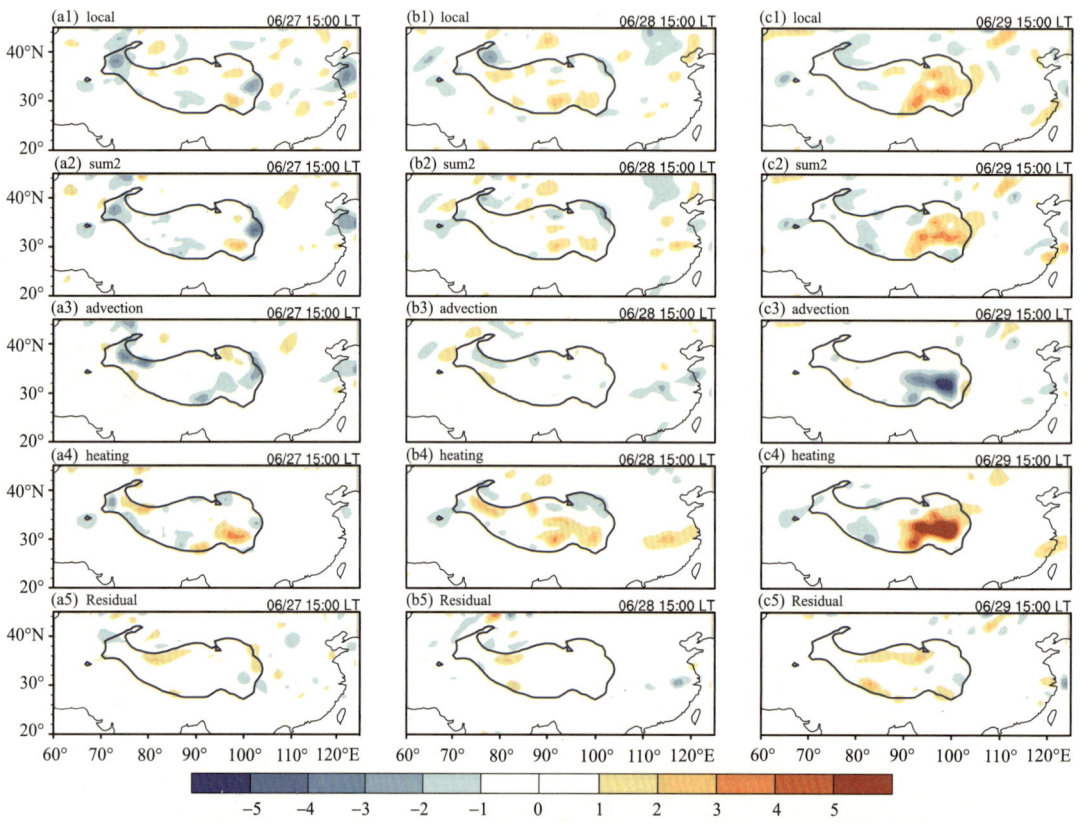

Fig. 9 Same as Fig. 8 but for the results from FAMIL2

The results presented in this section demonstrate that the surface PV budget equation (17) is well balanced both in the climate mean state and synoptic process. The contribution of diabatic heating with a negative peak in early morning and positive peak from evening to midnight dominates the diurnal cycle of the surface PV budget. The case study further shows that diabatic heating has an essential effect on the development and movement of the TPV center associated with positive PV generation.

6 Summary and discussion

6.1 Summary

The studies of Haynes and McIntyre (1987, 1990) and Hoskins (1991) demonstrated that the budget of globally integrated PV in the atmosphere is determined by the generation of PV at the Earth's surface. However, because of the complexity of terrain, wholly characterizing the features and budget of surface PV is a great challenge. The present study revealed the features of surface PV and provided a stepping stone to investigate the surface PV budget. An algorithm for evaluating the surface PV and its budget was derived, and the features and budget of surface PV based on the MERRA2 model level reanalysis and FAMIL2 model level data were diagnosed. The main conclusions drawn from the results of the study are as follows.

(1) Forms of the PV and PV budget equation are independent of coordinates; i.e., they are identical for all vertical coordinate systems.

(2) The surface PV is dominated by its vertical component over flat areas in both boreal winter and summer. However, the horizontal component of the surface PV is important over the slopes of the TP, being more than twice the vertical component over the steep southern slope of the TP. The prominent horizontal component of the surface PV is mainly attributed to the strong in-situ meridional gradient of the potential temperature. On steep terrain, neglecting the horizontal component of surface PV can lead to large errors in capturing the surface PV structure and associated dynamic processes. The results indicate that the complete expression of PV (Eq. (16)) can help reveal the fine-scale structure and dynamic effect of the surface PV.

(3) Surface PV budget analyses showed that the surface PV budget equation (17) is well balanced both in the climate mean state and synoptic process. The surface PV budget and the diabatic heating are characterized by a prominent diurnal cycle; the prominent diurnal cycle signal is stronger over the TP than in plain areas, indicating that the TP is an important area of surface PV variation. The contribution of diabatic heating that has a minimum in the early morning and maximum from evening to midnight dominates the diurnal cycle of the surface PV budget. The case study of the TPV further indicates that diabatic heating has an important effect on the development and movement of the TPV center associated with positive PV generation.

6.2 Discussion

To further reveal the important roles of surface PV and its budget in atmospheric general circulation, several issues are highlighted below for future study.

The lowest model levels of MERRA2 and FAMIL2 are at some distance (approximately 60 m and 30 m, respectively) from the Earth's surface. It has been demonstrated that the calculation of the surface PV is sensitive to the vertical resolution of the model near the surface layer. An improved surface layer presentation in the model is conducive to a better evaluation of the surface PV. Moreover, the present study did not explicitly calculate the friction term but instead incorporated the effects of the friction term into the residual term. It is noted that the residual term is the result of the combined effects of transient diffusion and friction, not just the friction term. Hoskins (1991) noted that there is a constraint in linking friction to diabatic heating, and this constraint links "westerlies" and "cooling" as well as "easterlies" and "heating" in an average sense. There are questions concerning the role of friction in the PV budget that need to be addressed in future work.

The diabatic heating is important to the surface PV budget both for the climate mean state and synoptic process. However, how diabatic heating originates, how it contributes to the formation of the TPV, and how it interacts with precipitation, especially on the diurnal timescale, are still unclear and require further study.

White et al. (2018) and Ren et al. (2019) showed the different topographic forcing have different impacts on the atmospheric circulation. Revealing the link between the surface PV in different regions and atmospheric circulation also requires further study.

References

AEBISCHER U, SCHÄR C, 1996. Low-level potential vorticity and cyclogenesis to the lee of the Alps[J]. Journal of the Atmospheric Sciences, 55(2): 186-207.

BANZON V, SMITH T M, CHIN T M, et al, 2016. A long-term record of blended satellite and in situ sea-surface temperature for climate monitoring, modeling and environmental studies[J]. Earth System Science Data, 8(1): 165-176.

BAO Q, WU X F, LI J X, et al, 2018. Outlook for El Niño and the indian ocean dipole in autumn-winter 2018-2019[J]. Chinese Science Bulletin, 63: 73-78.

CAO J, XU Q, 2011. Computing hydrostatic potential vorticity in terrain-following coordinates[J]. Monthly Weather Review, 139(9): 2955-2961.

CHEN B M, QIAN Z A, ZHANG L S, 1996. Numerical simulation of formation and development of vortices over the Qinghai-Xizang Plateau in summer[J]. Chinese Journal of Atmospheric Sciences, 20: 491-502.

EGGER J, HOINKA K P, SPENGLER T, 2015. Aspects of potential vorticity fluxes: Climatology and impermeability[J]. Journal of the Atmospheric Sciences, 72(8): 3257-3267.

ERTEL H, 1942. Ein neuer hydrodynamische wirbelsatz[J]. Meteorologische Zeitschrift Braunschweig, 59: 33-49.

FLAMANT C, RICHARD E, SCHAR C, et al, 2004. The wake south of the Alps: Dynamics and structure of the lee-side flow and secondary potential vorticity banners[J]. Quarterly Journal of the Royal Meteorological Society, 130(599): 1275-1303.

FLOHN H, 1957. Large-scale aspects of the "summer monsoon" in South and East Asia[J]. Journal of the Meteorological Society of Japan, 35A: 180-186.

HAYNES P H, MCINTYRE M E, 1987. On the evolution of vorticity and potential vorticity in the presence of diabatic heating and frictional or other forces[J]. Journal of the Atmospheric Sciences, 44(5): 828-841.

HAYNES P H, MCINTYRE M E, 1990. On the conservation and impermeability theorems for potential vorticity[J]. Journal of the Atmospheric Sciences, 47(16): 2021-2031.

HE B, BAO Q, WANG X, et al, 2019. CAS FGOALS-f3-L model datasets for CMIP6 historical atmospheric model intercomparison project simulation[J]. Advances in Atmospheric Sciences, 36(8): 771-778.

HE B, LIU Y M, WU G X, et al, 2020. CAS FGOALS-f3-L model datasets for CMIP6 gmmip tier-1 and tier-3 experiments [J]. Advances in Atmospheric Sciences, 37(1): 18-28.

HELD I M, SCHNEIDER T, 1999. The surface branch of the zonally averaged mass transport circulation in the troposphere[J]. Journal of the Atmospheric Sciences, 56(11): 1688-1697.

HOKE J E, ANTHES R A, 1976. The initialization of numerical models by a dynamic-initialization technique[J]. Monthly Weather Review, 104(12): 1551-1556.

HOSKINS B, 1991. Towards a PV-θ view of the general-circulation[J]. Tellus Series a-Dynamic Meteorology and Oceanography, 43(4): 27-35.

HOSKINS B, 1997. A potential vorticity view of synoptic development[J]. Meteorological Applications, 4: 325-334.

HOSKINS B, 2015. Potential vorticity and the PV perspective[J]. Advances in Atmospheric Sciences, 32(1): 2-9.

HOSKINS B, MCINTYRE M E, ROBERTSON A W, 1985. On the use and significance of isentropic potential vorticity maps[J]. Quarterly Journal of the Royal Meteorological Society, 111(470): 877-946.

LI G P, ZHAO B J, YANG J Q, 2002. A dynamical study of the role of surface sensible heating in the structure and intensification of the Tibetan Plateau vortices[J]. Chinese Journal of Atmospheric Sciences, 26: 519-525.

LI G P, LU H G, HUANG C H, et al, 2016. A climatology of the surface heat source on the Tibetan Plateau in summer and its impacts on the formation of the Tibetan Plateau vortex[J]. Chinese Journal of Atmospheric Sciences, 40: 131-141.

LI J, BAO Q, LIU Y, et al, 2019. Evaluation of FAMIL2 in simulating the climatology and seasonal-to-interannual variability of tropical cyclone characteristics[J]. Journal of Advances in Modeling Earth Systems, 11(4): 1117-1136.

LI L, ZHANG R, WEN M, 2014. Diurnal variation in the occurrence frequency of the tibetan plateau vortices[J]. Meteorology and Atmospheric Physics, 125(3-4): 135-144.

LI Z, FEI J F, HUANG X G, et al, 2017. A computational method of Ertel potential vorticity in terrain-following coordinates[J]. Acta Meteorologica Sinica, 75: 1011-1026.

LIN S J, 2004. A "vertically lagrangian" finite-volume dynamical core for global models[J]. Monthly Weather Review, 132(10): 2293-2307.

LIU Y M, HOSKINS B, BLACKBURN M, 2007. Impact of Tibetan orography and heating on the summer flow over Asia[J]. Journal of the Meteorological Society of Japan, 85B: 1-19.

LUCCHESI R, 2012. File specification for MERRA products[Z]. GMAO office note no. 1 (version 2.3), 87. https://gmao.Gsfc.Nasa.Gov/pubs/docs/lucchesi528.

LUO D, CHEN X, DAI A, et al, 2018a. Changes in atmospheric blocking circulations linked with winter Arctic warming: a new perspective[J]. Journal of Climate, 31(18): 7661-7678.

LUO D, CHEN X, FELDSTEIN S B, 2018b. Linear and nonlinear dynamics of North Atlantic Oscillations: A new thinking of symmetry breaking[J]. Journal of the Atmospheric Sciences, 75(6): 1955-1977.

LUO S W, HE M L, LIU X D, 1994. Study on the vortex of the Qinghai-Xizang (Tibet) Plateau in summer[J]. Science in China Series B-Chemistry, 37(5): 601-612.

MA T, LIU Y M, WU G X, et al, 2020. Potential vorticity diagnosis on the formation, development and eastward movement of a Tibetan plateau vortex and its influence on the downstream precipitation[J]. Chinese Journal of the Atmospheric Sciences, 44(3): 472-486.

MA T T, WU G X, LIU Y M, et al, 2019. Impact of surface potential vorticity density forcing over the Tibetan Plateau on the south China extreme precipitation in january 2008. Part I: Data analysis[J]. Journal of Meteorological Research, 33(3): 400-415.

ORTEGA S, WEBSTER P J, TOMA V, et al, 2018. The effect of potential vorticity fluxes on the circulation of the tropical upper troposphere[J]. Quarterly Journal of the Royal Meteorological Society, 144(712): 848-860.

PUTMAN W M, LIN S H, 2007. Finite-volume transport on various cubed-sphere grids[J]. Journal of Computational Physics, 227(1): 55-78.

QIAO Q M, ZHANG Y G, 1994. Synoptic Meteorology of the Tibetan Plateau and Its Effect on the Near Areas [M]. Beijing: China Meteorological Press: 251.

REN R C, WU G X, CAI M, et al, 2014. Progress in research of stratosphere-troposphere interactions: Application of isentropic potential vorticity dynamics and the effects of the Tibetan Plateau[J]. Journal of Meteorological Research, 28(5): 714-731.

REN R C, XIA X, RAO J, 2019. Topographic forcing from East Asia and North America in the northern winter stratosphere and their mutual interference[J]. Journal of Climate, 32(24): 8639-8658.

REYNOLDS R W, SMITH T M, LIU C, et al, 2007. Daily high-resolution-blended analyses for sea surface temperature[J]. Journal of Climate, 20(22): 5473-5496.

RIENECKER M M, SUAREZ M J, GELARO R, et al, 2011. MERRA: Nasa's modern-era retrospective analysis for research and applications[J]. Journal of Climate, 24(14): 3624-3648.

ROSSBY C G, 1940. Planetary flow patterns in the atmosphere[J]. Quarterly Journal of the Royal Meteorological Society, 66: 68-87.

SCHAR C, SPRENGER M, LÜTHI D, et al, 2003. Structure and dynamics of an alpine potential-vorticity banner[J].

Quarterly Journal of the Royal Meteorological Society, 129(588): 825-855.

SCHNEIDER T, 2005. Zonal momentum balance, potential vorticity dynamics, and mass fluxes on near-surface isentropes [J]. Journal of the Atmospheric Sciences, 62(6): 1884-1900.

WU G X, CAI Y P, TANG X J, 1995. Moist potential vorticity and slantwise vorticity development [J]. Acta Meteorologica Sinica, 53: 387-405.

WU G X, LIU Y M, WANG T M, et al, 2007. The influence of mechanical and thermal forcing by the Tibetan Plateau on Asian climate[J]. Journal of Hydrometeorology, 8(4): 770-789.

WU G X, LIU Y M, HE B, et al, 2012. Thermal controls on the Asian summer monsoon[J]. Scientific Reports, 2: 1-7.

WU G X, ZHUO H F, WANG Z Q, et al, 2016. Two types of summertime heating over the Asian large-scale orography and excitation of potential-vorticity forcing I. Over Tibetan Plateau[J]. Science China Earth Sciences, 59 (10): 1996-2008.

WU G X, LIU Y M, HE B, et al, 2018. Review of the impact of the Tibetan Plateau sensible heat driven air-pump on the Asian summer monsoon[J]. Chinese Journal of Atmospheric Sciences, 42: 488-504.

XU X, ZHAO T, SHI X, et al, 2015. A study of the role of the Tibetan Plateau's thermal forcing in modulating rainband and moisture transport in eastern China[J]. Acta Meteorologica Sinica, 73(1): 20-35.

YANAI M H, LI C F, SONG Z S, 1992. Seasonal heating of the Tibetan Plateau and its effects on the evolution of the Asian summer monsoon[J]. Journal of the Meteorological Society of Japan, 70(1B): 319-351.

YEH T C, LO S W, CHU P C, 1957. The wind structure and heat balance in the lower troposphere over Tibetan Plateau and its surroundings [J]. Acta Meteorologica Sinica, 28: 108-121.

YE D Z, GAO Y X, 1979. The meteorology of the Qinghai-Xizang (Tibet) Plateau [M]. Beijing: Science Press: 278.

YU J H, LIU Y M, MA T T, et al, 2019. Impact of surface potential vorticity density forcing over the Tibetan Plateau on the South China extreme precipitation in january 2008. Part II: Numerical simulation[J]. Journal of Meteorological Research, 33(3): 416-432.

ZHAO L, DING Y H, 2009. Potential vorticity analysis of cold air activities during the East Asian summer monsoon[J]. Chinese Journal of Atmospheric Sciences, 33(2): 359-374.

代表性论文 29

Quantification of Seasonal and Interannual Variations of the Tibetan Plateau Surface Thermodynamic Forcing Based on the Potential Vorticity

HE Bian[1,2], SHENG Chen[1,2], WU Guoxiong[1,2*], LIU Yimin[1,2], TANG Yiqiong[1,3]

(1 State Key Laboratory of Numerical Modeling for Atmospheric Sciences and Geophysical Fluid Dynamics (LASG), Institute of Atmospheric Physics (IAP), Chinese Academy of Sciences, Beijing 100029, China; 2 University of Chinese Academy of Sciences, Beijing 100029, China; 3 Nanjing University of Information Science and Technology, Nanjing 210044, China)

Abstract: In this study, a new index based on the potential vorticity (PV) framework is proposed for the quantification of the Tibetan Plateau (TP) surface thermodynamic and dynamic forcing. The results show that the derived TP surface PV (SPV) includes the topographical effect, near-surface absolute vorticity, and land-air potential temperature differences. The climatological annual cycle of the SPV suggests that the TP transitions from a cooling to a heating source in April. The SPV reaches a maximum from June to August, which is consistent with the evolution of the Asian summer monsoon precipitation. Further analysis suggests that the intensified SPV in the boreal summer results in a low-level cyclonic circulation anomaly associated with increased precipitation over the southeastern slope of the TP and South China and decreased precipitation over the Indian Ocean. In winter, the intensified SPV is associated with local cold air and divergence at the TP surface.

Key Points: The Tibetan Plateau surface potential vorticity index (TP-SPVI) includes the effects from orographic thermal and dynamical forcing.

The seasonal variation in the proposed TP-SPVI is consistent with the time evolution of the Asian summer monsoon precipitation.

The surface PV can be used as a metric for the identification of the TP surface forcing in reanalysis data and for model evaluation.

Plain Language Summary

The thermal forcing of Tibetan Plateau (TP) has long thought to be important in regulating the Asian summer monsoon. However, the surface sensible heat flux (SSHF) cannot represent the total thermodynamic forcing of the TP because the SSHF decreases when the monsoon precipitation intensifies

本文原载于 *Geophysical Research Letters*, 2022

during the prevailing Asian summer monsoon season. In this study, we propose a new index based on the potential vorticity (PV) framework, which can be used to quantify seasonal and interannual variations in the TP's surface thermodynamic and dynamic forcing. The results show that the proposed index can represent the TP surface-elevated heating effect during the boreal summer, which is consistent with those of numerical experiments performed in previous studies. The new index can be used as a metric for model evaluation and climate characterization in future studies to quantify the role of the TP in global climate change.

1 Introduction

The climatic effects of the Tibetan Plateau (TP) have been intensively studied since the mid-20th century. Here TP is referred to not only its platform, but also the imbedded major mountain ranges including the Kunlun, the Gangdises, and the Himalayas. Traditionally, the thermal effects of the TP are considered to be the main driving forces for the generation of a circulation structure over the TP during the boreal summer (Flohn, 1957; Yeh et al., 1957; Ye and Gao, 1979; Hoskins and Karoly, 1981; Chen and Trenberth, 1988; Wu et al., 1997, 2015; Wu and Zhang, 1998; Hsu and Liu, 2003; Duan and Wu, 2005; Liu et al., 2007), while the dynamical effect is also recognized as an important forcing in the formation of Asian climate (Queney et al., 1948; Charney and Eliassen, 1949; Bolin, 1950; Yeh, 1950; Liu et al., 2007; Chiang et al., 2015, 2017, 2020; Son et al., 2019, 2020, 2021). Theoretical and modeling studies have demonstrated that the relative importance of thermal forcing and mechanical forcing is very sensitive to the strength of the basic flow impinging upon the mountain (Held, 1983; Held and Ting, 1990; Held et al., 2002). The surface sensible heating is recognized as a crucial process based on which the TP surface thermal forcing drives atmospheric circulation, especially the heating on the southern slope of the TP (Wu et al., 2007).

The surface sensible heat flux (SSHF) over the TP has been calculated in many studies to identify the thermal forcing of the TP and its relationships with other climate systems (Luo and Yanai, 1984; Yanai et al., 1992; Zhang and Wu, 1999; Zhao et al., 2000; Ma et al., 2006, 2011; Yang et al., 2009; Duan et al., 2011, 2013, 2018; Zhu et al., 2012; Wang et al., 2014; Zhang et al., 2015, 2019; Han et al., 2017; Liu et al., 2020). However, the relationship between the summer SSHF over the TP and the Asian summer monsoon has been questioned in recent studies (Molnar et al., 1993; Boos and Kuang, 2010, 2013; Rajagopalan and Molnar, 2013). They found that the surface heating over the TP only correlates with summer monsoon rainfall in the early (May 20 to June 15) and late (September 1 to October 15) monsoon seasons and is insignificant during the main monsoon season (June 15 to August 31).

Therefore, more studies must be carried out to understand the correlation between the TP thermodynamic status and the Asian monsoon system, which is complex, highly nonlinear, and physically coupled. For example, the SSHF alone cannot represent the overall surface heating effect of the TP. In terms of seasonal variations in the Asian summer monsoon in the subtropical region (25°–40°N; Figs. 1 and S2 in the Online supplemental material, OSM), the mean precipitation in the monsoon season becomes stronger and two maxima appear (close to a longitude of 95° and 120°E, respectively) from mid-April to August over both the South and East Asian land (Figs 1a and S2b). The seasonal evolution of the precipitation is accompanied by weakened westerly wind south of the TP in early April (Fig. 1c) and intensified meridional wind over East Asia from May to August. However, the SSHF (Fig. 1b) over the Iranian Plateau (IP, ~45°–70°E) is strong (>60 W m^{-2}) from April to August. In the TP region, the

SSHF reaches a maximum from April to early June and then decreases when the monsoon precipitation increases from June to August. Theoretically, the SSHF can be expressed as follows:

$$\text{SSHF} = \rho C_p C_H u (T_g - T_a) \tag{1}$$

where ρ is the air density, C_p is the specific heat capacity at constant pressure, C_H is the bulk transfer coefficient for heat, u is the full wind speed near the surface, T_g is the ground temperature, and T_a is the air temperature near the surface. The variation in the SSHF can be determined based on the product of the wind and land-air temperature differences. During prevailing monsoon periods, both the westerly (Fig. 1c) and meridional winds over the TP are weak, whereas strong precipitation cools the surface and leads to a decrease in the land-air temperature difference. This explains the negative correlation between the SSHF and monsoon precipitation in midsummer, as identified by Rajagopalan and Molnar (2013) and Zhang et al. (2019).

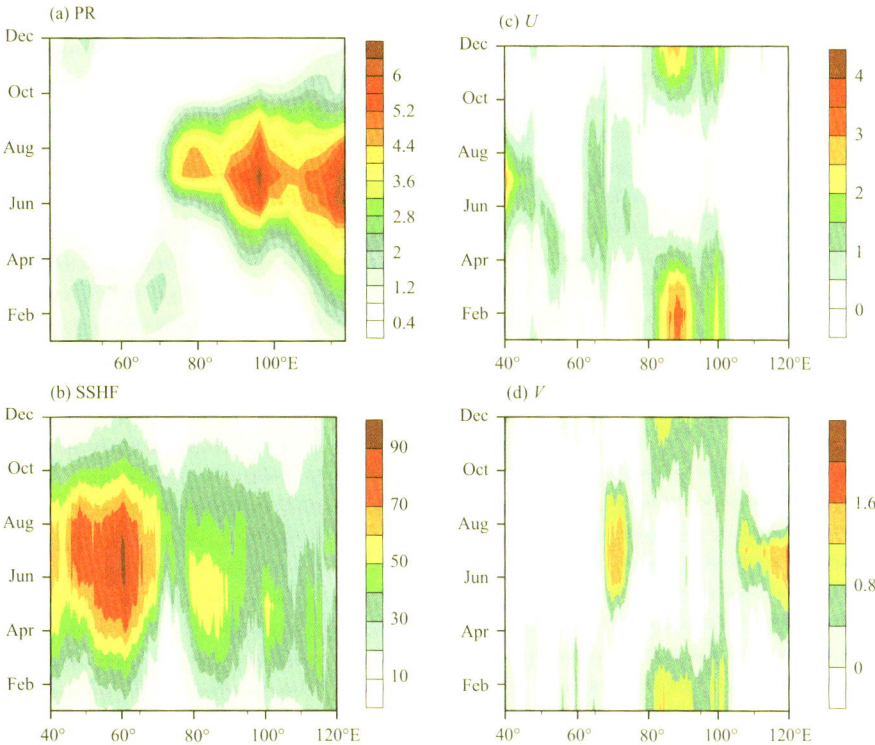

Figure 1 Climatological (1979—2019) annual cycle (25°—40°N mean) of the (a) precipitation (mm day^{-1}) from the GPCP, (b) surface sensible heating flux (W m^{-2}) at an elevation above 500 m, (c) near-surface zonal wind (m s^{-1}), and (d) meridional wind (m s^{-1}) from the ERA5 reanalysis

Although it remains unclear if the SSHF and Asian monsoon correlate during the boreal summer, the vertical diffusion heating over the TP was separated from other physical processes in a series of numerical experiments (Wu et al., 2012; He et al., 2015, 2019; He, 2017). The results of these experiments revealed that the TP surface sensible heating, especially the heating over the southern TP, indeed drives the generation of low-level circulation and surface air pumping over the TP. It remains unclear how this type of TP surface forcing in reanalysis datasets can be depicted to determine the role of TP surface thermodynamic forcing in Asian monsoon dynamics. In this study, we propose a new index, which can be used to determine the TP surface thermodynamic forcing based on the potential vorticity (PV) framework. The paper is structured as follows: The datasets and methods used in this study are introduced in Section 2. The features of the surface PV (SPV), both globally and in the TP region, are described in Section 3. The constructed TP thermodynamic forcing index and its correlation with the Asian monsoon system are presented in Section 4. The conclusions and discussion are provided in Section 5.

2 Datasets and Method

2.1 Datasets

The ECMWF Reanalysis v5 (ERA5) datasets from the European Center for Medium-Range Weather Forecasts (ECMWF; Hersbach et al., 2020) were used to determine the multilevel air temperature, specific humidity, and wind field. The observed precipitation data were adopted from the Global Precipitation Climatology Project (GPCP) monthly mean dataset (Adler et al., 2003). The study period was 1979 to 2019.

2.2 Method

The PV is a variable, which was proposed in the 1940s (Rossby, 1940; Ertel, 1942) and has been widely used for the analysis of synoptic events and climate change on isentropic surfaces (Thorpe, 1985; Hoskins et al., 1985; Haynes and McIntyre, 1987, 1990; Hoskins, 1991; Aebischer and Schar, 1996). In general, the three-dimensional PV can be expressed as:

$$\mathrm{PV} = \alpha \boldsymbol{\xi}_a \cdot \nabla_3 \theta \tag{2}$$

where α is specific volume, $\boldsymbol{\xi}_a$ is three-dimensional absolute vorticity, and θ is potential temperature. The PV of adiabatic and frictionless motions is conservative. Therefore, the changes in the PV can be used to identify processes related to diabatic heating and friction in a unified framework.

On Earth's surface, the application of traditional isentropic coordinates is not suitable because isentropic surfaces intersect with the surfaces of large mountainous areas. Therefore, a terrain-following coordinate model was adopted for the derivation of the SPV over orographic regions. We introduced the hydrostatic balance and σ coordinates:

$$\frac{\partial p}{\partial z} = -\rho g \tag{3}$$

and

$$\sigma = \frac{p}{p_s} \tag{4}$$

into Eq. (2). In a recent study (Sheng et al., 2021), each term in the three-dimensional PV was analyzed for different time scales. At seasonal mean time scales the vertical component product was identified to be the dominant term, whereas the horizontal component product was relatively small (see Fig. S3). Therefore, for simplicity, we used the vertical component product of the PV in the terrain-following coordinate model to estimate the TP surface thermodynamic forcing. The vertical PV product in the σ coordinate system can be simplified as follows:

$$\mathrm{PV}_\sigma = -\frac{g}{p_s}(f + \zeta_\sigma)\frac{\partial \theta}{\partial \sigma} \tag{5}$$

where ζ_σ denotes the relative vorticity in the σ coordinate system. On Earth's surface, Equation (5) can be expressed as:

$$\mathrm{PV}_{\sigma_1} = -\frac{g}{p_s}(f + \zeta_{\sigma_1})\frac{\theta_s - \theta_a}{1 - \sigma_1} \tag{6}$$

where σ_1 denotes the first σ level above the Earth's surface, which is the bottom level of the atmospheric model; ζ_{σ_1} is the relative vorticity at the σ_1 level; θ_s is the potential temperature at the Earth's surface; and θ_a is the air potential temperature at the σ_1 level. The distribution and variability of $PV_{\sigma1}$ are mainly determined by the orographic effect ($\frac{1}{p_s}$), surrounding circulation ($\zeta_{\sigma1}$), and vertical difference in the potential temperature ($\theta_s - \theta_a$), which is related to the vertical sensible heat flux. Therefore, the comparison of Equations (6) and (1) reveals that the vertical PV product is more suitable for the estimation of the surface forcing effect over the TP and adjacent regions than the SSHF. Because f is generally larger than ζ, the sign of $PV_{\sigma1}$ is determined by the land-air potential temperature difference ($\theta_s - \theta_a$) in the same hemisphere. In this study, we defined a new variable based on the SPV:

$$PV_{\sigma1}^* = -PV_{\sigma1} = -\left[-\frac{g}{p_s}(f+\zeta_{\sigma1})\frac{\theta_s-\theta_a}{1-\sigma_1}\right] \quad (7)$$

Thus, $PV_{\sigma1}^*$ is positive/negative when the land-air potential temperature difference ($\theta_s - \theta_a$) is positive/negative, which is related to the positive/negative heating gained from the Earth's surface.

3 SPV Distribution in Reanalysis Datasets

The seasonal and interannual variations in the SPV were analyzed. As noted in Section 2.2, the orographic elevation in terms of the surface pressure ($\frac{1}{p_s}$), absolute vorticity ($f+\zeta_{\sigma1}$), and vertical heating element ($\theta_s - \theta_a$) are the three key components that control the variation in distribution of the SPV. To understand the relative importance of these factors for the seasonal mean SPV, the global distributions of the SPV and the three components are shown in Fig. 2. In general, the June, July, and August (JJA) SPV (Fig. 2a) show positive values in the Northern Hemisphere and negative values in the Southern Hemisphere. The maximum SPV is located over the Tibetan and Iranian plateaus (TIPs) in Asia and the Rocky Mountains in North America. In December, January, and February (DJF; Fig. 2e), the SPV exhibits strong negative values over the TIPs and Rocky Mountains, which contrasts the JJA pattern. The SPV shows two strong anomalies over the northwestern Pacific and northwestern Atlantic oceans. These findings are consistent with the concept proposed in pioneering studies (Yeh et al., 1957; Ye and Gao, 1979) based on which the TP is a heat source in the boreal summer and cold source in the boreal winter and plays a dominant role in regulating climate change over East Asia. Thus, by analyzing Eq. (7), the change in the SPV can be considered as a measurement for the change in the TP thermal and dynamic forcing.

Figures 2b and 2f show the vertical heating elements ($\theta_s - \theta_a$) in JJA and DJF, respectively. Over Asia and North America, the distributions of the vertical heating elements ($\theta_s - \theta_a$) are very similar to the total SPV in both seasons. These patterns suggest that the land-air heat exchange is the most important factor enabling large-scale mountains to drive atmospheric circulation. This exchange is also the dominant PV source at the surface. The absolute vorticity term ($f+\zeta_{\sigma1}$) in JJA (Fig. 2c) and DJF (Fig. 2g) is similar. It is almost zonally symmetric at the global scale and increases with latitude. The absolute vorticity is negative in the Southern Hemisphere and positive in the Northern Hemisphere. This term can intensify the SPV at high latitudes in both hemispheres. The topographic elevation term is expressed as the reciprocal of the surface pressure and is shown in Figs. 2d and 2h for JJA and DJF, respectively. This term has a unique maximum over the TP and is almost twice that of other non-elevated areas, which indicates that the orographic elevation term in Eq. (7) can intensify the total SPV over the TP compared with other regions and that the TP is a PV source.

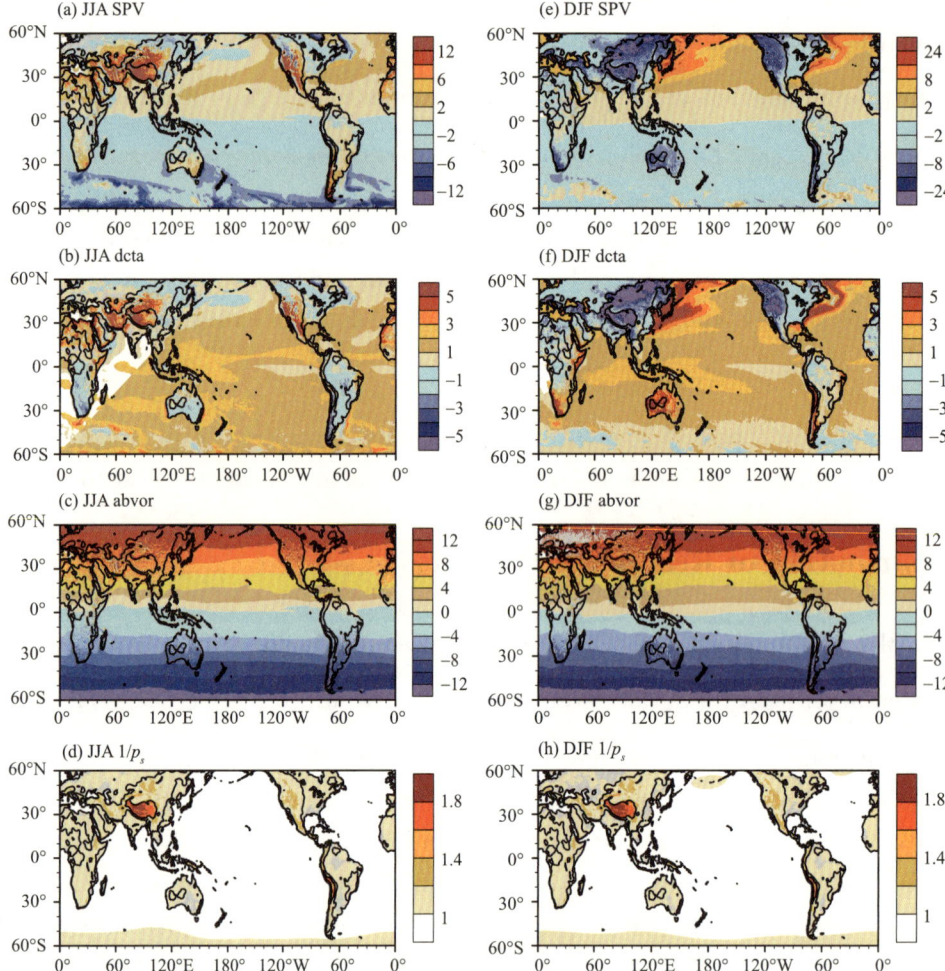

Figure 2 Climatological (1979—2019) spatial distribution of the SPV (PVU, 1 PVU=10^{-6} K m² kg⁻¹ s⁻¹) in (a) JJA and (e) DJF based on the ERA5 reanalysis. Land-air potential temperature difference ($\theta_s - \theta_a$) (K) in (b) JJA and (f) DJF. Absolute vorticity ($f + \zeta_{\sigma 1}$; 10^{-5} s⁻¹) in (c) JJA and (g) DJF. Orographic elevation effect ($\frac{1}{p_s}$; 10^{-3} hPa⁻¹) in (d) JJA and (h) DJF

The above-mentioned analysis provides a basis for estimating the intensity of the orographic surface forcing by quantifying the SPV over the TP region. Below, we further discuss the climatological annual cycle of the SPV over the TP region (Fig. 3) and compare it with the seasonal evolution of the SSHF and Asian monsoon system (Fig. 1). The annual cycle of the mean SPV averaged over the 25°—40°N band (Fig. 3) indicates that the TP (80°—100°E) changes from a negative PV source (cold source) to a positive PV source (heat source) from March to April. The SPV gradually intensifies and reaches its maximum from June to early August. Subsequently, it decreased from August to October and changed to a negative PV source again during late September, which is consistent with the seasonal evolution of the Asian summer monsoon precipitation (Figs. 1a and S2) and intensified meridional winds over East Asia from June to August (Fig. 1d). The positive PV is associated with the bend of isentropic surface in the lower troposphere and leads to cyclonic anomaly. It can contribute to the East Asian land precipitation. It's interesting that the IP is also a strong PV source throughout the year, except from November to January. The SPV over the IP also reaches its maximum from June to August, but the intensity is weaker than that over the TP. This implies that the TP and IP jointly form an important topographic thermodynamic forcing that could affect the variations of the Asian summer monsoon, which agrees with the results of previous studies (Wu et al., 2012, 2016, 2017; He et al., 2015; Liu et al., 2017).

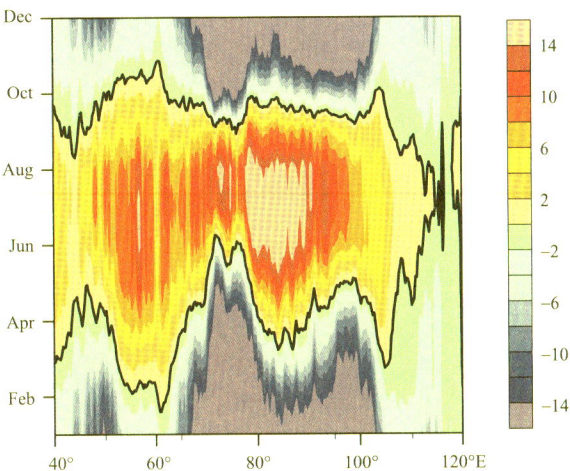

Figure 3 Climatological (1979—2014) annual cycle of the SPV (PVU) averaged over the (25°—40°N) zone with an elevation >500 m. The thick black line denotes the zero line of the SPV

4 Index for the Measurement of the TP Thermodynamic Forcing Based on the SPV

The analysis of seasonal and interannual SPV characteristics in the previous section suggests that the SPV can accurately depict the changes in the TP thermodynamic forcing. In the terrain-following coordinate, the SPV along the TP surface becomes a horizontally integrated SPV along the σ surface. Thus, the calculation of the total SPV over the TP surface on the σ surface can be used to determine the integrated intensity of the TP thermodynamic forcing. Based on Eq. (7), we defined an index for the estimation of the total TP surface thermodynamic forcing (TP-SPVI) on a σ surface:

$$I_{TP} = \iint PV^s_{\sigma 1} \, dxdy = \iint -\left[-\frac{g}{p_s}(f+\zeta_{\sigma 1})\frac{\theta_s-\theta_a}{1-\sigma_1}\right]dxdy \tag{8}$$

In this study, the TP region we focus mainly covers the area ranging from 25°—40°N and 70°—110°E, with an elevation above 500 m. It includes the southern boundary of the Himalayas and many large, interior mountain ranges. The evolution of the JJA mean I_{TP} from 1979 to 2019 is shown in Fig. 4a. The JJA I_{TP} ranges from ~39.5 potential vorticity unit (PVU) km² to 55 PVU km² and exhibits an interannual variability. The long-term variation in TP thermal forcing increases from 1979 to 1998 and decreases from 1998 to 2019. Decomposition analysis indicates that the potential temperature term dominates this variation (Fig. S4).

We further calculated the standard deviation of the I_{TP} and defined the strong/weak years based on the I_{TP} above/below the one-time standard deviation of the composite analysis. Among the 41 years, 7 years are strong (1995, 1998, 1999, 2000, 2006, 2012, and 2018) and 6 years are weak (1979, 1980, 1983, 1986, 1992, and 2019). To understand the corresponding monsoon circulation and precipitation changes associated with the I_{TP} intensity, we calculated the composites for the strong and weak years, respectively. The differences in the wind at 850 hPa and precipitation are shown in Fig. 4c together with the associated changes in the SPV (Fig. 4d). The SPV difference exhibits a positive pattern over the TP surface, with a maximum exceeding 4 PVU over the northwestern part of the TP (Fig. 4d). The associated changes in the Asian summer monsoon circulation (Fig. 4c) indicate a clear cyclonic pattern over the TP, with a positive precipitation difference over the southeastern edge of the TP and subtropical western Pacific. A negative precipitation difference can be mainly observed over the western part of India, the Indian Ocean, and the southern Indochina Peninsula. This pattern is very similar to the monsoon responses to the TP surface heat forcing derived from numerical simulations, as shown in Fig. 3b of Wu et al. (2012) and Fig. 2b of He et al. (2019). This result further confirms that our definition of the I_{TP} can capture the TP thermal forcing states during the boreal summer and the composite results are consistent with those of numerical experiments. Fig. 4b shows the linear regressions and correlations between the standardized I_{TP} and regional mean precipitation averaged over the area (15°—35°N, 85°—135°E; Fig. 4c). The regression coefficient was determined to be 0.47 (identical to the calculated Pearson correlation coefficient), implying that the intensified SPV over the TP can intensify the precipitation over the South and East Asian land.

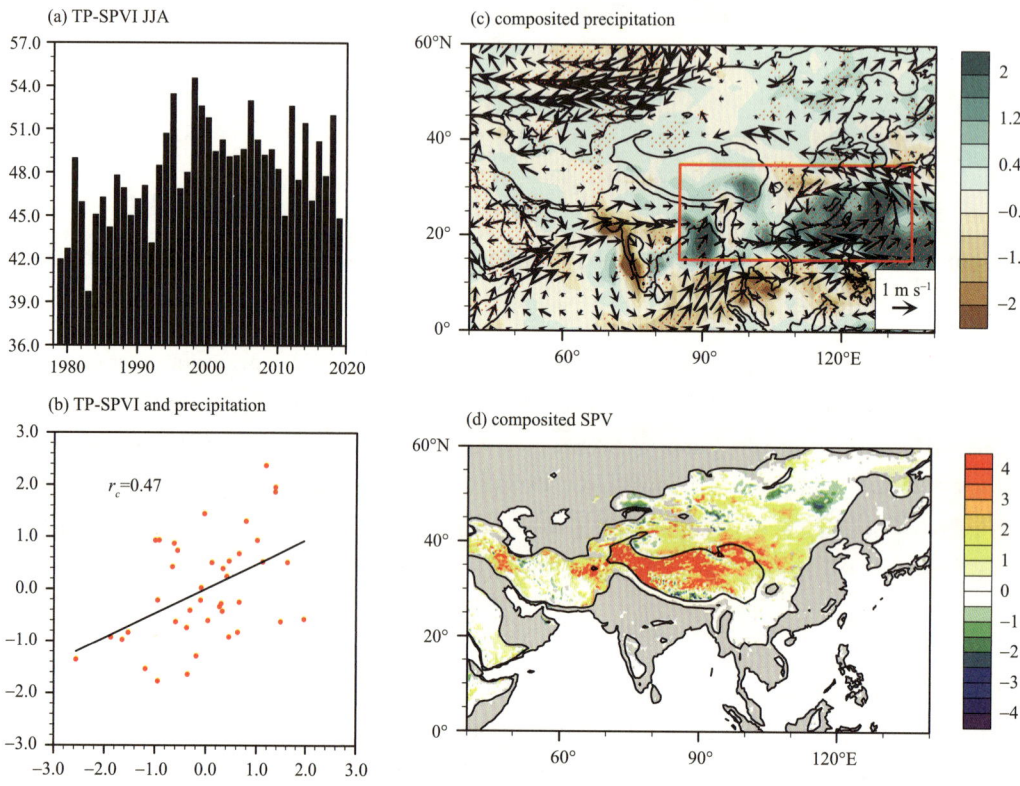

Figure 4 (a) Bar chart for the TP-SPVI from 1979 to 2019 (PVU km^{-2}). (b) Linear fit between the standardized JJA TP-SPVI (horizontal axis) and regional mean (15°—35°N, 85°—135°E) precipitation (vertical axis). The regression coefficient is 0.47 (identical to the calculated Pearson correlation coefficient). (c) Composite differences in the JJA precipitation (mm day^{-1}) and winds (m s^{-1}) at 850 hPa. The red dots are statistically significant at the 99% confidence level based on a Student's t-test. (d) Composite difference of JJA SPV (PVU). The thick black contours denote the topographic heights of 3000 and 500 m, respectively. The red rectangle denotes the area (15°—35°N, 85°—135°E)

Compared with the extensively studied effects of the TP thermodynamic forcing during the boreal summer, the effects of the TP thermodynamic forcing during the boreal winter have been rarely studied. To determine whether the new index proposed in this study is applicable to the boreal winter, we calculated the TP SPV and associated surface temperature and circulation. Figure 5a shows the evolution of the DJF mean I_{TP} values from 1979 to 2019. Based on the figure, the I_{TP} is negative in winter, suggesting that the TP is a cold source during the boreal winter, which agrees with previous results.

In addition, a composite analysis based on the one-time standard deviation of the I_{TP} was carried out to gain insights into the associated circulation and thermal structure related to the SPV changes in winter. The strong/weak years of the TP winter forcing are defined by the individual I_{TP} below or above one-time standard deviation. Among the 41 years, 7 years are strong (1983, 1992, 1998, 2001, 2007, 2013, and 2019) and 9 years are weak (1979, 1984, 1985, 2006, 2008, 2009, 2012, 2017, and 2018). The composite differences in the winter surface air temperature and near-surface winds at 10 m are shown in Fig. 5b. The cooling of the TP during winter is associated with a surface divergence anomaly over the TP, based on which the surface air is advected to both the northern and southern TP. The air flow anomaly is accompanied by surface air temperature changes, that is, mainly cooling on the TP above 3000 m, weak cooling over northern India, and warming over mid-high latitudes north of the TP. The above-mentioned analysis proves that the new scalar index represents the intensity of the surface temperature and wind anomaly over TP during the boreal winter. Finally, we also examined the composited precipitation associated with the changes in the SPV during the boreal winter (Fig. 5). The SPV correlates with precipitation changes over South China land and maritime continents over Western Pacific; however, the mechanism requires further studies.

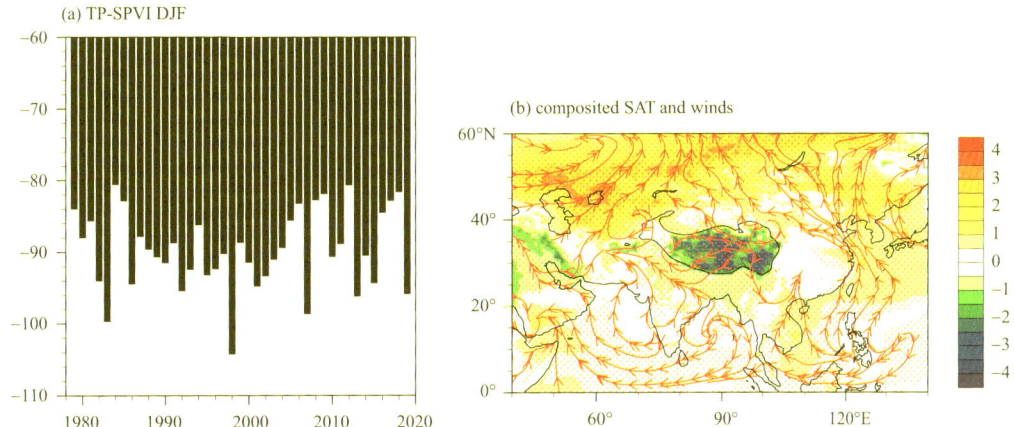

Figure 5 (a) Bar plots for the DJF mean TP-SPVI from 1979 to 2019 (PVU km^{-2}). (b) Composite differences in the surface air temperature (℃) and surface streamlines for the winds. The dotted region indicates statistical significance at the 99% confidence level based on a Student's t-test. The thick black contour denotes a topographic height of 3000 m

5 Conclusions and discussions

The SSHF has been used for a long time to estimate the TP surface heating status. However, the SSHF only partially represents the TP surface thermal forcing because the correlation between the SSHF over the TP and Asian summer monsoon precipitation is unclear. In this study, we constructed a new variable Surface PV (SPV) to measure the intensity of the TP thermodynamic forcing based on the PV framework. The SPV is mainly controlled by the near-surface vorticity, elevation, and potential temperature differences between the land and surface air associated with the SSHF. The annual cycle of the SPV suggests that the TP transitions from a cold source to a heat source, primarily in April, reaches its maximum from June to August, and transitions from a heat source to a cold source in late September, which is consistent with the development of the Asian monsoon system as well as previous results of numerical experiments of the effects of the TP surface heating. The above-mentioned analyses confirm that the SPV over the TP is representative of the mountain's thermodynamic as well as dynamic forcing and that the potential temperature difference term is dominant in the boreal summer. It can also be a measurement for the dynamical effect which may be important in different timescales. Therefore, the SPV can be used as a metric for model and climate evaluations to quantitatively estimate the role of the TP in global climate change.

Data Availability Statement

The ERA5 dataset is available at https://www.ecmwf.int/en/forecasts/datasets/reanalysis-datasets/era5. The GPCP precipitation dataset is available at http://www.esrl.noaa.gov/psd/data/gridded/data.gpcp.html.

References

ADLER R F, HUFFMAN G J, CHANG A, et al, 2003. The version-2 global precipitation climatology project (GPCP) monthly precipitation analysis (1979−present)[J]. Journal of Hydrometeorology, 4(6): 1147-1167.

AEBISCHER U, SCHÄR C, 1996. Low-level potential vorticity and cyclogenesis to the lee of the Alps[J]. Journal of the Atmospheric Sciences, 55(2): 186-207.

BOLIN B, 1950. On the influence of the earth's orography on the general character of the westerlies[J]. Tellus, 2(3): 184-195.

BOOS W R, KUANG Z, 2010. Dominant control of the South Asian monsoon by orographic insulation versus plateau heating[J]. Nature, 463(7278): 218-223.

BOOS W R, KUANG Z, 2013. Sensitivity of the South Asian monsoon to elevated and non-elevated heating[J]. Scientific Reports, 3.

CHARNEY J G, ELIASSEN A, 1949. A numerical method for predicting the perturbation of the middle latitude westerlies [J]. Tellus, 1: 38-55.

CHEN S C, TRENBERTH K E, 1988. Orographically forced planetary-waves in the Northern Hemisphere winter-steady-state model with wave-coupled lower boundary formulation[J]. Journal of the Atmospheric Sciences, 45(4): 657-680.

CHIANG J C H, FUNG I Y, WU C H, et al, 2015. Role of seasonal transitions and westerly jets in East Asian paleoclimate[J]. Quaternary Science Reviews, 108: 111-129.

CHIANG J C H, SWENSON L M, KONG W, 2017. Role of seasonal transitions and the westerlies in the interannual variability of the East Asian summer monsoon precipitation[J]. Geophysical Research Letters, 44(8): 3788-3795.

CHIANG J C H, KONG W, WU C H, et al, 2020. Origins of East Asian summer monsoon seasonality[J]. Journal of Climate, 33(18): 7945-7965.

DUAN A M, WU G X, 2005. Role of the Tibetan Plateau thermal forcing in the summer climate patterns over subtropical Asia[J]. Climate Dynamics, 24(7-8): 793-807.

DUAN A, LI F, WANG M, et al, 2011. Persistent weakening trend in the spring sensible heat source over the Tibetan Plateau and its impact on the Asian summer monsoon[J]. Journal of Climate, 24(21): 5671-5682.

DUAN A, WANG M, LEI Y, et al, 2013. Trends in summer rainfall over China associated with the Tibetan Plateau sensible heat source during 1980−2008[J]. Journal of Climate, 26(1): 261-275.

DUAN A M, LIU S F, ZHAO Y, et al, 2018. Atmospheric heat source/sink dataset over the Tibetan Plateau based on satellite and routine meteorological observations[J]. Big Earth Data, 2(2):179-189.

ERTEL H, 1942. Ein neuer hydrodynamische wirbelsatz[J]. Meteorologische Zeitschrift Braunschweig, 59: 33-49.

FLOHN H, 1957. Large-scale aspects of the "summer monsoon" in South and East Asia[J]. Journal of the Meteorological Society of Japan, 35A: 180-186.

HAN C, MA Y, CHEN X, et al, 2017. Trends of land surface heat fluxes on the Tibetan Plateau from 2001 to 2012[J]. International Journal of Climatology, 37(14): 4757-4767.

HAYNES P H, MCINTYRE M E, 1987. On the evolution of vorticity and potential vorticity in the presence of diabatic heating and frictional or other forces[J]. Journal of the Atmospheric Sciences, 44(5): 828-841.

HAYNES P H, MCINTYRE M E, 1990. On the conservation and impermeability theorems for potential vorticity[J]. Journal of the Atmospheric Sciences, 47(16): 2021-2031.

HE B, 2017. Influences of elevated heating effect by the himalaya on the changes in Asian summer monsoon[J]. Theoretical and Applied Climatology, 128(3-4): 905-917.

HE B, WU G, LIU Y, et al, 2015. Astronomical and hydrological perspective of mountain impacts on the Asian summer monsoon[J]. Scientific Reports, 5.

HE B, LIU Y M, WU G X, et al, 2019. The role of air-sea interactions in regulating the thermal effect of the Tibetan-Iranian Plateau on the Asian summer monsoon[J]. Climate Dynamics, 52(7-8): 4227-4245.

HELD I M, 1983. Stationary and quasi-stationary eddies in the extratropical troposphere: Theory[C]//HOSKINS B J, PEARCE R P. Large-Scale Dynamical Processes in the Atmosphere. London: Academic Press: 127-168.

HELD I M, TING M, 1990. Orograhpic versus thermal forcing of stationary waves—The importance of the mean low-level wind[J]. Journal of the Atmospheric Sciences, 47(4): 495-500.

HELD I M, TING M F, WANG H L, 2002. Northern winter stationary waves: Theory and modeling[J]. Journal of Climate, 15(16): 2125-2144.

HERSBACH H, BELL B, BERRISFORD P, et al, 2020. The era5 global reanalysis[J]. Quarterly Journal of the Royal Meteorological Society, 146(730): 1999-2049.

HOSKINS B, 1991. Towards a PV-θ view of the general-circulation[J]. Tellus Series a-Dynamic Meteorology and Oceanography, 43(4): 27-35.

HOSKINS B, KAROLY D J, 1981. The steady linear response of a spherical atmosphere to thermal and orographic forcing [J]. Journal of Atmospheric Sciences, 38(6): 1179-1196.

HOSKINS B, MCINTYRE M E, ROBERTSON A W, 1985. On the use and significance of isentropic potential vorticity maps[J]. Quarterly Journal of the Royal Meteorological Society, 111(470): 877-946.

HSU H H, LIU X, 2003. Relationship between the Tibetan Plateau heating and East Asian summer monsoon rainfall[J]. Geophysical Research Letters, 30(20): 4.

LIU Y M, HOSKINS B, BLACKBURN M, 2007. Impact of Tibetan orography and heating on the summer flow over Asia [J]. Journal of the Meteorological Society of Japan, 85B: 1-19.

LIU Y M, WANG Z Q, ZHUO H F, et al, 2017. Two types of summertime heating over Asian large-scale orography and excitation of potential-vorticity forcing II. Sensible heating over Tibetan-Iranian Plateau[J]. Science China-Earth Sciences, 60(4): 733-744.

LIU Y M, LU M M, YANG H J, et al, 2020. Land-atmosphere-ocean coupling associated with the Tibetan Plateau and its climate impacts[J]. National Science Review, 7(3): 534-552.

LUO H B, YANAI M, 1984. The large-scale circulation and heat-sources over the Tibetan Plateau and surrounding areas during the early summer of 1979. Part II. Heat and moisture budgets[J]. Monthly Weather Review, 112(5): 966-989.

MA Y, ZHONG L, SU Z, et al, 2006. Determination of regional distributions and seasonal variations of land surface heat fluxes from landsat-7 enhanced thematic mapper data over the central Tibetan Plateau area[J]. Journal of Geophysical Research-Atmospheres, 111(D10).

MA Y, ZHONG L, WANG B, et al, 2011. Determination of land surface heat fluxes over heterogeneous landscape of the Tibetan Plateau by using the modis and in situ data[J]. Atmospheric Chemistry and Physics, 11(20): 10461-10469.

MOLNAR P, ENGLAND P, MARTINOD J, 1993. Mantle dynamics, uplift of the Tibetan Plateau, and the Indian monsoon[J]. Reviews of Geophysics, 31(4): 357-396.

QUENEY P, 1948. The problem of air flow over mountains: A summary of theoretical studies[J]. Bulletin of the American Meteorological Society, 29: 16-29.

RAJAGOPALAN B, MOLNAR P, 2013. Signatures of Tibetan Plateau heating on Indian summer monsoon rainfall variability[J]. Journal of Geophysical Research-Atmospheres, 118(3): 1170-1178.

ROSSBY C G, 1940. Planetary flow patterns in the atmosphere[J]. Quarterly Journal of the Royal Meteorological Society,

66: 68-87.

SHENG C, WU G X, TANG Y Q, et al, 2021. Characteristics of the potential vorticity and its budget in the surface layer over the Tibetan Plateau[J]. International Journal of Climatology, 41: 439-455.

SON J H, SEO K H, WANG B, 2019. Dynamical control of the Tibetan Plateau on the East Asian summer monsoon[J]. Geophysical Research Letters, 46(13): 7672-7679.

SON J H, SEO K H, WANG B, 2020. How does the Tibetan Plateau dynamically affect downstream monsoon precipitation? [J]. Geophysical Research Letters, 47(23).

SON J H, KWON J I, HEO K Y, 2021. Weak upstream Westerly wind attracts Western North Pacific typhoon tracks to west[J]. Environmental Research Letters, 16(12):124041.

THORPE A J, 1985. Diagnosis of balanced vortex structure using potential vorticity[J]. Journal of the Atmospheric Sciences, 42(4): 397-406.

WANG Z, DUAN A, WU G, 2014. Time-lagged impact of spring sensible heat over the Tibetan Plateau on the summer rainfall anomaly in east China: Case studies using the WRF model[J]. Climate Dynamics, 42(11-12): 2885-2898.

WU G X, LI W J, GUO H, 1997. Sensible heat driven air-pump over the Tibetan Plateau and its impacts on the Asian summer monsoon[C]//YE D Z. Collection in Memory of Zhao Jiuzhang. Beijing: Science Press: 116-126.

WU G X, ZHANG Y S, 1998. Tibetan Plateau forcing and the timing of the monsoon onset over South Asia and the South China Sea[J]. Monthly Weather Review, 126(4): 913-927.

WU G X, LIU Y M, WANG T M, et al, 2007. The influence of mechanical and thermal forcing by the Tibetan Plateau on Asian climate[J]. Journal of Hydrometeorology, 8(4): 770-789.

WU G X, LIU Y M, HE B, et al, 2012. Thermal controls on the Asian summer monsoon[J]. Scientific Reports, 2: 7.

WU G X, DUAN A M, LIU Y M, et al, 2015. Tibetan Plateau climate dynamics: recent research progress and outlook[J]. National Science Review, 2(1): 100-116.

WU G X, ZHUO H F, WANG Z Q, et al, 2016. Two types of summertime heating over the Asian large-scale orography and excitation of potential-vorticity forcing I. Over Tibetan Plateau[J]. Science China Earth Sciences, 59(10): 1996-2008.

WU G, HE B, DUAN A, et al, 2017. Formation and variation of the atmospheric heat source over the Tibetan Plateau and its climate effects[J]. Advances in Atmospheric Sciences, 34(10): 1169-1184.

YANAI M H, LI C F, SONG Z S, 1992. Seasonal heating of the Tibetan Plateau and its effects on the evolution of the Asian summer monsoon[J]. Journal of the Meteorological Society of Japan, 70(1B): 319-351.

YANG K, QIN J, GUO X F, et al, 2009. Method development for estimating sensible heat flux over the Tibetan Plateau from CMA data[J]. Journal of Applied Meteorology and Climatology, 48(12): 2474-2486.

YEH T C, LO S W, CHU P C, 1957. The wind structure and heat balance in the lower troposphere over Tibetan Plateau and its surroundings [J]. Acta Meteorologica Sinica, 28: 108-121.

YE D Z, GAO Y X, 1979. The Meteorology of the Qinghai-Xizang (Tibet) Plateau [M]. Beijing: Science Press: 278.

YEH T G, 1950. The circulation of the high troposphere over China in the winter of 1945−1946[J]. Tellus, 2: 173-183.

ZHANG H, LI W, LI W, 2019. Influence of late springtime surface sensible heat flux anomalies over the Tibetan and Iranian Plateaus on the location of the South Asian high in early summer[J]. Advances in Atmospheric Sciences, 36(1): 93-103.

ZHANG Y, LI Z, LIU B, 2015. Interannual variability of surface sensible heating over the Tibetan Plateau in boreal spring and its influence on the onset time of the Indian summer monsoon[J]. Chinese Journal of Atmospheric Sciences, 39(6): 1059-1072.

ZHANG Y S, WU G X, 1999. Diagnostic investigations on the mechanism of the onset of Asian summer monsoon and abrupt seasonal transitions over the Northern Hemisphere part II: The role of surface sensible heating over Tibetan Plateau and surrounding regions[J]. Acta Meteorologica Sinica, 57: 56-73.

ZHAO P, CHEN L X, 2000. Study on climatic features of surface turbulent heat exchange coefficients and surface thermal forces over the Qinghai-Xizang Plateau[J]. Acta Meteorologica Sinica (14):13-29.

ZHU X, LIU Y, WU G, 2012. An assessment of summer sensible heat flux on the Tibetan Plateau from eight data sets[J]. Science China-Earth Sciences, 55(5): 779-786.

代表性论文 30

Linkage between Cross-Equatorial Potential Vorticity Flux and Surface Air Temperature over the Mid-High Latitudes of Eurasia during Boreal Spring

SHENG Chen[1,2], WU Guoxiong[1,2], HE Bian[1,2], LIU Yimin[1,2], MA Tingting[1]

(1 State Key Laboratory of Numerical Modeling for Atmospheric Sciences and Geophysical Fluid Dynamics (LASG), Institute of Atmospheric Physics, Chinese Academy of Sciences, Beijing 100029, China; 2 College of Earth and Planetary Sciences, University of Chinese Academy of Sciences, Beijing 100049, China)

Abstract: The source of potential vorticity (PV) for the global domain is located at the Earth's surface. PV in one hemisphere can exchange with the other through cross-equatorial PV flux (CEPVF). This study investigates the features of the climatic mean CEPVF, the connection in interannual CEPVF with the surface thermal characteristics, and the associated mechanism. Results indicate that the process of positive (negative) PV carried by a northerly (southerly) wind leads to the climatologically overwhelming negative CEPVF over almost the entire equatorial cross-section, while the change of the zonal circulation over the equator is predominately responsible for CEPVF variation. By introducing the concept of "PV circulation" (PVC), it is demonstrated that the interannual CEPVF over the equator is closely linked to the notable uniform anomalies of spring cold surface air temperature (SAT) over the mid-high latitudes of Eurasia by virtue of the PVC, the PV-θ mechanism, and the surface positive feedback. Further analysis reveals that equatorial sea surface temperature (SST) forcing, such as the El Niño-Southern Oscillation and tropical South Atlantic uniform SST, can directly drive anomalous CEPVF by changing the zonal circulation over the equator, thereby influencing SAT in the Northern Hemisphere. All results indicate that the equilibrium linkage between CEPVF and extratropical SAT is mainly a manifestation of the response of extratropical SAT to tropical forcing by virtue of PVC, and that the perspective of PVC can provide a reasonably direct and simple connection of the circulation and climate between the tropics and the mid-high latitudes.

Keywords: potential vorticity (PV), cross-equatorial PV flux (CEPVF), El Nino-Southern Oscillation (ENSO), surface air temperature (SAT), tropical South Atlantic

1 Introduction

Potential vorticity (PV) theory, with its mathematical elegance and completeness, has increasingly

本文原载于 *Climate Dynamics*, 2012, 38: 261-279

attracted the attention of dynamicists. Research on PV has deepened our understanding of various atmospheric phenomena, primarily because of its invertibility and conservation (e. g., Hoskins et al., 1985). Application of PV theory to synoptic meteorology has revealed the dynamic mechanisms of many complex weather events. Earlier related studies mostly focused on the redistribution of atmospheric interior PV (e. g., Hoskins et al., 1985), in which the impacts of PV redistribution on the general circulation (e. g., Hoskins et al., 2003; Hoskins, 2015; Liu et al., 2007; Luo et al., 2018a, 2018b; Ortega et al., 2018; Xie et al., 2020) and severe weather (e. g., Wu et al., 1995; Wu and Cai, 1997; Ma et al., 2022; Zhang et al., 2021a) are of primary interest.

Climate is the cumulative impact of weather over a certain period, e. g., month, season, year, decade or longer. Over long periods, energy generation and dissipation of the atmospheric system are prominent. Because the PV equation explicitly includes the impacts of diabatic heating and frictional dissipation on PV development, it is convenient and instructive to use the PV equation to study climate and its variations. With regard to climate, the features of the source and budget of the atmospheric PV have evoked great interests. For example, Haynes and McIntyre (1987, 1990) introduced the concept of PV density or PV substance, and proposed the impermeability theorem of PV that suggests PV is conserved over a closed isentropic surface. Based on the impermeability theorem, Hoskins (1991) proposed a three-fold division of the atmosphere: The Overworld, Middleworld, and Underworld. The "Overworld" is the region encompassed by isentropic surfaces that are everywhere above the tropopause. The "Middleworld" is the region with isentropic surfaces crossing the tropopause but not striking the Earth's surface. The "Underworld" is the region with isentropic surfaces intercepting the Earth's surface. He further illustrated that global atmospheric PV is changeable in the Underworld, but constant above the Underworld. Bretherton and Schär (1993) envisioned the impermeability theorem in terms of the so-called effective velocity, and proved that a particle of PV density or PV substance that moves with effective velocity always remains on the same isentropic surface. These findings imply that changes in globally integrated atmospheric PV depend solely on the PV flux on the Earth's surface (e. g., Sheng et al., 2021). Recently, a number of relevant data analysis regarding the source of Earth's surface PV has been published (e. g., Ma et al., 2019, 2022; Sheng et al., 2021, 2022; Zhang et al., 2021b).

The PV generated near the Earth's surface can be transported to the interior of the atmosphere, even from one hemisphere to the other. If the global atmospheric domain that is above the Underworld were covered by a lid of a potential temperature surface and divided into two hemispheric atmospheric domains by a vertical boundary along the equator, then because the PV flux cannot penetrate the upper lid, the hemispheric Earth's surface and vertical cross section along the equator become two effective boundaries through which the penetrating PV flux determines the changes in hemispheric PV. Over the long term, the integrated surface PV flux should be compensated by the integrated cross-equatorial PV flux (CEPVF). The CEPVF at the equatorial vertical cross section can therefore be considered as a monitor through which the behavior of the surface PV flux can be detected. Because surface PV flux is related to surface air temperature (SAT), diagnosing the variation of CEPVF might further help improve understanding of SAT variation. However, the features and climatological effects related to the integrated CEPVF have received little attention in the meteorological literature. The major objective of this study is to explore the features of the CEPVF and its connection to climate with focus on the SAT of the Northern Hemisphere (NH).

SAT is a vital basic variable in atmospheric science. Variations in SAT have pronounced effects on agriculture, socioeconomic development, and societal activities (Chen and Wu, 2018). Spring marks the transition from winter into summer, and it is the season for crops, vegetation recovery, and snowmelt (Chen et al., 2019). Any change in SAT over Eurasia during boreal spring could exert prominent influence on ecosystem recovery (Labat et al., 2004; Wang et al., 2011, Chen et al., 2019). In particular, SAT over Eurasia during boreal spring could connect atmospheric anomalies of the preceding winter and subsequent Asian summer monsoon activity (Ogi et al., 2003; Chen and Wu, 2018) by changing the land-sea thermal contrast (Liu and Yanai, 2001; D'Arrigo et al., 2006), in which the snow cover provides a memory effect (Ogi et al., 2003; Chen et al., 2016). Therefore, the linkage between CEPVF and SAT over the mid-high latitudes of Eurasia during boreal spring is a specific focus of this study.

The remainder of the paper is organized as follows. Section 2 presents the data, method and theory. Section 3 presents the CEPVF climatology. Section 4 analyzes the interannual relationship between the integrated CEPVF and SAT over the mid-high latitudes of Eurasia during boreal spring and its possible mechanism. Section 5 examines the possible forcing factors of the CEPVF. Finally, the summary and discussion are provided in Section 6.

2 Data, method, and theory

2.1 Data

This study uses monthly mean SAT data from the MERRA2 reanalysis product (Rienecker et al., 2011; Lucchesi, 2012; https://disc.gsfc.nasa.gov/datasets?project=MERRA-2) and sea surface temperature (SST) data from the COBE SST dataset (Ishii et al., 2005; https://psl.noaa.gov/data/gridded/data.cobe.html). The averaged SST anomalies of the NINO34 index (defined by the region: 5°S—5°N, 120°—170°W), which are used to represent the El Niño-Southern Oscillation (ENSO) condition, are obtained from the following web address: https://psl.noaa.gov/data/climateindices/list/.

PV and PV flux as defined in Eq. (1) and (3), respectively, involve the product of various variables and therefore the contribution of the transient process might not be represented properly in monthly mean data. To address this problem, we use 3-hourly instantaneous data on the pressure level obtained from MERRA2 to calculate PV and PV flux. The variables include air temperature, zonal wind, and meridional wind.

The research period of this study is 1980—2017. The horizontal resolution of all MERRA2 data is 1.25°×1.25°. The horizontal resolution of the SST data is 1°×1°. Climatic mean values calculated over December, January, and February (DJF), March, April, and May (MAM), June, July, and August (JJA), and September, October, and November (SON) are used to represent the conditions of boreal winter, spring, summer, and autumn, respectively.

2.2 Method

Besides the common correlation analysis, regression analysis and composite analysis, the partial correlation is also used in this study to investigate the correlation between the two variables under study by excluding the impact from a third variable. The partial correlation is calculated as follows:

$$r_{12.3} = \frac{r_{12} - r_{13}r_{23}}{\sqrt{(1-r_{13}^2)(1-r_{23}^2)}}$$

where r_{12} is the correlation coefficient between variables 1 and 2, r_{13} is the correlation coefficient between variables 1 and 3, and r_{23} is the correlation coefficient between variables 2 and 3. The coefficient $r_{12.3}$ is the partial correlation coefficient between variable 1 and variable 2 with variable 3 removed (Zar, 2010).

The linear trend and decadal variation (more than 9 years) of the data are removed to highlight interannual variability.

2.3 Theory

2.3.1 PV, PV budget equation, and PV flux

Here, W is defined as PV per unit volume, the PV density or PV substance. The expression of W in the general form (Haynes and McIntyre, 1987, 1990; Bretherton and Schär, 1993) is as follows:

$$W = \boldsymbol{\zeta}_a \cdot \nabla \theta \tag{1}$$

where $\boldsymbol{\zeta}_a$ is the 3D absolute vorticity vector, and θ is potential temperature. The budget equation for PV can be expressed in the conservation form as follows (Haynes and McIntyre, 1987, 1990):

$$\frac{\partial W}{\partial t} = -\nabla \cdot \boldsymbol{J} \tag{2}$$

with the "PV flux"

$$\boldsymbol{J} = \boldsymbol{V}_e W = \boldsymbol{V} W - \boldsymbol{\zeta}_a \dot{\theta} - \boldsymbol{F}_\zeta \theta \tag{3}$$

in which $\boldsymbol{V}$ is the 3D wind vector, $\dot{\theta} = \dfrac{\mathrm{d}\theta}{\mathrm{d}t}$ is diabatic heating rate, $\boldsymbol{F} = (F_x, F_y)$ is frictional force, $\boldsymbol{F}_\zeta$ is the vorticity of frictional force, and

$$\boldsymbol{V}_e = \boldsymbol{J}/W = \boldsymbol{V} - (\boldsymbol{\zeta}_a \dot{\theta} + \boldsymbol{F}_\zeta \theta)/W$$

is effective velocity. According to Bretherton and Schär (1993), because

$$\frac{\partial \theta}{\partial t} + \boldsymbol{V}_e \cdot \nabla \theta \equiv 0$$

the effective velocity $\boldsymbol{V}_e$ is parallel to θ surface. Therefore, following Eq. (3), the PV flux $\boldsymbol{J}$ is directed along isentropic surfaces without penetration, which represents an alternative way of envisioning the impermeability theorem.

Because the divergence of curl is zero, substituting Eq. (1) into Eq. (2) and choosing the form of $\boldsymbol{J}$ that satisfies Eq. (2) gives (Bretherton and Schär, 1993)

$$\frac{\partial \boldsymbol{\zeta}_a \theta}{\partial t} = -\boldsymbol{J} \tag{4}$$

By integrating Eq. (4) over the given period Δt and adopting the following definition:

$$\boldsymbol{J}_s = -\boldsymbol{\zeta}_a \theta \tag{5}$$

we obtain

$$\boldsymbol{J}_s = \int_{t_0}^{t_0+\Delta t} \boldsymbol{J} \, \mathrm{d}t + C = \int_{t_0}^{t_0+\Delta t} (\boldsymbol{V}_e W) \, \mathrm{d}t + C \tag{6a}$$

where C is a time-independent constant field. The flux $\boldsymbol{J}_s$ can be considered the temporally accumulated PV flux over a given period (Δt) with a different unit to PV flux $\boldsymbol{J}$. Equation (6a) implies that variation of $\boldsymbol{J}_s$ is parallel to variation of flux $\boldsymbol{J}$ averaged over the given period Δt. From Eq. (5), we can also obtain

$$W = -\nabla \cdot (\boldsymbol{J}_s) \tag{6b}$$

On the basis of Eq. (6b), and through mimicking the atmospheric convergence $C = -\nabla \cdot \boldsymbol{V}$, in which $\boldsymbol{V}$ is the atmospheric circulation, the accumulated PV flux $\boldsymbol{J}_s$ can be referred to as PV circulation (hereafter, PVC). From Eq. (6a), we can see PVC represents the cumulative effects of the transient PV flux.

From Eqs. (2) and (6b), we can also discern the difference between J_s and J. For a specific domain enclosed by boundary A, its gross PV is determined by the sum of the cross-A PVC J_s, whereas its gross PV generation is determined by the sum of the cross-A PV flux J. In simple terms, the convergence of J_s is PV itself (Eq. 6b), whereas the convergence of J is the local PV generation (Eq. (2)).

2.3.2 Gross PV budget in the atmosphere

We define a theta surface (θ_T) above the Underworld as the top of the atmosphere that is under investigation. Because flux J at θ_T is parallel to the θ_T surface, there is no flux J perpendicular to the θ_T surface (Bretherton and Schär, 1993; Schneider et al., 2003). Integrating the PV budget equation (Eq. (2)) globally from the Earth's surface to the θ_T surface and using Gauss's theorem, we obtain the following:

$$\frac{\partial}{\partial t}\iiint_{\text{global}} W \mathrm{d}v = -\iiint_{\text{global}} \nabla \cdot (J) \mathrm{d}v = -\iint_{\text{globalsurf}} J \cdot \mathrm{d}s \tag{7a}$$

If the domain is confined to the NH, then we have

$$\frac{\partial}{\partial t}\iiint_{\text{NH}} W \mathrm{d}v = -\iiint_{\text{NH}} \nabla \cdot (J) \mathrm{d}v = -\iint_{\text{NHsurf}} J \cdot \mathrm{d}s + \iint_{\text{EQ}} \text{CEPVF} \mathrm{d}s \tag{7b}$$

in which $\mathrm{d}s = \mathrm{d}x \mathrm{d}h$ with h ranging from the surface to θ_T, and

$$\text{CEPVF} = J \cdot j = J^y \tag{7c}$$

which is the meridional component of PV flux across the equatorial vertical section and represents cross-equatorial PV transport. The superscript indicates the vector component. Equation (7a) indicates that the budget of globally integrated PV is determined solely by the Earth's surface PV flux, which is consistent with previous studies (Haynes and McIntyre, 1987, 1990; Hoskins, 1991; Schneider et al., 2003).

Equation (7b) indicates that CEPVF also contributes to the PV budget in the NH. Because the PV budget is balanced in the NH in terms of the long-term mean (Eq. (7b)), the integrated CEPVF can be regarded as a monitor from which the integrated surface PV flux conditions can be detected. The surface PV flux is related to the surface thermal conditions (Eq. (3)), which means that CEPVF is closely linked to the surface thermal conditions.

2.3.3 PV flux and cross-equatorial PV flux in a pressure coordinate system

Most reanalysis data are archived in a pressure coordinate system. For hydrostatic large-scale motion, the pressure coordinate analogue of vorticity may be written as follows (Sheng et al., 2021):

$$\zeta_a = \frac{\partial v}{\partial p}i - \frac{\partial u}{\partial p}j - \left(f + \frac{\partial v}{\partial x} - \frac{\partial u}{\partial y}\right)k$$

where (u, v) is the horizontal wind vector, and the lowercase letter vectors (i, j, k) indicate unit vectors pointing eastward, northward, and downward, respectively.

Then, the component form of PVC (J_s) can be written as follows:

$$J_s = -\zeta_a \theta = (J_s^x, J_s^y, J_s^p) = -\frac{\partial v}{\partial p}\theta i + \frac{\partial u}{\partial p}\theta j + \left(f + \frac{\partial v}{\partial x} - \frac{\partial u}{\partial y}\right)\theta k \tag{8a}$$

in which

$$J_s^x = -\frac{\partial v}{\partial p}\theta, \quad J_s^y = \frac{\partial u}{\partial p}\theta, \quad J_s^p = (f + \zeta)\theta \tag{8b}$$

From Eqs. (6a), (7c), and (8b), we can obtain the following important relation:

$$\Delta \int_{t_0}^{t_0+\Delta t} \text{CEPVF} \, \mathrm{d}t = \Delta J_s^y = \Delta\left(\frac{\partial u}{\partial p}\theta\right) \tag{9}$$

where Δ indicates the temporal change or time variation. Equation (9) means that the temporal variation of CEPVF over a given period is determined substantially by the variation of vertical shear in the zonal wind weighted by potential temperature.

The PV flux $\boldsymbol{J}$ (Eq. 3) consists of advective flux ($\boldsymbol{J}_a$; the first term), heating flux (the second term), and friction flux (the third term). The effect of friction always tends to offset a change in PV but it cannot alter the sign of PV change. In simple terms, friction makes the eventual observed change more moderate than it would be otherwise. The effect of friction is usually confined in the surface layer, as is discussed in subsection 4.2.3, but it is neglectable in the free atmosphere. In the following analysis concerning the free atmosphere, we do not further investigate the effect of friction in relation to CEPVF.

Along the equator, the heating flux is smaller than the advective flux after integrating vertically over the equator (table not shown), and it tends to slightly offset the advective flux. Moreover, the correlation coefficient between the annual mean integrated advective flux and the sum of both the annual mean integrated advective and heating fluxes is as high as 0.92, passing the 0.01 significance level. This indicates that CEPVF is dominated by advective flux. Thus, we first consider only the advective flux in calculating CEPVF for the free atmosphere.

The meridional component of PV flux in Eq. (3) therefore gives

$$\text{CEPVF} = \boldsymbol{J}_a \cdot \boldsymbol{j} = J^y = vW \tag{10}$$

From Eq. (10), the vertically and zonally integrated CEPVF along the equator can be expressed as follows:

$$\{\text{CEPVF}\} = \int_{EQ} \int_{p_t}^{p_s} \text{CEPVF}\, dp\, dx = \int_{EQ} \int_{p_t}^{p_s} vW\, dp\, dx \tag{11}$$

From Eq. (11), a CEPVF index, namely CEPVFI, is defined as the normalized time series of $\{\text{CEPVF}\}$:

$$\text{CEPVFI} = \{\text{CEPVF}\}'/\sigma \tag{12}$$

where σ is the standard deviation of $\{\text{CEPVF}\}$ and $\{\text{CEPVF}\}'$ is the $\{\text{CEPVF}\}$ anomaly. A positive (negative) CEPVFI value corresponds to anomalous net northward (southward) CEPVF over the equatorial section. Since the climate mean CEPVF is negative (will be shown in Fig. 1), the positive (negative) anomalous CEPVF also implies less (more) southward CEPVF.

Because the 380 K isentropic surface is near the tropopause over the equator (Wilcox et al., 2012), and because the 100 hPa isobaric surface almost overlaps the 380 K isentropic surface at the equator, the 100 hPa isobaric surface can be considered the tropical tropopause at the equator. Calculation shows that the correlation between the CEPVFI when the upper integral boundary is set at 100 hPa and 380 K successively is as strong as 0.91, exceeding the 0.01 significance level (figure not shown). Furthermore, because the raw data are based on an isobaric surface, directly employing pressure coordinates can avoid extrapolation and interpolation errors. Thus, to focus on the signal in the troposphere in a pressure coordinate system, the upper integral boundary (p_t in Eq. 11) is chosen as 100 hPa in this study.

3 Climatology of cross-equatorial PV flux

Several important climatological features of CEPVF in different seasons are presented in Fig. 1. These features, which appear in all seasons, include two layers with strong southward CEPVF: One in the upper layer within 200—100 hPa and the other in lower layer within 1000—850 hPa; strong southward CEPVF occurring preferentially around the land in the lower layer; and negative CEPVF dominating over the entire equatorial section, which is the most interesting phenomenon.

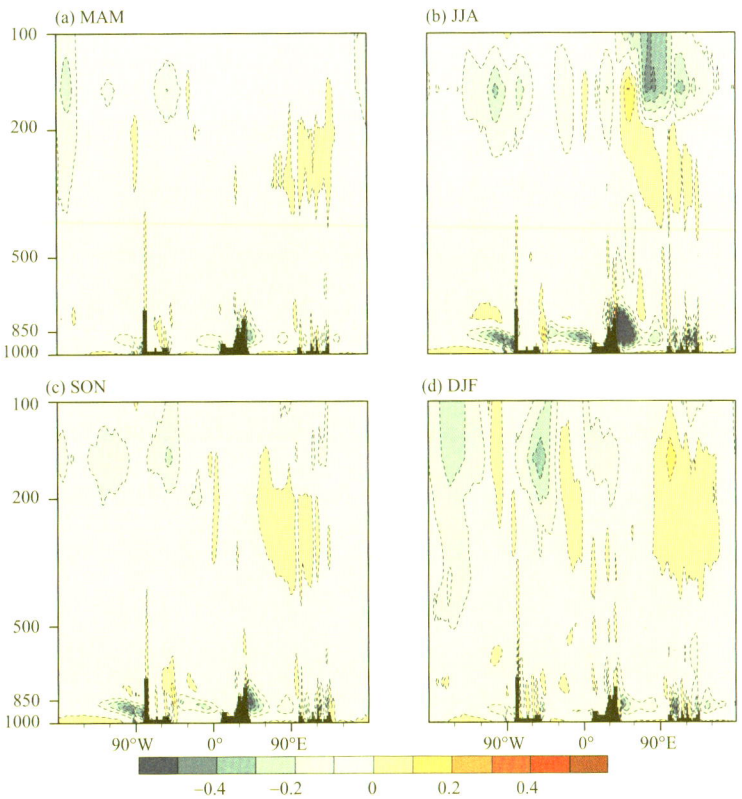

Fig. 1 Climatic mean distribution of cross-equatorial PV flux (CEPVF) along the equator in (a) MAM, (b) JJA, (c) SON, and (d) DJF. Unit: $10^{-7} kg^{-1}$ K m^2. Black shading indicates land

The CEPVF is linked with the meridional wind and PV. Zhao and Lu (2020) investigated the vertical structure of cross-equatorial flow (meridional wind). Their results clearly showed two layers in the distribution of cross-equatorial flow with absolute maxima at 200—100 and 1000—700 hPa. High PV in the atmosphere is also concentrated in both the upper layer (Hoskins et al., 1985) and the lower layer (Sheng et al., 2021; Zhao and Ding, 2009). Thus, the two layers of CEPVF (Fig. 1) do make sense because they are the product of the meridional wind and PV; however, it remains unclear why negative CEPVF is dominant over the entire equatorial section.

It might be conjectured that PV reverses sign across the equator with positive values to the north and negative values to the south. At the equator, a northerly wind ($v<0$) brings positive PV (PV>0) into the Southern Hemisphere (SH), while a southerly wind ($v>0$) brings negative PV (PV<0) into the NH. In both cases, CEPVF is always negative. Therefore, negative CEPVF is observed over almost the entire equatorial section. The equatorial cross section of the distribution of PV and the meridional wind is shown in Fig. 2. In all seasons, in the areas with red dots ($v>0$), PV is mostly negative (PV$<$ 0), whereas in the areas without dots ($v<0$), PV is mostly positive (PV>0). This configuration verifies the physical process, as documented above, that a northerly wind ($v<0$) brings positive PV (PV$>$ 0) into the SH, while a southerly wind ($v>0$) brings negative PV (PV<0) into the NH. This physical process can also be seen in Hoskins et al., (2020) and Hoskins and Yang (2021, see their Fig. 1c). Thus, these results indicate that it is this configuration that results in the overwhelming negative CEPVF over the equatorial section. However, certain areas of positive values do exist (Fig. 1) where the climatological meridional wind and PV have opposite signs, e.g., the region with positive values in the upper level shown in Figs. 1b and 1c. Because CEPVF is a combination of mean and transient eddy processes, these positive values imply that the transient eddy process also contributes to CEPVF, and thus the relevant physical implication deserves further study.

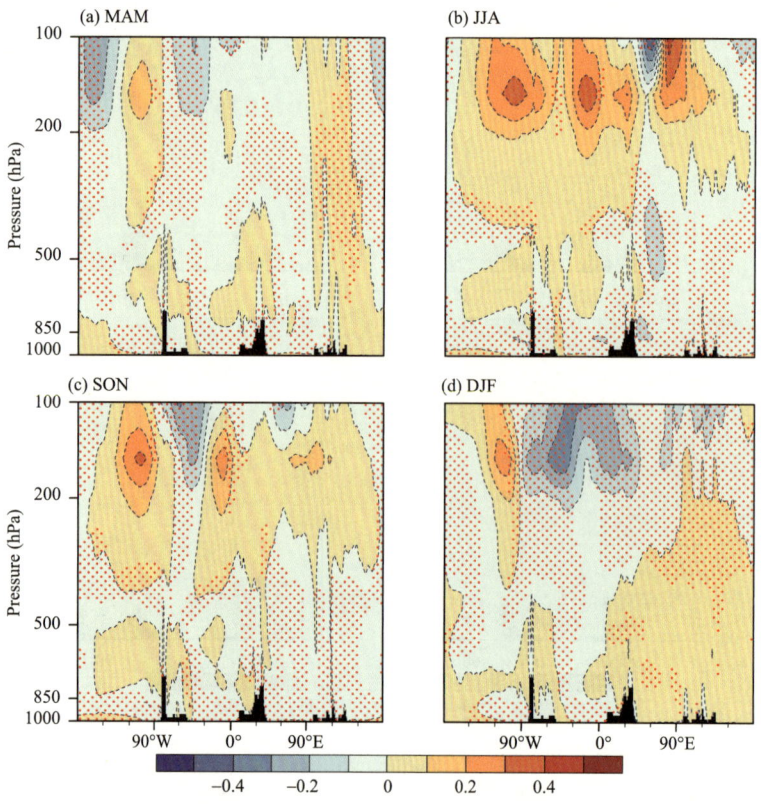

Fig. 2 Same as Fig. 1 but for PV per unit volume. Unit: 10^{-7} K m s kg^{-1}.
Red dots indicate southerly wind ($v>0$) and areas without dots indicate northerly wind ($v<0$)

The climatological annual cycle of the {CEPVF} and its standard deviation in each season are shown in Figs. 3a and 3b, respectively. The positive {CEPVF} component is very small and negative CEPVF dominates the {CEPVF} (Fig. 3a), which is consistent with Fig. 1. The negative {CEPVF} indicates that in terms of the climatological mean there is net PV transport from the NH to the SH. The

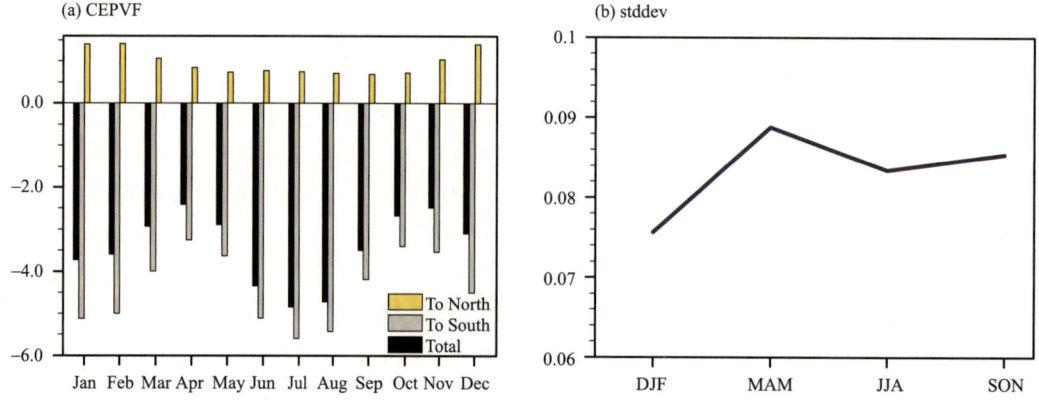

Fig. 3 (a) Climatological annual cycle of {CEPVF}. Unit: 10^4 K m Pa s^{-1}. Yellow, gray, and black bars indicate northward, southward, and total CEPVF along the equator, respectively. (b) Nondimensional {CEPVF} standard deviation in different seasons

{CEPVF} shows a semiannual cycle with absolute minima in April and November and absolute maxima in January and July. This semiannual cycle is consistent with the cycle of cross-equatorial mass exchange (Zhang et al., 2008). The value of the standard deviation of {CEPVF} peaks in MAM (Fig. 3b), reaches a trough in DJF, and is intermediate in JJA and SON. This means the strongest (weakest) interannual variability of {CEPVF} occurs in boreal spring (winter).

4 Cross-equatorial PV flux and SAT over Eurasia during boreal spring

As documented in subsection 2.3, because CEPVF is closely related to surface PV flux in the climatological mean state (Eq. (7b)), and because surface PV flux is related to surface SAT (Eq. (3)), there should be close linkage between CEPVF and SAT. This is shown to be true in data analyses, particularly in relation to boreal spring. In this section, we show this interannual connection between {CEPVF} and SAT over the mid-high latitudes of Eurasia during boreal spring when the interannual variability of {CEPVF} is the strongest (Fig. 3b) and discuss its possible mechanism.

4.1 Relationship between CEPVF and SAT

The variation of the CEPVFI (Eq. (12)) during boreal spring is shown in Fig. 4a. A positive (negative) value of the CEPVFI indicates net anomalous PV transport from the SH (NH) to the NH (SH). The CEPVFI shows strong interannual variability with the variation exceeding one standard deviation in 10 of the 38 years. A regression map of SAT against the CEPVFI over the mid-high latitudes of the Eurasian continent is presented in Fig. 4b, which shows a broad uniform negative pattern over the mid-high latitudes of Eurasia. Embedded within this broad uniform pattern are three significant negative centers: the Mediterranean region, the area southwest of Lake Baikal, and the far east of Russia. This finding indicates that a positive (negative) phase of the CEPVFI corresponds to cooling (warming) over Eurasia. This pattern bears close resemblance to the first empirical orthogonal function (EOF) mode of SAT over Eurasia during boreal spring that was revealed by Chen et al., (2016, see their Fig. 1a). Examination of the correlation between the CEPVFI and the time series of the first EOF mode of SAT indicates that the correlation coefficient reaches 0.36, exceeding the 0.05 significance level. This indicates that CEPVF is related closely to SAT over the mid-high latitudes of Eurasia during boreal spring. However, how CEPVF variation is related to SAT over the northern Eurasian continent remains unclear and requires further investigation.

4.2 Possible mechanism

4.2.1 CEPVF, general circulation, and meridional PVC at the equator

To understand the physical process via which the {CEPVF} is related to SAT, we first investigate the distribution of anomalous CEPVF along the equatorial section. The correlation between the CEPVFI and CEPVF along the equator is shown in Fig. 5a. It is evident that anomalous CEPVF over the equatorial Pacific is predominantly northward (southward) above (below) 700 hPa (Fig. 5a). Conversely, CEPVF over the equatorial Indian Ocean is southward (northward) in the upper (lower) layer.

The correlation map between the CEPVFI and meridional component of the PVC (J_s^y; Eq. (8b)) is shown in Fig. 5b. Because the variation of PV flux J over a given period is parallel to the variation of PVC (J_s), as stated in subsection 2.3.3, Fig. 5b bears marked resemblance to the vertical structure of

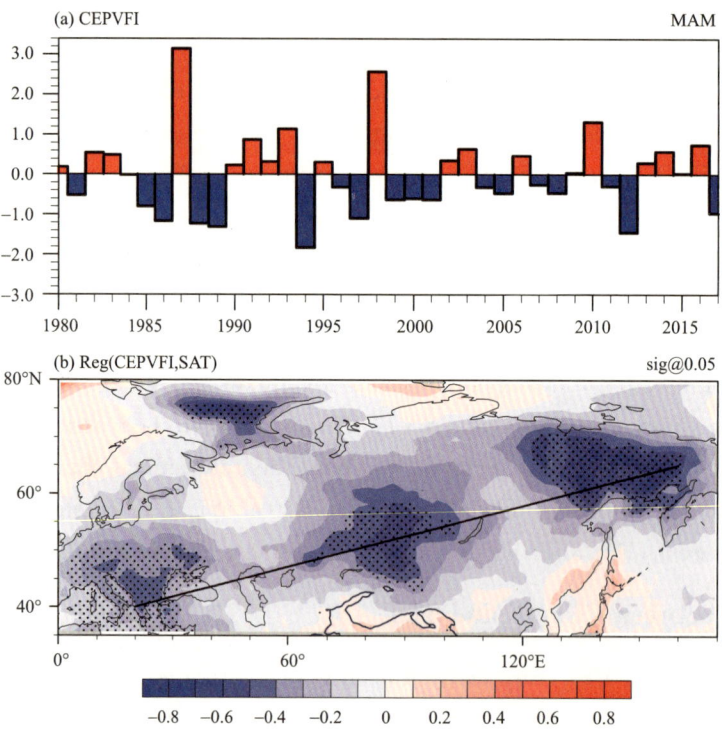

Fig. 4 (a) Normalized time series of spring CEPVFI. (b) Regression coefficients of SAT against the CEPVFI during boreal spring (shading, ℃). Areas exceeding the 0.05 significance level are highlighted by black dots. Three centers of negative SAT are evident along the black line

anomalous CEPVF shown in Fig. 5a, which is also in accordance with Eq. (9). The vertical structure with northward CEPVF in the upper level and compensatory southward CEPVF in the lower level over the equatorial Pacific is very clear (Fig. 5b). Moreover, the inverse vertical structure over the equatorial Indian Ocean is also prominent. This anomalous CEPVF pattern (Fig. 5b) revealed by the CEPVFI is consistent with the first EOF mode on J_s^y over the equatorial section, and the correlation coefficient of their time series reaches 0.68, passing the 0.01 significance level (figure not shown), which indicates that the CEPVFI could well capture the dominant mode of variation of CEPVF. A more interesting characteristic is that the structure shown in Fig. 5b is less chaotic than that in Fig. 5a. This is mainly because the PVC (J_s) is the time-integrated J (Eq. (6a)), which has the high-frequency information filtered out. It should be noted that the convergence of J represents the generation of W, whereas the convergence of J_s is equals to W itself. The similarity between Figs. 5a and 5b thus becomes natural. It should also be noted that certain differences remain between Figs. 5a and 5b. This is mainly because J is approximated to the advective PV flux, whereas J_s includes the total effect of advection, diabatic heating, and friction. Despite this difference, the similarity between Figs. 5a and 5b confirms that flux J and flux J_s are consistent in depicting the variation of CEPVF.

The nature of J_s^y is the vertical shear of the zonal wind weighted by potential temperature. From the definition of J_s^y (Eq. (8b)), the transitional area with the zero contour in Fig. 5b corresponds to the maximum or minimum zonal wind. As can be seen, the regions of maximum and minimum zonal wind are located in the lower layer near 700 hPa over the equatorial Pacific and Indian Ocean, respectively. Consequently, over the Pacific, the J_s^y is positive above the maximum zonal wind but negative below, whereas over the Indian Ocean, the J_s^y is negative above the minimum zonal wind but positive below.

The above results indicate that the variation of CEPVF is mainly related to the anomalous zonal circulation along the equator.

This result can be verified using the following linear temporal expansion of the meridional component of PVC (Eq. (8b)):

$$\Delta J_s^y = \Delta\left(\frac{\partial u}{\partial p}\theta\right) \approx \underbrace{\frac{\partial u}{\partial p} \cdot \Delta\theta}_{B} + \underbrace{\left(\Delta\frac{\partial u}{\partial p}\right) \cdot \theta}_{C}$$

A

which means that the temporal variations of CEPVF (term A) are induced by the variations of potential temperature (term B) and the zonal circulation (term C). The regression of terms A (red dashed line), B (green solid line), and C (blue solid line) on the CEPVFI along the equator at 300 and 850 hPa are shown in Figs. 5c and 5d, respectively. Because the regression of term B (green line) on CEPVFI is very small, it is multiplied by 100 for clarity in the figure. It is evident that term A along the equator matches well with term C at both upper (Fig. 5c) and lower (Fig. 5d) levels, further confirming that the temporal variation of CEPVF is determined substantially by changes of the zonal circulation along the equator.

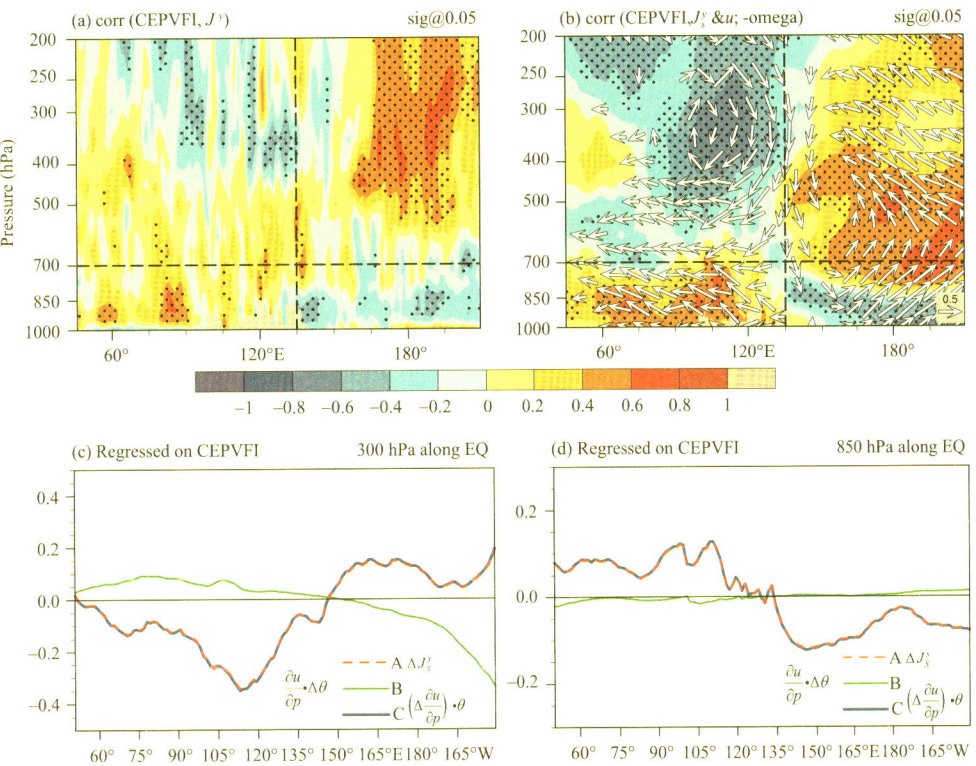

Fig. 5 Distribution of correlation coefficients between the CEPVFI and (a) J^y and (b) J_s^y (shading) and the latitudinal circulation (vectors) along the equatorial section during boreal spring. Vectors exceeding the 0.05 significance level are shown. Areas exceeding the 0.05 significance level are highlighted by black dots. (c) Regression of variation of J_s^y: term A (red dashed line), term B (green solid line), and term C (blue solid line) on CEPVFI along the equator at 300 hPa. Unit: 10^{-1} m K s^{-1} Pa^{-1}. The result regarding term B is multiplied by 100 for clarity. (d) Same as (c) except for 850 hPa

Wu and Meng (1998) found strong coupling between the monsoonal zonal circulation over the equatorial Indian Ocean and the Walker circulation over the Pacific. The coupling system operates very much like a pair of gears running on the equatorial Indian Ocean and Pacific (denoted as GIP), and the "gearing point" of the two cells is located near the Maritime Continent. When one cell rotates clockwise, the other rotates anticlockwise. The direction of this gearing with a clockwise (anticlockwise) Walker circulation over the Pacific and anticlockwise (clockwise) monsoonal zonal circulation over the equatorial Indian Ocean is defined as positive (negative). Wu and Meng (1998) demonstrated that the GIP is linked closely to ENSO. The circulation shown in Fig. 5b is in accordance with the negative GIP reported in their study, which suggests that CEPVF might be associated with ENSO.

4.2.2 CEPVF, PV, and PVC

Because the convergence of PVC (J_s) is directly related to PV itself (Eq. (6b)), whereas the convergence of PV flux (J) represents PV generation (Eq. (2)), for clear and intuitive illustration of the variation of PV itself we consider PVC in the following analysis. The distributions of the correlation coefficients between the CEPVFI and PV as well as horizontal PVC at different levels are presented in Fig. 6. Generally, in high-latitude regions, the PV distribution exhibits an equivalent barotropic structure, whereas in the tropics, both PV and horizontal PVC in the upper troposphere are out of phase with their counterparts in the lower troposphere. Corresponding to the positive phase of the CEPVFI, northward CEPVF can be seen clearly over the Pacific at 200 hPa (Fig. 6a). This northward CEPVF converges anomalous PV to the north of the equator, compensating the reduction of PV in situ. Meanwhile, southward CEPVF is also significant over the Maritime Continent and the Indian Ocean. The southward PVC over the Indian Ocean transfers PV toward the equator, corresponding to the formation of a belt of PVC divergence associated with negative PV over the north of the Tibetan Plateau (TP). This belt of divergence exudes PV that is conducive to the formation of positive PV over the mid-high latitudes of Eurasia (Eq. (6b)). It is further found that the broad pattern of positive PV bears close resemblance to the SAT pattern shown in Fig. 4b, suggesting connection between SAT anomalies and PV anomalies. The broad pattern of positive PV with three nodes over the Eurasian continent exhibits an equivalent barotropic structure that intrudes downward to 500 hPa (Fig. 6b).

The signal of horizontal PVC over the equator at 500 hPa is weaker than at 200 hPa, especially over the Indian Ocean. However, the three positive PV centers over Eurasia are very clear and the flux vector is significant (Fig. 6b). The horizontal PVC at 850 hPa (Fig. 6c) is opposite that in the upper level (Fig. 6a) along the equator, which is in accordance with Fig. 5b. Although the three positive PV centers are not clear over the mid-high latitudes of Eurasia at this lower level, the in situ divergence of the PVC vectors is prominent.

It is worth noting that near the equator over the Pacific sector PV is positive to the north and negative to the south at 850 hPa, and that this is reversed at 500 and 200 hPa (Fig. 6). This is because part of the PV ($-(f-\partial u/\partial y) \cdot \partial \theta/\partial p$) changes its sign across the equator, particularly if the maximum/minimum zonal wind is located at the equator. The intensified equatorial westerly flow in the lower troposphere and easterly flow above 600 hPa over the Pacific sector (Fig. 5b) thus contribute to the anomalous PV distribution there. To the west of the Maritime Continent, for the same reason, PV is negative to the north and positive to the south at 850 and 500 hPa, and the situation is reversed at 200 hPa (Fig. 6). It is because the equatorial westerly flow dominates below 400 hPa in this sector, whereas an easterly flow prevails aloft (Fig. 5b).

To further reveal the characteristics of the broad positive PV pattern in the mid-high latitudes over the Eurasian continent shown in Fig. 6a, the meridional vertical section crossing the central positive PV center (averaged over $70°-105°E$) embedded in this broad PV pattern is presented in Fig. 7. The corre-

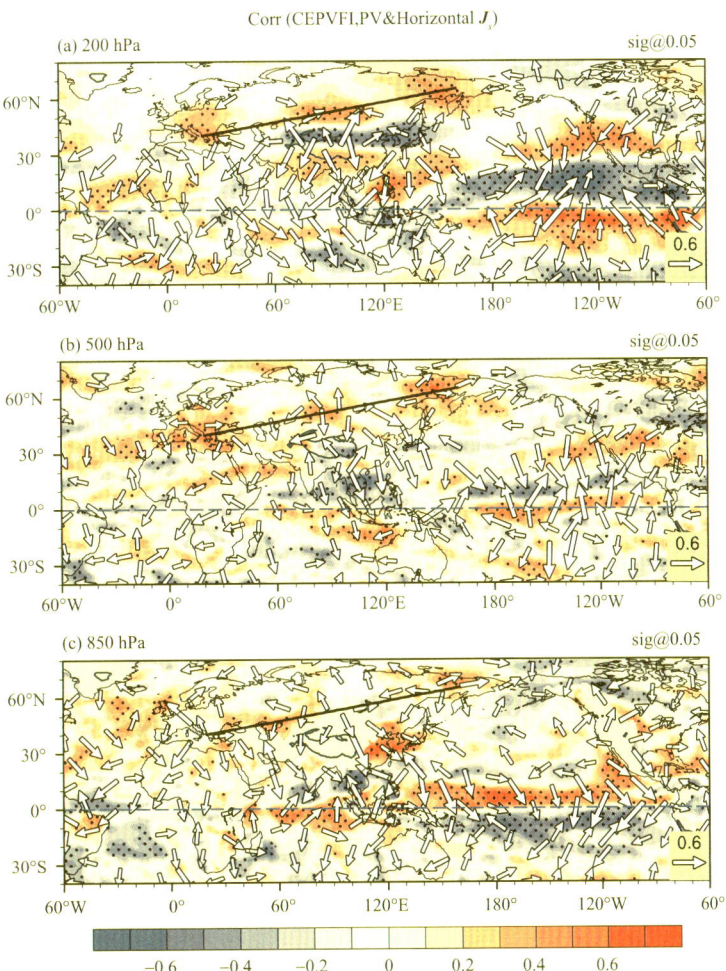

Fig. 6 Distribution of correlation coefficients between CEPVFI and PV (shading) and horizontal J_s (vectors) at (a) 200, (b) 500, and (c) 850 hPa during boreal spring. Areas exceeding the 0.05 significance level are highlighted by black dots. Vectors exceeding the 0.05 significance level are plotted. The blue line denotes the TP topographic boundary of 3000 m

lation coefficients between the CEPVFI and both PVC and PV are displayed. Because the positive vertical direction is downward in pressure coordinates, mimicking the vertical pressure velocity (ω), the vertical component of PVC (J_s^p) is multiplied by (-1) for intuitively plotting. Corresponding to the positive phase of the CEPVFI, in the tropical area below 300 hPa, positive PV exists to the south of the equator and negative PV exists to the north. This is because the prevailing equatorial zonal wind below 300 hPa in the Indian Ocean sector is easterly (Fig. 5b), resulting in cyclonic vorticity to the south of the equator and anticyclonic vorticity to the north. In both the NH and SH, the PVC is downward in the region of positive PV but upward in the region of negative PV. This is because, according to the definition (Eq. (8b)), the vertical component of PVC multiplied by (-1) ($-J_s^p = -(f+\zeta)\theta$) possesses the opposite sign to PV in a hydrostatic atmosphere.

A remarkable feature evident in Fig. 7 over the TP is the significant area of divergence (convergence) of PVC associated with negative (positive) PV in the upper troposphere above (below) 400 hPa. Diagnosis of the correlation between the CEPVFI and diabatic heating shows that in the area over the TP (28°—40°N, 80°—105°E), the area-averaged diabatic heating increases (decreases) with height below (above) 300 hPa (figure not shown). Positive and negative PV is then generated in the lower layer and upper layer, respectively. This explains at least partly why PV distribution is positive in the lower layer and negative in the upper layer over the TP. The divergence of PVC over the TP connects CEPVF to the SH and transports PV to high latitudes, which is conducive to the formation of a notable positive PV column over the mid-high latitudes of Eurasia (Eq. (6b)). Specifically, the divergence over the TP is related to the two gyres of PVC: one to its south and the other to its north. The southern one crosses the equator into the SH in the upper layer and returns to the NH in the lower layer, whereas the northern one moves northward into the higher latitudes, converging and intruding PV downward to the north of the TP, leading to increasing PV over the entire column in the mid-high latitudes of Eurasia. Finally, the positive PV column with an equivalent barotropic structure is formed over the mid-high latitudes of Eurasia.

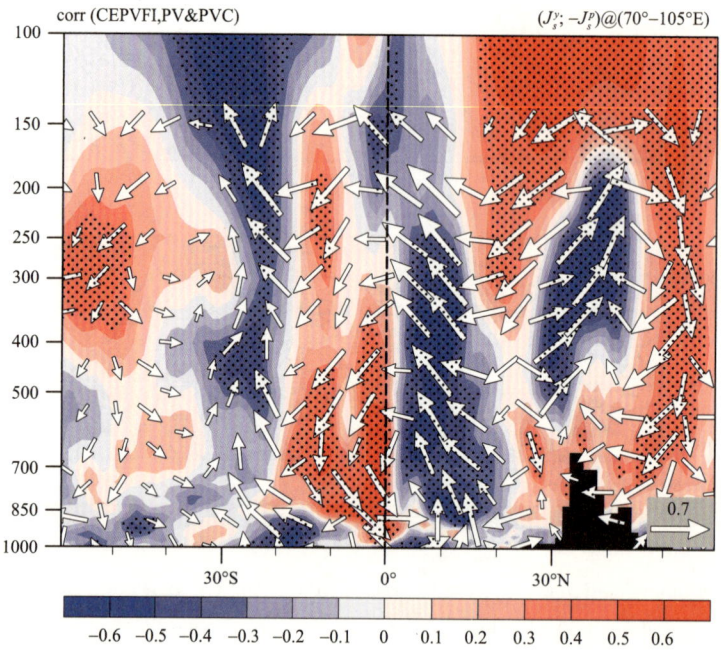

Fig. 7 Distribution of correlation coefficients between the CEPVFI and the 70°—105°E mean PV circulation (J_s^y; $-J_s^p$) (vectors) and PV (shading). Areas exceeding the 0.05 significance level are highlighted by black dots. Vectors exceeding the 0.05 significance level are plotted

4.2.3 PV, SAT, and surface feedback

To further reveal how the equivalent barotropic PV column is retained and how it is related to the in situ cold SAT, the vertical cross section of the correlation between the CEPVFI and both PV and potential temperature (along the black line in Fig. 4b) is shown in Fig. 8. Corresponding to the positive phase of the CEPVFI, three equivalent barotropic positive PV columns (Fig. 8) are located at positions coincident with the three cold SAT centers. Based on the PV-θ mechanism (Hoskins et al., 1985, 2003), isentropes in the troposphere will be vertically "sucked" toward a positive PV anomaly, which means the isentropes bow downward (upward) in the upper (lower) layer. As upward (downward) bowing of isentropes indicates a cold (warm) atmosphere, the three equivalent barotropic positive PV columns therefore lead to the broad uniform pattern of cold SAT shown in Fig. 4b.

As predicted by the PV-θ mechanism, an outstanding common feature (Fig. 8) is that a warm (cool) anomaly appears in each column above (below) 300 hPa. Recalling that the PV source is at the Earth's surface, at the boundary surface, the vertical PV transport due to flux $\boldsymbol{J}_s$ is $\boldsymbol{J}_s \cdot \boldsymbol{n} \approx \int_{t_0}^{t_0+\Delta t}(-f\dot{\theta}-F_\zeta\theta)\mathrm{d}t$, which depends on both heating and friction, where $\boldsymbol{n}$ is an upward unit vector. Because a cold surface favors a local anticyclone circulation (Thorpe, 1985; Hoskins et al., 1985), the atmosphere then exerts anticyclonic stress on the Earth, and the Earth reacts to the atmosphere by producing cyclonic torque, generating positive PV. Moreover, surface cooling over a cold surface also contributes to the generation of positive PV, which in return contributes to the formation of the lower part of the positive PV column over the cold surface, as presented in Fig. 8.

According to the thermal wind relation, a region with cold SAT and a cold air column above can generate a cyclonic vertical shear circulation. Conversely, a warm anomaly in the upper positive PV column should correspond to an anticyclonic vertical shear circulation. Consequently, a maximum anomalous cyclone circulation should exist in the middle column at around 300 hPa. Because the column is warm aloft and cold below, the static stability should also be maximized in the middle layer. Therefore, the positive PV can extend from the lower layer to the middle layer, as shown in Fig. 8. In return, the PV in the positive barotropic PV columns can further maintain the "warm aloft and cold below" thermal structure of the column due to the PV-θ mechanism. Consequently, the broad uniform pattern of cold SAT shown in Fig. 4b is retained at an equilibrium state attributable to positive feedback between the atmospheric circulation and surface cooling.

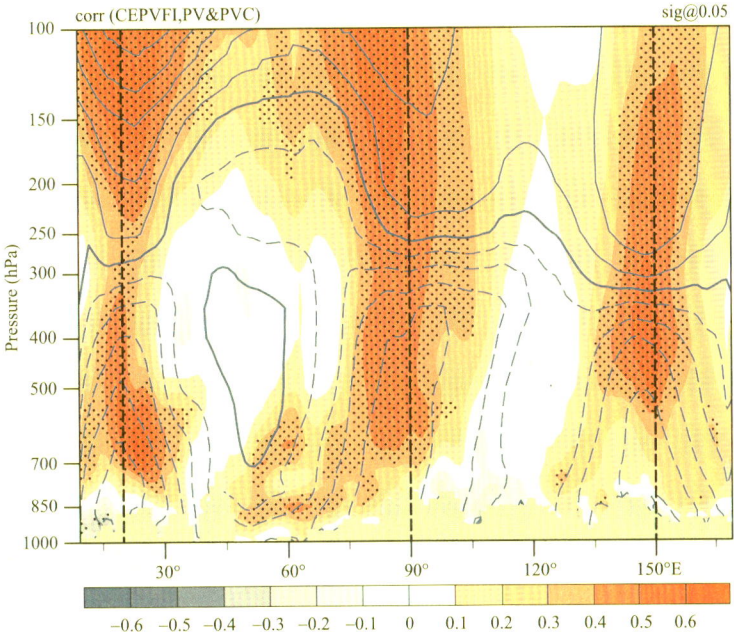

Fig. 8 Distribution of correlation coefficients between the CEPVFI and potential temperature (contours; solid (dashed) lines indicate positive (negative) values) and PV (shading) along the black line displayed in Fig. 1b during boreal spring. The black vertical dashed lines indicate the negative SAT centers displayed in Fig. 4b. Areas exceeding the 0.05 significance level are highlighted by black dots

The above discussion reveals that the significant connection between CEPVF and SAT over the mid-high latitudes of Eurasia is sustained via the PVC, the PV-θ mechanism, and the surface feedback involving friction, diabatic cooling, and the atmospheric circulation.

5 Possible forcing factors of the CEPVF

The anomalous CEPVF is related closely to the vertical distribution of anomalous zonal circulation over the equator (Eq. 9), and therefore any factors inducing the anomalous equatorial zonal circulation could drive the CEPVF. Because ENSO events have strong effect on the zonal circulation (Bjerknes 1969), ENSO as a strong signal of air-sea interaction along the equator could be a driver of CEPVF. The correlation coefficient between the spring CEPVFI and the preceding winter NINO34 index is as high as 0.62 (figure not shown), and the correlation coefficient between the spring CEPVFI and the current spring NINO34 index is as high as 0.67, as shown in Fig. 9a, both passing the 0.01 significance level and indicating a strong relation between ENSO and CEPVF. The correlation between NINO34 and the meridional component of the PVC (J_s^y) is shown in Fig. 9b. In the warm phase of ENSO (Fig. 9b), descending motion is observed over the Maritime Continent, and anomalous westerly and easterly winds are observed in the lower level over the equatorial Pacific and equatorial Indian Ocean, respectively, presenting a negative GIP pattern (Wu and Meng, 1998). The CEPVF induced by the warm phase of ENSO is very similar to the CEPVF revealed by the CEPVFI (Fig. 5b). This confirms that ENSO is an important forcing factor of CEPVF.

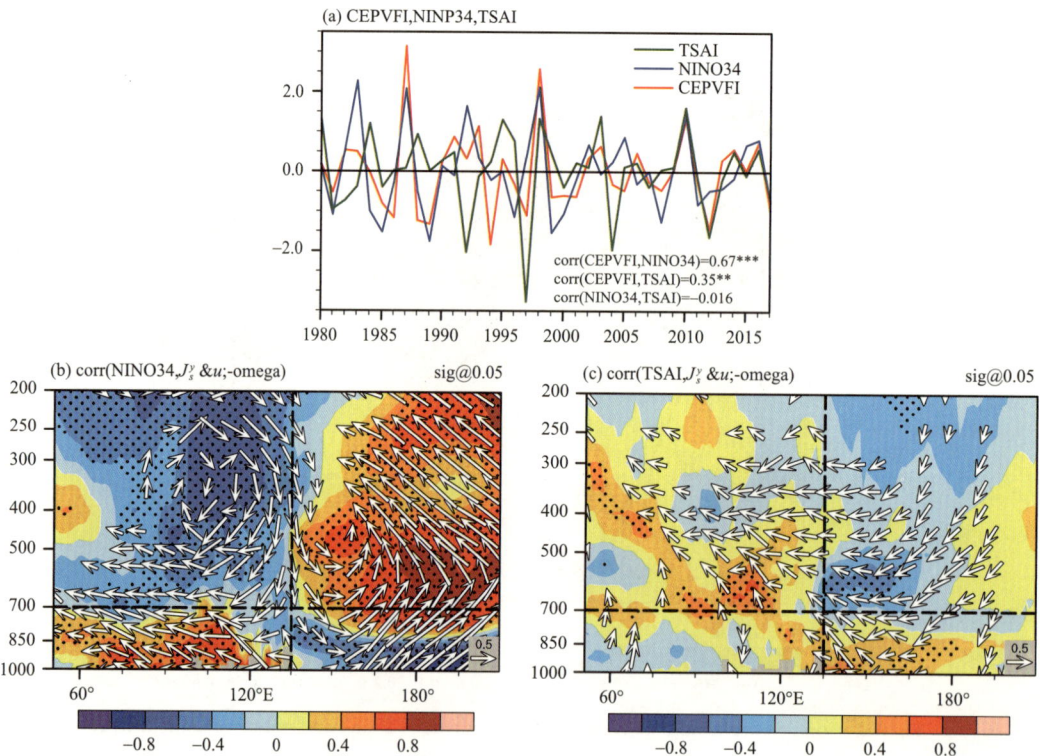

Fig. 9 (a) Normalized time series of the CEPVFI, NINO34 index, and TSAI during boreal spring. (b) Same as Fig. 5b but for the spring NINO34 index. (c) Same as Fig. 5b but for the TSAI index

In addition to ENSO, by employing the partial correlation between the CEPVFI and SST with the spring NINO34 index removed, we found another significant forcing signal of the tropical South Atlantic uniform SST mode (TSAM, Fig. 10), which is characterized as SST anomalies with the same sign over the region (30°S—0°, 40°W—10°E). This mode is a major EOF pattern for tropical South Atlantic variability (Huang et al., 2004). The time series of the area-averaged SST in the blue box in Fig. 10 is defined as the TSAM index (TSAI, shown in Fig. 9a) to represent tropical South Atlantic forcing. The correlation coefficient between the TSAI and the leading EOF time series of the SST in the blue box is as high as 0.99 (figure not shown), indicating that the following results are insensitive to the definition of the SST index. The correlation coefficient between the spring TSAI and the spring CEPVFI is 0.35, passing the 0.05 significance level. However, the correlation coefficient between the spring TSAI and the spring NINO34 index is as low as −0.016, failing the significance test (Fig. 9a). This implies that variation of SST in the tropical South Atlantic is unrelated to ENSO during boreal spring, which is consistent with both Chang et al. (2006) and Kucharski et al. (2007). The TSAM and ENSO therefore can be considered two independent forcing factors that drive CEPVF during boreal spring. As can be seen from Fig. 9c, a significant zonal circulation related to the TSAM is observed around the Maritime Continent, and the induced CEPVF that is concentrated around the Maritime Continent is significant.

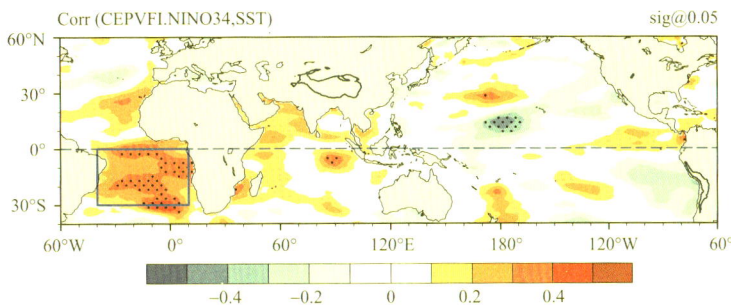

Fig. 10 Distribution of partial correlation coefficients between the CEPVFI and SST with the spring NINO34 index removed. The area covered by the blue box is (30°S−0°, 40°W−10°E). Areas exceeding the 0.05 significance level are highlighted by black dots. The solid blue line denotes the TP topographic boundary of 3000 m

To evaluate the relative roles of the two forcing factors, a regression equation is constructed as CEPVFI* = 0.68(spring NINO34) + 0.36(TSAI) through multiple regression analysis, as shown in Fig. 11a. The correlation between the CEPVFI and the CEPVFI* reconstructed by the spring NINO34 index and the TSAI is 0.75, passing the 0.01 significance level. The explained variance contributed by the NINO34 index and TSAI is approximately 45% and 13% of the total, respectively. Because the NINO34 index and TSAI are unrelated, the total variance explained by these two forcing factors is approximately 58%. These results indicate that the effects of ENSO and the TSAM largely dominate the variation of the CEPVFI.

Because the CEPVF-related zonal circulation induced by ENSO is different to that of the TSAM, the contributions of these two factors to the anomalous SAT should be different. To separate the contributions of ENSO and the TSAM to the anomalous SAT associated with the CEPVFI, Fig. 11 also presents the composite SAT differences between the strong positive (greater than 1 standard deviation) and strong negative (less than −1 standard deviation) NINO34 index (Fig. 11b) and TSAI (Fig. 11c). ENSO (Fig. 11b) is mainly responsible for the overall cooling over the mid-high latitudes of Eurasia. There are three evident cooling centers, i.e., over the Mediterranean region, northwest of Lake Baikal, and the far east of Russia, similar to the pattern of anomalous SAT induced by the CEPVFI shown in Fig. 4b, except that the center near Lake Baikal is shifted further northward. The TSAM (Fig. 11c) is mainly responsible for the cooling center to the south of Lake Baikal. The two independent forcing factors together largely contribute to the emergence of the wide-ranging cooling pattern related to the CEPVFI over the mid-latitudes of Eurasia.

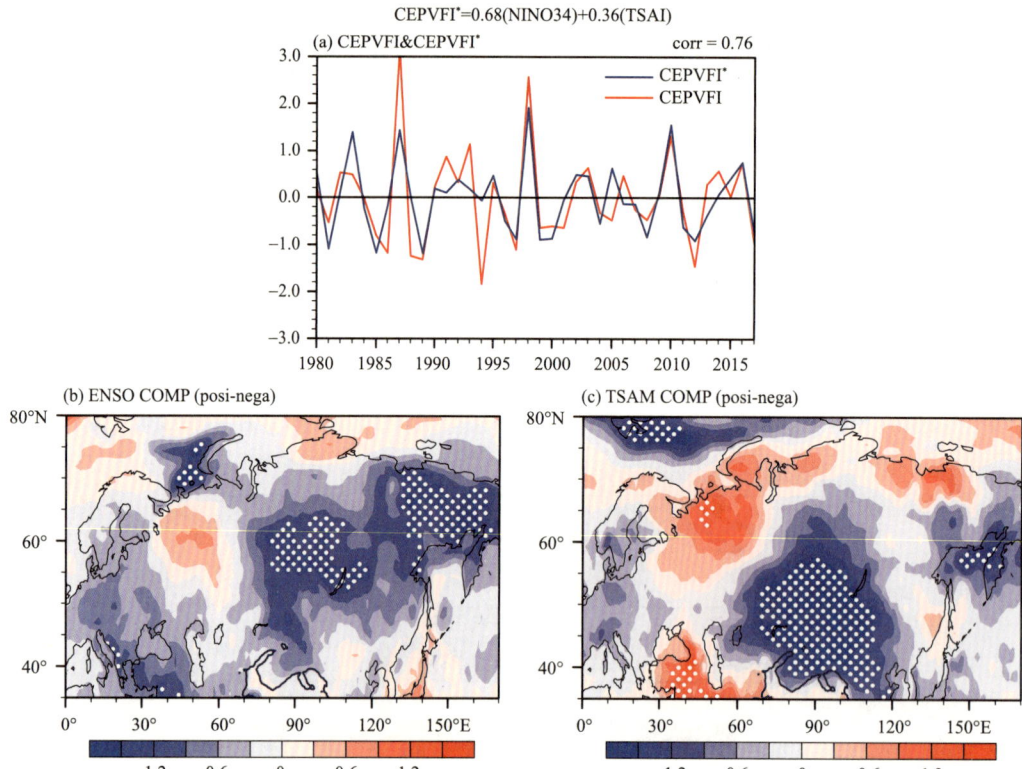

Fig. 11 (a) Normalized time series of the CEPVFI (red line) and CEPVFI* (blue line) during boreal spring. (b) Composite SAT differences between strong positive (greater than 1 standard deviation) and negative (less than −1 standard deviation) ENSO phases during boreal spring. (c) As in (b) but for the TSAI. Areas exceeding the 0.1 significance level are highlighted by white dots

6 Summary and discussion

6.1 Summary

The change in globally integrated atmospheric PV depends solely on PV flux on the Earth's surface, which is the lower boundary of the global domain. For the NH domain, the equatorial vertical section becomes another boundary, and the CEPVF becomes another boundary condition for the change of gross PV in the NH. Because the integrated surface PV flux associated with surface thermal conditions is balanced climatologically by the CEPVF (see Eqs. (3) and (7b)), the CEPVF is therefore related closely to the thermal conditions of the Earth's surface. The results of this study highlight that it is the atmospheric PVC (J_s, Eq. (8a)) that links the CEPVF to the tropical forcing and extratropical SAT response at the Earth's surface.

In addition to theoretical analysis, this study explored the climatological mean features of the CEPVF and focused on the correlation between CEPVF and SAT anomalies over the mid-high latitudes of Eurasia during boreal spring on the interannual time scale. The major results concerning the relationship between CEPVF and SAT are shown schematically in Fig. 12 and summarized as follows.

(1) Climatologically, overwhelming negative CEPVF largely dominates the entire equator, indicating that net PV is transported from the NH to the SH. We highlight that the physical process of a

northerly wind ($v<0$) bringing positive PV (PV>0) and a southerly wind ($v>0$) bringing negative PV (PV<0) dominates the distribution of the CEPVF.

(2) A CEPVF index, namely the CEPVFI, is defined as the normalized time series of the zonally and vertically integrated CEPVF over the equator. On the interannual time scale, the positive phase of the CEPVFI is closely related to the variation of SAT, particularly the significant cooling over the mid-high latitudes of Eurasia, through the PVC, PV-θ mechanism, and surface feedback, as indicated schematically in Fig. 12. The CEPVF over the equatorial Indian Ocean corresponds to the formation of a belt of divergence of PV over the north of the TP, which transfers PV toward the equator and contributes to the broad positive PV in the upper troposphere over the mid-high latitudes of Eurasia (Fig. 12a). The positive PV intrudes downward into the lower layer and forms three positive PV columns (Fig. 12b). Finally, owing to the PV-θ mechanism, the isentropes in the lower troposphere bow upward within these equivalent barotropic positive PV columns, leading to overall cold SAT over the mid-high latitudes of Eurasia (Figs. 12b and 12c). In return the cold surface and its cooling feedback to the atmosphere via surface friction and the diabatic process produce positive PV in the lower troposphere within the PV column. Through cooperation with the thermal wind relation, therefore, the cold SAT and the positive PV aloft are maintained at an equilibrium state.

(3) Because CEPVF is intrinsically proportional to a function of $-\frac{\partial u}{\partial p}$ (Eq. (9)), the variation of CEPVF can result from the change of the zonal circulation over the equator (Figs. 5c and 5d). Two independent forcing factors, namely ENSO and the TSAM, are identified as the main drivers of the CEPVF associated with the zonal circulation (Fig. 12c). Together, ENSO and the TSAM explain approximately 58% of the CEPVFI variance, of which ENSO accounts for approximately 45% and the TSAM accounts for approximately 13%. Owing to the differences in CEPVF induced by ENSO and the TSAM, the contributions of these two factors to the anomalous SAT differ. A warm ENSO is mainly responsible for the overall cold SAT over the mid-high latitudes of Eurasia, while a warm TSAM is mainly responsible for the cold SAT to the south of Lake Baikal. In combination, ENSO and the TSAM jointly explain a large proportion of the CEPVFI variance, and they influence a major part of the variation of SAT associated with the CEPVF during boreal spring.

Simplistically, with reference to Fig. 12d, we argue that once heating forcing (e. g. , ENSO) appears over the tropics (A), the zonal circulation along the equator will change (B). This process represents manifestation of the consensus that anomalous heating could induce anomalous zonal circulation (e. g. , Bjerknes, 1969). According to Eq. (9), anomalous zonal circulation (B) will induce anomalous CEPVF (C). Because PV is equal to the convergence of PVC (Eq. (6b)), changes in CEPVF (C) at the equatorial boundary of the NH will stimulate adjustment of PVC within the NH, leading to change in PV over the NH (D). Finally, in response to the PV-θ mechanism (Hoskins et al. , 1985, 2003) and owing to surface feedback associated with surface friction and diabatic cooling, anomalous cold SAT over the mid-high latitudes of Eurasia (E) is retained.

6.2 Discussion

On the basis of the results of this study, it may be speculated that the physical nature of the equilibrium linkage between CEPVF and extratropical SAT is mainly a manifestation of the response of extratropical SAT to tropical forcing that induces anomalous CEPVF. In this linkage, the CEPVF is much like a monitor through which the impacts of various tropical signals (including but not limited to ENSO and

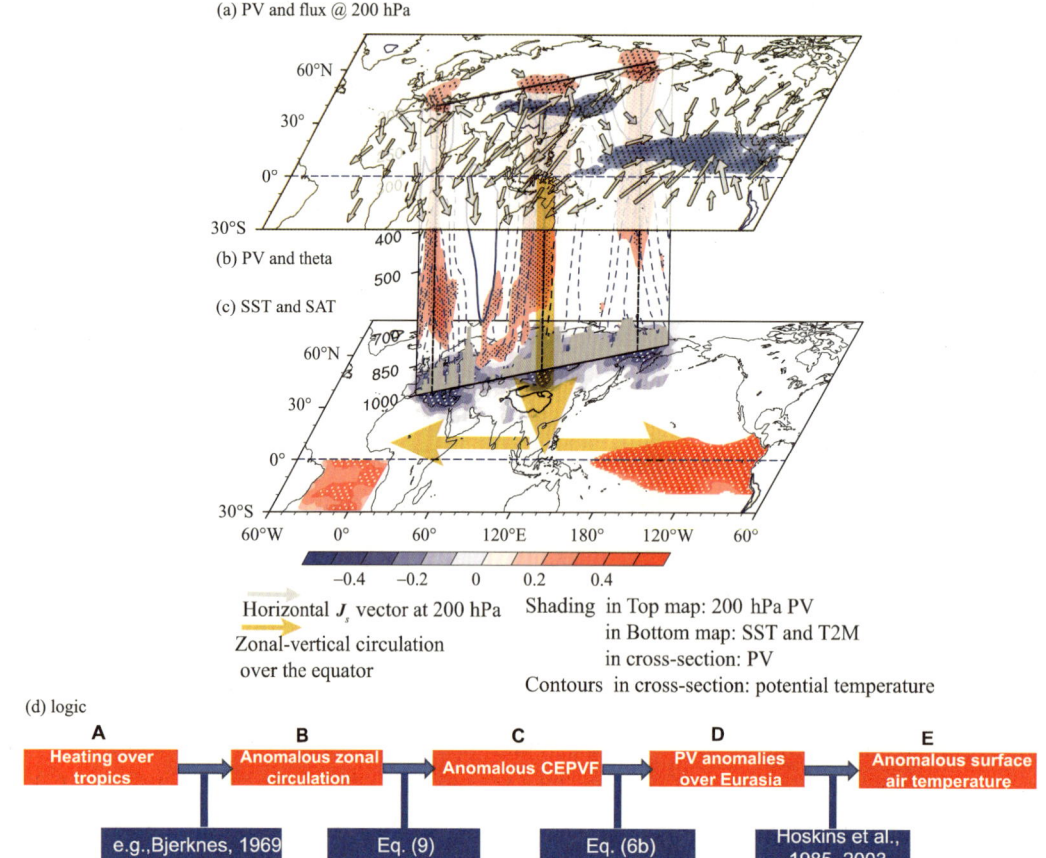

Fig. 12 Schematic showing the CEPVF influence on SAT over the mid-high latitudes of Eurasia during boreal spring. (a) PV (shading) and horizontal PV flux (vectors) at 200 hPa, (b) cross section of PV (shading) and potential temperature (contours), and (c) SST over oceans and SAT over land. Yellow vectors indicate the zonal circulation over the equator. (d) Logic diagram summarizing the findings of this study. The upper row with red boxes indicates physical nodes in this study, and the lower row with blue boxes indicates the foundation between these nodes

the TSAM) on extratropical SAT can be comprehensively detected. In this process, the PVC is like a bridge through which tropical forcing communicates with extratropical SAT. There are numerous frameworks through which to view the general circulation; however, the necessity to understand the behavior of the atmosphere demands that we seek an increasing array of diagnostic tools (Hoskins, 1991). This study sheds new light on the connection between the tropical and mid-high-latitude circulations. Owing to the existence of the easterly wind in the tropics, it is difficult to explain how tropical signals could cross the easterly wind belt and affect the extratropical circulation within the framework of Rossby waves. The perspective of PVC provides a relatively direct and simple connection between the tropics and the mid-high latitudes.

Although the tropical forcing factors that could drive the variability of zonal circulation associated with CEPVF anomalies are physically defensible (e.g., Bjerknes, 1969) and statistically significant, numerical sensitivity experiments should be conducted in future to further explore the detailed roles of ENSO and the TSAM in driving the zonal circulation associated with CEPVF anomalies. In addition to boreal spring, the climatological effects related to CEPVF in different seasons should be studied carefully to establish the connection between CEPVF and certain well-known extratropical forcing (e.g., the Arctic Oscillation, Arctic Sea ice, and South Asian high) and further deepen our insight into climate dynamics.

References

BJERKNES J, 1969. Atmospheric telecconnections from equatorial Pacific[J]. Monthly Weather Review, 97(3): 163-172.

BRETHERTON C S, SCHÄR C, 1993. Flux of potential vorticity substance—a simple derivation and a uniqueness property [J]. Journal of the Atmospheric Sciences, 50(12): 1834-1836.

CHANG P, FANG Y, SARAVANAN R, et al, 2006. The cause of the fragile relationship between the Pacific El Niño and the Atlantic Niño[J]. Nature, 443(7109): 324-328.

CHEN S, WU R, LIU Y, 2016. Dominant modes of interannual variability in Eurasian surface air temperature during boreal spring[J]. Journal of Climate, 29(3): 1109-1125.

CHEN S, WU R, 2018. Impacts of early autumn Arctic sea ice concentration on subsequent spring Eurasian surface air temperature variations[J]. Climate Dynamics, 51(7-8): 2523-2542.

CHEN S F, WU R G, CHEN W, 2019. Projections of climate changes over mid-high latitudes of Eurasia during boreal spring: Uncertainty due to internal variability[J]. Climate Dynamics, 53(9-10): 6309-6327.

D'ARRIGO R, WILSON R, LI J, 2006. Increased Eurasian-tropical temperature amplitude difference in recent centuries: Implications for the Asian monsoon[J]. Geophysical Research Letters, 33(22).

HAYNES P H, MCINTYRE M E, 1987. On the evolution of vorticity and potential vorticity in the presence of diabatic heating and frictional or other forces[J]. Journal of the Atmospheric Sciences, 44(5): 828-841.

HAYNES P H, MCINTYRE M E, 1990. On the conservation and impermeability theorems for potential vorticity[J]. Journal of the Atmospheric Sciences, 47(16): 2021-2031.

HOSKINS B, 1991. Towards a PV-θ view of the general-circulation[J]. Tellus-Series A: Dynamic Meteorology and Oceanography, 43(4): 27-35.

HOSKINS B, 2015. Potential vorticity and the PV perspective[J]. Advances in Atmospheric Sciences, 32(1): 2-9.

HOSKINS B, MCINTYRE M E, ROBERTSON A W, 1985. On the use and significance of isentropic potential vorticity maps[J]. Quarterly Journal of the Royal Meteorological Society, 111(470): 877-946.

HOSKINS B, PEDDER M, JONES D W, 2003. The omega equation and potential vorticity[J]. Quarterly Journal of the Royal Meteorological Society, 129(595): 3277-3303.

HOSKINS B, YANG G Y, FONSECA R M, 2020. The detailed dynamics of the June—August Hadley cell[J]. Quarterly Journal of the Royal Meteorological Society, 146(727): 557-575.

HOSKINS B, YANG G Y, 2021. The detailed dynamics of the Hadley cell. Part II: December—February[J]. Journal of Climate, 34(2): 805-823.

HUANG B H, SCHOPF P S, SHUKLA J, 2004. Intrinsic ocean-atmosphere variability of the tropical Atlantic ocean[J]. Journal of Climate, 17(11): 2058-2077.

ISHII M, SHOUJI A, SUGIMOTO S, et al, 2005. Objective analyses of sea-surface temperature and marine meteorological variables for the 20th century using ICOADS and the kobe collection[J]. International Journal of Climatology, 25(7): 865-879.

KUCHARSKI F, BRACCO A, YOO J H, et al, 2007. Low-frequency variability of the Indian monsoon-ENSO relationship and the tropical Atlantic: The "weakening" of the 1980s and 1990s[J]. Journal of Climate, 20(16): 4255-4266.

LABAT D, GODDERIS Y, PROBST J L, et al, 2004. Evidence for global runoff increase related to climate warming[J]. Advances in Water Resources, 27(6): 631-642.

LIU X D, YANAI M, 2001. Relationship between the Indian monsoon rainfall and the tropospheric temperature over the Eurasian continent[J]. Quarterly Journal of the Royal Meteorological Society, 127(573): 909-937.

LIU Y M, HOSKINS B, BLACKBURN M, 2007. Impact of Tibetan orography and heating on the summer flow over Asia[J]. Journal of the Meteorological Society of Japan, 85B: 1-19.

LUCCHESI R, 2012. File specification for merra products[Z]. Gmao office note no. 1 (version 2.3), 87. https//gmao.Gsfc.Nasa.Gov/pubs/docs/lucchesi528.

LUO D, CHEN X, FELDSTEIN S B, 2018a. Linear and nonlinear dynamics of North Atlantic Oscillations: A new thinking of symmetry breaking[J]. Journal of the Atmospheric Sciences, 75(6): 1955-1977.

LUO D, CHEN X, DAI A, et al, 2018b. Changes in atmospheric blocking circulations linked with winter arctic warming: A new perspective[J]. Journal of Climate, 31(18): 7661-7678.

MA T T, WU G X, LIU Y M, et al, 2019. Impact of surface potential vorticity density forcing over the Tibetan Plateau on the South China extreme precipitation in January 2008. Part I: Data analysis[J]. Journal of Meteorological Research, 33(3): 400-415.

MA T T, WU G, LIU Y, et al, 2022. Abnormal warm sea-surface temperature in the Indian Ocean, active potential vorticity over the Tibetan Plateau, and severe flooding along the Yangtze River in summer 2020[J]. Quarterly Journal of the Royal Meteorological Society.

OGI M, TACHIBANA Y, YAMAZAKI K, 2003. Impact of the wintertime North Atlantic Oscillation (NAO) on the summertime atmospheric circulation[J]. Geophysical Research Letters, 30(13): 1-4.

ORTEGA S, WEBSTER P J, TOMA V, et al, 2018. The effect of potential vorticity fluxes on the circulation of the tropical upper troposphere[J]. Quarterly Journal of the Royal Meteorological Society, 144(712): 848-860.

RIENECKER M M, SUAREZ M J, GELARO R, et al, 2011. Merra: Nasa's modern-era retrospective analysis for research and applications[J]. Journal of Climate, 24(14): 3624-3648.

SCHNEIDER T, HELD I M, GARNER S T, 2003. Boundary effects in potential vorticity dynamics[J]. Journal of the Atmospheric Sciences, 60(8): 1024-1040.

SHENG C, WU G X, TANG Y Q, et al, 2021. Characteristics of the potential vorticity and its budget in the surface layer over the Tibetan Plateau[J]. International Journal of Climatology, 41: 439-455.

SHENG C, HE B, WU G X, et al, 2022. Interannual influences of the surface potential vorticity forcing over the Tibetan Plateau on East Asian summer rainfall[J]. Advances in Atmospheric Sciences, 39(7): 1050-1061.

THORPE A J, 1985. Diagnosis of balanced vortex structure using potential vorticity[J]. Journal of the Atmospheric Sciences, 42(4): 397-406.

WANG X, PIAO S, CIAIS P, et al, 2011. Spring temperature change and its implication in the change of vegetation growth in North America from 1982 to 2006[J]. Proceedings of the National Academy of Sciences of the United States of America, 108(4): 1240-1245.

WILCOX L J, HOSKINS B J, SHINE K P, 2012. A global blended tropopause based on era data. Part I: Climatology[J]. Quarterly Journal of the Royal Meteorological Society, 138(664): 561-575.

WU G X, CAI Y P, TANG X J, 1995. Moist potential vorticity and slantwise vorticity development[J]. Acta Meteorologica Sinica, 53: 387-405.

WU G X, CAI Y P, 1997. Vertical wind shear and down-sliding slantwise vorticity development[J]. Chinese Journal of Atmospheric Sciences, 21: 273-282.

WU G X, MENG W, 1998. Gearing between the indo-monsoon circulation and the Pacific-Walker circulation and the ENSO. Part I: Data analyses[J]. Scientia Meteorologica Sinica, 22: 470-480.

XIE Y, WU G, LIU Y, et al, 2020. Eurasian cooling linked with Arctic warming: Insights from PV dynamics[J]. Journal of Climate, 33(7): 2627-2644.

ZAR J H, 2010. Biostatistical analysis[J]. Quarterly Review of Biology, 18: 797-799.

ZHANG G, MAO J, LIU Y, et al, 2021a. PV perspective of impacts on downstream extreme rainfall event of a Tibetan Plateau vortex collaborating with a southwest China vortex[J]. Advances in Atmospheric Sciences, 38(11): 1835-1851.

ZHANG G, MAO J, WU G, et al, 2021b. Impact of potential vorticity anomalies around the eastern Tibetan Plateau on quasi-biweekly oscillations of summer rainfall within and south of the Yangtze basin in 2016[J]. Climate Dynamics, 56(3-4): 813-835.

ZHANG Y, HUANG F, GONG X Q, 2008. The characteristics of the air mass exchange between the Northern and Southern Hemisphere[J]. Journal of Tropical Meteorology, 24: 74-80.

ZHAO L, DING Y H, 2009. Potential vorticity analysis of cold air activities during the East Asian summer monsoon[J]. Chinese Journal of Atmospheric Sciences, 33(2): 359-374.

ZHAO X, LU R, 2020. Vertical structure of interannual variability in cross-equatorial flows over the maritime continent and Indian ocean in boreal summer[J]. Advances in Atmospheric Sciences, 37(2): 173-186.

附录　吴国雄先生论文、著作列表

1979

[1] 徐国昌，陈美连，吴国雄. 甘肃省"4·22"特大沙尘暴分析[J]. 气象学报，1979，37(4)：26-35.

1983

[2] WU G X. The influence of large-scale orography upon the general circulation of the atmosphere [D]. London: The University of London, 1983.

1984

[3] WU G X. The nonlinear response of the atmosphere to large-scale mechanical and thermal forcing [J]. J Atmos Sci, 1984, 41: 2456-2476.

1985

[4] WU G X, CHEN S J. The effect of mechanical forcing on the formation of a meso-scale vortex[J]. Quarterly Journal of the Royal Meteorological Society, 1985, 111: 1049-1079.

[5] WU G X, CUBALSH U. The impact of the El Niño anomaly on the mean meridional circulation as simulated by a high-resolution model[J]. ECMWF Tech Memo, 1985, 105: 1-25.

[6] WU G X, LIU H Z. General circulation diagnostic package[J]. ECMWF Tech Memo, 1985, 96: 1-35.

1986

[7] 吴国雄. 国外大尺度动力学及中期数值天气预报的进展[J]. 南京气象学院学报，1986，3：305-313.

[8] 吴国雄. 欧洲中期天气预报中心的十年计划[J]. 气象科技，1986，2：1-10.

[9] 吴国雄. 中期数值天气预报的现状和展望[J]. 气象，1986，12(3)：2-7.

[10] 吴国雄，CUBASCH U. El Nino海温异常对纬向平均经圈环流及大气输送特征的影响[J]. 中国科学，1986，10：1109-1120.

[11] WU G X, WHITE A. A further study of the surface zonal flow predicted by an eddy flux parameterization scheme[J]. Quarterly Journal of the Royal Meteorological Society, 1986, 112: 1041-1056.

[12] WU G X. The effects of mechanical forcing on the mean meridional circulation of the atmosphere [J]. ECMWF Tech Memo, 1986, 112: 1-42.

1987

[13] 刘还珠，TIBALDI S，吴国雄. ECMWF预报模式在不同地形方案下的系统误差检验[J]. 气象，1987，13(3)：3-9.

[14] 吴国雄. "核冬天"理论及气溶胶的短波吸收作用[J]. 成都气象学院学报，1987，1：6-10.

[15] 吴国雄, 刘还珠. 全球大气环流时间平均统计图集[M]. 北京: 气象出版社, 1987.

[16] WU G X, CUBASCH U. The impact of El Niño SST anomaly on mean meridional circulation and transfer properties of the atmosphere[J]. Scientia Sinica, 1987, 30(5): 533-545.

[17] WU G X, TIBALDI S. The effects of mechanical forcing on the mean meridional circulation and transfer properties of the atmosphere[J]. Advances in Atmospheric Sciences, 1987, 4(01): 24-42.

1988

[18] 吴国雄. 大气的内外强迫源和西风指数的变化[J]. 气象, 1988, 14(8): 3-8.

[19] 吴国雄. 球面长波模式的设计和应用[J]. 气象学报, 1988, 46(2): 142-153.

[20] 吴国雄, 陈彪, 吴正贤. 不同波数域中干、湿空气的E-P剖面和余差环流[J]. 大气科学, 1988, 12(s1): 94-106.

[21] 吴国雄, TIBALDI S. 关于大气平均经圈环流的一种计算方案[J]. 中国科学, 1988, 4: 442-450.

[22] 吴国雄, TIBALDI S. 平均经圈环流在大气角动量和感热收支中的作用[J]. 大气科学, 1988, 12(1): 8-17.

[23] WU G X, CHEN B, WU Z X. Moist and dry Eliassen-Palm cross sections in different wave bands and residual circulation[J]. Chinese Journal of Atmospheric Sciences, 1988, 12(S1): 94-106.

[24] WU G X. Designation and application of a spherical long-wave spectral model[J]. Acta Meteorologica Sinica, 1988, 3(2): 132-144.

[25] WU G X. Formation of summer vortex on the eastern flank of Qinghai-Tibetan Plateau. Palmen Memorial on Extratropical Cyclones[J]. American Meteor Soc, 1988: 273-276.

[26] WU G X. Mountain torque and external forcing[J]. Advances in Atmospheric Sciences, 1988, 5: 141-148.

[27] WU G X, TIBALDI S. A scheme for evaluating the mean meridional circulation of the atmosphere[J]. Scientia Sinica, 1988, 31(10): 1235-1244.

[28] WU G X, TIBALDI S. Roles of the mean meridional circulation on atmospheric budgets of angular momentum and sensible heat[J]. Chinese Journal of Atmospheric Sciences, 1988, 12: 11-24.

1989

[29] WU G X, CHEN B. Non-acceleration theorem in a primitive Equation system: I. Acceleration of zonal meanflow[J]. Advances in Atmospheric Sciences, 1989, 6(1): 1-20.

1990

[30] 吴国雄. 大气水汽的输送和收支及其对副热带干旱的影响[J]. 大气科学, 1990, 14(1): 53-63.

[31] 吴国雄, 陈彪. E-P剖面的年变化和对流层西风加速[J]. 中国科学B, 1990, 7: 775-784.

[32] 吴国雄, 陈彪. 大气定常波传播的运动特征的时空变化[J]. 气象学报, 1990, 48(1): 34-45.

[33] 吴国雄, 陈彪. 原始方程系统中的无加速定理——II: 纬向平均温度的变化[J]. 大气科学, 1990, 14(2): 143-154.

[34] 吴国雄, 董步文. 大气平衡态的动力特征——I: 多平衡态的共面和非共面性质[J]. 大气科学, 1990, 14: 267-276.

[35] 吴国雄, 董步文. 大气平衡态的动力特征——II: 大气振荡及稳定平衡态的吸引机制[J]. 大气科学, 1990, 14: 385-394.

[36] 吴国雄, 邹晓蕾. 国外持续异常天气动力研究的进展. 旱涝气候研究进展[M]. 北京: 气象出版社, 1990.

[37] 吴正贤,李崇银,陈彪,等. 1982/83冬季El Nino期间大气环流异常分析[J]. 热带气象,1990,6(3):253-264.

[38] 杨伟愚,叶笃正,吴国雄. 夏季青藏高原气象学若干问题的研究[J]. 中国科学B,1990,10:1100-1111.

[39] WU G X. The impact of atmospheric water vapor transport and its budget on subtropical drought [J]. Chinese Journal of Atmospheric Sciences,1990,14:63-75.

[40] WU G X, CHEN B. Non-acceleration theorem in primitive equation system. Ⅱ. Variation of zonal mean temperature[J]. Chinese Journal of Atmospheric Sciences,1990,14:153-167.

[41] WU G X, DONG B W. Dynamical features of atmospheric equilibration. Part I. Co-surface and non-cosurface characters of multiple equilibria[J]. Chinese Journal of Atmospheric Sciences,1990,14:267-280.

[42] WU G X, DONG B W. Dynamical features of atmospheric equilibration. Part II. Mechanisms of atmospheric oscillation and attraction of steady equilibrium[J]. Chinese Journal of Atmospheric Sciences,1990,14:377-388.

1991

[43] 吴国雄. 大气环流动力学及其统计特征[C]//当代气候研究. 北京:气象出版社,1991.

[44] 邹晓蕾,吴国雄,叶笃正. 北半球两大地形下游冬季环流的动力分析——Ⅱ.行星波的垂直传播[J]. 气象学报,1991,49(3):257-268.

[45] 邹晓蕾,叶笃正,吴国雄. 北半球两大地形下游冬季环流的动力分析——Ⅰ.环流、遥相关和定常波的联系[J]. 气象学报,1991,49(2):129-140.

[46] BENZI R. 异常环流和阻塞[M]. 吴国雄,纪立人,译. 北京:气象出版社,1991.

[47] ZOU X L, YEH T C, WU G X. Analyses of the dynamic effects on winter circulation of the two main mountains in the Northern Hemisphere. Ⅰ. Relationship among general circulation, teleconnection and stationary waves[J]. Acta Meteorologica Sinica,1991,6(4):395-407.

1992

[48] 吴国雄. 海温异常对台风形成的影响[J]. 大气科学,1992,16(3):322-332.

[49] 杨伟愚,叶笃正,吴国雄. 夏季青藏高原热力场和环流场的诊断分析——Ⅰ:盛夏高原西部地区的水汽状况[J]. 大气科学,1992,16(1):41-51.

[50] 杨伟愚,叶笃正,吴国雄. 夏季青藏高原热力场和环流场的诊断分析——Ⅱ:环流场的主要特征及其大型垂直环流场[J]. 大气科学,1992,16(3):287-301.

[51] 杨伟愚,叶笃正,吴国雄. 夏季青藏高原热力场和环流场的诊断分析——Ⅲ:环流场稳定维持的物理机制[J]. 大气科学,1992,16(4):409-426.

[52] WU G X. A numerical study on the relationship between tropical-storm formation and ENSO [C]//LIGHTHILL S M J, ZHENG Z, HOLLAND G, et al. ICSU/WMO International Symposium on Tropical Cyclone Disasters. Beijing: Peking Univ Press, 1992: 163-169.

[53] WU G X, LAU N C. A GCM simulation of the relationship between tropical-storm formation and ENSO[J]. Monthly Weather Review,1992,120:958-977.

[54] WU G X, LIU H Z. Atmospheric precipitation in response to equatorial and tropical sea surface temperature anomalies[J]. Journal of the Atmospheric Sciences,1992,49(23):2236-2255.

[55] ZOU X L, WU G X, YEH T C. Analyses of the dynamic effects on winter circulation of the two main mountains in the Northern Hemisphere. Ⅱ. Vertical propagation of planetary waves[J]. Ac-

ta Meteorologica Sinica, 1992, 6(4): 408-420.

1993

[56] WU G X. The fourth international symposium on Asian monsoon held in Japan[J]. Advances in Atmospheric Sciences, 1993, 10: 128-134.

[57] WU G X. Variability of tropical-storm formation and ENSO[C]//YE D Z, MATSUNO T, ZENG Q C, et al. International Workshop on Climate Variability. Beijing: China Meteorological Press, 1993: 209-214.

[58] WU G X, CAI Y P. Modulation of atmospheric thermal and mechanical forcings and numerical modeling of mean meridional circulation[J]. Acta Meteorologica Sinica, 1993, 7: 412-422.

[59] WU G X, LIU H Z. Rainfall Pattern and tropical SSTA[C]//YE D Z, ZENG Q C, WU G X, et al. IVth International Summer Colloquium on Climate, Environment and Geophysical Fluid Dynamics. Beijing: China Meteorological Press, 1993: 166-176.

[60] YEH D C, ZENG Q C, WU G X, et al. Climate, Environment and Geophysical Fluid Dynamics-Proceedings of the IVth International Summer Colloquium and International Symposium for Young Scientists[C]. Beijing: China Meteorological Press, 1993.

1994

[61] 吴国雄,蔡雅萍. 大气热力强迫和动力强迫的调配及平均经圈环流的仿真模拟[J]. 气象学报, 1994, 52: 138-148.

[62] 吴国雄,刘辉,陈飞,等. 时变涡动输送和阻塞高压的形成——1980年夏季我国的南涝北旱[J]. 气象学报, 1994, 52: 308-320.

1995

[63] 刘辉,吴国雄,曾庆存. 北半球阻塞高压的维持Ⅰ: 准地转和Ertel位涡分析[J]. 气象学报, 1995, 53(2): 177-185.

[64] 刘辉,曾庆存,吴国雄. 北半球阻塞高压的维持Ⅱ: 瞬变扰动强迫和平均流位涡平流的形成[J]. 气象学报, 1995, 53(3): 337-348.

[65] 吴国雄. 大气中动能的双向转化和单向转化[J]. 大气科学, 1995, 19: 52-62.

[66] 吴国雄,蔡雅萍,唐晓菁. 湿位涡和倾斜涡度发展[J]. 气象学报, 1995, 53: 387-405.

[67] 吴国雄,刘还珠. 降水对热带海表温度异常的邻域响应——Ⅰ: 数值模拟[J]. 大气科学, 1995, 19(4): 422-434.

[68] 吴国雄,孙凤英,王晓春,等. 降水对热带海表温度异常的邻域响应——Ⅱ: 资料分析[J]. 大气科学, 1995, 19(6): 663-676.

[69] 吴国雄,薛纪善,王在志,等. 青藏高原化雪迟早的辐射效应对季节变化的影响[J]. 甘肃气象, 1995, 13(1): 1-8.

[70] 张绍晴,刘还珠,吴国雄,等. NWP模式中纬向平均环流系统误差的动力诊断[J]. 气象学报, 1995, 54: 569-579.

[71] PEIXOTO J P, OORT A H. 气候物理学[M]. 吴国雄,刘辉,译. 北京: 气象出版社, 1995.

[72] WU G X, LIU H, CHEN F, et al. Wave-mean flow interaction and formation of blocking- persistent anomalous weather in China in the summer of 1980[J]. Acta Meteorological Sinica, 1995, 9: 215-227.

[73] WU G X, ZHANG X H. Development of coupled models at LASG[C]//Atmospheric Ozone as a

Climate Gas. Berlin: Springer, 1995.

[74] ZHANG S Q, LIU H Z, WU G X. Diagnosis of NWP systematic forecast errors in zonal mean circulation[J]. Acta Meteorologica Sinica, 1995, 9: 288-301.

1996

[75] 黄荣辉, 郭其蕴, 吴国雄. 中国气候灾害的分布和变化[M]. 北京: 气象出版社, 1996.

[76] 王晓春, 吴国雄. 利用空间均匀网格对中国夏季降水异常区域特征的初步分析[J]. 气象学报, 1996, 54(3): 324-332.

[77] 吴国雄. 气候系统研究中的几个问题[C]//现代大气科学前沿与展望. 北京: 气象出版社, 1996.

[78] 吴国雄, 刘辉, 赵宇澄. 夏季东北亚阻塞高压形成的一种机制[C]//灾害性气候的过程及诊断. 北京: 气象出版社, 1996.

[79] 吴国雄, 王敬方. 热带和热带外海表温度异常与低空环流特征比较[J]. 气象学报, 1996, 54: 385-397.

[80] LIU H, JIN X Z, ZHANG X H, et al. A Coupling experiment of an atmosphere and an ocean model with a monthly anomaly exchange scheme[J]. Advances in Atmospheric Sciences, 1996, 13: 133-146.

[81] LIU H, WU G X, ZENG Q C. On the maintenance of blocking anticyclones of Northern Hemisphere part I: Quasi-geostrophic potential vorticity analysis[J]. Acta Meteorologica Sinica, 1996, 10(2): 142-147.

[82] WU G X, CAI Y P, TANG X J. Conservation of moist potential vorticity and down-sliding slantwise vorticity development[J]. Acta Meteorologica Sinica, 1996, 10(4): 399-418.

[83] WU G X, LIU H, ZHAO Y C, et al. A nine-layer atmospheric general circulation model and its performance[J]. Advances in Atmospheric Sciences, 1996, 13(1): 1-18.

[84] WU G X, ZHANG X H. Research in China on climate and its variability[J]. Theoretical and Applied Climatology, 1996, 55: 3-18.

[85] WU G X, ZHU B Z, GAO D Y. The impacts of Tibetan Plateau on local and regional climate [C]// Atmospheric Circulation to Global Change-Celebration of the 80th Birthday of Professor Ye Duzheng. Beijing: China Meteorological Press, 1996: 425-440.

1997

[86] 江灏, 王可丽, 吴国雄. 青藏高原地区地表涡度及其取值对大气长波辐射冷却的影响[J]. 高原气象, 1997, 16(3): 250-257.

[87] 王敬方, 吴国雄. 持续性东北冷夏的变化规律及相关特征[J]. 大气科学, 1997, 21(5): 523-532.

[88] 王晓春, 吴国雄. 中国夏季降水异常空间模与副热带高压的关系[J]. 大气科学, 1997, 21(2): 161-169.

[89] 吴国雄. 瞬变涡动输送在旱涝形成中的作用[C]//长江黄河流域旱涝规律和成因研究. 济南: 山东科学技术出版社, 1997.

[90] 吴国雄, 蔡雅萍. 风垂直切变和下滑倾斜涡度发展[J]. 大气科学, 1997, 21(3): 273-282.

[91] 吴国雄, 李伟平, 郭华, 等. 青藏高原感热气泵和亚洲夏季风[C]//叶笃正. 赵九章纪念文集. 北京: 科学出版社, 1997: 116-126.

[92] 吴国雄, 王敬方. 冬季中高纬 500 hPa 高度和海表温度异常特征及其现关分析[J]. 气象学报, 1997, 55: 11-21.

[93] 吴国雄, 张学洪, 刘辉. LASG 全球海洋-大气-陆面系统模式(GOALS/LASG)及其模拟研究[J].

应用气象学报，1997，8：15-28.

[94] LIU H, WU G X. Impacts of land surface on climate of July and onset of summer monsoon: A study with and AGCM plus SSIB[J]. Advances in Atmospheric Sciences, 1997, 14(3):273-282.

[95] WANG J F, WU G X. Evolution and characteristics of the persistent cold summer in northeast China[J]. Chinese Journal of Atmospheric Sciences, 1997, 21：295-305.

[96] WANG X C, WU G X. Regional Characteristics of summer precipitation anomalies over China[J]. Acta Meteorologica Sinica, 1997, 11：153-163.

[97] WANG X C, WU G X. The analysis of the relationship between the spatial modes of summer precipitation anomalies over China and the general circulation[J]. Chinese Journal of Atmospheric Sciences, 1997, 21：133-142.

[98] WU G X, LIU H Z. Vertical vorticity development owing to down-sliding at slantwise isentropic surface[J]. Dynamics of Atmospheres and Oceans, 1997, 27：715-743.

1998

[99] 孟文, 吴国雄. 赤道印度洋纬向季风环流和太平洋Walker环流的齿轮式耦合[J]. 中国学术期刊文摘(科技快报), 1998, 4(4)：484-487.

[100] 吴国雄. 海气相互作用研究进展——美国海气相互作用第九届学术会内容简介[J]. 气象, 1998, 24(7)：3-4.

[101] 吴国雄, 孟文. 赤道印度洋-太平洋地区海气系统的齿轮式耦合和ENSO事件——Ⅰ：资料分析[J]. 大气科学, 1998, 22(4)：470-480.

[102] 吴国雄, 孙岚, 刘辉, 等. 陆面感热和潜热输送对盛夏降水和副高分布的影响[C]//东亚季风和中国暴雨. 北京：气象出版社, 1998.

[103] 吴国雄, 王敬方. 夏季中高纬500 hPa高度和海表温度异常特征及其相关分析[J]. 气象学报, 1998, 56(1)：46-54.

[104] 吴国雄, 张永生. 青藏高原的热力和机械强迫作用以及亚洲季风的爆发——Ⅰ：爆发地点[J]. 大气科学, 1998, 22(6)：825-838.

[105] 张永生, 吴国雄. 关于亚洲夏季风爆发及北半球季节突变的物理机理的诊断分析——Ⅰ：季风爆发的阶段性特征[J]. 气象学报, 1998, 56：1-16.

[106] LIU H, ZHANG X H, WU G X. Cloud feedback on SST variability in the western equatorial Pacific in GOALS/LASG model[J]. Advances in Atmospheric Sciences, 1998, 15：410-423.

[107] WU G X, ZHANG Y S. Tibetan Plateau forcing and monsoon onset in South Asia and Southern China Sea[J]. Monthly Weather Review, 1998, 126：913-927.

[108] WU G X, ZHANG Y S. Tibetan Plateau forcing and the timing of the monsoon onset over South Asia and the South China Sea[J]. Monthly Weather Review, 1998, 126(4)：913-927.

[109] YEH T C, WU G X. The role of the heat source of the Tibetan Plateau in the general circulation[J]. Meteorology and Atmospheric Physics, 1998, 67：181-198.

1999

[110] 刘屹岷, 刘辉, 刘平, 等. 空间非均匀加热对副热带高压形成和变异的影响——Ⅱ：陆面感热与东太平洋副高[J]. 气象学报, 1999, 57(4)：385-396.

[111] 刘屹岷, 吴国雄, 刘辉, 等. 空间非均匀加热对副热带高压形成和变异的影响——Ⅲ：凝结潜热加热与南亚高压及西太平洋副高[J]. 气象学报, 1999, 57(5)：525-538.

[112] 刘屹岷, 吴国雄, 刘辉. 谱模式中负地形的处理与东亚副热带气候的模拟[J]. 大气科学, 1999, 23

(6):652-662.

[113] 孙菽芬,金继明,吴国雄. 用于GCM耦合的积雪模型的设计[J]. 气象学报,1999,57(3):38-45.

[114] 王东晓,兰健,吴国雄,等. 一个海洋环流模式伴随同化系统的初步试验[J]. 自然科学进展,1999,9(9):58-67.

[115] 王东晓,吴国雄,徐建军. 热带印度洋年代际海洋变率及其动力学解释[J]. 科学通报,1999,44(11):1226-1232.

[116] 吴国雄,刘还珠. 全型垂直涡度倾向方程和倾斜涡度发展[J]. 气象学报,1999,57(1):2-16.

[117] 吴国雄,刘屹岷,刘平. 空间非均匀加热对副热带高压带形成和变异的影响——Ⅰ:尺度分析[J]. 气象学报,1999(3):2-8.

[118] 吴国雄,张永生. 青藏高原的热力和机械强迫作用以及亚洲季风的爆发——Ⅱ:爆发时间[J]. 大气科学,1999,23(1):51-61.

[119] 张永生,吴国雄. 关于亚洲夏季风爆发及北半球季节突变的物理机理的诊断分析——Ⅱ:青藏高原及邻近地区地表感热加热的作用[J]. 气象学报,1999,57(1):57-74.

[120] 钟中,吴国雄,沙文钰. 东北太平洋辐合异常对西太平洋副高异常影响的数值试验[J]. 大气科学,1999,23(6):685-692.

[121] JIN J M, GAO X G, SOROOSHIAN S, et al. One-dimensional snow water and energy balance model for vegetated surfaces[J]. Hydrological Processes, 1999, 13(14-15):2467-2482.

[122] JIN J M, GAO X G, WU G X. Comparative analysis of physically based snowmelt models for climate simulations[J]. Journal of Climate, 1999, 12: 2643-2657.

[123] WANG D X, WU G X, XU J J. Interdecadal Variability in the Tropical Indian Ocean and its Dynamic Explanation[J]. Chinese Science Bulletin, 1999, 44: 1620-1626.

[124] WU G X. 1995—1998 China National Report on Meteorology and Atmospheric Sciences[M]. Beijing: China Meteorological Press, 1999.

[125] WU G X, SONG Z S, GAO D Y. The surface thermal feature of the Tibetan Plateau and its impacts on weather and climate[C]// China National Report on Meteorology and Atmospheric Sciences. Beijing: China Meteorological Press, 1999.

[126] XU J J, WU G X. Dynamic Features and Maintenance Mechanism of Asian Summer Monsoon Subsystem[J]. Advances in Atmospheric Sciences, 1999, 16(4):523-536.

2000

[127] 李伟平,林斌,吴国雄. 海温异常前兆指标的选取及其对我国东部夏季降水的回归分析[C]// 严重旱涝与低温的诊断分析和预测方法研究. 北京:气象出版社,2000.

[128] 李伟平,吴国雄,刘辉. 地表反照率的改变影响夏季北非副高的数值模拟[J]. 气象学报,2000,58(1):22-33.

[129] 刘平,吴国雄. 1998年夏季长江流域降水异常研究——热带环流Ⅱ:数值试验[C]// 严重旱涝与低温的诊断分析和预测方法研究. 北京:气象出版社,2000.

[130] 刘平,吴国雄,李伟平. 副热带高压带的三维结构特征[J]. 大气科学,2000,24(5):577-584.

[131] 刘平,吴国雄,刘还珠. 1998年夏季长江流域降水异常研究——热带环流Ⅰ:资料分析[C]// 严重旱涝与低温的诊断分析和预测方法研究. 北京:气象出版社,2000.

[132] 刘屹岷,吴国雄. 副热带高压研究回顾及对几个基本问题的再认识[J]. 气象学报,2000,58(4):500-512.

[133] 刘屹岷,吴国雄,颜金辉,等. 东亚季风区凝结潜热加热与南亚高压及西太平洋副热带高压[C]// 严重旱涝与低温的诊断分析和预测方法研究. 北京:气象出版社,2000.

[134] 毛江玉, 吴国雄. 赤道印度洋海温变化特征及其与大气环流的关系[C]//严重旱涝与低温的诊断分析和预测方法研究. 北京: 气象出版社, 2000.

[135] 孟文, 吴国雄. 赤道印度洋-太平洋地区海气系统的齿轮式耦合和ENSO事件Ⅱ. 数值模拟[J]. 大气科学, 2000, 24(1): 15-25.

[136] 孙岚, 吴国雄, 孙菽芬. 陆面过程对气候影响的数值模拟——SSiB与IAP/LASG L9R15 AGCM耦合及其模式性能[J]. 气象学报, 2000, 58(2): 179-193.

[137] 王宝灵, 谢金南, 吴国雄, 等. 青藏高原东北侧汛期降水若干问题研究[J]. 大气科学, 2000, 24(6): 775-784.

[138] 王东晓, 吴国雄, 朱江, 等. 大洋风生环流观测优化的伴随分析[J]. 中国科学, 2000, 30(1): 97-106.

[139] 吴国雄, 刘平, 刘屹岷, 等. 印度洋海温异常对西太副高的影响——大气中的两级热力适应[J]. 气象学报, 2000, 58(5): 513-522.

[140] 吴国雄, 刘屹岷. 热力适应、过流、频散和副高 Ⅰ. 热力适应和过流[J]. 大气科学, 2000, 24(4): 433-446.

[141] 吴国雄, 尉艺, 刘辉. 东亚持续强冬季风影响赤道海表温度初始异常的数值试验研究[J]. 气象学报, 2000, 55(6): 641-652.

[142] 颜金辉, 刘屹岷, 吴国雄. 中国区域降水与西太平洋副热带高压的关系[C]//严重旱涝与低温的诊断分析和预测方法研究. 北京: 气象出版社, 2000.

[143] 张琼, 吴国雄. 长江流域大范围旱涝与南亚高压及海温异常的关系[C]//严重旱涝与低温的诊断分析和预测方法研究. 北京: 气象出版社, 2000.

[144] 祝同文, 何金海, 吴国雄. 东亚季风指数及其与大尺度热力环流年际变化关系[J]. 气象学报, 2000, 58(4): 391-402.

[145] LI W P, LIU H, WU G X. A numerical simulation study of the impact of surface albedo on the summertime North Africa subtropical high[C]// IAP Global Ocean-Atmosphere-Land System Model. Beijing: Science Press, 2000.

[146] MENG W, WU G X. Gearing between the Indo-Monsoon circulation/Pacific-Walker circulation and ENSO, Part II: simulation[J]. Chinese Journal of Atmospheric Sciences, 2000, 23: 326-336.

[147] WANG D X, WU G X, ZHU J, et al. Analysis on observing optimization for the wind-driven circulation by an adjoint approach[J]. Science in China (Series D), 2000, 43: 243-252.

2001

[148] 黄立文, 吴国雄, 宇如聪, 等. 海洋风暴形成的一种动力学机制[J]. 气象学报, 2001, 59(6): 674-684.

[149] 李伟平, 吴国雄, 刘屹岷, 等. 青藏高原表面过程对夏季青藏高压的影响——数值实验[J]. 大气科学, 2001, 25: 809-816.

[150] 刘新, 吴国雄, 李伟平, 等. 夏季青藏高原加热和大尺度流场的热力适应[J]. 自然科学进展, 2001, 11(1): 33-39.

[151] 刘屹岷, 吴国雄, 宇如聪, 等. 热力适应、过流、频散和副高 Ⅱ. 水平非均匀加热与能量频散[J]. 大气科学, 2001, 24(4): 317-328.

[152] 孙岚, 吴国雄. 陆面蒸散对气候变化的影响[J]. 中国科学, 2001, 31(1): 59-69.

[153] 吴国雄. 海洋-大气-陆地相互作用与水分循环[C]//中国科学基金. 学科前沿与国家自然科学基金优先资助领域战略国际研讨会论文集. 北京: 高等教育出版社, 2001: 174-177.

[154] 吴国雄. 全型涡度方程和经典涡度方程比较[J]. 气象学报, 2001, 59(4): 385-392.

[155] 余晖, 吴国雄. 湿斜压性与热带气旋强度突变[J]. 气象学报, 2001, 59(4): 440-449.

[156] 张琼, 吴国雄. 长江流域大范围旱涝与南亚高压的关系[J]. 气象学报, 2001, 59(5): 569-577.

[157] LIU P, WASHINGTON W M, MEEHL G A, et al. Historical and future trends of the Sahara desert[J]. Geophysical Research Letters, 2001, 28: 2683-2686.

[158] LIU P, WU G X, SUN S F. Local meridional circulation and deserts[J]. Advances in Atmospheric Sciences, 2001, 18, 864-872.

[159] LIU X, WU G X, LI W P, et al. Thermal adaptation of the large-scale circulation to the summer heating over the Tibetan Plateau[J]. Progress in Natural Science, 2001, 11: 207-214.

[160] LIU Y M, WU G X, LIU H, et al. Dynamical effects of condensation heating on the subtropical anticyclones in the eastern hemisphere[J]. Climate Dynamics, 2001, 17: 327-338.

[161] SUN L, WU G X. Influence of land evapotranspiration on climate[J]. Science in China, 2001, 44: 838-846.

[162] SUN L, WU G X, SUN S F. Numerical simulations of land surface processes on climate-implementating of SSiB in IAP/LASG AGCM and its performance[J]. Acta Meteorologica Sinica, 2001, 15(2): 160-177.

[163] YU Y X, WU G X, WANG B L, et al. Water vapor content and mean transfer in the atmosphere over northwest China[J]. Acta Meteorologica Sinica, 2001, 15: 191-206.

2002

[164] 崔晓鹏, 吴国雄, 高守亭. 西大西洋锋面气旋过程的数值模拟和等熵分析[J]. 气象学报, 2002, 60(4): 385-399.

[165] 刘新, 李伟平, 吴国雄. 夏季青藏高原加热和北半球环流年际变化的相关分析[J]. 气象学报, 2002, 60(3): 267-277.

[166] 刘新, 吴国雄, 刘屹岷, 等. 青藏高原加热与亚洲环流季节变化和夏季风爆发[J]. 大气科学, 2002, 26(6): 781-793.

[167] 毛江玉, 吴国雄, 刘屹岷. 季节转换期间副热带高压带形态变异及其机制的研究Ⅰ:副热带高压结构的气候学特征[J]. 气象学报, 2002, 60(4): 400-408.

[168] 毛江玉, 吴国雄, 刘屹岷. 季节转换期间副热带高压带形态变异及其机制的研究Ⅱ:亚洲季风区季节转换指数[J]. 气象学报, 2002, 60(4): 409-420.

[169] 毛江玉, 吴国雄, 刘屹岷. 季节转换期间副热带高压带形态变异及其机制的研究Ⅲ:热力学诊断[J]. 气象学报, 2002, 60(4): 647-659.

[170] 王可丽, 吴国雄, 江灏, 等. 青藏高原云-辐射-加热效应和南亚夏季风——1985年与1987年对比分析[J]. 气象学报, 2002, 60(2): 173-179.

[171] 吴国雄, 丑纪范, 刘屹岷, 等. 副热带高压形成和变异的动力学问题[M]. 北京: 科学出版社, 2002.

[172] 吴国雄, 丑纪范, 刘屹岷. 关于夏季副热带高压形成和变化研究的进展. 大气科学发展战略[M]. 北京: 气象出版社, 2002.

[173] 吴国雄, 刘新, 张琼, 等. 青藏高原抬升加热气候效应研究的新进展[J]. 气候与环境研究, 2002, 7(2): 184-201.

[174] 吴国雄, 刘屹岷, 刘平, 等. 气候纬向平均副热带高压和Hadley环流的下沉支比较[J]. 气象学报, 2002, 61(5): 635-636.

[175] 张韬, 郭裕福, 吴国雄. 冷暖事件对大气能量循环和纬向平均环流影响的模拟研究[J]. 气象学报, 2002, 60(5): 513-526.

[176] 张韬, 吴国雄, 郭裕福. 海陆气全球耦合模式能量收支的误差[J]. 气象学报, 2002, 60(3): 278-288.

[177] 张韬, 吴国雄, 郭裕福. 一个具有高分辨率海洋分量的海气耦合模式[J]. 应用气象学报, 2002, 13(6): 688-695.

[178] 周兵, 徐海明, 吴国雄, 等. 云迹风资料同化对暴雨预报影响的数值模拟[J]. 气象学报, 2002, 60(3): 308-317.

[179] CHANG C P, WU G X, JOU B, et al. Selected Papaers of the Fourth Conference on East Asia and Western Pacific Meteorology and Climate[M]. New Jersey, London, Singapore, Hong Kong: World Scientific, 2002.

[180] KANG I S, WU G X, LIU Y M. Intercomparison of Atmospheric GCM Simulated Anomalies Associated with the 1997/98 El Niño[J]. Journal of Climate, 2002, 15: 2791-2805.

[181] LIU P, MEEHL G A, WU G X. Multi-model trends in the Sahara induced by increasing CO_2[J]. Geophysical Research Letters, 2002, 29: 1-28.

[182] LIU Y M, CHAN J C L, MAO J Y, et al. The role of Bay of Bengal convection in the onset of the 1998 South China Sea summer monsoon[J]. Monthly Weather Review, 2002, 130: 2731-2744.

[183] MAO J Y, WU G X, LIU Y M. Study on the variation in the configuration of subtropical anticyclone and its mechanism during seasonal transition—Part II: Seasonal transition indices over the Asian monsoon region[J]. Acta Meteorologica Sinica, 2002, 16: 17-32.

[184] MAO J Y, WU G X, LIU Y M. Study on the variation in the configuration of subtropical anticyclone and its mechanism during seasonal transition—Part III: Thermodynamica diagnoses[J]. Acta Meteorologica Sinica, 2002, 16: 33-50.

[185] WU G X, SUN Y M, LIU H, et al. Impacts of land surface processes on summer climate[C]// Selected Papers of the Fourth Conference on East Asia and Western Pacific Meteorology and Climate. New Jersey, London, Singapore, Hong Kong: World Scientific, 2002.

[186] ZHANG Q, WU G X, QIAN Y F. The bimodality of the 100 hPa South Asia high and its relationship to the climate anomaly over East Asia in summer[J]. Journal of the Meteorological Society of Japan, 2002, 80: 733-744.

[187] ZHANG T, GUO Y F, WU G X. Analysis of the zonal mean atmospheric climate state in IAP/LASG GOALS model simulations[J]. Advances in Atmospheric Sciences, 2002, 19(6): 1091-1102.

[188] ZHANG Y, LI T, WANG B, et al. Onset of the summer monsoon over the Indochina Peninsula: Climatology and interannual variations[J]. Journal of Climate, 2002, 15: 3206-3221.

2003

[189] 段安民, 刘屹岷, 吴国雄. 4—6月青藏高原热状况与盛夏东亚降水和大气环流的异常[J]. 中国科学D辑, 2003, 33(10): 997-1004.

[190] 段安民, 吴国雄. 7月青藏高原大气热源空间型及其与东亚大气环流和降水的相关研究[J]. 气象学报, 2003, 60(4): 447-455.

[191] 刘屹岷, 陈仲良, 毛江玉, 等. 孟加拉湾季风爆发对南海季风爆发的影响I: 个例分析[J]. 气象学报, 2003, 61(1): 1-9.

[192] 刘屹岷, 陈仲良, 吴国雄. 孟加拉湾季风爆发对南海季风爆发的影响II: 数值试验[J]. 气象学报, 2003, 61(1): 10-19.

[193] 毛江玉, 段安民, 刘屹岷, 等. 副高脊面反转与亚洲夏季风爆发可预测性分析[J]. 科学通报,

2003,48(增刊2):55-59.

[194] 任荣彩,吴国雄. 1998年夏季副热带高压的短期结构特征及形成机制[J]. 气象学报,2003,61(2):180-195.

[195] 吴国雄,丑纪范,刘屹岷,等. 副热带高压研究进展及展望[J]. 大气科学,2003,27(4):503-517.

[196] 张琼,刘平,吴国雄. 印度洋和南海海温与长江中下游旱涝[J]. 大气科学,2003,27(6):992-1006.

[197] 周兵,何金海,吴国雄,等. 东亚副热带季风特征及其指数的建立[J]. 大气科学,2003,27(1):123-135.

[198] CUI X P, GAO S T, WU G X. Moist potential vorticity and up-sliding slantwise vorticity development[J]. Chinese Physics Letters, 2003, 20(1):167-169.

[199] CUI X P, GAO S T, WU G X. Up-sliding slantwise vorticity development and the complete vorticity equation with mass forcing[J]. Advances in Atmospheric Sciences, 2003, 20(5):825-836.

[200] CUI X P, WU G X, GAO S T. An experiment study of frontal cyclones over the western Atlantic Ocean[J]. Acta Meteorologica Sinica, 2003, 17:321-336.

[201] HUANG Z C, WU G X, CHENG Z L. Research on causes and prevention methods of water damage to Sichuan-Tibet highway[J]. Bulletin of Soil and Water Conservation, 2003, 23(4):21-23,27.

[202] MAO J Y, WU G X, LIU Y M. Study on the variation in the configuration of subtropical anticyclone and its mechanism during seasonal transition—Part Ⅰ: Climatological features of subtropical high structure[J]. Acta Meteorologica Sinica, 2003, 17:274-286.

[203] WALISER D E, JIN K, KANG I S, et al. AGCM simulations of intraseasonal variability associated with the Asian summer monsoon[J]. Climate Dynamics, 2003, 21(5):423-446.

[204] WU G X, LIU Y M. Summertime quadruplet heating pattern in the subtropics and the associated atmospheric circulation[J]. Geophysical Research Letters, 2003,30(5).

[205] WU G X, YAO L K, YI Z J. Research on forming mechanism of inchoate cracks in cement concrete pavement[J]. Journal of Southwest Jiaotong University, 2003, 38(3):304-308.

[206] WU T W, LIU P, WANG Z Z, et al. The performance of atmospheric component model R42L9 of GOALS/LASG[J]. Advances in Atmospheric Sciences, 2003, 20(5):726-742.

[207] ZHANG T, WU G X, GUO Y F. Energy budget bias in global coupled ocean-atmosphere-land system model[J]. Acta Meteorologica Sinica, 2003, 17:287-306.

2004

[208] 段安民,毛江玉,吴国雄. 孟加拉湾季风爆发可预测性的分析和初步应用[J]. 高原气象,2004,23(1):18-25.

[209] 任荣彩,刘屹岷,吴国雄. 中高纬环流对1998年7月西太平洋副热带高压短期变化的影响机制[J]. 大气科学,2004,29(1):55-66.

[210] 吴国雄. 我国青藏高原气候动力学研究的近期进展[J]. 第四纪研究,2004,24(1):1-9.

[211] 吴国雄,刘屹岷,任荣彩,等. 定常态副热带高压与垂直运动的关系[J]. 气象学报,2004,62(5):587-597.

[212] 吴国雄,毛江玉,段安民,等. 青藏高原影响亚洲夏季气候研究的最新进展[J]. 气象学报,2004,62(5):528-540.

[213] 吴统文,吴国雄,王在志,等. GOALS/LASG模式对气候平均态的模拟[J]. 气象学报,2004,62(1):20-30.

[214] 吴统文,吴国雄,宇如聪. GOALS模式对热带太平洋ENSO年际变化特征的模拟评估[J]. 气象学

报,2004,62(2):154-166.

[215] DAI X G, WANG P, WU G X, et al. Teleconnection between Indian Monsoon and East Asian circulation[J]. Acta Meteorologica Sinica, 2004, 18(4): 397-410.

[216] DUAN A M, WU G X. Main heating modes over the Tibetan Plateau in July and the correlation patterns of circulation and precipitation over East Asia[J]. Acta Meteorologica Sinica, 2004, 18(2): 167-178.

[217] LIU Y M, WU G X, REN R C. Relationship between the subtropical anticyclone and diabatic heating[J]. Journal of Climate, 2004, 17: 682-698.

[218] LIU Y M, WU G X. Progress in the study on the formation of the summertime subtropical anticyclone[J]. Advances in Atmospheric Sciences, 2004, 21(3):322-342.

[219] MAO J Y, CHAN J, WU G X. Relationship between the onset of the South China Sea Summer monsoon and the structure of the Asian subtropical anticyclone[J]. Journal of the Meteorological Society of Japan, 2004, 82(3):845-859.

[220] WANG Z Z, WU G X, WU T W, et al. Simulation of Asian monsoon seasonal variations with climate model R42L9/LASG[J]. Advances in Atmospheric Sciences, 2004, 21(6):879-889.

[221] WU G X. Recent progress in the study of the Qinghai-Xizang Plateau climate dynamics in China[J]. Quaternary Sciences, 2004, 24(1):1-9.

[222] WU G X, LIU Y M, LIU P. Formation of the summertime subtropical anticyclone[C] // East Asian Monsoon. New Jersey, London, Singapore, Beijing, Shanghai, Hong Kong, Taipei, Chennai: World Scientific, 2004.

[223] WU G X, LIU Y M, MAO J Y, et al. Adaptation of the atmospheric circulation to thermal forcing over the Tibetan Plateau. observation, theory and modeling of atmospheric variability[C]//Selected Papers of Nanjing Institute of Meteorology Alumni in Commemoration of Professor Jijia Zhang. New Jersey, London, Singapore, Beijing, Shanghai, Hong Kong, Taipei, Chennai: World Scientific, 2004:92-114.

[224] WU T W, WANG Z Z, LIU Y M, et al. An evaluation of the effects of cloud parameterization in the R42L9 GCM[J]. Advances in Atmospheric Sciences, 2004, 21(2):153-162.

[225] WU T W, WU G X. An empirical formula to compute snow cover fraction in GCMs[J]. Advances in Atmospheric Sciences, 2004, 21(4):529-535.

2005

[226] 段安民,吴国雄.青藏高原气温的年际变率与大气环状波动模[J].气象学报,2005,63(5):790-798.

[227] 黄立文,吴国雄,宇如聪.中尺度海气相互作用对台风暴雨过程的影响[J].气象学报,2005,63(4):455-467.

[228] 梁潇云,刘屹岷,吴国雄.青藏高原隆升对春、夏季亚洲大气环流的影响[J].高原气象,2005,24(6):837-845.

[229] 梁潇云,刘屹岷,吴国雄.青藏高原对亚洲夏季风爆发位置及强度的影响[J].气象学报,2005,63(5):799-805.

[230] 毛江玉,吴国雄.1991年江淮梅雨与副热带高压的低频振荡[J].气象学报,2005,63(5):762-770.

[231] 王在志,吴国雄,刘平,等.全球海-陆-气耦合模式大气模式分量的发展及其气候模拟性能Ⅰ——水平分辨率的影响[J].热带气象学报,2005,21(3):225-237.

[232] 王在志,宇如聪,王鹏飞,等. 全球海-陆-气耦合模式大气模式分量的发展及其气候模拟性能 Ⅱ——垂直分辨率的提高及其影响[J]. 热带气象学报,2005,21(3):238-247.

[233] 吴国雄,刘屹岷,刘新,等. 青藏高原加热如何影响亚洲夏季的气候格局[J]. 大气科学,2005,29(1):47-56.

[234] 吴国雄,王军,刘新,等. 欧亚地形对不同季节大气环流影响的数值模拟研究[J]. 气象学报,2005,63(5):603-612.

[235] 赵兵科,姚秀萍,吴国雄. 2003年夏季淮河流域梅雨期西太平洋副高结构和活动特征及动力机制分析[J]. 大气科学,2005,29(5):771-779.

[236] 周天军,王在志,宇如聪,等. 基于LASG/IAP大气环流谱模式的气候系统模式[J]. 气象学报,2005,63(5):702-715.

[237] 周秀骥,吴国雄,丁一汇. 推动创新研究,繁荣大气科学——纪念《气象学报》创刊80周年[J]. 气象学报,2005,63(5):541-542.

[238] DUAN A M,LIU Y M,WU G X. Heating Status of the Tibetan Plateau from April to June and rainfall and atmospheric circulation anomaly over East Asia in midsummer[J]. Science in China (D),2005,48(2):250-257.

[239] DUAN A M,WU G X. Role of the Tibetan Plateau thermal forcing in the summer climate patterns over subtropical Asia[J]. Climate Dynamics,2005,24:793-807.

[240] DUAN A M,WU G X. Wave-mean flow interaction and its relationship with the atmospheric energy cycle with diabatic heating[J]. Science in China (D),2005,48(8):1293-1302.

[241] LIANG X Y,LIU Y M,WU G X. The role of land-sea distribution in the formation of the Asian summer monsoon[J]. Geophysical Research Letters,2005,32(3).

[242] XUE Y K,SUN S F,LAU K M,et al. Multiscale variability of the river runoff system in China and its long-term link to precipitation and sea surface temperature[J]. Journal of Hydrometeorology,2005,6(4):550-570.

[243] ZHANG T,WU G X,GUO Y F. The diabatic heating and the generation of available potential energy: Results from NCEP reanalysis[J]. Acta Meteorologica Sinica,2005,19(2):143-159.

2006

[244] 包庆,刘屹岷,周天军,等. LASG/IAP大气环流谱模式对陆面过程的敏感性试验[J]. 大气科学,2006,30(6):1077-1090.

[245] 段安民,吴国雄,刘屹岷. 定常条件下感热和地形影响的Rossby波[J]. 气象学报,2006,64(2):129-136.

[246] 段安民,吴国雄,张琼,等. 青藏高原气候变暖是温室气体排放加剧结果的新证据[J]. 科学通报,2006,51(8):989-992.

[247] 梁潇云,刘屹岷,吴国雄. 热带、副热带海陆分布与青藏高原在亚洲夏季风形成中的作用[J]. 地球物理学报,2006,49(4):983-992.

[248] 刘新,吴国雄,李伟平. 夏季青藏高原加热和环流场的日变化[J]. 地球科学进展,2006,21(12):69-78.

[249] 毛江玉,吴国雄. 青藏高原热状况和海温异常对亚洲季风季节转换年际变化的影响[J]. 地球物理学报,2006,49(5):1279-1287.

[250] 万日金,吴国雄. 江南春雨的气候成因机制研究[J]. 中国科学D辑,2006,36(10):936-950.

[251] 吴国雄,李建平,周天军,等. 影响我国短期气候异常的关键区——"亚印太交汇区"[J]. 地球科学进展,2006,21(11):1109-1118.

[252]张韬,吴国雄,郭裕福. GOALS模式中大气能量循环的诊断分析与不同版本计算结果的比较研究[J]. 大气科学, 2006, 30(1): 38-55.

[253]赵瑞霞,吴国雄. 黄河流域中上游水分收支以及再分析资料可用性分析[J]. 自然科学进展, 2006, 16(3): 316-324.

[254]周兵,吴国雄,梁潇云. 孟加拉湾深对流加热对东亚季风环流系统的影响[J]. 气象学报, 2006, 64(1): 48-56.

[255]DUAN A M, WU G X, ZHANG Q, et al. New proofs of the recent climate warming over the Tibetan Plateau as a result of the increasing greenhouse gases emissions[J]. Chinese Science Bulletin, 2006, 51(11): 1396-1400.

[256]FU Y F, LIU G S, WU G X, et al. Tower mast of precipitation over the central Tibetan Plateau summer[J]. Geophysical Research Letters, 2006, 33(5).

[257]GOSWAMI B N, WU G X, YASUNARI T. The annual cycle, intraseasonal oscillations, and roadblock to seasonal predictability of the Asian summer monsoon[J]. Journal of Climate, 2006, 19(20): 5078-5099.

[258]MAO J Y, WU G X. Intraseasonal variations of the Yangtze rainfall and its related atmospheric circulation features during the 1991 summer[J]. Climate Dynamics, 2006, 27(7-8): 815-830.

[259]REN F M, WU G X, DONG W J, et al. Changes in tropical cyclone precipitation over China[J]. Geophysical Research Letters, 2006, 33(20).

[260]WAN R J, WU G X. Mechanism of the Spring Persistent Rains over southeastern China[J]. Sciences in China D, 2007, 50: 130-144.

[261]WU G X, MAO J Y, DUAN A M, et al. Current progresses in study of impacts of the Tibetan Plateau on Asian summer climate[J]. Acta Meteorologica Sinica, 2006, 20(2): 144-158.

[262]YANAI M, WU G X. Effects of the Tibetan Plateau[C] // The Asian Monsoon. Chichester: Springer, 2006.

2007

[263]洪梅,张韧,吴国雄,等. 用遗传算法重构副热带高压特征指数的非线性动力模型[J]. 大气科学, 2007, 31: 346-352.

[264]刘新,王军,吴国雄,等. 欧亚地形对夏季南亚大气环流日变化影响的数值模拟研究[J]. 大气科学, 2007, 31(3): 389-399.

[265]刘屹岷,刘琨,吴国雄. 积云对流参数化方案对大气含水量及降水的影响[J]. 大气科学, 2007, 32(6): 1201-1211.

[266]任荣彩,刘屹岷,吴国雄. 1998年7月南亚高压影响西太平洋副热带高压短期变异的过程和机制[J]. 气象学报, 2007, 65(2): 183-197.

[267]任素玲,刘屹岷,吴国雄. 西太平洋副热带高压和台风相互作用的数值试验研究[J]. 气象学报, 2007, 65(3): 329-340.

[268]王在志,宇如聪,包庆,等. 大气环流模式(SAMIL)海气耦合前后性能的比较[J]. 大气科学, 2007, 32: 1-12.

[269]姚秀萍,吴国雄,刘屹岷,等. 热带对流层上空东风带扰动影响西太平洋副热带高压的个例分析[J]. 气象学报, 2007, 65(2): 198-207.

[270]姚秀萍,吴国雄,于玉斌,等. 与梅雨锋上低涡降水相伴的干侵入研究[J]. 中国科学D, 2007, 37(3): 417-428.

[271]赵瑞霞,吴国雄. 长江流域水分收支以及再分析资料可用性分析[J]. 气象学报, 2007, 65(3):

416-427.

[272] LIU Y M, BAO Q, DUAN A M, et al. Recent progress in the impact of the Tibetan plateau on climate in China[J]. Advances in Atmospheric Sciences, 2007, 24(6):1060-1076.

[273] MAO J Y, WU G X. Interannual Variability in the onset of the summer monsoon over the eastern Bay of Bengal[J]. Theoretical and Applied Climatology, 2007, 89: 155-170.

[274] WU G X, LIU Y M, WANG T M, et al. The influence of mechanical and thermal forcing by the Tibetan Plateau on Asian climate[J]. Journal of Hydrometeorology, 2007, 8(4):770-789.

[275] YAO X P, WU G X, ZHAO B K, et al. Research on the dry intrusion accompanying the low vortex precipitation[J]. Science in China Series D-Earth Sciences, 2007, 50(9):1396-1408.

[276] ZHAO B K, WU G X, YAO X P. Potential vorticity structure and inversion of the cyclogenesis over the Yangtze River and Huaihe River valleys[J]. Advances in Atmospheric Sciences, 2007, 24(1):44-54.

2008

[277] 包庆, WANG B, 刘屹岷, 等. 青藏高原增暖对东亚夏季风的影响——大气环流模式数值模拟研究[J]. 大气科学, 2008, 32(5):997-1005.

[278] 万日金, 王同美, 吴国雄. 江南春雨和南海副热带高压的时间演变及其与东亚夏季风环流和降水的关系[J]. 气象学报, 2008, 66(5):800-807.

[279] 万日金, 吴国雄. 江南春雨的时空分布[J]. 气象学报, 2008, 66(3): 310-319.

[280] 王同美, 吴国雄, 万日金. 青藏高原的热力和动力作用对亚洲季风环流的影响[J]. 高原气象, 2008, 27: 1-9.

[281] 王同美, 吴国雄. 南亚海陆热力差异及其对热带季风区环流的影响[J]. 热带气象学报, 2008, 27(1):37-43.

[282] 吴国雄, 刘屹岷, 宇婧婧, 等. 海陆分布对海气相互作用的调控和副热带高压的形成[J]. 大气科学, 2008, 32(4):720-740.

[283] 姚秀萍, 吴国雄, 刘还珠. 与2003年梅雨期西太副高东西向运动有关的热带上空东风带扰动的结构和演变特征[J]. 热带气象学报, 2008, 24(1):20-26.

[284] 赵兵科, 吴国雄, 姚秀萍. 2003年夏季梅雨期一次强气旋发展的位涡诊断分析[J]. 大气科学, 2008, 32(6): 1241-1255.

[285] 赵瑞霞, 吴国雄, 张宏. 夏季风期间长江流域的水汽输送状态及其年际变化[J]. 地球物理学报, 2008, 51(6): 1670-1681.

[286] DUAN A M, WU G X, LIANG X Y. Influence of the Tibetan Plateau on the summer climate patterns over Asia in the IAP/LASG SAMIL model[J]. Advances in Atmospheric Sciences, 2008, 25(4):518-528.

[287] DUAN A M, WU G X. Weakening trend in the atmospheric heat source over the Tibetan Plateau during recent decades. Part I: Observations[J]. Journal of Climate, 2008, 21(13):3149-3164.

[288] LAU K M, RAMANATHAN V, WU G X, et al. The joint aerosol—monsoon experiment: A new challenge for monsoon climate research[J]. Bulletin of the American Meteorological Society, 2008, 89(3): 369-384.

[289] REN S L, LIU Y M, WU G X. Interaction between typhoon and western Pacific subtropical anticyclone: data analyses and numerical experiments[J]. Acta Meteo Sinica, 2008, 22: 329-341.

[290] WAN R J, WANG T M, WU G X. Temporal variations of the spring persistent Rains and South China Sea sub-high and their correlations to the circulation and precipitation of the East Asian

summer monsoon[J]. Acta Meteorologica Sinica, 2008, 22(4):530-537.

[291] WANG B, BAO Q, HOSKINS B, et al. Tibetan Plateau warming and precipitation changes in East Asia[J]. Geophysical Research Letters, 2008, 35(14).

[292] WANG B, WU Z, LI J, et al. How to measure the strength of the East Asian summer monsoon [J]. Journal of Climate, 2008, 21(17):4449-4463.

[293] WANG T M, WU G X. Land-sea thermal contrast over South Asia and its influences on tropical monsoon circulation[J]. Journal of Tropical Meteorology, 2008, 14(1):77-80.

[294] WANG Z Z, MAO J Y, WU G X. The wavenumber-frequency characteristics of the tropical waves in an aqua-planet GCM[J]. Advances in Atmospheric Sciences, 2008, 25(4):541-554.

[295] YAO X P, WU G X, LIU H Z. Structural and evolution characteristics of the easterly vortex over the tropical region[J]. J Trop Meteor, 2008, 14(1):85-88.

[296] ZENG H L, JI J J, WU G X. An updated coupled model for land-atmosphere interaction. Part II: Simulations of biological processes[J]. Advances in Atmospheric Sciences, 2008, 25(4):632-640.

[297] ZENG H L, WANG Z Z, JI J J, et al. An updated coupled model for land-atmosphere interaction. Part I: simulations of physical processes[J]. Advances in Atmospheric Sciences, 2008, 25(4):619-631.

2009

[298] 李剑东, 刘屹岷, 孙治安, 等. 辐射和积云对流过程对大气辐射通量的影响[J]. 气象学报, 2009, 67:355-369.

[299] 孙长, 毛江玉, 吴国雄. 大气季节内振荡对夏季西北太平洋热带气旋群发性的影响[J]. 大气科学, 2009, 33(5):950-958.

[300] 王同美, 吴国雄, 宇婧婧. 春季青藏高原加热异常对亚洲热带环流和季风爆发的影响[J]. 热带气象学报, 2009, 25(s1):92-102.

[301] 张亚妮, 刘屹岷, 吴国雄. 线性准地转模型中副热带环流对潜热加热的定常响应Ⅰ. 基本性质及特征分析[J]. 大气科学, 2009, 33:868-878.

[302] 张亚妮, 刘屹岷, 吴国雄. 线性准地转模型中副热带环流对潜热加热的定常响应Ⅱ. 边界约束及风场与层结稳定度的自适应[J]. 大气科学, 2009, 33:891-902.

[303] DUAN A M, SUI C H, WU G X. Local air-sea interaction in intertropical convergence zone simulations[J]. Journal of Geophysical Research-Atmospheres, 2009, 114(22).

[304] DUAN A M, WU G X. Weakening trend in the atmospheric heat source over the Tibetan Plateau during recent decades. Part Ⅱ: Connection with climate warming[J]. Journal of Climate, 2009, 22(15):4197-4212.

[305] LI J D, LIU Y M, WU G X. Cloud radiative forcing in Asian monsoon region simulated by IPCC AR4 AMIP models[J]. Advances in Atmospheric Sciences, 2009, 26(5):923-939.

[306] MAO J Y, WU G X. Intraseasonal modulation of tropical cyclogenesis in the Western North Pacific: A case study[J]. Theoretical and Applied Climatology, 2009, 100:397-411.

[307] REN R C, WU G X, MING C, et al. Winter season stratospheric circulation in the SAMIL/LASG general circulation model[J]. Advances in Atmospheric Sciences, 2009, 26(3):451-464.

[308] WAN R J, WU G X. Temporal and spatial distributions of the spring persistent rains over southeastern China[J]. Acta Meteorologica Sinica, 2009, 23(5):598-608.

[309] WAN R J, ZHAO B K, WU G X. New evidences on the climatic causes of the formation of the spring persistent rains over southeastern China[J]. Advances in Atmospheric Sciences, 2009,

26(6):1081-1087.

[310] WU G X, LIU Y M, ZHU X, et al. Multi-scale forcing and the formation of subtropical desert and monsoon[J]. Annals Geophysics, 2009, 27: 3631-3644.

[311] YAO X P, WU G X, LIU Y M, et al. Case study on the impact of the vortex in the easterlies in the tropical upper troposphere on the western pacific subtropical anticyclone[J]. Acta Meteorologica Sinica, 2009, 23(3): 363-373.

2010

[312] 白莉娜, 任福民, 宋金杰, 等. 潜热通量异常对西北太平洋热带气旋活动影响的机理研究[J]. 海洋学报(中文版), 2010, 32(4): 32-40.

[313] 曾红玲, 季劲钧, 吴国雄. 全球植被分布对气候影响的数值试验[J]. 大气科学, 2010, 34(1): 1-11.

[314] 李剑东, 刘屹岷, 吴国雄. 积云对流和云物理过程调整对气候模拟的影响[J]. 大气科学, 2010, 34(5): 891-904.

[315] 刘琨, 刘屹岷, 吴国雄. SAMIL模式中Tiedtke积云对流方案对热带降水模拟的影响[J]. 大气科学, 2010, 34(1): 163-174.

[316] 吴国雄, 关月, 王同美, 等. 春季孟加拉湾涡旋形成及其对亚洲夏季风爆发的激发作用[J]. 中国科学, 2010, 40(11): 1459-1467.

[317] BAO Q, LIU Y M, SHI J C, et al. Comparisons of soil moisture datasets over the Tibetan Plateau and application to the simulation of Asia summer monsoon onset[J]. Advances in Atmospheric Sciences, 2010, 27(2): 303-314.

[318] BAO Q, WU G X, LIU Y M, et al. An introduction to the coupled model FGOALS1.1-s and its performance in East Asia[J]. Advances in Atmospheric Sciences, 2010, 27(5): 1131-1142.

[319] BAO Q, YANG J, LIU Y M, et al. Roles of Anomalous Tibetan Plateau warming on the severe 2008 winter storm in central-southern China[J]. Monthly Weather Review, 2010, 138(6): 2375-2384.

[320] LI J, WU G X. Atmospheric angular momentum transport and balance in the AGCM-SAMIL[J]. Advances in Atmospheric Sciences, 2010, 27(5): 1183-1192.

[321] LIU Y M, GUO L, WU G X, et al. Sensitivity of ITCZ configuration to cumulus convective parameterizations on an aqua planet[J]. Climate Dynamics, 2010, 34(2-3): 223-240.

[322] MAO J Y, SUN Z, WU G X. 20—50-day oscillation of summer Yangtze rainfall in response to intraseasonal variations in the subtropical high over the western North Pacific and South China Sea[J]. Climate Dynamics, 2010, 34(5): 747-761.

2011

[323] 宇婧婧, 刘屹岷, 吴国雄. 冬季青藏高原上空热状况的分析——I 气候平均[J]. 气象学报, 2011, 69(1): 79-88.

[324] 宇婧婧, 刘屹岷, 吴国雄. 冬季青藏高原上空热状况的分析——II 年际变化[J]. 气象学报, 2011, 69(1): 89-98.

[325] DUAN A M, LI F, WANG M R, et al. Persistent weakening trend in the spring sensible heat source over the Tibetan Plateau and its impact on the Asian summer monsoon[J]. Journal of Climate, 2011, 24(21): 5671-5682.

[326] MAO J Y, WU G X. Barotropic process contributing to the formation and growth of tropical cyclone Nargis[J]. Advances in Atmospheric Sciences, 2011, 28(3): 483-491.

[327] REN F M, LIANG J, WU G X, et al. Reliability analysis of climate change of tropical cyclone activity over the Western North Pacific[J]. Journal of Climate, 2011, 24(22):5887-5898.

[328] WANG X C, BAO Q, LIU K, et al. Features of rainfall and latent heating structure simulated by two convective parameterization schemes[J]. Science China-Earth Sciences, 2011, 54(11): 1779-1788.

[329] WENG H Y, WU G X, LIU Y M, et al. Anomalous summer climate in China influenced by the tropical Indo-Pacific Oceans[J]. Climate Dynamics, 2011, 36(3-4):769-782.

[330] WU G X, GUAN Y, WANG T M, et al. Vortex genesis over the Bay of Bengal in spring and its role in the onset of the Asian summer monsoon[J]. Science China-Earth Sciences, 2011, 54(1):1-9.

[331] YING M, CHEN B D, WU G X. Climate trends in tropical cyclone-induced wind and precipitation over mainland China[J]. Geophysical Research Letters, 2011, 38(01).

2012

[332] 王军, 包庆, 刘屹岷, 等. 大气环流模式SAMIL模拟的夏季全球加热场和东亚夏季风[J]. 大气科学, 2012, 36(1): 63-76.

[333] 王晓聪, 包庆, 刘琨, 等. 两种对流参数化方案下降水和潜热加热空间结构的模拟及其影响[J]. 中国科学, 2012, 42(4): 587-598.

[334] 应明, 吴国雄, 刘屹岷, 等. 海陆热力差异对热带气旋能量源汇及其季节变化的调控[J]. 中国科学, 2012, 42(9): 1329-1345.

[335] 赵瑞霞, 张宏, 吴国雄, 等. 中国东部初春水分循环季节推进过程的年代际突变[J]. 中国科学, 2012, 42: 434-446.

[336] 竺夏英, 刘屹岷, 吴国雄. 夏季青藏高原多种地表感热通量资料的评估[J]. 中国科学：地球科学, 2012, 42(1): 1-9.

[337] DUAN A M, WU G X, LIU Y M, et al. Weather and climate effects of the Tibetan Plateau[J]. Advances in Atmospheric Sciences, 2012, 29(5):978-992.

[338] LI J D, SUN Z A, LIU Y M, et al. A study on sulfate optical properties and direct radiative forcing using LASG-IAP general circulation model[J]. Advances in Atmospheric Sciences, 2012, 29(6):1185-1199.

[339] LI Q, REN R C, CAI M, et al. Attribution of the summer warming since 1970s in Indian Ocean Basin to the inter-decadal change in the seasonal timing of El Niño decay phase[J]. Geophysical Research Letters, 2012, 39(12).

[340] MAO J Y, WU G X. Diurnal variations of summer precipitation over the Asian monsoon region as revealed by TRMM satellite data[J]. Science China-Earth Sciences, 2012, 55(4):554-566.

[341] REN F M, BAI L N, WU G X, et al. A possible mechanism of the impact of atmosphere-ocean interaction on the activity of tropical cyclones affecting China[J]. Advances in Atmospheric Sciences, 2012, 29(4):661-674.

[342] REN R C, CAI M, XIANG C Y, et al. Observational evidence of the delayed response of stratospheric polar vortex variability to ENSO SST anomalies[J]. Climate Dynamics, 2012, 38(7-8): 1345-1358.

[343] WU G X, GUAN Y, LIU Y M, et al. Air-sea interaction and formation of the Asian summer monsoon onset vortex over the Bay of Bengal[J]. Climate Dynamics, 2012, 38(1-2):261-279.

[344] WU G X, LIU Y M, DONG B W, et al. Revisiting Asian monsoon formation and change

associated with Tibetan Plateau forcing: I. Formation[J]. Climate Dynamics, 2012, 39(5): 1169-1181.

[345] WU G X, LIU Y M, HE B, et al. Thermal Controls on the Asian Summer Monsoon[J]. Scientific Reports, 2012, 2.

[346] YING M, WU G X, LIU Y M, et al. Modulation of land-sea thermal contrast on the energy source and sink of tropical cyclone activity and its annual cycle[J]. Science China-Earth Sciences, 2012, 55(11):1855-1871.

[347] ZHOU L J, LIU Y M, BAO Q, et al. Computational performance of the high-resolution atmospheric model FAMIL[J]. Atmospheric and Oceanic Science Letters, 2012, 5:355-359.

[348] ZHU X Y, LIU Y M, WU G X. An assessment of summer sensible heat flux on the Tibetan Plateau from eight data sets[J]. Science China-Earth Sciences, 2012, 55(5):779-786.

2013

[349] 吴国雄,段安民,刘屹岷,等. 关于亚洲夏季风爆发的动力学研究的若干近期进展[J]. 大气科学, 2013, 37(2): 211-228.

[350] 吴国雄,段安民,张雪芹,等. 青藏高原极端天气气候变化及其环境效应[J]. 自然杂志, 2013, 35(3):167-171.

[351] 吴国雄,郑永骏,刘屹岷. 涡旋发展和移动的动力和热力问题 II:广义倾斜涡度发展[J]. 气象学报, 2013, 71(2): 198-208.

[352] 张亚妮,吴国雄,刘屹岷,等. 南亚高压的不稳定和位涡纬向非对称强迫对印度夏季风爆发的影响[J]. 中国科学 D, 2013, 43(1): 2072-2085.

[353] 郑永骏,吴国雄,刘屹岷. 涡旋发展和移动的动力和热力问题 I:PV-Q 观点[J]. 气象学报, 2013, 71(2): 185-197.

[354] BAO Q, LIN P F, ZHOU T J, et al. The Flexible Global ocean-atmosphere-land system model, spectral version 2: FGOALS-s2[J]. Advances in Atmospheric Sciences, 2013, 30(3):561-576.

[355] HE B, BAO Q, LI J D, et al. Influences of external forcing changes on the summer cooling trend over East Asia[J]. Climatic Change, 2013, 117(4):829-841.

[356] HU W T, DUAN A M, WU G X. Performance of FGOALS-s2 in simulating intraseasonal oscillation over the south Asian monsoon region[J]. Advances in Atmospheric Sciences, 2013, 30(3): 607-620.

[357] LI L J, LIN P F, YU Y Q, et al. The flexible global ocean-atmosphere-land system model, gridpoint version 2: FGOALS-g2[J]. Advances in Atmospheric Sciences, 2013, 30(3):543-560.

[358] LIU B Q, WU G X, MAO J Y, et al. Genesis of the South Asian high and its impact on the Asian summer monsoon onset[J]. Journal of Climate, 2013, 26(9):2976-2991.

[359] LIU Y M, HU J, HE B, et al. Seasonal evolution of subtropical anticyclones in the climate system model FGOALS-s2[J]. Advances in Atmospheric Sciences, 2013, 30(3):593-606.

[360] PAN W J, MAO J Y, WU G X. Characteristics and mechanism of the 10−20-day oscillation of spring rainfall over southern China[J]. Journal of Climate, 2013, 26(14):5072-5087.

[361] WANG J, BAO Q, NING Z, et al. Earth system model FGOALS-s2: Coupling a dynamic global vegetation and terrestrial carbon model with the physical climate system model[J]. Advances in Atmospheric Sciences, 2013, 30(6):1549-1559.

[362] WANG X C, LIU Y M, WU G X, et al. The application of flux-form semi-Lagrangian transport scheme in a spectral atmosphere model[J]. Advances in Atmospheric Sciences, 2013, 30(1):

89-100.

[363] WU G X, DUAN A M, LIU Y M, et al. Recent advances in the study on the dynamics of the Asian summer monsoon onset[J]. Chinese Journal of Atmospheric Sciences, 2013, 37(2):211-228.

[364] WU G X, REN S L, XU J M, et al. Impact of tropical cyclone development on the instability of South Asian high and the summer monsoon onset over Bay of Bengal[J]. Climate Dynamics, 2013, 41(9-10):2603-2616.

[365] WU G X, ZHENG Y J, LIU Y M. Dynamical and thermal problems in vortex development and movement. Part II: Generalized slantwise vorticity development[J]. Acta Meteorologica Sinica, 2013, 27(1):15-25.

[366] ZHANG Y N, WU G X, LIU Y M, et al. The effects of asymmetric potential vorticity forcing on the instability of South Asia high and indian summer monsoon onset[J]. Science China: Earth Sciences, 2013, 57: 337-350.

[367] ZHENG Y J, WU G X, LIU Y M. Dynamical and thermal problems in vortex development and movement. Part I: A PV-Q view[J]. Acta Meteorologica Sinica, 2013, 27(1):1-14.

2014

[368] 黄荣辉, 吴国雄, 陈文, 等. 大气科学和全球气候变化研究进展与前沿[M]. 北京:科学出版社, 2014.

[369] 任荣彩, 吴国雄, CAI M, 等. 平流层-对流层相互作用研究进展:等熵位涡理论的应用及青藏高原影响[J]. 气象学报, 2014, 72(5):853-868.

[370] 王子谦, 段安民, 吴国雄. 边界层参数化方案及海气耦合对WRF模拟东亚夏季风的影响[J]. 中国科学:地球科学, 2014, 44(3): 548-562.

[371] 吴国雄, 刘伯奇. 强迫和惯性不稳定对流发展在印度夏季风爆发过程中的作用[J]. 中国科学:地球科学, 2014, 44: 783-796.

[372] HOLBROOK N J, LI J P, COLLINS M, et al. Decadal climate variability and cross-scale interactions: ICCL 2013 expert assessment workshop[J]. Bulletin of the American Meteorological Society, 2014, 95(8): ES155-ES158.

[373] REN R C, WU G X, CAI M, et al. Progress in research of stratosphere-troposphere interactions: application of isentropic potential vorticity dynamics and the effects of the Tibetan Plateau[J]. Journal of Meteorological Research, 2014, 28(5):714-731.

[374] WANG Z Q, DUAN A M, WU G X. Impacts of boundary layer parameterization schemes and air-sea coupling on WRF simulation of the East Asian summer monsoon[J]. Science China-Earth Sciences, 2014, 57(7):1480-1493.

[375] WANG Z Q, DUAN A M, WU G X. Time-lagged impact of spring sensible heat over the Tibetan Plateau on the summer rainfall anomaly in East China: Case studies using the WRF model[J]. Climate Dynamics, 2014, 42(11-12):2885-2898.

[376] WU G X, LIU B Q. Roles of forced and inertially unstable convection development in the onset process of Indian summer monsoon[J]. Science China-Earth Sciences, 2014, 57(7):1438-1451.

[377] YU Y Y, REN R C, HU J G, et al. A mass budget analysis on the interannual variability of the polar surface pressure in the winter season[J]. Journal of the Atmospheric Sciences, 2014, 71(9): 3539-3553.

[378] ZHANG Y N, WU G X, LIU Y M, et al. The effects of asymmetric potential vorticity forcing on the instability of South Asia high and Indian summer monsoon onset[J]. Science China-Earth Sci-

ences, 2014, 57(2):337-350.

2015

[379] 吴国雄, 李占清, 符淙斌, 等. 气溶胶与东亚季风相互影响的研究进展[J]. 中国科学: 地球科学, 2015, 45: 1-19.

[380] WU G X, DUAN A M, LIU Y M, et al. Tibetan Plateau climate dynamics: Recent progress and outlook[J]. National Science Review, 2015, 2(1): 100-116.

[381] AN Z S, WU G X, LI J P, et al. Global monsoon dynamics and climate change[J]. Annual Review of Earth and Planetary Sciences, 2015, 43: 29-77.

[382] CUI Y F, DUAN A M, LIU Y M, et al. Interannual variability of the spring atmospheric heat source over the Tibetan Plateau forced by the North Atlantic SSTA[J]. Climate Dynamics, 2015, 45(5-6):1617-1634.

[383] CUTTER S L, ISMAIL ZADEH A, ALCANTARA AYALA I, et al. Pool knowledge to stem losses from disasters[J]. Nature, 2015, 522(7556):277-279.

[384] HE B, WU G X, LIU Y M, et al. Astronomical and hydrological perspective of mountain impacts on the Asian summer monsoon[J]. Scientific Reports, 2015, 5.

[385] HU W T, DUAN A M, WU G X. Impact of subdaily air-sea interaction on simulating intraseasonal oscillations over the tropical Asian monsoon region[J]. Journal of Climate, 2015, 28(3): 1057-1073.

[386] LI J Y, MAO J Y, WU G X. A case study of the impact of boreal summer intraseasonal oscillations on Yangtze rainfall[J]. Climate Dynamics, 2015, 44(9-10):2683-2702.

[387] LIU B Q, LIU Y M, WU G X, et al. Asian summer monsoon onset barrier and its formation mechanism[J]. Climate Dynamics, 2015, 45(3-4):711-726.

[388] LIU B Q, WU G X, REN R C. Influences of ENSO on the vertical coupling of atmospheric circulation during the onset of South Asian summer monsoon[J]. Climate Dynamics, 2015, 45(7-8): 1859-1875.

[389] WANG X C, LIU Y M, BAO Q, et al. Comparisons of GCM cloud cover parameterizations with cloud-resolving model explicit simulations[J]. Science China-Earth Sciences, 2015, 58(4): 604-614.

[390] WU G X, HE B, LIU Y M, et al. Location and variation of the summertime upper-troposphere temperature maximum over South Asia[J]. Climate Dynamics, 2015, 45(9-10):2757-2774.

[391] ZHOU L J, BAO Q, LIU Y M, et al. Global energy and water balance: Characteristics from finite-volume atmospheric model of the IAP/LASG (FAMIL1)[J]. Journal of Advances in Modeling Earth Systems, 2015, 7(1):1-20.

[392] ZHOU W L, JIN N, LIN Z, et al. From global change to future earth in China[J]. Advances in Climate Change Research, 2015, 6(2):92-100.

[393] WU G X, ZHENG Y J, LIU Y M. Severe weather diagnosis from the perspective of generalized slantwise vorticity development[C] // Dynamics and Predictability of Large-Scale High-Impact Weather and Climate Events. Cambridge, UK: Cambridge University Press, 2015: 16-36.

2016

[394] 任素玲, 刘屹岷, 吴国雄. 亚洲夏季风爆发前后西北太平洋和孟加拉湾热带气旋活动统计特征[J]. 气象学报, 2016, 74(6):837-849.

[395] 吴国雄，卓海峰，王子谦，等. 夏季亚洲大地形双加热及近对流层顶位涡强迫的激发 I：青藏高原主体加热[J]. 中国科学，2016，46(9)：1209-1222.

[396] 吴国雄，何编，刘屹岷，等. 青藏高原和亚洲夏季风动力学研究的新进展[J]. 大气科学，2016，40(1)：22-32.

[397] 吴胜刚，刘屹岷，邹晓蕾，等. WRF 模式对青藏高原南坡夏季降水的模拟分析[J]. 气象学报，2016，74(5)：744-756.

[398] 张亚妮，刘屹岷，吴国雄，等. 线性准地转模型中副热带感热加热强迫的定常波[J]. 气象学报，2016，74(6)：889-901.

[399] WU G X, LI Z Q, FU C B, et al. Advances in studying interactions between aerosols and monsoon in China[J]. Science China Earth Sciences, 2016, 59(1): 1-16.

[400] WU G X, LIU Y M. Impacts of the Tibetan Plateau on Asian climate[J]. Meteorological Monographs, 2016, 56(1): 7.1-7.29.

[401] WU G X, ZHUO H F, WANG Z Q, et al. Two types of summertime heating over the Asian large-scale orography and excitation of potential-vorticity forcing I. Over Tibetan Plateau[J]. Science China-Earth Sciences, 2016, 59(10): 1996-2008.

2017

[402] 孔晓宇，毛江玉，吴国雄. 2002 年夏季中高纬大气准双周振荡对华南降水的影响[J]. 大气科学，2017，41(6)：1204-1220.

[403] 刘屹岷，王子谦，卓海峰，等. 夏季亚洲大地形双加热及近对流层顶位涡强迫的激发 II：伊朗高原-青藏高原感热加热[J]. 中国科学（地球科学），2017，47(3)：354-366.

[404] CHEN X L, LIU Y M, WU G X. Understanding the surface temperature cold bias in CMIP5 AGCMs over the Tibetan Plateau[J]. Advances in Atmospheric Sciences, 2017, 34(12): 1447-1460.

[405] KONG X Y, MAO J Y, WU G X. Influence on the south China rainfall anomalies of the atmospheric quasi-biweekly oscillation in mid-high latitude during the Summer of 2002[J]. Chinese Journal of Atmospheric Sciences, 2017, 41(6): 1204-1220.

[406] LI J X, BAO Q, LIU Y M, et al. Evaluation of the computational performance of the finite-volume atmospheric model of the IAP/LASG (FAMIL) on a high-performance computer[J]. Atmospheric and Oceanic Science Letters, 2017, 10(4): 329-336.

[407] LIU Y M, WANG Z Q, ZHUO H F, et al. Two types of summertime heating over Asian large-scale orography and excitation of potential-vorticity forcing II. Sensible heating over Tibetan-Iranian Plateau[J]. Science China-Earth Sciences, 2017, 60(4): 733-744.

[408] REN R C, RAO J, WU G X, et al. Tracking the delayed response of the northern winter stratosphere to ENSO using multi reanalyses and model simulations[J]. Climate Dynamics, 2017, 48(9-10): 2859-2879.

[409] SUN S Y, REN R C, WU G X. Onset of the Bay of Bengal summer monsoon and the seasonal timing of ENSO's decay phase[J]. International Journal of Climatology, 2017, 37(14): 4938-4948.

[410] WU G X, HE B, DUAN A M, et al. Formation and variation of the atmospheric heat source over the Tibetan Plateau and its climate effects[J]. Advances in Atmospheric Sciences, 2017, 34(10): 1169-1184.

2018

[411] 马婷婷,吴国雄,刘屹岷,等. 青藏高原地表位涡密度强迫对我国2008年1月南方降水过程的影响——I 资料分析[J]. 气象学报,2018,76(6):870-886.

[412] 吴国雄,刘屹岷,何编,等. 青藏高原感热气泵影响亚洲夏季风的机制[J]. 大气科学,2018,42(3):488-504.

[413] 于佳卉,刘屹岷,马婷婷,等. 青藏高原地表位涡密度强迫对我国2008年1月南方降水过程的影响——Ⅱ 数值模拟[J]. 气象学报,2018,76(6):887-903.

[414] 于威,刘屹岷,杨修群,等. 青藏高原不同海拔地表感热的年际和年代际变化特征及其成因分析[J]. 高原气象,2018,37(5):1161-1176.

[415] LIU P, ZHU Y, ZHANG Q, et al. Climatology of tracked persistent maxima of 500-hPa geopotential height[J]. Climate Dynamics, 2018, 51(1): 701-717.

[416] ZHAO Y, DUAN A M, WU G X. Interannual variability of late-spring circulation and diabatic heating over the Tibetan Plateau associated with Indian Ocean forcing[J]. Advances in Atmospheric Sciences, 2018, 35(8):927-941.

2019

[417] 包庆,吴小飞,李矜霄,等. 2018—2019年秋冬季厄尔尼诺和印度洋偶极子的预测[J]. 科学通报,2019,64:73-78.

[418] HE B, LIU Y M, WU G X, et al. The role of air-sea interactions in regulating the thermal effect of the Tibetan-Iranian Plateau on the Asian summer monsoon[J]. Climate Dynamics, 2019, 52(7-8):4227-4245.

[419] LI J X, BAO Q, LIU Y M, et al. Evaluation of FAMIL2 in Simulating the Climatology and Seasonal-to-Interannual Variability of Tropical Cyclone Characteristics[J]. Journal of Advances in Modeling Earth Systems, 2019, 11(4):1117-1136.

[420] MA T T, WU G X, LIU Y M, et al. Impact of surface potential vorticity density forcing over the Tibetan Plateau on the south China extreme precipitation in January 2008. Part I: Data analysis[J]. Journal of Meteorological Research, 2019, 33(3):400-415.

[421] WANG L, BAO Q, WANG W C, et al. LASG global AGCM with a two-moment cloud microphysics scheme: energy balance and cloud radiative forcing characteristics[J]. Advances in Atmospheric Sciences, 2019, 36(7):697-710.

[422] WU G X, DUAN A M, LIU Y M. Atmospheric heating source over the Tibetan Plateau and its regional climate impact[C]// Oxford Research Encyclopedia of Climate Science. Oxford: Oxford University Press, 2019.

[423] YU J H, LIU Y M, MA T T, et al. Impact of surface potential vorticity density forcing over the Tibetan Plateau on the south China extreme precipitation in january 2008. Part Ⅱ: Numerical simulation[J]. Journal of Meteorological Research, 2019, 33(3):416-432.

2020

[424] 马婷,刘屹岷,吴国雄,等. 青藏高原低涡形成、发展和东移影响下游暴雨天气个例的位涡分析[J]. 大气科学,2020,44(3):472-486.

[425] 唐南军,任荣彩,吴国雄,等. 夏季青藏高原及周边上对流层水汽质量及其向平流层传输年际异常. I:水汽质量异常主导型[J]. 大气科学,2020,44(2):239-256.

[426] 唐南军, 任荣彩, 吴国雄. 青藏高原及周边UTLS水汽时空特征的多源资料对比[J]. 大气科学学报, 2020, 43(2):275-286.

[427] LIU S F, DUAN A, WU G. Asymmetrical response of the East Asian summer monsoon to the quadrennial oscillation of global sea surface temperature associated with the Tibetan Plateau thermal feedback[J]. Journal of Geophysical Research-Atmospheres, 2020, 125(20).

[428] LIU X L, LIU Y M, WANG X C, et al. Large-scale dynamics and moisture sources of the precipitation over the western Tibetan Plateau in boreal winter[J]. Journal of Geophysical Research-Atmospheres, 2020, 125(9).

[429] LIU Y M, LU M M, YANG H J, et al. Land-atmosphere-ocean coupling associated with the Tibetan Plateau and its climate impacts[J]. National Science Review, 2020, 7(3):534-552.

[430] MIAO H, WANG X C, LIU Y M, et al. An evaluation of cloud vertical structure in three reanalyses against CloudSat/cloud-aerosol lidar and infrared pathfinder satellite observations[J]. Atmospheric Science Letters, 2020, 20(7): e906.

[431] TANG Y Q, HE B, BAO Q, et al. The climate variability in global land precipitation in FGOALS-f3-L: A comparison between GMMIP and historical simulations[J]. Atmospheric and Oceanic Science Letters, 2020, 13(6):559-567.

[432] WU G X, LIU Y M, HE B. The nature of the thermal forcing of the Asian summer monsoon[C]// The Multiscale Global Monsoon System. World Scientific, 2020: 27-36.

[433] WU G X, MA T T, LIU Y M, et al. PV-Q perspective of cyclogenesis and vertical velocity development downstream of the Tibetan Plateau[J]. Journal of Geophysical Research-Atmospheres, 2020, 125(16).

[434] XIE Y K, WU G X, LIU Y M, et al. Eurasian cooling linked with Arctic warming: Insights from PV dynamics[J]. Journal of Climate, 2020, 33(7):2627-2644.

2021

[435] LI J X, BAO Q, LIU Y M, et al. Dynamical seasonal prediction of tropical cyclone activity using the FGOALS-f2 ensemble prediction system[J]. Weather and Forecasting, 2021, 36(5): 1759-1778.

[436] LI J X, BAO Q, LIU Y M, et al. Effect of horizontal resolution on the simulation of tropical cyclones in the Chinese Academy of Sciences FGOALS-f3 climate system model[J]. Geoscientific Model Development, 2021, 14(10):6113-6133.

[437] LIU X L, LU J H, LIU Y M, et al. Meridional tripole mode of winter precipitation over the Arctic and continental North Africa and Eurasia[J]. Journal of Climate, 2021, 34(24):9665-9678.

[438] MIAO H, WANG X C, LIU Y M, et al. A regime-based investigation into the errors of CMIP6 simulated cloud radiative effects using satellite observations[J]. Geophysical Research Letters, 2021, 48(18).

[439] SHENG C, WU G X, TANG Y Q, et al. Characteristics of the potential vorticity and its budget in the surface layer over the Tibetan Plateau[J]. International Journal of Climatology, 2021, 41(1):439-455.

[440] YU W, LIU Y M, YANG X Q, et al. Impact of North Atlantic SST and Tibetan Plateau forcing on seasonal transition of springtime South Asian monsoon circulation[J]. Climate Dynamics, 2021, 56(1-2):559-579.

[441] ZHANG G S, MAO J Y, LIU Y M, et al. PV Perspective of impacts on downstream extreme

rainfall event of a Tibetan Plateau vortex collaborating with a southwest China vortex[J]. Advances in Atmospheric Sciences, 2021, 38(11):1835-1851.

[442] ZHANG G S, MAO J Y, WU G X, et al. Impact of potential vorticity anomalies around the eastern Tibetan Plateau on quasi-biweekly oscillations of summer rainfall within and south of the Yangtze Basin in 2016[J]. Climate Dynamics, 2021, 56(3-4):813-835.

[443] ZHAO Y, DUAN A M, WU G X. Opposite Responses of the Indian Ocean to the thermal forcing of the Tibetan Plateau before and after the onset of the South Asian monsoon[J]. Journal of Climate, 2021, 34(20):8389-8408.

2022

[444] DUAN A M, PENG Y Z, LIU J P, et al. Sea ice loss of the Barents-Kara Sea enhances the winter warming over the Tibetan Plateau[J]. npj Climate and Atmospheric Science, 2022, 5(1).

[445] HE B, SHENG C, WU G X, et al. Quantification of seasonal and interannual variations of the Tibetan Plateau surface thermodynamic forcing based on the potential vorticity[J]. Geophysical Research Letters, 2022, 49(5).

[446] MA T T, WU G X, LIU Y M, et al. Abnormal warm sea-surface temperature in the Indian Ocean, active potential vorticity over the Tibetan Plateau, and severe flooding along the Yangtze River in summer 2020[J]. Quarterly Journal of the Royal Meteorological Society, 2022, 148(743):1001-1019.

[447] SHENG C, HE B, WU G X, et al. Interannual impact of the North Atlantic Tripole SST mode on the surface potential vorticity over the Tibetan Plateau during boreal summer[J]. Journal of Geophysical Research-Atmospheres, 2022, 127(9).

[448] SHENG C, HE B, WU G X, et al. Interannual influences of the surface potential vorticity forcing over the Tibetan Plateau on East Asian summer rainfall[J]. Advances in Atmospheric Sciences, 2022, 39(7):1050-1061.

[449] SHENG C, WU G X, HE B, et al. Linkage between cross-equatorial potential vorticity flux and surface air temperature over the mid-high latitudes of Eurasia during boreal spring[J]. Climate Dynamics, 2022.

[450] WANG J B, YANG S, LI Z N, et al. Optimal meridional positions of the Tibetan Plateau for intensifying the Asian summer monsoon[J]. Journal of Climate, 2022, 35(12):3861-3875.

[451] WU G X, TANG Y Q, HE B, et al. Potential vorticity perspective of the genesis of a Tibetan Plateau vortex in June 2016[J]. Climate Dynamics, 2022, 58(11-12):3351-3367.

[452] XIE Y K, WU G X, LIU Y M, et al. A dynamic and thermodynamic coupling view of the linkages between Eurasian cooling and Arctic warming[J]. Climate Dynamics, 2022, 58(9-10):2725-2744.

[453] YONG J L, REN S M, ZHANG S Q, et al. The behavior of moist potential vorticity in the interactions of binary typhoons Lekima and Krosa (2019) in with different high-resolution simulations[J]. Atmosphere, 2022, 13(2).

[454] YU W, LIU Y M, XU L L, et al. Potential impact of spring thermal forcing over the Tibetan Plateau on the following winter El Nino-Southern Oscillation[J]. Geophysical Research Letters, 2022, 49(6).

[455] ZHANG X Q, HE B, LIU Y M, et al. Evaluation of the seasonality and spatial aspects of the southern annular mode in CMIP6 models[J]. International Journal of Climatology, 2022, 42(7):3820-3837.